Contents

HUMAN MOLECULAR GENETICS 3

THIRD EDITION

Tom Strachan

BSc PhD, FMedSci

Professor of Human Molecular Genetics and Scientific Director, Institute of Human Genetics
University of Newcastle, Newcastle-upon-Tyne, UK

and

Andrew P. Read

MA PhD FRCPath FMedSci

Professor of Human Genetics
University of Manchester, Manchester UK

Garland Science
Taylor & Francis Group

LONDON AND NEW YORK

Vice President: Denise Schanck
Production Editor: Fran Kingston
Illustration: Touchmedia, Abingdon, UK
New Media Editor: Michael Morales
Copyeditor: Moira Vekony
Typesetter: Saxon Graphics, Derby, UK
Printer: Ajanta Offset Packaging, New Delhi, India

ISBN 0-8153-4184-9

Library of Congress Cataloging-in-Publication Data
Strachan, T.
Human molecular genetics / Tom Strachan, Andrew P. Read.--3rd ed.
    p. cm.
Includes bibliographical references and index. ISBN 0-8153-4184-9 (pbk.) -- ISBN 0-8153-4182-2 (hardback) -- ISBN 0-8153-4183-0 (diskette)
1. Human molecular genetics. I. Read, Andrew P., 1939- II. Title.
QH431.S787 2003
611'.01816--dc22
        2003017961

Published by Garland Science, a member of the Taylor & Francis Group, 29 West 35th Street, New York, NY 10001–2299, and 4 Park Square, Milton Park, Abingdon, Oxon OX14 4RN.

Distributed in the USA by
Fulfilment Center
Taylor & Francis
10650 Toebben Drive
Independence, KY 41051, USA
Toll Free Tel.: +1 800 634 7064; E-mail: taylorandfrancis@thomsonlearning.com

Distributed in Canada by
Taylor & Francis
74 Rolark Drive
Scarborough, Ontario M1R 4G2, Canada
Toll Free Tel.: +1 877 226 2237; E-mail: tal_fran@istar.ca

Distributed in the rest of the world by
Thomson Publishing Services
Cheriton House
North Way
Andover, Hampshire SP10 5BE, UK
Tel.: +44 (0)1264 332424; E-mail: salesorder.tandf@thomsonpublishingservices.co.uk

Printed in India

15 14 13 12 11 10 9 8 7 6 5 4 3 2 1

2D	Two-dimensional
2DGE	Two-dimensional gel electrophoresis
3'UTR	3'Untranslated region
5'UTR	5'Untranslated region
5-MeC	5-Methyl cytosine
A	Adenine
AAV	Adeno-associated virus
AcMNPV	Autographa californica nuclear polyhedrosis virus
ADAR	Adenosine deaminase acting on RNA
AID	Activation-induced deaminase
ALL	Acute lymphoblastoid leukemia
AMH	Anti-Mullerian hormone
AML	Acute myeloid leukemia
ARF	Alternative Reading Frame
ARMS	Amplification Refractory Mutation System
ARS	Autonomously replicating sequence
AS	Angelman syndrome
ASO	Allele-specific oligonucleotide
ASP	Affected sib pair
AT	Ataxia telangiectasia
ATCC	American Type Culture Collection
AVE	Anterior visceral endoderm
BAC	Bacterial artificial chromosome
BCR	Breakpoint cluster region
BER	Base excision repair
BMI	Body mass index
BOR	Branchio-oto-renal syndrome
bp	Base pairs
BrdU	Bromodeoxyuridine
C	Cytosine
CATH	Class Architecture Topology Homologous superfamily
CCC	Covalently closed circular
CCM	Chemical cleavage of mismatch
CD	Crohn's disease
cDNA	Complementary DNA
CDR	Complementarity-determining region
CEN	Centromere element
CF	Cystic fibrosis
CGH	Comparative genomic hybridization
CID	Chemically-induced dimerization
CIN	Chromosomal instability
cM	CentiMorgan
CNS	Central nervous system
C_ot	Product of DNA concentration and time
CREB	CRE-binding protein
CS	Cockayne syndrome
D	Displacement or Diversity
ddNTP	Dideoxynucleoside triphosphate
DGGE	Denaturing gradient gel electrophoresis
dHPLC	Denaturing high performance liquid chromatography
DIGE	Difference gel electrophoresis
DMD	Duchenne muscular dystrophy
DNA	Deoxyribonucleic acid
DNaseI	Deoxyribonuclease I
DOP-PCR	Degenerate oligonucleotide primed polymerase chain reaction
ds	Double-stranded
DS	Down syndrome
DZ	Dizygotic
EBV	Epstein–Barr virus
EC	Embryonal carcinoma
ECACC	European Collection of Cell Cultures
ECM	Extracellular matrix
EG	Embryonic germ
EGF	Epidermal growth factor
ELSI	Ethical legal and societal implications
EMS	Ethyl methylsulfonate
ENU	Ethyl nitrosurea
ER	Endoplasmic reticulum
ERCC	Excision repair cross-complementing
ERV	Endogenous retroviral sequence
ES	Embryonic stem
ESE	Exonic splice enhancer
ESI	Electrospray ionization
ESS	Exonic splice silencer
EST	Expressed sequence tags
EtBr	Ethidium bromide
ETDT	Extended TDT
FAP	Familial adenomatous polyposis
FISH	Fluorescence in situ hybridization
FITC	Fluorescein isothocyanate
FLAM	Free left Alu monomer
FRAM	Free right Alu monomer
FSSP	Fold classification based on Structure-Structure alignment of Proteins
G	Guanine
gcv	Ganciclovir
GDB	Genome database
GE	Genome equivalents
GFP	Green fluorescent protein
GO	Gene Ontology
GSS	Gerstmann–Sträussler–Scheinker
GST	Glutathione-S-transferase
H	Heavy
HAT	Histone acetyltransferase
HD	Huntington disease
HDAC	Histone deacetylase

HERV	Human endogenous retroviral sequence		MODY	Maturity onset diabetes of the young
HGDP	Human Genome Diversity Project		mRNA	Messenger RNA
HGP	Human Genome Project		MS	Mass spectrometry
HGT	Horizontal gene transfer		MS/MS	Tandem mass spectroscopy
HLA	Human leukocyte antigen		mtDNA	Mitochondrial DNA
HLH	Helix–loop–helix		MTOC	Microtubule-organizing center
HNPCC	Hereditary nonpolyposis colon cancer		MYr	Million years
HPLC	High pressure liquid chromatography		MZ	Monozygotic
HPRT	Hypoxanthine guanine phosphoribosyl transferase		NAS	Nonsense-associated altered splicing
HSC	Hematopoetic stem cell		NBS	Nijmegen breakage syndrome
HSCR	Hirschsprung disease		NCAM	Neural cell adhesion molecule
HSV-TK	Herpes simplex virus thymidine kinase		NER	Nucleotide excision repair
HTH	Helix–turn–helix		NF1	Neurofibromatosis type I
HUGO	Human Genome Organization		NMR	Nuclear magnetic resonance
HDV	Human delta virus		NOE	Nuclear Overhauser effect
IBD	Identity by descent or Inflammatory bowel disease		NPL	Nonparametric lod
			NRSE	Neural restrictive silencer element
IBS	Identity by state		NRSF	Neural restrictive silencer factor
ICAT	Isotope coded affinity tags		OD	Optical density
ICLC	Interlab Cell Line Collection		OI	Osteogenesis imperfecta
ICM	Inner cell mass		OLA	Oligonucleotide ligation assay
ICSI	Intracytoplasmic sperm injection		PAC	P1 artificial chromosome
Ig	Immunoglobulin		PCNA	Proliferating cell nuclear antigen
IM	Intermediate mesoderm		PCR	Polymerase chain reaction
IP_3	inositol 1,4,5-trisphosphate		PDB	Protein Databank
IPG	Immobilized pH gradient		PEG	Polyethylene glycol
IPTG	Isopropyl-thio-β-D-galactopyranoside		PFD	Polyostotic fibrous dysplasia
IRE	Iron-response element		PFGE	Pulsed field gel electrophoresis
ISCN	International System for Human Cytogenetic Nomenclature		PGC	Primordial germ cell
			Ph	Philadelphia
IVF	*In vitro* fertilization		PIC	Polymorphism information content
J	Joining		PIP_2	phosphatidyl inositol 4,5-bisphosphate
kb	Kilobases		PKD1	Adult polycystic kidney disease
KEGG	Kyoto Encyclopedia of Genes and Genomes		PKU	Phenylketonuria
L	Light		PM	Perfect match or Paraxial mesoderm
LCR	Locus control region		PMF	Peptide mass fingerprinting
LD	Linkage disequilibrium		PML	Promyelocytic leukemia
LINES	Long interspersed nuclear elements		PP_i	Pyrophosphate residue
LoH	Loss of heterozygosity		PSI-BLAST	Position-specific iterated BLAST
LPM	Lateral plate mesoderm		PTT	Protein truncation test
LTR	Long terminal repeat		Pu	Purine
m^7G	7-Methylguanosine		PWS	Prader–Willi syndrome
mAb	Monoclonal antibody		Py	Pyrimidine
MAD	Multi-wavelength anomalous dispersion		QTL	Quantitative trait locus
MALDI-TOF	Matrix-assisted laser desorption/ionization time-of-flight		RACE	Rapid amplification of cDNA ends
			REMI	Restriction enzyme-mediated integration
MS	mass spectrometry		RER	Rough endoplasmic Reticulum
MAPH	Multiplex amplifiable probe hybridization		REST	RE-1 silencing transcription factor
MAR	Matrix attachment region		RF	Replicative form
Mb	Mega base		RFLP	Restriction fragment length polymorphism
MCS	Multiple cloning site		RISC	RNA induced silencing complex
M-FISH	Multiplex FISH		RMSD	Root mean square deviation
MGSC	Mouse Genome Sequencing Consortium		RNA	Ribonucleic acid
MIN	Microsatellite instability		RNAi	RNA interference
MIR	Mammalian-wide interspersed repeat		RNase	Ribonucelase
miRNAs	MicroRNAs		RNP	Ribonucleoprotein
MM	Mismatch		rNTP	Ribonucleoside triphosphate
			RP	Retinitis pigmentosa

rRNA	Ribosomal RNA		TDT	Transmission disequilibrium test
RSP	Restriction site polymorphism		TF	Transcription factors
RT	Reverse transcriptase		TFR	Transferrin receptor
RT–PCR	Reverse transcriptase–polymerase chain reaction		TGF	Transforming growth factor
			TIGR	the Institute for Genome Research
SA	Splice acceptor		TK	Thymidine kinase
SAGE	Serial analysis of gene expression		T_m	Melting temperature
SAR	Scaffold attachment regions		TMP	Thymidine monophosphate
SCA1	Spinocerebellar ataxia type 1		TNF	Tumor necrosis factor
scFv	Single chain variable fragment		TOF	Time of flight
SCID	Severe combined immunodeficiency		TPA	Tissue plasminogen activator
SCOP	Structural Classification of Proteins		tRNA	Transfer RNA
SD	Splice donor		TS	Tumor suppressor
SDS	Sodium dodecyl sulfate		TTD	Trichothiodystrophy
SF1	Steroidogenic factor 1		U	Uracil
SINES	Short interspersed nuclear elements		UC	Ulcerative colitis
SIRAS	Single isomorphous replacement with anomalous scattering		UEC	Unequal crossover
			UESCE	Unequal sister chromatid exchange
siRNA	Short interfering RNA		UPD	Uniparental disomy
snoRNA	Small nucleolar RNA		UTR	Untranslated regions
SNP	Single nucleotide polymorphism		UV	Ultraviolet
snRNA	Small nuclear RNA		V	Variable
snRNP	Small nuclear ribonucleoprotein		VS	Varkud satellite
SCA	Spino-cerebellar ataxia		VCFS	Velocardiofacial syndrome
SRP	Signal recognition particle		VNTR	Variable number tandem repeat
ss	Single-stranded		VPC	Vector-producing cells
SSR	Simple sequence repeats		WAGR	Wilms tumor aniridia genital abnormalities mental retardation
SSRP	Simple sequence repeat polymorphism			
STAT	Signal transducers and activators of transcription		WLS	Williams syndrome
stRNA	Small temporal RNA		WS1	Waardenburg syndrome type 1
STRP	Short tandem repeat polymorphism		Xgal	5-bromo 4-chloro 3-indolyl β-D-galactopyranoside
STS	Sequence tagged site			
SV40	Simian virus 40		XP	Xeroderma pigmentosum
SVAS	Supravalvular aortic stenosis		X-SCID	X-linked Severe Combined Immunodeficiency Disease
T	Thymine			
TCR	T-cell receptor		YAC	Yeast artificial chromosome
TCS	Treacher Collins syndrome		ZPA	Zone of polarizing activity

Preface

Human Molecular Genetics has been revised and updated in the light of the discoveries following the Human Genome Project. As we enter the post genome era, we still believe that this book provides a bridge between elementary textbooks and the research literature, so that people with relatively little background in the subject can appreciate and read the latest research.

Human molecular genetics is a large subject. We have tried to make it more digestible by organizing the text into clearly demarcated color-coded sections, using statement headings, and identifying important new terms by bold typeface.

The first section (Chapters 1-7) covers basic material on DNA structure and function, chromosomes, cells and development, pedigree analysis and the basic techniques used in the laboratory. Part Two (Chapters 8-12) discusses the various genome sequencing projects and the insights they provide into the organization, expression, variation and evolution of our genome. In Part Three (Chapters 13-18) the focus is on mapping, identifying and diagnosing the genetic causes of mendelian and complex diseases and cancer. Finally, Part Four (Chapters 19-21) looks at the wider horizons of functional genomics, proteomics, bioinformatics, animal models and therapy.

We provide an extensive glossary, and three additional special glossaries within Chapters 5, 6 and 12. There are also two indices: a main one (marked by a green flash on the page border) and a disease index (marked by a red flash).

The 4 years since publication of the second edition of *Human Molecular Genetics* have been an eventful time. The draft human genome sequence appeared in 2001 and the 'finished' version in 2003. Readers familiar with the previous edition will notice many changes. There are new chapters on cells and development and on functional genomics. The sections on complex diseases have been completely rewritten and reorganized, as has the chapter on Genome Projects. Among many smaller changes are a new section on molecular phylogenetics (Chapter 12) and the introduction of ethics boxes to discuss some of the implications of the new knowledge. Additionally, virtually every page has been revised and updated to take account of the stunning developments of the past 4 years. The welcome move to full color production has meant a total revision of all the figures, with many completely new ones. As before, the figures (except for some taken from other publications) will be available for downloading from the publisher's website.

Some things have not changed. Our aim remains to explain the principles, and not to provide large numbers of facts. The facts are readily available, primarily through the internet, and we provide the necessary references. Scientific etiquette demands that in research-level reviews one uses references to credit the people who made the original discoveries. Here, however, the bibliographies at the end of each chapter have a more educational purpose; we often cite reviews rather than the first paper on a subject, and we have tried to choose references from easily accessible journals. We hope people who find no reference to their seminal paper will understand. Above all, as in previous editions, we try to convey the feel of fast-moving research, and hope that readers will end up sharing our excitement and enthusiasm for the continuing voyage of discovery into our genome.

As always, we are grateful to the many people who have commented on the previous edition, even if we have not always been able to incorporate their suggestions. For the present edition, we have appreciated advice and comments on chapters from many colleagues, notably Gavin Cuthbert, Ian Hampson, Mike Jackson, Ralf Kist, Chris Mathew, Heiko Peters, Nalin Thakker, Andy Wallace and John Wolstenholme. Richard Twyman deserves special thanks for substantial assistance with chapters 3, 19 and 20. It has been a pleasure to work with Jonathan Ray, Fran Kingston and the team at Garland Science/BIOS Scientific Publishers, and with Touchmedia who developed the full color illustrations. Lastly, we thank our long suffering families, especially Meryl, Alex, James and Gilly, and secretaries Anne, Kate, Leanne and Margaret for their support and for putting up with what has been at times quite a stressful time for all concerned.

Tom Strachan and Andrew Read

Supplementary Learning Aids

Supplementary material for students includes:

Human Molecular Genetics 3: Problems and Solutions

A companion text to Human Molecular Genetics 3, this self-assessment book contains multiple-choice questions, review questions and answers, and open-ended problems.

Supplementary materials available for Professors include

The Art of Human Molecular Genetics 3

A CD-ROM containing all the figures from the book for presentation purposes. The figures are available in JPEG and PowerPoint format and are free to those lecturers who adopt the text. Also available for purchase.

Garland Science Classwire ™

We are pleased to offer Garland Science Classwire ™ to adopters of Human Molecular Genetics 3, the website that allows you to:

▶ Access teaching resources provided by Garland Science

▶ Create a customized website for your course in minutes

▶ Communicate with your students online

▶ Build a continually expanding library of teaching resources

For **Human Molecular Genetics 3**, *online teaching resources include:*

▶ All of the images in the book in a downloadable, web-ready, as well as PowerPoint-ready, format

▶ Access to all other Garland Science resources held on Classwire

Adopters of Human Molecular Genetics 3 are entitled to unlimited use of the Garland Science Classwire service. You can view an online demonstration of Classwire ™ at www.classwire.com/garlandscience/demo1.html.

For more details please contact your local sales representative.

Classwire is a trademark of Chalkfree Inc.

Before we start –
Intelligent use of the Internet

Today's students and researchers do not need to be told to use the Internet. However, there are a few points particularly relevant to readers of this book that we would like to point out right at the start.

In this book we try to cover the principles of human molecular genetics, but we have not tried to list many facts. When we give facts, they are primarily there to illustrate principles. But principles without facts are rather sterile, and we expect you to look up facts, as necessary, primarily from the Internet. The genome projects, and all the associated research into human genetics, have produced a veritable tidal wave of data. Intelligent and discriminating use of the Internet is a key skill for any scientist, but perhaps especially for geneticists and students of genetics.

We have just run a Google search for "genetics" It produced 3 630 000 hits. Some of those sites are key resources, many are secondary, some are deliberately misleading and inaccurate. During the course of the book we recommend a number of sites for particular topics. We suggest the following core sites; we have chosen them because they are reliable, stable (not likely to change during the life of this book) and provide well-curated links to many other sites. With these as starting points, you should be able to evolve your own list of useful sites and profit sensibly from the astonishing riches that are out there on the Internet.

▶ General starting points for genetic data: http://www.ncbi.nlm.nih.gov; http://www.ebi.ac.uk/services/
▶ For genome data: www.ensembl.org; http://genome.cse.ucsc.edu
▶ For information on proteins: http://ca.expasy.org/;
▶ For information on any mendelian phenotype: http://www.ncbi.nlm.nih.gov/omim/
▶ Access to biomedical literature: http://www.ncbi.nlm.nih.gov/entrez/

PART ONE

The basics

CHAPTER ONE

DNA structure and gene expression

Chapter contents

1.1 Building blocks and chemical bonds in DNA, RNA and polypeptides

Molecular genetics is primarily concerned with the inter-relationship between the information macromolecules **DNA (deoxyribonucleic acid)** and **RNA (ribonucleic acid)** and how these molecules are used to synthesize **polypeptides**, the basic component of all proteins. In some viruses RNA is the hereditary material, but in all cells genetic information is stored in DNA molecules. Selected regions of the cellular DNA molecules serve as templates for synthesizing RNA molecules. The great majority of the RNA molecules are in turn used to specify the synthesis of polypeptides, either directly or by assisting at different stages in gene expression. As the vast majority of gene expression is dedicated to polypeptide synthesis, proteins represent the major functional end-point of DNA and account for the majority of the dry weight of a cell. The term **protein** was derived from the Greek *proteios*, meaning 'of the first rank' and reflects the important roles of proteins in diverse cellular functions, acting as enzymes, receptors, storage proteins, transport proteins, transcription factors, signaling molecules, hormones, and so on.

1.1.1 DNA, RNA and polypeptides are large polymers defined by a linear sequence of simple repeating units

In eukaryotes individual DNA molecules are found in the chromosomes of the nucleus, in mitochondria, and also in the chloroplasts of plant cells. They are large polymers, with a linear backbone of alternating sugar and phosphate residues. The sugar in DNA molecules is **deoxyribose**, a 5–carbon sugar, and successive sugar residues are linked by covalent phosphodiester bonds. Covalently attached to carbon atom number 1′ (one **prime**) of each sugar residue is a nitrogenous base. Four types of base are found: **adenine (A), cytosine (C), guanine (G) and thymine (T)** and they consist of heterocyclic rings of carbon and nitrogen atoms. They can be divided into two classes:

▶ **purines (A and G)** have two interlocked heterocyclic rings;

▶ **pyrimidines (C and T)** have one such ring.

A sugar with an attached base is called a **nucleoside**. A nucleoside with a phosphate group attached at carbon atom 5′ or 3′ constitutes a **nucleotide** *which is the basic repeat unit of a DNA strand (Figures 1.1 and 1.2)*. The composition of RNA molecules is similar to that of DNA molecules, but differs in that they contain **ribose** sugar residues in place of deoxyribose and **uracil (U)** instead of thymine (*Figures 1.1 and 1.2*).

Proteins are composed of one or more polypeptide molecules which may be modified by the addition of various carbohydrate side chains or other chemical groups. Like DNA and RNA, polypeptide molecules are polymers consisting of a linear sequence of repeating units, in this case **amino acids**. The latter consist of a positively charged amino group and a negatively charged carboxylic acid (carboxyl) group connected by a central carbon atom to which is attached an

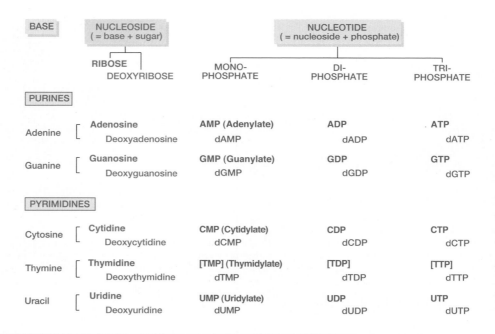

Figure 1.1: Common bases found in nucleic acids with corresponding nucleosides and nucleotides.

Note: (1) nucleotides with a single monophosphate are often described with names where the suffix -ine of the base is replaced by the suffix -ylate, as in adenylate, guanylate etc. (2) Brackets surrounding TMP, TDP and TTP indicate that they are not normally found.

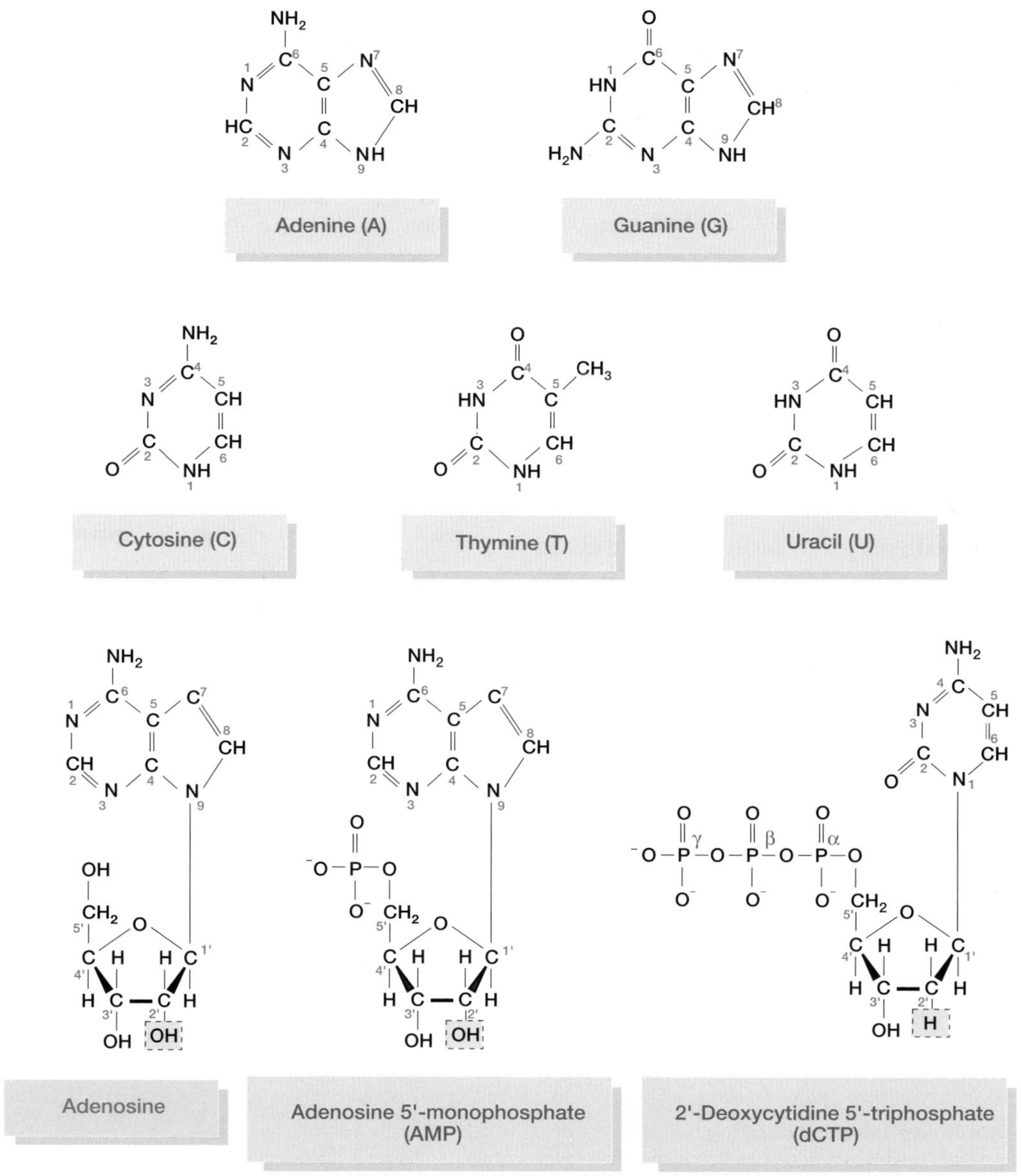

Figure 1.2: Structure of bases, nucleosides and nucleotides.

The bold lines at the bottom of the sugar rings are meant to indicate that the plane of the ring is set at an angle of 90° with respect to the plane of the corresponding base [i.e. if the plane of a base is represented as lying on the surface of the page, carbon atoms 2′ and 3′ (*three prime*) of the sugar can be viewed as projecting upwards out of the page and the oxygen atom as projecting down below the surface of the page]. *Note:* the numbering in deoxyribose and ribose sugars is confined to the five carbon atoms, numbered 1′ to 5′, but the numbering of the bases includes both carbon and nitrogen atoms which occur within the heterocyclic rings. The highlighted hydroxyl and hydrogen atoms connected to carbon 2′ indicate the essential difference between the ribose and deoxyribose sugar residues. Phosphate groups are denoted sequentially as α, β, γ etc., according to proximity to the sugar ring (see dCTP structure).

identifying side chain. The 20 different amino acids can be grouped into different classes depending on the nature of their side chains (*Figure 1.3*). Classification is based as follows:

▸ **basic** amino acids carry a side chain with a net positive charge; an amino (NH_2) group or histidine ring in the side chain acquires a H^+ ion at physiological pH;

▸ **acidic** amino acids carry a side chain with a net negative charge: a carboxyl group in the side chain loses a H^+ ion at physiological pH to form COO^-;

▸ **uncharged polar** amino acids are electrically neutral but carry side chains with polar electrical groups, which are distinguished by having fractional electrical charges (denoted as $\delta+$ or $\delta-$). For example, the hydrogen atom in both the *hydroxyl group* (–OH) and the *sulfhydryl group* (–SH) has a fractional positive charge while the oxygen/sulfur atoms have a fractional negative charge, leading to the following designations: ($-O^{\delta-}-H^{\delta+}$) and ($-S^{\delta-}-H^{\delta+}$);

▸ **nonpolar neutral** amino acids are **hydrophobic** (water-repelling). They often interact with one another and with other hydrophobic groups.

Polypeptides are formed by a condensation reaction between the amino group of one amino acid and the carboxyl group of the next to form *a repeating backbone (–NH–CHR–CO–)*, where the R side chains differ from one amino acid to another (see *Figure 1.21*).

1.1.2 Covalent bonds confer stability; weaker noncovalent bonds facilitate intermolecular associations and stabilize structure

The stability of the nucleic acid and protein polymers is primarily dependent on the strong **covalent bonds** that connect the constituent atoms of their linear backbones. In addition to covalent bonds, a number of weak **noncovalent** bonds (see *Table 1.1*) are important in interactions between these molecules and between groups within a single nucleic acid or protein molecule. Typically, such noncovalent bonds are weaker than covalent bonds by a factor of more than 10. Unlike covalent bonds, whose strength is determined only by the particular atoms involved, the strength of noncovalent bonds is also crucially dependent on their aqueous environment. The structure of water is particularly complex, with a rapidly changing network of noncovalent bonding occurring between the individual H_2O molecules. The predominant force in this structure is the **hydrogen bond**, a weak electrostatic bond formed between a partially positive hydrogen atom and a partially negative atom which, in the case of water molecules, is an oxygen atom.

Charged molecules are highly soluble in water. Because of the phosphate charges present in their component nucleotides, both DNA and RNA are negatively charged (*polyanions*). Depending on their amino acid composition, proteins may carry a net positive charge (**basic proteins**) or a net negative charge (**acidic proteins**). The hydrogen bonding potential of water molecules means that molecules with polar groups (including DNA, RNA and proteins) can form multiple interactions with the water molecules, leading to their solubilization. Thus, even electrically neutral proteins are often readily soluble if they contain an appreciable number of charged or neutral polar amino acids. In contrast, membrane-bound proteins are often characterized by a high content of hydrophobic amino acids which are thermodynamically more stable in the hydrophobic environment of a lipid membrane.

Unlike covalent bonds which require considerable energy input to break them, noncovalent bonds are constantly being made and broken at physiological temperatures. As a result, they readily permit *reversible (transient) molecular interactions*, which are essential for biological function. In the case of nucleic acids and proteins, they play a whole variety of key roles, ensuring faithful replication of DNA, transcription of

---

Table 1.1: Weak non-covalent bonding

Type of bond	Nature of bond
Hydrogen	Hydrogen bonds form when a *hydrogen atom* is sandwiched between two electron-attracting atoms, usually oxygen or nitrogen atoms. See Box 1.1 for examples of their importance in nucleic acid and protein structure and function
Ionic	Ionic interactions occur between *charged groups*. They can be very strong in crystals but in an aqueous environment, the charged groups are shielded by H_2O molecules and other ions in solution and so are quite weak. Nevertheless, they can be very important in biological function, as in the case of enzyme–substrate recognition
Van der Waals'	*Any* two atoms which are very close to each other show a weak attractive bonding interaction due to their fluctuating electrical charges (van der Waals' attraction) until they get extremely close, when they repel each other very strongly (van der Waals' repulsion). Although individually very weak, van der Waals' attractions can become important when there is a very good fit between the surfaces of two macromolecules
Hydrophobic forces	Water is a polar molecule. When hydrophobic molecules or chemical groups are placed in an aqueous environment they become forced together in order to minimize their disruptive effects on the complex network of hydrogen bonding between water molecules. Hydrophobic groups which are forced together in this way are said to be held together by *hydrophobic bonds*, even although the basis of their attraction is due to a common repulsion by water molecules

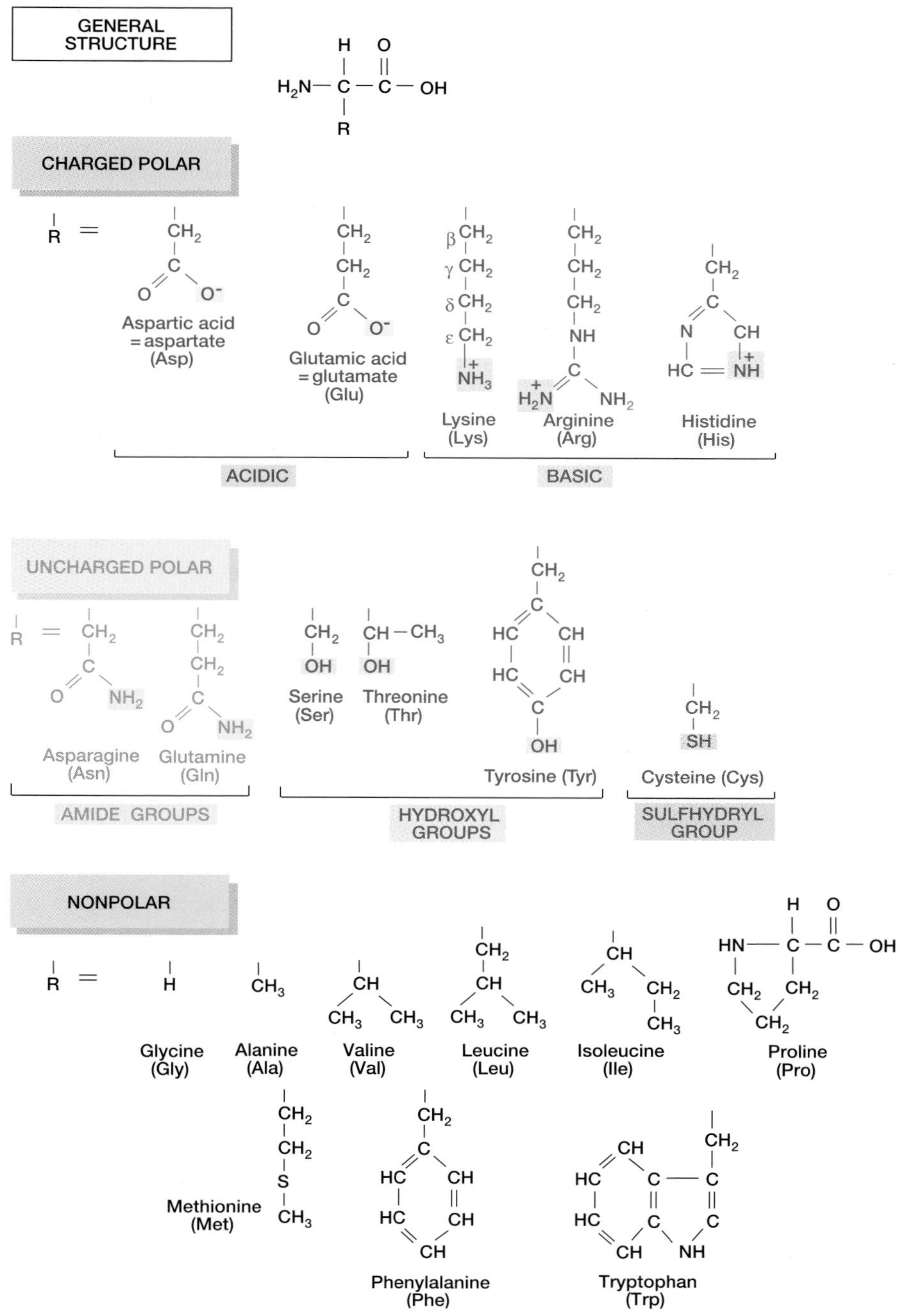

Figure 1.3: Structures of the 20 major amino acids.

Amino acids in a subclass (e.g. the acidic amino acids) are chemically very similar. Highlighted groups are polar chemical groups. The convention of numbering carbon atoms is to designate the central carbon atom as α and subsequent carbons of linear side chains as β, γ, δ etc. (see the example of the lysine side chain at top). Although, in general, polar amino acids are hydrophilic and nonpolar amino acids are hydrophobic, glycine (which has a very small side chain) and cysteine (whose sulfhydryl group is not so polar as a hydroxyl group) occupy intermediate positions on the hydrophilic–hydrophobic scale. *Note:* proline is unusual in that the side chain connects the nitrogen atom of the NH_2 group as well as the central carbon atom.

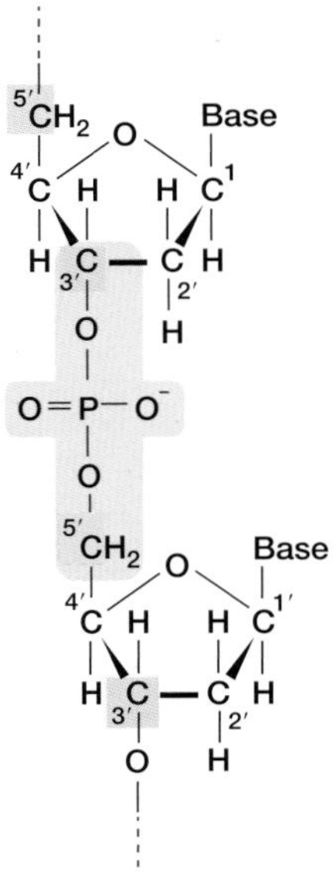

Figure 1.4: A 3′–5′ phosphodiester bond.

RNA, codon–anticodon recognition. Although individually weak, the combined action of numerous noncovalent bonds can make large contributions to the stability of the structure (**conformation**) of these molecules and so can be crucially important for specifying the shape of a macromolecule (see *Figure 1.7B* for the example of how intramolecular hydrogen bonding provides much of the shape of a transfer RNA molecule).

1.2 DNA structure and replication

1.2.1 The structure of DNA is an antiparallel double helix

As mentioned above, the linear backbone of a DNA molecule and of an RNA molecule consists of alternating sugar residues and phosphate groups. In each case, the bond linking an individual sugar residue to the neighboring sugar residues is a **3′,5′-phosphodiester bond**. This means that *a phosphate group links carbon atom 3′ of a sugar to carbon atom 5′ of the neighboring sugar* (*Figure 1.4*).

Whereas the RNA molecules within a cell normally exist as single molecules, the structure of DNA is a **double helix** in which two DNA molecules (***DNA strands***) are held together by weak hydrogen bonds to form a ***DNA duplex***. Hydrogen bonding occurs between laterally opposed bases, **base pairs** (bp), of the two strands of the DNA duplex according to **Watson–Crick rules**: *A specifically binds to T and C specifically binds to G* (*Figure 1.5*). As a result, the base composition of DNA from different cellular sources is not random: *the amount of adenine equals that of thymine, and the amount of cytosine equals that of guanine*. The base composition of DNA can therefore be specified unambiguously by quoting its %GC (= %G + %C) composition. For example, if a source of cellular DNA is quoted as being 42% GC, the base composition can be inferred to be: G, 21%; C, 21%; A, 29%; T, 29%.

DNA can adopt different types of helical structure. **A-DNA** and **B-DNA** are both *right-handed helices* (ones in which the helix spirals in a clockwise direction as it moves away from the observer). They have respectively 11 and 10 bp per turn. **Z-DNA** is a left-handed helix which has 12 bp per turn. Under physiological conditions, most of the DNA in a bacterial or eukaryotic genome is of the B-DNA form. Here each helical strand has a ***pitch*** (the distance occupied by a

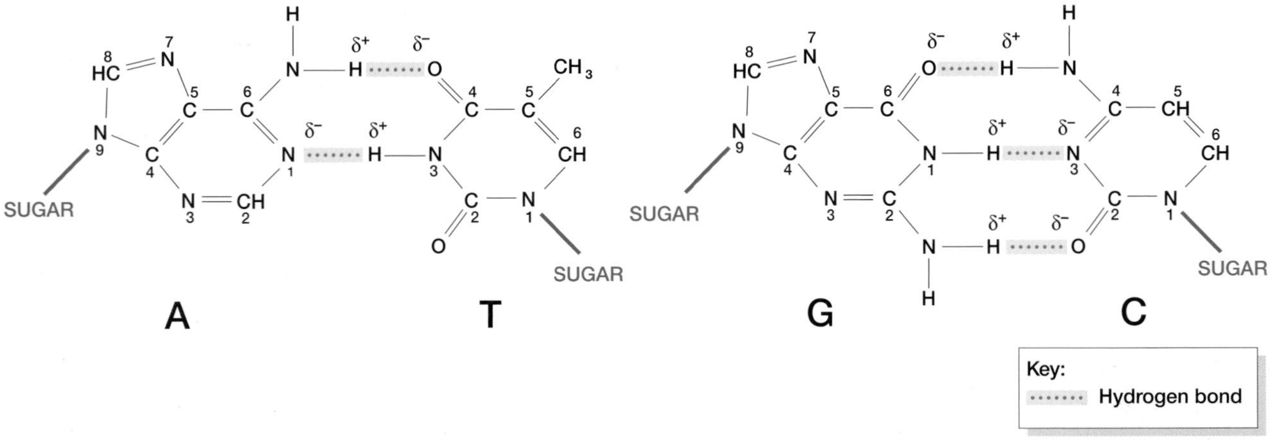

Figure 1.5: A–T base pairs have two connecting hydrogen bonds; G–C base pairs have three.

Fractional positive charges on hydrogen atoms and fractional negative charges on oxygen and nitrogen atoms are denoted by δ^+ and δ^- respectively.

single turn of the helix) of 3.4 nm. As the phosphodiester bonds link carbon atoms number 3′ and number 5′ of successive sugar residues, one end of each DNA strand, the so-called **5′ end**, will have a terminal sugar residue in which carbon atom number 5′ is not linked to a neighboring sugar residue (*Figure 1.6*). The other end is defined as the **3′ end** because of a similar absence of phosphodiester bonding at carbon atom number 3′ of the terminal sugar residue. The two strands of a DNA duplex are said to be **antiparallel** because they always associate (**anneal**) in such a way that the 5′→3′ direction of one DNA strand is the opposite to that of its partner (*Figure 1.6*).

Genetic information is encoded by the linear sequence of bases in the DNA strands (the **primary structure**). Consequently, two DNA strands of a DNA duplex are said to have **complementary** sequences (or to exhibit **base complementarity**) and the sequence of bases of one DNA strand can readily be inferred if the DNA sequence of its complementary strand is already known. It is usual, therefore, to describe a DNA sequence by writing the sequence of bases of one strand only, and in the 5′→3′ direction. This is the direction of synthesis of new DNA molecules during DNA replication, and also of transcription when RNA molecules are synthesized using DNA as a template (see below). However, when describing the sequence of a DNA region encompassing two neighboring bases (really a dinucleotide) on one DNA strand, it is usual to insert a 'p' to denote a connecting phosphodiester bond e.g. CpG means that a cytidine is covalently linked to a neighboring guanosine *on the same DNA strand*, while a CG base pair means a cytosine on one DNA strand is hydrogen-bonded to a guanine *on the complementary strand* (see *Figure 1.6*)].

Intermolecular hydrogen bonding also permits RNA–DNA duplexes and double-stranded RNA formation which are important requirements for gene expression (see *Box 1.1*). In addition, hydrogen bonding can occur between bases within a single DNA or RNA molecule. Sequences having closely positioned complementary *inverted repeats* are prone to forming **hairpin** structures or loops which are stabilized by hydrogen bonding between bases at the neck of the

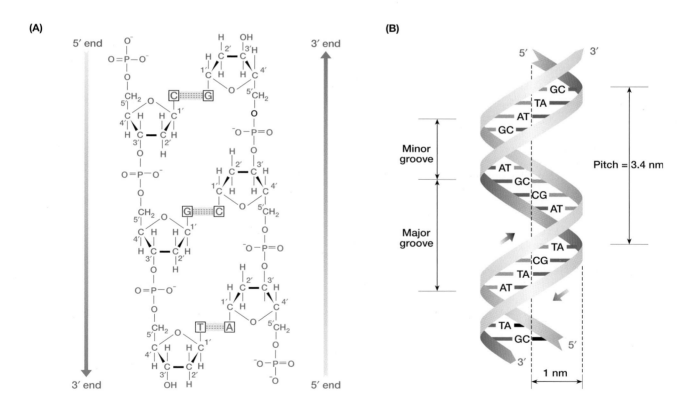

Figure 1.6: The structure of DNA is a double-stranded, antiparallel helix.

(A) *Antiparallel nature of the two DNA strands*. The two strands are **antiparallel** because they have opposite directions for linking of 3′ carbon atom to 5′ carbon atom. The structure shown is a double-stranded trinucleotide whose sequence can be represented as: 5′ pCpGpT–OH 3′ (DNA strand on left) / 5′ pApCpG–OH 3′ (DNA strand on right) (where p = phosphodiester bond and –OH = terminal OH group at 3′ end). This is normally abbreviated by deleting the 'p' *and* 'OH' symbols and giving the sequence on one strand only (e.g. the sequence could equally well be represented as 5′ CGT 3′ or 5′ ACG 3′). **(B)** *The double helical structure of DNA*. The two strands are wound round each other to form a *plectonemic* coil. The *pitch* of each helix represents the distance occupied by a single turn and in the B-DNA structure (see text) accommodates 10 nucleotides.

> **Box 1.1: Examples of the importance of hydrogen bonding in nucleic acids and proteins.**
>
> **Intermolecular hydrogen bonding in nucleic acids**
>
> This is important in permitting the formation of double-stranded nucleic acids as follows:
>
> ▶ *Double-stranded DNA*. The stability of the double helix is maintained by hydrogen bonding between A–T and C–G base pairs (Section 1.2.1 and *Figure 1.5*);
>
> ▶ *DNA–RNA duplexes*. These form naturally during RNA transcription and hydrogen bonding underpins the following types of base pairs: A–U; C–G and also A–T (bonding of A in the RNA strand with T in the DNA strand). See Section 1.3.3 and *Figure 1.12*;
>
> ▶ *Double-stranded RNA*. The genomes of some viruses consist of RNA–RNA duplexes, but in addition transient RNA–RNA duplexes form in all cells during RNA processing and gene expression. RNA splicing requires recognition of exon–intron boundaries following hydrogen bonding between the unspliced
>
> RNA transcripts and various small nuclear RNA molecules (Section 1.4.1). In addition, codon–anticodon recognition involves hydrogen bonding between two RNA molecules: mRNA and tRNA (Section 1.5.1 and *Figure 1.20*). *Hydrogen bonding between RNA molecules involves G–U base pairs as well as A–U and C–G base pairing* (see *Table 1.6*);
>
> **Intramolecular hydrogen bonding in nucleic acids**. This is important in providing secondary structure in both DNA and in RNA molecules, as in the case of formation of hairpins in DNA and the complex arms of tRNA (see Section 1.2.1 and *Figure 1.7*). *In the latter case, note that base-pairing includes G–U base pairs as well as A–U and G–C base pairs.*
>
> **Intramolecular hydrogen bonding in proteins**. Some fundamental units of protein secondary structure, such as the α-*helix*, the β-*pleated sheet* and the β-*turn* are largely defined by intrachain hydrogen bonding (Section 1.5.5 and *Figure 1.24*).

loop (*Figure 1.7A*; see also *Figure 9.6* for the example of microRNAs which originate by cleavage from *hairpin RNA* precursors). Such structural constraints, which are additional to those imposed by the primary structure, contribute to the **secondary structure** of the molecule. Certain RNA molecules, such as transfer RNA (tRNA), show particularly high degrees of secondary structure (*Figure 1.7B*).

Note: in the case of hydrogen bonding in RNA–RNA duplexes and also in intramolecular hydrogen bonding within an RNA molecule, *G–U base pairs are occasionally found in addition to A–U and C–G base pairs* (*Figure 1.7B*). This form of base pairing is not particularly stable, but does not significantly disrupt the RNA–RNA helix.

1.2.2 DNA replication is semi-conservative and synthesis of DNA strands is semi-discontinuous

During the process of DNA synthesis (**DNA replication**), the two DNA strands of each chromosome are unwound by a helicase and each DNA strand directs the synthesis of a complementary DNA strand. Two daughter DNA duplexes are formed, each of which is identical to the parent molecule (*Figure 1.8*). As each daughter DNA duplex contains *one strand from the parent molecule and one newly synthesized DNA strand*, the replication process is described as **semi-conservative**. The enzyme DNA polymerase catalyzes the synthesis of new DNA strands using the four deoxynucleoside triphosphates (dATP, dCTP, dGTP, dTTP) as nucleotide precursors.

DNA replication is initiated at specific points, which have been termed **origins of replication.** Starting from such an origin, the initiation of DNA replication results in a Y-shaped **replication fork**, where the parental DNA duplex bifurcates (splits) into two daughter DNA duplexes. The two strands of the parental DNA duplex are antiparallel, but act individually as templates for the synthesis of a complementary antiparallel daughter strand. It follows that the two daughter strands must

run in opposite directions (i.e. the direction of chain growth must be 5′→3′ for one daughter strand, the **leading strand**, but 3′→5′ for the other daughter strand, the **lagging strand**; *Figure 1.9*).

The reactions catalyzed by DNA polymerases involve addition of a dNMP residue to the free 3′ hydroxyl group of the growing DNA chain (the dNMP residue is provided by a dNTP precursor – the two distal β and γ phosphate residues are cleaved and the resulting pyrophosphate group is discarded). This requirement introduces an *asymmetry* into the DNA replication process: only the leading strand will have a free 3′ hydroxyl group at the point of bifurcation. This will permit sequential addition of nucleotides and continuous elongation in the same direction in which the replication fork moves.

The 5′→3′ direction of synthesis of the lagging strand is in the *opposite direction* to that in which the replication fork moves. As a result, synthesis has to be accomplished in steps, generating a progressive series of small, typically 100–1000 nucleotides long, fragments (**Okazaki fragments**). As only the leading strand is synthesized continuously, the synthesis of DNA strands is said to be **semi-discontinuous**. Each fragment of the lagging strand is synthesized in the 5′→3′ direction, which will be in the opposite direction to that in which the replication fork moves. Successively synthesized fragments are covalently joined at their ends using the enzyme DNA ligase. As a result, the lagging strand grows in the direction in which the replication fork moves.

1.2.3 The DNA replication machinery in mammalian cells is complex

As in the case of ribosomes, the DNA replication machinery is highly conserved and is basically similar from *Escherichia coli* to mammalian cells, with a variety of key different protein types (*Box 1.2*). However, the complexity is greater in mammalian cells, both in terms of numbers of different DNA polymerases,

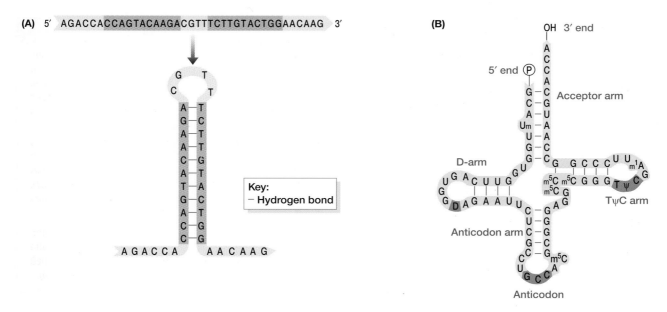

Figure 1.7: Intramolecular hydrogen bonding in DNA and RNA.

(A) *Formation of a double-stranded hairpin loop within a single DNA strand*. The highlighted sequences in the DNA strand above represent inverted repeat sequences which can hydrogen bond to form a hairpin structure (below). **(B)** *Transfer RNA (tRNA) has extensive secondary structure*. The example shown is a human tRNA^{Glu} gene. Minor nucleosides are: D, 5, 6-dihydrouridine; Ψ, pseudouridine (5-ribosyl uracil); m^{5}C, 5-methylcytidine; m^{1}A, 1-methyladenosine. The cloverleaf structure is stabilized by extensive intramolecular hydrogen bonding, mostly by Watson-Crick G–C and A–U base pairs, *but also with the occasional G–U base pair*. Four arms are recognized: the **acceptor arm** is the one to which an amino acid can be attached (at the 3′ end); the **TΨC arm** is defined by this trinucleotide; the **D arm** is named because it contains dihydrouridine residues; and the **anticodon arm** contains the anticodon trinucleotide in the center of the loop. The secondary structure of tRNAs is virtually invariant: there are always seven base pairs in the stem of the acceptor arm, five in the TΨC arm, five in the anticodon arm, and three or four in the D arm.

and numbers of constituent proteins and protein subunits. Most DNA polymerases in mammalian cells use an individual DNA strand as a template for synthesizing a complementary DNA strand and so are **DNA-directed DNA polymerases**. Unlike RNA polymerases, DNA polymerases *absolutely* require the 3′–hydroxyl end of a base-paired primer strand as a substrate for chain extension. Therefore, a previously synthesized **RNA primer** (which is synthesized by a **primase**) is required to provide the free 3-OH group that DNA polymerases need to start DNA synthesis.

There are more than 20 different types of DNA polymerase in mammalian cells (Friedberg *et al.*, 2000; 2002), which can be conveniently grouped into three major classes (see *Table 1.2*):

▶ **classical (high fidelity) DNA-directed DNA polymerases**. Of these, two are involved in standard chromosomal DNA replication, being specific for synthesizing DNA from the leading strand or lagging strand (DNA polymerases δ and α respectively), two are dedicated to DNA repair (β and ε)

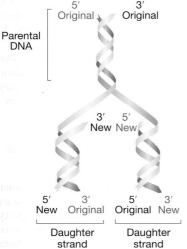

Figure 1.8: DNA replication is semi-conservative.

The parental DNA duplex consists of two complementary, antiparallel DNA strands which unwind and then individually act as templates for the synthesis of new complementary, antiparallel DNA strands. Each of the daughter DNA duplexes contains one original parental DNA strand and one new DNA strand, forming a DNA duplex which is structurally identical to the parental DNA duplex. **Note:** this figure shows the result of DNA duplication but not the way the process works (for which, see *Figure 1.9*).

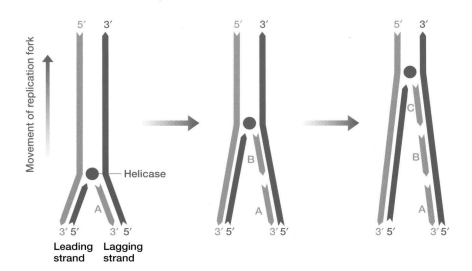

Figure 1.9: Asymmetry of strand synthesis during DNA replication.

The helicase unwinds the DNA duplex to allow the individual strands to be replicated. As replication proceeds, the 5′→3′ direction of synthesis of the leading strand is the same as that in which the replication fork is moving, and so synthesis can be continuous. The synthesis of the lagging strand, however, is in the direction opposite to that of movement of the replication fork. It needs to be synthesized in pieces (Okazaki fragments), first A, then B, then C which are subsequently sealed by the enzyme DNA ligase to form a continuous DNA strand. **Note:** the synthesis of these fragments is initiated using an *RNA* primer. In eukaryotic cells, the leading and lagging strands are synthesized, respectively, by DNA polymerases δ and α (see *Table 1.2*).

and one is dedicated to replicating and repairing mito-chondrial DNA (γ);

▶ *error-prone DNA-directed DNA polymerases.* A wide variety of different DNA polymerases have been identified with very low fidelity of DNA replication (e.g. the error rate for DNA polymerase ι (iota) is 20 000 times that for DNA polymerase ε. Some are known to be very highly expressed in immune system cells, and this observation plus

their low fidelity of DNA replication suggests involvement in hypermutation in B and T lymphocytes (see Section 10.6);

▶ *RNA-directed DNA polymerases.* Some DNA polymerases synthesize DNA by using an RNA template, and so have been described as *reverse transcriptases* (see Sections 1.2.4 and 1.3.1). They include an activity present in the enzyme *telomerase* which is responsible for replicating the ends of

Box 1.2: Major classes of proteins used in the DNA replication machinery.

Topoisomerases start the process of DNA unwinding by nicking (breaking) a single DNA strand. As a result, the tension holding the helix in its coiled and supercoiled form is released.

Helicases accomplish unwinding of the original double strand, once supercoiling has been eliminated by a topoisomerase.

DNA polymerases synthesize new DNA strands. DNA polymerases are complex aggregates of several different protein subunits, often including DNA *proof-reading* and *nuclease* activities so any wrongly incorporated bases can be identified and a local stretch of DNA containing the error can be cut away and then repaired. In cells, DNA typically replicates by new DNA synthesis from an existing DNA strand template, using *DNA-directed DNA polymerases*, and a wide variety of such DNA polymerases exist in mammalian cells. In more specialized situations, DNA can be synthesized in cells from RNA templates using *RNA-directed DNA polymerases* (also called *reverse transcriptases*) as in the case of synthesizing the ends of a

linear chromosome using the reverse transcriptase of the enzyme *telomerase*.

Primases. DNA polymerases find it difficult to start synthesizing *de novo* from a bare single-stranded template. Instead, a primer is required with a 3′ OH group onto which the polymerase can attach a dNTP. Primases attach a small RNA primer to the single-stranded DNA to act as a substitute 3′ OH for DNA polymerase to begin synthesizing from. This RNA primer is eventually removed by a ribonuclease and the gap is filled in using a DNA polymerase, then sealed using a DNA ligase.

Ligases catalyze the formation of a phosphodiester bond given an unattached but adjacent 3′ OH and 5′ phosphate.

Single-stranded binding proteins are important to maintain the stability of the replication fork. Single-stranded DNA is very labile, or unstable, so these proteins bind to it while it remains single stranded and keep it from being degraded.

Table 1.2: Major classes of mammalian DNA polymerase

(A) High fidelity (classical) DNA-directed DNA polymerases.

	DNA polymerase class				
	α	β	γ	δ	ε
Location	Nuclear	Nuclear	Mitochondrial	Nuclear	Nuclear
General DNA replication	Synthesis and priming of lagging strand	–	mtDNA replication	Synthesis of leading strand	–
$3' \rightarrow 5'$ exonuclease[a]	No	No	Yes	Yes	Yes
DNA repair function	–	By base excision	mtDNA repair	By nucleotide and base excision	By nucleotide and base excision

[a]Serves as a proof-reading activity.

(B) Low fidelity (error-prone) DNA-directed DNA polymerases.

DNA polymerase ζ (zeta)	Expressed in mutating B and T cells; involved in hypermutation?
DNA polymerase η (eta)	Mutates A and T nucleotides during hypermutation?
DNA polymerase ι (iota)	Very low fidelity of replication; thought to be involved in hypermutation
DNA polymerase μ (mu)	Highly expressed in B and T cells; thought to be involved in hypermutation

(C) RNA-directed DNA polymerases (reverse transcriptases).

Telomerase reverse transcriptase (Tert)	Replicates DNA at the ends of linear chromosomes
LINE-1/endogenous retrovirus reverse transcriptase	Occasionally converts mRNA and other RNA into cDNA which can integrate elsewhere into genome

linear chromosomes (Section 2.2.5) plus reverse transcriptases encoded by some highly repetitive DNA and endogenous retrovirus classes.

1.2.4 Viral genomes are frequently maintained by RNA replication rather than DNA replication

DNA is the hereditary material in all present-day cells and we are accustomed to thinking of **genomes** as the collective term for the hereditary DNA molecules of an organism or cell. Despite the ubiquity of DNA as the cellular hereditary material, many different classes of present-day viruses have an *RNA genome*. RNA molecules can undergo self-replication, but the *2′ OH group on the ribose residues of RNA makes the sugar–phosphate bonds rather chemically unstable*. In DNA the deoxyribose residues carry only hydrogen atoms at the 2′ position and so DNA is much more suited than RNA to being a stable carrier of genetic information. RNA replication is also error-prone with normal RNA replication error rates being about 10 000 times as high as that encountered during DNA replication.

Viruses have developed many different strategies to infect and subvert cells, and their genomes show extraordinary diversity (see *Table 1.3; Figure 1.10*). As a result of their high mutational load, viral RNA genomes are generally quite small but have the advantage of rapid mutation rates. Unlike

DNA viruses which generally replicate in the nucleus, RNA viruses usually replicate in the cytoplasm. Among some important exceptions to this general rule are a group of RNA viruses known as **retroviruses.** Retroviruses are unusual RNA viruses in that they replicate in the nucleus and also in that the RNA replicates via a DNA intermediate, using a *reverse transcriptase*, an RNA-directed DNA polymerase which, like RNA polymerases, has relatively high replication error rates. Following conversion of RNA into complementary DNA (cDNA), the retroviral cDNA integrates into host chromosomal DNA (see Section 21.5.3).

1.3 RNA transcription and gene expression

1.3.1 The flow of genetic information in cells is almost exclusively one way: DNA→RNA→protein

The expression of genetic information in all cells is very largely a one-way system: DNA specifies the synthesis of RNA and then RNA specifies the synthesis of polypeptides (which subsequently form proteins). Because of its universality, the DNA→RNA→polypeptide (protein) flow of genetic information has been described as the **central dogma** of molecular biology. Two successive steps are essential in all cellular organisms:

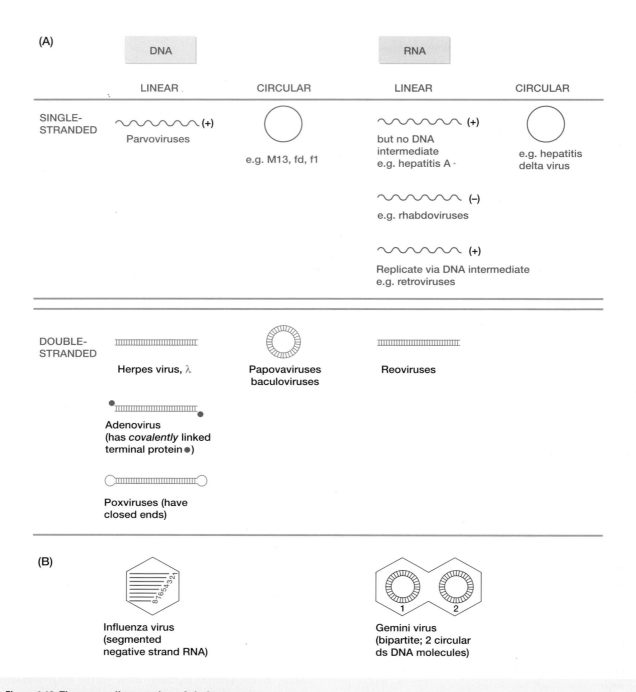

Figure 1.10: The extraordinary variety of viral genomes.

(A) *Strandedness and topology of viral genomes*. In the case of single-stranded viral genomes, the RNA used to make protein products may have the same sense as the genome which is therefore a *positive-strand genome* (+) or be the opposite sense (antisense) of the genome, which is described as a *negative-strand genome* (–). Some single-stranded (+) RNA viruses go through a DNA intermediate (***retroviruses***) and some double-stranded DNA viruses such as hepatitis B go through a replicative RNA form. **(B)** *Segmented and multipartite viral genomes*. Segmented genomes have a variety of different nucleic acid molecules specifying *monocistronic mRNAs*, as in the case of influenza viruses which have eight different negative single-strand RNA molecules. *Multipartite genomes* are a subset of segmented genomes where each of the different molecules is packaged into a separate virus particle.

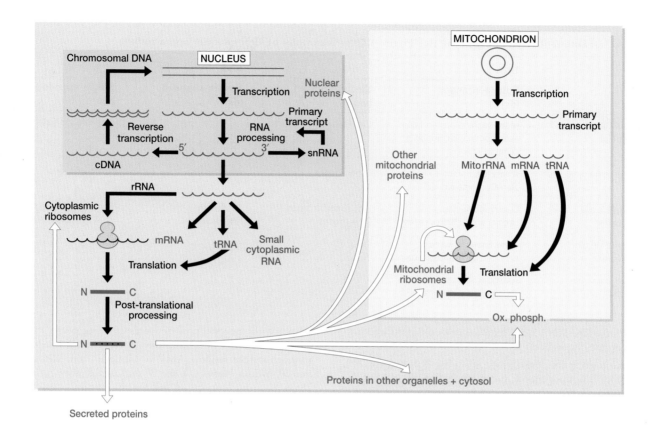

Figure 1.11: Gene expression in an animal cell.

Note: (i) very occasionally a small proportion of nuclear RNA molecules can be converted naturally to cDNA by virally encoded and cellular reverse transcriptases, and thereafter integrate into chromosomal DNA at diverse locations; (ii) the mitochondrion synthesizes its own rRNA and tRNA and a few proteins which are involved in the oxidative phosphorylation (Ox.phosph.) system. However, the proteins of mitochondrial ribosomes and the majority of the proteins in the mitochondrial oxidative phosphorylation system and other mitochondrial proteins are encoded by nuclear genes and translated on cytoplasmic ribosomes, before being imported into mitochondria.

1. **transcription.** This uses a ***DNA-directed RNA polymerase*** and occurs in the nucleus of eukaryotic cells and, to a limited extent, in mitochondria and chloroplasts, the only other organelles which have a genetic capacity in addition to the nucleus (see *Figure 1.11*);

2. **translation.** This occurs in **ribosomes**, large RNA–protein complexes which are found in the cytoplasm and also in mitochondria and chloroplasts. The RNA molecules which specify polypeptide are known as **messenger RNA (mRNA)**.

The expression of genetic information follows a **colinearity principle**: the *linear sequence* of nucleotides in DNA is decoded in groups of three nucleotides at a time (***base triplets***) to give a *linear sequence* of nucleotides in RNA which can be decoded in turn in groups of three nucleotides (***codons***) to give a *linear sequence* of amino acids in the polypeptide product.

Recently it has become clear that eukaryotic cells, including mammalian cells, contain nonviral chromosomal DNA sequences which encode cellular reverse transcriptases, such as members of the mammalian LINE-1 repetitive DNA family (see Section 9.5.2). Because some nonviral RNA sequences are known to act as templates for cellular DNA synthesis, the principle of unidirectional flow of genetic information in cells is no longer strictly valid.

1.3.2 Only a small fraction of the DNA in complex organisms is expressed to give a protein or RNA product

Only a small proportion of all the DNA in cells is ever transcribed. According to their needs, different cells transcribe different segments of the DNA (**transcription units**) which are discrete units, spaced irregularly along the DNA sequence. RNA polymerases use transcription units as templates to synthesize initially equivalent sized RNA molecules (***primary transcripts***) which are then modified to generate mature expression products. However, the great majority of the cellular DNA is never transcribed in any cell. Moreover, only

Table 1.3: Different classes of genome (see Figure 1.10 for examples of viral genome organization)

	DNA		RNA	
	Double-stranded (ds)	Single-stranded (ss)	Double-stranded (ds)	Single-stranded (ss)
Single circular molecule	Many bacteria and archaea; mitochondria; chloroplasts; some viruses	Some viruses	–	A very few viruses
Single linear molecule	A very few bacteria e.g. *Borrella*; some viruses	Some viruses	A few viruses	Some viruses
Multiple linear molecules	Eukaryotic nuclei; some segmented ds DNA viruses	Some segmented ss DNA viruses	Some segmented ds RNA viruses	Some segmented ss RNA viruses
Multiple circular molecules	Some bacteria; some bipartite and tripartite viruses	–	–	–
Mixed linear and circular	A very few bacteria e.g. *Agrobacterium tumefaciens*	–	–	–

a portion of the RNA made by transcription is translated into polypeptide. This is because:

▶ *some transcription units specify noncoding RNA*: the RNA product does not encode polypeptides like mRNA but has a different function. In addition to the well-established ribosomal RNA (rRNA) and transfer RNA (tRNA) we are now aware of a wide variety of noncoding RNAs of diverse functions (Section 9.2);

▶ *the primary transcript of transcription units specifying mRNA is subject to RNA processing*. As a result, much of the initial RNA sequence is discarded to give a much smaller mRNA (Section 1.4.1);

▶ *only a central part of the mature mRNA is translated*; sections of variable length at each end of the mRNA remain untranslated (Section 1.5.1).

In animal cells, DNA is found in both the nucleus and the mitochondria. The mitochondria have, however, only a very small fraction of the total cellular DNA and a very limited number of genes (see Section 9.1.2); the vast majority of the DNA of a cell is located in the chromosomes of the nucleus.

The fraction of **coding DNA** in the genomes of complex eukaryotes is rather small. This is partly a result of the noncoding nature of much of the sequence within genes. Another reason is that a considerable fraction of the genome of complex eukaryotes contains repeated sequences which are nonfunctional or which are not transcribed into RNA. The former include defective copies of functional genes (*pseudogenes* and *gene fragments*), and highly repetitive noncoding DNA.

1.3.3 During transcription genetic information in some DNA segments (genes) specifies RNA

RNA synthesis is accomplished using an RNA polymerase enzyme, with DNA as a template and ATP, CTP, GTP and UTP as RNA precursors. The RNA is synthesized as a single strand, with the direction of synthesis being $5' \rightarrow 3'$. Chain elongation occurs by adding the appropriate ribonucleoside monophosphate residue (AMP, CMP, GMP or UMP) to the free $3'$ hydroxyl group at the $3'$ end of the growing RNA chain. Such nucleotides are derived by splitting a pyrophosphate residue (PP_i) from the appropriate ribonucleoside triphosphate (rNTP) precursors. This means that the nucleotide at the extreme $5'$ end (*the initiator nucleotide*) *will differ from all others in the chain by carrying a $5'$ triphosphate group*.

Normally, only one of the two DNA strands acts as a template for RNA synthesis. During transcription, double-stranded DNA is unwound and the DNA strand which will act as a template for RNA synthesis forms a transient double-stranded **RNA–DNA hybrid** with the growing RNA chain. As the RNA transcript is complementary to this **template strand**, the transcript has the same $5' \rightarrow 3'$ direction and base sequence (except that U replaces T) as the opposite, nontemplate strand of the double helix. For this reason the nontemplate strand is often called the **sense strand**, and the template strand is often called the **antisense strand** (*Figure 1.12*). In documenting gene sequences it is customary to show only the DNA sequence of the sense strand. Orientation of sequences relative to a gene sequence normally refers to the sense strand. For example, the $5'$ end of a gene refers to sequences at the $5'$ end of the *sense* strand, and **upstream** or **downstream sequences** refer to sequences which flank the gene at the $5'$ or $3'$ ends, respectively, with reference to the *sense* strand.

In eukaryotic cells, three different RNA polymerase molecules are required to synthesize the different classes of RNA (*Table 1.4*). The vast majority of cellular genes encode polypeptides and are transcribed by RNA polymerase II. Increasing importance, however, is being paid to genes which encode RNA as their mature product: functional RNA molecules are now known to play a wide variety of roles, and catalytic functions have been ascribed to some of them (see Section 9.2.3).

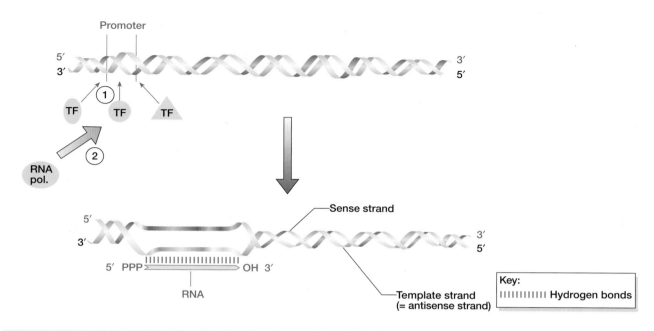

Figure 1.12: RNA is transcribed as a single strand which is complementary in base sequence to one strand (template strand) of a gene.

Various transcription factors (TF) are required to bind to a promoter sequence in the immediate vicinity of a gene (1), in order to subsequently position and guide the RNA polymerase that will transcribe the gene (2). Chain synthesis is initiated with a nucleoside triphosphate and chain elongation occurs by successive addition of nucleoside monophosphate residues provided by rNTPs to the 3′ OH. This means that the 5′ end will have a *triphosphate group*, which may subsequently undergo modification (e.g. by capping; see Section 1.4.2) and the 3′ end will have a free hydroxyl group. ***Note:*** the sequence of the RNA will normally be identical to the sense strand of the gene (*except U replaces T*) and complementary in sequence to the template strand.

1.3.4 *Cis*-acting regulatory elements and trans-acting transcription factors are required in eukaryotic gene expression

Eukaryotic RNA polymerases cannot initiate transcription by themselves. Instead, combinations of short sequence elements in the immediate vicinity of a gene act as recognition signals for **transcription factors** to bind to the DNA in order to guide and activate the polymerase. A major group of such short sequence elements is often clustered upstream of the coding sequence of a gene, where they collectively constitute the ***promoter***. After a number of general transcription factors bind to the promoter region, an RNA polymerase binds to the transcription factor complex and is activated to initiate the synthesis of RNA from a unique location. The transcription factors are said to be ***trans*-acting**, because they are synthesized by remotely located genes and are required to *migrate* to their sites of action. In contrast, the promoter elements are said to be ***cis*-acting** because their function is limited to the DNA duplex on which they reside (*Table 1.5*). Major functional groupings of *cis*-acting elements include:

▸ **promoters.** Common *cis*-acting elements include:
- **the TATA box**, often TATAAA or a variant and usually found at a position about 25 bp upstream (−25) from the transcriptional start site (*Figure 1.13*). It is typically found in genes which are actively transcribed by RNA

Table 1.4: The three classes of eukaryotic RNA polymerase

Class	Genes transcribed	Comments
I	28S rRNA; 18S rRNA; 5.8S rRNA	Localized in the nucleolus. A single primary transcript (45S rRNA) is cleaved to give the three rRNA classes listed
II	All genes that encode polypeptides; most snRNA genes	Polymerase II transcripts are unique in being subject to capping and polyadenylation
III	5S rRNA; tRNA genes; U6 snRNA; 7SL RNA; 7SK RNA; 7SM RNA; SiRNA	The promoter for some genes transcribed by RNA polymerase III (e.g. 5S rRNA, tRNA, 7SL RNA) is internal to the gene (see *Figure 1.13*) and for others (e.g. 7SK RNA) is located upstream

Table 1.5: Examples of cis-acting elements recognized by ubiquitous transcription factors

Cis element	DNA sequence is identical to, or a variant of	Associated *trans*-acting factors	Comments
GC box	GGGCGG	Spl	Spl factor is ubiquitous
TATA box	TATAAA	TFIID	TFIIA binds to the TFIID–TATA box complex to stabilize it
CAAT box	CCAAT	Many, e.g. C/EBP, CTF/NF1	Large family of *trans*-acting factors
TRE (TPA response element)	GTGAGT(A/C)A	AP-1 family, e.g. JUN/FOS	Large family of *trans*-acting factors
CRE (cAMP response element)	GTGACGT(A/C)A(A/G)	CREB/ATF family, e.g. ATF-1	Genes activated in response to cAMP

polymerase II but in a *restricted* sense, that is at a specific stage in the cell cycle (e.g. histones) or in specific cell types (e.g. β-globin). Mutation at the TATA element does not prevent initiation of transcription, but does cause the startpoint of transcription to be displaced from the normal position;

- **the GC box,** a variant of the consensus sequence GGGCGG, found in a variety of genes, many lacking a TATA box as in the case of *house-keeping genes* (Section 1.3.5). Although the GC box sequence is asymmetrical it appears to function in either orientation (*Figure 1.13*);

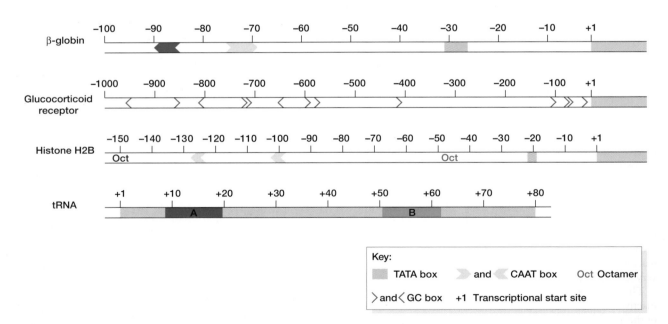

Key:
- ▮ TATA box
- ▷ and ◁ CAAT box
- Oct Octamer
- ⟩ and ⟨ GC box
- +1 Transcriptional start site

Figure 1.13: Eukaryotic promoters consist of a collection of conserved short sequence elements located at relatively constant distances from the transcription start site.

Alternative orientations for GC and CAAT box elements are indicated by chevron orientation: > means normal orientation; < means reverse orientation. The glucocorticoid receptor gene is unusual in possessing 13 upstream GC boxes (10 in the normal orientation; three in the reverse orientation). The tRNA genes are transcribed by RNA polymerase III and have an internal bipartite promoter comprising element A (usually within the nucleotides numbered +8 to +19 according to the standard tRNA nucleotide numbering system) and element B (usually between nucleotides +52 and +62). Specific transcription factors bind to these elements and then guide RNA polymerase III to start transcribing at +1.

- **the CAAT box,** often located at position −80. It is usually the strongest determinant of promoter efficiency. Like the GC box, it appears to be able to function in either orientation (*Figure 1.13*).

In addition to the above, more specific recognition elements are known to be recognized by tissue-restricted transcription factors (see *Table 10.3*).

▶ **enhancers** comprise groups of *cis*-acting short sequence elements, which can enhance the transcriptional activity of specific eukaryotic genes. However, unlike promoter elements whose positions relative to the transcriptional initiation site are relatively constant (*Figure 1.13*), enhancers are located a variable, and often considerable, distance from the transcriptional start site, and their function is independent of their orientation. They appear to bind gene regulatory proteins and, subsequently, the DNA between the promoter and enhancer loops out, allowing the proteins bound to the enhancer to interact with the transcription factors bound to the promoter, or with the RNA polymerase.

▶ **silencers** are regulatory elements which have similar properties to enhancers but *inhibit* the transcriptional activity of specific genes.

1.3.5 Tissue-specific gene expression involves selective activation of specific genes

The DNA content of a specific type of eukaryotic cell, a myocyte for example, is virtually identical to that of a lymphocyte, liver cell or any other type of nucleated cell from the same organism. What makes the different cell types different is that only a *proportion* of the genes in any one cell are significantly expressed and *the catalog of expressed genes varies between different cell types*. In some cells, particularly brain cells, a large number of different genes are expressed; in many other cell types, a large fraction of the genes is transcriptionally inactive.

Clearly, the genes that are expressed are the ones which define the functions of the cell. Some of the functions are general cell functions and are specified by so-called **housekeeping genes** (e.g. genes encoding histones, ribosomal proteins etc.). Other functions may be largely restricted to particular tissues or cell types (**tissue-specific gene expression**). Note, however, that even in the case of genes which show considerable tissue specificity in expression, some gene transcripts may be represented at very low levels in all cell types.

The distinction between transcriptionally active and inactive regions of DNA in a cell is reflected in the structure of the associated chromatin:

▶ *transcriptionally inactive chromatin* generally adopts a highly condensed conformation and is often associated with regions of the genome which undergo late replication during S phase of the cell cycle. It is associated with tight binding by the histone H1 molecule;

▶ *transcriptionally active chromatin* adopts a more open conformation and is often replicated early in S phase. It is marked by relatively weak binding by histone H1 mole-

cules and extensive acetylation of the four types of nucleosomal histones (i.e. histones H2A, H2B, H3 and H4; see Section 10.2.1).

Additionally, the promoter regions of vertebrate genes are generally characterized by absence of methylated cytosines (see below). Transcription factors can displace nucleosomes, and so the open conformation of transcriptionally active chromatin can be distinguished experimentally because it also affords access to nucleases: at very low concentrations, the enzyme deoxyribonuclease I (DNaseI) will digest long regions of nucleosome-free DNA. Although the regulatory regions may contain several sequence-specific binding proteins, the open chromatin structure is marked by the presence of *DNase I-hypersensitive sites* (see Section 10.5.2).

1.4 RNA processing

The RNA transcript of most eukaryotic genes undergoes a series of processing reactions. Often this involves removal of unwanted internal segments and rejoining of the remaining segments (*RNA splicing*). Additionally, in the case of RNA polymerase II transcripts, a specialized nucleotide linkage (*7-methylguanosine triphosphate*) is added to the 5′ end of the primary transcript (*capping*), and adenylate (AMP) residues are sequentially added to the 3′ end of mRNA to form a poly(A) tail (*polyadenylation*).

1.4.1 RNA splicing removes nonessential RNA sequences from the primary transcript

The linear sequence of a transcription unit dictates the synthesis of a correspondingly linear expression product, either a polypeptide or a mature noncoding RNA. However, for most vertebrate genes, only a small portion of the sequence of the gene is interpreted to give the final product. Instead, in the great majority of polypeptide-encoding genes and many of the genes specifying noncoding RNA, the genetic information comes in segments (**exons**) separated by intervening sequences which do not contribute genetic information that will be used to synthesize the final product (**introns**).

The initial transcription event involves the production of an RNA sequence complementary to the entire length of the gene, the so-called *primary transcript*. In the case of genes containing multiple exons, the primary transcript contains sequences complementary to both the exons and introns within the gene. Thereafter, however, the RNA transcript undergoes **RNA splicing**, a series of processing reactions whereby the intronic RNA segments are snipped out and discarded and the exonic RNA segments are joined end-to-end (spliced) to give a shorter RNA product (*Figure 1.14*). RNA splicing requires the nucleotide sequences at the exon/intron boundaries (**splice junctions**) to be recognized. In the vast majority of cases introns start with GT (*or GU at the RNA level*) and end with AG (the **GT–AG rule**; see *Figure 1.15*). Although the conserved GT (GU) and AG dinucleotides are crucially important for splicing, they are not by themselves sufficient to signal the presence of an intron. Comparisons of documented sequences have revealed that sequences immediately adjacent to the GT and AG di-

reaction is established by RNA–RNA base pairing between

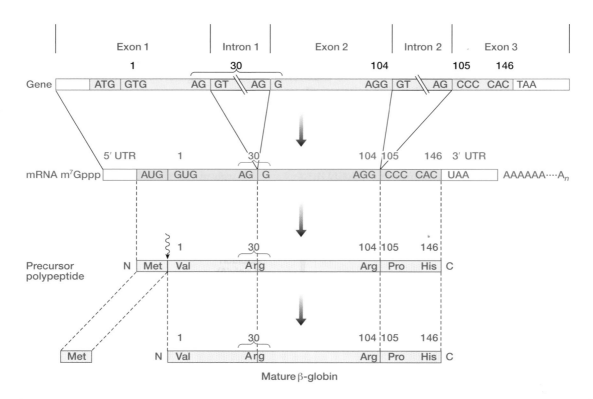

Figure 1.19: Expression of the human β-globin gene.

Exons 1 and 3 each contain noncoding sequences (shaded bars) at their extremities, which are transcribed and are present at the 5′ and 3′ ends of the β-globin mRNA, but are not translated to specify polypeptide synthesis. Such 5′ and 3′ untranslated regions (**5′ UTR** and **3′ UTR**), however, are thought to be important in ensuring high efficiency of translation (see text). The stop codon UAA represents the first three nucleotides of the 3′ untranslated region. **Note** that the initial translation product has 147 amino acids, but that the N-terminal methionine is removed by post-translational processing to generate the mature β-globin polypeptide. The first two bases of the codon specifying Arg30 are encoded by exon 1 and the third base is encoded by exon 2 (i.e. intron 1 separates the second and third bases of the codon, an example of a **phase 2 intron**; see *Box 12.2*. The second intron separates codons 104 and 105 and is an example of a **phase 0 intron**; see *Box 12.2*, and also *Figure 1.23* for an example of a **phase I intron**.

rRNA molecule: 28S rRNA, 5.8S rRNA and 5S rRNA, and about 50 ribosomal proteins. The 40S subunit contains a single 18S rRNA and over 30 ribosomal proteins. Ribosomes provide a structural framework for polypeptide synthesis *in which the RNA components are predominantly responsible for the catalytic function of the ribosome*; the protein components are thought to enhance the function of the rRNA molecules, and a surprising number of them do not appear to be essential for ribosome function.

The assembly of a new polypeptide from its constituent amino acids is governed by a **triplet genetic code**. Successive groups of three nucleotides (**codons**) in the linear mRNA sequence are decoded sequentially in order to specify individual amino acids. The decoding process is mediated by a collection of tRNA molecules, to each of which a specific amino acid has been covalently bound (at the free 3′ hydroxyl group of the tRNA; see *Figure 1.20*) by a specific amino acyl tRNA synthetase.

Different tRNA molecules bind different amino acids. Each tRNA has a specific trinucleotide sequence, called the **anti-**

codon, at a crucially important site located in the center of one arm of the tRNA (*Figure 1.7B*). This site provides the necessary specificity to interpret the genetic code: for an amino acid to be inserted in the growing polypeptide chain, the relevant codon of the mRNA molecule must be recognized via base pairing with a suitably complementary anticodon of the appropriate tRNA molecule (*Figure 1.20*).

One model of translation envisages that the 40S ribosomal subunit initially recognizes the 5′ cap via the participation of proteins that specifically bind to the cap. It then scans along the mRNA until it encounters the initiation codon, which is almost always AUG, specifying methionine (a few cases are known where ACG, CUG or GUG are used instead). Usually, though not always, the first AUG encountered will be the initiation codon. However, the AUG is recognized efficiently as an initiation codon only when it is embedded in a suitable *initiation codon recognition sequence*, the optimal being the sequence: **GCCPuCCAUGG**. The most important determinants in this sequence are the G following the AUG codon, and the purine (Pu), preferably A, preceding it by three

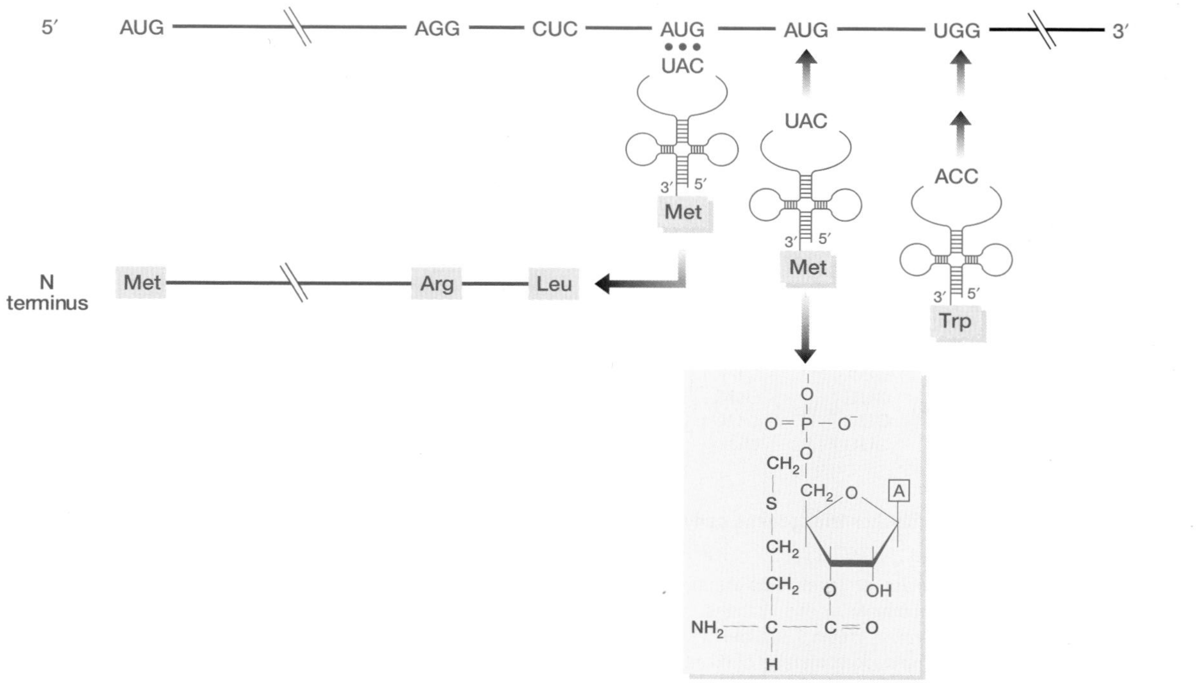

Figure 1.20: The genetic code is deciphered by codon–anticodon recognition.

The sequence of nucleotides in the mRNA sequence is interpreted from a translational start point (normally marked by the sequence AUG) and continues in the 5′→3′ direction until a stop codon is reached in that reading frame. Each codon in the mRNA is recognized by the complementary anticodon sequence of a tRNA molecule to which a specific amino acid is covalently bonded to the adenosine at the 3′ end (see insert).

nucleotides (Kozak, 1996). Subsequently, successive amino acids are incorporated into the growing polypeptide chain by a condensation reaction: the amino group of the incoming amino acid reacts with the carboxyl group of the last amino acid to be incorporated, resulting in a peptide bond between successive residues (*Figure 1.21*). This is catalyzed by a peptidyl transferase activity which resides in the RNA component of the large ribosomal subunit.

1.5.2 The genetic code is degenerate and not quite a universal code

The genetic code is a three-letter code. There are four possible bases to choose from at each of the three base positions in a codon. There are, therefore, $(4)^3 = 64$ possible codons, but there are only 20 major types of amino acid. As a result, the genetic code is a *degenerate code*: each amino acid is specified on average by about three different codons, although some amino acids, for example leucine and serine are specified by as many as six codons while others are much more poorly represented (*Figure 1.22*).

Although there are 64 codons, the corresponding number of tRNA molecules with different anticodons is less: just over 30 types of cytoplasmic tRNA and only 22 types of mitochondrial tRNA. The interpretation of all 64 codons on both cytoplasmic and mitochondrial ribosomes is possible because

the normal base pairing rules are relaxed when it comes to codon–anticodon recognition. The **wobble hypothesis** states that pairing of codon and anticodon follows the normal A–U and G–C rules for the first two base positions in a codon, but that exceptional 'wobbles' occur at the third position *and that G–U base pairs can also be used* (see *Table 1.6*).

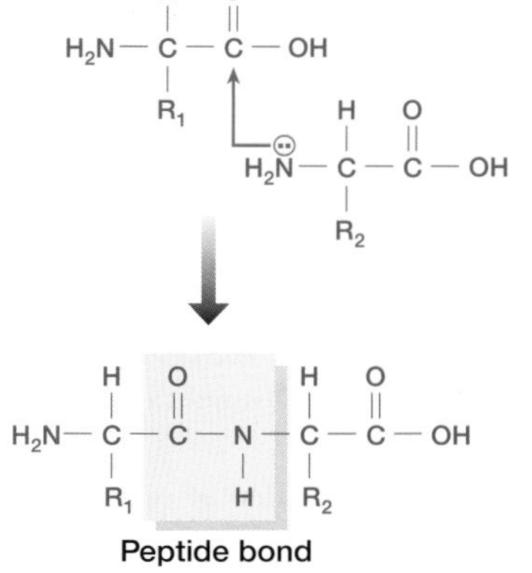

Figure 1.21: Polypeptides are synthesized by peptide bond formation between successive amino acids.

AAA, AAG } Lys
AAC, AAU } Asn
ACA, ACG, ACC, ACU } Thr
AGA } Arg (N)
AGG } STOP (M)
AGC, AGU } Ser
AUA { Ile (N) / Met (M)
AUG } Met
AUC, AUU } Ile

CAA, CAG } Gln
CAC, CAU } His
CCA, CCG, CCC, CCU } Pro
CGA, CGG, CGC, CGU } Arg
CUA, CUG, CUC, CUU } Leu

GAA, GAG } Glu
GAC, GAU } Asp
GCA, GCG, GCC, GCU } Ala
GGA, GGG, GGC, GGU } Gly
GUA, GUG, GUC, GUU } Val

UAA, UAG } STOP
UAC, UAU } Tyr
UCA, UCG, UCC, UCU } Ser
UGA { STOP (N) / Trp (M)
UGG } Trp
UGC, UGU } Cys
UUA, UUG } Leu
UUC, UUU } Phe

Figure 1.22: The nuclear and mitochondrial genetic codes are similar but not identical.

The four codons (shown in red) are interpreted differently in the nucleus and mitochondria of mammalian cells, with the mitochondrial interpretation given in green. Thus, the mitochondrial code has four stop codons instead of three (UAA, UAG, AGA, AGG), two Trp codons instead of one (UGA, UGG), four Arg codons instead of six (CGA, CGC, CGG, CGU), two Met codons instead of one (AUA, AUG) and two Ile codons instead of three (AUC, AUU). *Notes:* (1) the degeneracy of the genetic code most often involves the third base of the codon. Sometimes any base may be substituted (GGN = glycine, CCN = proline, etc., where N is any base). In other cases, any purine (Pu) or any pyrimidine (Py) will do (AAPu = lysine, AAPy = asparagine etc.). (2) signals in some mRNAs can lead to alternative interpretations of stop codons, such that UGA can specify a 21st amino acid, *selenocysteine*, and UAG can specify glutamine or a 22nd amino acid, *pyrrolysine* (Atkins and Gesteland, 2002).

Table 1.6: Codon–anticodon pairing admits relaxed base-pairing (wobbles) at the third base position of codons

Base at 5′ end of tRNA anticodon	Base recognized at 3′ end of mRNA codon
A	U only
C	G only
G (or I)*	C or U
U	A or G

* I = Inosine, a post-translationally modified form of adenine. See *Box 9.4* for more details of inosine and codon–anticodon recognition of cytoplasmic tRNA.

It is usual to describe the genetic code as a *universal code*, meaning that the same code is used throughout all life forms. This is not strictly true because of alternative interpretation of codons due to:

▶ *evolutionary divergence of organelle genetic codes.* Mitochondria and chloroplasts also have protein synthesis capacities, albeit limited. During evolution, these organelles have adopted slightly different genetic codes to that used in the case of nuclear genes. Thus, for example, translation of nuclear-encoded mRNA continues until a **termination codon** is encountered, which is one of three possibilities – *UAA, UAG, UGA* while in mammalian mitochondria there are four possibilities – UAA, UAG, AGA and AGG; see *Figure 1.22;*

▶ *context-dependent redefinition of codons.* Signals in some mRNAs, including a few types of nuclear-encoded mRNA, lead to redefinition of some codons. For example, in a wide variety of cells (including human cells), some nuclear-encoded mRNAs can alternatively interpret UGA as a 21st amino acid, *selenocysteine*, and UAG can be alter-

natively interpreted as glutamine (see Atkins and Gesteland, 2002).

The backbone of the primary translation product will therefore have at one end a methionine with a free amino group (the **N-terminal end**) and at the other end an amino acid with a free carboxyl group (the **C-terminal end**). Note that although codons are translated in a specific **translational reading frame**, overlapping genes are occasionally found in eukaryotes, in which different translational reading frames are used (see *Figure 9.3* for an example).

The predominant step in the control of translation is ribosome binding. In addition to the 5′ cap, the 5′ UTR (often < 200 bp) and 3′ UTR (usually very much longer than the 5′ UTR) both play critical roles in mRNA recruitment for translation. Several *cis*-acting elements that are involved in this process have been characterized and, in addition, a few *trans*-acting factors which bind to these elements have been identified. It is possible that the 5′ and 3′ UTR sequences interact to enhance translation. The 3′ UTR has a key role in translational regulation and signals for controlling translation, mRNA stability and localization have all been found in this region (see Wickens *et al.*, 1997; see also Section 10.2.6).

1.5.3 Post-translational modifications include chemical modifications of some amino acids and polypeptide cleavage

Primary translation products often undergo a variety of modification reactions, involving the addition of chemical groups which are attached covalently to the polypeptide chain at the translational and post-translational levels. This can involve simple chemical modification (hydroxylation, phosphorylation, etc.) of the side chains of single amino acids or the addition of different types of carbohydrate or lipid groups (see *Table 1.7*).

Protein modification by addition of carbohydrate groups

Glycoproteins contain oligosaccharides which are covalently attached to the side chains of certain amino acids. Few proteins in the cytosol are *glycosylated*, that is have attached carbohydrate, and those that are carry a single sugar residue, *N*-acetylglucosamine, covalently linked to a serine or threonine residue. By contrast, those proteins that are secreted from cells or exported to lysosomes, the Golgi apparatus or the plasma membrane are glycosylated. Oligosaccharide components of glycoproteins are largely preformed and added *en bloc* to polypeptides. Two major types of glycosylation are recognized:

▶ **N-glycosylation:** involves, in most cases, initial transfer of a common oligosaccharide sequence to the side chain NH_2 group of an asparagine residue within the endoplasmic reticulum (ER; see *Table 1.7*). Subsequent trimming of residues and replacement with different monosaccharides occurs in the Golgi apparatus;

▶ **O-glycosylation:** see *Table 1.7*.

Proteoglycans are proteins with attached *glycosaminoglycans* which usually contain disaccharide repeating units containing glucosamine or galactosamine. The most well-characterized proteoglycans are components of the extracellular matrix.

Protein modification by addition of lipid groups

Some proteins, notably membrane proteins, are modified by the addition of fatty acyl or prenyl groups which typically serve as membrane anchors. Examples of fatty acyl groups include the **myristoyl** group which is a C_{14} lipid that is found attached to a glycine residue at the extreme N terminus, and enables the modified protein to interact with a membrane receptor or the lipid bilayer of the membrane. Another fatty acyl group which serves as a membrane anchor is the C_{16} **palmitoyl** group, which becomes attached to the S atom of cysteine residues.

Prenyl groups typically become attached to cysteine residues close to the C terminus, and include **farnesyl** (C_{15}) groups and **geranylgeranyl** (C_{20}) groups. Many proteins that participate in signal transduction and protein targeting contain either a farnesyl or a geranylgeranyl unit at their C terminus.

Anchoring of a protein to the outer layer of the plasma membrane uses a different mechanism: the attachment of a **glycosylphosphatidyl inositol (GPI)** group. This complex glycolipid group contains a fatty acyl group that serves as the membrane anchor which is linked successively to a glycerophosphate unit, an oligosaccharide unit and finally through a phosphoethanolamine unit to the C terminus of the protein. The entire protein except for the GPI anchor is located in the extracellular space.

Post-translational cleavage

The primary translation product may also undergo internal cleavage to generate a smaller mature product. Occasionally the initiating methionine is cleaved from the primary translation product, as during the synthesis of β-globin (*Figure 1.19*). More substantial polypeptide cleavage is observed in the case of the maturation of many proteins, including plasma proteins, polypeptide hormones, neuropeptides, growth factors, etc. As described in the next section, cleavable signal sequences are often used to tag proteins that are destined to be

---

Table 1.7: Major types of modification of polypeptides

Type of modification (group added)	Target amino acids	Comments
Phosphorylation (PO_4^-)	Tyrosine, serine, threonine	Achieved by specific kinases. May be reversed by phosphatases
Methylation (CH_3)	Lysine	Achieved by methylases and undone by demethylases
Hydroxylation (OH)	Proline, lysine, aspartic acid	Hydroxyproline and hydroxylysine are particularly common in collagens
Acetylation (CH_3CO)	Lysine	Achieved by an acetylase and undone by deacetylase
Carboxylation (COOH)	Glutamate	Achieved by γ-carboxylase
Acetylation (CH_3CO)	Lysine	Achieved by an acetylase and undone by deacetylase
N-glycosylation (complex carbohydrate)	Asparagine, usually in the sequence: *Asn-X-(Ser/Thr)*	Takes place initially in the endoplasmic reticulum; X is any amino acid other than proline
O-glycosylation (complex carbohydrate)	Serine, threonine, hydroxylysine	Takes place in the Golgi apparatus; less common than *N*-glycosylation
GPI (glycolipid)	Aspartate at C terminus	Serves to anchor protein to *outer* layer of plasma membrane
Myristoylation (C_{14} fatty acyl group)	Glycine at N terminus (see text)	Serves as membrane anchor
Palmitoylation (C_{16} fatty acyl group)	Cysteine to form S-palmitoyl link	Serves as membrane anchor
Farnesylation (C_{15} prenyl group)	Cysteine at C terminus (see text)	Serves as membrane anchor
Geranylgeranylation (C_{20} prenyl group)	Cysteine at C terminus (see text)	Serves as membrane anchor

sent to a specific location. Additionally, in some cases, a single mRNA molecule may specify more than one functional polypeptide chain as a result of proteolytic cleavage of a large precursor polypeptide (*Figure 1.23*).

1.5.4 Protein secretion and intracellular export is controlled by specific localization signals or by chemical modifications

Proteins synthesized on mitochondrial ribosomes are required to function within the mitochondria. However, the numerous proteins that are synthesized on cytoplasmic ribosomes have diverse functions which may require them to be secreted from the cell where they were synthesized (as with hormones and other intercellular signaling molecules) or to be exported to specific intracellular compartments, such as the nucleus (histones, DNA and RNA polymerases, transcription factors, RNA-processing proteins, etc.), the mitochondrion (mitochondrial ribosomal proteins, many respiratory chain components, etc.), peroxisomes, and so on. To do this, a specific **localization signal** needs to be embedded in the structure of the polypeptide so that it can be sent to the correct address. Usually, the localization signal takes the form of a short peptide sequence. Often, but not always, this constitutes a so-called **signal sequence** (or **leader sequence**) which is removed from the protein by a specialized **signal peptidase** once the sorting process has been achieved.

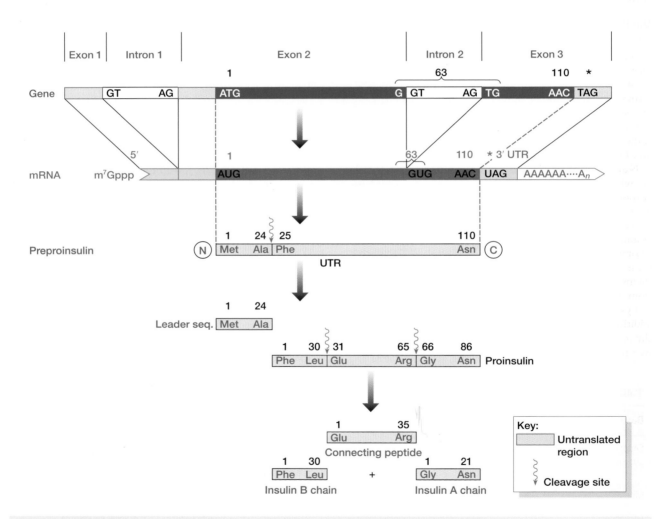

Figure 1.23: Insulin synthesis involves multiple post-translational cleavages of polypeptide precursors.

The first intron interrupts the 5′ untranslated region; the second intron interrupts base positions 1 and 2 of codon 63 and is classified as a phase I intron (see Box 12.2, and also *Figure 1.19* for other intron phases). The primary translation product, preproinsulin, has a **leader sequence** of 24 amino acids which is required for the protein to cross the cell membrane, and is discarded thereafter. The proinsulin precursor contains a central segment, the **connecting peptide**, which is thought to be important in maintaining the conformation of the A and B chain segments so that they can form disulfide bridges (see *Figure 1.25*).

Signals for export to the ER and extracellular space

In the case of secreted proteins the signal peptide comprises the first 20 or so amino acids at the N-terminal end and always includes a substantial number of hydrophobic amino acids (see *Table 1.8*). The signal sequence is guided to the ER by a **signal recognition particle (SRP)**, an RNA–protein complex consisting of a small cytoplasmic RNA species, 7SL RNA, and six specific proteins. The SRP complex binds both the growing polypeptide chain and the ribosome and directs them to an SRP receptor protein on the cytosolic side surface of the rough endoplasmic reticulum (RER) membrane. Thereafter a polypeptide can pass into the lumen of the ER, destined for export from the cell, unless there are additional hydrophobic segments which stop the transfer process, as in the case of **transmembrane proteins**.

Other signals

Like the ER signal, an N-terminal signal sequence is required to traverse the mitochondrial membranes, and is subsequently cleaved. Typically, a mitochondrial signal peptide has, in addition to many hydrophobic amino acids, several positively charged amino acids, usually spaced at intervals of about four amino acids. This structure is thought to form an *amphipathic α-helix*, a helical structure with charged amino acids on one surface and hydrophobic amino acids on another (see below and *Figure 1.24A*).

Nuclear localization signals can be located just about anywhere within the polypeptide sequence and typically consist of a stretch of four to eight positively charged amino acids, together with neighboring proline residues. Often, however, the signal is bipartite, and the positively charged amino acids are found in two blocks of two to four residues, separated by about 10 amino acids (see *Table 1.8*). Note that some nuclear proteins lack any nuclear localization sequences themselves but are transported into the nucleus with assistance from other nuclear proteins which do have appropriate signals.

Lysosomal proteins are targeted to the lysosome by the addition of a mannose 6-phosphate residue which is added in the *cis*-compartment of the Golgi apparatus and is recognized by a receptor protein in the *trans*-compartment of the Golgi.

1.5.5 Protein structure is highly varied and not easily predicted from the amino acid sequence

Proteins are composed of one or more polypeptides, each of which can be subject to post-translational modification. They can interact with specific **co-factors** (for example, divalent cations such as Ca^{2+}, Fe^{2+}, Cu^{2+}, Zn^{2+} or small molecules which are required for functional enzyme activity, e.g. NAD^+) or **ligands** (any molecule which a protein specifically binds), each of which can be powerful influences on the conformation of the protein. At least four different levels of structural organization have been distinguished for proteins (see *Table 1.9*).

Within a single polypeptide chain, there is ample scope for hydrogen bonding between different residues; irrespective of the side chains, the oxygen of a peptide bond's carbonyl (CO) group can hydrogen bond to the hydrogen of the NH group of another peptide bond. Fundamental structural units defined by hydrogen bonding between closely neighboring amino acid residues of a single polypeptide include:

▶ **the α-helix**. This involves formation of a rigid cylinder. The structure is dominated by hydrogen bonding between the carbonyl oxygen of a peptide bond with the hydrogen atom of the amino nitrogen of a peptide bond located four amino acids away (see *Figure 1.24*). Note that the DNA-binding domains of transcription factors are usually α-helical (see Section 10.2.4). An **amphipathic α-helix** has charged residues on one surface and hydrophobic ones on another (*Figure 1.24*). Identical α-helices with a repeating arrangement of nonpolar side chains can coil round each other to form a particular stable structure called a **coiled coil**. Long rod-like coiled coils are found in many fibrous proteins, such as the α-keratin fibers of skin, hair and nails or fibrinogen of the blood clot;

▶ **the β-pleated sheet**. This features hydrogen bond formation between opposed peptide bonds in parallel (often really antiparallel) segments of the same polypeptide chain (see *Figure 1.24*). β-pleated sheets form the core of most, but not all globular proteins;

Table 1.8: Examples of protein localization sequences

Destination of protein	Location and form of signal	Examples
Endoplasmic reticulum and secretion from cell	N-terminal peptide of 20 or so amino acids; very hydrophobic	Human insulin – 24 amino acid, highly hydrophobic signal peptide: N-Met-Ala-Leu-Trp-Met-Arg-Leu-Leu-Pro-Leu-Leu-Ala-Leu-Leu-Ala-Leu-Trp-Gly-Pro-Asp-Pro-Ala-Ala-Ala
Mitochondria	N-terminal peptide; α-helix with positively charged residues on one face and hydrophobic ones on the other	Human mitochondrial aldehyde dehydrogenase N-terminal 17 amino acids: N-Met-Leu-**Arg**-Ala-Ala-Ala-**Arg**-Phe-Gly-Pro-**Arg** -Leu-Gly-**Arg**-**Arg**-Leu-Leu
Nucleus	Internal sequence of amino acids; often a string of basic amino acids plus prolines; may be bipartite	SV40 T antigen – continuous: *Pro-Pro*-**Lys-Lys-Lys-Arg-Lys**-Val p53 – bipartite: **Lys-Arg**-Ala-Leu-*Pro*-Asn-Asn-Thr-Ser-Ser-Ser-*Pro*-Gln-*Pro*-**Lys-Lys-Lys**
Lysosome	Addition of mannose 6-phosphate residues	

Table 1.9: Levels of protein structure

Level	Definition	Comment
Primary	The linear sequence of amino acids in a polypeptide	Can vary enormously in length from a small peptide to thousands of amino acids long
Secondary	The path that a polypeptide backbone follows	May vary locally e.g. as α-helix or β-pleated sheet
Tertiary	The overall three dimensional structure of a polypeptide	Can vary enormously e.g. globular, rod-like, tube, coil, sheet etc.
Quaternary	The overall structure of a multimeric protein (= a combination of protein subunits)	Often stabilized by disulfide bridges and by binding to ligands etc.

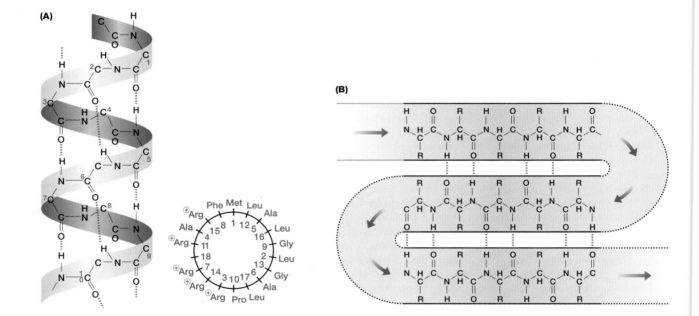

Figure 1.24: Regions of secondary structure in polypeptides are often dominated by intrachain hydrogen bonding.

(A) *Structure of an α-helix. Left:* only the backbone of the polypeptide is shown for clarity. The carbonyl (CO) oxygen of each peptide bond is hydrogen bonded to the hydrogen on the peptide bond amide group (NH) of the fourth amino acid away, so that the helix has 3.6 amino acids per turn. *Note:* for clarification purposes some bonds have been omitted. The side chains of each amino acid are located on the outside of the helix and there is almost no free space within the helix. *Right:* charged amino acids and hydrophobic amino acids are located on different surfaces in an amphipathic α-helix. The sequence shown is that for the 17-amino acid long signal peptide sequence for the mitochondrial aldehyde dehydrogenase (see *Table 1.8*). (B) *Structure of a β-pleated sheet. Note:* that hydrogen bonding occurs between the CO oxygen and NH hydrogen atoms of peptide bonds on adjacent parallel segments of the polypeptide backbone. The example shows a case of bonding between antiparallel segments of the polypeptide backbone (**antiparallel β-sheet**) and enforced abrupt change of direction between antiparallel segments is often accomplished using β-turns (see text). Arrows mark the direction from N terminus to C terminus. *Note:* parallel β-pleated sheets with the adjacent segments running in the same direction are also commonly found.

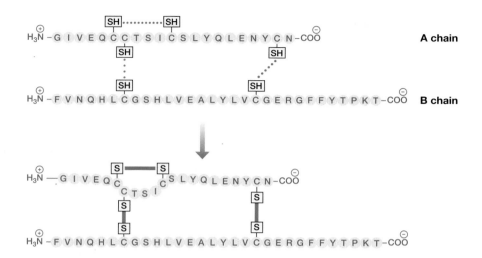

Figure 1.25: Intrachain and interchain disulfide bridges in human insulin.

The disulfide bridges (–S–S–) are formed by a condensation reaction between opposed sulfhydryl (–SH) groups of cysteine residues 6 and 11 of the A chain or between indicated residues of the different chains.

▶ **the β-turn.** Hydrogen bonding between the peptide bond CO group of amino acid residue n of a polypeptide with the peptide bond NH group of residue $n+3$ results in a hairpin turn. By permitting the polypeptide to reverse direction abruptly, compact globular shapes can be achieved. β-Turns are so named because they also often connect antiparallel strands in β-pleated sheets (see *Figure 1.24B*).

More complex structural motifs consisting of combinations of the above structural modules constitute **protein domains**; compact regions of a protein formed by folding back of the primary structure so that elements of secondary structure can be stacked next to each other. Such domains often represent functional units involved in binding other molecules. In addition, covalent **disulfide bridges** are often formed between the sulfhydryl (–SH) groups of pairs of cysteine residues occurring within the same polypeptide chain or on different polypeptide chains (see *Figure 1.25*).

Clearly, the tertiary or quaternary structure of proteins is determined by the primary amino acid sequence. However although secondary structure motifs such as α-helices, β-pleated sheets and β-turns can be predicted by analyzing the primary sequence, the overall three-dimensional structure cannot, at present, be accurately predicted. In addition to the structural complexity of simple polypeptides, many proteins are organized as complex aggregates of multiple polypeptide subunits.

Further reading

Alberts B, Johnson A, Lewis J, Raff M, Roberts K, Walter P (2001) *Molecular Biology of the Cell*, 4th Edn. Garland Publishing, New York.

Bell SP, Dutta A (2002) DNA replication in eukaryotic cells. *Annu. Rev. Biochem.* **71**, 307–331.

Big Picture Book of Viruses at: http://www.virology.net/Big_Virology /BVHomePage.html

Braden C, Tooze J (1999) *Introduction to Protein Structure*, 2nd Edn. Garland Science, New York.

Brow DA (2002) Allosteric cascade of spliceosome activation. *Annu. Rev. Genet.* **36**, 333–360.

Carradine CR, Drew HR (1997) *Understanding DNA. The molecule and how it works.* Academic Press, London.

Lodish H, Baltimore D, Berk A, Zipursky L, Matsudaira P, Darnell J (1995) *Molecular Cell Biology*, 3rd Edn. Scientific American Books, New York.

References

Atkins JF, Gesteland R (2002) The 22nd amino acid. *Science* **296**, 1409–1410.

Berget SM (1995) Exon recognition in vertebrate splicing. *J. Biol. Chem.* **270**, 2411–2414.

Fairbrother WG, Yeh RF, Sharp PA, Burge CB (2002) Predictive identification of exonic splicing enhancers in human genes. *Science* **297**, 1007–1113.

Friedberg EC, Feaver WJ, Gerlach VL (2000) The many faces of DNA polymerases: strategies for mutagenesis and for mutational avoidance. *Proc. Natl Acad. Sci. USA* **97**, 5681–5883.

Friedberg EC, Wagner R, Radman M (2002) Specialised DNA polymerases, cellular survival, and the genesis of mutations. *Science* **296**, 1627–1630.

Hastings ML, Krainer AR (2001) Pre-mRNA splicing in the new millennium. *Curr. Opin. Genet. Dev.* **13**, 302–309.

Kozak M (1996) Interpreting cDNA sequences: some insights from studies on translation. *Mamm. Genome* **7**, 563–574.

Lim LP, Burge CP (2001) A computational analysis of sequence features involved in recognition of short introns. *Proc. Natl Acad. Sci. USA* **98**, 11193–11198.

Staley JP, Guthrie C (1998) Mechanical devices of the spliceosome: motors, clocks, springs and things. *Cell* **92**, 315–326.

Tarn WY, Steitz JA (1997) Pre-mRNA splicing: the discovery of a new spliceosome doubles the challenge. *Trends Biochem. Sci.* **22**, 132–137.

Wickens M, Anderson P, Jackson RJ (1997) Life and death in the cytoplasm: messages from the 3′ end. *Curr. Biol.* **7**, 220–232.

CHAPTER TWO

Chromosome structure and function

Chapter contents

DNA functions in a context. Human DNA is structured into **chromosomes** and chromosomes function within cells which are derived by cell division from other cells. The process of cell division is, of course, a small component of the **cell cycle**, the process in which chromosomes and their constituent DNA molecules need to make perfect copies of themselves and then segregate into daughter cells. This needs to be very carefully orchestrated so that the daughter cells receive the correct set of chomosomes. In the usual form of cell division, **mitosis**, each daughter cell has the same number and types of chromosome as the parent cell. In addition, a specialized form of cell division, **meiosis**, occurs in certain cells of the testis and ovary and gives rise respectively to sperm and egg cells.

2.1 Ploidy and the cell cycle

The number of different chromosomes in any nucleated cell, the **chromosome set**, and the associated **DNA content** are designated **n** and **C** respectively. For humans n = 23 and C = approximately 3.5 pg (3.5×10^{-12} g). Cells may differ in the number of copies they have of the chromosome set (the **ploidy**). Sperm and egg cells carry a single chromosome set and are said to be **haploid** (n chromosomes; DNA content, C). Most human cells, however, carry two copies of the chromosome set and are **diploid** (2n chromosomes; DNA content = 2C). Almost all mammals are diploid (the red viscacha rat is an unusual exception – see Gallardo *et al.*, 1999), but among other organisms there are many examples of species that are normally haploid, **tetraploid** (4n) or **polyploid** (> 4n). **Triploidy** (3n) is less common because triploids have problems with meiosis (see below).

The diploid cells of our body are all derived ultimately from a single diploid cell, the **zygote**, by repeated rounds of mitotic cell division. Each round can be summarized as one turn of the **cell cycle** (*Figure 2.1*). This comprises a short stage of cell division, the **M phase** (mitosis; see *Figure 2.7*) and a long intervening **interphase**. Interphase can be divided into **S phase** (DNA synthesis), **G1 phase** (gap between M phase and S phase) and **G2 phase** (gap between S phase and M phase). From anaphase of mitosis right through until DNA duplication in S phase, a chromosome of a diploid cell contains a single DNA double helix and the total DNA content is 2C. G1 is the normal state of a cell, and the long-term end-state of non-dividing cells. Cells enter S phase only if they are committed to mitosis; nondividing cells remain in a modified G1 stage, sometimes called the **G0 phase**. The cell cycle diagram can give the impression that all the interesting action happens in S and M phases – but this is an illusion. A cell spends most of its life in G0 or G1 phase, and that is where the genome does most of its work.

A subset of the diploid body cells constitute the **germ-line** (see *Figure 2.9*). These give rise to specialized diploid cells in the ovary and testis that can divide by meiosis to produce haploid **gametes** (sperm and egg). In humans (n = 23) each gamete contains 22 **autosomes** (nonsex chromosomes) plus one sex chromosome. In eggs the sex chromosome is always an X; in sperm it may be an X or a Y. After fertilization the zygote is diploid (2n) with the chromosome

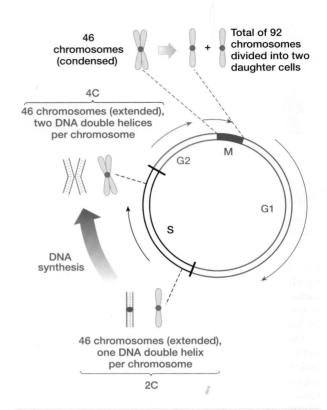

Figure 2.1: Human chromosomal DNA content during the cell cycle.

Interphase comprises G1 + S + G2. Chromosomes contain one DNA double helix from anaphase of mitosis right through until the DNA has duplicated in S phase. From this stage until the end of metaphase of mitosis, the chromosome consists of two chromatids each containing a DNA duplex, making two double helices per chromosome. The DNA content of a diploid cell before S phase is 2C (twice the DNA content of a haploid cell), while between S phase and mitosis it is 4C.

constitution 46,XX or 46,XY (*Figure 2.2*). The other cells of the body, apart from the germ-line, are known as **somatic cells**. Human somatic cells are usually diploid but some cells lack a nucleus and any chromosomes and are said to be **nulliploid**, and others are *naturally polyploid* as a result of multiple rounds of DNA replication without cell division (see Section 3.1.4).

2.2 Structure and function of chromosomes

Chromosomes as seen under the microscope and illustrated in textbooks are rather misleading because they represent an *unusual state* which occurs *briefly* in the cell cycle, during a part of the M phase as cells prepare to undergo cell division (metaphase). **Metaphase chromosomes** and **prometaphase chromosomes** are in a *highly condensed state* and have an *unusual two-chromatid structure* because at this stage the DNA

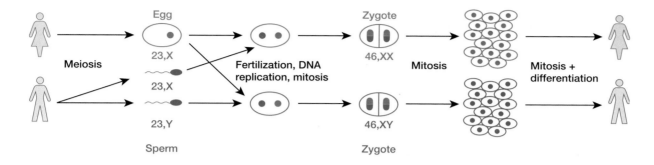

Figure 2.2: Human life, from a chromosomal viewpoint.

The haploid sperm and egg cells originate by meiosis from diploid precursors (see *Figure 2.9*). In the fertilized egg the sperm and egg chromosomes initially form separate male and female pronuclei. These combine during the first mitosis.

has replicated in readiness for cell division. Simply by being tightly packed into neat bundles, the chromosomes at metaphase are large enough to be seen with a light microscope but they are unrepresentative because the extreme packaging ensures that the genes are switched off.

The processes of cell division are fascinating in their own right, and errors in packaging or dividing up the genome have major medical consequences (Section 2.5). However, it is important to remember that for the vast majority of the cell cycle chromosomes have a quite different appearance. Throughout the long interphase stage, chromosomes are enormously extended and so much more diffuse than the metaphase chromosomes seen in *Figure 2.14*. Importantly, **an interphase chromosome** comprises only a *single chromatid and one DNA double helix*, and the extended structure allows genes to be expressed.

As functioning organelles, eukaryotic chromosomes seem to require only three classes of DNA sequence element: centromeres, telomeres and origins of replication. This simple requirement has been verified by the successful construction of artificial chromosomes in yeast: large foreign DNA fragments behave as autonomous chromosomes when joined to short sequences that specify a functional centromere, two telomeres and a replication origin (*Figure 5.17*). Recently **mammalian artificial chromosomes** have been constructed on similar principles (Huxley, 1997; Schindelhauer, 1999).

2.2.1 Packaging of DNA into chromosomes requires multiple hierarchies of DNA folding

In the cell the structure of each chromosome is highly ordered (Manuelidis, 1990). Even in the interphase nucleus the 2 nm DNA double helix is subject to at least two levels of coiling (*Figure 2.3*):

▶ the **nucleosome** is the most fundamental unit of packaging. It consists of a central core of eight **histone** proteins, small highly conserved basic (= positively charged) proteins of 102–135 amino acids. Each core comprises two molecules each of histones H2A, H2B, H3 and H4, around

which a stretch of 146 bp of double-stranded DNA is coiled in 1.75 turns. Adjacent nucleosomes are connected by a short length of spacer DNA. Electron micrographs of suitable preparations show a '*string of beads*' appearance. **Note:** sperm cells are different. Very tight packaging of DNA in the heads of sperm cells is achieved by replacing histones with an alternative class of small basic proteins known as ***protamines***;

▶ the string of beads, approximately 10 nm in diameter, is in turn coiled into a **chromatin fiber** of 30 nm diameter. The interphase chromosome seems to consist of these chromatin fibers, probably organized into long loops as described below.

During cell division, the chromosomes become ever more highly condensed. The DNA in a metaphase chromosome is compacted to about 1/10 000 of its stretched-out length. Loops of the 30 nm chromatin fiber, containing 20–100 kilobases (kb) of DNA per loop, are attached to a central **scaffold**. This consists of nonhistone acidic proteins, notably ***topoisomerase II***, an enzyme which has the interesting ability to pass one DNA double helix through another by cutting a gap and repairing it. Topoisomerase II and some other chromatin proteins are known to bind to AT-rich sequences, and the chromatin loops may be attached by stretches of several hundred base pairs of highly AT-rich (> 65%) DNA (**scaffold attachment regions**). In the chromatids of a metaphase chromosome the loop–scaffold complex is compacted yet further by coiling (see *Figure 2.3*).

2.2.2 Individual chromosomes occupy nonoverlapping territories in an interphase nucleus

We have long been aware of the high degree of organization in the cytoplasm, but the nucleus, too, has considerable substructure. In addition to the familiar ***nucleolus***, where rRNA is transcribed and ribosomal subunits are assembled, many other subnuclear compartments have recently been identified (*Cajal bodies, PML bodies, paraspeckles* etc.; see *Box 3.1*. The positioning of the chromosomes within the nucleus is also highly organized and specialized techniques have been

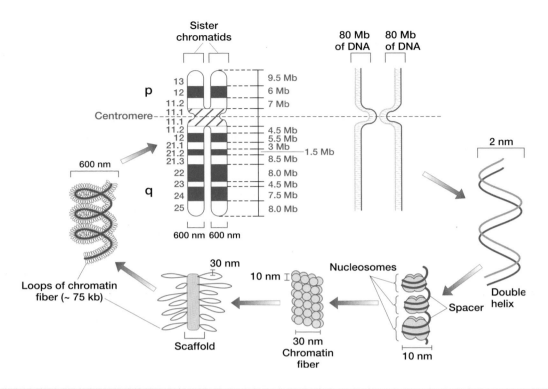

Figure 2.3: From DNA duplex to metaphase chromosome.

The figure shows human chromosome 17, as seen in a G-banded, 400-band preparation. The estimated packaging ratios (the degree of compaction of the linear DNA duplex) for human chromosomes are 1 : 6 for nucleosomes, 1 : 36 for the 30 nm fiber and > 1 : 10 000 for the metaphase chromosome. Presently, it is uncertain whether the DNA at the centromere of the metaphase chromosome has been delayed in its replication unlike the rest of the chromatid, or whether full DNA replication has occurred in the S phase and the constriction at the centromere is due to some other cause.

developed to follow the individual chromosome movements within living cells during interphase.

To some extent, mitosis places a restriction on chromosomal orientation during interphase. Just before the cell divides, microtubules attached to the centromere pull each chromosome towards one of the two poles of the *mitotic spindle* (the network of microtubules which position chromosomes during mitosis – see *Box 2.1*). During chromosome movement, the centromeres lead the way with the chromosome arms trailing behind, forming V-shapes. At the beginning of interphase the chromosomes tend to retain this so-called *Rabl orientation*, with the centromeres lined up facing one pole of the nucleus and the telomeres facing the opposite pole (in some cells where there is a short interphase, the chromosomes tend to adopt this orientation throughout interphase).

In mammals, the chromosomes of interphase cells are not usually found in the Rabl orientation, and the different centromeres are not so well aligned (although they do tend to cluster together at the periphery of the nucleus during the G1 phase before becoming much more dispersed during S phase). Nevertheless, although each chromosome is in a highly extended form, the chromosomes are not extensively entwined. Instead, they appear to occupy relatively small *non-overlapping* **chromosome territories** (*Figure 2.4* and Cremer and Cremer, 2001; Parada and Misteli 2002).

Although interphase chromosomes do not appear to have favored nuclear locations, chromosome positioning is nevertheless nonrandom. For example, the most gene-rich human chromosomes are known to concentrate at the center of the nucleus, whereas the more gene-poor chromosomes are located towards the nuclear envelope (*Figure 2.4*; see Boyle *et al.*, 2001 and Parada and Mistelli, 2002). Chromosome movements are probably restrained by interaction with the nuclear envelope and also by internal nuclear structures including the nucleolus (in the case of chromosomes containing ribosomal RNA genes).

2.2.3 Chromosomes as functioning organelles: the pivotal role of the centromere

Normal chromosomes have a single centromere that is seen under the microscope as the **primary constriction**, the region at which sister chromatids are joined. The centromere is essential for segregation during cell division. Chromosome fragments that lack a centromere (**acentric fragments**) do

Box 2.1: The mitotic spindle and its components.

The **mitotic spindle** (see *Figure* below) is formed from **microtubules** (polymers made from a repeating heterodimer of α-tubulin and β-tubulin) and microtubule-associated proteins. Each of the two spindle poles is defined by a **centrosome**, the major **microtubule organizing center**. The centrosome seeds the outward growth of polar microtubule fibers, with a **(–) end** at the centrosome end and the distal growing end defined as the **(+) end**.

Each centrosome is composed of a fibrous matrix (consisting of about 50 copies of a γ-tubulin ring complex in which is embedded a pair of **centrioles** (see *panel A* in figure below). Centrioles are short cylindrical structures composed of microtubules and associated proteins, and the two centrioles of the pair are always arranged at right angles to each other, forming an L shape (see *panel A* of the figure). At a certain point in the G1 stage the two centrioles in a pair separate and during S phase a daughter centriole begins to grow at the base of each mother centriole and at right angles to it until fully formed by G2. The two centriole pairs remain close together in a single centrosomal complex until the beginning of M phase when the complex splits in two and the two halves begin to separate. Each centrosome now develops its own radial array of microtubules (**aster**) and begin to migrate to opposite ends of the cell where they will form the **spindle poles** (see *panel C* in figure).

Three different forms of microtubule fiber are found in the spindle (see *panel C* in Figure below):

▶ **polar fibers** extend from the two poles of the spindle towards the equator. They develop at prophase while the nuclear membrane is still intact. ***Note:*** for the sake of clarity only a pair of such overlapping fibers is shown;

▶ **kinetochore fibers** do not develop until prometaphase. These fibers attach to the **kinetochore**, a large multiprotein structure attached to the centromere of each chromatid (*panel B*) and extend in the direction of the spindle poles;

▶ **astral fibers** form around each centrosome and extend to the periphery.

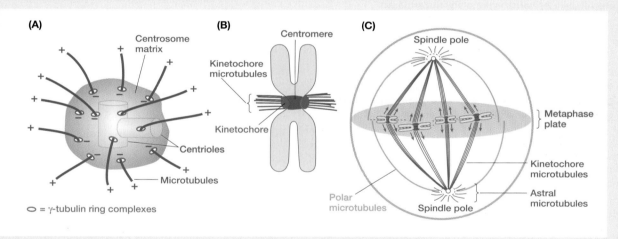

(A) *Centrosome structure;* (B) *kinetochore–centromere association;* (C) *mitotic spindle structure.*

not become attached to the ***mitotic spindle*** (see *Box 2.1* and *Figure 2.7*), and so fail to be included in the nuclei of either of the daughter cells.

During late prophase of mitosis, a pair of large multiprotein complexes known as ***kinetochores***, forms at each centromere, one attached to each sister chromatid. Microtubules attach to each kinetochore, linking the centromere of a chromosome and the two spindle poles (see *Box 2.1* and *Figure 2.7*). At anaphase, the kinetochore microtubules pull the two sister chromatids toward opposite poles of the spindle (*Figure 2.7*). Kinetochores play a central role in this process, by controlling assembly and disassembly of the attached microtubules and, through the presence of motor molecules, by ultimately driving chromosome movement.

Specific DNA sequences presumably specify the structure and function of centromeres. In simple eukaryotes, the sequences that specify centromere function are very short. For example, in the yeast *Saccharomyces cerevisiae* the centromere element (CEN) is about 110 bp long, consisting of two highly conserved flanking elements of 9 bp and 11 bp and a central AT-rich segment of about 80–90 bp (see *Figure 2.5*). The centromeres of such cells are interchangeable – a CEN fragment derived from one yeast chromosome can replace the centromere of another with no apparent consequence.

In mammals, the DNA of individual centromeres consists of hundreds of kilobases of repetitive DNA, some chromosome-specific and some nonspecific. A major component

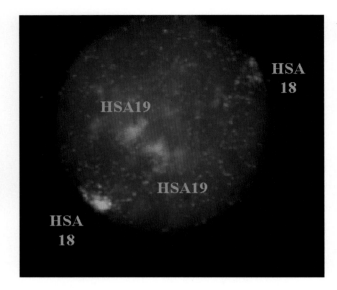

Figure 2.4: Individual chromosomes occupy distinct chromosome territories in the interphase nucleus.

The picture shows an example of ***chromosome painting*** (see Section 2.4.3) with two chromosome paints, one specific for HSA18 (= human chromosome 18 – green signal, peripheral location) and one for HSA19 (= human chromosome 19 – red signal, internal nuclear location) within an interphase nucleus. The nucleus appears blue as a result of counterstaining with DAPI, a fluorescent DNA-binding dye. Image kindly provided by Dr. Wendy Bickmore, MRC Human Genetics Unit, Edinburgh.

of human centromeric DNA is **α-satellite DNA**, a complex family of tandemly repeated DNA based on a 171 bp monomer – see Section 9.4.1 for a fuller description of human centromeric sequences). A variety of different proteins are known to associate with human centromeres, including CENP-B which binds directly to α-satellite DNA (see *Table 2.1*). Surprisingly, given that the chromosome segregation machinery is highly conserved across eukaryotes, centromeric DNA and associated proteins are evolving rapidly and have even been considered to be

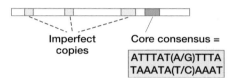

Centromere

TCACATGAT AGTGTACTA	80–90 bp > 90% (A+T)	TGATTTCCGAA ACTAAAGGCTT
I	II	III

Telomere

Tandem repeats based on the general formula
$(TG)_{1-3} TG_{2-3}/C_{2-3}A(CA)_{1-3}$

e.g.

5′....TGTGTGGGTGTGGTGTGTGTGG....3′
3′....ACACACCCACACCACACACACC....5′

Autonomous replicating sequence

Contains an 11-bp core consensus that is AT-rich, plus some imperfect copies of this sequence spanning an approximately 50-bp region of DNA

Imperfect copies   Core consensus =

ATTTAT(A/G)TTTA TAAATA(T/C)AAAT

Figure 2.5: The functional elements of a yeast chromosome.

responsible for the reproductive isolation of emerging species (Henikoff *et al.*, 2001).

2.2.4 Chromosomes as functioning organelles: origins of replication

The DNA in most diploid cells normally replicates only once per cell cycle. The initiation of replication is controlled by *cis*-acting sequences that lie close to the points at which DNA synthesis is initiated. Probably these are sites at which *trans*-acting proteins bind. Eukaryotic origins of replication have

Table 2.1: Major human centromere proteins

Name	Location	Features
CENP-A	Outer kinetochore plate	A centromere-specific histone H3-variant; essential for life being required to target CENP-C to kinetochore
CENP-B	Centromeric heterochromatin	Binds a 17 bp sequence found in α-satellite monomers (CENP-B box) ; needed for centromeric heterochromatin structure?
CENP-C	Inner kinetochore plate	Essential for proper centromere/kinetochore function
CENP-G	Inner kinetochore plate	Binds to a subset of α-satellite family sequences

Note: CENP-E (in the fibrous corona), CENP-F (outer kinetochore plate) proteins and the INCENP proteins (centric heterochromatin) are transiently associated with the centromere. See Craig *et al.* (1999) for further details.

been most comprehensively studied in yeast, where the presence of a putative replication origin can be tested by a genetic assay. To test the ability of a random fragment of yeast DNA to promote autonomous replication, it is incorporated into a bacterial plasmid together with a yeast gene that is essential for growth of yeast cells. This construct is used to transform a mutant yeast that lacks the essential gene. The transformed cells can only form colonies if the plasmid can replicate in yeast cells. However, the bacterial replication of origin in the plasmid does not function in yeast. The few plasmids that transform at high efficiency must therefore possess a sequence within the inserted yeast fragment that confers the ability to replicate extrachromosomally at high efficiency, a so-called **autonomously replicating sequence (ARS)** element.

ARS elements are thought to derive from authentic origins of replication and, in some cases, this has been confirmed by mapping a specific ARS element to a specific chromosomal location and demonstrating that DNA replication is indeed initiated at this location. Yeast ARS elements extend for only about 50 bp and consist of an AT-rich region which contains a conserved core consensus and some imperfect copies of this sequence (*Figure 2.5*). In addition, the ARS elements contain a binding site for a transcription factor and a multiprotein complex is known to bind to the origin.

The origins of replication of mammalian DNA have been much less well defined because of the absence of a genetic assay. Some initiation sites have been studied, but such studies have not been able to identify a unique origin of replication. This has led to speculation that replication can be initiated at multiple sites over regions tens of kilobases long (see Gilbert, 2001). Mammalian artificial chromosomes seem to work without specific ARS sequences being provided.

2.2.5 Chromosomes as functioning organelles: the telomeres

Telomere structure, function and evolution

Telomeres are specialized structures, comprising DNA and protein, which cap the ends of eukaryotic chromosomes. They have several likely functions:

▶ *maintaining structural integrity*. If a telomere is lost, the resulting chromosome end is unstable. It has a tendency either to fuse with the ends of other broken chromosomes, to be involved in recombination events or to be degraded. Telomere-binding proteins recognize the overhanging 3′ end of a telomere (see below) and can protect the terminal DNA *in vitro* and maybe *in vivo*;

▶ *ensuring complete DNA replication* (see section on telomerase below);

▶ *chromosome positioning*. Telomeres help to establish the three-dimensional architecture of the nucleus and / or chromosome pairing. In some cells, chromosome ends appear to be tethered to the nuclear envelope, suggesting that telomeres can help position chromosomes.

Eukaryotic telomeres consist of moderately long arrays of tandem repeats of a simple sequence which is TG-rich on one of the DNA strands and CA-rich on the complementary strand. In humans (and other animals) the repeating sequence is the hexanucleotide TTAGGG. The (TTAGGG)n array of a human telomere typically spans about 3–20 kb, beyond which (proceeding in the centromere direction) are about 100–300 kb of *telomere-associated repeats* before any unique sequence is encountered.

Unlike centromeres, the sequence of telomeres has been highly conserved in evolution – the simple sequence repeat is very similar in telomeres from different species, for example, TTGGGG in *Paramecium*, TAGGG in *Trypanosoma*, TTTAGGG in *Arabidopsis*, and TTAGGG in *Homo sapiens* (**Note**: in some species, however, there is some flexibility in the precise repeating unit as in the case of the yeast *S. cerevisiae*; see *Figure 2.5*). In addition, there is considerable conservation in the types of telomere-binding proteins (see Blackburn, 2001). The telomere-associated repeats are, however, not conserved in eukaryotes and their function is unknown.

Telomerase and the end-replication problem

During DNA replication, the lagging strand is synthesized in pieces because it must grow in a direction opposite to that of the 5′→3′ direction of DNA synthesis. A succession of 'back-stitching' syntheses are required to produce a series of DNA fragments whose ends are then sealed by DNA ligase (see *Figure 1.9* and sections 1.2.2; 1.2.3). Unlike RNA polymerases, DNA polymerases *absolutely* require a free 3′-OH group from a *double-stranded* nucleic acid from which to extend synthesis. This is achieved by employing an RNA polymerase to synthesize a complementary RNA primer to prime synthesis of each of the DNA fragments used to make the lagging strand. In these cases the RNA primer requires the presence of some DNA *ahead of* the sequence which is to be copied, to serve as its template. However, at the extreme end of a linear DNA molecule, there can never be such a template, and a different mechanism is required to solve the problem of replicating the ends of a linear DNA molecule.

The **end-replication problem** has been solved by extending the synthesis of the leading strand using a specialized form of *reverse transcriptase* (RNA-dependent DNA polymerase) provided by a specialized RNA-protein enzyme known as **telomerase**. Telomerase carries within its RNA component a short sequence near its 5′ end, CUAA **CCCUAA**C, with an internal hexanucleotide sequence (bold type) which is the antisense sequence of the human telomere repeat sequence (TTAGGG). This sequence will act as a template to prime extended DNA synthesis of telomeric DNA sequences on the leading strand. Further extension of the leading strand provides the necessary template for DNA polymerase α to complete synthesis of the lagging strand (*Figure 2.6*). This mechanism leaves the telomere itself with a protruding 3′ end which provides a single-stranded DNA target for binding by telomere-specific proteins such as human TRF2. However, the actual nature of the telomere sequence may not be important. The telomere length is known to be highly variable and is subject to genetic control (see Section 17.5.1).

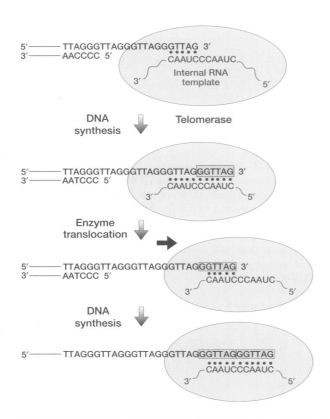

Figure 2.6: Telomerase extends the TG-rich strand of telomeres by DNA synthesis using an internal RNA template.

Note: newly synthesized hexanucleotides are shown in shading and the elongation mechanism is dependent on the RNA template containing an almost perfect tandem repeat as shown by underlining in the following sequence:
5' CUAACCCUAAC 3'.

2.2.6 Heterochromatin and euchromatin

In the interphase nucleus most of the chromatin exists in an extended state, dispersed through the nucleus and staining diffusely (**euchromatin**). However, some chromatin remains highly condensed throughout the cell cycle and forms dark-staining regions (**heterochromatin**). Genes located in euchromatin may or may not be expressed, depending on the cell type and its metabolic requirements, but genes that are located within heterochromatin, either naturally or as the result of a chromosomal rearrangement, are very unlikely to be expressed. There are two classes of heterochromatin:

▸ **constitutive heterochromatin** is always inactive and condensed. It consists largely of repetitive DNA and is found in and around the centromeres of chromosomes and in certain other regions (see *Figure 2.15* for location of constitutive heterochromatin on human chromosomes and *Table 9.2* for estimated amounts of DNA involved);

▸ **facultative heterochromatin** can exist in either a genetically active (decondensed) or an inactive (condensed)

form. Examples include mammalian X-chromosome inactivation (Section 10.5.6) or sex chromosome silencing during male meiosis. In the latter case both the X and the Y chromosome are inactivated during male meiosis (for a period of about 15 days in humans). They become condensed to form the **XY body** and are segregated into a special nuclear compartment.

In euchromatin, **G bands** show some of the properties of heterochromatin, but to a lesser degree (G bands are the dark chromosome bands resulting from positive staining with the Giemsa dye; the pale Giemsa negative bands are referred to as R bands – see Section 2.4.1). G band chromatin in metaphase chromosomes is more condensed than R band chromatin, and G bands are relatively poor in genes. The subset of R bands that are revealed by T-banding have a particularly high density of genes. Section 1.3.5 discusses the different structures of chromatin in transcriptionally active and inactive chromosomal regions.

2.3 Mitosis and meiosis are the two types of cell division

2.3.1 Mitosis is the normal form of cell division

As a person develops from an embryo, through fetus and infant to an adult, cell divisions are needed to generate the large numbers of cells required. Additionally, many cells have a limited life span, so there is a continuous requirement to generate new cells in the adult. All these cell divisions occur by mitosis. Mitosis is the normal process of cell division, from cleavage of the zygote to death of the person. In the lifetime of a human there may be something like 10^{17} mitotic divisions (Section 11.2.1).

The M phase of the cell cycle (*Figure 2.1*) consists of the various stages of nuclear division (prophase, prometaphase, metaphase, anaphase and telophase of mitosis), and cell division (**cytokinesis**), which overlaps the final stages of mitosis (*Figure 2.7*). In preparation for cell division, the previously highly extended chromosomes contract and condense so that, by **metaphase** of mitosis, they are readily visible under the microscope. Even though the DNA was replicated some time previously, it is only at prometaphase that individual chromosomes can be seen to comprise two **sister chromatids,** attached at the centromere.

The interaction between the different spindle fibers (see *Box 2.1*) pulls the chromosomes towards the center, and by metaphase each chromosome is aligned on the equatorial plane (**metaphase plate**). **Note:** during mitosis each chromosome in the diploid set behaves independently and *paternal and maternal homologs do not associate at all*. At anaphase the centromeres divide leading to physical separation of what were previously sister chromatids, and the pull by the spindle fibers ensures that the separated sister chromatids go to opposite poles (*Figures 2.7* and *2.8*). The DNA of the two sister chromatids is identical, barring any errors in DNA replication. Thus the effect of mitosis is to generate daughter cells that contain precisely the same set of DNA sequences.

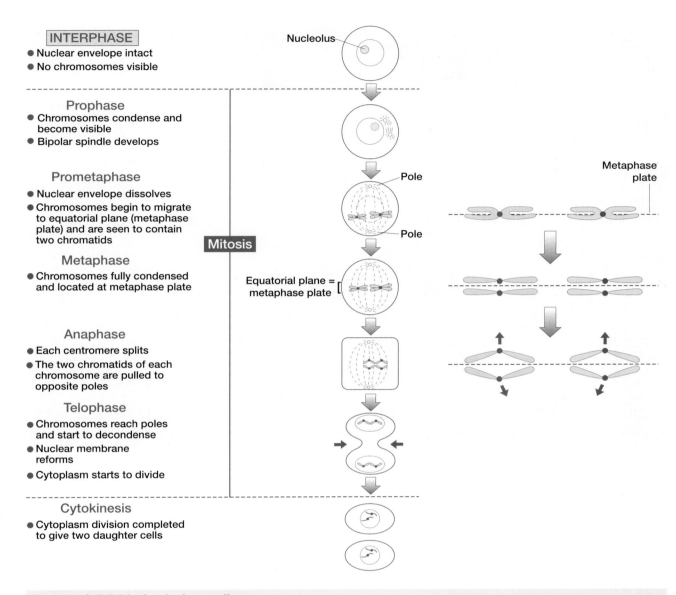

INTERPHASE
- Nuclear envelope intact
- No chromosomes visible

Prophase
- Chromosomes condense and become visible
- Bipolar spindle develops

Prometaphase
- Nuclear envelope dissolves
- Chromosomes begin to migrate to equatorial plane (metaphase plate) and are seen to contain two chromatids

Metaphase
- Chromosomes fully condensed and located at metaphase plate

Anaphase
- Each centromere splits
- The two chromatids of each chromosome are pulled to opposite poles

Telophase
- Chromosomes reach poles and start to decondense
- Nuclear membrane reforms
- Cytoplasm starts to divide

Cytokinesis
- Cytoplasm division completed to give two daughter cells

Figure 2.7: Cell division by mitosis: an outline.

2.3.2 Meiosis is a specialized form of cell division giving rise to sperm and egg cells

Primordial germ cells migrate into the embryonic gonad and engage in repeated rounds of mitosis to form **oogonia** in females and **spermatogonia** in males (***Note:*** this involves many more mitoses in males than in females – see *Box 11.4* – which may be a significant factor in explaining sex differences in mutation rate). Further growth and differentiation produces **primary oocytes** in the ovary and **primary spermatocytes** in the testis. These specialized diploid cells can undergo meiosis (*Figure 2.9*).

Meiosis involves two successive cell divisions but only one round of DNA replication, so the products are haploid. In males, the product is four spermatozoa; in females, however, there is ***asymmetric cell division*** because the cytoplasm divides unequally at each stage: the products of meiosis I (the first meiotic division) are a large **secondary oocyte** and a small cell (**polar body**). During meiosis II (the second meiotic division) the secondary oocyte then gives rise to the large mature egg cell and a second polar body.

There are two crucial differences between mitosis and meiosis (*Table 2.2*):

▶ the products of mitosis are diploid; the products of meiosis are haploid;

▶ the products of mitosis are genetically identical, the products of meiosis are genetically different.

Mitosis involves a single turn of the cell cycle (*Figure 2.1*). The DNA is replicated in S phase and the two copies are divided

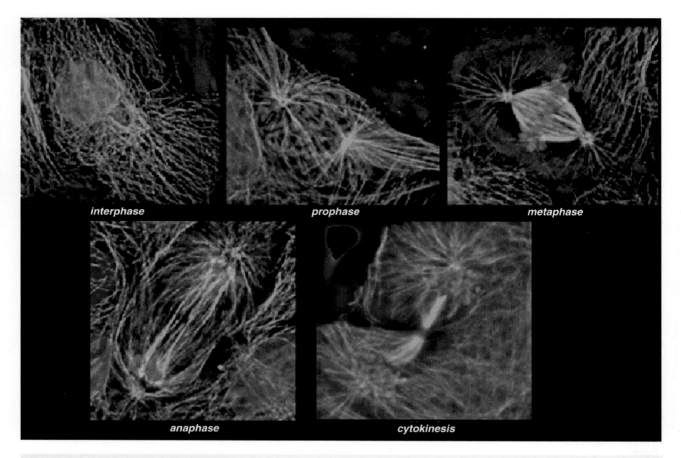

interphase   *prophase*   *metaphase*

anaphase   *cytokinesis*

Figure 2.8: A cell biologist's view of *Figure 2.7*.

The images are of HeLa cells, and were obtained using *deconvolution microscopy*. The DNA is stained with DAPI (false colored red), and the microtubules are stained with a β-tubulin antibody (false colored green). Image provided by William Earnshaw, University of Edinburgh. Reprinted from Pollard and Earnshaw (2002) *Cell Biology,* with permission from Elsevier.

Table 2.2: Mitosis and meiosis compared

	Mitosis	Meiosis
Location	All tissues	Only in testis and ovary
Products	Diploid somatic cells	Haploid sperm and egg cells
DNA replication and cell division	Normally one round of replication per cell division	Only one round of replication but two cell divisions
Extent of prophase	Short (~ 30 min in human cells)	Meiosis I is long and complex; can take years to complete
Pairing of homologs	None	Yes (in meiosis I)
Recombination	Rare and abnormal	Normally at least once in each chromosome arm
Relationship between daughter cells	Genetically identical	Different (recombination and independent assortment of homologs)

Box 2.3: Human chromosome nomenclature.

The **International System for Human Cytogenetic Nomenclature (ISCN)** is fixed by the Standing Committee on Human Cytogenetic Nomenclature (see Further reading). The basic terminology for banded chromosomes was decided at a meeting in Paris in 1971, and is often referred to as the Paris nomenclature.

Short arm locations are labeled **p** (*petit*) and long arms **q** (*queue*). Each chromosome arm is divided into regions labeled p1, p2, p3 etc., and q1, q2, q3, etc., counting outwards from the centromere. Regions are delimited by specific *landmarks,* which are consistent and distinct morphological features, such as the ends of the chromosome arms, the centromere and certain bands. Regions are divided into bands labeled p11 (one-one, not

eleven!), p12, p13 etc., sub-bands labeled p11.1, p11.2 etc. and sub-sub-bands for example p11.21, p11.22, in each case counting outwards from the centromere (*Figures 2.13, 2.15*).

Relative distance from the centromere is described by the words **proximal** and **distal**. Thus, proximal Xq means the segment of the long arm of the X that is *closest to the centromere*, while distal 2p means the portion of the short arm of chromosome 2 that is most *distant from the centromere*, and therefore closest to the telomere. Other common terms are as below.

When comparing human chromosomes with that of another species, the convention is to use the first letter of the genus name and the first two letters of the species name (e.g. HSA18 means human – **H**omo **sa**piens) chromosome 18.

The maximum resolution of metaphase FISH is several megabases. The use of the more extended prometaphase chromosomes can permit 1 Mb resolution but, because of problems with chromatin folding, two differentially labeled probe signals may appear to be side-by-side, unless they are separated by distances greater than 2 Mb. Recently, however, new variations have been developed, such as **fiber FISH** which involves artificial stretching of DNA or chromatin fibers, permitting very high resolution (see Heiskanen *et al.*, 1996). FISH on chromosomes from interphase nuclei (**interphase FISH;** see *Figure 2.17B*) can also provide high resolution analyses because the chromosomes are naturally very extended compared to metaphase or prometaphase chromosomes.

2.4.3 Chromosome painting, molecular karyotyping and comparative genome hybridization

Standard chromosome painting

A special application of FISH has been the use of DNA probes where the starting DNA is composed of a large collection of different DNA fragments from a single type of chromosome. Such probes can be prepared by combining all human DNA inserts in a *chromosome-specific DNA library* (see Section 8.3.2). Alternatively, human-specific fragments can be amplified from human *monochromosomal hybrid cells* (a type of human–rodent hybrid cell in which there is a full set of rodent chromosomes plus a single type of human chromosome (see *Box 8.4* for details of how the hybrid cells are made; the human sequences can be amplified *selectively* by *Alu-PCR*, a PCR method using primers derived from the Alu repeat sequence which is not found in rodent genomes; see *Box 5.1*).

The resulting hybridization signal represents the combined contributions of multiple labeled DNA clones deriving from many different loci spanning a whole chromosome, a *chromosome paint*, and so causes whole chromosomes to fluoresce (**chromosome painting**; see Ried *et al.*, 1998 and *Figure 2.16* for the basis of the method). *Figure 2.4* shows an example of chromosome painting in interphase nuclei, but most chromosome painting is carried out on metaphase chromosomes. An important application is in

defining *de novo* rearrangements and marker chromosomes in clinical and cancer cytogenetics (see *Figure 2.18* for an example). It is particularly helpful in cancer cytogenetics where tumor chromosome preparations are often of poor quality.

Molecular karyotyping and multiplex FISH (M-FISH)

Chromosome painting was initially limited by the relatively small number of differently colored fluorescent dyes (*fluorophores*) one could use to distinguish different chromosomes. To increase the number of different targets that can be detected, combinatorial labeling (labeling individual probes with more than one type of fluorophore) and ratio labeling (different ratios of the different fluorophores) have been used (see Lichter, 1997). The mixed colors are not detected by standard fluorescence microscopy using appropriate filters. Instead, automated digital image analysis is preferred by which various combinations of fluorophores are assigned artificial *pseudocolors*.

The above approaches have recently permitted all 24 different human chromosomes to be visualized simultaneously, a form of **molecular karyotyping.** The general method is known as **multiplex FISH (M-FISH)** and uses digital images acquired separately for each of five different fluorophores using a CCD (charge coupled device) camera. The images are analyzed by a software package which generates a composite image in which each chromosome is given a different *pseudocolor* depending on the fluorophore composition. Methods such as these are particularly applicable to analyzing tumor samples where complex chromosome rearrangements are particularly frequent (see *Figure 17.10* for an example).

Comparative genome hybridization (CGH)

A further extension of chromosome painting is **comparative genome hybridization** (CGH). CGH involves simultaneous painting of chromosomes in two different colors using as probes total DNA from two related sources, which are expected to show differences involving gain or loss of subchromosomal regions or even whole chromosomes. A common

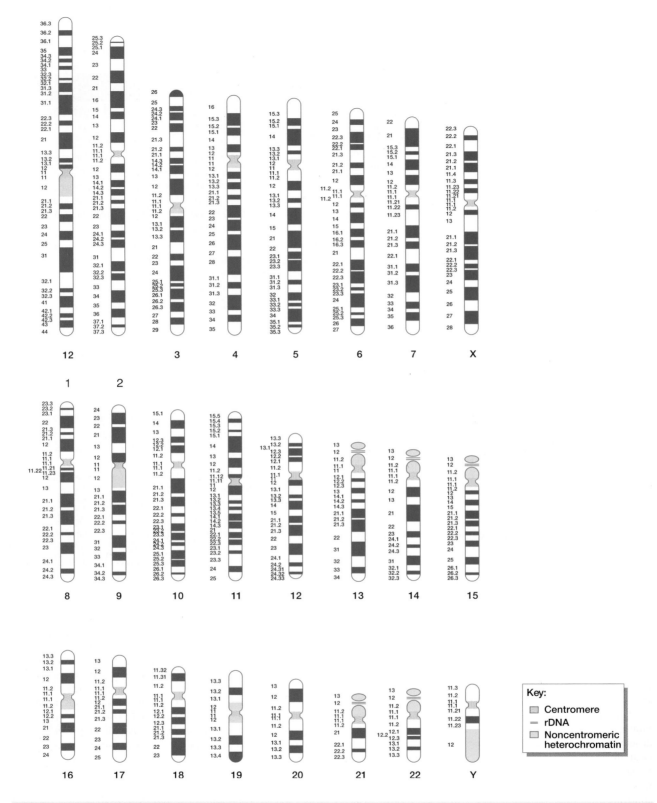

Figure 2.15: Banding pattern of human chromosomes.

This is a compilation of the best banding patterns that might be seen on each chromosome, and not a picture of how chromosomes appear in any one cell under the microscope. Chromosomes are numbered in order of size, except that 21 is actually smaller than 22. Arrays of repeated ribosomal DNA genes on the short arms of the acrocentric chromosomes 13, 14, 15, 21 and 22 often appear as thin stalks carrying knobs of chromatin (**satellites**). Constitutional heterochromatin occurs at centromeres, on much of the Y chromosome long arm, at secondary constrictions on 1q, 9q and 16q, and on the short arms of the acrocentric chromosomes.

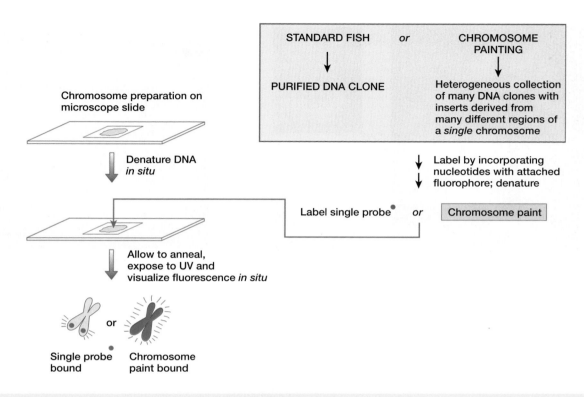

Figure 2.16: Basis of the chromosome FISH and chromosome painting methods.

Note: for simplicity the end result is shown for metaphase chromosomes only. See *Figure 2.17A* and *2.17B* for examples of metaphase FISH and interphase FISH respectively, and *Figures 2.18* and *2.4* for examples of chromosome painting on metaphase and interphase chromosomes respectively.

application is in analyzing tumor samples for evidence of regions of the genome which have been amplified, or where there has been subchromosomal loss. These are identified by comparing the ratios of the two color signals chromosome by chromosome (see *Figure 17.3* for an example).

2.5 Chromosome abnormalities

Chromosome abnormalities might be defined as changes resulting in a visible alteration of the chromosomes. How much can be seen depends on the technique used. The smallest loss or gain of material visible by traditional methods on standard cytogenetic preparations is about 4 Mb of DNA. However, FISH allows much smaller changes to be seen; the development of **molecular cytogenetics** has removed any clear dividing line between changes described as chromosomal abnormalities and changes thought of as molecular or DNA defects. An alternative definition of a chromosomal abnormality is an abnormality produced by specifically chromosomal mechanisms. Most chromosomal aberrations are produced by misrepair of broken chromosomes, by improper recombination or by malsegregation of chromosomes during mitosis or meiosis.

2.5.1 Types of chromosomal abnormality

Chromosomal abnormalities can be classified into two types according to the extent of their occurrence in cells of the body. A **constitutional abnormality** is present in *all* cells of the body. Where this occurs, the abnormality must have been present very early in development, most likely the result of an abnormal sperm or egg, or maybe abnormal fertilization or an abnormal event in the very early embryo. A **somatic** (or **acquired**) **abnormality** is present in only certain cells or tissues of an individual. As a result, an individual with a somatic abnormality is a **mosaic** (*Figure 4.10*), containing cells with two different chromosome constitutions, with both cell types deriving from the same zygote.

Chromosomal abnormalities, whether constitutional or somatic, mostly fall into two categories: numerical and structural abnormalities (see *Box 2.4*). Occasionally, abnormalities have been identified in which chromosomes have the correct number and structure, but represent unequal contributions from the two parents (Section 2.5.4). Perhaps unexpectedly, correct parental origin matters.

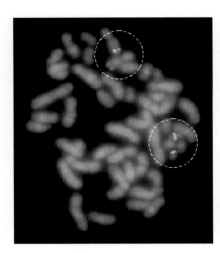
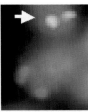

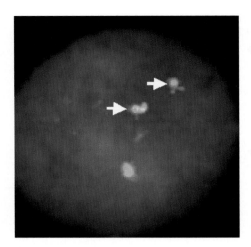

Figure 2.17: Two-color metaphase and interphase FISH to detect BCR-ABL rearrangement in chronic myeloid leukemia.

Most cases of chronic myeloid leukemia result from a t(9;22) reciprocal translocation generating gene fusion between the ABL oncogene at 9q34 and the BCR gene at 22q11 (see *Figure 17.4* for details). Hybridization with labeled probes for ABL (red signal) and BCR (green signal) is shown to *metaphase chromosomes from a CML patient* (on left and with areas within dashed circles expanded in the middle panels), and to *interphase chromosomes from the same patient* (right panel). The normal ABL and BCR genes give standard red and green signals respectively (*note:* metaphase signals from the two sister chromatids are visible at least in the case of the ABL gene). The white arrows show characteristic signals for the BCR-ABL fusion gene on the two translocation chromosomes [der(9) and der(22)]. These can be identified because of the very close positioning of the red and green signals, with overlapping red and green signals appearing orange–yellow. Image provided by Fiona Harding, Institute of Human Genetics, Newcastle Upon Tyne.

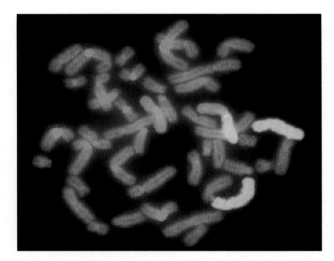

Figure 2.18: Chromosome painting can be used to define chromosome rearrangements.

In this case an abnormal X chromosome, with extra chromosomal material present on the short arm, identified by karyotyping a peripheral blood sample, was investigated using whole chromosome X paint (red) and whole chromosome 4 paint (green). This image confirms that the additional material present on the short arm of the abnormal X chromosome originated from chromosome 4. The background stain for the chromosomes is the blue DAPI stain. Image provided by Gareth Breese, Institute of Human Genetics, Newcastle Upon Tyne.

2.5.2 Numerical chromosomal abnormalities involve gain or loss of complete chromosomes

Three classes of numerical chromosomal abnormalities can be distinguished: polyploidy, aneuploidy and mixoploidy.

Polyploidy

One to three percent of recognized human pregnancies are **triploid**. The most usual cause is two sperm fertilizing a single egg (**dispermy**); sometimes the cause is a diploid gamete (*Figure 2.19*). Triploids very seldom survive to term, and the condition is not compatible with life. **Tetraploidy** is much rarer and always lethal. It is usually due to failure to complete the first zygotic division: the DNA has replicated to give a content of 4C, but cell division has not then taken place as normal. Although constitutional polyploidy is rare and lethal, all normal people have some polyploid cells (Section 3.1.4).

Aneuploidy

Euploidy means having *complete* chromosome sets (n, 2n, 3n etc.). **Aneuploidy** is the opposite: one or more individual chromosomes is present in an extra copy or is missing from a euploid set. **Trisomy** means having three copies of a particular chromosome in an otherwise diploid cell, for example trisomy 21 (47,XX+21 or 47,XY,+21) in Down syndrome. Monosomy is the corresponding lack of a chromosome, for example monosomy X (45,X) in Turner syndrome. Cancer cells often show extreme aneuploidy, with multiple chromosomal abnormalities (see *Figure 17.10*). Aneuploid cells arise through two main mechanisms:

▶ **nondisjunction:** failure of paired chromosomes to separate (disjoin) in anaphase of meiosis I, or failure of sister chromatids to disjoin at either meiosis II or at mitosis. Nondisjunction in meiosis produces gametes with 22 or 24 chromosomes, which after fertilization by a normal gamete make a trisomic or monosomic zygote. Nondisjunction in mitosis produces a mosaic.

▶ **anaphase lag:** failure of a chromosome or chromatid to be incorporated into one of the daughter nuclei following cell division, as a result of delayed movement (lagging) during anaphase. Chromosomes that do not enter a daughter cell nucleus are lost.

Mixoploidy (mosaicism and chimerism)

Mixoploidy means having two or more genetically different cell lineages within one individual. The genetically different cell populations can arise from the same zygote (**mosaicism**), or more rarely they can originate from different zygotes (**chimerism**). See Section 4.3.6 and *Figure 4.10* for a fuller explanation. Abnormalities that would be lethal in constitutional form may be compatible with life in mosaics.

Aneuploidy mosaics are common. For example, mosaicism resulting in a proportion of normal cells and a proportion of aneuploid (e.g. trisomic) cells can be ascribed to nondisjunction or chromosome lag occurring in one of the mitotic divisions of the early embryo (any monosomic cells that are formed usually die out).

Polyploidy mosaics (e.g. human diploid / triploid mosaics) are occasionally found. As gain or loss of a haploid set of chromosomes by mitotic nondisjunction is most unlikely, human diploid / triploid mosaics most probably arise by fusion of the second polar body with one of the cleavage nuclei of a normal diploid zygote.

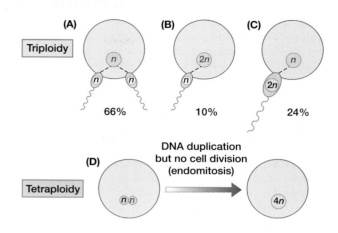

Figure 2.19: Origins of triploidy and tetraploidy.

Clinical consequences of numerical abnormalities

Having the wrong number of chromosomes has serious, usually lethal, consequences (*Table 2.4*). Even though the extra chromosome 21 in a man with Down syndrome is a perfectly normal chromosome, inherited from a normal parent, its presence causes multiple congenital abnormalities. Autosomal monosomies have even more catastrophic consequences than trisomies. These abnormalities must be the consequence of an imbalance in the levels of gene products encoded on different chromosomes. Normal development and function depend on innumerable interactions between gene products, including many that are encoded on different chromosomes. Altering the relative numbers of chromosomes will affect these interactions.

Having the wrong number of sex chromosomes has far fewer ill effects than having the wrong number of any autosome. 47,XXX and 47,XYY people often function

---

Box 2.4: Nomenclature of chromosome abnormalities.

Numerical abnormalities:

Triploidy	69,XXX, 69,XXY, 69,XYY
Trisomy	e.g. 47,XX,+21[a]
Monosomy	e.g. 45,X
Mosaicism	e.g. 47,XXX/ 46,XX

Structural abnormalities:

Deletion	e.g. 46,XY,**del**(4)(p16.3)[b]; 46,XX,**del**(5)(q13q33)[b]
Inversion	e.g. 46,XY,**inv**(11)(p11p15)
Duplication	e.g. 46,XX,**dup**(1)(q22q25)
Insertion	e.g. 46,XX,**ins**(2)(p13q21q31)[c]
Ring	e.g. 46,XY,**r**(7)(p22q36)
Marker	e.g. 47,XX,+**mar**[d]
Translocation, reciprocal	e.g. 46,XX,**t**(2;6)(q35;p21.3)[e]
Translocation, Robertsonian (gives rise to one **der**ivative chromosome)	e.g.45,XY,**der**(14;21)(q10;q10)[f] 46,XX,**der**(14;21)(q10;q10),+21[g]

Notes:

[a]Gain of a chromosome is indicated by +; loss of a chromosome by –.
[b]Terminal deletion (breakpoint at 4p16.3) and interstitial deletion (5q13–q33).
[c]A rearrangement of one copy of chromosome 2 by insertion of segment 2q21–q31 into a breakpoint at 2p13.
[d]Karyotype of a cell that contains a **marker chromosome** (an extra unidentified chromosome).
[e]A balanced reciprocal translocation with breakpoints in 2q35 and 6p21.3.
[f]A balanced carrier of a 14;21 Robertsonian translocation. q10 is not really a chromosome band, but indicates the centromere; **der** means **derivative chromosome** (used when one chromosome from a translocation is present).
[g]Translocation Down syndrome; a patient with one normal chromosome 14, a Robertsonian translocation 14;21 chromosome and two normal copies of chromosome 21.
This is a short nomenclature; a more complicated nomenclature is defined by the ISCN that allows complete description of any chromosome abnormality – see Further reading.

within the normal range; 47,XXY men have relatively minor problems compared to people with any autosomal trisomy, and even monosomy, in 45,X women, has remarkably few major consequences. In fact, since normal people can have either one or two X chromosomes, and either no or one Y, there must be special mechanisms that allow normal function with variable numbers of sex chromosomes. In the case of the Y chromosome, this is because it carries very few genes, whose only important function is to determine male sex. For the human X chromosome, the special mammalian mechanism of *X-chromosome inactivation* (Section 10.5.6) controls the level of X-encoded gene products independently of the number of X chromosomes present in the cell.

Autosomal monosomy is invariably lethal at the earliest stage of embryonic life. On every chromosome there are probably a few genes where reducing the level of gene product to 50% is incompatible with development. Also, while such a reduction is not obviously pathogenic for most genes (Section 16.4.2), it may have minor effects, and the combination of hundreds or thousands of these minor effects could be enough to disrupt normal development of the embryo. Trisomies make a smaller change than monosomies in relative levels of gene products, and their effects are somewhat less. Trisomic embryos survive longer than monosomic ones,

and trisomies 13, 18 and 21 are compatible with survival until birth. Interestingly, these three chromosomes seem to be relatively poor in genes (Section 9.1.4).

It is not so obvious why triploidy is lethal in humans and other animals. With three copies of every autosome, the dosage of autosomal genes is balanced and should not cause problems. Triploids are always sterile because triplets of chromosomes cannot pair and segregate correctly in meiosis, but many triploid plants are in all other respects healthy and vigorous. The lethality in animals is probably explained by imbalance between products encoded on the X chromosome and autosomes, for which X-chromosome inactivation is unable to compensate.

2.5.3 Structural chromosomal abnormalities result from misrepair of chromosome breaks or from malfunction of the recombination system

Chromosome breaks occur either as a result of damage to DNA (by radiation or chemicals, for example) or as part of the mechanism of recombination. In the G2 phase of the cell cycle (*Figure 2.1*) chromosomes consist of two chromatids. Breaks occurring at this stage are manifest as **chromatid breaks**, affecting only one of the two sister chromatids. Breaks

---

Table 2.4: Consequences of numerical chromosomal abnormalities.

Polyploidy

Triploidy (69,XXX, XXY or XYY)	1–3% of all conceptions; almost never live born; do not survive

Aneuploidy (autosomes)

Nullisomy (missing a pair of homologs)	Pre-implantation lethal
Monosomy (one chromosome missing)	Embryonic lethal
Trisomy (one extra chromosome)	Usually lethal at embryonic or fetal stages, but trisomy 13 (Patau syndrome) and trisomy 18 (Edwards syndrome) may survive to term and trisomy 21 (Down syndrome) may survive to age 40 or longer

Aneuploidy (sex chromosomes)

Additional sex chromosomes	(47, XXX; 47, XXY; 47, XYY) present relatively minor problems, with normal lifespan
Lacking a sex chromosome	45,X = Turner syndrome. About 99% of cases abort spontaneously; survivors are of normal intelligence but infertile and show minor physical signs. 45,Y = not viable

---

Table 2.5: Structural abnormalities resulting from misrepair of chromosome breaks or recombination between non-homologous chromosomes.

	One chromosome involved	Two chromosomes involved
One break	Terminal deletion (healed by adding telomere)	–
Two breaks	Interstitial deletion; Inversion;	Reciprocal translocation (*Figure 2.21*)
	Ring chromosome (*Figure 2.20*	Robertsonian translocation (*Figure 2.21*)
	Duplication or deletion by unequal sister-chromatid exchange (*Figure 11.7*)	Duplication or deletion by unequal recombination (*Figure 11.7*)
Three breaks	Various rearrangements, e.g. inversion with deletion, intrachromosomal insertion	Interchromosomal insertion (direct or inverted)

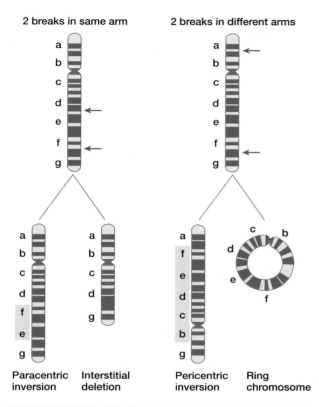

2 breaks in same arm

2 breaks in different arms

Paracentric
inversion

Interstitial
deletion

Pericentric
inversion

Ring
chromosome

Figure 2.20: Possible stable results of two breaks on a single chromosome.

occurring in G1 phase, if not repaired before S phase, appear later as chromosome breaks, affecting both chromatids. Cells have enzyme systems that recognize, and if possible repair, broken chromosome ends. Repairs can involve either joining two broken ends together, or adding a telomere to a broken end. Cell cycle checkpoint mechanisms (Section 18.7.3) normally prevent cells with unrepaired chromosome breaks from entering mitosis; if the damage cannot be repaired, the cell commits suicide (apoptosis).

Structural abnormalities arise when breaks are repaired incorrectly. Provided there are no free broken ends, it is possible to pass through the cell cycle checkpoints. Sometimes, therefore, the wrong broken ends get joined together. Any resulting **acentric chromosome** (lacking a centromere) or **dicentric chromosome** (possessing two centromeres) will not segregate stably in mitosis, and will eventually be lost. However, chromosomes with a single centromere can be stably propagated through successive rounds of mitosis, even if they are structurally abnormal. Meiotic recombination between mispaired chromosomes is a common cause of translocations, especially in spermatogenesis. *Table 2.5* summarizes the main stable structural abnormalities.

An additional rare class of structural abnormality not shown in *Table 2.5* is **isochromosomes**. These are symmetrical chromosomes consisting of either two long arms or two short arms of a particular chromosome. They are believed to arise from an abnormal U-type exchange between sister chromatids just next to the centromere of a chromosome. Human isochromosomes are rare except for i(Xq) and i(21q) which is an occasional cause of Down syndrome.

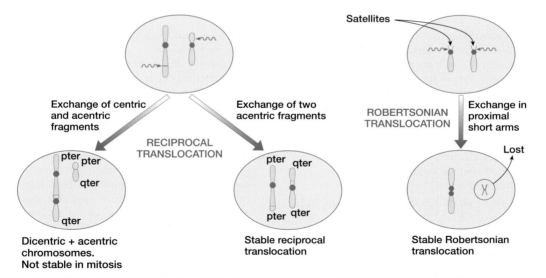

Figure 2.21: Origins of translocations.

Dicentric and acentric chromosomes are not stable through mitosis. Robertsonian translocations are produced by exchanges between the proximal short arms of the acrocentric chromosomes 13, 14, 15, 21 and 22 (arrays of repeated ribosomal DNA genes on the short arms of the acrocentric chromosomes 13, 14, 15, 21 and 22 often appear as thin stalks carrying knobs of chromatined – *satellites*). In a Robertsonian translocation both centromeres are present, but they function as one and the chromosome is stable. The small acentric fragment is lost, but this has no pathological consequences because it contains only repeated rDNA sequences, which are also present on the other acrocentric chromosomes.

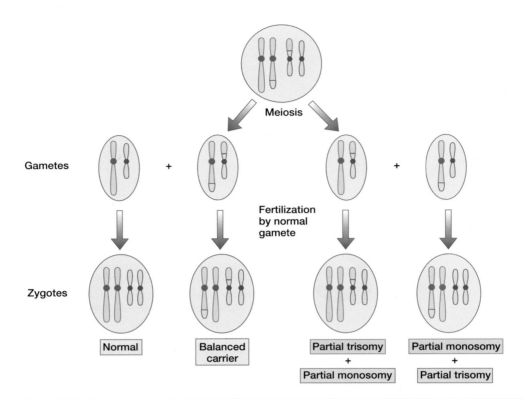

Figure 2.22: Results of meiosis in a carrier of a balanced reciprocal translocation.

Other modes of segregation are also possible, for example 3 : 1 segregation. The relative frequency of each possible gamete is not readily predicted. The risk of a carrier having a child with each of the possible outcomes depends on its frequency in the gametes and also on the likelihood of a conceptus with that abnormality developing to term. See the book by Gardner and Sutherland (Further reading) for discussion.

Structural chromosomal abnormalities are **balanced** if there is no net gain or loss of chromosomal material, and **unbalanced** if there is net gain or loss. In general, balanced abnormalities (inversions, balanced translocations) have no effect on the phenotype, although there are important exceptions to this:

▶ a chromosome break may disrupt an important gene;

▶ the break may affect expression of a gene even though it does not disrupt the coding sequence. It may separate a gene from a control element, or it may put the gene in an inappropriate chromatin environment, for example translocating a normally active gene into heterochromatin;

▶ balanced X-autosome translocations cause problems with X-inactivation (*Figure 14.10*).

Robertsonian translocations are sometimes called **centric fusions**, but this is misleading because in fact the breaks are in the proximal short arms. The translocation chromosome is really dicentric, but because the two centromeres are very close together they function as one, and the chromosome segregates regularly. The distal parts of the two short arms are lost as an acentric fragment. Short arms of acrocentric chromosomes contain only arrays of repeated ribosomal RNA genes, and the loss of two short arms has no phenotypic effect. Because there is no phenotypic effect, Robertsonian trans-

locations are regarded as balanced, even though in fact some material has been lost.

Unbalanced abnormalities can arise directly, through deletion or, rarely, duplication, or indirectly by malsegregation of chromosomes during meiosis in a carrier of a balanced abnormality. Carriers of balanced structural abnormalities can run into trouble during meiosis, if the structures of homologous pairs of chromosomes do not correspond:

▶ a carrier of a balanced reciprocal translocation can produce gametes that after fertilization give rise to an entirely normal child, a phenotypically normal balanced carrier, or various unbalanced karyotypes that always combine monosomy for part of one of the chromosomes with trisomy for part of the other (see *Figure 2.22*). It is not possible to make general statements about the relative frequencies of these outcomes. The size of any unbalanced segments depends on the position of the breakpoints. If the unbalanced segments are large, the fetus will probably abort spontaneously; an imbalance for smaller segments may result in live born abnormal babies;

▶ a carrier of a balanced Robertsonian translocation can produce gametes that after fertilization give rise to an entirely normal child, a phenotypically normal balanced

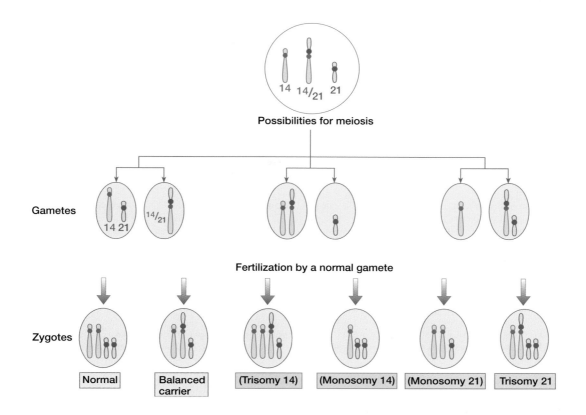

Figure 2.23: Results of meiosis in a carrier of a Robertsonian translocation.

Carriers are asymptomatic but often produce unbalanced gametes that can result in a monosomic or trisomic zygote. The bracketed monosomic and trisomic zygotes in this example would not develop to term.

carrier, or a conceptus with full trisomy or full monosomy for one of the chromosomes involved (*Figure 2.23*);

▶ a carrier of a pericentric inversion may produce unbalanced offspring because when the inverted and non-inverted homologs pair they form a loop so that matching segments pair along the whole length of the chromosomes. If a crossover occurs within the loop, the result is a chromosome carrying an unbalanced deletion and duplication. Paracentric chromosome inversions form similar loops, but any crossover within the loop generates an acentric or dicentric chromosome, which is unlikely to survive. For details of meiosis in carriers of inversions, see the book by Gardner and Sutherland (Further reading) or any other cytogenetics text.

2.5.4 Apparently normal chromosomal complements may be pathogenic if they have the wrong parental origin

The rare abnormalities described below demonstrate that it is not enough to have the correct number and structure of chromosomes; they must also have the correct parental origin. 46,XX conceptuses in which both genomes originate from the same parent (**uniparental diploidy**) never develop correctly. For some individual chromosomes, having both homologs derived from the same parent (**uniparental disomy**) also causes abnormality. A small number of genes are imprinted with their parental origin (Section 10.5.4) and are expressed differently according to the origin. It is assumed that the abnormalities of uniparental disomy and uniparental diploidy are caused by abnormal expression of such imprinted genes.

Uniparental diploidy is seen in **hydatidiform moles**, abnormal conceptuses with a 46,XX karyotype of exclusively paternal origin. Molar pregnancies show widespread hyperplasia of the trophoblast but no fetal parts, and have a significant risk of transformation into choriocarcinoma. Genetic marker studies show that most moles are homozygous at all loci, indicating that they arise by chromosome doubling from a single sperm. **Ovarian teratomas** are the result of maternal uniparental diploidy. These rare benign tumors of the ovary consist of disorganized embryonic tissues, without extra-embryonic membranes. They arise by activation of an unovulated oocyte.

Uniparental disomy (UPD), affecting a single pair of homologs, goes undiagnosed if the result is not abnormal, but is detected for chromosomes for which it produces characteristic syndromes (see *Box 16.6*). UPD can involve *isodisomy*, where both homologs are identical, or **heterodisomy**, where they are derived from both homologs in one parent. The usual cause is thought to be *trisomy rescue*: a conceptus that is trisomic and would otherwise die, occasionally loses one chromosome by mitotic nondisjunction or anaphase lag from a totipotent cell. The euploid progeny of this cell form the embryo, while all the aneuploid cells die. If each of the three copies has an equal chance of being lost, there will be a two in three chance of a single chromosome loss leading to the normal chromosome constitution and a one in three chance of uniparental disomy (either paternal or maternal). Uniparental isodisomy may possibly arise by selection pressure on a monosomic embryo to achieve euploidy by selective duplication of the monosomic chromosome.

Further reading

Choo KH (2001) Domain organization at the centromere and neocentromere. *Developmental Cell* **1**, 165–177.

Gardner RJM, Sutherland GR (1996) *Chromosome Abnormalities and Genetic Counseling,* 2nd Edn. OUP, Oxford [A thorough introduction to the nature, origin and consequences of human chromosomal abnormalities].

ISCN (1995) *An International System for Human Cytogenetic Nomenclature* (ed. F Mittelman). Karger, Basel.

Pollard TD, Earnshaw WC (2002) *Cell Biology.* Saunders, Philadelphia.

Rooney DE (2001) (ed.) *Human Cytogenetics: Constitutional Analysis,* 3rd Edn. Oxford University Press, Oxford [Detailed laboratory protocols].

Rooney DE (2001) (ed.) *Human Cytogenetics: Malignancy and Acquired Abnormalities,* 3rd Edn. Oxford University Press, Oxford [Detailed laboratory protocols].

Schinzel A (2001) *Catalogue of Unbalanced Chromosome Aberrations in Man.* Walter de Gruyter, Berlin.

Therman E, Susman M (1992) *Human Chromosomes: Structure, Behavior and Effects,* 3rd Edn. Springer, New York [An excellent compact introduction; emphasis on scientific bases rather than clinical implications].

Web-based cytogenetic resources. See compilations such as at: www.kumc.edu/gec/prof/cytogene.html

References

Blackburn EH (2001) Switching and signaling at the telomere. *Cell* **106**, 661–673.

Boyle S, Gilchrist S, Bridger JM, Mahy NL, Ellis JA, Bickmore WA (2001) The spatial organization of human chromosomes within the nuclei of normal and emerin-mutant cells. *Hum. Mol. Genet.* **10**, 211–219.

Craig JM, Earnshaw WC, Vagnarelli P (1999) Mammalian centromeres: DNA sequence, protein composition, and role in cell cycle progression. *Exp. Cell Res.* **246**, 249–262.

Craig JM, Bickmore WA (1993) Chromosome bands – flavours to savour. *Bioessays* **15**, 349–354.

Cremer T, Cremer C (2001) Chromosome territories, nuclear architecture and gene regulation in mammalian cells. *Nature Rev. Genet.* **2**, 292–301.

Cross I, Wolstenholme J (2001) An introduction to human chromosomes and their analysis. In: *Human Cytogenetics: Constitutional Analysis,* 3rd Edn (ed. DE Rooney) Oxford University Press, Oxford.

Gallardo MH, Bickham JW, Honeycutt RL, Ojeda RA, Kohler N (1999) Discovery of tetraploidy in a mammal. *Nature* **401**, 341.

Gilbert DM (2001) Making sense of eukaryotic replication origins. *Science* **294**, 96–100.

Heiskanen M, Peltonen L, Palotie A (1996) Visual mapping by high resolution FISH. *Trends Genet.* **12**, 379–384.

Henikoff S, Ahmad K, Malik HS (2001) The centromere paradox: stable inheritance with rapidly evolving DNA. *Science* **293**, 1098–1102.

Huxley C (1997) Mammalian artificial chromosomes and chromosome transgenics. *Trends Genet.* **13**, 345–347.

Manuelidis L (1990) A view of interphase chromosomes. *Science* **250**, 1533–1540.

Niimura Y, Gojobori T (2002) *In silico* chromosome staining: reconstruction of Giemsa bands from the whole human genome sequence. *Proc. Natl Acad. Sci. USA* **99**, 797–802.

Parada LA, Misteli T (2002) Chromosome positioning in the interphase nucleus. *Trends Cell Biol.* **12**, 425–432.

Saitoh Y, Laemmli UK (1994) Metaphase chromosome structure: bands arise from a differential folding path of the highly AT-rich scaffold. *Cell* **76**, 609–622.

Schindelhauer D (1999) Construction of mammalian artificial chromosomes: prospects for defining an optimal centromere. *Bioessays* **21**, 76–83.

Trask BJ (2002) Human cytogenetics: 46 chromosomes, 46 years and counting. *Nature Rev. Genet.* **3**, 769–778.

CHAPTER THREE

Cells and development

Chapter contents

With the exception of viruses, all life consists of **cells** – aqueous, membrane-bound compartments that interact with each other and the environment. Every cell arises either by the division or fusion of existing cells. Ultimately, there must be an unbroken chain of cells leading back to the first successful primordial cell that lived maybe 3.5 billion years ago. How that cell formed is an interesting question.

Some cells are independent unicellular organisms. Such organisms must perform all the activities that are necessary to sustain life, and must also be capable of reproduction. Therefore they are extremely sensitive to changes in their environment, they typically have very short life cycles and are thus suited to rapid proliferation. As a result, they can adapt quickly to changes around them – mutants that are more able to survive in a particular environment can flourish quickly. This has led to an enormous range of single-celled organisms which have evolved to fit different, sometimes extreme, environmental niches. However, the complexity of these organisms is always limited.

Multicellular organisms have comparatively greater longevity and changes in phenotype are correspondingly slow. Their success has been based on the partitioning of different functions into different cell types and the resulting availability of numerous forms of cell–cell interaction provides huge potential for functional complexity. Some unicellular organisms, such as the slime mold *Dictyostelium discoideum*, have a multicellular stage in their life cycle but exist predominantly as single cells. In contrast, animals and plants are multicellular throughout their vegetative existence, with specialized **germ cells** set aside to facilitate reproduction. Multicellular organisms can vary enormously in size, form and cell number, but in each case life begins with a single cell. The process of **development**, from single cell to mature organism, involves many rounds of cell division during which the cells must become increasingly specialized and organized in precise patterns. Their behavior and interactions with each other during development mold the overall morphology of the organism.

3.1 The structure and diversity of cells

3.1.1 Prokaryotes and eukaryotes represent the fundamental division of cellular life forms

Cells can be classified into broad taxonomic groups according to differences in their internal organization and major functional distinctions arising from early evolutionary divergence (see *Figure 3.1*). The major division of organisms into prokaryotes and eukaryotes is founded on fundamental differences in cell architecture (*Figure 3.2*).

Prokaryote cells have a simple internal organization, notably lacking organelles and intracellular compartments. There is no defined nucleus. The chromosomal DNA, which does not appear to be highly organized, exists as a nucleoprotein complex known as the **nucleoid**. Under the electron microscope prokaryotic cells appear relatively featureless. However, prokaryotes are far from primitive, since they have been through many more generations of evolution than ourselves. They comprise two kingdoms of life: **bacteria** (formerly called *eubacteria* to distinguish them from archae-

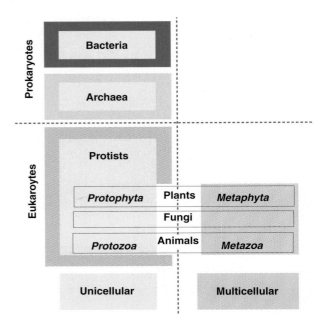

Figure 3.1: Classification of unicellular and multicellular organisms.

Note: protists include single-celled organisms that have been classified as animals (protozoa), plants (protophyta) and fungi (e.g. yeasts), but many protists are not classified into one of these groups

bacteria) and the **archaea** (a rather poorly understood group of organisms that superficially resemble bacteria and so were formerly termed *archaebacteria*). All prokaryotic organisms are unicellular. Bacteria are found in many environments and some are pathogenic to humans. Archaea are often found in extreme environments such as hot acid springs, but some species are found in more convivial locations along with eubacteria (e.g. in the guts of cows). The typical prokaryote genome is conventionally regarded as a single, circular chromosome containing less than 10 Mb of DNA. However, this view has been challenged recently as more prokaryote species have been characterized in detail and very diverse genome structures have been discovered. For example, prokaryotes have been identified that possess multiple circular or linear chromosomes, mixtures of circular and linear chromosomes, and genomes up to 30 Mb in size (e.g. *Bacillus megaterium*).

Eukaryote cells are thought to have first appeared about 1.5 billion years ago. They have a much more complex organization than their prokaryote counterparts, with internal membranes and membrane-bound **organelles** including a nucleus (see *Box 3.1*). There is only one kingdom of eukaryotic organisms – the **eukarya** – but this includes both unicellular species (e.g. yeast, protists, algae) and multicellular species (notably animals and plants). In all known eukaryotic organisms, the genome comprises two or more linear chromosomes that are contained within the nucleus. Each chromosome is a single, very long DNA molecule packaged in an elaborate and highly organized manner with histones and other proteins. The number and DNA content of the chromo-

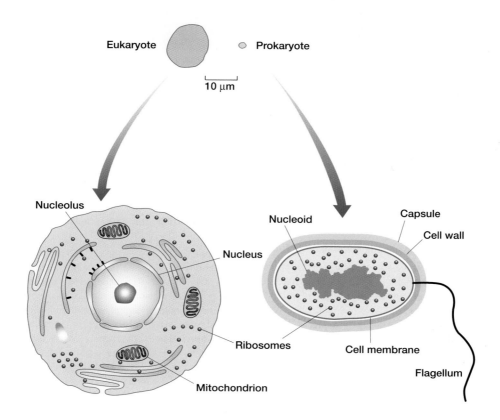

Figure 3.2: Prokaryotic and eukaryotic cell anatomy.

The eukaryotic cell shown in this figure is a generic animal cell. Examples of individual cell types are described in *Box 3.6*.

somes varies greatly between species, and although there is a tendency for the genome size to parallel the complexity of the organism, there are notable exceptions (see Section 3.1.4). In addition to the major organelles, the soluble portions of both the cytoplasm and the nucleoplasm are highly organized. In the nucleus, a variety of subnuclear structures have been identified which are arranged in the context of the **nuclear matrix** (see *Box 3.1*). Within the cytoplasm, there is a complicated network of different classes of protein filaments, collectively called the **cytoskeleton**, which determines cell shape and cell movement and which provides a framework for intracellular transport (see *Box 3.2*).

3.1.2 Cell size and shape can vary enormously, but rates of diffusion fix some upper limits

There are many factors that affect cell size (Su and O'Farrel, 1998; Saucedo and Edgar, 2002). Cells depend on diffusion to coordinate their metabolic activities, their interactions with the environment and their interactions with other cells. As they grow larger, the surface-to-volume ratio decreases. It is thought that the simple internal structure of prokaryotic cells limits their maximum size – typically bacterial cells are 1 μm in diameter. The complex internal membranes and compartmentalization of eukaryotic cells may be important in allowing them to grow larger. Nevertheless, metabolically

active internal regions are seldom more than 15–25 μm from the cell surface and this limits the cell size to approximately 50 μm. Indeed, the average diameter of cells in a multicellular organism lies within the range of 10–30 μm. Some specialized cells can grow much larger than this. Neurons, for example, can reach up to 1 m in length, although this reflects the projection of long and very thin axons while the cell body remains well within the parameters discussed above. Eggs are also very large cells. Mammalian eggs are about 100 μm in diameter but the eggs of other animals, which store nutrients required for development, can be much larger. The largest known cell is the ostrich egg, which can be up to 20 cm in length and has the volume of approximately 24 chicken eggs.

3.1.3 In multicellular organisms, there is a fundamental distinction between somatic cells and the germ line

In multicellular organisms, vegetative and reproductive functions are separated. A specialized population of **germ cells** is set aside to carry out reproductive functions while the role of the remaining **somatic cells** is to provide a vessel in which these reproductive cells are carried and a means to achieve reproduction (Wylie, 2000). In evolutionary terms, animals can be thought of as incubators and facilitators for the sperm and eggs that allow the species to reproduce. In plants and

Box 3.1: Intracellular organization of animal cells.

The diversity of organelles demonstrates the complexity of eukaryotic cells, but due to functional specialization the same organelles are not necessarily present in all cell types. Some human cells show extreme variations (e.g. mature red blood cells lack a nucleus and there are no organelles at all in terminally differentiated keratinocytes). In other cases, particular organelles are present in very few cell types and are required for specific functions. For example, epithelial cells lining the respiratory tract contain cilia to move particulate contaminants out of the lungs, while mature sperm are the only human cells to possess flagella. Most cells, however, contain a 'standard package' of organelles whose functions are discussed below.

NUCLEUS: the repository of the genetic material

The **nucleus** contains the vast majority (typically 99.5%) of the DNA of an animal cell, in the form of linear chromosomes. It is surrounded by a **nuclear envelope**, composed of two membranes separated by a narrow space and continuous with the endoplasmic reticulum (see below). The remainder of the cell is known as the **cytoplasm** and consists of various organelles, membranes and an aqueous compartment known as the **cytosol**. Communication between the nucleus and the cytoplasm occurs via **nuclear pores**, openings in the nuclear envelope which are surrounded by **nuclear pore complexes**. The latter are protein complexes which act as specific transporters of macromolecules between the nucleus and cytoplasm.

Within the nucleus, the chromosomes are arranged in a highly ordered way. This reflects the existence of a complex substructure known as the **nuclear matrix**, a network of proteins and RNA to which the chromosomes are attached via DNA sequences known as **matrix attachment regions (MARs)**. There are other discernible structures within the nucleus which suggest an intricate organization of different biochemical processes. The **nucleolus** is a discrete region where **ribosomal RNA (rRNA)** is synthesized and processed. The genes for rRNA, which are found predominantly as clusters on the short arms of chromosomes 13, 14, 15, 21 and 22 in human cells, are located in this organelle. The high RNA content gives the nucleolus its dense appearance in electron micrographs. Other **nuclear bodies** have been identified but are less well-characterized. For example, **Cajal bodies** (otherwise known as **coiled bodies** or **gems**) are thought to be sites of synthesis of **small nuclear ribonucleoprotein (snRNP)** particles and may also be involved in gene regulation. A number of transcription factors and cell cycle regulatory proteins have been shown to be located at foci within the nucleus, often associated with particular genetic loci. This suggests that chromatin is recruited to **transcription factories** at particular nuclear sites in contrast to the traditional view that transcription factors diffuse freely around the nucleus. One particularly interesting example is the **PML body**, which appears as a ring and is composed predominantly of the premyelocytic leukemia (PML) protein. In patients with acute premyelocytic leukemia, a translocation involving the PML gene

and another gene encoding a retinoic acid receptor generates a fusion protein that cannot form PML bodies in the normal way. Treatment with retinoic acid restores the ability of the cell to form PML bodies presumably by allowing the protein to become correctly localized, and also leads to cancer remission. Other nuclear bodies appear to be involved in processes downstream from transcription. These include **perichromatin fibrils** (sites of nascent RNA accumulation), **cleavage bodies** (sites of polyadenylation and cleavage) and **speckles** or **interchromatin granule clusters** (involved in RNA splicing).

MITOCHONDRIA: the power-houses of aerobic eukaryotic cells

A conspicuous feature in the cytoplasm of most eukaryotic cells, **mitochondria** are the organelles which are devoted to generating power. The mitochondrial **respiratory chain** is a series of five membrane-bound protein complexes, including various quinones, cytochromes and iron–sulfur proteins. These complexes are collectively responsible for **oxidative phosphorylation**, a reaction in which organic nutrients are oxidized by molecular oxygen and the resulting chemical energy is used to generate ATP. By subsequently diffusing to all parts of the cell the ATP can donate its stored energy (by hydrolysis of ATP to ADP) and the released energy is used to drive numerous cellular functions.

Although variable in size, mitochondria typically have a diameter of about 1 μm, similar to that of bacterial cells. Mitochondria have two membranes: a comparatively smooth outer membrane and a complex **inner mitochondrial membrane** which has a very large surface area because of numerous infoldings (*cristae*). The inner compartment, the **mitochondrial matrix**, is a very concentrated aqueous solution of many enzymes and chemical intermediates involved in energy metabolism.

Mitochondria (and the chloroplasts of plants cells) are the only other eukaryotic organelles which contain DNA. In addition, they have their own ribosomes which are distinct from the ribosomes found in the cytosol. The mitochondrial ribosomes are used to translate mRNA transcribed from mitochondrial DNA. This and other evidence suggests that mitochondria are the descendants of aerobic bacteria that lived symbiotically with the precursors of eukaryotic cells (Section 12.2.1). However, the mitochondrial DNA can specify only a very few of the mitochondrial functions and most mitochondrial proteins are encoded by genes in the nucleus. In the latter case, the mRNA is translated on ribosomes in the cytosol and the resulting proteins are then imported into mitochondria.

ENDOPLASMIC RETICULUM: a major site of protein and lipid synthesis

The **endoplasmic reticulum** consists of flattened, single membrane vesicles whose inner compartments, the *cisternae*, are interconnected to form channels throughout the cytoplasm. It is physically and functionally divided into two components. The **rough endoplasmic reticulum** is so-called because the surface

Box 3.1: *(continued)*

is studded with ribosomes which are larger than those found in the mitochondria, whereas the **smooth endoplasmic reticulum** lacks any adhering ribosomes. Proteins synthesized by the ribosomes which adhere to the rough endoplasmic reticulum cross the membrane of the endoplasmic reticulum and appear in the intracisternal space. From here, they are transported to the periphery of the cell where they will be incorporated into the plasma membrane, retrieved to the endoplasmic reticulum or secreted from the cell altogether. Many human proteins are *glycosylated* (have sugar residues added to them) and this process begins in the endoplasmic reticulum.

GOLGI APPARATUS: the machinery for secretion

The **Golgi complex** consists of flattened, single membrane vesicles which are often stacked. The primary functions of the Golgi apparatus are in the secretion of cell products, such as proteins, to the exterior and in helping to form the plasma membrane and the membranes of lysosomes. Small vesicles arise peripherally by a pinching-off process and in some of these vesicles secretory products are concentrated (**secretory vacuoles**). Glycoproteins arriving from the endoplasmic reticulum are further modified in the Golgi complex.

PEROXISOMES (MICROBODIES): specialists in dangerous chemistry

Peroxisomes (microbodies) are small single-membrane vesicles containing a variety of enzymes that use molecular oxygen to oxidize their substrates and generate hydrogen peroxide. The hydrogen peroxide is used by a major peroxisomal enzyme, catalase, to oxidize a wide range of compounds. Reactive oxygen species like peroxide and the superoxide radical are highly toxic to cells and must be carefully contained.

LYSOSOMES: intracellular digestive organs

Lysosomes are small vesicles, delineated by a single membrane, which contain hydrolytic enzymes such as ribonuclease and phosphatase. Lysosomes function in the digestion of materials brought into the cell by phagocytosis or pinocytosis, and help in the degradation of cell components following cell death.

PLASMA MEMBRANE (CELL MEMBRANE): guarded frontier of the cell

The **plasma membrane** is composed of a phospholipid bilayer in which the hydrophobic lipid groups are located in the interior. They are sandwiched by hydrophilic phosphate groups which on both sides are in contact with a polar aqueous environment, the exterior of the cell and the cytoplasm. As well as providing a generally protective barrier, the plasma membrane has a variety of important roles:

▶ it is *selectively permeable*, regulating transport of a variety of ions and small molecules into and out of the cell. It contains active transport systems for various ions such as Na^+, K^+ and Ca_2^+, and for various nutrients such as glucose and amino acids, as well as a number of important enzymes;

▶ it contains a variety of integral membrane proteins or proteins anchored to one face of the membrane which play important roles in cell–cell signaling.

CYTOSOL: a highly concentrated and structured aqueous solution

The **cytosol**, the aqueous component of the cytoplasm, makes up about half the volume of the cell and is the site of major metabolic activity including most protein synthesis and intermediary metabolism. As well as containing soluble components, the cytosol is very highly organized by a series of protein filaments, collectively called the **cytoskeleton**, which play a major role in cell movement, cell shape and intracellular transport (see *Box 3.2*).

CILIA and FLAGELLA: movers and shakers

Cilia and flagella are structures that extend from the plasma membrane and are used to facilitate movement. **Cilia** are small structures that beat backwards and forwards, or rotate. Unicellular organisms often achieve motility by the coordinated beating of thousands of cilia, but these structures are also present on fixed cells in animals where they are used to move streams of liquid. In humans, cilia are found on epithelial cells lining the respiratory tract (where they propel mucus and any particulate matter contained therein away from the lungs) and oviduct (where they provide a current to move the egg towards the uterus). Cilia are also present on a tiny embryonic structure called the **node** (see Section 3.7.5) where their rotation causes the perinodal fluid to move to one side of the embryo. This process is believed to be important in the specification of the embryo's left–right axis (*Box 3.7*). **Flagella** are larger structures that move in a whip-like fashion. Many unicellular organisms achieve motility with flagella, but unlike cilia there are generally only one or two flagella per cell. The only human cells to possess flagella are sperm, which use these organelles as a means of propulsion. Both cilia and flagella are composed of bundles of **microtubule filaments**, with a characteristic **'9 + 2'** (nine outer doublet microtubules surrounding two singlet microtubules in the center) or a **'9 + 0'** structure. These structures are attached to **basal bodies** under the plasma membrane, which arise by repeated duplication of centrioles during cell differentiation (see *Box 2.1*). Movement is produced by the sliding of the outer doublet microtubules over each other, which is controlled by the motor protein **dynein**. In humans, dynein deficiencies result in **immotile cilia syndrome**, which is characterized by recurrent respiratory infections, laterality defects and also results in infertility due to the failure of eggs to reach the uterus and the inability of sperm to whip their tails.

Box 3.2: The cytoskeleton: the key to cell movement and cell shape and a major framework for intracellular transport.

The cytoskeleton of eukaryotic cells is an internal scaffold of **protein filaments** which provides stability, generates the forces required for movement and changes in cell shape, facilitates the intracellular transport of organelles and also allows communication between the cell and its environment. There are three types of cytoskeletal filament: actin microfilaments, microtubules (made of tubulin) and intermediate filaments (made from a variety of different proteins, some of which are cell type specific). Actins and tubulins have been highly conserved during evolution. Microfilaments and microtubules are very dynamic structures and they are also *polarized* because of asymmetry during polymerization of the actin and tubulin subunits from which they are constructed. Thus, the polymers grow by attachment of new subunits at both ends but at different rates, resulting in a fast-growing end known as the **plus (+) end** and a slow-growing end known as the **minus (–) end**. Specific classes of **motor proteins** can bind to the polarized filaments and move steadily along them in one direction, powered by ATP hydrolysis.

Actin filaments (also known as **microfilaments**) can be arranged into parallel bundles or net-like lattices. They provide mechanical support in the form of stress fibers, allow controlled changes to cell shape (e.g. apical constrictions, constrictions during cell division), and facilitate cell movement by forming structures such as filopodia and lamellipodia (extensions to the cell that allow it to crawl along surfaces). Actin filaments are the basis of special structures such as microvilli, adhesion plaques and the acrosomal process of the sperm head. They consist of two-stranded helical polymers of the protein **actin** and have a diameter of about 7–8 nm. They interact with members of the **myosin superfamily** of motor proteins. In muscle cells the actin–myosin interaction forms contractile units which provide muscle cells with their contractile power. In nonmuscle cells, actins also have a more general cellular role, e.g. allowing membranes to invaginate, as in **endocytosis**.

Microtubules are long hollow cylinders about 25–30 nm in diameter and are much more rigid than actin filaments. They are constructed by assembling a series of polymers based on alternating α-tubulin and β-tubulin residues. Associated with microtubules are members of two superfamilies of motor protein, **dyneins** and **kinesins**, which are responsible for moving particular cargos along the microtubule tracks in the cell. This is how mitochondria, Golgi stacks and secretory vesicles, for example, reach their appropriate locations within the cell. The dyneins and kinesins move along microtubules in opposite directions.

Typically, the minus ends of individual microtubules in animal cells converge at a **microtubule-organizing centre (MTOC)**, located in a portion of the cytoplasm adjacent to the nucleus. A single, well-defined center of this type, known as the **centrosome** is found in most animal cells. During cell division, microtubules form the **mitotic spindle** and microtubule filaments attach to a protein assembly (the **kinetochore**) located at the centromeres of the duplicated chromosomes and ensure that the chromosomes are moved in an ordered way into the two daughter cells (*Box 2.1*). Microtubules also form the core of **cilia** and **flagella**, which are discussed in *Box 3.1*.

Intermediate filaments are 7–11 nm in diameter and have a mainly structural role. They are not polarized and have no associated motor proteins. The neurofilaments (specific to neural cells) and keratins (specific to epithelial cells) are intermediate filament proteins.

---

primitive animals, ordinary somatic cells can give rise to germ cells throughout the life of the organism. However, in most of the animals we understand in detail – insects, nematodes and vertebrates – the germ cells are set aside very early in development as a dedicated **germ line** and represent the sole source of gametes. The germ cells are the only cells in the body that are capable of meiosis, and are therefore the only cells that can give rise to haploid descendants that can take part in fertilization. The germ line cells in mammals originate from **primordial germ cells (PGCs)** which in the case of humans arise during the second week of development.

3.1.4 In multicellular organisms, no two cells carry exactly the same DNA sequence

As described in Section 2.1 the reference DNA content of cells, the **C value**, is provided by the haploid chromosome set (n), as in sperm and egg cells. This value varies widely for different organisms and the lack of a direct relationship between the value of C and the biological complexity is known as the **C value paradox**. For example, while most mammals have a haploid genome size in the range of about 2500–3500 Mb of DNA it is somewhat surprising to find that the haploid DNA content of a human cell is only 19% of that of an onion, 4% of that of some lilies and remarkably only 0.5% of that of the single-celled eukaryote *Amoeba dubia* (see *Table 3.1* and the database of genome sizes at http://www.cbs.dtu.dk/databases/DOGS).

Within a single individual there is also considerable variation in DNA content as a result of differences in the number of copies of the chromosome set (**ploidy**). Although most human somatic cells are diploid, there are exceptions. As discussed in *Box 3.1*, some terminally differentiated cells, such as red blood cells, keratinocytes and platelets, have no nuclei and are described as **nulliploid**. Other cells are **polyploid** as a result of DNA replication without cell division (**endomitosis**), or as a result of cell fusion. For example, the ploidy of hepatocytes ranges from 2 n to 8 n, that of cardiomyocytes (heart muscle cells) from 4 n to 8 n, and that of the giant megakaryocytes of the bone marrow from 16 n to 64 n. The latter cells individually give rise to thousands of nulliploid platelet cells (*Figure 3.3*). Polyploidy as a result of cell fusion results in some naturally occurring cells (e.g. muscle fibers) having multiple diploid nuclei, forming a **syncytium** (*Figure 3.3*).

As a result of inherited differences in the DNA sequence (mutations) the DNA of cells from different organisms can differ very considerably. The extent of the sequence differences is largely proportional to the evolutionary distance separating the two organisms. Thus, the DNA of a human cell is very closely related in sequence to that of a chimpanzee cell (98–99%) but increasingly more divergent compared to cells of a mouse, frog and fruit fly, respectively. DNA from cells of different individuals of the same species also shows mutational differences. There is approximately one change in every 1000 nucleotides when comparing the DNA from two unrelated humans. There are also differences in the DNA sequence of cells from a single individual. Such differences can arise in three ways.

▶ *Programmed differences in specialized cells*. Examples include sperm cells which bear an X or a Y chromosome and so carry different sets of sex chromosome DNA, and mature B and T lymphocytes which undergo *cell-specific DNA rearrangements* leading to cell-to-cell differences in the types of antibody or T-cell receptor produced.

▶ *Random mutation and instability of DNA.* The DNA of all cells is constantly subject to environmental damage, chemical degradation and repair, and this impacts on the accuracy of DNA replication. The small but significant error rate means that new mutations arise in the DNA sequence at every round of cell division. During development, each cell therefore builds up a unique profile of mutations and *no two cells of the same individual will have precisely the same DNA sequence*. Most of these mutations have no phenotypic effect at all, and even mutations occurring in essential genes have little impact on the organism as a whole when they occur in isolated cells (unless they cause that cell to proliferate; see Chapter 17). Only if a deleterious mutation occurs in the germ line, in stem cells or in somatic cells early in development, are the effects likely to be seen at the phenotypic level.

▶ *Chimerism and colonization*. Very occasionally, an individual may be represented by two or more clones of cells with very different DNA sequences, reflecting fusion of fraternal twin embryos early in development, or colo-

Table 3.1: The C-value paradox: genome size is not simply related to organism complexity

Organism	Genome size	Number of genes
Unicellular		
Escherichia coli	4.2 Mb	4300
Saccharomyces cerevisiae	13 Mb	6300
Amoeba dubia	670 000 Mb	?
Multicellular		
Caenorhabditis elegans	95 Mb	*ca.* 20 000
Drosophila melanogaster	165 Mb	*ca.* 13 000
Allium cepa (onion)	15 000 Mb	?
Mus musculus	3000 Mb	*ca.* 30 000
Homo sapiens	3300 Mb	*ca.* 30 000

nization of one embryo by cells derived from another (*Figure 4.10*). At a superficial level, such differences also exist where an individual has received a graft, transplant or transfusion.

3.1.5 Cells from multicellular organisms can be studied *in situ* or in culture

Single celled organisms can often be studied by culturing them on a medium that is representative of their normal environment. In the case of multicellular organisms, the environment of an individual cell may be difficult to reproduce. Therefore, cells from a multicellular organism can be studied either in the natural context of the entire organism or in an artificial environment outside it. Several different approaches can be used.

▶ Analysis *in situ* (as part of the original animal). For the simplest animals, such as *Caenorhabditis elegans*, it is possible to study individual cells in whole living specimens by

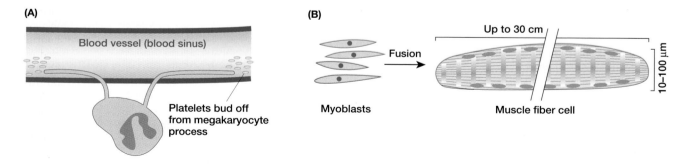

Figure 3.3: Some cells form by fragmentation or fusion of other cells.

Platelets are formed by budding from a giant megakaryocyte. They have no nucleus. Muscle cells are formed by fusion of large numbers of myoblast cells.

microscopy. The use of green fluorescent protein as a vital marker (*Box 20.4*) allows *in situ* studies to be carried out in living specimens of larger animals as long as the tissue under investigation is optically translucent. The embryos of many animals (e.g. *Drosophila*, zebrafish, mice) are translucent at early stages. For opaque specimens, it may be necessary to prepare tissue sections after treatment.

▶ As a **tissue explant**. Many studies in developmental biology are facilitated by investigating the behavior of particular tissue explants in isolation or in combination. This is used to test the effect of particular developmental molecules or to study the *inductive* effects of one cell population on another (see Section 3.4.2).

▶ As **primary cells**. A tissue explant can be dissociated, allowing individual cells to separate and grow individually in culture. This is not always a straightforward approach, since some cell types can be very difficult to grow and the effort required to maintain primary cells can be considerable.

▶ As a stable **cell line**. A stable cell line is often easy to grow in culture and so can be expanded easily and shipped to researchers throughout the world. Since the cells can be characterized for various diagnostic features, researchers can be confident that the results they obtain from a particular cell line can be referenced against or integrated with those of others working elsewhere on the same cell line.

There are many advantages to the study of tissue explants or cultured primary cells but they are difficult to handle and represent a nonpermanent resource. This is because most animal cells in culture will go through a certain number of divisions and then reach **senescence**, where they stop dividing and withdraw from the cell cycle, entering a quiescent state known as G_0 (Campisi, 1996). The number of rounds of division depends on the cell type, the species and the age of the source organism. For example, fetal human fibroblasts will divide approximately 60 times in culture, while fibroblasts taken from an adult human will undergo about half this number of divisions. There appears to be some relationship between the division potential and the longevity of the species. For example, the number of divisions achieved by human fetal fibroblasts (about 60 divisions, maximum lifespan 120 years) lies between the number achieved by fetal mouse fibroblasts (about 15 divisions, maximum lifespan 3 years) and fetal fibroblasts from the giant Galapagos tortoise (about 125 divisions, maximum lifespan 175 years).

The limitations of primary cells can be overcome by establishing a **permanent cell line**, which will divide in culture indefinitely. Many cell lines have been established from tumor explants, which have already lost certain growth restrictions. For example, the widely used **HeLa cell line** was derived from a cervical tumor removed from a patient called Henrietta Lacks in 1955. The problem with cell lines from tumor tissues is that they often have very different karyotypes compared to the original source tissue, resulting in phenotypic differences that become more pronounced with increasing passage number. Cell lines can also be obtained from primary cells by

persuading them to undergo **growth transformation (immortalization)** in culture. This process, which is analogous to the loss of growth restrictions seen in tumors, can occur spontaneously due to acquired mutations, or can be induced by irradiation, chemical mutagens or transforming viruses. More recently, cell lines have been established by transfection of primary cells with particular DNA sequences that are oncogenic. Thousands of cell lines, representing different human and animal tissues, are maintained in **cell banks** such as the American Type Culture Collection (ATCC), the European Collection of Cell Cultures (ECACC) and the Interlab Cell Line Collection (ICLC) (see *Table 3.2*; Stacey and Doyle, 2000).

An important application of cell lines is the maintenance of renewable genetic resources from particular human patients. This is required for a number of procedures, from quality control in tissue typing laboratories to linkage analysis and marker-based mapping. In this case, the phenotype of the cell is unimportant but the genotype must be accurately conserved. One of the most straightforward ways to achieve this is to make **lymphoblastoid cell lines** by transformation with Epstein–Barr virus (EBV), which remains *episomal* (*nonintegrated*) and therefore does not alter the endogenous genome. Human B lymphocytes have a receptor for EBV and once infected, they can become immortalized with a high rate of success to produce a cell line which can be propagated indefinitely in culture (or cryopreserved for future needs).

3.2 Cell interactions

3.2.1 Communication between cells involves the perception of signaling molecules by specific receptors

The survival and reproduction of individual cells in a multicellular organism is subordinate to the survival and reproduction of the organism as a whole. Somatic cells must cooperate with each other for the benefit of the organism. For this reason, cells within a multicellular organism need to communicate in order to coordinate and regulate physiological and biochemical functions. Cell communication is based on the production of **signaling molecules** by one population of cells and their recognition by **receptors** generally found on the surface of the responding cells (*Table 3.3*). Animal cells positively bristle with receptors for signaling molecules, and these receptors are connected to internal **signal transduction pathways** which allow the regulation of transcription factor activity, ultimately altering patterns of gene expression (see Twyman 2001 for an overview).

Signaling between cells can take place over different ranges. Most people are familiar with the concept of **endocrine signaling**, where a hormone released from an endocrine gland in one part of the body reaches a distant population of target cells by traveling through the blood stream. Endocrine signaling is important for the maintenance of homeostasis and also plays a major role in later development, when the vascular system has been established. In early development, however,

Table 3.2: Examples of human cell lines and their uses

ATCC code	Cell line	Origins and applications
Lymphoid cell lines		
CCL213	DAUDI	B-cell line, derived from Burkitt's lymphoma, used in cancer studies
CRL1432	NAMALWA	B-cell line, derived from Burkitt's lymphoma, used to manufacture interferons
TIB152	JURKAT E61	T-cell line, derived from acute lymphoblastic leukemia, produces large amounts of interleukin-2
CRL1942	SUP-T1	T-cell line, derived from T- lymphoblastic leukemia, supports HIV replication
Myeloid cell lines		
CCL240	HL-60	Derived from promyelocytic leukemia cells, used to study differentiation in the myeloid lineage
TIB202	THP	Derived from acute monocytic leukemia, used to study macrophage activation and maturation
Kidney cell lines		
CRL1573	293	Adenovirus transformed fetal kidney cells, used to study adenovirus
CRL2190	HK-2	Proximal tubular cell line transformed with E6/E7 genes from human papillomavirus, used as model system for proximal tubule studies
Liver cell lines		
HB8064	Hep3B	Produces many plasma proteins
HTB52	SK-HEP-1	Derived from liver adenocarcinoma, induces tumors in nude mice
Ovarian cell lines		
HTB75	Caov-3	From ovarian adenocarcinoma, produces mutant p53, used in cytokine and cancer studies
CRL1572	PA-1	From ovarian teratoma, used in developmental studies and drug testing

the most important signals travel over short distances. **Paracrine signaling** involves the release of a signaling molecule from one population of cells, and the diffusion of that molecule over a short distance to the responding cells. **Juxtracrine signaling** is based on the interaction of neighboring cells, and generally reflects the fact that the signal produced by one cell remains tethered to the plasma membrane. Cells also respond to molecules present in their immediate environment, the **extracellular matrix** (Section 3.2.4).

3.2.2 Activated receptors initiate signal transduction pathways that may involve enzyme cascades or second messengers, and result in the activation or inhibition of transcription factors

Once the receptor for a particular signaling protein has been activated, a chain of events is unleashed that eventually results in a change to the pattern of gene activity in the responding cell. The number and nature of links in the chain depends on the particular signaling molecule involved. Steroid and thyroid hormones and the developmental regulator retinoic acid are able to diffuse through the plasma membrane, so their receptors are located inside the cell. Once bound to their ligands, these receptors act directly as transcription factors to modulate the expression of downstream genes. Therefore, steroid-like molecules utilize a single-step signal transduction pathway (Tsai and O'Malley, 1994).

Most other signaling molecules remain outside the cell and bind to transmembrane receptors. Ligand binding to the extracellular domain of the receptor causes a conformational change in the internal domain which generally stimulates a latent enzyme activity. In the case of receptor kinases, this conformational change activates a kinase activity which is either integral to the receptor or integral to a protein that is associated with the receptor. This first causes the receptor to phosphorylate itself and then allows it to phosphorylate and hence activate other proteins inside the cell. These target proteins are themselves often kinases, which then phosphorylate proteins further downstream. Eventually, this cascade of kinase activity reaches a transcription factor, which is activated or inhibited as appropriate resulting in a change in gene expression. In some cases, this signaling cascade is short (e.g. the cytokine-regulated JAK-STAT pathway; Pellegrini and Dusanter-Fourt, 1997; Hou *et al.*, 2002; *Figure 3.4*) whereas in others there are many steps (e.g. the growth factor-regulated MAP kinase pathway; Robinson and Cobb, 1997; *Figure 3.5*).

For G-protein coupled receptors, signal transduction is achieved by the activation of second messengers (small molecules such as cAMP, calcium or lipids). Ligand binding causes the guanidine nucleotide binding protein (*G protein*) associated with the receptor to exchange GDP for GTP, which causes it to dissociate into α and $\beta\gamma$ units. Each of these units can then interact with downstream proteins to regulate second messenger levels (Bourne, 1997). Depending on the particular type of G-protein present, ligand binding may

Table 3.3: Important classes of signaling molecules and their receptors

	Receptor	Examples
SECRETED SIGNALS		
Various (peptides, proteins, small organic molecules, physical stimuli)	G-protein coupled receptor Single polypeptide. The central hydrophobic region forms a seven-pass transmembrane domain, the N-terminal domain binds the ligand, the internal C-terminal domain is associated with a guanidine nucleotide-binding protein	Receptors for adrenaline (epinephrine), serotonin, glucagon, follicle-stimulating hormone, histamine, opioids and neurokinins. Also olfaction receptors, taste receptors and the visual receptor rhodopsin
Growth factors, ephrins	Receptor tyrosine kinase Dimeric; N-terminal ligand-binding domain, internal kinase domain, spans membrane once	Receptors for insulin, fibroblast growth factors, epidermal growth factor, ephrins
Cytokines	Receptors with associated tyrosine kinase activity Oligomeric; N-terminal ligand-binding domain, internal domain associated with Janus kinase (JAK)	Receptors for growth hormone, prolactin, erythropoietin, colony stimulating factors, interleukins, interferons
Transforming growth factor-β family	Receptors with associated serine/threonine kinase activity Oligomeric; N-terminal ligand-binding domain, internal domain associated with SMAD proteins	Receptors for TGF-β proteins, bone morphogenetic proteins, Nodal, glial-derived neurotrophic factor
Hedgehog family	Patched	Receptors for Sonic hedgehog and Indian hedgehog in mammalian development
Wnt family	Frizzled	Receptors for Wnt family proteins in mammalian development
Steroid hormones, retinoids	Nuclear receptors Dimeric; reside in cytoplasm or nucleus, converted to transcription factors in association with ligand	Receptors for steroid hormones, thyroid hormones, vitamin D, retinoic acid
IMMOBILIZED SIGNALS		
Delta/Serrate family	Notch	Delta and Serrate in neural development

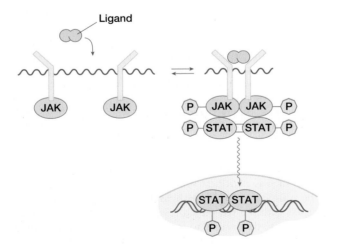

Figure 3.4: Cytokine receptors are dimeric or oligomeric, with each polypeptide spanning the membrane once.

The receptors possess no intrinsic tyrosine kinase activity, but they are constitutively associated with cytoplasmic tyrosine kinases of the Janus family (Janus kinases, JAKs). Ligand binding causes the receptors to dimerize. This results in reciprocal autotransphosphorylation of the associated JAKs, which become active and phosphorylate the receptor itself. The receptor is then able to recruit inactive transcription factors called STATs (signal transducers and activators of transcription) that bind to phosphotyrosine residues through their SH2 domains. The STATs are then phosphorylated by the JAKs, allowing them to dimerize and translocate to the nucleus, where they activate downstream genes. Redrawn from Twyman (2001) *Instant Notes in Developmental Biology*, published by BIOS Scientific Publishers.

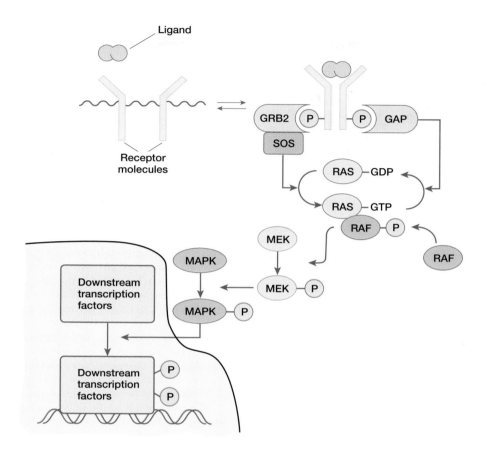

Figure 3.5: Growth factors act as ligands for receptor tyrosine kinases.

Ligand binding stimulates receptor dimerization and activates the cytoplasmic tyrosine kinase activity intrinsic to the receptor. The receptor can phosphorylate not only other proteins but also itself (autotransphosphorylation) resulting in the recruitment of proteins that bind specifically to phosphotyrosine residues, including enzymes that modulate the activity of Ras. Activated Ras recruits Raf to the membrane where its kinase activity is stimulated. Raf then activates MEK, which in turn activates MAP kinase. Both MEK and MAP kinase activate latent transcription factors therefore changing patterns of gene expression in the nucleus. Redrawn from Twyman (2001) *Instant Notes Developmental Biology*, published by BIOS Scientific Publishers.

Table 3.4: Cell junctions

Adherens junctions	Anchoring junctions in which the actin filaments of the cytoskeleton are connected to cadherins at the cell surface, therefore linking neighboring cells into a continuous adhesion belt
Desmosomes	Anchoring junctions in which the intermediate filaments of the cytoskeleton are connected to cadherins at the cell surface
Focal contacts	Anchoring junctions in which the actin filaments of the cytoskeleton are connected to integrins at the cell surface, providing punctate attachment sites to the extracellular matrix
Hemidesmosomes	Anchoring junctions in which the intermediate filaments of the cytoskeleton are connected to integrins at the cell surface, used to connect epithelial cells to the basal lamina
Gap junctions	These are pores that connect the cytosol of adjacent cells. They are formed by six identical connexin proteins that form a channel in each membrane. The attachment of connexin complexes on adjacent cells creates the gap junction
Tight junctions	Interconnected transmembrane proteins that seal epithelial cells into a continuous sheet. Depending on the density of the proteins in the junction, the barrier may be completely impermeable or selectively permeable to molecules of a particular size

stimulate or inhibit the enzyme adenylate cyclase, resulting in a change in the levels of intracellular cAMP (Houslay and Milligan, 1997). Other G proteins stimulate the production of lipids such as inositol-1,4,5-trisphosphate and diacylglycerol (Speigel *et al.*, 1996), or the release of calcium ions (Clapham, 1995). In turn, the second messengers activate downstream protein kinases such as cAMP-dependent protein kinase A and calcium-dependent protein kinase C, which go on to phosphorylate and therefore change the activity of particular transcription factors.

There is extensive cross-talk between the signaling pathways discussed above. For example, both protein kinase A and protein kinase C interact with the MAP kinase pathway, receptor tyrosine kinases can stimulate the JAK–STAT pathway and influence the levels of second messengers, and cytokine receptors can influence Ras activity. The response given by a particular cell depends on the sum of all signals arriving at its surface and is dependent on the particular receptors and signaling components that are present.

3.2.3 The organization of cells to form tissues requires cell adhesion

Cells become organized into tissues through the expression of **adhesion molecules** (Cunningham, 1995; Gumbiner, 1996). Essentially, adhesion molecules work in the same way as any other receptor–ligand interaction, although in this case the receptor and ligand are attached to the surfaces of adjacent cells and there may be hundreds of thousands of such molecules per cell, to ensure the binding is very strong. Cells may stick together directly, through the expression of complementary adhesion molecules on their surfaces, and/or they may form associations with the extracellular matrix. During development, changes in the expression of adhesion molecules allow cells to make and break connections with each other, which facilitates migration either individually or as sheets, resulting in large-scale rearrangements (McNeil, 2000; Irvine and Rauskolb, 2001). In the mature organism, adhesion interactions between cells are generally strengthened by the formation of specialized contact regions, known as **cell junctions** (Steinberg and McNutt, 1999; Perez-Moreno *et al.*, 2003; *Table 3.4*).

There are four major classes of cell adhesion molecule (Humphries and Newham, 1998; Hynes, 2002).

▶ **Cadherins**. These are calcium-dependent transmembrane adhesion proteins that recognize and bind to identical cadherins on the surface of other cells. The process of binding to identical molecules is known as **homophilic binding**. About 30 different cadherins are present in mammals, the most important of which are *E-cadherin* (restricted mainly to epithelial cells) and *N-cadherin* (expressed predominantly in the nervous system).

▶ **Ig-CAMs.** These are calcium-independent transmembrane adhesion molecules that have structural similarity to the immunoglobulin (Ig) family. For example, the *neural cell adhesion molecule (NCAM)* is expressed in the nervous system and in the developing somites.

▶ **Integrins**. These are calcium-dependent adhesion molecules that usually mediate cell-matrix interactions (see below), but certain leukocyte integrins are also involved in cell–cell adhesion.

▶ **Selectins**. These molecules are expressed by endothelial cells during an inflammatory response. They recognize carbohydrate residues on the surface of neutrophils and guide these cells to sites of inflammation.

3.2.4 The extracellular matrix provides a scaffold for all tissues in the body and is also an important source of signals that control cell behavior

The spaces between the cells of the body are not empty. The extracellular space is filled with a three-dimensional array of protein fibers embedded in a gel of complex carbohydrates called **glycosaminoglycans**. This material, the **extracellular material (ECM)**, is variable in composition, depending on what materials the local cells secrete into their immediate environment. Cells also influence the structure of the ECM by secreting modifying enzymes, such as proteases, and this can influence cell behavior (Streuli, 1999). In turn, the ECM provides a structural scaffold that plays an essential role in the maintenance of tissue integrity, and also guides cell behavior during development by interacting with membrane-spanning receptors (Giancotti and Ruoslahti, 1999; Bokel and Brown, 2002).

The components of the ECM can be divided into several groups (see *Table 3.5*):

▶ *structural proteins*, such as collagens and elastins;

▶ *adhesion proteins*, such as laminins and fibronectins;

▶ *proteoglycans*, which comprise a core protein associated with various glycosaminoglycans. Proteoglycans differ from glycoproteins in that sugar residues in the former represent up to 95% of the total weight of the molecule;

▶ the free glycosaminoglycan, *hyaluronic acid*.

The structural and adhesion proteins of the ECM play a major role in determining its function. For example, collagen-rich tissues are very strong (e.g. tendons) while elastin-rich tissues are very stretchy (e.g. blood vessels). The ECM proteins interact with cells via transmembrane receptors called **integrins**. These bind to ECM proteins on the outside of the cell and to actin microfilaments on the inside via bridging proteins such as talin and α-actinin. Such interactions allow cells to move by contracting actin filaments against the fixed structure of the ECM. It has also been shown that integrins activate internal signaling pathways, which suggests that the ECM can influence gene expression inside the cell and therefore modify cell behavior in response to specific ECM components. An important example of the role of the ECM in cell behavior is the maintenance of **cell polarity** by the basal lamina (see Dustin, 2002 for more examples). In the intestinal epithelium, the presence of the basal lamina on one face of the cell monolayer is responsible for the establishment and maintenance of a flat basal surface

Table 3.5: Molecular components of the extracellular matrix

Component	Distribution	Structure/function
Collagens	Many tissues	The collagens are large trimeric glycoproteins. There are several chemically distinct forms of collagen which vary in abundance in different tissues. The most abundant forms are collagens I, II and III, which tend to form fibrils stabilized by cross-links between lysine residues. They provide structural support, and influence cell differentiation and migration through interactions with integrins. Conversely, collagen IV forms lattices rather than fibrils and is an important component of the basal lamina. It interacts with cells indirectly, through association with laminin
Fibronectin	Many tissues	Fibronectin is a dimeric glycoprotein whose major function is to facilitate cell–matrix adhesion. Fibronectins have collagen- and heparin-binding domains, which help to organize the ECM. The fibronectins interact with cells via integrins and influence cell shape, movement and differentiation
Laminins	Epithelial tissues	Laminins are trimeric proteins comprising A, B1 and B2 subunits. They interact with collagen IV to form the basal lamina, and their major function is to facilitate the adhesion of cells to the basal lamina by interacting with integrins and other cell surface molecules. There are many tissue-specific forms of each laminin subunit
Entactin	Epithelial tissues	Associated stoichiometrically with laminin, and contains sequences that may allow interactions with cell surface integrins
Elastin	Many tissues, but abundant in blood vessels, skin and other structures that stretch and deform.	Elastin is a non-glycosylated protein that forms networks and sheets by cross-linking. The protein has a natural elastic recoil and provides tissues with the ability to regain shape after deformation
Tenascin	Embryo, adult nervous system	Tenascin is a hexameric glycoprotein that plays an important role in the control of cell migration. It has an adhesive effect on some cells and a repellent effect on others depending on the receptors displayed on the cell surface
Vitronectin	Blood and other tissues	This protein is generally associated with fibronectin, although not as widely distributed, and interacts with particular classes of integrins to facilitate cell–matrix adhesion
Proteoglycans	Many tissues	Very diverse family of molecules comprising a core protein, such as decorin or syndecan, associated with one or more glycosaminoglycans (e.g. chondroitin sulfate, dermatan sulfate, heparin, heparan sulfate). Often adopt higher order structure in the ECM. These molecules mediate cell adhesion, and can also bind growth factors and other bioactive molecules
Hyaluronic acid	Many tissues	The only glycosaminoglycan that is nonsulfated and not covalently attached to a protein to form a proteoglycan. Facilitates cell migration particularly during development and tissue repair

adjacent to the connective tissue and an apical surface characterized by a brush border.

The proteoglycan and glycosaminoglycan component of the ECM serves a number of functions. First, the molecules are very soluble, forming hydrated gels that act as cushions to protect tissues against compression. Where the proteoglycan content of the ECM is particularly high, the tissue is highly resistant to compression (e.g. cartilage). Proteoglycans can form complex superstructures in which individual proteoglycan molecules are arranged around a hyaluronic acid backbone. These complexes can act as biological reservoirs by storing active molecules such as growth factors. Indeed, proteoglycans may be essential for the diffusion of certain signaling molecules. For example, both the Hedgehog and Wingless signaling pathways are inhibited in the *Drosophila* mutant *sugarless*, which is unable to synthesize proteoglycans correctly (see Selleck, 2000). Finally, proteoglycans also help to mediate cell-matrix adhesion. This is achieved by binding to cell surface enzymes known as **glyco-syltransferases**, whose function is to add sugar residues to carbohydrate groups. In the absence of free sugar, the carbohydrate chain is held by the enzyme. When sugar is available, the reaction is completed and the carbohydrate chain is released. Such cycles of adhesion and release may be particularly important in the control of cell migration.

3.3 An overview of development

The term **development** as applied to animals refers to the process by which a single cell gives rise to a mature organism. It is often convenient to divide animal development into an **embryonic** stage, during which all the major organ systems are established, and a **post-embryonic** stage that (in the case of mammals) consists predominantly of growth and refinement. Developmental biologists tend to concentrate on the embryonic stage because this is where the most exciting and dramatic events occur, but this does not diminish the importance of post-embryonic development. Once the basic body plan is established, it is not clear when development stops. In humans, there is a continuum of post-embryonic growth and consolidation which is punctuated by birth and carries on for up to two decades afterwards, with some organs reaching maturity before others. It can be argued that human development ceases when the individual becomes sexually mature, but many tissues need to be replenished throughout life (e.g. skin, blood, intestinal epithelium) and in such cases development never really stops at all, only reaches an equilibrium. Some scientists even regard aging as a part of development, since it represents a natural part of the life cycle.

Development is a gradual process. The fertilized egg initially gives rise to a simple embryo with relatively crude features. As development proceeds, the number of cell types increases and the organization of those cells becomes more intricate. Complexity is achieved progressively. At the molecular level, development incorporates several different processes that affect the behavior of cells. These processes are inter-related and can occur separately or in combination in different parts of the embryo (Twyman, 2001):

▸ **cell proliferation,** by repeated cell division, which leads to an increase in cell number. In the mature organism, this is balanced by cell loss;

▸ **growth**, by the synthesis of macromolecules, which leads to an increase in overall size and biomass;

▸ **differentiation**, the process by which cells become structurally and functionally specialized;

▸ **pattern formation**, the process by which cells become organized, first to form the fundamental **body plan** of the organism and then the detailed structures of different organs and tissues;

▸ **morphogenesis**, or changes in overall form. Several underlying mechanisms may be involved in morphogenesis, including differential cell proliferation, selective cell–cell adhesion or cell–matrix adhesion, changes in cell shape and size, the selective use of programmed cell death and control over the symmetry and plane of cell division.

All of the above processes are controlled by genes, which specify where and when particular proteins are synthesized in the embryo and therefore how the different cells behave. While there are minor differences in DNA sequence among the cells in a multicellular organism, most cells at least contain the same genes. Therefore, for cells to diversify in the developing embryo, there must be *differential gene expression*. Gene expression is controlled by transcription factors, so development ultimately depends on which transcription factors are active in each cell (see, for example, Kuo *et al.*, 1992). As discussed in Section 3.2.2, the activity of transcription factors can be regulated by signaling between cells. Unraveling the signaling pathways and regulatory programs that control development is a major goal of biomedical research. In this chapter we will focus on vertebrate and particularly mammalian development. Our knowledge of early *human* development is fragmentary because access to samples for study is often restricted for ethical or practical reasons. As a result, much of the available information is derived from **animal models of development** (see *Box 3.3*).

3.4 The specialization of cells during development

3.4.1 Cell specialization involves an irreversible series of hierarchical decisions

Development can be likened to a stream running down the side of a mountain, with individual cells represented by leaves flowing with the water (*Figure 3.6*). The course of the water may branch many times before it reaches the bottom, but a leaf can only follow one path. As it approaches a branch point, the leaf must commit to one path or the other. Once that decision is made, the leaf cannot go back and choose a different path. The decision is irreversible.

After fertilization, the first few cell divisions which the mammalian zygote undergoes are symmetrical, giving daughter cells with the same developmental potential (or **potency**). The zygote and its immediate descendants are unspecialized and are said to be **totipotent**, because each cell retains the capacity to differentiate into all possible cells of the organism, including the extra-embryonic membranes. This is analogous to a leaf at the beginning of its mountain journey. As development proceeds cells become more specialized, and at the same time they become more *restricted* in their capacity to generate different types of descendent cells. The cells are forced to make decisions and choose alternative paths.

The first decision in the mammalian embryo is the choice between *inner cell mass* and *trophoblast*. The former becomes the embryo proper and the amnion while the latter becomes the chorion and the embryonic portion of the placenta. Cells of the inner cell mass are **pluripotent**, i.e. they can give rise to all of the cells of the embryo, and therefore of a whole animal, but they are no longer capable of giving rise to the extra-embryonic structures derived from the trophoblast. At any time up until this point, the potency of the embryonic cells is demonstrated by the ability of the embryo to form twins (*Box 3.4*).

Cells that have branched into the inner cell mass lineage are then faced with three choices. They can chose one of the three fundamental cell types of the embryo: the **ectoderm, mesoderm** or **endoderm**. Each of these **germ layers** can subsequently give rise to a specific and limited range of cell types but none of them is able to produce an entire embryo. They are said to be **multipotent**. Additionally, once cells are committed to one germ layer, they generally cannot produce cell types characteristic of the others, although recent evidence has suggested that stem cells might show more developmental plasticity than once assumed (see Section 3.4.6). Cells in the ectoderm lineage, for example, can give rise to epidermis, neural tissue or neural crest cells, but they cannot normally give rise to kidney cells (which derive from the mesoderm lineage) or liver cells (which derive from the endoderm lineage) (*Box 3.5*). Eventually, at the bottom of the mountain, progenitor cells arise which can give rise to only a single type of differentiated cell. These are described as **unipotent**.

Box 3.3: Animal models of development.

Research on development has focused on a relatively small number of **model organisms**, which are considered to be representative of the major taxonomic divisions of multicellular life. Some of these organisms were initially chosen for the ease with which they could be obtained and bred in captivity, while others were chosen because of specific experimental advantages they offered. The study of these organisms has shown that developmental genes, and indeed whole regulatory pathways, are very highly conserved. They have been instrumental in the identification and characterization of human genes involved in development. The importance of these species as experimental models is now reflected by their precedence among the genome projects that have been completed or are in progress (*Box 8.8*).

INVERTEBRATES

The major invertebrate models are the fruit fly (*Drosophila melanogaster*) and the nematode (*Caenorhabditis elegans*). Both these organisms are genetically amenable, and therefore suitable for large scale mutation screening, genetic analysis and genetic manipulation. The embryos of both species (as well as adult worms) are transparent and amenable to surgical manipulation. *Drosophila* was the first animal model to be studied in detail at the molecular level and many of the molecular mechanisms underpinning early embryonic development were established in this species. *Drosophila* also provides particularly useful models of neurogenesis and eye development. *Caenorhabditis* is remarkable for its stereotyped developmental program characterized by an almost invariant cell lineage and the availability of a complete wiring diagram of the nervous system. Lineage mutants provide a useful means to study cell memory in development. The nematode vulva is a well-established model of organogenesis. Other invertebrate models include mollusks, ascidians (tunicates) and annelids.

VERTEBRATES

Vertebrate model organisms serve as representative models of human development, both at the anatomic and molecular levels. The domestic chicken (*Gallus gallus*) and the African clawed frog (*Xenopus laevis*) are favored because both species produce robust embryos that develop outside the body and can be easily manipulated. However, these species are of little use for genetic analysis. In the case of *X. laevis*, this is primarily due to the long generation interval and the tetraploid genome, which makes genetic analysis difficult. There has been interest more recently in a related species, *X. tropicalis*, which produces smaller embryos but has a shorter generation interval and is diploid. The mouse (*Mus musculus*) is advantageous both in terms of its genetic amenability, its suitability for genetic manipulation (particularly gene targeting, which allows specific genes to be inactivated; see Chapter 20), and its closeness to humans. However, because the embryo must develop internally, surgical manipulation is difficult to carry out. The zebrafish (*Danio reiro*) combines genetic amenability with accessibility and therefore represents perhaps the most versatile of the vertebrate model organisms.

All vertebrate embryos pass through similar stages of development, including cleavage of a large egg, gastrulation, neurulation, somitogenesis and the formation of limb buds. They reach a **phylotypic stage** where the body plan of all vertebrates is much the same. Despite these similarities there are also major differences among the five vertebrate classes, predominantly in the cleavage and gastrulation stages, reflecting the alternative nutritional strategies. This is explained in more detail in Section 3.7.2. There are also differences in the methods used to specify the primary embryonic axes (Section 3.5.1), and in other developmental processes, such as sex-determination (*Box 3.9*).

Histology textbooks recognize over 200 different types of cell in the human body, encompassing a wide variety of functions (see *Box 3.6*). Some have a function which is limited to an organ system (e.g. hepatocyte, cardiomyocyte) while others may have more general functions e.g. fibroblasts. Some mature, specialized cells do not divide and are said to be *terminally differentiated*. Other cells divide actively and act as precursors of terminally differentiated cells. These types of cell are often distinguished by the suffix *–blast*, as in osteoblasts, chondroblasts, myoblasts etc. In some cases, the precursor cells are also capable of undergoing self-renewal, and these are known as **stem cells** (Section 3.4.3).

3.4.2 The choice between alternative fates may depend on lineage or position

A question often asked by developmental biologists is whether the **fate** of a cell (the range of cell types it can produce)

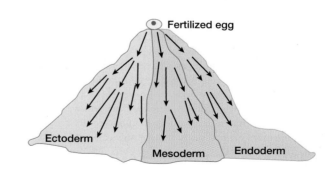

Figure 3.6: Developmental pathways can be thought of as branching streams running down the side of a mountain.

Box 3.4: Twinning in human embryos.

Approximately one in every 200 human pregnancies gives rise to twins. These are of two distinct types: fraternal (**dizygotic**) and identical (**monozygotic**). Fraternal twins result from the independent fertilization of two eggs and are no more closely related than any other siblings. Although developing in the same womb, the embryos have separate and independent sets of extra-embryonic membranes. Identical twins arise from the same fertilization event, and are produced by the division of the embryo while the cells are still totipotent or pluripotent. Early divisions, occurring during or prior to the morula stage, result in separate blastocysts which give rise to embryos shrouded by independent sets of extra-embryonic membranes. This represents about one-third of identical twins. In the remaining two-thirds, twinning occurs at the blastocyst stage, and this involves the division of the inner cell mass. The nature of the twinning reflects the exact stage at which the division occurs and how complete the division is. In most cases, the division occurs before day 9 of gestation, which is when the amnion is formed. Such twins share a common chorionic cavity but are surrounded by individual amnions. In a very small proportion of births, the division occurs after day 9 and the developing embryos are enclosed within a common amnion. Either through incomplete separation or subsequent fusion, these twins are occasionally conjoined.

---

depends on its **lineage** (i.e. the cells from which it is derived) or its **position** (i.e. the cells with which it is in contact). In the early development of vertebrate embryos, the position of a cell appears to be the most important determinant of cell fate. Cell fates are often specified by signals from nearby cells, a process termed **induction**. The formation of the neural plate in the *Xenopus* embryo provides a well-characterized example (Harland, 2000; Bainter *et al.*, 2001; *Figure 3.7*).

The neural plate arises from the surface ectoderm along the dorsal midline of the embryo while ectoderm on either side of the midline gives rise to epidermis. Initially, however, all the surface ectoderm is uncommitted or **naive** – that is, it is **competent** to give rise to either epidermis or neural plate. The signal for neural induction comes from the **notochord**, a mesodermal structure running along the craniocaudal axis of the embryo. Ectoderm cells above the notochord receive signals that promote neural development, while the surrounding ectoderm does not. However, if the notochord is removed, then all the ectoderm forms epidermis. Similarly, if an extra piece of notochord is grafted under ventral or lateral ectoderm (which would normally form epidermis), it forms neural plate tissue instead. Ectoderm removed from the presumptive neural plate region of the embryo and grafted elsewhere will form epidermis, while ectoderm removed from the ventral side of the embryo and grafted above the notochord will form neural plate. It is therefore clear that the fate of the ectoderm – epidermal or neural – depends on the position of the cells not their lineage, and is initially *reversible*. At this stage cell fate is said to be **specified**, which means it can still be altered by changing the environment of the cell. After a certain amount of time has elapsed, the fate of the ectoderm becomes fixed and can no longer be altered by grafting. At this stage, the cells are said to be **determined**, that is irreversibly committed to their fate. Determination generally means that the cell has initiated a molecular process that inevitably leads to differentiation. In some cases this is because a new transcription factor is synthesized and that transcription factor cannot be inactivated. In other cases, there may be modifications to the chromatin, which lock the pattern of gene expression in place. There may also be a loss of competence for induction. For example, ectoderm cells which are committed to becoming epidermis,

may stop synthesizing the receptor that responds to the signal emanating from the notochord.

There are fewer examples of cell fate specified by lineage in vertebrate embryos, but one explicit case is the behavior of stem cells, which we discuss below.

3.4.3 Stem cells are self-renewing progenitor cells

Even when an organism is fully grown, some cells, notably blood cells, epithelial cells in the skin and intestine, and sperm cells, need to be continuously manufactured. These are produced from self-renewing precursor cells, or **stem cells**. Stem cells can proliferate by normal *symmetrical cell division*, generating two similar daughter cells. As required, they can also undergo *asymmetric cell division*: one daughter cell has the same type of properties as the parent stem cell, but the other daughter cell has altered properties and becomes committed to producing a lineage of differentiated cells. The fate of the committed daughter cell is not influenced by its position, or by signals from other cells. The decision is *intrinsic* to the stem cell lineage. This type of nonconditional (autonomous) specification of cell fates results from the asymmetric distribution of regulatory molecules at cell division. In the case of neural stem cells, for example, there is asymmetric distribution of the proteins Notch-1 (concentrated at the apical pole) and Numb (concentrated at the basal pole). Division in the plane of the epithelial surface causes these determinants to be distributed equally, but a division at right angles to the epithelial surface causes them to be segregated in the daughter cells, which therefore develop differently (Kim and Schagat, 1996; *Figure 3.8*).

A useful example of stem cell lineages comes from blood cells. A single cell type, the **hematopoetic stem cell (HSC)**, can give rise to all blood cells (Morrison, *et al.*, 1995). This was dramatically verified by irradiating mice to ensure destruction of the bone marrow then grafting purified HSC cells from a different strain of mice whereupon the incoming cells were able to differentiate and re-populate the blood. HSCs are multipotent but the cell lineages to which they give rise become increasingly specialized, ultimately producing all of the terminally differentiated blood cells (*Figure 3.9*).

Box 3.5: Where our tissues come from – the developmental hierarchy in mammals.

Every type of human cell can be traced back through the developmental hierarchy to the fertilized egg. The chart below shows the developmental decisions that need to be made in order for the egg to give rise to each of these cell types, the journey down the mountain according to the analogy in section 3.4.1. Almost all cells derive from one of the three primary germ layers, the major exception being the primordial germ cells, which separate from the somatic lineage prior to gastrulation.

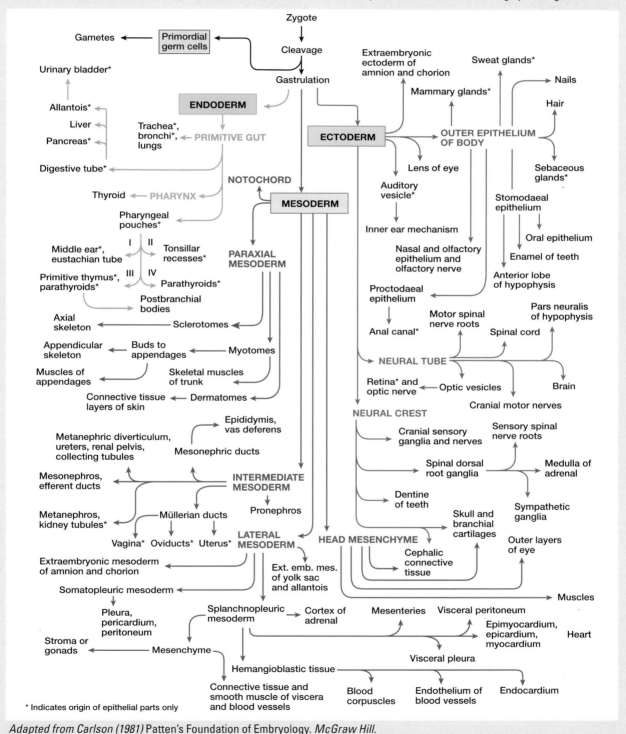

* Indicates origin of epithelial parts only

Adapted from Carlson (1981) Patten's Foundation of Embryology. McGraw Hill.

Box 3.6: The diversity of human cells.

Over 200 types of cells are described in histology textbooks. Here we illustrate the variety of size and form among some common cell types.

Ovum and **Sperm** (see *Figure 3.12* for characteristics, and *Box 11.4* for how they are generated from primordial germ cells).

Lymphocyte (*Figure 3.9*) – small round cell, 6–8 μm diameter. Very little cytoplasm. Typically 5000 per μl blood.

Erythrocyte (*Figure 3.9*) – flat biconcave disk 7.2 μm diameter. Erythrocytes have no nucleus, mitochondria or ribosomes. Metabolism is entirely by glycolysis. Lifespan 120 days. Typically 5 million per μl blood.

Megakaryocyte (*Figures 3.3 and 3.9*) – large bone marrow cell 35–150 μm diameter. Irregular lobulated nucleus containing 8–32 genomes, formed by endomitosis. Megakaryocytes fragment to form thousands of platelets.

Platelet (*Figures 3.3 and 3.9*) – 3–5 μm fragment of highly structured cytoplasm without nucleus. Lifespan 8 days. Typically 200 000 per μl blood.

Macrophage (*Figure 3.9*) – variable shape cell, wanders using pseudopodia. Specialized for engulfing particles by phagocytosis, contains many lysosomes to digest particles. Many macrophages fuse around large foreign bodies, forming giant tissue cells.

Epithelial cell – strongly adhesive cells that pack together to form epithelia with tight junctions (desmosomes) between cells. Some epithelial cells are specialized for ion transport, absorption or secretion.

Fibroblast – unspecialized cell of *connective tissue* capable of differentiating into cartilage, bone, fat and smooth muscle cells.

Hepatocyte – polyhedral cell 20–30 μm diameter, sometimes multinucleate. Rich in mitochondria and endoplasmic reticulum; contains lysosomes; may contain lipid droplets.

Muscle fiber cell (*Figure 3.3*) – multinucleated cell made by fusion of myoblasts. 10–100 μm diameter, may be several cm long. Nuclei lie around periphery; most of interior is occupied by 1–2 μm myofibrils, with many mitochondria in between.

Neuron – highly variable size and shape. Cell body 4–150 μm; usually many dendrites and one axon. One cell may have over 100 000 connections with other neurons. Axons of spinal cells innervating the feet are 1 m long.

Melanocyte (*Figure 3.9*) – epithelial cell with long branched processes that lie between keratinocytes and pass packages of pigment (melanosomes) into them. Typically 1500 per mm^2 of skin (regardless of skin color).

Keratinocyte – mature keratinocytes are scale-like structures full of keratin and devoid of nucleus or any organelles.

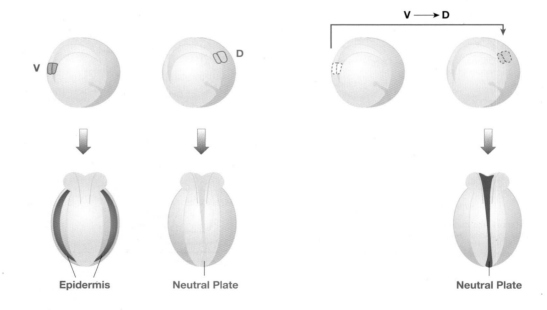

Figure 3.7: Grafting experiments in *Xenopus* show when ectoderm cells become committed to epidermal and neural cell fates.

On the left, ventral (V) ectoderm (green) gives rise to epidermis whereas dorsal (D) ectoderm (red) gives rise to the neural plate. On the right, if the ventral ectoderm is grafted onto the dorsal side of the embryo, it is respecified and is able to form the neural plate. This shows that the fate of the ectoderm is dependent on signals emanating from other cells in the vicinity of the neural plate, i.e. the cells of the axial mesoderm.

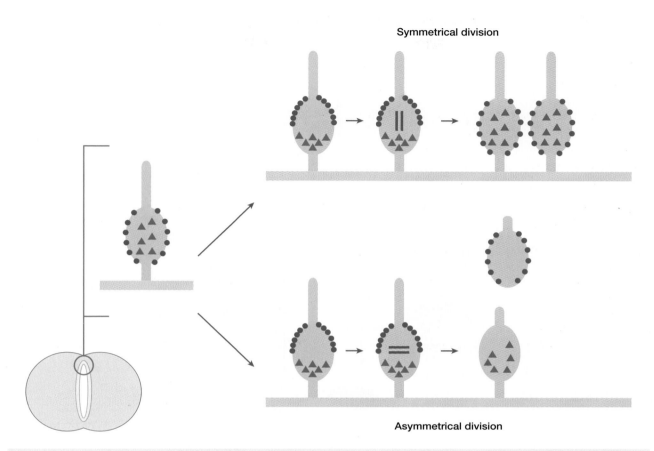

Figure 3.8: Cell fate in the descendants of neural stem cells depends on the plane of cell division, reflecting the asymmetric distribution of membrane-spanning proteins such as Notch and intracellular determinants such as Numb.

Symmetrical divisions occur in the plane of the neuroepithelium and result in the even distribution of Notch (green circles) and Numb (red triangles) between daughter cells. Asymmetrical divisions perpendicular to the neuroepithelium result in the formation of an apical neuronal progenitor and a basal replacement stem cell.

3.4.4 A variety of tissue stem cells are known to exist but much remains to be learned about them

A tissue stem cell (also sometimes described as a **somatic stem cell**) is an undifferentiated cell found among the differentiated cells in a tissue or organ (Verfaile, 2002). Being a stem cell, it can renew itself and differentiate to yield the major specialized cell types of the tissue or organ. Typically, there is a very small number of such cells in each tissue and although their precise origins are mostly unknown, they are thought to reside in specific areas of each tissue. In some cases they may remain *quiescent* (nondividing) for many years until activated by disease or injury, but in other cases where there is a need to replenish cells on a frequent basis (e.g. to replace epithelial cells of the skin and intestine) they are regularly active.

The first adult stem cells were found in the 1960s when bone marrow was found to contain at least two types of stem cell: *hematopoietic stem cells* which form all of the blood types and *bone marrow stromal cells*, a mixed population which generates bone, cartilage, fat and fibrous connective tissue. Since then, some surprising discoveries have been made and brain stem cells have been reported to generate all three of the major brain cell types: neurons (nerve cells) and the non-neuronal astrocytes and oligodendrocytes.

A variety of adult tissues has now been reported to contain stem cells (see *Table 3.6*), and some progress has been made in identifying specific locations for some tissue stem cells. One of the major difficulties in using tissue stem cells on an experimental basis has been that for the most part, they can remain in culture without differentiating for short periods only (unlike embryonic stem cells – see next section) and their differentiation potential is much more limited than that of embryonic stem cells. However, recent evidence has suggested that some adult stem cells may be more *plastic* in their developmental potential than previously thought. This will be explored in the following section.

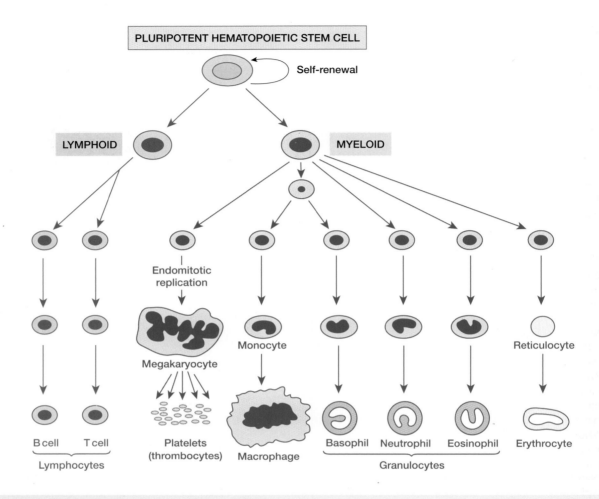

Figure 3.9: Commitment and differentiation in the blood cell lineage.

Table 3.6: Examples of adult stem cells

Stem cell type	Location	Differentiation into
Hematopoietic stem cells	Bone marrow	All blood cells (see *Figure 3.9*)
Bone marrow mesenchymal stem cells (stromal cells)	Bone marrow	Osteocytes (bone cells), chondrocytes (cartilage cells), adipocytes (fat cells) plus other kinds of connective tissue cells such as those in tendons
Neural stem cells	Brain	Neurons (nerve cells), astrocytes, oligodendrocytes
Intestinal epithelial stem cells	In deep crypts in the lining of the digestive tract	Absorptive cells, goblet cells, Paneth cells, enteroendocrine cells
Epidermal stem cells	Basal layer of epidermis	Keratinocytes
Follicular stem cells	Base of hair follicles	Hair follicle and epidermal cells

3.4.5 Embryonic stem (ES) cells have the potential to form any tissue

Cells from very early stages of development are comparatively undifferentiated. As their name suggests, **embryonic stem (ES) cells** are derived from embryos. In the early 1980s a significant breakthrough was made when successful culturing of cells from the ***inner cell mass*** of mouse blastocysts was achieved (Evans and Kaufman, 1981; Martin, 1981). ES cells are ***pluripotent*** and were soon shown to be able to give rise to healthy adult mice after being injected into blastocysts from a different mouse strain which were then transferred into the oviduct of a foster mother (the applications of this process are discussed in Chapter 20).

Mouse ES cells were cultured by transferring cells from the inner cell mass into a dish containing culture medium and coated on its inner surface with a layer of **feeder cells**. The inner cell mass (ICM) cells can grow and divide while attached to the sticky surface of the feeder cells which, as their name suggests, also release nutrients into the medium. After the ICM cells proliferate they are gently removed and *subcultured* by re-plating into several fresh culture dishes. After 6 months or so of repeated subculturing, the original 30 or so cells of the ICM can yield millions of ES cells. ES cells that have proliferated in cell culture for 6 or more months without differentiating, and which are pluripotent and appear genetically normal, are referred to as an **embryonic stem (ES) cell line**.

To test whether ES cells are likely to be pluripotent various experimental approaches can be followed, some leading to the production of desired types of differentiation product.

▶ *Spontaneous differentiation*. Normally the ES cells are grown under certain conditions which keep them in an undifferentiated state. Altering the culture conditions can allow the cells to adhere to each other to form cell clumps known as **embryoid bodies**, after which the cells begin to differentiate spontaneously to form cells of different types, but in a rather unpredictable fashion.

▶ *Directed differentiation*. The cells are manipulated (by changing the chemical composition of the culture medium etc.) so they will differentiate to form *specific* desired cell types.

▶ *Teratoma formation*. The cells are injected into an immunosuppressed mouse to test for the formation of a particular type of benign tumor known as a **teratoma**. Naturally occurring and experimentally induced teratomas typically contain a mixture of many differentiated or partly differentiated cell types – an indication that the ES cells are capable of differentiating into multiple cell types.

Note that although ES cells are termed stem cells, they are unlike tissue stem cells in that they do not divide asymmetrically to produce a replacement stem cell and a daughter cell ready for differentiation. ES cells divide *symmetrically* and all the daughter cells have the same potential to differentiate, depending on the experimental circumstances.

More recently, the successful culture of human ES cells has been reported (Thomson *et al.*, 1998). Human ES cells are derived from embryos that develop from eggs fertilized *in vitro* in an *in vitro fertilization* (IVF) clinic. The IVF procedure is intended to help couples who have difficulty in conceiving children, and normally results in an excess of fertilized eggs which may then be donated for research purposes. The embryos from which the ES cells are derived are typically 5-day-old or so blastocysts, hollow microscopic balls containing about 100 cells, of which about 30 constitute the cells of the ICM. The successful culturing of human ES cells was an important and exciting breakthrough because it raised the possibility of new therapeutic approaches and new avenues of research into cell differentiation. Disorders and severe injuries, where the pathogenesis results from loss of body cells, could be treated by **cell replacement strategies**,

and since ES cells can in principle be programmed to differentiate into cells of a specific type, expectations have been high (Chapter 21).

An alternative strategy is the isolation of **primordial germ cells**, the cells of the gonadal ridge which will normally develop into mature gametes (see figure in *Box 11.4*). These can be cultured *in vitro* and give rise to pluripotent **embryonic germ (EG) cells**, which behave in a very similar fashion to ES cells. Human EG cells, derived from the primordial germ cells or embryos and fetuses from 5 to 10 weeks old, were first cultured by Shamblott *et al.* (1998). See also Donovan and Gearhart, 2001.

3.4.6 The differentiation potential of tissue stem cells is controversial

In recent years there has been reassessment of the differentiation potential of tissue stem cells with several reports in the late 1990s that certain adult tissue stem cells appear to be pluripotent. The apparent ability of adult stem cells to differentiate into multiple rather different cell types is called **transdifferentiation** or **plasticity**. Reported examples include:

▶ hematopoietic stem cells differentiating into the three major types of brain cell (neurons, oligodendrocytes and astrocytes), skeletal muscle cells, cardiac muscle cells and hepatocytes (e.g. Petersen *et al.*, 1999);

▶ brain cells differentiating into blood and skeletal muscle cells or vice versa (e.g. Shih *et al.*, 2002; Clarke *et al.*, 2000; Bjornson *et al.*, 1999);

▶ bone marrow stromal cells differentiating into cardiac muscle cells and skeletal muscle cells (Orlic *et al.*, 2001).

The excitement that greeted these reports largely reflected the potential for **stem cell therapy**. In principle, any damaged tissue or organ might be renewed or even replaced using the most accessible stem cells, and these cells could even be manipulated in advance (see Weissman, 2000; Daley, 2002). For example, a patient's hematopoietic stem cells, which are easy to access, could be reprogrammed to produce brain cells replacing those lost in neurodegenerative disease or spinal cord injuries. However, doubts have been raised about the potential to exploit transdifferation therapeutically (see Medvinsky and Smith, 2003). See also Chapter 21 for general practical and ethical issues concerning using human stem cells.

3.5 Pattern formation in development

While differentiation gives rise to cells with specialized structures and functions, this does not constitute an organism unless the cells are organized in a useful way. Without organization, we might end up as an amorphous blob of heterogeneous tissue, with randomly distributed liver cells, neurons and skin cells unable to form recognizable tissues and organs. While there are many minor differences between individuals, all members of the same species tend to conform to the same basic body plan (Wolpert, 1996). In humans, and vertebrates in general, the body plan shows superficial bilateral symmetry with respect to the craniocaudal axis, while the other major axis (the dorsoventral axis) runs from back to belly. Inside the body, certain organs are placed asymmetrically, defining a left–right axis. The organs and tissues of the body are distributed in essentially the same way in relation to these axes in every individual, and this pattern emerges very early in development. Later, defined patterns emerge within particular organs. A good example is the formation of five fingers on each hand and five toes on each foot. More detailed patterns are generated by the arrangement of cells within tissues. During development, such patterns emerge gradually, with an initially crude embryo being progressively refined like a picture coming into sharp focus. In this section we discuss how pattern formation in the developing embryo is initiated, and the molecular mechanisms involved.

3.5.1 Emergence of the body plan is dependent on axis specification and polarization

Development begins with a single cell, which must somehow become polarized in order to produce an embryo with a head and tail, a back and front, and left and right sides. In many animals, the differentiation of the egg during gametogenesis involves the deposition of particular molecules at different intracellular sites, and these give the egg polarity. When the fertilized egg divides, these **determinants** are segregated into different daughter cells and the embryo thus becomes polarized. In other animals, the symmetry of the egg is broken by an external cue from the environment. In chickens, for example, the craniocaudal axis of the embryo is defined by gravity as the egg rotates on its way down the oviduct. In frogs, both mechanisms are used: there is pre-existing asymmetry in the egg defined by the distribution of maternal gene products, while the site of sperm entry provides another positional coordinate. In mammals the symmetry-breaking mechanism is unclear but probably also involves the sperm entry site (*Box 3.7*).

3.5.2 Homeotic mutations reveal the molecular basis of positional identity

Large scale mutagenesis screens in *Drosophila*, carried out in the 1980s, produced a small number of flies in which one body part developed with the likeness of another (a **homeotic transformation**). An example is the mutant *Antennapedia*, in which legs, instead of antennae, grow out of the head. What this reveals about pattern formation is that the **positional identity** of a cell, that is the information that tells each cell where it is in the embryo and therefore how to behave in order to generate a regionally-appropriate structure, is controlled by genes. These genes are known as **homeotic genes**.

Subsequent molecular analysis of the *Drosophila* homeotic mutants revealed two clusters of very similar genes (*Figure 3.10*), each encoding a transcription factor containing a conserved DNA-binding domain named the homeodomain. Remarkably, the genes were expressed in overlapping patterns along the head-to-tail axis of the fly, dividing the body into discrete zones. The particular combination of genes expressed in each zone appeared to establish a code that gave each cell along the axis a specific positional identity (see Morata, 1993; Lawrence and Morata, 1994). By manipulating the codes either by mutating one or more of the genes, or deliberately overexpressing them, it was possible to generate flies with specific body part transformations.

Very similar clusters of homeobox genes (*Hox* genes) are found in mammals. Humans, and mice, have four unlinked clusters of *Hox* genes which are expressed in overlapping patterns along the craniocaudal axis in a strikingly similar manner to that of flies (Krumlauf, 1994; Lumsden and Krumlauf, 1996; Burke, 2000; *Figure 3.10*). Furthermore, the general mechanism of *Hox* gene activity appears to be conserved, since knockout mutations and deliberate overexpression studies have achieved body part transformations involving mouse vertebrae. For example, targeted disruption of the *Hoxc8* gene produced mice with an extra pair of ribs due to the transformation of the first lumbar vertebra into the 13th thoracic vertebra (Le Mouellic *et al.*, 1992).

Two of the *Hox* gene clusters, *HoxA* and *HoxD*, are also expressed in overlapping patterns along the limbs. Mouse knockout and overexpression mutants for these genes have specific rearrangements of the limb segments. For example, mice with targeted disruptions of the *Hoxa11* and *Hoxd11* genes lack a radius and ulna (Davis *et al.*, 1995). Naturally occurring mutations in the human *HOXD13* gene are associated with a spectrum of hand defects, including polysyndactyly (Manouvrier-Hanu *et al.*, 1999).

3.5.3 Pattern formation often depends on signal gradients

Axis specification and polarization are important early events in development because cells in different parts of the embryo must behave differently, ultimately in terms of the gene products they synthesize, if they are to generate the appropriate body plan. A cell can only behave appropriately if it can tell where it is in relation to other cells. This, in turn, requires a reference framework. The major axes of the embryo provide a reference that allows the position of any cell to be absolutely and unambiguously defined.

How do cells become aware of their position along an axis and therefore behave accordingly? This question is pertinent because functionally equivalent cells at different positions are often required to produce different structures. Examples include the formation of different fingers from the same cell types in the developing hand, and the formation of different vertebrae (some with ribs and some without) from the same cell types in the somites.

In many developmental systems, the regionally-specific behavior of cells has been shown to depend on a signal gradient, which has different effects on equivalent target cells at different concentrations. Signaling molecules that work in this way are known as **morphogens**. In vertebrate embryos, the main craniocaudal axis of the body, as well as the antero-posterior and proximodistal axes of the limbs, are patterned using this mechanism (Wolpert, 1996; Ng *et al.*, 1999).

In the developing limbs (see Schwabe *et al.*, 1998; Niswander, 2003), there is a particular subset of cells at the posterior margin of each limb bud called the **zone of**

polarizing activity (ZPA) which is the source of a morphogen gradient (*Figure 3.11*). Cells nearest the ZPA form the smallest, most posterior digit of the hand or foot, while those furthest away form the thumb or great toe. The organizing ability of the ZPA can easily be demonstrated if a donor ZPA is grafted onto the anterior margin of a limb bud which already has its own ZPA. In this experiment, the limb becomes symmetrical, with posterior digits at both extremities. The morphogen in the developing limb appears to be the signaling protein **Sonic hedgehog** (Shh), although it is thought that the protein acts indirectly, since it is unable to diffuse more than a few cell widths away from its source. A bead soaked in Shh protein will substitute functionally for a ZPA, as will a bead soaked in retinoic acid, which is known to induce *Shh* gene expression. At the heart of the ZPA, all five of the distal *HoxD* genes (*Hoxd9–Hoxd13*) are expressed. However, as the strength of the signal diminishes, the *HoxD* genes are switched off one by one, until at the thumb-forming anterior margin of the limb bud, only *Hoxd9* remains switched on. In this way, signal gradients that specify the major embryonic axes are linked to the homeotic genes that control regional cell behavior.

As discussed above, *Hox* genes are expressed not only in the limbs but also along the major craniocaudal axis of the embryo. In this case, the source of the morphogen gradient that guides *Hox* gene expression is thought to be the *embryonic node*, and the morphogen itself is thought to be retinoic acid. The node secretes increasingly abundant amounts of retinoic acid as it regresses, such that posterior cells are exposed to larger amounts of the chemical than anterior cells, resulting in the progressive activation of more *Hox* genes in the posterior regions of the embryo.

3.6 Morphogenesis

Cell division, with progressive pattern formation and cell differentiation, would eventually yield an embryo with organized cell types, but that embryo would be a static ball of cells. Real embryos are dynamic structures, with cells and tissues undergoing constant interactions and rearrangements to generate structures and shapes. Cells form sheets, tubes, loose reticular masses and dense clumps. Cells migrate either individually or *en masse*. In some cases, such behavior is in response to the developmental program but in other cases, these processes drive development, bringing groups of cells together that would otherwise never come into contact. Several different mechanisms underlie morphogenesis, which are summarized in *Table 3.7* and discussed in more detail below (Hogan, 1999; Mathis and Nicolas, 2002; Peifer and McEwan, 2002; Lubarsky and Krasnow, 2003).

3.6.1 Morphogenesis can be driven by changes in cell shape and size

Orchestrated changes in cell shape can be brought about by reorganization of the cytoskeleton, and this can have a major impact on the structure of whole tissues. One of the landmark events in vertebrate development, the formation of the neural tube, is driven in part by changes in cell shape. Apical constrictions brought about by the local contraction of microfilaments cause some of the columnar cells of the neural plate to become wedge-shaped, allowing them to act as hinges. In combination with increased proliferation at the margins of the neural plate, this provides sufficient force for the entire neural plate to roll up into a tube (Schoenwolf and Smith, 1990).

Table 3.7: Morphogenetic processes in development and examples from model developmental systems

Process	Examples
Differential rates of cell proliferation	Selective outgrowth of vertebrate limb buds by proliferation of cells in the progress zone.
Alternative positioning and/or orientation of mitotic spindle	Different embryonic cleavage patterns in animals. Stereotyped cell divisions in nematodes.
Change of cell size	Cell expansion as adipocytes accumulate lipid droplets
Change of cell shape	Change from columnar to wedge-shaped cells during neural tube closure in birds and mammals
Cell fusion	Formation of trophoblast and myotubes in mammals
Cell death	Separation of digits in vertebrate limb bud. Selection of functional synapses in the mammalian nervous system
Gain of cell–cell adhesion	Condensation of cartilage mesenchyme in vertebrate limb bud
Loss of cell–cell adhesion	Delamination of cells from epiblast during gastrulation in mammals
Cell–matrix interaction	Migration of neural crest cells and germ cells. Axon migration
Loss of cell–matrix adhesion	Delamination of cells from basal layer of the epidermis

Reproduced from Twyman (2001) *Instant Notes in Developmental Biology*. © 2001 BIOS Scientific Publishers.

Box 3.7: Polarizing the mamalian embryo – signals and gene products.

THE DORSOVENTRAL AXIS

The first *overt* sign of asymmetry in the mammalian embryo is the segregation of the inner cell mass (ICM) to one side of the blastocyst (Lu *et al.*, 2001; Zernicka-Goetz, 2002). This defines the *embryonic–abembryonic axis*, with the ICM representing the embryonic pole. The embryonic face of the ICM is exposed to and in contact with the trophoblast while the abembryonic face is open to the blastocele. This difference in environment is sufficient to specify the first two distinct cell layers in the ICM – primitive ectoderm at the embryonic pole and primitive endoderm at the abembryonic pole. *This in turn defines the dorsoventral axis of the embryo.* It is not clear how the ICM becomes positioned asymmetrically in the blastocyst in the first place, but it is interesting to note that when the site of sperm entry is tracked using fluorescent beads, it is consistently localized to trophoblast cells at the embryonic–abembryonic border.

THE CRANIOCAUDAL AXIS

The position of sperm entry may also play an important role in defining the craniocaudal axis of the mammalian embryo, but it is still unclear how that axis becomes polarized (Beddington and Robertson, 1999; Lu *et al.*, 2001). Fertilization induces the second meiotic division, and the second polar body is generally extruded opposite the sperm entry site. This defines the *animal–vegetal axis* of the zygote, with the polar body at the animal pole. The subsequent cleavage divisions take place in the context of the animal–vegetal axis resulting in a blastocyst that shows bilateral symmetry aligned with the former animal–vegetal axis of the zygote. The axis of bilateral symmetry in the blastocyst predicts the alignment of the primitive streak, but not its orientation. The decision as to which end should form the head and which end should form the tail rests with a region of extra-embryonic tissue called the *anterior visceral endoderm (AVE)*. In mice, this is initially located at the tip of the egg cylinder. This rotates towards the future cranial pole of the craniocaudal axis just before gastrulation, and the embryonic node is established at the opposite extreme of the epiblast.

LEFT–RIGHT AXIS

The first overt sign of left–right asymmetry in the mammalian embryo is the looping of the heart tube. However, molecular asymmetry exists much earlier than this (Capdevila *et al.*, 2000). In the mammalian embryo a major determination step occurs during gastrulation when *rotation of cilia at the embryonic node results in a unidirectional flow of perinodal fluid that is required to specify the left–right axis.* Although the mechanism is unclear (Tabin and Vogan, 2003), the result is the activation of genes encoding the signaling molecules Nodal and Lefty-2 specifically on the left hand side of the embryo. These initiate signaling pathways that activate left-hand-specific transcription factors (e.g. Pitx2) and inhibit right-hand-specific transcription factors. On the right hand side of the embryo, where Lefty-2 and Nodal are absent, Pitx2 is not activated. Other proteins, such as Lefty-1, are expressed in the midline of the embryo, and establish a barrier that prevents the leaking of signals from one side of the embryo to the other. Mutations in the human *LEFTA*, *LEFTB* and *NODAL* genes are associated with a range of axis malformations including left–right reversal (situs inversus, situs ambiguous) and mirror image symmetry (isomerism). These malformations sometimes affect the whole body and sometimes only affect individual organs (heterotaxis). Mutations that affect subunits of the motor protein dynein, which is required for the unidirectional rotation of cilia, are also associated with laterality defects. Interestingly, these often occur along with recurrent respiratory infections and infertility, reflecting the immotility of cilia in other parts of the body (*Box 3.1*).

Overview of early axis formation in the mouse embryo from fertilization to the mid-streak stage (see facing page).

The animal pole of the animal–vegetal axis in the mouse embryo is defined as the point at which the second polar body is extruded just after fertilization. The embryonic–abembryonic axis in the blastocyst is orthogonal to the animal–vegetal axis and is defined by the location of the ICM, with the embryonic pole on the side of the blastocyst containing the ICM and the abembryonic pole on the side with the blastocele cavity. The P–D axis in the egg cylinder stage embryo is formed with the proximal pole located at the ectoplacental cone, and the distal pole at the bottom of the cup-shaped embryo. Before gastrulation, the P–D axis rotates 90° and is converted into the A–P axis. From Lu *et al.* (2001) *Curr. Opin. Genet. Dev.* **11**, 384–392. With permission from Elsevier.

Similar behavior within any flat sheet of cells will tend to cause that sheet to invaginate.

3.6.2 Major morphogenetic changes in the embryo result from differential cell affinity

Selective cell–cell adhesion and cell–matrix adhesion were described in Section 3.2.3 and 3.2.4 as mechanisms used to organize cells into tissues and maintain tissue boundaries. In development, regulating the synthesis of particular cell adhesion molecules allows cells to make and break contacts with each other and undergo very dynamic reorganization. Gastrulation, which is described in Section 3.7.5, is perhaps the most dramatic example of a morphogenetic process. The single sheet of epiblast turns in on itself and is converted into the three fundamental germ layers of the embryo, a process driven by a combination of changes in cell shape, selective cell proliferation and differences in cell affinity. Various different processes can result from altering the adhesive properties of cells (McNeil, 2000; Irvine and Rauskolb, 2001).

▶ **Migration**. The movement of an individual cell with respect to other cells in the embryo. Some cells, notably the neural crest and germ cells, undergo extensive migration during development and populate parts of the embryo that are very distant from their original locations.

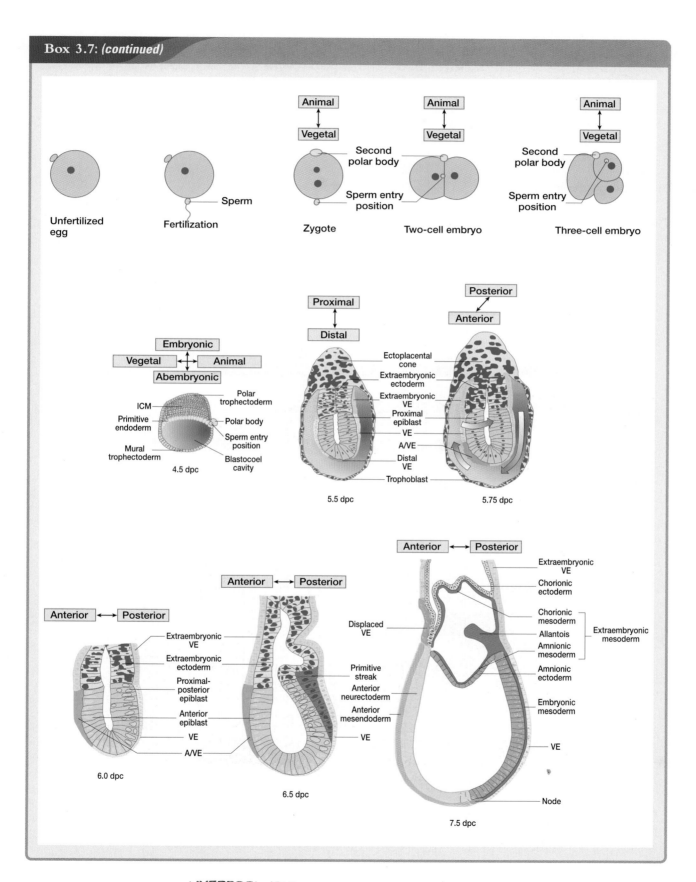

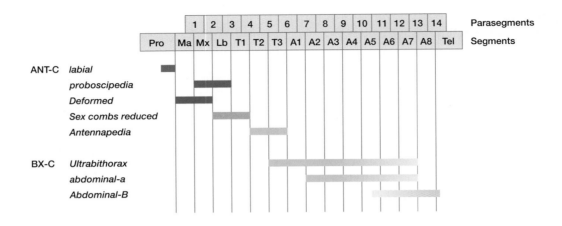

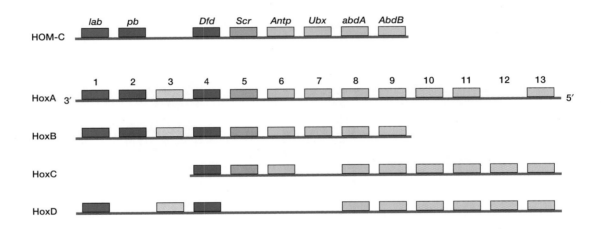

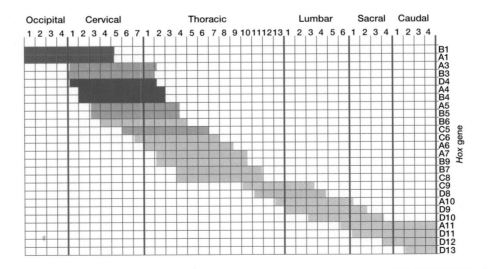

Figure 3.10: Comparison of the *Drosophila* and mouse HOM-C/*Hox* gene complexes and their expression domains along the anteroposterior axis of the embryo.

Redrawn from Twyman (2001) *Instant Notes in Developmental Biology*, published by BIOS Scientific Publishers.

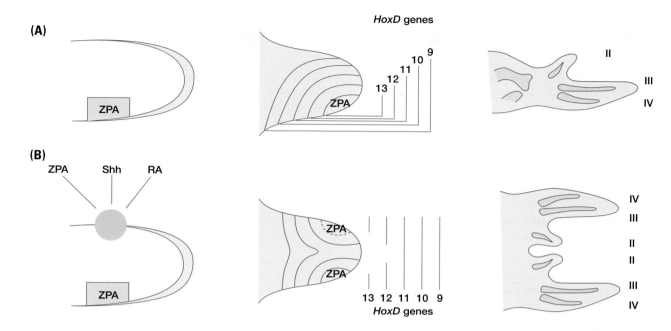

Figure 3.11: Grafting a second zone of polarizing activity to the anterior margin of the chick limb bud results in mirror image duplication of the digits

(A) A signal gradient established by Sonic hedgehog expression in the zone of polarizing activity (ZPA) generates a nested overlapping pattern of *HoxD* gene expression patterns in the developing limb bud, leading to the specification of five digits. **(B)** ZPA grafts establish an opposite gradient and result in a mirror image reversal of digit fates. The effect can be mimicked by beads coated in retinoic acid (RA) or sonic hedgehog protein (Shh). Redrawn from Twyman (2001) *Instant Notes in Developmental Biology*, published by BIOS Scientific Publishers.

▶ **Ingression.** The movement of a cell from the surface of an embryo into its interior.

▶ **Egression.** The movement of a cell from the interior of an embryo to the external surface.

▶ **Delamination.** The movement of cells out of an epithelial sheet, often to convert a single layer of cells into multiple layers. This is one of the major processes that underlie gastrulation in mammalian embryos. Cells can also delaminate from a basement membrane, as occurs in the development of the skin.

▶ **Intercalation.** The merging of cells from multiple cell layers into a single epithelial sheet.

▶ **Condensation.** The conversion of loosely packed mesenchyme cells into an epithelial structure. Sometimes called a **mesenchymal-to-epithelial transition**.

▶ **Dispersal.** The conversion of an epithelial structure into loose mesenchyme cells. An **epithelial-to-mesenchymal** transition.

▶ **Epiboly.** The spreading of a sheet of cells.

▶ **Involution.** The inward turning of an expanding sheet of cells so that the cells spread over the inside surface of the sheet and create a second layer.

3.6.3 Cell proliferation and programmed cell death (apoptosis) are important morphogenetic mechanisms

After an initial period of cleavage where all cells divide at much the same rate, cells in different parts of the embryo begin to divide at different rates. This can be used to generate new structures. For example, rapid cell division in selected regions of the lateral plate mesoderm gives rise to limb buds, while adjacent regions, dividing more slowly, do not form such structures. The plane of cell division is also important. For example, divisions perpendicular to the plane of an epithelial sheet will cause that sheet to expand by the incorporation of new cells. However, divisions in the same plane as the sheet will generate additional layers. If the cells are asymmetrical, as is the case for the neural stem cells discussed in Section 3.4.3, then the plane of cell division can influence the types of daughter cells that are produced. Furthermore, if the cleavage plane is not medial, cells of different sizes can be generated. This occurs in female gametogenesis, when meiotic division produces a massive egg containing most of the cytoplasm and vestigial *polar bodies* that are essentially waste vessels for the unwanted haploid chromosome set. Contrast this with male gametogenesis, where meiosis produces four equivalent spermatids (see figure in *Box 11.4*).

Programmed cell death (apoptosis) is another important morphogenetic mechanism, as it allows gaps to be introduced into the body plan (Vaux and Korsmeyer, 1999). The gaps between our fingers and toes are created by the death of interdigital cells in the hand and foot plates beginning at about 45 days gestation. In the mammalian nervous system, apoptosis is used to prune out the neurons with nonproductive connections, allowing the neuronal circuitry to be progressively refined. Remarkably, up to 50% of neurons are disposed of in this manner, and in the retina this can approach 80%.

3.7 Early human development: fertilization to gastrulation

3.7.1 Fertilization activates the egg and brings together the nuclei of sperm and egg to form a unique individual

Fertilization is the process whereby two sex cells (*gametes*) fuse together to create a new individual with genetic potentials derived from, but different to, both parents (Wassarman, 1999). The male gamete, the sperm cell, is a small cell with a greatly reduced cytoplasm and a haploid nucleus. The nucleus is highly condensed and transcriptionally inactive because the normal histones are replaced by a special class of packaging proteins known as **protamines**. At the posterior end is a long

flagellum which beats to provide propulsion and at the anterior end is the **acrosomal vesicle**, containing digestive enzymes (see *Figure 3.12*).

The female gamete, the egg cell (or **ovum**), is a large cell which contains material necessary for the beginning of growth and development (see *Figure 3.12*). The cytoplasm is extremely well endowed with very large numbers of mitochondria and ribosomes, and large amounts of DNA and RNA polymerase. There are also considerable amounts of proteins, RNAs, protective chemicals and morphogenetic factors. In many species, including birds, reptiles, fish, amphibians and insects, the egg contains a large or moderate amount of **yolk**, a collection of nutrients which is required to nourish the developing embryo before it can feed independently. Although mammalian embryos have a yolk sac, yolk is not required in mammalian eggs because the embryo is nourished by the placental blood supply.

Outside the plasma membrane is the *vitelline envelope*, which in mammals is a separate and thick *extracellular matrix* known as the **zona pellucida**. In mammals, too, the egg is surrounded by a layer of cells known as **cumulus cells** which serve to nurture the egg prior to, and just after, ovulation.

Human sperm cells have to migrate very considerable distances and out of the 280 million or so ejaculated into the vagina only about 200 reach the required part of the oviduct where fertilization takes place. Fertilization begins with

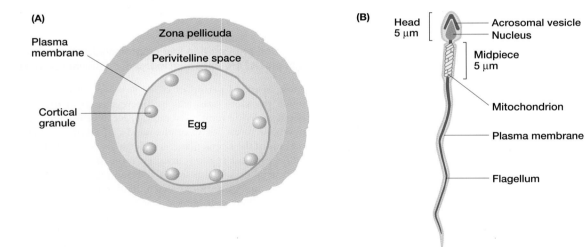

Figure 3.12: The specialized sex cells.

(A) *The ovum (egg)*. The mammalian ovum is a large cell, 120 μm in diameter, surrounded by an extracellular envelope, the *zona pellucida*, to which the sperm must first bind and which contains three glycoproteins, ZP-1, ZP-2 and ZP-3, that polymerize to form a gel. The first polar body (not shown here), the product of meiosis I, lies under the zona within the *perivitelline space*. At ovulation, oocytes are in metaphase II. Meiosis II is not completed until after fertilization. Fertilization triggers secretion of cortical granules which effectively inhibit further sperm from passing through the zona pellucida. **(B) *The sperm*.** This cell is much smaller than the ovum with a 5 μm head containing highly compacted DNA, a 5 μm cylindrical body, the midpiece, with many mitochondria, and a 50 μm tail. At the front, the *acrosome* contains enzymes that help the sperm to make a hole in the zona pellucida, allowing it to access and fertilize the egg. Semen typically contains 100 million sperm per ml. From Alberts *et al.* (2002) *Molecular Biology of the Cell*, 4th Edn., p. 1147. Copyright © 2002, Garland Science.

attachment of a sperm to the zona pellucida followed by release of enzymes from the acrosomal vesicle, which cause local digestion of the zona. The head of the sperm then fuses with the plasma membrane of the oocyte and the sperm nucleus passes into the cytoplasm. Within the oocyte, the haploid sets of sperm and egg chromosomes are initially separated from each other and constitute respectively the male and female **pronuclei**. They subsequently fuse to form a diploid nucleus. The fertilized oocyte is known as the **zygote**.

3.7.2 Cleavage partitions the zygote into many smaller cells

Cleavage is the developmental stage in which the zygote divides to form a number of smaller cells called **blastomeres**. The nature of the early cleavage divisions varies widely among different animal species, and in many insects (including *Drosophila*), the process does not even involve cell division (instead, the zygote nucleus undergoes a series of divisions in a common cytoplasm to generate a large, flattened multinucleated cell, the *syncytial blastoderm*). With some exceptions, the result of cleavage is usually a ball of cells, often surrounding a fluid-filled cavity called the **blastocele**.

In most invertebrates, the ball of cells resulting from cleavage is called a **blastula**, but in vertebrates the terminology varies. In amphibians and mammals, the term **morula** is used to describe the initial, loosely packed, ball of cells which results from early cleavage, and thereafter when the fluid-filled blastocele forms, the ball of cells is known as a

blastula in amphibians, but a **blastocyst** in mammals (see *Figure 3.13*). The situation is different in birds, fish, and reptiles, where the egg contains a lot of yolk which inhibits cell division. Here the cleavage is restricted to a flattened *blastodisc* at the periphery of the cell.

For many species (but *not* mammals – see below), cleavage divisions are rapid because there are no intervening G1 and G2 gap phases in the cell cycle between DNA replication and mitosis. In such cases there is no net growth of the embryo, and so as the cell number increases, the cell size decreases. Where this happens, the genome inherited from the zygote (the *zygotic genome*) is *transcriptionally inactive during cleavage*, and so cleavage is heavily dependent on maternally inherited gene products distributed in the egg cytoplasm. The maternal gene products regulate the cell cycle and determine the rate of cleavage, and the cleavage divisions are *synchronous*. This type of regulation is often referred to as *maternal genome* regulation and the maternal gene products are often referred to as *maternal determinants.*

Mammalian cleavage is exceptional in several ways:

▶ **early activation of the zygotic genome** – as early as the two-cell stage in some species. As a result, the cleavage divisions are controlled by the zygotic genome rather than by maternally inherited gene products, and they are slow divisions (because the cell cycles include the G1 and G2 gap phases), and asynchronous;

▶ **rotational cleavage** – the first cleavage plane is vertical, but in the second round of cell division, one of the cells cleaves vertically and the other horizontally – see *Figure 3.13*);

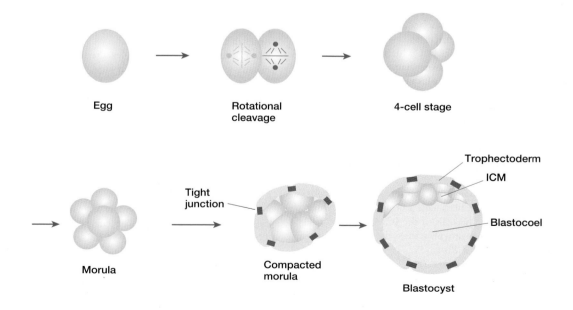

Egg   Rotational cleavage   4-cell stage

Trophectoderm
ICM
Blastocoel

Morula   Tight junction   Compacted morula   Blastocyst

Figure 3.13: Early development of the mammalian embryo, from fertilization to the formation of the blastocyst.

Redrawn from Twyman (2001) *Instant Notes Developmental Biology*, published by BIOS Scientific Publishers.

▶ **compaction** – the loosely associated blastomeres of the eight-cell embryo flatten against each other to maximize their contacts and form a tightly packed morula (**note:** compaction does occur in many nonmammalian embryos, including *Xenopus*, but compaction in mammals is strikingly overt). Compaction has the effect of introducing a degree of *cell polarity*. Before compaction the blastomeres are rounded cells with uniformly distributed microvilli, and the cell adhesion molecule E-cadherin is found wherever the cells are in contact with each other. After compaction, the situation is rather different: the microvilli become restricted to the *apical surface*, while E-cadherin becomes distributed over the *basolateral surfaces*. Now the cells form *tight junctions* with their neighbors and the cytoskeletal elements are reorganized to form an *apical band* (see *Table 3.4*).

At about the 16-cell morula stage in mammals, it becomes possible to discriminate between two types of cell: external polarized cells and internal nonpolar cells. As the population of nonpolar cells increases, the cells begin to communicate with each other through gap junctions. The distinction between the two types of cells is a fundamental one as will be explained in the next section.

3.7.3 Only a small percentage of the cells in the early mammalian embryo gives rise to the mature organism

In many animal models of development, the organism is formed from cells which have descended from all the cells of the early embryo. Mammals are rather different, however, and only a small minority of the cells of the early embryo give rise to the organism proper. This is so because much of early mammalian development is concerned with establishing tissues which act as a life support but which mostly do not contribute to the final organism, namely the **extra-embryonic membranes** and the **placenta** (see *Box 3.8*).

The two types of cells which can be distinguished by about the 16-cell stage in mammals, have different fates. The outer polarized cells comprise the **trophoblast** (or **trophecto-derm**) that will go on to form one of the four extra-embryonic membranes, the *chorion,* which provides the embryonic portion of the *placenta.* The inner nonpolar cells comprise the **embryoblast**. As the blastocele forms (at about the 32-cell stage in humans) the inner nonpolar cells congregate at one end of the blastocele to form an off-center **inner cell mass (ICM)**. The cells of the ICM will give rise to **all** the cells of the organism plus the other three extra-embryonic membranes (*Box 3.8*).

3.7.4 Implantation

After some time (day 5 of human development), the blastocyst *hatches*: an enzyme is released which bores a hole through the zona pellucida and the blastocyst squeezes out. The blastocyst is now free to interact directly with the uterine endometrium. Very soon after arriving in the uterus (day 6 of human devel-

opment), the blastocyst attaches tightly to the uterine epithelium (**implantation**). Trophoblast cells proliferate rapidly and differentiate into an inner layer of **cytotrophoblast** and an outer multinucleated cell layer, the **syncytiotrophoblast** which starts to invade the connective tissue of the uterus.

Even before implantation occurs, cells of the ICM begin to differentiate into distinct external and internal layers. The external cell layer is the **epiblast** (or **primitive ectoderm**). Epiblast cells will give rise to ectoderm, endoderm and mesoderm (and hence all of the tissues of the embryo), and, in addition, to the amnion, yolk sac and allantois. The internal cell layer, the **hypoblast** (or **primitive endoderm**) gives rise to the *extraembryonic mesoderm* that lines the primary yolk sac and the blastocele.

A fluid-filled cavity, the **amniotic cavity**, forms within the inner cell mass, enclosed by the *amnion*. The embryo, derived from part of the inner cell mass, now consists of distinct epiblast and hypoblast layers (and is known as the **bilaminar germ disc**). It is located between two fluid-filled cavities, the amniotic cavity on one side and the *yolk sac* on the other (*Figure 3.14*).

3.7.5 Gastrulation is a dynamic process whereby cells of the epiblast give rise to the three germ layers

As Lewis Wolpert has famously remarked 'It is not birth, marriage or death, but gastrulation, which is truly the important event in your life'. **Gastrulation** takes place during the third week of human development and is the first major morphogenetic process in development. During gastrulation, the orientation of the body is laid down, and the embryo is converted into a structure of three germ layers: **ectoderm**, **endoderm** and **mesoderm**, *all of which are derived from the epiblast*. The three germ layers are the progenitors of all the tissues of the organism (*Box 3.5*).

In mammals, birds and reptiles, the major structure characterizing gastrulation is a linear one, the **primitive streak**. This appears at about day 15 in human development, as a faint groove along the longitudinal midline of the now oval-shaped bilaminar germ disc. Over the course of the next day the *primitive groove* becomes deeper and elongates to occupy about half the length of the embryo. By day 16 a deep depression (the *primitive pit*), surrounded by a slight mound of epiblast (the *primitive node*) is evident at the end of the groove, near the center of the germ disc (see *Figure 3.14*).

The process of gastrulation is an *extremely dynamic* one, involving very rapid cell movements (Narasimha and Leptin, 2000; Myers *et al.*, 2002). At day 16 in human development the epiblast cells near the primitive streak begin to proliferate, flatten and lose their connections with one another. These flattened cells develop pseudopodia which allow them to migrate through the primitive streak into the space between the epiblast and the hypoblast (*Figure 3.14*). Some of the ingressing epiblast cells invade the hypoblast and displace its cells, leading eventually to complete replacement of the hypoblast by a new layer of cells, the definitive **endoderm**. Starting on day 16, some of the epiblast cells migrating

Box 3.8: Extra-embryonic membranes and the placenta.

Early mammalian development is primarily concerned with the formation of tissues which by and large* do not contribute to the final organism. They are the four **extra-embryonic membranes** (yolk sac, amnion, chorion and allantois) and the **placenta**, derived from a combination of embryonic tissue (the chorion) and maternal tissue. As well as protecting the embryo (and later the fetus), these life support systems are required to provide for its nutrition, respiration, and excretion.

(*The exceptions are: the dorsal part of the yolk sac, which is incorporated into the embryo as the precursor of the *primitive gut*, and the allantois which is represented in the adult as a fibrous cord, a residue of the umbilical cord.)

YOLK SAC. The most primitive of the four extra-embryonic membranes, the yolk sac is found in all amniotes and also in sharks, bony fishes and some amphibians. In bird embryos the yolk sac surrounds a nutritive *yolk mass* (the yellow part of the egg, consisting mostly of phospholipids). In many mammals (including humans and mice) the yolk sac does not contain any yolk. The yolk sac is generally important because:

▶ the *primordial germ cells*, which originate from the epiblast, migrate to the yolk sac prior to their colonization of the gonadal ridge;

▶ it is the source of the first blood cells of the conceptus and most of the first blood vessels (some of which extend themselves into the developing embryo).

The yolk sac originates from splanchnic (visceral) lateral plate mesoderm and endoderm.

AMNION. The amnion is found not just in amniotes (mammals, birds, reptiles) but also in primitive form in some bony fishes and in some amphibians. It is the innermost of the extra-embryonic membranes, remaining attached to and immediately surrounding the embryo. It contains amniotic fluid which bathes the embryo, thereby preventing the embryo from drying out during development, helping the embryo to float (and so reducing the effects of gravity on the body) and acting as a hydraulic cushion to protect the embryo from mechanical jolting etc). The amnion derives from ectoderm and somatic lateral plate mesoderm.

CHORION. Like the amnion, the chorion is found in amniotes and also in primitive form in some bony fishes and amphibians.

It too derives from ectoderm and somatic lateral plate mesoderm. In the embryos of birds, such as the chick, the chorion is pressed against the shell membrane but in mammalian embryos it is composed of *trophoblast* cells, which produce the enzymes that erode the lining of the uterus, helping the embryo to implant into the uterine wall. The chorion is also a source of hormones (chorionic gonadotropin) that influence the uterus as well as other systems. In all these cases, the chorion serves as a surface for respiratory exchange. In placental mammals, the chorion provides the fetal component of the *placenta* (see below).

ALLANTOIS. The most evolutionarily recent of the extra-embryonic membranes, the allantois is only found in amniotes (which possibly should be renamed as *allantoisotes*). It acts as a waste (urine) storage system in birds, reptiles and indeed in most amniotes but not in placental mammals. The allantois derives from an outward bulging of the floor of the hindgut and so is composed of endoderm and splanchnic lateral plate mesoderm. Although prominent in some mammals, in others (including humans) it is vestigial (does not now have any function) except that its blood vessels give rise to the *umbilical cord*.

PLACENTA. The placenta is a feature only found in placental mammals, and is derived partly from the conceptus and partly from the uterine wall. It develops after implantation, when the embryo induces a response in the neighboring maternal endometrium changing it to become a nutrient-packed highly vascular tissue called the **decidua**. During the second and third weeks of development, the trophoblast tissue becomes vacuolated and these vacuoles connect to nearby maternal capillaries, rapidly filling with blood. As the chorion forms, it projects outgrowths known as chorionic villi into the vacuoles, bringing the maternal and embryonic blood supplies into close contact. At the end of three weeks, the chorion has differentiated fully and contains a vascular system that is connected to the embryo. Exchange of nutrients and waste products occurs over the chorionic villi. Initially, the embryo is completely surrounded by the decidua, but as it grows and expands into the uterus, the overlying decidual tissue (decidua capsularis) thins out and then disintegrates. The mature placenta is derived completely from the underlying decidua basalis.

---

through the primitive streak diverge into the space between the epiblast and the nascent definitive endoderm to form a third layer, the **intraembryonic mesoderm.** When the intraembryonic mesoderm and definitive endoderm have formed, the residual epiblast is now described as the **ectoderm** and the new three-layered structure is referred to as the **trilaminar germ disc** (*Figure 3.14*).

The ingressing mesoderm cells migrate in different directions, some laterally and cranially, while others are deposited on the midline. The cells that migrate through the primitive pit and come to rest on the midline form two structures:

▶ the **prechordal plate,** a compact mass of mesoderm cranial to the primitive pit. The prechordal plate will

induce important cranial midline structures such as the brain;

▶ the **notochordal process**, a hollow tube which sprouts from the primitive pit and grows in length as cells proliferating in the region of the primitive node add on to its proximal end, and as the primitive streak regresses. By day 20 of human development the notochordal process is completely formed, but then transforms from a hollow tube to a solid rod, the **notochord.** The notochord will later induce formation of components of the nervous system (see next section).

The ingressing mesoderm cells which migrate laterally condense into rod- and sheet-like structures on either side

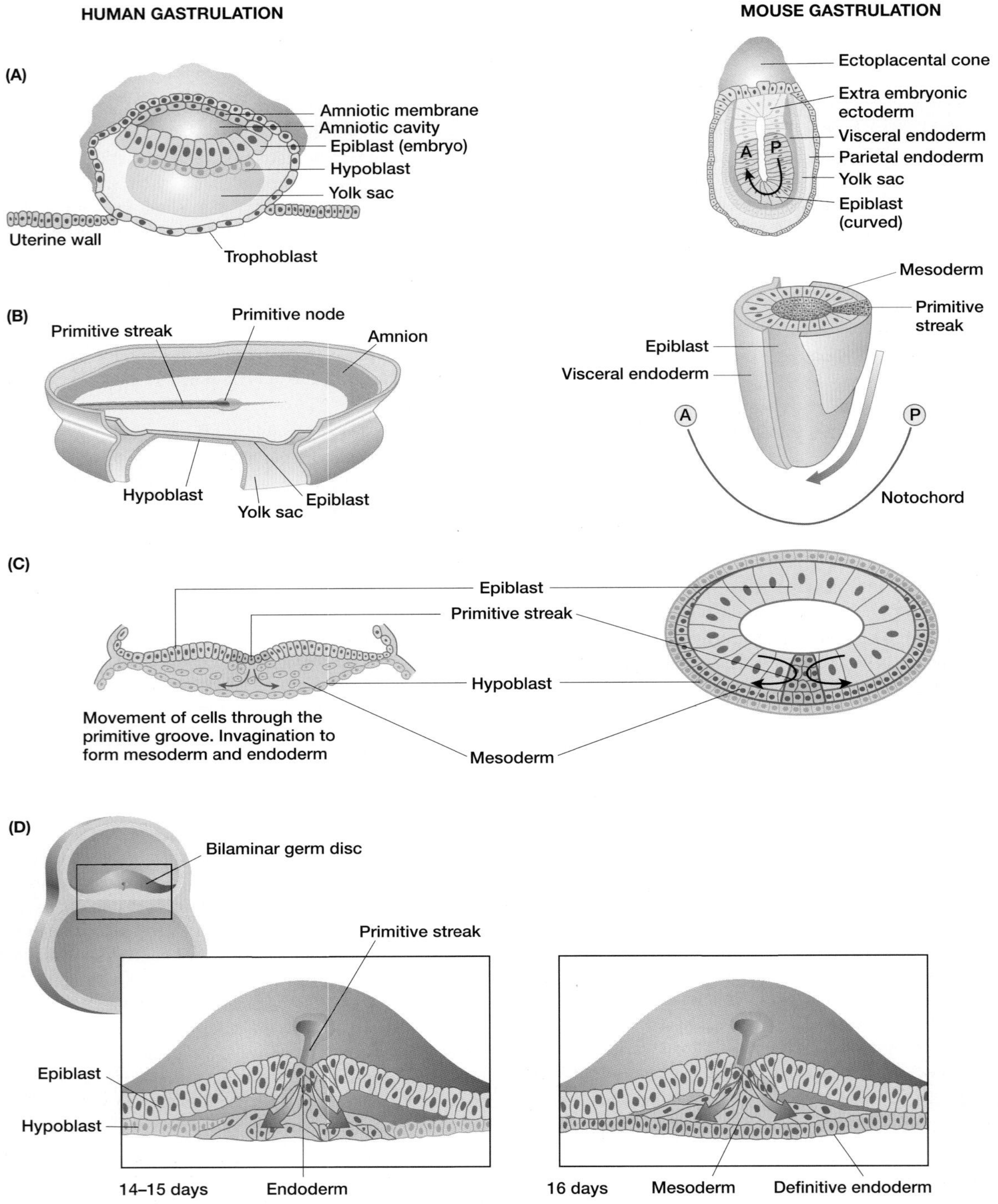

HUMAN GASTRULATION

(A)
- Amniotic membrane
- Amniotic cavity
- Epiblast (embryo)
- Hypoblast
- Yolk sac
- Uterine wall
- Trophoblast

(B)
- Primitive streak
- Primitive node
- Amnion
- Hypoblast
- Yolk sac
- Epiblast

(C)
- Epiblast
- Primitive streak
- Hypoblast
- Mesoderm

Movement of cells through the primitive groove. Invagination to form mesoderm and endoderm

(D)
- Bilaminar germ disc
- Primitive streak
- Epiblast
- Hypoblast
- 14–15 days
- Endoderm
- 16 days
- Mesoderm
- Definitive endoderm

MOUSE GASTRULATION

- Ectoplacental cone
- Extra embryonic ectoderm
- Visceral endoderm
- Parietal endoderm
- Yolk sac
- Epiblast (curved)
- A P
- Mesoderm
- Primitive streak
- Epiblast
- Visceral endoderm
- A P
- Notochord
- Epiblast
- Primitive streak
- Hypoblast
- Mesoderm

Figure 3.14: Gastrulation in humans and other primates involves the reorganization of a flat bilaminar germ disc to form a trilaminar embryo through the programmed delamination of cells from the epiblast.

(A–C) While the outcome of gastrulation is similar in all mammals, there may be major differences in the details of morphogenesis, particularly in the way extra-embryonic structures are formed and used. Gastrulation in the mouse is shown on the right as a comparison. In this species, the flat germ disc is replaced by a cup-shaped egg cylinder, and the cells move 'out' of the primitive streak onto the surface of the embryo rather than into it. The future anteroposterior axis is wrapped around the base of the cup. (D) After 14–15 days of human development, ingressing mesodermal cells invade the hypoblast and displace its cells, leading to formation of the embryonic endoderm. On day 16 some of the ingressing epiblast cells diverge into the space between the epiblast and the nascent endoderm to form embryonic mesoderm. **Note** that whereas the epiblast gives rise to all three germ layers (ectoderm, endoderm and mesoderm), the hypoblast gives rise to the extraembryonic endoderm that lines the yolk sac. Human embryos redrawn from Larsen (2002) *Human Embryology* 3rd Edn, with permission from Elsevier. Mouse embryos redrawn from Twyman (2001) *Instant Notes Developmental Biology*, Published by BIOS Scientific Publishers.

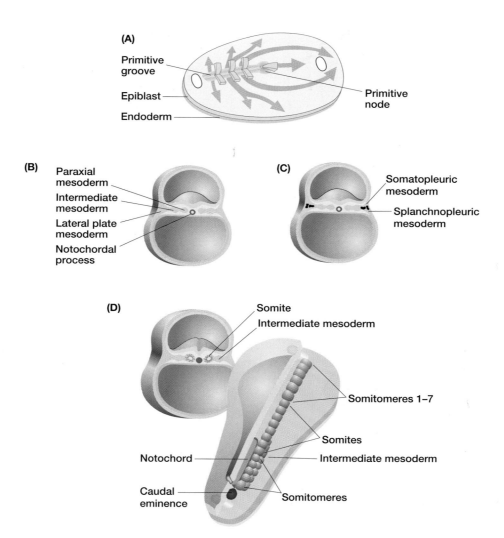

Figure 3.15: Migration paths and fates of the mesoderm cells which ingress during gastrulation.

(A) *Early mesoderm migration*. Epiblast cells ingressing through the primitive node migrate cranially to form the *prechordal plate,* a compact mass of mesoderm just cranial to the primitive node and the *notochordal process*, a dense midline *tube* which will later form a *solid rod*, the *notochord*. Epiblast cells ingressing through the primitive groove migrate to form the lateral mesoderm flanking the midline. **(B)** *Early differentation of lateral mesoderm*. The mesoderm immediately flanking the notochordal process forms cylindrical condensations, the *paraxial mesoderm*. The neighboring, less pronounced cylindrical condensations are the *intermediate mesoderm*. The rest of the lateral mesoderm forms a flattened sheet, the *lateral plate mesoderm*. **(C)** *Differentiation of lateral plate mesoderm (LPM)*. Vacuoles form in the LPM which splits into two layers, a ventral layer, the *splanchnopleuric mesoderm*, gives rise to the mesothelial covering of the visceral organs and the dorsal *somatopleuric mesoderm* which gives rise to the inner lining of the body walls and to most of the dermis. **(D)** *Somitomere formation*. The paraxial mesoderm forms a series of rounded whorl-like structures, *somitomeres*. Except for somitomeres 1–7 (see text), somitomeres will give rise to *somites*, blocks of segmental mesoderm which establish the segmental organization of the body. In this diagram of a 21 day human embryo six central somitomeres have already differentiated into somites. Adapted from Larsen (2002) with permission from Churchill Livingstone Publishers.

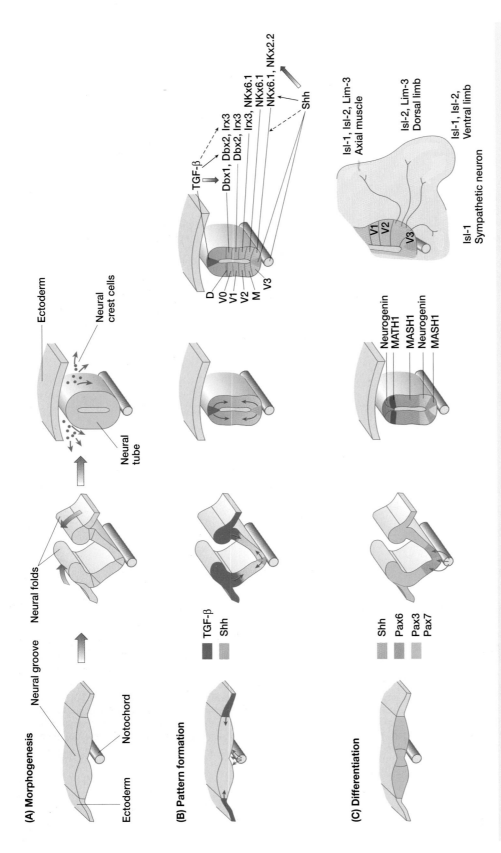

Figure 3.16: The nervous system as an example of the coordination of different developmental processes during organogenesis.

(A) *Morphogenesis*. The change from a flat neural plate to a closed neural tube is caused by the formation of hinge points where cells become apically constricted, and the proliferation of ectoderm at the margins of the neural plate pushing the sides together. As the neural folds come together, neural crest cells delaminate and begin to migrate away. (B) *Pattern formation*. Dorsoventral pattern formation involves a competition between secreted signals from the notochord (Sonic hedgehog) and ectoderm (initially BMP4, but later other members of the TGF-β superfamily). As the neural plate folds up, the signals are expressed in the neural tube itself. The result is the activation of different transcription factors in different zones resulting in the regional specification of different neuronal subtypes (D = dorsal; V0, V1 etc = ventral subtypes; M = motor neurons). (C) *Differentiation*. The genes that control neurogenesis in the developing nervous system (the proneural and neurogenic genes) are also expressed in discrete zones along the dorsoventral axis, possibly under the control of *Pax3*, *Pax6* and *Pax7* whose expression domains are defined in part by the Sonic hedgehog signal emanating from the notochord. The expression domains of the proneural genes *neuongenin*, *MATH1* and *MASH1* are shown, while those of the neurogenic genes *Notch*, *Delta* and *Serrate* are not. Neural development is characterized by the expression of the transcription factor NeuroD. Then different classes of neurons begin to express distinct groups of transcription factors. Those of the LIM homeodomain family are especially important in the determination of subneuronal fates by regulating the behavior of projected axons.

of the notochord. There are three main structures (see *Figure 3.15*).

▶ The **paraxial mesoderm (PM)**, a pair of cylindrical condensations lying immediately adjacent to and flanking the notochord. The PM first develops into a series of whorl-like structures known as **somitomeres** which form in a cranial–caudal sequence through the third and fourth weeks of human development. The first seven *cranial* somitomeres will eventually go on to form the striated muscles of the face, jaw and throat, but the other somitomeres develop further into discrete *blocks* of segmental mesoderm known as **somites** (see *Figure 3.15*). Cervical, thoracic, lumbar and sacral somites will establish the segmental organization of the body by giving rise to most of the axial skeleton (including the vertebral column), the voluntary muscles, and part of the dermis of the skin.

▶ The **intermediate mesoderm (IM)**, a pair of less pronounced cylindrical condensations, just lateral to the paraxial mesoderm. The IM will develop into the urinary system and parts of the genital system.

▶ The **lateral plate mesoderm (LPM)**, the remainder of the lateral mesoderm, forming a flattened sheet. Starting on day 17 of human development, the LPM splits into two layers:

Box 3.9: Sex determination: genes and the environment in development.

Humans, like all mammals, are **sexually dimorphic**, i.e. there are separate male and female sexes. The decision between male and female development is made at conception, when the sperm delivers either an X chromosome or a Y chromosome to the egg, which always contains an X chromosome (Schafer and Goodfellow, 1996). The only exceptions to this general rule occur when errors in meiosis produce gametes with missing or extra sex chromosomes, resulting in individuals with *sex-chromosome aneuploidies* (Section 2.5.2). Although the sex of the human embryo is established at conception, sexual differentiation does not begin to occur until the embryo is about 5 weeks old. There are two forms of sexual differentiation, one involving **primary sexual characteristics** (the development of the gonad and the choice between sperm and egg development) and one involving **secondary sex characteristics** (the sex-specific structures of the urogenital system and the external genitalia). Primary sex characteristics are dependent on the genotype of the embryo whereas secondary sex characteristics are dependent on signals from the environment, mediated by hormonal signaling.

Male development depends on the presence or absence of the Y chromosome, which contains a critical male-determining gene called *SRY* (sex-determining region of the Y chromosome). *SRY* encodes a transcription factor that activates downstream genes required for testis development. The testis then produces sex hormones which are required for the development of male secondary sex characteristics. For a long time, it was thought that female gonad development was a 'default state' and that the *SRY* gene was sufficient to switch the indifferent embryonic gonad from female to male differentiation. This was supported by several lines of evidence – rare XX males often have a small fragment of the Y chromosome, including *SRY*, translocated onto the tip of one of their X chromosomes, and genetically female mice transgenic for the mouse *Sry* gene develop as males. However, more recent studies suggest that genes on the X chromosome and autosomes are involved in the positive regulation of ovarian development. The overexpression of such genes, which include *DAX* and *WNT4A*, can feminize XY individuals even if they possess a functional *SRY* gene.

Early gamete development appears to be controlled more by the environment than the genotype of the germ cells. Female *primordial germ cells (PGCs)* introduced into the testis will begin to differentiate into sperm, and male PGCs introduced into the ovary will begin to differentiate into oocytes. This may reflect the regulation of the cell cycle, since PGCs entering the testis arrest prior to meiosis while those entering the ovary commence meiosis immediately. Therefore, PGCs of either sex that colonize somatic tissue outside the gonad begin to differentiate into oocytes since there is no signal to arrest the cell cycle. In all of these unusual situations, however, functional gametes are not produced. Differentiation aborts at a relatively late stage, presumably because the genotype of the germ cells themselves also plays a critical role in gamete development.

Unlike the situation with primary sex characteristics, it appears that female secondary sex characteristics *are* the default state. One of the genes regulated by SRY is *NR5A1*, which encodes another transcription factor called steroidogenic factor 1 **(SF1)**. This activates genes required for the production of male sex hormones, including *HSD17B3* (encoding hydroxysteroid-17-β-dehydrogenase 3, which is required for testosterone synthesis) and *AMH* (the gene encoding anti-Mullerian hormone). Both hormones play important roles in the differentiation of the male urogenital system. AMH, for example, causes the *Mullerian ducts* (which become the fallopian tubes and uterus in females) to break down. Mutations inhibiting the production, distribution, elimination or perception of such hormones produce feminized XY individuals. For example, androgen insensitivity syndrome results from defects in the testosterone receptor, which prevent the body responding to the hormone even if it is produced at normal levels. XY individuals with this disease appear outwardly as normal females, but due to the effects of SRY and AMH they possess undescended testes instead of ovaries, and they lack a uterus and fallopian tubes. Mutations that lead to the overproduction of male sex hormones in females have the opposite effect, i.e. virilization of XX individuals. Occasionally, this occurs in developing male/female fraternal twins, when the female twin is exposed to male hormones from her brother. The CYP19 enzyme converts androgens to estrogens so mutations that increase its activity can result in the feminization of males while those decreasing or abolishing its activity can lead to the virilization of females.

- the *splanchnopleuric mesoderm*, the layer adjacent to the endoderm, gives rise to the mesothelial covering of the visceral organs;
- the *somatopleuric mesoderm*, the layer adjacent to the ectoderm, gives rise to the inner lining of the body wall, part of the limbs and most of the dermis of the skin.

3.8 Neural development

As described above, gastrulation results in a remarkable set of changes in the embryo converting it from a two-layered structure to a three-layered one. The development of the embryo is now programmed towards organizing tissues into the precursors of the many organs and systems contained in the adult. We focus on the early development of the nervous system as an example of organogenesis because it shows how the processes of differentiation, pattern formation and morphogenesis are exquisitely coordinated.

3.8.1 The nervous system develops after the ectoderm is induced to differentiate by the underlying mesoderm

The development of the nervous system marks the onset of organogenesis, and begins at the end of the third week of human development. The initiating event is the induction of the overlying ectoderm by the axial mesoderm (Wilson and Edlund, 2001). Within the axial mesoderm, cells of the prechordal plate and the cranial portion of the notochordal plate transmit signals to overlying ectoderm cells causing them to differentiate into a thick plate of neuroepithelial cells (**neurectoderm**). The resulting **neural plate** appears at day 18 of human development but grows rapidly and changes proportions over the next 2 days (*Figure 3.16*).

At the start of the fourth week, a process known as **neurulation** results in conversion of the neural plate into a **neural tube**, the precursor of the brain and spinal cord. The neural plate begins to crease ventrally along its midline to form a depression called the neural groove. This is thought to develop in response to inductive signals from the closely apposed notochord. Thick neural folds rotate around the neural groove and meet dorsally, initially at a mid-point along the craniocaudal axis (*Figure 3.16A*). Closure of the neural tube proceeds in a zipper-like fashion and bidirectionally. Where closure is incomplete, there is an anterior neuropore and a posterior neuropore. Occasionally, there is a partial failure of neural tube closure resulting in conditions such as spina bifida.

During neurulation, a specific population of cells arising along the lateral margins of the neural folds detaches from the neural plate and migrates to many specific locations within the body. This highly versatile group of cells, termed the **neural crest**, gives rise to part of the peripheral nervous system, melanocytes, some bone and muscle, the retina and other structures (Garcia-Castro and Bonner-Fraser, 1999; Knecht and Bonner-Fraser, 2002; *Box 3.5*).

3.8.2 Pattern formation in the neural tube involves the coordinated expression of genes along two axes

As soon as the neural plate is formed, three large cranial vesicles (the future brain) become visible as well as a narrower caudal section that will form the spinal cord. This cranio-caudal polarity reflects the *regional specificity of neural induction*, that is the signals coming from the axial mesoderm contain positional information which cause the overlying ectoderm to form neural tissue specific for different parts of the axis. The precise nature of the signal in amniotes is not understood (Stern, 2002). In *Xenopus*, there is a well-established model in which neural development is prevented by members of the bone morphogenetic protein (BMP) family of signaling proteins and neural induction is initiated by BMP antagonists. However, the mechanism in birds and mammals appears to be more complex and is an area of active research. It seems likely that a general neuralizing signal is released from the mesoderm which induces the formation of neural plate that is anterior in character. This must be 'posteriorized' by another signal that originates in the caudal region of the embryo. Further molecules, secreted specifically by the anterior mesoderm, are required to form the head.

Whatever the underlying mechanism, the signals activate different sets of transcription factors along the axis, and these confer positional identities on the cells and regulate their behavior. In the forebrain and midbrain, transcription factors of the Emx and Otx families are expressed. In the hindbrain and spinal cord, positional identities are controlled by the *Hox* genes. In the hindbrain it appears that the positional values of cells are fixed at the neural plate stage and that migrating neural crest cells with this positional information impose positional identities on the surrounding tissues. Conversely, positional identity in the spinal cord is imposed by signals from the surrounding paraxial mesoderm. This can be shown by transplanting cells to different parts of the axis. Cranial neural crest cells behave according to their lineage when moved to a new position, that is they do the same things as they would in their original position. Trunk crest cells, on the other hand, behave according to their new position – they do the same things as their neighbors.

The developing nervous system is a good model of pattern formation and differentiation, since various cell types (different classes of neurons and glia) arise along the dorsoventral axis. This process is controlled by a hierarchy of genetic regulators (Lumsden and Krumlauf, 1996; Tanabe and Jessel, 1996; Eislen 1999; *Figure 3.16B,C*).

Dorsoventral polarity is generated by opposing sets of signals originating from the notochord and the adjacent ectoderm. The notochord secretes the signaling molecule Sonic hedgehog, which has ventralizing activity, while the ectoderm secretes several members of the transforming growth factor (TGF)-β superfamily, including BMP4, BMP7 and a protein appropriately named Dorsalin. As the neural plate begins to fold, these same signals begin to be expressed in the extreme ventral and dorsal regions of the neural tube itself – the floor plate and roof plate respectively. The opposing signals have opposite effects on the activation of various homeodomain class transcription factors. As shown in *Figure 3.16B*, this divides the neural tube into discrete dorsoventral zones, or transcription factor domains, which later become the regional centers of different classes of neurons. For example, the zone defined by the expression of *Nkx6.1* alone becomes the region populated by motor neurons, which are residents of the ventral third of the neural tube.

3.8.3 Neuronal differentiation involves the combinatorial activity of transcription factors

Neurons do not arise uniformly in the neural ectoderm but are restricted to specific regions demarcated by the expression of **proneural genes** such as *neurogenin*, *MASH1* and *MATH1* (*Figure 3.16C*), which encode transcription factors of the bHLH family. Proneural gene expression confers on cells the ability to form neuroblasts, but not all proneural cells can adopt this fate. Instead, there is competition between the cells involving the expression of **neurogenic genes** such as *Notch* and *Delta*. The successful cells form neuroblasts and inhibit the surrounding cells from doing the same. Therefore, neuroblasts arise in a precise spacing pattern. This pattern forming process is termed **lateral inhibition**. The neurons then begin to differentiate according to their position with respect to the dorsoventral and craniocaudal axes of the nervous system, which is defined by the combination of transcription factors as discussed in the previous section (Panchision and McKay, 2002). Further diversification is controlled by refinements in the expression patterns of these transcription factors. For example, all motor neurons initially express two transcription factors of the LIM homeodomain family: Islet-1 and Islet-2.

Table 3.8: A selection of human developmental genes that have been associated with specific disease phenotypes. Several major classes of gene products are considered: signaling proteins, receptors, transcription factors, structural proteins and enzymes.

Gene	Product and normal function	Associated disorder
Secreted signaling proteins		
SHH	Sonic hedgehog Signaling protein responsible for patterning the neural tube, somites, gut and limb buds	Holoprosencephaly, a disorder in which the developing forebrain fails to separate into left and right hemispheres. In severe cases there is a single brain ventricle and cyclopia. In very mild cases, the disease may manifest as a single central incisor
EDN3	Endothelin 3, a peptide hormone that regulates vasoconstriction, required during development for the differentiation of neural crest derivatives	Hirschsprung disease, a disorder of neural crest differentiation in which the enteric ganglia are not formed, causes megacolon (chronic constipation and intestinal obstruction) due to absence of peristaltic movements
GH1	Growth hormone 1, a polypeptide hormone expressed in the pituitary gland that is responsible for regulating growth	Pituitary dwarfism, a form of dwarfism that can be treated by administration of purified or recombinant growth hormone during childhood
LEFTB	Lefty2, a Nodal-related signaling protein expressed specifically on the left-hand side of the early embryo, helps to establish left–right asymmetry	Laterality defects – situs inversus, isomerism, or heterotaxis, due to the failure of left-right axis specification
Receptors		
FGFR3	A receptor for fibroblast growth factors, expressed particularly strongly in cartilage and nervous system. Has a major role in bone development	Achondroplasia, the most common form of short-limb dwarfism Crouzon syndrome and craniosynostosis, severe craniofacial abnormalities
EDNRB	Endothelin B receptor, found on neural crest cells and required for their differentiation into melanocytes and enteric ganglia in the gut	Hirschsprung disease (see *EDN3*, above)
KIT	KIT is a receptor tyrosine kinase, so called because it was originally found as an oncogene in a feline sarcoma virus. The receptor is widely expressed but has particularly important roles in the development of the blood cell lineage, melanocytes and germ cells	Piebaldism, characterized by congenital patches of white skin and hair due to the failure of melanocyte proliferation

Table 3.8: continued

Gene	Product and normal function	Associated disorder
GHR	Growth hormone receptor, transduces signals from growth hormone (see above)	Laron dwarfism, a form of dwarfism where the serum levels of growth hormone are normal and which is unresponsive to growth hormone therapy. Also known as growth hormone insensitivity syndrome
Transcription factors		
HOXD13	Transcription factor involved in pattern formation. Confers positional information along the craniocaudal axis and in limb development	Polysyndactyly (extra fused digits) caused by the mis-specification of cell types and the consequent formation of abnormal bone structures in the distal regions of the limb
PAX6	Transcription factor with multiple roles in eye development	Eye defects ranging from the mild ectopia pupillae (off-center pupils) to aniridia (partial or complete absence of the iris, in combination with malformations of the lens and anterior chamber, and corneal degeneration)
TBX5	Transcription factor expressed specifically in the forelimb region of the developing embryo, has an important role in establishing limb identity (arm vs leg development)	Holt–Oram syndrome, a disorder of the upper limbs which also affects heart development
SRY	Transcription factor expressed in gonadal ridge of male embryos, and required to initiate male sexual differentiation. Found on the Y-chromosome	Male sex reversal (outward female appearance in XY individuals) often associated with complete gonadal dysgenesis
Structural proteins		
COL6A1	Alpha subunit of collagen VI, major component of microfibrils in the extracellular matrix that provide structural rigidity to tissues.	Bethlem myopathy, involving the contraction of joints (particularly the elbows and ankles), is due to the lack of collagen VI microfibrils. Overexpression may contribute to the heart defects in Down syndrome.
ELN	Elastin, a protein of the extracellular matrix that allows tissues to return to their original shape after deformation	Cutis laxa, loose and baggy skin resulting from permanent deformation can be caused by abnormal and malfunctioning elastin Aortic stenosis, caused by the deformation of the aorta
LAMA3	Alpha 3 subunit of laminin 5, a component of the basal membrane of the skin which plays an important role in keratinocyte differentiation	Junctional epidermolysis bullosa gravis, a skin blistering disorder where the basal cells detach from the basement membrane
USH2A	Extracellular matrix protein required for eye and inner ear development	Type II Usher's syndrome, characterized by severe deafness from birth and the onset of retinitis pigmentosa in late teens
Enzymes		
WRN	A helicase probably involved in the repair of double-stranded DNA breaks	Werner syndrome, a disease of premature aging
CYP11B1	Steroid 11-beta-hydroxylase, required for the synthesis of aldosterone (a hormone that acts on the kidneys and regulates mineral and water balance)	Congenital adrenal hyperplasia, involving rapid childhood growth but premature termination of bone elongation resulting in short adult stature
HSD17B3	17 beta-hydroxysteroid dehydrogenase, required for the synthesis of testosterone	Male pseudohermaphroditism, where boys are born with unambiguous external female appearance, but virilize during puberty
EXT1	Glycosyltransferase required for the synthesis of a heparan sulfate, an important constituent of the extracellular matrix	Hereditary multiple exostoses, a disorder in which cartilaginous lumps develop near the ends of the bones particularly in the limbs but occasionally also in the ribs and shoulders

Later, only those motor neurons projecting their axons to the ventral limb muscles express these LIM transcription factors and no others. Neurons expressing Isl-1, Isl-2 and a third transcription factor, Lim-3, project their axons to the axial muscles of the body wall. Neurons expressing Isl-2 and Lim-1 project their axons to dorsal limb muscles, while those expressing Isl-1 alone project their axons to sympathetic ganglia (*Figure 3.16C*). The transcription factors determine the particular combinations of receptors expressed on the axon growth cones and therefore determine the response of growing axons to different physical and chemical cues (chemoattractants etc.) (Tessier-Lavigne and Goodmans, 1996). Once the first axons reach their targets, further axons can find their way by growing along existing axon paths – a process termed *fasciculation*.

3.9 Conservation of developmental pathways

3.9.1 Many human diseases are caused by the failure of normal developmental processes

The most dramatic human diseases involve striking morphological abnormalities that are due to disruptions of the general processes of differentiation, pattern formation and morphogenesis. An example is holoprosencephaly, a failure in the normal process of forebrain development, which in its severest form gives rise to individuals with a single eye (cyclopia) and no nose. As is the case for other human diseases, the phenotype can be influenced by both genetic and environmental factors. In some cases, it is clear that specific mutations have caused this defect, for example mutations in the *SHH* gene which encodes the signaling protein Sonic hedgehog. Otherwise, the abnormality can be traced to an environmental cause, such as limited cholesterol intake in the maternal diet. The relative roles of genes and the environment on a developmental process are clearly illustrated by *sex-determination*, which is discussed in *Box 3.9*.

While some developmental abnormalities can be traced to the use of drugs (chemicals that are known to have *teratogenic* effects include alcohol, certain antibiotics, thalidomide, retinoic acid and illegal drugs such as cocaine and heroin), many are the result of mutations in specific genes. As stated at the beginning of this chapter, some of the most important developmental genes are regulatory in nature, encoding either transcription factors or components of signaling pathways. Genes encoding structural components of the cell or extracellular matrix, and even metabolic enzymes, also play an important role in development and reveal important disease phenotypes. Some examples are presented in *Table 3.8*.

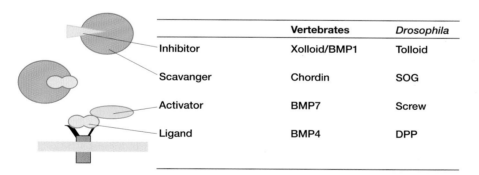

	Vertebrates	*Drosophila*
Inhibitor	Xolloid/BMP1	Tolloid
Scavanger	Chordin	SOG
Activator	BMP7	Screw
Ligand	BMP4	DPP

	Humans	*Drosophila*	*Caenorhabditis*
Ligand	EGF	BOSS	LIN-3
Receptor	EGFR	Sevenless	LET-23
SH2/SH3 adaptor	GRB2	Drk	SEM-5
G-protein	Ras	Ras1	LET-60
GTPase activator and nucleotide exchanger	GAP/GNRP	Gap1/SOS	Gap-1

Figure 3.17: Evolutionary conservation of developmental pathways.

(A) The BMP4/Chordin pathway that underlies neural induction in *Drosophila* and vertebrates. **(B)** The growth factor signaling pathway, which has diverse roles in vertebrates, flies and worms.

3.9.2 Developmental processes are highly conserved at both the single gene level and the level of complete pathways

Where genes have been shown to cause developmental diseases, there is often a remarkable degree of evolutionary conservation among animals. This conservation applies not only to the genes, but also to the entire pathways in which they are involved (Pires-daSilva and Sommer, 2003; *Figure 3.17*).

Conserved molecular pathways are often used for similar processes in very distantly related species. For example, as stated briefly above, neural induction in *Xenopus* embryos involves a battle between the opposing effects of BMP4, which favors ventral and lateral fates, and dorsalizing (neuralizing) factors such as Chordin, Noggin and Follistatin. The *Drosophila* orthologs of BMP4 and Chordin are proteins called Decapentaplegic (Dpp) and Short gastrulation (Sog) respectively. Remarkably, these proteins play out equivalent roles in the formation of the *Drosophila* nervous system. Indeed the relationship goes further. The activity of Dpp in *Drosophila* is enhanced by the protein Tolloid (Tol) which degrades Sog. In *Xenopus*, the role of Tol is played by its ortholog Xolloid (Xol), and in zebrafish the equivalent molecule is BMP1. Both Xolloid and BMP1 degrade Chordin. There is also a cross-phylum pairing of *Xenopus* BMP7 and the *Drosophila* protein Screw, accessory proteins that are necessary for BMP4/Dpp activity.

In other cases, the same developmental pathway is used for very different purposes in different species. In mammals, epidermal growth factor (EGF) is used to promote the proliferation of epidermal cells. This growth factor binds to the EGF receptor, which is a receptor tyrosine kinase, and initiates the Ras-Raf-MAP kinase cascade (Section 3.2.1). In *Drosophila*, the same pathway is used to promote differentiation of one of the eight photoreceptor cell types during eye development. In the nematode, the same pathway is used to stimulate the division and differentiation of vulval cells.

The best example of evolutionary conservation involves the homeobox-containing genes, which appear to be present in all animals and to carry out very similar functions. The fundamental role of these genes in pattern formation is demonstrated by the ability of orthologous genes from very different species to substitute for each other. For example, human *HOX* and *OTX* genes have been introduced into *Drosophila* lines in which the orthologous genes have been mutated, and this has led to a complete rescue of the mutant phenotype.

However, there are also significant differences between species, even those that are closely related. Given the extensive similarity between humans and mice, it is perhaps surprising that the process of gastrulation should be so different (*Figure 3.14*). Indeed there are very great differences among the vertebrate embryos prior to gastrulation, reflecting different strategies for nutrient acquisition. Sex determination mechanisms are also very diverse (Morrish and Sinclair, 2002). Not all mammals use the human model of XY sex determination (Graves, 2002) and many reptiles dispense with the use of heteromorphic sex chromosomes altogether by relying on the temperature of the environment to specify the sex of the embryo.

Further reading

Arias AM, Stewart A (2002) *Molecular Principles of Animal Development*. Oxford University Press, Oxford.

Alberts B, Johnson A, Lewis J, Raff M, Roberts K, Walter P (2002) *Molecular Biology of the Cell*, 4th Edn. Garland Science, New York.

Gilbert SF (2003) *Developmental Biology*, 7th Edn. Sinauer Associates, Sunderland.

Twyman RM (2001) *Instant Notes in Developmental Biology*. BIOS Scientific Publishers, Oxford.

Wolpert L (2002) *Principles of Development*, 2nd Edn. Oxford University Press, Oxford.

References

Bainter JJ, Boos A, Kroll KL (2001) Neural induction takes a transcriptional twist. *Dev. Dynamics* **222**, 315–327.

Beddington RSP, Robertson EJ (1999) Axis development and early asymmetry in mammals. *Cell* **96**, 195–209.

Bjornson CRR, Rietz RL, Reynolds BA *et al*. (1999) Turning brain into blood: a hematopoietic fate adopted by adult neural stem cells in vivo. *Science* **283**, 354–357.

Bokel C, Brown NH (2002) Integrins in development: Moving on, responding to, and sticking to the extracellular matrix. *Dev. Cell* **3**, 311–321.

Bourne HR (1997) How receptors talk to trimeric G proteins. *Curr. Opin. Cell Biol.* **9**, 134–142.

Burke AC (2000) *Hox* genes and the global patterning of the somitic mesoderm. *Curr. Top. Dev. Biol.* **47**, 155–181.

Campisi J (1996) Replicative senescence: an old live's tale? *Cell* **84**, 497–500.

Capdevila J, Vogan KJ, Tabin CJ, Belmonte JCI (2000) Mechanisms of left-right determination in vertebrates. *Cell* **101**, 9–21.

Clapham DE (1995) Calcium signaling. *Cell* **80**, 259–268.

Clarke DL, Johansson CB, Wilbertz J *et al*. (2000) Generalized potential of adult neural stem cells. *Science* **288**, 1660–1663.

Cunningham BA (1995) Cell adhesion molecules as morphoregulators. *Curr. Opin. Cell Biol.* **7**, 628–633.

Daley GQ (2002) Prospects for stem cell therapeutics: myths and medicines. *Curr. Opin. Genet. Dev.* **12**, 607–613.

Davis AP, Witte DP, Hsieh-Li HM *et al*. (1995) Absence of radius and ulna in mice lacking *Hoxa-11* and *Hoxd-11*. *Nature* **375**, 791–795.

Donovan PJ, Gearhart J (2001) The end of the beginning for pluripotent stem cells. *Nature* **414**, 92–97.

Dustin ML (2002) Shmoos, rafts, and uropods – The many facets of cell polarity. *Cell* **110**, 13–18.

Eislen JS (1999) Patterning motoneurons in the vertebrate nervous system. *Trends Neurosci.* **22**, 321–326.

Evans MJ, Kaufman MH (1981) Establishment in culture of pluripotential cells from mouse embryos. *Nature* **292**, 154–156.

Garcia-Castro M, Bonner-Fraser M (1999) Induction and differentiation of the neural crest. *Curr. Opin. Cell Biol.* **11**, 695–698.

Giancotti FG, Ruoslahti E (1999) Transduction – Integrin signalling. *Science* **285**, 1028–1032.

Graves JAM (2002) The rise and fall of *SRY*. *Trends Genet.* **18**, 259–264.

Gumbiner BM (1996) Cell adhesion: the molecular basis of tissue architecture and morphogenesis. *Cell* **84**, 345–357.

Harland R (2000) Neural induction. *Curr. Opin. Genet. Dev.* **10**, 357–362.

Hogan BLM (1999) Morphogenesis. *Cell* **96**, 225–233.

Hou SX, Zheng ZY, Chen X *et al.* (2002) The JAK/STAT pathway in model organisms: Emerging roles in cell movement. *Dev. Cell* **3**, 765–778.

Houslay MD, Milligan G (1997) Tailoring cAMP-signalling responses through isoform multiplicity. *Trends Biochem. Sci.* **22**, 217–224.

Humphries MJ, Newham P (1998) The structure of cell-adhesion molecules. *Trends Cell. Biol.* **8**, 78–83.

Hynes RO (2002) Integrins: Bidirectional, allosteric signaling machines. *Cell* **110**, 673–687.

Irvine KD, Rauskolb C (2001) Boundaries in development: formation and function. *Annu. Rev. Cell Dev. Biol.* **17**, 189–214.

Kim H, Schagat T (1996) Neuroblasts: a model for the asymmetric division of cells. *Trends Genet.* **13**, 33–39.

Knecht AK, Bonner-Fraser M (2002) Induction of the neural crest: A multigene process. *Nature Rev. Genet.* **3**, 453–461.

Krumlauf R (1994) *Hox* genes in vertebrate development. *Cell* **78**, 191–201.

Kuo CJ, Conley PB, Chen L *et al.* (1992) A transcriptional hierarchy involved in mammalian cell type specification *Nature* **355**, 457–461.

Larsen W (2002) *Human Embryology*, 3rd Edn. Churchill Livingstone, New York.

Lawrence PA, Morata G (1994) Homeobox genes: their function in *Drosophila* segmentation and pattern formation. *Cell* **78**, 181–189.

Le Mouellic H, Lallemand Y, Brulet P (1992) Homeosis in the mouse induced by a null mutation in the *Hox 3.1* gene. *Cell* **69**, 251–264.

Lu CC, Brennan J, Robertson EJ (2001) From fertilization to gastrulation: axis formation in the mouse embryo. *Curr. Opin. Genet. Dev.* **11**, 384–392.

Lubarsky B, Krasnow MA (2003) Tube morphogenesis: Making and shaping biological tubes. *Cell* **112**, 19–28.

Lumsden A, Krumlauf R (1996) Patterning the vertebrate neuraxis. *Science* **274**, 1109–1115.

Manouvrier-Hanu S, Holder-Espinasse M, Lyonnet S (1999) Genetics of limb anomalies in humans. *Trends Genet.* **15**, 409–417.

Martin GR (1981) Isolation of a pluripotent cell line from early mouse embryos cultured in medium conditioned by teratocarcinoma stem cells. *Proc. Natl Acad. Sci. USA* **78**, 7634–7638.

Mathis L, Nicolas J-F (2002) Cellular patterning of the vertebrate embryo. *Trends Genet.* **18**, 627–635.

McNeill H (2000) Sticking together and sorting things out: adhesion as a force in development. *Nature Rev. Genet.* **1**, 100–108.

Medvinsky A, Smith A (2003) Stem cells: Fusion brings down barriers. *Nature* **422**, 823–825.

Morata G (1993) Homeotic genes of *Drosophila*. *Curr. Opin. Genet. Dev.* **3**, 606–613.

Morrish BC, Sinclair AH (2002) Vertebrate sex determination: many means to an end. *Reproduction* **124**, 447–457.

Morrison SJ, Uchida N, Weissman IL (1995) The biology of hematopoietic stem cells. *Ann. Rev. Cell Biol.* **11**, 35–71.

Myers DC, Sepich DS, Solnica-Krezel L (2002) Convergence and extension in vertebrate gastrulae: cell movements according to or in search of identity? *Trends Genet.* **18**, 447–455.

Narasimha M, Leptin M (2000) Cell movements during gastrulation: come in and be induced. *Trends Cell Biol.* **10**, 169–172.

Ng JK, Tamura K, Buscher D *et al.* (1999) Molecular and cellular basis of pattern formation during vertebrate limb development. *Curr. Top. Dev. Biol.* **41**, 37–66.

Niswander L (2003) Pattern formation: Old models out on a limb. *Nature Rev. Genet.* **4**, 133–143.

Orlic D, Kajstura J, Chimenti S, Jakoniuk I *et al.* (2001) Bone marrow cells regenerate infarcted myocardium. *Nature* **410**, 701–705.

Panchision DM, McKay RDG (2002) The control of neural stem cells by morphogenetic signals. *Curr. Opin. Genet. Dev.* **12**, 478–487.

Peifer M, McEwen DG (2002) The ballet of morphogenesis: Unveiling the hidden choreographers. *Cell* **109**, 271–274.

Pellegrini S, Dusanter-Fourt I (1997) The structure, regulation and function of the Janus kinases (JAKs) and the signal transducers and activators of transcription (STATs). *Eur. J. Biochem.* **248**, 615–633.

Perez-Moreno M, Jamora C, Fuchs E (2003) Sticky business: Orchestrating cellular signals at adherens junctions. *Cell* **112**, 535–548.

Petersen BE, Bowen WC, Patrene KD *et al.* (1999) Bone marrow as a potential source of hepatic oval cells. *Science* **284**, 1168–1170.

Pires-daSilva A, Sommer RJ (2003) The evolution of signalling pathways in animal development. *Nature Rev. Genet.* **4**, 39–49.

Robinson MJ, Cobb MH (1977) Mitogen activated kinase pathways. *Curr. Opin. Cell Biol.* **9**, 180–186.

Saucedo LJ, Edgar BA (2002) Why size matters: altering cell size. *Curr. Opin. Genet. Dev.* **12**, 565–571.

Schafer AJ, Goodfellow PN (1996) Sex determination in humans. *Bioessays* **18**, 955–963.

Schoenwolf GC, Smith JL (1990) Mechanisms of neurulation: traditional viewpoint and recent advances. *Development* **109**, 243–270.

Schwabe JWR, Rodriguez-Esteban C, Belmonte JCI (1998) Limbs are moving: Where are they going? *Trends Genet.* **14**, 229–235.

Selleck SB (2000) Proteoglycans and pattern formation: sugar biochemistry meets developmental genetics. *Trends Genet.* **16**, 206–212.

Shamblott MJ, Axelman J, Wang SP, Bugg EM *et al.* (1998) Derivation of pluripotent stem cells from cultured human primordial germ cells. *Proc. Natl Acad. Sci. USA* **95**, 13726–13731.

Shih CC, Mamelak A, LeBon T, Forman SJ (2002) Hematopoietic potential of neural stem cells. *Nature Med.* **8**, 535–536.

Speigel S, Foster D, Kolesnick R (1996) Signal transduction through lipid second messengers. *Curr. Opin. Cell Biol.* **8**, 159–167.

Stacey G, Doyle A (2000) Cell banks: A service to animal cell technology. In: *Encyclopedia of Animal Cell Technology* (ed Spier R). Wiley, New York, pp. 293–320.

Steinberg MS, McNutt PM (1999) Cadherins and their connections: adhesion junctions have broader functions. *Curr. Opin. Cell Biol.* **11**, 554–560.

Stern CD (2002) Induction and initial patterning of the nervous system – the chick embryo enters the scene. *Curr. Opin. Genet. Dev.* **12**, 447–451.

Streuli C (1999) Extracellular matrix remodelling and cellular differentiation. *Curr. Opin. Cell Biol.* **11**, 634–640.

Su TT, O'Farrell PH (1998) Size control: cell proliferation does not equal growth. *Curr. Biol.* **19**, R687–R689.

Tabin CJ, Vogan KJ (2003) A two-cilia model for vertebrate left-right axis specification. *Genes Dev.* **17**, 1–6.

Tanabe Y, Jessel TM (1996) Diversity and pattern in the developing spinal cord. *Science* **274**, 1115–1123.

Tessier-Lavigne M, Goodmans CS (1996) The molecular biology of axon guidance. *Science* **274**, 1123–1133.

Thomson JA, Itskovitz-Elder J, Shapiro SS *et al.* (1998) Embryonic stem cell lines derived from human blastocysts. *Science* **282**, 1145–1147.

Tsai MJ, O'Malley BW (1994) Molecular mechanisms of action of steroid/thyroid receptor superfamily members. *Annu. Rev. Biochem.* **63**, 451–486.

Twyman RM (2001) Signal transduction in development. In: *Instant Notes in Developmental Biology*. BIOS Scientific Publishers, Oxford.

Vaux DL, Korsmeyer SJ (1999) Cell death in development. *Cell* **96**, 245–254.

Verfaile CM (2002) Adult stem cells: assessing the case for pluripotency. *Trends Cell. Biol.* **12**, 502–508.

Wassarman PM (1999) Mammalian fertilization: Molecular aspects of gamete adhesion, exocytosis, and fusion. *Cell* **96**, 175–183.

Weissman IL (2000) Translating stem and progenitor cell biology to the clinic: barriers and opportunities. *Science* **287**, 1442–1446.

Wilson SI, Edlund T (2001) Neural induction: toward a unifying mechanism. *Nature Neurosci.* **4(Suppl)**, 1161–1168.

Wolpert L (1996) One hundred years of positional information. *Trends Genet.* **12**, 359–364.

Wylie C (2000) Germ cells. *Curr. Opin. Genet. Dev.* **10**, 410–413.

Zernicka-Goetz M (2002) Patterning the embryo: the first spatial decisions in the life of a mouse. *Development* **129**, 815–829.

CHAPTER FOUR

Genes in pedigrees and populations

Chapter contents

4.1 Monogenic versus multifactorial inheritance

The simplest genetic characters are those whose presence or absence depends on the **genotype** at a single **locus**. That is not to say that the character itself is programmed by only one pair of genes: expression of any human character is likely to require a large number of genes and environmental factors. However, sometimes a particular genotype at one locus is both necessary and sufficient for the character to be expressed, given the normal human genetic and environmental background. Such characters are called **Mendelian**. Mendelian characters can be recognized by the characteristic pedigree patterns they give (Section 4.2). Over 6000 Mendelian characters are known in man. As described in the 'Before we start – intelligent use of the Internet', the essential starting point for acquiring information on any such character, whether pathological or nonpathological, is the OMIM database (www.ncbi.nlm.nih.gov/Omim/).

Most human characters are governed by genes at more than one locus. The further away a character is from the primary gene action, the less likely is it to show a simple Mendelian pedigree pattern. DNA sequence variants are virtually always cleanly Mendelian – which is their major attraction as genetic markers (Section 13.2). Protein variants (electrophoretic mobility or enzyme activity) are usually Mendelian but can depend on more than one locus because of post-translational modification (Section 1.5.3). The failure or malfunction of a developmental pathway that results in a birth defect is likely to involve a complex balance of factors. Thus the common birth defects (cleft palate, spina bifida, congenital heart disease etc.) are rarely Mendelian. Behavioral traits like IQ test performance or schizophrenia are still less likely to be Mendelian – but they may still be genetically determined to a greater or lesser extent.

Non-Mendelian characters may depend on two, three or many genetic loci, with greater or smaller contributions from environmental factors. We use **multifactorial** here as a catch-all term covering all these possibilities. More specifically, the genetic determination may involve a small number of loci (**oligogenic**) or many loci each of individually small effect (**polygenic**); or there may be a single major locus with a polygenic background. For **dichotomous characters** (characters that you either have or don't have, like extra fingers) the underlying loci are envisaged as **susceptibility genes**, while for **quantitative** or **continuous characters** (height, weight etc.) they are seen as **quantitative trait loci (QTLs)**. Any of these characters may tend to run in families, but the pedigree patterns are not Mendelian and do not admit of simple analysis.

4.2 Mendelian pedigree patterns

4.2.1 Dominance and recessiveness are properties of characters, not genes

A character is **dominant** if it is manifest in the **heterozygote** and **recessive** if not. Note that dominance and recessiveness are properties of characters, not genes. Thus sickle cell anemia is recessive because only HbS **homozygotes** manifest it, but sickling trait, which is the phenotype of HbS heterozygotes, is dominant. Most human dominant syndromes are known only in heterozygotes. Sometimes homozygotes have been described, born from matings of two heterozygous affected people, and often the homozygotes are much more severely affected. Examples are achondroplasia (short-limbed dwarfism, MIM 100800) and Type 1 Waardenburg syndrome (deafness with pigmentary abnormalities; MIM 193500). Nevertheless we describe achondroplasia and Waardenburg syndrome as dominant because these terms describe **phenotypes** seen in heterozygotes. In experimental organisms, where this uncertainty does not exist, geneticists tend to use the term **semi-dominant** when the heterozygote has an intermediate phenotype, reserving 'dominant' for conditions where the homozygote is indistinguishable from the heterozygote – Huntington disease (adult-onset progressive neurological deterioration, MIM 143100) for example. The question of dominance has been well reviewed by Wilkie (1994). Males are **hemizygous** for loci on the X and Y chromosomes, where they have only a single copy of each gene, so the question of dominance or recessiveness does not arise in males for X- or Y-linked characters.

4.2.2 There are five basic Mendelian pedigree patterns

Figure 4.1 shows the symbols used for pedigree drawing and *Box 4.1* summarizes the main factors of each pattern. Mendelian characters may be determined by loci on an **autosome** or on the X or Y **sex chromosomes**. Autosomal

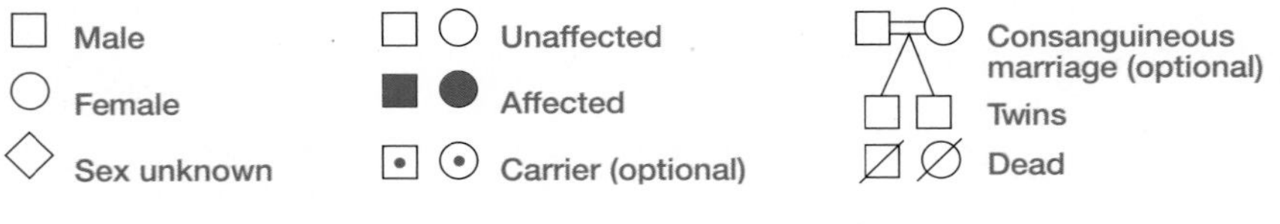

□ Male	□ ○ Unaffected	⊟○ Consanguineous marriage (optional)
○ Female	■ ● Affected	△△ Twins
◇ Sex unknown	⊡ ⊙ Carrier (optional)	⊘ ⊘ Dead

Figure 4.1: Main symbols used in pedigrees.

Generations are usually labeled in Roman numerals, and individuals within each generation in Arabic numerals; III-7 or III$_7$ is the seventh person from the left (unless explicitly numbered otherwise) in generation III. An arrow ↗ can be used to indicate the proband or propositus (female: proposita) through whom the family was ascertained.

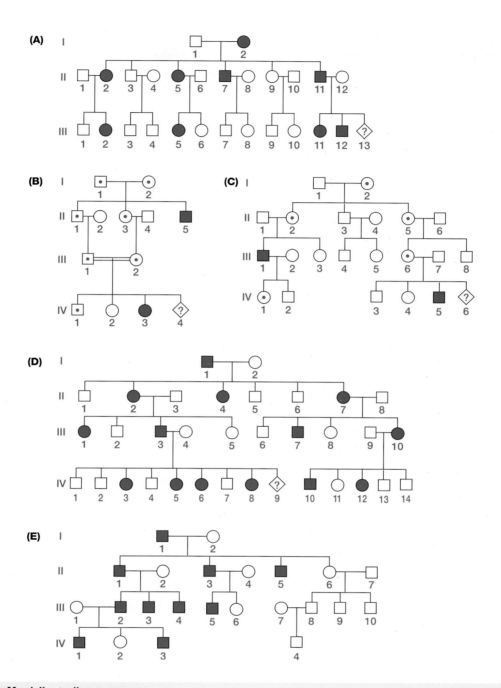

Figure 4.2: Basic Mendelian pedigree patterns.

(A) Autosomal dominant; **(B)** autosomal recessive; **(C)** X-linked recessive; **(D)** X-linked dominant; **(E)** Y-linked. The risk for the individuals marked with a query are **(A)** 1 in 2; **(B)** 1 in 4; **(C)** 1 in 2 males or 1 in 4 of all offspring; **(D)** negligibly low for males, 100% for females. See Section 4.3 and *Figure 4.5* for complications to these basic patterns.

characters in both sexes, and X–linked characters in females, can be dominant or recessive. Nobody has two genetically different Y chromosomes (in the rare XYY males, the two Y chromosomes are duplicates). Thus there are five archetypal Mendelian pedigree patterns (*Figure 4.2*). Special considerations apply to X- and Y-linked conditions as described below,

so that in practice the important Mendelian pedigree patterns are autosomal dominant, autosomal recessive and X-linked (dominant or recessive). These basic patterns are subject to various complications discussed in Section 4.3 (below), and illustrated in *Figure 4.5*.

Box 4.1: Characteristics of the Mendelian patterns of inheritance.

Autosomal dominant inheritance (*Figure 4.2A*):

an affected person usually has at least one affected parent (for exceptions see *Figure 4.4*);

affects either sex;

transmitted by either sex;

a child of an affected x unaffected mating has a 50% chance of being affected (this assumes the affected person is heterozygous, which is usually true for rare conditions).

Autosomal recessive inheritance (*Figure 4.2B*):

affected people are usually born to unaffected parents;

parents of affected people are usually asymptomatic carriers;

there is an increased incidence of parental consanguinity;

affects either sex;

after the birth of an affected child, each subsequent child has a 25% chance of being affected (assuming both parents are phenotypically normal carriers).

X-linked recessive inheritance (*Figure 4.2C*):

affects mainly males;

affected males are usually born to unaffected parents; the mother is normally an asymptomatic carrier and may have affected male relatives;

females may be affected if the father is affected and the mother is a carrier, or occasionally as a result of nonrandom X-inactivation (*Section 4.2.2*);

there is no male-to-male transmission in the pedigree (but matings of an affected male and carrier female can give the *appearance* of male to male transmission, see *Figure 4.5G*).

X-linked dominant inheritance (*Figure 4.2D*):

affects either sex, but more females than males;

females are often more mildly and more variably affected than males;

the child of an affected female, regardless of its sex, has a 50% chance of being affected;

for an affected male, all his daughters but none of his sons are affected.

Y-linked inheritance (*Figure 4.2E*):

affects only males.

affected males always have an affected father (unless there is a new mutation).

all sons of an affected man are affected.

X-inactivation (lyonization) blurs the distinction between dominant and recessive X-linked conditions.

Carriers of 'recessive' X-linked conditions often manifest some signs of the condition, while compared to affected males, heterozygotes for 'dominant' X-linked conditions are usually more mildly and variably affected. This is a consequence of **X-inactivation**. As described in Section 10.5.6, mammals compensate for the unequal numbers of X chromosomes in males and females by permanently inactivating all X chromosomes except one in each somatic cell. XY males keep their single X active, whilst XX females inactivate one X (chosen at random) in each cell. Inactivation takes place early in embryonic life, and once a cell has chosen which X to inactivate, that choice is transmitted clonally to all its daughter cells.

A female heterozygous for an X-linked condition (dominant or recessive), is a *mosaic* (see Section 4.3.6). Each cell expresses either the normal or the abnormal allele, but not both. Where the phenotype depends on a circulating product, as in hemophilia (failure of blood to clot, MIM 306700, 306900), there is an averaging effect between the normal and abnormal cells. Female carriers have an intermediate phenotype, and are usually clinically unaffected but biochemically abnormal. Where the phenotype is a localized property of individual cells, as in hypohidrotic ectodermal dysplasia (missing sweat glands, abnormal teeth and hair; MIM 305100) female carriers show patches of normal and abnormal tissue. Occasional **manifesting heterozygotes** are seen for X-linked recessive conditions. These women may be quite severely affected because by bad luck most cells in some critical tissue have inactivated the normal X.

The Y chromosome carries relatively few genes

No known human character, apart from maleness itself, gives the stereotypical Y-linked pedigree of *Figure 4.2E* (claims for 'porcupine men' and 'hairy ears' are dubious, see MIM 146600 and 425500 respectively). Since normal females lack all Y-linked genes, any such genes must code either for nonessential characters or for male-specific functions. Some genes exist as functional copies on both the Y and the X; they might prove an exception to this argument, but they would not give a classical Y-linked pedigree pattern. Interstitial deletions of Yq are an important cause of male infertility, but of course infertile males will not produce pedigrees like *Figure 4.2E*. Jobling and Tyler-Smith (2000) and Skaletsky *et al.* (2003) summarize the gene content of the Y chromosome and its possible involvement in disease.

Genes located in the Xp–Yp pairing region show pseudoautosomal inheritance

As mentioned in Section 2.3.3, the distal 2.6 Mb of Xp and Yp are homologous and are subject to crossing over in meiosis. Thus the few genes in these regions segregate in a **'pseudoautosomal'** and not a sex-linked pattern.

4.2.3 The mode of inheritance can rarely be defined unambiguously in a single pedigree

Given the limited size of human families, it is rarely possible to be completely certain of the mode of inheritance of a character simply by inspecting a single pedigree. In experimental animals one would set up a test cross and check for a 1 : 2 or 1 : 4 ratio. In human pedigrees the proportion of affected children is not a very reliable indicator. Mostly this is because the numbers are too small, but in addition, the way in which the family was ascertained can bias the ratio of affected to unaffected children. For recessive conditions, the proportion of affected children often seems to be greater than 1 in 4. This is because families are normally ascertained when they have an affected child; families where both parents are carriers but, by good fortune, nobody is affected are systematically missed. These **biases of ascertainment**, and the ways of correcting them, are discussed in Section 15.2.1.

For many of the rarer conditions, the stated mode of inheritance is no more than an informed guess. Assigning modes of inheritance is important, because that is the basis of the risk estimates used in genetic counseling. However, it is important to recognize that the modes of inheritance are often working hypotheses rather than established fact. OMIM uses an asterisk to denote entries with relatively well-established modes of inheritance. Only when a cloned copy of the gene is available can the inheritance be defined with certainty.

4.2.4 One gene–one enzyme does not imply one gene–one syndrome

Pedigree patterns provide the essential entry point into human genetics, but they are only a starting point for defining genes. It would be a serious error to imagine that the 6000 or so known Mendelian characters define 6000 DNA coding sequences. This would be an unjustified extension of the **one gene–one enzyme hypothesis** of Beadle and Tatum. Back in the 1940s this hypothesis allowed a major leap forward in understanding how genes determine phenotypes. Since then it has been extended: some genes encode nontranslated RNAs, some proteins are not enzymes, and many proteins contain several separately encoded polypeptide chains. But even with these extensions, Beadle and Tatum's hypothesis cannot be used to imply a one-to-one correspondence between entries in the OMIM catalogue and DNA transcription units.

The genes of classical genetics are abstract entities. Any character that is determined at a single chromosomal location will segregate in a Mendelian pattern – but the determinant may not be a gene in the molecular geneticist's sense of the word. Fascio-scapulo-humeral muscular dystrophy (severe but nonlethal weakness of certain muscle groups; MIM 158900) is caused by small deletions of sequences at 4q35 – but nobody, at the time of writing, has managed to find a relevant protein-coding sequence at that location, despite intensive searching and sequencing. The 'gene' for Charcot–Marie–Tooth disease type 1A (motor and sensory neuropathy; MIM 118220) turned out to be a 1.5 Mb tandem duplication on chromosome 17p11.2 (Section 16.5.2). These examples are unusual – most OMIM entries probably do describe the consequences of mutations affecting a single transcription unit. However, there is still no one-to-one correspondence between phenotypes and transcription units because of three types of heterogeneity:

▶ **locus heterogeneity** is where the same clinical phenotype can result from mutations at any one of several different loci;

▶ **allelic heterogeneity** is where many different mutations within a given gene can be seen in different patients with a certain genetic condition (explored more fully in Chapter 16). Many diseases show both locus and allelic heterogeneity;

▶ **clinical heterogeneity** is used here to describe the situation when mutations in the same gene produce two or more different diseases. Section 16.6 gives examples.

Locus heterogeneity is common in syndromes that result from failure of a complex pathway

Hearing loss provides good examples of locus heterogeneity. When two people with autosomal recessive profound congenital hearing loss marry, as they often do, the children most often have normal hearing (*Figure 4.3*). It is easy to see that many different genes would be needed to construct so exquisite a machine as the cochlear hair cell, and a defect in any of those genes could lead to deafness. The children have normal hearing whenever the parents carry mutations in

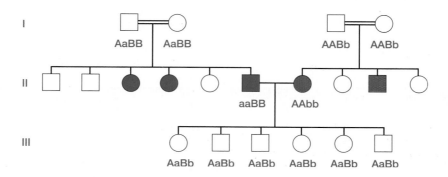

Figure 4.3: Complementation: parents with autosomal recessive profound hearing loss often have children with normal hearing.

II₆ and II₇ are offspring of unaffected but consanguineous parents, and each has affected sibs, making it likely that each has autosomal recessive hearing loss. All their children are unaffected, showing that II₆ and II₇ have nonallelic mutations.

different genes. This is an example of **complementation** (*Box 4.2*). Such locus heterogeneity is only to be expected in conditions like deafness, blindness or mental retardation, where a rather general pathway has failed; but even with more specific pathologies, multiple loci are very frequent. A striking example is Usher syndrome, an autosomal recessive combination of hearing loss and progressive blindness (retinitis pigmentosa), which can be caused by mutations at 10 or more unlinked loci (Hereditary Hearing Loss Homepage, www.uia.ac.be/dnalab/hhh/). OMIM has separate entries for known examples of locus heterogeneity (defined by linkage or mutation analysis), but there must be many undetected examples still contained within single entries.

Allelic series are a cause of clinical heterogeneity

Sometimes several apparently distinct human phenotypes turn out to be all caused by different allelic mutations at the same locus. The difference may be one of degree: mutations that partially inactivate the dystrophin gene produce Becker muscular dystrophy, while mutations that completely inactivate the same gene produce the similar but more severe Duchenne muscular dystrophy (lethal muscle wasting; MIM 310200). Other times the difference is qualitative: inactivation of the androgen receptor gene causes androgen insensitivity (46,XY embryos develop as females; MIM 313700), but expansion of a run of glutamine codons within the same gene causes a very different disease, spinobulbar muscular atrophy or Kennedy disease (MIM 313200). These and other genotype–phenotype correlations are discussed in more depth in Chapter 16.

4.2.5 Mitochondrial inheritance gives a recognizable matrilineal pedigree pattern

In addition to the mutations in genes carried on the nuclear chromosomes, mitochondrial mutations are a significant cause of human genetic disease (see the MITOMAP database for details; www.mitomap.org). The mitochondrial genome (Section 9.1.2) is small but highly mutable compared to nuclear DNA, probably because mitochondrial DNA replication is more error-prone and the number of replications is much higher. Mitochondrially encoded diseases have two unusual features, **matrilineal inheritance** and frequent **heteroplasmy**.

Inheritance is matrilineal, because sperm do not contribute mitochondria to the zygote (this assertion rests on limited evidence; however, paternally-derived mitochondrial variants have almost never been detected in children). Thus a mitochondrially inherited condition can affect both sexes, but is passed on only by affected mothers (*Figure 4.4*), giving a recognizable pedigree pattern.

Cells contain many mitochondrial genomes. In some patients with a mitochondrial disease, every mitochondrial genome carries the causative mutation (**homoplasmy**), but in other cases a mixed population of normal and mutant genomes is seen within each cell (heteroplasmy). Unlike nuclear genetic mosaicism, which must arise postzygotically (Section 4.3.6), mitochondrial heteroplasmy can be transmitted from heteroplasmic mother to heteroplasmic child. In such cases the proportion of abnormal mitochondrial genomes can vary remarkably between the mother and child, suggesting that a surprisingly small number of maternal mitochondrial DNA molecules give rise to all the mitochondrial DNA of the child (see Section 11.4.2). The complicated molecular pathology of mitochondrial diseases is discussed in Section 16.6.6.

4.3 Complications to the basic Mendelian pedigree patterns

In real life various complications often disguise a basic mendelian pattern. *Figure 4.5* shows a number of common complications.

4.3.1 Common recessive conditions can give a pseudo-dominant pedigree pattern

If a character is common in the population, there is a high chance that it may be brought into the pedigree independently by two or more people. A common recessive character like blood group O may be seen in successive generations because of repeated marriages of group O people with heterozygotes. This produces a pattern resembling dominant inheritance (*Figure 4.5A*). Thus the classic pedigree patterns are best seen with rare characters.

4.3.2 Failure of a dominant condition to manifest is called nonpenetrance

With dominant conditions, **nonpenetrance** is a frequent complication. The **penetrance** of a character, for a given

---

Box 4.2: The complementation test to discover whether two recessive characters are determined by allelic genes.

	one locus	two loci
Parental cross	$a_1a_1 \times a_2a_2$	aaBB × AAbb
Offspring	a_1a_2	AaBb
Phenotype	mutant	wild-type

Animals homozygous for the two characters are crossed and the phenotype of the offspring observed. If both animals carry mutations at the same locus the progeny will not have a wild-type allele, and so will be phenotypically abnormal. If there are two different loci the progeny are heterozygous for each of the two recessive characters, and therefore phenotypically normal. Very occasionally alleles at the same locus can complement each other (interallelic complementation).

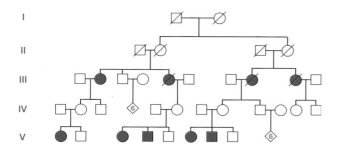

Figure 4.4: Pedigree of a mitochondrial disease.

A typical pedigree pattern, showing mitochondrially-determined hearing loss (family reported by Prezant *et al.*, 1993). Note the incomplete penetrance.

genotype, is defined as the probability that a person who has the genotype will manifest the character. By definition, a dominant character is manifest in a heterozygous person, and so should show 100% penetrance. Nevertheless, many human characters, while generally showing dominant inheritance, occasionally skip a generation. In *Figure 4.5B*, II$_2$ has an affected parent and an affected child, and almost certainly carries the mutant gene, but is phenotypically normal. This would be described as a case of nonpenetrance.

There is no mystery about nonpenetrance – indeed, 100% penetrance is the more surprising phenomenon. Very often the presence or absence of a character depends, in the main and in normal circumstances, on the genotype at one locus, but an unusual genetic background, a particular lifestyle or maybe just chance means that the occasional person may fail to manifest the character. Nonpenetrance is a major pitfall in genetic counseling. It would be an unwise counselor who, knowing the condition in *Figure 4.5B* was dominant and seeing III$_7$ was free of signs, told her that she had no risk of having affected children. One of the jobs of genetic counselors is to know the usual degree of penetrance of each dominant syndrome.

Frequently, of course, a character depends on many factors and does not show a Mendelian pedigree pattern even if entirely genetic. There is a continuum of characters from fully penetrant Mendelian to multifactorial (*Figure 4.6*), with increasing influence of other genetic loci and/or the environment. No logical break separates imperfectly penetrant Mendelian from multifactorial characters; it is a question of which is the most useful description to apply.

Age-related penetrance in late-onset diseases

A particularly important case of reduced penetrance is seen with late-onset diseases. Genetic conditions are not necessarily congenital (present at birth). The genotype is fixed at conception, but the phenotype may not manifest until adult life. In such cases the penetrance is age-related. Huntington disease (progressive neurodegeneration; MIM 143100) is a well-known example (*Figure 4.7*). Delayed onset might be caused by slow accumulation of a noxious substance, by slow tissue death or by inability to repair some form of environmental damage. Hereditary cancers are caused by a second

mutation affecting a cell of a person who already carries one mutation in a tumor suppressor gene (Chapter 17). Depending on the disease, the penetrance may become 100% if the person lives long enough, or there may be people who carry the gene but who will never develop symptoms no matter how long they live. Age-of-onset curves such as *Figure 4.7* are important tools in genetic counseling, because they enable the geneticist to estimate the chance that an at-risk but asymptomatic person will subsequently develop the disease.

4.3.3 Many conditions show variable expression

Related to nonpenetrance is the **variable expression** frequently seen in dominant conditions. *Figure 4.5C* shows an example from a family with Waardenburg syndrome. Different family members show different features of the syndrome. The cause is the same as with nonpenetrance: other genes, environmental factors or pure chance have some influence on development of the symptoms. Nonpenetrance and variable expression are typically problems with dominant, rather than recessive, characters. Partly this reflects the difficulty of spotting nonpenetrant cases in a typical recessive pedigree. However, as a general rule, recessive conditions are less variable than dominant ones, probably because the phenotype of a heterozygote involves a balance between the effects of the two alleles, so that the outcome is likely to be more sensitive to outside influence than the phenotype of a homozygote. However, both nonpenetrance and variable expression are occasionally seen in recessive conditions.

These complications are much more conspicuous in humans than in plants or other animals. Laboratory animals and crop plants are far more genetically uniform than humans. What we see in human genetics is typical of a wild population. Nevertheless, mouse geneticists are familiar with the way expression of a mutant can change when it is bred onto a different genetic background – an important consideration when studying mouse models of human diseases.

Anticipation is a special type of variable expression

Anticipation describes the tendency of some variable dominant conditions to become more severe (or have earlier onset) in successive generations. Until recently, most geneticists were skeptical that this ever really happened. The problem is that true anticipation is very easily mimicked by random variations in severity. A family comes to clinical attention when a severely affected child is born. Investigating the history, the geneticist notes that one of the parents is affected, but only mildly. This looks like anticipation, but may actually be just a bias of ascertainment. Had the parent been severely affected, he or she would most likely never have become a parent; had the child been mildly affected, the family would not have come to notice. Given the lack of any plausible mechanism for anticipation, and the statistical problems of demonstrating it in the face of these biases, most geneticists were unwilling to consider anticipation seriously – until molecular developments obliged them to do so.

Anticipation suddenly became respectable, even fashionable, with the discovery of unstable expanding

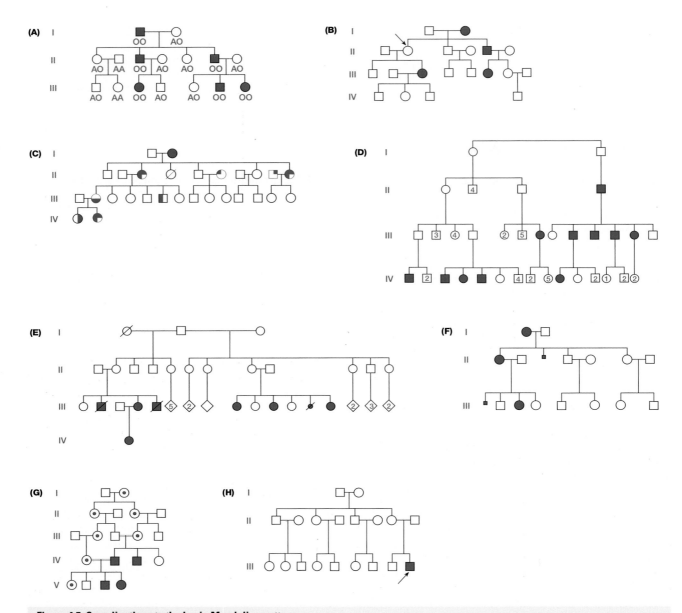

Figure 4.5: Complications to the basic Mendelian patterns.

(A) A common recessive, such as blood group O, can give the appearance of a dominant pattern. (B) Autosomal dominant inheritance with nonpenetrance in II2. (C) Autosomal dominant inheritance with variable expression: in this family with Waardenburg syndrome, shading of first quadrant = hearing loss; second quadrant = different colored eyes; third quadrant = white forelock; fourth quadrant = premature graying of hair. (D) Genetic imprinting: in this family autosomal dominant glomus tumors manifest only when the gene is inherited from the father (family reported by Heutink *et al.,* 1992). (E) Genetic imprinting: in this family autosomal dominant Beckwith–Wiedemann syndrome manifests only when the gene is inherited from the mother (family reported by Viljoen and Ramesar, 1992). (F) X-linked dominant incontinentia pigmenti. Affected males abort spontaneously (small squares). (G) An X-linked recessive pedigree where inbreeding gives an affected female and apparent male-to-male transmission. (H) A new autosomal dominant mutation, mimicking an autosomal or X-linked recessive pattern.

trinucleotide repeats in Fragile-X syndrome (mental retardation with various physical signs; MIM 309550), and later in myotonic dystrophy (a very variable multi-system disease with characteristic muscular dysfunction; MIM 160900) and Huntington disease. Severity or age of onset of these diseases correlates with the repeat length, and the repeat length tends to grow as the gene is transmitted down the generations (see Section 16.6.4). These conditions show true anticipation. Now once again we see claims for anticipation being made for many diseases, and it is important to bear in mind that the old objection about bias of ascertainment remains valid. To be credible, a claim of anticipation requires careful statistical backing, not just clinical impression.

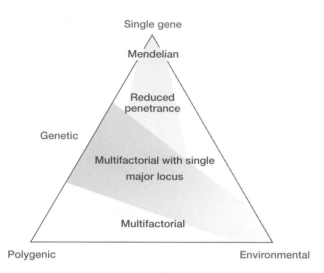

Figure 4.6: The spectrum of human characters.

Few characters are purely Mendelian, purely polygenic or purely environmental. Most depend on some mix of major and minor genetic determinants, together with environmental influences. The mix of factors determining any given character could be represented by a point located somewhere within the triangle.

4.3.4 For imprinted genes, expression depends on parental origin

Certain human characters are autosomal dominant, affect both sexes and are transmitted by parents of either sex, but manifest only when inherited from a parent of one particular sex. For example there are families with autosomal dominant glomus tumors that are expressed only in men or women who inherit the gene from their father (*Figure 4.5D*), while Beckwith–Wiedemann syndrome (exomphalos, macroglossia, overgrowth; MIM 130650) is sometimes dominant but expressed only in children who inherit it from their mother (*Figure 4.5E*). These parental sex effects are evidence of **imprinting**, a poorly understood phenomenon whereby certain genes are somehow marked ('imprinted') with their parental origin. The many questions that surround the mechanism and evolutionary purpose of imprinting are discussed in Section 10.5.4 and a particularly striking clinical example is described in *Box 16.6*.

4.3.5 Male lethality may complicate X-linked pedigrees

For some X-linked dominant conditions, absence of the normal allele is lethal before birth. Thus affected males are not born, and we see a condition that affects only females, who pass it on to half their daughters but none of their sons (*Figure 4.5F*). There may be a history of miscarriages, but families are rarely big enough to prove that the number of sons is only half the number of daughters. An example is incontinentia pigmenti (linear skin defects following defined patterns known as Blaschko's lines, often accompanied by neurological or skeletal problems; MIM 308310).

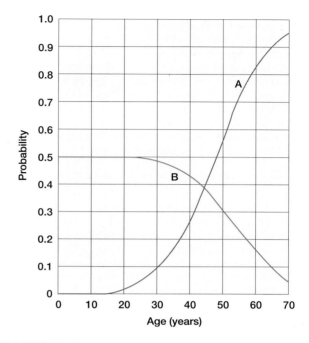

Figure 4.7: Age of onset curve for Huntington disease.

Curve A: probability that an individual carrying the disease gene will have developed symptoms by a given age. Curve B: risk that a healthy child of an affected parent carries the disease gene at a given age. From Harper (2001). Practical Genetic Counselling, 5th edition, Hodder Arnold. Reproduced by permission of Hodder Arnold.

4.3.6 New mutations often complicate pedigree interpretation, and can lead to mosaicism

Many cases of severe dominant or X-linked genetic disease are the result of fresh mutations, striking without warning in a family with no previous history of the disease. A fully penetrant lethal dominant would necessarily always occur by fresh mutation, and the parents would never be affected – an example is thanatophoric dysplasia (severe shortening of long bones and abnormal fusion of cranial sutures; MIM 187600). For a nonlethal but severe dominant condition a similar argument applies, but to a lesser degree. If the disease prevents most affected people from reproducing, but if nevertheless fresh cases of the disease keep occurring, many or most of these must be caused by new mutations. Serious X-linked recessives also show a significant proportion of fresh mutations, because the gene is exposed to natural selection whenever it is in a male. Autosomal recessive pedigrees, on the other hand, are not significantly affected – a mutant allele can propagate for many generations in asymptomatic carriers, and we can safely assume that the parents of an affected child are both carriers.

On the assumption that, averaged over time, new mutations exactly replace the disease genes lost through natural selection, their frequency in the population depends on a simple relationship (described in Section 4.5.2) between the rate at which natural selection is removing disadvantageous genes and the rate at which new mutation is creating them. The general mechanisms that affect the population frequency of alleles are discussed in Chapter 11, *Box 11.2*.

When a normal couple with no relevant family history have a child with severe abnormalities (*Figure 4.5H*), deciding the mode of inheritance and recurrence risk can be very difficult; the problem might be autosomal recessive, autosomal dominant with a new mutation, X-linked recessive (if the child is male) or nongenetic. A further complication is introduced by germinal mosaicism (see below).

Mosaics have two (or more) genetically different cell lines

We have seen that in serious autosomal dominant and X-linked diseases, where affected people have few or no children, the disease genes are maintained in the population by recurrent mutation. A common assumption is that an entirely normal person produces a single mutant gamete. However, this is not necessarily what happens. Unless there is something special about the mutational process, such that it can happen only during gametogenesis, mutations may arise at any time during postzygotic life. Postzygotic mutations produce **mosaics** with two (or more) genetically distinct cell lines.

Mosaicism can affect somatic and/or germ line tissues. Postzygotic mutations are not merely frequent, they are inevitable. Human mutation rates are typically 10^{-7} per gene per cell generation, and our bodies contain perhaps 10^{13} cells. It follows that every one of us must be a mosaic for innumerable genetic diseases. Indeed, as Professor John Edwards memorably remarked, a normal man may well produce the whole of the OMIM catalogue in every ejaculate. This should cause no anxiety. If a cell in your finger mutates to the Huntington disease genotype, or a cell in your ear picks up a cystic fibrosis mutation, there are absolutely no consequences for you or your family. Only if a somatic mutation results in the emergence of a substantial clone of mutant cells is there a risk to the whole organism. This can happen in two ways:

▶ the mutation causes abnormal proliferation of a cell that would normally replicate slowly or not at all, thus generating a clone of mutant cells. This is how cancer happens, and this whole topic is discussed in detail in Chapter 17.

▶ the mutation occurs in an early embryo, affecting a cell which is the progenitor of a significant fraction of the whole organism. In that case the mosaic individual may show clinical signs of disease.

Mutations occurring in a parent's germ line can cause *de novo* inherited disease in a child. An early germ-line mutation can produce a person who harbors a large clone of mutant germ-line cells (**germinal** – or **gonadal** – **mosaicism**). As a result, a normal couple with no previous family history may produce more than one child with the same serious dominant disease. The pedigree mimics recessive inheritance. Even if the correct mode of inheritance is realized, it is very difficult to calculate a recurrence risk to use in counseling the parents (van der Meulen *et al.*, 1995). Usually an empiric risk (Section 4.4.4) is quoted. *Figure 4.8* shows an example of the uncertainty that mosaicism introduces into counseling, in this case in an X-linked disease.

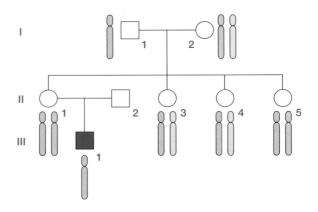

Figure 4.8: A new mutation in X-linked recessive Duchenne muscular dystrophy.

The three grandparental X chromosomes were distinguished using genetic markers, and are shown in blue, pink and brown (ignoring recombination). III_1 has the grandpaternal X, which has acquired a mutation at some point in the pedigree. There are four possible points at which this could have happened:

if III_1 carries a new mutation, the recurrence risk for all family members is very low;

if II_1 is a germinal mosaic, there is a significant risk (but hard to quantify) for her future children, but not for her sisters;

if II_1 was the result of a single mutant sperm, she has the standard recurrence risk for X-linked recessives, but her sisters are free of risk;

if I_1 was a germinal mosaic all the sisters have a significant risk, which is hard to quantify.

Molecular studies can be a great help in these cases. Sometimes it is possible to demonstrate directly that a normal father is producing a proportion of mutant sperm. Direct testing of the germ line is not possible in women, but other accessible tissues such as fibroblasts or hair roots can be examined for evidence of mosaicism. A negative result on somatic tissues does not rule out germ-line mosaicism, but a positive result, in conjunction with an affected child, proves it (*Figure 4.9*).

Chimeras contain cells from two separate zygotes in a single organism

Mosaics start life as a single fertilized egg. **Chimeras** on the other hand are the result of fusion of two zygotes to form a single embryo (the reverse of twinning), or alternatively of limited colonization of one twin by cells from a nonidentical co-twin (*Figure 4.10*). Chimerism is proved by the presence in pooled tissue samples of too many parental alleles at several loci (if just one locus were involved, one would suspect mosaicism for a single mutation). Blood-grouping centers occasionally discover chimeras among normal donors, and some intersex patients turn out to be XX/XY chimeras. Strain *et al.* (1998), for example, describe a 46,XY/46,XX boy who was the result of two embryos amalgamating after an *in vitro* fertilization in which three embryos were transferred to the mother.

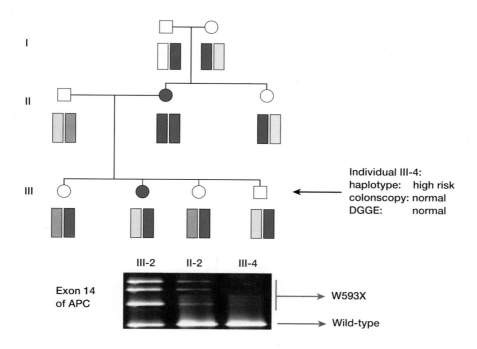

Individual III-4:
haplotype: high risk
colonscopy: normal
DGGE: normal

Figure 4.9: Germline and somatic mosaicism in a dominant disease.

Individuals II-2 and III-2 suffer from familial adenomatous polyposis, a dominantly inherited form of colorectal cancer that maps to chromosome 5 (MIM 175100, see Section 17.5.3). The parents of II-2 are unaffected. Denaturing gradient gel electrophoresis (see Section 18.3.2) demonstrated a mutation, W593X, in exon 14 of the *APC* gene in III-2 (seen as the upper bands in track 1 of the gel). For II-2 the gel shows the mutant bands, but only very weakly, showing that the blood of this person is mosaic for the mutation. The mutation is absent in III-4, even though studies with linked markers (color-coded chromosomes – see Section 18.5) showed that he inherited the high-risk (blue) chromosome from his mother. Clinical examination (colonoscopy) confirmed that III-4 was free of the disease. II-2 must be a germline as well as a somatic mosaic for the mutation. Example and gel courtesy of Professor Bert Bakker, Leiden.

4.4 Genetics of multifactorial characters: the polygenic threshold theory

4.4.1 Some history

By the time Mendel's work was rediscovered in 1900, a rival school of genetics was well established in the UK and elsewhere. Francis Galton, the remarkable and eccentric cousin of Charles Darwin, devoted much of his vast talent to systematizing the study of human variation. Starting with an article on 'Hereditary Talent and Character' published the same year, 1865, as Mendel's paper (and expanded in 1869 to a book, *Hereditary Genius*), he spent many years investigating family resemblances. Galton was devoted to quantifying observations and applying statistical analysis. His Anthropometric Laboratory, established in London in 1884, recorded from his subjects (who paid him threepence for the privilege) their weight, sitting and standing height, arm span, breathing capacity, strength of pull and of squeeze, force of blow, reaction time, keenness of sight and hearing, color discrimination and judgements of length. In one of the first applications of statistics he compared physical attributes of parents and children, and established the degree of correlation between relatives. By 1900 he had established a large body of

knowledge about the inheritance of such attributes, and a tradition (**biometrics**) of their investigation.

When Mendel's work was rediscovered, a controversy arose. Biometricians accepted that Mendelian genes might explain a few rare abnormalities or curious quirks, but they pointed out that most of the characters likely to be important in evolution (body size, build, strength, skill in catching prey or finding food) were **continuous** or **quantitative characters** and not amenable to Mendelian analysis. We all have these characters, only to different degrees, so you cannot define their inheritance by drawing pedigrees and marking in the people who have them. Mendelian analysis requires **dichotomous characters** that you either have or don't have. A controversy, heated at times, ran on between mendelians and biometricians until 1918. That year saw a seminal paper by R.A. Fisher demonstrating that characters governed by a large number of independent Mendelian factors (**polygenic** characters) would display precisely the continuous nature, quantitative variation and family correlations described by the biometricians. Later Falconer extended this model to cover dichotomous characters. Fisher's and Falconer's analyses created a unified theoretical basis for human genetics. The following sections set out their ideas, in a nonmathematical form. A more rigorous treatment can be

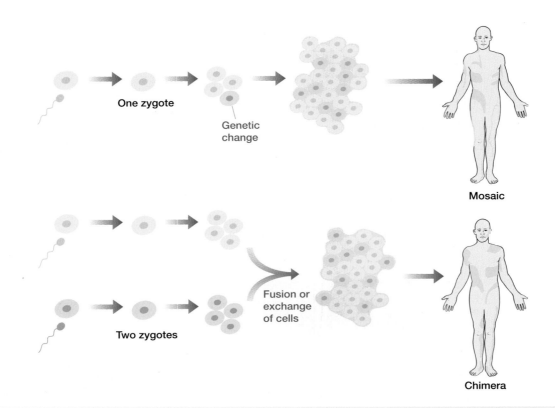

Figure 4.10: Mosaics and chimeras.

Mosaics have two or more genetically different cell lines derived from a single zygote. The genetic change indicated may be a gene mutation, a numerical or structural chromosomal change, or in the special case of lyonization, X-inactivation. A chimera is derived from two zygotes, which are usually both normal but genetically distinct.

found in any textbook of quantitative genetics, for example Falconer and Mackay, 1996 (see Further reading).

4.4.2 Polygenic theory of quantitative traits

Any variable character that depends on the additive action of a large number of individually small independent causes will show a **Normal (Gaussian) distribution** in the population. *Figure 4.11* gives a highly simplified illustration of this. We suppose the character depends on alleles at a single locus, then at two loci, then at three. As more loci are included we see two consequences:

▶ the simple one-to-one relationship between genotype and phenotype disappears. Except for the extreme phenotypes, it is not possible to infer the genotype from the phenotype;

▶ as the number of loci increases, the distribution looks increasingly like a Gaussian curve. Addition of a little environmental variation would smooth out the three-locus distribution into a good Gaussian curve.

A more sophisticated treatment, allowing dominance and varying gene frequencies, leads to the same conclusions. Since relatives share genes, their phenotypes are correlated, and Fisher's 1918 paper predicted the size of the correlation for different relationships.

A much misunderstood feature, both of biometric data and of polygenic theory, is **regression to the mean**. Imagine, for the sake of example only, that variation in IQ were entirely genetically determined. *Figure 4.12* shows how in our simplified two-locus model, for each class of mothers, the average IQ of their children is half way between the mother's value and the population mean. This is regression to the mean – but its implications are often misinterpreted (see *Box 4.3*). Note however a hidden assumption in this simple model: that there is random mating. For each class of mothers, the average IQ of their husbands is assumed to be 100. Thus the average IQ of the children is the mid-parental IQ, as common sense would suggest. In the real world, highly intelligent women tend to marry men of above average intelligence (**assortative mating**) and we would not expect regression half way to the population mean, even if IQ were a purely genetic character.

A second assumption of our simplified model is that there is no dominance: each person's phenotype is the sum of the contribution of each allele at each relevant locus. If we allow dominance, the effect of some of a parent's genes will be masked by dominant alleles and invisible in their phenotype, but they can still be passed on and can affect the child's phenotype. Given dominance, the expectation for the child is no longer the mid-parental value. Our best guess about the likely phenotypic effect of the masked recessive alleles is

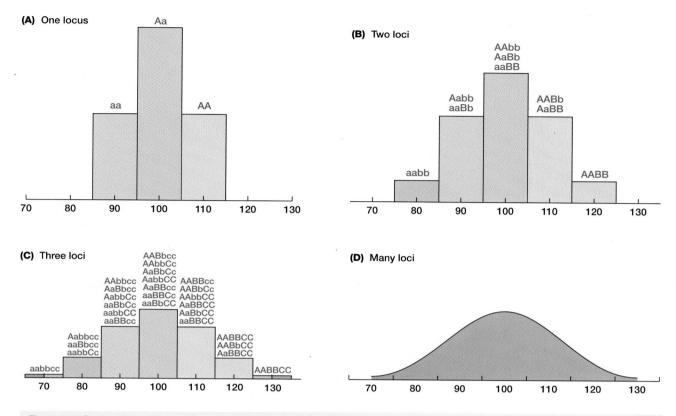

Figure 4.11: Successive approximations to a Gaussian distribution.

The charts show the distribution in the population of a hypothetical character that has a mean value of 100 units. The character is determined by the additive (co-dominant) effects of alleles. Each upper case allele adds 5 units to the value, and each lower case allele subtracts 5 units. All allele frequencies are 0.5. **(A)** The character is determined by a single locus; **(B)** two loci; **(C)** three loci: addition of a minor amount of 'random' (environmental or polygenic) variation produces the Gaussian curve **(D)**.

obtained by looking at the rest of the population. Therefore, the child's expected phenotype will be displaced from the mid-parental value towards the population mean. How far it will be displaced depends on how important dominance is in determining the phenotype.

The heritability is the proportion of variance due to additive genetic effects

Gaussian curves are specified by only two parameters, the mean and the variance (or the standard deviation, which is the square root of the variance). Variances have the useful property of being additive when they are due to independent causes (*Box 4.4*). Thus the overall variance of the phenotype V_P is the sum of the variances due to the individual causes of variation – the environmental variance V_E and the genetic variance V_G. V_G can in turn be broken down to a variance V_A due to simply additive genetic effects and an extra term V_D due to dominance effects. The **heritability** (h^2) of a trait is the proportion of the total variance that is genetic, that is V_G/V_P. For animal breeders interested in breeding cows with higher milk yields, this is an important measure of how far a breeding program can create a herd in which the average animal

resembles today's best. Strictly, V_G/V_P is the broad heritability. Dominance variance cannot be fixed by breeding, so the selection response is determined by the narrow heritability, V_A/V_P. Heritabilities of human traits are often estimated as part of segregation analysis (see Section 15.2 and Table 15.4). However, it should be borne in mind that for many human behavioral traits, the simple partitioning of variance into environmental and genetic components is not applicable. We give our children both their genes and their environment. Genetic disadvantage and social disadvantage tend to go together, so genetic and environmental factors are often correlated. If genetic and environmental factors are not independent, V_P does not equal $V_G + V_E$; there are additional interaction variances. A proliferation of variances can rapidly reduce the explanatory power of the models, and in general this has been a difficult area in which to work.

The term 'heritability' is often misunderstood. Heritability is quite different from the mode of inheritance. The mode of inheritance (autosomal dominant, polygenic, etc.) is a fixed property of a trait, but heritability is not. 'Heritability of IQ' is shorthand for 'heritability of variations in IQ'. Contrast the two questions:

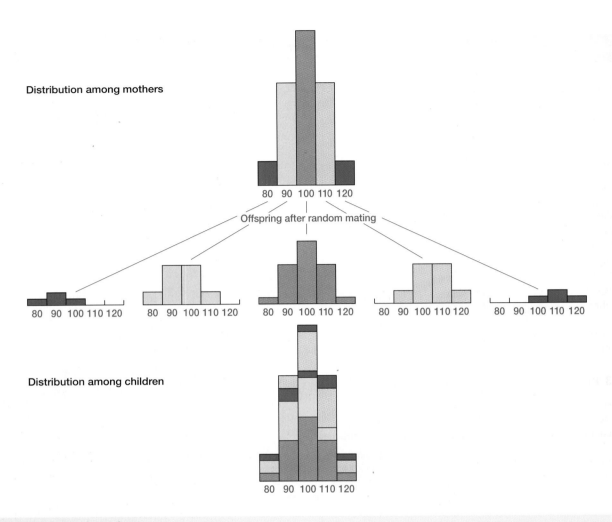

Figure 4.12: Regression to the mean.

The same character as in *Figure 4.11B*: mean 100, determined by co-dominant alleles A, a, B and b at two loci, all gene frequencies = 0.5. Top: distribution in a series of mothers. Middle: distributions in children of each class of mothers, assuming random mating. Bottom: summed distribution in the children. *Note* that: (a) the distribution in the children is the same as the distribution in the mothers; (b) for each class of mothers, the mean for their children is halfway between the mothers' value and the population mean (100); and (c) for each class of children (bottom), the mean for their mothers is half way between the children's value and the population mean.

Box 4.3: Two common misconceptions about regression to the mean.

(1) After a few generations everybody will be exactly the same.

(2) If a character shows regression to the mean, it must be genetic.

Figure 4.12 shows that the first of these beliefs is wrong. In a simple genetic model:

the overall distribution is the same in each generation;

regression works both ways: for each class of children, the average for their mothers is half way between the children's value and the population mean. This may sound paradoxical but it can be confirmed by inspecting, for example, the right-hand column of the bottom histogram in *Figure 4.12* (children of IQ

120). One-quarter of their mothers have IQ 120, half 110 and one-quarter 100, making an average of 110.

Regarding the second of these beliefs, regression to the mean is not a genetic mechanism but a purely statistical phenomenon. Whether the determinants of IQ are genetic, environmental or any mix of the two, if we take an exceptional group of mothers (for example, those with an IQ of 120), then these mothers must have had an exceptional set of determinants. If we take a second group who share half those determinants (their children, their sibs or either of their parents), the average phenotype in this second group will deviate from the population mean by half as much. Genetics provides the figure of one half, but not the principle of regression.

Box 4.4: Partitioning of variance.

Variance of phenotype (V_P) = Genetic variance (V_G) + Environmental variance (V_E)

V_G = variance due to additive genetic effects (V_A) + variance due to dominant effects (V_D)

$V_P = V_A + V_D + V_E$

Heritability (broad) = V_G / V_P

Heritability (narrow) = V_A / V_P

▶ *to what extent is IQ genetic?* This is a meaningless question;

▶ *How much of the differences in IQ between people in a particular country at a particular time are caused by their genetic differences, and how much by their different environments and life histories?* This is a meaningful question, even if difficult to answer.

In different social circumstances, the heritability of IQ will differ. The more equal a society is, the higher the heritability of IQ should be. If everybody has equal opportunities, a number of the environmental differences between people have been removed. Therefore more of the remaining differences in IQ will be due to the genetic differences between people.

4.4.3 Polygenic theory of discontinuous characters

Most of the classical 'polygenic' continuously variable characters like height or weight are of little interest to medical geneticists (although we discuss obesity; see Section 15.6.8). Much more interesting are the innumerable diseases and malformations that tend to run in families, but that do not show Mendelian pedigree patterns. A major conceptual

tool in non-Mendelian genetics was provided by Falconer's extension of polygenic theory to dichotomous or discontinuous characters (those you either have or do not have).

Falconer postulated an underlying continuously variable **susceptibility**. You may or may not have a cleft palate, but every embryo has a certain susceptibility to cleft palate. The susceptibility may be low or high; it is polygenic and follows a Gaussian distribution in the population. Together with the polygenic susceptibility, Falconer postulated the existence of a **threshold**. Embryos whose susceptibility exceeds a critical threshold value develop cleft palate; those whose susceptibility is below the threshold, even if only just below, avoid cleft palate. Stripped of mathematical subtlety, the model can be represented as in *Figure 4.13*. The threshold can be imagined as the neutral point of the balance. Changing the balance of factors tips the phenotype one way or the other.

For cleft palate, a polygenic threshold model seems intuitively reasonable (Fraser, 1980). All embryos start with a cleft palate. During early development the palatal shelves must become horizontal and fuse together. They must do this within a specific developmental window of time. Many different genetic and environmental factors influence embryonic development, so it seems reasonable that susceptibility should be polygenic. Whether the palatal shelves meet and fuse with ample time to spare, or whether they only just manage to fuse in time, is unimportant – if they fuse then a normal palate forms, and if they do not fuse then a cleft palate results. Thus there is a natural threshold superimposed on a continuously variable process.

Falconer's threshold theory helps explain how recurrence risks vary in families. Affected people must have an unfortunate combination of high-susceptibility alleles. Their relatives who share genes with them will also, on average, have a raised susceptibility, the divergence from the population mean depending on the proportion of shared genes. Thus polygenic threshold characters tend to run in families (*Figure 4.14*). Parents who have had several affected children may just have been unlucky, but on average they will have more high-risk alleles than parents with only one affected child. The threshold is fixed, but the average susceptibility, and hence the recurrence risk, rises with an increasing number of previous affected children.

Many supposed threshold conditions have different incidences in the two sexes. This implies sex-specific thresholds. Congenital pyloric stenosis, for example, is five times more common in boys than girls. The threshold must be higher for girls than boys, therefore relatives of an affected girl have a higher average susceptibility than relatives of an affected boy (*Figure 4.15*). The recurrence risk is correspondingly higher, although in each case a baby's risk of being affected is five times higher if it is a boy (*Table 4.1*).

Figure 4.13: Multifactorial determination of a disease or malformation.

The angels and devils can represent any combination of genetic and environmental factors. Adding an extra devil or removing an angel can tip the balance, without that particular factor being *the* cause of the disease. Courtesy of Professor R.S.W. Smithells.

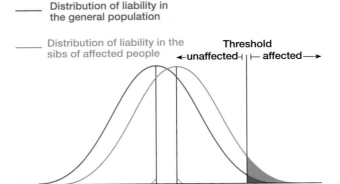

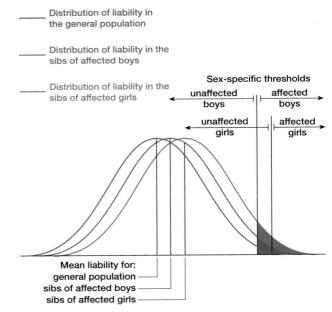

Figure 4.14: Falconer's polygenic threshold model for dichotomous non-Mendelian characters.

Liability to the condition is polygenic and normally distributed (green curve). People whose liability is above a certain threshold value (the balance point in *Figure 4.13*) are affected. Their sibs (lilac curve) have a higher average liability than the population mean and a greater proportion of them have liability exceeding the threshold. Therefore the condition tends to run in families.

4.4.4 Counseling in non-Mendelian conditions uses empiric risks

In genetic counseling for non-Mendelian conditions, risks are not derived from polygenic theory; they are **empiric risks** obtained through population surveys such as those in *Table 4.1*). This is fundamentally different from Mendelian conditions, where the 1 in 2, 1 in 4, and so on, risks come from theory. The effect of family history is also quite different. If a couple are both carriers of cystic fibrosis, the risk of their next child being affected is 1 in 4. This remains true regardless how many affected or normal children they have already produced (summed up in the tag 'chance has no memory'). If they have had a baby with neural tube defect, the recurrence risk is about

Figure 4.15: A polygenic character with sex-specific thresholds.

If a non-Mendelian dichotomous character affects predominantly males, this is accommodated in multifactorial threshold theory by postulating a lower threshold for males than females. It follows that recurrence risks are higher for relatives of affected females, but the majority of those recurrent cases will be male. See *Table 4.1* for an example of data fitting this interpretation.

1 in 25 in the UK – but if they have already had two affected babies, the recurrence risk is about 1 in 12 (perhaps summed up by 'to him that hath shall be given'). It is not that having a second affected baby has caused their recurrence risk to increase, but it has enabled us to recognize them as a couple who always had been at particularly high risk. A cynic would say it involves the counselor being wise after the event – but the practise accords with our understanding based on threshold theory, as well as with epidemiological data, and it represents the best we can offer in an imperfect state of knowledge.

Table 4.1: Recurrence risks for pyloric stenosis.

Relatives of	Sons	Daughters	Brothers	Sisters
Male proband	19/296 (6.42%)	7/274 (2.55%)	5/230 (2.17%)	5/242 (2.07%)
Female proband	14/61 (22.95%)	7/62 (11.48%)	11/101 (10.89%)	9/101 (8.91%)

More boys than girls are affected, but the recurrence risk is higher for relatives of an affected girl. Data fit a polygenic threshold model with sex-specific thresholds (*Figure 4.15*). Data from Fuhrmann and Vogel (1976).

4.5 Factors affecting gene frequencies

4.5.1 There can be a simple relation between gene frequencies and genotype frequencies

A thought experiment: picking genes from the gene pool

Over a whole population there may be many different alleles at a particular locus, although each individual person has just two alleles, which may be identical or different. We can imagine a **gene pool**, consisting of all alleles at the A locus in the population. The **gene frequency** of allele A_1 is the proportion of all A alleles in the gene pool which are A_1. Consider two alleles, A_1 and A_2 at the A locus. Let their gene frequencies be p and q respectively (p and q are each between 0 and 1). Let us perform a thought experiment:

▶ pick an allele at random from the gene pool. There is a chance p that it is A_1 and a chance q that it is A_2;

▶ pick a second allele at random. Again the chance of picking A_1 is p and the chance of picking A_2 is q (we assume the gene pool is sufficiently large that removing the first allele has not significantly changed the gene frequencies in the remaining pool). It follows that:

• the chance that both alleles were A_1 is p^2;

• the chance that both alleles were A_2 is q^2;

• the chance that the first allele was A_1 and the second A_2 is pq. The chance that the first was A_2 and the second A_1 is qp. Overall, the chance of picking one A_1 and one A_2 allele is 2pq.

The Hardy–Weinberg distribution

If we pick a person at random from the population, this is equivalent to picking two genes at random from the gene pool. The chance the person is A_1A_1 is p^2, the chance they are A_1A_2 is 2pq, and the chance they are A_2A_2 is q^2. This simple relationship between gene frequencies and genotype frequencies (the **Hardy–Weinberg distribution**, see *Box 4.5*) holds whenever a person's two genes are drawn independently and at random from the gene pool. A_1 and A_2 may be the only alleles at the locus (in which case p + q = 1) or there may be other alleles and other genotypes (p + q < 1). For X-linked loci males, being hemizygous (only one allele) are A_1 or A_2 with frequencies p and q respectively, while females can be A_1A_1, A_1A_2 or A_2A_2 (*see Box 4.5*).

Limitations of the Hardy–Weinberg distribution

These simple calculations break down if the underlying assumption, that a person's two genes are picked independently from the gene pool, is violated. In particular, there is a problem if there has not been random mating. Assortative mating can take several forms, but the most generally important is **inbreeding**. If you marry a relative you are marrying somebody whose genes resemble your own. This increases the likelihood of your children being homozygous and decreases the likelihood that they will be heterozygous. Rare recessive conditions are strongly associated with parental consanguinity, and Hardy–Weinberg calculations that ignore this will overestimate the carrier frequency in the population at large.

Use of the Hardy–Weinberg distribution in genetic counseling

Gene frequencies or genotype frequencies are essential inputs into many forms of genetic analysis, such as linkage analysis (Section 13.3) and segregation analysis (Section 15.2), and they have a particular importance in calculating genetic risks. *Box 4.6* gives examples.

4.5.2 Genotype frequencies can be used (with caution) to calculate mutation rates

Mutant genes are being created by fresh mutation and being removed by natural selection (Box 11.2). For a given level of selection we can calculate the mutation rate that would be required to replace the genes lost by selection. If we assume that there is an equilibrium in the population between the rates of loss and of replacement, the calculation tells us the present mutation rate. We can define the **coefficient of selection (s)** as the relative chance of reproductive failure of a genotype due to selection (the fittest type in the population has s = 0, a genetic lethal has s = 1).

▶ **For an autosomal recessive** condition, a proportion q^2 of the population is affected. The loss of disease genes each generation is sq^2. This is balanced by mutation at the rate of $\mu(1 - q^2)$ where μ is the mutation rate per gene per generation. At equilibrium $sq^2 = \mu(1 - q^2)$, or approximately (if q is small) $\mu = sq^2$.

▶ For a rare **autosomal dominant** condition homozygotes are excessively rare. Heterozygotes occur with frequency 2pq (frequency of disease gene = p). Only half the genes

Box 4.5: Hardy–Weinberg equilibrium genotype frequencies for allele frequencies p (A_1) and q (A_2).

	Autosomal locus			X-linked locus				
				Males		Females		
Genotype	A_1A_1	A_1A_2	A_2A_2	A_1	A_2	A_1A_1	A_1A_2	A_2A_2
Frequency	p^2	2pq	q^2	p	q	p^2	2pq	q^2

Note that these genotype frequencies will be seen whether or not A_1 and A_2 are the only alleles at the locus.

Box 4.6: The Hardy–Weinberg distribution can be used (with caution) to calculate carrier frequencies and simple risks for counseling.

An autosomal recessive condition affects 1 newborn in 10 000. What is the expected frequency of carriers?

Phenotypes:	Unaffected		Affected
Genotypes:	AA	Aa	aa
Frequencies:	p^2	$2pq$	$q^2 = 1/10\,000$

q^2 is 10^{-4}, and therefore $q = 10^{-2}$ or 1/100.

1 in 100 genes at the A locus are a, 99/100 are A.

The carrier frequency, $2pq$, is $2 \times 99/100 \times 1/100$, very nearly 1 in 50. This assumes that the frequency of the condition has not been raised by inbreeding.

If a parent of a child affected by the above condition remarries, what is the risk of producing an affected child in the new marriage?

To produce an affected child, both parents must be carriers, and the risk is then 1 in 4. Thus the overall risk is:

(the parent's carrier risk) × (the new spouse's carrier risk) × 1/4

= 1 × 1/50 × 1/4

= 1/200

This assumes there is no family history of the same disease in the new spouse's family.

X-linked red-green color blindness affects 1 in 12 British males. What proportion of females will be carriers and what proportion will be affected?

	Males		Females		
Genotypes:	A_1	A_2	A_1A_1	A_1A_2	A_2A_2
Frequencies:	p	q = 1/12	p^2	$2pq$	q^2

q = 1/12, therefore p = 11/12

$2pq = 2 \times 1/12 \times 11/12 = 22/144$

$q^2 = 1$ in 144. Thus this single-locus model predicts that 15% of females will be carriers and 0.7% will be affected.

Box 4.7: Mutation-selection equilibrium.

Autosomal recessive condition*	$\mu = sq^2$	or	$\mu = F(1-f)$
Autosomal dominant condition	$\mu = sp$	or	$\mu = \frac{1}{2}F(1-f)$
X-linked recessive condition	$\mu = sq/3$	or	$\mu = 1/3 F(1-f)$

    μ = mutation rate per gene per generation

    p,q = gene frequencies

    s = coefficient of selection

    f = biological fitness = 1–s

    F = frequency of condition in the population

*This formula gives a seriously wrong estimate of the mutation rate if there is heterozygote advantage, see *Box 4.8*.

---

lost through their reproductive failure are the disease allele, so the rate of gene loss is very nearly sp. Again this is balanced by a rate of new mutation of μq^2, which is approximately μ if q is almost 1. Thus $\mu = \mathbf{sp}$.

▶ For an **X-linked recessive** disease the rate of gene loss through affected males is sq. This is balanced by a mutation rate 3μ, since all X chromosomes in the population are available for mutation, but only the one-third of X chromosomes which are in males are exposed to selection. Thus $\mu = \mathbf{sq/3}$.

These results are summarized in *Box 4.7*. Estimates derived using them can be compared with the general expectation, from studies in many organisms, that mutation rates are typically 10^{-5}–10^{-7} per gene per generation.

4.5.3 Heterozygote advantage can be much more important than recurrent mutation for determining the frequency of a recessive disease

The formula $\mu = sq^2$ gives an unexpectedly high mutation rate for some autosomal recessive conditions. Consider cystic fibrosis (CF), for example. Until very recently, virtually nobody with CF lived long enough to reproduce, therefore s = 1. CF affects about one birth in 2000 in the UK. Thus $q^2 = 1/2000$, and the formula gives $\mu = 5 \times 10^{-4}$. This would be a strikingly high mutation rate for any gene – but there is evidence that new CF mutations are in fact very rare. This follows from the uneven ethnic distribution of CF and the existence of strong linkage disequilibrium (Section 13.5.2).

The missing factor is **heterozygote advantage**. CF carriers have, or had in the past, some reproductive advantage over normal homozygotes. There has been debate over what this advantage might be. The CF gene encodes a membrane chloride channel, which is required by *Salmonella typhi* for it to enter epithelial cells, so maybe heterozygotes are relatively resistant to typhoid fever (Pier *et al.*, 1998). Whatever the cause of the heterozygote advantage, if s_1 and s_2 are the coefficients of selection against the AA and aa genotypes respectively, then an equilibrium is established (ignoring recurrent mutation) when the ratio of the gene frequencies of A and a, p/q, is s_2/s_1. *Box 4.8* illustrates the calculation for cystic fibrosis, and shows that a heterozygote advantage too small to observe in population surveys can have a major effect on gene frequencies.

Box 4.8: Selection in favor of heterozygotes for CF.

For CF, the disease frequency in the UK is about one in 2000 births.

Phenotypes:	Unaffected		Affected
Genotypes:	AA	Aa	aa
Frequencies:	p^2	$2pq$	$q^2 = 1/2000$

q^2 is 5×10^{-4}, therefore $q = 0.022$ and $p = 1-q = 0.978$

$p/q = 0.978 / 0.022 = 43.72 = s_2 / s_1$

If $s_2 = 1$ (affected homozygotes never reproduce), $s_1 = 0.023$

The present CF gene frequency will be maintained, even without fresh mutations, if Aa heterozygotes have on average 2.3% more surviving children than AA homozygotes.

It is worth remembering that the medically important Mendelian diseases are those that are both common and serious. They must all have some or other special trick to remain common in the face of selection. The trick may be an exceptionally high mutation rate (Duchenne muscular dystrophy), or propagation of nonpathological premutations (Fragile X), or onset of symptoms after reproductive age (Huntington disease) – but for common serious recessive conditions it is most often heterozygote advantage.

Further reading

Falconer DS, Mackay TFC (1996) *Introduction to Quantitative Genetics.* Longman, Harlow.

Forrest DW (1974) *Francis Galton: The Life and Work of a Victorian Genius.* Elek, London.

References

Fisher RA (1918). The correlation between relatives under the supposition of mendelian inheritance. *Trans. Roy. Soc.* **52**, 399–433.

Fuhrmann W, Vogel F (1976) *Genetic Counselling.* Springer, New York.

Fraser FC (1980) The William Allan Memorial Award Address: Evolution of a palatable multifactorial threshold model. *Am. J. Hum. Genet.* **32**, 796–813.

Harper PS (2001) *Practical Genetic Counselling,* 5th Edn. Arnold, London.

Heutink P, van der Mey AG, Sandkuijl LA *et al.* (1992) A gene subject to genomic imprinting and responsible for hereditary paragangliomas maps to chromosome 11q23-qter. *Hum. Molec. Genet.* **1**, 7–10.

Jobling MA, Tyler-Smith C (2000) New uses for new haplotypes: the human Y chromosome, disease and selection. *Trends Genet.* **16**, 356–362.

Pier GB, Grout M, Zaidi T *et al.* (1998) *Salmonella typhi* uses CFTR to enter intestinal epithelial cells. *Nature* **393**, 79–82.

Prezant TR, Agapian JV, Bowman MC *et al.* (1993) Mitochondrial ribosomal RNA mutation associated with both antibiotic induced and non syndromic deafness. *Nature Genet.* **4**, 289–294.

Skaletsky H, Kuroda-Kawaguchi T, Minx PJ *et al.* (2003) The male-specific region of the human Y chromosome is a mosaic of discrete sequence classes. *Nature* **423**, 825–837.

Strain L, Dean JCS, Hamilton MPR, Bonthron DT (1998) A true hermaphrodite chimera resulting from embryo amalgamation after in vitro fertilization. *N. Engl J. Med.* **338**, 166–169.

Van der Meulen MA, van der Meulen MJP, te Meerman GJ (1995) Recurrence risk for germinal mosaics revisited. *J. Med. Genet.* **32**, 102–104.

Viljoen D, Ramesar R (1992) Evidence for paternal imprinting in familial Beckwith–Wiedemann syndrome. *J. Med. Genet.* **29**, 221–225.

Wilkie AOM (1994). The molecular basis of dominance. *J. Med. Genet.* **31**, 89–98.

CHAPTER FIVE

Amplifying DNA: PCR and cell–based DNA cloning

Chapter contents

5.1 The importance of DNA cloning

The fundamentals of current DNA technology are very largely based on two quite different approaches to studying specific DNA sequences within a complex DNA population (see *Figure 5.1*):

▶ **DNA cloning**. The desired DNA sequence or fragment must be *selectively amplified* to produce large numbers of identical copies, resulting in purification of the desired product. Thereafter, its structure and function can be comprehensively studied;

▶ **molecular hybridization**. The fragment of interest is not amplified or purified in any way; instead, it is *specifically detected* within a complex mixture of many different sequences.

Before DNA cloning, our knowledge of DNA was extremely limited. DNA cloning technology changed all that and revolutionized the study of genetics. To understand why that was, one should appreciate the tremendous size and complexity of DNA sequences (compared to, say, protein sequences). Individual nuclear DNA molecules contain hundreds of millions of nucleotides. When DNA is isolated from cells using standard methods, these huge molecules are fragmented by shear forces, generating complex mixtures of still very large DNA fragments (typically 50–100 kb long).

Given the complexity of the DNA isolated from the cells of a typical eukaryote, or even prokaryote, the challenge was how to analyze it. What was needed was some method of fraction-ating the DNA so that different subpopulations of DNA sequences could be purified. For many eukaryotes, one early approach had been to separate different populations of DNA sequence by centrifugation. Ultracentrifugation in equilibrium density gradients (e.g. in CsCl density gradients) typically fractionates the DNA from a eukaryotic cell into a major band (the **bulk DNA**) and several minor bands. The buoyant densities of the minor DNA bands are different from that of the bulk DNA (and from each other) because they are composed of tandemly repetitive **satellite DNA** sequences whose base composition is significantly different to that of the bulk DNA. The satellite DNA sequences were found to be involved in specific aspects of chromosome structure and function. Although valuable and interesting, the purified satellite DNAs were a minor component of the genome and did not contain genes. *What DNA cloning offered was a general method for purifying and studying any DNA sequence.*

DNA cloning requires selective amplification of a specific DNA component (the target DNA) occurring within a large complex starting DNA population, which may be the total genomic DNA within a particular tissue (or cell type), or **complementary DNA (cDNA)** prepared from the total RNA of a particular tissue. Amplification is achieved using a DNA polymerase and can be carried out *in vitro*, or within cells.

In vitro DNA cloning

Here the desired target DNA is selected using oligonucleotide primers which will bind specifically to this sequence. Once the sequence-specific primers are bound to the target, a heat-

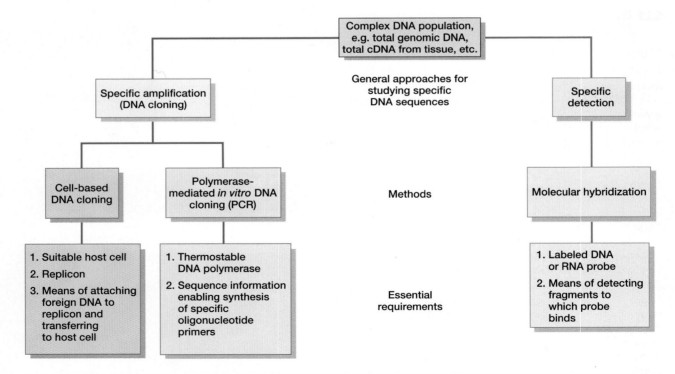

Figure 5.1: General approaches for studying specific DNA sequences in complex DNA populations.

stable DNA polymerase can generate additional copies of the target sequence. The new DNA copies of the target in turn serve as templates to make more copies still in a **chain reaction** which has been named the **polymerase chain reaction (PCR)**.

Cell-based DNA cloning

The target DNA molecules are attached to a single type of **replicon** molecule (one which is capable of *independent* DNA replication within a suitable host cell) and transferred into suitable host cells. The hybrid target–replicon molecules typically undergo many rounds of DNA replication (***DNA amplification***), after which the amplified target DNA molecules can be separated from other contents of the cells and also from the replicon molecules. *The key to cell-based DNA cloning is that when the different target–replicon molecules are transferred into cells (**transformation**), each cell typically takes up only one type of target DNA-replicon molecule, and so all the descendants of the cell contain exactly the same type of target molecules (**DNA clones**).* By growing up large numbers of cells starting from an isolated cell colony it is possible to prepare pure populations of the desired target DNA.

5.2 PCR: basic features and applications

PCR has revolutionized molecular genetics by permitting *rapid* cloning and analysis of DNA. Since the first reports describing this new technology in the mid 1980s there have been numerous applications in both basic and clinical research.

5.2.1 Principles of basic PCR and reverse transcriptase (RT) PCR

Choice of starting nucleic acid

Usually, PCR is designed to permit *selective amplification* of a specific target DNA sequence(s) within a heterogeneous collection of DNA sequences. Often the starting DNA is total genomic DNA from a particular tissue or cultured cells, in which case the target DNA is typically a tiny fraction of the starting DNA. For example, if the intention is to amplify the 1.6 kb β-globin gene from human DNA (haploid genome size = 3300 Mb), the target : starting DNA ratio is 1.6 kb : 3300 Mb or 1 : 2 000 000. Many PCR reactions involve amplification of even smaller targets, such as single exons which average 140 bp, and so represent less than 0.000 005% of a starting population of human genomic DNA.

In the case of coding DNA target sequences, the starting DNA may often be total cDNA prepared by isolating RNA from a suitable tissue or cell line and then converting it into DNA using the enzyme Reverse transcriptase (**RT–PCR**). Depending on the extent to which the target sequence is expressed as transcripts in the original RNA population there may be significant enrichment of the target sequence (when compared to its representation in genomic DNA). However, under optimal conditions RT–PCR can sometimes be used to amplify target sequences from tissues where there is only a basal transcription level.

Choice of primers

To permit selective amplification, some prior DNA sequence information from the target sequences is required. The information is used to design two oligonucleotide primers (**amplimers**), optimally about 18–25 nucleotides long, which are specific for sequences flanking the target sequence (that is, their base sequences are represented perfectly in the sequences flanking the target sequence).

For most PCR reactions, the goal is to amplify a single DNA sequence and so it is important to reduce the chance of the primers binding to other locations in the DNA than the desired ones. It is important, therefore, to avoid repetitive DNA sequences and a variety of other considerations need to be factored into primer design:

- *base composition.* The GC content should be between 40 and 60% with an even distribution of all four nucleotides;

- *melting temperature* (T_m; see Section 6.2.1 for definition). The calculated T_m values for two primers used together should not differ by > 5°C and the T_m of the amplification product should not differ from those of the primers by > 10°C;

- *3′ terminal sequences.* The 3′ sequence of one primer should not be complementary in sequence to any region of the other primer in the same reaction. **Note**: correct base matching at the 3′ end of the primer is critical and can be exploited to ensure that even a specific allele can be amplified but not others, providing the basis of ***allele-specific PCR*** (see *Box 5.1*);

- *self-complementary sequences.* Inverted repeats or any self-complementary sequences > 3 bp in length are to be avoided.

Primer design is aided by various commercial software programs and by freeware programs such as those compiled at http://www.hgmp.mrc.ac.uk/GenomeWeb/nucprimer.html.

Cyclical nature and exponential amplification

PCR consists of a series of cycles of three successive reactions:

- **denaturation**, typically set to be at about 93–95°C for human genomic DNA;

- **primer annealing** at temperatures usually from about 50°C to 70°C depending on the *melting temperature* of the expected duplex (the reannealing temperature is typically set to be about 5°C below the calculated melting temperature);

- **DNA synthesis**, typically at about 70–75°C. The DNA polymerase used is heat-stable (it needs to elongate efficiently at 70–75°C and should not be adversely affected by the denaturation steps). In the presence of a suitably heat-stable DNA polymerase and DNA precursors (the four deoxynucleoside triphosphates, dATP, dCTP, dGTP and dTTP), the primers initiate the synthesis of new DNA strands which are complementary to the individual DNA strands of the target DNA segment.

The orientation of the primers is deliberately chosen so that the direction of new strand synthesis occurring from one

Box 5.1: A glossary of PCR methods.

▶ **Allele-specific PCR**. Designed to amplify a DNA sequence while excluding the possibility of amplifying other alleles. Based on the requirement for precise base matching between the 3′ end of a PCR primer and the target DNA. See *Figure 5.4*.

▶ **Alu-PCR**. A PCR method carried out using a primer specific for the Alu repeat, a sequence which occurs about once every 3 kb in human DNA. When neighboring Alu repeats are orientated in opposite directions, a single type of Alu primer can amplify the intervening sequence (e.g. primers A or B in *Figure A* opposite).

▶ **Anchored PCR**. Uses a sequence-specific primer and a *universal primer* for amplifying sequences adjacent to a known sequence. The universal primer recognizes and binds to a common sequence that is artificially added to all of the different DNA molecules in the starting population e.g. by covalently attaching a *double-stranded* **oligonucleotide linker***.

▶ **Differential display-PCR**. A form of RT–PCR for comparing mRNA populations expressed from two related sources of cells in order to pick out differentially expressed genes (see *Figure 7.16*).

▶ **DOP–PCR (degenerate oligonucleotide-primed PCR)**. Uses partially **degenerate oligonucleotide primers** (sets of oligonucleotide sequences that have been synthesized in parallel to have the same base at certain nucleotide positions, while differing at other positions) to amplify a variety of related target DNAs.

▶ **Hot-start PCR**. A way of increasing the specificity of a PCR reaction. Mixing of all PCR reagents prior to an initial heat denaturation step allows more opportunity for nonspecific binding of primer sequences. To reduce this possibility, one or more components of the PCR are physically separated until the first denaturation step.

▶ **Inverse PCR**. Another way of accessing DNA that is immediately adjacent to a known sequence (cf. anchored PCR). In this case, the starting DNA population is digested with a restriction nuclease, diluted to low DNA concentration and then treated with DNA ligase to encourage formation of circular DNA molecules by intramolecular ligation. The PCR primers are positioned so as to bind to a known DNA sequence and then initiate new DNA synthesis *in a direction leading away from the known sequence and towards the unknown adjacent sequence*, leading to amplification of the unknown sequence. See *Figure B* opposite (X and Y are uncharacterized sequences flanking a known sequence).

▶ **Island-rescue PCR**. A specialized method for amplifying sequences at CpG islands.

▶ **Linker-primed PCR (ligation adaptor PCR)**. A form of *indiscriminate* amplification. A complex starting DNA is digested with a restriction nuclease to produce multiple types of fragment with the same type of overhanging end.

A *double-stranded* **oligonucleotide linker*** of known sequence and with similar overhanging ends is ligated to the fragments. PCR is then conducted using primers which are *specific for the linker sequences* to amplify all fragments of the DNA source flanked by linker molecules.

▶ **Nested primer PCR**. A way of increasing the specificity of a PCR reaction. The products of an initial amplification reaction are diluted and used as the starting DNA source for a second reaction in which a different set of primers is used, corresponding to sequences located close, but *internal*, to those used in the first reaction.

▶ **Quantitative PCR**. See Real-time PCR.

▶ **RACE-PCR**. A form of **anchor-primed PCR** (see above) for **r**apid **a**mplification of **c**DNA **e**nds (see *Figure 7.12*).

▶ **Real-time PCR**. A form of *quantitative* PCR using a fluorescence-detecting thermocycler machine to amplify specific nucleic acid sequences *and simultaneously* measure their concentrations. There are two major research applications: (i) to *quantify* gene expression (and to confirm differential expression of genes detected by microarray hybridization analyses); and (ii) to screen for mutations and single nucleotide polymorphisms. In analytical labs it is also used to measure the abundance of DNA or RNA sequences in clinical and industrial samples.

▶ **RT–PCR (reverse transcriptase PCR)**. PCR where the starting population is mRNA and an initial reverse transcriptase step to produce cDNA is required.

▶ **Touch-down PCR**. A way of increasing the specificity of a PCR reaction. Most thermal cyclers can be programmed to perform runs in which the annealing temparature is lowered incrementally during the PCR cycling from an initial value above the expected T_m to a value below the T_m. By keeping the stringency of hybridization initially very high, the formation of spurious products is discouraged, allowing the expected sequence to predominate.

▶ **Whole genome PCR**. Indiscriminate PCR by using comprehensively degenerate primers or by attaching oligonucleotide linkers to a complex DNA population then using linker-specific oligonucleotide primers to amplify all sequences.

***Note:** a double-stranded **oligonucleotide linker (adaptor)** is constructed by designing two oligonucleotides to be complementary in sequence. The two oligonucleotides are chemically synthesized in separate reactions and once purified, they are allowed to base pair to each other to form the desired double-stranded sequence. Oligonucleotide linkers are ligated (linked) to a DNA sequence to permit: (i) PCR using an oligonucleotide specific for the linker; or (ii) addition of desirable features e.g. restriction sites to help cloning (see *Figure 5.10*). Complex **polylinkers** containing several restriction sites are routinely inserted into cloning vectors (see Section 5.3.5).

Box 5.1: *(continued)*

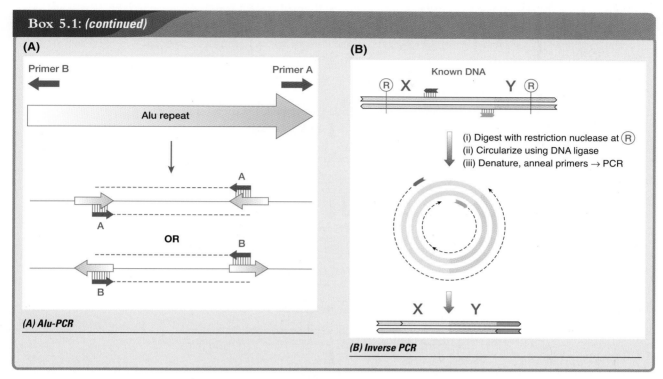

(A)

Primer B

Primer A

Alu repeat

A

A

OR

B

B

(A) Alu-PCR

(B)

Known DNA

Ⓡ X   Y Ⓡ

(i) Digest with restriction nuclease at Ⓡ
(ii) Circularize using DNA ligase
(iii) Denature, anneal primers → PCR

X   Y

(B) Inverse PCR

primer is towards the other primer binding site. As a result, the newly synthesized strands can in turn serve as templates for new DNA synthesis, causing a *chain reaction* with an exponential increase in product (*Figure 5.2*).

The requirement for a heat-stable enzyme drove researchers to isolate DNA polymerases from microorganisms whose natural habitat is hot springs. An early and widely used example was Taq polymerase that is obtained from *Thermus aquaticus* and is thermostable up to 94°C, with an optimum working temperature of 80°C. However, Taq polymerase lacks an associated 3′→5′ exonuclease activity to provide a *proof-reading function* and so copying errors (incorporation of the wrong base) can be frequent compared to that occurring within cells. As a result, alternative enzymes with a 3′→5′ proofreading exonuclease activity, such as Pfu polymerase from *Pyrococcus furiosus,* are now widely used (Cline *et al.,* 1996).

5.2.2 PCR has two major limitations: short sizes and low yields of products

A clear disadvantage of PCR as a DNA cloning method is the size range of the amplification products. Unlike cell-based DNA cloning where the upper size limit of cloned DNA sequences can approach 2 Mb, reported PCR products are typically in the 0–5 kb size range, often at the lower end of this scale. Although small segments of DNA can usually be amplified easily by PCR, it becomes increasingly more difficult to obtain efficient amplification as the desired product length increases. **Long-range PCR** protocols have been developed, however, leading to products which are tens of kilobases in length, such as a 42 kb product from the bacterio-

phage λ (see Cheng *et al.,* 1994). The modified conditions often involve a mix of two types of heat-stable polymerases in an effort to provide optimal levels of DNA polymerase and 3′→5′ exonuclease activity which serves as a *proofreading mechanism.*

The amount of material that can be cloned in a single PCR reaction is also a limitation, and it is time-consuming and expensive to repeat the same PCR reaction many times to achieve large quantities of the desired DNA. In addition, the PCR product may not be in a suitable form that will permit some subsequent studies. As a result, it is often convenient to clone the PCR product in a cell-based cloning system in order to obtain large quantities of the desired DNA and to permit a variety of analyses.

Various plasmid cloning systems are used to propagate PCR-cloned DNA in bacterial cells. Once cloned, the insert can be cut out using suitable restriction nucleases and transferred into other plasmids which may have specialized uses in permitting expression to give an RNA product, or to provide large quantities of a protein. Several thermostable polymerases including Taq polymerase have a terminal deoxynucleotidyl transferase activity which selectively modifies PCR-generated fragments by adding a single nucleotide, generally adenine, to the 3′ ends of amplified DNA fragments.

The resulting overhangs can make it difficult to clone PCR products and a variety of approaches are commonly used to facilitate cloning, including the use of vectors with overhanging T residues in their cloning site polylinker (*Figure 5.3*), and the use of 'polishing' enzymes such as T4 polymerase or Pfu polymerase which can remove the overhanging single nucleotides. Alternatively, PCR primers can be modified by designing an approximately 10 nucleotide extension

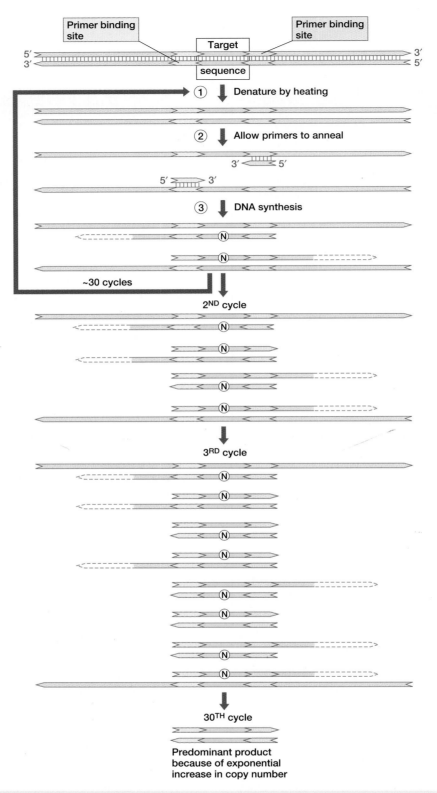

Figure 5.2: PCR is an *in vitro* method for amplifying DNA sequences using defined oligonucleotide primers.

After identifying a sequence to be amplified (the ***target sequence***), oligonucleotide primers are designed to be complementary to DNA sequences located on opposite DNA strands and flanking the target sequence. PCR consists of cycles of denaturation, annealing of primers and then DNA synthesis in which the primers are incorporated into the newly synthesized DNA strands. The first cycle will result in two ***new DNA strands (N)*** whose 5′ ends are fixed by the position of the oligonucleotide primer but whose 3′ ends are variable (indicated by dotted lines). After the second cycle, the four new strands consist of two products with variable 3′ ends, just as in the first cycle, but now two strands of fixed length (with both 5′ and 3′ends) defined by the primer sequences. After the third cycle, six of the eight new strands are of the desired fixed length and after 30 cycles or so there is a massive increase (amplification) of this type of product.

containing a suitable restriction site at their 5′ end. The nucleotide extension does not base-pair to target DNA during amplification but afterwards the amplified product can be digested with the appropriate restriction enzyme to generate overhanging ends for cloning into a suitable vector (see Section 5.5.3).

5.2.3 General applications of PCR

Because of its simplicity, PCR is a popular technique with a wide range of applications which depend on essentially three major advantages of the method:

▶ it is very *rapid* and easy to perform;

▶ it is very *sensitive*, enabling amplification from minute amounts of target DNA – even the DNA from a single cell (Li *et al.*, 1988). Various applications have been found in diagnosis, in genetic linkage analysis (including single-sperm typing) and in forensic science (trace tissues left by an individual – such as a single hair or discarded skin cells – can be PCR-typed using highly polymorphic DNA markers to identify the individual of origin);

▶ it is very *robust*, and it is often possible to amplify DNA from tissues or cells which are badly degraded, or embedded in some medium that makes it difficult to isolate DNA by the standard methods. Short sequences can be amplified from small amounts of degraded DNA extracted from decomposed tissues at archaeological/historical sites, and successful PCR amplification is possible from formalin-fixed tissue samples (with important advantages to molecular pathology and, in some cases, genetic linkage studies).

The many applications of PCR and the need to optimize efficiency and specificity have prompted a wide variety of PCR approaches (*Box 5.1*). Because of the crucial importance of primer specificity, various modifications are often used to reduce the chances of nonspecific primer binding (as, for example by using hot-start PCR, nested primers, touch-down PCR; see *Box 5.1*).

The crucial dependence of correct base pairing at the extreme 3′ end of bound primers has allowed methods to be developed which permit distinction between alleles that differ at just a single nucleotide (**allele-specific PCR**). In the popular ***ARMS (amplification refractory mutation system)*** method, primers are designed with their 3′ end nucleotides designed to base-pair with the variable nucleotide which distinguishes the two alleles, and with the remaining primer sequence designed to be complementary to the sequence immediately adjacent to the variable nucleotide. Under suitable experimental conditions amplification will not take place where the 3′ end nucleotide is not perfectly base-paired thereby distinguishing the two alleles (*Figure 5.4*).

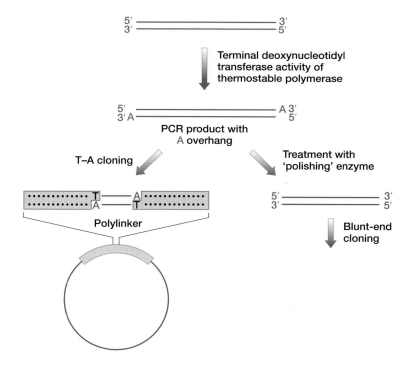

Figure 5.3: Cloning of PCR products in bacterial cells.

PCR products frequently have an overhanging adenosine at their 3′ ends (see text). The T-A cloning system has a polylinker system with complementary thymine overhangs to facilitate cloning. An alternative is to trim back the adenine overhangs using a suitable 'polishing' enzyme, which leaves the fragment blunt-ended.

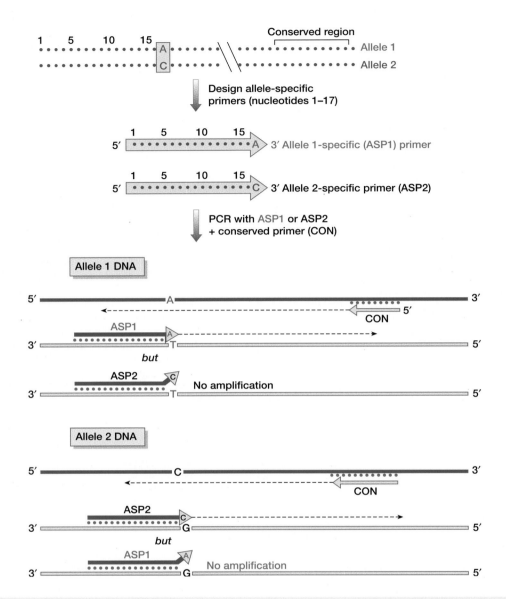

Figure 5.4: Allele-specific PCR using the ARMS system is dependent on perfect base-pairing of the 3′ end nucleotide of primers.

The allele-specific oligonucleotide primers ASP1 and ASP2 are designed to be identical to the sequence of the two alleles over a region preceding the position of the variant nucleotide, up to and terminating in the variant nucleotide itself. ASP1 will bind perfectly to the complementary strand of the allele 1 sequence, permitting amplification with the conserved primer. However, the 3′-terminal C of the ASP2 primer mismatches with the T of the allele 1 sequence, making amplification impossible. Similarly ASP2 can bind perfectly to allele 2 and initiate amplification, unlike ASP1.

5.2.4 Some PCR reactions are designed to permit multiple amplification products and to amplify previously uncharacterized sequences

The standard PCR/RT–PCR reactions are based on the need for very specific primer binding in order to allow *selective* amplification of a desired *known* target sequence. Sometimes, however, it is desirable to design PCR reactions to amplify DNA sequences for which there is only limited or no prior sequence information.

Amplifying new members of a DNA family by DOP–PCR

Previously uncharacterized DNA sequences can sometimes be cloned using PCR if they are members of a gene or repetitive DNA family at least one of whose members has previously been characterized. For example, many new members of the mammalian *Wnt* gene family were isolated for the first time after noting that the products of the first few *Wnt* genes had very similar amino acid sequences within a particular conserved domain. This allowed the design of primers for this sequence based on *degenerate oligonucleotides*, with in each case a mix of different oligonucleotides representing various permutations of amino acids. ***Degenerate oligonucleotide primed PCR (DOP–PCR)*** can allow a variety of different but closely related genes, including novel genes, to be amplified at the same time and then fractionated and purified by cell-based DNA cloning.

Indiscriminate amplification

If the source of DNA is precious and in very limiting quantities, it is possible to use PCR to amplify all of the DNA by covalently attaching *double-stranded* **linker oligonucleotides** to the extremities of all of the DNA sequences in the starting population. The linker oligonucleotides are prepared by individually synthesizing two oligodeoxyribonucleotides which are designed to be complementary in sequence and to be able to base-pair to form a double-stranded DNA sequence with an overhanging end. The DNA to be amplified is digested with a restriction nuclease which produces similar overhanging ends so that the linkers can be covalently attached (ligated). Linker-specific primers can permit amplification of target DNA molecules with linkers at both ends. As a result, all of the DNA sequences in a starting sequence can be amplified simultaneously, permitting ***whole genome amplification*** in the case of genomic DNA. An alternative approach is to use comprehensively degenerate oligonucleotides as primers so that the primers can bind to very many binding sites.

Using a known sequence to amplify a neighbouring uncharacterized DNA sequence

Various ingenious modifications have been designed to progress from a known starting DNA sequence to an uncharacterized neighboring sequence, both in genomic DNA and in cDNA. Examples include ***anchored PCR, inverse PCR, RACE-PCR*** (*Box 5.1*).

5.3 Principles of cell-based DNA cloning

5.3.1 An overview of cell-based DNA cloning

Cell-based DNA cloning was first developed in the early 1970s. It was made possible by the discovery of ***type II***

restriction endonucleases, bacterial enzymes which are capable of cleaving DNA at all locations which contain a small, *specific* recognition sequence, usually 4–8 bp long. These enzymes normally serve to protect bacteria from invading bacteriophages by selectively cleaving the foreign DNA (see *Box 5.2* and the next section). The big advantage they offered to molecular geneticists was a way of cutting DNA into defined fragments that could be joined easily to other similarly cut DNA fragments.

The essence of cell-based DNA cloning is to use restriction nucleases to cut DNA molecules in a starting DNA population (the ***target DNA***) into pieces of manageable size, then attach them to a **replicon** (any sequence capable of independent DNA replication) and transfer the resulting hybrid molecules (**recombinant DNA**) into a suitable host cell which is then allowed to proliferate by cell division. Because the replicon can replicate inside the cells (often to high copy numbers), so does the attached target DNA, resulting in a form of *cell-based DNA amplification*.

There are four essential steps in cell-based DNA cloning (see *Figure 5.5*):

▶ **construction of recombinant DNA molecules**. Suitably sized DNA fragments are typically produced by cutting with a restriction nuclease followed by covalent attachment (**ligation**) of the target DNA fragments to a single type of ***replicon*** molecule. The construction of recombinant DNA is facilitated by ensuring that the target DNA and replicon molecules are cut with restriction nucleases which produce the same types of end before joining the target DNA fragments to replicon molecules using the enzyme DNA ligase;

▶ **transformation.** The recombinant DNA molecules are transferred into host cells (often bacterial or yeast cells) in which the chosen replicon can undergo DNA replication independently of the host cell chromosome(s). Replicons used in cell cloning are often referred to as **vector molecules** because they help to transport the passenger target molecules into cells and then aid their replication within the cells;

Box 5.2: Restriction endonucleases and modification–restriction systems.

Bacteriophages that are liberated from a bacterium of a particular strain can infect other bacteria of the same strain but not those of a different strain. This is because the phage DNA has the same **modification** pattern as the DNA of bacterial strains it can infect; the phage is **'restricted'** to that strain of bacteria. The restriction is not an absolute one: some phages can escape restriction and can acquire the modification pattern of the new host.

The basis of modification–restriction systems is now known to involve two types of enzyme activity:

▶ a sequence-specific **DNA methylase** activity provides the basis of the modification pattern;

▶ a sequence-specific **restriction endonuclease** activity underpins the restriction phenomenon: it cleaves phage DNA whose methylation pattern is different from that of the host cell DNA.

The bacterial strain possesses a *DNA methylase activity with the same sequence specificity as the corresponding restriction nuclease activity*. As a result, cellular restriction endonucleases will not cleave the appropriately methylated host cell DNA but may cleave incoming phage DNA, if not methylated appropriately.

Note: some plasmids and bacteriophages possess genes for modification and restriction systems, and can confer this specificity on a host cell.

▶ **selective propagation of cell clones** involves two stages. Initially the transformed cells are plated out by spreading on an agar surface in order to encourage the growth of *well-separated cell colonies*. The cells in a single cell colony are all identical (since they descended from a single cell) and are described as **cell clones**. Subsequently, *individual colonies* can be picked from a plate and the cells are allowed to undergo a second stage of growth in liquid culture;

▶ **isolation of recombinant DNA clones** by harvesting expanded cell cultures and selectively isolating the recombinant DNA.

5.3.2 Restriction endonucleases enable the target DNA to be cut into manageable pieces which can be joined to similarly cut vector molecules

Type II restriction nucleases

The recognition sequences for the vast majority of type II restriction endonucleases are normally **palindromes** (the sequence of bases is the same on both strands when read in the $5' \rightarrow 3'$ direction, as a result of a twofold axis of symmetry). Depending on the location of the cleavage sites produced by a restriction nuclease the resulting **restriction fragments** may have:

▶ **blunt ends** (the cleavage points occur exactly on the axis of symmetry);

▶ **overhanging ends** (the cleavage points do not fall on the symmetry axis, so that the resulting restriction fragments possess so-called *5' overhangs* or *3' overhangs* (see *Table 5.1*). **Note:** the two overhanging ends of each fragment are complementary in base sequence, and will have a tendency to associate with each other (or with any other similarly complementary overhang) by forming base pairs. *As a result of their tendency to stick to other ends of the same type*, overhanging ends of this type are often described as **sticky ends**.

Type II restriction nucleases offered two great advantages for DNA cloning and DNA analysis:

▶ *a defined set of starting DNA fragments*. When DNA is isolated from tissues and cultured cells the unpackaging of the huge DNA molecules and inevitable physical shearing results in heterogeneous populations of DNA fragments of randomly different lengths. Using restriction endonucleases it was possible to convert this hugely heterogeneous population of broken DNA fragments into sets of *restriction fragments* of *defined lengths*;

▶ *a way of artificially recombining DNA molecules*. Restriction fragments with the same types of sticky end can be easily joined together using a **DNA ligase**. To construct a recombinant DNA, therefore, one could cut the replicon (**vector**) molecule and the target DNA with the same type of restriction nuclease, or with restriction nucleases producing the same types of sticky end.

The termini of restriction fragments which have the same type of overhanging ends can associate in a variety of different ways, either *intra*molecularly (**cyclization**), or between molecules to form linear **concatemers** or circular compound molecules. Intermolecular reactions occur most readily at high DNA concentrations. At very low DNA concentrations, however, individual termini on different molecules have less opportunity of making contact with each other, and

---

Table 5.1: Examples of commonly used restriction endonucleases (see also Table 6.3 for rare-cutters).

Enzyme	Source	Sequence cut; N = A, C, G or T	Restriction fragment with the following ends
Producing blunt ends			
AluI	Arthrobacter luteus	A G↓C T	5' C T--------------A G 3'
		T C↑G A	3' G A--------------T C 5'
Producing 5' overhangs			
EcoRI	Escherichia coli R factor	G↓A A T T C	5' A A T T C--------------G 3'
		C T T A A↑G	3' G--------------C T T A A 5'
Producing 3' overhangs			
PstI	Providencia stuartii	C T G C A↓G	5' G--------------C T G C A 3'
		G↑A C G T C	3' A C G T C --------------G 5'
Recognising non-palindromic sequence			
MnlI	Moraxella nonliquefaciens	C C T C N N N N N N N↓	5' ---------C C T C N N N N N N N 3'
		G G A G N N N N N N↑N	3' N---------G G A G N N N N N N 5'
Recognising bipartite recognition sequence			
BstXI		C C A N N N N N↓N T G G	5' N T G G----C C A N N N N N 3'
		G G T N↑N N N N N A C C	3' N N N N N A C C----G G T N

Note: names are normally derived from the first letter of the genus and the first two letters of the species name, e.g. PstI is the first restriction nuclease to have been isolated from *Providencia stuartii,* – see http://rebase.neb.com/rebase/rebase.html for the REBASE database of restriction nucleases.

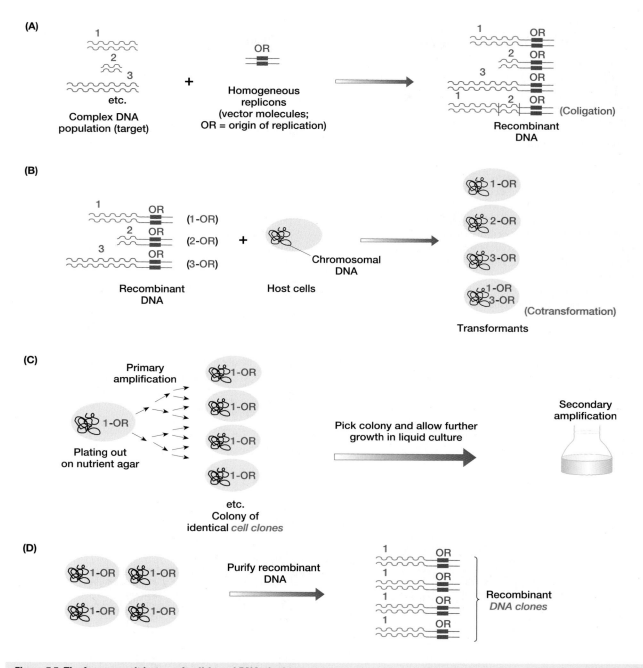

Figure 5.5: The four essential steps of cell-based DNA cloning.

(A) *Formation of recombinant DNA*. *Note* that, in addition to simple vector–target ligation products, co-ligation events may occur whereby two unrelated target DNA sequences may be ligated in a single product (e.g. sequences 1 plus 2 in the bottom example). OR, origin of replication. **(B)** *Transformation*. This is a key step in DNA cloning because cells normally take up only one foreign DNA molecule. *Note* that occasionally, however, co-transformation events are observed, such as the illustrated cell at bottom which has been transformed by two different DNA molecules (a recombinant molecule containing sequence 1 and a recombinant molecule containing sequence 3). **(C)** *Amplification to produce numerous cell clones*. This is another key step. After plating out the transformed cells, *individual* clone colonies can be separated on a dish and then *individually picked* into a secondary amplification step to ensure clone homogeneity. **(D)** *Isolation of recombinant DNA clones*.

intramolecular cyclization is favored. Generally, ligation reactions are designed to promote the formation of recombinant DNA (by ligating target DNA to vector DNA), although vector cyclization, vector–vector concatemers and target DNA–target DNA ligation are also possible (see *Figure 5.6*). To achieve this, the vector molecules are often treated so as to prevent or minimize their ability to undergo cyclization.

Simple vectors for cloning in bacterial cells

During cell-based DNA cloning target DNA fragments must be able to replicate inside cells. As they lack a functional origin of replication, they need to be attached to a replicon (vector) which can replicate within the host cell, independently of the host cell chromosomes. The vector may have an origin of replication that originates from either a natural extrachromosomal replicon or, in some cases, a chromosomal replicon (as in the case of yeast artificial chromosomes, see Section 5.4.4).

The most frequently used host cells are modified bacterial or fungal host cells. Bacterial cell hosts are particularly widely used because of their capacity for rapid cell division. They have a single circular double-stranded chromosome with a single origin of replication. Replication of the host chromosome subsequently triggers cell division so that each of the two resulting daughter cells contains a single chromosome like their parent cell (i.e. the copy number is maintained at one copy per cell). However, the replication of extrachromosomal replicons is not constrained in this way: many such replicons go through several cycles of replication during the cell cycle and can reach high copy numbers. As a result, large amounts of target DNA can be produced by being co-replicated with a replicon. There are two general classes of extra chromosomal replicon:

▶ **plasmids** – small circular double-stranded DNA molecules which individually contain very few genes. Their existence is intracellular, being vertically distributed to daughter cells following host cell division, but can be transferred horizontally to neighboring cells during bacterial conjugation events. Natural examples include plasmids which carry the sex factor (F) and those which carry drug-resistance genes;

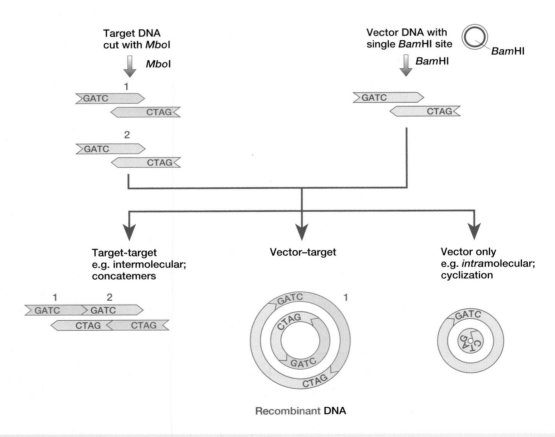

Figure 5.6: Cohesive termini can associate intramolecularly and intermolecularly.

Note: only some of the possible outcomes are shown. For example, vector molecules may also form intermolecular concatemers, multimers can undergo cyclization and co-ligation events can involve two different target sequences being included with a vector molecule in the same recombinant DNA molecule (see *Figure 5.5A*). The tendency towards cyclization of individual molecules is more pronounced when the DNA is at low concentration and the chances of collision between different molecules with complementary sticky ends is reduced.

▶ **bacteriophages** – viruses which infect bacterial cells. DNA-containing bacteriophages often have genomes containing double-stranded DNA which may be circular or linear. Unlike plasmids, they can exist extracellularly. The mature virus particle (**virion**) has its genome encased in a protein coat so as to facilitate adsorption and entry into a new host cell.

5.3.3 Introducing recombinant DNA into recipient cells provides a method for fractionating a complex starting DNA population

The plasma membrane of cells is selectively permeable and does not normally admit large molecules such as long DNA fragments. However, cells can be treated in certain ways (e.g. by exposure to certain high ionic strength salts, short electric shocks, etc.) so that the permeability properties of the plasma membranes are altered. As a result, a fraction of the cells become **competent**, meaning that they are capable of taking up foreign DNA from the extracellular environment.

Only a small percentage of the cells will take up the foreign DNA (**DNA transformation**). *However, those that do take up foreign DNA, will often take up only a single molecule* (which can, however, subsequently replicate many times within a cell). This is the basis of the critical fractionation step in cell-based DNA cloning: the population of transformed cells can be thought of as a sorting office in which the complex mixture of DNA fragments is sorted by depositing individual DNA molecules into individual recipient cells (*Figure 5.7*).

Because circular DNA (even nicked circular DNA) transforms much more efficiently than linear DNA, most of the cell transformants will contain cyclized products rather than linear recombinant DNA concatemers and, if an effort has been made to suppress vector cyclization (e.g. by dephosphorylation), most of the transformants will contain recombinant DNA. Note, however, that **cotransformation** events (the occurrence of more than one type of introduced DNA molecule within a cell clone, see *Figure 5.5B*) may be comparatively common in some cloning systems.

The transformed cells are allowed to multiply. In the case of cloning using plasmid vectors and a bacterial cell host, a solution containing the transformed cells is simply spread over the surface of nutrient agar in a petri dish (**plating out**). This usually results in the formation of **bacterial colonies** which consist of **cell clones** (identical progeny of a single ancestral cell). Picking an individual colony into a tube for subsequent growth in liquid culture permits a secondary expansion in the number of cells which can be scaled up to provide very large yields of cell clones, all identical to an ancestral single cell (*Figure 5.7*). If the original cell contained a single type of foreign DNA fragment attached to a replicon, then so will the descendants, resulting in a huge amplification in the amount of the specific foreign fragment. Expanded cultures representing cell clones derived from a single cell can then be processed to recover the recombinant DNA.

To recover recombinant DNA selectively from lysed cells, physical differences between host cell DNA and recombinant DNA are exploited. In the case of bacterial cells, the double-stranded bacterial chromosome is circular like any plasmids containing introduced foreign DNA but very much larger in size. As a result, it is prone to nicking and shearing during cell lysis and subsequent DNA extraction, generating linear DNA fragments with free ends. After subjecting the isolated DNA to a denaturation step, for example by alkaline treatment, the linearized host cell DNA readily denatures, but the strands of *covalently closed circular (CCC)* plasmid DNA are unable to separate, and when allowed to renature, the two strands reassociate to form native **superhelical** molecules or so-called **supercoiled DNA** (*Figure 5.8*). The denatured host cell DNA precipitates out of solution, leaving behind the CCC plasmid DNA.

If required, further purification is possible by *equilibrium density gradient centrifugation (isopycnic centrifugation)*: the partially purified DNA is centrifuged to equilibrium in a solution of cesium chloride containing ethidium bromide (EtBr). EtBr binds DNA by *intercalating* between the base pairs, thereby causing the DNA helix to unwind. Unlike chromosomal DNA, a CCC plasmid DNA has no free ends and can only unwind to a limited extent, which limits the amount of EtBr it can bind. EtBr–DNA complexes are denser when they contain less EtBr, so CCC plasmid DNA will band at a lower position in the cesium chloride gradient than either chromosomal DNA or plasmid circles that are open, enabling separation of the recombinant DNA from host cell DNA. The resulting recombinant DNA molecules will normally be identical to each other (representing a single target DNA fragment) and are referred to as **DNA clones**.

5.3.4 DNA libraries are a comprehensive set of DNA clones representing a complex starting DNA population

The first attempts at cloning human DNA fragments in bacterial cells concentrated on target sequences which were highly abundant in a particular starting DNA population. For example, nucleated human cells contain much the same set of DNA sequences, but the mRNA populations can be quite different. Although the mRNA population in each cell is complex, some cells are particularly devoted to synthesizing a specific type of protein and so they have a few predominant mRNA species (e.g. much of the mRNA made in erythrocytes consists of α- and β-globin mRNA). The enzyme **reverse transcriptase** (RT; RNA-dependent DNA polymerase) can be used to make a cDNA copy that is complementary in base sequence to the mRNA. Hence, cDNA from erythrocytes is greatly enriched in globin cDNA, facilitating its isolation.

Modern DNA cloning approaches offer the possibility of making comprehensive collections of DNA clones (**DNA libraries**) from extremely complex starting DNA populations (such as total human genomic DNA). This approach enables DNA sequences which are very rare in the starting population to be represented in a library of DNA clones, whence they can be isolated individually by selecting a suitable host cell colony and amplifying it. Two basic varieties of this method have popularly been undertaken, depending on the nature of the starting DNA: genomic DNA libraries and cDNA libraries.

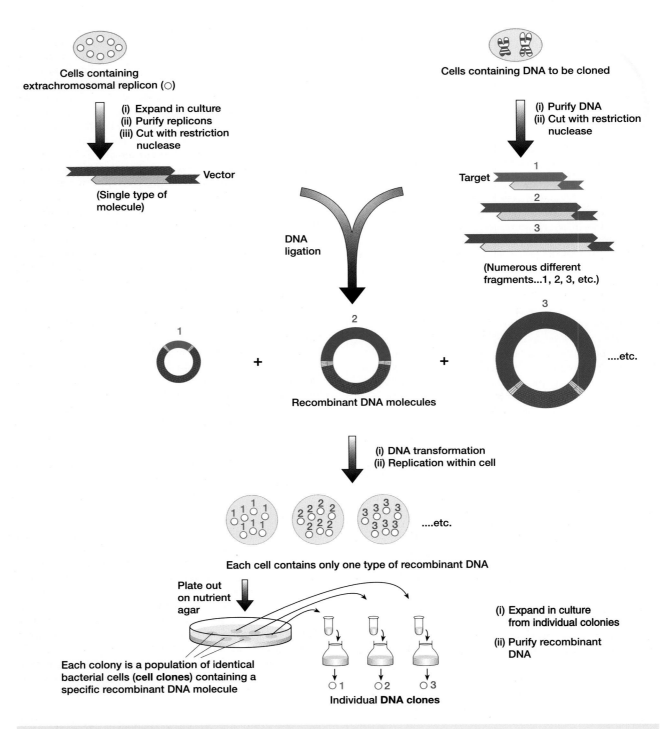

Figure 5.7: DNA cloning in bacterial cells.

The example illustrates cloning of genomic DNA but could equally be applied to cloning cDNA.

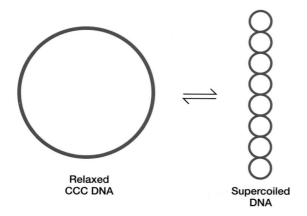

Figure 5.8: Covalently closed circular (CCC) DNA and DNA supercoiling.

The CCC DNA is shown on left in schematic form (no attempt is made to show the double helical structure). Unless one of the two DNA strands becomes nicked, the twisting of the double helix cannot be relaxed and the tension induced causes spontaneous formation of a **supercoiled** structure shown on the right. Nicking of either DNA strand, however, relieves the tension by permitting rotation at the free end.

Newly constructed libraries are said to be ***unamplified***, although this is a misleading term because the initially transformed cells have been amplified to form separated cell colonies. Often, cell colony formation is allowed to proceed on top of membranes that are overlaid on to the surface of nutrient agar in sterile culture dishes. Copies of the library can then be made by ***replica plating*** on to a similar sized membrane prior to overlaying on to a nutrient agar surface and colony growth. More recently, individually picked cell colonies have been spotted in gridded arrays on to suitable membranes or into the wells of microtiter dishes where they can be stored for long periods at −70°C in the presence of a cell-stabilizing medium such as glycerol.

For multiple distribution, ***amplified libraries*** are required. The cells from representative primary filters are washed off into cell culture medium, diluted and stabilized by the presence of glycerol, or some alternative stabilizing agent. Individual aliquots can then be plated out at a later stage to regenerate the library. This additional amplification step, however, may result in distortion of the original representation of cell clones because during the amplification stage there may be differential rates of growth of different colonies.

Genomic DNA libraries

In the case of complex eukaryotes, such as mammals, all nucleated cells have essentially the same DNA content, and it is often convenient to prepare a genomic library from easily accessible cells, such as white blood cells. The starting material is genomic DNA which has been fragmented in some way, usually by digesting with a restriction endonuclease.

Typically, the genomic DNA is digested with a 4-bp cutter, such as *Mbo*I which recognizes the sequence GATC. This sequence will occur about every 280 bp on average in human genomic DNA, and so there will be few DNA sequences which lack a recognition site for this enzyme. Complete digestion of the starting DNA with *Mbo*I would produce very small fragments. Instead, **partial restriction digestion** is carried out (low enzyme concentration, short time of incubation, etc.), and so cleavage will occur at only a small number of the potential restriction sites.

Partial restriction digestion not only produces desirably large fragments for cloning, but importantly, it also allows *random DNA fragmentation*. Thus, for a specific sequence location, the pattern of cutting will be different on different copies of the same starting DNA sequence (*Figure 5.9*). Such random fragmentation ensures that the library will contain as much representation as possible of the starting DNA. Additionally, it has the advantage that it produces *clones with overlapping inserts*. As a result, after characterization of the insert of one clone, attempts can be made to access clones from the same general region by identifying those with inserts showing some similarities to that of the original clone (see *Box 8.5*).

The **complexity** (number of independent DNA clones) of a genomic DNA library can be defined in terms of **genome equivalents (GE)**. A genome equivalent of 1, a so-called ***one-fold library***, is obtained when the number of independent clones = genome size/average insert size. For example, for a human genomic DNA library with an average insert size of 40 kb, 1 GE = 3000 Mb/40 kb = 75 000 independent clones. A library such as this which has 300 000 clones is sometimes termed a fourfold library because it has 4 GE. Because of sampling variation, however, the number of GEs must be considerably greater than 1 to have a high chance of including any particular sequence within that library. Consequently, attempts are normally made to prepare complex (> 4 GE) libraries.

cDNA libraries

As gene expression can vary in different cells and at different stages of development, the starting material for making cDNA libraries is usually total RNA from a specific tissue or specific developmental stage of embryogenesis. As the vast majority of mRNA is polyadenylated, poly(A)$^+$ mRNA is selected by specific binding to a complementary oligo(dT) or poly(U) sequence connected to a solid Sepharose or cellulose matrix. The isolated poly(A)$^+$ mRNA can then be converted, using reverse transcriptase, to a double-stranded cDNA copy. To assist cloning, double-stranded ***oligonucleotide linkers (adaptors)*** which contain suitable restriction sites are ligated to each end of the cDNA (*Figure 5.10*).

5.3.5 Recombinant screening is often achieved by insertional inactivation of a marker gene

An essential requirement for cell-based DNA cloning systems is a method of detecting cells containing the appropriate vector molecule, and within this group the subset containing recombinant DNA. *Generalized screening* for recombinants is useful for screening DNA libraries made from DNA populations which have not been so well characterized. Increasingly,

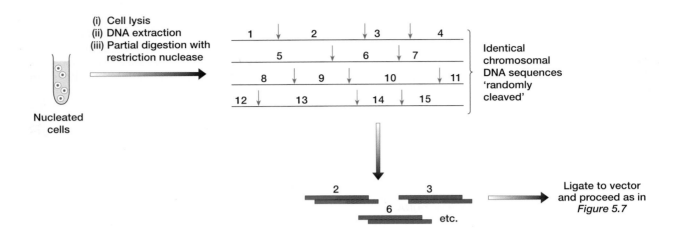

Figure 5.9: Making a genomic DNA library.

All nucleated cells of an individual will have the same genomic DNA content so that any easily accessible cells (e.g. white blood cells) can be used as source material. Because DNA is extracted from numerous cells with identical DNA molecules, the isolated DNA will contain large numbers of identical DNA sequences. However, partial digestion with a restriction endonuclease will cleave the DNA at only a small subset of the available restriction sites, and the pattern will differ between individual molecules, resulting in almost random cleavage. This will generate a series of restriction fragments which, if they derive from the same locus, may share some common DNA sequence (e.g. fragment 6 partially overlaps fragments 2 and 3, as shown, and also fragments 9 and 10, and 13 and 14).

however, *directed screening* is carried out to *test specifically for the presence of some previously known sequence or one that is closely related to a known sequence.*

Screening for cells transformed by vector molecules

Identification of cells containing the vector molecule requires engineering or selection of the vector molecule to contain a suitable marker gene whose expression provides a means of identifying cells containing it. Two popularly used marker gene systems are based on:

▸ **antibiotic resistance genes.** The host cell strain that is used is chosen to be one that is sensitive to a particular antibiotic, often ampicillin, tetracycline or chloramphenicol. The corresponding vector has been engineered to contain a gene which confers resistance to the antibiotic. After transformation, cells are plated on agar containing the antibiotic to rescue cells transformed by the vector;

▸ **β-galactosidase gene complementation.** The host cell is a mutant which contains a fragment of the β-galactosidase gene but cannot make any functional β-galactosidase. The vector is engineered to contain a different fragment of the β-galactosidase gene. After transformation, *functional complementation* occurs: the host cell and vector-encoded β-galactosidase fragments are able to combine so as to produce an active enzyme. Functional β-galactosidase activity is assayed by conversion of a colorless substrate, Xgal (5-bromo, 4-chloro, 3-indolyl β-D-galactopyranoside), to a blue product.

Generalized recombinant screening

Generalized screens for recombinant DNA usually rely on **insertional inactivation:** the vector is designed to contain

some marker gene which confers some phenotype that can be easily scored and contains within it unique restriction sites to enable foreign DNA to be inserted into the marker gene, thereby inactivating it and changing the phenotype. To achieve this, it is usual to modify the marker gene by inserting within it a **multiple cloning site polylinker**, a double-stranded oligonucleotide designed to contain recognition sequences for specific restriction nucleases (pre-existing restriction sites for these enzymes will be deleted from the vector if necessary to ensure the presence of unique cloning sites).

Because the polylinker is both short (ca. 30 bp) and a multiple of three nucleotides in length (maintaining the reading frame of the marker gene) it does not affect expression of the marker gene. However, when a foreign DNA fragment is then cloned into the polylinker, the marker gene will now have a large insertion which will inactivate it. Popularly used systems include:

▸ **β-galactosidase-based screens.** In the case of a marker β-galactosidase gene, insertional inactivation results in colorless cells in the presence of Xgal, while cells containing nonrecombinant vector are blue;

▸ **suppressor tRNA-based screens.** Suppressor tRNA genes are mutant tRNA genes carrying an altered anticodon sequence that is complementary to one of the normal termination codons: UAA (*ochre*), UAG (*amber*) or UGA (*opal*). In response to the relevant stop codon the suppressor tRNA inserts an amino acid (see *Box 5.3*). The host cell typically carries a defective marker gene designed to have a premature stop codon, resulting in a phenotype that can be easily scored. If the vector carries a suitable suppressor tRNA gene to suppress the marker gene mutation, the wild-type phenotype will be restored. Cloning of foreign DNA into

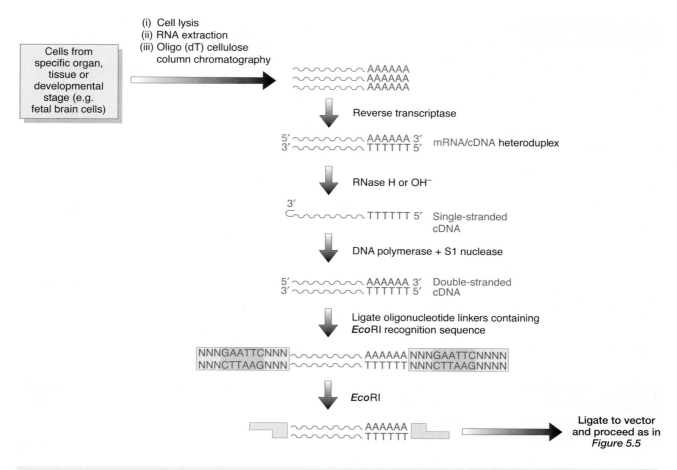

Figure 5.10: Making a cDNA library.

The reverse transcriptase step often uses an oligo(dT) primer to prime synthesis of the cDNA strand. More recently, mixtures of random oligonucleotide primers have been used instead to provide a more normal representation of sequences. RNase H will specifically digest RNA that is bound to DNA in an RNA–DNA hybrid. The 3′ end of the resulting single-stranded cDNA has a tendency to loop back to form a short hairpin. This can be used to prime second-strand synthesis by DNA polymerase and the resulting short loop connecting the two strands can then be cleaved by the S1 nuclease which specifically cleaves regions of DNA that are single stranded.

the suppressor tRNA gene results in insertional inactivation, restoring the mutant phenotype.

Directed recombinant screening by hybridization and PCR

DNA library screening is usually carried out directionally by testing cells for the presence of a previously characterized DNA sequence, or of one that is related to a known DNA sequence. For example, a DNA library with very large clone inserts could be screened to retrieve a very large recombinant clone for functional analyses, or a DNA library from a poorly studied species could be screened for sequences related to a well known human gene. This can be achieved by DNA hybridization-based screening or by PCR:

▶ *hybridization-based screening.* A specific DNA clone of interest is labeled in some way then used as a hybridization probe to identify cell colonies containing the sequence of interest (see Section 6.4.1);

▶ PCR-based screening. Once a DNA sequence is known it is usually possible to design a *specific PCR assay* to test for its presence. The sequence is then said to have been *tagged* (since it can always be recognized by the specific PCR assay). If the sequence occurs at a unique *site* (location) within the genome of interest, then that site is said to be a **sequence tagged site (STS**; see *Box 5.4*). Sequence tagged sites have been very useful in providing rough physical maps of genomes (see Section 8.3.2) but the same principle can be used in library screening. In this case thousands of individual cell clones are stored and recombinant DNA isolated from each one of them is individually deposited into the wells of multiple microtitre dishes. Pools of clones from different groups of wells – arranged in different hierarchies – can be tested for the presence of a specific STS to identify a well which contains the desired DNA, and thereafter the original cell clone (see *Figure 5.11*).

Box 5.3: Nonsense suppressor mutations.

Base changes in the anticodon of a tRNA may enable it to insert an amino acid in response to a stop codon. Glutamate has two codons, GAG and GAA, which are recognized by two different tRNA molecules. The tRNA^{Glu} at top carries a CUC anticodon which recognizes the glutamate codon GAG. Mutation of the tRNA gene can produce a mutant tRNA^{Glu} which has a C→A change at the 3′ base in the anticodon. This mutant tRNA can now recognize the *amber* stop codon UAG and by inserting a glutamate it *suppresses* the amber stop signal. The bottom example illustrates a similar (C→A) mutation applied to the 3′ base at the anticodon of the other tRNA^{Glu} this time generating a mutant tRNA which can suppress the effect of an *ochre* stop codon.

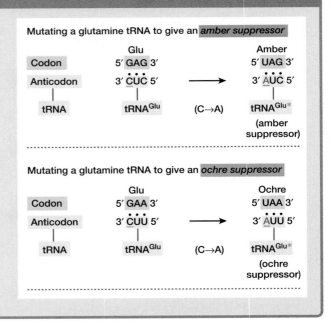

Box 5.4: The importance of sequence tagged sites (STSs).

Sequence tagged sites are important mapping tools simply because the presence of that sequence can very conveniently be assayed by PCR. Most STSs are nonpolymorphic and in genomes that have been sequenced the unique subchromosomal location for each STS is known precisely. An example of how an STS is developed from a DNA sequence is shown below.

In a complex genome such as the human genome, the chances of a 16-nucleotide-long primer binding by chance to a related but different sequence other than the intended target is not insignificant. However, the chances of both primers binding to unintended related sequences which just happen to be *both in close proximity and also in a suitable orientation* is normally very low. The specificity of the reaction can be assayed simply by size-fractionating the amplification products on an agarose gel. If there is a single strong PCR product of the expected approximate size (141 bp in the above example), there is an excellent chance that the assay is *specific* for the intended target sequence, thus defining the STS.

```
1                                         40
5′ CCCAGCGGGCCCGCGGCGCAGGGGCCCGGCGGGGCCCTGG

               PRIMER A
41        5′ CAGTGAGCATCAGATA ---➤ 3′        80
GGCCGCCCGGCAGTGAGCATCAGATACAGAACCTAGACGA

81                                        120
ACCTAGGACCAGTACCTACAAGGTACTCTAGATGATCTAT

121                                       160
ACTGAGGATCCTATTCAGATCCTAGGTACCACACTGATTA

161                                       200
AGGATACTAGCTATACGGACATGGCATTACACCCCCGGGG 3′
               ●●●●●●●●●●●●●●●●
       ◀---- 3′ TGCCTGTACCGTAATG 5′
               PRIMER B
```

5.4 Cloning systems for amplifying different sized fragments

Cell-based DNA cloning has been used widely as a tool for producing quantities of pure DNA for physical characterization and functional studies of individual genes, gene clusters or other DNA sequences of interest. However, the size of different DNA sequences of interest can vary enormously (e.g. human gene sizes are known to vary between 0.1 kb and 2 Mb). The first cell-based cloning systems to be developed could clone only rather small DNA fragments. Recently, however, there have been rapid developments in cloning systems that permit cloning of very large DNA fragments (see *Table 5.2*).

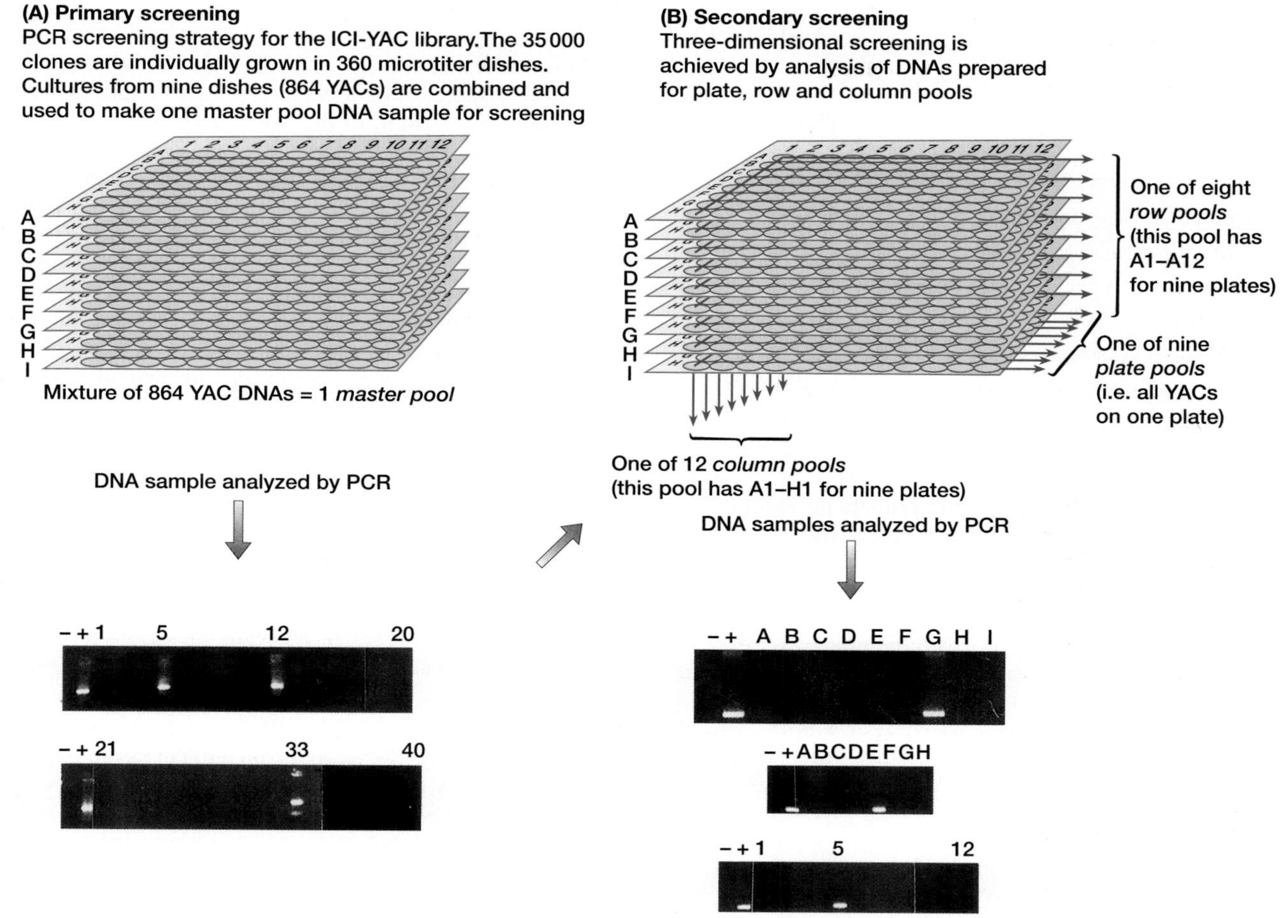

(A) Primary screening
PCR screening strategy for the ICI-YAC library. The 35 000 clones are individually grown in 360 microtiter dishes. Cultures from nine dishes (864 YACs) are combined and used to make one master pool DNA sample for screening

(B) Secondary screening
Three-dimensional screening is achieved by analysis of DNAs prepared for plate, row and column pools

Mixture of 864 YAC DNAs = 1 *master pool*

One of eight *row pools* (this pool has A1–A12 for nine plates)

One of nine *plate pools* (i.e. all YACs on one plate)

DNA sample analyzed by PCR

One of 12 *column pools* (this pool has A1–H1 for nine plates)

DNA samples analyzed by PCR

Figure 5.11: PCR-based library screening.

The example illustrates screening of a human YAC library. Approximately 35 000 individual clones were individually deposited into the 96 wells of 360 microtiter dishes. To facilitate screening, a total of 40 **master pools** were generated by combining all 864 clones in sets of nine microtiter dishes (plates A–I). Modified from Jones *et al.* (1994) *Genomics*, **24**, pp. 266–275 with permission from Academic Press Inc. **(A)** *Primary screening* involves PCR assay of the 40 master pools. In this example, three master pools were positive when referenced against positive (+) and negative (–) controls: pools 5, 12, 33. **(B)** *Secondary screening* identifies single YACs by assaying different subsets of the 864 YACs in a positive master pool, in this case master pool 12. Three-dimensional screening of each of nine **plate pools** (of 96 YACs each), eight **row pools** (of 106 YACs each), 12 **column pools** (of 72 YACs each) and identified a positive YAC in plate 12G (top panel), row E (middle panel), column 5 (bottom panel). The example here involved screening for YACs containing an anonymous X chromosome sequence. From Jones *et al.* (1994) *Genomics* **24**(1), 266–275, with permission from Elsevier. Photos were kindly provided by Dr Sandie Herrell, University of Newcastle upon Tyne.

5.4.1 Standard plasmid vectors provide a simple way of cloning small DNA fragments in bacterial (and simple eukaryotic) cells

In order to adapt natural plasmid molecules as cloning vectors, several modifications are normally made:

▶ insertion of an ***antibiotic resistance gene*** (to enable screening for the presence of the vector – see Section 5.3.5);

▶ insertion of a ***marker gene*** containing within it a ***multiple cloning site polylinker*** (to enable screening for recombinants – Section 5.3.5).

As an example, the plasmid vector pUC19 contains a polylinker with unique cloning sites for multiple restriction nucleases and an ampicillin resistance gene to permit identification of transformed cells (*Figure 5.12*). In addition, selection for recombinants is achieved by insertional inactivation of a component of the β-galactosidase gene, a complementary portion of this gene being provided by using a specially modified *Escherichia coli* host cell.

Table 5.2: Sizes of inserted DNA commonly obtained with different cloning vectors.

Cloning vector	Size of insert
Standard high copy number plasmid vectors	0–5 kb
Bacteriophage λ insertion vectors	0–10 kb
Bacteriophage λ replacement vectors	9–23 kb
Cosmid vectors	30–44 kb
Bacteriophage P1	70–100 kb
PAC (P1 artificial chromosome) vectors	130–150 kb
BAC (bacterial artificial chromosome) vectors	up to 300 kb
YAC (yeast artificial chromosome) vectors	0.2–2.0 Mb

5.4.2 Lambda and cosmid vectors provide an efficient means of cloning moderately large DNA fragments in bacterial cells

The major disadvantage of plasmid vectors is that their capacity for accepting large DNA fragments is severely limited: most inserts are a few kilobases in length and inserts larger than 5–10 kb are very rare. Additionally, standard methods of transformation of bacterial cells with plasmid vectors are relatively inefficient. To address these difficulties, attention was focused at an early stage on the possibility of using bacteriophage **lambda** as a cloning vector. The wild-type λ virus particle (**virion**) contains a genome of close to 50 kb of linear double-stranded DNA packaged within a protein coat and has evolved a highly efficient mechanism of infecting *E. coli* cells.

After the λ virion attaches to the bacterial cell, the coat protein is discarded and the λ DNA is injected into the cell. At the extreme termini of the λ DNA are overhanging 5′ ends which are 12 nucleotides long and complementary in base sequence. Because these large 5′ overhangs can base-pair, they are effectively sticky ends, similar to, but more cohesive than, the small sticky ends generated by some restriction nucleases (see Section 5.3.2). Such cohesive properties are recognized in the name given to this sequence – the **cos sequence**. Once inside the bacterial cell, the cos sequences base-pair, and sealing of nicks by cellular ligases results in the formation of a double-stranded circular DNA. Thereafter the λ DNA can enter two alternative pathways (*Figure 5.13*):

▶ **the lytic cycle.** The λ DNA replicates, initially bidirectionally, and subsequently by a rolling circle model which generates linear multimers of the unit length. Coat proteins are synthesized and the λ multimers are snipped at the cos sites to generate unit lengths of λ genome which are packaged within the protein coats. Some of the λ gene products lyse the host cell, allowing the virions to escape and infect new cells;

▶ **the lysogenic state.** The λ genome possesses a gene *att* which has a homolog in the *E. coli* chromosome. Apposition of the two *att* genes can result in recombination between the λ and *E. coli* genomes and subsequent integration of the λ DNA within the *E. coli* chromosome. In this state, the λ DNA is described as a **provirus** and the host cell as a **lysogen** because, although the λ DNA can

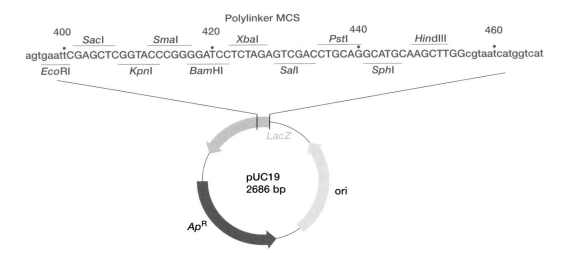

Figure 5.12: Map of plasmid vector pUC19.

The origin of replication (ori) was derived originally from a ColE1-like plasmid, pMB1. The ampicillin resistance gene (*Ap*^R) permits selection for cells containing the vector molecule. A portion of the *lacZ* gene is included and is expressed to give an amino-terminal fragment of β-galactosidase. This is *complemented* by a mutant *lacZ* gene in the host cell: the products of the vector and host cell *lacZ* sequences, although individually inactive, can associate to form a functional product. The 54-bp polylinker multiple cloning site (capital letters) is inserted into the vector *lacZ* (lower case letters) component in such a way as to preserve the reading frame and functional expression. However, cloning of an insert into the multiple cloning site (MCS) will cause insertional inactivation, and absence of β-galactosidase activity.

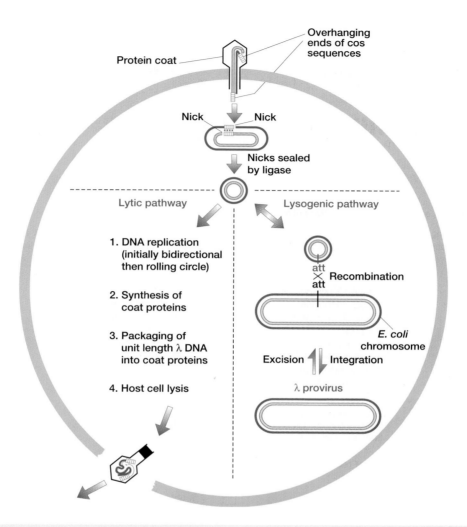

Figure 5.13: Phage λ can enter both lytic and lysogenic pathways.

remain stably integrated for long periods, it has the capacity for excision from the host chromosome and entry into the lytic cycle (*Figure 5.13*). Genes required for lysogenic function are located in a central segment of the λ genome (*Figure 5.14*).

The decision to enter the lytic cycle or the lysogenic state is controlled by two regulatory genes, *cI* and *cro*. These two genes are mutually antagonistic: in the lytic state the *cro* protein dominates, leading to repression of *cI*, whereas in the lysogenic state the *cI* repressor dominates and suppresses transcription of other λ genes including *cro*. In normally growing host cells, the lysogenic state is favored and the λ genome replicates along with the host chromosomal DNA. Damage to host cells favors a transition to the lytic cycle, enabling the virus to escape the damaged cell and infect new cells.

In order to design suitable cloning vectors based on λ, it was necessary to design a system whereby foreign DNA could be attached to the λ replicon *in vitro* and for the resultant recombinant DNA to be able to transform *E. coli* cells at high efficiency. The latter requirement was achieved by developing an *in vitro* **packaging system** which mimicked the way in which wild-type λ DNA is packaged in a protein coat, resulting in high infection efficiency (*Figure 5.15*).

Several major types of cloning vector that have been developed by modifying phage λ, or utilizing the size selection imposed by cos sequences, are described in the following sections.

▶ **Replacement λ vectors.** Only DNA molecules from 37 to 52 kb in length can be stably packaged into the λ particle. The central segment of the λ genome contains genes that are required for the lysogenic cycle but are not essential for lytic function. As a result, it can be removed and replaced by a foreign DNA fragment. Using this strategy, it is possible to clone foreign DNA up to 23 kb in length, and such vectors are normally used for making genomic DNA libraries;

▶ **insertion λ vectors.** Lambda vectors used for making cDNA libraries do not require a large insert capacity (most cDNAs are < 5 kb long). Design of insertion vectors often involves modification of the λ genome to permit insertional cloning into the *cI* gene;

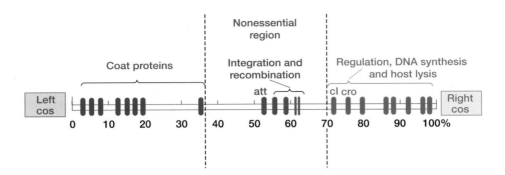

Figure 5.14: Map of the λ genome, showing positions of genes (vertical bars).

In λ replacement vectors, the nonessential region is removed by restriction endonuclease digestion, leaving a left λ arm and a right λ arm. A foreign DNA fragment can be ligated to the two arms in place of the original 'stuffer' fragment, providing maximal insert sizes of over 20 kb.

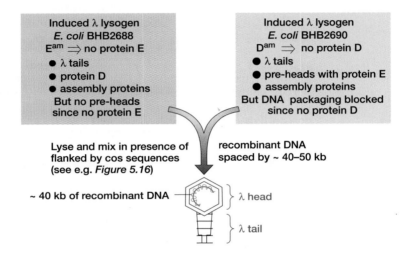

Figure 5.15: *In vitro* DNA packaging in a phage λ protein coat can be performed using a mixed lysate of two mutated λ lysogens.

Normal *in vivo* packaging of λ DNA involves first making **pre-heads**, structures composed of the major capsid protein encoded by gene *E*. A unit length of λ DNA is inserted in the pre-head, the unit length being prepared by cleavage at neighboring cos sites. A minor capsid protein D is then inserted in the pre-heads to complete head maturation, and the products of other genes serve as **assembly proteins**, ensuring joining of the completed tails to the completed heads. A defect in producing protein E, resulting from an amber mutation introduced into gene *E* (*E*^am), prevents pre-heads being formed by BHB2688. An amber mutation in gene *D* (*D*^am) prevents maturation of the pre-heads, with enclosed DNA, into complete heads. The components of the BHB2688/BHB2690 mixed lysate, however, complement each other's deficiency and provide all the products for correct packaging.

▶ *cosmid vectors* contain **cos** sequences inserted into a small plas**mid** vector. Large (~30–44 kb) foreign DNA fragments can be cloned using such vectors in an *in vitro* packaging reaction because the total size of the cosmid vector is often about 8 kb (*Figure 5.16*).

5.4.3 Large DNA fragments can be cloned in bacterial cells using vectors based on bacteriophage P1 and F factor plasmids

Bacterial artificial chromosome (BAC) vectors

Many vectors used for DNA cloning in bacterial cells are based on high to medium copy number replicons. The high copy number results in a high yield of DNA: each cell in which a vector molecule is propagated will have several to many copies of the vector molecule. An important disadvantage is that such vectors often show structural instability of inserts, resulting in deletion or rearrangement of portions of the cloned DNA. Such instability is particularly common in the case of DNA inserts of eukaryotic origin where repetitive sequences occur frequently. As a result, it is difficult to clone and maintain intact large DNA in bacterial cells.

In order to overcome this limitation, attention has recently been focused on vectors based on low copy number replicons, such as the *E. coli* fertility plasmid, the F-factor. This plasmid contains two genes, *parA* and *parB*, which maintain the copy

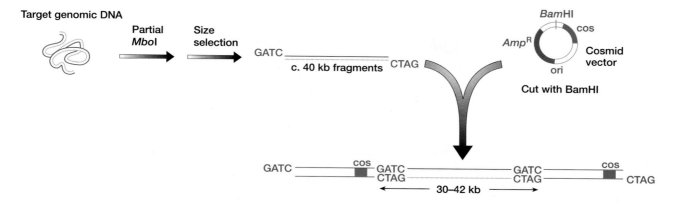

Figure 5.16: Ligation to cleaved cosmid vector molecules can produce vector–target concatemers, resulting in a large exogenous DNA fragment flanked by cos sequences.

number of the F-factor at 1–2 per *E. coli* cell. Vectors based on the F-factor system are able to accept large foreign DNA fragments (> 300 kb). The resulting recombinants can be transferred with considerable efficiency into bacterial cells using **electroporation** (a method of exposing cells to high voltages in order to relax the selective permeability of their plasma membranes). However, because the resulting **bacterial artificial chromosomes (BACs)** contain a low copy number replicon, only low yields of recombinant DNA can be recovered from the host cells.

Bacteriophage P1 vectors and P1 artificial chromosomes (PACs)

Certain bacteriophages have relatively large genomes, thereby affording the potential for developing vectors that can accommodate large foreign DNA fragments. One such is ***bacteriophage P1*** which, like phage λ, packages its genome in a protein coat; 110–115 kb of linear DNA is packaged in the P1 protein coat. P1 cloning vectors have therefore been designed in which components of P1 are included in a circular plasmid.

The P1 plasmid vector can be cleaved to generate two vector arms to which up to 100 kb of foreign DNA can be ligated and packaged into a P1 protein coat *in vitro*. The recombinant P1 phage can be allowed to adsorb to a suitable host, following which the recombinant P1 DNA is injected into the cell, circularizes and can be amplified (Sternberg, 1992). An improvement on the size range of inserts accepted by the basic P1 cloning system has been the use of bacteriophage T4 *in vitro* packaging systems with P1 vectors which enables the recovery of inserts up to 122 kb in size. More recently, features of the P1 and F-factor systems have been combined to produce cloning systems (Iouannou *et al.*, 1994).

5.4.4 Yeast artificial chromosomes (YACs) enable cloning of megabase fragments

The most popularly used system for cloning very large DNA fragments involves the construction of **yeast artificial chromosomes (YACs**; see Schlessinger, 1990). Certain eukaryotic

sequences, notably those with repeated sequence organizations, are difficult, or impossible, to propagate in bacterial cells which do not have such types of DNA organization, but would be anticipated to be tolerated in yeast cells which are eukaryotic cells. However, the main advantage offered by YACs has been the ability to clone very large DNA fragments. Such a development proceeded from the realization that the great bulk of the DNA in a chromosome is not required for normal chromosome function. As detailed in *Figure 2.5*, the essential functional components of yeast chromosomes are threefold:

▶ ***centromeres*** are required for disjunction of sister chromatids in mitosis and of homologous chromosomes at the first meiotic division;

▶ ***telomeres*** are required for complete replication of linear molecules and for protection of the ends of the chromosome from nuclease attack;

▶ ***autonomous replicating sequence*** elements are required for autonomous replication of the chromosomal DNA and are thought to act as specific replication origins.

In each case, the DNA segment necessary for functional activity *in vivo* in yeast is limited to at most a few hundred base pairs of DNA (*Figure 2.5*). As a result, it became possible to envisage a novel cloning system based on the use of *chromosomal replicons* (autonomous replicating sequence elements) as an alternative to extrachromosomal replicons (those found in plasmids and bacteriophages) and involving the construction of an *artificial chromosome*.

To make a YAC, it is simply necessary to combine four short sequences that can function in yeast cells: two telomeres, one centromere and one ARS element, together with a suitably sized foreign DNA fragment to give a linear DNA molecule in which the telomere sequences are correctly positioned at the termini (*Figure 5.17*). The resulting construct cannot be transfected directly into yeast cells. Instead, yeast cells have to be treated in such a way as to remove the external cell walls. The resulting yeast **spheroplasts** can accept exogenous fragments but are osmotically unstable and need to be embedded in agar. The overall transformation efficiency is very low and

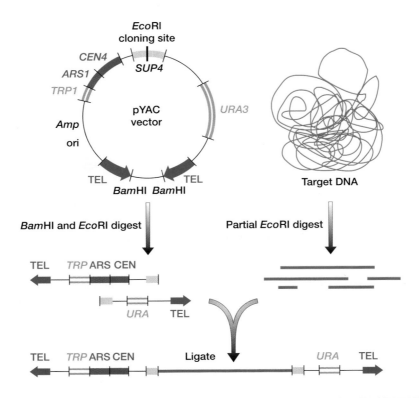

Figure 5.17: Making YACs.

Vector DNA sequences include: CEN4, centromere sequence; TEL, telomere sequences; ARS1, autonomous replicating sequence; *Amp*, gene conferring ampicillin resistance; ori, origin of replication for propagation in an *E. coli* host. The vector is used with a specialized yeast host cell, AB1380, which is red colored because it carries an ochre mutation in a gene, *ade-2*, involved in adenine metabolism, resulting in accumulation of a red pigment. However, the vector carries the *SUP4* gene, a ***suppressor tRNA gene*** (see *Box 5.3*) which overcomes the effect of the *ade-2* ochre mutation and restores wild-type activity, resulting in colorless colonies. The host cells are also designed to have recessive *trp1* and *ura3* alleles which can be complemented by the corresponding *TRP1* and *URA3* alleles in the vector, providing a selection system for identifying cells containing the YAC vector. Cloning of a foreign DNA fragment into the *SUP4* gene causes insertional inactivation of the suppressor gene function, restoring the mutant (red color) phenotype.

the yield of cloned DNA is low (about one copy per cell). Nevertheless, the capacity to clone large exogenous DNA fragments (up to 2 Mb) has made YACs a vital tool in physical mapping (see Section 8.3.2).

5.5 Cloning systems for producing single-stranded and mutagenized DNA

Single-stranded DNA clones are useful for several applications including **DNA sequencing** (because the sequences obtained are clearer and easier to read) and **site-directed mutagenesis** (where a specific site in a cloned DNA needs to be altered in a precise, pre-determined way). Site-directed mutagenesis can be designed to create specific nucleotide substitutions, deletions, and so on which can help identify key amino acid residues or other functionally important sequences if a functional assay is already available for the DNA sequence of interest.

5.5.1 Single-stranded DNA for use in DNA sequencing is obtained using M13 or phagemid vectors or by linear PCR amplification

Single-stranded recombinant DNA clones are usually used for DNA sequencing as templates using vectors based on certain bacteriophages which naturally adopt a single-stranded DNA form at some stage in their life cycle. Because the vector sequence is already known, it is convenient to use a single vector-specific sequencing primer which is complementary to a sequence in the vector adjacent to the cloning site.

M13 vectors

M13 is a *filamentous bacteriophage* which can infect certain *E. coli* strains. Its 6.4-kb circular single-stranded genome is enclosed in a protein coat, forming a long filamentous structure. After adsorption to the bacterium, the M13 genome enters the bacterial cell, and is converted to a double-stranded form, the *replicative form* (**RF**) which serves as a template for making numerous copies of the genome. After a certain time,

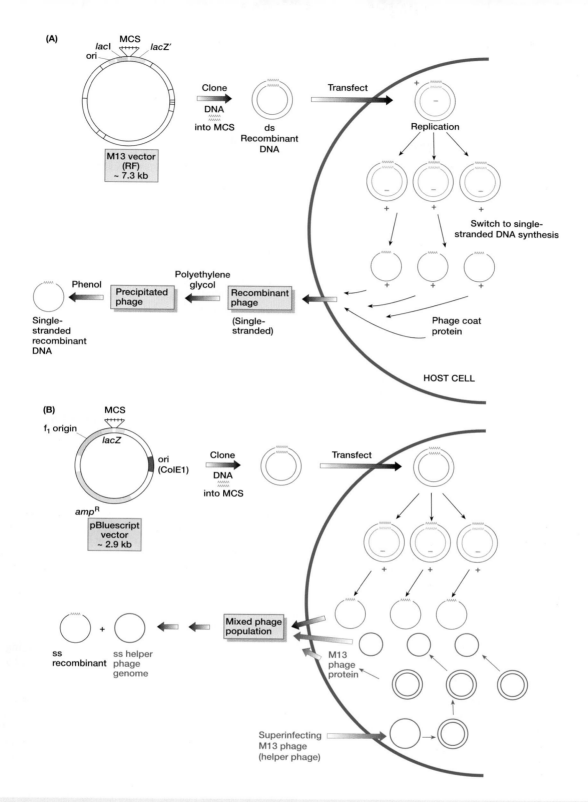

Figure 5.18: Producing single-stranded recombinant DNA using M13 and phagemid vectors.

(A) *M13 vectors*. M13 vectors are replicative forms (RF) of M13 derivatives containing a nonfunctional component of the *lacZ* β-galactosidase system which can be complemented in function by the presence of a complementary *lacZ* component in the *E. coli* JM series. The double-stranded M13 recombinant DNA enters the normal cycle of DNA replication to generate numerous copies of the genome, prior to a switch to production of single-stranded DNA (+ strand only). Mature recombinant phage exit from the cell without lysis.
(B) *Phagemid vectors*. The pBluescript series of plasmid vectors contain two origins of replication: a normal one from *Co*/E1 and a second from phage f1 which, in the presence of a filamentous phage genome, will specify production of single-stranded DNA. Superinfection of transformed cells with M13 phage results in two types of phage-like particles released from the cells: the original superinfecting phage and the plasmid recombinants within a phage protein coat. Sequencing primers specific for the phagemid vector are used to obtain unambiguous sequences.

a phage-encoded product switches DNA synthesis towards production of single strands which migrate to the cell membrane. Here they are enclosed in a protein coat, and hundreds of mature phage particles are extruded from the infected cell without cell lysis. M13 vectors are based on the RF with a multiple cloning site for accepting foreign inserts of limited size. The latter can be transfected into suitable strains of *E. coli*. After a certain period, phage particles are harvested and stripped of their protein coats to release single-stranded recombinant DNA for direct use as templates in DNA sequencing reactions (*Figure 5.18A*).

Phagemid vectors

A small segment of the genome of a filamentous bacterio**phage**, such as M13 (or the related filamentous phages fd or f1) can be inserted into a plas**mid** to form a hybrid vector known as a **phagemid**. The selected phage sequences contain all the *cis*-acting elements required for DNA replication and assembly into phage particles. They permit successful cloning of inserts several kilobases long (unlike M13 vectors in which such inserts tend to be unstable). Following transformation of a suitable *E. coli* strain with a recombinant phagemid, the bacterial cells are *superinfected* with a filamentous **helper phage**, such as f1, which is required to provide the coat protein. Phage particles secreted from the superinfected cells will be a mixture of helper phage and recombinant phagemids (*Figure 5.18B*). The mixed single-stranded DNA population can be used directly for DNA sequencing because the primer for initiating DNA strand synthesis is designed to bind specifically to a sequence of the phagemid vector adjacent to the cloning site. Commonly used phagemid vectors include the pEMBL series of plasmids and the pBluescript family (see *Figure 5.18B*).

Linear PCR amplification

A form of PCR-based sequencing known as *cycle sequencing* (*Box 7.1*) uses a modified PCR reaction to generate single-stranded templates for sequencing. This is achieved by using only a single primer so that a single-stranded product accumulates but in a linear fashion rather than the exponential amplification seen in normal PCR.

5.5.2 Oligonucleotide mismatch mutagenesis can create a predetermined single nucleotide change in any cloned gene

Many *in vitro* assays of gene function aim to gain information on the importance of individual amino acids in the encoded polypeptide. This may be relevant when attempting to assess whether a particular missense mutation found in a known disease gene is pathogenic, or just generally in trying to evaluate the contribution of a specific amino acid to the biological function of a protein.

A popular general approach involves cloning the gene or cDNA into an M13 or phagemid vector to recover single-stranded recombinant DNA (see previous section). A mutagenic oligonucleotide primer is then designed whose sequence is perfectly complementary to the gene sequence in the region to be mutated except for a single difference: *at the intended mutation site it bears a base that is complementary to the desired mutant nucleotide rather than the original.* The mutagenic oligonucleotide is then allowed to prime new DNA synthesis to create a complementary full-length sequence containing the desired mutation. The newly formed *heteroduplex* is used to transform cells, and the desired mutant genes can be identified by screening for the mutation (see *Figure 5.19*).

Other small-scale mutations can also be introduced in addition to single nucleotide substitutions. For example, it is possible to introduce a three-nucleotide deletion that will result in removal of a single amino acid from the encoded polypeptide, or an insertion that adds a new amino acid. Provided the mutagenic oligonucleotide is long enough, it will be able to bind specifically to the gene template even if there is a considerable central mismatch. Still larger mutations can be introduced by using **cassette mutagenesis** in which case a specific region of the original sequence of the original gene is deleted and replaced by oligonucleotide cassettes (Bedwell *et al.*, 1989).

5.5.3 PCR-based mutagenesis includes coupling of desired sequences or chemical groups to a target sequence and site-specific mutagenesis

Site-directed mutagenesis by PCR has become increasingly popular and various strategies have been devised to enable base substitutions, deletions and insertions (see below and Newton and Graham, 1997). In addition to producing specific predetermined mutations in a target DNA, a form of mutagenesis known as 5′ add-on mutagenesis permits addition of a desired sequence or chemical group in much the same way as can be achieved using ligation of oligonucleotide linkers.

5′ add-on mutagenesis is a commonly used practice in which a new sequence or chemical group is added to the 5′ end of a PCR product by designing primers which have the desired specific sequence for the 3′ part of the primer while the 5′ part of the primer contains the novel sequence or a sequence with an attached chemical group. The extra 5′ sequence does *not* participate in the first annealing step of the PCR reaction (only the 3′ part of the primer is specific for the target sequence), but it subsequently becomes incorporated into the amplified product, thereby generating a recombinant product (*Figure 5.20A*). Various popular alternatives for the extra 5′ sequence include: (i) a suitable restriction site, which may facilitate subsequent cell-based DNA cloning; (ii) a functional component, for example a promoter sequence for driving expression; (iii) a modified nucleotide containing a reporter group or labeled group, such as a biotinylated nucleotide or fluorophore.

Mismatched primer mutagenesis. The primer is designed to be only partially complementary to the target site but in such a way that it will still bind specifically to the target. Inevitably this means that the mutation is introduced close to the extreme end of the PCR product. As illustrated in *Fig. 18.8*, this approach may be exploited to introduce an artificial diagnostic restriction site that permits screening for a known mutation. Mutations can also be introduced at any point within a chosen sequence using mismatched primers. Two mutagenic reactions are designed in which the two separate PCR products have partially overlapping sequences containing the mutation.

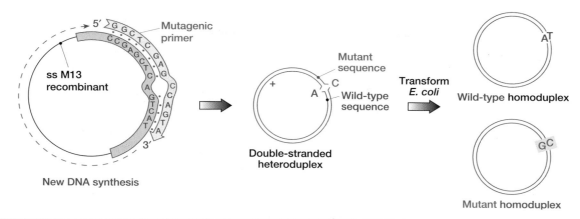

Figure 5.19: Oligonucleotide mismatch mutagenesis can create a desired point mutation at a unique predetermined site within a cloned DNA molecule.

The figure illustrates only one of many different methods of cell-based oligonucleotide mismatch mutagenesis. The example illustrates the use of a mutagenic oligonucleotide to direct a single nucleotide substitution in a gene. The gene is cloned into M13 in order to generate a single-stranded recombinant DNA. An oligonucleotide primer is designed to be complementary in sequence to a portion of the gene sequence encompassing the nucleotide to be mutated (A) and containing the desired noncomplementary base at that position (C, not T). Despite the internal mismatch, annealing of the mutagenic primer is possible, and second-strand synthesis can be extended by DNA polymerase and the gap sealed by DNA ligase. The resulting heteroduplex can be transformed into E. coli, whereupon two populations of recombinants can be recovered: wild-type and mutant homoduplexes. The latter can be identified by molecular hybridization (by using the mutagenic primer as an allele-specific oligonucleotide probe; see Figure 6.11) or by PCR-based allele-specific amplification methods (see Figure 5.4).

The denatured products are combined to generate a larger product with the mutation in a more central location (Higuchi, 1990; see *Figure 5.20B*).

5.6 Cloning systems designed to express genes

The cloning systems described in Section 5.4 are used when the aim is simply to amplify the introduced DNA to obtain sufficient quantities for a variety of subsequent structural and functional studies. However, in the case of gene sequences, there are many circumstances where instead of just amplifying and propagating the cloned DNA, it is desirable to be able to *express* the gene in some way (**expression cloning**). In each case appropriate expression signals need to be provided by the cloning system. Expression cloning can be conducted using PCR-based systems, but is usually carried out using cell-based cloning systems. A wide variety of cloning systems can be used, depending on:

▶ **type of expression product.** For some purposes, it may be sufficient to be able to obtain an RNA product. Examples include generation of antisense RNA probes (*riboprobes* − see *Figure 6.3*) for use in tissue *in situ* hybridization studies, or generation of antisense RNA for inhibiting or destroying expression of specific genes, either in functional studies or for therapeutic purposes (Sections 20.2.6 and 21.7.5). In many cases, however, a protein product is desired;

▶ **type of environment.** Sometimes, it may be sufficient to express the product *in vitro*. Often, however, it may be

desirable to be able to express the product in a particular cellular system which may be a highly defined prokaryotic or eukaryotic cell line;

▶ **purpose of expression system.** The expression system may be designed simply for investigating expression. In many cases, however, the purpose may be to retrieve large quantities of an expression product, as in the need to generate large quantities of a specific protein to assist subsequent crystallography studies or attempts to raise specific antibodies against the protein.

Many different expression cloning vectors have been designed to be used in different host cell systems, ranging from bacterial cells to mammalian cells, with specific vectors being engineered to be useful in specific host cell types.

5.6.1 Large amounts of protein can be produced by expression cloning in bacterial cells

Cloning of eukaryotic cDNA in an expression vector is often required to produce medically relevant compounds or proteins for basic follow-up research studies, for example structural studies. Usually in these cases, the cDNA is simply designed to provide the genetic information specifying the protein sequence and the expression signals are provided externally by stitching suitably strong promoters, regulatory elements, and so on into the expression vector. As the expression system is based on recombinant DNA and also quite often involves bringing into existence artificial fusion proteins or proteins modified by adding certain peptide tags to them, the resulting proteins are sometimes described as **recombinant proteins**.

(A)

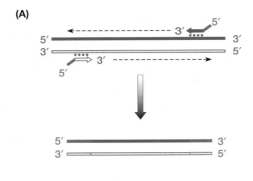

(B)

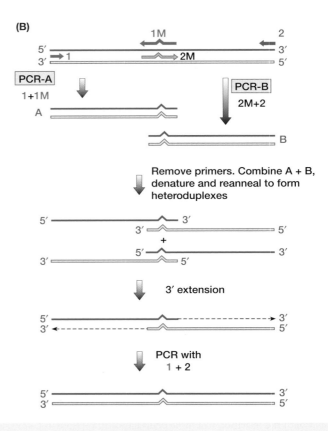

Figure 5.20: PCR mutagenesis.

(A) *5′ add-on mutagenesis*. Primers can be modified at the 5′ end to introduce, for example, a labeled group (e.g. *Figure 7.11*), a sequence containing a suitable restriction site or a phage promoter to drive gene expression. **(B)** *Site-specific mutagenesis*. The mutagenesis shown can result in an amplified product with a specific pre-determined mutation located in a central segment. PCR reactions A and B are envisaged as amplifying overlapping segments of DNA containing an introduced mutation (by deliberate base mismatching using a mutant primer: 1M or 2M). After the two products are combined, denatured and allowed to reanneal, the DNA polymerase can extend the 3′ end of heteroduplexes with recessed 3′ ends. Thereafter, a full-length product with the introduced mutation in a central segment can be amplified by using the outer primers 1 and 2 only.

Bacterial cells have the advantage that they grow rapidly and can be expanded easily in culture to very large culture volumes and the well-studied *E. coli* has been a favorite host cell for expression of heterologous (foreign) proteins (Baneyx, 1999). A wide variety of proteins can be expressed using expression systems based on bacteriophage **T7 RNA polymerase**, which is capable of producing complete transcripts from almost any coding sequence.

As the production of very large amounts of a *heterologous* protein can be detrimental, and even toxic, to host cell growth, it is advantageous to have the expression controlled by an **inducible promoter**. As *E. coli* RNA polymerase does not recognize the T7 promoter, DNA cloned in the vector will remain largely unexpressed in the absence of T7 RNA polymerase. The normal host RNA polymerase will not recognize the T7 promoter but for cloning purposes a strain is used which has an inducible T7 RNA polymerase within the bacterial chromosome. For example, when using the **pET vector** cloning systems (*Figure 5.21A*), the host is modified to contain a T7 RNA polymerase regulated by a lac promoter and so can be

induced when desired using the *lac* inducer isopropyl-thio-β-D-galactopyranoside (IPTG). As a result, transformed cells can be selected and grown up in large quantities without expression; thereafter, IPTG can be added to induce expression and the cells can be harvested shortly afterwards.

Although bacteria have many advantages for heterologous protein expression, there are various types of limitation:

▶ *post-translational processing*. The absence of normal glycosylation or phosphorylation patterns for certain eukaryotic proteins produced in bacterial cells means that the proteins become unstable, or show limited or no biological activity;

▶ *protein length*. Many eukaryotic proteins, notably some mammalian proteins, are very much larger than bacterial proteins and cannot easily be synthesized in *E. coli*;

▶ **protein folding and solubility.** Often, overexpression leads to the production of *inclusion bodies* – insoluble aggregates of misfolded protein. The inclusion bodies can easily be purified, but the expressed protein can usually

(A)

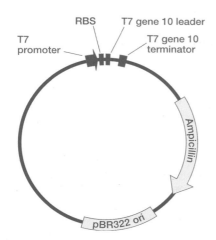

Figure 5.21: Examples of bacterial expression vectors.

(A) *pET-3 bacterial expression vectors*. High-level expression is possible using the bacteriophage T7 promoter. The vector also contains a sequence encoding the T7 gene *10* leader peptide and a T7 gene *10* terminator (T7 ter). Cloning is possible in either the *Nde*I or the *Bam*H1 cloning site. If the latter, a fusion protein is made containing 13 N-terminal amino acids from *T7* gene *10* with the T7 leader sequence ensuring high-level translation . Expression is induced by adding IPTG which induces expression of a T7 RNA polymerase gene located within the modified bacterial chromosome. RBS, Ribosome binding site. **(B) *pGEX-4T gene fusion vectors*.** The multiple cloning site is located so that a *fusion protein* is produced following transcription from the tac promoter, comprising an N-terminal component of GST and a C-terminal component of the desired protein. Commencing at the *Eco*RI the multiple cloning sites for pGEX-4T-1, -2 and -3 are arranged so that all three amino acid reading frames are possible. The GST-fusion protein can be purified easily on a glutathione affinity purification column such as glutathione sepharose 4B, and the desired protein can be cleaved at the thrombin cleavage site.

(B)

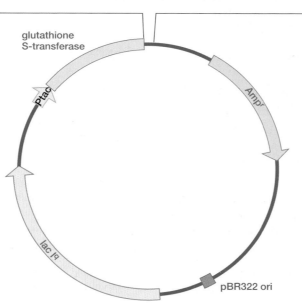

only be solubilized using strongly denaturing conditions and a major problem is then how to achieve efficient folding *in vitro*.

Efforts to increase yield and solubility have often involved the production of **fusion proteins** where the desired protein is coupled to an endogenous protein, for example, maltose-binding protein, thioredoxin, ubiquitin, and so on. It is also common practice for protein expression vectors to be modified so that an **affinity tag** is added, assisting purification of the recombinant by *affinity chromatography*. Two favorite systems are:

▶ *GST-glutathione affinity.* Glutathione-S-transferase (GST) is a small protein with a very high affinity for its substrate glutathione. A GST-fusion protein can be generated (see *Figure 5.21B)* and purified by selective binding to a column containing glutathione.

▶ *polyhistidine–nickel ion affinity.* The side chains of certain amino acids such as histidine have a very high affinity for some metal ions. The expression vectors in this case usually lead to coupling of an affinity tag of six consecutive histidine residues (see *Figure 5.23*). The side chains of the $(His)_6$ tag bind selectively and strongly to nickel ions assisting purification by affinity chromatography using a nickel–nitrilotriacetic acid matrix.

Expression libraries

Plasmid vectors are often used in expression cloning because they are easy to work with, and if the object is to express a *specific* gene of interest they are the tools of choice. However, sometimes the object is to produce a large number of different recombinants as a resource for expression, an **expression library**. In such cases, it is often advantageous to use modified bacteriophage λ vectors because large numbers of recombinants can be screened comparatively easily.

Expression libraries are typically constructed by cloning cDNA from a tissue of interest using a λ vector such as λgt11 or λZAP. Filters containing individual phage-infected bacterial colonies can be screened by exposure to antibody. Positively reacting bacteria can then be propagated to isolate the cDNA clone, and the isolated cDNA clone can, in turn, be used to screen a genomic library to identify the cognate gene.

5.6.2 Phage display is a form of expression cloning in which proteins are expressed on bacterial cell surfaces

Phage display is a form of expression cloning of foreign genes using phage (see Clackson and Wells, 1994). Genetic engineering techniques are used to insert foreign DNA fragments into a suitable phage coat protein gene. The modified gene can then be expressed as a *fusion protein* which is incorporated into the virion and displayed on the surface of the phage which, however, retains infectivity (**fusion phage**). If an antibody is available for a specific protein, phage displaying that protein can be selected by preferential binding to the antibody: affinity purification of virions bearing a target determinant can be achieved from a 10^8-fold excess of phage

not bearing the determinant, using even minute quantities of the relevant antibody.

Initially, phage display involved the use of filamentous phages such as fd, f1, M13, where the foreign gene was incorporated into a gene specifying a minor coat protein such as the gene III protein (see *Figure 5.22*). Several useful applications have been devised:

▶ *antibody engineering.* Phage display is proving a powerful alternative source of constructing antibodies, including humanized antibodies, bypassing immunization and even hybridoma technology (see Winter *et al.*, 1994);

▶ *general protein engineering.* Phage display is a powerful adjunct to random mutagenesis programs as a way of selecting for desired variants from a library of mutants;

▶ *studying protein–protein interactions.* This involves a library-based method that can be used to identify proteins that interact with a given protein. In the same way that antibodies can be used in affinity screening, a desired protein (or any other molecule to which a protein can bind) is used as the selective agent. The protein can select fusion phage that display any other proteins that significantly bind to it.

5.6.3 Eukaryotic gene expression is carried out with greater fidelity in eukaryotic cell lines

The biological properties of many eukaryotic proteins synthesized in bacteria may not be very representative of the native molecules because of lack of normal post-translational processing and incorrect or inefficient protein folding. While bacterial expression systems have the great advantage of very high level protein expression, the disadvantages have prompted the alternative use of eukaryotic cell hosts for recombinant protein expression, including insect and mammalian cells. In addition to hosting protein expression mammalian cells are also often used as a way of screening the effects of *in vitro* manipulations on *transcriptional and post-transcriptional control sequences*.

When using animal cell lines as host cells, consideration has to be given to how the foreign DNA should be transported into the host cells (see *Box 5.5 and Table 5.3*) and the duration of the expression. In the latter case, two types of expression system are possible:

▶ **transient expression.** The DNA carried by the expression vector is intended to remain as an *independent* genetic element within the transfected cells, a so-called **episome**, rather than integrate into the chromosomes of the host cell. Expression of the **transgene** (any gene which has been introduced into animal or plant cells) reaches a maximal level about 2–3 days following transfection of the expression vector into the mammalian cell line but thereafter expression diminishes rapidly due to cell death or loss of the expression construct;

▶ **stable expression.** The DNA carried by the expression vector is designed to *integrate* into the host cell chromosomes. This can take over a month to establish but once established, expression products should be found in all cells provided the transgene is capable of being expressed.

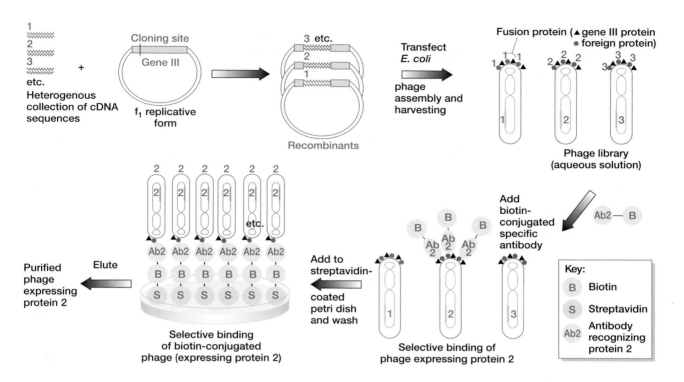

Figure 5.22: Phage display.

Foreign cDNA is cloned into phage vectors in order to express foreign proteins on the phage surface. Here the DNA is inserted into gene III of phage f1 (or M13, fd), which encodes a minor phage coat protein. The cloning site is located at the region specifying the extreme N-terminal sequence of the gene III protein. A phage expression library is produced following transfection of *E. coli*, phage assembly, extrusion from the cells and phage harvesting (see also *Figure 5.18*). Recombinants with in-frame inserts may often be expressed to give a fusion protein in which the N-terminal component consists of a foreign protein sequence. An antibody specific for one of the foreign proteins can bind specifically to the phage which displays the sequence, leading to its purification. Such *affinity purification* permits identification of cDNA sequences encoding an uncharacterized protein of interest (see Parmley and Smith, 1988).

Transient high-level protein expression in insect cells using baculovirus

Baculovirus gene expression is a popular method for producing large quantities of recombinant proteins in insect host cells, the protein yields being higher than in mammalian expression systems, while the costs are lower. In most cases, post-translational processing of eukaryotic proteins expressed in insect cells is similar to the protein processing which occurs in mammalian cells, and expression of very large proteins is possible. As a result, the proteins produced in insect cells have comparable biological activities and immunological reactivities to proteins expressed in mammalian cells.

The baculovirus usually used for protein expression is *Autographa californica* nuclear polyhedrosis virus (AcMNPV) which can be propagated in certain insect cell lines. The viral polyhedron protein is transcribed at high rates and although essential for viral propagation in its normal habitat, it is not needed in culture. As a result, its coding sequence can be replaced by that for a foreign protein. The cloning vector is designed to express the heterologous protein from the powerful polyhedrin promoter resulting in expression levels accounting for over 30% of the total cell protein.

Transient expression in mammalian cells

Expressing mammalian proteins in mammalian cells has the obvious advantage that correct protein folding and post-translational modification is not an issue and it is possible to analyze downstream signals and cellular effects. Stable expression systems in mammalian cells (typically based on chromosome-integrated plasmid sequences) have delivered kilograms of complex proteins in industrial-scale **bioreactors** but typically have required large investments in time, resources and equipment. As an alternative, large-scale transient expression systems have been developed for producing recombinant proteins in mammalian cells (Wurm and Bernard, 1999).

In addition to delivering protein expression, some mammalian cell lines have found major applications in screening the effects of *in vitro* manipulations upon transcriptional and post-transcriptional *control sequences*. A good example is provided by **COS cells**, stable lines of African green monkey kidney cells which were derived from the SV40-permissive simian cell-line, CV-1. When CV-1 cells are infected with SV40, the normal SV40 lytic cycle ensues. However, Gluzman (1981) was able to transform CV-1 cells after a segment of the SV40 genome

Box 5.5: Transferring genes into cultured animal cells.

A variety of methods can be used to transfer genes into human and animal cells, but can be grouped into two classes:

▶ **transduction**. This describes *virus-mediated* gene transfer. Certain animal DNA and RNA viruses naturally infect human and mammalian cells. Modifications of these viruses enable their use as vectors to transfer exogenous genes into suitable target cells at high efficiency. They offer both transient expression systems offering high copy number expression, as in the case of adenovirus vectors, and stable expression systems such as those based on retroviruses, a class of RNA virus whose natural lifecycle involves making cDNA copies which integrate into host cell chromosomes (see *Table 5.3*).

▶ **transfection**. This describes nonviral mediated gene transfer. Note that the term *transfection* is analogous to the process of *transformation* in bacteria. The latter term was not applied to the process of transferring genes into animal cells because of its association with an altered phenotype and unrestrained growth (see below). Transgenes can be transferred by different means:

- *using nonreplicating plasmid vectors*

- *using plasmid vectors with viral replicons*. Often an SV40 replicon is used (as with COS cells; see Section 5.6.3), or Epstein–Barr virus (human host cells) or bovine papilloma virus (mouse host cells).

▶ Various transfection methods are available, including the use of:

- *calcium phosphate*. The calcium phosphate and DNA form co-precipitates on the surface of the target cells. The high concentration of the DNA on the plasma membrane may increase the efficiency of transfection.

- **liposomes** (artificial lipid vesicles) to which the DNA binds. Liposomes can form spontaneously in aqueous solution following artificial mixing of phospholipid molecules. The liposomes can fuse with the plasma membrane and so permit access to the cell interior (see *Figure* below);

- **electroporation**. A popular method whereby an electric shock is used to cause temporary membrane depolarization in the target cells, thereby assisting passage of large DNA molecules.

Whether by transduction or transfection, genes which have been transferred into animal (or plant) cells are known as **transgenes**. Transgenes can have different fates as follows:

- *episomal transgenes*. Transgenes which do not integrate into host cell chromosomes may be maintained in the nucleus in an extrachromosomal state (**episome**). If the transgene is linked to a vector with an origin of replication that functions in the host cell, it can be amplified, sometimes to quite high copy number. If the transgene is not coupled to such an origin of replication, it persists for just a short time before it is diluted (as the host cells divide) and degraded;

- *integrated transgenes*. In some cases, the transgenes can integrate into the chromosomes of the host cell and be stably inherited. This state is often described as **stable transformation (note**: the strict usage of the term *transformation* signifies that the phenotype of the cell is altered so that the cell acquires unrestrained growth characteristics) and the resulting cell is described as a *cell line*. Integration is a very inefficient process, and so the rare stably transformed cells must be isolated from the background of nontransformed cells by selection for some marker (see text).

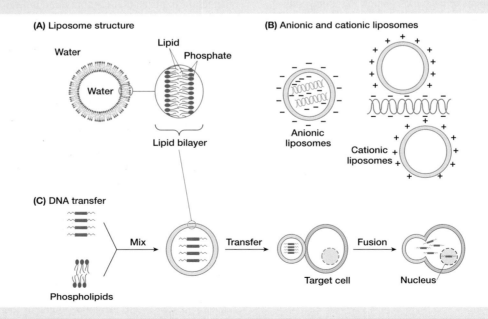

(A) Liposome structure

Water · Lipid · Phosphate · Water · Lipid bilayer

(B) Anionic and cationic liposomes

Anionic liposomes · Cationic liposomes

(C) DNA transfer

DNA · Phospholipids · Mix · Transfer · Target cell · Fusion · Nucleus

Table 5.3: Common viral vector systems for expression in mammalian cells

Vector system based on	Host range and location	Other comments
Adenovirus	Broad mammalian host range; generally nuclear *episomal*	Insert size up to 8 kb only; high level expression; high titres of recombinant virus
Adeno-associated virus	Broad mammalian host range; wild type virus *integrates* into specific site in human chromosome 19	Insert size up to 4.5 kb only; requires adenovirus for packaging; stable expression
Epstein–Barr	Retained as a nuclear *episome* in human, monkey and dog, but not generally in rodent	
Herpes simplex	Broad mammalian host range; *lytic*	Insert size up to 150 kb; recombinant viruses are made replication-deficient by deletion of one of the relevant viral genes
Papilloma	Retained as a nuclear *episome* in rodent (BPV) or human and monkey (HPV)	Used to study gene regulation and for high level transgene expression
Polyoma	Broad mammalian host range; can *integrate*	Replicates best in mouse cells; used to study gene regulation and for high level transgene expression
Retrovirus	Variable host range but some have broad mammalian host range; *integrates* as cDNA copy into host chromosomes	Insert size limited to max. of 8.5 kb; low titres of recombinant virus; stable expression
SV40	Broad mammalian host range; can *integrate* but is *episomal* in presence of SV40 origin of replication plus large T antigen	
Vaccinia	Broad mammalian host range; *lytic*	Used primarily for transgene overexpression

containing a mutant origin of replication integrated into the chromosomes of CV-1.

The ensuing COS cells (**C**V1 with defective **o**rigin of **S**V40) *constitutively* (stably) express the SV40-encoded **large T antigen**, the only viral protein which is required for activating the SV40 origin of replication. By expressing the large SV40 T antigen in a stable way, COS cells permit *any* introduced circular DNA with a functional SV40 origin of replication to replicate independently of the host cell chromosomes, with no clear size limitation. When transient expression vectors are

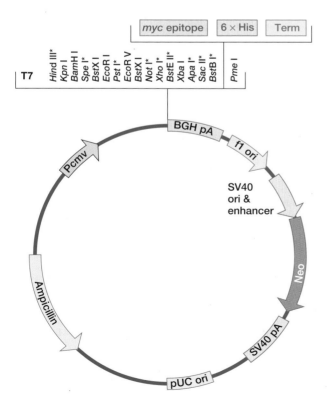

Figure 5.23: A mammalian expression vector, pcDNA3.1/*myc*-HIS.

The Invitrogen pcDNA series of plasmid expression vectors offer high level constitutive expression in mammalian cells. Cloned cDNA inserts can be transcribed from the strong cytomegalovirus (P_{CMV}) promoter (which ensures high level expression) with a bovine growth hormone polyadenylation sequence element (BGHpA) enabling generation of a defined 3′ end to the mRNA produced from the insert. A *neo* gene marker (regulated by a SV40 promoter/enhancer and poly A sequence) permits selection by growth in G418. The polylinker at top contains a multiple cloning site, followed by sequences specifying six consecutive histidine residues (6×His; which facilitates recombinant protein purification), a *myc* epitope tag (which allows antibody screening of the recombinant protein using an antibody specific for this sequence) and finally translation termination signals. Components for propagation in *E. coli* are indicated in yellow including a permissive origin of replication from plasmid ColE1 (pUCori), an ampicillin-resistance gene (Amp) and an f1 origin providing an option for producing single-stranded recombinants (Section 5.5.1).

transfected into COS cells, permanent cell lines do not result because massive vector replication makes the cells inviable. Even though only a low proportion of cells are successfully transfected, the amplification of the introduced DNA to high copy numbers in those cells compensates for the low take-up rate.

Stable expression in mammalian cells

Transgenes can stably integrate into host chromosomal DNA but the process is very inefficient and so the rare stably transformed cells must be isolated from the background of nontransformed cells by *selection* for some marker. Two broad approaches have been used:

▶ *functional complementation of mutant host cells*. The host cells are genetically deficient in some way but the original function can be restored by an *endogenous marker*. The transgene and the marker can be transferred as separate molecules by a process known as **co-transformation**. An example is the use of cells which are genetically deficient in thymidine kinase (Tk⁻) and a *Tk* gene marker. TK is required to convert thymidine into thymidine monophosphate (TMP) but TMP can also be synthesized by enzymatic conversion from dUMP. The drug *aminopterin* blocks the dUMP→TMP reaction and so in the presence

of this drug cells cannot grow unless they have a source of thymidine and a functional *Tk* gene. Selection for Tk⁺ cells is usually achieved in **HAT medium** (hypoxanthine, aminopterin, thymidine);

▶ *use of dominant selectable markers*. The major disadvantage of endogenous markers is that they can only be used with mutant host cell lines in which the corresponding gene is nonfunctional. As a result, endogenous markers have largely been superseded by **dominant selectable markers** which confer a phenotype that is entirely novel to the cell and hence can be used in any cell type. Markers of this type are usually drug-resistance genes of bacterial origin which can confer resistance to drugs that are known to affect eukaryotic and bacterial cells. For example, the aminoglycoside antibiotics (which include *neomycin* and *G418*) are inhibitors of protein synthesis in both bacterial and eukaryotic cells. The neomycin phosphotransferase (**neo**) gene confers resistance to neomycin, G418 etc. and so cells which have been transformed with the *neo* gene can be selected by growth in G418 etc. See *Figure 5.23* for an example of a mammalian expression vector with a *neo* marker.

Further reading

Colosimo A, Goncz KK, Holmes AR et al. (2000) Transfer and expression of foreign genes in mammalian cells. *Biotechniques* **29**, 314–331.

Higgins SJ, Hames BD (1999) *Protein Expression. A Practical Approach.* Oxford University Press, Oxford.

Ling MM, Robinson BH (1997) Approaches to DNA mutagenesis: an overview. *Anal. Biochem.* **254**, 157–178.

McPherson MJ, Møller SG (2000) *PCR: The Basics.* BIOS Scientific Publishers, Oxford.

Old RW, Twyman RM, Primrose SB (2001) *Principles of Gene Manipulation*, 6th Edn. Blackwell Scientific Publications Ltd, Oxford.

REBASE database of restriction nucleases at *http://rebase.neb.com/rebase/rebase.html*

Sambrook J, Russell D (2001) *Molecular Cloning: A Laboratory Manual*, 3rd Edn. Cold Spring Harbor Laboratory Press, Cold Spring Harbor, New York.

References

Baneyx F (1999) Recombinant protein expression in *E. coli. Curr. Opin. Biotechnol.* **10**, 411–421.

Bedwell DM, Strobel SA, Yun K, Jongeward GD, Emr SD (1989) Sequence and structural requirements of a mitochondrial protein import signal defined by saturation cassette mutagenesis. *Mol. Cell. Biol.* **9**, 1014–1025.

Cheng S, Fockler C, Barnes WM, Higuchi R (1994) Effective amplification of long targets from cloned inserts and human genomic DNA. *Proc. Natl Acad. Sci. USA* **91**, 5695–5699.

Clackson T, Wells JA (1994) In vitro selection from protein and peptide libraries. *Trends Biotechnol.* **12**, 173–184.

Cline J, Braman JC, Hogrefe HH (1996) PCR fidelity of Pfu DNA polymerase and other thermostable DNA polymerases. *Nucl. Acids Res.* **24**, 3546–3551.

Gluzman Y (1981) SV40-transformed simian cells support the replication of early SV40 mutants. *Cell* **23**, 175–182.

Higuchi R (1990) Recombinant PCR. In: *PCR Protocols. A Guide to Methods and Applications* (eds MA Innis, DH Gelfand, JJ Sninsky, TJ White). Academic Press, San Diego, pp. 177–183.

Iouannou PA, Amemiya CT, Garnes J, Kroisel PM, Shizuya H, Chen C, Batzer MA, de Jong P (1994) A new bacteriophage P1-derived vector for the propagation of large human DNA fragments. *Nature Genet.* **6**, 84–89.

Li HH, Gyllensten UB, Cui XF, Saiki RK, Erlich HA, Arnheim N (1988) Amplification and analysis of DNA sequences in single human sperm and diploid cells. *Nature* **335**, 414–417.

Newton CR, Graham A. (1997) PCR. 2nd Edn. Springer-Verlag, New York.

Parmley SF, Smith GP (1998) Antibody-selectable filamentous fd phage vectors: affinity purification of target genes. *Gene* **73**, 305–318.

Schlessinger D (1990) Yeast artificial chromosomes: tools for mapping and analysis of complex genomes. *Trends Genet.* **6**, 248–258.

Sternberg N (1992) Cloning high molecular weight DNA fragments by the bacteriophage P1 system. *Trends Genet.* **8**, 11–16.

Winter G, Griffiths AD, Hawkins RE, Hoogenboom HR (1994) Making antibodies by phage display technology. *Ann. Rev. Immunol.* **12**, 433–455.

Wurm F, Bernard A (1999) Large-scale transient expression in mammalian cells for recombinant protein production. *Curr. Opin. Biotechnol.* **10**, 156–159.

CHAPTER SIX

Nucleic acid hybridization: principles and applications

Chapter contents

Nucleic acid hybridization is a fundamental tool in molecular genetics which takes advantage of the ability of individual single-stranded nucleic acid molecules to form double-stranded molecules (that is, to **hybridize** to each other). For this to happen, the interacting single-stranded molecules must have a sufficiently high degree of *base complementarity*. Standard nucleic acid hybridization assays involve using a labeled nucleic acid *probe* to identify related DNA or RNA molecules (that is, ones with a significantly high degree of sequence similarity) within a complex mixture of unlabeled nucleic acid molecules, the *target* nucleic acid (**note:** *target* has a rather different usage in DNA cloning when it typically refers to specific DNA fragments which one intends to amplify by cloning).

6.1 Preparation of nucleic acid probes

In standard nucleic acid hybridization assays the probe is **labeled** in some way. Nucleic acid probes may be made as single-stranded or double-stranded molecules (see *Figure 6.1*), *but the working probe must be in the form of single strands.*

Conventional DNA probes are isolated by cell-based DNA cloning or via PCR. In both cases probes are usually double-stranded to begin with. DNA cloned within cells may range in size from 0.1 kb to hundreds of kilobases, but DNA cloned by PCR is usually less than 1 kb in length. The probes are usually labeled by incorporating labeled dNTPs during an *in vitro* DNA synthesis reaction.

RNA probes derive from single-stranded RNA molecules typically a few hundred bp to several kilobases long. They can conveniently be generated from DNA which has been cloned in a specialized plasmid vector containing a phage *promoter* sequence immediately adjacent to the multiple cloning site. An RNA synthesis reaction is performed using the relevant phage RNA polymerase and the four rNTPs, at least one of which is labeled. Specific labeled RNA transcripts can then be generated from the cloned insert.

Oligonucleotide probes are single-stranded and very short (typically 15–50 nucleotides long). They are made by chemical synthesis (unlike other probes which originate from DNA cloning). Mononucleotides are added, one at a time, to a starting mononucleotide, conventionally the 3′ end nucleotide, which is initially bound to a solid support. Generally, oligonucleotide probes are designed with a specific sequence chosen in response to prior information about the target DNA. Sometimes, however, **degenerate oligonucleotides** are used as probes, involving labeling of a collection of related oligonucleotides which are synthesized in parallel, having been designed to be identical at certain nucleotide positions but different at others. Oligonucleotide probes are often labeled by incorporating a ^{32}P atom or other labeled group at the 5′ end.

6.1.1 Nucleic acids can conveniently be labeled *in vitro* by incorporation of modified nucleotides

DNA and RNA can be labeled *in vivo*, by supplying labeled deoxynucleotides to tissue culture cells, but this procedure is of limited general use. A much more versatile method involves *in vitro* **labeling**: the purified DNA, RNA or oligonucleotide is labeled *in vitro* by using a suitable enzyme to incorporate

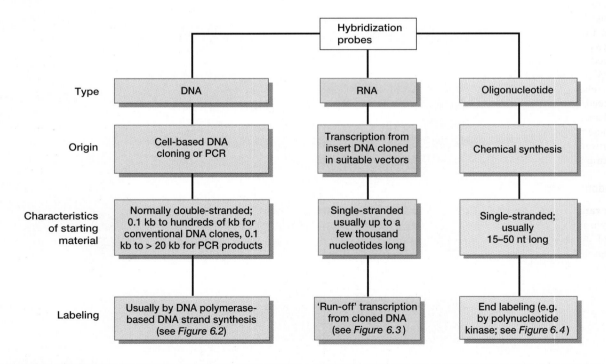

Figure 6.1: Origin and characteristics of nucleic acid hybridization probes.

labeled nucleotides. Two major types of procedure have been widely used:

▶ **strand synthesis labeling** – the standard labeling method, in which DNA or RNA polymerase is used to make labeled DNA or RNA copies of a starting DNA. The *in vitro* DNA or RNA synthesis reaction requires that at least one of the four nucleotide precursors carries a labeled group. Labeling of DNA is normally done by one of three methods: nick-translation; random primed labeling or PCR-mediated labeling. Labeling of RNA is achieved using an *in vitro* transcription system;

▶ **end-labeling** – a more specialized procedure in which a labeled group is added to one or a few terminal nucleotides only. End labeling is useful for labeling single-stranded oligonucleotides (see below) and in *restriction mapping*. Inevitably, because only one or a very few labeled groups are incorporated, the **specific activity** of the labeled nucleic acid (the amount of label incorporated divided by the total mass) is much less than that for probes in which there has been incorporation of multiple labeled nucleotides along the length of the molecule.

Labeling DNA by nick translation

The **nick-translation** procedure involves introducing single-strand breaks (*nicks*) in the DNA, leaving exposed 3′ hydroxyl termini and 5′ phosphate termini. The nicking can be achieved by adding a suitable endonuclease such as pancreatic DNase I. The exposed nick can then serve as a start point for introducing new nucleotides at the 3′ hydroxyl side of the nick using the DNA polymerase activity of *E. coli* DNA polymerase I at the same time as existing nucleotides are removed from the other side of the nick by the 5′→3′ exonuclease activity of the same enzyme. As a result, the nick will be moved progressively along the DNA ('translated') in the 5′→3′ direction (see *Figure 6.2A*). If the reaction is carried out at a relatively low temperature (about 15°C), the reaction proceeds no further than one complete renewal of the existing nucleotide sequence. Although there is no net DNA synthesis at these temperatures, the synthesis reaction allows the incorporation of labeled nucleotides in place of the previously existing unlabeled ones.

Random primed DNA labeling

The **random primed DNA labeling** method (sometimes known as **oligolabeling**; see Feinberg and Vogelstein, 1983) is based on hybridization of a mixture of all possible hexanucleotides: the starting DNA is denatured and then cooled slowly so that the individual hexanucleotides can bind to suitably complementary sequences within the DNA strands. Synthesis of new complementary DNA strands is primed by the bound hexanucleotides and is catalyzed by the **Klenow subunit** of *E. coli* DNA polymerase I (which contains the polymerase activity in the absence of an associated 5′→3′ exonuclease activity). DNA synthesis occurs in the presence of the four dNTPs, at least one of which has a labeled group (see *Figure 6.2B*). This method produces labeled DNAs of high

specific activity. Because all sequence combinations are represented in the hexanucleotide mixture, binding of primer to template DNA occurs in a random manner, and labeling is uniform across the length of the DNA.

DNA labeling by PCR

The standard PCR reaction can be modified to include one or more labeled nucleotide precursors which become incorporated into the PCR product throughout its length.

RNA labeling

RNA probes (**riboprobes**) can be obtained by *in vitro* transcription of insert DNA cloned in a suitable plasmid vector with a phage promoter. For example, the plasmid vector pSP64 contains the bacteriophage SP6 promoter sequence immediately adjacent to a multiple cloning site. The SP6 RNA polymerase can then be used to initiate transcription *from a specific start point* in the SP6 promoter sequence, transcribing through any DNA sequence that has been inserted into the multiple cloning site. By using a mix of NTPs, at least one of which is labeled, high specific activity radiolabeled transcripts can be generated (*Figure 6.3*). Bacteriophage T3 and T7 promoter/RNA polymerase systems are also used commonly for generating riboprobes. Labeled sense and antisense riboprobes can be generated from any gene cloned in such vectors (the gene can be cloned in either of the two orientations) and are widely used in tissue *in situ* hybridization (Section 6.3.4).

End-labeling

Single-stranded oligonucleotides are usually labeled using polynucleotide kinase (**kinase end-labeling**). Typically, the label is provided in the form of a ^{32}P at the γ-phosphate position of ATP and the polynucleotide catalyses an exchange reaction with the 5′-terminal phosphates (see *Figure 6.4A*). Larger DNA fragments can also be end-labeled, but often by alternative methods, including:

▶ **fill-in end-labeling** (*Figure 6.4B*) – the DNA is treated with a suitable restriction enzyme that generates an overhanging 5′ end and a polymerase activity is used to add labeled complementary nucleotides to 'fill in' the recessed ends. This is usually achieved using the **Klenow subunit** of *E. coli* DNA polymerase I (see above). As required, the labeled fragments can be cleaved internally with another restriction nuclease to produce two fragments labeled at one end each which can be size fractionated;

▶ **primer-mediated 5′ end-labeling** – a simple PCR method in which a primer is used with a labeled group attached to its 5′ end. As PCR proceeds the primer with its 5′ end-label is incorporated into the PCR product.

6.1.2 Nucleic acids can be labeled by isotopic and nonisotopic methods

Isotopic labeling and detection

Traditionally, nucleic acids have been labeled by incorporating nucleotides containing radioisotopes. Such **radiolabeled**

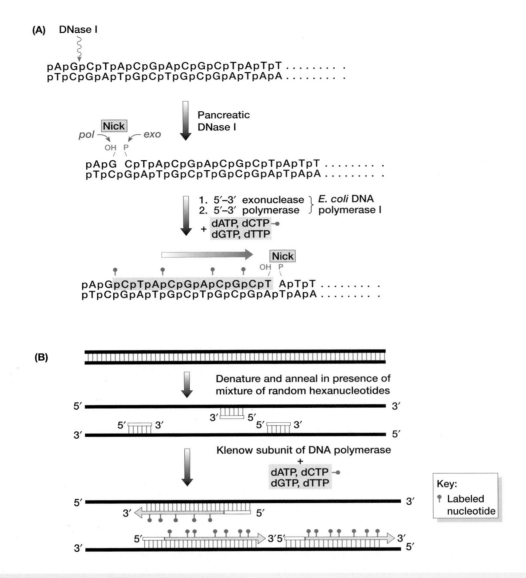

Figure 6.2: DNA labeling by *in vitro* DNA strand synthesis.

(A) *Nick translation*. Pancreatic DNase I introduces single-stranded nicks by cleaving internal phosphodiester bonds (p), generating a 5′ phosphate group and a 3′ hydroxyl terminus. Addition of the multisubunit enzyme *E. coli* DNA polymerase I contributes two enzyme activities: (i) a 5′→3′ exonuclease attacks the exposed 5′ termini of a nick and sequentially removes nucleotides in the 5′→3′ direction; (ii) a DNA polymerase adds new nucleotides to the exposed 3′ hydroxyl group, continuing in the 5′→3′ direction, thereby replacing nucleotides removed by the exonuclease and causing lateral displacement (translation) of the nick. **(B)** *Random primed labeling*. The Klenow subunit of *E. coli* DNA polymerase I can synthesize new radiolabeled DNA strands using as a template separated strands of DNA, and random hexanucleotide primers.

probes contain nucleotides with a radioisotope (often ^{32}P, ^{33}P, ^{35}S or ^{3}H), which can be detected specifically in solution or, much more commonly, within a solid specimen (**autoradiography;** see *Box 6.1*).

The intensity of an autoradiographic signal is dependent on the intensity of the radiation emitted by the radioisotope, and the time of exposure, which may often be long (1 or more days, or even weeks in some applications). ^{32}P has been used widely in nucleic acid hybridization assays because it emits high-energy β-particles which afford a high degree of sensitivity of detection. It has the disadvantage, however, that it is relatively unstable (see *Table 6.1*).

The high energy of ^{32}P β-particles emission can be a disadvantage under circumstances when fine physical resolution is required to interpret autoradiographic images unambiguously. As a result, radionuclides which emit less energetic β-particles

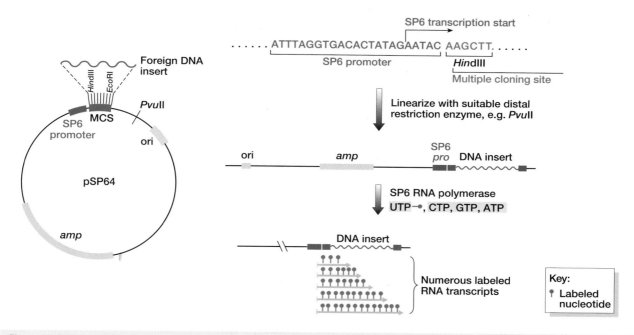

Figure 6.3: Riboprobes (RNA probes) are generated by run-off transcription from cloned DNA inserts in specialized plasmid vectors.

The plasmid vector pSP64 contains a promoter sequence for phage SP6 RNA polymerase linked to the multiple cloning site (MCS) in addition to an origin of replication (ori) and ampicillin resistance gene (*amp*). After cloning a suitable DNA fragment in one of the 11 unique restriction sites of the MCS, the purified recombinant DNA is linearized by cutting with a restriction enzyme at a unique restriction site just distal to the insert DNA (*Pvu* II in this example). Thereafter labeled insert-specific RNA transcripts can be generated using SP6 RNA polymerase and a cocktail of NTPs, at least one of which is labeled (UTP in this case).

Box 6.1: Principles of autoradiography.

Autoradiography is a procedure for localizing and recording a radiolabeled compound within a solid sample, which involves the production of an image in a photographic emulsion. In molecular genetic applications, the solid sample often consists of size-fractionated DNA or protein samples that are embedded within a dried gel, fixed to the surface of a dried nylon membrane or nitrocellulose filter, or located within fixed chromatin or tissue samples mounted on a glass slide. The photographic emulsions consist of silver halide crystals in suspension in a clear gelatinous phase. Following passage through the emulsion of a β-particle or a γ-ray emitted by a radionuclide, the Ag$^+$ ions are converted to Ag atoms. The resulting latent image can then be converted to a visible image once the image is **developed**, an amplification process in which entire silver halide crystals are reduced to give metallic silver. The **fixing** process results in removal of any unexposed silver halide crystals, giving an autoradiographic image which provides a two-dimensional representation of the distribution of the radiolabel in the original sample.

Direct autoradiography involves placing the sample in intimate contact with an X-ray film, a plastic sheet with a coating of photographic emulsion; the radioactive emissions from the sample produce dark areas on the developed film. This method is best suited to detection of weak to medium strength β-emitting radionuclides (e.g. ^{3}H, ^{35}S, etc.). However, it is not suited to high-energy β-particles (e.g. from ^{32}P): such emissions pass through the film, resulting in the wasting of the majority of the energy. **Indirect autoradiography** is a modification in which the emitted energy is converted to light by a suitable chemical (**scintillator** or **fluor**). One popular approach uses **intensifying screens**, sheets of a solid inorganic scintillator which are placed behind the film in the case of samples emitting high-energy radiation, such as ^{32}P. Those emissions which pass through the photographic emulsion are absorbed by the screen and converted to light. By effectively superimposing a photographic emission upon the direct autoradiographic emission, the image is intensified.

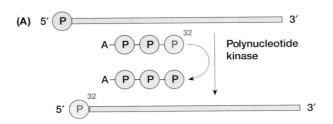

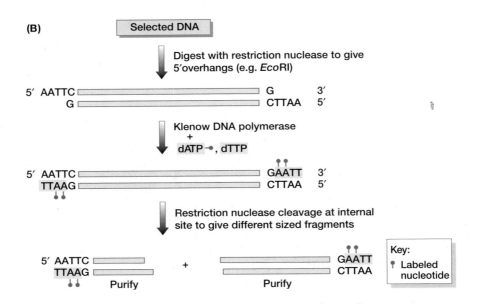

Figure 6.4: End-labeling of nucleic acids.

(A) *Kinase end-labeling of oligonucleotides.* The 5'-terminal phosphate of the oligonucleotide is replaced in an exchange reaction by the ^{32}P-labeled γ-phosphate of [γ-^{32}P]ATP. The same procedure can be used to label the two 5' termini of double-stranded DNA. (B) *Fill-in end-labeling.* The DNA of interest is cleaved with a suitable restriction nuclease to generate 5' overhangs. The overhangs act as a primer for Klenow DNA polymerase to incorporate labeled nucleotides complementary to the overhang. Fragments labeled at one end only can be generated by internal cleavage with a suitable restriction enzyme to generate two differently sized fragments which can easily be size-fractionated.

Table 6.1: Characteristics of radioisotopes commonly used for labeling DNA and RNA probes.

Radioisotope	Half-life	Decay type	Energy of emission
^{3}H	12.4 years	β$^-$	0.019 MeV
^{32}P	14.3 days	β$^-$	1.710 MeV
^{33}P	25.5 days	β$^-$	0.248 MeV
^{35}S	87.4 days	β$^-$	0.167 MeV

have been preferred in certain procedures, for example ^{35}S and ^{33}P in the case of DNA sequencing and tissue *in situ* hybridization and ^{3}H for chromosome *in situ* hybridization. ^{35}S and ^{33}P have moderate half-lives. ^{3}H has a very long half-life but is disadvantaged because of the comparatively low energy of the β-particles which it emits, necessitating very long exposure times.

^{32}P-labeled and ^{33}P-labeled nucleotides used in DNA strand synthesis labeling reactions have the radioisotope at the α-phosphate position, because the β- and γ-phosphates from dNTP precursors are not incorporated into the growing DNA chain. Kinase-mediated end-labeling, however, uses [γ-^{32}P]ATP (*Figure 6.4A*). In the case of ^{35}S-labeled nucleotides which are incorporated during the synthesis of DNA or RNA strands, the NTP or dNTP carries a ^{35}S isotope in place of the O$^-$ of the α-phosphate group. ^{3}H-labeled nucleotides carry the radioisotope at several positions. Specific detection of molecules carrying a radioisotope is most often performed by autoradiography (see *Box 6.1*).

Nonisotopic labeling and detection

Nonisotopic labeling systems involve the use of nonradioactive probes and are finding widespread applications in a variety of different areas (Kricka, 1992). Two major types of nonradioactive labeling are conducted:

▸ **direct nonisotopic labeling** – a nucleotide containing an attached labeled group is incorporated. Often such systems involve incorporation of modified nucleotides containing a **fluorophore**, a chemical group which can fluoresce when exposed to light of a certain wavelength (see *Figure 6.5* and *Box 6.2*).

▸ **indirect nonisotopic labeling** usually features chemical coupling of a modified **reporter molecule** to a nucleotide precursor. After incorporation into DNA, the reporter groups can be specifically bound by an **affinity molecule**, a protein or other ligand which has a very high affinity for the reporter group. Conjugated to the affinity molecule is a **marker** molecule or group which can be detected in a suitable assay (*Figure 6.6*). The reporter molecules on modified nucleotides need to protrude sufficiently far from the nucleic acid backbone so as to facilitate their detection by the affinity molecule and so a long carbon atom **spacer** is required to separate the nucleotide from the reporter group.

Two indirect nonisotopic labeling systems are widely used:

▸ the **biotin–streptavidin** system utilizes the extremely high affinity of two ligands: **biotin** (a naturally occurring vitamin) which acts as the reporter, and the bacterial protein **streptavidin**, which is the affinity molecule. Biotin and streptavidin bind together extremely tightly with an affinity constant of 10^{-14}, one of the strongest known in biology. Biotinylated probes can be made easily by including a suitable biotinylated nucleotide in the labeling reaction (see *Figure 6.7*).

▸ **digoxigenin** is a plant steroid (obtained from *Digitalis* plants) to which a specific antibody has been raised. The digoxigenin-specific antibody permits detection of nucleic acid molecules which have incorporated nucleotides containing the digoxigenin reporter group (see *Figure 6.7*).

A variety of different marker groups or molecules can be conjugated to affinity molecules such as streptavidin or the digoxigenin-specific antibody. They include various **fluorophores** (*Box 6.2*), or enzymes such as alkaline phosphatase and peroxidase which can permit detection via colorimetric assays or chemical luminescence assays etc.

6.2 Principles of nucleic acid hybridization

Nucleic acid hybridization is a fundamental technique in molecular genetics. A glossary of relevant terms is provided in *Box 6.3* to assist readers who may be unfamiliar with the terminology.

6.2.1 Nucleic acid hybridization is a method for identifying closely related molecules within two nucleic acid populations

Definition and rationale

The usual purpose of **nucleic acid hybridization** is to gain some information about an *imperfectly understood* and typically complex population of nucleic acids (the **target**). This is achieved by using a *known* population of nucleic acid molecules as a **probe** in order to identify closely related nucleic acids within the target.

The *specificity* of the interaction between probe and target comes from **base complementarity**, because both populations are treated in such a way as to ensure that all the nucleic acids sequences present are *single stranded*. Thus, if either the probe or the target is initially double-stranded, the individual strands must be separated (**denatured**), generally by heating or by alkaline treatment. After mixing single strands of probe with single strands of target, strands with complementary base sequences are allowed to **reassociate** (= **reanneal**) to form double-stranded nucleic acids. When this happens two types of product are formed:

▸ **homoduplexes** – complementary strands within the probe, or within the target reanneal to regenerate double-stranded molecules *originally found* in the probe or target populations;

▸ **heteroduplexes** – a single-stranded nucleic acid within the probe base-pairs with a complementary strand in the target population. The double-stranded nucleic acid formed here is a *new* combination of nucleic acid strands. Heteroduplexes define the *usefulness* of a nucleic acid hybridization assay because the whole point of the hybridization assay is to use the known probe to identify related nucleic acid fragments in the target (*Figure 6.8*).

Probes are often homogeneous nucleic acid populations (e.g. a specific cloned DNA or chemically synthesized oligonucleotide) and are usually labeled and in solution. By

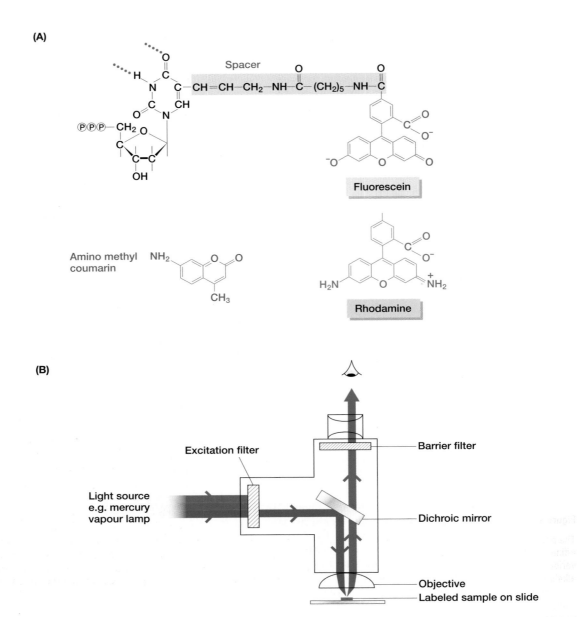

Figure 6.5: Fluorescence microscopy and structure of common fluorophores.

(A) *Structure of fluorophores*. The example on top shows fluorescein–dUTP. The fluorescein group is linked to the 5′ carbon atom of the uridine by a spacer group so that when the modified nucleotide is incorporated into DNA, the fluorescein group is readily accessible. Below is the structure of rhodamine from which a variety of fluorophores has been derived. **(B)** *Fluorescence microscopy*. The excitation filter is a color barrier filter which in this example is selected to let through only blue light. The transmitted blue light is of an appropriate wavelength to be reflected by the dichroic (beam-splitting) mirror onto the labeled sample which then fluoresces and emits light of a longer wavelength, green light in this case. The longer wavelength of the emitted green light means that it passes straight through the dichroic mirror. The light subsequently passes through a second color barrier filter which blocks unwanted fluorescent signals leaving the desired green fluorescence emission to pass through to the eyepiece of the microscope. A second beam-splitting device can also permit the light to be recorded in a CCD camera.

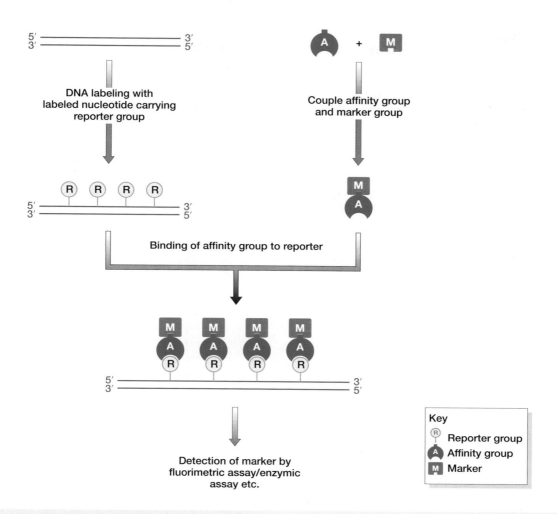

Figure 6.6: General principles of indirect nonisotopic labeling.

The protein recognizing the reporter group is often a specific antibody, as in the digoxigenin system, or any other ligand that has a very high affinity for a specific group, such as streptavidin in the case of using biotin as the reporter (see *Figure 6.7*). The marker can be detected in various ways. If it carries a specific fluorescent dye, it can be detected in a fluorimetric assay. Alternatively, it can be an enzyme such as alkaline phosphatase which can be coupled to an enzyme assay, yielding a product that can be measured colorimetrically.

contrast, target nucleic acid populations are typically unlabeled complex populations and are usually bound to a solid support. However, some important hybridization assays use unlabeled probes bound to a solid support which are used to interrogate labeled target DNA populations in solution.

Melting temperature and hybridization stringency

Denaturation of double-stranded probe DNA is generally achieved by heating a solution of the labeled DNA to a temperature that is high enough to break the hydrogen bonds holding the two complementary DNA strands together. The energy required to separate two perfectly complementary DNA strands is dependent on a number of factors, notably:

▶ **strand length** – long homoduplexes contain a large number of hydrogen bonds and require more energy to separate them; because the labeling procedure typically results in

short DNA probes, this effect is negligible above an original (i.e. prior to labeling) length of 500 bp;

▶ **base composition** – GC base pairs have one more hydrogen bond than AT base pairs. Strands with a high % GC composition are thus more difficult to separate than those with a low % GC composition;

▶ **chemical environment** – the presence of monovalent cations (e.g. Na^+ ions) stabilizes the duplex, whereas some strongly polar molecules such as **formamide** ($H-CO-NH_3^+$) and **urea** ($H-_3N^+-CO-NH_3^+$) act as chemical *denaturants*: they destabilize the duplex by disrupting the hydrogen bonds between base pairs.

A useful measure of the stability of a nucleic acid duplex is the **melting temperature (T_m)**. This is the temperature corresponding to the mid-point in the observed transition from double-stranded form to single-stranded form. Conveniently,

Box 6.2: Fluorescence labeling and detection systems.

Fluorescence labeling of nucleic acids was developed in the 1980s and has proved to be extremely valuable in many different applications including chromosome *in situ* hybridization, tissue *in situ* hybridization and automated DNA sequencing. The fluorescent labels can be used in direct labeling of nucleic acids by incorporating a modified nucleotide (often 2′ deoxyuridine 5′ triphosphate) containing an appropriate **fluorophore**, a chemical group which fluoresces when exposed to a specific wavelength of light. Popular fluorophores used in direct labeling include: *fluorescein*, a pale green fluorescent dye; *rhodamine*, a red fluorescent dye; and *amino methyl coumarin*, a blue fluorescent dye (see *Figure 6.5A*). Alternatively, indirect labeling systems can be used whereby modified nucleotides containing a reporter group (such as biotin or digoxigenin) are incorporated into the nucleic acid (see *Figure 6.6*) and then the reporter group is specifically bound by an affinity molecule (such as streptavidin or a digoxigenin-specific antibody) to which is attached a fluorophore e.g. *amino methylcoumarin acetic acid* (*AMCA*), *fluorescein isothiocyanate* (*FITC*) or other fluorescein derivatives, and *tetramethylrhodamine isothiocyanate* (*TRITC*) or other rhodamine derivatives.

Detection of the fluorophore in direct or indirect labeling systems is accomplished by passing a beam of light from a suitable light source (e.g. a mercury vapor lamp in fluorescence microscopy,

an argon laser in automated DNA sequencing) through an appropriate color filter (*excitation filter*) designed to transmit light at the desired **excitation wavelength**. In fluorescence microscopy systems this light is reflected onto the fluorescently labeled sample on a microscope slide using a *dichroic mirror,* one which reflects light of certain wavelengths while allowing light of other wavelengths to pass straight through. The light then excites the fluorophore to fluoresce and as it does so, it emits light at a slightly longer wavelength, the **emission wavelength**. The light emitted by the fluorophore passes back up and straight through the dichroic mirror, through an appropriate barrier filter and is then transmitted into the eye piece of the microscope, and can also be captured using a suitable CCD (*charged coupled device*) camera. Maximum emission and excitation wavelengths for common fluorophores are as indicated below:

Fluorophore	Color	Excitation max. (nm)	Emission max. (nm)
AMCA	Blue	399	446
Fluorescein	Green	494	523
CY3	Red	552	565
Rhodamine	Red	555	580
Texas Red	Red	590	615

this transition can be followed by measuring the **optical density (OD)** of the DNA. The bases of the nucleic acids absorb 260 nm ultraviolet (UV) light strongly. However, the adsorption by double-stranded DNA is considerably less than that of the free nucleotides. This difference, the so-called **hypochromic effect**, is due to interactions between the electron systems of adjacent bases, arising from the way in which adjacent bases are stacked in parallel in a double helix. If duplex DNA is gradually heated, therefore, there will be an increase in the light absorbed at 260 nm (the OD_{260}) towards the value characteristic of the free bases. The temperature at which there is a mid-point in the optical density shift is then taken as the T_m (see *Figure 6.9*).

For mammalian genomes, with a base composition of about 40% GC, the DNA denatures with a T_m of about 87°C under approximately physiological conditions. The T_m of perfect hybrids formed by DNA, RNA or oligonucleotide probes can be determined according to the formulae in *Table 6.2*. Often, hybridization conditions are chosen so as to promote heteroduplex formation and the hybridization temperature is often as much as 25°C below the T_m. However, after the hybridization and removal of excess probe, hybridization washes may be conducted under more stringent conditions so as to disrupt all duplexes other than those between very closely related sequences.

Probe–target heteroduplexes are most stable thermo-dynamically when the region of duplex formation contains perfect base matching. Mismatches between the two strands of a heteroduplex reduce the T_m: for normal DNA probes, each

1% of mismatching reduces the T_m by approximately 1°C. Although probe–target heteroduplexes are usually not as stable as reannealed probe homoduplexes, a considerable degree of mismatching can be tolerated if the overall region of base complementarity is long (> 100 bp; see *Figure 6.10*).

Increasing the concentration of NaCl and reducing the temperature reduces the **hybridization stringency**, and enhances the stability of mismatched heteroduplexes. This means that comparatively diverged members of a multigene family or other repetitive DNA family can be identified by hybridization using a specific family member as a probe. Additionally, a gene sequence from one species can be used as a probe to identify homologs in other comparatively diverged species, provided the sequence is reasonably conserved during evolution.

Conditions can also be chosen to maximize hybridization stringency (e.g. lowering the concentration of NaCl and increasing the temperature), so as to encourage dissociation (denaturation) of mismatched heteroduplexes. If the region of base complementarity is small, as with oligonucleotide probes (typically 15–20 nucleotides) hybridization conditions can be chosen such that a single mismatch renders a heteroduplex unstable (see Section 6.3.1).

6.2.2 The kinetics of DNA reassociation are defined by the product of DNA concentration and time (C_0t)

When double-stranded DNA is denatured (by heat, for example) and the complementary single strands are allowed to

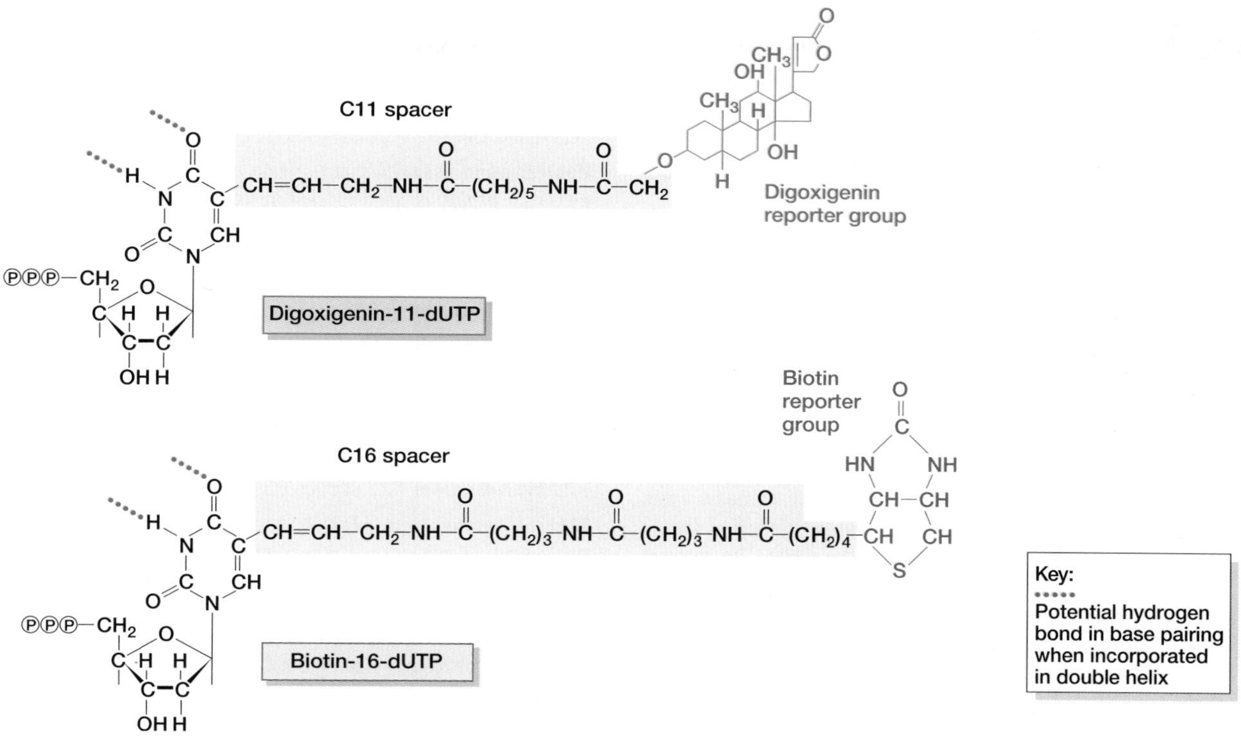

Figure 6.7: Structure of digoxigenin- and biotin-modified nucleotides.

Note that the digoxigenin and biotin groups in these examples are linked to the 5′ carbon atom of the uridine of dUTP by spacer groups consisting respectively of a total of 11 carbon atoms (digoxigenin-11-UTP) or 16 carbon atoms (biotin-16-dUTP). The digoxigenin and biotin groups are ***reporter groups***: after incorporation into a nucleic acid they are bound by specific ligands containing an attached marker such as a fluorophore.

reassociate to form double-stranded DNA, the speed at which the complementary strands reassociate will depend on the starting concentration of the DNA. If there is a high concentration of the complementary DNA sequences, the time taken for any one single-stranded DNA molecule to find a complementary strand and form a duplex will be reduced. **Reassociation kinetics** is the term used to measure the speed at which complementary single-stranded molecules are able to find each other and form duplexes. There are two major parameters: the starting concentration (C_o) of the specific DNA sequence in moles of nucleotides per liter and the reaction time (t) in seconds. Since the rate of reassociation is proportional to C_o and to t, the $C_o t$ value (often loosely referred to as the **cot value**) is a useful measure. The $C_o t$ value will also vary depending on the temperature of reassociation and the concentration of monovalent cations. As a result, it is usual to use fixed reference values: a reassociation temperature of 65°C and an [Na$^+$] concentration of 0.3 M NaCl.

Most hybridization assays use an excess of target nucleic acid over probe in order to encourage probe–target formation. This is so because the probe is usually *homogenous*, often consisting of a single type of cloned DNA molecule or

RNA molecule, but the target nucleic acid is typically *heterogeneous*, comprising for example genomic DNA or total cellular RNA. In the latter case the concentration of any one sequence may be very low, thereby causing the rate of reassociation to be slow. For example, if a Southern blot uses a cloned β-globin gene as a probe to identify complementary sequences in human genomic DNA, the latter will be present in very low concentration (the β-globin gene is an example of a **single copy** sequence and in this case represents only 0.00005% of human genomic DNA). It is therefore necessary to use several micrograms of target DNA to drive the reaction. By contrast, certain other sequences are highly repeated in genomic DNA (see Chapter 9), and this greatly elevated DNA concentration results in a comparatively rapid reassociation time.

Because the amount of target nucleic acid bound by a probe depends on the copy number of the recognized sequence, hybridization signal intensity is proportional to the copy number of the recognized sequence: single copy genes give weak hybridization signals, highly repetitive DNA sequences give very strong signals. If a particular probe is heterogeneous and contains a low copy sequence of interest (such as a specific gene), mixed with a highly

Box 6.3: A glossary of nucleic acid hybridization (for individual methods see *Box 6.4*).

Allele-specific oligonucleotide (ASO) hybridization. Uses short ca. 20 nucleotide-long oligonucleotide probes specific for individual alleles under *stringent* hybridization conditions where a single base mismatch precludes successful hybridization. See *Figure 6.11*.

Anneal. If two single-stranded nucleic acids share sufficient *base complementarity* they will form a double-stranded DNA duplex. To anneal means to allow hydrogen bonds to form between two single strands and so has the opposite meaning to that of *denature*.

Base complementarity. The degree to which the sequences of two single-stranded nucleic acids can form a DNA duplex by Watson–Crick base pairing (Section 1.2.1).

Competition hybridization (= suppression hybridization). Hybridization reaction in which a probe which contains some repetitive DNA sequences undergoes a *prehybridization step* to block access to repetitive sequences in the probe. The probe is first denatured and allowed to reassociate in the presence of an unlabeled DNA population that has been enriched in repetitive DNA: as a result the repetitive elements within the probe are effectively removed by annealing to complementary repetitive DNA sequences, leaving available only the nonrepetitive sequences.

Denature. To separate the individual strands of a double-stranded DNA duplex by breaking the hydrogen bonds between them (by heating them or by treating them with a chemical denaturant, such as formamide).

Fluorescence *in situ* hybridization (FISH). Any *in situ hybridization* reaction in which nucleic acids are labeled by attaching chemical groups which can fluoresce under certain wavelengths.

Heteroduplex. A double-stranded DNA formed when two single-stranded sequences with partial base complementarity are allowed to anneal. Heteroduplexes can arise in different ways:

▶ *allelic heteroduplexes*. Denaturing then cooling a *single* diploid DNA sample, or a mix of DNA samples from *different individuals* will allow complementary strands from two alleles which differ slightly in DNA sequence to form almost perfectly matched duplexes;

▶ *paralogous heteroduplexes*. Denaturing then cooling a single DNA sample from a complex eukaryotic cell can also give rise to reassociation between *nonallelic related DNA sequences* such as closely related different members of a gene family or of a noncoding repetitive DNA family;

▶ *interspecific heteroduplexes*. Denatured DNA samples from species with reasonably closely related genes, e.g. humans and mice, are mixed and single strands allowed to reanneal.

Homoduplex. A double-stranded DNA formed when two single-stranded sequences with perfect base complementarity are allowed to anneal.

Hybridization assay. A test reaction in which a known nucleic acid (the *probe*) is used to search for related sequences in a heterogeneous collection of DNA sequences (the *target*). In a standard hybridization assay the probe is labeled and usually in solution while the target is unlabeled and usually bound to a solid support; in reverse hybridization the opposite is true.

***In situ* hybridization.** A hybridization reaction in which a probe is hybridized to nucleic acids from cells or chromosomes which have been mounted on a glass slide, e.g. denatured DNA from metaphase chromosomes (*chromosome in situ hybridization*; Section 2.4.2) or RNA prepared from sectioned tissue (*tissue in situ hybridization*). See Section 6.3.4.

Melting temperature (T_m). A measure of the stability of a nucleic acid duplex, it is the temperature corresponding to the mid-point in the observed transition from double-stranded to single-stranded forms. Conveniently, this transition can be followed by measuring the optical density of the DNA at a wavelength of 260 nm. See *Figure 6.9*.

Probe. A *known* nucleic acid population used to query a complex heterogeneous nucleic acid population in a hybridization assay. May originally be double stranded but to work as a probe has to be denatured to give single strands. Usually the probe is labeled and in solution, but see *reverse hybridization* assays.

Reassociate. To re-anneal after a prior denaturation step.

Reassociation kinetics. The rates at which complementary DNA strands reassociate. Highly repetitive DNA strands reassociate rapidly; single copy sequences reassociate slowly.

Reverse hybridization assay. An assay in which the target is labeled and usually in solution while the probe is unlabeled and usually bound to a solid support. Examples include reverse dot-blot assays and microarray hybridization.

Subtraction hybridization. A hybridization-based method of identifying nucleic acid sequences known to be present in one population (the *plus* population) but absent in another closely related population (the *minus* population). Hybridization is designed to have the (–) population in vast excess and so act as the *driver DNA*, ensuring that all related sequences in the (+) population are removed as heteroduplexes, leaving the desired sequences that are found only in the (+) population.

Target. A complex heterogeneous nucleic acid population which is queried in a hybridization assay by a known, often homogeneous, nucleic acid. Usually unlabeled and bound to a solid support but in *reverse hybridization* assays targets are labeled and in solution.

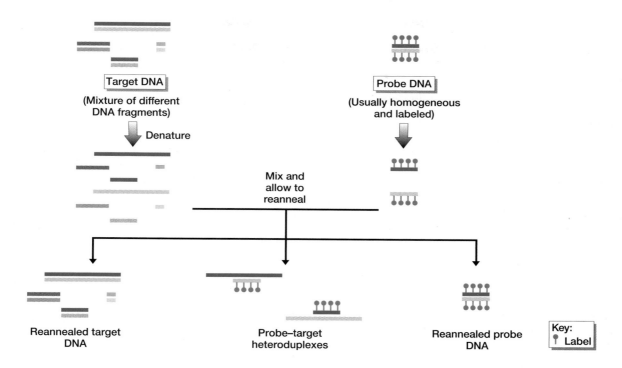

Figure 6.8: A nucleic acid hybridization assay requires the formation of heteroduplexes between labeled single-stranded nucleic acid probes and complementary sequences within a target nucleic acid.

The probe is envisaged to be strongly related in sequence to a central segment of one of the many types of nucleic acid molecule in the target. Mixing of denatured probe and denatured target will result in reannealed probe–probe homoduplexes (bottom right) and target–target homoduplexes (bottom left) but also in **heteroduplexes** formed between probe DNA and any target DNA molecules that are significantly related in sequence (bottom center). If a method is available for removing the probe DNA that is not bound to target DNA, the heteroduplexes can easily be identified by methods that can detect the label.

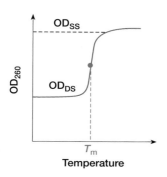

Figure 6.9: Denaturation of DNA results in an increase in optical density.

OD_{SS} and OD_{DS} indicate the optical density of single-stranded and double-stranded DNA respectively. The difference between them represents the hypochromic effect (see text).

abundant repetitive DNA sequence, the weak hybridization signal obtained with low copy sequence will be completely masked by the strong signal from the repetitive DNA. This effect can, however, be overcome by *competition hybridization* (see *Box 6.3*).

6.2.3 A wide variety of nucleic acid hybridization assays can be used

Early experiments in nucleic acid hybridization utilized *solution hybridization*, involving mixing of aqueous solutions of probe and target nucleic acids. However, the very low concentration of single copy sequences in complex genomes meant that reassociation times were inevitably slow. One widely used way of increasing the reassociation speed is to artificially increase the overall DNA concentration in aqueous solution by abstracting water molecules (e.g. by adding high concentrations of *polyethylene glycol*).

An alternative to solution hybridization which facilitated detection of reassociated molecules involved immobilizing the target DNA on a solid support, such as a membrane made of nitrocellulose or nylon, to both of which single-stranded DNA binds readily. Attachment of labeled probe to the immobilized target DNA can then be followed by removing the

Table 6.2: Equations for calculating T_m.

Hybrids	T_m (°C)
DNA–DNA	$81.5 + 16.6(\log_{10}[Na^+]^a) + 0.41(\%GC^b) - 500/L^c$
DNA–RNA or RNA–RNA	$79.8 + 18.5(\log_{10}[Na^+]^a) + 0.58(\%GC^b) + 11.8\,(\%GC^b)^2 - 820/L^c$
oligo–DNA or oligo–RNA[d]:	
For < 20 nucleotides	$2\,(l_n)$
For 20–35 nucleotides	$22 + 1.46\,(l_n)$

[a] Or for other monovalent cation, but only accurate in the 0.01–0.4 M range.
[b] Only accurate for 30–75% GC.
[c] L = length of duplex in base pairs.
[d] oligo = oligonucleotide: l_n = effective length of primer = 2 × (no. of G + C) + (no. of A + T).

Note: for each 1% formamide, the T_m is reduced by about 0.6°C, while the presence of 6 M urea reduces the T_m by about 30°C.

solution containing unbound probe DNA, extensive washing of the membrane and drying in preparation for detection.

This is the basis of the *standard* nucleic acid hybridization assays currently in use. More recently, however, **reverse hybridization assays** have also become popular. In these cases, the probe population is unlabeled and fixed to the solid support, while the target nucleic acid is labeled and present in aqueous solution. Note, therefore, that the distinction between probe and target is not based primarily on which is the labeled population and which is the unlabeled population. Instead, the important consideration is that the target DNA should be the complex, *imperfectly understood* population which the probe (whose molecular identity is known) attempts to query. Depending on the nature and form of the probe and target, a very wide variety of nucleic acid hybridization assays can be devised (*Box 6.4*).

6.3 Nucleic acid hybridization assays using cloned DNA probes to screen uncloned nucleic acid populations

Numerous applications in molecular genetics involve taking an individual DNA clone and using it as a hybridization probe to screen for the presence of related sequences within a complex target of uncloned DNA or RNA. Sometimes the assay is restricted to simply checking for the presence or absence of sequences related to the probe. In other cases, useful information can be obtained regarding the size of the complementary sequences, their subchromosomal location or their locations within specific tissues or groups of cells.

6.3.1 Dot-blot hybridization, a rapid screening method, often employs allele-specific oligonucleotide probes

The general procedure of **dot-blotting** involves taking an aqueous solution of target DNA, for example total human genomic DNA, and simply spotting it on to a nitrocellulose or nylon membrane then allowing it to dry. The variant technique of **slot-blotting** involves pipeting the DNA through an individual slot in a suitable template. In both methods the target DNA sequences are denatured, either by previously exposing to heat, or by exposure of the filter containing them to alkali. The denatured target DNA sequences now

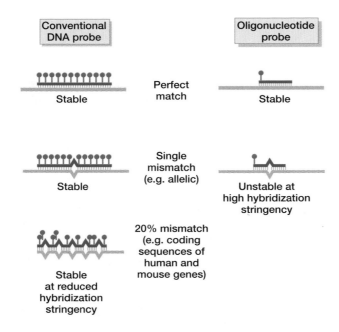

Figure 6.10: Nucleic acid hybridization can identify target sequences considerably diverged from a conventional nucleic acid probe, or identical to an oligonucleotide probe.

Box 6.4: Standard and reverse nucleic acid hybridization assays.

STANDARD ASSAYS	Labeled probe in solution	Unlabeled target bound to solid support
Dot blot (*Figure 6.11*)	Any labeled DNA or RNA but often an oligonucleotide	Complex DNA or RNA population; not size-fractionated, but spotted directly onto membrane
Southern blot (*Figure 6.12 for method; Figures 7.9, 18.12B, 18.12C for examples*)	Any	Often complex genomic DNA (but may be individual DNA clones); digested with restriction nuclease and size fractionated, then transferred to membrane
Northern blot (*Figure 6.13*)	Any	Complex RNA population (e.g. total cellular RNA or poly A$^+$ RNA) which has been size-fractionated, then transferred to membrane
Chromosome *in situ* hybridization (*e.g. Figures 2.16–2.18, 14.12.*)	Usually a labeled genomic clone	DNA within chromosomes (often metaphase) of lysed cells on a microscope slide
Tissue *in situ* hybridization (*Figure 6.15*)	Usually a labeled antisense riboprobe or oligonucleotide	RNA within cells of fixed tissue sections on a microscope slide
Colony blot (*Figure 6.16*)	Any	Cell colonies separated after plating out on agar then transferred to membrane
Plaque lift	Any	Phage-infected bacterial colonies separated after plating out on agar and transferred to membrane
Gridded clone hybridization assay (*Figure 6.17*)	Any	Clones robotically spotted onto membrane in geometric arrays

REVERSE ASSAYS	Labeled target in solution	Unlabeled probes bound to solid support
Reverse dot blot	Complex DNA	Oligonucleotides spotted onto membrane
DNA microarray or microarray of presynthesized oligonucleotides (*Figure 6.19*)	Complex DNA	DNA clones or oligonucleotides robotically spotted onto microscope slide
Oligonucleotide microarray (*Figures 17.18*)	Complex DNA	Oligonucleotides synthesized on glass

immobilized on the membrane are exposed to a solution containing single-stranded labeled probe sequences (the label is often ^{32}P to optimize detection). After allowing sufficient time for probe–target heteroduplex formation, the probe solution is decanted, and the membrane is washed to remove excess probe that may have become nonspecifically bound to the filter. It is then dried and exposed to an autoradiographic film.

A useful application of dot-blotting involves distinguishing between alleles that differ by even a single nucleotide substitution. To do this **allele-specific oligonucleotide (ASO)** probes are constructed from sequences spanning the variant nucleotide site. ASO probes are typically 15–20 nucleotides long and are normally used under hybridization conditions at which the DNA duplex between probe and target is stable *only if there is perfect base complementarity between them*: a single mismatch between probe and target sequence is sufficient to render the short heteroduplex unstable (*Figure 6.10*). Typically, this involves designing the oligonucleotides so that the single nucleotide difference between alleles occurs in a central segment of the oligonucleotide sequence, thereby maximizing the

thermodynamic instability of a mismatched duplex. Such discrimination can be employed for a variety of research and diagnostic purposes. Although ASOs can be used in conventional Southern blot hybridization (see below), it is more convenient to use them in dot-blot assays (see *Figure 6.11*).

Another method of ASO dot blotting uses a **reverse dot blotting** approach. This means that the oligonucleotide probes are not labeled and are fixed on a filter or membrane whereas the target DNA is labeled and provided in solution. Positive binding of labeled target DNA to a specific oligonucleotide on the membrane is taken to mean that the target has that specific sequence. This approach, and related DNA microarray methods, have many diagnostic applications.

6.3.2 Southern and Northern blot hybridizations detect nucleic acids that have been size-fractionated by gel electrophoresis

Southern blot hybridization

In this procedure, the target DNA is digested with one or more restriction endonucleases whose recognition sequences

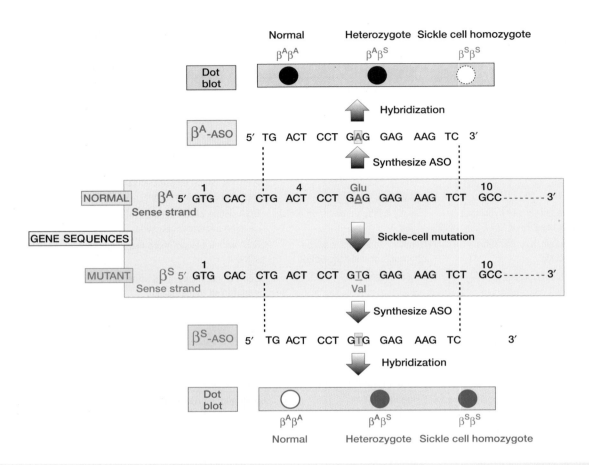

Figure 6.11: Allele-specific oligonucleotide (ASO) dot-blot hybridization can identify individuals with the sickle cell mutation.

The schematic dot blot at the top shows the result of probing with an ASO specific for the normal β-globin allele (β^A-ASO; shown immediately below). The results are positive (filled circle) for normal individuals and for heterozygotes but negative for sickle cell homozygotes (dashed, unfilled circle). The dot blot at the bottom shows the result of probing with an ASO specific for the sickle cell β-globin allele (β^S-ASO; shown immediately above), and in this case the results are positive for the sickle cell homozygotes and heterozygotes but negative for normal individuals. The β^A-ASO and β^S-ASO were designed to be 19 nucleotides long in this case chosen from codons 3 to 9 of respectively the sense β^A and β^S globin gene sequences surrounding the sickle cell mutation site. The latter is a single nucleotide substitution (A→T) at codon 6 in the β-globin gene, resulting in a G<u>A</u>G (Glu)→G<u>T</u>G (Val) substitution (see middle sequences).

are often 4–6 bp in length, generating fragments that are several hundred or thousands of base pairs in length. The restriction fragments are size-fractionated by agarose gel electrophoresis, denatured and transferred to a nitrocellulose or nylon membrane for hybridization (*Figure 6.12*). During the electrophoresis DNA fragments, which are negatively charged because of the phosphate groups, are repelled from the negative electrode towards the positive electrode, and sieve through the porous gel. Smaller DNA fragments move faster. For fragments between 0.1 and 30 kb long, the migration speed depends on fragment length, but scarcely at all on the base composition. Thus, fragments in this size range are fractionated by size in a conventional agarose gel electrophoresis system.

An important application of Southern blot hybridization in mammalian genetics is to use a probe to identify related sequences which may belong to the *same genome* (other members of a family of evolutionarily related genes or DNA sequences) or to *other genomes* (e.g. an *orthologous* gene, that is a direct equivalent of the gene used as a probe). Once a newly isolated probe is demonstrated to be related to other uncharacterized sequences, attempts can then be made to isolate the other members of the family by screening *genomic DNA libraries* (Section 5.3.4). Additionally, screening can also be conducted on genomic DNA samples from different species (a **zooblot**) to identify sequences conserved between different species (see *Figure 7.9*). Because coding sequences are comparatively highly conserved, this is one way of identifying coding DNA.

Northern blot hybridization

Northern blot hybridization is a variant of Southern blotting in which the target nucleic acid is undigested RNA instead of

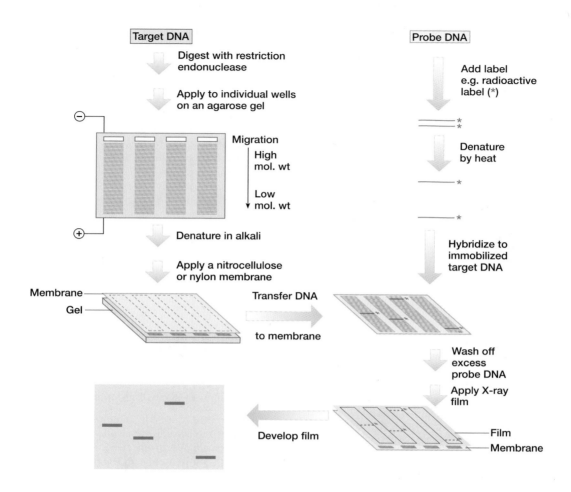

Figure 6.12: Southern blot hybridization detects target DNA fragments that have been size-fractionated by gel electrophoresis.

DNA. A principal use of this method is to obtain information on the expression patterns of specific genes. Once a gene has been cloned, it can be used as a probe and hybridized against a Northern blot containing, in different lanes, samples of RNA isolated from a variety of different tissues (see *Figure 6.13*). The data obtained can provide information on the range of cell types in which the gene is expressed, and the relative abundance of transcripts. Additionally, by revealing transcripts of different sizes, it may provide evidence for different isoforms (e.g. resulting from alternative promoters, splice sites, polyadenylation sites etc.).

6.3.3 Pulsed field gel electrophoresis extends Southern hybridization to include detection of very large DNA molecules

Standard agarose gel electrophoresis can resolve DNA fragments in a limited size range, from 100 bp to about 30 kb as a result of a *sieving effect*: the DNA molecules pass through pores in the agarose gel and small molecules are able to migrate more quickly through the pores. Above a certain size of DNA fragment, however, the sieving effect is no longer effective and the resolution of DNA fragments above 40 kb is extremely limited. Because many mammalian genes and other functional sequence units are very large, an alternative electrophoresis method is required for separating very large fragments of DNA.

Pulsed field gel electrophoresis (PFGE) is a more recent modification of agarose gel electrophoresis which can resolve DNA fragments in a size range from about 20 kb to several Mb in length. The very large DNA molecules contained in mammalian chromosomes – typically hundreds of Mb in length – cannot be size-separated by this method but specialized restriction nucleases are known to cleave vertebrate DNA rather infrequently producing large restriction fragments which can be size separated by PFGE. These enzymes, sometimes known as *rare-cutter restriction endonucleases* often recognize GC-rich recognition sequences which contain one or more CpG dinucleotides. Because the CpG dinucleotide occurs at low frequencies in vertebrate DNA, human DNA and other vertebrate DNAs have

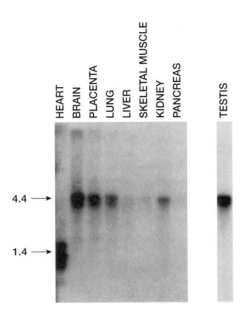

HEART BRAIN PLACENTA LUNG LIVER SKELETAL MUSCLE KIDNEY PANCREAS TESTIS

4.4 →

1.4 →

Figure 6.13: Northern blot hybridization is used to evaluate the gross expression patterns of a gene.

Northern blotting involves size-fractionation of samples of total RNA [or purified poly(A)$^+$ mRNA], transfer to a membrane and hybridization with a suitable labeled nucleic acid probe. The example shows the use of a labeled cDNA probe from the *FMR1* (fragile-X mental retardation syndrome) gene. Highest levels are detected in the brain and testis (4.4 kb), with decreasing expression in the placenta, lung and kidney respectively. Multiple smaller transcripts are present in the heart. Reproduced from Hinds *et al.* (1993) with permission from Nature Publishing Group.

comparatively few recognition sequences for restriction enzymes which cut at sequences containing CpGs (*Table 6.3*).

Conventionally prepared genomic DNA is not suitable for PFGE, because the procedures involved in lysing the cells and purifying the DNA result in shear forces causing considerable fragmentation of the DNA. Instead, the DNA is isolated in such a way as to minimize artificial breakage of the large molecules, and then is digested with appropriate rare-cutter restriction endonucleases. To prepare high molecular weight DNA, samples of cells, for example white blood cells, are mixed with molten agarose and then transferred into wells in a block-former and allowed to cool. As a result, the cells become entrapped in solid agarose blocks (*Figure 6.14*). The agarose blocks are removed and incubated with hydrolytic enzymes which diffuse through the small pores in the agarose and digest cellular components, but leave the high molecular weight chromosomal DNA virtually intact. Individual blocks containing purified high molecular weight DNA can then be incubated in a buffer containing a rare-cutter restriction endonuclease.

To size-separate the large restriction fragments contained within the agarose blocks, the blocks are placed in wells at one

end of an agarose gel contained within a PFGE apparatus. As in conventional gel electrophoresis the negatively charged DNA is repelled from the negative electrode and migrates in the electric field. However, during a PFGE run, *the relative orientation of the gel and the electric field is periodically altered, typically by setting a switch to deliver brief pulses of power, alternatively activating two differently oriented fields* (*Figure 6.14*). Variants of the technique use a single electric field but with periodic reversals of the polarity (*field inversion gel electrophoresis*), or periodic rotation of the gel or electrodes.

In common to each of the variant methods is the principle of a *discontinuous electric field* so that the DNA molecules are intermittently forced to change their conformation and direction of migration during their passage through the gel. The time taken for a DNA molecule to alter its conformation and re-orient itself in the direction of the new electric field is strictly size dependent; as a result, DNA fragments up to several Mb in size can be fractionated efficiently, including intact DNA from whole yeast chromosomes (Schwartz and Cantor, 1984).

6.3.4 In *in situ* hybridization probes are hybridized to denatured DNA of a chromosome preparation or RNA of a tissue section fixed on a glass slide

Chromosome *in situ* hybridization

A simple procedure for mapping genes and other DNA sequences is to hybridize a suitable labeled DNA probe against chromosomal DNA that has been denatured *in situ*. To do this, an air-dried microscope slide chromosome preparation is made (often metaphase or prometaphase chromosomes from peripheral blood lymphocytes or lymphoblastoid cell lines). Treatment with ribonuclease (RNase) and proteinase K results in partially purified chromosomal DNA, which is denatured by exposure to formamide. The denatured DNA is then available for *in situ* hybridization with an added solution containing a labeled nucleic acid probe, overlaid with a coverslip. Depending on the particular technique that is used, chromosome banding of the chromosomes can be arranged either before or after the hybridization step. As a result, the signal obtained after removal of excess probe can be correlated with the chromosome band pattern in order to identify a map location for the DNA sequences recognized by the probe. Chromosome *in situ* hybridization has been revolutionized by the use of *fluorescence in situ hybridization (FISH)* techniques (see Section 2.4.2).

Tissue *in situ* hybridization

In this procedure, a labeled probe is hybridized against RNA in tissue sections (Wilkinson, 1998). Tissue sections are made from either paraffin-embedded or frozen tissue using a *cryostat*, and then mounted on to glass slides. A hybridization mix including the probe is applied to the section on the slide and covered with a glass coverslip. Typically, the hybridization mix has formamide at a concentration of 50% in order to reduce

Table 6.3: Examples of 'rare-cutter' restriction endonucleases

Enzyme	Source	Sequence cut; CG = CpG; N = A, C, G or T	Average expected fragment size (kb) in human DNA[a]
SmaI	*Serratia marcescens*	CCCGGG	78
BssHII	*Bacillus stearothermophilus*	GCGCGC	390
Sac II	*Streptomyces lividans*	CCGCGG	390
SfiI	*Streptomyces fimbriatus*	GGCCNNNNNGGCC	400
NotI	*Norcadia otitidis-caviarum*	GCGGCCGC	9766

[a]Assuming 40% G + C, and a CpG frequency 20% of that expected.

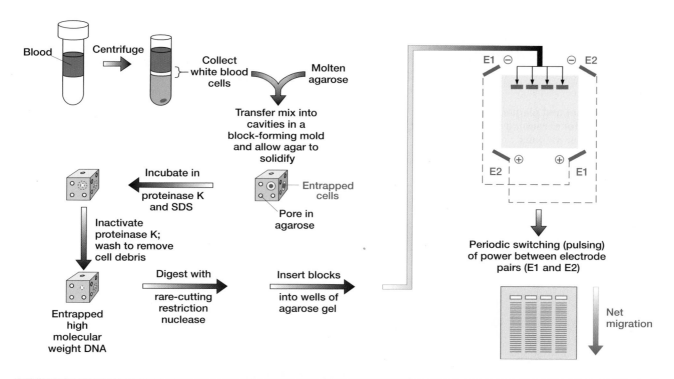

Figure 6.14: Fractionation of high molecular weight DNA from blood cells by pulsed field gel electrophoresis.

the hybridization temperature and minimize evaporation problems.

Although double-stranded cDNAs have been used as probes, single-stranded *complementary RNA* probes (**ribo-probes**) are preferred: the sensitivity of initially single-stranded probes is generally higher than that of double-stranded probes, presumably because a proportion of the denatured double-stranded probe renatures to form probe homoduplexes. cRNA riboprobes that are complementary to the mRNA of a gene are known as **antisense riboprobes** and can be obtained by cloning a gene in the

reverse orientation in a suitable vector such as pSP64 (*Figure 6.3*). In such cases, the phage polymerase will synthesize labeled transcripts from the opposite DNA strand to that which is normally transcribed *in vivo*. Useful controls for such reactions include *sense riboprobes* which should not hybridize to mRNA except in rare occurrences where both DNA strands of a gene are transcribed.

Labeling of probes is performed using either selected radioisotopes, notably ^{35}S, or by nonisotopic labeling. In the former case, the hybridized probe is visualized using auto-radiographic procedures. The localization of the silver grains is often visualized using only **dark-field microscopy** (direct light is not allowed to reach the objective; instead, the illuminating rays of light are directed from the side so that only scattered light enters the microscopic lenses and the signal appears as an illuminated object against a black background). However **bright-field microscopy** (where the image is obtained by direct transmission of light through the sample) provides better signal detection (see *Figure 6.15*). Fluorescence labeling is a popular nonisotopic labeling approach and detection is accomplished by fluorescence microscopy (see *Box 6.2, Figure 6.5*).

6.4 Hybridization assays using cloned target DNA and microarrays

Some of the technologies described in the preceding section (e.g. Southern blot-hybridization and dot-blot hybridization) are also used to study cloned DNA as well as uncloned DNA. The techniques described in the next two sections, however, are dedicated to analyzing cloned DNA. In addition, the very recently developed and very powerful microarray technologies are described in a third section.

6.4.1 Colony blot and plaque lift hybridization are methods for screening separated bacterial colonies or plaques

As described in Section 5.3.5 colonies of bacteria or other suitable host cells which contain recombinant DNA can

generally be selected or identified by the ability of the insert to inactivate a marker vector gene (e.g. β-galactosidase, or an antibiotic-resistance gene). However, if the desired recombinant DNA contains a DNA sequence that is closely related to an available nucleic acid probe, it can be specifically detected by hybridization. In the case of bacterial cells used to propagate plasmid recombinants, the cell colonies are allowed to grow on an agar surface and then transferred by surface contact to a nitrocellulose or nylon membrane, a process known as **colony blotting** (see *Figure 6.16*). Alternatively the cell mixture is spread out on a nitrocellulose or nylon membrane placed on top of a nutrient agar surface, and colonies are allowed to form directly on top of the membrane. In either approach, the membrane is then exposed to alkali to denature the DNA prior to hybridizing with a labeled nucleic acid probe.

After hybridization, the probe solution is removed, and the filter is washed extensively, dried and submitted to autoradiography using X-ray film. The position of strong radioactive signals is related back to a master plate containing the original pattern of colonies, in order to identify colonies containing DNA related to the probe. These can then be individually picked and amplified in culture prior to DNA extraction and purification of the recombinant DNA.

A similar process is possible when using phage vectors. The plaques which are formed following lysis of bacterial cells by phage will contain residual phage particles. A nitrocellulose or nylon membrane is placed on top of the agar plate in the same way as above, and when removed from the plate will

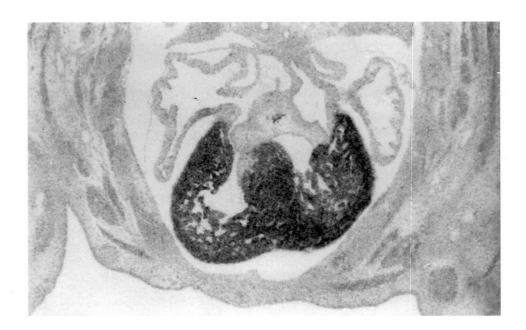

Figure 6.15: Tissue *in situ* hybridization provides high resolution gene expression patterns.

The example shows the pattern of hybridization produced using a ^{35}S-labeled β-myosin heavy chain antisense riboprobe against a transverse section of tissue from a 13-day embryonic mouse. The dark areas represent strong labeling, notably in the ventricles of the heart. Kindly supplied by Dr David Wilson, University of Newcastle upon Tyne, UK.

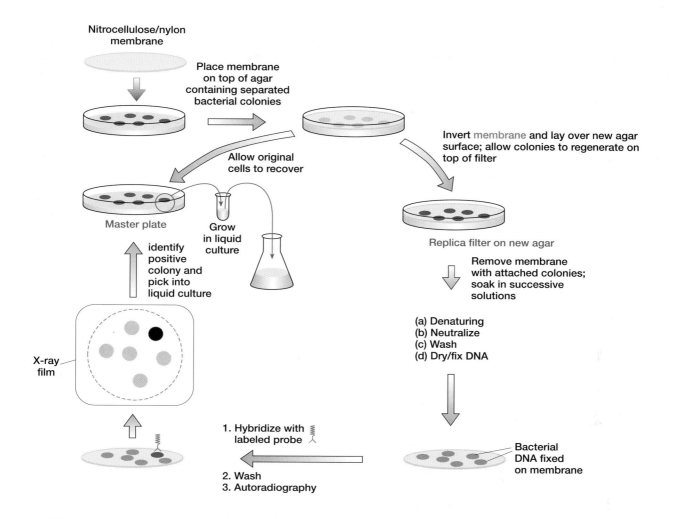

Figure 6.16: Colony blot hybridization involves replicating colonies on to a durable membrane prior to hybridization with a labeled nucleic acid probe.

This method is popularly used to identify colonies containing recombinant DNA, should a suitable labeled probe be available.

constitute a faithful copy of the phage material in the plaques, a so called **plaque-lift**. Subsequent processing of the filter is identical to the scheme in *Figure 6.16*.

6.4.2 Gridded high density arrays of transformed cell clones or DNA clones has greatly increased the efficiency of DNA library screening

Once it became possible to create complex DNA libraries, more efficient methods of clone screening were required. Rather than simply plate out cell colonies on a cell culture dish and transfer them to a membrane in standard colony blotting, it was preferable to pick individual colonies and transfer them onto large membranes in the format of a *high-density gridded array*.

The process of generating arrays was enormously simplified by the application of **robotic gridding devices** which could perform the necessary spotting automatically

by pipeting from clones arranged in microtitre dishes into pre-determined linear co-ordinates on a membrane. The resulting high-density clone filters permitted rapid and efficient library screening (see *Figure 6.17*) and could be copied and distributed to numerous laboratories throughout the world.

6.4.3 DNA microarray technology has enormously extended the power of nucleic acid hybridization

Recently developed **DNA microarrays** have provided a scale-up in hybridization assay technology because of their huge capacity for miniaturization and automation (Schena *et al.*, 1998). Although reminiscent of the filter-based arrays, microarray construction involves quite different procedures. The surfaces involved have typically been chemically-treated glass microscope slides rather than porous membranes,

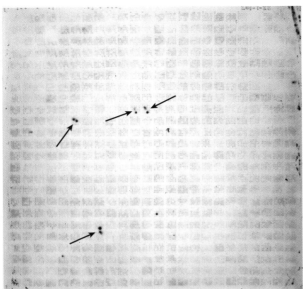

Figure 6.17: Gridded clone hybridization filters have facilitated physical mapping of the human genome.

The figure illustrates an autoradiograph of a membrane containing human YAC clones (i.e. total DNA from individual yeast clones containing human YACs). The membrane contains a total of 17 664 clones which had been gridded in arrays of a unit grid of 6 × 6 clones. The hybridization signals include weak signals from all clones by using a ^{35}S-labeled probe of total yeast DNA plus strongly hybridizing signals obtained with a ^{32}P-labeled unique sex chromosome probe (*DXYS646*). Original photo from Dr Mark Ross, Sanger Centre, Cambridge. Reproduced from Ross and Stanton (1995) In: *Current Protocols in Human Genetics*, Vol. 1, this material is used by permission of Wiley-Liss, Inc., a subsidiary of John Wiley & Sons, Inc.

although more recently there has been a move towards more porous substrates such as nitrocellulose-coated glass surfaces in an effort to bind more DNA and increase sensitivity. There are two quite distinct types of microarray technology according to differences in how the nucleic acid samples are generated and delivered to the microarrays:

▶ **microarrays of pre-synthesized nucleic acids.** Here the nucleic acids present in the microarrays have previously been synthesized (often they consist of collections of different DNA clones but they could in principle be collections of previously synthesized oligonucleotides). The construction of the microarray in this case means that individual DNA clones or oligonucleotides are spotted at individual locations on the surface of a microscope slide, specified by precise x,y co-ordinates in a miniaturized grid (*Figure 6.18A*). The high precision in delivering the samples requires quite sophisticated robotic contact printing devices but detailed instructions for making this type of device have been posted at http://cmgm.stanford.edu/pbrown/mguide/;

▶ **microarrays of oligonucleotides synthesized *in situ*.** This approach has been pioneered by the Affymetrix company and typically involves a combination of photolithography technology from the semi-conductor industry with the chemistry of oligonucleotide synthesis. In this case, many thousands of different oligonucleotides are *assembled in situ* on the surface of a glass slide, in a series of sequential synthesis steps involving adding one nucleotide at a time. The process requires covalent coupling of mononucleotides to a linker molecule which terminates with a photolabile protecting group (*Figure 6.18B*).

A photomask is used to determine which positions react with light: an opening in the photomask at a particular position will admit light from an external light source, destroying the photolabile protecting groups. A chemical coupling reaction is then used to add a particular type of nucleotide to the new deprotected site, and the process is repeated using a different mask, but those positions hidden by the mask will be protected from the light. In this way, *specific* oligonucleotide sequences can be constructed at predetermined locations depending on the arrangement of holes cuts in the photolithograph used for that synthesis step. The similarity with the production of silicon chips has led to the popular use of the term **DNA chip** for microarrays of this type. Because of the complex technology required to produce them, DNA chips need to be purchased from specialist companies.

As in the case of reverse dot-blotting (Section 6.3.1), the DNA microarray technologies operate on a *reverse* nucleic acid hybridization approach. The **probe** is the set of unlabeled nucleic acids fixed to the microarray. Although complex, the probe is the *known* quantity and it is used to query the **target** which although consisting of labeled nucleic acids in solution, is derived from sources *about which information is desired*.

Once a microarray has been constructed or bought, and a source of nucleic acid to be investigated has been isolated, the hybridization reaction can be carried out. The target is labeled with a **fluorophore**, and allowed to come into contact with the microarray, enabling probe–target heteroduplexes to form, after which hybridization washes minimize nonspecifically bound label. Most microarray hybridizations use two fluorophores, usually **Cy3** (green channel excitation) and **Cy5** (red channel excitation).

Following hybridization, bound fluorescent label is detected using a high resolution **laser scanner** and the scanning process involves acquiring an image for both fluorophores to build a ratio image. The final hybridization pattern is obtained by analyzing the signal emitted from each spot on the array using **digital imaging software** which converts the signal into one of a palette of colors according to its intensity (*Figure 6.19*).

Although the technology for establishing DNA microarrays was only developed very recently, already there have been numerous important applications and their impact on future biomedical research and diagnostic approaches is expected to be profound. Principal applications include:

(A)

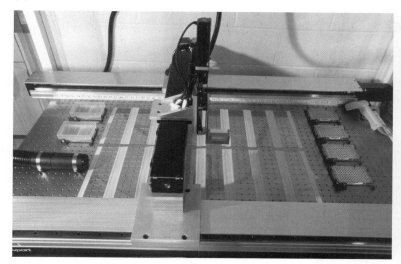

(B)

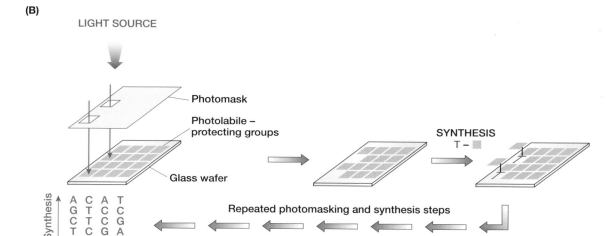

Figure 6.18: Construction of DNA and oligonucleotide microarrays.

(A) *Robotic spotting for construction of DNA microarrays*. Left: a microarray robot, with a table configuration which contains 160 slides with four microtitre plates, two wash stations and the dryer. Right: a laser scanner showing the optical table, power supplies for the lasers and photomultiplier tube cooling, the Ludi stage and lenses (see Cheung *et al.*, 1999 for more details). The microspotting of samples by robots can be performed by physical contact between spotting pins and the solid surface (of a microscope slide) or by an *ink-jetting* approach as is used in standard printing (the sample is loaded into a miniature nozzle equipped with a piezoelectric fitting and an electric current is used to expel a precise amount of liquid from the jet onto the substrate). Images kindly supplied by Aldo Massimi, Raju Kucherlapati, and Geoffrey Childs at the Albert Einstein College of Medicine. Reprinted from Cheung *et al.* (1999) *Nature Genet.* **21** (Suppl.), 15–19, with permission from Nature Publishing Group. **(B)** *Construction of an oligonucleotide microarray* by combining photolithography and *in situ* synthesis of oligonucleotides. Oligonucleotides are synthesized *in situ* in sequential steps starting from a 3′ mononucleotide which is anchored to the surface of a glass wafer. The photolithography entails modifying the glass wafer with photolabile protecting groups which can be eliminated when exposed to light and the use of carefully constructed *photomasks* which allow light to pass through onto carefully selected spatial co-ordinates. For those areas of the wafer which receive light passing through the photomask, the removal of the photolabile protective groups permits a new synthesis step. In this example thymidine is shown being coupled together with a protective photolabile group.

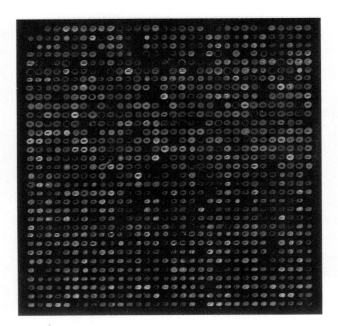

Figure 6.19: Gene expression profiling by hybridization to a cDNA clone microarray.

A 1.0-cm^2 DNA microarray containing 1046 cDNA clones from a whole human blood cell cDNA library was hybridized with a fluorescent labeled cDNA probe prepared by reverse transcription from human bone marrow mRNA. Confocal laser scans of the fluorescence label are represented in a pseudocolor rainbow scale to indicate rough quantification of expression levels: background levels being violet, then progressing through indigo, blue–green, yellow, orange to red, the most abundant expression. The parallel format of the array allows precise comparisons and differential expression measurements. Image kindly provided by Dr. Mark Schena, Stanford. Reproduced from Strachan *et al.* (1997) *Nature Genetics* **16**, 126–132, with permission from Nature Publishing Group.

▸ *expression screening*. The focus of most current microarray-based studies is the monitoring of RNA expression levels (Granjeaud *et al.*, 1999), which can be done by using either cDNA clone microarrays (*Figure 6.19*) or gene-specific oligonucleotide microarrays (usually constructed by *in situ* oligonucleotide synthesis; see *Figure 17.18* and *19.6*, and Section 19.3.3);

▸ *DNA variation screening*. Oligonucleotide microarrays are required for this general purpose and several applica-

tions have been devised. The re-sequencing of the human mitochondrial genome by using DNA microarray was a successful test of the power of this technology for assessing large-scale sequence variation in individuals (see *Figure 7.4*). There is also huge potential for assaying for mutations in known disease genes. In addition, there have been vigorous efforts to identify and catalog human single nucleotide polymorphism (SNP) markers.

Further reading

Molecular Probes. *Handbook of Fluorescent Probes and Research Products* at http://www.molecularprobes.com/handbook/

Sambrook J, Russell D (2001) *Molecular Cloning: A Laboratory Manual,* 3rd Edn. Cold Spring Harbor Laboratory Press, Cold Spring Harbor, NY.

Schena M (1999) *Microarrays: A Practical Approach.* Oxford University Press, Oxford.

Schena M (2002) *Microarray analysis.* John Wiley and Sons, New York.

Various authors (1999) The Chipping Forecast. *Nature Genet.* **21** (Suppl.), 1–60.

Various authors (2002) The Chipping Forecast II. *Nature Genet.* **32** (Suppl.), 465–552.

References

Cheung VG, Morley M, Aguilar F, Massimi A, Kucherlapati R, Childs G (1999) Making and reading microarrays. *Nature Genet.* **21** (Suppl.), 15–19.

Feinberg AP, Vogelstein B (1983) A technique for radiolabeling DNA restriction endonuclease fragments to high specific activity. *Anal. Biochem.* **132**, 6–13.

Granjeaud S, Bertucci F, Jordan BR (1999) Expression profiling: DNA arrays in many guises. *BioEssays* **21**, 781–790.

Hinds HL, Ashley CT, Sutcliffe JS *et al.* (1993) Tissue specific expression of FMR-1 provides evidence for a functional role in fragile X syndrome. *Nature Genet.* **3**, 36–43.

Kricka LJ (1992) *Nonisotopic DNA Probing Techniques*. Academic Press, San Diego, CA.

Ross MT, Stanton VPJ (1995) Screening large-insert libraries by hybridization. In: *Current Protocols in Human Genetics*, Vol. 1.

(eds NJ Dracopoli, JL Haines, BR Korf, CC Morton, CE Seideman, DT Moir, D Smith). John Wiley & Sons, New York, pp. 5.6.1–5.6.30.

Schena M, Heller RA, Theriault TP, Konrad K, Lachenmeier E, Davis RW (1998) Microarrays: biotechnology's discovery platform for functional genomics. *Trends Biotechnol.* **16**, 301–306.

Schwartz DC, Cantor CR (1984) Separation of yeast chromosome-sized DNAs by pulsed field gradient gel electrophoresis. *Cell* **37**, 67–75.

Strachan T, Abitbol M, Davidson D, Beckmann JS (1997) A new dimension for the Human Genome Project: towards comprehensive expression maps. *Nature Genet.* **16**, 126–132.

Wilkinson D (1998) In Situ *Hybridization: A Practical Approach*. 2nd Edn. IRL Press, Oxford.

CHAPTER SEVEN

Analyzing DNA and gene structure, variation and expression

Chapter contents

7.1 Sequencing and genotyping DNA

Individuals within a species carry genetic differences, many of which can have important consequences. Thus, while it has been clearly pivotal to commit so much effort to genome projects we will continue to be interested in mapping and sequencing the DNA of individuals within a species, notably our own species, and in typing DNA variation (**genotyping**).

While traditional methods of large-scale physical mapping and sequencing will of course be useful for obtaining new genome sequences, they will increasingly be overtaken by more powerful, fully automated methods of establishing DNA structure, directly or in scanning for genetic variation. Oligonucleotide-based microarray hybridization can, for example, permit *re-sequencing* of the DNA of individuals once a reference sequence has been established.

7.1.1 Standard DNA sequencing involves enzymatic DNA synthesis using base-specific dideoxynucleotide chain terminators

Chemical methods for DNA sequencing (Maxam and Gilbert, 1980) continue to have some applications (e.g. sequencing of oligonucleotides; see Section 7.2.4). The vast majority of current DNA sequencing, however, uses an enzymatic method first developed by Fred Sanger, winning him a second Nobel prize (the first was for developing protein sequencing). In the **dideoxy sequencing** method the DNA is provided in a *single-stranded* form (see *Box 7.1*) and acts as a template for making a new complementary DNA strand *in vitro* using a suitable DNA polymerase. There are four parallel reactions, each containing the four dNTPs, plus a small proportion of one of the four analogous **dideoxynucleotides** (**ddNTPs**) which will serve as a base-specific chain terminator.

Box 7.1: Producing single-stranded DNA sequencing templates.

Single-stranded templates for DNA sequencing can be provided by:

▶ producing single-stranded *recombinant DNA* using specialized cloning vectors such as M13 or phagemids, a widely used approach – see Section 5.5.1 and *Figure 5.18*. Here synthesis of the complementary DNA is primed by a **universal sequencing primer** that is complementary to the vector sequence flanking the cloning site and so can be used to prime new strand synthesis from any single-stranded recombinant produced using that vector (see *Figure* at bottom left);

▶ *cycle sequencing* (also called *linear amplification sequencing*). Like the standard PCR reaction, cycle sequencing uses a thermostable DNA polymerase and a temperature cycling format of denaturation, annealing and DNA synthesis. The difference is that cycle sequencing employs only *one* primer and includes a ddNTP chain terminator in the reaction. The use of only a single primer means that unlike the exponential increase in product during standard PCR reactions, the product accumulates linearly (see *Figure* at bottom right).

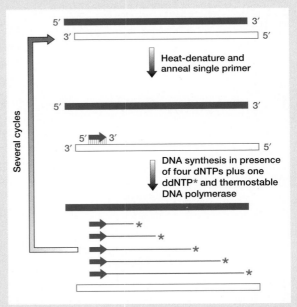

A universal sequencing primer can be used to sequence many different template DNAs.

Cycle sequencing involves linear amplification using a single primer to initiate DNA synthesis.

The ddNTPs are closely related to the normal dNTPs; they differ only in that *they lack a hydroxyl group at the 3′ carbon position as well as the 2′ carbon (Figure 7.1)*. A dideoxynucleotide can be incorporated into the growing DNA chain by forming a phosphodiester bond between its 5′ carbon atom and the 3′ carbon of the previously incorporated nucleotide. However, since ddNTPs lack a 3′ hydroxyl group, any ddNTP that is incorporated into a growing DNA chain cannot participate in phosphodiester bonding at its 3′ carbon atom, thereby causing abrupt termination of chain synthesis.

By ensuring that one of the four dNTPs or the primer is labeled with a distinctive radioisotope group or fluorophore, the growing DNA strand becomes labeled. By setting the concentration of the ddNTP to be very much lower than that of its normal dNTP analog, there will be competition between a specific ddNTP molecule and an excess of analogous dNTP molecules for inclusion in the growing DNA chain – if a dNTP is included, chain extension continues, but occasionally a ddNTP will be incorporated, causing chain termination. Each reaction is therefore a *partial reaction* because chain termination will occur *randomly* at one of the possible choices for a specific type of base *in any one DNA strand*.

Because the DNA in a DNA sequencing reaction is normally a *population* of identical molecules, each one of the four base-specific reactions will generate a *collection* of labeled DNA fragments. The synthesized DNA fragments in each of the four reactions will cover a range of different sizes. They have a common 5′ end *but variable 3′ ends* (the 5′ end is defined by the sequencing primer; the 3′ ends are variable because the insertion of the selected ddNTP occurs *randomly* at one of the many different positions that will accept that specific base (see *Figure 7.2*).

Fragments that differ in size by even a single nucleotide can be size-fractionated on a ***denaturing polyacrylamide gel***

(a gel which contains a high concentration of a denaturing agent – such as 8M urea – which ensures that migrating DNA remains single stranded). In former times DNA sequencing reactions used radioisotope labeling such as ^{35}S or ^{33}P and after exposing the dried sequencing gel to an X-ray film, the sequence was read manually by following successive bands on the autoradiograph. This cumbersome approach has been overtaken by automated DNA sequencing methods.

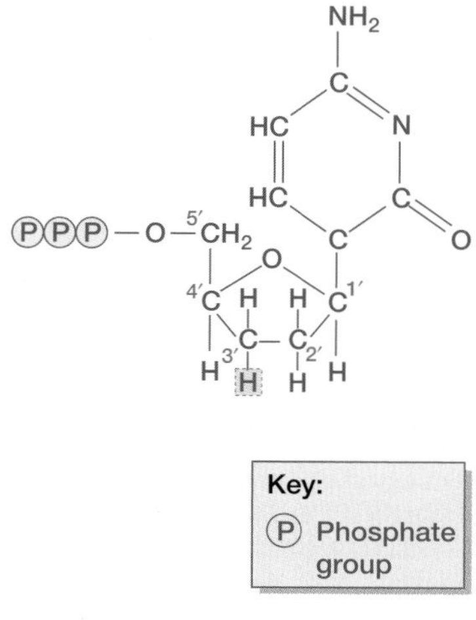

7.1.2 Automated DNA sequencing and microarray-based re-sequencing

Automated DNA sequencing

Automated DNA sequencing uses ***fluorescence labeling***: the DNA is labeled by incorporating a primer or dNTP which carries a ***fluorophore*** (a chemical group capable of fluorescing – see Section 6.1.2). The use of different fluorophores in the four base-specific reactions means that, unlike conventional DNA sequencing, all four reactions can be loaded into a single lane. During electrophoresis, a monitor detects and records the fluorescence signal as the DNA passes through a fixed point in the gel (*Figure 7.3A*). This allows an output in the form of intensity profiles for each of the differently colored fluorophores while simultaneously storing the information electronically (*Figure 7.3B*). If the DNA sequence is known to be coding sequence, the output can immediately be translated in different reading frames to infer a polypeptide sequence.

While electrophoresis in slab acrylamide gels is used in many automated DNA sequencers, high throughput DNA sequencing often now uses **capillary sequencers** where DNA samples migrate through very thin long glass capillary tubes which have been filled with gel (see Meldrum, 2000). By avoiding the need to cast large gels, a higher degree of automation can be achieved.

Re-sequencing using microarray hybridization

Once sequences have been established, it is possible in principle to use them as *reference* sequences to help design oligonucleotides for ***hybridization-based DNA sequencing***. This involves synthesizing oligonucleotides *in situ* on a glass surface to act as a probe for hybridizing to the test DNA to be sequenced (see Section 6.4.3 for the principle of oligonucleotide microarrays). Re-sequencing by microarray hybridization can be done in a highly automated way with a throughput far exceeding that of automated DNA sequencing. A test case has involved re-sequencing of the 16.5 kb mitochondrial DNA (mtDNA) (Chee *et al.*, 1966; see *Figure 7.4*), but very large-scale sequencing by microarray hybridization poses formidable technical challenges.

7.1.3 Basic genotyping of restriction site polymorphisms and variable number of tandem repeat polymorphisms

There are two basic types of polymorphism which are amenable to genotyping using simple methods: restriction site

Figure 7.1: Structure of a dideoxynucleotide, 2′,3′ dideoxy CTP (ddCTP).

Note: the hydroxyl group attached to carbon 3′ in normal nucleotides is replaced by a hydrogen atom (shown by shading).

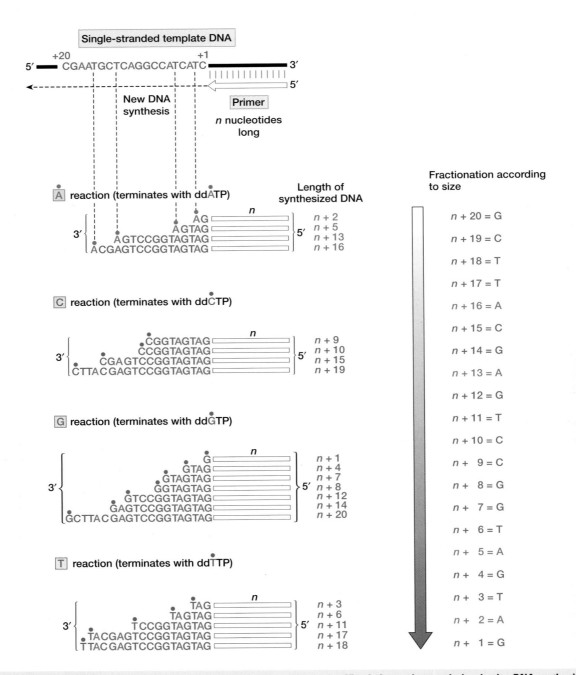

Figure 7.2: Dideoxy DNA sequencing relies on *random* incorporation of base-specific chain terminators during *in vitro* DNA synthesis.

The sequencing primer binds specifically to a region 3' of the desired DNA sequence and primes synthesis of a complementary DNA strand in the indicated direction. Four parallel base-specific reactions are carried out, each with all four dNTPs and with one ddNTP. Competition for incorporation into the growing DNA chain between a ddNTP and its normal dNTP analog results in a population of fragments of different lengths. The fragments will have a common 5' end (defined by the sequencing primer) but variable 3' ends, depending on where a dideoxynucleotide (shown with a filled circle above) has been inserted. For example, in the A-specific reaction chain, extension occurs until a ddA nucleotide (shown as A with a filled red circle above) is incorporated. This will lead to a population of DNA fragments of lengths $n+2$, $n+5$, $n+13$, $n+16$ nucleotides, etc., which if size fractionated as shown on the right will produce the following sequence from $n=1$ to $n+20$, GATGATGGCCTGAGCATTCG, which is the reverse complement of the sequence shown at top left.

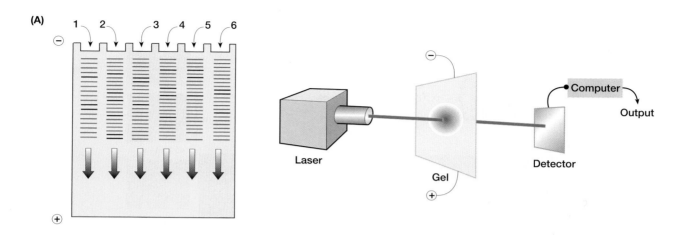

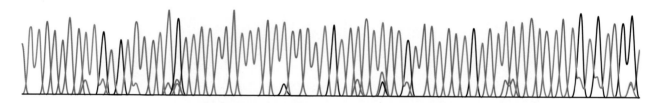

Figure 7.3: Automated DNA sequencing using fluorescent primers.

(A) *Principles of automated DNA sequencing*. All four reaction products are loaded into single lanes of the electrophoresis gel or single gel capillaries. Four separate fluorescent dyes are used as labels for the base-specific reactions (the label can be incorporated by being attached to a base-specific ddNTP, or by being attached to the primer and having four sets of primers corresponding to the four reactions). During the electrophoresis run, a laser beam is focused at a specific constant position on the gel. As the individual DNA fragments migrate past this position, the laser causes the dyes to fluoresce. Maximum fluorescence occurs at different wavelengths for the four dyes, the information is recorded electronically and the interpreted sequence is stored in a computer database. **(B) *Example of DNA sequence output*.** This shows a typical output of sequence data as a succession of dye-specific (and therefore base-specific) intensity profiles. The example illustrated shows a cDNA sequence from the recently identified human polyhomeotic gene, *PHC3* (Tonkin *et al.*, 2002). Data provided by Dr. Emma Tonkin, Institute of Human Genetics, University of Newcastle upon Tyne, UK.

polymorphisms (RSPs) and variable number of tandem repeat (VNTR) polymorphisms. Various derivative polymorphisms are described in *Box 7.2*.

Genotyping RSPs

Single nucleotide polymorphisms (SNPs) are often used as markers in studying susceptibility to common disease, and a variety of automated *high throughput genotyping* methods have been devised to detect them (see *Box 18.2*). A subset of SNPs cause a loss or gain of a restriction site (**restriction site polymorphisms** or **RSPs**) and they are often typed on a small scale using simple genotyping methods.

In the past RSPs would be typed by Southern hybridization using a nearby probe to detect the altered restriction fragment, and when this type of assay was used, RSPs were described as **restriction fragment length polymorphisms (RFLPs**; see *Figure 7.5A* for the general principle). Now, however, it is simpler to use a PCR-based

method to type an RSP: primers are designed from sequences flanking the polymorphic restriction site and the amplified product is cut with the appropriate restriction enzyme and size-fractionated by agarose gel electrophoresis (*Figure 7.6*).

Genotyping VNTR polymorphisms

Microsatellite polymorphism is a type of VNTR where the array size and the size of the tandem repeats are short (see *Box 7.2* for alternative names), and so it is convenient to carry out typing by PCR. Primers are designed from sequences known to flank a specific microsatellite locus, permitting PCR amplification of alleles whose sizes differ by integral repeat units (*Figure 7.7*). The PCR products can then be size-fractionated by polyacrylamide gel electrophoresis. The PCR normally includes a radioactive or fluorescent nucleotide precursor which becomes incorporated into the small PCR products and facilitates their detection. To ensure adequate size fractionation of alleles, the PCR products are denatured prior to

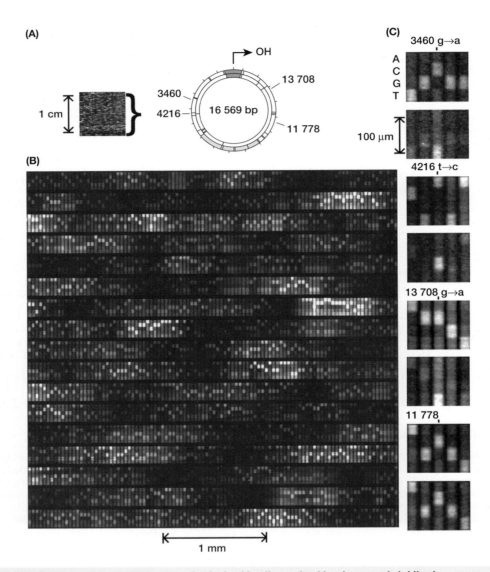

Figure 7.4: Re-sequencing of the mitochondrial genome by single chip-oligonucleotide microarray hybridization.

(A) An image of the array hybridized to 16.6 kb of mitochondrial target RNA (L strand) plus a map of the mtDNA genome. **(B)** A *portion* of the hybridization pattern magnified. **(C)** The ability of the array to detect and read single-base differences in a 16.6-kb sample is illustrated. Two different target sequences were hybridized in parallel to different chips. The hybridization patterns are compared for four different positions in the sequence. The top panel of each pair shows the hybridization of a reference sequence and the lower panel shows the pattern generated by a sample from a patient with Leber's hereditary optic neuropathy. Three known pathogenic mutations, at nucleotides 3460, 4216, and 13 708, are clearly detected. For comparison, the fourth panel in the set shows a region around position 11 778 that is identical in the two samples. Reprinted with permission from Chee *et al.* (1996) *Science* **274**, 610–614, American Association for the Advancement of Science.

electrophoresis. An example of the use of a (CA)$_n$ microsatellite is shown in *Figure 7.8*.

Minisatellite VNTR polymorphisms are often typed by Southern hybridization. DNA is digested with a restriction enzyme(s) known to cut at sites closely flanking the relevant VNTR sequence. This produces a restriction fragment containing the VNTR plus neighboring unique sequence DNA. A probe from the latter sequence can detect length variation as an RFLP (see *Figure 7.5B*).

7.2 Identifying genes in cloned DNA and establishing their structure

Identifying genes within cloned DNA relies on assaying for gene-specific properties. Two major features permit the DNA of genes to be distinguished from DNA that does not have a coding function: (i) high overall evolutionary *conservation*, and (ii) *expression* to give RNA transcripts. In the vast majority of cases the RNA transcripts undergo splicing and are translated

---

Box 7.2: Common classes of DNA polymorphism which are amenable to simple genotyping methods.

As detailed in Chapter 11, DNA variation can occur at different levels. Occasionally large changes are seen such as duplications, insertions, deletions and transpositions of kilobase- and even megabase-sized stretches of DNA, some of which are disease-associated and some not. However, the most frequent changes to the DNA sequence involve single nucleotide substitutions, insertions and deletions. They are usually not associated with disease unless they alter a coding sequence or an important regulatory sequence. A DNA variation is described as a *polymorphism* if it is sufficiently common in a population so that it cannot be explained by recurring mutation (see Section 11.1). Polymorphisms which can be genotyped simply fall into two basic classes:

▸ **SNP (single nucleotide polymorphism)**. An SNP is a polymorphism arising by change of a single nucleotide. Generally assayed by DNA sequencing (Sections 7.1.1; 7.1.2), or by primer extension assays (*Box 18.2*);

▸ **VNTR (variable number of tandem repeats) polymorphism.** A polymorphism which arises because of instability in an array of tandem repeats causing the number of repeat units to change. This is a general term which includes microsatellite VNTR polymorphism and minisatellite VNTRs (see below), but is often used loosely to mean the ministatellite class.

Some derivative subclasses of the above classes of polymorphism are:

▸ **RSP (restriction site polymorphism)**. An RSP is a subset of SNPs where the change in nucleotide causes either loss or gain of a restriction site. This can be typed by PCR or alternatively by Southern hybridization, in which case the polymorphism can also be described as an RFLP (see below);

▸ **RFLP (restriction fragment length polymorphism)**. A polymorphism which causes a change in the length of a restriction fragment and is typed by Southern hybridization. It can arise in two ways:

  • as a result of an RSP (see above);

  • as a result of length variation of a restriction fragment containing a moderately long VNTR array of tandem repeats. The flanking restriction sites are not changed, but the length between them can expand or contract depending on the number of repeat units.

▸ **microsatellite VNTR** [also called **short tandem repeat polymorphism (STRP)** or **simple sequence repeat polymorphism (SSRP)**]. A type of VNTR where the array is small (usually less than 100 bp) and the repeat unit is small, usually 1–4 nucleotides long – see also Section 9.4.3;

▸ **minisatellite VNTR** (confusingly, often shortened to VNTR). Here the array size is moderately long and the repeat unit is often from 9 to 65 bp long – see also Section 9.4.2.

---

to give polypeptides (and so are distinguished by having long *open reading frames – ORFs*). In addition, *vertebrate* genes are often associated with **CpG islands** (see *Box 9.3*. These features have permitted a variety of different methods for identifying genes in cloned vertebrate DNA (Monaco, 1994).

Routine methods for gene identification

Once genomic DNA clones were available, genes have traditionally been defined by using simple methods as a first resort. To test for evidence of expression, standard approaches include screening cDNA libraries (Section 5.3.5); carrying out RT–PCR (Section 5.2.1); and hybridizing test probes against Northern blots (*Figure 6.13*). Later on, expression assays are usually extended to *in situ* hybridization assays against RNA in tissue sections (*Figure 6.15*). The alternative approach of looking for evolutionary conservation used to rely heavily on **zooblots** to identify sequences that have been strongly conserved across a range of species (see *Figure 7.9*). More recently, homology searching of sequence databases has been an important way of identifying genes: if a test sequence is closely related to some other coding sequence there is a good chance that it is also a coding sequence (see *Box 7.3*).

In addition to the routine gene identification methods, two more specialized procedures have been used: exon trapping (an artificial RNA splicing assay) and cDNA selection.

7.2.1 Exon trapping identifies expressed sequences by using an artificial RNA splicing assay

RNA splicing involves fusion of exonic sequences at the RNA level and excision of intronic sequences. Spliceosomes are able to accomplish this *in vivo* by recognizing certain sequences at exon–intron boundaries: a *splice donor* sequence at the junction between an exon and its downstream (3′) intron, and a *splice acceptor* sequence at the junction between an exon and its upstream (5′) intron (see *Figure 1.15*). A cosmid or other suitable genomic DNA clone containing an internal exon flanked by intronic sequences will therefore contain functional splice donor and acceptor sequences.

Exons can be identified in cloned genomic DNA by subcloning the DNA into a suitable expression vector and transfecting into an appropriate eukaryotic cell line in which the insert DNA is transcribed into RNA and the RNA transcript undergoes RNA splicing. Such techniques are known as **exon trapping** (often called **exon amplification** if a PCR reaction is used to recover the exons from a cDNA copy of the spliced RNA). For example, in the method of Church *et al.* (1994), the DNA is subcloned into a plasmid expression vector pSPL3 (*Figure 7.10A*) which contains an **artificial minigene** that can be expressed in a suitable host cell. The minigene consists of: a segment of the simian virus 40 (SV40)

(A)

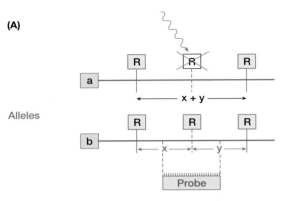

Alleles

Assay
  (i) digest with restriction nuclease R
  (ii) size fractionate on gel
  (iii) hybridize labeled probe

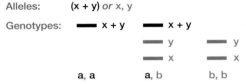

(B)

Alleles
vary in
number
of tandem
repeats

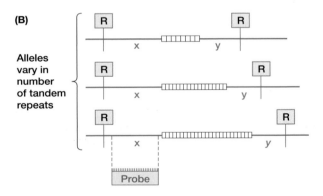

Assay: (as in A above)

Allele sizes: x + y + (n × repeats) where n is variable

Figure 7.5: Typing an RFLP by a hybridization based-assay.

(A) *Typing an RSP-type RFLP*. This is the usual type of RFLP and is caused by a minor alteration to the DNA sequence causing loss (or gain) of a restriction site. This type of polymorphism is more easily typed by a PCR assay – see *Figure 7.6*. **(B) *Typing a VNTR-type RFLP*.** A hybridization assay is only used if the expansion and contraction of the VNTR involves significant length changes. Otherwise, a PCR-type assay is used instead.

---

genome which contains an origin of replication plus a powerful promoter sequence; two splicing-competent exons separated by an intron which contains a multiple cloning site; and an SV40 polyadenylation site.

The recombinant DNA is transfected into **COS cells** which, as explained in Section 5.6.3, allow any circular DNA

that contain an SV40 origin of replication to replicate independently of the cellular DNA. Transcription from the SV40 promoter results in an RNA transcript which normally splices to include the two exons of the minigene. If the DNA cloned into the intervening intron contains a functional exon, however, the foreign exons can be spliced to the exons present in the vector's minigene. After making a cDNA copy using reverse transcriptase, PCR reactions using primers specific for vector exon sequences should distinguish between normal splicing and splicing involving exons in the insert DNA (see *Figure 7.10B*).

7.2.2 cDNA selection identifies expressed sequences in genomic clones by heteroduplex formation

The **cDNA selection** method entails hybridizing a complex cloned DNA, such as the insert of a YAC, to a complex mixture of cDNAs, such as the inserts from all cDNA clones in a cDNA library (Lovett, 1994). The principle underlying the technique is that *cognate* cDNAs corresponding to genes found within the YAC will bind preferentially to the YAC DNA; several rounds of hybridization should lead to a huge enrichment of the desired cDNA sequences, enabling the identification of the corresponding genes. Considerable blocking of repetitive DNA sequences is required.

Early approaches used immobilized YACs, but more modern approaches have used a solution hybridization reaction and biotin–streptavidin capture methods (see *Figure 7.11*). Like all expression-based systems, the method depends on appropriate levels of gene expression (the cognate cDNAs should not be too rare in the starting population). Additionally, genes containing very short exons may be missed because the heteroduplexes formed with cognate cDNAs may not be sufficiently stable. Another problem is that cDNAs may bind to pseudogenes which show a high degree of homology to the cognate functional genes.

7.2.3 Achieving full-length cDNA sequences: overlapping clone sets, and RACE-PCR amplification

An early priority in defining gene structure is to obtain a *full-length* cDNA sequence and define translational initiation and termination sites and polyadenylation site(s).

Defining overlapping clone sets

To obtain a full length cDNA sequence, an initial approach has been to screen a variety of different cDNA libraries and then map the extent of overlap between the inserts of positive clones (by sequencing or by PCR/hybridization-based mapping). As a result, a series of overlapping cDNA clones can be established, a **cDNA clone contig** (for an example, see the contig for the cystic fibrosis cDNA in Riordan *et al.*, 1989). Full or selected clone sequencing will define a consensus sequence which may provide a full-length cDNA sequence.

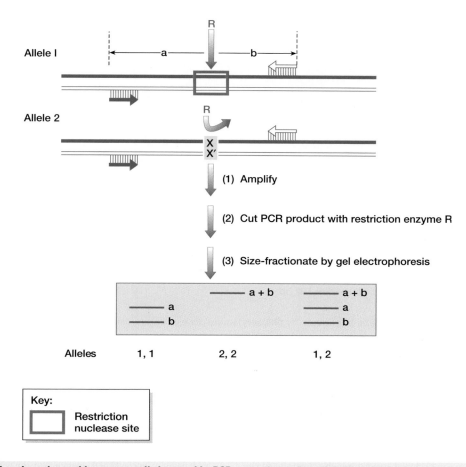

Figure 7.6: Restriction site polymorphisms can easily be typed by PCR as an alternative to laborious RFLP assays.

Alleles 1 and 2 are distinguished by a polymorphism which alters the nucleotide sequence of a specific restriction site for restriction nuclease R. Allele 1 possesses the site, but allele 2 has an altered nucleotide(s) X, X' and so lacks it. PCR primers can be designed simply from sequences flanking the restriction site to produce a short product. Digestion of the PCR product with enzyme R and size-fractionation can result in simple typing for the two alleles.

RACE–PCR to extend short cDNAs

RT–PCR can help define transcripts but a variant of RT–PCR which is particularly effective in extending short cDNA sequences to achieve a full length cDNA is the **RACE (rapid amplification of cDNA ends)** technique (Frohman *et al.*, 1988). RACE–PCR is an anchor PCR modification of RT–PCR. Its rationale is to amplify sequences between a single previously characterized region in the mRNA (cDNA) and an *anchor sequence* that is coupled to the 5' or the 3' end. A primer is designed from the known internal sequence and the second primer is selected from the relevant anchor sequence (see *Figure 7.12*).

7.2.4 Mapping transcription start sites and defining exon–intron boundaries

Important regulatory sequences are often located close to transcription start sites. Although 5' RACE–PCR can permit rescue of sequences corresponding to the 5' end of an mRNA

(and therefore, the transcriptional start site), two major methods are preferentially used to define the transcriptional start site: nuclease S1 protection and primer extension. Exon–intron structure can be determined by referencing the cDNA sequence against sequences of cognate genomic DNA clones. Subsequently, attempts may be made to complete gene characterization at the genomic level by sequencing of promoter regions, 5' and 3' flanking sequences and intron sequences.

Nuclease S1 protection

The endonuclease S1 is an enzyme from the mold *Aspergillus oryzae* which cleaves single-stranded RNA and DNA but not double-stranded molecules. In order to map the transcription start site for a gene, a genomic DNA clone suspected of containing the start site is required. The DNA clone is then digested with a suitable restriction endonuclease to generate a fragment that is expected to contain the transcription start site. As shown in *Figure 7.13A*, hybridization to the cognate

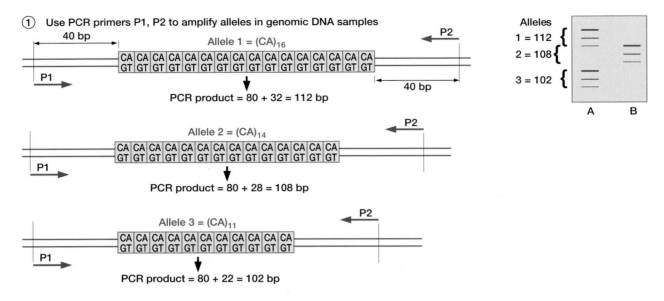

Figure 7.7: PCR is used to type short tandem repeat polymorphisms (STRPs).

The example illustrates typing of a microsatellite marker, in this case a (CA)/(TG) dinucleotide repeat polymorphism which has three alleles as a result of variation in the number of the (CA)/(TG) repeats. On the autoradiograph each allele is represented by a major upper band and two minor 'shadow bands' (see *Figure 7.8*). Individuals A and B have genotypes (in brackets) as follows: A (1,3); B (2,2).

mRNA and S1 nuclease digestion defines the distance of the transcription start site from the unlabeled end of the restriction fragment. If more precise localization is required, the labeled DNA fragment in the heteroduplex can be sequenced by a chemical method of DNA sequencing (see Maxam and Gilbert, 1980). Note that, in much the same way, nuclease S1 mapping can also be used to map other boundaries between coding and noncoding DNA such as exon–intron boundaries (see below) and the 3′ end of a transcript.

Primer extension

The method is very similar to that for nuclease S1 protection. In this case, the chosen restriction fragment must be shorter than the mRNA and the overhang is filled in using reverse transcriptase (*Figure 7.13B*). As with nuclease S1 mapping, a more accurate location for the transcription initiation site is possible by using the chemical sequencing method of Maxam and Gilbert (1980) to sequence the labeled DNA strand.

Determining exon–intron organization

Not all human genes have introns (see *Table 9.5*), but when present they are often large when compared to exons. The presence of introns can usually be inferred from comparative mapping of cognate genomic and cDNA clones. Once the full cDNA sequence has been established, sequencing primers can be designed as required from various segments of the cDNA

and used in *cycle DNA sequencing* with denatured genomic clones as the DNA sequencing templates (see *Box 7.1*). The sequences obtained should cross an exon–intron boundary, unless the exon is very large, in which case additional sequencing primers may be required. Of course, when a genome has been sequenced, all that is required is to cross-reference cDNA clones against the available genomic sequence.

7.3 Studying gene expression

7.3.1 Principles of expression screening

Expression screening can be done at different levels and using a variety of different technologies. The target is either RNA transcripts or proteins. Protein expression is normally tracked by using highly specific antibodies, but RNA transcripts can be followed by several different types of approach. Usually these involve molecular hybridization with a specific antisense nucleic acid probe or some variant of RT–PCR. However, some alternative, ingenious methods have been devised such as **SAGE** (*serial analysis of gene expression*) which tracks large numbers of individual transcripts by following representative short sequence tags (see Section 19.3.2). Important parameters in gene expression are the source of the material under study, the expression resolution and the throughput (see *Figure 7.14*).

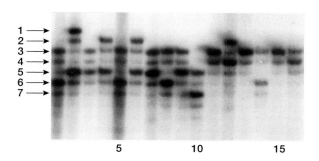

5   10   15

Figure 7.8: Example of typing for a CA repeat.

The example illustrated shows typing of members of a large family with the (CA)/(TG) marker *D17S800*. Arrows to the left mark the top (main) band seen in different alleles 1–7. **Note** that individual alleles show a strong upper band followed by two lower 'shadow bands', one of intermediate intensity immediately underneath the strong upper band, and one that is very faint and is located immediately below the first shadow band. For the indicated individuals, the genotypes (in brackets) are as follows: 1 (3,6); 2 (1,5); 3 (3,5); 4 (2,5); 5 (3,6); 6 (2,5); 7 (3,5); 8 (3,6); 9 (3,5); 10 (5,7); 11 (3,3); 12 (2,4); 13 (3,3); 14 (3,6); 15 (3,3); 16 (3,4). **Note** also that in the latter case, the middle band is particularly intense because it contains both the main band for allele 4 plus the major shadow band for allele 3. Slipped strand mispairing (see Section 11.3.1) is thought to be the major mechanism responsible for producing shadow bands at tandem dinucleotide repeats (Hauge and Litt, 1993).

Source of material under study

The material studied can vary widely. Frequently, crude RNA/cDNA or protein extracts are prepared, but in other cases expression is sampled in tissue sections or even whole embryos which have been fixed in such a way as to preserve the original *in vivo* morphology. Expression can also be studied in live cells in tissue culture (but always there is the question of how representative they are in relation to the same type of cells *in vivo*). In living experimental organisms where the tissues are optically transparent gene expression can be tracked with the aid of fluorescent tags.

New **laser capture microdissection** methods entail using lasers to microdissect tissue to produce pure cell populations from sources such as tissue biopsies and stained tissue, and even single cells (see Schutze and Lahr, 1998; Simone *et al.*, 1998). Such developments will allow a variety of gene expression analyses to be focused on single cells, or on homogeneous cell populations which will be more representative of the *in vivo* state than cell lines.

Gene expression resolution

Some methods are designed simply to track the gross expression of a gene in RNA extracts or protein extracts. Such *low resolution* expression patterns are usually attempted as a first-pass approach. In addition to being able to sample expression in different tissues, they may provide useful information on product size and on possible isoforms. In contrast, *high resolution* expression can be obtained using methods which track expression patterns within a cell, or within groups

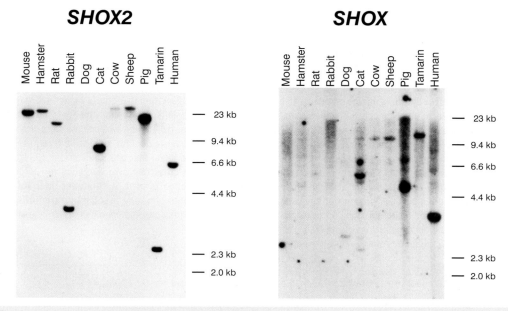

Figure 7.9: Zooblot hybridization identifies evolutionarily conserved sequences.

Some genes show extraordinary conservation across species; some others less so, including the one illustrated here. The human *SHOX* gene is located in the ***major pseudoautosomal region*** at the tip of Xp/Yp (*Figure 12.15*). It is a locus for some short stature syndromes which are characterized by skeletal abnormalities, and may be an important contributor to Turner syndrome. Although conserved in a variety of different mammals, rodents lack it (see right panel, lanes on left), as a result of gene deletion in the evolutionary past, presumably at an early stage in the lineage leading to rodents. The recent sequencing of the mouse genome confirmed the absence of a *SHOX* homolog. The related autosomal *SHOX2* gene is much more highly conserved (left panel). Reproduced from Clement-Jones *et al.* (2000) *Human Molec. Genet.* **9**, 696, with permission from Oxford University Press.

Box 7.3: Database homology searching.

Powerful computer programs have been devised to permit searching of nucleic acid and protein sequence databases for significant sequence matching (**sequence homology**) with a test sequence under investigation. Popularly used programs are the different **BLAST** and **FASTA** programs (see Ginsburg, 1994 and *Table* below).

Program	Compares
FASTA	A nucleotide sequence against a nucleotide sequence database, or an amino acid sequence against a protein sequence database
TFASTA	An amino acid sequence against a nucleotide sequence database translated in all six reading frames
BLASTN	A nucleotide sequence against a nucleotide sequence database
BLASTX	A nucleotide sequence translated in all six reading frames against a protein sequence database
EST BLAST	A cDNA/EST sequence against cDNA/EST sequence databases
BLASTP	An amino acid sequence against a protein sequence database
TBLASTN	An amino acid sequence against a nucleotide sequence database translated in all six reading frames

Note: as the design of comparable programs such as FASTA and BLASTN is different, they may give different results (see Ginsburg, 1994). All of the above programs are accessible through the Internet from various centers, such as the US National Center for Biotechnology Information (http://www.ncbi.nih.gov/) and the European Bioinformatics Institute (http://www.ebi.ac.uk/).

Programs such as BLAST and FASTA use algorithms to identify optimal sequence alignments and typically display the output as a series of pair-wise comparisons between the test sequence (*query sequence*) and each related sequence which the program identifies in the database (*subject sequences*).

Different approaches can be taken to calculate the optimal sequence alignments. For example, in nucleotide sequence alignments the algorithm devised by Needleman and Wunsch (1970) seeks to maximize the number of matched nucleotides. In contrast, other programs such as that of Waterman *et al.* (1976) the object is to minimize the number of mismatches. Pair-wise comparisons of sequence alignments are comparatively simple when the test sequences are very closely matched and have similar, preferably identical, lengths. When the two sequences that are being matched are significantly different from each other, and especially when there are clear differences in length due to deletions/insertions, considerable effort may be necessary to calculate the optimal alignment (see panel immediately below).

```
GATATTATCACTGGAGCCTGGCAGGAGCT          GATATTATCACTGGAGCCTGGCAGGAGCT
*** **** *********** *******     OR    *** **** ********** * *******
GATTTTATGACTGGAGCCTGA-AGGAGCT          GATTTTATGACTGGAGCCT-GAAGGAGCT
```

The difficulty in sequence alignments. Here the two nucleotide sequences are clearly related but at the sequence **GGC** shown at top, there is uncertainty as to the best alignment with the corresponding sequence **GA** in the bottom sequence.

If the nucleotide sequence under investigation is a coding sequence, then nucleotide sequence alignments can be aided by parallel amino acid sequence alignments using the assumed translational reading frame for the coding sequence. This is so because there are 20 different amino acids but only four different nucleotides. Pair-wise alignments of amino acid sequences may also be aided by taking into account the *chemical subclasses* of amino acids. *Conservative substitutions* are nucleotide changes which result in an amino acid change but where the new amino acid is *chemically related* to the replaced amino acid and typically belongs to the same subclass (Box 11.3). As a result, algorithms used to compare amino acid sequences

typically use a **scoring matrix** in which pairs of scores are arranged in a 20×20 matrix where higher scores are accorded to identical amino acids and to ones which are of similar character (e.g. isoleucine and leucine) and lower scores are given to amino acids that are of different character (e.g. isoleucine and aspartate; see Henikoff and Henikoff, 1992). The typical output gives two overall results for percent sequence relatedness, often termed **% sequence identity** (matching of identical residues only) and **% sequence similarity** (matching of both identical residues and ones that are chemically related; see panel below).

```
Score = 52.8 bits (125), Expect = 9e-08
Identities = 39/120 (32%), Positives = 57/120 (47%), Gaps = 9/120 (7%)

Query:   1  AKLLIKHDSNIGIPDVEGKIPLHWAANHKDPSAVHTVRCILDAAPTESLLNWQDYEGRTP  60
            A+LL++HD++       G  PLH A +H +     + V+ +L +      W  Y   TP
Sbjct: 548  AELLLEHDAHPNAAGKNGLTPLHVAVHHNN---LDIVKLLLPRGGSPHSPAWNGY---TP  601
Query:  61  LHFAVADGNLTVVDVLTSY-ESCNITSYDNLFRTPLHWAALLGHAQIVHLLLERNKSGTI  119
            LH A    +V   L Y  S N S    + TPLH AA  GH ++V LLL +   +G +
Sbjct: 602  LHIAAKQNQIEVARSLLQYGGSANAESVQGV--TPLHLAAQEGHTEMVALLLSKQANGNL  659
```

Sequence identity and sequence similarity. The BLASTP output here resulted from querying the Swiss-prot protein database with a query sequence of amino acids 165–283 of the newly identified inversin protein. The subject sequence shown here is a mouse erythrocyte ankyrin sequence. The program considers not just *sequence identity* (39 of the 120 positions, or 32%, have identical residues in the two sequences; shown as **red** letters), but also *sequence similarity* (here indicated as 'positives') whereby an additional 19 positions have chemically similar amino acids (shown as **+**).

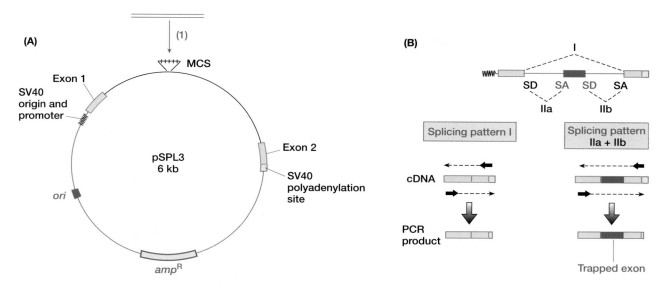

1. Clone genomic DNA fragment into multiple cloning site (MCS)
2. Transfect into COS cells
3. Expression from SV40 promoter → RNA product
4. Isolate RNA and use as template for making cDNA
5. Amplify in PCR reaction using primers specific for exons in vector

Figure 7.10: Exon trapping using the pSPL3 vector.

(A) The pSPL3 plasmid vector. This shuttle vector can be propagated in *E. coli* (using the ori origin of replication and selection for ampicillin resistance) and also in monkey COS cells (using the functional SV40 origin of replication) (Church *et al.*, 1994). The pSPL3 vector contains a minigene (in black): transcription occurs from the SV40 promoter and the RNA undergoes splicing under control of the host cell's RNA splicing machinery, resulting in fusion of the two vector exon sequences. **(B) Splicing patterns**. The normal splicing pattern which is seen when only the vector exons are present is indicated by splicing pattern I. If a genomic DNA fragment cloned into pSPL3 contains an exon with functional splice donor (SD) and splice acceptor (SA) sequences, a different splicing pattern (IIa + IIb) may occur. The two splicing patterns can be distinguished at the cDNA level by using various vector-specific PCR primers, and size-fractionation on gels can lead to recovery of the amplified exon from genomic DNA.

of cells and tissues which are spatially organized in a manner representative of the normal *in vivo* organization.

Gene expression throughput

Some methods are designed to obtain expression data for only one or a very small number of genes at a time (*low throughput* expression). Other methods can simultaneously track the expression of many genes at a time, and in some cases where a genome project has identified all the genes in an organism it has been possible to conduct **whole genome expression screening** (see Section 19.3).

7.3.2 Hybridization-based gene expression analyses: from single gene analyses to whole genome expression screening

Traditional hybridization-based expression screening has been low throughput, focused on analyzing RNA transcripts from one or only a few genes at a time, but the resolution can vary from low to high. Recently, however, microarray-based gene expression analyses have ushered in a new very high throughput type of low resolution expression analysis, allowing simultaneous sampling of RNA transcripts from thousands of genes at a time.

Northern blot hybridization

This approach affords low resolution expression patterns by hybridizing a gene or cDNA probe to total RNA or poly(A)$^+$ RNA extracts prepared from different tissues or cell lines. Because the RNA is size-fractionated on a gel, it is possible to estimate the size of transcripts. The presence of multiple hybridization bands in one lane may indicate the presence of differently sized isoforms (see *Figure 6.13*).

Tissue *in situ* hybridization

High resolution spatial expression patterns of RNA in tissues and groups of cells are normally obtained by **tissue *in situ* hybridization**. Usually, tissues are frozen or embedded in wax then cut using a microtome to give very thin sections (e.g. 5 microns thick) which are mounted on a microscope slide. Hybridization of a suitable gene-specific probe to the tissue on the slide can then give detailed expression images

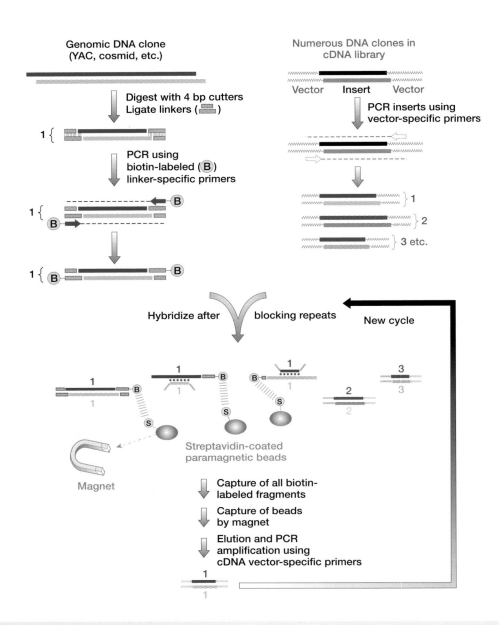

Figure 7.11: cDNA selection using magnetic bead capture.

The method relies on heteroduplex formation between single strands of a single genomic DNA clone (numbered 1) and of a complex cDNA population, such as the inserts of a cDNA library (numbered 1, 2, 3, etc.). The genomic DNA strands are labeled with a biotin group (attached to PCR primers which are incorporated during amplification). The hybridization reaction will favor heteroduplex formation involving those cDNA clones *cognate* with the genomic DNA clone. In this example, genomic DNA clone 1 and cDNA clone 1 are envisaged to be cognate, i.e. contain common sequences, allowing opposite sense strands to bond together, giving a heteroduplex. Hybridization products with a biotin group (including genomic DNA–cDNA heteroduplexes) will bind to streptavidin-coated paramagnetic beads and can be removed from other reaction components by a magnet. The separated beads can then be treated to elute the biotin-containing molecules and, by using PCR primers specific for the vector sequences flanking the cDNA, the bound cDNA can be amplified. This population is submitted to further hybridization cycles to enrich for the desired cDNA.

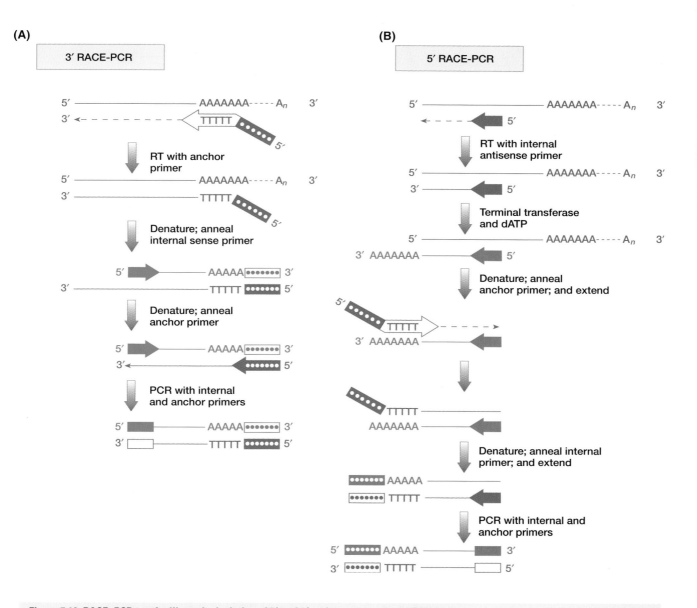

Figure 7.12: RACE–PCR can facilitate the isolation of 5′ and 3′ end sequences from cDNA.

A preliminary step in RACE–PCR involves the introduction of a specific end sequence by a form of 5′-**add-on mutagenesis** (see Section 5.5.3). **(A)** *3′ RACE–PCR.* A starting antisense primer is used with a specific 5′ extension sequence (*anchor sequence*, often > 15 nucleotides long) which becomes incorporated into the cDNA transcript at the reverse transcriptase step. An internal sense primer is then used to generate a short second strand ending in a sequence complementary to the original anchor sequence. Thereafter, PCR is initiated using the internal sense primer and an anchor sequence primer. **(B)** *5′ RACE–PCR.* Here an internal antisense primer is used to prime synthesis from a mRNA template (red) of a partial first cDNA strand (black). A poly(dA) is added to the 3′ end of the cDNA using terminal transferase. Second-strand synthesis is primed using a sense primer with a specific extension (anchor) sequence. This strand is used as a template for a further synthesis step using the internal primer in order to produce a complementary copy of the anchor sequence. PCR can then be accomplished using internal and anchor sequence primers.

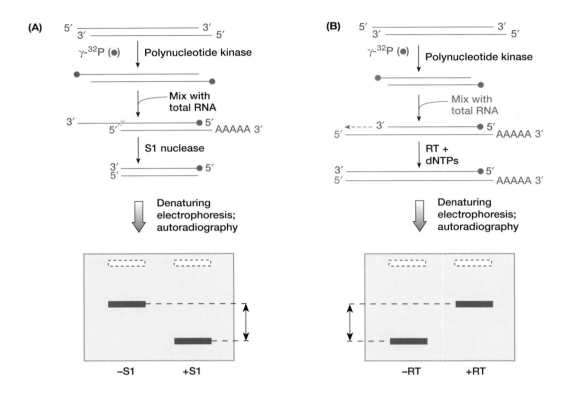

Figure 7.13: The transcriptional initiation site can be mapped by nuclease S1 protection or primer extension assays.

(A) *Nuclease S1 protection assay*. A restriction fragment from the 5′ end of a cloned gene is suspected of containing the transcription initiation site. It is end-labeled at the 5′ ends, then denatured and mixed with total RNA from cells in which the relevant gene is thought to be expressed. The cognate mRNA can hybridize to the antisense DNA strand to form an RNA–DNA heteroduplex. Subsequent treatment with nuclease S1 results in progressive cleavage of the overhanging 3′ DNA sequence until the point at which the DNA is hybridized to the 5′ end of the mRNA. Size-fractionation on a denaturing electrophoresis gel can identify the size difference between the original DNA and the DNA after nuclease S1 treatment. **(B)** *Primer extension assay*. Here, the restriction fragment suspected of containing the transcriptional initiation site is deliberately chosen to be small. Hybridization with a cognate mRNA will leave the mRNA with an overhanging 5′ end. The DNA can serve as a primer for reverse transcriptase (RT) to extend its 3′ end until the 5′ end of the mRNA is reached. The size increase after reverse transcriptase treatment (+RT) compared with before treatment (−RT) maps the transcription initiation site. More precise mapping is possible in both methods by sequencing the DNA following treatment with S1 or RT.

representative of the distribution of the RNA in the tissue of origin (see *Figure 6.15*). Often, the tissues used include embryonic tissues which have the advantage that their miniature size permits expression screening of many tissues in a single section.

Whole mount *in situ* hybridization

An extension of tissue *in situ* hybridization is to study expression in a *whole embryo*. **Whole mount *in situ* hybridization** is a popular method for tracking expression during development in whole embryos from model vertebrate organisms. Because of the ethical and practical difficulties in carrying out equivalent human gene analyses, there has been considerable reliance on extrapolating from analyses carried out on mouse embryos (see *Figure 7.15*). The relatively high amount of tissue available means the method is a relatively sensitive one and automation of the technique has enhanced its popularity.

Single cell gene expression profiling

Using suitably labeled probes specific RNA sequences can be tracked within *single cells* to identify sites of RNA processing, transport and cytoplasmic localization. By using quantitative fluorescence *in situ* hybridization (FISH) and digital imaging microscopy, it has even been possible to visualize *single RNA transcripts in situ* (Femino *et al.*, 1998). A further refinement uses combinations of different types of oligonucleotide probe labeled at multiple sites with one of a variety of spectrally distinct fluorophores. This has allowed transcripts from multiple genes to be tracked simultaneously (Levsky *et al.*, 2002).

Large-scale expression screening using microarrays

Gene expression screens have been transformed by the capacity to prepare high density oligonucleotides or cDNA clone **microarrays** on glass surfaces (see Section 6.4.3 for the general procedures). In the case of some genomes which have been completely sequenced, it affords the possibility of **whole**

	RESOLUTION	THROUGHPUT	EXAMPLES
RNA	High	Low	Tissue *in situ* hybridization Cellular *in situ* hybridization
	Low	Low	Northern blot hybridization RNA dot blot hybridization RT-PCR Ribonuclease protection assay
	Low	High	DNA microarray hybridization Differential display Serial analysis of gene expression (SAGE)
PROTEIN	High	Low	Immunocytochemistry Immunofluorescence microscopy
	Low	Low	Immunoblotting (western blotting)
	Low	High	2-D gel electrophoresis Mass spectrometry

Figure 7.14: Expression mapping can be conducted at different levels.

Throughput refers to the number of genes/proteins that can be studied at a time.

genome expression screening whereby the expression of every single gene in an organism can be monitored simultaneously (see Section 19.3).

7.3.3 PCR-based gene expression analyses: RT–PCR and mRNA differential display

As described in Section 5.2, the great advantages of PCR are its speed, sensitivity and simplicity. Although it is not so suited to providing spatial patterns of expression (in the way that tissue *in situ* hybridization, for example, does), it can provide rapid, gross patterns of expression which may be valuable.

Conventional RT–PCR

Reverse transcriptase–PCR (RT–PCR; see Section 5.2.1 for the basic principle) can provide rough quantification of the expression of a particular gene (useful in the case of cell types or tissues that are not easy to access in great quantity, e.g. early stage pre-implantation human embryos; see Daniels *et al.*, 1995). The extreme sensitivity of PCR means that RT–PCR can also be used to study expression in single cells. In addition, RT–PCR can be useful for identifying and studying different *isoforms* of an RNA transcript. For example, different mRNA isoforms may be produced by alternative splicing and can be identified when exon-specific primers identify extra amplification products in addition to the expected products (for an example, see Pykett *et al.*, 1994).

mRNA differential display

By using *partially degenerate* PCR primers (primers where the choice of base at some positions is deliberately flexible), it is

possible to devise modified forms of RT–PCR which can track the expression of many genes simultaneously. One such technique is **mRNA differential display**. It uses a *modified* oligo(dT) primer which at its 3′ end has a different single nucleotide (that is, A, C or G *but not T*) or a different dinucleotide (e.g. CA). As a result, it will bind to the poly(A) tail of a *subset of mRNAs* (Liang *et al.*, 1993). For example, if the oligonucleotide TTTTTTTTTTTT**CA** (= T_{11}CA) is used as a primer, it will preferentially prime cDNA synthesis from those mRNAs where the dinucleotide **TG** precedes the poly (A) tail.

The upstream primer is usually an arbitrary short sequence (often 10 nucleotides long) but, because of mismatching, especially at the 59 end, it can bind to many more sites than expected for a decamer). The resulting amplification patterns are deliberately designed to produce a complex ladder of bands when size-fractionated in a long polyacrylamide gel (*Figure 7.16*).

Unlike DNA microarray screens, mRNA differential display is a way of simultaneously monitoring the expression of multiple genes where the identities of the genes being tracked *have not previously been established*. Instead of screening the expression of known genes, therefore, it is simply an expression scanning method. Its main use has been in *comparative gene expression* studies to identify how gene expression alters when two sources are compared (e.g. the comparison of two types of cells at different physiological or developmental stages). This can allow identification of a small subset of genes whose expression patterns are different between the cell types.

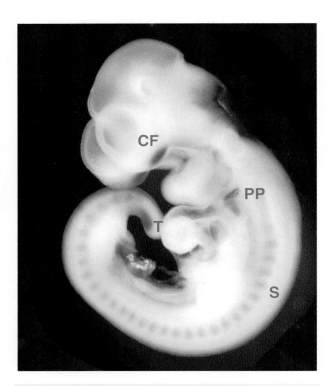

Figure 7.15: Whole mount *in situ* hybridization.

The example shows *Pax9* expression in an E9.5 (9.5 embryonic day = 9.5 days post coitum) mouse embryo. Expression is evident in the craniofacial region (CF; in what will develop into nasal mesenchyme), pharyngeal pouches (PP), somites (S) and tailbud (T). Image kindly provided by Dr. Heiko Peters, Institute of Human Genetics, University of Newcastle upon Tyne, UK.

7.3.4 Protein expression screens typically use highly specific antibodies

Because of their exquisite diversity and sensitivity in detecting proteins, antibodies have numerous applications in research, and their therapeutic potential is considerable (see Section 21.3.4). Traditionally, antibodies have been isolated by immunizing animals, but increasingly genetically engineered antibodies are being used (see *Box 7.4*). Antibodies have been used to detect proteins by different methods, and different labeling systems can be used.

Antibody labeling and detection systems

Antibodies can be labeled in different ways and, as with nucleic acid labeling and detection, antibodies can be used in either direct or indirect detection systems. In ***direct detection*** methods, the purified antibody is labeled appropriately with a reporter molecule (e.g. fluorescein, rhodamine, biotin etc.; see also Section 6.1.2) and then used directly to bind the target protein. In ***indirect detection*** systems, the primary antibody is used as an intermediate molecule and is not linked directly to a labeled group.

Once bound to its target, the primary antibody is in turn bound by a secondary reagent which is conjugated to a reporter group. A common secondary reagent is **protein A**, a protein found in the cell wall of *Staphylococcus aureus*. For unknown reasons protein A binds strongly to sites in the second and third constant regions of the Fc portion of Ig heavy chain). Alternatively one can use a ***secondary antibody*** (an antibody raised against a primary antibody).

Typical reporter groups may be a fluorochrome (Section 6.1.2), an enzyme (e.g. horseradish peroxidase, alkaline phosphatase, b-galactosidase etc.) or colloidal gold. Direct detection systems have the disadvantage that a large variety of primary antibodies may need to be conjugated to reporter molecules, whereas the indirect systems offer the use of readily available commercial affinity-purified secondary antibodies. Labeling-detection systems for use with specific methods of tracking protein expression are outlined in *Table 7.1*.

Immunoblotting (Western blotting)

This method is designed to survey gross protein expression using cell extracts which are fractionated according to size. Usually this is achieved by ***one-dimensional SDS–PAGE***, a form of poly*a*crylamide *g*el *e*lectrophoresis in which the mixture of extracted proteins is first dissolved in a solution of sodium dodecyl sulfate (SDS), an anionic detergent that disrupts nearly all noncovalent interactions in native proteins. Mercaptoethanol or dithiothreitol is also added to reduce disulfide bonds. Following electrophoresis, the fractionated proteins can be visualized by staining with a suitable dye (e.g. Coomassie blue) or a silver stain.

Two-dimensional PAGE gels may also be used: the first dimension involves *isoelectric focusing*, that is separation according to charge in a pH gradient, and the second dimension, at right angles to the first, involves size-fractionation by SDS–PAGE (see Stryer, 1995). In this case the fractionated proteins are transferred ('blotted') to a sheet of nitrocellulose and then exposed to a specific antibody (see *Figure 7.17*).

Immunocytochemistry (immunohistochemistry)

This technique is concerned with studying the overall expression pattern at the protein level, within a tissue or other multicellular structure. It can therefore be regarded as the protein equivalent of the tissue *in situ* hybridization methods used to screen RNA expression. As in the latter case, tissues are typically either frozen or embedded in wax and then cut into very thin sections with a microtome before being mounted on a slide. A suitably specific antibody is allowed to bind to the protein in the tissue section and can produce expression data that can be related to histological staining of neighboring tissue sections (*Figure 7.18*).

Immunofluorescence microscopy

This method is used when investigating the *subcellular location* for a protein of interest. A suitable fluorescent dye, such as fluorescein or rhodamine, is coupled to the desired antibody, enabling the relevant protein to be localized within the cell by using fluorescence microscopy (see *Figure 6.5B*).

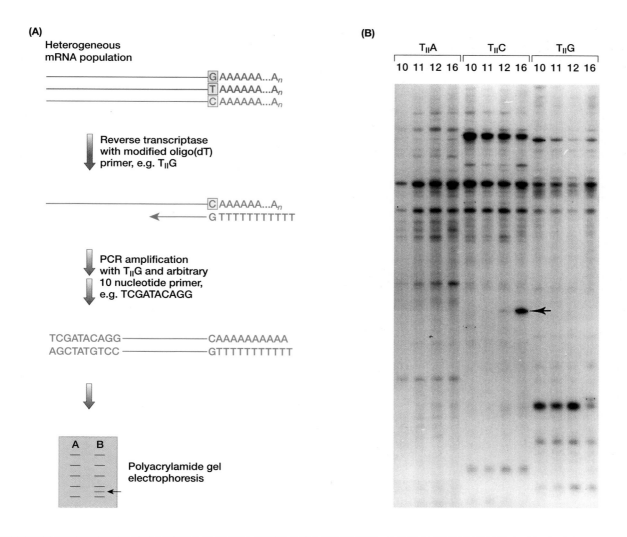

Figure 7.16: mRNA differential display is a modified RT–PCR method for multiplex gene expression scanning.

(A) *Schematic representation*. Total RNA from two or more cell types is reverse transcribed using a modified oligo (dT) primer (in this example, TTTTTTTTTTTG or $T_{11}G$ for short, which should prime cDNA synthesis preferentially from mRNA sequences where a C precedes the poly (A) tail. Amplification is carried out using an arbitrary primer and products are size-fractionated on a polyacrylamide gel. Differences in amplification bands between the RNA sources that are being compared (A and B) indicates differential expression. (B) *An application*: identifying genes differentially expressed at different stages of mouse heart development (embryonic days 10, 11, 12 and 16). Three sets of reaction conditions were used, with a $T_{11}G$ primer in each case and one of three different arbitrary primers AP1, AP2 and AP3. The figure shows a section of the gel where several bands can be seen to change in intensity at the different developmental stages. A particularly prominent change (shown by the arrow) turned out to be β-globin. The photograph was kindly provided by Andy Curtis and David Wilson, University of Newcastle upon Tyne, UK.

Box 7.4: Obtaining antibodies.

TRADITIONAL METHODS OF OBTAINING ANTIBODIES

Antibodies to human gene products have traditionally been obtained by repeatedly injecting suitable animals (e.g. rodents, rabbits, goats, etc.) with a suitable **immunogen**. Two types of immunogen are commonly used:

▸ **synthetic peptides**. The amino acid sequence (as inferred from the known cDNA sequence) is inspected and a synthetic peptide (often 20–50 amino acids long) is designed. The idea is that, when conjugated to a suitable molecule (e.g. keyhole limpet hemocyanin), the peptide will adopt a conformation that resembles that of the corresponding segment of the native polypeptide. This approach is relatively simple, but success in generating suitably specific antibodies is far from assured and difficult to predict.

▸ **fusion proteins**. An alternative approach is to insert a suitable cDNA sequence into a modified bacterial gene contained within an appropriate expression cloning vector. The rationale is that a hybrid mRNA will be produced which will be translated to give a fusion protein with an N-terminal region derived from the bacterial gene and the remainder derived from the inserted gene (see *Figure*). The N-terminal bacterial sequence is often designed to be quite short, but may nevertheless confer some advantages. For example it can provide a signal sequence to ensure secretion of the fusion protein into the extracellular medium, thereby simplifying its purification, and it may protect the foreign protein from being degraded within the bacterium. Because the fusion protein contains most or all of the desired polypeptide sequence, the probability of raising specific antibodies may be reasonably high.

If the animal's immune system has responded, specific antibodies should be secreted into the serum. The antibody-rich serum (**antiserum**) which is collected contains a heterogeneous mixture of antibodies, each produced by a different B lymphocyte [*because immunoglobulin gene rearrangements are cell-specific as well as cell type (B-lymphocyte)-specific*, see Section 10.6]. The different antibodies recognize different parts (**epitopes**) of the immunogen (**polyclonal antisera**). A homogeneous preparation of antibodies can be prepared, however, by propagating a *clone* of cells (originally derived from a single B lymphocyte).

Because B cells have a limited life span in culture, it is preferable to establish an *immortal* cell line: antibody-producing cells are fused with cells derived from an immortal B-cell tumor. From the resulting heterogeneous mixture of hybrid cells, those hybrids that have both the ability to make a particular antibody and the ability to multiply indefinitely in culture are selected. Such **hybridomas** are propagated as individual clones, each of which can provide a permanent and stable source of a single type of **monoclonal antibody (mAb)**.

The above methods of raising antibodies are not always guaranteed to produce suitably specific antibodies. An alternative approach for tracking the subcellular expression of a protein of interest is to use a previously obtained antibody to track it as a result of binding to an artificially coupled epitope (**epitope tagging**). In this procedure a recombinant DNA construct is generated by coupling a sequence that encodes an epitope for which a previously obtained antibody is available, to the coding sequence of the protein of interest in much the same way as in the *Figure* except that in this case the vector system is designed to be expressed in mammalian cells or other cells in which it is intended to investigate expression. Expression of this construct in the desired cells can be monitored by using the antibody specific for the epitope tag to track the protein. Commonly used epitope tags are shown below.

Sequence of tag	Origin	Location	mAb
DYKDDDDK	Synthetic FLAG	N, C terminal	Anti-FLAG M1
EQKLISEEDL	Human c-Myc	N, C terminal	9E10
MASMTGGQQMG	T7 gene 10	N terminal	T7.Tag Ab
QPELAPEDPED	HSV protein D	C terminal	HSV.Tag Ab
RPKPQQFFGLM	Substance P	C terminal	NC1/34
YPYDVPDYA	influenza HA1	N, C terminal	12CA5

Flag[Q33], HSV, Herpes Simplex Virus.

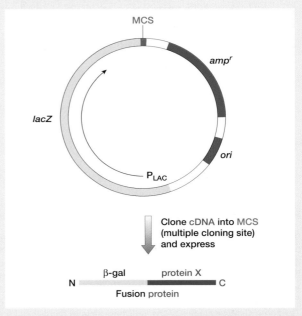

Fusion proteins are often designed as immunogens for raising antibodies.

GENETICALLY ENGINEERED ANTIBODIES

Antibodies generated by the above classical approaches originate from animals. Once the various immunoglobulin genes had been cloned, however, DNA cutting and ligation technology could be used to generate new antibodies including both **partially humanized antibodies** and **fully human antibodies** (see Section 21.3.4). For example, transgenic mice have been engineered to contain human immmunoglobulin loci permitting the *in vivo* production of fully human antibodies.

Box 7.4: *(continued)*

Novel approaches can bypass the need for hybridoma technology, and immunization altogether. The powerful **phage display technology** permits the construction of a virtually limitless repertoire of human antibodies with specificities against both foreign and self-antigens (see Winter *et al.*, 1994). The essence of this method is that the gene segments encoding antibody heavy and light chain variable sequences are cloned and expressed on the surface of a filamentous bacteriophage, and rare phage are selected from a complex population by binding to an antigen of interest (see Section 19.4.6 for a fuller explanation).

The plasmid vector shown here has an origin of replication (ori) and an ampicillin resistance gene (amp^R) which is designed for growth in *E. coli*. The multiple cloning site (MCS) is located immediately adjacent to a *lacZ* gene which can encode β-galactosidase with transcription occurring from the *lacZ* promoter (P_{lac}) in the direction shown by the arrow. A cDNA sequence from a gene of interest (gene X) is cloned in a suitable orientation into the MCS. Expression from the *lacZ* promoter will result in a β-galactosidase–X fusion protein which can be produced in large quantities and used as an immunogen in order to provoke the production of antibodies to protein X. A popular alternative is to use **GST fusion proteins**, where glutathione-S-transferase is coupled to the protein of interest and the fusion protein can be purified easily by affinity chromatography using glutathione–agarose columns.

Table 7.1: Antibody labeling–detection for tracking protein expression

Label	Detection method	Application
Iodine-125	X-ray film	Immunoblotting
Enzyme	Chromogenic substrate detected by eye	Immunoblotting; Immunocytochemistry
Biotin	Avidin or streptavidin coupled to various labels	Immunoblotting; Immunocytochemistry
Fluorochrome	Fluorescence microscopy (*Figure 6.5B*)	Immunocytochemistry; Immunofluorescence microscopy

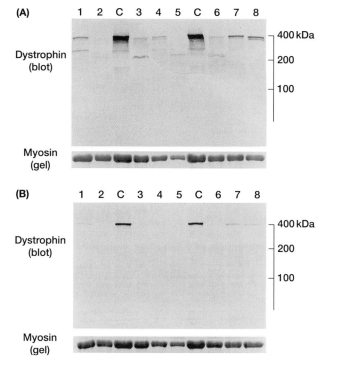

Figure 7.17: Immunoblotting (Western blotting) detects proteins that have been size-fractionated on an electrophoresis gel.

Immunoblotting involves detection of polypeptides after size-fractionation in a polyacrylamide gel and transfer ('blotting') to a membrane. This example illustrates its application in detecting dystrophin using two antibodies. The Dy4/6D3 antibody is specific for the rod domain and was generated by using a fusion protein immunogen (see *Box 7.4*). The Dy6/C5 antibody is specific for the C-terminal region and was generated by using a synthetic peptide immunogen. Reproduced from Nicholson *et al.* (1993) with permission from the BMJ Publishing Group. The photograph was kindly provided by Louise Anderson (formerly Nicholson), University of Newcastle upon Tyne, UK.

Ultrastructural studies

Higher resolution still of the intracellular localization of a gene product or other molecule is possible using electron microscopy. The antibody is typically labeled with an electron-dense particle, such as colloidal gold spheres.

7.3.5 Autofluorescent protein tags provided a powerful way of tracking subcellular localization of proteins

Green fluorescent protein (GFP) is a 238-amino acid protein originally identified in the jellyfish *Aequoria victoria*. Similar proteins are expressed in many jellyfish and appear to be responsible for the green light that they emit, being stimulated by energy obtained following oxidation of luciferin or another photoprotein (See Tsien, 1998). When the GFP gene was cloned and transfected into target cells in culture, expression of GFP in heterologous cells was also marked by emission of the green fluorescent light. This means that GFP is an *autofluorescent protein*: by itself it can act as a functional fluorophore. It can therefore serve as a unique reporter since it does not require other agents such as antibodies, cofactors, enzyme substrates, and so on. As a result, GFP can readily be followed by conventional and confocal fluorescence microscopy, and has became a popular tool for tracking gene expression in animals (see Section 20.3.1).

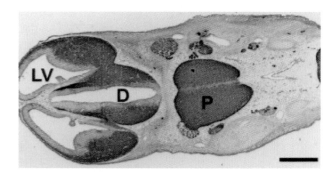

Figure 7.18: Immunocytochemistry.

In this example, β-tubulin expression was screened in a transverse section of the brain of a 12.5-embryonic day mouse. The antibody detection system used identifies expression ultimately as a brown color reaction based on horseradish peroxidase/3,3′ diaminobenzidine. The underlying histology was revealed by counterstaining with Toluidine blue. Abbreviations: LV, lateral ventricle; D, diencephalon; P, pons. Figure kindly provided by Steve Lisgo, Institute of Human Genetics, University of Newcastle upon Tyne, UK.

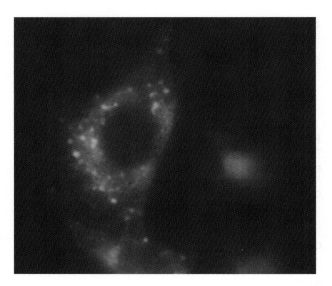

Figure 7.19: Tagging with the green fluorescent protein provides a powerful way of tracking protein expression.

This example shows a live transiently transfected HeLa cell expressing a GFP-tagged Batten disease protein. *CLN3* was cloned into the GFP expression vector, pEGFP-N1, so that a fusion protein consisting of the Batten disease protein with a GFP sequence coupled to its C terminus (a *GFP tag*) was produced. This cell is an example of a small proportion of HeLa cells expressing CLN3p/GFP in a vesicular punctate pattern distributed throughout the cytoplasm. These and other analyses indicate that the Batten disease protein is a Golgi integral membrane protein. Reproduced from Kremmidiotis *et al.* (1999) *Hum. Mol. Genet.* **8**, 523–531, with permission from Oxford University Press.

The most successful and popular applications have used GFP as a tag in a fusion protein where it is coupled to the protein whose expression is to be tracked. In such cases, the principal aim is to investigate the subcellular localization of the protein under investigation. GFP itself is not specifically localized within cells: in most cell types the fluorescence of GFP appears to be homogeneously spread all over the nucleus, cytoplasm and distal cell processes. Genetic engineering can be used to produce vectors containing a GFP coding sequence into which a coding sequence for an uncharacterized protein, X, can be cloned. The resulting GFP–X fusion construct can be transfected into suitable target cells such as cultured mammalian cells and expression of the GFP–X fusion protein can be monitored to track the subcellular location of the protein. See *Figure 7.9* for an example of tracking expression of the protein produced by CLN3, the gene associated with Batten disease, a neurodegenerative disease of childhood.

Further reading

Sambrook J, Russell D (2001) *Molecular Cloning: A Laboratory Manual,* 3rd Edn. Cold Spring Harbor Laboratory Press, Cold Spring Harbor, NY.

References

Chee M, Yang R, Hubbell E *et al.* (1996) Accessing genetic information with high-density arrays. *Science* **274**, 610–614.

Church DM, Stotler CJ, Rutter JL, Murrell JR, Trofatter JA, Buckler AJ (1994) Isolation of genes from complex sources of mammalian genomic DNA using exon amplification. *Nature Genet.* **6**, 98–105.

Clement-Jones M, Schiller S, Rao E *et al.* (2000) The short stature homeobox gene *SHOX* is involved in skeletal abnormalities in Turner Syndrome. *Hum. Molec. Genet.* **9**, 695–702.

Daniels R, Kinis T, Serhal P, Monk M (1995) Expression of myotonin protein kinase gene in preimplantation human embryos. *Hum. Molec. Genet.* **4**, 389–393.

Femino AM, Fay FS, Fogarty K, Singer RH (1998) Visualization of single RNA transcripts *in situ. Science* **280**, 585–590.

Frohman MA, Dush MK, Martin GR (1988) Rapid production of full length cDNAs from rare transcripts: amplification using a single gene-specific oligonucleotide primer. *Proc. Natl Acad. Sci. USA* **85**, 8998–9002.

Ginsburg M (1994) In: *Guide to Human Genome Computing* (ed. MJ Bishop), pp. 215–248. Academic Press, New York.

Hauge Y, Litt M (1993) A study of the origin of 'shadow bands' seen when typing dinucleotide repeat polymorphisms by the PCR. *Hum. Mol. Genet.* **2**, 411–415.

Henikoff S, Henikoff JG (1992) Amino acid substitution matrices from protein blocks. *Proc. Natl Acad. Sci. USA* **89**, 10915–10919.

Kremmidiotis G, Lensink IL, Bilton RL, Woollatt E, Chataway TK, Sutherland GR, Callen DF (1999) The Batten disease gene product (CLN3p) is a Golgi integral membrane protein. *Hum. Mol. Genet.* **8**, 523–531.

Levsky JM, Shenoy SM, Pezo RC, Singer RH (2002) Single cell gene expression profiling. *Science* **297**, 836–840.

Liang P, Averboukh L, Pardee AB (1993) Distribution and cloning of eukaryotic mRNAs by means of differential display: refinements and optimization. *Nucleic Acids Res.* **21**, 3269–3275.

Lovett M (1994) Fishing for complements: finding genes by direct selection. *Trends Genet.* **10**, 352–357.

Maxam AM, Gilbert W (1980) Sequencing end labeled DNA with base-specific chemical cleavages. *Methods Enzymol.* **65**, 499–560.

Meldrum D (2000) Automation for genomics, part two: sequencers, microarrays and future trends. *Genome Res.* **10**, 1288–1303.

Monaco AP (1994) Isolation of genes from cloned DNA. *Curr. Opin. Genet. Dev.* **4**, 360–365.

Needleman SB, Wunsch CD (1970) A general method applicable to the search of similarities in the amino acid sequences of two proteins. *J. Mol. Biol.* **48**, 443–453.

Pykett MJ, Murphy M, Harnish PR, George DL (1994) The neurofibromatosis 2 (NF2) tumor suppressor gene encodes multiple alternatively spliced transcripts. *Hum. Mol. Genet.* **3**, 559–564.

Riordan JR, Rommens JM, Kerem B *et al.* (1989) Identification of the cystic fibrosis gene: cloning and characterization of complementary DNA. *Science* **245**, 1066–1073.

Schutze K, Lahr G (1998) Identification of expressed genes by laser-manipulated manipulation of single cells. *Nature Biotechnol.* **16**, 737–742.

Simone NL, Bonner RF, Gillespie JW, Emmert-Buck MR, Liotta LA (1998) Laser-capture microdissection: opening the microscopic frontier to molecular analysis. *Trends Genet.* **16**, 272–276.

Stryer L (1995) *Biochemistry*, 4th Edn. W.H. Freeman & Co., New York.

Tonkin E, Hagan DM, Li W, Strachan T (2002) Identification and characterisation of novel mammalian homologs of *Drosophila* polyhomeotic permits new insights into relationships between members of the polyhomeotic family. *Hum. Genet.* **111**, 435–442.

Tsien RY (1998) Green fluorescent protein. *Ann. Rev. Biochem.* **67**, 509–554.

Waterman MS, Smith TF, Beyer WA (1976) Some biological sequence metrics. *Adv. Math.* **20**, 367–387.

Winter G, Griffiths AD, Hawkins RE, Hoogenboom HR (1994) Making antibodies by phage display technology. *Annu. Rev. Immunol.* **12**, 433–455.

PART TWO

The human genome and its relationship to other genomes

CHAPTER EIGHT

Genome projects and model organisms

Chapter contents

8.1 The ground-breaking importance of genome projects

8.1.1 Genome projects prepared the way for systematic studies of the Universe within

After many centuries of investigation we have built up an approximate understanding of at least the more accessible parts of our external Universe. The scale is impressive, and some concepts are certainly outside our normal experience (such as the 13 dimensions which are now thought to exist!). However, there is also a largely unexplored Universe *within us*, and it too is of impressive scale (the complexity of the human brain is a useful example: about 10^{11} neurons and somewhere in the region of 10^{15} interconnections).

Until recently, exploratory voyages into our internal universe have been modest and limited in scale. The application of microscopy to the study of cells and subcellular structures provided one major route into this internal world, to be followed by pioneering advances in biochemistry and then molecular and cellular biology. Now, at the start of a new millennium, we are poised to move from these initial investigations onto an altogether higher plane. Now it will be possible to conduct a serious and systematic exploration of our internal universe.

The catalyst which paved the way for this new phase of discovery was the **Human Genome Project (HGP)**, a truly international endeavor. Officially inaugurated in 1990, it was biology's first 'big project', with a projected time-span of 15 years. The HGP and other genome projects sought for the first time to know the precise chemical instructions which define living organisms, the complete genome sequences (*genome* means the total set of *different* DNA molecules – see the genomics glossary in *Box 8.1*). For many scientists, this was biology's equivalent of the periodic table: all matter can be reduced to a periodic table of *elements*, but at a higher level, every living thing can be reduced to a periodic table of genes.

The goals of the Human Genome Project (HGP)

The major *rationale* of the HGP was to acquire fundamental information concerning our genetic make-up which would further our basic scientific understanding of human genetics and of the role of various genes in health and disease. Since the small (16.5 kb) human mitochondrial DNA (mtDNA) sequence had been published in 1981 (see Section 9.1.2), the primary goal of the HGP was to sequence the vastly larger (ca. 3000 Mb) nuclear genome. As a first step towards achieving this, there was a need for high resolution human genetic maps which could then be used as a *scaffold* (framework) for constructing high resolution physical maps, culminating in the ultimate physical map, the complete sequence of the human genome.

In addition to the primary goal of mapping and sequencing the human nuclear genome, the HGP envisaged at its outset a series of ancillary projects:

▸ *development of appropriate technologies and tools.* This would include developments in genetic and physical mapping approaches, in DNA sequencing technology, in database design and construction, in informatics for sequence analysis and so on;

▸ *genome projects for five model organisms*: the bacterium *E. coli*, the yeast *Saccharomyces cerevisiae*, the roundworm *Caenorhabditis elegans*, the fruit fly *Drosophila melanogaster*, and the mouse. Here, the object was two-fold: to provide much-needed information for these model organisms; and to provide test-cases for implementing and refining the various technologies and tools which would be needed for the HGP;

▸ *ethical, legal and societal implications (ELSI).* A significant fraction of the total expenditure was devoted to this important, but easy to overlook, component.

By 2003 the sequencing goals for both the human genome and the five initial model organism genomes had been achieved, and the sequences were available through the internet. During the course of the HGP, other genome projects were initiated for a wide variety of other model organisms and by May 2003 the genome sequences of 140 other organisms had been determined (see Section 8.4). Additional ancillary studies have been concerned with the extent of *sequence variation* within the human genome.

8.1.2 The medical and scientific benefits of the genome projects are expected to be enormous

For many human biologists and geneticists, the HGP was an exciting, historic mission. A major justification was the anticipated medical benefits (see Collins and McKusick, 2001; Subramanian *et al.*, 2001; van Ommen, 2002). For inherited disorders where a major single gene is causative, comprehensive prenatal/presymptomatic diagnosis of disorders will be possible in individuals judged to be at risk of carrying a disease gene. Additionally, information on gene structure will be used to explore how individual genes function and how they are regulated. Such information will provide sorely needed explanations for biological processes in humans. It would also be expected to provide a framework for developing new therapies for diseases, and to extend the current therapeutic approaches. Wide-scale application of mutation screening would be expected to usher in a radical change in the approach to medical care, moving more from one of treating advanced disease to one of *preventing disease*, based on the identification of individual risk (**personalized medicine**).

Exciting though such possibilities are, however, there may be unexpected difficulties in understanding precisely and comprehensively how some genes function and are regulated (cautionary precedents are the slow progress in predicting protein structure from the amino acid sequence, and the lack of understanding of the precise ways in which the regulation of globin gene expression is coordinated, decades after the relevant sequences have been obtained). In addition, the single gene disorders which should be the easiest targets for developing novel therapies are very rare; the most common disorders are multifactorial and present considerable challenges. So although the data collected in the Human Genome

Box 8.1: A genomics glossary.

centiMorgan (cM). A unit of distance in a *genetic map* (see below). In the human genome IcM corresponds roughly to a physical map distance of 1 Mb.

centiRay (cR). A unit of map distance in a *radiation hybrid map* (see below).

Clone. DNA clones are populations of identical DNA molecules which have been purified by cell-based cloning methods (Section 5.3.1) or by PCR (Section 5.2.1).

Contig. A series of DNA clones which have been shown to contain insert DNA molecules that derive from neighboring and *overlapping* regions of a chromosome – see *Box 8.5*.

DNA marker. A general term for a DNA sequence which has been, or can be, placed on a genetic map (in the case of *polymorphic markers* – see below) or on a physical map (in the case of all markers).

CpG island. Short stretch of GC-rich DNA, often < 1 kb, containing frequent unmethylated CpG dinucleotides. CpG islands tend to mark the 5′ ends of genes – see *Box 9.3*.

DNA library. A collection of DNA clones which is meant to collectively represent a starting population of DNA. For a *genomic DNA library*, the starting DNA is the total DNA from a given cell population (which shows little variation between different cell types). In the case of a *cDNA library*, the starting DNA is *cDNA* prepared using reverse transcriptase from single-stranded RNA from a specific tissue (with very considerable variation in the cDNA of different tissues). See Sections 5.3.4 and 5.3.5 for how libraries are made and screened.

EST (expressed sequence tag). An expressed *STS (sequence tagged site*; see below) obtained by randomly selecting a cDNA clone for sequencing and designing specific primers for specifically PCR amplifying the corresponding fragment from genomic DNA.

Genetic map. A map which relies on tracing the inheritance of phenotypes and/or polymorphic markers, through generations. Polymorphic loci are positioned relative to one another on the basis of the frequency with which they recombine during meiosis. The unit of distance is *1 centiMorgan (1 cM)* which denotes a 1% chance of recombination.

Genome. The collective name for the *different DNA molecules* found in the cells of a particular species. In humans, the genome comprises 25 different DNA molecules: a single type of mitochondrial DNA and 24 different nuclear DNA molecules (see Section 9.1.1). Because the amount of DNA in the nucleus is so large, however, the term genome is often loosely used to mean the set of *nuclear DNA* molecules (more accurately termed the *nuclear genome*; mitochondrial DNA is often described as the *mitochondrial genome*).

Hybrid cell mapping. Human DNA markers can be assigned to a specific chromosomal or subchromosomal location by using panels of different hybrid cells containing a full complement of chromosomes from a rodent species (hamster or mouse) and a variable subset of human chromosomes, or of fragments of human chromosomes broken by exposure to X-rays (*radiation hybrids*; see *Box 8.4*).

Microsatellite marker. A type of DNA marker which is commonly used, largely because markers of this type can be very polymorphic. See *Figures 7.7 and 7.8*.

Physical map. A map which provides information on the *linear structure* of DNA molecules. The most detailed physical map is the nucleotide sequence.

Polymorphic markers. Polymorphic (genetic) markers are DNA sequences which show variation between individuals and which are used in constructing genetic maps by following how alleles segregate in large families. Markers may be located within coding sequences or other gene components but are mostly located in noncoding DNA. Commonly used markers are *microsatellites* and *SNPs*, although in the past *RFLPs* were used and even protein polymorphisms.

Radiation hybrid (RH) map. A genome map in which STSs are positioned relative to one another according to the frequency with which they are separated by radiation-induced chromosome breaks. The frequency is assayed by analyzing a panel of **hybrid cell** (human–hamster) lines which contain different patterns of human chromosome fragments initially generated by exposure to X-rays. The unit of map distance is **1 centiRay (1 cR)**, denoting a 1% chance of a break occurring between two loci.

RFLP (restriction fragment length polymorphism). A type of DNA marker used widely in the past but rather infrequently in modern times because they are often not very polymorphic and are not so easy to type. See Section 7.1.3.

SNP (single nucleotide polymorphism). SNPs provide a type of DNA marker which is increasingly being used. They occur very frequently in DNA and can be typed very easily by automated methods, allowing very large numbers of samples to be analyzed at a time. See Section 7.1.3 and *Box 18.2* for how SNPs are typed.

STS (sequence tagged site). Any short (usually < 500 bp) sequence which is uniquely represented in a genome and for which primers have been designed enabling specific PCR amplification of that sequence (see *Box 5.4*). STSs were often designed by randomly sequencing the ends of genomic clones and so were often nonpolymorphic, but a subset of STSs are known to be polymorphic, including *microsatellite markers* (see above).

Project will inevitably be of medical value, some of the most important medical applications may take some considerable time to be developed.

As we move into the **post-genome era** huge international efforts are focusing on how the human genome sequence can specify a person, and how the DNA of other organisms is related to us and to their biologies. The sequence information obtained by studying the structure of genomes (conventional **genomics**) is paving the way for other large-scale approaches to investigate the function of genomes (*functional genomics*) and how different genomes relate to each other (*comparative genomics*). These topics are covered in Chapter 19 and Section 12.3.

8.2 Background and organization of the Human Genome Project

8.2.1 DNA polymorphisms and new DNA cloning technologies paved the way for sequencing our genome

Since the mid 1950s we have had a very rough physical map of the human genome — a cytogenetic map based on distinguishing the chromosomes by size and shape. To progress towards very detailed physical maps needed new approaches. A major problem was that the initial goal of obtaining a human genetic map – which could act as a *scaffold* for building detailed physical maps – seemed impossible. Instead, human geneticists watched enviously as **classical genetic maps** were constructed decades ago for *Drosophila* and mouse, and then continuously refined.

Classical genetic maps are based on **genes**. They are constructed by crossing different mutants in order to determine whether the two gene loci are linked or not. A classical human genetic map could never be achieved, however, because the frequency of mating between two individuals suffering from different genetic disorders is vanishingly small. Without a genetic map to provide anchor points, it was difficult to imagine how detailed physical maps of all the chromosomes could be made.

A turning point was the growing realization in the late 1970s that much – indeed, the great majority – of the sequence variation in our genome occurred *outside of genes, and could be assayed*. The variation that had been studied up until then was focused on protein polymorphisms. Only a few protein markers could be studied because coding DNA is a very small fraction (2%) of the genome and not so prone to variation (because it is functionally important and so highly conserved during evolution). By contrast, the vast majority of the > 98% of our DNA which is noncoding is not well conserved and is very susceptible to changes in DNA sequence.

In the late 1970s methods became available to assay *DNA variation* for the first time (by screening for *restriction fragment length polymorphisms*, or *RFLPs* – Box 8.1). At last, the idea of constructing a comprehensive, nonclassical human genetic map became a possibility (Botstein *et al.,* 1980). From now on human geneticists would build increasingly detailed genetic linkage maps using DNA markers which were randomly scat-

tered throughout our genome. By testing to see if specific alleles from two or more markers segregated together in family studies, the DNA markers could be allocated to a particular **linkage group.** Individual linkage groups could in turn be assigned to specific chromosomes by physical mapping of one or more of the constituent markers (e.g. by labeling a marker and hybridizing it to a metaphase chromosome preparation – see *Figures 2.16* and *2.17* – or by using panels of *hybrid cells* – see section 8.3.2).

Another important requirement was the development of powerful DNA cloning technologies which allowed the assembly of clones containing large pieces of DNA (*large insert DNA libraries*). The inserts of the clones had been generated by randomly fragmenting the genome but they could be tested to see which particular DNA markers they contained and whether they shared markers with inserts in other clones. If they did, the clones often had *overlapping* insert DNAs, and it became possible to produce ordered maps of DNA clones according to overlap between the DNA inserts, so called *clone contigs*.

8.2.2 The Human Genome Project was mainly conducted in large genome centers with high-throughput sequencing capacity

Organization of the Human Genome Project

While the U.S. Human Genome Project provided the initial momentum to the publicly funded Human Genome Project, several other countries quickly developed their own Human Genome Projects. Centers in the UK and France were quick to make their mark, and more recently considerable contributions were made by centers in some other countries, notably Japan and Germany. In order to co-ordinate the different national efforts, the **Human Genome Organization (HUGO)** was established in 1988 with a remit of facilitating exchange of research resources, encouraging public debate

Figure 8.1: Large-scale DNA sequencing at the Wellcome Trust Sanger Institute.

The Wellcome Trust Sanger Institute at Hinxton, UK has been the single biggest contributor to the publicly funded Human Genome Project. Data can be accessed at http://www.sanger.ac.uk

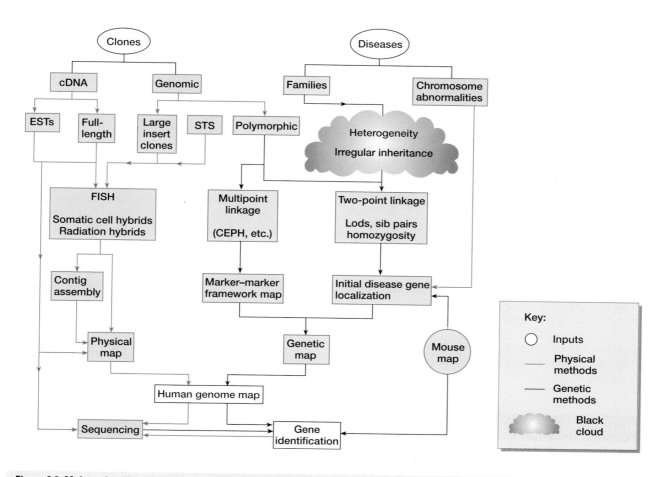

Figure 8.2: Major scientific strategies and approaches used in the Human Genome Project.

The major scientific thrust of the Human Genome Project began with the isolation of human genomic and cDNA clones (by cell-based cloning or PCR-based cloning). These were then used to construct high resolution genetic and physical maps prior to obtaining the ultimate physical map, the complete nucleotide sequence of the 3000-Mb nuclear genome. Inevitably, the project interacted with research on mapping and identifying human disease genes. In addition, ancillary projects included studying genetic variation (the Human Genome Diversity Project – Section 8.3.7), genome projects for model organisms (Section 8.4) and research on ethical, legal and social implications. The data produced were channeled into mapping and sequence databases permitting rapid electronic access and data analysis. EST, expressed sequence tag; STS, sequence tagged site.

and advising on the implications of human genome research (McKusick, 1989).

Because of the large scale involved, much of the technology required to sequence the human genome was concentrated in a few very large **genome centers** with industrial scale sequencing capacities (*Figure 8.1*). Automated fluorescence labeling-based DNA sequencing became the norm and the advent of *capillary-based DNA sequencing* (Section 7.1.2) provided a much needed boost to high-throughput sequencing. For the publicly funded HGP, most of the sequence was contributed by five large centers, the Wellcome Trust Sanger Institute in the UK and four centers in the USA located at: The Whitehead Institute/Massachussetts Institute of Technology in Massachusetts, Washington University, the DoE Joint Genome Institute and Baylor College of Medicine. Interacting with these and some other large centers was a

world-wide network of small laboratories mostly attempting to map and identify disease genes and typically focusing on very specific subchromosomal regions (see *Figure 8.2*).

Communication between the network of genome centers and interacting laboratories was – and continues to be – very largely based on electronic communication. The need to manage and store the huge amount of sequencing data that was being produced quickly led to the development of large **electronic databases** which, at least for the publicly funded mapping and sequencing efforts, are freely accessible through the internet. Analyses can then be conducted from remote computer terminals throughout the world. Depending on the source of input data, two types of database were relevant:

▶ *central repositories for storing globally produced mapping and sequence data*. Universal DNA and protein **sequence data-bases** were established decades before the onset of the genome projects but specialized *species-specific* mapping databases were established more recently, such as the **Genome database** (**GDB**) a mapping database aimed specifically at storing *human* mapping data (**note:** there is a specific nomenclature for naming DNA segments and genes in different species – see Further Reading and see *Box 8.2* for the human nomenclature);

▶ *databases for storing locally produced data.* In order to improve their own efficiency, the big genome mapping and sequencing centers have stored data produced in their own laboratories in dedicated databases. Unlike data input, data access is freely available through the Web from publicly funded genome centers.

8.3 How the human genome was mapped and sequenced

The official Human Genome Project was meant to have been a 15 year project spanning 1990–2005, but progress was faster than expected. Genetic maps were developed ahead of the original schedule and the final stage of large-scale DNA sequencing was facilitated by developments in automated fluorescence-based DNA sequencing, and by added momentum resulting from competition with a private company, Celera. By 2003, an essentially finished sequence

was available to all through the internet. See *Box 8.3* for a timeline of some of the major milestones.

8.3.1 The first useful human genetic maps were based on microsatellite markers

The realization in the early 1980s that a comprehensive human genetic map was now attainable sparked serious efforts to construct one. The first such map was reported in 1987 and was based mostly on **restriction fragment length polymorphisms** (**RFLPs**). Despite this heroic achievement, the RFLP map had serious limitations: the average spacing between the markers was considerable, and more significantly RFLP markers are not very informative markers (only two alleles) and are not so easy to type.

Thereafter, attention switched to making maps with **microsatellite markers** which have the advantage of being highly informative, easy to type and dispersed throughout the genome (see Section 9.4.3 and *Figures 7.7* and *7.8*). Within another 5 years researchers at the Généthon laboratory in France reported the first microsatellite-based linkage map of the human genome (Weissenbach *et al.*, 1992), and 2 years later an international consortium published a further improved map, mostly based on microsatellites with a high marker density, approximately one marker every centiMorgan (cM) (Murray *et al.*, 1994).

Box 8.2: Human gene and DNA segment nomenclature.

The nomenclature used is decided by the HUGO nomenclature committee. Genes and pseudogenes are allocated symbols of usually two to six characters: Pseudogene sequences are indicated by a P following the relevant gene symbol. For anonymous DNA sequences, the convention is to use D (= DNA) followed by 1–22, X or Y to denote the chromosomal location, then S for a unique segment, Z for a chromosome-specific repetitive DNA family or F for a multilocus DNA family, and finally a serial number. The letter E following the number for an anonymous DNA sequence indicates that the sequence is known to be expressed.

Symbol	Interpretation
CRYB1	Gene for crystallin β polypeptide 1
GAPD	Gene for glyceraldehyde-3-phosphate dehydrogenase
GAPDL7	GAPD-like gene 7, functional status unknown
GAPDP1	GAPD pseudogene 1
AK1	Gene for adenylate kinase, locus 1
AK2	Gene for adenylate kinase, locus 2
PGK1*2	Second allele at *PGK1* locus
B3P42	Breakpoint number 42 on chromosome 3
DYS29	Unique DNA segment, number 29 on the Y chromosome
D3S2550E	Unique DNA segment number 2550 on chromosome 3, known to be expressed
D11Z3	Chromosome 11-specific repetitive DNA family number 3
DXYS6X	DNA segment found on the X chromosome, with a known homolog on the Y chromosome, and representing the 6th XY homolog pair to be classified
DXYS44Y	DNA segment found on the Y chromosome, with a known homolog on the X chromosome, 44th XY homolog pair
D12F3S1	DNA segment on chromosome 12, first member of multilocus family 3
DXF3S2	DNA segment on chromosome X, second member of multilocus family 3
FRA16A	Fragile site A on chromosome 16

Box 8.3: Major milestones in mapping and sequencing the human genome.

1956: The first physical map of the human genome is determined – light microscopy of stained tissue reveals that our cells contain 46 chromosomes, with a total of 24 different types of chromosome.

1977: Fred Sanger and colleagues in Cambridge, UK publish the dideoxy DNA sequencing method, which is still the basis of current DNA sequencing technology more than a quarter of a century later.

1980: Botstein *et al.* (1980) propose that a human genetic map can be constructed using a set of random DNA markers (RFLPs).

1981: Fred Sanger and colleagues publish the complete sequence of human mitochondrial DNA (see Section 9.1.2).

1984: Workshop in Alta, Utah partially sponsored by the U.S. Department of Energy (DoE) to evaluate methods of mutation detection and characterization and project future technologies. A principal conclusion is that an enormously large, complex and expensive sequencing program is required to permit high efficiency mutation detection.

1987: The U.S. Department of Energy Report on a Human Genome Initiative envisages three major objectives: generation of refined physical maps of human chromosomes; development of support technologies and facilities for human genome research; and expansion of communication networks and of computational and database capacities.

1988: The U.S. National Institutes of Health (NIH) sets up an Office of Human Genome Research (later re-named the National Center for Human Genome Research) to co-ordinate NIH genome activities in co-operation with other U.S. organizations. The **Human Genome Organization (HUGO)** is established in the same year to co-ordinate international efforts and has a remit of facilitating exchange of research resources, encouraging public debate and advising on the implications of human genome research (McKusick, 1989).

1990: Official launch of the **Human Genome Project (HGP)** following implementation of a $3 billion 15-year project in the U.S.

1991: The Genome Database (GDB), a repository for human DNA mapping data, is established.

1992: Jean Weissenbach and colleagues at the Généthon laboratory, France publish the first comprehensive human genetic linkage map based on microsatellite markers (Weissenbach *et al.*, 1992).

1993: Daniel Cohen and colleagues at the Généthon laboratory, France publish a first generation physical map of the human genome, based on large insert DNA clones (Cohen *et al.*, 1993).

1995: Eric Lander and colleagues at the Whitehead Institute/Massachusetts Institute of Technology publish the first detailed physical map of the human genome, based on *sequence tagged sites* (Hudson *et al.*, 1995).

1998: *GeneMap '98*, the first reasonably comprehensive map of *gene-based markers*, is published by an international consortium led by researchers at the Sanger Centre, UK (Deloukas *et al.*, 1998).

1999: An international consortium led by researchers at the Sanger Centre, UK publishes the first complete sequence of a human chromosome, chromosome 22 (Dunham *et al.*, 1999).

2001: Publication of rough working drafts of the sequence of the human genome (comprising roughly 90% of the total sequence) by an international consortium of publicly funded researchers and by a private company, Celera (International Human Gene Sequencing Consortium, 2001; Venter *et al.*, 2001).

2003: Completion of the sequence of the human genome.

The high resolution human genetic map published in 1994 met the first major scientific goal of the Human Genome Project and now permitted an important *scaffold* for developing detailed physical maps of all the chromosomes. From now on the major focus of the HGP would be on developing and refining physical maps leading to the ultimate physical map, the complete DNA sequence of each chromosome (see section 8.3.2). However, genetic mapping of the human genome has continued, in two major ways:

▶ *refinement of microsatellite maps*. Currently, the most detailed map is the one constructed by the deCODE genetics group in Iceland. It involved typing 5136 microsatellite markers in 146 families, with a total of 1257 meiotic events (Kong *et al.*, 2002).

▶ *development of single nucleotide polymorphism (SNP) maps* (Section 8.3.7)

8.3.2 The first high resolution physical maps of the human genome were based on clone contigs and STS landmarks

General approaches in physical mapping of the human genome

Although different types of marker were used in constructing the different human genetic maps, there was a *common underlying principle* – the markers were typed in members of a variety of multigeneration families, and the data were fed into a computer to check for markers with co-segregating alleles. Physical mapping is different because many different types of map are possible (see *Table 8.1*). The first physical map of the human genome was based on chromosome banding and was obtained more than 40 years ago (see *Figure 2.14* for a modern example of a chromosome banding map).

Table 8.1: Different types of physical map can be used to map the human nuclear genome.

Type of map	Examples/methodology	Resolution
Cytogenetic	Chromosome banding maps	An average band has several Mb of DNA
Chromosome breakpoint maps	Somatic cell hybrid panels containing human chromosome fragments derived from natural, translocation or deletion chromosomes	Distance between adjacent chromosomal breakpoints on a chromosome is usually several Mb
	Monochromosomal radiation hybrid (RH) maps	Distance between breakpoints is often many Mb
	Whole genome RH maps	Resolution can be as high as 0.5 Mb
Restriction map	Rare-cutter restriction maps, e.g. NotI maps	Several hundred kb for rare-cutter restriction maps
Clone contig map	Overlapping YAC clones	Average YAC insert has several hundred kb of DNA
	Overlapping cosmid clones	Average cosmid insert is 40 kb
STS (sequence-tagged site) map	Requires prior sequence information from ordered clones so that STSs can be ordered	Less than 1 kb possible, but standard STS maps have resolutions in tens of kb
EST (expressed sequence tag) map	Requires cDNA sequencing then mapping cDNAs back to other physical maps	Average resolution in the human nuclear genome is ~90 kb
DNA sequence map	Complete nucleotide sequence of chromosomal DNA	1 bp

Although the resolution is coarse, cytogenetic maps have provided a very useful general framework for ordering human DNA sequences by *in situ* hybridization and defined cytogenetic breakpoints have enabled additional mapping tools. **Long range restriction maps** have been generated, too, by using *rare-cutting restriction nucleases*, for example to produce a *Not*I restriction map of 21q (Ichikawa *et al.*, 1993). However, in the context of the HGP, the first high resolution physical map of the human genome was made possible by making **libraries of genomic DNA clones** using cell-based DNA cloning (see Section 5.3.4 for the general principle). Once available, the libraries could be screened to identify individual clones which could then be grouped into *sets of clones with inserts originating from the same chromosome and subchromosomal region*. Eventually, it would be possible to organize the clones into groups which spanned large subchromosomal regions and eventually whole chromosomes.

To prepare libraries of genomic DNA clones it was usual to start with a permanent (lymphoblastoid) cell line to provide a continuously renewable source of a homogeneous cell population. After isolating genomic DNA from the cell line by standard methods, the DNA is cut with a restriction nuclease and cloned into a suitable vector to prepare the genomic DNA library of choice. Early attempts usually used lambda and cosmid vectors to generate libraries with inserts in the 15–40 kb range (Section 5.4.2). Clone inserts could be screened (initially by hybridization with a previously isolated small cDNA clone, but more recently by PCR) and then readily mapped to a subchromosomal region by **chromosome in situ hybridization** (see *Figures 2.16, 2.17*; **note:** mapping of the much shorter cDNA clones by hybridizing them to metaphase chromosomes was technically much more difficult than mapping large genomic clones, and was not generally attempted).

In the early human genomic DNA libraries the vast majority of the clones were both *anonymous* (because the identity of their insert DNA was not known) and *unmapped*. Gradually more and more different cDNA clones were characterized, enabling the corresponding (*cognate*) genomic DNA clones to be identified and then mapped to subchromosomal regions (cDNA clones were easier to characterize because they had short inserts which could be sequenced relatively quickly, and the libraries were less complex, with particular types of clone often predominating because of differential gene expression, e.g. globin transcripts in blood samples).

To aid mapping of human genomic DNA clones, additional approaches were developed. They included:

▶ *enriching the starting DNA.* Instead of using whole genome DNA, individual chromosomes were purified by *flow cytometry* using the same principles as used to fractionate cells in a FACS cell sorter. By collecting sufficient numbers of a particular type of chromosome, **chromosome-specific DNA libraries** were generated (Davies *et al.*, 1981). Additional **chromosome microdissection** procedures enabled DNA libraries to be made from DNA isolated from a specific subchromosomal region (Ludecke *et al.*, 1989);

▶ *hybrid cell mapping.* After sequencing a short bit at the end of a genomic clone, a PCR assay could be devised for this specific sequence and then used to type a panel of hybrid cells which were designed to lack certain human chromosomes or subchromosomal regions (see *Box 8.4*);

▶ *genetic linkage* to a DNA fragment which had previously been placed on the physical map using one of the various mapping techniques described above.

Box 8.4: Hybrid cell mapping.

BASIC PRINCIPLES OF SOMATIC CELL HYBRIDS

Under certain experimental conditions, cultured cells from different species can be induced to fuse together, generating **somatic cell hybrids**. For human mapping purposes, hybrid cells are typically constructed by fusing human cells with rodent (mouse or hamster) cells. The initial fusion products are described as **heterokaryons** because the cells contain both a human nucleus and a rodent nucleus. Eventually, heterokaryons proceed to mitosis, and the two nuclear envelopes dissolve. Thereafter, the human and rodent chromosomes are brought together in a single nucleus. Such hybrid cells are unstable. For unknown reasons, most human chromosomes fail to replicate in subsequent rounds of cell division, and are lost. This gives rise eventually to a variety of more or less stable hybrid cell lines, each with the full set of rodent chromosomes plus a few types of human chromosome (see *Figure, upper panel*). The loss of the human chromosomes occurs essentially at random, but can be controlled by selection.

Panels of hybrid cells with different subsets of human chromosomes can be used to map a human gene or DNA sequence to a *specific human chromosome*. It is most efficient, however, to use panels of **monochromosomal hybrids** (cells containing just a single type of human chromosome), collectively representing all 24 types of human chromosome (Cuthbert *et al.,* 1995). To make a monochromosomal hybrid, donor human cells are exposed to colcemid, causing the chromosome set to become partitioned into discrete subnuclear packets (***micronuclei***). Subsequent centrifugation can result in the formation of ***microcells***, consisting of a single micronucleus with a thin rim of cytoplasm, surrounded by an intact plasma membrane. The microcells are fused with recipient rodent cells (**microcell fusion**) to generate hybrids, some with a single human chromosome (see, for example, Warburton *et al.*, 1990).

RADIATION HYBRIDS

A human gene or DNA sequence can also be mapped to a *subchromosomal location* using panels of **radiation hybrids**, hybrid cells which contain *fragments* of human chromosomes integrated within a full set of rodent chromosomes. The human chromosome fragments are generated by subjecting donor cells to a lethal dose of radiation causing chromosome breaks (the average fragment size is a function of the radiation dose). After irradiation the donor cells are fused with recipient rodent cells and a selection system is used to pick out recipient cells that have taken up some of the donor chromosome fragments (see Walter *et al.*, 1994).

Previously, panels of **monochromosomal radiation hybrids** were used where the donor cell line was a monochromosomal hybrid: in this case fragments from the single type of human chromosome plus fragments from the broken rodent chromosomes would integrate into the rodent chromosome set of the recipient cell (Cox *et al.*, 1990). They have been superseded by panels of **whole-genome radiation hybrids** where the donor is an irradiated normal human diploid cell (Walter *et al.* 1994). For any one hybrid, only a *proportion* of the

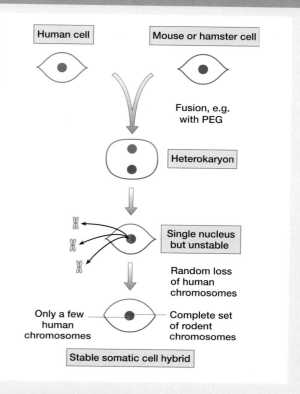

Principle of somatic cell hybrids.

pieces of the broken human chromosomes will integrate, and so different hybrids will have different fractions of the human genome integrated into the recipient rodent chromosome set.

A **radiation hybrid (RH) map** can be constructed by typing a panel of hybrids with a set of human DNA markers. Because the individual hybrids have different but overlapping subsets of the human genome, the different markers will be present in some hybrids but absent in others. Although the pattern of fragment integration is mostly random, individual markers will give related typing patterns if they are located as close neighbors on a particular chromosome. The principle of a radiation hybrid map is reminiscent of meiotic linkage analysis (*Chapter 13*): the nearer together two DNA sequences are on a chromosome, the lower the probability that they will be separated by the chance occurrence of a breakpoint between them. The frequency of chromosome breakage between two markers can be defined by a value θ, analogous to the recombination frequency in meiotic mapping. θ varies from 0 (the two markers are never separated) to 1.0 (the two markers are always broken apart).

As in meiotic mapping, θ underestimates the distance between markers that are far apart on the same chromosome, in this case because a cell can take up two markers on separate fragments. A more accurate estimate is provided by an **RH mapping function**, $D = -\ln(1-\theta)$, which is analogous to the Haldane mapping function used in meiotic linkage analysis (*Section 13.1*).

Box 8.4: *(continued)*

D is measured in **centiRays** (**cR**). D is dependent on the dosage of radiation, so it is referenced against the number of rads. For example, a distance of 1 cR_{8000} between two markers represents a 1% frequency of breakage between them after exposure to 8000 rad of X-rays.

Two radiation hybrid panels have been particularly important in the human genome project. The **Genebridge 4** panel consists of 93 human–hamster radiation hybrids with an average human fragment size of 25 Mb and a 32% retention of any particular human sequence in each hybrid. Laboratories can map any

unknown STS by scoring the 93 Genebridge hybrids and comparing the pattern with patterns of previously mapped markers held on a central server (*Figure, lower panel*). A second human–hamster panel, the **Stanford G3 panel**, was made using a higher dose of radiation, so that the average human fragment size is smaller. The 83 hybrids in G3 average 16% retention of the human genome, with an average fragment size of 2.4 Mb. Thus G3 can be used for finer mapping. The impressive results of large-scale use of these panels can be accessed at http:www.ncbi.nlm.nih.gov/genemap98/ (Deloukas *et al.*, 1998).

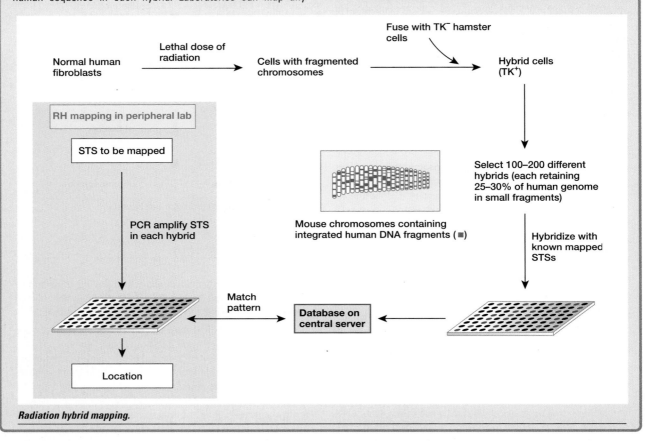

Radiation hybrid mapping.

A YAC-based physical map of the human genome

At the official beginning of the Human Genome Project in 1990, the available genomic DNA libraries contained inserts up to 40 kb in length (cosmid clones), which were anonymous in the vast majority of cases, and also very largely unmapped. Because of the very large size of the human genome, a cosmid library with an average insert size of about 40 kb would need to have several hundreds of thousands of different clones to ensure a high probability that approaching 100% of the genome was represented in the library. Screening these complex libraries to isolate individual clones and organizing the clones into related sets were daunting tasks.

To reduce the problem of screening huge numbers of clones and to make it easier to organize them into related sets,

cloning systems offering very large insert sizes were attractive. Novel methods were developed to achieve this aim by making *artificial* eukaryotic chromosomes. The chromosome system was based on linear yeast chromosomes where it had been known that only very small sequences were indispensable for chromosome function. By purifying these sequences and then combining them with large human DNA fragments it was possible to make hybrid molecules containing megabase-sized human inserts but which still behaved as chromosomes in yeast cells, so-called **yeast artificial chromosomes (YACs)** – see Section 5.4.4 for the details.

YAC libraries with an average insert size of, say, 1 Mb would require ca. 12 000–15 000 clones only to be reasonably representative of the human genome, and would have the advantage of enabling large genes (plus gene clusters or other functional segments) to be contained in a single clone. Taking this approach, Daniel Cohen and colleagues at the CEPH lab in Paris constructed a YAC-based map of the human genome, the first reasonably detailed high resolution physical map (Cohen *et al.*, 1993). An updated YAC map covering perhaps 75% of the human genome and consisting of 225 contigs with an average size of 10 Mb was subsequently published by the same group (Chumakov *et al.*, 1995).

The underlying principle in YAC maps (and all other clone-based physical maps), is to *order the clones* in the library on the basis of the subchromosomal region of origin for the insert DNAs. The map is built by defining sets of clones where the insert DNA derives from a *common subchromosomal region* and where the insert DNA of any one clone *overlaps* the insert DNA of some other clones in the clone set. This means that the relevant subchromosomal region is represented by a linear array of partially overlapping clones without leaving any gaps. Such a *contig*uous set of cloned DNA sequences is called a **clone contig** (see *Box 8.5*).

A high resolution STS (sequence tagged site) map of the human genome

The accuracy of clone contig maps is crucially dependent on the extent to which the clone insert DNA is a true representation of the original genomic sequence. Although the human YAC map was an impressive achievement, there were considerable areas of the genome not represented in the map and there was a major inherent limitation: the insert DNAs were often not faithful representations of the genomic DNA. The large YAC inserts are prone to rearrangements (including loss of internal sequences) and there was a substantial problem with *chimerism* (where a single transformed cell contains two or more pieces of human DNA from noncontiguous portions of the genome, often from different chromosomes, as a result of *co-ligation* or *co-transformation* – see respectively *Figure 5.5A* and *Figure 5.5B*).

To insure against possible problems due to infidelity of clone inserts, HGP physical mapping strategies had also emphasized the need to develop maps based on **sequence tagged sites** (**STSs** – see *Box 5.4*). By having a sufficiently high density of STS landmarks, the problem of insert instability in YAC libraries could be side-stepped: the large number of STSs would mean that the physical coverage of any problem region could rapidly be restored by STS typing of other kinds of clones (BACs, P1, PACs, etc.). Taking this approach, Eric Lander and colleagues at the Whitehead Institute of Biomedical Research and Massachusetts Institute of Technology reported the landmark achievement of a human STS map with over 15 000 STSs and an average spacing of just less than 200 kb (Hudson *et al.*, 1995).

The human STS map was an integrated physical map in which STSs had been used to type: (a) a panel of human *radiation hybrid cells* (see *Box 8.4*); and (b) the CEPH YAC library. The STS markers were of two types, nonpolymorphic and polymorphic. *Nonpolymorphic* STS markers included STSs derived by sequencing genomic clones at random and then developing PCR primers from nonrepetitive regions, and

STSs selected from cDNA sequences (using primers corresponding to sequences not separated by an intron), so-called *expressed sequence tags* or **ESTs** – see Section 8.3.4. Polymorphic STS markers mostly consisted of microsatellite markers, the majority of which had been used by the Généthon team in human genetic mapping.

The STS map published by Hudson *et al.* (1995) provided an extensive physical framework for the human genome, and because it contained over 2400 ESTs it also provided an embryonic *human gene map*. From now on the focus of the human genome project would move in two directions: creating high resolution gene (transcript) maps; and using the existing STS map to provide a framework for constructing clone contigs using *bacterial artificial chromosome clones (BACs)*.

8.3.3 The final stage of the Human Genome Project was crucially dependent on BAC/PAC clone contigs

Sequencing strategies

Because YAC inserts are often not faithful representations of the original starting DNA, second generation clone contig maps of the human genome were required to provide the cloned human DNA that would be sequenced. **BACs** and **P1 artificial chromosomes (PACs)** were selected as the cloning systems because although their insert sizes (100–250 kb) are much smaller than those of YACs this disadvantage is more than outweighed by the greater insert fidelity (see Section 5.4.3). A variety of different human BAC, and to a lesser extent PAC libraries, were therefore used as the physical template for sequencing.

The sequencing strategy used in the publicly funded Human Genome Project was based on *hierarchical* shotgun sequencing (in contrast Celera used a *whole genome* shotgun sequencing strategy; see *Figure 8.3*). Shotgun sequencing means that a starting DNA is randomly cleaved into small fragments, typically by sonication followed by end-repair, and the resulting fragments are cloned into a vector from which single-stranded recombinant DNA can be prepared with ease and used directly in sequencing – see *Box 7.1*. In the hierarchical shotgun sequencing approach, the DNA which is submitted for shotgun sequencing is the purified inserts of individual BAC clones which have been placed accurately on a physical map; by contrast, whole genome shotgun sequencing involves shotgun sequencing directly from the isolated genomic DNA (*Figure 8.3*).

The basis of the sequencing methodology used in human genome sequencing was the dideoxysequencing method which Fred Sanger and colleagues had developed a quarter of a century ago. Although the underlying method had not changed, various improvements in efficiency were introduced by automating various aspects of sequence generation and data analysis. The development of fluorescence labeling-based automated DNA sequencers and subsequently *capillary sequencers* (see Section 7.1.2) enabled much higher sequencing throughputs. Various dedicated computer programs helped with sequence interpretation and assembly, notably **PHRED**

Box 8.5: Physical mapping by building clone contigs.

The construction of genomic DNA libraries involves carefully controlled *partial restriction digestion* of the DNA to be cloned. The enzyme typically used to produce restriction fragments from the starting DNA is *Mbo*I which recognizes a 4-bp recognition sequence (↓GATC) and would be expected to cut DNA once every 300 bp or so, on average. Instead by exposing the starting DNA to very low concentrations of enzyme and for short periods of time, only a very small number of the total restriction sites available will be cut. For example, when making BAC libraries, the desired cloning fragment size is about 200 kb, and for this to be achieved less than 1% of the *Mbo*I sites available for cleavage should be cut.

The starting DNA is usually obtained from millions of diploid cells and so each of the original 23 types of human DNA molecule (corresponding to the 23 different types of chromosome) will be represented by millions of identical copies of the original pair of chromosomal DNA molecules. However, when submitted to partial restriction digestion, the DNA molecules are *cleaved randomly*. Thus for any one subchromosomal region, the million or so relevant molecules will show different patterns of restriction site cleavage (see *Figure, top panel*).

As a result of the partial restriction digestion, nonidentical fragments are generated which possess overlapping sequences derived from the same subchromosomal region. Although the individual fragments will be cloned into different host cells, it is possible to identify clones with *overlapping inserts* by screening for similarities between the inserts of different clones, and ultimately define *a set of clones with overlapping inserts such that all the original DNA sequence on the chromosomal region is represented, without gaps, within the inserts of this group of clones*, a so-called **clone contig** (see figure, middle panel).

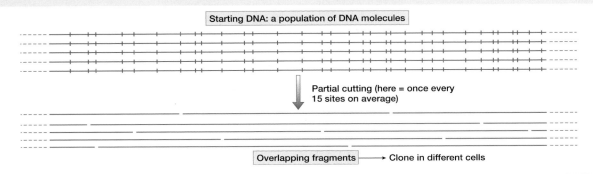

Upper panel. *Generating overlapping DNA fragments by partial restriction cutting.*

Middle panel. *A clone contig – a series of clones with partially overlapping DNA inserts.*

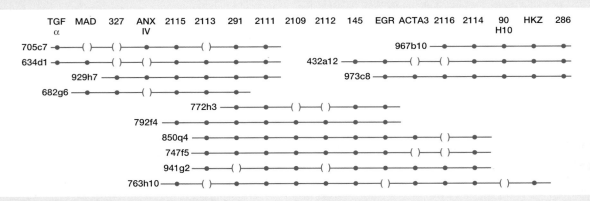

Lower panel: *Example of an STS contig on chromosome 2.*

Markers at top include STSs from genes (TGFα, MAD etc.) as well as anonymous markers (D2S327, D2S2115, etc. which are abbreviated here by removing the prefix D2S). Clones are YACs and brackets indicate absence of expected STS most likely as a result of internal YAC rearrangements.

(analyzes raw sequence traces and provides a *quality score* at each base position to indicate the degree of confidence that the assigned base call is correct), and **PHRAP** (assembles raw sequences into sequence contigs by scanning for overlapping sequences shared by two or more independent shotgun clones).

The problem with repetitive DNA

Assembling the individual clone sequences to identify overlaps (and hence reconstruct the sequence of the genome) is crucially dependent on an important assumption: that the overlapping sequences are *uniquely represented* within the genome. However, a very large fraction (ca. 50%) of the human genome is composed of repetitive DNA

(Section 9.4 and 9.5). Classes of highly repetitive interspersed DNA classes, for example LINE-1 and Alu repeats were well known and so were avoided wherever possible when seeking evidence for significant overlaps between the sequences of clones.

Despite the above precautions, areas that were very rich in *known* repetitive sequences would prove problematic, while another concern was previously unidentified *low copy number repeats* (a very real concern since it was subsequently found that a sizeable fraction of the genome has undergone *segmental duplication*, with sequences spanning tens of kilobases and sometimes hundreds of kilobases present in two or more regions of the genome – see Section 12.2.5). At least in the publicly funded HGP, the hierarchical shotgun cloning

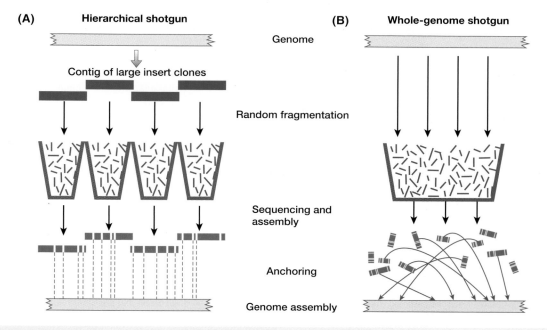

Figure 8.3: Different shotgun sequencing strategies for sequencing the human genome.

(A) *Hierarchical shotgun sequencing*. Human genomic DNA is fragmented by partial restriction digestion and the resulting large restriction fragments are cloned into BAC vectors to generate a BAC library. BAC clones are organized into large contigs by typing all clones with STS markers to identify clones with overlapping inserts. The inserts of selected BAC clones are shotgun cloned and sequenced. The sequenced fragments of a BAC are then assembled to give the BAC sequence and the full BAC sequences are integrated to remove overlaps.
(B) *Whole genome shotgun sequencing*. Here isolated genomic DNA is submitted directly to shotgun cloning and sequencing, and the sequenced pieces are assembled into large contigs spanning megabases. Adapted from Waterston *et al.* (2002) *Proc. Natl Acad. Sci. USA* **99**, p. 3713, with permission from the National Academy of Sciences, USA.

strategy facilitated sequence assembly, whereas the whole genome shotgun cloning strategy used by the private company Celera made immense demands on computing power, and critics have argued that this strategy *if used in isolation* was bound to fail because of the complexity and high repeat content in the human genome (see *Box 8.6*).

8.3.4 The first high density human gene maps were based on EST markers

At the outset of the HGP there was much debate over whether to go for an all-out assault (indiscriminate sequencing of all three billion bases), or whether to focus initially just on the very small fraction which represented

---

Box 8.6: Co-operation, competition and controversy in the genome projects.

Cooperation versus competition in publicly funded genome projects

Because of their scale the genome projects are major undertakings. In many cases, there have been laudable examples of *co-operation*: sharing of resources between different centers and labs, and agreed subdivision of different project tasks etc. The yeast genome project was an excellent example, involving highly organized co-operation between different European centers which subsequently extended into co-operation on functional analyses. In other cases, tensions have been evident as a result of fierce *competition* between different laboratories and wasteful duplication of effort, as in the *E. coli* genome project. Here a competitive race between American and Japanese groups developed, leading to a very close finish: the American lab deposited its sequence in GenBank 1 week before the Japanese group.

Tensions between the public and private sectors: gene patenting

The different priorities of the public and private sectors have provided various points of tension in genome projects. An early area of dispute concerned *gene patents*. This issue first appeared in 1991 when the U.S. NIH applied for patents for more than 7000 fragments of human brain cDNA clones whose sequences had been established as part of an EST mapping exercise led by Dr. Craig Venter. This attempt met with widespread opposition from the scientific community, especially since nothing was known about the functions of the expressed sequences. Under pressure, the U.S. Patents Office rejected the applications. An important new question – *who owns the human genome?* – had been raised for the first time (see Thomas *et al.*, 1996), and the idea of commercial monopoly of what is quite simply our genetic heritage appeared alarming and offensive to many people.

Dr. Venter subsequently left NIH to set up a new commercially backed institute, the Institute of Genome Research and, adopting a factory-style approach to EST sequencing, quickly compiled the world's largest human gene databank. In April 1994 the drug company SmithKline Beecham invested £80 million for an exclusive stake in Venter's database and announced to scientists that they could have access to it only if they agreed to concede first rights to any patentable discovery. Again the prospect of a corporation trying to monopolize control of a large part of the expressed human genome alarmed the scientific community. Many felt that a case could be made for patenting, after identifying the function of a gene, but not before.

Thousands of patents have been awarded for human DNA sequences where the sequence has been associated with some functional feature (see Thomas *et al.*, 1996) but in October 1998 the U.S. patent office awarded the first patent for an EST sequence (to Incyte Pharmaceuticals). As described by Knoppers (1999), the issue of patentability of the human genome continues to be problematic.

Tensions between the public and private sectors: genome sequencing

Privately-funded genome sequencing has also been controversial. The Celera human and mouse genome sequencing efforts were established in aggressive competition with the longer-established publicly funded projects. Celera audaciously claimed in 1999 that that their **whole genome shotgun strategy** (see Text and *Figure 8.3*) could produce a draft human genome sequence in 2 years, rapidly overtaking the slower clone-by-clone mapping and sequencing strategy of the publicly funded HGP.

As it happened, the HGP and Celera reports announcing the achievement of draft human genome sequences were published simultaneously (International Human Genome Sequencing Consortium, 2001; Venter *et al.*, 2001). However, this was never an equal race because at all stages Celera declined to make their data readily and freely available (external access to their data was denied and even when completed required expensive subscription charges, and remains restricted). By stunning contrast, the publicly funded laboratories had committed themselves to making new sequence data immediately and widely available (posting updates every 24 hours on the Internet).

Celera, like everyone else, had continuous and unfettered access to the DNA sequence data coming out of the publicly funded HGP, and unashamedly captured huge blocks of the publicly available human genome sequence data, re-processed it, and fed the resulting data into its own compilation. As a result, the Celera sequencing effort extensively parasitized on the publicly generated data. Given that the Celera strategy relied on the more difficult whole genome shotgun strategy, there has been huge skepticism as to whether the human genome sequence reported by Venter *et al.* (2001) had been in any way a vindication of the whole genome shotgun strategy. Instead, Waterston *et al.* (2002) have provided data indicating that the Celera sequence reported by Venter *et al.* (2001) was not an independent sequence of the human genome at all (since so much of the Celera sequence data had resulted from capturing and re-packaging large quantities of the HGP sequence).

coding DNA sequence, by far the most interesting and medically relevant part. Supporters of whole genome sequencing eventually won the argument by emphasizing that finding all genes could be difficult (some genes may be very restricted in expression) and that some noncoding DNA is functionally important, for example in the case of regulatory elements and sequences important for chromosome function. However, in the early years of the project, competing commercial interests made gene-finding a priority.

An initial approach involved large-scale sequencing of short sequences at the 3′ untranslated regions (UTRs) of cDNA clones which had been selected at random from a variety of human cDNA libraries (Adams *et al.*, 1991). These short sequences were described as **expressed sequence tags** (**ESTs**) because, just as for the more general *sequence tagged sites,* the sequence permitted a specific PCR assay to be designed for the expressed sequence (genomic sequences specifying 3′ UTR sequences are less commonly separated by introns than is coding DNA and so PCR primers from a 3′ UTR EST can often amplify the specific sequence in a *genomic DNA* sample). Later on, sequencing was extended to 5′ ends of cDNA clones and sequences were eventually obtained from the ends of hundreds of thousands of human cDNA clones.

After large numbers of human ESTs had been obtained (by both privately and publicly funded research groups), the next task was to start placing them on physical maps of the human genome. This involved typing YAC contigs for the presence of individual ESTs, or screening panels of human–hamster *radiation hybrids* (see Section 8.3.2, and also *Box 8.4* for the principle of radiation hybrids). A rough chromosomal distribution of human genes had previously been obtained by hybridization studies (*Figure 8.4*), but placing large numbers of ESTs on physical maps was the first systematic approach to constructing a **human gene map**.

The mapping data was integrated relative to the human genetic map and then cross-referenced to cytogenetic band maps of the chromosomes. In order to define a *nonredundant gene set*, attempts were then made to integrate mapping information on the different ESTs, as in the case of the **UniGene** system (*http://www.ncbi.nlm. nih.gov/UniGene/*). The first reasonably comprehensive human gene maps were reported by Schuler *et al.* (1996) and Deloukas *et al. (*1998). In the latter case, map positions for what was estimated (incorrectly) to be 30 000 human genes were reported and at the time this was thought to represent somewhat less than half of the total human gene catalog. However, sequencing of the human genome provided an unexpected surprise (see next section).

8.3.5 The draft human genome sequence suggested 30 000-35 000 human genes, but getting a precise total is difficult

Before the human genome had been sequenced, predictions for the total number of human genes (which were based on extrapolation from a variety of limited datasets) were mostly in the range of 60 000–100 000 genes. After the draft sequences were reported in 2001, the revised estimate was surprising low, perhaps only about 30 000–35 000 genes.

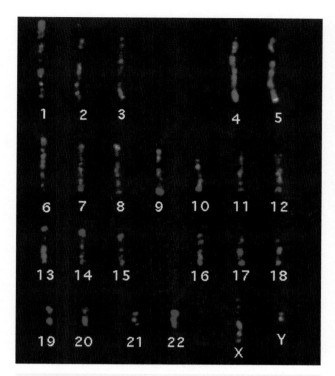

Figure 8.4: An early map of genome-wide human gene distribution.

Most human genes are associated with *CpG islands* (see *Box 9.3*). A purified human CpG island fraction was labeled with a Texas red stain and hybridized to human metaphase chromosomes (Craig and Bickmore, 1994). Late replicating chromosomal regions (mostly transcriptionally inactive), are shown in green [as a result of incorporation of fluorescein isothocyanate (FITC)-labeled bromodeoxyuridine (BrdU)]. Yellow regions (overlap of red and green signals), denote late-replicating regions rich in genes (or strictly, CpG islands). Early replicating gene-poor regions of the genome are invisible (since there is no counterstain), as are centromeres (where the anti-BrdU cannot get access). High gene density (shown by the red color from the labeled CpG island fraction) is found on certain chromosomes (e.g. chromosome 22) while others (e.g. chromosomes 4, 18, X, Y) are gene-poor. Adapted from Craig and Bickmore (1994) *Nature Genet.* **7**, 376–381, Fig. 1, with permission from Nature Publishing Group.

With scarcely 50% more genes than in *C. elegans*, a 1-mm long worm with only 959 somatic cells, another blow had been delivered to human pride, perhaps even more surprising than the long-established *C-value paradox* (cellular DNA content is not always related to functional complexity; some types of ameba have vastly more DNA per cell than we do – see *Table 3.1*).

Difficulties in establishing the precise number of human genes

Even by 2003, after essentially all of the human genome had been sequenced, the number of human genes is still not known with confidence, although more recent estimates

(Ensembl build 29) suggest a total close to 30 000 or perhaps even below 30 000 (see human–mouse comparisons in the paper by the Mouse Genome Sequencing Consortium, 2002). To understand the difficulty in estimating the precise gene number, it is instructive to consider how genes are defined. As detailed in Section 7.2, there are only two essential gene-specific criteria: transcription into RNA and evidence of evolutionarily conserved sequences. To identify all genes experimentally on the grounds of either criterion is, however, very difficult at the practical level, because:

▶ although searching for associated transcript sequences is often rewarding (see Camargo *et al.*, 2002), genes that are expressed at low levels and/or at unusual cellular locations and stages of development may not be well represented in available cDNA libraries;

▶ genes encoding untranslated RNAs may be difficult to identify in the absence of a sizeable open reading frame.

As a result of the above, genes are often missed using experimental methods alone, and when the draft human genome sequences were published, experimental support existed for perhaps only 11 000 genes; the others were predicted by computer (*in silico* **analyses**). Computer-based programs which are used to identify genes (see Zhang, 2002) compare a test sequence against other known gene sequences (homology screening) and seek to identify exons (exon prediction programs):

▶ *homology searches against sequence databases.* For any test sequence, the nucleotide sequence can be matched against all nucleotide sequences in available databases, and possible translated polypeptide sequences can be matched against all known protein sequences (see *Table 8.2* and *Box 7.3*). As more and more sequences pour into the sequence databases, this is becoming a particularly effective way of identifying sequences which have been highly conserved during evolution, which is indicative of conserved function;

▶ *exon prediction programs.* Some programs use only information about the input sequence such as the popular **GENSCAN** program (see Burge and Karlin, 1997). Others rely heavily on confirmatory homology-based screening. As yet, however, even the best such programs, such as GENSCAN have been only moderately successful in identifying exons when tested against genes whose exon organizations had previously been established. Cross-species comparisons have provided another important way of identifying exons, because unlike most genome sequences exons tend to be evolutionarily well conserved (see e.g. Batzoglou *et al.*, 2000);

▶ *integrated gene-finding software packages.* Various software packages have been produced which use general sequence homology-based database searching programs together with programs designed to identify gene-associated motifs and exons. Usually the output is in a graphical format. Popular packages are **NIX** (http://www.hgmp.mrc.ac.uk/Registered/Webapp/nix/; see *Figure 8.6* for an output using this software package) and the **Genotator** program (see http://www.fruitfly.org/~nomi/genotator/genotator-paper.html).

Notwithstanding the value of computer-based gene and exon predictions, problems continue with estimating human gene number, involving both over- and underprediction. *Overprediction* can occur, for example, when clusters of what appeared to be separate genes turn out to be different parts of a single very large gene, or a result of artifactual cDNAs [because of the way that cDNA libraries are constructed, some apparent unspliced cDNAs are artifacts due to priming of complementary strands by oligo(dA) tracts in genomic DNA]. *Underprediction* can occur because of the difficulty in finding genes with very small exons and restricted expression, and in finding and confirming genes encoding untranslated RNA.

Table 8.2: Major electronic databases which serve as repositories of nucleotide or protein sequences.

Database type	Database	Location	URL
NUCLEOTIDE SEQUENCE	GenBank	Maintained by the US National Center for Biotechnology Information (NCBI) at the US National Institutes of Health (NIH)	http://www.ncbi.nlm.nih.gov
	EMBL	Maintained by the European Bioinformatics Institute (EBI) at Hinxton, near Cambridge, UK	http://www.ebi.ac.uk
	DDBJ	Maintained by the National Institute of Genetics, Japan at Mishima, Shizuoka	http://www.ddbj.nig.ac.jp/
PROTEIN SEQUENCE	SWISS-PROT	Protein sequences with high quality annotation. Maintained collaboratively by the Swiss Institute of Bioinformatics, Geneva and the European Bioinformatics Institute (EBI) at Hinxton, UK	http://ca.expasy.org/sprot http://www.ebi.ac.uk/swissprot/
	TREMBL	Translations of coding sequences from the EMBL database which have not yet been deposited in Swiss-Prot	http://www.ebi.ac.uk/trembl/ http://ca.expasy.org/sprot
	PIR	Maintained collaboratively by the US National Biomedical Research Foundation (NBRF), Georgetown; the Japan International Protein Information Database in Japan (JIPID) and the Munich Information Center for Protein Sequences (MIPS). Available at numerous locations world-wide	http://pir.georgetown.edu/ http://www.ddbj.nig.ac.jp/ http://mips.gsf.de/ http://www.hgmp.mrc.ac.uk/

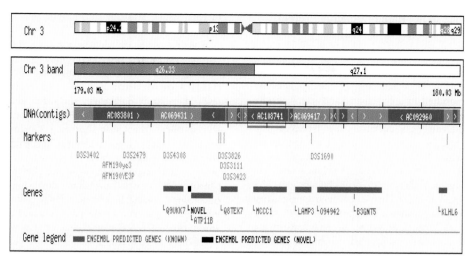

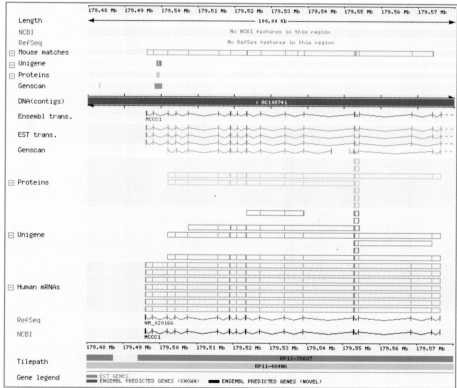

Figure 8.5: Graphical interfaces at the Ensembl genome browser allow ready navigation through individual chromosomes in assembled genome sequences.

The Ensembl genome browser (http://www.ensembl.org) was developed out of a collaboration between the Wellcome Trust Sanger Institute and the European Bioinformatics Institute, and allows browsing by chromosome for a number of assembled genome sequences, including human and mouse. After human chromosome 3 had been initially selected, a subchromosomal region spanning the 3q26.33–3q27.1 border was selected by clicking on a chromosome ideogram (shown by the red open box in the top panel). The selected region is shown in the middle panel in expanded form, defined as the 1-Mb region from 179.03 Mb to 180.03 Mb. The red open box in the middle panel is a 100-kb segment (179.48 Mb to 179.58 Mb) and is shown in expanded view in the lower panel, illustrating gene, exon, RNA, protein structure etc.

8.3.6 The final stages of the Human Genome Project: gene annotation and gene ontology

As human genome sequencing reached the final phase, strenuous efforts were devoted to developing new software which would permit the huge amounts of human genome information to be searched in a systematic and user-friendly way. The clever design of **genome browsers** such as *Ensembl* (developed out of a collaboration between the Wellcome Trust Sanger Institute and the European Bioinformatics Institute) provides graphical interfaces to portray genome information for individual chromosomes and subchromosomal regions. Users can quickly navigate the sequence of a selected human chromosome moving from large scale to nucleotide scale, identifying genes, and associated exons, RNAs, and proteins. Central to the navigation are click-over tools which enable a myriad of connections to other databases and programs (including direct links to other assembled genome sequences, notably the mouse genome), allowing users to follow through on requested information (see http://www.ensembl.org/ and *Figure 8.5*). As more and more information is developed regarding genes and other functional units, more informative and precise **gene anno-**

tation will be available in frequent, periodical updates of the genome browsers.

Another major development has been in **gene ontology**, where systematic and hierarchical vocabularies are being developed to define gene function. The **Gene Ontology (GO) Consortium** has involved collaboration between researchers studying human and other genomes to develop a common system of defining gene function which can apply to all organisms (the Gene Ontology Consortium, 2000; 2001; see also http://www.geneontology.org/). The use of a common vocabulary will allow searching across multiple proteins and species for common characteristics. The Gene Ontology Consortium has developed three separate ontologies – *biological process, cellular component* and *molecular function* – to describe gene products and these allow for the annotation of molecular characteristics across species. Each vocabulary is structured as directed acyclic graphs, wherein any term may have more than one parent as well as zero, one, or more children. This makes attempts to describe biology much richer than would be possible with a hierarchical graph. Currently the GO vocabulary consists of more than 11 000 terms, which will, in time, all have strict definitions for their usage.

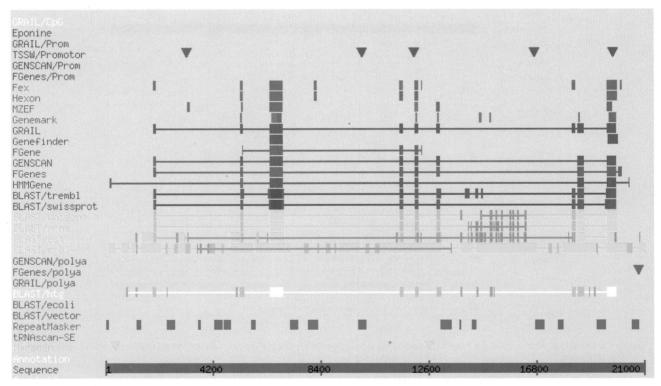

Figure 8.6: Gene finding by computer-based analysis of genomic sequences.

This example shows *NIX analysis* (see text) of a PAC sequence from a region of chromosome 12q24.1 encompassing the Darier's disease locus. The nucleotide size of the region is illustrated by the bar at the bottom. Analyses include the use of programs to scan for gene-associated motifs such as promoter sequences (green inverted triangles at top), and polyadenylation sites (ochre-colored inverted triangles) and various exon prediction programs (GRAIL, GENSCAN etc.). Significant homologies to other sequences at the nucleotide level and at the protein level are indicated by the boxes for the various BLAST programs. Data provided by Dr. Victor Ruiz-Perez and Simon Carter, University of Newcastle upon Tyne, UK.

8.3.7 Analyses of human genome sequence variation are important for anthropological and medical research

At the outset, the Human Genome Project was conceived as a project to obtain the nucleotide sequence of a collection of cloned human DNA fragments collectively amounting to one or a very few haploid genomes. What it did not consider was the genetic diversity of humans. Information on human genetic diversity has subsequently been considered desirable in different contexts:

▶ *anthropology*. The information should be of help in anthropological and historical research in tracking human origins, prehistoric population movements and social structure;

▶ *forensic analysis.* The accuracy of DNA-based tests which are carried out to identify people or confirm close biological relationships (*DNA profiling*) is dependent, in part, on knowing how the diagnostic DNA markers vary from one population to the next;

▶ *medical research.* Common diseases are *multifactorial*, and it can be frustratingly difficult to identify the underlying genes (see Chapter 15). A huge effort has recently been devoted to identifying genetic variants which can explain differences between human populations with respect to increased susceptibility to particular diseases, or indeed comparative resistance to certain diseases.

The Human Genome Diversity Project

The idea of a global effort to study human genome sequence diversity, the **Human Genome Diversity Project (HGDP)**, was proposed by Cavalli-Sforza *et al.* (1991). The emphasis was predominantly focused on the need to collect DNA samples from a large number of ethnic groups. However, although supported by HUGO this project has been in considerable difficulties (see Greely, 2001). From the outset it has been dogged by a conspicuous lack of funding. As long as the primary aim of this project was simply to find markers for ethnic groups and to trace the origins of human migrations and ancestral lineages, the case for large scale funding was not so persuasive.

The HDGP has also been beset by controversy. In some cases, researchers have visited isolated human populations on the margins of survival, obtained samples quickly without taking time to explain their significance and then left with little or no subsequent communication. While proponents refer to the need to safeguard our cultural heritage, critics have witheringly used terms such as *helicopter genetics* to refer to the insensitive quickly-in, quickly-out approach often used to obtain samples.

Single nucleotide polymorphism (SNP) maps

While highly polymorphic microsatellite markers are distributed throughout the genome, they are not particularly amenable to automated typing and occur only once every 30 kb or so. *Single nucleotide polymorphisms* (*SNPs*) are not very polymorphic (typically only two alleles), but they can be typed very easily by automation, and they occur very frequently in the genome, on average about once every kilobase in the human genome (see Sections 7.1.3 and *Box 18.2* for how they are typed). As a result, human **SNP maps** have found important applications in constructing even higher resolution genetic maps which were required for investigating chromosomal regions expected to contain common disease genes.

The *International SNP Consortium Ltd* was inaugurated in 1999 as a not-for-profit partnership between the Wellcome Trust and about a dozen private, mostly pharmaceutical, companies. Within 2 years it had delivered a human SNP map with a total of 1.42 million SNPs, equating to about one SNP every two kilobases (International SNP map working group, 2001). Now strenuous efforts are being applied to map genes for common disease using high density SNP maps and much commercial interest is also being devoted to applications in *pharmacogenomics*.

8.3.8 Without proper safeguards, the Human Genome Project could lead to discrimination against carriers of disease genes and to a resurgence of eugenics

Any major scientific advance carries with it the fear of exploitation. The HGP is no exception, and perceived benefits of the project can also have a downside. When we know all the human genes and can detect large numbers of disease-associated mutations, there will be enormous benefit in targeting prevention of disease to those individuals who can be shown to carry disease genes. However, the same information could also be used to discriminate against such individuals. For example, there is widespread anxiety that insurance companies could insist on large-scale genetic screening tests for the presence of genes that confer susceptibility to common disorders such as diabetes, cardiovascular disease, cancers and various mental disorders. Perfectly healthy individuals who happen to be identified as carrying such disease-associated alleles may then be refused life or medical insurance. Clearly such discrimination is practiced on a small scale at the moment; what is alarming to many is the prospect of discrimination against a very large percentage of the individuals in our society. It is also important to preserve people's right not to know. A fundamental ethical principle in all genetic counseling and genetic testing is that genetic information should be generated only in response to an explicit request from a fully informed adult patient.

Another troublesome area is the question of biological determinism and whether comprehensive knowledge of human genes could foster a revival of *eugenics*, the application of selective breeding or other genetic techniques to 'improve' human qualities (Garver and Garver, 1994). In the past, negative eugenic movements in some countries (including the USA and Germany) discriminated against individuals who were adjudged to be inferior in some way, notably by forcing them to be sterilized. The possibility also exists of a preoccupation with *genetic enhancement* to positively select for heritable qualities that are judged to be desirable (see Ethics Box 3 in Chapter 21).

In recognition of the above problems, the U.S. Human Genome Project has devoted considerable resources to support **ethical, legal and social impact (ELSI) research projects** (see e.g. http://www.nhgri.nih.gov/ELSI).

8.4 Genome projects for model organisms

Mapping the human genome was not the only scientific focus of the Human Genome Project; right at the outset, the value of sequencing genomes of five, key **model organisms** was clearly recognized, and since then, the list has become quite a substantial one. It encompasses a wide variety of single-celled microbial organisms as well as various multicellular model organisms, with many being particularly suited to genetic analysis. In part, the sequencing of smaller genomes was also considered as a pilot for large-scale sequencing of the human genome. By May 2003 the genomes of over 140 organisms had been completed (or very nearly so; see http://wit.integratedgenomics.com/GOLD/; http://www2.ebi.ac.uk/genomes/).

8.4.1 There is a huge diversity of prokaryotic genome projects

The diversity of prokaryotic genome sequencing projects

A variety of prokaryotic models have long been established (see *Box 8.7*). Prokaryotic genomes are typically small (often only one or a few megabases) they are particularly amenable to comparatively rapid sequencing, resulting in the rapid development of many **prokaryotic genome projects**. By May 2003 the genomes of a total of 122 different prokaryotes (16 archaea and 106 bacteria) had been sequenced, and those of another 342 (23 archaeal, 319 bacterial) were in progress (see also Doolittle, 2002).

The first prokaryotic genome to be completed (in 1995) was the 1.83-Mb *Haemophilus influenzae* genome. This was a landmark: the first time that the genome of a free-living organism had been sequenced (see Tang *et al.*, 1997). Subsequently, a variety of other *firsts* were achieved: the genome of the *smallest autonomous self-replicating entity* (***Mycoplasma genitalium*** in 1995), the first archaeal genome (***Methanococcus jannaschii*** in 1996), and then the important achievement of the complete sequence of the 4.6-Mb ***E. coli* genome** (see Pennisi, 1997).

The list of prokaryotic organisms whose genomes have been sequenced reveals different priorities. In some cases, the driving force was to understand evolutionary relationships between different organisms, as in the case of the archaeal genomes (see Olsen and Woese, 1997) and in the case of *Mycoplasma genitalium* it was to understand what constitutes a *minimal genome*, this being the smallest known cellular genome (which is now known to have only 470 genes). In other cases, as in *E. coli* and *Bacillus subtilis*, the priority was simply to further basic research on popular experimental organisms. For many researchers, the big prize has been *E. coli*, the most-intensively studied bacterium. Surprisingly, given the huge amount of prior investigation, almost 40% of the initially identified 4288 genes had no known function and became the subject of intensive investigation. For many other organisms, however, the primary motivation for genome sequencing has been their medical relevance.

Disease-related prokaryotic genome projects

In some cases prokaryotes were selected for genome sequencing because of their known *associations* with chronic diseases (see Danesh *et al.*, 1997), or because they were known to be *causative agents* of disease (see *Table 8.3*). In addition to having a more complete understanding of these organisms, the new information can be expected to lead to more sensitive diagnostic tools and new targets for establishing drugs/vaccines.

8.4.2 The *S. cerevisiae* genome project was the first of many successful protist genome projects

Protists are single-celled eukaryotes, a term which encompasses unicellular animals (**protozoa**), unicellular fungi, and unicellular plants. A variety of protist genome projects have been undertaken, in some cases motivated by the need to understand basic model organisms; in other cases as a means of combating disease caused by pathogenic protozoa.

The *Saccharomyces cerevisiae* genome project

The budding yeast **Saccharomyces cerevisiae** is a unicellular fungus and has long been a favorite eukaryotic model organism, partly because it is so amenable to genetic analyses (*Box 8.7*). Its 16 chromosomes were sequenced by European and American consortia and the complete sequence was reported by Goffeau *et al.* (1996). This represented another milestone in biology: the first complete sequence for a eukaryotic cell. The data indicate that yeast genes are closely clustered, being spaced on average once every 2 kb. Of the 6340 genes, about 7% specify untranslated RNA. Although one of the most intensely studied organisms, 60% of its genes had no experimentally determined function when the sequence was first reported. However, a sizeable fraction of yeast genes could be shown to have an identifiable mammalian homolog and so in only about 25% of yeast genes was there no clue whatsoever to their function (Botstein *et al.*, 1997). The successful conclusion of this project has now opened up large-scale functional analyses (see Chapter 19).

The *Schizosaccharomyces pombe* genome project

The fission yeast is another unicellular fungus which has long been a favorite model organism (see *Box 8.7*). The complete 13.8-Mb sequence was published by Wood *et al.* (2002) and revealed a total of 4824 protein-encoding genes. The data show considerable differences between the *S. cerevisiae* and *S. pombe* genomes, including hundreds of genes present in *S. pombe* but apparently absent in *S. cerevisiae* and vice versa, differences in the number of introns (4700 in *S. pombe* compared to only 275 in *S. cerevisiae*) and in transposable elements (very few in *S. pombe* when referenced against *S. cerevisiae*) etc.

Box 8.7: Model unicellular organisms.

A variety of unicellular organisms are particularly suited to genetic and biochemical analyses and offer important advantages of extremely rapid generation times, easy large-scale culture etc. Although very distantly related to us in evolutionary terms, they are advantageous to study for a variety of scientific and medical reasons. Benefits for human geneticists and medical researchers include new insights into a wide variety of research areas, some with clear medical applications, including:

▶ *gene function and cellular processes:* a variety of crucially important core cellular activities have been extraordinarily conserved during evolution. We can gain insights into how diverse mammalian genes function and the nature of crucial cellular processes such as ribosome biogenesis, cell cycle control, membrane transport, polymerase function etc. by studying equivalent genes and processes in unicellular organisms;

▶ *pathogenesis:* many single-celled organisms are known to cause disease, and over the years medicine has often struggled to find appropriate effective drugs or other treatments. By determining the complete genomes of these pathogenic organisms and studying the exact molecular basis of how disease is caused, new insights could be expected in how to deal with pathogenic microorganisms;

▶ *evolution:* sequence analysis provides the most useful way of understanding how organisms are related to each other. Thus, while rRNA sequence analysis provided the breakthrough that was to establish the fundamental division into the three major kingdoms of life: archaea, bacteria, and eukaryotes (*Box 12.4*), full comparison of genome sequences will undoubtedly provide additional important insights. See also *Figure 12.22*.

BACTERIA

A variety of normally *nonpathogenic* bacteria have long been favorite model organisms, notably the rod-shaped **Escherichia coli**, which lives in the gut of humans and other vertebrates (in a symbiotic relationship: their contribution is to synthesize vitamin K and B-complex vitamins which we gratefully absorb). Because of intensive studies over decades we have built up more knowledge of *E. coli* than of any other type of cell and most of our understanding of the fundamental mechanisms of life, including DNA replication, transcription, protein synthesis etc. has come from studies of this organism.

In addition, of course, a variety of *pathogenic* bacteria are responsible for diseases of varying severity. Among these are strains of *E. coli* causing human infections including meningitis, septicemia, urinary tract infections, and intestinal infections. An important example is **E. coli O157:H7** which as a result of acquiring a particular bacteriophage sequence in the past, has been genetically modified to produce a toxin. In some cases the toxin results in heavy bleeding which can be fatal in young children or in the elderly; in other cases kidney failure can result. A variety of genome projects have been launched in a bid to obtain the complete sequences of pathogenic micro- organisms (see *Table 8.3*).

ARCHAEA

Archaea are prokaryotes which superficially resemble bacteria in form but which diverged from bacteria at a very early stage in evolution. They were initially found in unusual, often extreme environments – at very high temperatures in hot springs and near rift vents in the deep sea, in waters of extreme pH or salinity, in oxygen-deficient muds of marshes, in petroleum deposits deep underground, and at the bottom of oceans. Now, however, they are known to inhabit more familiar environments such as soils and lakes and have been found thriving inside the digestive tracts of cows, termites, and marine life where they produce methane.

The metabolic and energy conversion systems of archaea resemble those of bacteria but the systems used to handle and process genetic information (DNA replication, transcription, translation) are more closely aligned to those of eukaryotes than to their bacterial counterparts. Archaea are *not* known to be associated with disease, and the major interest in studying their genomes is based on their position as a quite separate kingdom from other life forms, and the resulting interest in knowing more about how they evolved to be so different.

YEASTS (UNICELLULAR FUNGI)

Yeasts are single-celled fungi, and so are eukaryotes. Common on plant leaves and flowers, soil and salt water, they are also found on the skin surface and in the intestinal tracts of warm-blooded animals, where they may live symbiotically or as parasites. They typically replicate by *budding* rather than binary fission: the cytoplasm and dividing nucleus from the parent cell is initially a continuum with the *bud*, or daughter yeast, before a new cell wall is deposited to separate the two.

Yeasts have been valuable model organisms because a variety of key molecules are known to have been strongly conserved from yeasts to mammals. Importantly, the human versions of a number of yeast *cell cycle* and *DNA repair* genes have been found to be directly involved in human cell division. Malfunctions can lead to cancer or birth defects. The normal activity of these genes is most easily studied in the yeast cells, which are readily manipulated in the laboratory. This provides important insights into their mechanisms in humans, which helps direct experiments in more complicated cell types. Studying the control of cell division in yeast is thus very relevant to human health and understanding many clinical disorders. Usually yeasts are not associated with disease, but some *pathogenic yeasts* – in particular *Candida* species – constitute a common health problem.

▶ **Saccharomyces cerevisiae** is a budding yeast which has long been of importance in baking and brewing. Because it is simple and easy to grow, it has been a favorite subject of basic research and one of the most extensively studied of eukaryotes. Partly because of a high frequency of *nonhomologous recombination*, it has been very amenable to genetic analyses. It has been used as a model to dissect

continued on next page

various aspects of cell biology, including cell cycle control, protein trafficking and transcriptional regulation.

▶ **Schizosaccharomyces pombe** is a fission yeast and has a rapid generation time (2–4 hours). It has only more recently been extensively studied, primarily as a model of cell cycle control (there is a distinct G2 phase to the cycle) and differentiation. It is only distantly related to *S. cerevisiae*, and in some aspects of chromosome structure and RNA processing it is more closely related to higher eukaryotes than is *S. cerevisiae*.

▶ **Candida albicans** is naturally present in the mouth of healthy people but can cause irritating, sometimes debilitating infections in individuals whose immune system is not as active as normal, causing vaginal and oral thrush, diaper rash etc.

PROTOZOA (UNICELLULAR ANIMALS)

Protozoa are a large group of unicellular animals (nonphotosynthetic), including: **amebae, flagellates** and **ciliates** (which move by the aid of pseudopodia, flagella and cilia respectively), and other organisms with complex life-cycles. They are of interest to biomedical researchers as models of various facets of cell and developmental biology, and also because many are parasites associated with disease. The latter may be hosts for pathogenic bacteria (causing diseases such as Legionnaire's disease, salmonellosis, tuberculosis etc.), or cause disease directly, as in the case of:

▶ **Entamoeba histolytica**: a parasitic ameba causing severe gastrointestinal disease;

▶ **Trypanosomes**: parasitic flagellates causing the tropical fever known as sleeping sickness;

▶ **Giardia**: a parasitic flagellate causing severe diarrhea;

▶ **Plasmodium**: causing malaria;

▶ **Toxoplasma**: causing gastrointestinal disease and damage to internal organs.

As models of cell and developmental biology, most interest has been focused on **Dictyostelium discoideum**, a so-called *social*

ameba: although growth as separate, independent cells is typical, the cells can interact to form multicellular structures when challenged by adverse conditions such as starvation. Up to 100 000 cells signal to each other by releasing the chemoattractant cAMP and aggregate together by *chemotaxis* to form a mound that is surrounded by an extracellular matrix. This mechanism for generating a multicellular organism differs radically from the early steps of metazoan embryogenesis. However, subsequent processes depend on cell–cell communication in both *Dictyostelium* and metazoans. The organism is uniquely suited for studies of cytokinesis, motility, phagocytosis, chemotaxis, signal transduction, and aspects of development such as cell sorting, pattern formation, and cell-type determination. Many of these cellular behaviors and biochemical mechanisms are either absent or less accessible in other model organisms.

Ciliates have not been as extensively investigated as the other protozoan classes. The one which has received most attention has been **Tetrahymena**, a freshwater organism that commonly inhabits streams, lakes and ponds, and a genome project for **Tetrahymena thermophila** is under active consideration (see Turkewitz *et al.* (2002). Tetrahymena has large cells (40–50 μm along the anterior-posterior axis) and like other ciliates, the cells have a striking variety of highly complex and specialized cell structures. As is typical of ciliates, the nuclear apparatus of *Tetrahymena* is composed of two structurally and functionally differentiated types of nuclei, a phenomenon known as *nuclear dimorphism*. The *micronucleus* is the germline, i.e. the store of genetic information for the sexual progeny. It is diploid and contains five pairs of chromosomes. The *macronucleus* (MAC) is the somatic nucleus, i.e. the nucleus is actively expressed during vegetative multiplication and no known MAC DNA is transmitted to the sexual progeny. *Tetrahymena* is a well-established model for cellular and developmental biology, with major interest revolving round cell motility, developmentally programmed DNA rearrangements, regulated secretion, phagocytosis and telomere maintenance and function.

The *Plasmodium falciparum* genome project

The report by Gardner *et al.* (2002) of sequencing of the *P. falciparum* genome was another landmark: the first time that a genome of a eukaryotic parasite had been sequenced. *P. falciparum* is the deadliest of the malaria parasites and the simultaneous reporting of the complete genome of its host, the mosquito *Anopheles gambiae* (see Section 8.4.4) offers a new dimension in the battle against malaria, an often lethal condition which kills 1 million people each year, mostly in sub-Saharan Africa.

Other protist genome projects

The other protist genome projects include genome projects for various other **pathogenic protozoa** involved in human

parasitic infections. In most cases the genome sizes are substantial, typically in the 30–90 Mb range (see *Table 8.3*). In addition, genome projects have been developed for other organisms that have been well-studied and are amenable to biochemical/genetic analyses, including the molds **Aspergillus nidulans** and **Neurospora crassa**.

8.4.3 The *Caenorhabditis elegans* genome project was the first animal genome project to be completed

The *Caenorhabditis elegans* genome project

Although a simple organism, only about 1 mm long, *C. elegans* had been regarded as an important model of development and was also useful for modeling other processes relevant to

Table 8.3: Germ wars: examples of genome projects for pathogenic microorganisms (see Further Reading for internet sources)

Organism	Genome size (no. of chromosomes)	Associated disease
Bacteria		
Bacillus anthracis	4.5 Mb (1)	Anthrax
Bordetella pertussis	3.88 Mb (1)	Whooping cough
Borrelia burgdorferi	0.95 Mb (1)	Lime disease
Chlamydia pneumoniae	1.0 Mb (1)	Respiratory disease; coronary heart disease
Chlamydia trachomatis	1.7 Mb (1)	Trachoma, a major cause of blindness
Clostridium difficile	4.4 Mb (1)	Antibiotic-associated diarrhea; pseudomembranous colitis
Helicobacter pylori	1.67 Mb (1)	Peptic ulcers
Mycobacterium leprae	2.8 Mb (1)	Leprosy
Mycobacterium tuberculosis	4.4 Mb (1)	Tuberculosis
Rickettsia prowazekii	1.1 Mb (1)	Typhus
Salmonella typhi	4.5 Mb (1)	Typhoid fever
Treponema pallidum	1.1 Mb (1)	Syphilis
Vibrio cholerae	2.5 Mb (1)	Cholera
Yersinia pestis	4.38 Mb (1)	Plague
Protozoa		
Leishmania major	33.6 Mb (36)	Leishmaniasis
Plasmodium falciparum	23 Mb (14)	Malaria
Trypanosoma brucei	54 Mb (22)[a]	African trypanosomiasis (sleeping sickness)
Trypanosoma cruzi	87 Mb (>42)	American trypanosomiasis (Chagas disease)

[a]Organized as 11 pairs.

human cells (see *Box 8.8*). Because of its large genome size (nearly 100 Mb), the *C. elegans* genome project was also viewed as the major pilot model for large-scale sequencing of the human genome. The successful genome project was carried out by the Wellcome Trust Sanger Institute and Washington University School of Medicine (*C. elegans* Sequencing Consortium, 1999). It was another landmark achievement, providing for the first time, the genetic instructions for a multicellular animal (**metazoan**).

The *C. elegans* genome project initially reported a total of close to 19 000 polypeptide-encoding genes and over 1000 genes encoding untranslated RNA molecules, giving an average spacing of one gene every 5 kb. A surprisingly high number of the genes appear to occur as part of **operons** where individual genes are transcribed as part of large multigenic RNA transcripts. At the time of the initial report, comparison with published sequences from elsewhere revealed that about one in three of the recently identified *C. elegans* genes showed similarities to previously known genes, and 12 000 of the polypeptide-encoding genes were of unknown function, most being predicted genes without experimental confirmation.

For the reported 19 000 polypeptide-encoding genes, supporting experimental evidence was available in only a little over 9000 cases; the remaining close to 10 000 predicted genes were identified solely by computer-based analyses of the sequence. Subsequent analyses to test these genes for evidence of corresponding RNA transcripts suggested that at least 80% of the computer-predicted genes were authentic, leading Reboul *et al.* (2001) to conclude that there are at least 17 300 genes in *C. elegans*. Now large-scale efforts are being made to

investigate specific gene function and large-scale chemical mutagenesis programs are seeking to produce large numbers of mutant phenotypes.

8.4.4 Metazoan genome projects are mostly focusing on models of development and disease

The success of the *C. elegans* sequencing project ushered in a new and confident era. To add to the few longer-standing animal genome projects (e.g. *D. melanogaster*), a plethora of new projects were developed with different motivations:

▶ basic research – for example the desire to fully understand valuable models of development;

▶ commercial – for example genome projects for farm animals;

▶ medical – for example to understand disease models and the pathogenesis caused by parasitic nematodes or borne by disease vectors such as mosquitoes, the carriers of malaria.

The *D. melanogaster* genome project

This was initially conducted largely as a collaboration between the University of California at Berkeley laboratory and a consortium of European laboratories, but subsequently with major participation by a private company, Celera. The sequence reported by Adams *et al.* (2000) was not the complete sequence, of the 165-Mb genome, but that of nearly all of the approximately 120-Mb euchromatic portion in which the vast majority of the genes were known to reside. The release 3.0 version issued in August/September 2002 has eliminated most of the gaps in the euchromatic sequence (http://www.fruitfly.org/sequence/ index.html).

Box 8.8: Model multicellular animals for understanding development, disease and gene function.

A wide range of multicellular animal models have been used for understanding basic processes of developmental and cell biology, for understanding gene function, and as disease models. They range from invertebrate worms and flies to different species of fish, frogs, birds and mammals. See Hedges (2002) and see *Figures 12.23* and *12.24* for phylogenics.

CAENORHABDITIS ELEGANS (ROUNDWORM)

C. elegans is a *nematode* or *roundworm* (as opposed to flatworms and segmented worms). Roundworms outnumber all other complex creatures on the planet and are found almost everywhere in the temperate world, flourishing in soil. They may be free living (such as *C. elegans*) or parasitic. Roundworms infect a billion humans, spread diseases including river blindness and elephantiasis, and devour crops.

C. elegans is 1 mm long. There are two sexes: male (XO); and hermaphrodite (XX), a modified female. The hermaphrodite is the predominant sex. By producing both sperm and eggs it can self-fertilize, resulting in homozygosity for alleles. The male develops through occasional loss of one of the two hermaphrodite X chromosomes, and hermaphrodites will mate preferentially with males, as available. The cell lineage is *invariant*, and there are precisely 959 somatic cells in the adult hermaphrodite and 1031 in the adult male.

C. elegans is an important model of development, can be cultured easily in the lab (on agar plates feeding on bacteria, or in liquid culture) and is very amenable to genetic analyses. In the latter case, in addition to the standard methods of knocking out gene function at the DNA level, transient inactivation of expression can be achieved for specific genes by **RNA interference (RNAi) technology**. In this approach an investigator injects double-stranded RNA for the gene of interest into the oocyte. The double-stranded RNA inactivates expression of the homologous gene and resulting mutant phenotypes can be examined for clues to gene function (see Section 20.2.6).

Several features make *C. elegans* a good model organism for developmental biology and related studies:

▶ *expression studies.* Because *C. elegans* is transparent throughout its life cycle, the *green fluorescent protein* gene (**GFP**; see *Box 20.4*) can be linked to a *C. elegans* gene and used to find out where the gene is expressed in the worm;

▶ *lineage studies.* The transparency of *C. elegans* means every cell can be seen and followed during development. As a result, and because of the invariance of the cell lineage, the exact lineage of every cell in *C. elegans* is known—information which is unknown in all other multicellular organisms;

▶ *nervous system.* There is a complete wiring diagram of the nervous system: all 302 neurons and the connections between them are known. *C. elegans* also possesses genes for most of the known molecular components of vertebrate brains. While many scientists believe that the human brain is so complex that we can never hope to fully understand it, understanding the simple nervous system of *C. elegans* will provide enormous insights and knowing all the *C. elegans* genes will be important in understanding its nervous system;

▶ *aging.* This is readily studied since the worm develops from one cell into the fully grown form within 3 days and survives for only 2 weeks. Numerous mutations have been identified in *C. elegans* which result in major extensions of life span, with some mutants living more than five times as long as wild-type worms;

▶ *apoptosis.* Genes involved in **apoptosis** (programmed cell death) are often highly conserved and the pattern of apoptosis is well understood in *C. elegans* development (a total of 1090 cells develop initially in the hermaphrodite but 131 cells are programmed to die during development).

DROSOPHILA MELANOGASTER

The fruit fly is so-called because of its penchant for rotting fruit. It has a short life cycle, is particularly amenable to sophisticated genetic analyses, and has been studied extensively over many decades. During this time a large number of mutants have been systematically analyzed and a vast amount of information has been built up concerning gene function (see Perrimon, 1998 for a summary of new advances in studying gene function). A variety of features and approaches have assisted genetic mapping and functional analyses:

▶ **polytene chromosomes**. These are *interphase* chromosomes 2 mm long found in the salivary gland cells in the larval stages. They are distinctive because they are produced by repeated replication without separation into daughter nuclei, resulting in a set of 1024 copies of the normally single DNA duplex, arranged side-by-side like drinking straws in a box. Because of this parallel DNA amplification, polytene chromosomes are unique among interphase chromosomes in being visible under the light microscope. As a result of their extended conformation, they allow precise localization (tens of kilobases) of chromosome breakpoints and precise localization of DNA clones by *in situ* hybridization;

▶ the **P-element**. This *Drosophila transposable element* permits several types of experimental manipulation, including *mutagenesis* (see Spradling *et al.*, 1995) and *transgenesis* (see Section 20.2.2). Unequal recombination between adjacent P element inserts can also produce precise deletions;

▶ spatially and temporally restricted expression of transgenes is possible using the **GAL4-UAS** system of conditional gene expression. Large-scale mutagenesis screens are possible and *RNAi technology* (see above) has recently been used to transiently inactivate specific genes. In many cases (perhaps two-thirds of the 12 000 *Drosophila* genes) loss of function does not result in a mutant phenotype, but transgene misexpression often gives clues to gene function by producing dominant/dominant negative phenotypes. One generation screens for suppressors/enhancers of dominant mutant phenotypes can identify interacting genes. The yeast flp-frt recombinase system can be used to induce mitotic clones and so form homozygous patches permitting observation of phenotypes of lethal recessive mutations at late stages of development. Mitotic recombination can also be used in a one-generation screen to score mutant phenotypes in clones and recover lethal mutations that affect late development.

Although *Drosophila* is an invertebrate, and the number of *Drosophila* genes is perhaps only half of the number of human genes, there are nevertheless some remarkable similarities between human and *Drosophila* genes. Much of the difference in gene number is due to gene duplication events resulting in larger gene families in humans and a large proportion of *Drosophila* genes have human homologs, including genes underlying many genetic disorders and cancer genes. The level of homology has permitted successful electronic screening to identify human cDNAs corresponding to mutant *Drosophila* genes (Banfi *et al.*, 1996). Many of the highly conserved genes play important roles in early development which is comparatively well understood in *Drosophila* and many of the relevant pathways in early development and in some other crucial cellular processes are essentially conserved from *Drosophila* to mammals. As a result, *Drosophila* can be used as a model system of exploring gene function and interacting partners of genes which have direct relevance to human systems.

FISHES

Different fishes have been used as models, notably the zebrafish which is an excellent model of development and an increasingly important disease model, the pufferfish because of its very compact genome, and more recently, Medaka, another important model of development. Comparative genomic approaches in these three species permit an important window on vertebrate genome evolution.

▶ **Zebrafish (*Brachydanio rerio*)**

The zebrafish is a small freshwater fish that originated in rivers in India and is now common as an aquarium fish throughout the world. It has a short generation time and large numbers of eggs are produced at each mating. It is a principal model of vertebrate *development*: fertilization is external so that all aspects of development are accessible, and the embryo is transparent, facilitating identification of developmental mutants. Genes that are important in vertebrate development are often very highly conserved and so human developmental control genes normally have easily identifiable orthologs in zebrafish. Large scale mutagenesis screens have produced a large number of valuable developmental mutants and some of these have been used to model human disorders (see *Box 20.6*). **RNA interference technology** (see above) is also being used to inactivate specific genes.

▶ **The pufferfish** e.g. ***Takifugu rubripes rubripes*** (see below) and ***Tetraodon nigroviridis*** (see next page).

The value has mostly been in *comparative genomics* (see Section 12.3). The pufferfish has an extremely compact genome with much the same number of genes as mammals

continued on next page

Box 8.8: Model multicellular animals for understanding development, disease and gene function *(continued)*

compressed into a genome only about one-seventh of the size of the human or mouse genomes. Conservation of exons and important regulatory sequences aids identification of human equivalents by comparative mapping (see Clark, 1999).

▶ **Medaka**

The medaka (*Oryzias latipes*) is a small egg-laying freshwater fish predominantly found in Japan. It is an increasingly important model of development which is distantly related to zebrafish (the two separated from a common ancestor perhaps more than 110 million years ago). Like the zebrafish it is well-suited to genetic and embryological analyses with a short generation time (2–3 months), inbred strains, genetic maps, transgenesis, enhancer trapping and availability of stem cells (Wittbrodt *et al.*, 2002). The phenotypes recovered from developmental screens in medaka and in zebrafish indicate a nonoverlapping spectrum of embryonic-lethal phenotypes.

THE CHICK

The chick has several advantages as a *model of development*. Like mammals, birds are amniotes (the embryo has amniotic membranes) and their development closely resembles that of mammals. However, while a mammalian embryo depends on the mother for its nutrition (with exchange occurring across the placenta), avian embryos do not have a placenta and so are

self-developing systems. Because the chick embryo develops outside the body, it is accessible at all stages in development. In addition to being obtained easily, the chick embryo has the advantage of being large, and relatively translucent, allowing delicate microsurgical manipulations to be performed easily. As such, it offers an excellent system in which molecular studies can be combined with classical embryology (see Brown *et al.*, 2003).

Popular experimental manipulations of chick embryos include: surgical manipulations and tissue grafting; retrovirus-mediated gene transfer; electroporation of developing embryos; and embryo culture. In addition, the chick offers a unique system for the *study of cellular processes*: the **DT40 cell line**. Chicken DT40 cells continue to diversify their immunoglobulin genes and uniquely among available vertebrate cell lines, DT40 cells can carry out homologous recombination at a rate approaching that of illegitimate recombination. As a result, targeted gene deletion and mutation analyses can be carried out in the cultured cell line with relative convenience. Rapid advances are being made in chicken transgenics, embryonic stem (ES) cell technology and cryopreservation of sperm, blastodisc cells, primordial germ cells and ES cells. These newer technologies will permit the creation of chick-based systems for modeling human disease.

XENOPUS (AFRICAN CLAWED FROG)

This African frog gets its name (*Xenopus* = strange foot) from the sharp claws on the toes of its large, strong, webbed hind feet. It has long been a favorite model of embryonic development and cell biology: all developmental stages are accessible and the comparatively large size of the eggs and very early embryos facilitates micromanipulation, including microinjections (of mRNA, antibodies and antisense oligonucleotides), cell grafting and labeling experiments. Powerful methods for generating transgenic embryos were developed in the mid 1990s (see Beck and Slack, 2001) and impressive mutagenesis screens are being planned for *X. tropicalis*.

Photo from Amaya *et al.* (1998) Trends Genet. **14**, 253–255, © Elsevier.

▶ *Xenopus laevis (pictured on left, above)*

X. leavis has a longish generation time of 1–2 years, and can be induced to produce 300–1000 eggs at a time which are large (1–1.3 mm). It has been an important model for establishing the mechanisms for early fate decisions, patterning of the basic body plan and organogenesis. Contributions in cell biology and biochemistry include seminal work on chromosome replication, chromatin and nuclear assembly, cell cycle components, cytoskeletal elements, and signaling pathways. *X. laevis* has an *allotetraploid genome* (a very significant number of genes are duplicated as a result of a genome duplication event occurring maybe 30 million years ago, although many of the originally duplicated genes have since been lost) and a major disadvantage is that it has not been so easy to carry out genetic analyses.

▶ *Xenopus tropicalis (pictured on right, above)*

The smaller *X. tropicalis* has a comparatively short generation time (< 5 months), and can be induced to produce 1000–3000 eggs at a time, albeit smaller than those of *X. laevis*. It is evolutionarily closely related to *X. laevis* but has a diploid genome and is more amenable to genetic analyses.

THE MOUSE

The model organism considered most relevant to the Human Genome Project (see Meisler, 1996). It is the mammalian species with the most highly developed genetics and it is an extensively used model of mammalian development. Its small size and short generation time have allowed large-scale mutagenesis programs and extensive genetic crosses, and various features aid in mapping genes and phenotypes (see *Box 14.2*). Because of the comparatively high level of sequence conservation between human and mouse coding sequences, almost all human genes have an easily identifiable mouse homolog. Large chromosomal segments are conserved between mouse and man (*Figures 14.7, 14.8*) and so if a region of the mouse genome is mapped to high resolution, the information can be used to make predictions about the orthologous region of the human genome (and vice versa). This is particularly relevant to medical research because orthologous mouse and human mutants often show similar phenotypes so that positional cloning of a disease gene in one species may have considerable relevance to the other species (Section 14.3.5). The ability to construct mice with predetermined genetic modifications to the germline (by transgenic technology and gene targeting in embryonic stem cells) has been a powerful tool in studying gene expression and function and in creating mouse models of human disease. See Chapter 20 for details.

THE RAT

Rats, being considerably larger than mice, have for many years been the mammal of choice for physiological, neurological, pharmacological and biochemical analysis. They may also provide genetic model systems for complex human vascular and neurological disorders, such as hypertension and epilepsy (for various reasons, there are no mouse models for such diseases). Genetic analysis in laboratory rats, however, is much less advanced than in mice, partly because of the relatively high cost of rat breeding programs and the current difficulty in modifying the rat germline by gene targeting. Recently, however, high resolution genetic and physical maps have been constructed, and the genome sequence is now available.

The initial report of the *Drosophila* genome sequence reported 13 601 polypeptide-encoding genes, with a gene density of about one per 9 kb. The low gene number was a surprise when referenced against the initially reported 19 000 polypeptide-encoding genes of *C. elegans*, a simpler organism with perhaps about only 10% of the number of cells of *D. melanogaster*. In order to relate *D. melanogaster* to more distantly related *Drosophila* species other genome projects have recently been initiated, notably the *D. pseudoobscura* project.

The *Anopheles gambiae* genome project

Anopheles gambiae is the mosquito which spreads the deadly malaria parasite, *Plasmodium falciparum*. With funding mainly provided by the U.S. NIH–NIAID and the French Ministry of Research, the *Anopheles gambiae* genome was sequenced by a collaboration between Celera Genomics, the French national sequencing center Génoscope, and the Institute for Genome Research (TIGR), in association with several university research groups (Holt *et al.,* 2002). Its sequence was reported simultaneously with that of *Plasmodium falciparum* (see Section 8.4.2), offering a new dimension in combating malaria.

The mouse (*Mus musculus*) genome project

Because of various features the mouse provides the model genome which is most relevant to the Human Genome Project (see *Box 8.8*), and was expected to have essentially the same number of genes. The publicly funded mouse genome sequencing effort began in 1999 and had initially planned to deliver a draft sequence by 2005. However, when the private company Celera then announced its intention to sequence the mouse genome in 2 years, concerted attempts were made to accelerate the public sequencing effort. Towards that end, the **Mouse Genome Sequencing Consortium (MGSC)** was formed by the U.S. National Institutes of Health and the U.K.'s Wellcome Trust in association with three private companies (GlaxoSmithKline, Merck Genome Research Institute and Affymetrix, Inc.). The MGSC sequencing effort was delegated to three of the major centers involved in the human genome project: the Wellcome Trust Sanger Institute, the Whitehead Institute/MIT and Baylor College of Medicine.

In April 2001 Celera announced that their rival mouse sequencing project had delivered a sixfold coverage of the mouse genome, involving sequences from three mouse strains: 129X1/SvJ, DBA/2J and A/J. The Celera sequence data were not freely available; instead, access was restricted to those prepared to pay expensive subscription fees. In May 2002, the MSGC announced that they had completed a sevenfold coverage of the C57BL/6J mouse genome which was 96% complete and released it on the internet immediately, before publishing a conventional journal article in December 2002 (Mouse Genome Sequencing Consortium, 2002).

Annotated versions of the mouse genome sequence data became available as a result of the **Ensembl project**, a collaboration between the European Bioinformatics Institute and the Wellcome Trust Sanger Institute (http://www.ensembl.org/Mus_musculus), and through the University of California at Santa Cruz (http://genome.ucsc.edu/cgi-bin/hgGateway?db=mm2). The data from the initial draft sequence suggested that the mouse had much the same number of genes as in humans (ca. 30 000 or somewhat fewer than 30 000 – see Mouse Genome Sequencing Consortium, 2002), but with some interesting differences (see Section 12.4.1). As well as providing important data on the mouse genes and the structure of the mouse genome, the data have also been valuable in defining sequence variation and in *comparative genomics*. Comparison with the human genome sequence, for example, has been very helpful in defining highly conserved sequences (not just coding DNA but also regulatory and other sequences and in understanding genome evolution (Section 12.3.2).

Other metazoan genome projects (see *Table 8.4*)

▶ ***Bird genome projects.*** The front runner is the **chick** because of its importance as a model of development (*Box 8.8*).

▶ ***Fish genome projects.*** By the end of 2002, most of the genomic sequence had been obtained for two species of **pufferfish**, a model of a compact vertebrate genome, and the **zebrafish**, a model of development and an increasingly important model of disease and gene function (see *Table 8.4* and *Box 8.8*).

▶ ***Frog genome projects.*** As detailed in *Box 8.8*, **Xenopus** is an excellent model of development and a commitment has been made to sequence the genome of *Xenopus tropicalis*, which is more genetically tractable than the more popularly studied *Xenopus laevis*.

▶ ***Fly genome projects.*** In addition to *Drosophila* and mosquito projects (Section 8.4.4), a genome project has also been initiated for the **honey bee** which is of interest because of: (1) its powerful social instincts and unique behavioral traits (useful to neurobiologists); (2) its relevance to human health (the potentially serious consequences of bee stings, and as a model for antibiotic resistance, immunity, allergic reaction, etc.); (3) its importance to the agricultural community as a pollinator.

▶ ***Worm genome projects.*** In addition to projects for *free living* nematodes (the completed *C. elegans* project – Section 8.4.3 – and the more recent project to sequence a distantly related species, *C. briggsae* – see Section 12.3.2), a major collaborative project has been developed between the Wellcome Trust Sanger Institute and the Universities of Washington and Edinburgh to sequence hundreds of thousands of ESTs from about 20 species of **parasitic nematodes.** They include *Ascaris lumbricoides* which is the most common nematode parasite of humans, infecting an estimated 1.47 billion individuals (pathology can result from pneumonia caused by the worm's migration through the lungs, blocking of the gastrointestinal tract or the bile or pancreatic duct – see http://nematode.net).

▶ ***Mammalian genome projects.*** Projects have recently been initiated to sequence the genomes of:
  - the ***chimpanzee,*** our closest relative. The interest in sequencing the genome of the chimpanzee arises from its evolutionarily very close relationship to humans. The information may provide valuable insights into human

diseases where counterparts in chimpanzees are rare or appear to be because there are a number of medical conditions that affect humans but not chimpanzees (e.g. HIV progression to AIDS, *Plasmodium falciparum*-induced malaria) – see Cyranoski (2002) and Olson and Varki, (2003);

- the **rat**, a valuable model for cardiovascular, psychological and physiological research. A draft genome sequence was announced by Baylor College of Medicine and deposited in GenBank in November 2002. The sequence covers 90% of the genome (2560 Mb of the estimated total of 2800 Mb; see http://www.hgsc.bcm.tmc.edu/projects/rat/);

- the **cow**, **sheep** and **pig** – valuable farmyard animals;

- the **dog**, a valuable disease model (dogs suffer many of the same diseases as humans). The dog is also important model for the genetics of behavior, and is used extensively in pharmaceutical research.

▶ **Simple marine animal genome projects.**

- The **sea urchin** has been an important model system for many years in the study of basic biology, particularly in

developmental biology. It is a *deuterostome* and so is comparatively closely related to vertebrates and humans (see *Figure 12.23*). There is a large body of information about gene expression, and the sea urchin is a useful model for learning how pathways of genes and proteins regulate growth and development. A genome project has recently been initiated for *Strongylocentrotus purpuratus*.

- **Ascidians** (also called **tunicates,** and commonly known as **sea squirts**) are a group of marine animals that spend most of their lives attached to docks, rocks or the undersides of boats. Belonging to the phylum urochordata, they are actually more closely related to vertebrates than to most other invertebrate animals. *Ciona intestinalis* has the smallest genome of any experimentally manipulatable *chordate*, and provides a good system for exploring the evolutionary origins of the chordate lineage. A draft sequence of the *C. intestinalis* genome was reported by Dehal *et al.* (2002). The 117-Mb genome has 16 000 or so genes and comparative genomics analyses with sequenced vertebrate genomes is providing important information on the evolution of chordates and vertebrates.

Table 8.4: Important metazoan (multicellular animal) genome projects (see Further reading for electronic reference sources)

Animal class	Species (common name)	Genome size	Co-ordinating center[a]
Ascidians	*Ciona intestinalis* (sea squirt)	200 Mb	US DOE Joint Genome Institute
Birds	*Gallus gallus* (chick)	1200 Mb	Washington University
Fish	*Brachydanio rerio* (zebrafish)	1900 Mb	Wellcome Trust Sanger Institute
	Fugu rubripes (pufferfish)	400 Mb	International Fugu Genome Consortium
	Tetraodon nigroviridis (pufferfish)	350 Mb	Genoscope, Paris
Frogs	*Xenopus tropicalis*	1700 Mb	US DoE Joint Genome Institute
Insects	*Apis mellifera* (honey bee)	270 Mb	Baylor College of Medicine
	Aedes aegypti (yellow fever mosquito)	780 Mb	TIGR (The Institute for Genome Research)
	Anopheles gambiae (malaria mosquito)	278 Mb	International (but principally Celera & Genoscope)
	Drosophila melanogaster (fruitfly; euchrom. region)	165 Mb	Celera/Drosophila Genome Center/Howard Hughes Medical Institute
	Drosophila pseudoobscura (euchrom.region)	125 Mb	Baylor College of Medicine
Mammals	*Bos taurus* (cow)	3000 Mb	Baylor College of Medicine
	Canis familiaris (dog)	2800 Mb	Whitehead/MIT
	Mus musculus (mouse)	2500 Mb	Mouse genome sequencing consortium/Celera
	Pan troglodytes (common chimp)	3500 Mb	Whitehead/MIT and Washington University
	Rattus norvegicus (rat)	3100 Mb	Baylor College of Medicine
Roundworms	*Caenorhabditis elegans*	100 Mb	Wellcome Trust Sanger Institute/Washington University
	Caenorhabditis briggsae	80 Mb	Washington University
Sea urchins	*Strongylocentrotus purpuratus*	800 Mb	Baylor College of Medicine (likely)

[a]Web addresses for the major sequencing centers involved are as follows:

Baylor College of Medicine Genome Sequencing Center	http://hgsc.bcm.tmc.edu/
Celera	http://www.celera.com/
DoE Joint Genome Institute	http://www.jgi.doe.gov/
Genoscope	http://www.genoscope.cns.fr/
TIGR (The Institute for Genome Research)	http://www.tigr.org/
Washington University Genome Sequencing Center	http://genome.wustl.edu/
Wellcome Trust Sanger Institute	http://www.sanger.ac.uk/
Whitehead Institute/MIT Center for Genome Research	http://www-genome.wi.mit.edu/

Further reading

Borsani G, Ballabio A, Banfi S (1998) A practical guide to orient yourself in the labyrinth of genome databases. *Hum. Mol. Genet.* **7**, 1641–1648.

Database issue of Nucleic Acid Research (2003) *Nucl. Acid Res.* **31**, 1–516

Hedges SB (2002) The origin and evolution of model organisms. *Nature Rev. Genet.* **3**, 838–849.

Human Genome *Nature* Issue (15 February 2001). *Nature* **409**, 813–958 (papers are available electronically at the Nature Genome Gateway at http://www.nature.com/genomics/human/).

Human Genome *Science* Issue (16 February 2001). *Science* **291**, 1177–1351 (papers are available electronically at http://www.sciencemag.org/content/vol291/issue5507/index.shtml).

Mouse Genome *Nature* Issue (5 December 2002). *Nature* **420**, 509–590 (papers are available electronically at http://www.nature.com/nature/mousegenome/index.html).

User's Guide to the Human Genome. *Nature Genetics Supplement,* September 2002 (available electronically through the Nature Genome Gateway at http://www.nature.com/genomics/).

Wilkie T (1993) *Perilous Knowledge: the Human Genome Project and its Implications.* Faber and Faber, New York.

Electronic information on the Gene Ontology Consortium at http://www.geneontology.org/

Electronic information on the Human Genome Project (and related projects) can be found at many locations. Useful web sites include the following sites:
- the U.S. National Human Genome Research Institute (NGHRI) at http://www.nhgri.nih.gov/
- *the U.S. National Center for Biotechnology Information (NCBI) at http://www.ncbi.nlm.nih.gov/*
- the Genome Web maintained at the UK Human Genome Mapping Resource Centre at http://www.hgmp.mrc.ac.uk/GenomeWeb/
- The Nature Genome Gateway at http://www.nature.com/genomics/

Electronic information on Genome Projects for microorganisms and model organisms can be found at a variety of sites, including:
- The TIGR microbial genome database at http://www.tigr.org/tdb/mdb/mdbcomplete.html and http://www.tigr.org/tdb/mdb/mdbinprogress.html
- The European Bioinformatics parasite genome webpage at http://www.ebi.ac.uk/parasites/paratable.html
- The Parasite Genomes Web Site at http://www.rna.ucla.edu/par/
- The U.S. national Human Genome Research Institute's web site at http://www.genome.gov/Pages/Research/Sequencing/Proposals/

Electronic information on Model Organisms includes:
- the NIH model organism site at http://www.nih.gov/science/models
- the Model Organisms Virtual Library at http://www.ceolas.org/VL/mo/
- the NCBI Model Organisms site at http://www.ncbi.nlm.nih.gov/About/model/
- the Model and other organisms glossary at http://www.genomicglossaries.com/content/model_organisms_glossary.asp
- the Tree of Life web project at http://tolweb.org/tree/

Electronic summary lists for all Genome Projects can be found at a variety of sites, including:
- A list of completed and ongoing genome projects compiled by Integrated Genomics at http://ergo.integratedgenomics.com/GOLD/
- a list of completed genomes at the European Bioinformatics Institute at http://www2.ebi.ac.uk/genomes/

Human Genome Browsers and Integrated Databases include:
- Ensembl at http://www.ensembl.org
- NCBI Map Viewer at http://www.ncbi.nlm.nih.gov/cgi-bin/Entrez/map-search
- UCSC Genome Browser at http://genome.ucsc.edu
- The HOWDY integrated database at http://www-alis.tokyo.jst.go.jp/HOWDY

References

Adams MD, Kelley JM, Gocayne JD et al. (1991) Complementary DNA sequencing: expressed sequence tags and Human Genome Project. *Science* **252**, 1651–1656.

Adams MD, Celniker SE, Holt RA et al. (2000) The genome sequence of *Drosophila melanogaster. Science* **287**, 2185–2195.

Banfi S, Borsani G, Rossi E et al. (1996) Identification and mapping of human cDNAs homologous to Drosophila mutant genes through EST database searching. *Nature Genet.* **13**, 167–174.

Batzoglou S, Pachter L, Mesirov JP, Berger B, Lander ES (2000) Human and mouse gene structure: comparative analysis and application to exon prediction. *Genome Res.* **10**, 950–958.

Beck CW, Slack JM (2001) An amphibian with ambition: a new role for Xenopus in the 21st century. *Genome Biol.* **2**, 1029.1–1029.5.

Botstein D, White RL, Skolnick M, Davis RW (1980) Construction of a genetic linkage map in man using restriction fragment length polymorphism. *Am. J. Hum. Genet.* **32**, 314–331.

Botstein D, Chervitz SA, Cherry JM (1997) Yeast as a model organism. *Science* **277**, 1259–1260.

Brown WRA, Hubbard SJ, Tickle C, Wilson SA (2003). The chicken as a model for large scale analysis of vertebrate gene function. *Nature Rev. Genet.* **4**, 87–98.

Burge C, Karlin S (1997) Prediction of complete gene structures in human genomic DNA. *J. Mol. Biol.* **268**, 78–94.

C. elegans Sequencing Consortium (1998) Genome sequence of the nematode *C. elegans*: a platform for investigating biology. *Science* **282**, 2012–2017.

Camargo AA, de Souza SJ, Brentani RR, Simpson AJG (2002) Human gene discovery through experimental definition of transcribed regions of the human genome. *Curr. Opin. Chem. Biol.* **6**, 13–16.

Cavalli-Sforza LL, Wilson AC, Cantor CR, Cook-Deegan RM, King MC (1991) Call for a worldwide survey of human genetic diversity: a vanishing opportunity for the Human Genome Project. *Genomics* **11**, 490–491.

Chumakov IM, Rigault P, Le Gall I *et al.* (1995) A YAC contig map of the human genome. *Nature.***377**, 175–297

Clark MS (1999) Comparative genomics: the key to understanding the Human Genome Project. *Bioessays* **21**, 121–130.

Cohen D, Chumakov I, Weissenbach J (1993) A first generation physical map of the human genome. *Nature* **366**, 698–701.

Collins FS, McKusick VA (2001) Implications of the human genome project for medical science. *J. Am. Med. Assoc.* **285**, 540–544.

Collins F, Guyer MS, Chakravarti A (1997) Variations on a theme: cataloging human DNA sequence variation. *Science* **262**, 43–46.

Cox DR, Burmeister M, Proce ER, Kim S, Myers RM (1990) Radiation hybrid mapping: a somatic cell genetic method for constructing high-resolution maps of mammalian chromosomes. *Science,* **250**, 245–250.

Craig JM, Bickmore WA (1994) The distribution of CpG islands in mammalian chromosomes. *Nature Genet.* **7**, 376–381.

Cuthbert AP, Trott DA, Ekong RM *et al.* (1995) Construction and characterization of a highly stable human:rodent monochromosomal hybrid panel for genetic complementation and genome mapping studies. *Cytogenet. Cell Genet.* **71**, 68–76.

Cyranoski D (2002) Almost human. *Nature* **418**, 910–912.

Danesh J, Newton R, Beral V (1997) A human germ project? *Nature* **389**, 21–24.

Davies KE, Young BD, Elles RG, Hill ME, Williamson R (1981) Cloning of a representative genomic library of the human X chromosome after sorting by flow cytometry. *Nature* **293**, 374–376.

Dehal P, Satou Y, Campbell RK (2002) The draft genome of *Ciona intestinalis*: insights into chordate and vertebrate origins. *Science* **298**, 2157–2167.

Deloukas P, Schuler GD, Gyapay G *et al.* (1998) A physical map of 30,000 human genes. *Science* **282**, 744–746.

Doolittle RF (2002) Microbial genomes multiply. *Nature* **416**, 697–700.

Dunham I, Shimizu N, Roe BA *et al.* (1999) The DNA sequence of human chromosome 22. *Nature* **402**, 489–495.

Gardner MJ, Shallom SJ, Carlton JM *et al.* (2002) Genome sequence of the human malaria parasite *Plasmodium falciparum. Nature* **419**, 498–511.

Garver KL, Garver B (1994) The Human Genome Project and eugenic concerns. *Am. J. Hum. Genet.* **54**, 148–158.

Gene Ontology Consortium (2000) Gene ontology: tool for the unification of biology. *Nature Genet.* **25**, 25–29.

Gene Ontology Consortium (2001) Creating the gene ontology resource: design and implementation. *Genome Res.* **11**, 1425–1433.

Goffeau A, Barrell BG, Bussey H *et al.* (1996) Life with 6000 genes. *Science* **274**, 546–567.

Greely HT (2001) Human genome diversity: what about the other human genome project? *Nature Rev. Genet.* **2,** 222–227.

Hedges SB (2002) The origin and evolution of model organisms. *Nature Rev. Genet.* **3**, 838–849.

Holt RA, Subramaniam GM, Halpern A *et al.* (2002) The genome sequence of the malaria mosquito *Anopheles gambiae. Science* **298**, 129–149.

Hudson TJ, Stein LD, Gerety SS *et al.* (1995) An STS-based map of the human genome. *Science* **270**, 1945–1954.

Ichikawa H, Hosoda F, Arai Y, Shimizu K, Ohira M, Ohki M (1993) A *Not*I restriction map of the entire long arm of human chromosome 21. *Nature Genet.* **4**, 361–365.

International Human Genome Sequencing Consortium (2001) Initial sequencing and analysis of the human genome. *Nature* **409**, 860–921.

International SNP map working group (2001) *Nature* **409**, 928–933.

Knoppers BM (1999) Status, sale and patenting of human genetic material: an international survey. *Nature Genet.* **22**, 23–25.

Kong A, Gudbjartsson DF, Sainz J *et al.* (2002) A high-resolution recombination map of the human genome. *Nature Genet.* **31**, 241–247.

Ludecke HJ, Senger G, Claussen U, Horsthemke B (1989) Cloning defined regions of the human genome by microdissection of banded chromosomes and enzymatic amplification. *Nature* **338**, 348–350.

McKusick V (1989) HUGO News. The Human Genome Organization: History, Purposes, and Membership. *Genomics* **5**, 385–387.

Meisler MH (1996) The role of the laboratory mouse in the human genome project. *Am. J. Hum. Genet.* **59**, 764–771.

Mouse Genome Sequencing Consortium (2002) Initial sequencing and comparative analysis of the mouse genome. *Nature* **420**, 520–562.

Murray JC, Buetow K, Weber JL *et al.* (1994) A comprehensive human linkage map with centimorgan density. *Science* **265**, 2049–2054.

Olsen GJ, Woese CR (1997) Archaeal genomics: an overview. *Cell* **89**, 991–994.

Olson MV, Varki A (2003) Sequencing the chimpanzee genome: insights into human evolution and disease. *Nature Rev. Genet.* **4**, 20–28.

Pennisi E (1997) Laboratory workhorse decoded. *Science* **277**, 1432–1434.

Perrimon N (1998) New advances in *Drosophila* provide opportunities to study gene functions. *Proc. Natl Acad. Sci. USA* **95**, 9716–9717.

Reboul J, Vaglio P, Tzellas N *et al.* (2001) Open-reading-frame sequence tags (OSTs) support the existence of at least 17,300 genes in *C. elegans. Nature Genet.* **27**, 332–336.

Schuler GD, Boguski MS, Stewart EA *et al.* (1996) A gene map of the human genome. *Science* **274**, 540–546.

Spradling AC, Stern DM, Kiss I, Roote J, Laverty J, Rubin GM (1995) Gene disruptions using P transposable elements: an integral component of the *Drosophila* genome project. *Proc. Natl Acad. Sci. USA* **92**, 10824–10830.

Subramanian G, Adams MD, Venter JC, Broder S (2001) Implications of the human genome for understanding human biology and medicine. *J. Am. Med. Assoc.* **286**, 2296–2307.

Tang CM, Hood DW, Moxon ER (1997) *Haemophilus* influence: the impact of whole genome sequencing on microbiology. *Trends Genet.* **13**, 399–404.

Thomas SM, Davies ARW, Birtwistle NJ, Crowther SM, Burke JF (1996) Ownership of the human genome. *Nature* **380**, 387–388.

Turkewitz AP, Orias E, Kapler G (2002) Functional genomics: the coming of age for *Tetrahymena thermophila. Trends Genet.* **18**, 35–40.

van Ommen GJ (2002) The Human Genome Project and the future of diagnostics, treatment and prevention. *J. Inherit. Metab. Dis.* **25,** 183–188.

Venter JC, Adams MD, Myers EW *et al.* (2001) The sequence of the human genome. *Science.* **291**, 1304–1351.

Walter MA, Spillett DJ, Thomas P, Weissenbach J, Goodfellow PN (1994) A method for constructing radiation hybrid maps of whole genomes. *Nat. Genet.* **7**, 22–28.

Waterston R, Lander ES, Sulston J (2002) On the sequencing of the human genome. *Proc. Natl Acad. Sci. USA* **99**, 3712–3716.

Weissenbach J, Gyapay G, Dib C, Vignal A, Morissette J, Millasseau P, Vaysseix G, Lathrop M (1992) A second generation linkage map of the human genome. *Nature* **359**, 794–801.

Wittbrodt J, Shima A, Schartl M (2002) Medaka–a model organism from the far East. *Nature Rev. Genet.* **3**, 53–64

Wood V, Gwilliam R, Rajandream MA *et al.* (2002) The genome sequence of *Schizosaccharomyces pombe. Nature* **415**, 871–880.

Zhang MQ (2002) computational prediction of eurkaryotic protein-coding genes. *Nature Rev. Genet.* **3**, 698–709.

CHAPTER NINE

Organization of the human genome

Chapter contents

9.1 General organization of the human genome

9.1.1 An overview of the human genome

The **human genome** is the term used to describe the total genetic information (DNA content) in human cells. It really comprises *two* genomes: a complex *nuclear genome* with about 30 000 genes, and a very simple *mitochondrial genome* with 37 genes (*Figure 9.1*). The nuclear genome provides the great bulk of essential genetic information, most of which specifies polypeptide synthesis on cytoplasmic ribosomes.

Mitochondria possess their own ribosomes and the very few polypeptide-encoding genes in the mitochondrial genome produce mRNAs which are translated on the mitochondrial ribosomes. However, the mitochondrial genome specifies only a very small portion of the specific mitochondrial functions; the bulk of the mitochondrial polypeptides are encoded by nuclear genes and are synthesized on cytoplasmic ribosomes, before being imported into the mitochondria.

Human–mouse comparisons have shown that less than 5% of the genome is strongly conserved, including the 1.5% devoted to coding DNA and a somewhat higher percentage including conserved sequences within untranslated sequences, regulatory elements and so on (Mouse Genome Sequencing Consortium, 2002; Dermitzakis *et al.*, 2002). The majority of the coding DNA is used to make mRNA and hence polypeptides but a significant minority (at least 5% and probably close to 10%) of human genes specifies noncoding

(= untranslated) RNA (*RNA genes*). A variety of novel RNA genes have recently been identified, forcing a reassessment of RNA function.

The coding sequences frequently belong to families of related sequences (*DNA sequence families*) which may be organized into clusters on one or more chromosomes or be dispersed. Such duplicated sequences have arisen by various *gene duplication* mechanisms which have occurred during evolution. Sequencing of the genome provided the first genome-wide assessment of duplication, and revealed a very significant amount of primate-specific *segmental duplication* (very closely related blocks of sequences are found on different chromosomes or in different regions of a single chromosome as a result of very recent duplications – Section 12.2.5 and *Figure 12.13;* see Bailey *et al.*, 2002).

The mechanisms giving rise to duplicated genes also give rise to nonfunctional gene-related sequences, including *pseudogenes* and *gene fragments* (Section 9.3.6). There are numerous defective copies of RNA genes scattered through the genome, and for some polypeptide-encoding genes, too, many related pseudogenes are also found: analyses of the finished chromosome 21 and 22 sequences predicts a total of about 20 000 pseudogenes in the genome (Harrison *et al.*, 2002; Collins *et al.*, 2003).

As in other complex genomes, a very large component of the human genome is made up of *noncoding DNA*. A sizeable component is organized in *tandem* head-to-tail (→→→→) repeats, but the majority consists of interspersed repeats which have originated from RNA transcripts by *retrotransposition*

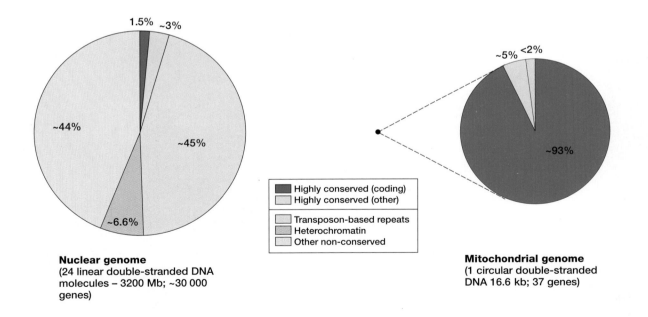

Nuclear genome
(24 linear double-stranded DNA molecules – 3200 Mb; ~30 000 genes)

Highly conserved (coding)
Highly conserved (other)

Transposon-based repeats
Heterochromatin
Other non-conserved

Mitochondrial genome
(1 circular double-stranded DNA 16.6 kb; 37 genes)

Figure 9.1: Organization of the human genome.

The dot in the middle represents the relative size of the mitochondrial genome using the same scale as for the nuclear genome. *Note* also the profound difference between the two genomes in the extent of highly conserved DNA (coding sequence, regulatory sequence etc.) and the fraction of highly repetitive noncoding DNA.

(cellular reverse transcriptases can copy RNA transcripts to make natural cDNA which can integrate elsewhere in the genome).

9.1.2 The mitochondrial genome consists of a small circular DNA duplex which is densely packed with genetic information

General structure and inheritance of the mitochondrial genome

The human mitochondrial genome is defined by a single type of circular double-stranded DNA whose complete nucleotide sequence has been established (Anderson *et al.*, 1981; see also the Mitomap mitochondrial genome database at http://www.mitomap.org/). It is 16 569 bp in length and is 44% (G + C). The two DNA strands have significantly different base compositions: the **heavy (H) strand** is rich in guanines, the **light (L) strand** is rich in cytosines. Although the mitochondrial DNA is principally double stranded, a small section shows

a *triple-DNA strand* structure due to the repetitive synthesis of a short segment of heavy strand DNA, **7S DNA** [see *Figure 9.2*, and Clayton (1992) for a general review of transcription and replication of animal mitochondrial DNAs]. Human cells typically contain thousands of copies of the double-stranded mitochondrial DNA molecule, but the number can vary considerably in different cell types (see *Box 9.1*).

During zygote formation, a sperm cell contributes its nuclear genome, but not its mitochondrial genome, to the egg cell. Consequently, the mitochondrial genome of the zygote is usually determined exclusively by that originally found in the unfertilized egg. The mitochondrial genome is therefore *maternally inherited*: males and females both inherit their mitochondria from their mother but males do not transmit their mitochondria to subsequent generations. Thus mitochondrially encoded genes or DNA variants give the pedigree pattern shown in *Figure 4.4*. During mitotic cell division, the mitochondrial DNA molecules of the dividing cell segregate in a purely random way to the two daughter cells.

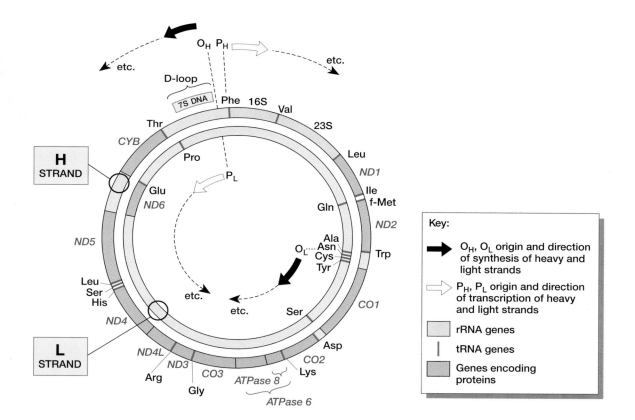

Figure 9.2: The human mitochondrial genome.

The **D loop** has a triple-stranded structure owing to duplicate synthesis of a stretch of heavy (H) strand. Transcription of the heavy strand originates from two closely spaced promoters in the D loop region (grouped for the sake of clarity as P_H). Transcription from the P_H promoters runs clockwise round the circle; but anticlockwise from the light strand promoter P_L. In both cases, the large primary transcripts are cleaved to generate RNAs for individual genes. The genes all lack introns and are closely clustered with one case of overlapping genes: the *ATPase 8* gene partly overlaps the *ATPase 6* gene (see *Figure 9.3*). Other polypeptide-encoding genes specify seven NADH dehydrogenase subunits (*ND4L* and *ND1–ND6*); three cytochrome *c* oxidase subunits (*CO1–CO3*) and cytochrome *b* (*CYB*).

Box 9.1: Genome copy number variation in human cells.

Textbooks often state that the cells of an organism show little variation in their DNA content, and this is undoubtedly true when referenced against RNA or protein content. Nevertheless, there can be appreciable differences in both mtDNA and nuclear DNA content in different cell types.

▶ **Mitochondrial genome copy number variation**. Certain cells (e.g. terminally differentiated skin cells) lack any mitochondria and so have no mtDNA. The mtDNA copy number in other somatic cells varies but typically is in the 1000–10 000 range (e.g. lymphocytes have about 1000 mtDNA molecules). The gametes are unusual: sperm cells have a few hundred copies of mtDNA and oocytes have perhaps 100 000 copies, accounting for over 30% of the oocyte DNA.

▶ **Nuclear genome copy number variation**. Nucleated (diploid) cells show little variation in DNA content, but

differential **ploidy** means for some cells there are substantial differences in DNA content: *nulliploidy* – cells have no DNA at all, as in the case of many types of terminally differentiated cells e.g. erythrocytes (which lack a nucleus) and terminally differentiated skin cells (which lack organelles); *haploidy* – there is half the DNA content of diploid cells in egg and sperm cells; *polyploidy* – some cells naturally have many copies of the normal chromosome set as a result of **endomitotic replication** (where cells undergo several rounds of DNA duplication but without any cell division; e.g. regenerating cells of the liver and other tissues are naturally tetraploid, and the giant megakaryocytes of the bone marrow can contain up to 16 times the amount of DNA in diploid cells), or as a result of **syncytial cell fusion** (e.g. muscle fiber cells form by fusion of multiple cells producing single cells with multiple nuclei – see *Figure 3.3*).

Mitochondrial genes

The human mitochondrial genome contains 37 genes. For 28 of the genes the heavy strand is the sense strand; for the other nine, the light strand is the sense strand (*Figure 9.2*). Of the 37 genes, a total of 24 specify a mature RNA product: 22 mitochondrial tRNA molecules and two mitochondrial rRNA molecules, a 23S rRNA (a component of the large subunit of mitochondrial ribosomes) and a 16S rRNA (a component of the small subunit of the mitochondrial ribosomes). The remaining 13 genes encode polypeptides which are synthesized on mitochondrial ribosomes.

Each of the 13 polypeptides encoded by the mitochondrial genome is a subunit of one of the mitochondrial **respiratory complexes**, the multichain enzymes of **oxidative phosphorylation** which are engaged in the production of ATP. There is, however, a total of about 100 different polypeptide subunits in the mitochondrial oxidative phosphorylation system, and so the vast majority are encoded by nuclear genes (see *Box 9.2*). All other mitochondrial proteins are encoded by the nuclear genome and are translated on cytoplasmic ribosomes before being imported into the mitochondria (*Box 9.2; Figure 1.11*).

The mitochondrial genetic code

The mitochondrial genetic code is used to decode the heavy and light chain transcripts to give a total of only 13 polypetides. This very small functional load has allowed the mitochondrial genetic code to *drift* from the 'universal' genetic code (which is retained for nuclear genes because of the need to conserve the functions of 30 000 or so genes). There are 60 mitochondrial sense codons, one fewer than in the nuclear genetic code, and four stop codons. Two of the four stop codons, UAA and UAG, also serve as stop codons in the nuclear genetic code, but the other two are AGA and AGG which specify arginine in the nuclear genetic code (see *Figure 1.22*). UGA encodes tryptophan rather than serving as a stop codon and AUA specifies methionine not isoleucine.

The mitochondrial genome encodes all the rRNA and tRNA molecules it needs for synthesizing proteins but relies on nuclear-encoded genes to provide all other components (such as the protein components of mitochondrial ribosomes, amino acyl tRNA synthetases etc.). As there are only 22 different types of human mitochondrial tRNA, individual tRNA molecules need to be able to interpret several different codons. This is possible because of **third base wobble** in codon interpretation. Eight of the 22 tRNA molecules have anticodons which are each able to recognize families of four codons *differing only at the third base*, and 14 recognize pairs of codons which are identical at the first two base positions and share either a purine or a pyrimidine at the third base. Between them, therefore, the 22 mitochondrial tRNA molecules can recognize a total of 60 codons [$(8 \times 4) + (14 \times 2)$].

In addition to their differences in genetic capacity and different genetic codes, the mitochondrial and nuclear genomes differ in many other aspects of their organization and expression (*Table 9.1*).

Coding and noncoding DNA

Unlike its nuclear counterpart, the human mitochondrial genome is extremely compact: approximately 93% of the DNA sequence represents coding sequence. All 37 mitochondrial genes lack introns and they are tightly packed (on average one per 0.45 kb). The coding sequences of some genes (notably those encoding the sixth and eighth subunits of the mitochondrial ATPase) show some overlap (*Figures 9.2 and 9.3*) and, in most other cases, the coding sequences of neighboring genes are contiguous or separated by one or two noncoding bases. Some genes even lack termination codons; to overcome this deficiency, UAA codons have to be introduced *at the post-transcriptional level* (Anderson *et al.*, 1981; see legend to *Figure 9.3*).

The only significant region lacking any known coding DNA is the **displacement (D) loop region**. This is the

Box 9.2: The limited autonomy of the mitochondrial genome.

Mitochondrial component	Encoded by mitochondrial genome	Encoded by nuclear genome
Components of oxidative phosphorylation system	**13 subunits**	**> 80 subunits**
I NADH dehydrogenase	7 subunits	> 41 subunits
II Succinate CoQ reductase	0 subunits	4 subunits
III Cytochrome *b–c*1 complex	1 subunit	10 subunits
IV Cytochrome *c* oxidase complex	3 subunits	10 subunits
V ATP synthase complex	2 subunits	14 subunits
Components of protein synthesis apparatus	**24**	**ca. 80**
rRNA components	2 rRNAs	None
tRNA components	22 tRNAs	None
Ribosomal proteins	None	ca. 80
Other mitochondrial proteins	**None**	**All** (e.g. mitochondrial DNA and RNA polymerases plus numerous other enzymes, structural and transport proteins etc.)

Table 9.1: The human nuclear and mitochondrial genomes

	Nuclear genome	Mitochondrial genome
Size	3200 Mb	16.6 kb
No. of different DNA molecules	23 (in XX cells) or 24 (in XY cells); all linear	One circular DNA molecule
Total no. of DNA molecules per cell	46 in diploid cells, but varies according to ploidy	Often several thousands (but variable – see *Box 9.1*)
Associated protein	Several classes of histone and nonhistone protein	Largely free of protein
No. of genes	~ 30 000–35 000	37
Gene density	~ 1/100 kb	1/0.45 kb
Repetitive DNA	Over 50% of genome, see *Figure 9.1*	Very little
Transcription	The great bulk of genes are transcribed individually (*monocistronic transcription units*)	Co-transcription of multiple genes from both the heavy and the light strands (*polycistronic transcription units*)
Introns	Found in most genes	Absent
% of coding DNA	~ 1.5%	~ 93%
Codon usage	See *Figure 1.22*	See *Figure 1.22*
Recombination	At least once for each pair of homologs at meiosis	Not evident
Inheritance	Mendelian for sequences on X and autosomes; paternal for sequences on Y	Exclusively maternal

region in which a triple-stranded DNA structure is generated by duplicate synthesis of a short piece of the H-strand DNA, known as 7S DNA (see *Figure 9.2*). The replication of both the H and L strands is unidirectional and starts at specific origins. In the former case, the origin is in the D loop and only after about two-thirds of the daughter H strand has been synthesized (by using the L strand as a template and displacing the old H strand) does the origin for L strand replication become exposed. Thereafter, replication of the L strand proceeds in the opposite direction, using the H strand as a template (*Figure 9.2*). The D loop also contains the predominant promoter for transcription of both the H and L strands. Unlike transcription of nuclear genes, in which individual genes are almost always transcribed separately using individual

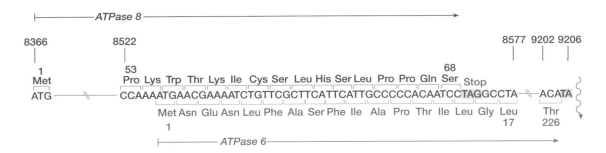

Figure 9.3: The genes for mitochondrial ATPase subunits 6 and 8 are partially overlapping and translated in different reading frames.

Note: the overlapping genes share a common sense strand, the H strand. Coding sequence co-ordinates are as follows: *ATPase* subunit 8, 8366–8569; *ATPase* subunit 6, 8527–9204. The C terminus of the *ATPase 6* subunit gene is defined by the post-transcriptional introduction of a UAA codon: following transcription the RNA is cleaved after position 9206 and polyadenylated, resulting in a UAA codon where the first two nucleotides are derived ultimately from the TA at positions 9205–9206 and the third nucleotide is the first A of the poly(A) tail. Other human genes are known to be overlapping but are often transcribed from opposite strands.

promoters, transcription of the mitochondrial DNA starts from the promoters in the D loop region and continues, in opposing directions for the two different strands, round the circle to generate large multigenic transcripts (see *Figure 9.2*). The mature RNAs are subsequently generated by cleavage of the multigenic transcripts.

9.1.3 The nuclear genome consists of 24 different DNA molecules corresponding to the 24 different human chromosomes

Size and structure of human chromosomes

The nucleus of a human cell typically contains more than 99% of the cellular DNA (except for some specialized cells, notably the oocyte – see *Box 9.1*). The nuclear genome is distributed between 24 different types of linear double-stranded DNA molecule, each of which has histones and other nonhistone proteins bound to it, constituting a chromosome. The 24 different chromosomes (22 types of autosome and two sex chromosomes, X and Y) can easily be differentiated by chromosome banding techniques (*Figure 2.15*), and have been classified into groups largely according to size and, to some extent, centromere position (*Table 2.3*).

The DNA selected for sequencing in the Human Genome Project was not the total nuclear genome, but the ***euchromatic portion***, comprising close to 3000 Mb. There is also over 200 kb of ***constitutive heterochromatin***, permanently condensed and transcriptionally inactive regions, giving a total genome size of the order of 3200 Mb. The average size of a human chromosome is therefore about 140 Mb but with considerable size variation between chromosomes, and variable amounts of constitutive heterochromatin (*Table 9.2*). The latter comprises approximately 3 Mb segments at each centromere plus large components on several chromosomes, including: the short arms of the acrocentric chromosomes 13, 14, 15, 21, and 22; the long arm of the Y chromosome; and large regions of the long arms of chromosomes 1, 9, and 16 (corresponding to *secondary chromosome constrictions* – see *Table 9.2* and *Figure 2.15* for a graphical view).

Base composition in the human nuclear genome

The draft human genome sequences (International Human Genome Sequencing Consortium, 2001; Venter *et al.*, 2001) suggest a genome-wide average of 41% GC for the euchromatic component. The base composition does, however, vary considerably between chromosomes from 38% GC for chromosomes 4 and 13 up to 49% for chromosome 19. It also varies considerably along the lengths of chromosomes. For example, the average GC content on chromosome 17q is 50% for the distal 10.3 Mb but drops to 38% for the adjacent 3.9 Mb. There are regions of less than 300 kb with even wider swings in GC content, for example, from 33.1% to 59.3%.

There is a clear correlation between GC composition and the extent of Giemsa staining during chromosome banding. For example, 98% of large-insert clones mapping to the darkest G-bands are in 200-kb regions of low GC content (average 37%), whereas more than 80% of clones mapping to the lightest G-bands are in regions of high GC content (average 45%). Analysis of the data does not, however, support the existence of strict ***isochores***, which have been defined as compositionally homogenous large-scale regions and classified into five or so groups on the basis of differing %GC composition (International Human Genome Sequencing Consortium, 2001).

The proportion of some combinations of nucleotides can vary considerably. Like other vertebrate nuclear genomes, for example, the human nuclear genome has a conspicuous shortage of the dinucleotide **CpG** (that is, neighboring cytosine and guanine residues on the same DNA strand in the 5′→3′ direction; 'p' denotes phosphodiester bond). Taking the overall average figure of 41% GC the individual base frequencies are C = G = 0.205, and so the *expected frequency* for the dinucleotide CpG is $(0.205)^2 = 0.042$. However, the *observed CpG frequency* is approximately one-fifth of this. Despite the general CpG depletion, certain small regions of transcriptionally active DNA have the expected CpG density and, significantly, are unmethylated (**CpG islands**; see *Box 9.3*).

Table 9.2: DNA content of human chromosomes

Chromosome	Total amount of DNA (Mb)	Amount of heterochromatin (Mb)	Chromosome	Total amount of DNA (Mb)	Amount of heterochromatin (Mb)
1	279	30	13	118	16
2	251	3	14	107	16
3	221	3	15	100	17
4	197	3	16	104	15
5	198	3	17	88	3
6	176	3	18	86	3
7	163	3	19	72	3
8	148	3	20	66	3
9	140	22	21	45	11
10	143	3	22	48	13
11	148	3	X	163	3
12	142	3	Y	51	27

Data abstracted from the International Human Genome Sequence Consortium (2001). Using these figures the size of the total human genome is 3289 Mb, but this figure (and the total amounts for individual chromosomes) is known to include some artefactual duplication and a more realistic value may be ~ 3200 Mb.

9.1.4 The human genome contains about 30 000—35 000 unevenly distributed genes but precise numbers are uncertain

Human gene number

The total number of genes in the human genome is now thought to be in the 30 000–35 000 range. As all but 37 of these genes are located in the nuclear genome, this gives a rough estimate of ca. 1400 genes per chromosome on average. The great majority of the genes are polypeptide-encoding but a significant minority (at least 5%, and probably about 10%) specify untranslated RNA molecules (Section 9.2).

The International Human Genome Sequencing Consortium (2001) and Venter *et al.* (2001) had estimated respectively 30 000–40 000 genes and 26 000–38 000 genes, but favored predictions nearer the *lower* ends of these ranges. Such estimates were much lower than previous ones based on incomplete data sets (see Section 8.3.5), but there remains considerable uncertainty about the precise gene number. There are general difficulties in identifying genes in the first place. When the draft genome sequences were published in 2001, about 11 000 or so genes could be identified with confidence; many thousands of others were predicted by computer-based analyses of the sequence. Computer-based prediction of polypeptide-encoding genes has been very helpful, but is not always so reliable (false positives and inaccuracy in identifying genuine exons – see Zhang, 2002). Computer-based prediction of RNA genes is particularly poor, see Section 8.3.5.

The comparatively low number of human genes was a surprise. After all, the very simple, 1 mm long roundworm, *Caenorhabditis elegans* (which consists of only 959 somatic cells and which has a genome only 1/30 of the size of the human genome), had previously been shown to contain 19 099 polypeptide-encoding genes and over 1000 RNA genes (*C. elegans* sequencing consortium, 1998). Genome complexity might not always parallel biological complexity (*Drosophila melanogaster* has substantially fewer genes than the simpler *C. elegans*), but sequenced invertebrate genomes (e.g. insects, roundworm, sea squirt) are tending to show of the order of 14 000–20 000 genes while the vertebrates (human, mouse, pufferfish etc.) have tended towards a figure of about 30 000–35 000 (see *Table 12.4*). The unexpected low gene number has also been rationalized both on the basis of the very large increase in transcriptional complexity one might expect as gene number moves from, say, 20 000 to 30 000 (Claverie, 2001), and because of additional complexity one can expect because of the increased frequency of alternative splicing in complex genomes (Maniatis and Tasic, 2002; see Section 10.3.2).

Human gene distribution

Human genes are not evenly distributed on the chromosomes. The constitutive heterochromatin regions are devoid of genes but even within the euchromatic portion of the genome gene density can vary substantially between chromosomal regions and also between whole chromosomes. The first generalized

Box 9.3: DNA methylation and CpG islands

DNA methylation is likely to have different biological roles. For some species, such as the yeast *S. cerevisiae* and the roundworm *C. elegans*, it does not appear to occur at all; in many others it plays important roles. In bacteria DNA methylation is largely restricted to a proportion of adenine and cytosine residues and appears to act as a host defense mechanism: host cell restriction endonucleases recognize and cleave invading (unmethylated) phage DNA at specific recognition sequences but the same sequences in host DNA are specifically methylated and so are protected from cleavage (see *Box 5.2*).

Where it occurs in metazoans (multicellular animals) DNA methylation often involves methylation of a proportion of *cytosine* residues, giving **5-methylcytosine (C^m)**. In *D. melanogaster* the amount of DNA methylation is very low and most of the 5-methylcytosine is found in CpT dinucleotides (C^{m}pT). In other animals the dinucleotide **CpG** is a common target for cytosine methylation by specific cytosine methyltransferases, forming C^{m}pG (see *Figure, panel A*). The genomes of most invertebrates – other than *Drosophila* – have moderately high levels of C^{m}pG which is concentrated in large domains of methylated DNA separated by similarly large domains of unmethylated DNA (***mosaic methylation***).

Vertebrates have the highest levels of 5-methylcytosine in the animal kingdom and here the methylation is *dispersed* throughout the genome. The DNA methylation is known to have important consequences for gene expression and allows particular gene expression patterns to be stably transmitted to daughter cells (Section 10.4.2). It has also been suggested to provide a form of host-defense against transposons (Section 10.4.3). Although methylation is dispersed throughout vertebrate genomes, only a small percentage of cytosines is methylated (about 3% in human DNA, mostly as C^{m}pG but with a small percentage as C^{m}pNpG, where *N* = any nucleotide).

5-methylcytosine is chemically unstable and is prone to *deamination*, resulting in thymine—see *Figure, panel A*. The other bases are also prone to deamination (e.g. unmethylated cytosine is prone to deamination to give uracil). Over evolutionarily long periods of time the number of CpG dinucleotides in vertebrate DNA has gradually fallen because of the slow but steady conversion of **C**p**G** to **T**p**G** (and to Cp**A** on the complementary strand). Although the overall frequency of CpG in the vertebrate genome is low, there are small stretches of unmethylated DNA which are characterized by having the *normal, expected CpG frequency*. Such islands of *normal* CpG density (**CpG islands**) are comparatively GC-rich (typically over

(A) *The cytosine in CpG dinucleotides is a target for methylation at the 5 carbon giving 5-methylcytosine.*

The latter spontaneously deaminates to give thymine (T) which is inefficiently recognized by the DNA repair system and so tends to persist (deamination of unmethylated cytosine, however, gives uracil which is recognized by the DNA repair system). **(B)** The vertebrate CpG dinucleotide is gradually being replaced by TpG and CpA.

50% GC) and extend over hundreds of nucleotides, often marking the 5′ ends of genes. When the draft human genome sequence is filtered to remove high copy number repetitive noncoding DNA sequences, about 30 000 CpG islands could be identified (International Human Genome Sequencing Consortium, 2001).

insight into whole genome gene distribution was obtained after hybridizing purified **CpG island** fractions of the genome to metaphase chromosomes. On this basis, it was concluded that gene density must be high in subtelomeric regions, and that some chromosomes (e.g. 19 and 22) are gene-rich while others (e.g. X, 18) are gene-poor (*Figure 8.4*). The predictions of differential CpG island density and differential gene density were subsequently confirmed when draft sequences encompassing about 90% of the genome were reported (International Human Genome Sequencing Consortium, 2001).

The difference in %GC between Giemsa pale and dark bands also reflects differential gene densities because GC-rich chromosomes (e.g. chromosome 19) and regions (e.g. pale G bands) are also comparatively rich in genes. For example, the gene-rich human leukocyte antigen (HLA) complex (180 genes in a span of 4 Mb) is located within the pale 6p21.3 band, while a full 2.4 Mb of DNA which appears to be devoted almost exclusively to a single mammoth gene, the dystrophin gene, lies within a dark G band.

9.2 Organization, distribution and function of human RNA genes

Although the great majority of human genes encode polypeptides (Section 9.3), a significant minority specify noncoding (= untranslated) RNA molecules as their end product and so are described as **RNA genes** (Eddy, 2001; Huttenhofer *et al.*, 2002; Storz, 2002; see also the noncoding RNA database at http://biobases.ibch.poznan.pl/ncRNA/). The mitochondrial genome is exceptional in that 65% (24/37) of the genes specify mature RNA molecules but even in the case of the nuclear genome there are probably about 3000 RNA genes, accounting for close to 10% of the total gene number (*Figure 9.4*).

It is likely that current estimates of RNA gene number are conservative (because of the difficulty in identifying RNA genes in sequenced DNA; see Section 8.3.5). Comprehensive analyses of mouse transcripts (Section 9.2.3), and microarray-based analyses of human chromosome 21 and 22 transcripts (Kapranov *et al.*, 2002) have been interpreted to suggest many more transcripts than expected from the predicted gene counts. In addition to the RNA genes there are many related pseudogene/gene fragments, especially in the case of RNA genes transcribed by RNA polymerase III.

In common with other cellular genomes most of the known RNA genes are devoted to making molecules which assist in the general process of gene expression (*Figure 9.4*). Some, notably the rRNA and tRNA families, are involved in mRNA translation. Many other RNA families are involved in *RNA maturation*, involving both *cleavage* and *base-specific modification* of other RNA molecules (mRNA, rRNA, tRNA and other RNA species). In addition, a substantial number of other RNA genes, belonging to different RNA classes have more recently been identified. Many have, or are expected to have, important regulatory roles, emphasizing the very considerable functional diversity of RNA molecules (*Table 9.3*; Section 9.2.3).

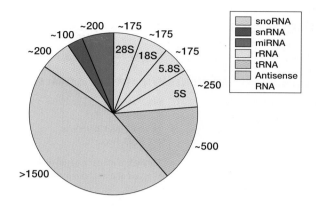

Figure 9.4: Human RNA genes according to class.

The best estimate (as of mid 2003) is a total of over 3000 human RNA genes distributed between the different classes shown. ***Note:*** (i) for operational reasons (see text) the draft human genome sequences excluded the rRNA gene clusters and given numbers are estimated from other data; (ii) because of the difficulty in identifying RNA genes (see Section 8.3.5) the number of some categories of small RNA, such as the miRNAs may be considerable under-estimates; (iii) the predicted number of antisense RNA genes is based on data from Collins *et al.* (2003) and is supported from equivalent analyses in mouse (by the FANTOM Consortium and the RIKEN Genome Exploration Research Group Phase I & II Team (2002)).

9.2.1 A total of about 1200 human genes encode rRNA or tRNA and are mostly organized into large gene clusters

Ribosomal RNA (rRNA) genes

There are approximately 700–800 human rRNA genes, mostly organized in tandemly repeated clusters, and many related pseudogenes. Homogeneous multigene families that occur in tandem arrays are under-represented in the draft human genome sequence (owing to the selection of restriction enzymes used in constructing the BAC libraries and the decision to delay the sequencing of BACs with low complexity fingerprints indicative of tandemly repeated DNA). As a result, the precise number of rRNA genes could not be inferred from the draft human genome sequences.

In addition to the 16S and 23S mitochondrial rRNA molecules, there are four types of cytoplasmic tRNA, three associated with the large ribosome subunit (28S, 5.8S and 5S rRNAs), and one with the small ribosome subunit (18S rRNA). Of these, the 28S, 5.8S and 18S rRNAs are encoded by a ***single transcription unit*** (see *Figure 10.2*) which is organized into five clusters, each with 30–40 tandem repeats, located on the short arms of human chromosomes 13, 14, 15, 21, and 22.

The 5S rDNA genes also occur in tandem arrays, the largest of which is on chromosome 1q41-42, close to the telomere. There are 200–300 true 5S genes in these arrays but there appear to be many dispersed pseudogenes. The major rationale

Table 9.3: Functional diversity of human RNA

Class of RNA	Examples	Function
(A) MAJOR RNA CLASSES INVOLVED IN ASSISTING GENERAL GENE EXPRESSION		
Ribosomal RNA (rRNA)	16S rRNA	Component of small mitochondrial ribosomal subunit (*Figure 9.2*)
	23S rRNA	Component of large mitochondrial ribosomal subunit (*Figure 9.2*)
	28S, 5.8S, and 5S rRNAs	Components of large cytoplasmic ribosomal subunit (*Figure 10.2*)
	18S rRNA	Component of small cytoplasmic ribosomal subunit (*Figure 10.2*)
Transfer RNA (tRNA)	22 types of mitochondrial tRNA	Binding to codons in mitochondrial mRNA (*Figure 9.2*)
	49 types of cytoplasmic tRNA	Binding to codons in cytoplasmic mRNA (*Figure 9.4*)
Small nuclear RNA (snRNA) (involved in RNA splicing)	Many, including:	
	U1, U2, U4 and U6 snRNAs	Components of major spliceosome
	U5 snRNA	Component of both major and minor spliceosomes
	U4acat, U6acat, U11, and U12 snRNAs	Components of minor spliceosome
	U7 snRNA	Histone mRNA transcriptional termination
Small nucleolar RNA (snoRNA) (involved in RNA modification and processing)	Over 100 different types:	
	ca. 80 C/D box snoRNAs	Site-specific methylation of the 2′ OH group of rRNA
	ca. 15 H/ACA snoRNAs	Site-specific rRNA modification by formation of pseudouridine.
	U3 and U8 snoRNAs	rRNA processing
(B) OTHER RNA CLASSES (see also the Noncoding RNAs Database at http://biobases.ibch.poznan.pl/ncRNA/)		
MicroRNA	At least 200 classes likely	Very small (~22 ntds) regulatory RNA molecules (Section 9.2.3)
X-chromosome inactivation associated	*XIST* RNA	*See* Section 10.5.6
	TSIX RNA	*See* Section 10.5.6
Imprinting-associated	Many e.g. *H19* RNA	*See Figure 10.24 for some examples*
Nervous system-specific	e.g. BC200 RNA	?
Antisense RNA	Probably 1500 or so types	e.g. to *HOXA11*, *MSX1* etc. (see *Figure 10.24*)
Others	Telomerase RNA	Component of telomerase (Section 2.2.5)
	PCA3 RNA	Prostate cancer antigen-3
	PCGEM1 RNA	Highly over-expressed in prostate cancer
	SRA1 RNA	Specific co-activator of several steroid receptors
	TTY2 RNA	Testis-specific family
	7SK RNA	Negative transcriptional regulator of RNA polymerase II elongation
	7SL RNA	Component of signal recognition particle for transporting proteins

for the repetition of cytoplasmic rRNA genes is based on gene dosage: by having a comparatively large number of these genes, the cell can satisfy the huge demand for cytoplasmic ribosomes needed for protein synthesis.

Transfer RNA (tRNA) genes

In addition to the 22 mitochondrial tRNA genes, the draft human genome sequence published in 2001 revealed a total of 497 nuclear genes encoding cytoplasmic tRNA molecules and 324 tRNA-derived putative pseudogenes. Humans therefore, appear to have fewer genes specifying cytoplasmic tRNA than the worm (584), but more than the fly (284). In metazoans tRNA gene number is not so related to organismal complexity, but rather to special demands for tRNA abundance in certain tissues or stages of embryonic development (the frog *Xenopus laevis*, for example, has large oocytes which

must each be loaded with as much as 40 ng of tRNA; the high tRNA demand is met by having thousands of tRNA genes).

The 497 cytoplasmic tRNA genes can be grouped into 49 families according to their anticodon specificities. Although the universal genetic code provides 61 different sense codons which need to be recognized by anticodons in tRNA molecules, *wobble* at the third base position of codons means that when the third base position in a codon is a pyrimidine (U or C), a single anticodon can base pair with the two alternative codons. The choice of anticodon to interpret alternative human codons follows general rules for eukaryotic cytoplasmic tRNA (*Box 9.4*). On this basis one would predict a total of 46 different classes of human tRNA but despite the generality of third base wobble, three pairs of such codons – AU(U/C), UA(U/C), AA(U/C) – appear to be served by two anticodons each, and there are therefore an extra three tRNA classes (see *Figure 9.5*). There is only a very

---

Box 9.4: Anticodon specificity of eukaryotic cytoplasmic tRNAs.

As in the case of interpreting mitochondrial codons (Section 9.1.2) *third base wobble* means that there is no 1:1 correspondence between codons in cytoplasmic mRNA and the tRNA anticodons which recognize them. In this case alternative codons which differ in having a C or a U at the third base position can be recognized by a single anticodon. The decoding rules for cytoplasmic mRNA codons are:

▶ *codons in 'two-codon boxes'* (codons ending with U/C which encode a different amino acid from those ending with A/G). Here the U/C wobble position is typically decoded by a G at the 5′ base position in the tRNA anticodon. Thus, there is no tRNA with an **A**AA anticodon to match the UU**U** codon for Phe, but the **G**AA anticodon can recognize both UU**U** and UU**C** codons in the mRNA (see *Figure 9.5*);

▶ *nonglycine codons in 'four-codon boxes'* (codons where U, C, A and G in the wobble position all encode the same amino acid, but not glycine). Here the U/C wobble position is decoded by an **inosine (I)** at the 5′ position in the anticodon (*inosine is produced by post-transcriptional modification of an adenine: the amino group at carbon 6 of adenosine is replaced by a C=O carbonyl group*). For example, the GU**U**

and GU**C** codons of the four-codon valine box are decoded by a tRNA with an anticodon of **A**AC, which is no doubt modified to **I**AC. In addition to base pairing with C and U, inosine can also base pair with A (and so the **I**AC anticodon can recognize each of GU**U**, GU**C**, GU**A**). To avoid possible translational misreading, tRNAs with inosine at the 5′ base of the anticodon cannot be utilized in two-codon boxes;

▶ *glycine codons*. GGU and GGC codons are decoded by a **G**CC anticodon, rather than the expected **I**CC anticodon.

Only 16 anticodons are required to decode the 32 codons ending in a pyrimidine. The minimum set of anticodons is therefore 61 (the number of different sense codons) minus 16 = 45. However, in addition a specialized tRNA carries an anticodon to the codon UGA (which normally functions as a stop codon). Under conditions of high selenium, this tRNA will decode UGA in only a very small number of cases to insert the 21st amino acid, **selenocysteine**, in a select group of '**selenoproteins**' (all of which have redox activities; in mammals thioredoxin reductase and glutathione peroxidase are among the most widely found selenoproteins).

---

rough correlation of human tRNA gene number with amino acid frequency (*Table 9.4*).

Although the tRNA genes appear to be dispersed throughout the human genome (they are found on all chromosomes except chromosome 22 and the Y chromosome), there is striking clustering. More than half of them (280 out of 497) reside on either chromosome 6 (which contains 140 tRNA genes – including almost all the different types of tRNA gene – in a region of only 4 Mb at 6p2), or on chromosome 1 (where many of the Asn and Glu tRNA genes are loosely clustered). In addition, many of the other tRNA genes are clustered, for example 18 of the 30 Cys tRNAs are found in a 0.5-Mb stretch of chromosome 7.

9.2.2 Small nuclear RNA and small nucleolar RNA are encoded by mostly dispersed, moderately large gene families

In addition to rRNA and tRNA two other major RNA classes are involved in assisting general gene expression: small nuclear RNA (snRNA) and small nucleolar RNA (snoRNA). They are encoded by families of close to 100 genes (snRNA) or somewhat exceeding 100 genes (snoRNA) which, although dispersed, do show some clustering of subfamilies.

Small nuclear RNA (snRNA) genes

The heterogeneous collection of **small nuclear RNA (snRNA)** molecules include many which are uridine-rich and are named accordingly, for example U3 snRNA means the third uridine-rich small nuclear RNA to be classified. Several are **spliceosomal RNAs**, required for functioning of

the major and minor spliceosomes (see *Table 9.3*) and are encoded by a family of over 80 genes. More than 70 of these genes specify snRNAs used in the major spliceosome. They include 44 identified genes specifying U6 snRNA and 16 specifying U1 snRNA.

There is some evidence for clustering particularly in the case of the U1 and U2 snRNA families, but because of the way BACs were selected for obtaining the draft genome sequence (see above) there is under-representation in the draft sequence. The U2 RNA genes were previously known to be located at the *RNU2* locus, a tandem array of nearly identical 6.1-kb units at 17q21–q22 which is highly variable in the number of repeat units (from six to more than 30 repeats). The U1 RNA genes are clustered with about 30 copies at the *RNU1* locus at 1p36.1, but this cluster is thought to be loose and irregularly organized. There are a large number of related nonfunctional sequences (pseudogenes, gene fragments etc): for example 1135 U6 snRNA-related sequences were identified in the draft sequence (International Human Genome Sequencing Consortium, 2001).

Small nucleolar RNA (snoRNA) genes

A large family of **small nucleolar RNAs (snoRNAs)** are mostly employed in the nucleolus to direct or *guide* site-specific base modifications in rRNA (Smith and Steitz, 1997; Filipowicz, 2000), but are also known to carry out base modifications on other stable RNAs including U6 snRNA. There are two subfamilies. **C/D box snoRNAs** are mostly involved in guiding *site-specific* 2′-O-ribose methylations (there are 105–107 varieties of this methylation in rRNA). **H/ACA snoRNAs** are mostly involved in guiding *site-specific pseudouridylations* (where uridine is isomerized to give

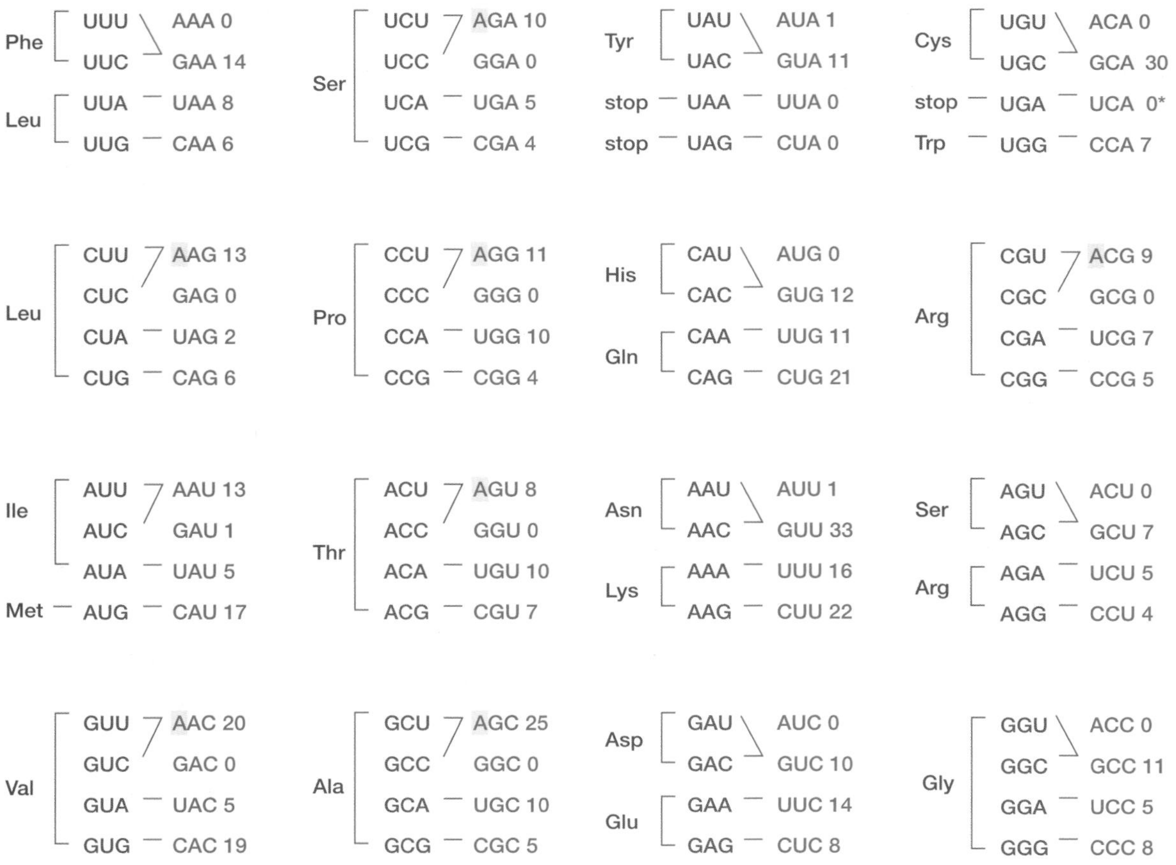

Amino acid	Codon	Anticodon	No. genes
Phe	UUU	AAA	0
Phe	UUC	GAA	14
Leu	UUA	UAA	8
Leu	UUG	CAA	6
Ser	UCU	AGA	10
Ser	UCC	GGA	0
Ser	UCA	UGA	5
Ser	UCG	CGA	4
Tyr	UAU	AUA	1
Tyr	UAC	GUA	11
stop	UAA	UUA	0
stop	UAG	CUA	0
Cys	UGU	ACA	0
Cys	UGC	GCA	30
stop	UGA	UCA	0*
Trp	UGG	CCA	7
Leu	CUU	AAG	13
Leu	CUC	GAG	0
Leu	CUA	UAG	2
Leu	CUG	CAG	6
Pro	CCU	AGG	11
Pro	CCC	GGG	0
Pro	CCA	UGG	10
Pro	CCG	CGG	4
His	CAU	AUG	0
His	CAC	GUG	12
Gln	CAA	UUG	11
Gln	CAG	CUG	21
Arg	CGU	ACG	9
Arg	CGC	GCG	0
Arg	CGA	UCG	7
Arg	CGG	CCG	5
Ile	AUU	AAU	13
Ile	AUC	GAU	1
Ile	AUA	UAU	5
Met	AUG	CAU	17
Thr	ACU	AGU	8
Thr	ACC	GGU	0
Thr	ACA	UGU	10
Thr	ACG	CGU	7
Asn	AAU	AUU	1
Asn	AAC	GUU	33
Lys	AAA	UUU	16
Lys	AAG	CUU	22
Ser	AGU	ACU	0
Ser	AGC	GCU	7
Arg	AGA	UCU	5
Arg	AGG	CCU	4
Val	GUU	AAC	20
Val	GUC	GAC	0
Val	GUA	UAC	5
Val	GUG	CAC	19
Ala	GCU	AGC	25
Ala	GCC	GGC	0
Ala	GCA	UGC	10
Ala	GCG	CGC	5
Asp	GAU	AUC	0
Asp	GAC	GUC	10
Glu	GAA	UUC	14
Glu	GAG	CUC	8
Gly	GGU	ACC	0
Gly	GGC	GCC	11
Gly	GGA	UCC	5
Gly	GGG	CCC	8

Figure 9.5: Numbers of human tRNA genes classified according to anticodon.

The codons are connected by lines to the (unmodified) anticodons on the right. Joined lines in a V shape link alternative codons ending in a U or C which can be decoded by a single anticodon because of third base wobble. The number next to each anticodon is the number of human genes encoding tRNAs with that anticodon. Thus, for example, at the top left it can be seen that the UUU phenylalanine codon is not decoded by an AAA anticodon as there are no tRNA genes carrying such an anticodon. The shaded adenines are almost certainly modified as *inosines* (see *Box 9.4*). **Note:** (i) despite the provision of third base wobble, single genes appear to encode tRNAs with anticodons which might not have been expected to be needed (AUA, AUU and GAU); (ii) the asterisk next to the anticodon UCA signifies that there is an unusual tRNA carrying this anticodon which can occasionally interpret a small subset of UGA codons as selenocysteine instead of stop—see *Box 9.4*). Modified from International Human Genome Sequencing Consortium (2001) *Nature* **409**, 860–921 with permission from the Nature Publishing Group.

pseudouridine, the most common modified base; 95 different pseudouridylations are required for rRNA).

Single snoRNAs specify one, or at most two such modifications. SnoRNA genes are often found within the introns of other genes and although most snoRNA genes appear to be single-copy and dispersed, some large clusters are known, including two which are found within the large *SNURF–SNRPN* transcription unit on 15q. The latter genes are paternally imprinted and expressed in the brain, and have been considered to play an important role in Prader–Willi syndrome (see *Figure 10.24* and references therein).

9.2.3 MicroRNAs and other novel regulatory RNAs are challenging preconceptions on the extent of RNA function

Textbooks generally emphasize the importance of proteins as end points of gene expression because of the wide variety in their functions and their importance in regulating gene expression. RNA molecules have generally been seen as less important (the Celera draft genome sequence paper by Venter *et al.*, 2001 did not present any analyses on human RNA genes!). In recent years, however, a variety of discoveries are forcing a radical re-appraisal of RNA function. The realization in 1982 that some RNA molecules could have a catalytic

Table 9.4: Distribution of human cytoplasmic tRNA gene families genes according to amino acid specified

Amino acid	Frequency*	Number of corresponding tRNA genes	Amino acid	Frequency*	Number of corresponding tRNA genes
Alanine	7.06%	40	Lysine	5.65%	38
Arginine	5.69%	30	Methionine	2.23%	17
Aspartate	4.78%	10	Phenylalanine	3.75%	14
Asparagine	3.58%	34	Proline	6.10%	25
Cysteine	2.25%	30	Selenocysteine	<0.01%	1
Glutamine	4.63%	32	Serine	8.00%	26
Glutamate	6.93%	22	Threonine	5.31%	25
Glycine	6.62%	24	Tryptophan	1.30%	7
Histidine	2.56%	12	Tyrosine	2.76%	12
Isoleucine	4.43%	19	Valine	6.12%	44
Leucine	9.95%	35			

* human proteome-wide average frequency

function (and so function as a **ribozyme**) has led to identification of catalytic functions in several other types of RNA. They include both rRNA (recent X-ray crystallography data indicate that peptide bond formation is catalyzed by rRNA, *not* protein– Nissen *et al.*, 2000) and also snRNA (Valadkhan and Manley, 2001).

A variety of other key RNA molecules have been well-studied including the *telomerase RNA* (see Section 2.2.5) and the *SRP RNA* (also called 7SL RNA) of the *signal recognition particle* (the ribonucleoprotein complex which recognizes the signal sequence on proteins destined for export, affording the proteins passage across the cell membrane). More recently, an exciting variety of novel human RNA molecules have been identified with known or expected regulatory roles. This is a continuing process but the most comprehensive analysis of mammalian transcripts so far (from the mouse genome) suggests that a very significant percentage of transcripts will be noncoding RNAs (FANTOM Consortium and the RIKEN Genome Exploration Research Group Phase I & II Team, 2002)

MicroRNAs: novel small regulatory RNA molecules

MicroRNAs (miRNAs) are very small (approximately 22-nucleotides long) RNA molecules which can function as antisense regulators of other genes (Ambros, 2001; Gottesman, 2002). They derive from larger, ~ 70-nucleotide-long precursors containing an *inverted repeat* which permits double-stranded *hairpin RNA* formation. Such hairpin precursor RNAs are cleaved by a type of ribonuclease III (specific for double-stranded RNA) known as *dicer*[a]. The first such sequences to be described in animals, the lin-4 and let-7 RNAs, were identified as *small temporal RNA* (*stRNA*) by genetic analyses in *C. elegans*. Both lin-4 and let-7 RNAs are developmentally regulated and also control various developmental programs themselves. They act as antisense regulators by binding to complementary sequences in the 3′ UTR of mRNA from target genes, inhibiting translation and so repressing the synthesis of the target gene proteins (*Figure 9.6A*). Other miRNAs, for example in plants, have also been shown to be developmental regulators, and this finding plus the strong evolutionary conservation of miRNA has led to the expectation of similar functions for mammalian miRNAs.

Various approaches have been taken to identify mammalian miRNAs (see Gottesman, 2002 for a summary), resulting in the recent identification of novel miRNAs in both humans and mice (Lagos-Quintana *et al.*, 2001, 2002; Mourelatos *et al.*, 2002). By the time of writing (mid 2003), 200 or so human miRNA genes have been estimated. Although dispersed on many chromosomes there is evidence for some clustering, notably a cluster of at least seven miRNA genes within a 0.8-kb region on chromosome 13 (Mourelatos *et al.*, 2002). On the expectation that mammalian miRNAs have important regulatory functions active efforts are currently being made to identify target genes.

---

[a]*Note:* this type of ribonuclease activity is thought to form part of a conserved *genetic surveillance system* that can degrade a specific mRNA in response to the presence of double-stranded RNA corresponding to the specific mRNA. It can be utilized in certain experimental analyses, known as *RNA interference* (*RNAi*) to specifically inactivate target genes within cells by cleaving long double-stranded RNA produced from artificially introduced transgenes to give ca. 22-nucleotide-long antisense RNA molecules (known as *siRNA = short interfering RNA*). The siRNA molecules can base-pair with mRNA from the endogenous gene which corresponds to the introduced transgene, thereby specifically inhibiting expression (see Section 20.2.6).

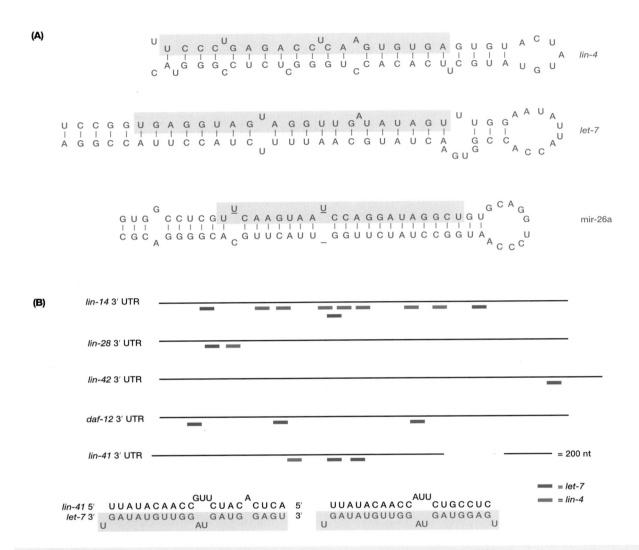

Figure 9.6: MicroRNAs are very short (21–22 nucleotides) RNAs which can function as antisense regulators.

(A) *miRNA precursor structure.* The *C. elegans lin-4* and *let-7* stRNAs belong to the miRNA class, showing significant similarity to mammalian miRNAs such as the human mir-26a miRNA. Mature microRNA (miRNA) sequences (shaded) are derived from a precursor with inverted repeats (which form a hairpin RNA by intramolecular hydrogen bonding). Cleavage of the hairpin RNA is carried out by a type of RNase III nuclease known as dicer. Sometimes two different miRNAs are derived from the same precursor. **(B) *Antisense regulation by lin-4 and let-7 RNAs.*** mRNA from target genes regulated by the *lin-4* and *let-7* RNAs have regions in their 3′ UTRs which show very significant complementarity to these miRNAs. Base pairing is usually not perfect as shown in the case of predicted base-pairing between the *let-7* RNA and its two target sequences in the *lin-41* 3′ UTR sequence. From Banerjee and Slack (2002) *Bioessays* **24**, 119–129, reprinted with permission of Wiley-Liss, Inc., a subsidiary of John Wiley & Sons, Inc.

Genes encoding moderate- to large-sized regulatory RNA molecules

An increasing number of genes specify moderate- to large-sized noncoding RNAs with known or expected regulatory functions (many but not all can be likened to '*noncoding mRNA*' molecules because they are transcribed by RNA polymerase II and undergo capping and polyadenylation[b]). They include genes which specify ***7SK RNA***, a negative transcriptional regulator of RNA polymerase II elongation (Yang *et al.*, 2001); ***SRA1 (steroid receptor activator) RNA,*** which serves as a specific co-activator of several steroid receptors

(Lanz *et al.*, 1999); and ***XIST,*** which is central to *X-chromosome inactivation* (Section 10.5.6).

Several moderate to large-sized regulatory **antisense RNAs** are also known, including the ***TSIX*** antisense transcript which regulates *XIST*, a variety of antisense transcripts which regulate imprinted genes (Section 10.5.5; see *Figure 10.24* for some examples), and a large number of other anti-

---

[b]***Note:*** by way of comparison the 28S, 18S and 5.8S RNAs are transcribed by RNA polymerase I; 5S rRNA, tRNA, snoRNA and miRNA are transcribed by RNA polymerase III; snRNAs are a surprising mix: some are transcribed by RNA polymerase III, and some by RNA polymerase II).

sense transcripts (Lehner *et al.*, 2002). Although the total number of such antisense transcripts in the human genome is not accurately known, a recent re-evaluation of the genes on human chromosome 22 has identified 16 possible antisense RNA genes suggesting that there may be about 1500 or so antisense RNA genes in the human genome (Collins *et al.*, 2003). In support of this estimate several hundred mouse antisense RNAs have been predicted, using the most conservative estimates, from a very comprehensive study by the FANTOM Consortium and the RIKEN Genome Exploration Research Group Phase I & II Team (2002).

9.3 Organization, distribution and function of human polypeptide-encoding genes

9.3.1 Human genes show enormous variation in size and internal organization

Size diversity

Genes in simple organisms such as bacteria are comparatively similar in size, and usually very short. In complex organisms variation in gene size is considerable, notably in the human

genome (*Figure 9.7*). The enormous size of some human genes means that transcription can be time consuming, taking about 16 hours for the 2.4–Mb dystrophin gene (Tennyson *et al.*, 1995). Although there is a direct correlation between gene and product sizes, there are some striking anomalies. For example, apolipoprotein B has 4563 amino acids and is encoded by a 45-kb gene but the largest protein encoded by the giant 2.4–Mb dystrophin gene has only 3685 amino acids.

Diversity in exon-intron organization

A very small minority of human genes lack introns (see *Table 9.5*), and are generally small in size. For those that do possess introns, there is an inverse correlation between gene size and fraction of coding DNA (*Figure 9.7*). This does not arise because exons in large genes are smaller than those in small genes: the average exon size in human genes is less than 200 bp and, although very large exons are known (see *Box 9.5*), exon size is comparatively independent of gene length (*Table 9.6*). Instead, there is huge variation in intron lengths, and large genes tend to have very large introns (the collagen type 7 and titin genes are rather remarkable exceptions – see *Table 9.6*). Transcription of long introns is, however, costly in time and

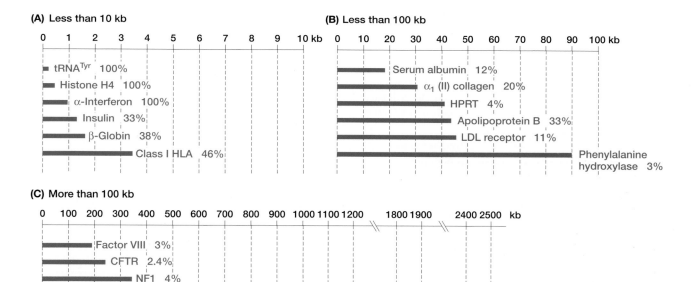

Figure 9.7: Human genes vary enormously in size and exon content.

Exon content is shown as a percentage of the lengths of indicated genes. ***Note*** the generally inverse relationship between gene length and percentage of exon content. Asterisks emphasize that the lengths given for the indicated Ig heavy chain and light chain loci correspond to the *germline organizations*. (Ig and T-cell receptor genes have unique organizations, requiring cell-specific somatic rearrangements in order to be expressed in B or T lymphocytes respectively – see Section 10.6.) CFTR, cystic fibrosis transmembrane regulator; HPRT, hypoxanthine phosphoribosyl transferase; NF1, neurofibromatosis type 1.

Table 9.5: Examples of human genes with uninterrupted coding sequences

For more detailed lists, see http://exppc01.uni-muenster.de/expath/frames.htm

All 37 mitochondrial genes

Many RNA genes (especially genes encoding small RNAs, e.g. most tRNA genes, but also some large RNAs e.g. *XIST* RNA)

Retrogenes (see *Table 9.11*)

Interferons

Histone genes

Many ribonuclease genes

Heat shock protein genes

Many G protein coupled receptors

Some genes with HMG boxes (e.g. *SRY*, many *SOX* genes)

Various neurotransmitter receptor and hormone receptor genes, e.g. dopamine D1 and D5 receptors, 5-HT$_{1B}$ serotonin receptor, angiotensin II type 1 receptor, formyl peptide receptor, bradykinin B2 receptor, α2 adrenergic receptor

energy and natural selection favors short introns in highly expressed genes (Castillo-Davis *et al.*, 2002).

Diversity in repetitive DNA content

Genes often have repetitive DNA components within noncoding introns and flanking sequences, but in addition repetitive DNA sequences are found to different extents in coding DNA. Tandem repetition of **microsatellite sequences** (short sequence motifs – see Section 9.4.3) is common and may simply reflect statistically expected frequencies for certain base compositions. Tandem repetition of sequences encoding known or assumed **protein domains** is also quite common, and may be functionally advantageous in some cases by providing a more available biological target. In some cases, the sequence homology between the repeats can be very high; in other cases it may be rather low (see *Table 9.7*).

9.3.2 Functionally similar genes are occasionally clustered in the human genome, but are more often dispersed over different chromosomes

As seen in Section 9.2, some families of RNA genes are clustered. In the case of polypeptide-encoding gene families, genes encoding identical products or ones which are very closely related in sequence are often found in one or more clusters which may be dispersed on several chromosomes. However, some families of functional genes encoding products which have conserved *components* only (domains, significant motifs etc.) are often dispersed in the genome. Genes encoding functionally related products which do not show a very significant sequence homology are typically dispersed.

Functionally identical genes

A very few human polypeptides are known to be encoded by two or more *identical* gene copies. Often, these are encoded by recently duplicated genes in a gene cluster, for example the duplicated α-globin genes. In addition, very occasionally some genes on *different chromosomes* encode identical polypeptides. Examples include:

▶ *histone genes.* The NHGRI histone sequence database lists a total of 86 different histone sequences distributed over 10 different chromosomes, albeit with two large clusters on 6p (http://genome.nhgri.nih.gov/histones/chrmap.shtml) but some subfamily members are identical although encoded by genes on different chromosomes;

▶ *ubiquitin genes.* The 76 amino acid **ubiquitin** is a highly conserved protein which plays a major role in protein degradation and cellular stress response. Human ubiquitin genes are found at different loci distributed on several chromosomes. Some are found as a series of tandem full-length coding sequence repeats which undergo co-transcription (*polycistronic transcription units*). Others are monomers (but fused to ribosomal protein genes, constituting *bicistronic transcription units*) – see Nei *et al.*, 2000; Section 9.3.3.

Functionally similar genes

A large fraction of human genes are members of gene families where individual genes are *closely related but not identical* in sequence. In many such cases the genes are clustered and have arisen by tandem gene duplication, as in the case of the different members of each of the α-globin and β-globin gene clusters (see *Figure 9.11*). Genes which encode clearly related products but which are located on different chromosomes are generally less related, as in the case of the α-globin and β-globin genes. However, in the case of the *HOX* homeobox gene family, which consists of clusters of approximately 10 genes on each of four chromosomes, individual genes on different chromosomes may be more related to each other than they are to members of the same gene cluster (*Figure 12.9*). In addition to the above, genes encoding closely related tissue-specific isoforms, or subcellular compartment-specific isozymes are often located on different chromosomes (see *Table 9.8*).

Functionally related genes

Some genes encode products which may not be so closely related in sequence, but are clearly functionally related. The

Box 9.5: Human genome and human gene statistics.

Genome size	~ 3200 Mb
Nuclear genome	~ 3200 Mb
Mitochondrial genome	37 kb
Euchromatic component	~ 2900–3000 Mb
Constitutive heterochromatin	> 200 Mb (*Table 9.2; Figure 2.15*)
Highly conserved fraction	> 100 Mb (>3%)
Coding DNA	~ 50 Mb (~1.5%)
Other (regulatory etc.)	~ 100 Mb (3%)
Segmentally duplicated DNA	>150 Mb (>5%)
Noncoding repetitive DNA	> 50% of genome
Transposon-based repeats	~ 1400 Mb (~ 43%; see *Table 9.15*)
Gene number	~ 30 000–35 000
Nuclear genome	~ 30 000–35 000 (Section 9.1.3).
Mitochondrial genome	37 (Section 9.1.2).
Per chromosome	Average of ~ 1400; but depends on chromosome length and type (see *Figure 8.4*); ~ 60 per band in a 550-band chromosome preparation
Polypeptide-encoding genes	~ 30 000 but considerable uncertainty
RNA genes	~ 3000, but some uncertainty (see *Figure 9.4*)
Pseudogenes	~20 000
Gene density	~ 1/100 kb in nuclear genome; 1/0.45 kb in mitochondrial genome
Gene size (genomic extent)	Average = 27 kb, but enormous variation (see *Figure 9.7*).
Intergenic distance	Average = ca. 75 kb in nuclear genome.
CpG island number	~ 30 000 (in genome sequence filtered to remove noncoding repeats)
Exon number	Average = 9. Generally correlated with gene length, but wide variation.
Largest number	363 (in the titin gene)
Smallest number	1 (that is, no introns – *Table 9.5*)
Exon size	Average = 122 bp for internal exons with comparatively little length variation, but 3′ exons can be considerably longer (Zhang, 1998).
Largest exons	Many kb long, e.g. exon 26 of the apoB gene (*APOB*) is 7.6 kb
Smallest exons	< 10 bp
Intron size	Enormous variation; strong direct correlation with gene size (see *Table 9.6*):
Largest introns	Hundreds of kb e.g. intron 8 of the human *WWOX* gene is ~ 800 kb.
Smallest introns	Tens of bp
mRNA size	Average of about 2.6 kb, but considerable variation (titin mRNA is > 115 kb long!)
5′ UTR	Average of about 0.2–0.3 kb
3′ UTR	Average of about 0.77 kb but likely to be an underestimate because of under-reporting of long 3′ UTRs
Noncoding RNA size	Highly variable; from ~ 21–22 nucleotides (microRNA) to many kb e.g. *XIST* (17 kb)
Polypeptide size	Average of about 500–550 amino acids
Largest polypeptide	Titin: 38 138 codons in titin gene (but significant length variation)
Smallest polypeptides	Tens of amino acids e.g. various small hormones etc.

Table 9.6: Average sizes of exons and introns in human genes

Gene product	Size of gene (kb)	Number of exons	Average size of exon (bp)	Average size of intron (bp)
tRNA^{tyr}	0.1	2	50	20
Insulin	1.4	3	155	480
β-Globin	1.6	3	150	490
Class I HLA	3.5	8	187	260
Serum albumin	18	14	137	1100
Type VII collagen	31	118	77	190
Complement C3	41	29	122	900
Phenylalanine hydroxylase	90	26	96	3500
Factor VIII	186	26	375	7100
CFTR (cystic fibrosis)	250	27	227	9100
Titin	283	363	315	466
Dystrophin	2400	79	180	30 770

Table 9.7: Examples of large-scale intragenic repetitive coding DNA

Gene product	Size of encoded repeat in amino acids	No. of copies	Nucleotide sequence homology between copies
Involucrin	10	59	High homology for central 39 repeats
Apolipoprotein?(a)	114 = kringle 4-like repeat[a]	37	High homology; 24 of the repeats are identical in sequence
Plasminogen	~ 75–80	5	Low homology but conserved protein domains (kringles[a])
Collagen	18	57	Low homology but conserved amino acid motifs based on $(Gly-X-Y)_6$
Serum albumin	195	3	Low homology
Proline-rich protein genes	16–21	5	Low homology
Tropomyosin α-chain	42	7	Low homology
Immunoglobulin ε-chain, C region	108	4	Low homology
Dystrophin	109	24	Low homology

[a]A kringle is a cysteine-rich sequence that contains three internal disulfide bridges and forms a pretzel-shaped structure.

products may be subunits of the same protein or macromolecular structure, components of the same metabolic or developmental pathway, or may be required to specifically bind to each other as in the case of ligands and their relevant receptors. In almost all such cases, the genes are not clustered and are usually found on different chromosomes (see *Table 9.8* for some examples).

9.3.3 Overlapping genes, genes-within-genes and polycistronic transcription units are occasionally found in the human genome

Bidirectional gene organization and partially overlapping genes

Simple genomes have high gene densities (roughly one per 0.5 kb, 1 kb, and 2 kb for the genomes of human mito-chondria, *E. coli*, and *S. cerevisiae* respectively) and often show examples of partially overlapping genes. Different reading frames may be used, sometimes from a common sense strand (see *Figure 9.3*). The genes of complex organisms are much less clustered (only one gene per 100 kb in the human nuclear genome) and overlapping genes are not so common. Occasionally, however, very closely neighboring genes are found, some with their 5′ ends separated by a few hundred nucleotides and transcribed from opposite strands. Such a **bidirectional gene organization** is often found in the case of DNA repair genes, for example, and may provide for common regulation of the gene pair (Adachi and Lieber, 2002).

Partially overlapping genes in the complex nuclear genomes of mammals are rare and, where they do occur, the overlapping genes are usually transcribed from the two different DNA strands. Strong gene clustering occurs in GC-

Table 9.8: **Distribution of genes encoding functionally related products**

Genes which encode	Organization	Examples
The same product	Often clustered but may also be on different chromosomes	The two α-globin genes on 11p (*Figure 9.11*); genes encoding rRNA (*Figure 10.2*); some histone subfamilies (see http://genome.nhgri.nih.gov/histones/chrmap.shtml)
Tissue-specific protein isoforms or isozymes	Sometimes clustered; sometimes non-syntenic	Clustering of pancreatic and salivary amylase genes (1p21); nonsynteny of α-actin genes expressed in skeletal (1p) and cardiac (15q) muscle
Isozymes in different cellular compartments	Usually nonsyntenic	Cytoplasmic (c) and mitochondrial (m) isozymes for various enzymes e.g. aldehyde dehydrogenase (c)-9q and aldehyde dehydrogenase (m)-12q; thymidine kinase (c)-17q and thymidine kinase (m)-16q
Enzymes in same metabolic pathway	Usually nonsyntenic	Genes encoding enzymes in steroidogenesis: steroid 11-hydroxylase-8q; steroid 17-hydroxylase-10q; steroid 21-hydroxylase-6p
Subunits of the same protein	Usually nonsyntenic	α-globin-16p and β-globin-11p; ferritin heavy chain-11q and ferritin light chain-22q
Interacting components of signaling pathway	Usually nonsyntenic	JAK1–1p; STAT1–2q
Ligand plus associated receptor	Usually nonsyntenic	Insulin-11p and insulin receptor-19p; interferon β-9p; interferon β receptor-21q

rich subchromosomal regions and regions of particularly high gene density often show some examples of overlapping genes. For example, the class III region of the HLA complex at 6p21.3 has an average gene density of about one gene per 15 kb, and is known to contain several examples of overlapping genes (*Figure 9.8A*).

Genes-within-genes

The small nucleolar RNA (snoRNA) genes are unusual in that the *majority* of them are located within other genes, often ones which encode a ribosome-associated protein or a nucleolar protein. Possibly this arrangement has been maintained to permit co-ordinate production of protein and RNA components of the ribosome (Tycowski *et al.*, 1993). In addition to the snoRNA genes some other genes, including several polypeptide-encoding genes, are located within the introns of larger genes. Illustrative examples are: the *NF1* (neurofibromatosis type I) gene (three small internal genes transcribed from the opposite strand – see *Figure 9.8B*); the *F8* (blood clotting factor VIII) gene (two internal genes transcribed in opposite directions – see *Figure 11.20*) and the *RB1* (retinoblastoma susceptibility) gene (one internal gene transcribed from the opposite strand – see *Figure 9.19*).

Polycistronic transcription units

Polycistronic (= multigenic) transcription units are common in the simple genomes of bacteria and are also quite frequently found in *C. elegans*. The simple human mitochondrial genome (Section 9.1.2) and the major rRNA gene clusters (*Figure 10.2*) provide two major examples of polycistronic transcription units in the human genome. In addition, some rare examples are known of polypeptide-encoding **bicistronic transcription units** in the nuclear genome: transcription

starts from one gene and continues though a neighboring downstream gene to give a precursor protein which is cleaved to give different proteins.

The A and B chains of insulin can be considered to derive from a bicistronic transcription unit (*Figure 1.23*), but they are intimately related functionally. Sometimes, however, bicistronic transcription units generate functionally distinct proteins. The *UBA52* and *UBA80* genes, for example, generate ubiquitin and a ribosomal protein, S27a or L40, respectively. Other ubiquitin genes are organized as tandem full coding sequence repeats which form polycistronic transcription units (see Nei *et al.*, 2000). There are no introns in ubiquitin genes but in other bicistronic transcription units splicing is required to link transcripts of exons from one gene onto those of a downstream gene. The *SNURF–SNRPN* transcription unit provides an example of two polypeptides encoded by different exons (Gray *et al.*, 1999) but is also employed to make untranslated RNA transcripts which are paternally imprinted – see *Figure 10.24*.

9.3.4 Polypeptide-encoding gene families can be classified according to the degree and extent of sequence relatedness in family members

A large percentage of actively expressed human genes, encoding both noncoding RNA and polypeptides, are members of **DNA sequence families** which show a high degree of sequence similarity. However, the *extent* of sequence sharing and the organization of family members can vary widely. Many family members may be nonfunctional (*pseudogenes* and *gene fragments* – see below) and rapidly accumulate sequence differences, leading to marked sequence divergence.

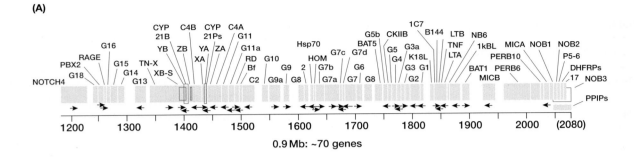

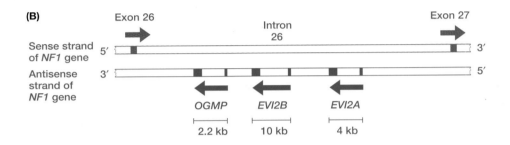

Figure 9.8: Overlapping genes and genes within genes.

(A) Overlapping genes. Genes in the class III region of the HLA complex are tightly packed and overlapping in some cases. **(B) Genes within genes.** Intron 26 of the gene for neurofibromatosis type I (NF1) contains three internal two-exon genes each transcribed from the opposing strand to that used to transcribe the *NF1* gene. Genes are: *OGMP,* oligodendrocyte myelin glycoprotein; *EVI2A* and *EVI2B*, human homologs of murine genes thought to be involved in leukemogenesis, and located at e̲cotropic v̲iral i̲ntegration sites.

Classical gene families

Members of classical gene families exhibit a high degree of sequence homology over most of the gene length or, at least, the coding DNA component. Examples include histone gene families (histones are strongly conserved and subfamily members are virtually identical), and the α-globin and β-globin gene families (members of an individual family show a high degree of sequence similarity).

Gene families encoding products with large, highly conserved domains

In some gene families there is particularly pronounced homology within specific strongly conserved regions of the genes; the corresponding sequence similarity between the remaining portion of the coding sequence in the different genes may be quite low. Often such families encode transcription factors that play important roles in early development, and the conserved sequence encodes a protein domain which is required to bind specifically to the DNA of selected target genes (see *Table 9.9*).

Gene families encoding products with short conserved amino acid motifs

The members of some gene families may not be very obviously related at the DNA sequence level, but nevertheless encode gene products that are characterized by a common general function and the presence of very short conserved

sequence motifs, such as the **DEAD box,** the sequence Asp–Glu–Ala–Asp (DEAD in the one-letter amino acid code), or the **WD repeat** (tryptophan–aspartate) – see *Figure 9.9.*

Gene superfamilies

Members of a **gene superfamily** are much more distantly related in evolutionary terms than those in a classical or conserved domain/motif gene family. They encode products that are functionally related in a *general* sense, and show only very weak sequence homology over a large segment, without very significant conserved amino acid motifs. Instead, there may be some evidence for general common *structural* features, and a general related function. Illustrative examples are:

▶ the *immunoglobulin superfamily* (*Figure 9.10*) – a very large family encompassing immunoglobulin (Ig) genes, T-cell receptor genes, HLA genes, and many others. The genes encode products which are considerably diverged at the sequence level but which function in the immune system and contain Ig-like domains;

▶ the *globin superfamil*y – a small family which includes not only the members of the α-globin and β-globin gene families (*Figure 9.11*) which function in oxygen transport and storage in blood, but also equivalent genes which encode muscle and brain globins, myoglobin and neuroglobin, respectively (see *Figure 12.4*).

Table 9.9: Examples of human genes with sequence motifs which encode highly conserved domains

Gene family	Number of genes	Sequence motif/ domain
Homeobox genes	38 *HOX* genes (see *Figure 12.9*) plus 214 orphan homeobox genes	*Homeobox* specifies a *homeodomain* of ca. 60 amino acids. A wide variety of different subclasses have been defined
PAX genes	9	*Paired box* encodes a *paired domain* of ~ 128 amino acids; *PAX* genes often have in addition a type of homeodomain known as a *paired-type homeodomain*
SOX genes	18	*SRY*-like *HMG box* which encodes a domain of ca. 69 amino acids
TBX genes	18	*T-Box* which encodes a domain of ca. 170 amino acids
Forkhead domain genes	49	The *forkhead domain* is ~ 110 amino acids long
POU domain genes	24	The *POU domain* is ~ 150 amino acids long

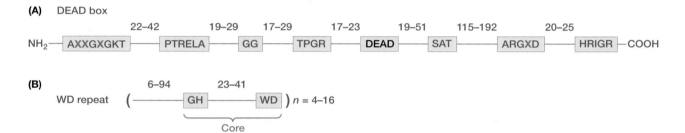

(A) DEAD box

(B) WD repeat

Figure 9.9: Some gene families are defined by functionally related gene products bearing very short conserved amino acid motifs.

(A) *Motifs in the DEAD box family*. This gene family encodes products implicated in cellular processes involving alteration of RNA secondary structure, such as translation initiation and splicing. Eight very highly conserved amino acid motifs are evident, including the DEAD box (Asp–Glu–Ala–Asp). Numbers refer to frequently found size ranges for intervening amino acid sequences (see Schmid and Linder, 1992). X = any amino acid. See **inside front cover** for the one-letter amino acid code. **(B)** *WD repeat family*. This gene family encodes products that are involved in a variety of regulatory functions, such as regulation of cell division, transcription, transmembrane signaling, mRNA modification, etc. The gene products are characterized by 4–16 tandem WD repeats consisting of about 44–60 amino acids each containing a core sequence of fixed length beginning with a GH (Gly–His) dipeptide and terminating in the dipeptide WD (Trp–Asp) preceded by a variable length sequence (see Smith *et al.*, 1999).

▶ the *G protein–coupled receptor superfamily* – a very large and diverse family of receptors that mediate ligand-induced signaling between the extracellular and intracellular environments via interaction with intracellular G proteins. They share a common structure of seven α-helix transmembrane segments, but typically have low (< 40%) sequence similarity to each other.

9.3.5 Genes in human gene families may be organized into small clusters, widely dispersed or both

Human gene families may be classified into those showing evidence of close gene clustering, and those that are dispersed over several different chromosomal locations. This classification is somewhat arbitrary, however, since some gene families consist of multiple gene clusters at different chromosomal locations (see *Table 9.10*), and others such as the histone gene family (http://genome.nhgri.nih.gov/histones/

chrmap. shtml) may be dominated by one or two large clusters but also have several dispersed *orphan genes*.

Gene families organized in a single cluster

Genes in an individual gene cluster are thought to arise by *tandem gene duplication* events (*Figure 12.3*). Different organizations are evident:

▶ *tandem gene organization*. The genes are highly related to each other in terms of both sequence and function, although certain family members may be nonfunctional. There are very few examples for polypeptide-encoding genes (the polyubiquitin genes are one striking example), but several RNA gene families (rRNA, U2 snRNA) show this organization;

▶ *close clustering*. The genes are not quite tandemly repeated; instead, they are closely clustered, and may be regulated by a single *locus control region* – see the example of the α- and β-globin gene clusters in *Figure 9.11*. The individual genes

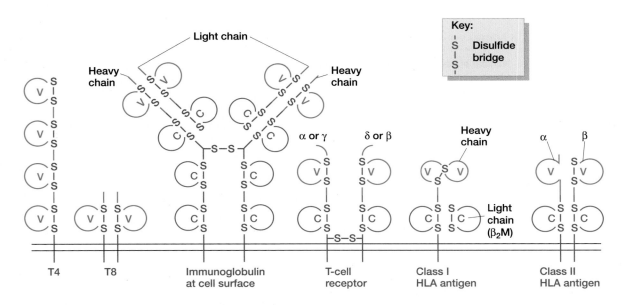

Figure 9.10: Members of the Ig superfamily are surface proteins with similar types of domain structure.

A few examples of the very large Ig superfamily are illustrated. Many members are dimers consisting of extracellular variable domains (V) located at the N-terminal ends and constant (C) domains, located at the C-terminal (membrane-proximal) ends. The light chain of class I HLA antigens, β_2-microglobulin, has a single constant domain and does not span the membrane. It associates with the transmembrane heavy chain which has two variable and one constant domain, giving an overall structure similar to that of the class II HLA antigens.

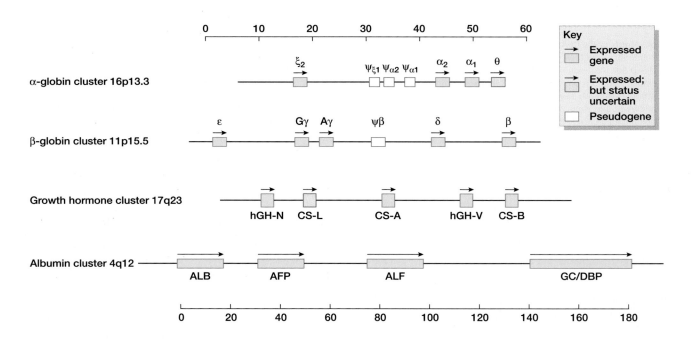

Figure 9.11: Examples of human clustered gene families.

Genes in a cluster are closely related in sequence and typically transcribed from the same strand. The functional status of the θ-globin and *CS-L* genes is uncertain. The scales at top (globin and growth hormone clusters) and bottom (albumin cluster) are in kilobases.

Table 9.10: **Examples of clustered and interspersed multigene families**

Family	Copy no.	Organization	Chromosome location (s)
(A) CLUSTERED GENE FAMILIES			
Single cluster gene families			
Growth hormone gene cluster	5	Clustered within 67 kb; one conventional pseudogene	17q22–24
α-Globin gene cluster	7	Clustered over ~ 50 kb (see *Figure 9.11*)	16p13.3
Class I HLA heavy chain genes	~ 20	Clustered over 2 Mb (see *Figure 9.12*)	6p21.3
Multiple cluster gene families			
HOX genes	38	Organized in four clusters on 2p, 7, 12, 17 (see *Figure 12.9*)	
Histone gene family	61	Modest-sized clusters at a few locations; two large clusters on chromosome 6	Many
Olfactory receptor gene family	> 900	About 25 large clusters scattered throughout the genome	Many
(B) INTERSPERSED GENE FAMILIES			
Pyruvate dehydrogenase	2	One intron-containing gene and one testis-expressed *retrogene*	Xp22; 4q22-q23
Aldolase	5	Three functional genes and two pseudogenes on five different chromosomes	Many
PAX	9	All nine are functional genes	Many
NF1 (Neurofibromatosis type I)	> 12	One functional gene at 17q; others are defective *non-processed* DNA copies (*Figure 9.13*)	Many; mostly pericentromeric
Ferritin heavy chain	> 15	One functional gene on chromosome 11; most are *processed* pseudogenes	Many

usually show a high degree of sequence and functional identity to each other, but many family members may be *pseudogenes* (Section 9.3.6);

▶ *compound clusters*. In other clustered gene families, however, the physical relationship between genes in a cluster may be less close and a cluster of related genes may also contain within it genes that are unrelated in sequence and function, constituting a **compound gene cluster**. For example, the HLA complex on 6p21.3 is dominated by families of genes which encode class I and II HLA antigens and various serum complement factors, but individual family members may be separated by functionally unrelated genes such as members of the steroid 21-hydroxylase gene family, etc.

Gene families organized in multiple gene clusters

Some gene families are organized in multiple clusters. Occasionally, the clusters may be closely related on the same chromosome as a result of recent duplication, for example the inverted clusters containing the *SMN1* and *SMN2* genes associated with spinal muscular atrophy (see Frugier *et al.*, 2002). More often, however, they are distributed over two or more chromosomal locations. Different organizations are evident. Some families show comparatively high similarity between genes on different clusters; some less so. An outstanding example is the *olfactory receptor gene family* which encodes a diverse repertoire of receptors allowing us to discriminate thousands of different odors. The > 900 members of this family are organized in large clusters at more than 25 different chromosomal locations, representing all chromosomes other than chromosome 20 and the Y chromosome (Glusman *et al.*, 2001).

Sequence homology is generally greater within a cluster than between clusters (compare, for example, members of the α-globin cluster on 16p, with those of the β-globin cluster on 11p – see *Figure 12.4*), but occasionally because of strong functional selection genes in different clusters may be more related to each other than those in a single cluster, as in the case of *HOX* genes (*Figure 12.9*).

Dispersed gene families

Members of some families are dispersed at two or more different chromosomal locations. The genes at the different locations are usually quite divergent in sequence unless gene duplication occurred relatively recently, or there has been considerable selection pressure to maintain sequence conservation. Family members may have originated from:

▶ *different genomes*. The original mitochondrial genome may have derived from an aerobic bacterium with subsequent transfer of many of the original bacterial genes to the nuclear genome. As a result, the nuclear genome contains duplicated genes, encoding *cytoplasm-specific* and *mitochondrial-specific* isoforms for certain enzymes and other key metabolic products (see *Table 9.8* for some examples).

▶ *ancient gene/genome duplication events*. Families of this type typically contain only a few members, as in the case of the PAX gene family, and appear to have evolved by a combination of gene duplication and/or genome duplication events over a long period of evolutionary time. Usually all or many of the family members are functional, and significant sequence homology between the gene products may be restricted to crucially important domains e.g the paired domain of PAX gene products.

▶ *by retrotransposition events.* Some gene families have expanded comparatively recently in evolutionary terms by a process whereby RNA transcribed from one or a small number of functional genes is converted by cellular reverse transcriptase into *natural cDNA* which then becomes integrated elsewhere in the chromosomes. Most such copies are nonfunctional but some gene families have a functional intron–containing gene and a functional processed gene copy (see next section).

9.3.6 Pseudogenes, truncated gene copies and gene fragments are commonly found in multigene families

Families of polypeptide-encoding genes (and RNA genes) are frequently characterized by *defective copies* (**pseudogenes**) of essentially all the sequence of a functional gene (or at least its coding sequence), or of portions of it, e.g. *truncated copies* lacking the 5′ or 3′ ends or *internal fragments*, in some cases a single exon. A large variety of different classes are found. The following examples are meant to be illustrative of the types of defective gene copies found in different types of gene family.

Nonprocessed pseudogenes in a gene cluster.

Individual gene clusters often have defective gene copies which have been copied *at the level of genomic DNA* by tandem *gene duplication*. The copies can contain sequence corresponding to exons, introns and promoter regions of the func-

tional genes (**nonprocessed pseudogenes**), but are usually recognized to be defective by the presence of inappropriate termination codons in sequences corresponding to exons. Classical examples are found in the α-globin and β-globin clusters (see *Figure 9.11*).

Truncated genes and internal gene fragments in a gene cluster

The class I HLA gene family at 6p21.3 is a classical example of a gene cluster which is characterized by nonprocessed pseudogenes, truncated gene copies and gene fragments. Although the number of class I *HLA* genes can vary on different chromosome 6s, comprehensive analysis of one of these identified 17 family members clustered over about 2 Mb, comprising: six expressed genes, four conventional full-length pseudogenes, five truncated gene copies, and two small internal gene fragments (Geraghty *et al.*, 1992; see *Figure 9.12*). The family would have originated by tandem gene duplications and the fragmented gene copies arose by *unequal crossover* or *unequal sister chromatid exchange* (Section 11.3.2).

Nonprocessed pseudogenes in a dispersed gene family

Two illustrative examples are sequences related to the *NF1* (neurofibromatosis type I) and the *PKD1* (adult polycystic kidney disease) genes. These genes are located respectively at 17q11.2, close to the centromere (*pericentromeric*), and at 16p13.3, close to the telomere (*subtelomeric*). Human pericentromeric regions are typically composed of sequences which

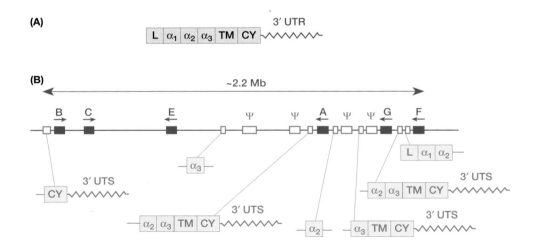

Figure 9.12: Clustered gene families often contain nonprocessed pseudogenes and truncated genes or gene fragments: example of the class I HLA gene family.

(A) *Structure of a class I HLA heavy chain mRNA.* The full-length mRNA contains a polypeptide-encoding sequence; blocks represent different domains as follows: L, leader sequence; α_1, α_2, α_3, extracellular domains; TM, transmembrane sequence; CY, cytoplasmic tail); and a 3′-untranslated sequence (3′-UTS). The three extracellular domains α_1–α_3 are each encoded essentially by a single exon. The very small 5′-UTS is not shown. **(B)** *The class I HLA heavy chain gene cluster.* The cluster is located at 6p21.3 and comprises about 20 genes. They include six expressed genes (blue), four full-length nonprocessed pseudogenes (Ψ), and a variety of partial gene copies (small, open red boxes). Some of the latter are truncated at the 5′ end (e.g. the one next to *HLA-B*), some are truncated at the 3′ end (e.g. the one next to *HLA-F*) and some contain single exons (e.g. the one next to *HLA-E*).

have been copied recently during evolution and which are located on several chromosomes. Subtelomeric regions are also comparatively unstable and prone to duplication (Eichler, 2001; Mefford and Trask, 2002). They make a substantial contribution to the primate-specific *segmental duplication* which accounts for over 150 Mb of the human genome, although the instability effect seems to be partly chromosome specific (see Bailey *et al.*, 2002; Section 12.2.5).

In the case of the *NF1* gene at least 11 nonprocessed pseudogene/gene fragment copies (containing sequences resembling *NF1* introns as well as exons) are distributed over seven different chromosomes, nine being located at pericentromeric regions (Regnier *et al.*, 1997; see *Figure 9.13*). The *PKD1* gene has 46 exons spanning 50 kb. A truncated 5′ gene copy comprising approximately 70% of the gene (exons 1 to 34 plus intervening introns) has been faithfully replicated at least three times and inserted into a more proximal location at 16p13.1 (the European Polycystic Kidney Disease Consortium, 1994).

Processed pseudogenes in a dispersed polypeptide-encoding gene family

Interspersed gene families often have defective gene copies which contain sequences corresponding to the exons of a functional gene (but not the introns) and they usually contain at one end an oligo(dA)/(dT) sequence. Such **processed pseudogenes** have been copied *at the cDNA level* by retrotransposition (see also Section 9.5.1). Cellular reverse transcriptases transcribe mRNA into *natural* cDNA which can then integrate into chromosomal DNA (*Figure 9.14*) most likely with the assistance of the LINE1 transposition machinery (see Section 9.5.2). Processed pseudogenes can be quite prolific. For example, there are 79 proteins in cytoplasmic ribosomes and a family of 95 ribosomal protein genes (16 are duplicated) but a staggering 2090 processed pseudogenes of this type have been identified in the nuclear genome (Zhang, 2002).

Processed pseudogenes are typically not expressed (because they lack a promoter sequence) but some examples are known of *expressed processed genes*. Here the natural cDNA has integrated into a chromosomal DNA site which happens, by chance, to be adjacent to a promoter which can drive expression of the processed gene copy. Selection pressure may ensure continued expression of the processed gene copy which is then considered a **retrogene**. A variety of intronless retrogenes are known to have testis-specific expression patterns and often they are autosomal homologs of an intron-containing X-linked gene (see *Table 9.11*). The selection

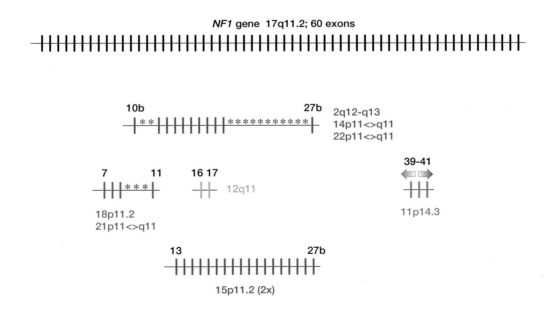

Figure 9.13: Dispersed nonprocessed pseudogenes originating from the pericentromeric *NF1* (neurofibromatosis type I) gene.

Exons are represented as thin vertical boxes. For convenience, the introns in the *NF1* gene are represented as having equal lengths. Although the *NF1* gene has 60 exons, the exon numbering goes from 1 to 49 with some neighboring exons assigned the same number, but distinguishing letters (e.g. exons 10a, 10b, and 10c). Highly homologous pseudogene copies of the *NF1* gene are found at eight or more other genome locations, mostly in pericentromeric regions. In each case, the pseudogene comprises a copy of a portion of the full-length gene only, with some exons and intervening introns. Rearrangements of the pseudogenes are apparent. Some have caused deletion of exons and introns (shown by asterisks). One has involved an inversion so that an exon-39 copy is inverted compared to neighboring exon copies. Data kindly provided by Dr. Nick Thomas and Dr. Meena Upadhyaya, University of Wales College of Medicine.

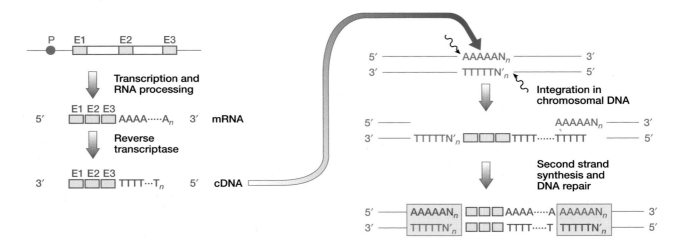

Figure 9.14: Processed pseudogenes and retrogenes originate by reverse transcription from RNA transcripts.

The reverse transcriptase function could be provided by LINE-1 repeats. The model for integration shown in the figure is only one of several possibilities. Here integration is envisaged at staggered breaks (indicated by curly arrows) in A-rich sequences, but could be assisted by the LINE1 endonuclease (Section 9.5.2). If the A-rich sequence is included in a 5′ overhang, it could form a hybrid with the distal end of the poly(T) of the cDNA, facilitating second-strand synthesis. Because of the staggered breaks during integration, the inserted sequence will be flanked by short direct repeats (boxed sequences). E1–E3 represent exons. P = promoter. Transposed copies do not carry a promoter and so normally they will not be expressed and will acquire deleterious mutations (***processed pseudogenes***). However, some processed copies of polypeptide-encoding genes are functional (***retrogenes***), having integrated at a location adjacent to a functional promoter and having been subject to selection pressure to conserve function (*Table 9.11*).

Table 9.11: **Examples of human intronless retrogenes and their parental intron–containing homologs**

(for more information, see http://exppc01.uni-muenster.de/expath/frames.htm)

Retrogene	Intron-containing homolog	Product
GK2 at 4q13	*GK1* at Xp21.3	Glycerol kinase
PDHA2 at 4q22-q23	*PDHA1* at Xp22	Pyruvate dehydrogenase
PGK2 at 6p12.3	*PGK* at Xq13	Phosphoglycerate kinase
TAF1L at 9p13.3	*TAF1* at Xq13.1	TATA box binding protein associated factor, 250 kDA
MYCL1 at 1p34.2	*MYCL2* at Xq22-q23	Homolog of v-myc oncogene
GLUD1 at 10q23.3	*GLUD2* at Xq25	Glutamate dehydrogenase
SNAIL1L1 at 2q33-q37	*SNAIL1* at 20q13	Snail-related developmental regulator

pressure here may be the requirement for expression during male meiosis when autosomal genes are transcriptionally active but when *both* the X and the Y chromosomes are silenced, being condensed to form heterochromatin. However, some functional retrogenes are copies of non X–linked genes, such as a copy of the *SNAIL1* developmental regulator (Locascio *et al.*, 2002).

Processed pseudogenes in an RNA-encoding gene family

Although the size of some interspersed polypeptide–encoding gene families testifies to the success of retrotransposition as a mechanism for generating processed gene copies, the really successful (in terms of high copy number) retrotranspositions have been performed from RNA polymerase III transcripts. For example, the ***Alu repeat*** family (see Section 9.5.3) is considered to have arisen as processed pseudogenes copied from the gene encoding SRP RNA (also called 7SL RNA), a component of the *signal recognition particle*. Genes such as this, which are transcribed by RNA polymerase III, often contain an *internal promoter* (*Figure 10.4*) which facilitates the expression of newly transposed copies in permissive regions of the genome.

9.3.7 Human proteome classification has begun but the precise functions of many human proteins remain uncertain

The sequencing of the human genome has provided valuable information on the predicted set of human proteins, the **human proteome**. For many genes, protein functions had previously been assigned, but analysis of the large number of novel genes has permitted extension of the previous classifications. Various databases are dedicated to recording sequence features which are shared by multiple proteins and are indicative of common or related functions (although not all proteins can be assigned to the available categories because some proteins do not appear to share sequence with others). Commonly used databases include the **InterPro database** maintained by the European Bioinformatics Institute and the **Pfam database** maintained at the Wellcome Trust Sanger Institute (see Further reading). Categories include:

▶ **protein families** (on the basis of general functional similarity) – see *Table 9.12*;

▶ **protein domains** – see *Table 9.13*. The huge number of zinc fingers testifies to their importance in a wide variety of DNA–protein interactions;

▶ **protein repeats** – the most common are the G-protein beta WD-40 repeat (~400 protein matches) and the ankyrin repeat (> 260 protein matches).

Functional classification has also begun, and the **Gene Ontology (GO) consortium** has defined functional classification categories according to which cellular component the protein operates in, its molecular function and the overall biological process to which it contributes (Section 8.3.6). Of course, this is an ongoing process: the functions of many human genes remain to be determined, using various approaches. An initial classification of human proteins according to molecular function and biological process is illustrated in *Figure 9.15*.

9.4 Tandemly repeated noncoding DNA

Highly repeated noncoding human DNA often occurs in *arrays* (or *blocks*) of tandem repeats of a sequence which may be a simple one (1–10 nucleotides), or a moderately complex one (tens to hundreds of nucleotides). Individual arrays can occur at a few or many different chromosomal locations. According to the *array size* three major subclasses can be defined: *satellite DNA, minisatellite DNA* and *microsatellite DNA* (*Table 9.14*). Satellite DNA is transcriptionally inactive as is the vast majority of minisatellite DNA, but in the case of microsatellite DNA a significant percentage (albeit a very small one) is located in coding DNA.

9.4.1 Satellite DNA consists of very long arrays of tandem repeats and can be separated from bulk DNA by density gradient centrifugation

Human satellite DNA is comprised of very large arrays of tandemly repeated DNA. The repeat unit may be a simple sequence (only a few nucleotides long) or a moderately complex sequence (*Table 9.14*; see Singer, 1982). Satellite DNA makes up most of the heterochromatic regions of the genome, and is notably found in the vicinity of the centromeres (**pericentromeric heterochromatin**). When the repeat unit is very short, the base composition of the repeat units, and so the overall base composition of satellite DNA, may diverge substantially from that of the total genomic DNA. As a result, three human satellite DNAs, satellites I, II and III, have been able to be separated from the bulk DNA by

---

Table 9.12: The top 15 protein families in the human proteome

Data obtained in January 2003 from the InterPro database maintained by the European Bioinformatic Institute at http://www.ebi.ac.uk/proteome/).

InterPro reference	Protein family name	Proteins matched
IPR000276	Rhodopsin-like G protein-coupled receptor	826
IPR000719	Protein kinase	688
IPR001909	KRAB box (Kruppel-associated box)	314
IPR001806	Ras GTPase superfamily	192
IPR005821	Ion transport protein	149
IPR000387	Tyrosine specific protein phosphatase and dual specificity protein phosphatase	139
IPR001254	Serine protease, trypsin family	128
IPR000379	Esterase/lipase/thioesterase, active site	112
IPR007114	Major facilitator superfamily (MFS)	100
IPR001993	Mitochondrial substrate carrier	86
IPR001664	Intermediate filament protein	85
IPR001128	Cytochrome P450	84

Table 9.13: The top 15 protein domains in the human proteome

Data obtained in January 2003 from the InterPro database at http://www.ebi.ac.uk/proteome/

InterPro reference	Name of protein domain	Total number in proteome
IPR007087	Zinc finger, C2H2 type	28654
IPR002126	Cadherin	4131
IPR006209	Epidermal growth factor (EGF)-like domain	3107
IPR003006	Immunoglobulin/major histocompatibility complex	2387
IPR002048	Calcium-binding EF-hand	1885
IPR001452	SH3 domain	1815
IPR003961	Fibronectin, type III	1812
IPR000504	RNA-binding region RNP-1 (RNA recognition motif)	1783
IPR001356	Homeobox	1435
IPR002965	Proline-rich extensin	1229
IPR001478	PDZ/DHR/GLGF domain	1143
IPR001841	Zinc finger, RING	1132
IPR001849	Plekstrin-like	1061
IPR000210	BTB/POZ domain	494
IPR005225	Small GTP-binding protein domain	189

buoyant density gradient centrifugation. Each satellite class includes a number of different tandemly repeated DNA sequence families (*satellite subfamilies*), some of which are shared between different classes. Satellites II and III mostly contain simple sequence repeats but higher order structure is also evident.

Alphoid DNA and centromeric heterochromatin

Other types of satellite DNA sequence cannot easily be resolved by density gradient centrifugation. They were first identified by digestion of genomic DNA with a restriction endonuclease which typically has a single recognition site in the basic repeat unit. In addition to the basic repeat unit size (monomer), such enzymes will produce a characteristic pattern of multimers of the unit length because of occasional random loss of the restriction site in some of the repeats (Singer, 1982). **Alpha satellite** (or **alphoid DNA**) consists of tandem repeats of a 171–bp repeat unit and makes up the bulk of the centromeric heterochromatin. High sequence divergence between individual members of the alphoid DNA family means that there are specific subfamilies for each of the human chromosomes (Choo *et al.*, 1991).

The precise function of satellite DNA remains unclear (see Csink and Henikoff, 1998; Henikoff *et al.*, 2001). The centromeric DNA of human chromosomes largely consists of various families of satellite DNA (see *Figure 9.16*). Of these, only the α–satellite is known to be present on all chromosomes, and its repeat units often contain a binding site for a specific centromere protein, CENP-B. Cloned α–satellite arrays have been shown to seed *de novo* centromeres in human cells, indicating that α–satellite plays an important role in centromere function (Grimes and Cook, 1998).

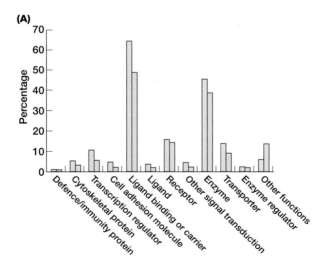

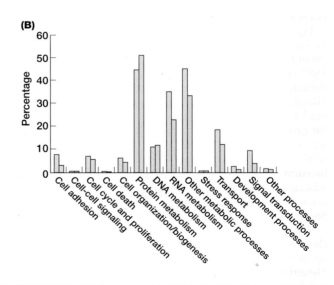

Figure 9.15: A classification of human and mouse proteins based on their molecular functions and the biological processes they are involved in.

Gene ontology (GO) terms have been grouped into approximately a dozen categories falling within the larger ontologies of **(A)** *molecular function* and **(B)** *biological process*. Blue bars: mouse proteins; red bars: human proteins. Modified from the Mouse Genome Sequencing Consortium (2002) *Nature* **420**, 520–562 with permission from Nature Publishing Group.

Table 9.14: Major classes of tandemly repeated human DNA

Class	Size of repeat unit (bp)	Major chromosomal location(s); transcriptional status
Satellite DNA (arrays often within *100 kb to several Mb* size range	5–171	Especially at centromeres; not transcribed
α (alphoid DNA)	171	Centromeric heterochromatin of all chromosomes
β (*Sau*3A family)	68	Notably the centromeric heterochromatin of 1, 9, 13, 14, 15, 21, 22 and Y
Satellite 1 (AT-rich)	25–48	Centromeric heterochromatin of most chromosomes and other heterochromatic regions
Satellites 2 and 3	5	Most, possibly all, chromosomes
Minisatellite DNA (arrays often within the *0.1–20 kb* range)	9–64	At or close to telomeres of all chromosomes; vast majority not transcribed
Telomeric family	6	All telomeres
Hypervariable family	9–64	All chromosomes, often near telomeres
Microsatellite DNA (= *simple sequence repeats, SSR*) (arrays typically *< 100 bp*)	12	Dispersed throughout all chromosomes; some small arrays of very simple sequence

9.4.2 Minisatellite DNA is composed of moderately sized arrays of tandem repeats and is often located at or close to telomeres

Minisatellite DNA comprises a collection of moderately sized arrays of tandemly repeated DNA sequences which are dispersed over considerable portions of the nuclear genome (*Table 9.14*). Like satellite DNA sequences, they are not normally transcribed (but see below).

Hypervariable minisatellite DNA sequences are highly polymorphic and are organized in over 1000 arrays (from 0.1 to 20 kb long) of short tandem repeats (Jeffreys, 1987). The repeat units in different hypervariable arrays vary considerably in size, but share a common core sequence, GGGCAGGAXG (where X = any nucleotide), which is similar in size and in G content to the *chi* sequence, a signal for generalized recombination in *E. coli*. While many of the arrays are found near the telomeres, several hypervariable minisatellite DNA sequences occur at other chromosomal locations. Although the great majority of hypervariable minisatellite DNA sequences are not transcribed, some rare cases are known to be expressed (e.g. the *MUC1* locus – Swallow *et al.*, 1987).

The significance of hypervariable minisatellite DNA is not clear, although it has been reported to be a 'hotspot' for homologous recombination in human cells (Wahls *et al.*, 1990). Nevertheless it has found many applications. Various individual loci have been characterized and used as genetic markers, although the preferential localization in subtelomeric regions has limited their use for genome–wide linkage studies. A major application has been in *DNA fingerprinting*, in which a single DNA probe which contains the common core sequence can hybridize simultaneously to multiple minisatellite DNA loci on all chromosomes, resulting in a complex individual-specific hybridization pattern (see Section 18.7.1).

Another major family of minisatellite DNA sequences is found at the termini of chromosomes, the telomeres. The principal constituent of a human chromosome's **telomeric DNA** is 3–20 kb of tandem hexanucleotide repeat units, especially TTAGGG, which are added by a specialized enzyme, *telomerase*. By acting as buffers to protect the ends of the chromosomes from degradation and loss and by providing a mechanism for replicating the ends of the linear DNA of chromosomes, these simple repeats are directly responsible for telomere function (*Figure 2.6*; Section 2.2.5).

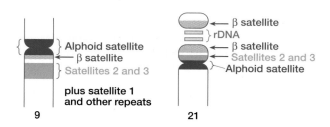

Figure 9.16: Satellite DNA organization at centromeres.

The locations of different classes of satellite DNA are shown for chromosome 9 and for chromosome 21 (one of the five autosomal acrocentric chromosomes). The illustration in this case is redrawn from Tyler-Smith and Willard (1993) *Curr. Opin. Genet. Dev.* **3**, 390–397, with permission from Elsevier.

9.4.3 Microsatellite DNA consists of short arrays of simple tandem repeats and is dispersed throughout the human genome

Microsatellite DNA, also called **simple sequence repeats (SSR),** are small arrays of tandem repeats of a simple sequence (usually less than 10 bp). They are interspersed throughout the genome, accounting for over 60 Mb (2% of the genome), and are thought to have arisen mostly by *replication slippage (Figure 11.5)*. Arrays of dinucleotide repeats are the most common type, accounting for about 0.5% of the genome. CA/TG repeats are very common (1 per 36 kb), and are often highly polymorphic *(Figures 7.7, 7.8)*. AT/TA (1 per 50 kb) and AG/CT (1 per 125 kb) repeats are also quite common but CG/GC repeats are very rare (1 per 10 Mb), because the CpG dinucleotide is prone to methylation and subsequent de-amination (Section 9.1.3).

Of the mononucleotide repeats, runs of A and of T are very common (see *Figure 9.19* for intragenic examples); runs of G and of C are very much rarer. Individual classes of trinucleotide and tetranucleotide tandem repeats are comparatively rare, but are often highly polymorphic and increasingly have been investigated to develop highly polymorphic markers. See Tables 14 and 15 of the International Human Genome Sequencing Consortium (2001) for further information.

The significance of microsatellite DNA is not known. Alternating purine–pyrimidine repeats, such as tandem repeats of the dinucleotide pair CA/TG, are capable of adopting an altered DNA conformation, *Z*-DNA, *in vitro*, but there is little evidence that they do so in the cell. Although microsatellite DNA has generally been identified in intergenic DNA or within the introns of genes, a few examples have been recorded within the coding sequences of genes and are often mutation hotspots because they are prone to replication slippage (see *Figure 11.14* for some examples) and in some limited cases to unstable expansion (Sections 11.5.2 and 16.6.4).

9.5 Interspersed repetitive noncoding DNA

9.5.1 Transposon-derived repeats make up > 40% of the human genome and mostly arose through RNA intermediates

Almost all of the interspersed repetitive noncoding DNA in the human genome is derived from **transposable elements** (also called **transposons**), mobile DNA sequences which can migrate to different regions of the genome (Smit, 1996; Prak and Kazazian, 2000). Close to 45% of the genome can be recognized as belonging to this class (International Human Genome Sequencing Consortium, 2001; Li *et al.*, 2001), but much of the remaining 'unique' DNA must also be derived from ancient transposon copies that have diverged too far to be recognized as such. Often dismissed as junk DNA in the past, there is increasing evidence that such transposons may be valuable to mammalian cells (see Dennis, 2002).

In humans and other mammals there are four major classes of transposon, but only a tiny minority are actively transposing. They can be organized into two groups according to the method of transposition:

▶ **retrotransposons** (also abbreviated to **retroposons**). Here the copying mechanism uses *reverse transcriptase* to make cDNA copies of RNA transcripts, resembling the way in which processed pseudogenes and retrogenes are generated (Section 9.3.6). *Replicative* (or *copy*) *transposition* ensures that a copy is made of an existing sequence after which the copy migrates and inserts elsewhere in the genome. Three mammalian transposon classes fall into this group: long interspersed nuclear elements (LINES); short interspersed nuclear elements (SINES) and retrovirus-like elements containing long terminal repeats;

▶ **DNA transposons.** Members of this fourth class of transposon migrate by **conservative transposition.** There is no copying of the sequence; instead, the sequence is excised and then re-inserted elsewhere in the genome (a 'cut and paste' mechanism).

According to their ability to transpose independently or not, transposable elements can be *autonomous* or *nonautonomous (Figure 9.17)*. Of the four classes of transposable element, LINES and SINES predominate and are described in more detail in Sections 9.5.2 and 9.5.3 respectively. The other two classes are briefly described here.

Human LTR transposons

LTR transposons include autonomous and nonautonomous retrovirus-like elements which are flanked by **long terminal (direct) repeats (LTRs)** containing necessary transcriptional regulatory elements. The *autonomous* members are known as **endogenous retroviral sequences** (or **ERV**), and they contain **gag** and **pol** genes, which encode a protease, reverse transcriptase, RNAse H and integrase. There are three major classes of human ERV (HERV) accounting for a total of about 4.6% of the human genome *(Table 9.15)*. Very many are defective and transposition has been extremely rare during the last several million years. The very small HERV-K group, however, shows conservation of intact retroviral genes (Lower *et al.*, 1996), and some members of the HERV-K10 subfamily have undergone transposition comparatively recently during evolution. *Nonautonomous* retroviral elements lack the *pol* gene and often also the *gag* gene (the internal sequence having been lost by homologous recombination between the flanking LTRs). The MaLR family of such elements account for almost 4% or so of the genome *(Table 9.15)*.

Human DNA transposon fossils

DNA transposons have terminal *inverted* repeats and encode a **transposase** which regulates transposition. They account for close to 3% of the human genome and can be grouped into different classes which can be subdivided into many families with independent origins (see Smit, 1996 and the RepBase database of repeat sequences at http://www.girinst.org/). There are two major human families, MER1 and MER2 plus a variety of less frequent families *(Table 9.15)*. Virtually all the resident human DNA transposon sequences are no longer active and so are *transposon fossils*. DNA transposons tend to have short life spans within a species, unlike some of the other transposable elements such as LINES (Section 9.5.2). Quite a few functional human genes, however, appear to have originated from DNA

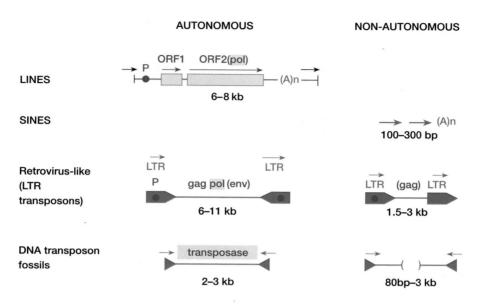

Figure 9.17: Mammalian transposon families.

Only a small proportion of members of any of the above families may be capable of transposing; many have lost such capacity by acquiring inactivating mutations and many are short truncated copies. See *Figures 9.18* and *9.19* for the typical structures of some human transposable elements. Modified from International Human Genome Sequencing Consortium (2001) *Nature* **409**, 860–921 with permission from the Nature Publishing Group.

Table 9.15: Major classes and families of interspersed repetitive DNA in the human genome (excluding the Y chromosome)

Class	Family	No. of copies	Fraction of genome (%)
SINE	Alu family	~ 1 200 000	10.7
	MIR	~ 450 000	2.5
	MIR3	~ 85 000	0.4
LINE	LINE-1 family	~ 600 000	17.3
	LINE-2 family	~ 370 000	3.3
	LINE-3 family	~ 44 000	0.3
LTR elements	ERV families	~ 240 000	4.7
	MaLR	~ 285 000	3.8
DNA transposon	MER-1 (Charlie)	~ 213 000	1.4
	MER-2 (Tigger)	~ 68 000	1.0
	Others	~ 60 000	0.4

Data from International Human Genome Sequencing Consortium, 2001; Mouse Genome Sequencing Consortium, 2002.

transposons, notably genes encoding the RAG1 and RAG2 recombinases and the major centromere-binding protein CENPB (see also Jurka and Kapitonov, 1999; Smit, 1999; International Human Genome Sequencing Consortium, 2001).

9.5.2 Some human LINE1 elements are actively transposing and enable transposition of SINES, processed pseudogenes and retrogenes

Long interspersed nuclear elements (LINEs) have been very successful transposable elements and they have a long evolutionary history. As autonomous transposable elements, they can encode the necessary products to ensure retrotransposition, including the essential reverse trancriptase. Human LINES consist of three distantly related families: LINE1 (L1), LINE2 and LINE3, collectively comprising about 20% of the genome (*Table 9.15*). They are primarily located in euchromatic regions and are located preferentially in the dark AT-rich G bands (Giemsa positive) of metaphase chromosomes (Korenberg and Rykowski, 1988). Of the three human families, LINE1 (or L1) is the only family which is still actively transposing, and it is predominant, making up about 17% of the genome. It is the most important human transposable element and is also found in other mammals, including mice (Ostertag and Kazazian, 2001).

The full-length **LINE1 (L1) element** is about 6.1 kb long and encodes two proteins: an RNA-binding protein; and a protein with both endonuclease and reverse transcriptase activities (*Figure 9.18A*). Unusually an *internal promoter* is located within the 5′ UTR and so copies of full-length transcripts carry with them their own promoter which can be used following integration in a permissive region of the genome. After translation, the LINE1 RNA assembles with its own encoded proteins and moves to the nucleus. The endonuclease cuts a DNA duplex on one strand leaving a free 3′ OH group which serves as a primer for reverse transcription from the 3′ end of the LINE RNA. The endonuclease's preferred cleavage site is TTTT↓A; hence the preference for integrating into AT-rich regions. AT-rich DNA is poor in genes and so their tendency to integrate into AT-rich DNA means that LINES impose a lower mutational burden, making it easier for their host to accommodate them.

During integration, the reverse transcription often fails to proceed to the 5′ end, resulting in truncated, nonfunctional insertions. Accordingly, most LINE-derived repeats are short, with an average size of 900 bp for all LINE1 copies, and only about one in 100 copies are full length. The LINE1 machinery is responsible for most of the reverse transcription in the genome, allowing retrotransposition of the nonautonomous SINEs and the creation of processed pseudogenes and retrogenes (Esnault *et al.*, 2000; Section 9.3.6). Of the 6000 or so full-length LINE1 sequences, about 60–100 are still capable of transposing, and occasionally cause disease by disrupting gene function following insertion into an important conserved sequence (Section 11.5.6).

9.5.3 Alu repeats occur more than once every 3 kb in the human genome and may be subject to positive selection

Short interspersed nuclear elements (SINEs) are about 100–400 bp long and have been very successful in colonizing mammalian genomes, resulting in a variety of high copy number families. Some human SINES are primate specific

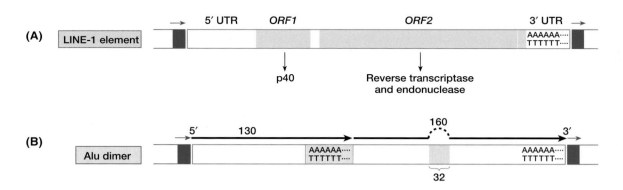

Figure 9.18: The human LINE1 and Alu repeat elements.

(A) *The LINE1 element*. The 6.1 kb LINE1 (L1) element has two open reading frames: a 1-kb ORF1 which encodes an RNA-binding protein and a 4-kb ORF2 which specifies a protein with both endonuclease and reverse transcriptase activities. An internal promoter lies within a region of untranslated DNA preceding ORF1 (conventionally called the 5′-UTR) while at the other end there is an (A)$_n$/(T)$_n$ sequence, often described as the 3′ poly(A) tail. The LINE1 endonuclease cuts (↓) one strand of a DNA duplex, preferably within the sequence TTTT↓A and the reverse transcriptase uses the released 3′-OH end to prime cDNA synthesis. New insertion sites are flanked by a small target site duplication of 7–20 bp. **(B)** *The Alu repeat*. The consensus standard Alu dimer is shown with two similar repeats terminating in an (A)$_n$/(T)$_n$ like sequence. They have different sizes because of the insertion of a 32-bp element within the larger repeat. Alu monomers also exist in the human genome, as do various truncated copies of both monomers and dimers.

such as the Alu family; others are not restricted to primates, being also found in marsupials and monotremes, and they have been described as MIR (**m**ammalian-wide **i**nterspersed **r**epeat) families (*Table 9.15*). SINES do not encode any proteins and are not autonomous. LINES and SINES share sequences at their 3' end and SINES have been shown to be mobilized by neighboring partner LINES (Kajikawa and Okada, 2002). By parasitizing on the LINE element transposition machinery, SINES can attain very high copy numbers.

Mammalian SINES have originated from copies of tRNA (in many cases), or from SRP(7SL) RNA, as in the case of both the **Alu repeat** (Ullu and Tschudi, 1984) and the mouse **B1 repeat** (see Section 12.4.1). Genes encoding tRNA and SRP RNA are transcribed by RNA polymerase III and are unusual in that they have *internal promoters (Figure 10.4)*. However, the internal polymerase III promoter carried by Alu repeats is not sufficient for active transcription *in vivo* and appropriate flanking sequences are required for its activation. Following integration, therefore, a newly transposed Alu copy will become inactive unless it fortuitously lands in a region which enables the promoter to be active.

The Alu repeat is the most abundant sequence in the human genome, occurring on average more than once every 3 kb (International Human Genome Sequencing Consortium, 2001; Li *et al.*, 2001). There is a series of Alu subfamilies of different evolutionary ages with only about 5000 copies having integrated into the genome in the past 5 million or so years since divergence of humans and African apes (see Batzer and Deininger, 2002). The full-length Alu repeat is about 280 bp long and consists of two tandem repeats, each approximately 120 bp in length followed by a short sequence which is rich in A residues on one strand and T residues on the complementary strand. However, there is asymmetry between the tandem repeats: one repeat contains an internal 32-bp sequence lacking in the other (*Figure 9.18B*). Monomers, containing only one of the two tandem repeats, and various truncated versions of dimers and monomers are also common, giving a genome-wide average of 230 bp.

Alu repeats have a relatively high GC content and, although dispersed mainly throughout the euchromatic regions of the genome, are preferentially located in the GC-rich and gene-rich *R chromosome bands*, in striking contrast to the preferential location of LINES in AT-rich DNA (Korenberg and Rykowski, 1988). When located within genes, however, they are, like LINE1 elements, confined to introns and the

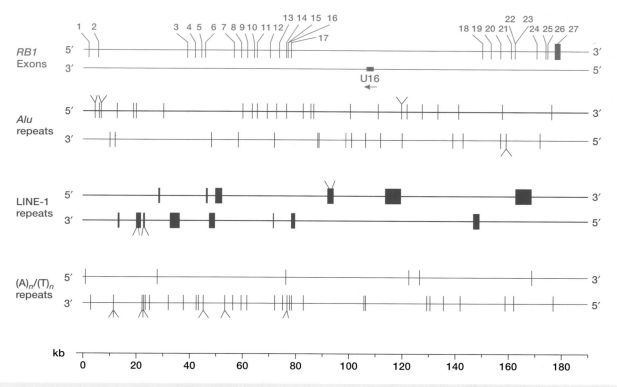

Figure 9.19: Location of Alu, LINE-1 and (A)$_n$/(T)$_n$ repeats within the human retinoblastoma susceptibility gene, *RB1*.

The 72-kb intron 17 contains a G protein-coupled receptor gene, U16, which is actively transcribed in the opposite direction to the *RB1* gene. The top line (5'→3') of each pair shows the repeat elements orientated in the *RB1* sense direction; the bottom line (3'→5') shows them in the antisense orientation. There are 46 Alu repeats and 17 LINE-1 elements (some closely clustered shown by divergent lines), all located within introns. Only two of the LINE-1 elements approach the full 6.1 kb length. The (A)$_n$/(T)$_n$ sequences (*n* = 12 or greater) indicated are only those located outside the interspersed repeats. No examples were found for (C)$_n$/(G)$_n$ for *n* = 12 or greater. Redrawn from Toguchida *et al.* (1993) *Genomics* **17**, 535–543 with permission from Elsevier.

untranslated regions (*Figure 9.19*). Despite the tendency to be located in GC-rich DNA, *newly transposing* Alu repeats show a preference for AT-rich DNA resembling that of LINEs, but progressively older Alus show a progressively stronger bias towards GC-rich DNA (International Human Genome Sequencing Consortium, 2001).

The bias in the overall distribution of Alus towards GC-rich (and, accordingly, gene-rich) regions must result from strong selection pressure. It suggests that Alu repeats are not just genome parasites, but are making a useful contribution to cells containing them. Some Alu sequences are known to be actively transcribed and may have been recruited to a useful function.

The *BCYRN1* gene which encodes a neural small cytoplasmic RNA, BC200, arose from an Alu monomer and is one of the few Alu sequences which are transcriptionally active under *normal* circumstances (Martignetti and Brosius, 1993). In many species SINEs are transcribed under conditions of stress, and the resulting RNAs bind a specific protein kinase (PKR) and block its ability to inhibit protein translation. SINE RNAs would thus promote protein translation under stress. Possibly a general function for SINES (Schmid, 1998) lies, therefore, in regulating protein translation (SINE RNA can be quickly transcribed in large quantities from thousands of elements and can function without protein translation).

Further reading

Human Genome *Nature* Issue (15 February 2001). *Nature* **409**, 813–958 (papers are available electronically via the Nature Genome Gateway at http://www.nature.com/genomics/human/)

Human Genome *Science* Issue (16 February 2001). *Science* **291**, 1177–1351 (papers are available electronically at http://www.sciencemag.org/content/vol291/issue5507/index.shtml)

InterPro proteome analysis database at http://www.ebi.ac.uk/proteome/

MITOMAP human mitochondrial genome database at http://www.mitomap.org

Mouse Genome *Nature* Issue (5 December 2002). *Nature* **420**, 447–590 (papers are available electronically via the Nature Genome Gateway at http://www.nature.com/nature/mousegenome/index.html)

NCBI guide to on-line information resources on the human genome at http://www.ncbi.nlm.nih.gov/genome/guide/human/

Noncoding RNAs Database at http://biobases.ibch.poznan.pl/ncRNA/

Pfam protein domain family database at http://www.sanger.ac.uk/Software/Pfam/

Repeat Sequence Database at http://www.girinst.org

References

Adachi N, Lieber MR (2002) Bidirectional gene organization: a common architectural feature of the human genome. *Cell* **109**, 807–809.

Ambros V (2001) microRNAs: tiny regulators with great potential. *Cell* **107**, 823–826.

Anderson S, Bankier AT, Barrell BG *et al.* (1981) Sequence and organization of the human mitochondrial genome. *Nature* **290**, 457–465.

Bailey JA, Gu Z, Clark RA *et al.* (2002) Recent segmental duplications in the human genome. *Science* **297**, 1003–1007.

Batzer MA, Deininger PL (2002) Alu repeats and human genetic diversity. *Nature Rev. Genet.* **3**, 370–378.

***C. elegans* sequencing consortium** (1998) Genome sequence of the nematode C. elegans: a platform for investigating biology. *Science* **282**, 2012–2018.

Castillo-Davis CI, Mekhedov SL, Hartl DL, Koonin EV, Kondrashov FA (2002) Selection for short introns in highly expressed genes. *Nature Genet.* **31**, 415–418.

Choo KH, Vissel B, Nagy A, Earle E, Kalitsis P (1991) A survey of the genomic distribution of alpha satellite DNA on all the human chromosomes, and derivation of a new consensus sequence. *Nucl. Acids Res.* **19**, 1179–1182.

Claverie JM (2001) Gene number. What if there are only 30,000 human genes? *Science* **291**, 1255–1257.

Clayton DA (1992) Transcription and replication of animal mitochondrial DNAs. *Int. Rev. Cytol.* **141**, 217–232.

Collins JE, Goward ME, Cole CG *et al.* (2003) Reevaluating human gene annotation: a second-generation analysis of chromosome 22. *Genome Res.* **13**, 27–36.

Craig JM, Bickmore WA (1994) The distribution of CpG islands in mammalian chromosomes. *Nature Genet.* **7**, 376–381.

Csink AK, Henikoff S (1998) Something from nothing: the evolution and utility of satellite repeats. *Trends Genet.* **14**, 200–204.

Dennis C (2002) A forage in the junkyard. *Nature* **420**, 458–459.

Dermitzakis ET, Reymond A, Lyle R *et al.* (2002) Numerous potentially functional but non-genic conserved sequences on human chromosome 21. *Nature* **420**, 578–582.

Eddy SR (2001) Noncoding RNA genes and the modern RNA world. *Nature Rev. Genet.* **2**, 919–929.

Eichler EE (2001) Recent duplication, domain accretion and the dynamic mutation of the human genome. *Trends Genet.* **17**, 661–669.

Esnault C, Maestre J, Heidmann T (2000). Human LINE retrotransposons generate processed pseudogenes. *Nature Genet.* **24**, 363–367.

European Polycystic Kidney Disease Consortium (1994) The polycystic kidney disease 1 gene encodes a 14 kb transcript and lies within a duplicated region on chromosome **16**. *Cell* **77**, 881–894.

FANTOM Consortium and the RIKEN Genome Exploration Research Group Phase I & II Team (2002) Analysis of the mouse transcriptome based on functional annotation of 60,770 full-length cDNAs. *Nature* **420**, 563–573.

Filipowicz W (2000) Imprinted expression of small nucleolar RNAs in brain: time for RNomics. *Proc. Natl Acad. Sci. USA*, **97**, 14035–14037.

Frugier T, Nicole S, Cifuentes-Diaz C, Melki J (2002) The molecular bases of spinal muscular atrophy. *Curr. Opin. Genet. Dev.* **12**, 294–298.

Geraghty DE, Koller BH, Pei J, Hansen JA (1992) Examination of four HLA class I pseudogenes. Common events in the evolution of HLA genes and pseudogenes. *J. Immunol.* **149**, 1947–1956.

Gottesman S (2002) Stealth regulation: biological circuits with small RNA switches. *Genes. Dev.* **16**, 2829–2842.

Glusman G, Yanai I, Rubin I, Lancet D (2001) The complete human olfactory subgenome. *Genome Res.* **11**, 685–702.

Gray TA, Saitoh S, Nicholls RD (1999) An imprinted mammalian bicistronic transcript encodes two independent proteins. *Proc. Natl Acad. Sci. USA* **96**, 5616–5621.

Grimes B, Cooke H (1998) Engineering mammalian chromosomes. *Hum. Mol. Genet.* **7**, 1635–1640.

Harrison PM, Hegyi H, Balasubramanian S *et al.* (2002) Molecular fossils in the human genome: identification and analysis of the pseudogenes in chromosomes 21 and 22. *Genome Res.* **12**, 272–280.

Henikoff S, Ahmad K, Malik HS (2001) The centromere paradox: stable inheritance with rapidly evolving DNA. *Science* **293**, 1098–1102.

Huttenhofer A, Brosius J, Bachellerie JP (2002) RNomics: identification and function of small, non messenger RNAs. *Curr. Opin. Chem. Biol.* **6**, 835–843.

International Human Genome Sequencing Consortium (2001) Initial sequencing and analysis of the human genome. *Nature* **409**, 860–921.

Jeffreys AJ (1987) Highly variable minisatellites and DNA fingerprints. *Biochem. Soc. Trans.* **15**, 309–317

Jurka J, Kapitonov VV (1999) Sectorial mutagenesis by transposable elements. *Genetica* **107**, 239–248.

Kajikawa M, Okada N (2002) LINEs mobilize SINEs in the eel through a shared 3' sequence. *Cell* **111**, 433–444.

Kapranov P, Cawley SE, Drenkow J *et al.* (2002) Large-scale transcriptional activity in chromosomes 21 and 22. *Science* **296**, 916–919.

Korenberg JR, Rykowski MC (1988) Human genome organization: Alu, lines, and the molecular structure of metaphase chromosome bands. *Cell* **53**, 391–400.

Lagos-Quintana M, Rauhut R, Lendeckel W, Tuschl T (2001) Identification of novel genes coding for small expressed RNAs. *Science* **294**, 853–858.

Lagos-Quintana M, Rauhut R, Yalcin A, Meyer J, Lendeckel W, Tuschl T (2002) Identification of tissue-specific microRNAs from mouse. *Curr. Biol.* **12**, 735–739.

Lanz RB, McKenna NJ, Onate SA *et al.* (1999) A steroid receptor coactivator, SRA, functions as an RNA and is present in an SRC-1 complex. *Cell* **97**, 17–27.

Lehner B, Williams G, Campbell RD, Sanderson CM (2002) Antisense transcripts in the human genome. *Trends Genet.* **18**, 63–65.

Li W-H, Gu Z, Wang H, Nekrutenko A (2001) Evolutionary analyses of the human genome. *Nature* **409**, 847–849.

Locascio A, Vega S, de Frutos CA, Manzanares M, Nieto MA (2002) Biological potential of a functional human SNAIL retrogene. *J. Biol. Chem.* **277**, 38803–38809.

Lower R, Lower J, Kurth R (1996) The viruses in all of us: characteristics and biological significance of human endogenous retrovirus sequences. *Proc. Natl Acad. Sci. USA* **93**, 5177–5184.

Maniatis T, Tasic B (2002) Alternative pre-mRNA splicing and proteome expansion in metazoans. *Nature* **418**, 236–243.

Martignetti JA, Brosius J (1993) BC200 RNA: a neural RNA polymerase III product encoded by a monomeric Alu element. *Proc. Natl Acad. Sci. USA* **90**, 11563–11567.

Mefford HC, Trask BJ (2002) The complex structure and dynamic evolution of human subtelomeres. *Nature Rev. Genet.* **3**, 91–102.

Mourelatos Z, Dostie J, Paushkin S *et al.* (2002) miRNPs: a novel class of ribonucleoproteins containing numerous microRNAs. *Genes Dev.* **16**, 720–728.

Mouse Genome Sequencing Consortium (2002) Initial sequencing and comparative analysis of the mouse genome. *Nature* **420**, 520–562.

Nei M, Rogozin IB, Piontkivska H (2000) Purifying selection and birth-and-death evolution in the ubiquitin gene family. *Proc. Natl Acad. Sci. USA* **97**, 10866–10871.

Nissen P, Hansen J, Ban N, Moore PB, Steitz TA (2000) The structural basis of ribosome activity in peptide bond synthesis. *Science* **289**, 920–930.

Ostertag EM, Kazazian HH Jr (2001) Biology of mammalian L1 retrotransposons. *Annu. Rev. Genet.* **35**, 501–538.

Prak EL, Kazazian HH Jr (2000) Mobile elements and the human genome. *Nature Rev. Genet.* **1**, 134–144.

Regnier V, Meddeb M, Lecointre G *et al.* (1997) Emergence and scattering of multiple neurofibromatosis (NF1)-related sequences during hominoid evolution suggest a process of pericentromeric interchromosomal transposition. *Hum. Mol. Genet.* **6**, 9–16.

Schmid CW (1998) Does SINE evolution preclude Alu function? *Nucl. Acids Res.* **26**, 4541–4550.

Schmid SR, Linder P (1992) D-E-A-D protein family of putative RNA helicases. *Mol. Microbiol.* **6**, 283–291.

Singer MF (1982) Highly repeated sequences in mammalian genomes. *Int. Rev. Cytol.* **76**, 67–112.

Smit AF (1996) The origin of interspersed repeats in the human genome. *Curr. Opin. Genet. Dev.* **6**, 743–748.

Smit, AF (1999) Interspersed repeats and other mementos of transposable elements in mammalian genomes. *Curr. Opin. Genet. Dev.* **9**, 657–663.

Smith CM, Steitz JA (1997) Sno storm in the nucleolus: new roles for myriad small RNPs. *Cell* **89**, 669–672.

Smith TF, Gaitatzes C, Saxena K, Neer EJ (1999) The WD-repeat: a common architecture for diverse functions. *Trends Biochem. Sci.* **24**, 181–184.

Storz G (2002) An expanding universe of noncoding RNAs. *Science* **296**, 1260–1263.

Swallow DM, Gendler S, Griffiths B, Corney G, Taylor-Papadimitriou J, Bramwell ME (1987) The human tumour-associated epithelial mucins are coded by an expressed hypervariable gene locus PUM. *Nature* **328**, 82–84.

Tennyson CN, Klamut HJ, Worton RG (1995) The human dystrophin gene requires 16 hours to be transcribed and is co-transcriptionally spliced. *Nature Genet.* **9**, 184–190.

Toguchida J, McGee TL, Paterson JC, Eagle JR, Tucker S, Yandell DW, Dryja TP (1993) Complete genomic sequence of the human retinoblastoma susceptibility gene. *Genomics* **17**, 535–543.

Tycowski KT, Shu M-D, Steitz JA (1993) A small nucleolar RNA is processed from an intron of the human gene encoding ribosomal protein S3. *Genes Dev.* **7**, 1176–1190.

Tyler-Smith C, Willard HF (1993) Mammalian chromosome structure. *Curr. Opin. Genet. Dev.* **3**, 390–397.

Ullu E, Tschudi C (1984) Alu sequences are processed 7SL RNA genes. *Nature* **312**, 171–172.

Valadkhan S, Manley JL (2001) Splicing-related catalysis by protein-free snRNAs. *Nature* **413**, 701–707.

Venter JC, Adams MD, Myers EW *et al.* (2001) The sequence of the human genome. *Science* **291**, 1304–1351.

Wahls WP, Wallace LJ, Moore PD (1990) Hypervariable minisatellite DNA is a hotspot for homologous recombination in human cells. *Cell* **60**, 95–103.

Yang Z, Zhu Q, Luo K, Zhou Q (2001) The 7SK small nuclear RNA inhibits the CDK9/cyclin T1 kinase to control transcription. *Nature* **414**, 317–322.

Young JM, Friedman C, Williams EM, Ross JA, Tonnes-Priddy L, Trask BJ (2002) Different evolutionary processes shaped the mouse and human olfactory receptor gene families *Hum. Mol. Genet.* **11**, 535–546.

Zhang MQ (1998) Statistical features of human exons and their flanking regions. *Hum. Mol. Genet.* **7**, 919–932.

Zhang MQ (2002) Computational prediction of eukaryotic protein-coding genes. *Nature Rev. Genet.* **3**, 698–709.

Zhang Z, Harrison P, Gerstein M (2002) Identification and analysis of over 2000 ribosomal protein pseudogenes in the human genome. *Genome Res.* **12**, 1466–1482.

CHAPTER TEN

Human gene expression

Chapter contents

10.1 An overview of gene expression in human cells

The control mechanisms used to regulate human gene expression may be more complex than those in lower eukaryotes but many of the same basic principles apply. Mammals are particularly complex multicellular organisms and so it is perhaps unsurprising that some of the mechanisms controlling mammalian gene expression are not used in bacteria or in some other eukaryotes. Expression is restricted spatially and temporally (*Box 10.1*). Although simplistic, it is convenient to consider three broad levels at which gene regulation can operate:

▶ *transcriptional regulation of gene expression.* Primary control of gene regulation in eukaryotes occurs at the level of initiation of transcription, through the core promoter of a gene, and at the level of recruitment and processivity of the relevant RNA polymerase. Gene expression is initiated by binding of transcription factors to the promoter. Basal transcription levels can be modulated by binding of protein factors to other regulatory regions in nearby flanking or intronic sequences;

▶ *post–transcriptional regulation of gene expression.* This includes mechanisms operating at the level of RNA processing, mRNA transport, translation, mRNA stability, protein processing, protein targeting, protein stability and so on. At the level of RNA processing, different mechanisms such as RNA splicing (which is really a *co-transcriptional mechanism*) enable a gene to generate a variety of different gene products (*isoforms*). Translation regulation often involves recognition of regulatory sequences in untranslated regions by *trans*-acting protein factors;

▶ *epigenetic mechanisms and long range control of gene expression.* Genetic factors depend on changes in DNA

Box 10.1: Spatial and temporal restriction of gene expression in mammalian cells.

SPATIAL RESTRICTION OF GENE EXPRESSION

Housekeeping genes need to be expressed in essentially all types of nucleated cells because they encode a key product that is required to fulfill a general function in all cells, e.g. protein synthesis, energy production, etc. Many eukaryotic genes, however, show much more restricted **tissue-specific gene expression** patterns. Spatial restriction of gene expression can occur at different levels:

▶ *multiple organ/tissue pattern*. In some cases, a gene may be performing a similar type of role in different organ systems. A variety of genes which play a crucial role in early development may be involved in regulating target genes in several different organ systems e.g. the sonic hedgehog gene is expressed in various parts of the developing nervous system, in developing limbs, and elsewhere. In other cases a gene can encode different variants (*isoforms*) in different tissues by using tissue-specific promoters or tissue-specific alternative splicing. In some cases these may have different functions (Section 10.3.2);

▶ *specific tissue, cell lineage or cell type*. Some genes have a function which is appropriate for a particular cell type or cell lineage, as in the case of the β-globin gene which is expressed in erythroid cells;

▶ *individual cells*. Some specialized genes produce different products in *individual* cells belonging to the same cell type. For example, the different B lymphocytes in a person express different (*cell-specific*) antibody molecules, the different T cells produce different T cell receptors, and individual olfactory neurons produce different olfactory receptors. *Note:* there may also be variation in expression between different cells of the same type in a tissue as a result of *randomized monoallelic expression* (as in the case of *X-chromosome inactivation*—Section 10.5.6);

▶ *intracellular distribution*. The proteins of different genes are transported to different intracellular (or extracellular) locations. In some cases, different isoforms of the same protein may be sent to different intracellular locations (Section 10.3.2). In addition, gene control mechanisms are required to send mRNA for some genes to different intracellular locations (Section 10.2.6).

TEMPORAL RESTRICTION OF GENE EXPRESSION

▶ *Cell cycle stage*. In addition to general large-scale gene silencing as the chromosomes condense at mitosis, some genes are only expressed at specific times in the cell cycle. For example, many histone genes are expressed only at the S (DNA synthesis) phase, and the expression of various cell cycle regulators is programmed to occur at specific cell cycle stages.

▶ *Developmental stage*. At the very earliest stages of development transcription does not occur; instead cells rely on previously synthesized RNA. Later in development some genes may be expressed transiently at specific stages. Some gene families contain members that are expressed at different developmental stages as in the case of globin genes (*Figures 10.22, 10.23*).

▶ *Differentiation stage*. As cells differentiate, their genomes are modified resulting in altered gene expression patterns. In some terminally differentiated cells, transcription does not occur. The genome modifications that result in the progression to a nucleated adult somatic cell used to be thought to be irreversible until the birth of Dolly, the cloned sheep (Section 20.2.2).

▶ *Inducible expression*. Some genes are activated in response to environmental cues or extracellular signaling from other cells (Section 10.2.5). Such gene expression is easily reversed if the inducing factor is removed.

sequence. Additional changes that are heritable (from cell to daughter cell, or from parent to child) *but that do not depend on changes in genome sequence* are described as **epigenetic.** DNA methylation is one of a very few epigenetic mechanisms known to operate in mammalian cells, where it plays a very important part in gene regulation (Bird, 2002). In addition to acting as a general method of maintaining repression of transcription, it also is crucially involved in mechanisms operating on some genes to ensure that only one of the two parentally inherited alleles is normally expressed (**monoallelic expression**), even although the nucleotide sequence of the nonexpressed allele may be identical to that of the expressed allele. Epigenetic mechanisms often result in the perpetuation of altered chromatin conformation over long distances.

Table 10.1 provides an overview of the different types of mechanism known to be involved in regulating expression of human genes.

10.2 Control of gene expression by binding of *trans*-acting protein factors to *cis*-acting regulatory sequences in DNA and RNA

Much of the control of gene expression (whether it occurs at the level of transcription initiation, RNA processing, translation, RNA transport etc.) involves the binding of protein factors to regulatory nucleic acid sequences. The latter can be DNA sequences found in the vicinity of the gene or even within it, or can be transcript sequences at the level of precursor RNA or mRNA. As the protein factors engaged in regulating gene expression are themselves encoded by distantly located genes, *they are required to migrate to their site of action*, and so are called **trans-acting** factors. In contrast, regulatory sequences to which they bind are typically **cis-acting** because they are on the *same* DNA or RNA molecule as the gene or RNA transcript that is being regulated.

Table 10.1: Overview of the regulation of gene expression in human cells

Selective expression mechanism	Examples
Transcriptional	
Histone modification, chromatin remodeling	See Section 10.2.1
Binding of tissue-specific transcription factors to *cis*-acting elements of a single gene	See *Table 10.3*
Direct binding of hormones, growth factors or intermediates to response elements in inducible transcription elements	cAMP response elements, steroid hormone response elements etc. (see *Table 10.4*; *Figure 10.10*)
Use of alternative promoters in a single gene	See *Figure 10.14* for dystrophin gene; *Figure 10.20* for *Dnmt1* gene
Post-transcriptional	
Alternative splicing	Section 10.3.2; *Figures 10.15 and 10.16*
Alternative polyadenylation	Section 10.3.2; *Figure 10.15E*
Tissue-specific RNA editing	Section 10.3.3; *Figure 10.17*
Translational control mechanisms	Section 10.2.6; *Figure 10.13*; *Figure 9.6*.
Epigenetic mechanisms/long-range control by chromatin structure	
Allelic exclusion	DNA rearrangements in B and T lymphocytes which produce cell-specific immunoglobulins and T-cell receptors (Section 10.6.1)
	Inactivation by the *XIST* gene product of many genes on the one X chromosome on which it is expressed in female cells (Section 10.5.6)
	Random allelic exclusion by unknown mechanisms e.g. *IL-2*, *IL-4*, *PAX5* etc. (Section 10.5.3; *Box 10.4*)
	Imprinting of certain genes (Sections 10.5.4, 10.5.5)
Long range control by chromatin structure	Competition for enhancers or silencers (e.g. in globin expression; see Section 10.5.2. and Section 10.5.5)
	Position effects (Section 10.5.1)
Cell position-dependent short-range signaling	Section 10.4.1

Control by DNA–protein binding

In eukaryotic cells a major gene expression control point is at the level of transcription *initiation*. Chromatin is a highly organized and densely packed structure which does not easily afford access to RNA polymerases and so *chromatin remodeling* enzymes are required to alter the folding and basic structure of chromatin, making it a more open and active structure that will enable transcription to take place. Three different types of RNA polymerase are known to transcribe different classes of genes (see *Table 1.4*), and in each case are large enzymes, consisting of 8–14 subunits.

The RNA polymerases transcribe genes following binding of proteins (*transcription factors*) to specific regulatory DNA sequences within the gene or in its vicinity. The transcription factors may be of two kinds:

▶ *general transcription factors* are required for transcription of most promoters for a specific RNA polymerase class (that is, the general transcription factors are mostly unique to a polymerase class although at least one factor, the **TATA box–binding protein** is common to all three polymerase classes). In the case of polypeptide-encoding genes, for example, a particular polymerase, RNA polymerase II, works in concert with associated general transcription factors to produce a basal level of transcription;

▶ *specialized transcription factors* can modulate basal expression levels and include **tissue-specific transcription factors** and factors which are concerned with the transcription of specific gene sets. Proteins which bind to specific DNA sequences and cause stimulation of transcription are known as **activators** (or **trans-activators**). Those with an antagonistic effect of silencing transcriptional activity are known as **repressors.**

In addition to transcription factors which can activate or repress gene expression following DNA binding, a wide variety of regulatory proteins affect gene expression but *do not bind to DNA itself, but rather to the transcription factors themselves*. Such proteins are known as **coactivators** or **corepressors** (depending on whether they bind to transcriptional activators or repressors – see Lemon and Tjian, 2000).

Control by RNA–protein binding

In addition to transcription factors, RNA-binding proteins are used to regulate gene expression. The most well-studied examples involve binding to regulatory sequences in the untranslated sequences of mRNA, permitting translational control of gene expression. In addition, specific RNA–protein binding interactions are expected to be involved in the control of gene expression at the level of differential RNA processing too, as in the case of binding of SR and HnRNP proteins to pre-mRNA in order to modulate the choice of exons in splicing. The latter mechanisms are considered separately in Section 10.3 in order to illustrate the tremendous complexity of expression mechanisms that can be used to decode single genes, and the significance of the huge numbers of isoforms that can be produced as a result.

10.2.1 Histone modification and chromatin remodeling facilitate access to chromatin by DNA-binding factors

Transcriptionally inactive DNA is organized in a condensed chromatin structure where there is a tight association between the core histones and the DNA. The dense packing of nucleosomes can deny access to a variety of different proteins which are required to interact with DNA, including not just proteins involved in gene expression, but also factors required for DNA replication, DNA repair and so forth. Local chromatin structure can be reversibly changed from a condensed to a more accessible conformation by histone modification and chromatin remodeling.

Histone modification

Histones are organized in nucleosomes in such a way that their tails (notably the N-terminal ends) protrude from the histone octamer. The exposed N-terminal tails of the four core histones are very highly conserved in sequence, and perform crucial functions in regulating chromatin structure. Histones H3 and H4 in particular are well known to contain certain amino acids which are subject to modification (see Goll and Bestor, 2002 and *Figure 10.1*). The overall resulting pattern of histone residue modifications in specific regions of the chromatin has been suggested to form a **histone code** which dictates a particular biological outcome (see Berger, 2002; Turner, 2002).

Histone acetylation is the most well investigated of the histone modifications (which also include methylation, phosphorylation and ubiquitinylation). Various **histone acetyltransferases (HATs)** are known to catalyze the addition of acetyl (CH_3CO-) groups to the $\epsilon-NH_3+$ amino group on the side chains of up to 13 of the 30 lysine residues at the exposed N-terminal tails of a histone octamer. HATs function as *transcriptional co-activators*. In part, this may be due to the loss of positive charges when the charged lysine side chains are modified, resulting in reduced affinity between histones and DNA. The modified histones are also targets for chromatin remodeling machines (see below). The net effect is that RNA polymerase and transcription factors find it easier to access the promoter region. **Histone deacetylases (HDACs)** have the opposite effect of removing acetyl groups, and promoting transcriptional repression. HDACs are thought to be recruited as part of a corepressor complex *in response to DNA methylation* (Section 10.4.3; see *Figure 10.21*).

Histone methylation is carried out by different classes of methyltransferases targeting certain arginine and certain lysine residues in histones. Histone arginine methylation is known to be involved in transcriptional activation but histone lysine methylation can be a signal for transcriptional repression, as in methylation of H3K9 (histone 3, lysine at position 9) which may be induced following deacetylation of the same residue (Section 10.4.2; see *Figure 10.21*).

Chromatin remodeling

ATP-dependent **chromatin remodeling complexes** use ATP hydrolysis to change temporarily the structure of

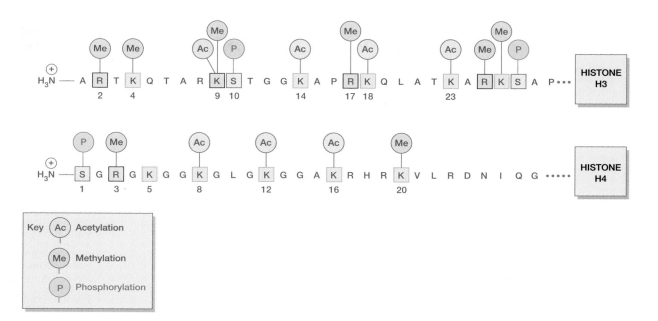

Figure 10.1: Histone H3 and H4 modifications.

Modifications of N-terminal residues of histones 3 and 4 (which are part of the exposed tails when in the histone octamer) are shown. Note that H3K9 (the lysine residue at position number 9 of histone H3) can be acetylated and also methylated. Acetylation is associated with transcriptionally active chromatin but if the region of chromatin is methylated at CpG, proteins which bind to CpG can recruit histone deacetylases which will lead to removal of the acetyl residues, and the H3K9 residue then becomes methylated by a histone methyltransferase bound to the CpG-binding proteins. The methylated H3K9 residue is then a target for binding by proteins which induce chromatin to condense (see *Figure 10.21*).

nucleosomes (see Narlikar *et al.*, 2002). By remodeling chromatin, various proteins may be afforded more ready access to DNA. Even after the remodeling complex has dissociated from the DNA, the nucleosomes can remain for some time in a remodeled state in which the DNA–histone contacts have been loosened. In some cases, the remodeling can involve altered *positioning* of the nucleosomes, causing them to slide along the DNA. As a result of nucleosome sliding, sequences previously wrapped round histone octamers become more accessible. In addition, some remodeling complexes can restore the transcriptionally inactive state.

Chromatin remodeling complexes come in different varieties, each containing multiple (typically > 10) subunits. In addition to carrying ATPase domains, they exhibit other domains which interact with modified histones, *permitting concerted action between histone modification and chromatin remodeling.* They include the **bromodomains** of the SWI/SNF family, which are known to interact with acetylated lysine residues in histones, and the **chromodomains** of the Mi-2 family which interact with methylated lysines (see Berger, 2002; Narlikar *et al.*, 2002).

10.2.2 Ubiquitous transcription factors are required for transcription by RNA polymerases I and III

RNA polymerases I and III in eukaryotic cells are dedicated to transcribing genes to give RNA molecules (rRNA, tRNA etc.) which assist in expression of the polypeptide-encoding

genes. The transcribed genes are housekeeping genes as rRNA and tRNA are required in essentially all cells to assist in protein synthesis. As a result, ubiquitous transcription factors are required to assist RNA polymerases I and III.

Transcription by RNA polymerase I

RNA polymerase I is confined to the nucleolus and is devoted to transcription of the 18S, 5.8S and 28S rRNA genes. The latter are consecutively organized on a common 13-kb *polycistronic transcription unit* (see *Figure 10.2*). A compound unit of the 13-kb transcription unit and an adjacent 27-kb nontranscribed *spacer* is tandemly repeated about 30–40 times on the short arms of each of the five human acrocentric chromosomes, at the *nucleolar organizer regions.* The resulting five clusters of rRNA genes, each about 1.5 Mb long, are referred to as **ribosomal DNA** or **rDNA**.

Initiation of transcription of the 28S, 5.8S and 18S rRNA genes follows binding of two transcription factors to a core promoter element at the transcription initiation site and an *upstream control element* located over 100 nucleotides upstream. One of the transcription factors, **UBF (*upstream binding factor*)**, is a homodimer and its identical subunits may bind first to the core promoter element and upstream control element, bringing them together so that they can be bound by the second factor, SL1 (*selectivity factor 1*; known in mouse as TIF-1B; see *Figure 10.3*). The bound transcription factors subsequently recruit RNA polymerase I to form an initiation complex.

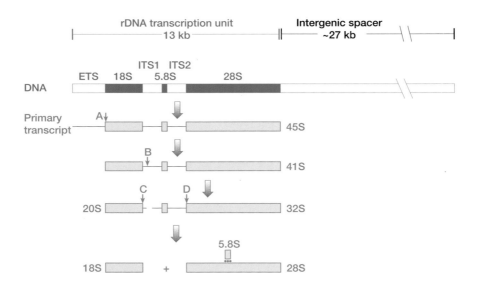

Figure 10.2: The major human rRNA species are synthesized by cleavage from a common 13-kb transcription unit which is part of a 40-kb tandemly repeated unit.

Small arrows indicated by letters A–D signify positions of endonuclease cleavage of RNA precursors. Cleavage of the 41S precursor at B generates two products: 20S + 32S. Following cleavage of the 32S precursor at D, and excision of the small 5.8S rRNA, hydrogen bonding takes places between the 5.8S rRNA and a complementary central segment of the 28S rRNA. The approximately 6 kb of RNA sequence originating from the external and internal transcribed spacer units (ETS, ITS1 and ITS2) are degraded in the nucleus. S is the sedimentation coefficient, a measure of size.

The primary transcript expressed from the single 13-kb transcription unit is a 45S precursor rRNA which undergoes a variety of cleavage reactions and base-specific modifications (carried out by a large number of different types of **small nucleolar RNA**) to generate the mature 28S, 5.8S and 18S rRNA species (see *Figure 10.2*). Thus, these genes differ from the vast majority of nuclear genes, which are transcribed individually. Instead, rDNA transcription resembles mtDNA transcription (Section 9.1.2; *Figure 9.2*): both result in multigenic transcripts which yield functionally related products. This unusual use of polygenic primary *transcripts* is no different in principle, however, from the way in which a single primary *translation product* is occasionally cleaved to generate two or more functionally related polypeptides (see the example of human insulin in *Figure 1.23*).

Transcription by RNA polymerase III

RNA polymerase III is also involved in transcription of a variety of housekeeping genes, encoding various small stable RNA molecules such as 5S rRNA, tRNA molecules, 7SL RNA and some of the snRNA molecules needed for RNA splicing. These genes are characterized by promoters that lie *within* the coding sequence of the gene, rather than upstream of it. A **bipartite promoter** is found in tRNA genes, consisting of two well conserved sequences, the **A box** and the **B box**, but the promoter of the 5S rRNA gene consists of a single element, the **C box**. In each case, transcription by RNA polymerase III is thought to proceed by binding of ubiquitous transcription factors to the promoter elements,

followed by subsequent binding of other factors and finally recruitment of the polymerase (*Figure 10.4*).

10.2.3 Transcription by RNA polymerase II requires complex sets of *cis*-acting regulatory sequences and tissue-specific transcription factors

RNA polymerase II is responsible for transcribing all genes which encode polypeptides and also certain species of snRNA gene. RNA polymerase II (like RNA polymerases I and III) is dependent on auxiliary general transcription factors, and several are known which can be complex in structure. For example, TFIID consists of the **TATA box–binding protein, TBP** (which is also found in association with RNA polymerases I and III) plus various TBP-associated factors, or **TAF proteins** (see also *Table 10.2*). The complex of polymerase and general transcription factors is known as the **basal transcription apparatus** and is all that is required to *initiate* transcription. Genes are constitutively expressed at a minimum rate determined by the *core promoter* (see below) unless the rate of transcription is increased or switched off by additional positive or negative regulatory elements (which may be located some distance away *or be intrinsic components of the promoter region itself*).

Some of the genes encoding polypeptides are *housekeeping genes*, but unlike the genes transcribed by RNA polymerases I and III, a large percentage of genes transcribed by RNA polymerase II show tissue-restricted or tissue-specific expression patterns. Since the DNA in different nucleated cells of an

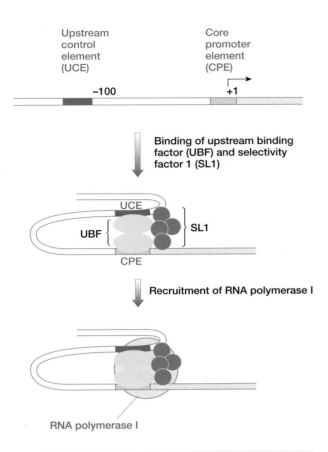

Figure 10.3: Initiation of transcription by RNA polymerase I.

One possible model envisages initial binding of the two identical subunits of the upstream binding factor to the upstream control element and the core promoter element. This forces these two sequences to come into close proximity, enabling their subsequent binding by the selectivity factor 1 (SL1; four subunits). The stabilized structure permits binding of other factors (not shown) and subsequently RNA polymerase I.

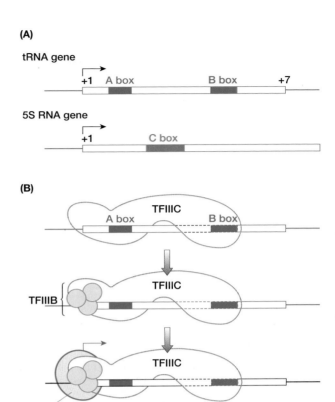

Figure 10.4: tRNA and 5S rRNA genes have promoters located within the coding sequence.

(A) *Positions of promoter elements in tRNA and 5S rRNA genes.* The promoter elements A and B in the tRNA genes are located in the sequences specifying the D loop and the TψCG loops respectively (see tRNA structure in *Figure 1.7B*). **(B) *Initiation of transcription of a tRNA gene.*** Binding of the TFIIIC transcription factor to the promoter elements permits subsequent binding of the trimeric TFIIIB factor to the sequence immediately upstream of the transcription start site. In response to binding of the TFIIIB factor, RNA polymerase III binds and initiates transcription. In the case of the 5S rRNA genes, a similar mechanism occurs but in this case an additional transcription factor, TFIIIA, is required to bind to the C box, and the bound TFIIIA factor permits subsequent binding of TFIIIB followed by recruitment of TFIIIC and RNA polymerase III as in the case of tRNA genes.

individual is essentially identical, the identity of a cell, whether it be a hepatocyte or a T lymphocyte for instance, is very largely defined by the proteins made by the cell. In addition to general ubiquitous transcription factors, therefore, tissue-specific or tissue-restricted transcription factors regulate the expression of many genes which encode polypeptides, by recognizing and binding specific *cis*-acting sequence elements.

Partly because of the large size of mammalian nuclear genomes and also because of the general need for more sophisticated control systems imposed by having very large numbers of interacting genes, control elements in eukaryotic cells are quite elaborate. Often, regulation of expression of individual human genes is controlled by several sets of *cis*-acting regulatory elements. While the individual regulatory elements may be composed of multiple short sequence elements (typically 4–8 nucleotides long) distributed over a few hundred base pairs, the different classes of regulatory

element which modulate the expression of a single gene may be located at considerable distances. A variety of different types of *cis*-acting elements can be recognized, including promoters, enhancers, silencers, boundary elements (insulators) and response elements (see *Box 10.2* and *Figures 10.5, 10.6*).

Tissue specificity and developmental stage specificity of gene expression are often conferred by enhancer and silencer sequences and a variety of *cis*-acting sequences have been identified which are specifically recognized by **tissue-**

Table 10.2: Subunit composition of eukaryotic RNA polymerase II and associated general transcription factors

Factor	Number of subunits	Functions
Polymerase II	12	Catalyzes RNA synthesis
TFIID	12	Recognizes the core promoter and supplies a scaffold upon which the rest of the general transcriptional machinery can assemble. Consists of the TATA-binding protein (TBP) and several TBP-associated factors (TAFs)
TFIIB	1	Binds TBP, selects start site and recruits polymerase II
TFIIA	3	Stablizes binding of TFIIB and TBP
TFIIF	2	Binds polymerase II and TFIIB
TFIIE	2	Recruits TFIIH which contains a *helicase* and a *protein kinase*
TFIIH	9	Unwinds DNA at promoter; phosphorylates the C-terminal domain of polymerase II, resulting in a conformational change. The activated polymerase disengages from the general transcription factors and acquires new proteins to help it transcribe over long distances without dissociating from the DNA

Note: the nomenclature for general transcription factors uses the common prefix TF (transcription factor) followed by a Roman numeral for the associated RNA polymerase.

specific transcription factors. For example, specific expression in erythroid cells is often signaled by one of two sequences: TGACTCAG (or its reverse complement CTGAGTCA; both are recognized by the erythroid-specific transcription factor NF-E2); or the sequence [A/T]GATA[A/G] (or its reverse complement, both being recognized by the GATA series of transcription factors. See *Figure 10.7* for examples, and *Table 10.3* for other tissue-specific *cis*-acting elements).

In addition to actively promoting tissue-specific transcription, some *cis*-acting silencer elements confer tissue or developmental stage specificity *by blocking expression in all but the desired tissue*. For example, the neural restrictive silencer element (NRSE) represses expression of several genes in all tissues other than neural tissues (Schoenherr *et al.*, 1996). A transcription factor that binds to the NRSE and which is variously called NRSF (neural restrictive silencer factor) or REST (RE-1 silencing transcription factor) is ubiquitously expressed in non-neural tissue and neuronal precursors during early development but subsequently it is specifically *not expressed* in more mature (postmitotic) neurons. NRSF/REST appears to be able to recruit a co-repressor, co-REST, which appears to serve to recruit molecular machinery which can induce transcriptional silencing across chromosomal regions containing neuronally expressed genes (Lunyak *et al.*, 2002).

10.2.4 Transcription factors contain conserved structural motifs that permit DNA binding

Transcription factors recognize and bind a short nucleotide sequence, usually as a result of extensive complementarity between the surface of the protein and surface features of the double helix in the region of binding. Although the *individual* interactions between the amino acids and nucleotides are weak (usually hydrogen bonds, ionic bonds and hydrophobic interactions), the region of DNA–protein binding is typically characterized by about 20 such contacts, which *collectively* ensure that the binding is strong and specific. In human and other eukaryotic transcription factors, two distinct functions can often be identified and located in different parts of the protein:

▶ an **activation domain** activates transcription of the target genes once the transcription factor has bound to it. Activation domains are thought to stimulate transcription by interacting with basal transcription factors so as to assist the formation of the transcription complex on the promoter. Although not so well studied as DNA-binding domains, some are known to be rich in aspartate and glutamate residues (*acidic activation domains*); others are rich in proline or glutamate;

▶ a **DNA-binding domain** permits specific binding of the transcription factor to its target genes. Several conserved structural motifs have been identified which are common to many different transcription factors with quite different specificities, including the leucine zipper, helix–loop–helix, helix–turn–helix, and zinc finger motifs which are described below. Each of the motifs uses α-helices (or occasionally β-sheets; see *Figure 1.24*) to bind to the major groove of DNA. Clearly, although the motifs in general provide the basis for DNA binding, the precise collection of sequence elements in the DNA-binding domain will provide the basis for the required sequence-specific recognition. Most transcription factors bind to DNA as *homodimers*, with the DNA-binding region of the protein usually distinct from the region responsible for forming dimers.

The leucine zipper motif

The **leucine zipper** is a helical stretch of amino acids rich in leucine residues (typically occurring once every seven amino acid residues, i.e. once every two turns of the helix – see *Figure*

Box 10.2: Classes of *cis*-acting sequence elements involved in regulating transcription of polypeptide-encoding genes.

PROMOTERS are combinations of short sequence elements (usually located in the immediate upstream region of the gene – often within 200 bp of the transcription start site) which serve to *initiate* transcription. They can be subdivided into different components.

▶ The **core promoter** directs the basal transcription complex to initiate transcription of the gene. In the absence of additional regulatory elements it permits constitutive expression of the gene, but at very low (basal) levels. Core promoter elements are typically located very close to the transcription initiation site, at about nucleotide positions –45 to +40 (see Butler and Kadonaga, 2002). As illustrated in *Figure 10.5*, they include: the *TATA box* located at position ca. –25, surrounded by GC-rich sequences and recognized by the *TATA-binding protein* subunit of TFIID); the *BRE sequence* located immediately upstream of the TATA element at around –35 and recognized by the TFIIB component; the *Inr (initiator) sequence* located at the start site of transcription and bound by TFIID; the *DPE* or ***D**ownstream **P**romoter **E**lement*, located at about position +30 relative to transcription and recognized by TFIID.

▶ The **proximal promoter region** is the sequence located immediately upstream of the core promoter, usually from –50 to –200 bp (promoter elements found further upstream would be said to map to the *distal promoter region*). Additional ***non-core promoter elements*** are typically located in the proximal promoter region (although they can be found in the core promoter region too). They include: **GC box**es (also called ***Sp1 boxes***, the consensus sequence is GGGCGG which is often found in multiple copies within 100 bp of the transcription initiation site, and is bound by the ubiquitous Sp1 transcription factor); **CCAAT boxes** typically located at position –75. CCAAT boxes are recognized by CTF (CCAAT-binding Transcription Factor; also known as NF-I for Nuclear Factor I) and by CBF (CCAAT box-Binding Factor; also known as NF-Y for Nuclear Factor Y). ***Note*** that CCAAT and GC boxes serve to *modulate* the basal transcription of the core promoter and operate as *enhancer* sequences (see next section), while *silencer elements* (see below) may also be integral components of the promoter.

ENHANCERS are positive transcriptional control elements which are particularly prevalent in the cells of complex eukaryotes such as mammals but which are absent or very poorly represented in simple eukaryotes such as yeast (see Martin, 2001). They serve to increase the basal level of transcription which is initiated through the core promoter elements. Their functions, unlike those of the core promoter, are independent of both their orientation and, to some extent, their distance from the genes they regulate. Enhancer elements can be distantly located from the genes they regulate (e.g. the β-globin *locus control region* – *Figure 10.23*). Enhancers often contain within a span of only 200–300 bp, several elements recognized by ubiquitous transcription factors plus several recognized by tissue-specific transcription factors (*Figure 10.7*). In addition, some enhancer elements may be integral components of promoters, as in the case of the CCAAT and GC boxes (see above).

SILENCERS serve to reduce transcription levels. Although less well-studied, two classes have been distinguished: ***classical silencers*** (also called *silencer elements*) are position-independent elements that direct an active transcriptional repression mechanism; ***negative regulatory elements*** are position-dependent elements that result in a passive repression mechanism (see Ogbourne and Antalis, 1998). Where studied in human genes, silencer elements have been reported in various positions: close to the promoter, some distance upstream and also within introns. However, the evidence for such sequences often relies on *in vitro* DNA-binding studies and their significance *in vivo* is still uncertain.

BOUNDARY ELEMENTS (INSULATORS) are regions of DNA, often spanning from 0.5 kb to 3 kb, which function to block (or *insulate*) the spreading of the influence of agents that have a positive effect on transcription (enhancers) or a negative one (silencers, heterochromatin-like repressive effects). See Bell *et al.* (2001).

RESPONSE ELEMENTS modulate transcription in response to specific external stimuli. They are usually located a short distance upstream of the promoter elements (often within 1 kb of the transcription start site). A variety of such elements respond to specific hormones (e.g. retinoic acid or steroid hormones such as glucocorticoids) or to intracellular second messengers such as cyclic AMP (see Section 10.2.5 and *Table 10.4*).

10.8), which readily forms a dimer. Each monomer unit consists of an amphipathic α-helix (hydrophobic side groups of the constituent amino acids face one way; polar groups face the other way, see *Figure 1.24*). The two α-helices of the individual monomers join together over a short distance to form a *coiled-coil* (see Section 1.5.5) with the predominant interactions occurring between opposed hydrophobic amino acids of the individual monomers. Beyond this region the two α-helices separate, so that the overall dimer is a Y-shaped structure. The dimer is thought to grip the double helix much like a clothes peg grips a clothes line (see *Figure 10.9*). In addition to forming homodimers, leucine zipper proteins can occasionally form heterodimers depending on the compatibility of the hydrophobic surfaces of the two different monomers. Such heterodimer formation provides an important combinatorial control mechanism in gene regulation.

The helix–loop–helix motif

The **helix–loop–helix (HLH) motif** is related to the leucine zipper and should be distinguished from the *helix–turn–helix (HTH)* motif described in the next section. It consists of two α-helices, one short and one long, connected by a flexible loop. Unlike the short turn in the HTH motif, the loop in the HLH motif is flexible enough to permit folding back so that the two helices can pack against each

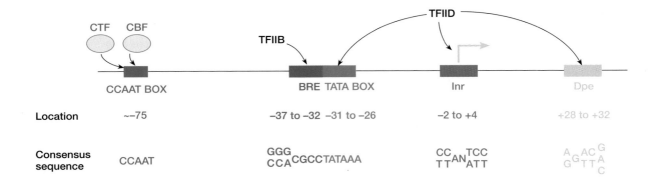

Figure 10.5: Conserved locations in complex eukaryotes for regulatory promoter elements bound by ubiquitous transcription factors.

Note: the core promoter of individual genes need not contain all elements. For example, many promoters lack a TATA box and use instead the functionally analogous initiator (INR) element. GC boxes are usually found in promoters too but their locations are more variable (see *Figure 10.6* and *Figure 1.13* for some examples). BRE, TFIIB recognition element; DPE, downstream promoter element. Adapted from Butler and Kadonga (2002) *Genes Dev.* **16**, 2583–2592 with permission from Cold Spring Harbor Laboratory Press.

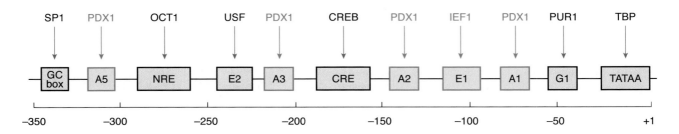

Figure 10.6: The human insulin gene promoter contains a variety of sequence elements recognized by ubiquitous and tissue-specific transcription factors.

Arrows indicate binding of transcription factors (top row) to regulatory sequence elements (boxed) present upstream of the human insulin gene. Ubiquitous or widely expressed transcription factors are shown in black; those shown in red are specific for pancreatic beta cells. The PDX1 transcription factor binds to four sequence motifs of the form C(C/T)TAATG which are present in the insulin promoter (A1, A2, A3, A5). Abbreviations: CRE, cAMP response element; NRE, negative regulatory element.

Table 10.3: **Examples of *cis-acting* sequences recognized by tissue-restricted and tissue-specific transcription factors**

Consensus binding sequence	Transcription factor	Expression pattern
(A/T)GATA(A/G)	GATA-1,-2, etc.	Erythroid cells
TGACTCAG	NF-E2	Erythroid cells
GTTAATNATTAAC (= PE element)	HNF-1	Liver, kidney, stomach, intestine, spleen
T(G/A)TTTG(C/T)	HNF-5	Liver
ATGCAAAT	POU2F2 (OTF-2)	Lymphoid cells
GCCTGCAGGC	Ker1	Keratinocytes
(C/T)TAAAAATAA(C/T)3	MBF-1	Myocytes
(C/T)TA(A/T)AAATA(A/G)	MEF-2	Myocytes
CAACTGAC	MyoD	Myoblasts + myotubes
C/A)A(C/A)AG	TCF-1	T cells

TCGACCCTCTGGAACCTATCAGGGACCACAGTCAGCCAGGCAAGCACATC
← GATA-1

TGCCCAAGCCAAGGGTGGAGGCATGCAGCTGTGGGGGTCTGTGAAAACAC
← CACC box

GATA-1 →   NF-E2 →
TTGAGGGAGCAGATAACTGGGCCAACCATGACTCAGTGCTTCTGGAGGCC

AACAGGACTGCTGAGTCATCCTGTGGGGGTGGAGGTGGGACAAGGGAAAG
← NF-E2   ← CACC box

GATA-1 →
GGGTGAATGGTACTGCTGATTACAACCTCTGGTGCTGCCTCCCCCTCCTG
← CACC box

TTTATCTGAGAGGGAAGGCCATGCCCAAAGTGTTCACAGCCAGGCTTCAG
← GATA-1

Figure 10.7: The HS-40 α-globin regulatory site contains many recognition elements for erythroid-specific transcription factors.

Note that the HS-40 site appears to be a *locus control region* for the α-globin gene cluster (Section 10.5.2).

other, that is the two helices lie in planes that are parallel to each other, in contrast to the two helices in the HTH motif (*Figure 10.8*). The HLH motif mediates both DNA binding and protein dimer formation (see *Figure 10.9*) and it permits occasional heterodimer formation. In the latter case, however, heterodimers form between a full-length HLH protein and a truncated HLH protein which lacks the full length of the α-helix necessary to bind to the DNA. The resulting heterodimer is unable to bind DNA tightly. As a result, HLH heterodimers are thought to act as a control mechanism, by enabling *inactivation* of specific gene regulatory proteins.

The helix–turn–helix motif

The **HTH motif** is a common motif found in homeoboxes, and a number of other transcription factors. It consists of two short α-helices separated by a short amino acid sequence which induces a turn, so that the two α-helices are orientated differently (i.e. the two helices do not lie in the same plane, unlike those in the HLH motif; *Figure 10.8*). The structure is very similar to the DNA-binding motif of several bacteriophage regulatory proteins such as the λ cro protein whose binding to DNA has been intensively studied by X-ray crystallography. In the case of both the λ cro protein and eukaryotic HTH motifs, it is thought that while the HTH motif in general mediates DNA binding, the more C-terminal

helix acts as a specific **recognition helix** because it fits into the major groove of the DNA (*Figure 10.9*), controlling the precise DNA sequence which is recognized.

The zinc finger motif

The **zinc finger motif** involves binding of a zinc ion by four conserved amino acids so as to form a loop (finger), a structure which is often tandemly repeated. Although several different forms exist, common forms involve binding of a Zn^{2+} ion by two conserved cysteine residues and two conserved histidine residues, or by four conserved cysteine residues. The resulting structure may then consist of an α-helix and a β-sheet held together by co-ordination with the Zn^{2+} ion, or of two α-helices. In either case, the primary contact with the DNA is made by an α-helix binding to the major groove. The so-called C2H2 (Cys_2/His_2) zinc finger typically comprises about 23 amino acids with neighboring fingers separated by a stretch of about seven or eight amino acids (*Figure 10.8*).

10.2.5 A variety of mechanisms permit transcriptional regulation of gene expression in response to external stimuli

In eukaryotic cells gene expression can be altered in a semi-permanent way as cells differentiate, or in a temporary, easily reversible way in response to extracellular signals (**inducible gene expression**). Environmental cues such as the extracellular concentrations of certain ions and small nutrient molecules, temperature shock and so on can result in dramatic alteration of gene expression patterns in cells exposed to changes in these parameters. In complex multicellular animals there are also fundamental requirements for cells to communicate with each other and different modes of *cell signaling* are possible (Sections 3.2.1, 3.2.2). In some cases, alteration of gene expression is conducted at the translational level which can offer certain advantages. In other cases, gene expression is altered by modulating transcription.

Transcriptional regulation in response to cell signaling can take different forms, *but the end-point is always the same*: a previously inactive transcription factor is specifically activated and then subsequently binds to specific regulatory sequences located in the promoters of target genes, thereby modulating their transcription. In the case of transcription regulated by signaling molecules or their intermediaries, such regulatory sequences are often referred to as **response elements** (see *Table 10.4*).

Ligand-inducible transcription factors

Small hydrophobic hormones and morphogens such as steroid hormones, thyroxine and retinoic acid are able to diffuse through the plasma membrane of the target cell and to bind intracellular receptors in the cytoplasm or nucleus. These receptors (often called *hormone nuclear receptors*) are inducible transcription factors: following binding of the homologous ligand, the receptor protein associates with a specific DNA response element located in the promoter regions of perhaps 50–100 target genes and activates their transcription.

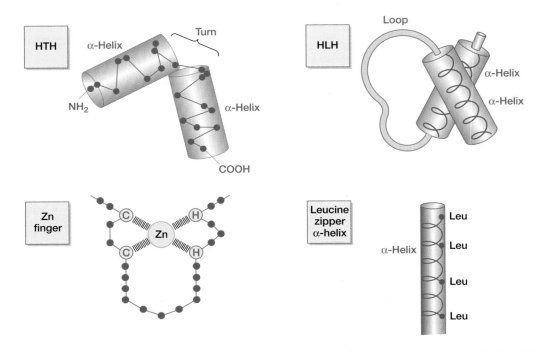

Figure 10.8: Structural motifs commonly found in transcription factors and DNA-binding proteins.

Abbreviations: HTH, helix–turn–helix; HLH, helix–loop–helix. **Note** that the leucine zipper monomer is *amphipathic* [i.e. has hydrophobic residues (leucines) consistently on one face of the helix, see Figure 1.24]. Two such helices can align with their hydrophobic faces in opposition to form a coiled-coil structure.

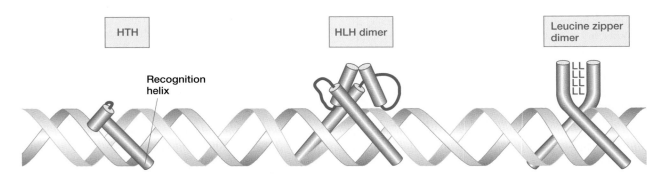

Figure 10.9: Binding of conserved structural motifs in transcription factors to the double helix.

Note: the individual monomers of the helix–loop–helix (HLH) dimer and the leucine zipper dimer are colored differently to permit distinction, but may be identical (homodimers). HLH heterodimers and leucine zipper heterodimers may provide a higher level of regulation (see text).

Although thyroxine and retinoic acid are structurally and biosynthetically unrelated to the steroid hormones, their receptors belong to a common nuclear receptor superfamily. Two conserved domains characterize the family: a centrally located ***DNA-binding domain*** of about 68 amino acids, and an approximately 240 amino acid ***ligand-binding domain*** located close to the C terminus (*Figure 10.10*). The DNA-binding domain contains zinc fingers and binds as a dimer with each monomer recognizing one of two hexanucleotides in the response element. The two hexanucleotides are either inverted repeats or direct repeats which are typically separated by three or five nucleotides (*Figure 10.10*). In the absence of the ligand, the receptor is inactivated by direct repression of the DNA-binding domain function by the ligand-binding domain, or by binding to an inhibitory protein, as in the case of the glucocorticoid receptor (*Figure 10.11*).

Table 10.4: Examples of response elements in inducible gene expression

Consensus Response Element (R.E.)	Response to	Protein factor which recognizes R.E.
(T/G)(T/A)CGTCA	cAMP	CREB (also called ATF)
CC(A/T)(A/T)(A/T)(A/T)(A/T)(A/T)GG	Serum growth factor	Serum response factor
TTNCNNNAAA	Interferon-gamma	Stat-1
TGCGCCCGCC	Heavy metals	Mep-1
TGAGTCAG	Phorbol esters	AP1
CTNGAATNTTCTAGA	Heat shock	HSP70, etc.

Note: see also hormone response elements in *Figure 10.10*.

Activation of transcription factors by signal transduction

Unlike lipid-soluble hormones or morphogens, hydrophilic signaling molecules such as polypeptide hormones, cannot diffuse through the plasma membrane. Instead, they bind to a specific receptor on the cell surface. After binding of the ligand molecule, the receptor undergoes a conformational change and becomes activated in such a way that it passes on the signal via other molecules within the cell (*signal transduction* – see Section 3.2.1).

Many cell surface receptors have a kinase activity or can activate intracellular kinases (see Table 3.3), and signal transduction pathways are often characterized by complex regulatory interplay between *kinases* and *phosphatases* which can activate or repress intermediates by phosphorylation/ dephosphorylation. In many cases, the phosphorylation or dephosphorylation induces an altered conformation. In the case of activation of a signaling molecule, the altered conformation often means that a signaling factor is no longer inhibited by some repressor sequence present in an inhibitory protein to which it is bound, or in a domain or sequence motif within its own structure.

In terms of transcriptional activation, two general mechanisms permit rapid transmission of signals from cell-surface receptors to the nucleus, both involving *protein phosphorylation*: protein kinases are activated and then translocated from the cytoplasm to the nucleus where they phosphorylate target transcription factors; inactive transcription factors stored in the cytoplasm are activated by phosphorylation and translocated into the nucleus. The following two sections provide examples to illustrate the above two mechanisms (see also Karin and Hunter, 1995).

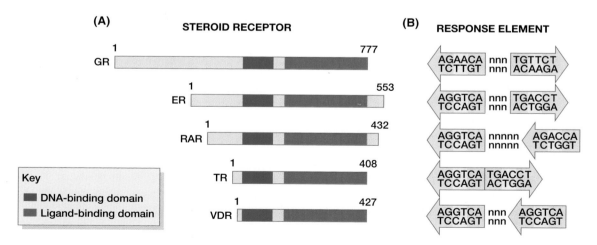

Figure 10.10: Steroid receptors and the respective response elements.

(A) *Structure of members of the nuclear receptor superfamily*. Numbers refer to protein size in amino acids. ER, estrogen receptor; GR, gluocorticoid receptor; PR, progesterone receptor; RAR, retinoic acid receptor; TR, thyroxine receptor; VDR, vitamin D receptor.
(B) *Response elements. Note:* (i) that the response elements are often perfect inverted hexanucleotide repeats, but that the response elements for retinoic acid and vitamin D$_3$ are imperfect direct hexanucleotide repeats; (ii) the hexanucleotides all have the general sequence AGNNCA with the central two nucleotides (shown by shading) conferring specificity and belonging to one of three classes: GT, AC or AA.

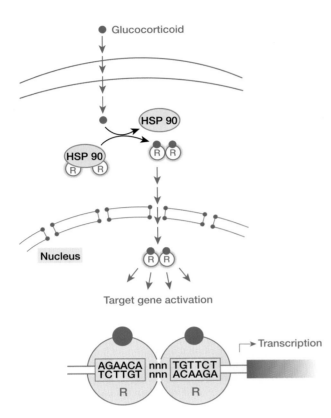

Figure 10.11: Transcriptional regulation by glucocorticoids.

The glucocorticoid receptor is normally inactivated by being bound to an inhibitor protein, Hsp90. Binding of glucocorticoids to the glucocorticoid receptor releases Hsp90, the receptor dimerizes and then activates selected genes which have a glucocorticoid response element in their promoter (see *Figure 10.10*).

Hormonal signaling through the cyclic AMP pathway

Cyclic AMP is an important **second messenger** (see *Table 10.5*) which acts in response to a variety of hormones and other signaling molecules. It is synthesized from ATP by a membrane bound enzyme, adenylate cyclase. Hormones which activate adenylate cyclase bind to a cell surface receptor which is of the *G protein-coupled receptor* class. Binding of the hormone to the receptor promotes the interaction of the receptor with a G protein which consists of three subunits, α, β and γ. Following this interaction the α subunit of the G protein is activated, causing it to dissociate and stimulate adenylate cyclase.

The increase in intracellular cAMP produced by activated adenylate cyclase can then activate the transcription of specific target sequences that contain a *cAMP response element* or CRE. This function of cAMP is mediated by the enzyme *protein kinase A*. Cyclic AMP binds to protein kinase and activates it by permitting release of the two catalytically active subunits which then enter the nucleus and phosphorylate a specific transcription factor, *CREB* (CRE-binding protein). Activated CREB then activates transcription of genes with the cAMP response element (*Figure 10.12A*).

Activation of NF-κB via tumor necrosis factor signaling

NF-κB is a transcription factor which is involved in a variety of aspects of the immune response. In its inactive state, NF–κB is retained in the cytoplasm where it is complexed with an inhibitory subunit, *IκB*. However, the latter can be targeted for degradation following phosphorylation. The consequent destruction of IκB permits *NF-κB* to translocate to the nucleus and activate its various target genes. IκB phosphorylation is achieved by a kinase cascade after tumor necrosis factor (TNF) binds to a specific cell surface TNF receptor, causing activation of TNF receptor associated factor 2 (TRAF2 – see *Figure 10.12B*).

10.2.6 Translational control of gene expression can involve recognition of UTR regulatory sequences by RNA-binding proteins

Protein synthesis is the major ultimate step of gene expression and an important control point for regulation. A complex series of proteins is involved in translation initiation in eukaryotes and different pathways for initiating translation have been uncovered (see Dever, 2002). The choice of the initiating AUG (methionine) codon is a key one but for some genes alternative choices are made for the start codon in a mRNA, yielding isoforms which vary in their N-terminal sequence – see the example of the Wilms tumor gene, *WT1* in *Figure 10.16A*. The significance of these isoforms is not well appreciated.

Table 10.5: Examples of secondary messengers in cell signaling

Secondary messenger	Characteristics
Cyclic AMP (cAMP)	Produced from ATP by adenylate cyclase. Effects are usually mediated through protein kinase A. See example of activation of CREB factor (*Figure 10.12A*)
Cyclic GMP (cGMP)	Produced from GTP by guanylate cyclase. Best-characterized role is in visual reception in the vertebrate eye
Phospholipids/Ca^{2+}	Activated downstream of G protein-coupled receptors and protein tyrosine kinases. Hydrolysis of phosphatidylinositol 4,5-bis-phosphate (PIP$_2$) yields diacylglycerol and inositol 1,4,5-trisphosphate (IP$_3$) which activate protein kinase C and mobilize Ca^{++} from intracellular stores

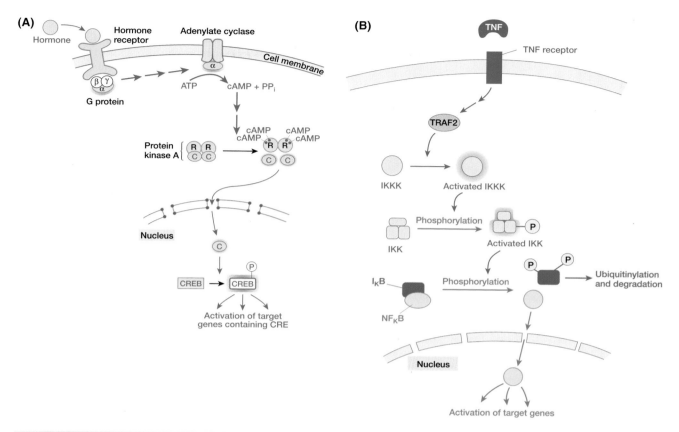

Figure 10.12: Selected target genes can be actively expressed in response to extracellular stimuli by signal transduction from activated cell surface receptors.

(A) *Activation of a protein kinase and translocation to the nucleus:* hormonal signaling through the cyclic AMP–protein kinase A signal transduction pathway. Binding of hormone to a specific cell surface receptor promotes the interaction of the receptor with a G protein. The activated G protein α subunit dissociates from the receptor and stimulates the membrane-bound adenylate cyclase to synthesize cAMP. The latter binds to the regulatory subunits of protein kinase A, enabling release of the catalytic subunits (c) which migrate to the nucleus and activate the transcription factor CREB (CRE-binding protein) by phosphorylation. Activated CREB binds to cAMP response elements in the promoters of target genes. **(B)** *Activation of a cytoplasmic transcription factor (NF-κB) and translocation to the nucleus.* Binding of TNF causes a change in its membrane receptor, allowing it to recruit a number of intracellular signaling proteins. They in turn recruit and activate IKKK, a kinase which phosphorylates IKB, the IκB kinase. IKB is normally bound to NFκB but the addition of phosphate groups to IKB marks it for ubiquitinylation and degradation. The released NFκB has an exposed nuclear localization sequence which now allows it to migrate to the nucleus where it activates a set of target genes.

An increasing number of eukaryotic and mammalian mRNA species have also been shown to contain regulatory sequences in their untranslated sequences, most frequently at the 3′ end (see Wickens *et al.*, 1997). Several eukaryotic and mammalian RNA-binding proteins have also been identified and shown to bind to specific regulatory sequences present in untranslated sequences, thereby providing the basis for translational control of gene expression (Siomi and Dreyfuss, 1997). A variety of different **RNA-binding domains** have been identified and they include elements which have previously been associated with DNA-binding properties of transcription factors such as zinc fingers and homeodomains (Siomi and Dreyfuss, 1997).

RNA trafficking

The interaction between *cis*-acting regulatory elements in RNA and *trans*-acting RNA-binding proteins can be envisaged to alter RNA structure in various ways: facilitating or hindering interactions with other *trans*-acting factors; altering higher order RNA structure; bringing together initially remote RNA sequences; or providing localization or targeting signals for transport of RNA molecules to specific intracellular locations (**RNA trafficking**). Numerous eukaryotic and mammalian mRNAs are known to be transported as ribonucleoprotein (RNP) particles to specific locations within some types of cells, notably of the nervous system (see Hazelrigg, 1998). For example, tau mRNA is localized to the proximal portions of axons rather than to dendrites, where many mRNA molecules are located in mature neurons, and myelin basic protein mRNA is transported with the aid of kinesin to the processes of oligodendrocytes.

RNA trafficking may provide a more efficient way to localize proteins than simply transporting them: a single mRNA can give rise to many different protein molecules, assuming that it can engage with ribosomes. Different sequential steps have been envisaged: initial translational repression, transport within the cell, localization (to the specific subcellular destination) and then localization-dependent translation. Recently, key regulatory sequences which are required for various steps in this process have been identified in the untranslated sequences, predominantly the 3′ UTR of many mRNA species (see Wickens *et al.*, 2002).

Translational control of gene expression in response to external stimuli

Translational control of gene expression can permit a more rapid response to altered environmental stimuli than the alternative of activating transcription. Iron metabolism provides two useful examples. Increased iron levels stimulate the synthesis of the iron-binding protein, ferritin, without any corresponding increase in the amount of ferritin mRNA. Conversely, decreased iron levels stimulate the production of transferrin receptor (*TFR*) without any effect on the production of transferrin receptor mRNA. The 5′-UTR of

both ferritin heavy chain mRNA and light chain mRNA contain a single **iron–response element (IRE)**, a specific *cis*-acting regulatory sequence which forms a hairpin structure. Several such IRE sequences are also found in the 3′ UTR of the transferrin receptor mRNA (see Klausner *et al.*, 1993). Regulation is exerted by binding of IREs by a specific IRE-binding protein which is activated at low iron levels (see *Figure 10.13*).

Translational control of gene expression during early development

Gene expression during oocyte maturation and the earliest embryonic stages is regulated at the level of translation, not transcription (see De Moor and Richter, 2001). Following fertilization of a human oocyte, no mRNA is made initially until the four to eight cell stage when *zygotic transcription* is activated, that is, transcription of the genes present in the zygote. Before this time, cell functions are specified by *maternal mRNA* that was previously synthesized during oogenesis.

Extrapolation from studies in model organisms would suggest that a variety of mRNAs are stored in oocytes in an inactive form, characterized by having short oligo(A) tails. Such mRNAs were previously subject to deadenylation and the resulting short oligo(A) tail means that they cannot be translated. Subsequently, at fertilization or later in development, the stored inactive mRNA species can be activated by **cytoplasmic polyadenylation**, restoring the normal size

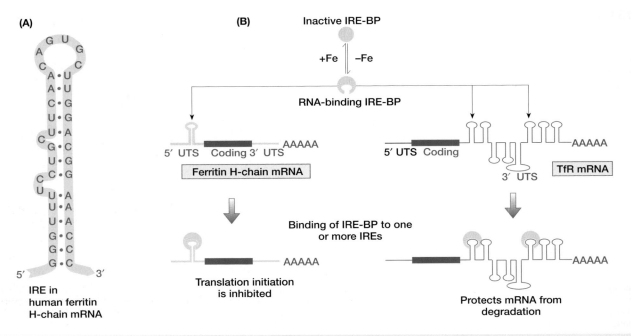

Figure 10.13: The IRE-binding protein regulates the production of ferritin heavy chain and transferrin receptor by binding to iron-response elements (IREs) in the 5′- or 3′-untranslated regions.

(A) Structure of the IRE in the 5′-UTR of the ferritin heavy chain. **(B)** Binding of the IRE-binding protein to ferritin and transferrin receptor mRNAs has contrasting effects on protein synthesis.

poly(A) tail. The same type of poly(A) polymerase activity is used as in standard polyadenylation of newly formed mRNA (which occurs in the nucleus), but in addition to the AAUAAA signal, the mRNA needs to have a uridine-rich upstream *cytoplasmic polyadenylation element* (see Wahle and Kuhn, 1997).

Two other mechanisms which regulate translation of some mRNAs during development are **translational masking** (whereby RNA-binding proteins can recognize and bind specific sequences in the 3′ UTRs of the mRNAs, thereby repressing translation – see Gray and Wickens, 1998) and **antisense regulation**. In the latter case, some mRNAs are known to be regulated by a complementary RNA sequence, as in the case of *microRNAs*, very small RNAs which have been shown in some cases to regulate genes during development by binding to complementary sequences in their 3′ UTRs (see *Figure 9.6* and Bannerjee and Slack, 2002; Pasquinelli and Ruvkun, 2002).

10.3 Alternative transcription and processing of individual genes

In addition to the control that is exerted in selecting specific *genes* (or their transcripts) for activation or repression, control mechanisms can also select between specific *alternative* transcripts of a single gene. Differential promoter usage or differential RNA processing events can result in a large number of different isoforms and these and other mechanisms have challenged the classical definition of a gene.

10.3.1 The use of alternative promoters can generate tissue-specific isoforms

Several individual mammalian genes are known to have two or more **alternative promoters**, which can result in alternative expression products (*isoforms*) with different properties

(see Ayoubi and van de Ven, 1996). Often the individual promoters drive transcription from alternative versions of a first exon which is then spliced in each case to a common set of downstream exons. In addition, however, some alternative promoters are located in more distal portions of a gene (*alternative internal promoters*) and drive expression of truncated products, as in the case of some promoters in the dystrophin gene (see below). The isoforms can provide:

▶ tissue-specificity (a frequent occurrence because different promoters can contain different regulatory elements; see the example of the human dystrophin gene below);

▶ developmental stage specificity (e.g. the insulin-like growth factor II gene);

▶ differential subcellular localization (e.g. soluble and membrane-bound isoforms);

▶ differential functional capacity (as in the case of the progesterone receptor);

▶ sex-specific gene regulation (see the case of the *Dnmt1* methyltransferase gene (Section 10.4.2; *Figure 10.20*)).

One of the most celebrated examples of differential promoter usage in humans concerns the giant dystrophin gene which comprises a total of more than 79 exons distributed over about 2.4 Mb of DNA in Xp21. At least seven different alternative promoters can be used. Three of the alternative promoters are located near the conventional start site and comprise a brain cortex-specific promoter, a muscle-specific promoter located 100 kb downstream, and a promoter which is used in Purkinje cells of the cerebellum and located a further 100 kb downstream (see *Figure 10.14*). Usage of these promoters results in large isoforms with a molecular weight of 427 kDa (referred to as Dp427 where Dp = Dystrophin protein and often given a suffix to indicate tissue specificity, for example Dp427m, to indicate the muscle specific isoform). The three Dp427 isoforms differ in their extreme N-terminal amino acid sequence as a result of using three different alternatives for exon

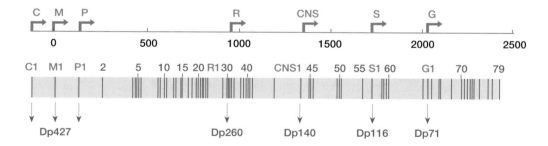

Figure 10.14: At least seven distinct promoters can be used to generate tissue- and cell type-specific expression of the dystrophin gene.

The positions of the seven alternative promoters are illustrated at the top: C, cortical; M, muscle; P, Purkinje; R, retinal (+ brain + cardiac muscle); CNS, central nervous system (+ kidney); S, Schwann cell; G, general (almost ubiquitously expressed, but undetectable in fully differentiated skeletal muscle). The approximate positions of the exons are illustrated below. *Note:* each promoter uses its own first exon (in red: C1, M1, P1, R1, CNS1, S1 and G1) together with downstream exons (in blue). The internal promoters are located immediately upstream of indicated exons as follows: R, exon 30; CNS, exon 45; S, exon 56; G, exon 63. The full-length C-, M- and P-dystrophins are about 427 kDa (Dp427). The four internal promoters R, CNS, S and G generate progressively smaller isoforms: Dp260, Dp140, Dp116, and Dp71. Alternative splicing is known to occur, notably at the 3′ end; for more information, see http://www.dmd.nl/isoforms.html.

1. In addition to the alternative promoters encoding the conventional large isoforms, at least four other alternative internal promoters can be used, generating smaller isoforms (*Figure 10.14*).

10.3.2 Human genes are prone to alternative splicing and alternative polyadenylation

The use of alternative promoters typically involves using alternative exons at the beginning of transcription, but other mechanisms, notably alternative splicing contribute to wide-scale usage of alternative exons. Unexpectedly low numbers of genes in some comparatively complex organisms (*Drosophila* and humans have respectively about 0.7 × and 1.5 × the number of genes as in a simple 1-mm-long worm, *C. elegans*) has suggested that biological complexity could be very dependent on alternative expression of genes (Maniatis and Tasic, 2002; Roberts and Smith, 2002).

Alternative splicing: prevalence and patterns

At least 50% (and quite possibly a much higher percentage) of human genes undergo **alternative splicing**, whereby different exon combinations are represented in transcripts from the same gene during RNA processing. For many genes numerous isoforms can in principle be generated at the RNA level, but it is often not clear how many of the possible alternative transcripts are biologically important (although some clearly are important – see below). Various classes of alternative splicing can occur, resulting in alternative combinations of *coding exons* and of *noncoding exons*, and in **exon length variants** which share some common sequence (see *Figure 10.15*). There are several consequences for polypeptide-encoding genes:

▸ **different protein isoforms**. This can be brought about by alternative combinations of coding exons or variant coding exons resulting in amino acid differences. Sometimes proteins can be produced which lack a whole functionally

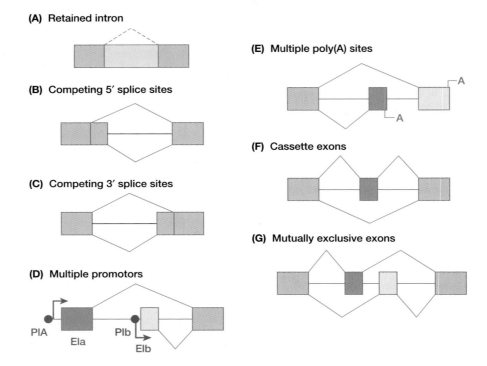

Figure 10.15: Types of alternative splicing event.

Internal *cassette exons* (F) can be included or omitted (skipped) independently of other exons (**exon skipping**) while *mutually exclusive exons* (G) occur in arrays of two or more exons, only one of which can be selected at a time for inclusion in the mature RNA. Illustrative examples of some of these events are as follows: cassette exons, exon 5 of the *WT1* gene (*Figure 10.16A*); exon length variants due to competing 5′ splice sites (B), variants of exon 9 in the *WT1* gene (*Figure 10.16A*); mutually exclusive exons, variants for exons 4, 6, 9 or 17 of the *Dscam* gene (*Figure 10.16B*). The choice of alternative promoters also introduces alternative 5′ exons (see *Figure 10.14*). Adapted from Roberts and Smith (2002) *Curr. Opin. Chem. Biol.* **6**, 375–383 with permission from Elsevier. *Note* that in addition to alternative splicing, occasional examples of **trans-splicing** have also been found in the human genome. Whereas alternative splicing brings together sequences transcribed from different combinations of exons within a *single* transcription unit on a *single DNA strand*, *trans*-splicing brings together sequences transcribed from exons belonging to different transcription units on *different DNA strand*s (see Finta and Zaphiropoulos, 2002; Maniatis and Tasic, 2002).

important domain or an important localization signal, with important functional consequences (see *Box 10.3*);

▶ *different untranslated sequences.* Alternative combinations of noncoding exons and variant noncoding exons result in different 5′ or 3′ untranslated sequences, and sometimes different polyadenylation sites (see *Figure 10.15*). The functional significance of differing untranslated sequences is often not clear, but for some genes it is a striking phenomenon – e.g. growth hormone receptor mRNA exhibits at least eight different 5′ UTR sequences as a result of alternative splicing (Pekhletsky *et al.*, 1992).

Alternative splicing can result in large numbers of potential isoforms, including up to a staggering 38 016 protein isoforms for the *Drosophila Dscam* gene (Schmucker *et al.*, 2000; see *Figure 10.16B*). Some of these alternatives have been shown to be developmentally and tissue-specifically regulated. In some well-studied genes, alternative splice forms have been shown to be extraordinarily well conserved, such as the +KTS and −KTS isoforms of the Wilms tumor *WT1* gene (*Figure 10.16A; Box 10.3*) which have been conserved in distantly related organisms such as the pufferfish.

Alternative splicing: regulation

The best understood model system for understanding the regulation of splicing is the sex determination pathway in *Drosophila* which also controls gene dosage. Alternative splicing is used in each branch of this pathway to control the expression of transcriptional regulators or chromatin-associated proteins that influence transcription, and both positive and negative control of splicing is evident (see Lopez, 1998). In mammalian cells candidate splice regulators are the **SR family** of RNA-binding proteins [which have a distinctive C-terminal domain rich in serine (S)-arginine (R) dipeptides] and some HnRNP (heterogeneous nuclear ribonucleoprotein particle) proteins. These proteins are known to promote various steps in assembly of spliceosomes and they are also known to bind to **splicing enhancer sequences**, regulatory sequences which can enhance splice site recognition (see Blencowe 2000; Caceres and Kornblihtt, 2002; Fairbrother *et al.* 2002).

Alternative polyadenylation

The usage of alternative polyadenylation signals is also quite common in human mRNA, and different types of **alternative polyadenylation** have been identified (see Edwards-Gilbert *et al.*, 1997). In many genes two or more polyadenylation signals are found in the 3′ UTR and the alternatively polyadenylated transcripts can show tissue specificity; in other cases, alternative polyadenylation signals may be brought into play following alternative splicing.

10.3.3 RNA editing is a rare form of processing whereby base-specific changes are introduced into RNA

RNA editing is a form of post-transcriptional processing which can involve enzyme-mediated insertion or deletion of nucleotides or substitution of single nucleotides *at the RNA level*. Insertion or deletion RNA editing appears to be a peculiar property of gene expression in mitochondria of kinetoplastid protozoa (such as trypanosomes) and slime molds. **Substitution RNA editing** is frequently employed in some systems, such as the mitochondria and chloroplasts of vascular plants where individual mRNAs may undergo multiple C→U or U→C editing events.

In mammals there is no evidence for insertion or deletion RNA editing, but substitution editing has been observed in a limited number of genes (see Gerber and Keller, 2001). The RNA editing which is known to occur mostly involves *deamination* (removal of amino groups) from very select cytosine or adenine residues (catalyzed by a family of RNA-dependent deaminases), but *transamination* (modification by acquisition of an amino group) can also occur, as in the case of U→C editing in the Wilms tumor gene (see *Figure 10.16A*). The two classes of deamination-based RNA editing known are:

Box 10.3: Alternative splicing can alter the functional properties of a protein

The following, far from exhaustive, list is merely meant to illustrate some ways in which the biological properties of a protein can be altered as a result of alternative splicing. See Lopez (1998) and Graveley (2001) for further information.

▶ *Tissue-specific isoforms* e.g. tropomyosin and calcitonin isoforms (the latter includes thyroid-expressed calcitonin and the neural calcitonin gene-related peptide).

▶ *Membrane-bound and soluble isoforms.* Protein localization can be regulated by generating soluble forms of numerous membrane receptors e.g. class I and II HLA, IgM, CD8, growth hormone receptor, IL-4, IL-5, IL-7, IL, erythropoietin, G-CSF, G-MCSF, LIF (leukemia inhibitory factor), and the FAS apoptosis-signaling receptor.

▶ *Alternative intracellular localization.* A useful example is provided by the *WT1* Wilms tumor gene which specifies a protein with four zinc fingers at its C terminal and has up to 24 different isoforms. Differential splicing can lead to inclusion or omission of a sequence specifying three amino acids, KTS which governs nuclear localization (see *Figure 10.16A*).

▶ *Altered function.* The +KTS and −KTS isoforms of the *WT1* gene product also differ in their ability to bind to specific DNA sequences in target genes. The former are thought to have a role in binding splicing factors; the latter may have a more general role in binding domains that harbor general transcription factors. Other examples include: transcription factor isoforms (which activate/repress transcription depending on the nature of the domains that are included or excluded from the protein product – see Lopez, 1998); and apoptosis-promoting and apoptosis-inducing isoforms of various genes, such as the Ich-1 (caspase 2) gene.

▶ **C→U editing.** This occurs in a very few genes only, notably the human apolipoprotein gene *APOB* where it has been well studied. In the liver the *APOB* gene encodes a 14.1-kb mRNA transcript and a 4536 amino acid product, apoB100. However, in the intestine a specific cytosine deaminase, APOBEC1, converts a single cytosine at nucleotide 6666 to uridine, thereby generating a premature stop codon. The truncated (7 kb) mRNA encodes a product, apoB48, identical in sequence to the first 2152 amino acids of apoB100 (*Figure 10.17*). The *activation-induced deaminase (AID)*, an enzyme involved in immunoglobulin DNA recombination and mutation, shows considerable similarity to APOBEC1 but appears to deaminate *deoxy*cytidines (Section 10.6);

▶ **A→I editing.** This is carried out by members of the *ADAR family* of deaminases (**a**denosine **d**eaminase **a**cting on **R**NA) on mRNA encoding some ligand-gated ion channels including glutamate receptors and related proteins. An adenosine is deaminated to give **inosine (I)**, a base not normally present in mRNA (the amino group at carbon 6 of adenosine is replaced by a C=O carbonyl group). Inosine behaves like guanosine: it base–pairs preferentially with cytosine and when present in a codon is translated during protein synthesis as if it were G. In the case of the glutamate receptor B gene, for example, RNA editing replaces a CAG (glutamine) codon by CIG which is translated as if it were CGG, giving arginine. This type of editing which brings about a Gln→Arg is often referred to as **Q/R editing** (after the single letter code for the two amino acids involved).

Although its full significance remains uncertain, RNA editing is important for survival in mammals (for example, knockouts of individual mouse ADAR genes can give severe phenotypes). Perhaps it arose in evolution to correct mistakes made at the genome level, but it can also provide a form of regulation of expression and additional ways of creating protein diversity.

10.4 Differential gene expression: origins through asymmetry and perpetuation through epigenetic mechanisms such as DNA methylation

The concept of tissue specificity of human gene expression is long established. What is much less clear is how such patterns get laid down initially. Since the DNA content of all nucleated cells in an organism is virtually identical, genetic mechanisms cannot explain how differential gene expression first develops in cells. To explain this, C. H. Waddington invoked **epigenetic mechanisms** of gene control during development. Genetic mechanisms explain heritable states (characters) which result from changes in DNA sequence (mutations), but epigenetic mechanisms describe heritable states *which do not depend on DNA sequence.* In recent times, a variety of epigenetic mecha-

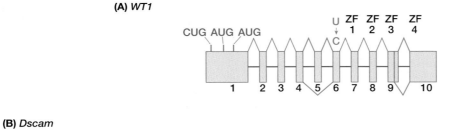

(A) WT1

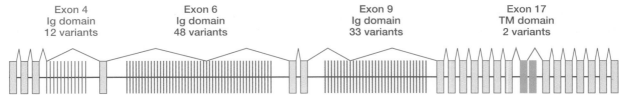

(B) Dscam

Figure 10.16: Functional differences and massive potential complexity due to alternative splicing in respectively the *WT1* Wilms tumor gene and the *Drosophila Dscam* gene.

(A) *WT1* splicing. Twenty-four isoforms are possible because of a combination of alternative usage of three different initiation codons in exon 1, a U→C RNA editing substitution in exon 6, and two alternative splicing events: variable omission (skipping) of exon 5 and also length variation for exon 9 (due to competing 5′ splice sites for intron 9). The exon 9 variation results in inclusion or omission of a KTS peptide. The +KTS isoforms are specifically localized to spliceosomal sites in the nucleus and are thought to have a role in binding splicing factors while the −KTS isoforms are more generally distributed in the nucleoplasm and may have a more general role in binding domains that harbor general transcription factors (Larsson *et al.*, 1995). **(B) *Dscam* splicing.** A total of 38 016 (12 × 48 × 33 × 2) possible isoforms can be generated by selecting from mutually exclusive variants for each of exons 4, 6, and 9 which encode immunoglobulin-like domains and for exon 17 which encodes a transmembrane region (see Schmucker *et al.*, 2000). Adapted from Roberts and Smith (2002) *Curr. Opin. Chem. Biol.* **6**, 375–383 with permission from Elsevier Science.

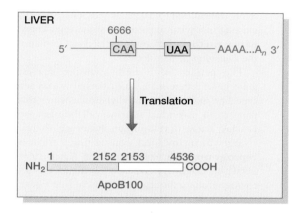

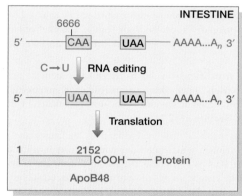

Figure 10.17: Tissue-specific RNA editing during processing of the human apolipoprotein B gene (APOB).

In the liver codon 2153 (CAA) at nucleotide positions 6666–6668 of the *APOB* mRNA specifies glutamine. In the intestine, however, C→U RNA editing at position 6666 causes replacement of the CAA codon by a stop codon, **U**AA, resulting in a shorter product, ApoB48.

nisms have been identified to operate in the cells of vertebrates, including ones which can perpetuate particular states of gene expression in somatic cell lineages.

10.4.1 Selective gene expression in cells of mammalian embryos most likely develops in response to short range cell–cell signaling events

In order to explain subsequent tissue-, cell- and developmental stage-specific patterns of expression, some mechanism is required to set up an asymmetry or axis in the fertilized egg cell or in very early development. In *Drosophila*, the egg is inherently asymmetrical because of transfer of gene products from asymmetrically sited nurse cells. The embryo develops initially as a multinucleate *syncytium* (effectively one big cell) and regionalization depends on the response of individual nuclei to long-range gradients of regulatory molecules. In mammals, however, the egg cell is relatively small and early embryonic development creates an apparently symmetrical aggregate of individual cells. Nevertheless, development becomes asymmetric.

The generation of asymmetry in mammalian cells could derive from early positional clues. Some aspects of early development are inherently asymmetrical including the point of entry of the sperm during fertilization, the attachment of the embryo to the uterine wall during implantation and the location of cells with respect to their neighbors. As the embryo develops into a ball of cells, and later on as more complex structures develop, individual cells will vary in the number of cell neighbors available. Direct cell–cell signaling or short range intercellular signaling events can provide a means of identifying cell position, and triggering differential gene expression. For example, if an intercellular signaling molecule has a range of, say, one cell diameter, then the cells at the outside of the *blastula* (Section 3.7.2; *Figure 3.13*) will receive different signals from those surrounded by neighbors on all sides, and the different positional cues may be translated into differential gene expression. As particular cell systems develop during, for example, organogenesis (mostly accomplished between the 4th and 9th embryonic weeks), particular cell type growth or differentiation factors may then induce the expression of developmental stage- and/or tissue-specific transcription factors.

10.4.2 DNA methylation is an important epigenetic factor in perpetuating gene repression in vertebrate cells

Once differential expression patterns have been set up, epigenetic mechanisms can ensure that they are stably inherited when cells divide, providing a form of **cell memory** which is transmitted through cell lineages. The epigenetic mechanisms can ensure stable inheritance of a transcriptionally activated state ('open' chromatin conformation) for some target genes or genome regions, or alternatively can organize the chromatin of some genome regions to adopt a highly condensed, transcriptionally inactive form. At least two and possibly three different epigenetic mechanisms are now considered to operate in animal development:

▶ **DNA methylation** – an epigenetic mechanism in which chromatin is organized into closed, transcriptionally inactive states (see below);

▶ **Polycomb–trithorax** gene regulation. The Polycomb group of repressors and the trithorax group of activators maintain the correct expression of several key developmental regulators (including the homeotic genes) by changing the structure of chromatin, either into a 'closed' (transcriptionally repressed) or 'open' (transcriptionally active) conformations – see Mahmoudi and Verrijzer, 2001).

▶ **histone modification** – a likely third epigenetic mechanism which may be responsible for perpetuating expression states at specific genomic locations (Turner, 2002; see Section 10.2.1).

DNA methylation is now recognized as an important epigenetic mechanism which interacts with histone modification (see Section 10.4.3) to permit the stable transmission from a diploid cell to daughter cells of chromatin states which repress gene expression (see Bird, 2002). However, the precise function of DNA methylation in eukaryotes is still imperfectly understood and clearly shows species differences (see *Box 9.3*). Vertebrate cytosine methyltransferases recognize a CpG

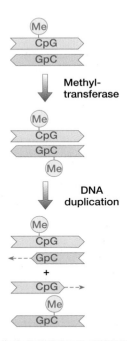

Figure 10.18: CpG methylation is perpetuated by a requirement for the specific methyl transferase to recognize a *hemimethylated* target sequence.

The sequence CpG has dyad symmetry. Following methylation of a **hemi-methylated** target (methylated on one strand only), the two methylated strands will separate at DNA duplication and act as templates for the synthesis of two unmethylated daughter strands. The resulting daughter duplexes will now provide new hemi-methylated targets for continuing the same pattern of methylation.

target sequence but unlike bacterial methylases, they show a strong preference for recognizing a *hemi-methylated* DNA target (one that is already methylated on one strand only). The sequence CpG shows *dyad symmetry* and so, following DNA replication, the newly synthesized DNA strands will receive the same CpG methylation pattern as the parental DNA (*Figure 10.18*). As a result, the CpG methylation pattern can be stably transmitted to daughter cells. The perpetuation of a pre-existing methylation pattern is sometimes known as **maintenance methylation** and is carried out in mammalian cells by the *Dnmt1 methyltransferase*.

The pattern of 5-methylcytosine distribution in the genome of differentiated somatic cells varies according to cell type but maintenance methylation ensures that methylation patterns in individual somatic cell lineages are quite stable. During early development, however, there are dramatic changes in methylation, constituting a form of **epigenetic reprogramming** (see Razin and Kafri, 1994; Reik *et al.*, 2001; Li, 2002). There are two major types of epigenetic reprogramming during development (see *Figure 10.19*).

Reprogramming in germ cells

The *primordial germ cells* of the embryo (the cells from which the gametes will ultimately derive) start off with highly methylated DNA, but then progressive demethylation occurs during development. By the time the primordial germ cells have entered the gonads the demethylation is largely complete and will be finalized shortly afterwards. However, after gonadal differentiation, and as the germ cells begin to develop, *de novo* methylation occurs. This leads to substantial methylation of the DNA of mammalian sperm and egg cells. The sperm genome is more heavily methylated than the egg's genome, and *sex-specific differences* in methylation patterns are evident, notably at imprinted loci (see Merteneit *et al.*, 1998 for references).

Reprogramming in the early embryo

The genome of the fertilized oocyte is an aggregate of the sperm and egg genomes and so it and the very early embryo are substantially methylated with methylation differences at paternal and maternal alleles of many genes. Later on, at the morula and early blastula stages in the pre-implantation embryo, **genome–wide demethylation** occurs. Later still, at the pre-gastrulation stage, widespread **de novo methylation** is carried out. However, the extent of this methylation varies in different cell lineages: **somatic cell lineages** are heavily methylated; **trophoblast–derived lineages** (giving rise to the placenta, yolk sac etc.) are undermethylated; **early primordial germ cells** are spared; their genomic DNA remains very largely unmethylated until after gonadal differentiation (as described above).

The contribution of the different methyltransferases to *de novo* methylation is still unresolved. The *Dnmt3a* and *Dnmt3b methyltransferases* are strong candidates but are not by themselves sufficient. Instead, it appears likely that they interact with Dnmt1 methyltransferase. The *Dnmt1* gene is highly expressed in male germ cells, mature oocytes and in the early embryo, and expression is subject to sex-specific regulation as

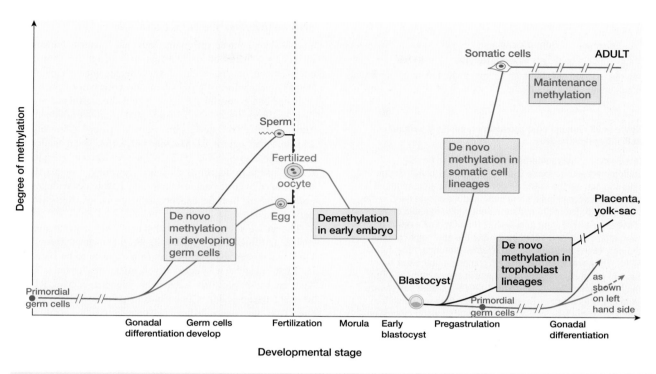

Figure 10.19: Changes in DNA methylation during mammalian development.

Developmental stages for gametogenesis and early embryo development are expanded for clarity; those for later development are contracted, as indicated by double slashes. **Note** the very rapid changes in DNA methylation during: (i) **gametogenesis** – *de novo* methylation gives rise to substantially methylated genomes in the sperm and egg (albeit with differences in both the overall level of methylation and the *pattern of methylation in these genomes* – see text); and in (ii) **the early embryo** where a wave of genome-wide demethylation occurs at the preimplantation stage (morula and early blastula), and is succeeded shortly afterwards by large-scale *de novo* methylation beginning at the pre-gastrulation stage. The latter is particularly pronounced in somatic lineages, and to a lesser extent in trophoblast lineages giving rise to placenta and yolk sac, *but does not occur in the primordial germ cells* (the cells of the embryo which will eventually give rise to sperm and egg cells).

a result of using oocyte-specific and spermatocyte-specific promoters (see *Figure 10.20*)

10.4.3 Animal DNA methylation may provide defense against transposons as well as regulating gene expression

Although not all eukaryotes appear to be subject to DNA methylation, its function in animal cells does appear to be critically important, and targeted knock-out of the cytosine methyltransferase gene in mice results in embryonic lethality. The precise function of DNA methylation in animal cells, however, remains unclear. Current views have focused in particular on two aspects of animal cells: the genome size (animals have comparatively large genomes with large numbers of genes, and also large numbers of highly repetitive DNA families belonging to the transposon class); and *the mode of development* (especially the variation in terms of life-span and rate of cell turnover). Two quite contrasting views regarding the primary function of DNA methylation in animal cells have been the subject of much controversy: the **host defense model** and the **gene regulation model.**

Host defense as a primary function for DNA methylation

Like the restriction-modification function of DNA methylation in bacteria (see *Box 5.2*), the host defense model envisages that the primary function of DNA methylation in animal cells is to confer a form of genome protection, but in this case checking the spread of transposons (Yoder *et al.*, 1997). About 45% of the DNA sequence in the human genome can be classified as belonging to transposon families and a small fraction of these sequences in the human genome is known to be actively transposing (Section 9.5). Transposon families in the human and other genomes are known to be heavily methylated (about 90% of the 5-methylcytosines are thought to be located in retrotransposon families) and so DNA methylation has been viewed as a mechanism for repressing such transposition, which if left unchecked could be expected to be damaging to cells. However, recently obtained data from an invertebrate chordate, *Ciona intestinalis*, appear to be inconsistent with the genome defense model: multiple copies of an apparently active retrotransposon and a large fraction of highly repeated SINEs were predominantly unmethylated, while genes, by contrast, appeared to be methylated (Simmen *et al.*, 1999).

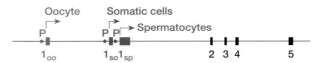

Figure 10.20: Sex-specific promoters regulate the *Dnmt1* methyl transferase gene.

The *Dnmt1* methyltransferase gene appears to be the predominant maintenance DNA methyltransferase in mouse cells and may be the major *de novo* methyltransferase too. It is highly expressed in male germ cells, mature oocytes and in the early embryo. There are five exons but with a variable choice between three different possibilities for exon 1 as a result of using alternative promoters (much like the alternative promoter choices for the different p427 dystrophin isoforms – see *Figure 10.14*). They are: **exon 1so** (used in <u>so</u>matic cells); **exon 1o** (used in oocytes); and **exon 1sp** (used in <u>sp</u>ermatocytes). The oocyte-specific exon is associated with the production of very large amounts of active Dmnt1 methyltransferase, which is truncated at the N terminus and sequestered in the cytoplasm during the later stages of growth. The spermatocyte-specific exon interferes with translation and prevents production of Dmnt1 during the crossing over stage of male meiosis. See Mertineit *et al.* (1998).

Gene regulation as the primary function for DNA methylation

DNA methylation in vertebrates has been viewed as a mechanism for silencing transcription and may constitute a default position. DNA sequences which are transcriptionally active require to be unmethylated (at least at the promoter regions). While DNA methylation in invertebrates may serve to repress transposons and other repeated sequence families, it may have acquired a special role in vertebrates as a mechanism for regulating expression of endogenous genes and reducing transcriptional noise (by silencing of a large fraction of genes whose activity is not required in a cell).

The counter argument is that the methylation status of the 5′ regions of tissue-specific genes cannot be correlated with expression in different tissues, and that the role of methylation in gene expression is in specialized biological functions resulting from mechanisms (e.g. imprinting etc.) which use *allele-specific* gene expression (Walsh and Bestor, 1999).

DNA methylation and gene expression

The DNA of transcriptionally active and inactive chromatin differs in a number of features including the degree of compaction and the extent of its methylation (see *Table 10.6*). While methylation of CpG islands downstream of promoters does not block continued transcription through these regions (Jones, 1999), there is no doubt that methylated promoter regions are correlated with transcriptional silencing. In addition, the extent of **histone acetylation** is an important factor (see also Section 10.2.1). Specific histone acetyltransferases add acetyl groups to lysine residues close to the N terminus of histone proteins, inducing a more open chromatin conformation; histone deacetylation promotes repression of gene expression.

The processes of DNA methylation and histone modification are linked (see Li, 2002). Repression at methylated CpG sequences in promoter regions appears to be mediated by proteins which specifically bind to methylated CpG. Two of these proteins have been identified, MeCP1 and MeCP2 (**me**thylated **C**pG-binding **p**roteins 1 and 2), and the latter has been shown to be essential for embryonic development and to function as a transcriptional repressor. MeCP2 silences gene expression partly by recruiting HDAC activity, resulting in chromatin remodeling. Removal of the acetyl group from H3K9 (histone 3, lysine 9 residue) is followed by MeCP2-aided methylation of H3K9 and the resulting histone methylation is a signal for recruiting proteins such as HP1 which cause the chromatin to become condensed (see Fuks *et al.*, 2002 and references therein and *Figure 10.21*).

10.5 Long range control of gene expression and imprinting

10.5.1 Chromatin structure may exert long-range control over gene expression

Unlike bacterial genes, eukaryotic genes are usually *individually transcribed*. Promoters and related upstream elements typically control expression of a single gene with a transcription start point located within 1 kb of the control element. Some *cis*-acting elements, however, exert *long-range control* over a much larger chromosomal region and there is increasing evidence for co-ordinate regulation of gene clusters.

Table 10.6: Features associated with transcriptionally active and inactive chromatin

Feature	Transcriptionally active chromatin	Transcriptionally inactive chromatin
Chromatin structure	Open, extended conformation	Highly condensed conformation; particularly apparent in heterochromatin (both facultative heterochromatin and constitutive heterochromatin)
DNA methylation	Relatively unmethylated, especially at promoter regions	Methylated, including at promoter regions
Histone acetylation	Acetylated histones	Deacetylated histones

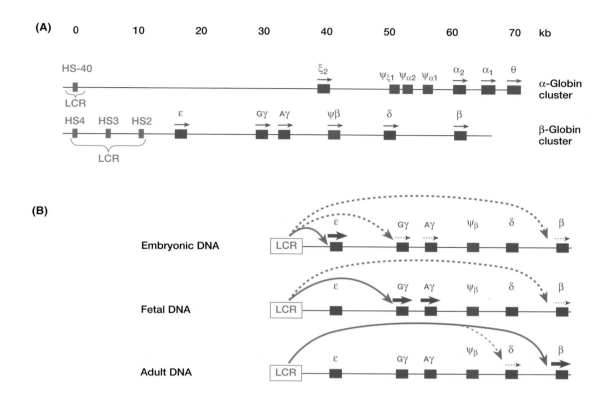

Figure 10.23: Gene expression in the α- and β-globin gene clusters can be controlled by common locus control regions.

(A) *Organization of the human α- and β-globin gene clusters.* The locus control regions (LCRs) consist of one or more erythroid-specific DNase I-hypersensitive sites (HS-40, etc.) located upstream of the cluster. Arrows mark the direction of transcription of expressed genes. The functional status of the θ-globin gene is uncertain: it is expressed, but θ-globin is not incorporated into any hemoglobin molecule.
(B) *Proposed regulation of gene expression by the β-globin LCR.* The strong red arrows indicate a powerful enhancer effect by the LCR on the indicated genes, resulting in a high expression level; dotted red arrows indicate correspondingly weak effects. There has been some controversy regarding this model – see text.

relevant gene is said to exhibit **functional hemizygosity**: only one half of the maximum gene product is normally obtained *even although the sequences of both parental alleles are perfectly consistent with normal gene expression and may even be identical*. In some cases the allelic exclusion may be a property of select cells or tissues while in other cells of the same individual both alleles may be expressed normally.

Although initially considered a rarity, **monoallelic expression** of biallelic genes has been demonstrated for a growing number of human genes. A variety of different expression mechanisms can be involved and two broad classes of mechanism are involved (see Chess, 1998; Ohlsson *et al.*, 1998):

▶ *allelic exclusion according to parent of origin* (*imprinting*). In some cases, the choice of which of the two inherited copies is expressed is *not random*. This means that for some genes the allele whose expression is repressed is always the paternally inherited allele; in others it is always the maternally inherited allele (Section 10.5.4);

▶ *allelic exclusion independent of parent of origin*. Here the decision as to which of the two alleles is repressed is

initially made randomly, but afterwards that pattern of allelic exclusion is transmitted stably to daughter cells following cell division. A variety of different mechanisms may be involved (*Box 10.4*).

10.5.4 Genomic imprinting involves differences in the expression of alleles according to parent of origin

Various observations in mammals have suggested that the maternal and paternal genomes in an individual are not equivalent (*Box 10.5*). In addition to genetic differences between the DNA of the sperm genome and the oocyte genome, there are also epigenetic differences. A major difference is in both the total amount of DNA methylation (the sperm genome is more extensively methylated than the oocyte genome) and the pattern of DNA methylation in specific DNA sequence classes. For example, LINE1 sequences are highly methylated in sperm cells but only partially methylated in the oocyte (see Razin and Kafri, 1994; Yoder *et al.*, 1997). At some individual gene loci, too, there are major differences between the extent of methylation of paternal and maternal alleles. For example,

Box 10.4: Mechanisms resulting in monoallelic expression from biallelic genes in human cells.

Mechanism	Examples of relevant genes and cellular location of monoallelic expression
A. Allelic exclusion according to parent of origin	
Genomic imprinting	A small number of genes (see Further Reading for details of the Imprinted Gene Catalogue). Cellular locations depend on where an individual gene is expressed but note that some imprinted genes show monoallelic expression in some cell types but biallelic expression in others (*Table 10.7*)
B. Random allelic exclusion (independent of parent of origin)	
Allelic exclusion due to X-inactivation	Confined to certain X-linked genes in females only. Expression of the allele from the noninactivated X chromosome only in cells in which genes are expressed (Section 10.5.6)
Allelic exclusion following programmed DNA rearrangement	Immunoglobulin gene expression in B lymphocytes; T cell receptor gene expression in T lymphocytes (Section 10.6.3)
Allelic exclusion by unknown mechanism	Olfactory receptor genes in neurons; NK cell receptor genes; certain interleukin genes(*IL2, IL4*); *XIST* (in cells of the early female embryo); *PAX5* (in mature B cells and early progenitors)

Box 10.5: The nonequivalence of the maternal and paternal genomes

In addition to the obvious X/Y sex chromosome difference, non-equivalence between paternally and maternally inherited autosomes and X chromosomes is indicated from the observations listed below.

EXPERIMENTALLY INDUCED UNIPARENTAL DIPLOIDY IN MICE

The male pronucleus of a fertilized mouse oocyte can be removed and replaced by a second female pronucleus to generate a **gynogenote** (sometimes called a **parthenogenote**; all 38 chromosomes are of maternal origin). If, instead, the female pronucleus were to be replaced by a second male pronucleus an **androgenote** is formed. Despite having normal diploid chromosomes, such embryos fail to develop and die prior to mid-gestation. Gynogenotes show severe deficiencies in extra-embryonic structures but a relatively normal embryo; in contrast, in androgenotes the embryo is more severely affected than the extraembryonic structures (see Bestor, 1998 for references).

NATURALLY OCCURRING UNIPARENTAL DIPLOIDY IN HUMANS (see also Section 2.5.4)

Human uniparental conceptuses are not uncommon. Androgenetic conceptuses develop as **hydatidiform moles** which consist of masses of hydropic chorionic villi and other placental structures, but lack embryonic tissues. Gynogenetic conceptuses give rise to dermoid cysts which develop into **ovarian teratomas**, consisting of a mass of well-differentiated but highly disorganized adult tissues, often including bone, tooth, cartilage, skin, and other tissues but usually lacking any extraembryonic structures.

Triploid abortuses may be considered to represent a combination of a diploid genome inherited from one parent and a normal haploid genome from the other. The phenotype is different depending on which parent contributes the diploid genome.

UNIPARENTAL DISOMY (see also Section 2.5.4)

Some conceptuses have a normal 46,XX or 46,XY karyotype but may have inherited two copies of the same chromosome from just one of the two parents. This may result in abnormal phenotypes which are different according to parental origin of the relevant chromosome. For example, 46,XX or 46,XY individuals who inherit both copies of chromosome 15 from their father develop Angelman syndrome; if both copies of chromosome 15 are maternally inherited, Prader–Willi syndrome results (see *Box 16.6*).

SUBCHROMOSOMAL MUTATIONS CAUSING DIFFERENTIAL ABNORMAL PHENOTYPES ACCORDING TO PARENT OF ORIGIN

▶ Deletion of certain chromosomal regions produces a different phenotype when on the maternal or paternal chromosome. The best example is deletion of 15q12, which on the paternal chromosome produces Prader–Willi syndrome and on the maternal chromosome produces Angelman syndrome (*see Box 16.6*).

▶ Certain human characters are autosomal dominant but manifest only when inherited from one parent. In some families glomus tumors are inherited as an autosomal dominant character, but expressed only in people who inherit the gene from their father. Beckwith–Wiedemann syndrome (MIM 130650) is sometimes dominant but expressed only by people who inherit it from their mother. Example pedigrees are shown in *Figures 4.5D, 4.5 E*.

▶ Allele loss in many cancers (Chapter 17) preferentially involves the paternal allele.

the paternal allele of the *H19* gene is heavily methylated; the maternal allele is undermethylated.

As suggested by the observations in *Box 10.5*, differences between the paternal and maternal genomes lead to differences in expression between paternal and maternal alleles. **Genomic imprinting** (also called *gametic* or *parental imprinting*) in mammals describes the situation where there is nonequivalence in expression of alleles at certain gene loci, dependent on the parent of origin (Reik *et al.*, 2001; Sleutels and Barlow, 2002). In all (or at least some) of the tissues where the gene is expressed, the expression of either the paternally inherited allele or the maternally inherited allele, is consistently repressed, resulting in monoallelic expression. The same pattern of monoallelic expression can be faithfully transmitted to daughter cells following cell division. However, as the nucleotide sequence of the allele whose expression is repressed may be perfectly consistent with gene expression (and may even be identical to that of the expressed allele), this is an epigenetic phenomenon, not a genetic one.

Prevalence and evolution of imprinting

Most human genes are not subject to imprinting, otherwise we would not see so many simple Mendelian characters. Systematic surveys have been made to identify imprinted chromosomal regions in the mouse. Unlike in humans, all the mouse chromosomes are acrocentric and Robertsonian translocations can permit crosses to be set up which produce offspring having both copies of one particular chromosome derived from a single parent (**uniparental disomy, UPD** – Section 2.5.4). These reveal that UPD for some chromosomes has no phenotypic effect; for others it produces abnormal phenotypes. The abnormal phenotypes are sometimes complementary for different parental origins, for example overgrowth is often seen in maternal UPD and growth retardation in paternal UPD. For some chromosomes, UPD is lethal.

Further dissection at the chromosomal and genetic level shows that imprinting is a property of a limited number of individual genes or small chromosomal regions. Currently, a total of about 60 genes are known to be imprinted in humans and mice, including genes which specify polypeptides and also genes encoding functional noncoding RNA (see Tycko and Morrison, 2002). The known physiological functions of such genes can show considerable variation, but suggestive evidence indicates a considerable number of genes controlling growth and neurobehavioral traits. Two major clusters of imprinted genes are known in the human genome: a 1-Mb region at 11p15.5 (encompassing the Beckwith–Wiedemann syndrome region) containing at least eight imprinted genes (see Maher and Reik, 2000); and a 2.2-Mb cluster at 15q11-q13 region (encompassing the Prader–Willi and Angelman syndrome regions) and containing more than 10 imprinted genes (Meguro *et al.*, 2001; *Figure 10.24*).

The great majority of known imprinted genes are autosomal. However, the *XIST* gene which has a major role in establishing X chromosome inactivation (see next section) may be considered an example of an imprinted X-linked gene since expression of the maternally inherited allele is preferentially repressed in trophoblast. An imprinted X-linked gene which affects cognitive function has also been suggested from differential behavior patterns in Turner syndrome. Girls with Turner syndrome lack a Y chromosome but have only one X chromosome. If the X chromosome is inherited from the mother, socially disruptive behavior is common, but if inherited from the father, the girl shows behavior closer to normal (Skuse *et al.*, 1997).

Imprinting is known to occur in seed plants, some insects and mammals. No major imprinting effect, as judged by phenotype, has been observed in some model organisms such as *Drosophila, C. elegans* and the zebrafish, although the potential for imprinting may exist in *Drosophila*. Mammals are unusual in the way in which embryos are totally dependent on flow of nutrients from the maternal placenta. As many imprinted genes are involved in regulating fetal growth, one explanation envisages **parental genome conflict**: the paternal genome propagates itself best by creating an embryo which aggressively removes nutrients from the mother; the maternal genome suppresses this to protect the mother and spare some resources for future offspring. As seen in cases of uniparental diploidy (see *Box 10.5*), paternal genes are preferentially expressed in the trophoblast and extra-embryonic membranes, while maternal genes are preferentially expressed in the embryo.

10.5.5 The mechanism of genomic imprinting is unclear but a key component appears to be DNA methylation

To confirm imprinting of a gene it is necessary to identify an individual who is heterozygous for a sequence variant present in the mature mRNA; mRNA from different tissues can then be checked for monoallelic or biallelic expression, and the origin of each allele determined by typing the parents. For some genes, this type of analysis has shown that imprinting is confined to only certain tissues or to certain stages of development (see *Table 10.7*). Thus, imprinting allows an extra level of control of gene expression, but it is not possible to compress its functioning into a simple uniform story.

Imprinted genes are usually found to be organized as gene clusters, harboring **imprint control elements,** *cis*-acting regulatory elements which act over long distances. Within a single such cluster it is usual to find some genes which show preferential paternal expression in close proximity to other genes which show preferential maternal expression (see *Figure 10.24* for examples). Major imprint control elements have been found to be restricted to small DNA regions known as **imprinting centers**. In the case of the Prader–Willi syndrome (PWS)/Angelman syndrome (AS) region at 15q11-q13 an imprinting center has been defined upstream of, and extending into, the 5′ end of the *SNURF-SNRPN* gene. It has two imprint control elements: the **PWS-SRO** element at the promoter and 5′ end of *SNURF-SNRPN* which is responsible for establishing and maintaining the paternal imprint; and the **AS-SRO** element located about 35 kb upstream of *SNURF-SNRPN*, which is responsible for the maternal imprint (see Perk *et al.*, 2002).

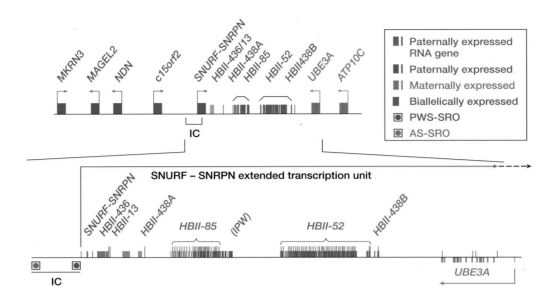

Figure 10.24: The Prader–Willi syndrome/Angelman syndrome-associated imprinted gene cluster at 15q11-q13.

Arrows show direction of transcription. Imprinted polypeptide-encoding genes include *UBE3A* and *ATP10C,* which are preferentially expressed from maternal chromosome 15 and *MKRN3, NDN,* and *MAGEL2,* preferentially expressed from paternal chromosome 15.
In addition, the complex *SNURF/SNRPN* transcription unit (> 148 exons; and extending more than 460 kb to overlap the *UBE3A* gene and possibly the *ATP10C* gene) is imprinted. It encodes two proteins (the SNURF protein; encoded by exons 1–3 and the SNRPN spliceosomal protein encoded by exons 4–10) plus some RNA transcripts (not shown here, but known in the literature as *IPW, PAR5, UBE3AS* etc.) and may regulate the *UBE3A* and *ATP10C* genes as an antisense RNA regulator. In addition to this complexity a total of 79 snoRNA sequences are located within the introns of the *SNURF/SNRPN* transcription unit, comprising one copy each of *HBII-436, HBII-13, HBII-437* (a likely pseudogene), two identical but separated copies of *HBII-438,* 27 copies of *HBII-85* and 47 copies of *HBII-52* (shown as large vertical lines compared to the small vertical lines for the *SNURF/SNRPN* exons – see Runte *et al.,* 2001). Almost all of these copies (except *HBII-437*) are paternally expressed, especially in brain, and they may have a direct role in Prader–Willi syndrome (Gallagher *et al.,* 2002). IC, imprinting center. Adapted from Runte *et al.* (2001) *Hum. Mol. Genet.* **10** (23): 2687–2700, by permission of Oxford University Press.

Imprinted gene clusters usually also contain genes specifying untranslated RNAs whose expression often correlates with repression of nearby polypeptide-encoding genes. The PWS-AS cluster provides many examples but other examples are known such as the *H19* gene in 11p15.5. The functions of imprinted RNA genes in proximity to imprinted polypeptide-encoding genes, although by and large unknown,

are expected to be regulatory ones, and direct evidence is available in at least one case: the mouse *Air* gene has been shown to regulate the imprinted *Igf2r* (insulin growth factor II receptor gene – see Rougeulle and Heard, 2002; Sleutels *et al.,* 2002).

The above observations suggest that some mechanism must be able to distinguish between maternally and paternally

Table 10.7: Examples of tissue and developmental stage regulation of imprinted genes in mammals

Gene	Repressed allele	Differences in expression patterns
IGF2 (insulin-like growth factor 2)	Maternal	Imprinted in many tissues but biallelic expression in brain, adult liver, chondrocytes etc.
PEG1/MEST	Maternal	Imprinted in fetal tissue but biallelically expressed in adult blood
UBE3A (ubiquitin protein ligase 3)	Paternal	Imprinted exclusively in brain; biallelically expressed in other tissues
KvLQT1 (potassium channel)	Paternal	Imprinted in several tissues but biallelically expressed in heart
WT1 (Wilms tumor gene)	Paternal	Frequently imprinted in cells of placenta and brain but biallelic expression in kidney

inherited alleles: as chromosomes pass through the male and female germlines they must acquire some *imprint* to signal a difference between paternal and maternal alleles in the developing organism. A key component, at least in maintaining the imprinted status, is allele-specific DNA methylation: all imprinted genes are characterized by CG-rich regions of differential methylation, and the imprinting of several imprinted genes has been shown to be disrupted in mutant mice deficient in the *Dnmt1* cytosine methyltransferase gene which is the major maintenance methylase.

Intriguingly, *Dnmt1* is known to have sex-specific exons (Section 10.4.2; *Figure 10.20*). In oocytes this results in an oocyte-specific N-terminal truncated protein product which conceivably could specifically methylate the maternal alleles of genes such as the insulin-like growth factor II receptor. The spermatocyte-specific exon of *Dnmt1* interferes with translation of *Dnmt1* mRNA, and it is less clear how paternal-specific patterns of methylation could be acquired. During development, the imprint would be expected to be stably inherited at least for many rounds of DNA duplication (but see below). Clearly, there must also be a mechanism for erasing the imprints during the germ line, as required when, for example, a man passes on an allele which he had inherited from his mother (see *Figure 10.25*). Demethylation occurring in the early embryo leaving the primordial germ cells essentially unmethylated (see *Figure 10.19*), is a way in which this could be achieved.

10.5.6 X chromosome inactivation in mammals involves very long range *cis*-acting repression of gene expression

Nature of X chromosome inactivation

X chromosome inactivation is a process that occurs in all mammals, resulting in selective inactivation of alleles on one of the two X chromosomes in females (Lyon, 1999). It provides a mechanism of **dosage compensation** which overcomes sex differences in the expected ratio of autosomal (A) gene dosage to X chromosome gene dosage. Males with a single X chromosome have one allele only for X-linked genes and so are *constitutionally hemizygous* for X-linked genes. There is therefore a sex difference in the ratio of A : X gene dosage (2 : 1 in males but 1 : 1 in females). The dosage ratios are important because the products of X-linked genes need to interact with the products of autosomal genes in a variety of important metabolic and developmental pathways, and there is tight regulation of the amounts of product for key *dosage-sensitive genes*.

To counteract the sex difference in A : X gene dosage, a compensatory adjustment is made in the cells of most female mammals: one of the two parental X chromosomes is inactivated. X inactivation involves modification of chromatin structure, resulting in a condensed, heterochromatinized structure, the **Barr body** (which can be seen to lie along the inside of the nuclear envelope in female cells). Most genes

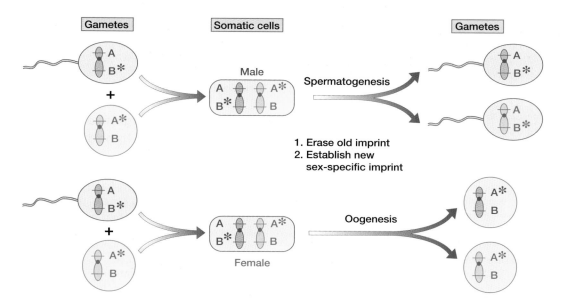

Figure 10.25: Genomic (gametic) imprinting requires erasure of the imprint in the germ line.

The diagram illustrates the fate of a chromosome carrying two genes, A and B, which are subject to imprinting: A is imprinted in the female germline, B is imprinted in the male germline, as indicated by asterisks. As a result, in diploid somatic cells A is imprinted when present on a maternally inherited chromosome and B is imprinted when present on a paternally inherited chromosome. An individual chromosome may pass through the male and female germlines in successive generations: a man may transmit a chromosome inherited from his mother and a woman can transmit a chromosome inherited from her father, as indicated by the gametes in the right panel. As a result, there must be a mechanism whereby the old imprint is erased from the germline prior to establishing a new sex-specific imprint.

on the inactivated X chromosome are subject to some mechanism which causes them to become transcriptionally inactive. By inactivating one of the two parental X chromosomes, therefore, female mammals become **functionally hemizygous** for most X-linked genes. Not all genes on the inactivated X chromosome are inactivated; the few which escape X-inactivation include those where there is a functional homolog on the Y chromosome, and some genes where gene dosage does not seem to be important (see *Section 12.2.8* for examples of genes that escape X-inactivation).

Timing of X chromosome inactivation

At very early stages in development both X chromosomes are active, but X inactivation is initiated as cells begin to differentiate from totipotent or pluripotent lineages, occurring at the late blastula stage in mice, and most likely also in humans. In each cell that will give rise to the female fetus, one of the two parental X chromosomes is selected for inactivation, but the decision to inactivate the paternally inherited X chromosome (X^p) or the maternal X (X^m) is usually random, and so *varies from cell to cell* [**note:** trophoblast cells and marsupial mammals are different: the X^p chromosome is preferentially inactivated, an example of tissue-restricted imprinting, and in individuals with an X:autosome translocation, the outcome is that the normal X chromosome is consistently inactivated]. The X inactivation pattern needs to be erased from generation to generation and, perhaps as part of this mechanism, the X chromosome is known to be inactivated transiently during gametogenesis in both males and females.

Once a progenitor cell in the early embryo has committed to inactivating the X^p or the X^m chromosome, the inactivation pattern shows **clonal inheritance**: all cell descendants within the resulting cell lineage carry the same X inactivation pattern as the progenitor cell (see *Figure 10.26A*). This means that *all female mammals are mosaics*, comprising mixtures of cell lines in which the paternal X is inactivated and cell lines where the maternally inherited X is inactivated. This is graphically illustrated by the mosaic coat color pattern of the calico cat (*Figure 10.26B*).

Mechanism of X chromosome inactivation

The process of X chromosome inactivation is complex, and distinct molecular mechanisms are involved in initiation of inactivation and maintenance of the inactivation (see Avner and Heard, 2001, Brockdorff, 2002). Two important *cis-acting* elements have been defined genetically:

▶ the **X-inactivation center** (*Xic*), controls the initiation and propagation of X-inactivation. The presence of *Xic* is essential for X inactivation: an X chromosome which carries *Xic* can undergo inactivation but one which lacks it cannot. *Xic* is also important for chromosome 'counting'. In rare individuals with an abnormal number of X chromosomes (45,X; 47,XXX; 47,XXY etc), a single X chromosome remains active no matter how many are present. By contrast, in triploid individuals either one or two X chromosomes remain active and in tetraploids two X chromosomes remain active. Some kind of *counting mechanism* ensures that one X chromosome remains active for

every two sets of autosomes. *Xic* function is dependent on a *XIST*, a gene specifying a functional noncoding RNA (see below) but while *XIST* (*Xist*) is essential for *initiating* X chromosome inactivation, it is not required for *maintaining* the inactivation state;

▶ the **X-controlling element** (*Xce*), which affects the choice of which X chromosome remains active and which is inactivated. Skewing of X inactivation (from the normal 50 : 50 of inactive to active Xs) can occur in females who are heterozygous for *Xce* alleles (an X chromosome with a strong *Xce* allele is more likely to be active than one with a weak *Xce* allele). Xce is distinct from *Xic* and *XIST* and maps telomeric to them, but the molecular basis remains unclear.

After the X inactivation center was mapped to human Xq13, analyses of this region uncovered an extraordinary gene: **XIST** (called **Xist** in rodents). *XIST* shows monoallelic expression and is uniquely expressed from the inactive X chromosome. Its primary transcript undergoes splicing and polyadenylation to generate a 17-kb mature noncoding RNA. It appears to be the primary signal for spreading the transcriptionally inactive state along the X chromosome from which it is synthesized, and somehow *cis*-limited spreading of this RNA product acts so as to coat the inactivated X chromosome over very long distances. The RNA is then thought to recruit some protein factors which organize the chromatin into a closed transcriptionally inactive conformation (Brockdorff, 2002).

Another extraordinary gene, **TSIX** (called **TsiX** in rodents) has a transcription unit overlapping all of the *XIST* gene, but on the antisense strand. This partner gene is expressed in undifferentiated embryonic stem cells and early embryos, and has been proposed to control *XIST* expression in *cis* at the onset of X inactivation (see Avner and Heard, 2001; Brockdorff, 2002).

10.6 The unique organization and expression of Ig and TCR genes

As shown in *Figure 9.10*, immunoglobulins (Igs) and T cell receptors (TCRs) are *heterodimers*. There are three Ig genes in human cells: one, *IGH*, specifies the heavy chain and two, *IGK* and *IGL* encode respectively the alternative kappa (κ) and lambda (λ) light chains. The four human TCR genes encode respectively the four different varieties of TCR chain: α and β which form the common αβ heterodimer plus γ and δ which form the rarer γδ heterodimer.

The organization and expression of these seven genes is in many ways quite different from that of other genes. This is so because each person needs to produce a huge number (of the order of about 2.5×10^7 different varieties) of Igs and TCRs manufactured by B and T lymphocytes respectively. These cells are the major agents of the *adaptive immune system* and they need to recognize a myriad of foreign antigens and so be capable of mounting a defense against a large number of different pathogens. An individual B or T cell is, however, **monospecific**: it produces a single type of Ig or TCR heterodimer *with a unique antigen-binding site, and so is specific*

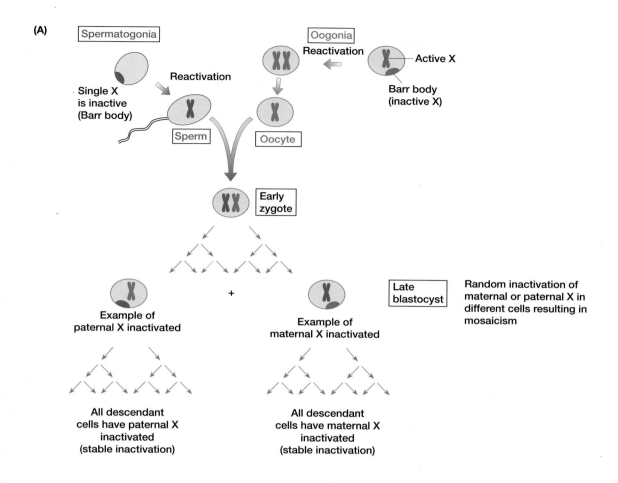

Figure 10.26: X chromosome inactivation in mammals.

(A) *The process of X inactivation*. In the early XX female zygote, both X chromosomes are active but around the late blastocyst stage a choice is made randomly in each cell to inactivate either the paternal or the maternal X. The choice that a cell makes is preserved in all its descendants. An adult XX female has clonal populations of cells with the paternal or maternal X inactivated. The inactive X is reactivated in oocytes some time before meiosis. During spermatogenesis both the X and Y chromosomes are transiently inactivated. Adapted from Migeon (1994) *Trends Genet.* **10**, 230–235, with permission from Elsevier Trends Journals. **(B) *The calico (tortoiseshell and white) cat*.** This cat (which is always female) has an X chromosome with a gene for black coat color and another X chromosome with a gene specifying orange coat color. Differently colored coat patches result because of clonal X-inactivation. The white coat patches are due to a separate gene

for a *particular* antigen. It is the *population* of different B and T cells in any one individual that enables the synthesis of so many different types of these molecules. By providing a large repertoire of Igs and TCRs, the possibilities for being able to recognize and bind very many different types of foreign antigen are greatly increased.

The Ig and TCR genes in most cells are inactive but extraordinary shuffling of these genes occurs in B and T cells respectively in order to activate them. When this happens, new coding combinations are created enabling a single individual to produce an extraordinary variety of Igs and TCRs. Four major DNA shuffling mechanisms are known to apply in vertebrates:

▶ *VDJ or VJ recombination, a widespread mechanism*. The variable regions of both the Igs and TCRs are specified by two or three types of gene segment, always a V (variable region) and a J (joining region) gene segment, and in some cases also a D (diversity region) gene segment. In germline DNA, each of these gene segments is present in multiple copies (see *Table 10.8* and Section 10.6.1). However, the DNA of mature B cells and T cells undergoes *cell-specific* DNA recombination so as to bring together a specific combination of a V, D and J segment, or of a V and J segment, creating respectively a **VDJ exon** or a **VJ exon** (see section 10.6.1);

▶ *somatic hypermutation, a major mechanism in humans and mice*. Additional Ig diversity results when point mutations are introduced into the V region of VJ and VDJ exons due to *error-prone DNA repair*;

▶ *gene conversion, a major mechanism in rabbits and chickens.* Ig diversity can also result when stretches of nucleotide sequence are copied from upstream *V pseudogene (ΨV) segments* into the sequence of V gene segments in VJ and VDJ exons;

▶ *class-switch recombination, a widespread mechanism*. The constant region of Igs is specified by transcription units consisting of several exons. In the case of Ig heavy chains, several different transcription units specify the different classes of heavy chain, such as Cμ (IgM), Cδ (IgD), Cγ (IgG) etc. During maturation of B cells intrachromatid recombination within the heavy chain gene brings specific types of constant region transcription unit in proximity to the VDJ exon (Section 10.6.2).

Very recently, it has been shown that the somatic hypermutation, gene conversion and class-switch recombination events (which can all occur in a single species, although some mechanisms are particularly prevalent in particular species) are all controlled by a single gene which encodes **activation-induced deoxycytidine deaminase** (AID; see Petersen-Mahrt *et al.*, 2002).

10.6.1 DNA rearrangements in B and T cells generate *cell-specific* exons encoding Ig and TCR variable regions

The genes which encode the three Ig chains (one heavy chain and two types of light chain) and the four TCR chains (α, β, γ, and δ) are located on different chromosomes and show an extraordinary organization. In each case the variable region is specified by two or three different gene segments which are present in numerous different copies that are sequentially repeated in germ-line DNA (see *Table 10.8* and *Figure 10.27*). The gene segments comprise:

▶ **V (variable region) gene segments.** V gene segments encode most of the variable region;

▶ **J (joining region) gene segments.** J gene segments encode the *joining region,* a small part at the C-terminal end of the variable region;

▶ **D (diversity region) gene segments.** D gene segments encode a small *diversity region* near the C-terminal ends of the variable region of Ig heavy chains, TCRβ chains and TCRδ chains.

The unique arrangement of gene segments in the Ig and TCR gene clusters reflects the very unusual way in which B and T lymphocytes undergo somatic recombinations to *activate* previously nonfunctional Ig or TCR genes and then express functional products from them. They do this by bringing together specific combinations of V + D + J gene segments (in the case of the genes encoding the Ig heavy chain, TCRβ and TCRδ) or of V + J gene segments (in the case of the other four genes). Once assembled, the V + J or V + D + J

Table 10.8: Human Ig and TCR genes

Gene	Location	Number of V, D, J gene segments			Number of C transcription units
		V	D	J	
IGH	14q32.3	123–129[a]	27	9	11
IGK	2p12	76	0	5	1
IGL	22q11	70–71[a]	0	7–11[a]	7–11[a]
TRA	14q11.2	49 (+5)[b]	0	61	1
TRB	7q34	64–67[a]	2	14	2
TRG	7p15-p14	12–15[a]	0	5	2
TRD	14q11.2	1 (+5)[b]	3	4	1

Note: numbers include nonfunctional (pseudogene) sequences—see *Figure 10.27* for an example from the *IGH* gene.

[a]Number varies on different haplotypes. [b]Five V gene segments are shared between the neighboring *TRA* and *TRD* genes. For pictorial representations see under individual gene names in various databases such as the Atlas of Genetics and Cytogenetics in Oncology and Haematology at www.infobiogen.fr/services/chromcancer/Genes/Geneliste.html

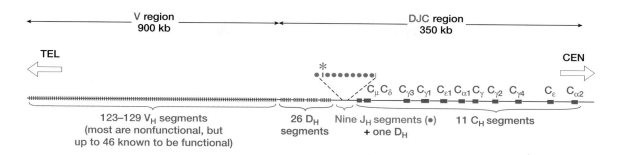

Figure 10.27: Multiple gene segments in the *IGH* Ig heavy chain gene at 14q32.

The gene spans 1250 kb of 14q32.3 from telomeric end (left) to centromeric end (right). There are 123–129 V_H gene segments (depending on the haplotype), of which the majority are known to be nonfunctional pseudogene segments, 27 D_H gene segments, nine J_H gene segments, and 11 C transcription units (*each of which contains several exons*). For further details, see http://www.infobiogen.fr/ services/chromcancer/Genes/IgHID40.html

units behave as *functional exons* (see below). For each of the seven genes, *the choice of which of the many V, (D) or J gene segments is brought together varies from lymphocyte to lymphocyte* and so the newly created **VJ exons** and **VDJ exons** are *cell-specific* (see *Figure 10.28*). As a result, individual B and T cells produce different Igs and TCRs. In a sense, therefore, every individual is a mosaic with respect to the organization of the Ig and TCR genes in B and T lymphocytes, *and even identical twins will diverge genetically.*

Once a VDJ or VJ exon has been assembled, a functional Ig or TCR gene has been created. Now the new VDJ or VJ exon will provide the first exon for this gene and the downstream exons are provided by the exons in the neighboring C transcription unit. In the case of the Ig heavy chain gene there are several different functional C transcription units with different biological properties. Initially, however, splicing involves the VDJ exon and the exons of the neighboring Cμ and Cδ transcription units, and alternative splicing can result in μ or δ chain synthesis which become incorporated into IgM and IgD immunoglobulins (*Figure 10.28*). As the B cell matures, however, subsequent somatic recombinations result in joining of the previously assembled VDJ exon to different C transcription units to produce γ, α, or ε heavy chains to be incorporated into IgG, IgA, or IgE (the heavy chain *class switch* – see text; *Figure 10.29*).

The genetic mechanisms leading to the production of functional VJ and VDJ exons often involve large scale deletions of the sequences separating the selected gene segments (most likely by intrachromatid recombination, in much the same way as in *Figure 10.29B*) and in some cases inversions of the intervening sequence. Conserved **recombination signal sequences** flank the 3′ ends of each V and J segment and both the 5′ and 3′ ends of each D gene segment. They enable joining of V to J, or D to J followed by V to DJ, but never V to V or D to D etc.

The recombination events are directed by a **V(D)J recombinase**, a complex which contains two lymphocyte-specific proteins **RAG1** and **RAG2**, as well as enzymes that help repair damaged DNA in all our cells. RAG1 and RAG2 make double-stranded breaks at the recombination signal sequences, and the breaks are repaired by stitching together appropriate V,

D and J gene segments while excluding intervening sequences. Although site-specific recombination is normally precise, the recombination that occurs at Ig and TCR genes in B and T cells respectively is deliberately not precise. Instead, a variable number of nucleotides is often lost from the ends of the recombining gene segments and one or more randomly chosen nucleotides may also be inserted, creating **junctional diversification**. This increases the diversity of variable region coding sequences enormously.

10.6.2 Heavy chain class switching involves joining of a single VDJ exon to alternative constant region transcription units

Although a B cell produces only one type of Ig molecule, the *class* (or **isotype**) of the heavy chain can change during development: within a single lineage descendants of an original cell can produce an Ig with the same antigen-binding site as before but using a different class of heavy chain (**class switching** or **isotype switching**). Such switching involves differential joining of the same VDJ exon that was brought together by two successive somatic recombinations (see *Figure 10.28*) to alternative constant region transcription units.

Class switching involves an intrachromatid recombination which results in joining of a VDJ exon to a more distal constant region transcription unit (**VDJ–C joining**). It involves the following progression:

▶ *(i) initial synthesis of IgM only by immature naive B cells.* This occurs early in B cell development because RNA splicing brings together sequence transcribed from the VDJ exon and the exons of the neighboring Cμ transcription unit (*Figure 10.29A*);

▶ *(ii) later synthesis of both IgM and IgD by mature naive B cells.* Later in B cell development but at a stage when the B cells are still **immunologically naive** (that is, they have not yet been exposed to foreign antigen), a *partial* class switch occurs and the B cells now make IgD as well as IgM. This occurs because now additional, alternative RNA splicing can also bring together sequence transcribed from

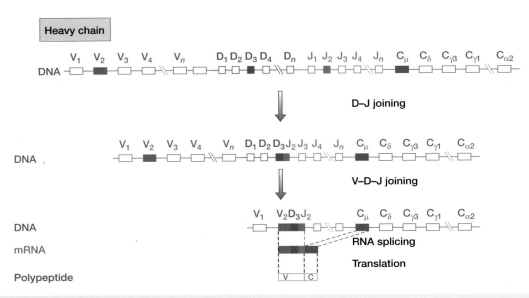

Figure 10.28: Cell-specific VDJ recombination as a prelude to making an Ig heavy chain.

Two sequential somatic recombinations produce first D–J joining, then a mature VDJ exon. In this particular example, the second out of 129 different V segments (V_2) is fused to the third D region segment (D_3) and the second J region segment (J_2) to produce a functional $V_2D_3J_2$ exon but the choice is cell-specific so that a neighboring B lymphocyte may have a functional $V_{129}D_{17}J_1$ exon for example. Once the VDJ exon has assembled the gene can be transcribed using the VDJ exon as the first exon, with the subsequent exons provided by the nearest C transcription units. To begin with, the $C\mu$ and $C\delta$ transcription units are nearest and the first types of Ig heavy chain made are the μ chain and then μ plus δ, the heavy chains characteristic of IgM and IgD respectively. As the B cell matures, however, subsequent somatic recombinations result in joining of the previously assembled VDJ exon to different C transcription units (heavy chain **class switch**, see text and *Figure 10.29*).

the VDJ exon and the exons of the neighboring Cδ transcription unit (*Figure 10.29A*). Most of the IgM and IgD production is devoted to making membrane bound receptors but exposure to foreign antigen will trigger secretion of soluble IgM antibody in a weak **primary immune response**;

▶ *(iii) synthesis of IgG, IgE or IgA by mature B cells*. After exposure to foreign antigens, B cells develop processes for refining the affinity of Ig binding (**affinity maturation**) so that they can respond more effectively to foreign antigen on a future occasion (when they will be able to secrete large amounts of soluble antibody with a very high affinity for the foreign antigen in a powerful **secondary immune response**). Affinity maturation is aided by somatic hyper-mutation/gene conversion events and so on. In the mature B cell class switching now occurs by a different mechanism and involves a recombination event enabling joining at the DNA level of the same VDJ exon to either a Cγ, Cε or Cα transcription unit. The mechanism involves deletion of the intervening sequence by intrachromatid recombination (*Figure 10.29B*). The Ig molecules secreted by mature B cells can bind to different cell types which have specialized receptors for the tail portion (Fc) of the soluble antibody. Such **Fc receptors** can selectively bind different types of Ig:

- **IgG** – in addition to activating the **complement system** IgG can be specifically bound by Fc receptors on *macrophages* and *neutrophils*, powerful phagocytic cells;

- **IgA** – a type of Fc receptor unique to secretory epithelia transports IgA antibodies so that they constitute the principal antibody class in secretions, including saliva, tears, milk and respiratory and intestinal secretions;

- **IgE** – Fc receptors on *mast cells* and *basophils* bind IgE with very high affinity. The bound IgE molecules then function as passively acquired antigen receptors. Subsequent binding of antigen will trigger the mast cell or basophil to secrete a variety of cytokines and histamine. In addition, mast cells will secrete factors which attract and activate *eosinophils* which also have Fc receptors that bind IgE molecules and which can kill various types of parasites.

10.6.3 The monospecificity of Igs and TCRs is due to allelic and light chain exclusion

There are three functional Ig genes in human cells (*IGH* encoding the heavy chain, and two genes, *IGK* and *IGL* which encode the functionally interchangeable κ and λ light chains), and because these occur on both maternal and paternal homologs, six genes are potentially available for making Ig chains. However, an *individual* B cell is **monospecific**: it produces only one type of Ig molecule with a single type of heavy chain and a single type of light chain. This is so for two reasons:

▶ **allelic exclusion**. A light chain or a heavy chain can be synthesized from a maternal chromosome or a paternal

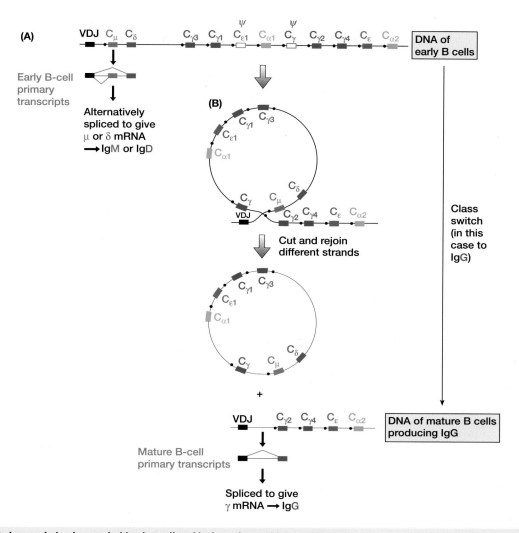

Figure 10.29: Ig heavy chain class switching is mediated by intrachromatid recombination.

(A) *Early (partial) switch to IgD*. The initial heavy chain class is IgM because RNA splicing brings together sequences transcribed from the VDJ exon and the exons of the neighboring *Cμ transcription unit*. As naive B cells mature, however, alternative RNA splicing brings together sequences from the VDJ exon and exons of the Cδ transcription unit, leading to the additional production of IgD. **(B) *Late switch to IgA, IgG and IgE*.** Here class switching occurs by intrachromatid recombination, where the same VDJ exon is brought close to the initially more distal constant region transcription units: a Cα, Cγ (as illustrated here) or Cε transcription unit. The new VDJ–C combination is expressed to give IgA, IgG or IgE.

chromosome in any one B cell, but *not from both parental homologs*. As a result, there is **monoallelic expression** at the heavy chain gene locus in B cells. This phenomenon also applies to TCR gene clusters;

▸ **light chain exclusion**. A light chain synthesized in a single B cell may be a κ chain, or a λ chain, but never both. As a result of this requirement plus that of allelic exclusion, there is monoallelic expression at one of the two functional light chain gene clusters *and no expression at the other*.

The decision of which of the two heavy chain alleles to use to make a heavy chain and which of the four possible light chain genes should be used to make a light chain appears to be random. Most likely, in each B-cell precursor, productive DNA rearrangements are attempted at all six Ig genes but the chances of productive arrangements in more than one light chain gene or more than one heavy chain gene may not be high. Additionally, however, there appears to be some kind of **negative feedback regulation**: a functional rearrangement at one of the heavy chain alleles suppresses rearrangements occurring in the other allele, and a functional rearrangement at any one of the four genes capable of encoding a light chain suppresses rearrangements occurring in the other three.

Further reading

Brivanou AH, Darnewell Jr. JE (2002) Signal transduction and the control of gene expression. *Science* **295**, 813–818.

Gellert M (2002) V(D)J recombination: RAG proteins, repair factors and regulation. *Annu. Rev. Biochem.* **71**, 101–132.

Imprinted Gene Catalogue at http://cancer.otago.ac.nz/IGC/Web/home.html

Janeway CA, Travers P, Walport M, Capra JD (2001) *Immunobiology. The Immune System in Health and Disease.* 5th Edn. Garland Publishing. Available with animations at http://www.blink.uk.com/immunoanimations/

Latchman D (1998) *Gene regulation. A Eukaryotic Perspective.* Stanley Thornes (Publishers) Ltd., Cheltenham.

Orphanides G, Reinberg D (2002) A unified theory of gene expression. *Cell* **108**, 439–451.

Palacois IM, St. Johnson D (2001) Getting the message across: the intracellular localization of mRNAs in higher eukaryotes. *Annu. Rev. Cell Dev. Biol.* **17**, 569–614.

Plath K, Mylnarczyk-Evans S, Nusinow DA, Panning B (2002) *XIST* RNA and the mechanism of X chromosome inactivation. *Annu. Rev. Genet.* **36**, 233–278.

Sandberg K, Mulroney SE (eds) (2002) *RNA Binding Proteins: New Concepts in Gene Regulation.* Kluwer Academic Publishers, Boston.

Turner BM, Turner BS (2002) *Chromatin and Gene Regulation: Mechanisms in Epigenetics.* Blackwell Science, Oxford.

Travers A (1993) *DNA–Protein Interactions.* Chapman & Hall, London.

van Driel R, Otte AP (1997) *Nuclear Organization, Chromatin Structure and Gene Expression.* Oxford University Press, Oxford.

References

Avner P, Heard E (2001) X-chromosome inactivation: counting, choice and initiation. *Nature Rev. Genet.* **2**, 59–67.

Ayoubi TA, Van De Ven WJ (1996) Regulation of gene expression by alternative promoters. *FASEB J* **10**, 453–460.

Bell AC, West AG, Felsenfeld G (2001) Insulators and boundaries: versatile regulatory elements in the eukaryotic genome. *Science* **291**, 447–450.

Bannerjee D, Slack F (2002) Control of developmental timing by small temporal RNAs: a paradigm for RNA-mediated regulation of gene expression. *BioEssays* **24**, 119–128.

Berger SL (2002) Histone modifications in transcriptional regulation. *Curr. Opin. Genet. Dev.* **12**, 142–148.

Bestor TH (1998) Cytosine methylation and the unequal developmental potentials of the oocyte and sperm genomes. *Am. J. Hum. Genet.* **62**, 1269–1273.

Bird A (2002) DNA methylation patterns and epigenetic memory. *Genes Dev.* **16**, 6–21.

Blencowe BJ (2000) Exonic splicing enhancers: mechanism of action, diversity and role in human genetic diseases. *Trends Biochem. Sci.* **25**, 106–110.

Brockdorff N (2002) X chromosome inactivation: closing in on proteins that bind *Xist* RNA. *Trends Genet.* **18**, 352–358.

Bulger M, Groudine M (1999) Looping versus linking: toward a model for long-distance gene activation. *Genes Dev.* **13**, 2465–2477.

Bulger M, Sawado T, Schubeler D, Groudine M (2002) ChIPs of the β-globin locus: unravelling gene regulation within an active domain. *Curr. Opin. Genet. Dev.* **12**, 170–177.

Butler JE, Kadonaga JT (2002) The RNA polymerase II core promoter: a key component in the regulation of gene expression. *Genes Dev.* **16**, 2583–2592.

Caceres JF, Kornblihtt AR (2002) Alternative splicing: multiple control mechanisms and involvement in human disease. *Trends Genet.* **18**, 186–193.

Chess A (1998) Expansion of the allelic exclusion principle. *Science* **279**, 2067–2068.

De Moor CH, Richter JD (2001) Translational control in vertebrate development. *Int. Rev. Cytol.* **203**, 567–608.

Dever TE (2002) Gene-specific regulation by general translation factors. *Cell* **108**, 545–556.

Edwards-Gilbert G, Veraldi KL, Milcarek C (1997) Alternative poly (A) site selection in complex transcription units: means to an end? *Nucl. Acid Res.* **13**, 2547–2561.

Fairbrother WG, Yeh RF, Sharp PA, Burge CB (2002) Predictive identification of exonic splicing enhancers in human genes. *Science* **297**, 1007–1013.

Finta C, Zaphiropoulos PG (2002) Intergenic mRNA molecules resulting from trans-splicing. *J. Biol. Chem.* **277**, 5882–5890.

Fuks F, Hurd PJ, Wolf D, Nan X, Bird AP, Kouzarides T (2002) The methyl-CpG-binding protein MeCP2 links DNA methylation to histone methylation. *J. Biol. Chem* **278**, 4035–4040.

Gallagher RC, Pils B, Albalwi M, Francke U (2002) Evidence for the role of PWCR1/HBII-85 C/D box small nucleolar RNAs in Prader-Willi syndrome. *Am. J. Hum. Genet.* **71**, 669–678.

Gerber AP, Keller W (2001) RNA editing by base deamination: more enzymes, more targets, new mysteries. *Trends Biochem. Sci.* **26**, 376–384.

Goll MG, Bestor TH (2002) Histone modification and replacement in chromatin activation. *Genes Dev.* **16**, 1739–1742.

Graveley BR (2001) Alternative splicing: increasing diversity in the proteomic world. *Trends Genet.* **17**, 100–107.

Gray NK, Wickens M (1998) Control of translation initiation in animals. *Annu. Rev. Cell Dev. Biol.* **14**, 399–458.

Hazelrigg T (1998) The destinies and destinations of RNAs. *Cell* **95**, 451–460.

Jones PA (1999) The DNA methylation paradox. *Trends Genet.* **15**, 34–37.

Karin M, Hunter T (1995) Transcriptional control by protein phosphorylation: signal transmission from the cell surface to the nucleus. *Curr. Biol.* **5**, 747–757.

Klausner RD, Rouault TA, Harford JB (1993) Regulating the fate of mRNA: the control of cellular iron metabolism. *Cell* **72**, 19–28.

Kleinjan D-J, van Heyningen V (1998) Position effects in human genetic disease. **7**, 1611–1618.

Larsson SH, Charlieu JP, Miyagawa K et al. (1995) Subnuclear localization of WT1 in splicing or transcription factor domains is regulated by alternative splicing. *Cell* **81**, 391–401.

Lemon B, Tjian R (2000) Orchestrated response: a symphony of transcription factors for gene control. *Genes Dev.* **14**, 2551–2569.

Li E (2002) Chromatin modification and epigenetic reprogramming in mammalian development. *Nature Rev. Genet.* **3**, 662–673.

Lopez AJ (1998) Alternative splicing of pre-mRNA: developmental consequences and mechanisms of regulation. *Annu. Rev. Genet.* **32**, 279–305.

Lunyak VV, Burgess R, Prefontaine GG et al. (2002) Corepressor-dependent silencing of chromosomal regions encoding neuronal genes. *Science* **298**, 1747–1752.

Lyon MF (1999) X-chromosome inactivation. *Curr. Biol.* **9**, R235–R237.

Maher E, Reik W (2000) Beckwith-Wiedemann syndrome: imprinting in clusters revisited. *J. Clin. Invest.* **105**, 247–252.

Mahmoudi T, Verrijzer CP (2001) Chromatin silencing and activation by Polycomb and trithorax group proteins. *Oncogene* **20**, 3055–3066.

Maniatis T, Tasic B (2002) Alternative pre-mRNA splicing and proteome expansion in metazoans. *Nature* **418**, 236–243.

Martin DIK (2001) Transcriptional enhancers – on/off gene regulation as an adaptation to silencing in higher eukaryotic nuclei. *Trends Genet.* **17**, 444–448.

Meguro M, Kashiwagi A, Mitsuya K, Nakao M, Kondo I, Saitoh S, Oshimura M (2001) A novel maternally expressed gene, ATP10C, encodes a putative aminophospholipid translocase associated with Angelman syndrome. *Nature Genet.* **28**, 19–20.

Merteneit C, Yoder JA, Taketo T, Laird DW, Trasier JM, Bestor TH (1998) Sex-specific exons control DNA methyltransferase in mammalian germ cells. *Development* **125**, 889–897.

Migeon BR (1994) X-chromosome inactivation: molecular mechanisms and genetic consequences. *Trends Genet.* **10**, 230–235.

Narlikar GJ, Fan H-Y, Kingston RE (2002) Co-operation between complexes that regulate chromatin structure and transcription. *Cell* **108**, 475–487.

Ogbourne S, Antalis TM (1998) Transcriptional control and the role of silencers in transcriptional regulation in eukaryotes. *Biochem. J.* **331**, 1–14.

Ohlsson R, Tycko B, Sapienza C (1998) Monoallelic expression: 'there can only be one'. *Trends Genet.* **14**, 435–438.

Pasquinelli AE, Ruvkun G (2002) Control of developmental timing by microRNAs and their targets. *Annu. Rev. Cell. Dev. Biol.* **18**, 495–513.

Pekhletsky RI, Chernov BK, Rubtsov PM (1992) Variants of the 5′-untranslated sequence of human growth hormone receptor mRNA. *Mol. Cell. Endocrinol.* **90**, 103–109.

Perk J, Makedonski K, Lande L, Cedar H, Razin A, Shemer R (2002) The imprinting mechanism of the Prader-Willi/Angelman regional control center. *EMBO J.* **21**, 5807–5814.

Petersen-Mahrt SK, Harris RS, Neuberger MS (2002) AID mutates *E. coli* suggesting a DNA deamination mechanism for antibody diversification. *Nature* **418**, 99–10.

Razin A, Kafri T (1994) DNA methylation from embryo to adult. *Prog. Nucl. Acid Res. Mol. Biol.* **48**, 53–81.

Reik W, Dean W, Walter J (2001) Epigenetic reprogramming in mammalian development. *Science* **293**, 1089–1093.

Roberts GC, Smith CWJ (2002) Alternative splicing: combinatorial output from the genome. *Curr. Opin. Chem. Biol.* **6**, 375–383.

Rougeulle C, Heard E (2002) Antisense RNA in imprinting: spreading silence through *Air. Trends Genet.* **18**, 434–437.

Runte M, Huttenhofer A, Gross S, Keifmann M, Horsthemke B, Buiting K (2001) The IC-SNURF-SNRPN transcript serves as a host for multiple small nucleolar RNA species and as an antisense RNA for UBE3A. *Hum. Mol. Genet.* **10**, 2687–2700.

Schmucker D, Clemens JC, Shu H et al. (2000) *Drosophila* Dscam is an axon guidance receptor exhibiting extraordinary molecular diversity. *Cell* **101**, 671–684.

Schoenherr CJ, Paquette AJ, Anderson DJ (1996) Identification of potential target genes for the neuron-restrictive silencer factor. *Proc. Natl Acad. Sci. USA* **93**, 9881–9886.

Simmen MW, Leitgeb S, Charlton J, Jones SJM, Harris B, Clark VH, Bird AP (1999) Nonmethylated transposable elements and methylated genes in a chordate genome. *Science* **283**, 1164–1167.

Siomi H, Dreyfuss G (1997) RNA-binding proteins as regulators of gene expression. *Curr. Opin. Genet. Dev.* **7**, 345–353.

Skuse DH, James RS, Bishop DV et al. (1997) Evidence from Turner's syndrome of an imprinted X-linked locus affecting cognitive function. *Nature* **387**, 705–708.

Sleutels F, Barlow DP (2002) The origins of genomic imprinting in mammals. *Adv. Genet.* **46**, 119–163.

Sleutels F, Zwart R, Barlow DP (2002) The non-coding *Air* RNA is required for silencing autosomal imprinted genes. *Nature* **415**, 810–813.

Turner BM (2002) Cellular memory and the histone code. *Cell* **111**, 285–291.

Tycko B, Morison IM (2002) Physiological functions of imprinted genes. *J. Cell Physiol.* **192**, 245–258.

Wahle E, Kuhn (1997) The mechanism of 3′ cleavage and polyadenylation of 3′ eukaryotic pre-mRNA. *Prog. Nucl. Acid Res. Mol. Biol.* **57**, 41–71.

Walsh CP, Bestor TH (1999) Cytosine methylation and mammalian development. *Genes Dev.* **13**, 26–34.

Wickens M, Anderson P, Jackson RJ (1997) Life and death in the cytoplasm: messages from the 39′ end. *Curr. Opin. Genet. Dev.* **7**, 233–241.

Wickens M, Bernstein DS, Kimble J, Parker RA (2002) PUF family portrait: 3′ UTR regulation as a way of life. *Trends Genet.* **18**, 150–157.

Yoder JA, Walsh CP, Bestor TH (1997) Cytosine methylation and the ecology of intragenomic parasites. *Trends Genet.* **13**, 335–340.

CHAPTER ELEVEN

Instability of the human genome: mutation and DNA repair

Chapter contents

11.1 An overview of mutation, polymorphism, and DNA repair

As in other genomes, the DNA of the human genome is not a static entity. Instead, it is subject to a variety of different types of heritable change (mutation). Large-scale chromosome abnormalities involve loss or gain of chromosomes, or breakage and rejoining of chromatids (see Section 2.5). Smaller scale mutations can be grouped into different mutation classes according to the effect on the DNA sequence as follows:

▶ **base substitutions** – involve replacement of usually a single base; in rare cases several clustered bases may be replaced simultaneously as a result of a form of gene conversion (Section 11.3.3);

▶ **deletions** – one or more nucleotides are eliminated from a sequence;

▶ **insertions** – one or more nucleotides are inserted into a sequence.

Mutations can also be categorized on the basis of whether they involve a single DNA sequence (*simple mutations* – Section 11.2) or whether they involve *exchanges* between two allelic or nonallelic sequences (Section 11.3). The three mutation classes listed above can each arise by simple mutations or by sequence exchanges.

New mutations arise in single individuals, either in somatic cells or in the germline (Section 3.1.3). If a germline mutation does not seriously impair an individual's ability to have offspring who can transmit the mutation, it can spread to other members of a (sexual) population. Allelic sequence variation has traditionally been described as a DNA polymorphism if more than one variant (allele) at a locus occurs in a human population with a frequency greater than 0.01 (a frequency that is higher than could be maintained simply by recurring mutation). There is, however, a variety of different types of polymorphism ranging from single nucleotide changes to very large scale changes (*Box 11.1*)

For human genomic DNA the **mean heterozygosity** (also termed *average nucleotide diversity*) is of the order of 0.08%: approximately 1 out of 1250 bases differs on average between allelic sequences. This figure was initially estimated by averaging data from a limited number of single locus sequencing studies (see Przeworski *et al.*, 2000). Heterozygosity values do vary widely for different loci and certain genes are exceptionally polymorphic, notably some HLA genes (see *Figure 12.29*), but analysis of the recently available global human single nucleotide polymorphism (SNP) datasets has confirmed the 0.08% mean heterozygosity value (Reich *et al.*, 2002). Because, however, mutation rates are comparatively low, the vast majority of the differences between allelic sequences within an individual are inherited, rather than resulting from *de novo* mutations.

Mutations are the raw fuel that drives evolution, but they can also be pathogenic (Sections 11.4, 11.5). They can be the direct cause of a phenotypic abnormality or they can result in increased susceptibility to disease. The usually low level of mutation may therefore be viewed as a balance between permitting occasional evolutionary novelty at the expense of causing disease or death in a proportion of the members of a species.

Mutations often arise as copying errors during DNA replication. Although the fidelity of DNA replication *in vivo* is normally extremely high, misincorporation occurs at a low frequency, dependent on the relative free energies of correctly and incorrectly paired bases. Very minor changes in helix geometry can stabilize G–T base pairs (with two hydrogen bonds; *note the frequent occurrence of G–U base pairing in RNA*, see *Figure 1.7B*). To reduce the error rate of misincorporation many DNA polymerases contain an integral $3' \rightarrow 5'$ exonuclease which can serve as a **proofreading activity** (see *Table 1.2*). When an incorrect base is inserted during DNA synthesis, DNA synthesis does not proceed. Instead, the $3' \rightarrow 5'$ exonuclease activity removes one nucleotide at a time from the $3'$ hydroxyl terminus until a correctly base-paired terminus is obtained, enabling DNA synthesis to proceed again. Even then, however, the size of the human genome makes huge demands on the fidelity of any DNA polymerase: a sequence of six billion nucleotides needs to be replicated accurately every single time a human cell divides.

DNA is also subject to significant spontaneous chemical attack in the cell and to damage caused by exposure to natural ionizing radiation and to reactive metabolites. In order to minimize the mutation rate, therefore, it is necessary to have effective **DNA repair** systems which identify and correct many abnormalities in the DNA sequence (Section 11.6). In addition, errors that arise in the mRNA sequence during gene expression are subject to **RNA surveillance** mechanisms which ensure removal of mRNAs which have inappropriate termination codons (Section 11.4.4).

11.2 Simple mutations

11.2.1 Mutations due to errors in DNA replication and repair are frequent

Mutations can be induced in our DNA by exposure to a variety of mutagens occurring in our external environment or to mutagens generated in the intracellular environment. In the case of radiation-induced mutation, for example, Dubrova *et al.* (1996, 2002) reported that the normal germline mutation rate for hypervariable minisatellite loci was doubled as a consequence of heavy exposure to the radioactive fallout from the Chernobyl accident. However, under normal circumstances by far the greatest source of mutations is from *endogenous mutation*, including spontaneous errors in DNA replication and repair.

During an average human lifetime in the order of 10^{17} cell divisions can be estimated to take place: about 2×10^{14} divisions are required to generate the approximately 10^{14} cells in the adult, and additional mitoses are required to permit cell renewal in the case of certain cell types, notably epithelial cells (see Cairns, 1975). As each cell division requires the incorporation of 6×10^9 new nucleotides, error-free DNA replication in an average lifetime would require a DNA replication repair process whose accuracy was great enough that the correct

Box 11.1: Classes of genetic polymorphism and sequence variation.

Differences in individual phenotypes are largely due to genetic variation and a variety of different types of genetic polymorphisms and large-scale sequence variation are known. Ultimately, however, the effect on the phenotype is expressed at the protein or RNA levels. Changes in polypeptide-encoding DNA can lead to amino acid changes causing **protein polymorphisms**. In addition to DNA and protein polymorphisms which have been studied for many years are poorly understood *quantitative* differences in expression at the transcript level. Allelic human gene **expression variation** may be caused by changes in regulatory DNA, including sequences which regulate transcription and splicing. This type of sequence variation may often underlie the susceptibility to common diseases but quantitative methods to explore allelic variation in human gene expression have only very recently been developed (Yan *et al.*, 2002). Common classes of **DNA polymorphism** and large-scale sequence variation are described below.

Single nucleotide polymorphism (SNP)

As the name suggests, this involves a single nucleotide. For most SNPs a single nucleotide is substituted by a different nucleotide (**nucleotide substitution**), but the term also encompasses changes involving nucleotide <u>in</u>sertions or <u>del</u>etions (simple **indel polymorphisms**). Typically, SNPs have only two alleles. Some SNPs cause changes in restriction sites (**restriction site polymorphism**; Section 7.1.3). Since coding DNA accounts for only about 1.5% of the human genome, most SNPs are found in noncoding DNA, such as within introns and intergenic sequences. However, the chromosomal location of SNPs is far from uniform: large chromosomal regions with very few SNPs are often found adjacent to large regions containing many SNPs (see Section 11.2.6). For typing of SNPs, see Section 7.1.3.

Simple variable number of tandem repeat (VNTR) polymorphism

VNTR polymorphism traditionally describes alleles at loci containing tandemly repeated runs of a *simple sequence*. It encompasses two classes: microsatellites = **SSR** (**simple sequence repeat**) **polymorphism** where the simple sequence is from one to several nucleotides long and the array length ranges from less than 10 to over 100 nucleotides; and **minisatellite DNA polymorphism** which involves arrays often spanning hundreds of nucleotides and consisting of tandem repeats of a sequence of between 9 and several tens of nucleotides long. VNTR loci often have multiple alleles and some hypervariable minisatellites show extraordinary variation (e.g. the MS32 locus has a heterozygosity value of 0.975).

Both types of polymorphism are very rarely found in polypeptide-encoding DNA. Exceptions include very rare nonframeshifting SSR polymorphisms and the very occasional expressed minisatellite polymorphism e.g. the MUC1 locus at 1q21 is known to encode a highly polymorphic glycoprotein found in several epithelial tissues and body fluids as a result of extensive variation in minisatellite-encoded repeats (Swallow *et al.*, 1987). In addition, some of these polymorphisms may be located very close to genes and can affect their expression. A notable example is the *INS* VNTR, a polymorphic minisatellite located

596 bp upstream of the insulin gene translation initiation site and consisting of a variable number of tandem repeats based on the consensus sequence ACAGGGGTGTGGGG. Different alleles appear to confer differential susceptibility to type I diabetes, probably because of differential effects on the expression of the insulin gene (Section 15.6.4).

Transposon repeat polymorphism

Close to 45% of the human genome is composed of transposon-based repeats (Section 9.5.1). The vast majority are no longer active, but some LINE1, Alu and LTR-based transposons are still actively transposing. As a result of evolutionarily recent transpositions some loci are polymorphic. For example, many members of the Yb9, Yc1 and Yc2 Alu subfamilies have inserted in the human genome so recently that about a one-third of the analyzed elements are polymorphic for the presence/absence of the Alu repeat in diverse human populations (Roy-Engel *et al.*, 2001).

Large-scale VNTR polymorphism

Large-scale tandem repeats are prone to copy number variation as a result of unequal crossover or unequal sister chromatid exchanges resulting in a type of *large-scale* VNTR polymorphism. Examples include α-satellite repeats at centromeres and various tandemly repeated RNA genes such as rRNA genes and the U2 snRNAs at the *RNU2* locus at 17q21–q22 (from 6 to over 30 repeats of nearly identical 6.1-kb units). Some polypeptide-encoding gene clusters include large tandem repeats which are prone to this type of polymorphism e.g. the 21-hydroxylase/complement C4 cluster (Section 11.5.3).

Inversion polymorphism

The sequencing of the euchromatic portion of the human genome revealed many examples of low to moderate copy number sequences with very high (>95%) sequence homology between the related sequences. For some low copy number repeats, very high sequence homology extends over tens to hundreds of kb as a result of evolutionarily recent (primate-specific) **segmental duplication** – see Section 12.2.5. Such sequences can predispose to unequal recombination causing translocations and large-scale deletions, duplications and inversions. Although the deletions and some duplications are often associated with disease and typically extend over megabase (but sub-cytogenetic) intervals, large-scale inversion polymorphisms may also occur which do not directly contribute to disease (Section 11.5.5).

Chromosomal polymorphisms and large scale sequence variants

Some polymorphisms involve very large changes to noncoding DNA such that alleles can be distinguished by traditional cytogenetic analyses. C-banding (which identifies hetero-chromatin; see *Box 2.2*) often reveals size variation for specific blocks of heterochromatin as in the case of chromosomes 9, 16 and the Y. A large pericentromeric inversion on chromosome 9 is frequently found in the normal population. Occasional inversions with breakpoints apparently outside heterochromatin sequences can also be quite common in the normal population.

nucleotide was inserted on the growing DNA strands on each of about 6×10^{26} occasions.

Such a level of DNA replication fidelity is impossible to sustain; indeed, the observed fidelity of replication of DNA polymerases is very much less than this and uncorrected replication errors occur with a frequency of about 10^{-9}–10^{-11} per incorporated nucleotide (see Cooper *et al.*, 2000). As the coding DNA of an average human gene is about 1.65 kb, coding DNA mutations will occur spontaneously with an average frequency of about 1.65×10^{-6}–1.65×10^{-8} per gene per cell division. Thus, during the approximately 10^{16} mitoses undergone in an average human lifetime, each gene will be a locus for about 10^{8}–10^{10} mutations (but for any one gene, only a tiny minority of cells will carry a mutation). In many cases, a deleterious gene mutation in a somatic cell will be inconsequential: the mutation may cause lethality for that single cell, but will not have consequences for other cells. However, in some cases, the mutation may lead to an inappropriate continuation of cell division, causing cancer (see Chapter 17).

11.2.2 The frequency of individual base substitutions is nonrandom according to substitution class

Base substitutions are among the most common mutations and can be grouped into two classes. **Transitions** are substitutions of a pyrimidine by a pyrimidine (C↔T), or of a purine by a purine (A↔G). **Transversions** are substitutions of a pyrimidine by a purine or of a purine by a pyrimidine. When one base is substituted by another, there are always two possible choices for transversion, but only one choice for a transition (see *Figure 11.1*). One might, therefore, expect transversions to be twice as frequent as transitions.

Because the substitution of alleles in a population takes thousands or even millions of years to complete, nucleotide substitutions cannot be observed directly. Instead, they are always inferred from pairwise comparisons of DNA molecules that share a common origin, such as orthologs in different species. When this is done, the transition rate in mammalian genomes is found to be unexpectedly higher than transversion rates. For example, Collins and Jukes (1994) compared 337 pairs of human and rodent orthologs and found that the transition rate exceeded the transversion rate by a ratio of 1.4 : 1 for substitutions which did not lead to an altered amino acid, and by a ratio of more than 2 : 1 for those that did result in an amino acid change.

Figure 11.1: Transversions are theoretically expected to be twice as frequent as transitions.

Red arrows, transversions; black arrows, transitions.

Transitions may be favored over transversions in coding DNA because they usually result in a more conserved polypeptide sequence (see below). In both coding and noncoding vertebrate DNA the excess of transitions over transversions is at least partly due to the comparatively high frequency of C→T transitions, resulting from instability of cytosine residues occurring in the CpG dinucleotide. In such dinucleotides the cytosine is often methylated at the carbon atom 5 and 5-methylcytosines are susceptible to spontaneous deamination to give thymine (*Box 9.3*). Presumably as a result of this, the CpG dinucleotide is a hotspot for mutation in vertebrate genomes: its mutation rate is about 8.5 times higher than that of the average dinucleotide (see Cooper *et al.*, 2000) and CpG→TpG transitions are the most common type of pathogenic point mutations. Other factors favoring transitions over transversions are likely to include differential repair of mispaired bases by the sequence-dependent proofreading activities of the relevant DNA polymerases.

11.2.3 The frequency and spectrum of mutations in coding DNA differs from that in noncoding DNA

Many mutations are generated essentially randomly in the DNA of individuals. As a result, coding DNA and noncoding DNA are about equally susceptible to mutation. Clearly, however, the major consequences of mutation are largely restricted to the approximately 1.5% of the DNA in the human genome which is known to be coding DNA and to the other 3% or so of highly conserved sequences (including regulatory sequences etc.). Mutations that occur within coding DNA can be grouped into two classes:

▶ **synonymous (silent) mutations** do not change the sequence of the gene product. This applies only to polypeptide-encoding DNA. A synonymous mutation causes a codon change but does not result in an altered amino acid because of the *degeneracy* of the genetic code. *Note:* some mutations which appear to be silent may not be because they affect splicing (by activating a cryptic splice site or by altering an exonic splice enhancer sequence – Section 11.4.3);

▶ **nonsynonymous mutations** change the sequence of the gene product, which may be a polypeptide or functional noncoding (= untranslated) RNA.

True silent mutations in mammalian genomes are thought to be effectively **neutral mutations** (conferring no advantage or disadvantage to the organism in whose genome they arise). In contrast, nonsynonymous mutations can be grouped into three classes: those having a deleterious effect; those with no effect; and those with a beneficial effect (e.g. improved gene function or gene–gene interaction). Most new nonsynonymous mutations are likely to have a deleterious effect on gene expression and so can result in disease or lethality. However, the population frequency of this type of mutation is very much reduced because of *natural selection* (see Box 11.2). The overall mutation rate in coding DNA is much less than that in noncoding DNA, therefore, and coding DNA sequences (and important regulatory sequences etc.) show a relatively high degree of evolutionary conservation.

Box 11.2: Mechanisms that affect the population frequency of alleles.

Individuals within a population differ from each other, mostly as a result of inherited genetic variation. The frequency of any mutant allele in a population is dependent on a number of factors, including natural selection, random genetic drift and sequence exchanges between nonallelic sequences.

NATURAL SELECTION

Natural selection is the process whereby some of the inherited genetic variation will result in differences between individuals regarding their ability to survive and reproduce successfully. The differential reproduction is due to differences between individuals in their capacity to engage in reproduction (affected by parameters such as mortality, health and mating success) and to produce healthy offspring (differences in fertility, fecundity and viability of the offspring). *The* **fitness** *of an organism* is a measure of the individual's ability to survive and to reproduce successfully. In the simplest models, the fitness of an individual is considered to be determined solely by its genetic make-up, and all loci are imagined to contribute independently to the fitness of an individual, so that each locus can be treated separately. As a result, one can also talk about *the fitness of a genotype.*

The great majority of new nonsynonymous mutations in coding DNA reduce the fitness of their carriers. They are therefore selected against and removed from the population (**negative** or **purifying selection**). Occasionally, a new mutation may be as fit as the best allele in the population; such a mutation is selectively **neutral**. Very rarely, a new mutation confers a selective advantage and increases the fitness of its carrier. Such a mutation will be subjected to **positive** (or **advantageous**) **selection**, which would be expected to foster its spread through a population. If we consider a locus with two alleles that have different fitnesses, the heterozygote may have a fitness intermediate between the two types of homozygote. The mode of selection in this case is **codominant** and the selection will be directional, resulting in an increase of the advantageous allele. In some cases, however, a new mutation may not be advantageous in homozygotes, but only in heterozygotes (**heterozygote advantage**). This situation, in which the heterozygote has a higher fitness than both the mutant homozygote *and the normal homozygote*, is a form of balancing selection known as **overdominant selection** (see the example of cystic fibrosis in *Box 4.8*).

RANDOM GENETIC DRIFT

Changes in allele frequency can occur simply by chance (***random genetic drift***). Even if all the individuals in a population had exactly the same fitness so that natural selection could not operate, allele frequencies would nevertheless change because of *random sampling of gametes*. Sampling occurs because only a small proportion of the available gametes in any generation is ever passed on to the next generation (not all individuals in a population will reproduce because of circumstances or choice). Even if there were to be no excess of gametes (so that every single individual contributes two gametes to the next generation)

sampling would still occur because heterozygotes can produce two types of gamete bearing different alleles but the two gametes passed to the next generation may by chance carry the same allele.

Genetic drift has little effect in large populations. This is so because although only a small fraction of the total gametes is transmitted, *enough* gametes are transmitted so that they are by and large *statistically representative of all the gametes in the population*. However, when population sizes are small (e.g. as a result of geographical isolation) the number of gametes that are transmitted to the next generation will be proportionately smaller and random genetic drift can cause considerable changes in allele frequencies (see Figure). In the absence of new mutation and other factors affecting allele frequency, such as selection, alleles subject to random genetic drift will eventually reach

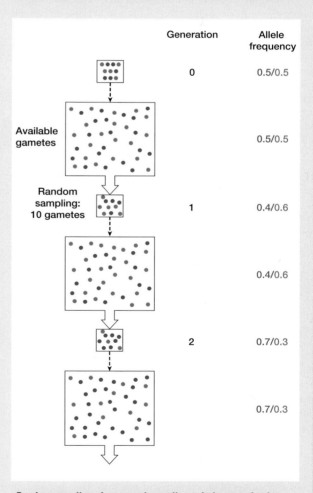

Random sampling of gametes in small populations can lead to considerable changes in allele frequencies.

Modified from Bodmer and Cavalli-Sforza (1976) *Genetics, Evolution and Man.*

Box 11.2: Mechanisms that affect the population frequency of alleles (continued).

fixation (the point at which the allele frequency in the population is 0 or 100%).

INTERLOCUS SEQUENCE EXCHANGE

Individual genes in some gene families encode essentially the same product, but there may be sequence exchanges occurring between the different gene copies. For example, human 5.8S rRNA, 18S rRNA and 28S rRNA are encoded by tandemly repeated transcription units (see *Figure 10.2*) and are particularly prone to sequence exchanges between the different

repeats. Simply as a result of the sequence exchanges between different repeats, one type of repeat can increase in population frequency (see *Figure 11.8* for an illustration of the general principle). In such cases, where multiple loci produce essentially identical products, all the genes can effectively be considered as the equivalent of alleles, although not in the Mendelian sense (which normally allows a maximum of two alleles in a diploid cell). The frequency of a specific repeat ('allele') can therefore be determined in part by the frequency with which it engages in sequence exchanges.

Selection pressure (the constraints imposed by natural selection) reduces both the overall frequency of surviving mutations in coding DNA and the spectrum of mutations seen. For example, deletions/insertions of one or several nucleotides are frequent in noncoding DNA but are conspicuously absent from coding DNA. This is so because often such mutations will cause a shift in the translational reading frame (**frameshift mutation**), introducing a premature termination codon and causing loss of gene expression. Even if insertions/deletions do not cause a frameshift mutation, they can often affect gene function, for example, as a result of removing a key coding sequence. Instead, coding DNA is marked by a comparatively high frequency of nonrandom base substitution occurring at locations which lead to minimal effects on gene expression (see next section).

11.2.4 The location of base substitutions in coding DNA is nonrandom

Nucleotide substitutions occurring in noncoding DNA usually have no net effect on gene expression. Exceptions include some changes in promoter elements or some other DNA sequence that regulates gene expression, or which is required for splicing (*Figure 1.15*). Substitutions occurring in coding DNA sequences which specify polypeptides show a very nonrandom pattern of substitutions because of the need to conserve polypeptide sequence and biological function. In principle, base substitutions can be grouped into three classes, depending on their effect on coding potential (see *Box 11.3*).

The different classes of base substitution listed in *Box 11.3* show differential tendencies to be located at the first, second or third base positions of codons. Because of the design of the genetic code, different degrees of degeneracy characterize different sites. Base positions in amino acid-specifying codons can be grouped into three classes:

▶ **nondegenerate sites** are base positions where all three possible substitutions are nonsynonymous. They include the first base position of all but eight codons, the second base position of all codons and the third base position of two codons, AUG and UGG (see *Figure 11.2*). Taking into account the observed codon frequencies in human genes, they comprise about 65% of the base positions in human codons. The base substitution rate at nondegenerate sites is

very low, consistent with a strong conservative selection pressure to avoid amino acid changes (see below);

▶ **fourfold degenerate sites** are base positions in which all three possible substitutions are synonymous and are found at the third base position of several codons (see *Figure 11.2*). They comprise about 16% of the base positions in human codons. The substitution rate at fourfold sites is very similar to that within introns and pseudogenes, consistent with the assumption that synonymous substitutions are selectively neutral (Section 11.2.5);

▶ **twofold degenerate sites** are base positions in which one of the three possible substitutions is synonymous. They are often found at the third base positions of codons, but also at the first base position in eight codons (see *Figure 11.2*). They comprise about 19% of the base positions in human codons. As expected, the substitution rate for twofold degenerate sites is intermediate: only one out of the three possible substitutions, a transition, maintains the same amino acid. The other two possible substitutions are transversions which, because of the way in which the genetic code has evolved, are often conservative substitutions. For example, at the third base position of the glutamate codon GAA, an A→G transition is silent, while the two transversions (A→C; A→T) result in replacement by a closely similar amino acid, aspartate.

The design of the genetic code and the degree to which one amino acid is functionally similar to another affect the relative *amino acid mutabilities*. Certain amino acids may play key roles which cannot be substituted easily by others. For example, cysteine is often involved in disulfide bonding which can play a crucially important role in establishing the conformation of a polypeptide (see *Figure 1.25*). As no other amino acid has a side chain with a sulfhydryl group, there is strong selection pressure to conserve cysteine residues at many locations, and cysteine is among the least mutable of the amino acids (*Table 11.1*). In contrast, certain other amino acids such as serine and threonine have very similar side chains, and substitutions at both the first base position of codons (A<u>C</u>X→U<u>C</u>X; where X = any nucleotide) and second base positions (A<u>C</u>Py→A<u>G</u>Py; where Py = pyrimidine) can result in serine→threonine substitutions. Presumably as a result, serine and threonine are among the most mutable of the amino acids (*Table 11.1*).

Box 11.3: Classes of single base substitution in polypeptide-encoding DNA.

From time to time a single nucleotide substitution within a coding exon alters RNA splicing, causing defective gene expression. This may happen in various ways: by activating a cryptic splice site within an exon (*Figure 11.12*); by affecting nucleotides in the splice donor and acceptor sequences immediately adjacent to the conserved GT and AG signals (*Figure 1.15*); or by altering internal sequences which regulate splicing, notably *exonic splice enhancers* (Section 11.4.3). Other single base substitutions can be classified into synonymous (silent) substitutions, or nonsynonymous substitutions where a codon is mutated to cause a change of amino acid sequence.

SYNONYMOUS (SILENT) SUBSTITUTIONS

The substitution results in a new codon but specifying the same amino acid. They are the most frequently observed change in coding DNA (because they are almost always neutral mutations and are not subject to selection pressure). Silent substitutions mostly occur at the third base position of a codon because third base wobble means that the altered codon often specifies the same amino acid as before. Occasionally, however, substitution at the first base position is involved, as in the case of some leucine codons (CUA⇔UUA, CUG⇔UUG) and some arginine ones (AGA⇔CGA, AGG⇔CGG).

NONSENSE MUTATIONS

These represent a form of nonsynonymous substitution where a codon specifying an amino acid is replaced by a stop codon. Because such mutations are almost always associated with a dramatic reduction in gene function, selection pressure ensures that they are normally rare. The average human polypeptide is specified by about 500–550 codons, a size which would be expected to harbor over 20 termination codons, if there were no functional constraints.

MISSENSE MUTATIONS

These are nonsynonymous substitutions where the altered codon specifies a different amino acid. They can be classified into two subgroups:

▶ *conservative substitutions* result in replacement of an amino acid by another that is chemically similar to it. Often, the effect of such substitutions on protein function is minimal because the side chain of the new amino acid may be functionally similar to that of the amino acid it replaces (see footnote below). To minimize the effect of nucleotide substitution, the genetic code appears to have evolved so that *codons specifying related amino acids are themselves related*. For example, the Asp (GAC, GAT) and Glu (GAA, GAG) codon pairs ensure that third base wobble in a GAX codon (where X is any nucleotide) has a minimal effect. However, some first codon position changes can also be conservative, e.g. CUX (Leu)⇔GUX (Val);

▶ *nonconservative substitutions* result in replacement of one amino acid by another with a dissimilar side chain (see table below). Sometimes a charge difference is introduced; other changes may involve replacement of polar side chains by nonpolar ones and vice versa. Base substitutions at the first and second codon positions can often result in nonconservative substitutions, e.g. CGX (Arg)→GGX (Gly), CCX (Pro), CUX (Leu) or CAX (Gln/His), etc (where X = any nucleotide).

Physicochemical distances between pairs of amino acids.

Arg	Leu	Pro	Thr	Ala	Val	Gly	Ile	Phe	Tyr	Cys	His	Gln	Asn	Lys	Asp	Glu	Met	Trp	
110	145	74	58	99	124	56	142	155	144	112	89	68	46	121	65	80	135	177	Ser
	102	103	71	112	96	125	97	97	77	180	29	43	86	26	96	54	91	101	Arg
		98	92	96	32	138	5	22	36	198	99	99	113	107	172	138	15	61	Leu
			38	27	68	42	95	114	110	169	77	76	91	103	108	93	87	147	Pro
				58	69	59	89	103	92	149	47	42	65	78	85	65	81	128	Thr
					64	60	94	113	112	195	86	91	111	106	126	107	84	148	Ala
						109	29	50	55	192	84	96	133	97	152	121	21	88	Val
							135	153	147	159	98	87	80	127	94	98	127	184	Gly
								21	33	198	94	109	149	102	168	134	10	61	Ile
									22	205	100	116	158	102	177	140	28	40	Phe
										194	83	99	143	85	160	122	36	37	Tyr
											174	154	139	202	154	170	196	215	Cys
												24	68	32	81	40	87	115	His
													46	53	61	29	101	130	Gln
														94	23	42	95	174	Asn
															101	56	95	110	Lys
																45	160	181	Asp
																	126	152	Glu
																		67	Met

Quantifying the degree of similarity between two amino acids is based on properties such as polarity, molecular volume and chemical composition, with large numbers signifying greater dissimilarity. In the table above from Grantham (1974) the most similar pairs are: Leu⇔Ile (5), Met⇔Ile (10), and Met⇔Leu (15); the most dissimilar pairs are Cys⇔Trp (215), Cys⇔Phe (205), and Cys⇔Lys (202).

UUU	Phe	17.1	UCU	Ser	14.7	UAU	Tyr	12.1	UGU	Cys	10.1
UUC	Phe	20.4	UCC	Ser	17.5	UAC	Tyr	15.5	UGC	Cys	12.4
UUA	Leu	7.3	UCA	Ser	11.9	(UAA	STOP)		(UGA	STOP)	
UUG	Leu	12.7	UCG	Ser	4.5	(UAG | STOP)		UGG	Trp	13.0
CUU	Leu	12.9	CCU	Pro	17.3	CAU	His	10.6	CGU	Arg	4.7
CUC	Leu	19.5	CCC	Pro	20.0	CAC	His	15.0	CGC	Arg	10.8
CUA	Leu	7.0	CCA	Pro	16.7	CAA	Gln	11.9	CGA	Arg	6.3
CUG	Leu	40.1	CCG	Pro	7.0	CAG	Gln	34.4	CGG	Arg	11.8
AUU	Ile	15.8	ACU	Thr	12.9	AAU	Asn	16.7	AGU	Ser	12.0
AUC	Ile	21.3	ACC	Thr	19.1	AAC	Asn	19.3	AGC	Ser	19.4
AUA	Ile	7.2	ACA	Thr	14.9	AAA	Lys	24.0	AGA	Arg	11.7
AUG	Met	22.3	ACG	Thr	6.2	AAG	Lys	32.5	AGG	Arg	11.6
GUU	Val	10.9	GCU	Ala	18.6	GAU	Asp	22.1	GGU	Gly	10.8
GUC	Val	14.6	GCC	Ala	28.4	GAC	Asp	25.7	GGC	Gly	22.6
GUA	Val	7.0	GCA	Ala	16.0	GAA	Glu	29.0	GGA	Gly	16.4
GUG	Val	28.7	GCG	Ala	7.6	GAG	Glu	40.3	GGG	Gly	16.4

Key

N Nondegenerate site
N Two-fold degenerate site
N Four-fold degenerate site

Figure 11.2: Codon frequencies in human genes and locations of nondegenerate, twofold and fourfold degenerate sites.

Observed codon frequencies are given as values out of 1000 (e.g. UUU = 17.1/1000 or 0.0171). They were derived from the Codon Usage database (http://www.kazusa.or.jp/codon/) and involved sampling of 21 930 294 codons from GenBank Release 131.0 (15 August 2002). *Note:* although eight of the 61 first base positions are twofold degenerate, about 96% of all possible substitutions at the first base position are nonsynonymous. Of the substitutions at the second base position, 100% are nonsynonymous and at the third base position, about 33%.

11.2.5 Substitution rates vary considerably between different genes and between different gene components

The neutral substitution rate

The recent availability of draft sequences for the human and mouse genomes has enabled genome-wide analyses of sequence divergence and substitution rates (Mouse Genome Sequencing Consortium, 2002). As a reference point the neutral substitution rate was estimated from alignments of nonfunctional DNA. This was achieved by identifying ancestral repeat sequences (noncoding transposon-based repeats which were judged to have inserted into the genomic DNA of the common ancestor and become fixed before the divergence of humans and mice. A total of 165 Mb of ancestral repeat sequences could be identified (orthologous sequences were determined by alignment of adjacent nonrepetitive DNA).

The overall sequence identity for ancestral repeat sequences was estimated as 66.7%. The sequence identity at fourfold degenerate sites (see previous section), another set of potentially functionless sequences, was 67%. These remarkably similar values have led to estimates of 0.46–0.47 substitutions per site for the neutral substitution rate (Mouse Genome Sequencing Consortium, 2002). Taking into account differences in mutation rate in the lineages leading to modern humans and mice (Section 11.2.6), this has been estimated to reflect a neutral substitution rate of about 2×10^{-9} per site per year in the human lineage.

The substitution rate in different gene components

Comparisons of over 14 000 pairs of orthologous human and mouse gene sequences have also emphasized differences in substitution rate within different gene components (Mouse Genome Sequencing Consortium, 2002; see *Figure 11.3* for a visual representation based on a subset of the data). As expected coding regions are the most conserved (85% sequence identity, or 0.165 substitutions per nucleotide site) but intron sequences are much less conserved (68.6%

Table 11.1: Relative amino acid mutability

(Ala = 100; data from Collins and Jukes, 1994)

Thr	116	Asp	84
Ser	114	Lys	77
His	107	Glu	76
Asn	107	Pro	67
Met	102	Leu	58
Ala	100	Gly	57
Gln	99	Phe	55
Val	98	Tyr	53
Arg	94	Trp	31
Ile	92	Cys	29

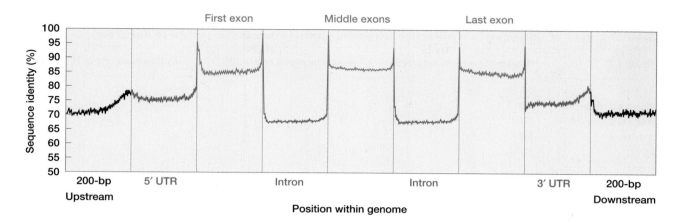

Figure 11.3: Variation in human–mouse sequence conservation across a typical gene.

The rate of nucleotide substitution varies in different components of a gene, as revealed by alignment of over 3000 human RefSeq mRNAs and upstream/downstream genomic sequences against orthologous mouse sequences. Adapted from the Mouse Genome Sequence Consortium (2002) *Nature* **420**, 520–562, Figure 25A with permission from Nature Publishing Group.

sequence identity which is close to the overall 69.1% sequence identity for genome-wide sequence comparisons). The untranslated regions show intermediate conservation (5′ UTR, 75.9%; 3′ UTR, 74.7%) as do the flanking 200 bp upstream (promoter region − 73.9%) and downstream 200 bp (70.9%).

The substitution rate in different genes and domains

Different genes are conserved to different extents. Generally, the substitution rate at synonymous sites is relatively conserved but the nonsynonymous substitution rate can vary markedly. If we use K_S to denote the number of substitutions per synonymous site and K_A to represent the number of substitutions per nonsynonymous site, genes which are subject to *negative (purifying) selection* will have $K_A \ll K_S$. Genes subject to positive selection for amino acid replacement can also show $K_A < K_S$ because the positive selection may be operating in only a few crucially important sites and is offset by purifying selection at a large number of other sites. When over 12 000 orthologous human and mouse gene pairs were examined by the Mouse Genome Sequencing Consortium (2002) the median value of K_A/K_S was found to be 0.115.

At one extreme are proteins whose sequences are extremely highly conserved because they are required for crucial cell functions, specifying proteins such as actins, calmodulin, histones, ribosomal proteins, ubiquitin and so on. For example, the ubiquitin proteins of humans, mouse and *Drosophila* show 100% sequence identity, and comparison with the yeast ubiquitin reveals 96.1% sequence identity. These genes are not especially protected from mutation, because the rate of synonymous codon substitution is typical of that for many protein-encoding genes. Instead, what distinguishes them is the extremely low rate of nonsynonymous codon substitution compared with other genes (see *Table 11.2* for some examples).

The genes which show the highest rates of nonsynonymous codon substitution include many which are implicated in

mammalian defense and immune response systems (*Table 11.2*), as do six out of the eight common protein domains associated with the highest K_A/K_S values (Mouse Genome Sequencing Consortium, 2002 − Table 13). The high K_A/K_S ratio in these cases indicates that they are either under reduced purifying selection, increased positive selection or both. Increased positive selection could reflect a competitive struggle between a mammalian host and its pathogens where each is under strong pressure to respond to innovations in the other genome. The same six out of eight domains with the highest K_A/K_S values are found in secreted proteins and secreted domains have considerably higher K_A/K_S ratios than nuclear and cytoplasmic domains (*Table 11.3*). Catalytic domains seem to have comparatively low K_A/K_S ratios.

11.2.6 The substitution rate can vary in different chromosomal regions and in different lineages

Substitution rate and heterozygosity in different chromosomal regions

For various reasons the substitution rate in the mitochondrial genome is much higher than in the nuclear genome DNA (Section 11.4.2), but the latter shows very considerable regional variation. The draft human and mouse genome sequences offered an ideal opportunity to assess this. Using aligned ancestral repeat sequences (see previous section) and aligned fourfold degenerate sites the Mouse Genome Sequencing Consortium (2002) calculated the apparent neutral substitution rate for about 2500 overlapping 5-Mb windows across the human genome. Regional variation was evident when the average rates on different chromosomes were compared. The substitution rate is lowest on the X chromosome and there are significant differences between the autosomes (*Figure 11.4A*). Since the X chromosome spends two-thirds of its time in females the low substitution rate may reflect differences in the number of germ cell divisions in males and females (*Box 11.4*)

Table 11.2: Rates of synonymous and nonsynonymous substitutions in mammalian protein-coding genes

Gene	Number of codons compared	Nonsynonymous rate ($\times 10^9$)	Synonymous rate ($\times 10^9$)
Actin α	376	0.01	2.92
Ribosomal protein S14	150	0.02	2.16
Ribosomal protein S17	134	0.06	2.69
Aldolase A	363	0.09	2.78
HPRT	217	0.12	1.57
Insulin	51	0.20	3.03
α-Globin	141	0.56	4.38
β-Globin	146	0.78	2.58
Albumin	590	0.92	5.16
Ig V_H	100	1.10	4.76
Growth hormone	189	1.34	3.79
Ig κ	106	2.03	5.56
Interferon-β_1	159	2.38	5.33
Interferon-γ	136	3.06	5.50

Data from human–rodent comparisons extracted from Table 4.1 of Grauer and Li (2000).

Subchromosomal variation in the substitution rate (and also in the rates of deletion and insertions) is also very evident, as in the case of human chromosome 22 (*Figure 11.4B*). In part this reflects the base composition: the substitution rate does seem to be higher in regions of extremely high or extremely low G + C content but the relationship is complex. Using the deCode Genetics high resolution recombination map of the human genome (Kong *et al.*, 2002) as a reference, a correlation was also found between the substitution rate and the recombination rate and also single nucleotide polymorphism SNP density (see *Figure 11.4C*). While the average heterozygosity in the human genome is of the order of 1 out of 1250 bases (Reich *et al.*, 2002), SNP density can vary enormously in different parts of the genome and when averaged across windows of 200 kb, the

rates of heterozygosity show up to a 10-fold variation (International SNP Map Working Group, 2001). At a finer resolution sizeable regions of the human genome spanning over tens of thousands of base pairs appear to have intrinsically high and low rates of sequence variation. For these regions only a small proportion (up to 25%) of the variation appears to be due to the local mutation rate; the biggest contributor is shared genealogical history (Reich *et al.*, 2002).

Human versus mouse substitution rates

The draft human and mouse sequences have also permitted analysis of how substitution rates can vary in different lineages. Since synonymous substitutions have been considered to be effectively neutral from the point of view of selective

Table 11.3: Sequence conservation and substitution rates for orthologous genes and domain-encoding DNA in humans and mice

Orthologous region	Amino acid identity (%)	K_A	K_S	K_A/K_S
Full length protein	78.5	0.071	0.602	0.115
Domain-containing protein regions	93.5	0.032	0.601	0.061
Domain-free protein regions	71.1	0.090	0.586	0.155
All predicted domains	95.1	0.024	0.627	0.062
Catalytic domains	96.6	0.015	0.578	0.033
Non-catalytic domains	94.9	0.026	0.635	0.068
Nuclear domains	98.6	0.008	0.655	0.050
Secreted domains	88.9	0.058	0.694	0.091
Cytoplasmic domains	96.7	0.015	0.587	0.041

K_A, Nonsynonymous substitution rate; K_S, synonymous substitution rate. Data abstracted from Table 12 of the Mouse Genome Sequencing Consortium (2002) *Nature* **420**, 520–562, with permission from Nature Publishing Group.

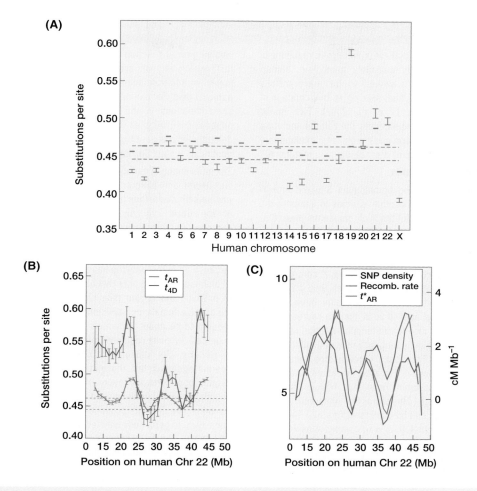

Figure 11.4: Variation in substitution rates between human chromosomes and subchromosomal regions.

Variation in the estimated number of substitutions per site within fourfold degenerate sites (blue; t_{4D}) and in ancestral repeat sites (red; t_{AR}) is shown for different human chromosomes **(A)**, and along human chromosome 22 **(B)**. Dashed lines in (A) indicate genome-wide averages (note the conspicuously high number of substitutions in fourfold degenerate sites in human chromosome 19 but low substitution rates on the X chromosome) **(C)**. Correlation of the substitution rate in ancestral repeats (t_{AR}) across chromosome 22 with SNP density and recombination rate. Adapted from the Mouse Genome Sequence Consortium (2002) *Nature* **420**, 520–562, Figures 29A and 30A with permission from Nature Publishing Group.

constraints, the concept of a constant **molecular clock** (whereby a given gene or gene product undergoes a constant rate of molecular evolution) was suggested over 30 years ago. Since then, however, various lines of evidence have argued against a constant molecular clock (see Ayala, 1999; and *Table 11.2* which shows apparent differences in the rate of synonymous codon substitutions as well as in the case of nonsynonymous codon substitutions).

To assess the relative rates of nucleotide substitutions in two lineages leading to present-day species A and B, many studies have used a **relative rate test** with a distantly related reference species C (known to have branched off earlier in evolution, before the A–B split). Pairwise comparisons of orthologs in A and C, and in B and C are used to calculate the K value, the frequency of synonymous substitutions. The K_{AC} and K_{BC} values then provide a measure of the relative rates of mutation in the lineages leading to species A and to species B.

Tests of this type have suggested that mutation rate in lineages leading to the primates is lower than that in rodent lineages, and lower still in the lineage leading to modern day humans (e.g. Wu and Li, 1985; Li and Tanimura, 1987). However, more recent studies (e.g. Kumar and Subramanian, 2002) have challenged this view, often on the basis of perceived inaccuracies in mammalian phylogeny.

To look for possible differences in mutation rates in the human and mouse lineages, the Mouse Genome Sequencing Consortium (2002) investigated the divergence of 18 subfamilies of human and mouse ancestral repeats which were active shortly before human–mouse divergence. Thousands of ubiquitously distributed copies for each of the subfamilies were compared within each genome to assess divergence from a consensus sequence, and so regional variation in substitution rates could be minimized. As can be seen in the representative examples in *Table 11.4* the least diverged ancestral repeats in

Box 11.4: Sex differences in mutation rate and the question of male-driven evolution.

Since Haldane first observed that most mutations resulting in hemophilia were generated in the male germ line, human mutations have been often assumed to be preferentially inherited through the paternal line, leading to the concept of **male-driven evolution** (see Li *et al.*, 2002). Two major approaches have been taken to estimate the relative mutation rates in the male and female germ lines: molecular evolutionary methods and direct observation of disease-causing mutations.

Molecular evolutionary methods

Usually this involves comparing homologous X-linked and Y-linked genes with orthologs in another species to estimate the rate of synonymous mutations for the X chromosome gene (K_{SX}) and the Y chromosome gene (K_{SY}). Unlike autosomes, the sex chromosomes spend different amounts of time in the two sexes. Y chromosome sequences located outside the pseudoautosomal region spend all their time in males; X chromosome sequences spend on average two-thirds of their time in females and one-third of their time in males (females have two X chromosomes; males have one). If α is used to represent the ratio of the male mutation rate to the female mutation rate, this is equivalent to setting the male mutation rate at α relative to a female mutation rate of 1. For most Y chromosome sequences, the mutation rate is therefore α. For X chromosome sequences it is $2/3 \times 1$ (the female mutation rate) plus $1/3 \times \alpha$ (the male mutation rate) $= 2/3 + \alpha/3$. Therefore the observed K_{SX}/K_{SY} ratio $= (2/3 + \alpha/3)/\alpha$ and so can be used to estimate α. This often results in estimates of about 4–6 for α in comparisons involving primates (see Li *et al.*, 2002, Makova and Li, 2002) although some studies have given values closer to 2, a value not so different from equivalent comparisons in rodents where there are less germ cell divisions. Possibly, therefore, mutations are not so heavily dependent on replication errors as was previously thought.

Direct observation of disease-causing mutations

Clearly, the mutations assessed in this case are a special subset. Samples are analyzed from a patient with a *de novo* mutation and the two parents (the parent who passed on the faulty chromosome is typically identified by typing for markers closely flanking the disease gene). In most cases the mutation will have been transmitted through the germ line, but if only a blood sample is typed, the mutations may include some **post-zygotic mutations** (not present in sperm or egg but arising following formation of the zygote).

The available data point to a generally large bias in favor of paternal mutations, at least in the case of simple point mutations (see Crow, 2000; Li *et al.*, 2002 and *Table 11.5*). The most striking sex biases in mutation rates are unusual, however, in that they involve gain-of-function mutations in genes (FGFR2, FGFR3 and RET) which encode proteins belonging to the same protein family. In addition, some types of mutation do not show such paternal bias. For example, the great majority of *de novo* large-scale deletions in the dystrophin gene appear to arise in oogenesis (Grimm *et al.*, 1994; see also below) and large deletions encompassing the neurofibromin gene often occur in oogenesis (Lopez Correa *et al.*, 2000).

Sex differences in mutation rates may be due to different factors (see Hurst and Ellegren, 1998; Li *et al.*, 2002). In the case of **premeiotic mutations** (inherited mutations which occur in one of the germ cell divisions preceding meiosis) a major contributory factor is likely to be the large sex difference in the number of human germ cell divisions. In females, the number of cell divisions from zygote to fertilized oocyte is constant because all of the oocytes have been formed by the fifth month of development and only two further cell divisions are required to produce the zygote (see Figure, panel A). The number of successive female cell divisions from zygote to mature egg is thought to be about 24, which is close to the 30–31 estimated male cell divisions required from zygote to stem spermatogonia at puberty. Six subsequent cell divisions are required for spermatogenesis but thereafter the spermatogenesis cycle occurs approximately every 16 days or 23 cycles per year (Figure, panel B). As illustrated in the Figure, the number of cell divisions required to produce sperm is age dependent. By puberty male germ cells will have undergone 30 mitotic divisions and sperm can be produced following a further six divisions. Thereafter sperm can continuously be produced because the stem cells undergo one division every 16 days (= 23 divisions per year). So if puberty occurs at, say 15 years, the number of germ cell divisions required to produce sperm in a man aged n years = $36 + [23 \times (n-15)]$, or about 265 divisions for a 25 year old man and about 840 divisions for a 50 year old man.

Errors in DNA replication/repair during germ cell divisions can be expected to provide the great majority of simple point mutations, and so one might then expect that the male mutation rate would be substantially greater than that of the female, and that a paternal age effect would be notable (see e.g. Crow, 2000). However, some more complex classes of inherited mutations may be more disposed to occur at meiosis (**meiotic mutations**) rather than at the many germ cell divisions preceding meiosis. For example, large deletions may often arise by unequal crossover at meiosis (Section 11.3.2). In the case of large-scale deletions in X-linked genes such as dystrophin which do not have a Y chromosome homolog, a significant bias towards meiotic mutation would mean that most deletions would arise in oogenesis.

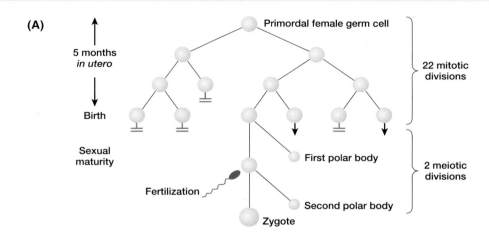

(A)

5 months
in utero

Birth

Sexual
maturity

Primordal female germ cell

22 mitotic
divisions

First polar body

Fertilization

Second polar body

Zygote

2 meiotic
divisions

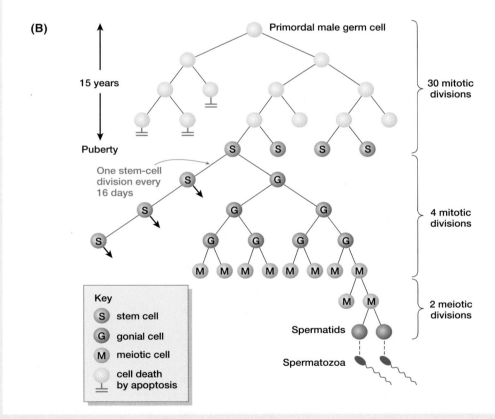

(B)

15 years

Puberty

One stem-cell
division every
16 days

Primordal male germ cell

30 mitotic
divisions

4 mitotic
divisions

2 meiotic
divisions

Spermatids

Spermatozoa

Key

S stem cell

G gonial cell

M meiotic cell

  cell death
  by apoptosis

The sex difference in germ cell divisions.

(A) Human oogenesis occurs only during fetal life and ceases by the time of birth. The total number of cell divisions is thought to be around 24. **(B)** Human spermatogenesis continues through adult life and sperm cells originate from self-renewing stem cells via spermatogonia. The number of germ cell divisions is therefore very age dependent. Modified from Vogel and Motulsky (1996) *Human Genetics. Problems and Approaches*, 3rd Edn, with permission from Springer Verlag. © 1996, Springer-Verlag.

the human subfamilies show about 16% divergence from a consensus sequence (or about 0.17 substitutions per site) whereas repeats in each of the mouse subfamilies have diverged by at least 26–27% (about 0.34 substitutions per site). Assuming that human–mouse divergence took place about 75 Myr (million years) ago the average substitution rates would have been 2.2×10^{-9} in the human lineage and 4.5×10^{-9} in the mouse lineage. These figures are the *average* rates ever since

human–mouse divergence and the current substitution rate per year in the mouse genome is thought to be much higher (see Mouse Genome Sequencing Consortium, 2002). The same comparisons also allowed an estimate of the rate of occurrence of small (< 50 bp) insertions and deletions. Both species showed a net loss of nucleotides (the ratios of deleted bases : inserted bases were in the 2 : 1 to 3 : 1 range), but the overall loss was at least twice as high in mouse.

Table 11.4: Ancestral repeats have diverged more rapidly in the mouse lineage than in the human lineage

Subfamily	Class of substitution	Mouse		Human		Adjusted ratio
		Divergence rate	Substitution rate	Divergence rate	Substitution rate	
L1MA6	LINE1	0.28	0.35	0.16	0.184	1.98
L1MA7	LINE1	0.28	0.35	0.16	0.181	1.96
L1MA8	LINE1	0.27	0.34	0.15	0.172	1.96
L1MA9	LINE1	0.28	0.35	0.18	0.201	1.86
MLT1A	MaLR	0.31	0.39	0.21	0.242	1.73
MLT1A0	MaLR	0.30	0.38	0.19	0.219	1.80
MLT1A1	MaLR	0.29	0.37	0.19	0.214	1.78
MER20	DNA	0.29	0.37	0.19	0.222	1.76
MER33	DNA	0.27	0.33	0.18	0.211	1.63
Tigger6a	DNA	0.29	0.37	0.18	0.211	1.85

Abstracted from the Mouse Genome Sequencing Consortium (2002) *Nature* **420**, 520–562. Data are shown for a representative 10 out of 18 subfamilies of ancestral repeats from Table 6 of the Nature paper.

Table 11.5: A bias towards paternally inherited mutations causing human disease

Disease	Gene	Number of paternal mutations	Number of maternal mutations	Ratio of paternal/ maternal mutations (α)
X-linked dominant				
Pelizaeus–Merzbacher disease	*PLP*	4	1	4
Rett syndrome	*MECP2*	27	2	13.5
Autosomal dominant				
Achondroplasia	*FGFR3*	40	0	Inf.
Apert syndrome	*FGFR2*	57	0	Inf.
Crouzon and Pfeiffer syndromes	*FGFR2*	22	0	Inf.
Denys–Drash syndrome	*WT1*	2	0	Inf.
Hirschsprung disease	*RET*	0	3	0
Multiple endocrine neoplasia 2A	*RET*	10	0	Inf.
Multiple endocrine neoplasia 2B	*RET*	25	0	Inf.
Neurofibromatosis type 2	*NF2*	13	10	1.3
Von Hippel–Lindau disease	*VHL*	4	3	1.3
Total		204	19	10

Data abstracted from Li *et al.* (2002) *Curr. Opin. Genet. Dev.* **12**, 650–656 with permission from Elsevier.

11.3 Genetic mechanisms which result in sequence exchanges between repeats

In addition to very frequent simple mutations, there are several mutation classes which involve *sequence exchange* between allelic or nonallelic sequences, often involving repeated sequences. For example, tandemly repetitive DNA is prone to deletion/insertion polymorphism whereby different alleles vary in the number of integral copies of the tandem repeat. Such *variable number tandem repeat (VNTR) polymorphisms* can occur in the case of repeated units that are very short (microsatellites); intermediate (minisatellites) or large. Different genetic mechanisms can account for VNTR polymorphism depending on the size of the repeating unit (see the following two sections). In addition, interspersed repeats can also predispose to deletions/duplications by a variety of different genetic mechanisms. These are discussed particularly in the context of disease mutations and are therefore presented in Section 11.4.

11.3.1 Replication slippage can cause VNTR polymorphism at short tandem repeats (microsatellites)

Germline mutation rates at microsatellite loci vary but are often in the range of 10^{-3} to 10^{-4} per locus per generation (see Ellegren, 2000). Novel length alleles at (CA)/(TG) microsatellites and at tetranucleotide marker loci are known to be formed without exchange of flanking markers. This means that they are not generated by unequal crossover (see below). Instead, as new mutant alleles have been observed to differ by a single repeat unit from the originating parental allele, the most likely mechanism to explain length variation is a form of exchange of sequence information which commences by *slipped strand mispairing*. This occurs when the normal pairing between the two complementary strands of a double helix is altered by staggering of the repeats on the two strands, leading to incorrect pairing of repeats. Although slipped strand mispairing can be envisaged to occur in nonreplicating DNA, replicating DNA may offer more opportunity for slippage and hence the mechanism is often also called **replication slippage** or **polymerase slippage** (see *Figure 11.5*). In addition to mispairing between tandem repeats, slippage replication has been envisaged to generate large deletions and duplications by mispairing between *noncontiguous repeats* and has been suggested to be a major mechanism for DNA sequence and genome evolution (Levinson and Gutman, 1987; see also Dover, 1995). The pathogenic potential of short tandem repeats is considerable (Sections 11.5.1, 11.5.2).

11.3.2 Large units of tandemly repeated DNA are prone to insertion/deletion as a result of unequal crossover or unequal sister chromatid exchanges

Homologous recombination describes recombination (**crossover**) occurring at meiosis or, rarely, mitosis between identical or very similar DNA sequences, and usually involves breakage of nonsister chromatids of a pair of homologs and rejoining of the fragments to generate new recombinant strands. **Sister chromatid exchange** is an analogous type of sequence exchange involving breakage of individual sister chromatids and rejoining fragments that initially were on different chromatids of the same chromosome. Both homologous recombination and sister chromatid exchange normally involve *equal* exchanges – cleavage and rejoining of the chromatids occurs at the same position on each chromatid. As a result, the exchanges occur between allelic sequences and at corresponding positions within alleles. In the case of intragenic equal crossover between two alleles, a new allele can result which is a **fusion gene** (or **hybrid gene**), comprising a terminal fragment from one allele and the remaining sequence of the second allele (*Figure 11.6*). However, equal sister chromatid exchanges cannot normally produce genetic variation because sister chromatids have identical DNA sequences.

Unequal crossover (UEC) is a form of *nonallelic homologous recombination* in which the crossover takes place between nonallelic sequences on *nonsister chromatids* of a pair of homologs (*Figure 11.7*). Often the sequences at which crossover takes place show very considerable sequence homology which presumably stabilizes mispairing of the chromosomes. The analogous exchange between *sister chromatids* is called **unequal sister chromatid exchange** (**UESCE**; see *Figure 11.7*).

Both UEC and UESCE occur predominantly in regions of the genome where there are tandem repeats of a moderate to large sized sequence with high homology between the repeats (e.g. in rDNA clusters, complex satellite DNA etc.). In such cases, the very high degree of sequence homology between the different repeats can facilitate abnormal pairing of nonallelic repeats on nonsister chromatids or sister chromatids – the chromatids mispair with one chromatid out of register with the other by an integral number of repeat units. If chromosome breakage and rejoining occurs while the chromatids are mispaired in this way, there will be a reciprocal sequence exchange causing insertion on one chromatid, and an equal sized deletion on the other (which may be pathogenic). Such exchanges can also lead to *concerted evolution* by causing a particular sequence variant to spread through an array of tandem repeats, resulting in homogenization of the repeat units (see *Figure 11.8*).

While UEC and UESCE are particularly common in tandemly repeated DNA, they can also be initiated by mispairing between repeats which are separated by a considerable amount of intervening sequence. For example, mispairing of nonallelic Alu repeats or other interspersed repeats can occasionally occur and can cause the formation of a tandemly duplicated locus from an originally single-copy locus (*Figure 11.9*).

11.3.3 Gene conversion events may be relatively frequent in tandemly repetitive DNA

Gene conversion describes a *nonreciprocal* transfer of sequence information between a pair of nonallelic DNA sequences (*interlocus gene conversion*) or allelic sequences (*interallelic gene*

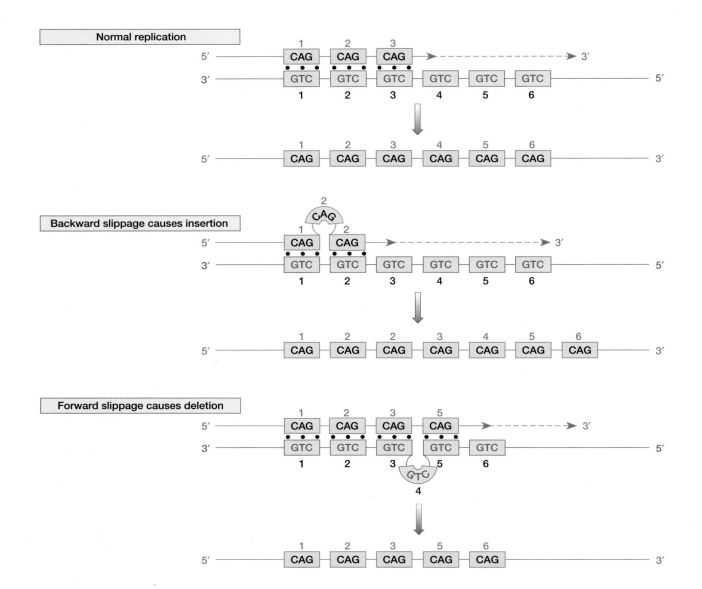

Figure 11.5: Slipped strand mispairing during DNA replication can cause insertions or deletions.

Short tandem repeats are thought to be particularly prone to slipped strand mispairing (= mispairing of the complementary DNA strands of a single DNA double helix). The examples show how slipped strand mispairing can occur during replication, with the lower strand representing a parental DNA strand and the upper strand representing the newly synthesized complementary strand. In such cases, slippage involves a region of nonpairing (shown as a bubble) containing one or more repeats of the newly synthesized strand (backward slippage) or of the parental strand (forward slippage), causing, respectively, an insertion or a deletion on the newly synthesized strand. **Note:** it is conceivable that slipped strand mispairing can also cause insertions/deletions in nonreplicating DNA. In such cases, two regions of nonpairing are required, one containing repeats from one DNA strand and the other containing repeats from the complementary strand (see Levinson and Gutman, 1987). A mismatch repair enzyme might then introduce an insertion or deletion by 'correcting' the non-pairing.

conversion). One of the pair of interacting sequences, the **donor**, remains unchanged. The other DNA sequence, the **acceptor**, is changed by having some or all of its sequence replaced by a sequence copied from the donor sequence (*Figure 11.10*). The sequence exchange is therefore a directional one; the acceptor sequence is modified by the donor sequence, but not the other way round.

One possible mechanism for gene conversion envisages formation of a heteroduplex between a DNA strand from the donor gene and a complementary strand from the acceptor gene. Following heteroduplex formation, conversion of an acceptor gene segment may occur by **mismatch repair** – DNA repair enzymes recognize that the two strands of the heteroduplex are not perfectly matched and 'correct' the

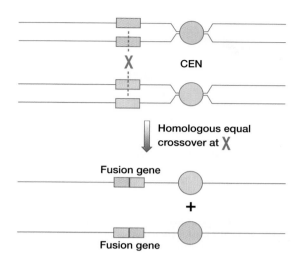

Figure 11.6: Homologous equal crossover can result in fusion genes.

The example shows how intragenic equal crossover occurring between alleles on nonsister chromatids can generate novel fusion genes composed of adjacent segments from the two alleles. Note that similar exchanges between genes on sister chromatids do not result in genetic novelty because the gene sequences on the interacting sister chromatids would be expected to be identical.

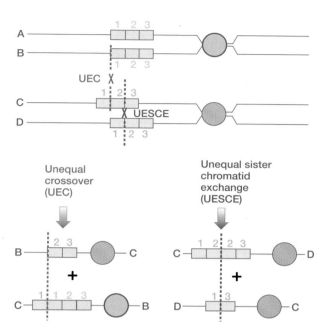

Figure 11.7: Unequal crossover and unequal sister chromatid exchange cause insertions and deletions.

The examples illustrate unequal pairing of chromatids within a tandemly repeated array. Unequal crossover involves unequal pairing of *nonsister chromatids* followed by chromatid breakage and rejoining. Unequal sister chromatid exchange involves unequal pairing of sister chromatids followed by chromatid breakage and rejoining. For the sake of simplicity, the breakages of the chromatids are shown to occur between repeats, but of course breaks can occur within repeats. Note: both types of exchange are reciprocal – one of the participating chromatids loses some DNA, while the other gains some.

DNA sequence of the acceptor strand to make it perfectly complementary in the converted region to the sequence of the donor gene strand (see *Figure 11.10*).

Gene conversion has been well-described in fungi where all four products of meiosis can be recovered and studied (tetrad analysis). In humans and mammals it is not possible to do this. Gene conversion cannot be demonstrated unambiguously in higher organisms because it can never be distinguished from double crossover events, for example (although double crossovers occurring in very close proximity would normally be expected to be extremely unlikely). Nevertheless, there are numerous instances in mammalian genomes where an allele at one locus shows a pattern of mutations which strongly resembles those found in alleles at another locus of the same species, suggesting gene conversion-like exchanges between loci.

Although simple comparisons of two sequences may be suggestive, the evidence for gene conversion is most compelling when a new mutant allele can be compared directly with its progenitor sequence. Certain highly mutable loci lend themselves to this type of analysis. In particular, some hypervariable minisatellite loci have high germline mutation rates (often 1% or more per gamete) and individual repeats often show nucleotide differences so that repeat subclasses can be recognized. Germline mutations can be studied by detecting and characterizing mutant minisatellite alleles in individual gametes. To do this, PCR analysis has been conducted on multiple dilute aliquots of DNA isolated from the sperm of an individual (**small pool PCR**), where each

aliquot is calibrated to contain a few, perhaps 100, input molecules (Jeffreys *et al.*, 1994).

The PCR products recovered from individual pools are typed to identify any new mutations that result in a novel allele whose length is sufficiently different as to be distinguishable from the progenitor allele. Analyses of the patterns of germline mutation at three such loci have failed to identify exchanges of flanking markers and have shown that most mutations occurring at these loci are polar, involving the preferential gain of a few repeats at one end of a tandem repeat array. There is a bias towards gain of repeats and evidence was obtained for nonreciprocal sequence exchange between alleles, suggesting interallelic gene conversion (Jeffreys *et al.*, 1994). Evidence for interlocus gene conversion has also been obtained in human genes, notably the steroid 21-hydroxylase gene (see Section 11.5.3).

11.4 Pathogenic mutations

Deleterious mutations typically affect gene expression, either by altering a coding sequence directly or by altering intragenic

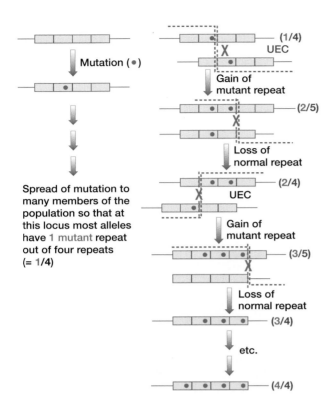

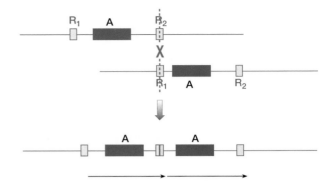

Figure 11.8: Unequal crossover in a tandem repeat array can result in sequence homogenization.

Note that the initial spread of the novel sequence variant to the same position in the chromosomes of other members of a sexual population can result by random genetic drift (see *Box 11.2*). Once the mutation has achieved a reasonable population frequency (left panel) it can spread to other positions within the array (right panel). This can occur by successive gain of mutant repeats as a result of unequal crossover (or unequal sister chromatid exchanges) and occasional loss of normal repeats. Eventually the mutant repeat can replace the original repeat sequence at all positions within the array, leading to sequence homogenization for the mutant repeat. Such sequence homogenization is thought to result in species-specific concerted evolution for repetitive DNA sequences. UEC, unequal crossover.

Figure 11.9: Tandem gene duplication can result from unequal crossover or unequal sister chromatid exchange, facilitated by short interspersed repeats.

The double arrow at the bottom indicates the extent of the tandem gene duplication of a segment containing gene A and flanking sequences. Original mispairing of chromatids could be facilitated by a high degree of sequence homology between nonallelic short repeats (R_1, R_2). Note that the same mechanism can result in large scale deletions (see *Figure 16.3*).

11.4.1 There is a high deleterious mutation rate in hominids

The mutation rate for *neutral mutations* (those which are neither detrimental nor advantageous for the organism carrying them) is easy to estimate by first deriving the rate of change of some presumed neutral sequence (Section 11.2.5). The **deleterious mutation rate**, by contrast, has been more difficult to measure and no convincing estimate existed for any vertebrate until a study reported by Eyre-Walker and Keightley (1999). They investigated amino acid changes in 46 proteins occurring in the human ancestral line after its divergence from the chimpanzee. If all nonsynonymous substitutions were neutral, 231 new substitutions would have been expected in their sample of 46 genes (given an average neutral mutation rate of 0.0056 nonsynonymous substitutions per nucleotide and a total of 41 471 nucleotides investigated). Instead, only 143 nonsynonymous substitutions were observed; the remaining 88 such substitutions that would have been expected were inferred to have been removed by natural selection because they had been deleterious.

Eyre-Walker and Keightley (1999) estimated a deleterious rate of 1.6 mutations per person per generation on the assumption of 60 000 human genes. Recalculating on the basis of the more recent estimate of 30 000 genes with an average coding sequence of 1.6 kb gives a deleterious rate of about 0.84 out of 2.2 coding DNA mutations per person per generation. Coding DNA accounts for less than 1.5% of the human genome and on the basis of an estimated average mutation rate of about 2.5×10^{-8} mutations per nucleotide site, the total number of mutations occurring in our diploid genome has been calculated to about 175 per generation (Nachman and Crowell, 2000).

or extragenic sequences which are important for gene expression (see *Table 16.1* for different ways in which gene expression can be altered by mutation). The great majority of recorded pathogenic mutations have been identified in coding sequence (mostly nonsynonymous substitutions, nonsense mutations and frameshifting insertions/deletions). Because of its relatively high mutability, the CpG dinucleotide is often located at hotspots for pathogenic mutation in coding DNA (see Cooper *et al.*, 2000). Other hotspots include tandem repeats within coding DNA (see below). Of the noncoding intragenic mutations, splice site mutations and mutations in conserved elements of the untranslated sequences are important. Extragenic noncoding sequence mutations include mutations in the promoter and other control elements.

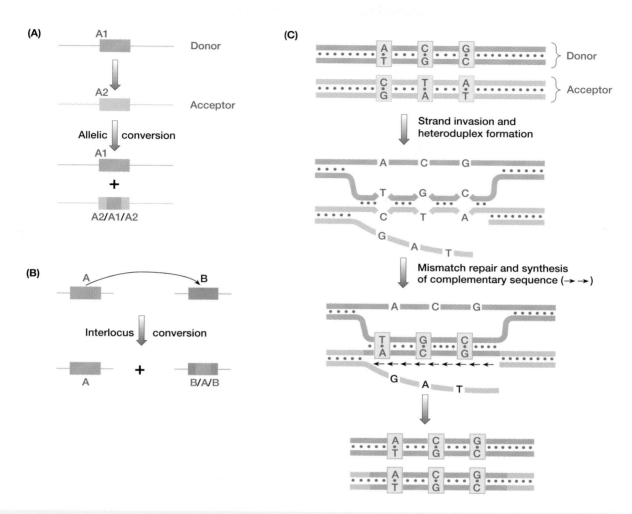

Figure 11.10: Gene conversion involves nonreciprocal sequence exchanges.

(A) *Interallelic gene conversion.* Note the nonreciprocal nature of the sequence exchange – the donor sequence is not altered but the acceptor sequence is altered by incorporating sequence copied from the donor sequence. **(B)** *Interlocus gene conversion.* This is facilitated by a high degree of sequence homology between nonallelic sequences, as in the case of tandem repeats. **(C)** *Mismatch repair of a heteroduplex.* This is one of several possible models to explain gene conversion. The model envisages invasion by one strand of the donor sequence to form a heteroduplex with the complementary strand of the acceptor sequence, thereby displacing the other strand of the acceptor. Mismatch repair enzymes recognize the mispaired bases in the heteroduplex and 'correct' the mismatches so that the acceptor sequence is 'converted' to be perfectly complementary in sequence to the donor strand. Subsequent replication of the acceptor strand and sealing of nicks completes the conversion.

11.4.2 The mitochondrial genome is a hotspot for pathogenic mutations

Because of the very large size of the human nuclear genome, most mutations occur in nuclear DNA sequences. By comparison, the mitochondrial genome is a small target for mutation (about 1/200 000 of the size of the nuclear genome). Unlike nuclear genes, mitochondrial genes are present in thousands of copies in each human somatic cell. Some cells, such as brain and muscle cells, have particularly high oxidative phosphorylation requirements and so more mitochondria. The oocyte is exceptional, having about 100 000 mtDNA molecules, many more than in somatic cells.

In normal individuals, ~99.9% of the mtDNA molecules are identical (**homoplasmy**). However, if a new mutation arises and spreads in the mtDNA population, there will be two significantly frequent mtDNA genotypes (**heteroplasmy**). Given that a mutation in mitochondrial DNA must arise on a single mtDNA molecule, one might intuitively expect that the chances of a single mtDNA mutation becoming fixed would be very low and the mutation rate correspondingly low. On these grounds, one could anticipate that the proportion of clinical disease due to pathogenic mutation in the mitochondrial genome should be extremely low. Instead, the frequency of 'mitochondrial disorders' is rather high (Section 16.6.6) and the mitochondrial genome can be considered to be a mutation hotspot. This is also true for animal mitochondrial DNA in general where mutations have been reported to be fixed at a rate about 10 times greater than in equivalent sequences in the nuclear genome (Brown *et al.*, 1979).

There are several explanations for the high pathogenic impact and mutability of mtDNA. Ninety-three percent of the mtDNA is coding DNA (but only 1.6% of nuclear DNA). Reactive oxygen intermediates produced by the respiratory chain are thought to cause substantial oxidative damage to mtDNA (which, unlike nuclear DNA, is not protected by histones). The mtDNA also has to undergo many more rounds of replication than chromosomal DNA. Mitochondria lack an adequate DNA-repair mechanism and although several well characterized mtDNA repair systems are now known, some frequent mutations cannot be repaired, including thymidine dimers (Section 11.6).

The question of how mtDNA mutations become *fixed* has required the explanation of a developmental **mtDNA bottleneck**. During oogenesis there is considerable fluctuation in both the number of germ cells and the number of mitochondria per cell. Thus in the 3-week female fetus there are of the order of 50 primordial germ cells, each with about 10 mitochondria per cell but rapid expansion in cell number subsequently occurs so that by the ninth week the fetus will have over half a million oogonia each with 200 mitochondria. The bottleneck is thought to occur at a very early stage between the primordial germ cell stage and the oogonia stage when a very small number of primordial germ cells migrate to the gonad, and mutant mtDNA molecules may become fixed by random drift and possibly by selection (see e.g. Jenuth *et al.*, 1996; Chinnery *et al.*, 2000).

11.4.3 Most splicing mutations alter a conserved sequence needed for normal splicing, but some occur in sequences not normally required for splicing

Many genes naturally undergo alternative forms of RNA splicing. In addition, mutations can sometimes produce an aberrant form of RNA splicing which is pathogenic. Sometimes this results in the sequences of whole exons being excluded from the mature RNA (*exon skipping*; see below) or retention of whole introns. On other occasions, the abnormal splicing pattern may exclude part of a normal exon or result in new exonic sequences. Point mutations which alter a conserved sequence that is normally required for RNA splicing are comparatively common. Occasionally, however, aberrant splicing of a gene can be induced by mutation of other sequence elements which resemble *splice donor* (= 5′ splice site) or *splice acceptor* (= 3′ splice site) sequences but which are not normally involved in splicing.

Mutations altering sequences that are important for splicing

Conserved sequences important for splicing include: the essentially invariant *GT* and *AG* dinucleotides, located respectively at the start (5′) and end (3′) of an intron; the immedi-

ately adjacent intronic and exonic sequences at the *splice donor* and *splice acceptor* sequences, including the *polypyrimidine tract* which precedes the end of an intron; and the *splice branch site* (*Figure 1.15*). In addition, splicing is known to be regulated through *splice enhancer sequences* (positive regulation) and *splice silencer sequences* (negative regulation) which occur within both exons and introns.

Exonic splice enhancer (ESE) sequences are discrete but degenerate sequences of about 6–8 nucleotides long which are known to bind splice regulatory proteins (Blencowe, 2000; Section 10.3.2). They are present in most if not all exons (both constitutive and alternatively spliced) and a set of 10 ESE motifs have recently been predicted for human exons, some important for regulating splice donor or splice acceptor recognition (Fairbrother *et al.*, 2002). **Exonic splice silencer (ESS)** sequences and other splice silencers are much less well understood (Fairbrother and Chasin, 2000).

Mutations which alter sequences important for splicing may have different consequences, as follows:

▶ *intron retention* – due to complete failure in splicing. This is more likely when an intron is small and the neighboring sequence lacks alternative legitimate splice sites or *cryptic splice sites* (sequences which resemble the consensus splice site sequences but which are not normally used by the splicing apparatus; see *Figure 11.11A*). Splicing is required for efficient export of mRNA from intron-containing genes (Luo and Reed, 1999) and so retention of intronic sequence in mRNA usually means that the mRNA is retained in the nucleus to avoid contact with the translation machinery (if translated, there would be a possibility of inappropriate amino acids or a frameshift in the translational reading frame);

▶ *exon skipping* – the splicing apparatus uses an *alternative* legitimate splice site. Mutation of a splice donor sequence often results in skipping of the upstream exon; mutation of the splice acceptor often results in skipping of the downstream exon (*Figure 11.11A*). Note, however, that other outcomes are possible, too, including use of alternative exonic or intronic cryptic splice sites (see below), and so the outcome is not always easily predictable – e.g. see Takahara *et al.* (2002). When an exon is skipped, various outcomes are possible. If the number of nucleotides in the exon is not divisible by three, a frameshift will introduce a premature termination codon, often resulting in an unstable RNA transcript and no polypeptide. If exon skipping does not cause a frameshift, the absence of the normally encoded amino acids will often result in a nonfunctional or abnormal polypeptide depending on the importance of these amino acids to protein function and/or structure.

Mutations of sequences not normally important for RNA splicing

Cryptic (or *latent*) **splice sites** coincidentally resemble the sequences of authentic splice sites but are not normally used in splicing, unless: (i) a mutation directly alters the sequence so that the splicing apparatus now recognizes it as a normal splice

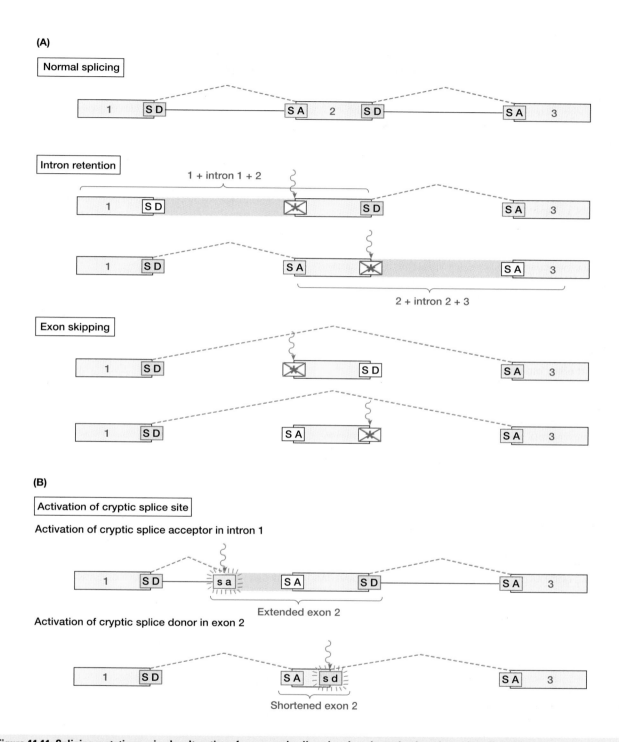

Figure 11.11: Splicing mutations arise by alteration of conserved splice signals or by activation of cryptic splice sites.

(A) *Alteration of conserved splice signals.* Mutation at splice donor or splice acceptor sequences (see *Figure 1.15* for consensus sequences) can result in: (a) intron retention where there is failure of splicing and an intervening intron sequence is not excised; (b) exon skipping where the spliceosome brings together the splice donor and splice acceptor sites of non-neighboring exons. ***Note:*** a splice site mutation may occasionally cause indirect activation of an alternative cryptic splice site (not normally used in splicing), in preference to using another legitimate splice site – see e.g. Takahara *et al.,* 2002). **(B) *Direct activation of cryptic splice sites.*** A mutation can directly activate a cryptic splice site by changing its sequence so that it becomes more like the consensus splice donor or acceptor sequence. The *altered* cryptic splice site can now be recognized and used by the spliceosome. See *Figures 11.12* and *11.13* for examples of activation of an exonic and an intronic cryptic splice site, respectively.

site (*direct activation*); (ii) a mutation occurs in an authentic splice site, causing a splice donor or splice acceptor to be faulty, in which case the splicing apparatus scans for other possible alternatives and selects a cryptic splice site (*indirect activation*; see the section on *exon skipping* immediately above). Because individual splice donor and splice acceptor sequences often show some variation from the consensus sequences shown in *Figure 1.15*, cryptic splice sites occur frequently within genes. If the altered mRNA is translated, the use of an *intronic* cryptic splice site will introduce new amino acids; using an *exonic* cryptic splice site will result in a deletion of coding DNA (*Figure 11.11B*).

See *Figures 11.12* and *11.13* respectively for worked examples of activation of a cryptic splice donor within an exon, and a cryptic splice acceptor within an intron. The former is a cautionary reminder that apparently silent mutations may yet be pathogenic. Note that in some cases mutations which occur within exons but not at cryptic splice sites can also induce skipping of that exon (see next section).

11.4.4 Mutations that introduce a premature termination codon often result in unstable mRNA but other outcomes are possible

Several different classes of mutation can introduce a premature termination codon (*chain-terminating mutations*). Nonsense mutations produce a premature termination codon simply by substituting a normal codon with a stop codon. Frameshifting insertions and deletions usually also introduce a premature termination codon not too far downstream of the mutation site. This happens because there is no selection pressure to avoid stop codons in the other translational reading frames and so, given established nucleotide frequencies, at least one stop

codon is usually encountered within a stretch of 100 nucleotides downstream of the mutation site. A variety of splice site mutations too can introduce a premature termination codon, for example by skipping of a single exon containing a number of nucleotides that cannot be divided by three. There are several possible consequences for gene expression for chain-terminating mutations:

▶ **unstable mRNA.** This is by far the most frequent consequence. An mRNA carrying a premature termination codon at least 50 nucleotides upstream of the last splice junction is usually rapidly degraded *in vivo* by a form of *RNA surveillance* known as **nonsense-mediated mRNA decay (NMD)** – see Lykke-Andersen *et al.* (2001); Maquat (2002). This can avoid the potentially lethal consequences of producing a truncated polypeptide which could interfere with vital cell functions;

▶ **truncated polypeptide.** NMD ensures that truncated polypeptides are a rare outcome *in vivo*, but it is dependent on splicing. Accordingly, nonsense mutations in one of the 5% or so of human genes which lack introns (*Table 9.5*) can result in truncated polypeptides. The effect of truncated polypeptides may be difficult to predict and will depend among other things on the extent of the truncation, the stability of the polypeptide product and its ability to interfere with expression of normal alleles;

▶ **exon skipping.** For a small subset of nonsense mutations potentially harmful effects on gene expression may be mitigated by a process termed **nonsense-associated altered splicing (NAS)** – see Wang *et al.* (2002). Here the normal splicing pattern is altered, e.g. by exon skipping, so that the stop codon is bypassed and a stable mRNA is produced lacking the mutation. The existence of NAS has

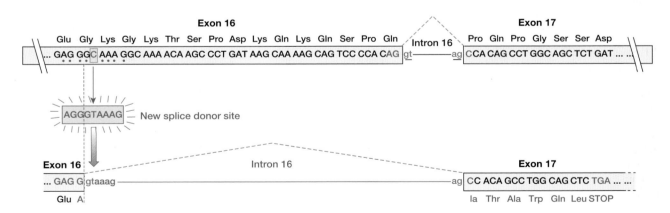

Figure 11.12: When a silent mutation is not silent.

This example shows a mutation that was identified in a LGMD2A limb girdle muscular dystrophy patient. The mutation was found in the calpain 3 gene, a known locus for this form of muscular dystrophy, but occurred at the third base position of a codon and appeared to be a silent mutation. It would lead to replacement of one glycine codon (GGC) by another glycine codon (GGT). However, the mutation is believed nevertheless to be pathogenic. The substitution results in activation of a cryptic splice donor sequence (AGGGCAAAAG) within exon 16 resulting in aberrant splicing with the loss of coding sequence from exon 16 and the introduction of a frameshift. See Richard and Beckmann (1995). ***Note:*** another possibility for pathogenic synonymous mutations are ones which cause their effect by mutating an exonic splice enhancer sequence (Section 11.4.3).

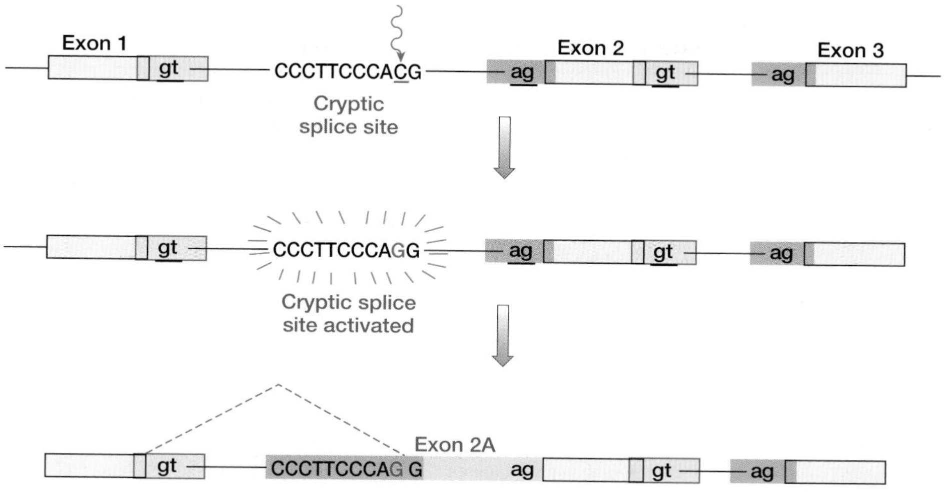

Figure 11.13: Mutations can cause abnormal RNA splicing by activation of cryptic splice sites.

Activation of a cryptic splice acceptor sequence located within an intron (compare *Figure 11.12* which illustrates activation of a cryptic splice donor site within an exon). A mutation can result in the alteration of a sequence which is not important for RNA splicing so as to create a new, alternative splice site. In the example illustrated, the mutation is envisaged to change a single nucleotide in intron 1. The nucleotide happens to occur within a cryptic splice site sequence that is closely related to the splice acceptor consensus sequence (see *Figure 1.15*), but does not have the conserved AG dinucleotide. The mutation overcomes this difference, activating the cryptic splice site so that it competes with the natural splice acceptor site. If it is used by the splicing apparatus, a novel exon, exon 2A, results, which contains additional sequence which may or may not result in a frameshift.

been interpreted by some to mean that there must be some nuclear translation mechanism which allows codons to be read *before* the mRNA is exported to the cytoplasm. However, an alternative explanation for most if not all, cases of NAS is that the nonsense mutation alters an ***exonic splice enhancer*** (see Section 11.4.3; and Maquat, 2002 for alternative views on how NAS is activated).

11.5 The pathogenic potential of repeated sequences

The human genome has a very high proportion of repetitive DNA sequences (Sections 9.5 and 9.6), which are prone to copy number variation and sequence exchanges (*Table 11.6*). A reduction in repeat number can often result in a pathogenic deletion, but expansion by sequence duplication can be pathogenic too (see Mazzarella and Schlessinger, 1998). Certain chromosomal regions, notably the subtelomeric and pericentromeric regions, harbor large tracts of duplicated DNA and instability of such regions can predispose to disease – see Eichler, 1998). Interspersed repeats can also cause pathogenic mutations by a variety of different mechanisms (*Table 11.6*).

11.5.1 Slipped strand mispairing of short tandem repeats predisposes to pathogenic deletions and frameshifting insertions

Insertions and deletions in coding DNA are rare because they usually introduce a translational frameshift. However, occa-

sionally, a series of tandem repeats of a small number of nucleotides occurs by chance in the coding sequence for a polypeptide. Such repeats, like microsatellite loci, are comparatively prone to mutation by slipped strand mispairing. As a result, the copy number of tandem repeats is liable to fluctuate, introducing a deletion or an insertion of one or more repeat units. If the mutation occurs in polypeptide-encoding DNA, a resulting deletion will often have a profound effect on gene expression. Frameshifting deletions will normally result in abolition of gene expression and can be common.

Even if the deletion does not produce a frameshift, deletions of one or more amino acids can still be pathogenic (*Figure 11.14*). Small frameshifting insertions will also be expected to lead to loss of gene expression and often the insertion is a tandem repeat of sequences flanking it. However, nonframeshifting insertions would often not be expected to be pathogenic, unless the insertion occurs in a critically important region, destabilizing an essential structure or impeding gene function in some way.

11.5.2 Unstable expansion of short tandem repeats can cause a variety of diseases but the mutational mechanism is not well understood

Certain short tandem repeats within or in the immediate vicinity of a gene can expand to considerable lengths and affect gene expression, causing disease. On occasions a modestly expanded repeat which causes disease may be perfectly stable and be propagated without change in size

Table 11.6: Repeated DNA sequences often contribute to pathogenesis

Type of repeated DNA	Type of mutation	Mechanism and examples
Tandem repeats		
Very short repeats within genes	Deletion	Slipped strand mispairing (see *Figure 11.5*). Examples in *Figure 11.14*
	Frameshifting insertion	Slipped strand mispairing
	Triplet repeat expansion	Initially by slipped strand mispairing?; subsequent large-scale expansion by unknown mechanism
Moderate sized intragenic repeats	Intragenic deletion	UEC/UESCE[a] (see *Figure 11.7*)
	Partial or total gene deletion UEC/UESCE[a] (*Figure 11.7*)	Examples in *Figure 11.16*
Large tandem repeats containing whole genes	Alteration of gene sequence Gene conversion (*Figure 11.10*)	Examples in *Figures 11.16 and 11.17*
Interspersed repeats		
Short direct repeats	Deletion	Slipped strand mispairing or intrachromatid recombination?
Interspersed repeat elements (e.g. Alu repeats)	Deletion/duplication	UEC/UESCE[a] (see Section 11.5.4)
Inverted repeats	Inversion	Intrachromatid exchange, e.g. Factor VIII (see *Figure 11.20*)
Low copy number long repeats	Large scale deletion/duplication	UEC/UESCE[a] (see *Table 11.7* and *Figure 11.19*)
Active transposable elements	Intragenic insertion by retrotransposons	Retrotransposition. For examples, see Section 11.5.6

[a] UEC–unequal crossover; UESCE–unequal sister chromatid exchange

through several generations. In other cases, however, the expanded repeat is *unstable*. The discovery that human disease can be caused by large-scale expansion of highly unstable trinucleotide repeats was quite unexpected (studies in other organisms had not revealed precedents for such a phenomenon), but the list of human examples is now considerable (see *Table 16.6*).

Although 64 possible trinucleotide sequences are possible, when allowance is made for cyclic permutations $(CAG)_n = (AGC)_n = (GCA)_n$ and reading from either strand $[5'(CAG)_n$ on one strand $= 5'(CTG)_n$ on the other], there are only 10 different trinucleotide repeats at the level of genomic DNA (*Figure 11.15*). In addition to unstable *triplet repeat* expansion, the majority of disease alleles at the cystatin B gene which cause progressive myoclonus epilepsy involve expansions of a 12-nucleotide repeat $(C)_4G(C)_4GCG$ (Lalioti *et al.*, 1997). As detailed in *Table 16.6*, genes containing unstable expanding short tandem repeats fall into two major classes according to the size of the expansion and its location as follows:

▶ *modest* $(CAG)_n$ *expansions resulting in polyglutamine tracts.* The CAG codon specifies glutamate. The stable, nonpathogenic range is 10–30 repeats; unstable pathogenic alleles often have in the range of 40–200 repeats. The expanded polyglutamine tract causes the protein to aggregate within certain cells and kill them;

▶ *very large noncoding repeat expansions.* Various types of repeat (e.g. CGG, CCG, CTG, GAA) found in noncoding sequence (the promoter, or the untranslated regions or intronic sequences) undergo very large expansions. The expansions inhibit expression of closely neighboring genes, causing loss of function. Stable, nonpathogenic alleles typically have 5–50 repeats; unstable pathogenic alleles have several hundreds or thousands of copies (see *Table 16.6*). **Note:** some very large expansions of short noncoding tandem repeats simply affect chromosome structure, causing **fragile sites**, but without causing disease (presumably because no important gene is located nearby). Examples include FRAXF and FRA16A (both due to CCG expansion).

In each case, repeats below a certain threshold length are stable in mitosis and meiosis, but become extremely unstable above the threshold length. The unstable repeats are virtually never transmitted unchanged from parent to child. Both expansions and contractions can occur, but there is a bias towards expansion. The average size change often depends on the sex of the transmitting parent, as well as the length of the repeat.

The nature of the expansion mechanism is still not clearly understood (see Djian, 1998; Sinden *et al.*, 2002). Slipped strand mispairing (see *Figure 11.5*) has been considered to be a likely component of the expansion mechanism, given the observation that interrupted repeats appear to be stable and

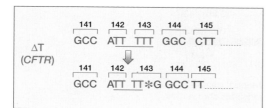

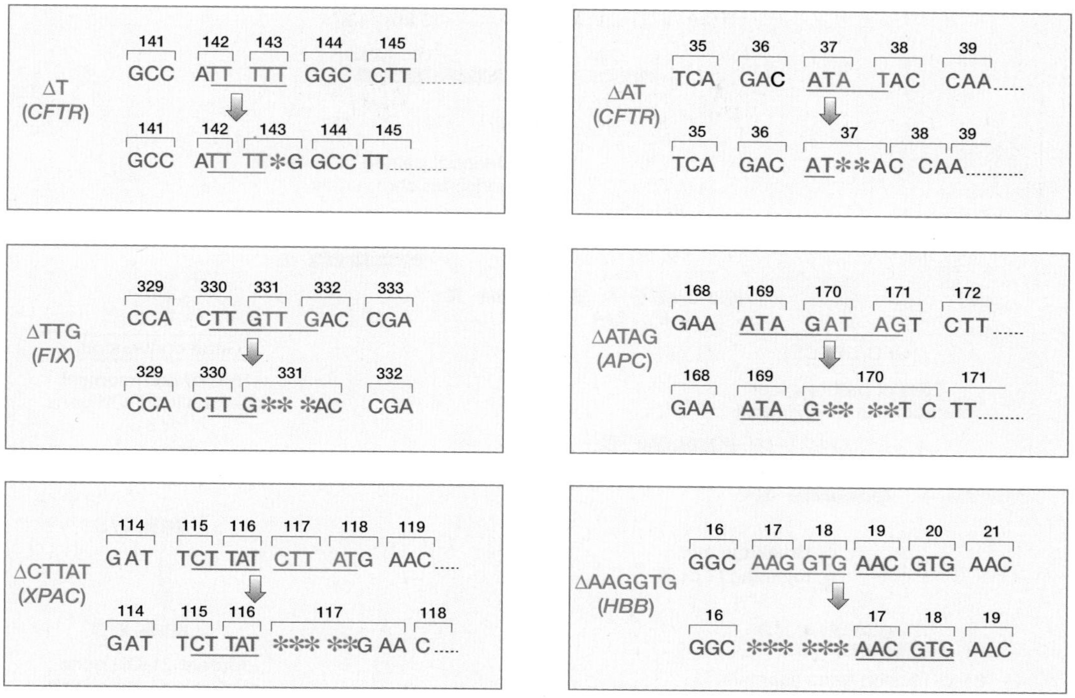

Figure 11.14: Short tandem repeats are deletion/insertion hotspots.

The six deletions illustrated are examples of pathogenic deletions occurring at tandemly repeated units of from 1 to 6 bp. The deletions of 3 and 6 bp do not cause frameshifts, and pathogenesis is thought to be due to removal of one or two amino acids that are critically important for polypeptide function. ***Note:*** in the case of the 6-bp deletion the original tandem repeat is not a perfect one. Genes (and associated diseases) are: *CFTR*, cystic fibrosis transmembrane regulator; *FIX*, factor IX (hemophilia B); *APC*, adenomatous polyposis coli; *XPAC*, xeroderma pigmentosa complementation group C; *HBB*, β-globin (β-thalassemia). Though not illustrated here, small insertions are often tandem repeats of sequences flanking them.

only homogeneous repeats are unstable. For example, In spinocerebellar ataxia type 1, 123/126 normal sized CAG repeats were interrupted by one or two CAT triplets, while 30/30 expanded (CAG)*n* alleles contained no interruption (Chung *et al.*, 1993). But why the changes in length are biased towards expansion remains unclear. Understanding of unstable repeat expansion is progressing rapidly; the reader is advised to consult a recent review for more information.

11.5.3 Tandemly repeated and clustered gene families may be prone to pathogenic unequal crossover and gene conversion-like events

Many human and mammalian gene clusters contain nonfunctional pseudogenes which may be closely related to functional gene members. Interlocus sequence exchanges between pseudogenes and functional genes can result in disease by removing or altering some or all of the sequence of a functional gene. For example, unequal crossover (or unequal sister chromatid exchange) between a functional gene and a related pseudogene can result in deletion of the functional gene or the formation of fusion genes containing a segment derived from the pseudogene. Alternatively, the pseudogene can act as

a donor sequence in gene conversion events and introduce deleterious mutations into the functional gene.

The classical example of pathogenesis due to gene–pseudogene exchanges is steroid 21-hydroxylase deficiency, where over 95% of pathogenic mutations arise as a result of sequence exchanges between the functional 21-hydroxylase gene, *CYP21B*, and a very closely related pseudogene, *CYP21A*. The two genes occur on tandemly repeated DNA segments approximately 30 kb long which also contain other duplicated genes, notably the complement C4 genes, *C4A*

AAC/GTT	AGG/CCT
AAG/CTT	ATC/GAT
AAT/ATT	
ACC/GGT	CAG/CTG
ACG/CGT	CCG/CGG
ACT/AGT	

Figure 11.15: The 10 possible trinucleotide repeats.

Both DNA strands are shown. All other trinucleotide repeats are cyclic permutations of one or another of these (see text).

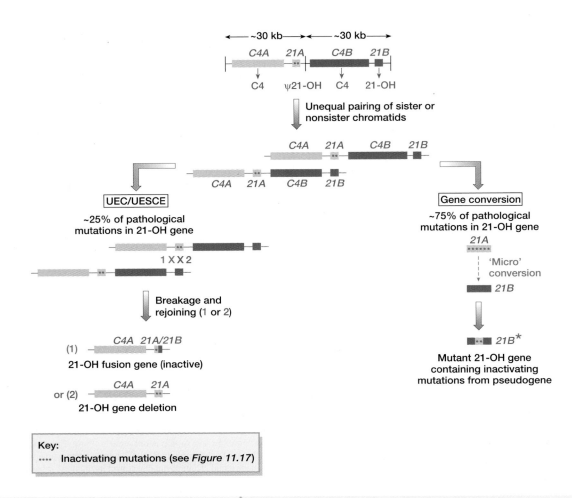

Figure 11.16: Almost all 21-hydroxylase gene mutations are due to sequence exchange with a closely related pseudogene.

The duplicated complement C4 genes and steroid 21-hydroxylase genes are located on tandem 30-kb repeats which show about 97% sequence identity. Both the *C4A* and *C4B* genes are expressed to give complement C4 products; the *CYP21B* gene (*21B*) encodes a 21-hydroxylase product, but the *CYP21A* (*21A*) gene is a pseudogene. About 25% of pathological mutations at the 21-hydroxylase locus involve a 30-kb deletion resulting from unequal crossover (UEC) or unequal sister chromatid exchange (UESCE). The remaining mutations are point mutations where small-scale gene conversion of the *CYP21B* gene occurs – a small segment of the *CYP21A* gene containing deleterious mutations is copied and inserted into the *CYP21B* gene replacing a short segment of the original sequence (see *Figure 11.10C* for one possible mechanism). Possibly gene conversion events are, like UEC and UESCE, primed by unequal pairing of the tandem repeats on sister or nonsister chromatids.

and *C4B*. Large pathogenic deletions uniformly result in removal of about 30 kb of DNA (corresponding to one repeat unit length) containing *CYP21B* gene sequence (see *Figure 11.16*). Nonpathogenic deletions of the same length are also found because sometimes the *C4A* + *CYP21A* repeat unit is deleted (and so the functional *CYP21B* gene is retained), and the C4/21-OH genes show a form of large-scale VNTR polymorphism locus with different alleles containing 1, 2, 3 or 4 of the 30-kb units (Collier *et al.*, 1989).

Virtually all of the 75% of pathogenic point mutations are copied from deleterious mutations in the pseudogene, suggesting a gene conversion mechanism (see *Figures 11.16* and *11.17*). Analysis of one such mutation which arose *de novo* suggests that the conversion tract is a maximum of 390 bp

(Collier *et al.*, 1993). Gene conversion events are also found in the duplicated C4 genes, both of which are normally expressed. A likely priming event for conversions in the *CYP21–C4* gene cluster is unequal pairing of chromatids so that a *CYP21A–C4A* unit pairs with a *CYP21B–C4B* unit (see *Figure 11.17*).

11.5.4 Interspersed repeats often predispose to large deletions and duplications

Short direct repeats

In several cases, the endpoints of deletions are marked by very short direct repeats. For example, the breakpoints in numerous pathogenic deletions in mtDNA occur at perfect or almost

Location of mutation	Normal 21-OH gene sequence (CYP21B)	21-OH pseudogene sequence (CYP21A)	Mutant 21-OH gene sequence
Intron 2	CCCACCTCC	CCCAGCTCC	CCCAGCTCC
Exon 3 (codons 110–112)	GGA GAC TAC TC Gly Asp Tyr Ser	G(.........)TC	G(.........)TC Val
Exon 4 (codon 172)	ATC ATC TGT Ile Ile Cys	ATC AAC TGT	ATC AAC TGT Ile Asn Cys
Exon 6 (codons 235–238)	ATC GTG GAG ATG Ile Val Glu Met	AAC GAG GAG AAG	AAC GAG GAG AAG Asn Glu Glu Lys
Exon 7 (codon 281)	CAC GTG CAC His Val His	CAC TTG CAC	CAC TTG CAC His Leu His
Exon 8 (codon 318)	CAC CAG GAG His Gln Glu	CTG TAG GAG	CTG TAG GAG Leu STOP
Exon 8 (codon 356)	CTG CGG CCC Leu Arg Pro	CTG TGG CCC	CTG TGG CCC Leu Trp Pro

Figure 11.17: Pathogenic point mutations in the steroid 21-hydroxylase gene originate by copying sequences from the 21-hydroxylase pseudogene.

The copying is thought to involve a gene conversion-like mechanism (see *Figures 11.16* and *11.10C*).

perfect short direct repeats. Of these, the most common is a deletion of 4977 bp which has been found in multiple patients with Kearns–Sayre syndrome, an encephalomyopathy characterized by external ophthalmoplegia, ptosis, ataxia and cataract. The deletion results in elimination of the intervening sequence between two perfect 13-bp repeats and loss of the sequence of one of the repeats (*Figure 11.18*). The mitochondrial genome is recombination deficient and Shoffner *et al.* (1989) have postulated that such deletions arise by a replication slippage mechanism, similar to that occurring at short tandem repeats (*Figure 11.5*). Partial duplications of the mitochondrial genome are also distinctive features of certain diseases, notably Kearns–Sayre syndrome. The ends of the duplicated sequences, like those of the common deletions, are often marked by short direct repeats, and the mechanisms of duplication and deletion appear to be closely related (see Poulton and Holt, 1994).

The Alu repeat as a recombination hotspot

Some large-scale deletions and insertions may be generated by pairing of nonallelic interspersed repeats, followed by breakage and rejoining of chromatid fragments. For example, the Alu repeat occurs approximately once every 3 kb and mispairing between such repeats has been suggested to be a frequent cause of deletions and duplications. Some large genes have many internal Alu sequences in their introns or untranslated sequences, making them liable to frequent internal deletions and duplications. For example, the Alu repeat occurs once every 1.5 kb in the 45-kb low density lipoprotein receptor gene. A very high frequency of pathogenic deletions in this gene are likely to involve an Alu repeat, usually at both endpoints, and occasional pathogenic intragenic duplications also involve Alu repeats (see Hobbs *et al.*, 1990). Such observations have suggested a general role for Alu sequences in promoting recombination and recombination-like events. Initial gene duplications in the evolution of clustered multigene families may often have involved an unequal crossover event between Alu repeats or other dispersed repetitive elements. It should be noted, however, that some Alu-rich genes do not appear to be loci for frequent Alu-mediated recombination.

Low copy number long repeats

The sequencing of the euchromatic portion of the human genome revealed very many examples of interspersed low copy number sequences exhibiting very high sequence homology (often > 95% sequence identity) often extending over tens to hundreds of kilobases. Repeated sequences of this type have typically undergone evolutionarily recent, primate-specific duplications termed **segmental duplication** (Section 12.2.5). The very high sequence homology over such long regions predisposes to unequal recombination between the repeats (also called **duplicons**). When present on different chromosomes such closely related repeats may possibly predispose to translocation. When present on the same chromosome they often predispose to unequal (= nonallelic) homologous recombination which can cause large-scale deletions and duplications.

Duplicon-mediated deletions, and to a lesser extent duplications, which extend over megabase intervals are often pathogenic (Stankiewicz and Lupski, 2002; see *Table 11.7* and *Figure 11.19*), causing loss of gene function or inappropriately high gene dosage for dosage-sensitive genes contained within the intervals. Such rearrangements are usually not resolved by standard cytogenetic analyses and are often classified (from a *chromosomal* point of view) as

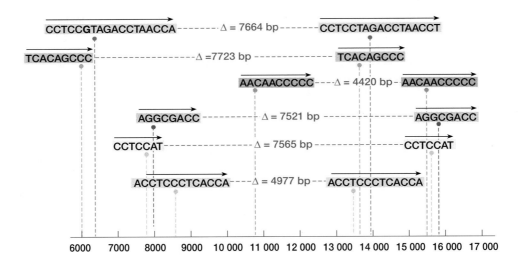

Figure 11.18: Short direct repeats mark the endpoints of many pathogenic deletions in the mitochondrial genome.

Note: as mitochondria are recombination-deficient one likely mechanism to explain the deletions is slipped strand mispairing (see text).

microdeletions and microduplications. Because of the widespread prevalence of segmental duplication in the human genome, the contribution of duplicons to pathogenesis may be very considerable (Stankiewicz and Lupski, 2002).

11.5.5 Pathogenic inversions can be produced by intrachromatid recombination between inverted repeats

Occasionally, clustered inverted repeats with a high degree of sequence identity may be located within or close to a gene. The high degree of sequence similarity between inverted repeats may predispose to pairing of the repeats by a mech-

anism that involves a chromatid bending back upon itself. Subsequent chromatid breakage at the mispaired repeats and rejoining can then result in an inversion, in much the same way as the natural mechanism used for the production of some immunoglobulin κ light chains.

The classic example of pathogenic inversions is a mutation which accounts for more than 40% of cases of severe hemophilia A. Intron 22 of the factor VIII gene, *F8*, contains a CpG island from which two internal genes are transcribed: *F8A* in the opposite direction to the host gene *F8*, and *F8B* in the same direction as *F8* (see *Figure 11.20*). *F8A* belongs to a gene family with two other closely related members located several hundred kilobases upstream of the *F8* gene and transcribed in the

Table 11.7: **Low copy number repeats can predispose to large-scale pathogenic deletions and duplications**

Trait/syndrome	Gene(s)	Chromosome location	Rearrangement type	Size (kb)	Duplicon size (kb)
Male infertility (AZFa)	*DBY, USP9Y*	Yq11.2	Del	800	10
Male infertility (AZFc)	*RBMY, DAZ?*	Yq11.2	Del	3500	229
Williams Beuren syndrome	*ELN, GTF2I?*	7q11.23	Del	1600	320
Prader–Willi syndrome	?	15q12pat	Del	3500	500
Angelman syndrome	*UBE3A*	15q12mat	Del	3500	500
Smith Magenis syndrome	?	17p11.2	Del	3700	200
DiGeorge/VCF syndrome	*TBX1,?*	22q11.2	Del	3000/1500	225–400
Peripheral neuropathy (CMT1A)	*PMP22*	17p12	Dup	1400	24
Peripheral neuropathy (HNPP)	*PMP22*	17p12	Del	1400	24

Abstracted from Stankiewicz and Lupski (2002). *Curr. Opin. Genet. Dev.* **12**, 312–319 with permission from Elsevier.

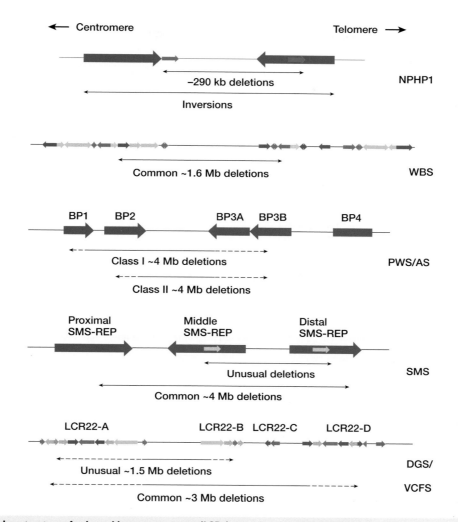

Figure 11.19: Complex structure of selected low-copy repeats (LCRs) associated with human disease

Note the complex structure of LCRs consisting of both direct repeats (arrowheads in same direction) and inverted repeats (arrowheads in opposite directions). Disease abbreviations are: NPHP1, familial juvenile nephronophthisis 1; WBS, Williams–Beuren syndrome: PWS/AS, Prader–Willi syndrome/Angelman syndrome; SMS, Smith–Magenis syndrome; DGS/VFCS, DiGeorge syndrome/velocardiofacial syndrome. Reproduced from Stankiewicz and Lupski (2002) *Trends Genet.* **18**, 74–82 with permission from Elsevier.

opposite direction to *F8A*. As a result, the region between the *F8A* gene and the other two members is susceptible to inversions – the *F8A* gene can pair with either of the other two members on the same chromatid, and subsequent chromatid breakage and rejoining in the region of the paired repeats results in an inversion which disrupts the factor VIII gene (Lakich *et al.*, 1993, see *Figure 11.20*).

Several examples of duplicons arising though segmental duplication (Section 12.2.5) and other low to moderate copy number repeats also predispose to inversions but the inversions are not directly pathogenic because the breakpoints do not result in aberrant gene expression (unlike the example in the Factor VIII gene). Instead large-scale ***inversion polymorphisms*** may be generated as in the case of duplicons at the Williams–Beuren locus (Osborne *et al.*, 2001) and highly homologous olfactory receptor repeats (Giglio *et al.*, 2001).

11.5.6 DNA sequence transposition is not uncommon and can cause disease

A proportion of moderately and highly repeated interspersed elements are capable of transposition via an RNA intermediate (Section 9.5). Defective gene expression due to DNA transposition is comparatively rare and represents only a small component of molecular pathology. However, several examples have been recorded of genetic deficiency due to insertional inactivation by retrotransposons. For example, in one study, hemophilia A was found to arise in two out of 140 unrelated patients as a result of a *de novo* insertion of a LINE-1 (Kpn) repeat into an exon of the factor VIII gene. Other instances are known of insertional inactivation by an actively transposing Alu element. Additionally, a number of other examples have been recorded of pathogenesis due to intra-

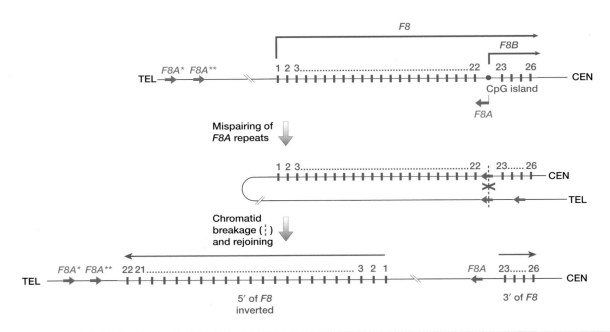

Figure 11.20: Inversions disrupting the factor VIII gene result from intrachromatid recombination between inverted repeats.

Intron 22 of the factor VIII gene (*F8*) contains a CpG island from which two internal genes are transcribed: *F8B* in the same direction (a novel exon is spliced onto exons 23–26 of *F8*) and *F8A* gene which is transcribed from the opposite strand. *F8A** and *F8A*** are sequences very closely related to *F8A* but located about 500 kb upstream and transcribed from the opposite strand. The high degree of sequence identity between the three members of the *F8A* gene family means that pairing of the *F8A* gene with one of the other two members on the same chromatid can occur by looping back of the chromatid. Subsequent chromatid breakage and rejoining can result in an inversion of the region between the *F8A* gene and the other paired family member, resulting in disruption of the factor VIII gene (see Lakich *et al.*, 1993).

genic insertion of undefined DNA sequences. See Kazazian (1998) for references.

11.6 DNA repair

DNA in cells suffers a wide range of damage, partly in response to extracellular agents, but largely as a result of endogenous mechanisms, including spontaneous chemical hydrolysis, intracellular interaction with reactive oxygen groups and errors in replication and recombination.

Extracellular agents causing DNA damage

▸ *Ionizing radiation* – gamma rays and X-rays can cause single- or double-strand breaks in DNA.

▸ *Ultraviolet light* – especially the UV-C rays (~ 260 nm) that are absorbed strongly by DNA but also longer-wavelength UV-B rays that penetrates the ozone shield. UV causes *cross-linking* between adjacent thymines on a DNA strand to form a stable chemical dimer (*Figure 11.21C*).

▸ *Environmental chemicals* – includes hydrocarbons (e.g. some of the ones found in cigarette smoke), some plant and microbial products (e.g. the aflatoxins produced in moldy peanuts) and chemicals used in cancer chemotherapy. *Alkylating agents* can transfer an *alkyl group* (an aliphatic hydrocarbon such as a methyl group) onto bases and can cause cross-linking between bases on different DNA strands or within a strand.

Endogenous mechanisms causing DNA damage

▸ *Depurination* (*Figure 11.21A*). Approximately 5000 adenine or guanines are lost every day from each nucleated human cell by spontaneous fission of the base-sugar link.

▸ *Deamination* (*Figure 11.21A*). About 100 cytosines spontaneously deaminate per day in each nucleated human cell to produce uracil (which preferentially base pairs with adenine, causing the DNA replication machinery to insert an A when it encounters U on the template strand). Less frequently adenines spontaneously deaminate to give hypoxanthine.

▸ *Reactive oxygen* species. Reactive oxygen species include the superoxide anion, O_2^-, which is both an ion and a *free radical* (a cluster of atoms one of which contains an unpaired electron in its outermost shell of electrons; this is an extremely unstable configuration, and radicals quickly react with other molecules or radicals to achieve the stable configuration of four pairs of electrons in their outermost shell). Reactive oxygen species can be formed by the effect of ionizing radiation on some cellular molecules but also occur as an unavoidable by-product of cellular respiration (some electrons passing 'down' the respiratory chain are diverted from the main path and go directly to reduce oxygen molecules to the superoxide anion). In the cell reactive oxygen species attack purine and pyrimidine rings.

▶ *Mistakes in DNA replication.* Incorrect *proofreading* results in incorporation of mismatched bases e.g. uracil is often incorrectly inserted instead of thymine into DNA.

▶ *Mistakes in replication or recombination* cause strand breaks to be left in DNA.

All these lesions must be repaired if the cell is to survive. DNA repair seldom involves simply undoing the change that caused the damage (*direct repair*). Almost always a stretch of DNA containing the damaged nucleotide(s) is excised and the gap filled by resynthesis (*excision repair*). The importance of effective DNA repair systems is highlighted by the approximately 130 human genes participating in DNA repair (see Wood *et al.*, 2001 and supplement at http://www.cgal.icnet.uk/DNA_Repair_Genes.html#DR) and by severe diseases affecting people with deficient repair systems (see below).

11.6.1 DNA repair usually involves cutting out and resynthesizing a whole area of DNA surrounding the damage

To cope with all these forms of damage, human cells are capable of at least five types of DNA repair (for reviews, see the October 1995 issue of *Trends in Biochemical Sciences*; and Lindahl and Wood, 1999).

Direct repair – reverses the DNA damage

Three genes have been implicated in this infrequently used mechanism. Of these, the most well characterized encodes the O^6-methylguanine-DNA methyltransferase which is able to remove methyl groups from guanines which have been incorrectly methylated. (*Note:* in bacteria thymine dimers can be removed in a photoreactivation reaction that depends on visible light and an enzyme, photolyase, but although mammals possess enzymes related to photolyase, they use them for a quite different purpose, to control their circadian clock; Van der Horst *et al.*, 1999.)

Base excision repair (BER) – uses glycosidase enzymes to remove abnormal bases (*Figure 11.22A*)

BER corrects much the commonest type of DNA damage (of the order of 20 000 altered bases in each nucleated cell in our body each day). We have at least eight genes encoding different **DNA glycosylases**, each responsible for identifying and removing a specific kind of base damage (see *Table 11.8*). After base removal an endonuclease, **AP endonuclease**, and a phosphodiesterase cut the sugar–phosphate backbone at the position of the missing base and remove the sugar–phosphate residue. The gap is filled by resynthesis using a DNA polymerase, and the remaining nick is sealed by DNA ligase III. The same process is used to repair spontaneous depurination.

Nucleotide excision repair (NER) – removes thymine dimers and large chemical adducts (*Figure 11.22B*)

NER differs from BER in using different enzymes and even where there is only a single abnormal base to correct, its nucleotide is removed along with many other adjacent nucleotides; that is, NER removes a large 'patch' around the damage. *Figure 11.22* illustrates the process. Defects in nucleotide excision repair cause the autosomal recessive disease xeroderma pigmentosum (XP; Lambert *et al.*, 1998). Seven complementation groups, XPA–XPG, have been defined by cell fusion studies. XP patients are exceedingly sensitive to UV light. Sun-exposed skin develops thousands of freckles, many of which progress to skin cancer.

Post-replication repair (homologous recombination) is required to correct double-strand breaks (Haber, 1999)

The usual mechanism is a gene conversion-like process (*recombinational repair*), where a single strand from the homologous chromosome invades the damaged DNA. Alternatively, broken ends are rejoined regardless of their sequence, a desperate measure that is likely to cause mutations. The eukaryotic machinery for recombination repair is less well defined than the excision repair systems. Human genes involved in this pathway include *NBS* (mutated in Nijmegen breakage syndrome; see Section 13.5.2), *BLM* (mutated in Bloom syndrome; MIM 210900) and the *BRCA2* and *BRCA1* breast cancer susceptibility genes (Section 17.5.1).

Mismatch repair – corrects mismatched base pairs caused by mistakes in DNA replication

Cells deficient in mismatch repair have mutation rates 100–1000 times higher than normal, with a particular tendency to replication slippage in homopolymeric runs (*Figure 11.5*). In humans the mechanisms involve at least five proteins and defects cause hereditary nonpolyposis colon cancer (Section 17.5.3 and *Figure 17.12*).

All these systems, except for direct repair, require exo- and endonucleases, helicases, polymerases and ligases, usually acting in multiprotein complexes that have some components in common. Sorting out the individual pathways has been greatly aided by the very strong conservation of repair mechanisms across the whole spectrum of life. Not only the reaction mechanisms but also the protein structures and gene sequences are often conserved from *E. coli* to man. A downside of the conservation is a confusing gene nomenclature, referring sometimes to human diseases (*XPA* etc.), sometimes to yeast mutants (*RAD* genes) and sometimes to mammalian cell complementation systems (ERCC – excision repair cross-complementing): for example *XPD*, *ERCC2* and *RAD3* are the same gene in man, mouse and yeast. Generally, eukaryotes have multiple systems corresponding to each single system in *E. coli*, so that, for example, nucleotide excision repair requires six proteins in *E. coli* but at least 30 in mammals.

11.6.2 DNA repair systems share components and processes with the transcription and recombination machinery

As well as sharing components with each other, many repair systems share components with the machinery for DNA replication, transcription and recombination. DNA

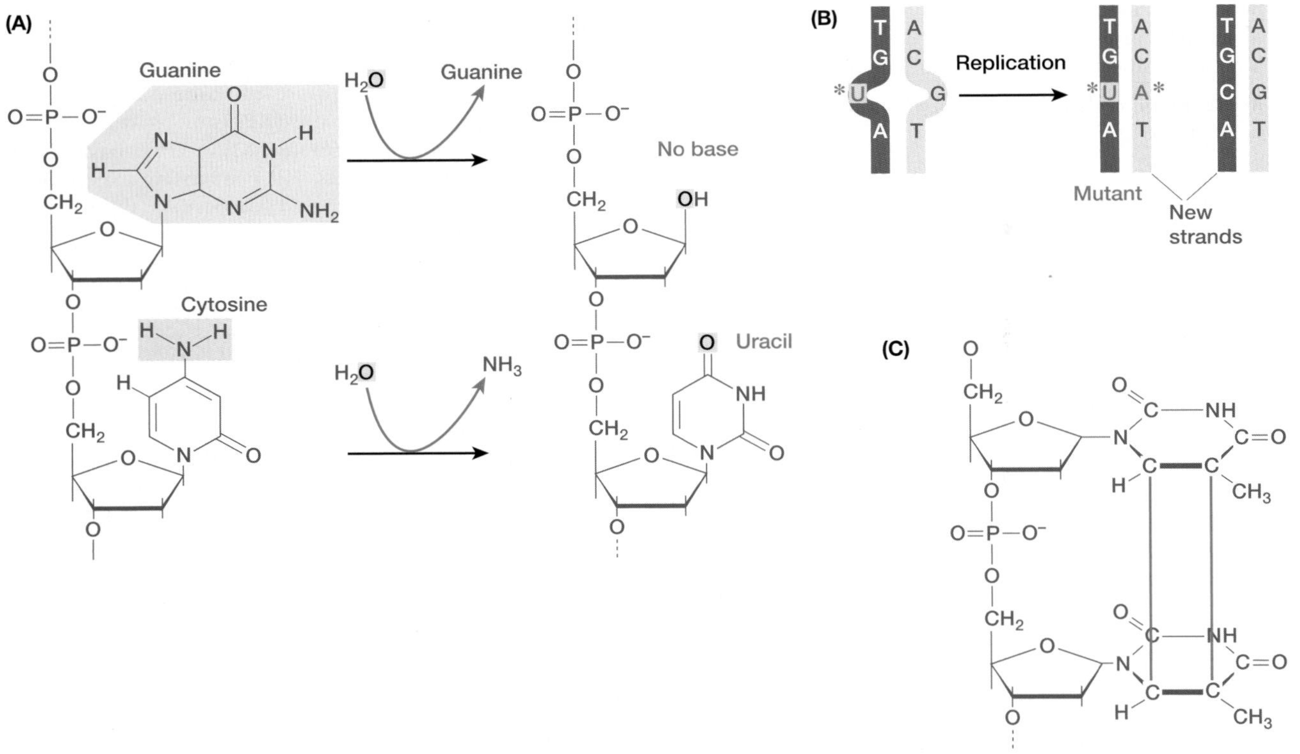

Figure 11.21: DNA damage by chemical modifications of nucleotides.

(A) *Examples of depurination (above) and deamination (below)*. **(B)** *How chemical modification causes a mutation*. Here a uracil has been produced by deamination of a cytosine residue. If not repaired, DNA replication will insert an adenine on the complementary strand, and so the net effect is that of a C→T transition. Repair is possible by the base excision repair pathway (*Figure 11.21A*). The net effect of a depurination is a deletion of the single nucleotide whose base was removed. **(C)** *A thymidine dimer* – covalent bonds (red) link carbons of neighboring thymine bases. Thymidine dimers can be repaired by the nucleotide excision repair pathway (*Figure 11.21B*).

Table 11.8: **Human DNA glycosylases (see Wood *et al.*, 2001)**

Gene	Enzyme	Major altered base released
UNG	Uracil N-glycosylase	U
SMUG1	Single-strand selective monofunctional uracil DNA glycosylase	U
MBD4	Methyl-CpG binding domain protein 4	U or T opposite G at CpG sequences
TDG	Thymine DNA glycosylase	U, T or ethenoC opposite G
OGG1	8-oxoguanine DNA-glycosylase 1	8-oxoG opposite C
MYH	MutY homolog	A opposite 8-oxoG
NTHL1 (NTH1)	nth endonuclease III-like 1	Ring-saturated or fragmented pyrimidines
MPG	N-methylpurine-DNA glycosylase	3-meA, ethenoA, hypoxanthine

polymerases and ligases are required for both DNA replication and resynthesis after excision of a defect. The recombination machinery is involved in double-strand break repair. The link with transcription is particularly intriguing (Lehmann, 1995). The general transcription factor TFIIH is a multiprotein complex that includes the XPB and XPD proteins. TFIIH exists in two forms. One form is concerned with general transcription and the other with repair, probably specific repair of transcriptionally active DNA. This system is deficient in two rare diseases, Cockayne syndrome (CS; MIM 216400) and

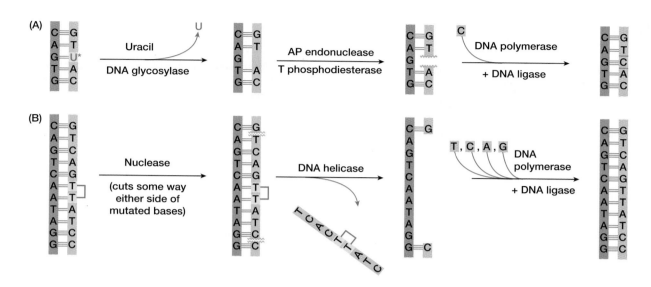

Figure 11.22: Base and nucleotide excision repair pathways.

Sugar-phosphate backbones are shown as vertical shaded boxes and hydrogen bonding by red lines.
(A) *Base excision repair*. Here a specific DNA glycosylase removes a uracil following deamination of a cytosine (*Figure 11.21A*) and the residual sugar-phosphate moiety is then removed by the sequential action of AP endonuclease and a phosphodiesterase. DNA polymerase inserts a deoxycytidine (dCMP) and DNA ligase seals the gap. **(B) *Nucleotide excision repair*.** Here a pyrimidine dimer is recognized as a bulky lesion; a nuclease within a multienzyme complex makes cuts some distance away on either side of the strand containing the mutation, and a helicase removes the intervening segment to leave a large gap (which in reality may be 20 nucleotides or more rather than the small gap shown here). The gap is repaired by sequential insertion of dNMP residues by DNA polymerase followed by sealing by DNA ligase.

trichothiodystrophy (TTD; MIM 601675). Clinically, and in cell biology, CS and TTD both overlap XP, and in some cases the same genes are responsible, but CS and TTD patients have developmental defects that presumably reflect defective transcription, and they do not have the cancer susceptibility of XP patients.

11.6.3 Hypersensitivity to agents that damage DNA is often the result of an impaired cellular response to DNA damage, rather than defective DNA repair

Many human diseases that involve hypersensitivity to DNA-damaging agents, or a high level of cellular DNA damage, are not caused by defects in the DNA repair systems themselves, but by a defective cellular response to DNA damage. Normal cells react to DNA damage by stalling progress through the cell cycle at a checkpoint until the damage has been repaired, or triggering apoptosis if the damage is unrepairable. Part of the machinery for doing this involves the ATM protein. The role of ATM is described in Section 17.5.1. Briefly, it senses DNA damage and relays the signal to the p53 protein, the 'guardian of the genome'. People with no functional ATM have ataxia telangiectasia (MIM 208900; Lambert *et al.* 1998). Their cells are hypersensitive to radiation, and they have chromosomal instability and a high risk of malignancy, but the DNA repair machinery itself is intact. Fanconi anemia (see MIM 227650) is another heterogeneous group of diseases (at least five complementation groups) that result in defective responses to DNA damage, without having specific defects in DNA repair.

Further reading

Bamshad M, Wooding SP (2003) Signatures of natural selection in the human genome. *Nature Rev. Genet.* **4**, 99–111.

Cooper DN, Krawczak M (1993) *Human Gene Mutation.* BIOS Scientific Publishers, Oxford.

Graur D, Li W-H (2000) *Fundamentals of Molecular Evolution*, 2nd Edn. Sinauer Associates, Sunderland, MA.

Li W-H (1997) *Molecular Evolution.* Sinauer Associates Sunderland, MA.

Marnett LJ, Plastras JP (2001) Endogenous DNA damage and mutation. *Trends Genet.* **17**, 214–221.

Nickoloff JA, Hoekstra MF (eds) (1998) *DNA Damage and Repair. Vol. 2: DNA Repair in Higher Eukaryotes.* Humana Press, Totowa, New Jersey.

TIBS October 1995 issue on DNA repair (1995) *Trends Biochem. Sci.* **20**, 381–440.

Electronic resources and mutation databases

Codon Usage Database at http://www.kazusa.or.jp/codon/

Human Gene Mutation Database at http://archive.uwcm.ac.uk/uwcm/mg/hgmd0.html

Human DNA Repair Gene List at http://www.cgal.icnet.uk/DNA_Repair_Genes.html

HGV base (human genome variation database) at http://hgvbase.cgb.ki.se

Locus-specific mutation databases – see compilations at various centers e.g. within the Human Gene Mutation database at http://archive.uwcm.ac.uk/uwcm/mg/docs/oth_mut.html

Nomenclature for the description of sequence variations at http://www.dmd.nl/mutnomen.html

SNP databases – see compilations at various centers e.g. the UK HGMP Resource Centre at http://www.hgmp.mrc.ac.uk/GenomeWeb/human-gen-db-mutation.html

References

Ayala FJ (1999) Molecular clock mirages. *Bioessays* **21**, 71–75.

Blencowe BJ (2000) Exonic splicing enhancers: mechanism of action, diversity and role in human genetic diseases. *Trends Biochem. Sci.* **25**, 106–110.

Bodmer W, Cavalli-Sftorza LL (1976) *Genetics, Evolution and Man.* Freeman, San Francisco.

Brown WM, George M Jr, Wilson AC (1979) Rapid evolution of animal mitochondrial DNA. *Proc. Natl Acad. Sci. USA* **76**, 1967–1971.

Cairns J (1975) Mutation selection and the natural history of cancer. *Nature* **255**, 197–200.

Chinnery PF, Thorburn DR, Samuels DC et al. (2000) The inheritance of mitochondrial DNA heteroplasmy: random drift, selection or both? *Trends Genet.* **16**, 500–505.

Chung MY, Ranum LPW, Duvick IA, Servadio A, Zoghbi HY, Orr HT (1993) Evidence for a mechanism predisposing to intergenerational CAG repeat instability in spinocerebellar ataxia type 1. *Nature Genet.* **5**, 254–258.

Collier S, Sinnott PJ, Dyer PA, Price DA, Harris R, Strachan T (1989) Pulsed field gel electrophoresis identifies a high degree of variability in the number of tandem 21-hydroxylase and complement C4 gene repeats in 21-hydroxylase deficiency haplotypes. *EMBO J.* **8**, 1393–1402.

Collier PS, Tassabehji M, Sinnott PJ, Strachan T (1993) A de novo pathological point mutation at the 21-hydroxylase locus: implications for gene conversion in the human genome. *Nature Genet.* **3**, 260–265 and *Nature Genet.* **4**, 101.

Collins DW, Jukes TH (1994) Rates of transition and transversion in coding sequences since the human-rodent divergence. *Genomics* **20**, 386–396.

Cooper DN, Krawczak M, Antonorakis SE (2000) The nature and mechanisms of human gene mutation. In: *The Metabolic and Molecular Bases of Inherited Disease*, Vol. 1, 8th Edn (eds CR Scriver, AL Beaudet, WS Sly, D Valle). McGraw-Hill, New York.

Crow JF (2000) The origins, patterns and implications of human spontaneous mutation. *Nature Rev. Genet.* **1**, 40–47.

Djian P (1998) Evolution of simple repeats in DNA and their relation to human disease. *Cell* **94**, 155–160.

Dover GA (1995) Slippery DNA runs on and on and on. *Nature Genet.* **10**, 254–256.

Dubrova YE, Nesterov VN, Krouchinsky NG, Ostapenko VN, Neumann R, Neil DL, Jeffreys AJ (1996) Human minisatellite mutation rate after the Chernobyl accident. *Nature* **380**, 683–686.

Dubrova YE, Bersimbaev RI, Djansugurova LB et al. (2002) Nuclear weapons tests and human germline mutation rates. *Science* **295**, 1037.

Eichler EE (1998) Masquerading repeats: paralogous pitfalls of the human genome. *Genome Res.* **8**, 758–762.

Ellegren H (2000) Microsatellite mutations in the germline. *Trends Genet.* **16**, 551–558.

Eyre-Walker A, Keightley PD (1999) High genomic deleterious mutation rates in hominids. *Nature* **397**, 344–347.

Fairbrother WG, Chasin LA (2000) Human genomic sequences that inhibit splicing. *Mol. Cell Biol.* **20**, 6816–6825.

Fairbrother WG, Yeh RF, Sharp PA, Burge CB (2002) Predictive identification of exonic splicing enhancers in human genes. *Science* **297**, 1007–1013.

Giglio S, Broman KW, Matsumoto N et al. (2001) Olfactory receptor-gene clusters, genomic-inversion polymorphisms and common chromosome rearrangements. *Am. J. Hum. Genet.* **68**, 874–883.

Grantham R (1974) Amino acid difference formula to help explain protein evolution. *Science* **185**, 862–864.

Grimm T, Meng G, Liechti-Gallati S, Bettecken T, Muller CR, Muller B (1994) On the origin of deletions and point mutations in Duchenne muscular dystrophy: most deletions arise in oogenesis and most point mutations result from events in spermatogenesis. *J. Med. Genet.* **31**, 183–186.

Haber JA (1999) Gatekeepers of recombination. *Nature* **398**, 665–667.

Hobbs HH, Russell DW, Brown MS, Golding JL (1990) The LDL receptor locus in familial hypercholesterolaemia: mutational analysis of a membrane protein. *Annu. Rev. Genet.* **24**, 133–170.

Hurst LD, Ellegren H (1998) Sex biases in the mutation rate. *Trends Genet.* **14**, 446–452.

International SNP Mapping Working Group (2001) A map of human genome sequence variation containing 1.42 million single nucleotide polymorphisms. *Nature* **409**, 928–933.

Jeffreys A, Tamaki K, MacLeod A, Monckton DG, Neil DL, Armour JAL (1994) Complex gene conversion events in germline mutation at human microsatellites. *Nature Genet.* **6**, 136–145.

Jenuth JP, Peterson AC, Fu K, Shoubridge EA (1996) Random genetic drift in the female germline explains the rapid segregation of mammalian mitochondrial DNA. *Nature Genet.* **14**, 146–150.

Kazazian HH Jr (1998) Mobile elements and disease. *Curr. Opin. Genet. Dev.* **8**, 343–350.

Kong A, Gudbjartsson DF, Sainz J et al. (2002) A high-resolution recombination map of the human genome. *Nature Genet.* **31**, 241–247.

Kumar S, Subramanian S (2002) Mutation rates in mammalian genomes. *Proc. Natl Acad. Sci. USA* **99**, 803–808.

Lakich D, Kazazian Jr HH, Antonarakis SE, Gitschier J (1993) Inversions disrupting the factor VIII gene are a common cause of severe haemophilia A. *Nature Genet.* **5**, 236–241.

Lalioti MD, Scott HS, Buresi C et al. (1997) Dodecamer repeat expansion in cystatin B gene in progressive myoclonus epilepsy. *Nature* **386**, 847–851.

Lambert WC, Kuo H-R, Lambert MW (1998) Xeroderma pigmentosum and related disorders. In: *Principles of Molecular Medicine* (ed. Jameson JP). Humana Press, Totowa, New Jersey.

Lehmann AR (1995) Nucleotide excision repair and the link with transcription. *Trends Biochem. Sci.* **20**, 402–405.

Levinson G, Gutman GA (1987) Slipped strand mispairing: a major mechanism for DNA sequence evolution. *Mol. Biol. Evol.* **4**, 203–221.

Li WH, Tanimura M (1987) The molecular clock runs more slowly in man than in apes and monkeys. *Nature* **326**, 93–96.

Li WH, Yi S, Makova K (2002) Male driven evolution. *Curr. Opin. Genet. Dev.* **12**, 650–656

Lindahl T, Wood RD (1999) Quality control by DNA repair. *Science* **286**, 1897–1905.

López Correa C, Brems H, Lázaro C, Marynen P, Legius E (2000) Unequal meiotic crossover: a frequent cause of *NF1* microdeletions. *Am. J. Hum. Genet.* **66**, 1969–1974.

Luo MJ, Reed R (1999) Splicing is required for rapid and efficient mRNA export in metazoans. *Proc. Natl Acad. Sci. USA* **96**, 14937–14942.

Lykke-Andersen J, Shu M-D, Steitz JA (2001) Communication of the position of exon–exon junctions to the mRNA surveillance machinery by the protein RNPS1. *Science* **293**, 1836–1839 (see also the preceding paper in that issue).

Makova KD, Li WH (2002) Strong male-driven evolution of DNA sequences in humans and apes. *Nature* **416**, 624–626.

Maquat LE (2002) NASty effects on fibrillin pre-mRNA splicing: another case of ESE does it, but proposals for translation-dependent splice site choice live on. *Genes Dev.* **16**, 1743–1753.

Mazzarella R, Schlessinger D (1998) Pathological consequences of sequence duplications in the human genome. *Genome Res.* **8**, 1007–1021.

Mouse Genome Sequencing Consortium (2002) Initial sequencing and comparative analysis of the mouse genome. *Nature* **420**, 520–562.

Nachman MW, Crowell SL (2000) Estimate of the mutation rate per nucleotide in humans. *Genetics* **156**, 297–304.

Osborne LR, Li M, Pober B et al. (2001) A 1.5 million-base pair inversion polymorphism in families with Williams–Beuren syndrome. *Nature Genet.* **29**, 321–325.

Poulton J, Holt IJ (1994) Mitochondrial DNA: does more lead to less? *Nature Genet.* **8**, 313–315.

Przeworski M, Hudson RR, Di Rienzo A (2000) Adjusting the focus on human variation. *Trends Genet.* **16**, 296–302.

Reich DE, Schaffner SF, Daly MJ et al. (2002) Human genome sequence variation and the influence of gene history, mutation and recombination. *Nature Genet.* **32**, 135–142.

Richard I, Beckmann JS (1995) How neutral are synonymous codon mutations? *Nature Genet.* **10**, 259.

Roy-Engel AM, Carroll ML, Vogel E et al. (2001) Alu insertion polymorphisms for the study of human genomic diversity. *Genetics* **159**, 279–290.

Shoffner JM, Lott MT, Voljavec AS, Soueidan SA, Costigan DA, Wallace DC (1989) Spontaneous Kearns-Sayre/chronic external ophthalmoplegia plus syndrome associated with a mitochondrial DNA deletion: a slip-replication model and metabolic therapy. *Proc. Natl Acad. Sci. USA* **86**, 7952–7956.

Sinden RR, Potaman VN, Oussatcheva E, Pearson CE, Lyubchenko YL, Shlyakhtenko LS (2002) Triplet DNA structures and human genetic disease: dynamic mutations from dynamic DNA. *J. Biosci.* **27**, 53–65.

Stankiewicz P, Lupski JR (2002) Molecular-evolutionary mechanisms for genomic disorders. *Curr. Opin. Genet. Dev.* **12**, 312–319.

Swallow DM, Gendler S, Griffiths B (1987) The hypervariable gene locus PUM, which codes for the tumour associated epithelial mucins, is located on chromosome 1, within the region 1q21-24. *Ann. Hum. Genet.* **51**, 289–294.

Takahara K, Schwarze U, Imamura Y et al. (2002) Order of intron removal influences multiple splice outcomes, including a two-exon skip, in a COL5A1 acceptor-site mutation that results in abnormal pro-alpha1(V) N-propeptides and Ehlers–Danlos syndrome type I. *Am. J. Hum. Genet.* **71**, 451–465.

Van der Horst GTJ, Muijtjens M, Kobayashi K et al. (1999) Mammalian Cry1 and Cry2 are essential for maintenance of circadian rhythms. *Nature* **398**, 627–630.

Vogel F, Motulsky AG (1996) *Human Genetics. Problems and Approaches*, 3rd Edn. Springer Verlag, Berlin.

Wang J, Chang YF, Hamilton JI, Wilkinson MF (2002) Nonsense-associated altered splicing: a frame-dependent response distinct from nonsense-mediated decay. *Mol. Cell* **10**, 951–957.

Wood RD, Mitchell M, Sgouros J, Lindahl T (2001) Human DNA repair genes. *Science* **291**, 1284–1291.

Wu CI, Li WH (1985) Evidence for higher rates of nucleotide substitution in rodents than in man. *Proc. Natl Acad. Sci. USA* **82**, 1741–1745.

Yan H, Yuan W, Velculescu VE, Vogelstein B, Kinzler KW (2002) Allelic variation in human gene expression. *Science* **297**, 1143.

CHAPTER TWELVE

Our place in the tree of life

Chapter contents

Supporters of Theodosius Dobzhansky's view that "nothing in biology makes sense except in the light of evolution" will have been encouraged by the torrents of data flowing out of the different genome projects. Sequence comparisons across whole genomes are now the mainstay of a new discipline, **comparative genomics**, and they are beginning to provide powerful new insights into our place in the tree of life.

The evolutionary origins of the human genome (and all genomes) are, of course, as old as life itself but the present chapter is not intended as an overview of molecular evolutionary genetics *per se*. Many fascinating areas are not covered here, therefore, such as the early evolution of the genetic code (e.g. Knight and Landweber, 2000) or the idea that RNA used to be the primary information molecule before being superseded by DNA (e.g. Joyce, 2002).

Instead, the present chapter is meant to focus on how comparative analyses of present day genomes have shed light on the evolutionary origin of human DNA and human genes. Much of the data is derived from comparisons of mammalian and other animal genomes, although comparison with more distant genomes is occasionally used to explain certain **footprints of evolution**, as in the origin of introns and mitochondrial DNA. The comparative data allow insights into our uniqueness when compared with mammalian models, notably the mouse (an important model for understanding early human development and also human disease) and primates, our closest living relatives. A final section considers the recent origins and relatedness of human populations.

12.1 Evolution of gene structure and duplicated genes

Eukaryotic genes and proteins are typically larger and more complex than those from simple organisms. The larger gene sizes in eukaryotes is principally due to the presence of large introns, and the average intron size often reflects genome (and biological) complexity. Eukaryotic coding sequences are also typically longer than those of prokaryotes as a result of different mechanisms. *Intragenic duplication* allows coding sequence lengths to expand and often to diversify. *Intergenic recombination* brings together different combinations of **protein domains** (discrete structural or functional modules). The near ubiquity of introns in the genes of complex genomes is thought to reflect their importance in allowing coding sequences to be expanded and modified during evolution.

12.1.1 Spliceosomal introns probably originated from group II introns and first appeared in early eukaryotic cells

Following the discovery of split genes in 1977, the significance of **spliceosomal introns** (see *Box 12.1*) has been intensely debated. The introns found in complex genomes are generally large compared with those in other species and the intron sequences are not so well conserved. Nevertheless, introns contain functionally important sequences involved in gene regulation and the sequences of some short introns have been

considerably conserved in evolution. Highly expressed genes often have very short introns (possibly because of selection for rapid mRNA processing). Whatever function is proposed for introns (e.g. permitting recombination to enable evolutionary novelty as in Section 12.1.3), it cannot be a general one: a small minority of genes in complex organisms lack introns (see *Table 9.5*).

The evolution of spliceosomal introns has been a controversial issue (see Logsdon, 1998; Lynch and Richardson, 2002). It is now largely accepted that spliceosomal introns did not originate before eukaryotic cells appeared (none of the hundreds of sequenced prokaryotic genomes harbors the signature of current or past introns in protein-coding genes). Instead, they most likely first appeared very early on in the evolution of eukaryotic cells: the only potentially early diverging eukaryotic group in which they have not been found are the parabasalids such as *Trichomonas* which, however, have some of the necessary splicing machinery.

Spliceosomal introns probably originated from self-splicing group II introns which are processed by similar mechanisms (see *Box 12.1*; Lynch and Richardson, 2002). Spliceosomal introns cannot carry out splicing by themselves, requiring five separate snRNA molecules and many proteins, but group II introns are normally cohesive, *self-splicing* introns. However, some functional group II introns are known to have fragmented into different components, somewhat reminiscent of the dispersed spliceosomal processing system. Some group II introns can also encode their own reverse transcriptase and act as mobile elements. Spliceosomes may have originated, therefore, from a group II intron within an early organelle (transfer of sequences from organelle DNA to nuclear DNA is well established; see *Box 12.4*). The subsequent fragmentation into different components may have taken place over an extended period.

Since they first appeared spliceosomal introns have periodically integrated into genes (and in some cases, been eliminated) during evolution. Some clearly have ancient origins. For example, the positioning of the two major introns in the the globin superfamily has been very well conserved suggesting that the two introns integrated into these positions in an ancestral globin gene probably more than 800 MYr (million years) ago (*Figure 12.1*). Exceptionally high conservation of intron positions can also be found in distantly related orthologs. For example, the human Huntington's disease gene has 67 exons spanning 170 kb. The equivalent gene in pufferfish (which diverged from humans over 400 MYr ago) is only 23 kb long, but also has 67 exons with almost perfect conservation of intron positions (Baxendale *et al.*, 1995). Some other spliceosomal introns, however, appear to be of more recent evolutionary origin. For example, intron location in numerous individual gene families (e.g. actins, myosins, tubulins) are not well conserved.

12.1.2 Complex genes can evolve by intragenic duplication, often as a result of exon duplication

Like other eukaryotic genes, human genes often show evidence of intragenic DNA duplication which can be

---

Box 12.1: Intron groups.

Introns are heterogeneous entities with different functional capacities and notable structural differences, including enormous length differences (unlike exons which appear to be much more homogeneous in length; see *Table 9.6*). Depending on the extent to which they rely on extrinsic factors to engage in RNA splicing and on the nature of the splicing reaction, they can be classified into different **intron groups** as follows.

▶ **Spliceosomal introns** are the conventional introns of eukaryotic cells. They are transcribed into RNA in the primary transcript and are excised at the RNA level during RNA processing by *spliceosomes*. Only a few short sequences appear to be important for gene function [those at, or close to, the splice junctions and at the branch site (*Figure 1.15*) plus splice enhancers and silencers (Section 11.4.3)]. As a result, spliceosomal introns can tolerate large insertions and can be very long (sometimes over 1 Mb). Spliceosomal introns are likely to have arisen comparatively recently in evolution, and may have evolved from group II introns (see Section 12.1.1). By tolerating the insertion of mobile elements, they facilitated *exon shuffling*.

▶ **Group I and II introns** have significant secondary structure and are *self-splicing introns* (they can catalyze their own excision without the need for a spliceosome). They are found in both bacteria and eukaryotes, but are very restricted in their distribution, being found primarily in rRNA and tRNA genes and in a few protein-coding genes found in some types of mitochondria, chloroplasts and bacteriophages. Both groups may also act as mobile elements, and mobile group II introns encode a reverse transcriptase-like activity, which is strikingly similar to that of LINE-1 elements. Group I and II introns differ in the identity of conserved splicing signals and in the nature of the splicing reaction (e.g. the group I introns are the only intron class, which requires a free guanine nucleoside – see Bonen and Vogel, 2001).

▶ **Archaeal introns** have been found only in tRNA and rRNA genes in *archaea*. They have no conserved internal structure and, unlike group I and group II introns, are not self-splicing. Although they require proteins for the splicing mechanism, they do not, unlike spliceosomal introns, require *trans*-acting RNA molecules for the splicing reaction.

---

substantial. For example, many genes are known to encode polypeptides whose sequences are completely or largely composed of large repeats, with sequence homology between the repeats being very high in some cases (see *Table 9.7*). By repeating a previously designed domain larger polypeptides can be constructed with a variety of evolutionary advantages. The human ubiquitin-encoding genes, *UBB* and *UBC*, encode long multimeric proteins consisting of tandem repeats of a whole ubiquitin protein unit which are cleaved to generate multiple ubiquitin copies (three in the case of *UBB*; nine in the case of *UBC*).

Aside from the unusual case of ubiquitin genes, intragenic duplication often involves some form of **exon duplication** so that a particular protein domain is duplicated (*Figure 12.2*). About 10% of genes in humans, *Drosophila melanogaster* and *Caenorhabditis elegans* have duplicated exons (Letunic *et al.*, 2002) and several advantages can be envisaged:

▶ **structural extension**. Repeating domains may be particularly advantageous in the case of proteins that have a major structural role. An illustrative example is provided by the 41 exons of the *COL1A1* gene which encode the part of α1(I) collagen that forms a triple helix; each exon encodes essentially an integral number of copies (one to three) of an 18-amino acid motif which itself is composed of six tandem repeats of the structure Gly–X–Y where X and Y are variable amino acids;

▶ **diversity through domain divergence**. In most cases, intragenic duplication events have been followed by substantial nucleotide sequence divergence between the different repeat units. Such divergence presumably provides the opportunity of acquiring different, though related, functions. Sometimes the degree of sequence divergence between the repeats is such that the repeated structure may

not be obvious at the sequence level as in the case of immunoglobulin domains (*Figure 9.10*).

▶ *diversity through alternative splicing* (see e.g. Letunic *et al.*, 2002). One type of alternative splicing produces different isoforms by selecting one exon sequence from a group of duplicated exons to be included in the spliced product. As the sequences of the duplicated exons may show small differences this type of alternative splicing produces a series of related isoforms. Many human genes undergo this type of splicing but the *Drosophila Dscam* gene is the most celebrated example (see *Figure 10.16*).

Intragenic exon duplication can be explained by a variety of mechanisms including unequal crossover, or *unequal sister chromatid exchange* (see Long, 2001). In order to avoid frameshifts in the translational reading frame, duplication is confined to **symmetrical exons** or *symmetrical exon groups* (*Box 12.2*).

12.1.3 Exon shuffling can bring together new combinations of protein domains

Only a few thousand conserved protein domains are known throughout Nature but for metazoan species many proteins contain domains found in another protein (Li *et al.*, 2001). Fibronectin, a large extracellular matrix protein, contains multiple repeated domains encoded by individual exons or pairs of exons and is a good example of classical exon duplication. One of the repeated domains, now known as the fibronectin type I domain, was subsequently found in tissue plasminogen activator. Like fibronectin, tissue plasminogen activator also contains other domains. They include an epidermal growth factor (EGF)-like domain characteristic of the EGF precursor, and two kringle domains which have been found in other polypeptides such as prourokinase and plasminogen etc. (*Figure 12.2*).

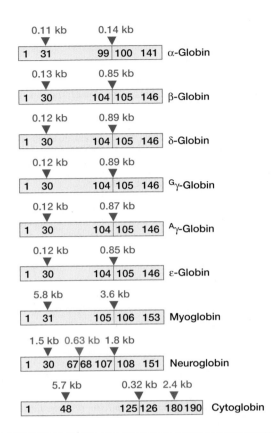

Figure 12.1: Ancient intron insertion is suggested from conservation of intron position in the globin superfamily.

Boxes represent the mature polypeptides. Numbers contained within the boxes are the amino acid positions. Note the generally strong conservation of intron positions suggesting that an ancestral globin gene had two introns (shown in blue) perhaps more than 800 Myr ago (see *Figure 12.4*). The additional introns in the recently identified neuroglobin sequence (red) and cytoglobin sequences (purple) may have been inserted some very considerable time ago. Note that the expansion of the cytoglobin protein included an expansion of the N-terminal region by about 17 amino acids and the C-terminal region by about 23 amino acids (see Pesce *et al.*, 2002).

Not infrequently protein domains are encoded by individual *symmetrical exons* or *symmetrical exon groups* (*Box 12.2*). Such observations have suggested the possibility of **exon shuffling** between genes: exons or exon groups encoding whole domains are copied and inserted into other genes (see Patthy, 1999; Kaessmann *et al.*, 2002). The mechanism of exon shuffling most likely involves LINE element-assisted retrotransposition (see Section 12.1.6).

12.1.4 Gene duplication has played a crucially important role in the evolution of multicellular organisms

Mutation is the motor of evolution. In prokaryotes short doubling times and large populations allow mutations which confer a survival advantage (in response to changes in the

environment) to quickly become established. In multicellular animals, however, changes in important genes can often be harmful and usually there is strong conservative *selection pressure* to maintain the sequence of a gene. To propagate mutations in complex organisms, therefore, there is a need for **gene duplication**.

For each case of gene duplication, one of the two gene copies is surplus to requirement and so can diverge rapidly (because of the absence of selection pressure to conserve function). Even if there is no problem with gene dosage effects, an additional gene copy often acquires deleterious mutations and degenerates into a nonfunctional **pseudogene**, or is lost by DNA turnover processes (which can happen over evolutionary timescales when a DNA sequence is not important). In some cases, however, the diverged gene copy mutates to produce a functional product with altered properties that may be selectively advantageous and conserved (*Figure 12.3*). Alterations in coding sequences could take some time to be established, but alterations in regulatory sequences may quickly provide novel expression characteristics. The resulting gene product could become expressed in different tissues or at different stages of development and subsequently adapt to the new environment (see the example of globin genes – Section 12.1.5).

Duplicated genes are common in complex genomes and they can arise quite rapidly on an evolutionary timescale: an average gene undergoes duplication about once every 100 MYr (Lynch and Conery, 2000). They originate by a variety of different mechanisms. Some cause limited duplications, often involving single genes and are considered in Sections 12.1.5 and 12.1.6. Others bring about large-scale and whole genome duplications and are discussed in the context of chromosome and genome evolution (Section 12.2). The principal mechanisms are described in *Box 12.3*.

12.1.5 The globin superfamily has evolved by a process of gene duplications, gene conversions, and gene loss/inactivation

Evolution of the five globin gene classes

Globins are porphyrin-containing proteins that bind oxygen reversibly and so are important in the respiratory system. They fall into four functional classes (see *Table 12.1*): the well-studied hemoglobin and myoglobin, and the more recently identified neuroglobin and cytoglobins (see Pesce *et al.*, 2002). Hemoglobins transport oxygen in the blood and are heterotetramers consisting of two identical globin chains encoded by members of the *α-globin gene family* on chromosome 16, and two encoded by *β-glsobin gene family* members on chromosome 11. Myoglobin is encoded by a gene on chromosome 22. It is monomeric and found in muscle cells where it facilitates the diffusion of oxygen to the mitochondria. The neuroglobin gene on chromosome 14 encodes a monomer and is expressed in the brain where it may increase the availability of oxygen to the brain tissues (Burmester *et al.*, 2000). The recently identified cytoglobin gene is ubiquitously expressed; its function is unknown.

Sequence homology between the globin polypeptides suggest that this family has been shaped by a series of gene

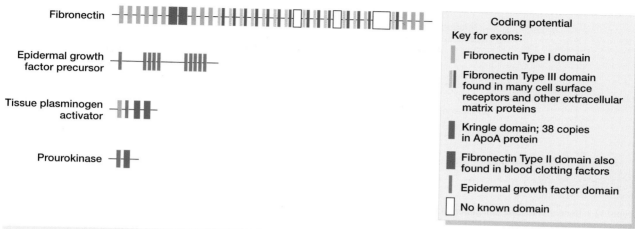

Figure 12.2: Exon duplication and exon shuffling.

Conspicuous exon duplication is found in the human fibronectin (*FN1*) and epidermal growth factor (*EGF*) genes. The *FN1* gene has 12 copies of an exon encoding the fibronectin type I domain and 15 copies of a pair of exons, which together specify the fibronection type III domain; the *EGF* gene has 9 copies of an exon specifying an EGF domain. Exon shuffling has meant that exons encoding these domains are found in many other genes such as those encoding tissue plasminogen activator and prourokinase. See *Figure 12.7* for a possible mechanism for exon shuffling.

Box 12.2: Symmetrical exons and intron phases.

Not all exons contain coding DNA. **Noncoding exons** occur because introns are sometimes located within untranslated sequences, such as the 5'UTR and 3'UTR sequences of genes encoding polypeptides. Introns, which split coding DNA into different coding exons, allow diversification by *exon duplication* and *exon shuffling*. Three **intron phases** can be distinguished depending on the point of insertion within coding DNA (see *Figure*): *phase 0* (between the third base of a codon and the first base of the next codon); *phase 1* (between the first and second bases of one codon); *phase 2* (between the second and third bases of a codon).

Coding exons can be classified according to the phases of the introns flanking them. *Symmetrical exons* (with a total number of nucleotides exactly divisible by three) will be flanked by introns of the same phase and so may be classified as: 0-0, 1-1 or 2-2 according to the phase of the flanking introns. *Nonsymmetrical exons* (where the number of nucleotides is not exactly divisible by 3) can be classified as 0-1, 0-2, 1-0 etc. Making a copy of an exon by exon duplication and exon shuffling is often limited to symmetrical exons, or to groups of neighboring nonsymmetrical exons which would not result in a frameshift if duplicated or copied into another gene, e.g. neighboring 0-1 and 1-0 exons.

Examples of intron phases in the β-globin and insulin genes.

Numbers above genes refer to codon/amino acid positions. *HBB* – human β-globin gene. *INS* – human insulin gene. **Note:** Exon 2 of the β-globin gene would be classified as a *nonsymmetrical* 2-0 exon because it is flanked by introns of different phases, in this case a phase 2 intron upstream and a phase 0 intron downstream.

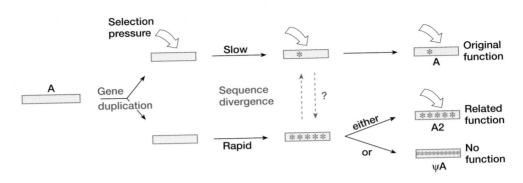

Figure 12.3: Gene duplication can generate different functional gene varieties but often often results in formation of a pseudogene.
Duplication of gene A results in two equivalent gene copies. Selection pressure need be applied to only one gene copy (top) to maintain the presence of the original functional gene product. The other copy (bottom), will continue to be expressed but, in the absence of selection pressure to conserve its sequence, will accumulate mutations (red asterisks) relatively rapidly. It may acquire deleterious mutations and become a nonfunctional pseudogene (ψA) which will eventually be eliminated from the genome. In some cases, however, the mutational differences may lead to a different expression pattern or other property that is selectively advantageous (A2).

Table 12.1: Classes of globin and human globin genes

Protein	Function	Structure	Polypeptide(s)	Gene(s)	Gene location
Hemoglobin* (Hb)	O$_2$ transport in blood. Portland, Gower 1 and Gower 2 are early embryonic forms. At about 8 weeks of gestation, the fetal liver synthesizes predominantly HbF and a little HbA. 70% of Hb in newborns is HbF, but by the end of year this has declined to 1% as HbA predominates. HbA2 is a minority Hb in children and accounts for < 3% of the adult Hb	Heterotetramer *Portland* ($\zeta_2\gamma_2$) *Gower 1* ($\zeta_2\epsilon_2$) *Gower 2* ($\alpha_2\epsilon_2$) *HbF* = fetal Hb ($\alpha_2\gamma_2$) *HbA* adult Hb ($\alpha_2\beta_2$) *HbA2* ($\alpha_2\delta_2$)	Alpha-globin Zeta (ζ) – globin Beta-globin Gamma-globin Delta-globin Epsilon (ϵ) – globin	*HBA1,HBA2* *HBZ* *HBB* *HBG1,HBG2* *HBD* *HBE1*	16p13.3, α-globin cluster 16p13.3, α-globin cluster 11p15.5, β-globin cluster 11p15.5, β-globin cluster 11p15.5, β-globin cluster 11p15.5, β-globin cluster
Myoglobin	O$_2$ transport and storage in muscle	Monomer	Myoglobin	*MB*	22q13.1
Neuroglobin	Expressed predominantly in nervous system where it supplies oxygen; may be multifunctional	Monomer	Neuroglobin	*NGB*	14q24
Cytoglobin	Expressed in almost all tissues; exact function unknown	Monomer	Cytoglobin	*CYGB*	17q25.3

Note: The theta globin polypeptide synthesized by the *HBQ1* gene in the α-globin cluster (see *Figure 9.11*) does not get incorporated into a Hb molecule and its function, if any, is unknown.

duplication events, some ancient and some more recent. Thus, globin polypeptides from the different chromosomal locations are generally distantly related to each other suggesting ancient gene duplications (although the α-globin and β-globin families are clearly more closely related to each other than to myoglobin, neuroglobin or cytoglobin).

The degree of sequence homology between the different globins plus the observation of generally conserved intron positions (*Figure 12.1*) suggests that the four families originated from a single ancestral globin gene by a series of duplications from around 800 MYr to 450 MYr ago (*Figure 12.4*).

Box 12.3: Gene duplication mechanisms and paralogy

Gene duplication in metazoan genomes means that **homologous genes** (or **homologs** – genes with significant sequence identity suggesting a close evolutionary relationship) can be one of two types (see *Figure*). **Orthologs** are genes present in different genomes which are directly related through descent from a common ancestor. **Paralogs** are genes present in a single genome as a result of gene duplication. A variety of different mechanisms can result in gene duplication.

▶ **Tandem gene duplication**. Single genes can undergo tandem duplication as a result of unequal crossover events or unequal sister chromatid exchanges (*Figure 11.9*). Clustered human gene families typically originate by sequential tandem duplications (e.g., the α-globin and β-globin gene clusters – see Section 12.1.5). The same mechanisms can give rise to larger-scale duplications, involving segments containing several genes, but very large-scale tandem duplications are relatively rare, partly because of the higher likelihood of gene dosage effects, etc.

▶ **Retrotransposition-mediated gene duplication** (= *duplicative transposition*). This is likely to be an important contributor. Although it often results in intronless copies of an intron-containing gene *it can also sometimes result in copying of intron sequences* (Section 12.1.6).

▶ **Horizontal** (= *lateral*) **gene transfer**. This means ancient gene transfers between different genomes (Brown, 2003). The International Human Gene Sequencing Consortium (2001) suggested that hundreds of human genes likely originated by horizontal gene transfer from bacteria at some point in the vertebrate lineage. Although this is now known to be incorrect (see *Box 12.4*), at least some human nuclear genes appear to have originated by ancient horizontal gene transfer following the acquisition of a proto-mitochondrial genome (Section 12.2.1).

▶ **Segmental duplication**. A significant proportion of the human genome consists of closely related sequence blocks at different genomic locations showing >90% sequence identity over segments spanning from kilobase lengths to hundreds of kilobases. This type of duplication appears to have occurred very recently during evolution. See Section 12.2.5.

▶ **Polyploidy**. Whole-genome duplication has the attraction that it immediately offers gene copies for mutation to work on without incurring problems due to differences in gene dosage. The genomes of several species have clearly undergone polyploidy but the extent to which this has shaped the evolution of human genes has been controversial (Section 12.2.3).

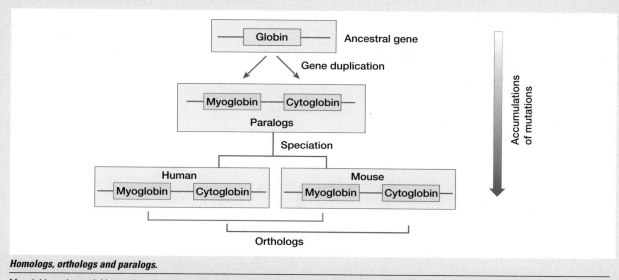

Homologs, orthologs and paralogs.

Myoglobin and cytoglobin are thought to have originated from a gene duplication occurring about 500 MYr ago. See *Figure 12.4* for the relationships with the other globin genes.

Thereafter, the ancestral α– and β–globin genes underwent a series of additional duplications, some occurring very recently (see *Figure 12.4*). For example, the two human α-globin genes *HBA1* and *HBA2* encode identical products, and the products of the two γ-globin genes, *HBG1* and *HBG2*, differ by a single amino acid. In other cases, the duplicated genes within a cluster are clearly more diverged in sequence, presumably because the relevant duplication events occurred some time ago. Some duplications gave rise to conventional pseudogenes. Maybe in the future one each of the two human α-globin genes and the two γ-globin genes will degenerate into pseudogenes.

Evolution of mammalian β-globin gene clusters

Comparative sequence analyses of the β–globin gene clusters in different mammals reveals somewhat different gene organizations. Different types of single gene duplication have occurred in different lineages, and there is also evidence of gene

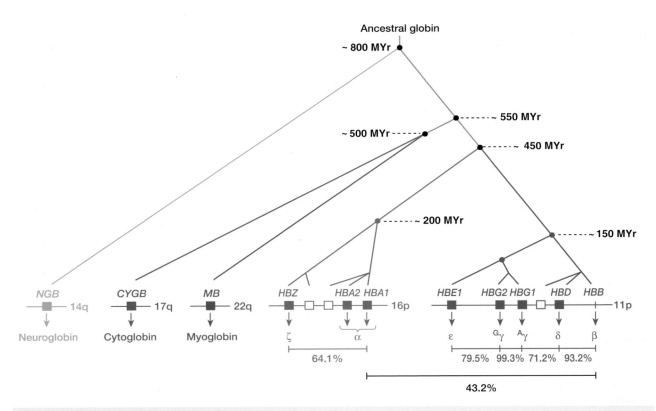

Figure 12.4: Evolution of the globin superfamily.

Globins encoded within a gene cluster show a greater degree of sequence identity than those encoded on the different chromosomes (where values of sequence identity are significantly less than 30% except for comparisons between the α-globin and β-globin gene clusters). The α- and β-globin gene clusters have later diversified by tandem gene duplications. Some of the duplications have been very recent: the *HBA1* and *HBA2* genes encode identical α-globins; *HBG1* and *HBG2* encode γ-globins that differ by a single amino acid.

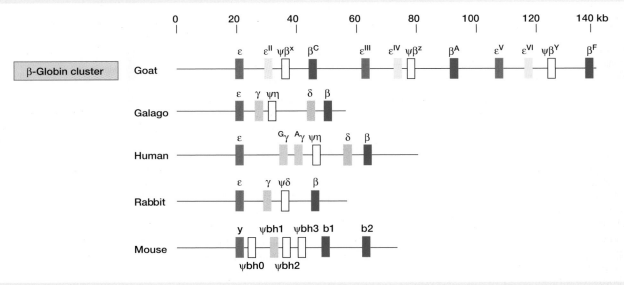

Figure 12.5: The β-globin gene cluster illustrates the considerable differences in organization of orthologous mammalian gene families.

The large number of genes in the goat β-globin cluster reflects a tandem triplication event. Open boxes denote pseudogenes. Redrawn from Hardison and Miller (1993) *Mol. Biol. Evol.* **10**, 73–102. © 1993, Oxford University Press.

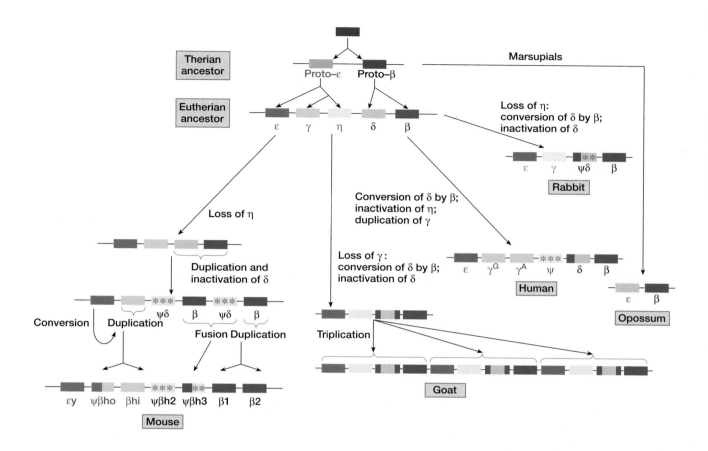

Figure 12.6: The evolution of the mammalian β-globin gene cluster has involved frequent gene duplications, conversions and gene loss or inactivation.

Note that, in addition to gene duplication and gene loss events, there are frequent examples where the sequence of one gene shows evidence of having been copied from the sequence of another gene. This is loosely described as conversion in this figure, but may have involved mechanisms other than gene conversion in some cases. For example, the conversion of δ by β in the lineage leading to rabbits could have involved an unequal crossover event of the type that most likely resulted in the production of the ψh3 gene in the mouse lineage. Redrawn from Tagle *et al.* (1992) *Genomics* **13**, 741–760 with permission from Elsevier.

loss, gene fusion and *gene conversion* (nonreciprocal sequence exchanges where a sequence copied from a donor gene is used to replace some sequence of an acceptor gene – see *Figure 11.10* for the mechanism). For example, humans have one β-globin and one ε-globin gene but two γ-globin genes; the mouse has two β-globin genes, two ε-like globin genes but only one γ-globin gene (*Figure 12.5*). In the lineage leading to present day goats early loss of a γ-globin gene was followed by large-scale triplication (presumably as a result of two successive unequal recombination events, the first of which produced a duplication). *Gene conversion* appears to have been frequent. For example, conversion of δ-globin by β-globin has been apparent in both the human and goat lineages (*Figure 12.6*).

The above types of event can all be explained by simple sequence exchanges occurring in this cluster.

As marsupials such as the opossum have only two genes in this cluster, a β-globin gene and an ε-globin gene, there is the possibility that the current β-globin gene cluster arose from a single globin gene which underwent an initial duplication generating a proto β-like globin gene and a proto ε-like globin gene which then underwent additional duplications to generate the five types of gene in this cluster (*Figure 12.6*). The initial perfect sequence identity of tandemly duplicated sequences makes them liable to further sequence exchanges by unequal crossover (which can give rise to further gene duplication, gene loss and fusion genes) and gene conversion. Eventually, however, duplicated sequences will diverge unless conservative selection pressure maintains the sequence and so although there is considerable sequence homology between the individual 1.6 kb gene sequences, there is much less between the longer flanking sequences.

Evolution of differential globin gene expression

Tandem gene duplication normally means that both the regulatory sequences and coding sequences are duplicated. Subsequent sequence divergence in *cis*-acting regulatory regions may lead to altered expression, both spatially (e.g. in different tissues) and temporally (e.g. at different stages in development). This may be a very powerful evolutionary advantage of gene duplication. As the regulatory sequences are often comprised of very small sequence elements, mutation can quickly result in divergent expression. By then expressing copies of essentially the same gene in different spatial or temporal compartments, it is possible for the gene copies to acquire different functions: the different environments will provide differential selection pressure causing the coding sequences to diverge in an effort to adapt optimally to the different environments.

The globin superfamily provides some examples. Ancient gene duplications of an ancestral globin gene generated copies which underwent divergence in regulatory sequences, causing them to be expressed in different tissues (e.g. blood, muscle, nervous system). The gene copies adapted to their environments, leading to very considerable sequence divergence of the products, forming respectively hemoglobin, myoglobin and neuroglobin – each still having a core function of binding oxygen. Subsequent refinements in the globins making up hemoglobin introduced different varieties of globin, which became expressed at particular developmental stages and maybe became more specialized. Maybe, for example, the ε-, ζ- and γ-globin chains are especially suited to binding oxygen in the comparatively oxygen-poor environment of early development, whereas the α- and β-globin chains may be the preferred polypeptides in the environment of adult tissues.

12.1.6 Retrotransposition can permit exon shuffling and is an important contributor to gene evolution

Retrotransposition, the process whereby a natural cDNA copy is made from an RNA transcript and then inserted into a new chromosomal location, has been a very powerful agent in sculpting the human genome. Over 40% of the human genome consists of retrotransposon-derived repeats and a small percentage of these repeats are actively transposing (Section 9.5.1). The retrotransposition machinery has made an important contribution to gene evolution by making copies of exons and shuffling them from one genome location to another, and by providing a mechanism for some genes to be duplicated.

Exon shuffling by LINE-mediated transduction

LINE1 (L1) elements belong to the non-LTR class of retrotransposons and can transpose autonomously (Section 9.5.2). Under experimental conditions, LINE1 elements have been shown to insert into the intron of a gene, make a copy of a downstream exon and then to transpose this exon copy into another gene (Moran *et al.*, 1999). This is possible because the LINE1 retrotransposition machinery has a weak specificity for its own 3′ end, causing it to overlook the regulatory signals at that end. After a LINE1 element inserts into a gene, transcription of the LINE1 repeat often bypasses its own weak poly(A) sequence and uses instead a downstream poly(A) signal from the host gene. In so doing it can make a copy of a host exon which can be stitched into another gene after another retrotransposition event (**LINE1-mediated 3′**

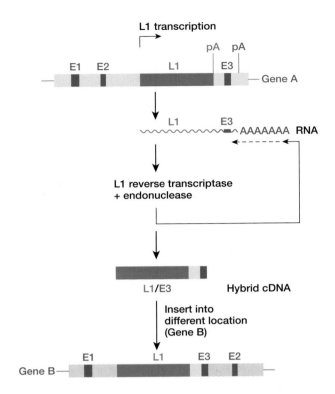

Figure 12.7: Exon shuffling between genes can be mediated by transposable elements.

The LINE1 (L1) sequence family contains members that actively transpose in the human genome. LINE1 elements have weak poly(A) signals and so transcription can continue past such a signal until another nearby poly(A) signal is reached, as in the case of gene A at top. The resulting RNA copy can contain a transcript not just of LINE1 sequences but also of a downstream exon (in this case E3). The LINE1 reverse transcriptase complex can then act on the extended poly(A) sequence to produce a cDNA copy that contains both LINE1 and E3 sequences. Subsequent transposition into a new chromosomal location may lead to insertion of exon 3 into a different gene (gene B). See Moran *et al.* (1999).

transduction; see *Figure 12.7*). This mechanism may therefore be the basis of the *exon shuffling* that appears to have been important in gene evolution (Section 12.1.3).

Gene duplication by retrotransposition

Retrotransposition of gene sequences has been a common event in the shaping of complex genomes. Copying of a spliced RNA from an intron-containing gene means that intron-containing sequences are removed, and when the RNA that is copied is a mRNA, the copied sequence will lack any promoter sequences. As a result, retrotransposition of copied mRNA sequences typically results in the formation of inactive **processed pseudogenes** (*Figure 9.14*). Occasionally, however, such a cDNA copy will integrate close to a functional promoter and may be subject to selection pressure to maintain gene function, becoming a **retrogene**. The classic examples are autosomal processed copies of X-linked genes which produce proteins that are functionally required during spermatogenesis. During spermatogenesis, the X and Y chromosomes condense to form the transcriptionally inactive **XY body**, but an autosomal gene copy can provide the necessary gene product. Retrogenes derived from non-X-linked genes are also known – see Section 9.3.6; *Table 9.11*.

Until very recently, it was thought that gene duplication by retrotransposition was limited to copying exonic sequences only. However, analyses of a chimeric new gene (Courseaux and Nahon, 2001) have shown that intronic sequences can be copied by an indirect route in certain situations. This is possible when the antisense strand of an intron-containing gene produces a mature RNA transcript containing sequence complementary to the intron sequences of the gene transcribed from the other strand. As antisense RNA transcripts are now known to be quite frequent in the human genome retrotransposed copies may sometimes lead to duplication of intron-containing genes.

12.2 Evolution of chromosomes and genomes

12.2.1 The mitochondrial genome may have originated following endocytosis of a prokaryotic cell by a eukaryotic cell precursor

In addition to the nucleus, mitochondria have a genome, as do the chloroplasts of plant cells. The organization and expression of mitochondrial and chloroplast genomes shows considerable similarities to that of prokaryotic cells (see below), suggesting that eukaryotic cells originated after a eukaryotic cell precursor (**protoeukaryote**) engulfed some type of prokaryotic cell (the **symbiont**). Such a process is thought to have conferred a selective advantage for the resulting new cell and has been termed **endosymbiosis** (*Figure 12.8*).

The endosymbiont hypothesis imagines that the genome of the engulfed prokaryotic cell gave rise to the present day mitochondrial genome. Present day prokaryotic cells (bacteria and archaea) typically contain one or a few megabases of DNA and contain several hundreds or thousands of genes, but mitochondrial genomes are much smaller. For example, the human mitochondrial genome is about 16 kb with only 37 genes, and the vast majority of mitochondrial proteins and functions are specified by nuclear genes (see Section 9.1.2). It is likely, therefore, that many of the genes originally present in the engulfed cell were transferred to the genome of the host cell by **horizontal** or **lateral gene transfer** (Doolittle, 1998). Horizontal gene transfers may have been extensive during the early stages of cellular evolution.

The identities of the cells involved in the endocytosis giving rise to mitochondria has been the subject of some debate. Initial hypotheses imagined that the host cell had some eukaryote characteristics, but more recent hypotheses have considered an archaeal cell as the host. The **hydrogen hypothesis** proposes that eukaryotes arose by engulfment of a hydrogen-producing α-proteobacterium by an anaerobic, hydrogen-dependent archaeal-like host. A variant, the **syntrophic hypothesis**, suggests that the hydrogen-producing symbiont was a δ-proteobacterium. The archaeal-like host was envisaged to have been strictly *autotrophic* (able to synthesize its organic compounds from simple molecules from the environment, rather than rely as *heterotrophs* do on ingesting organic compounds synthesized by other organisms). Subsequently, however, to avoid pointless cycling of metabolites in its cytoplasm the host lost its autotrophic pathway and an irreversible heterotroph emerged containing ancestral mitochondria but no longer dependent on hydrogen. More efficient oxygen-based respiration was then adopted by many such organisms and aerobic mitochondria evolved.

The hydrogen/syntrophic hypotheses were founded on observations of certain eukaryotes that lack mitochondria:

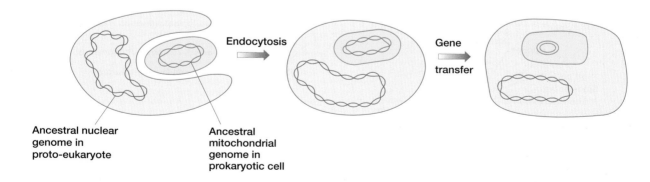

Ancestral nuclear genome in proto-eukaryote

Ancestral mitochondrial genome in prokaryotic cell

Figure 12.8: The human mitochondrial genome probably originated following endocytosis of a prokaryotic cell by a eukaryotic precursor cell.

In this particular example, the endocytosing cell does not have a nucleus, but some models have imagined endocytosis by a protoeukaryotic cell with a nucleus. Following endocytosis, genes in the genome of the prokaryote are imagined to have been transferred to the precursor of the nuclear genome, leaving a much reduced mitochondrial genome. See Doolittle (1998) for one possible mechanism of gene transfer.

Box 12.4: The universal tree of life and horizontal gene transfer.

Molecular phylogenetics (the classification of organisms according to relatedness of proteins or nucleic acids) produced a major shock in the late 1970s: rRNA analyses revealed that a group of methane-producing bacteria were very different from other bacteria. The ***archaebacteria*** – as they were first termed – also seemed to have some unusual cellular features, and they showed a close relationship to eukaryotes in many aspects of information transfer (DNA replication, DNA repair, transcription, translation, etc.). As a result, there was growing acceptance of the need to replace the previous division of life into prokaryotes and eukaryotes by an alternative division into three ***domains***.

▶ Bacteria – the commonly encountered prokaryotes, which have traditionally been well-studied (e.g. gram-negative and gram-positive bacteria, cyanobacteria etc.).

▶ Archaea – prokaryotes which resemble eukaryotes in their information transfer processes. They have often been isolated from extreme environments (e.g., hot springs, very high salt concentration, extremes of pH etc.), but are also known to occur in more usual habitats, including soils and lakes, and they have been found thriving inside the digestive tracts of cows, termites, and marine life where they produce methane.

▶ Eukaryotes.

Of these, the bacteria were imagined to have diverged first from the *universal last common ancestor* (see *Figure*). Mitochondria and chloroplasts are thought to have originated by endocytosis of certain prokaryotic cells by eukaryotic cell precursors, and subsequent transfer of many of the genes in the endocytosed genome to the host genome, a form of **horizontal gene transfer (HGT)**.

While HGT appears to have played a crucial role in genome evolution in the past, several recent examples of naturally occurring HGT are known, notably between different bacterial species (e.g. plasmid and phage-mediated transfer of pathogenic and antibiotic resistance genes). HGT between eukaryotic genomes can also occur (e.g. between different *Drosophila* species via DNA transposons). The International Human Genome Sequencing Consortium (2001) also came to the startling conclusion that some hundreds of human genes had evolved by HGT from bacteria at some point in the vertebrate lineage. The human genes had shown clear homologies to bacterial genes without appearing to be found in some invertebrate (*D. melanogaster, C. elegans*) genomes. The conclusion is, however, now known to be incorrect because extended phylogenetic analyses identified homologs in other invertebrates, and so the human genes can be explained in terms of descent through common ancestry – the absence of homologs in some species can be explained by gene loss in some lineages (see Brown, 2003 for references).

The Universal Tree of Life shown in this figure is far from being universally accepted, and it has been called into question by studies of various other protein sequence datasets. Inevitably, phylogenetic classification is difficult if common horizontal gene transfer has to be taken into account, since different gene sets can give conflicting phylogenetic results. As a result, ***whole genome approaches to tree construction*** are now being applied. This field is a dynamic one and readers are advised to consult recent reviews.

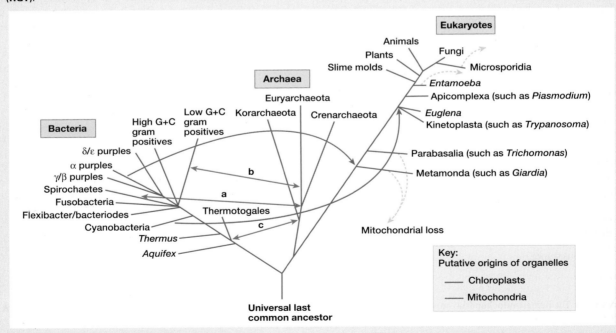

The universal tree of life based on rRNA phylogeny.

As well as being involved in formation of the mitochondrial and chloroplast genomes, horizontal gene transfer has been observed in other cases, e.g. between spirochetes and archaea (arrow a); low G+C gram positive bacteria and archaea (arrow b); and between thermophilic bacteria and archaea (arrow c). Reproduced from Brown (2003) *Nature Rev. Genet.* **4**, 121–132 with permission from the Nature Publishing Group.

Giardia, parabasalids such as *Trichomonas,* Entamoeba and a few ciliates and fungi. The most accepted phylogenetic classification places *Giardia* and *Trichomonas* as the organisms that diverged earliest from the eukaryotic lineage (*Box 12.4; Figure 12.22*). The parabasalids, as well as some amitochondrial ciliates and fungi, have specialized organelles called **hydrogenosomes**, which ferment pyruvate and produce hydrogen. The hydrogenosome lacks any genome that might help trace its origins; however, phylogenetic analysis of hydrogenosome-targeted proteins indicates a possible shared evolutionary path with mitochondria, and the hydrogenosome itself may be a highly derived mitochondrion. The mitochondria-lacking eukaryotes also possess bacterial-like metabolic enzymes (in addition to other known bacterial-like genes). It is perhaps significant, too, that whereas most eukaryotes use histones to compact their nuclear DNA, the only prokaryotes that have histones and nucleosomes are the Euryarchaeota, the division of the Archaea that includes the hydrogen-consuming methanogens. See Brown (2003) for further details and additional hypotheses.

12.2.2 Reduced selection pressure caused the mitochondrial genetic code to diverge

The human mitochondrial genetic code is slightly different from the 'universal' genetic code used in cellular genomes and plant mitochondrial genomes (*Figure 1.22*). Although identical to the genetic codes of other mammalian mitochondrial genomes, it also shows some differences to the nonuniversal genetic code in the mitochondria of other eukaryotes, such as *Drosophila* and yeast cells.

The mitochondrial genetic code is believed to have diverged as a result of reduced selection pressure. The genome of the endocytosed prokaryote genome, which gave rise to the mitochondrial genome would originally have contained at least several hundreds and possibly thousands of genes. Like all large genomes, it would have been subject to strong conservative selection pressure to maintain the universal genetic code (slight alterations in the code could result in lack of gene function).

Subsequently, genes were transferred from the precursor mitochondrial genome to the nuclear genome, possibly by successive processes of organelle lysis, incorporation of the DNA into the nuclear genome, loss of organelle copies and fixation of gene loss by genetic drift (see Doolittle, 1998). As the coding potential steadily diminished by gene transfer to the nuclear genome, there would have been progressively less selection pressure to conserve the original genetic code.

Eventually a severely depleted genome resulted (only 13 genes in the human mitochondrial genome encode polypeptides). Because only a tiny number of polypeptides would be involved, selection pressure to maintain the universal genetic code would have been relaxed and a certain degree of drift from the normal codon interpretations could be tolerated without provoking disastrous consequences. It is also likely that the codons that have been altered (see *Figure 1.22*) have not been used extensively in locations where amino acid substitutions would have been deleterious.

Of course, the process of mitochondrial genome depletion was a slow one that continued through metazoan evolution. As a result, different organisms show differences in the genes retained by the mitochondrial genome (and so the mitochondrial genetic code differs in some species). The mitochondrial genomes of plants, however, retained comparatively many genes and so no drift from the universal genetic code was possible there.

12.2.3 The evolution of vertebrate genomes may have involved whole genome duplication

Genome duplication resulting in **polyploidy** is an effective way of increasing genome size and versatility. Gene copies are simultaneously available for all genes, avoiding problems of differential gene dosage. Several species are naturally polyploid or are known to be degenerate polyploids, including the majority of flowering plants, many fishes, the yeast *Saccharomyces cerevisiae, Xenopus laevis,* and at least one mammal, the tetraploid red viscacha rat (see Otto and Whitton, 2000; Wolfe, 2001). Constitutional tetraploidy is rare and lethal in humans, but humans and other diploid organisms have some naturally polyploid cells as a result of sequential chromosome duplications by mitosis without intervening cell division, or by cell fusions (Section 3.1.4).

Since several species have clearly undergone genome duplication, polyploidization events are clearly not rare. This observation plus comparisons of gene characteristics in vertebrates and invertebrates has suggested the possibility that all vertebrates have undergone at least one round of genome duplication during evolution. Following genome duplication, and a transient tetraploid state, subsequent large-scale chromosome rearrangements could have been expected to cause chromosome divergence and restore diploidy, but now with twice the number of chromosomes.

The relaxation of selection pressure on many of the gene copies following genome duplication would result in the great majority acquiring deleterious mutations and becoming pseudogenes. Later on, because of the lack of any selection pressure to conserve them, the defective gene copies could be expected to be eliminated from the genome by various DNA turnover mechanisms (see Lynch and Conery, 2000). Some considerable time after a genome duplication event, therefore, the evidence for previous genome-wide duplication could be sparse; only the very few duplicated genes which retained functional activity would be expected to be maintained.

The initial proposals for ancient tetraploidization events during vertebrate evolution envisaged two rounds of genome duplication, the **2R hypothesis** (see Wolfe, 2001). Much of the evidence to support the 2R hypothesis was based on the observations that crucially important genes and gene clusters in invertebrates such as *Drosophila,* have 3–4 equivalents in vertebrates. Important examples include the four classical *Hox* gene clusters (*Figure 12.9*), the four paraHox gene clusters (human chromosomes 13q13-14, Xq13-q22, 4q11-12, 5q31-33) and the MHC clusters (on human chromosomes 1q21-q25, 6p21.3-p22.2, 9q33-q34, 19p13.1-p13.4 – see Abi-Rached *et al.,* 2002). In each of these cases there is a single

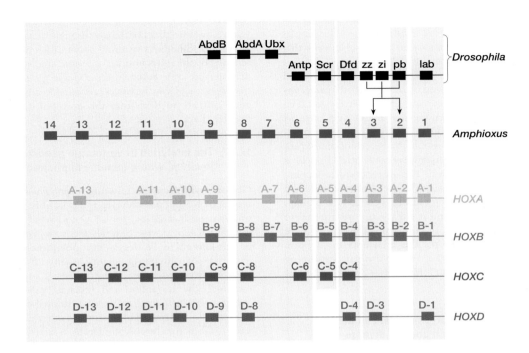

Figure 12.9: The organization of _Hox_ gene clusters in mammals and _Amphioxus_ suggests the possibility of one or two rounds of ancestral genome duplication.

Humans and other mammals have four clusters containing 9–11 classical _Hox_ genes, but _Amphioxus_, the invertebrate considered to be most closely related to vertebrates, has 14 such genes. The linear order of the genes in a cluster is thought to dictate the temporal order in which they are expressed during development and also their anterior limits of expression along the anterior–posterior axis (see _Figure 3.10_). Shaded boxes indicate paralogous groups consisting of genes with very similar expression patterns and presumably similar functions. Although the four classical mammalian _Hox_ clusters show strong general conservation of gene order, the paralogy groups indicate that there has presumably been gene loss from what might have been an ancestral cluster with 13 or 14 genes. A rearrangement in the _Drosophila_ lineage led to separation of the genes into two subclusters.

gene cluster in _Amphioxus_, the closest invertebrate relative of the vertebrates.

Following the genome projects, it has become apparent that vertebrates have around 30 000–35 000 genes, approximately twice the gene number found in invertebrates rather than the widely expected fourfold difference. A recent comprehensive analysis of the draft human genome sequence by McLysaght _et al._ (2002) has suggested that it has many more duplicated genes than would be expected by chance. About 25% of human genes have clearly related _paralogs_ and comparisons with orthologs in _D. melanogaster_ and _C. elegans_ indicate that a burst of gene duplication activity took place about 350–600 MYr ago, consistent with at least one round of whole genome duplication.

12.2.4 There have been numerous major chromosome rearrangements during the evolution of mammalian genomes

Large-scale chromosome rearrangements have been frequent during mammalian genome evolution and chromosome evolution can be uncoupled from phenotype evolution. A classic example is provided by two species of muntjac (a type

of small deer). The Chinese muntjac and the Indian muntjac (_Figure 12.10_) are so closely related that they can mate and can produce offspring, which, however, are not viable. The Chinese muntjac has 46 chromosomes, but as a result of various chromosome fusion events the Indian muntjac has only 6 chromosomes in females and 7 in males.

While the above example is an extraordinary one, significant chromosome rearrangements occur regularly during mammalian evolution. Inversions appear to be particularly frequent; translocations somewhat less so. Centromeres (which are composed of rapidly evolving sequence) can change positions. When human chromosomes are compared with those of our nearest living relatives, the great apes, there are very strong similarities in chromosome banding patterns (Yunis and Prakash, 1982). The most frequent rearrangements have been inversions (including both pericentric and paracentric inversions) and in addition there have been some translocations (see Section 12.4.2 and see _Figures 12.27, 12.28_).

While it is simple to identify orthologous chromosomes in very closely related species, chromosome comparisons between more distantly related species generally shows that only small segments of chromosomes are conserved. Conservation of linear gene order (**conservation of**

Figure 12.10: The Chinese and Indian muntjacs.

The Chinese muntjac (*Muntiacus reevesi;* left panel) and the Indian muntjac (*Muntiacus muntjak;* right panel) are very closely related but have very different karyotypes (see text).

synteny) is therefore often limited to small chromosome segments. To assess conservation of synteny, previous approaches mapped orthologous genes to chromosomes in the two species but the genome projects have been able to deliver more detailed conservation of synteny maps. See *Figure 12.11* for human–mouse conservation of synteny.

The 342 or so chromosome segments shared by humans and mice mean that *on average* conservation of synteny extends to somewhat less than 10 Mb, and the sequences of most individual human and mouse chromosomes have orthologs on a variety of different chromosomes in the other species. The X chromosome is a notable exception: the vast majority of

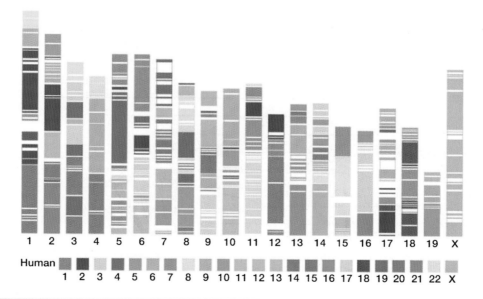

Figure 12.11: Human–mouse conservation of synteny is generally limited to small chromosome segments.

Segments and blocks >300 kb in size with conserved synteny in human chromosomes are superimposed on the 20 mouse chromosomes (top). Each color corresponds to a particular *human* chromosome. The 342 segments are separated from each other by thin white lines within the 217 blocks of consistent color. The X chromosomes are represented as single, reciprocal syntenic blocks, and human chromosomes 17 and 20 correspond entirely to a portion of mouse chromosomes 11 and 2 (but with extensive rearrangements into at least 16 segments in the former case). Other chromosomes show evidence of much more extensive interchromosomal rearrangement. Figure reproduced from the Mouse Genome Sequencing Consortium (2002) *Nature* **420**, 520–562 with permission from the Nature Publishing Group.

sequences on the human X chromosome have orthologs on the mouse X chromosome (because mammalian X-inactivation has evolved to ensure an effective 2 : 1 gene dosage ratio for autosomal : X-linked genes – see Section 12.2.8). Even in the case of the X chromosome, however, there have been numerous inversions, scrambling gene order between humans and mice, and analysis of the lengths and numbers of conserved synteny segments on the autosomes fits with a process of random chromosome breakage (Mouse Genome Sequencing Consortium, 2002).

12.2.5 Segmental duplication in primate lineages and the evolutionary instability of pericentromeric and subtelomeric sequences

One of the major surprises that came out of sequencing the human genome was the extent of evolutionarily recent **segmental duplication**, an important form of subgenomic duplication which can lead to chromosome rearrangements. About 5% of the human genome sequence is now thought to be represented in interspersed duplicated copies with > 90% sequence identity extending over segments from 1 kb up to several hundreds of kb in length (Bailey *et al.*, 2002; Samonte and Eichler, 2002; see *Figure 12.12*). They include:

▶ *intrachromosomal duplications* which tend to occur in euchromatic regions. When the sequence identity exceeds 95% and extends over 10 kb or more, these sequence elements often predispose to large-scale deletions, duplications and inversions many of which are associated with disease (Sections 11.5.4, 11.5.5);

▶ *interchromosomal duplications* which tend to be located in pericentromeric or subtelomeric regions. The duplications may involve genes or portions of genes, which can give rise

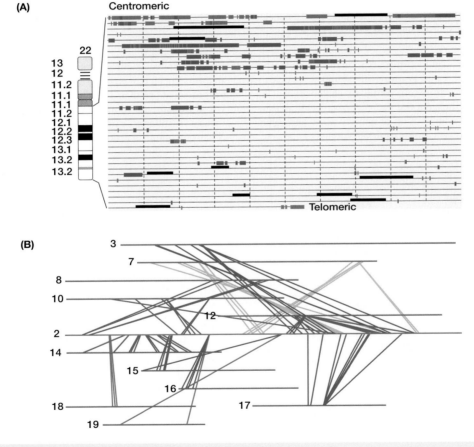

Figure 12.12: Examples of recent segmental duplications in the human genome.

(A) *Segmental duplications on 22q.* Horizontal lines represents successive 1 Mb sequences from centromeric end (top left) to telomeric end (bottom right). Black bars denote sequence gaps. Sequences involved in interchromosomal duplications (red) are mostly restricted to the most centromeric and telomeric regions. Intrachromosomal duplications are shown in blue. Adapted from Eichler (2001) *Trends Genet.* **11**, 661–669 with permission from Elsevier. **(B)** *Interchromosomal homologies for human chromosome 2 sequences.* Chromosome 2 is shown as a red horizontal bar in the middle. Eleven other chromosomes (green horizontal bars) contain segments which are highly homologous to chromosome 2 sequences (linked by colored vertical lines). Reproduced from Venter *et al.* (2001) *Science* **291**, 1304–1351 with permission from the American Association for the Advancement of Science.

to dispersed nonfunctional gene copies (*Figure 9.13*), chimeric transcripts, and possibly new genes. Primate interchromosomal duplications appear to be a little less frequent than intrachromosomal (Samonte and Eichler, 2002).

Screening of the mouse genome for sequences closely related to the segmentally duplicated human sequences typically reveals single copy sequences. Human segmental duplications therefore originated comparatively recently: segmental duplication events appear to have been ongoing in the primate lineage for the last 40 MYr or so, with possibly the majority occurring in the last 12 MYr (Samonte and Eichler, 2002). Segmental duplication is not, however, primate-specific: unrelated recent pericentromeric duplications have occurred in the mouse (Thomas *et al.*, 2003). The mechanism for segmental duplication is not well understood, but generally results in mosaic structures combining sequence modules from a different variety of chromosomal locations. Certain regions of the genome, notably pericentromeric regions seem to readily accept copies of sequences from other genome regions and exchange them with other such regions (see *Figure 12.13*).

12.2.6 The human X and Y chromosomes exhibit substantial regions of sequence homology, including common pseudoautosomal regions

In mammals, pairs of homologous autosomal chromosomes are structurally virtually identical (*homomorphic*); chromosome pairing at meiosis is presumed to be facilitated by the high degree of sequence identity between homologs, albeit by a mechanism that is not understood. By contrast, the X and Y chromosomes of humans and other mammalian species are *heteromorphic*. The human X chromosome is a submetacentric

chromosome which contains over 160 Mb of DNA, whereas the Y is acrocentric and is much smaller (containing about 50 Mb of DNA). The human X chromosome contains numerous important genes; in marked contrast, the Y chromosome has only about 50 genes (*Table 12.2*) and the bulk of the Y is composed of genetically inert constitutive heterochromatin. Note, however, that many of the genes in the nonrecombining portion of the Y chromosome are involved in spermatogenesis/sex determination (*Table 12.2*) and that the X chromosome too appears to be enriched in sex- and reproduction-related genes (e.g. Wang *et al.*, 2001).

Despite being morphologically distinct, the X and Y chromosomes show substantial regions of homology, including a variety of Xp–Yq and Xq–Yp homologies, as well as Xp–Yp and Xq–Yq homologies. These homologies have allowed a variety of X–Y gene pairs to be identified (see *Figure 12.14*). The existence of such homologies suggests that the two chromosomes have evolved from an ancestral homomorphic pair of chromosomes. Clearly, the two chromosomes have subsequently undergone substantial divergence, and sequences that are physically close on one chromosome may have very widely spaced counterparts on the other. The X and Y are also able to pair during male meiosis and so can exchange sequences just as autosomal homologues do. However, the meiotic exchanges are very limited in extent, being confined to small **pseudoautosomal regions** at the chromosome tips (such sequences are therefore not X-linked or Y-linked, hence the term *pseudoautosomal*). In humans there are two pseudoautosomal regions).

▶ The **major pseudoautosomal region** (**PAR1**) extends over 2.6 Mb at the extreme tips of the short arms of the X and Y and is known to contain at least 13 genes (Ried *et al.*, 1998; Gianfrancesco *et al.*, 2001). It is the site of an *obligate crossover* during male meiosis which is thought to be

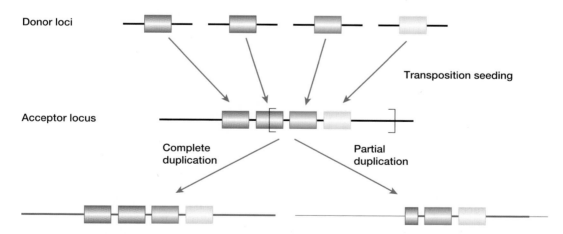

Figure 12.13: Model of segmental duplication.

Moderately long to large sequence copies (1–200 kb segments) originate by duplication of sequences from certain regions in the genome (donor loci) and transpose, integrating elsewhere in the genome (*duplicative transposition*). As shown, different donor sequences from disparate regions can insert into common acceptor regions. The different duplicative transposition events occur independently over time, creating larger blocks of duplicated sequence that are *mosaic* in structure. Portions of this mosaic structure can in turn be duplicated and copied to other regions of the genome. Rearrangements (deletions and inversions) subsequently alter the structure of these regions. Reproduced from Samonte and Eichler (2002) *Nature Rev. Genet.* **3**, 65–72, with permission from the Nature Publishing Group.

Table 12.2: Functional classification of human Y-chromosome genes

Gene category	Genes	Known/putative function(s)	Expression specificity	Multiple copies on Y?	Has active X homolog?	X homolog inactivated in female?
Pseudoautosomal	Many*	As diverse as autosomal genes	Diverse	No	Yes	Yes (except SYBL1, HSPRY3)
NRY class 1	RPSRY, ZFY, USP9Y, DBY, UTY, TB4Y, SMCY, EIF1AY	Housekeeping	Broad	No	Yes	No
NRY class 2	TTY1, TSPY PRY, TTY2, CDY, XKRY, DAZ, BPY2	Spermatogenesis	Testis	Yes	No	N/A
NRY class 3	SRY	Male determination	Testis	No	Yes	Yes
	RBMY	Spermatogenesis	Testis	Yes	Yes	Yes
	AMELY	Tooth development	Tooth bud	No	Yes	Maybe
	VCY	Unknown	Testis	Yes	Yes	N/A
	PCDHY	Unknown	Brain	No	Yes	No

(NRY, non-recombining region of the Y chromosome; N/A, not applicable).

*At least 17, of which 13 are in the major pseudoautosomal region (see *Figure 12.15*).

required for correct meiotic segregation. This very small region is evolutionarily unstable (see below) and contains highly recombinogenic sequences (the sex-averaged recombination frequency is 28% which, for a region of only 2.6 Mb, is approximately 10 times the normal recombination frequency). The high figure is, of course, mostly due to the obligatory crossover in male meiosis resulting in a crossover frequency approaching 50%. The boundary between the major pseudoautosomal region and the sex-specific region has been shown to map within the *XG* blood group gene, with the *SRY* male determinant gene occurring only about 5 kb from the PAR1 boundary on the Y chromosome (*Figure 12.15*).

▶ The **minor pseudoautosomal region** (**PAR2**) spans 330 kb at the extreme tips of the long arms of the X and Y and contains only four genes (Ciccodicola *et al.*, 2000; Charchar *et al.*, 2003). Crossover between the X and Y in this region is not so frequent as in PAR1, and is neither necessary nor sufficient for successful male meiosis.

12.2.7 Human sex chromosomes evolved from autosomes and diverged due to periodic regional suppression of recombination

Distinct sex chromosomes have been independently developed in many animals with disparate evolutionary lineages, including not only mammals, but birds (where the females are ZW, the *heterogametic sex*, and the males are ZZ, the *homogametic sex*), and certain species of fish, reptiles and insects. In each case, it is thought that the different sex chromosomes started off as virtually identical autosomes, except that one of them happened to evolve a major sex-determining locus (the

SRY locus in humans, located 5 kb from the PAR1 boundary on the Y). Subsequent evolution resulted in the two chromosomes becoming increasingly dissimilar until, in many species, one sex chromosome (the Y in mammals, the W in birds) was reduced to a small chromosome, rich in repetitive sequences but with only a very few functional genes. There would appear to be evolutionary pressure to adopt the strategy of having two structurally and functionally different sex chromosomes.

The autosomal origins and plasticity of pseudoautosomal regions

The pseudoautosomal regions have not been well-conserved in evolution. There is no equivalent to PAR2 in mouse and even in some primates, and known mouse orthologs of PAR2 genes are either autosomal or map close to the centromere of the X. There are also very significant species differences in PAR1. In humans the PAR1 boundary occurs within the *XG* gene, which does not appear to have an ortholog in mouse (*Figure 12.15*). The equivalent mouse pseudoautosomal region (PAR), only 0.7 Mb long and located at the tip of the *long arm* of the X, shows very little sequence homology to human PAR1 (Perry *et al.*, 2001). The mouse PAR boundary lies within the *Mid1* gene (formerly *Fxy*), whose human ortholog *MID1* is located more proximally within the X chromosome-specific region. The steroid sulfatase gene, *Sts,* is the only other gene known to be located in the mouse PAR but the human homolog is located about 3.5 Mb proximal to PAR1 on the X chromosome. Three of the human PAR1 genes have autosomal mouse orthologs and other human PAR1 genes have autosomal orthologs in some other mammals. The PAR1 region has therefore been imagined to have evolved by repeated addition of autosomal segments onto the

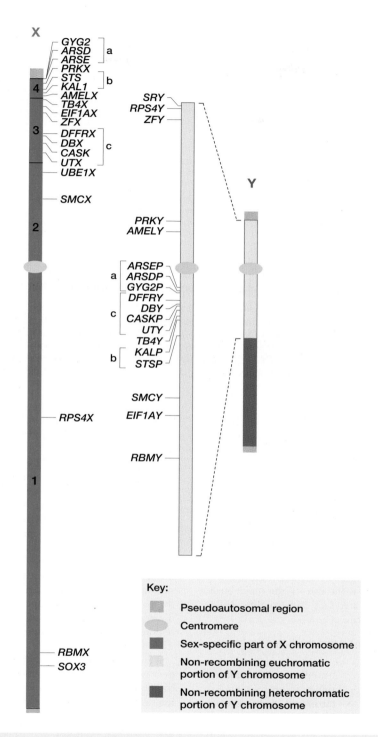

Figure 12.14: The human X and Y chromosomes show several regions of homology indicative of a common evolutionary origin.

The pseudoautosomal regions at the tips of Xp and Yp are identical (see e.g. *Figure 12.15*), as are those at the tips of Xq and Yq. The remaining nonrecombining regions show several clearly homologous XY gene pairs plus the *SOX3-SRY* pair (which have an HMG domain in common). Numbers 1 to 4 on the X chromosome correspond to different 'evolutionary strata' (see text). Some of the Y chromosome homologs have degenerated into pseudogenes (symbol terminates in a P e.g. *ARSEP, ARSDP* etc.). Reproduced from Lahn and Page (1999) *Science* **286**, 964–967, with permission from the American Association for the Advancement of Science.

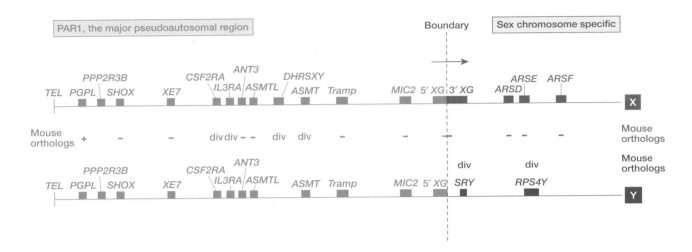

Figure 12.15: Organization and evolutionary instability of the major human pseudoautosomal region (PAR1).

The 2.6 Mb PAR1 region is common to the tips of Xp (top left) and Yp (bottom left; TEL = telomere) and contains at least 13 genes. Adjacent to PAR1 are the long sex-specific parts of the X and Y for which the genes immediately adjacent to PAR1 are shown to the right of the boundary line. The PAR1 boundary occurs within the *XG* blood group gene on the X chromosome. On the Y chromosome, there is a truncated *XG* gene homolog: the promoter and first few exons are present in the pseudoautosomal region, but thereafter there are unrelated Y chromosome-specific sequences, e.g *SRY, RPS4Y*. Genes within PAR1 and in the neighboring sex-specific regions (which were part of a former pseudoautosomal region) have been poorly conserved during evolution, and mouse orthologs are often undetectable (–) or highly diverged (div.).

pseudoautosomal region of one of the sex chromosomes, before being recombined onto the other sex chromosome (Graves *et al.*, 1998).

PAR1 and neighboring regions are thought to be comparatively unstable regions. Frequent DNA exchanges result in a high incidence of gene fusions, exon duplications and *exon shuffling* (see Ried *et al.*, 1998). Many of the PAR1 genes and genes in the nearby sex-specific regions (which were previously pseudoautosomal regions – see below) do not appear to have detectable orthologs in the mouse (by analysis of the available mouse genome sequence or by hybridization assay). Those that do have orthologs are often undergoing extremely rapid sequence divergence as in the case of the major male determinant, *SRY*, which is located only 5 kb from the PAR1 boundary, and the steroid sulfatase gene, *STS*.

Autosomal origins for human Xp and evolutionary strata on the X chromosome

Comparison of genes in distantly related mammals indicates that much of the short arm of the human X chromosome has recently been acquired by X–autosomal translocation. Mammals have been classified into two subclasses **prototheria (**the monotremes or egg-laying mammals) and **theria** which in turn are subdivided into two subclasses: **metatheria** (marsupials) and **eutheria**, a group which includes placental mammals. Many eutherian X-linked genes are found to be X-linked in marsupials. However, genes mapping to a large part of human Xp (distal to Xp11.3) have orthologs on autosomes of both marsupials and monotremes. Because the prototherian divergence pre-dated the metatherian–eutherian split (see *Figure 12.24*), the simplest

explanation is that at least one large autosomal region was translocated to the X chromosome early in the eutherian lineage.

Outside the shared *PAR1* and *PAR2* regions, there are still about 20 X-Y gene pairs which are relics of the extensive sequence identity that once existed between the ancestral X and Y chromosomes. For the X–Y pairs, most of the genes on the X are located on Xp but the equivalent sequences seem to be found throughout the euchromatic portion of the Y (*Figure 12.14*). There is, however, a clear regional difference in the sequence divergence of the X–Y pairs. By taking K_S (the mean number of synonymous substitutions per synonymous site – see Section 11.2.5) as a measure of sequence divergence, the X–Y pairs can be grouped into four classes of sequence divergence (and therefore of age) that correspond to linear regions along the X chromosome (Lahn and Page, 1999; see *Table 12.3*).

The data in *Table 12.3* indicate that the ages of the X–Y gene pairs decreases in stepwise fashion as one proceeds along the nonrecombining portion of the X, from the end of the long arm (distal Xq) to the end of the short arm (distal Xp). At least four *evolutionary strata* are evident, corresponding to the four sets of genes in *Figure 12.14* and recombination between the X and the Y has been thought to be suppressed regionally, beginning with stratum 1 at about 240–320 MYr ago (shortly after divergence of the lineages leading to mammals and birds (see *Figure 12.24* for phylogenetic information). Thereafter, suppression of recombination spread in discrete steps up to strata 2, then strata 3 and finally strata 4. The most likely way in which recombination was suppressed would have been by inversions (which are known to be able to suppress recombi-

Table 12.3: Homologous X- and Y-linked genes can be grouped into four categories of sequence divergence which correspond to the location of the X-linked gene on the X chromosome (see *Figure 12.14*)

X-Y gene pair (*-pseudogene)	K_s	DNA divergence (%)	X-Y gene pair (*-pseudogene)	K_s	DNA divergence (%)
Group 1			**Group 2**		
RPS4X/Y	0.97	18	UBE1X/Y	0.58	16
RBMX/Y	0.94	29	SMCX/Y	0.52	17
SOX3/SRY	1.25	28			
Group 3			**Group 4**		
TB4X/Y	0.29	7	GYG2/GYG2P*	0.11	7
EIF1AX/Y	0.32	9	ARSD/ARSDP*	0.09	7
ZFX/Y	0.23	7	ARSE/ARSEP*	0.05	4
DFFRX/Y	0.33	11	PRKX/Y	0.07	5
DBX/Y	0.36	12	STS/STSP*	0.12	11
CASK/CASKP*	0.24	15	KALI/KALP*	0.07	6
UTX/Y	0.26	12	AMELX/Y	0.07	7

nation over broad regions in mammals), and which appeared to have occurred on the Y chromosome (e.g. a Y-specific inversion would explain why the *PAR1* boundary crosses a gene that is intact on the X chromosome but disrupted on the Y – see *Figure 12.15*). A sequence of likely events to explain how the sex chromosomes evolved is illustrated in *Figure 12.16*.

12.2.8 Sex chromosome differentiation results in progressive Y chromosome degeneration and X chromosome inactivation

The human Y chromosome could be on its way to extinction but there is still a future for males!

The evolution of sex determination systems has been crucial for the development of complex multicellular organisms because of the genetic novelty afforded by recombination. After a major sex-determining locus has been established during evolution it is essential to suppress recombination in the region containing that locus (in order to maintain the sex differences). Unlike the human X chromosome which can recombine *throughout its length* with a partner X chromosome in female meiosis, the human Y chromosome has been viewed as an essentially asexual (non-recombining) chromosome.

Population genetics predicts that a nonrecombining chromosome should degenerate by a process known as **Muller's ratchet**. If the mutation rate is reasonably high the absence of recombination means that harmful mutations can gradually accumulate in genes on that chromosome over long evolutionary time scales (there is no possibility of crossover to pick up instead an allelic sequence lacking the harmful mutation). Mutant alleles may drift to fixation as Y chromosomes with fewer mutants are lost by chance, or they may 'hitchhike' along with a favorable allele in a region protected from recombination.

Once mutations accumulate in the nonrecombining Y and cause loss of gene function, there is no selective pressure to retain the relevant DNA segment. DNA turnover mechanisms will ensure that the chromosome gradually, but inexorably, contracts by a series of deletions. As a result, the human Y chromosome may be heading towards extinction. Maleness will continue, however, by switching to an alternative sex determination system. Most likely it will be conferred simply by X : autosome gene dosage ratio, and XO individuals will be male (as in the case of *Drosophila*).

The necessary development of X chromosome inactivation

The evolution of the mammalian sex determination system is also inextricably interwoven with the evolution of the X-inactivation mechanism for dosage compensation (Section 10.5.6; see Ellis, 1998). In response to large-scale destruction of Y chromosome sequences, there would have been pressure to increase gene expression on the X chromosome. However, this would lead to excessive X chromosome gene expression in females which could cause reduced fitness. As a result a form of gene dosage compensation evolved whereby a single X chromosome was selected to be inactivated in female cells (**X-inactivation**).

The rationale for X chromosome inactivation is to act as a dosage compensation mechanism for those X chromosome genes which do *not* have homologs on the Y chromosome. However, a small minority of human X-linked genes do have *functional* homologs on the Y chromosome. Because these genes will not show sex differences in gene dosage, they would be expected to escape X-inactivation. All *PAR1* genes tested escape X-inactivation. *PAR2* genes are different. The two most telomeric genes, *IL9R* and *CXYorf1* escape inactivation, but the two proximal genes, *SYBL1* and *HSPRY3*, are both X-inactivated. The apparent discrepancy is due to a compensatory **Y-inactivation** mechanism for these genes: when present on the Y chromosome, *SYBL1* and *HSPRY3* are both methylated and not expressed.

In addition to the genes in the pseudoautosomal regions, perhaps about one-fifth of the total genes on the X chromosome escape inactivation (see Carrel *et al.*, 1999). Escaping

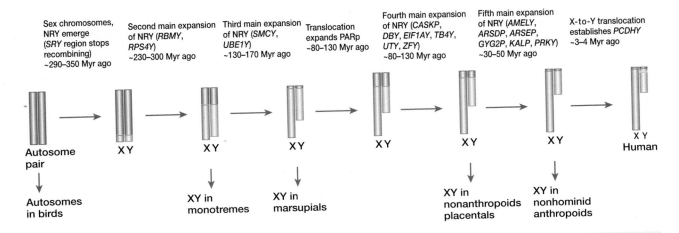

Figure 12.16: Human sex chromosome evolution.

The figure shows the overall shrinkage of the Y chromosome and blockwise expansion of its nonrecombining region (NRY), probably as a result of successive large-scale inversions. Evolutionarily new NRY genes are placed in parentheses, and phylogenetic branches are indicated by arrows. Green regions are freely recombining. Red regions are X-chromosome specific. Blue regions are Y-specific (NRY). The yellow region represents *PCDHX/Y* (protocadherin X/Y)-containing sequence that has translocated from the X to the NRY (some other likely translocations are omitted for simplicity). The diagram is not drawn to scale and centromeres are omitted, as their locations are uncertain for many evolutionary stages. Abbreviations: PAR1, major pseudoautosomal region. Adapted from Lahn *et al.* (2001) *Nature Rev. Genet.* **2**, 207–216, with permission from the Nature Publishing Group.

genes tend to be concentrated in clusters mainly on Xp, and many of them appear to derive from recent autosomal additions to the sex chromosomes. Genes which do not have a functional homolog on the Y but yet escape inactivation may be ones where 2 : 1 dosage differences are not a problem. Species differences for the nonpseudoautosomal expression patterns can also occur. For example, the human nonpseudoautosomal genes *ZFX, RPS4X* and *UBE1* all escape inactivation, but the murine homologs *Zfx, Rps4* and *Ube1X* (which unlike the human *UBE1* gene has a homolog on the Y) are all subject to X inactivation. For those X-linked genes where gene dosage needs to be tightly controlled, X-inactivation appears to have evolved as an adaptation for the decay of a homologous Y-linked gene (Jegalian and Page, 1998).

12.3 Molecular phylogenetics and comparative genomics

Molecular phylogenetics describes the use of nucleic acid or protein comparisons in order to establish the evolutionary relationships between organisms or populations. This, of course, complements classical phylogenetic approaches which have been based on anatomical and morphological features of living organisms and information gleaned from the fossil record.

Until recently when molecular evolutionary biologists compared DNA sequences (or the inferred protein sequences) they typically used sequences from at most a very few genomic locations. Phylogenies based on very small sequence datasets can, however, be misleading and give differing results. For example, comparisons of proteins involved in transcription, translation, DNA replication and repair emphasize

similarities between archaea and eukaryotes, whereas comparisons of proteins involved in cellular metabolism group archaea with bacteria. When efforts were made to get a broader perspective, they relied on comparative genetic maps (O'Brien *et al.*, 1999) or comparative karyotyping. The genome projects have of course changed all that. They have allowed us views from the all-embracing perspective of whole genome sequences, and a new science has been born – **comparative genomics**.

The sequence information that has been streaming out of genome projects provides much more definitive and representative DNA profiles on which to infer phylogenies. By determining the sequences from a whole series of genomes (Chapter 8) we are gaining much more profound insights into how genomes have been shaped during evolution. Of course, the data also help us to understand how current genomes relate to each other, and aid the identification of important conserved sequences.

12.3.1 Molecular phylogenetics uses sequence alignments to construct evolutionary trees

To construct an evolutionary tree it is necessary to compare nucleic acid sequences (or occasionally the inferred protein sequences; nucleic acid sequences are more informative, but in comparisons of very distantly related genes protein sequences have often been used). If two or more sequences show a sufficient degree of similarity (**sequence homology**) they can be assumed to be derived from a common ancestral sequence. Sequence alignments can then be used to derive quantitative scores describing the extent of relationship between the sequences.

Comparing sequences of equal fixed length is usually straightforward if there is a reasonably high sequence homology. Often, however, the nucleic acid sequences that are compared will have previously undergone deletions or insertions, and so rigorous mathematical approaches to sequence alignment are needed. Complementary algorithms have been devised. The **similarity approach** of Needleman and Wunsch (1970) seeks to maximize the number of matched nucleotides; the **distance approach** of Waterman *et al.* (1976) aims to minimize the number of mismatches. Computer programs such as **Clustal** carry out multiple sequence alignments (Jeanmougin *et al.*, 1998).

Once the sequences have been aligned, **evolutionary trees** can be constructed. They are most commonly represented as diagrams which use combinations of lines (*branches*) and nodes. The different organisms (sequences) which are compared are located at external nodes but are connected via branches to *interior nodes* (intersections between the branches) which represent ancestral forms for two or more organisms. A **rooted tree** (or **cladogram**) infers the existence of a common ancestor (represented as the trunk or *root* of the tree) and indicates the *direction* of the evolutionary process (*Figure 12.17*). The root of the tree may be determined by comparing sequences against an *out group* sequence (one which is clearly related in evolution but is distantly related to the sequences under study). An **unrooted tree** does not infer a common ancestor and shows only the evolutionary relationships between the organisms. Note that the number of possible rooted trees is usually much higher than the number of unrooted trees.

Constructing evolutionary trees

A variety of different approaches are used for constructing evolutionary trees, but many use a **distance matrix** method. The first step calculates the **evolutionary distance** between all pairs of sequences in the dataset and arranges them in a table (matrix). This may be expressed as the number of nucleotide differences or amino acid replacements between the two sequences, or the number of nucleotide (amino acid) differences per nucleotide (amino acid) site (see *Figure 12.18A* for an example).

Having calculated the matrix of pairwise differences, the next step is to link the sequences according to the evolutionary distance between them. For example, in one approach the two members of pair of sequences which have the smallest distance score are connected with a root in between them. The mean of the distances from each member of this pair to a third node is used for the next step of the distances matrix and the process is repeated until all sequences have been placed in the tree. This always results in a rooted tree but variant methods such as the *neighbor relation* method can create unrooted trees. The critical assumption is of a constant *molecular clock* (the mutation rate is constant in different lineages) which may often not be correct (Section 11.2.6). The **neighbor joining method** is a variant which does not require that all lineages have diverged by equal amounts per unit time. It is, therefore, especially suited for datasets comprising lineages with largely varying rates of evolution (see *Figure 12.31* for an example of a tree constructed by this method).

Alternatives to matrix distance methods include maximum parsimony and maximum likelihood methods. **Maximum parsimony** methods seek to use the minimum number of evolutionary steps. They consider all the possible evolutionary trees that could explain the observed sequence relationships but then select those that require the fewest changes. **Maximum likelihood** methods create all possible trees and then use statistics to evaluate which tree is likely. For a small number of sequences this may be possible but for a large number of sequences the number of generated trees becomes so large that *heuristics* (methods which produce an answer in a computable length of time but for which the answer may not be optimal) are used to select a subset of trees to create.

Assessing the accuracy of an evolutionary tree

Once an evolutionary tree has been derived, statistical methods can be used to gain a measure of its reliability. A popular method is **bootstrapping**, a form of Monte-Carlo simulation. Typically, a subsample of the data is removed and replaced by a randomly generated equivalent data set and the resulting *pseudosequence* is analyzed to see if the suggested evolutionary pattern is still favored (*Figure 12.18B*). The resampling *randomizes* the data but if there is a clear relationship between two sequences, randomization will not erase it. If on the other hand a node connecting the two sequences in the original tree is spurious, it may disappear upon randomization *because the randomization process changes the frequencies of the individual sites*. Bootstrapping often involves re-sampling subsets of data 1000 times. The highest **bootstrap value** is often given as a percentage so that a value of 100 means that the simulations fully support the original interpretation. Values of 95–100 indicate a high level of confidence in a predicted node. Bootstrap values less than 95 do not mean that the original grouping of sequences is wrong, but that the available data do not provide convincing support. See *Figure 12.31* for some examples.

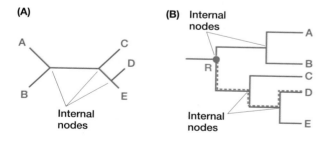

(A)

(B) Internal nodes

Internal nodes

Internal nodes

Figure 12.17: Unrooted and rooted evolutionary trees.

(A) *An unrooted tree*. The tree has five *external nodes* (A, B, C, D and E), which are linked by lines (*branches*) that intersect at *internal nodes*. Such a tree specifies only the relationships between the organisms under study, but does not define the evolutionary path. **(B)** *A rooted tree*. From one particular internal node, *the root* (shown as R in red), there is a unique evolutionary path that leads to any other node, such as the path to D (dashed red line).

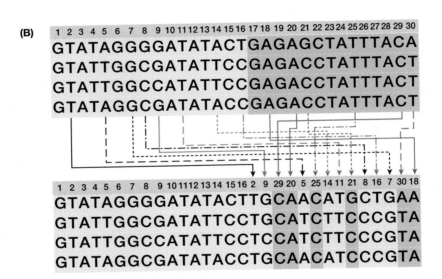

Figure 12.18: Constructing an evolutionary tree by the distance matrix method and verifying it by bootstrapping.

(A) *Tree construction*. Multiple sequence alignments are used to calculate the evolutionary distances between pairs of sequences. Sequences 1 and 2 differ at 6/30 (= 0.20) nucleotide positions; 3 and 4 differ at 3/30 (= 0.10) positions. These and other pairwise difference values are entered into a distance matrix and computer programs use the data to assess the sequence relationships and construct an evolutionary tree. Distance matrix methods need to include statistical estimates of the amount of multiple substitutions that may have occurred (a simple A→ T substitution may actually have resulted by successive A→ C→ T substitution). **(B)** *Bootstrap analysis*. From the original dataset, a subset (here sites 17–30) is selected to be discarded, but the remainder (sites 1–16) is retained. Original sites 17–30 are replaced by an equally sized dataset of *randomly chosen sites* from the original 30 sites producing a new *pseudosequence alignment* where some of the original sites are repeated (2, 9, 5, 14, 11 etc.) and some are absent (17, 21, 23 etc). The process is repeated a 1000 or so times to generate 1000 pseudosequence alignments and evolutionary trees are constructed in each case, and referenced against the original to give a **bootstrap value** (see text).

12.3.2 New computer programs align large scale and whole genome sequences, aiding evolutionary analyses and identification of conserved sequences

The sequence data flowing out of genome projects is, of course, only the start of a long campaign to work out what they mean. Because of the vast sizes of the datasets, new computer programs have been needed to enable large-scale and whole genome sequence alignments. The new programs have been invaluable in assisting identification of well-conserved sequences and in understanding evolutionary relationships.

Large-scale comparative sequence analyses for identifying conserved sequences

A principal challenge for genome projects is to identify RNA genes, regulatory sequences and other functionally important sequences that do not make proteins. Analysis of the sequence of a complex genome (low percentage of coding DNA) is not easy because computer programs cannot easily predict RNA genes and regulatory sequences (polypeptide-encoding genes are easier — there are long open reading frames and large amounts of sequence data on genes and their expression products to compare against). An alternative is to seek out highly conserved sequences by comparing different genome

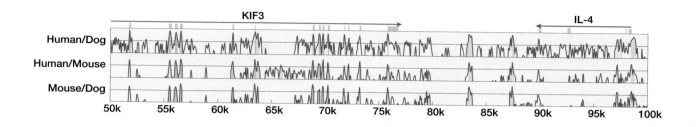

Figure 12.19: VISTA sequence alignment of a 50 kb genomic sequence containing human *KIF3* and *IL4* gene sequences with orthologous sequences in dog (D) and mouse (M).

Conserved sequences are shown relative to their positions in the human genome (horizontal axes), and their percent identities (50%–100%) are indicated on the vertical axes. The locations of coding exons (blue rectangles: at top) and the 3′-UTR of *KIF3* (turquoise rectangle) are shown above the profile. Horizontal arrows indicate the direction of transcription for each gene. Peaks of highly conserved sequences include coding sequences (blue) and also noncoding (red) sequences of unknown function. Adapted from Dubchak *et al.* (2000) *Genome Res.* **10**, 1304–1306 with permission from Cold Spring Harbor Laboratory Press.

sequences over large intervals. Different computer programs, such as **VISTA** (http://www-gsd.lbl.gov/vista) and **Pipmaker** (http://bio.cse.psu.edu/pipmaker/) allow large-scale sequence alignments and express the results in graphical format (see *Figure 12.19* for an example).

As a result of comparative mammalian analyses (mostly human–mouse) the amount of highly conserved sequence in the human genome is now thought to be around 5%, of which only about 30% (~1.5% of the total genome) is thought to be polypeptide-encoding DNA. Many of the analyses have compared human and mouse sequences but some regulatory sequences that are involved in primate-specific gene expression may require comparisons with a range of primate sequences.

The genomes of model organisms are also being compared with genome sequences of distantly related species of the same genus. For example, the nematodes *C. elegans* and *C. briggsae* diverged from a common ancestor 100 MYr ago and both genome sequences can now be interrogated via ENSEMBL at http://www.ensembl.org/). A *Drosophila pseudoobscura* genome project (http://www.hgsc.bcm.tmc.edu/projects/drosophila/) will permit comparison with the *D. melanogaster* genome sequence.

Whole mammalian genome sequence alignments

Complex genomes contain a high proportion of neutrally evolving sequence. Because of the lack of selection pressure, such sequences diverge rapidly. To gain greater insights into genome evolution, computer programs need to be able to align neutrally evolving sequence over at least a large proportion of the genomes being compared. This has been achieved for the human and mouse genomes using a modification of the **BLASTZ** program (Schwartz *et al.*, 2003) and *whole human–mouse genome sequence alignments* are available together with illustrations of their usefulness at http://bio.cse.psu.edu/genome/hummus/.

The whole genome alignments have shown that about 40% of the human genome sequence aligns to that of the mouse, and that the highly repetitive DNA sequences can be partitioned into two classes. ***Lineage-specific repeats*** were intro-

duced by transposition after divergence of humans and mice from a common ancestor (within the last 80–100 MYr, approximately). ***Ancient conserved repeats*** have been retained from the common ancestor and orthologous human and mouse repeats can be identified and aligned even although they are > 80–100 MYr old. The data also revealed differences in the ratio of the two repeat classes. In humans 24.4% of the genome (or about 53% of the interspersed repeats) are lineage-specific but in mouse they account for 32.4% of the genome, or close to 85% of the interspersed repeats (about 39% of the mouse genome). The ancestral repeats have largely been replaced in the mouse genome and make up only about 5% of the genome; in humans 22% of the genome is composed of ancestral repeats. The difference reflects the difference in nucleotide substitution rates in the two genomes (Section 11.2.6).

12.3.3 Gene number is generally proportional to biological complexity

Genome sequencing has revealed some surprises, and at first sight gene number might not seem to parallel biological complexity as well as expected. Who would have predicted that we have only about 1.5 times the number of genes found in a 1-mm long nematode worm containing only about 1000 cells, or that this tiny worm appears to have thousands more genes than the much more complex fruitfly? Nevertheless, if we consider sequenced genomes there is a clear trend: vertebrate genomes have roughly twice the number of genes present in invertebrate genomes, which in turn have many more genes than unicellular organisms (*Table 12.4*).

Metazoan gene number generally increases according to specialization but while gene duplication has provided a major boost to developing complex metazoans, some species e.g. *D. melanogaster* have a surprisingly low number of duplicated genes. It is important to consider that genes can also be lost from lineages during evolution.

As can be seen in *Table 12.4*, gene density decreases as one moves from simple genomes to complex metazoan genomes. This reflects the emergence of introns in early eukaryotes and

Table 12.4: Gene number and density in simple and complex genomes

	UNICELLULAR GENOMES				MULTICELLULAR GENOMES		
Genome	**Genome size [range]**	**Gene number**	**Gene density**	**Genome**	**Genome size (or number of Mb sequenced*)**	**Gene number**	**Gene density**
PROKARYOTES				**Invertebrate**			
Bacterial (n = 97)	3.10 Mb [0.58–9.11 Mb]	Average = 2840	1 per 1.09 kb	Ascidian (*C. intestinalis*)	117 Mb*	16000	~1 per 7 kb
				Nematode (*C. elegans*)	97 Mb	19000	~1 per 5 kb
Archaeal (n = 16)	2.23 Mb [1.66-5.75 Mb]	Average = 2200	1 per 1.02 kb	Fruitfly (*D. melanogaster*)	123 Mb*	14000	~1 per 9 kb
				Mosquito (*A. gambiae*)	278 Mb	14000	~1 per 20 kb
UNICELLULAR EUKARYOTES							
Microsporidians				**Plant**			
E. cuniculi	2.9 Mb**	2000**	1 per 1.45 kb	Thale cress (*A. thaliana*)	115 Mb*	25500	~1 per 4.5 kb
Unicellular fungi							
S. cerevisiae	14 Mb	6300	1 per 2.2 kb				
S. pombe	14 Mb	4800	1 per 2.9 kb				
				Vertebrate			
				Pufferfish (*T. rubripes*)	365 Mb	31000	~1 per 12 kb
Protozoans				Mouse (*M. musculus*)	2500 Mb*	~30000?	~1 per 80 kb
P. falciparum	23 Mb	5300	1 per 4.3 kb	Human	2900 Mb*	~30000	~1 per 100 kb
P. Y. yoeli	23 Mb	5900	1 per 3.9 kb				

* – does not represent the full genome size because excludes highly repetitive tandem repeats which are difficult to clone and sequence, but the total gene number.

**– small genome size and small gene number reflect the obligate intracellular status of microsporidians.

an increasing tendency towards accumulating repetitive DNA within introns and integenic regions as genomes become more complex. Thus. for example, about 45% of the human genome is composed of transposon-based repeats but the corresponding value for mouse is 37% and the equivalent values in *D. melanogaster* and *C. elegans* are considerably lower.

12.3.4 The extent of progressive protein specialization is being revealed by proteome comparisons

When the genome sequences of extensively-studied model organisms such as *E. coli* and the yeast *S. cerevisiae* were first determined, it was surprising to find that a significant number of genes were identified whose functions were unknown. In the case of the draft human genome sequences close to one half had no known function, an important reminder that there is a long way to go before we really begin to understand our genome. Of the remainder, close to one half are involved in signal transduction or nucleic acid binding (*Figure 12.20*).

Comparisons of the proteins predicted from the human genome sequence against those predicted from sequenced genomes of various model organisms has provided some insights into gene deployment during evolution (*Figure 12.21* and *Table 12.5*). Approximately 20% of human proteins show sequence homology to widely distributed proteins found in both eukaryotes and prokaryotes. About 1% of the proteins did not have a homolog in other animal genomes tested. However, at that time there were very few database entries for non-human primates and so the 1% or so may turn out to be primate-specific rather than human-specific (at most, only a

tiny number of human genes are expected to be human-specific).

The availability of complete genomes from different species also allows a form of **comparative proteomics** in which all known or predicted protein sequences encoded by the different genomes are interrogated for the presence of specific protein domains, motifs or other subsequence component. The leading resource is **InterPro** (Integrated resource of Protein Families, Domains and Sites), an amalgam of various protein sequence databases maintained at the European Bioinformatics Institute (http://www.ebi.ac.uk/interpro/). The InterPro data are provided in various forms, including lists cataloging the frequencies of specific protein families, protein domains and protein repeats within the proteomes of sequenced genomes (see http://www.ebi.ac.uk/proteome/).

When the top 25 human InterPro entries (based on prevalence) are compared across different types of eukaryote genomes, several categories are absent not just from the unicellular yeast *S. cerevisiae* (11/25), but also from the plant *A. thaliana* (6/25) and to a lesser extent from invertebrate animals (3/25 from *C. elegans* and 2/25 from *D. melanogaster* – see *Table 12.5*). In part this depends on vertebrate specialization, such as immune system genes. Olfactory receptor genes are clearly important in mice and humans and although there are no direct counterparts listed in the other organisms, the top InterPro entry for *C. elegans* is the nematode 7TM chemoreceptor which may function as an olfactory receptor. There are also major species differences in the frequency rankings of InterPro entries, even in the case of humans and mice, as in the case of the prevalence of

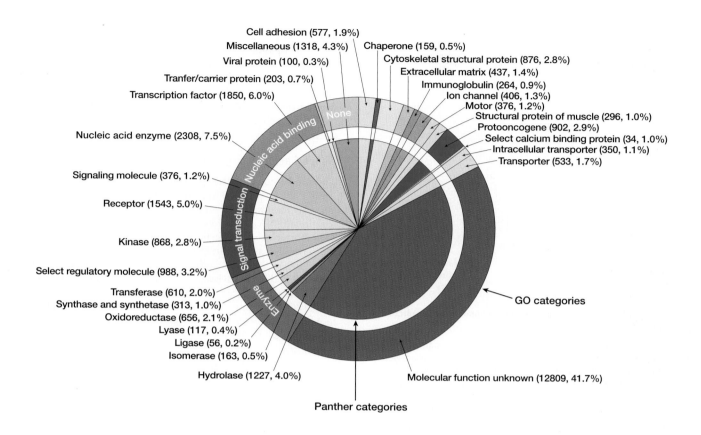

Cell adhesion (577, 1.9%)
Miscellaneous (1318, 4.3%)
Viral protein (100, 0.3%)
Tranfer/carrier protein (203, 0.7%)
Transcription factor (1850, 6.0%)

Chaperone (159, 0.5%)
Cytoskeletal structural protein (876, 2.8%)
Extracellular matrix (437, 1.4%)
Immunoglobulin (264, 0.9%)
Ion channel (406, 1.3%)
Motor (376, 1.2%)
Structural protein of muscle (296, 1.0%)
Protooncogene (902, 2.9%)
Select calcium binding protein (34, 1.0%)
Intracellular transporter (350, 1.1%)
Transporter (533, 1.7%)

Nucleic acid enzyme (2308, 7.5%)

Signaling molecule (376, 1.2%)

Receptor (1543, 5.0%)

Kinase (868, 2.8%)

Select regulatory molecule (988, 3.2%)

Transferase (610, 2.0%)
Synthase and synthetase (313, 1.0%)
Oxidoreductase (656, 2.1%)
Lyase (117, 0.4%)
Ligase (56, 0.2%)
Isomerase (163, 0.5%)
Hydrolase (1227, 4.0%)

None

Nucleic acid binding

Signal transduction

Enzyme

GO categories

Molecular function unknown (12809, 41.7%)

Panther categories

Figure 12.20: A preliminary functional classification of human polypeptide-encoding genes.

Known or predicted functions for 26 383 human polypeptide-encoding genes. Classification is according to the GO molecular function categories as shown in the outer circle (Gene Ontology classification – see Section 8.3.6) or to Celera's Panther molecular function categories (inner circle). Reproduced from Venter *et al.* (2001) *Science* **291**, 1304–1351, with permission from the American Association for the Advancement of Science.

olfactory receptors, C2H2-type zinc fingers and KRAB boxes (domains found in C2H2-type zinc finger proteins), and rhodopsin-type G protein-coupled receptors (*Table 12.5*).

12.4 What makes us human?

Where have we come from? What makes us human? Fundamental questions such as these are ones which will be best answered once we have complete genomes, gene lists and gene product inventories for living things. Genome projects started to deliver genome sequences in 1995 and the first 10–20 or so years of the new millennium should see an explosion of sequenced genomes, detailed gene annotations and comprehensive information on proteomes and RNA products. Already, comparative analyses of the available data have begun to give us a wealth of molecular detail on what nucleic acid sequences and inferred proteins we share with other species and how we differ from them.

Comparisons can of course be done at different levels and humans can be placed in a variety of phylogenetic groups. In an evolutionary series of progressively more specialized groups we are eukaryotes, metazoans, bilaterians, coelomates, deuterostomes, chordates, craniates, vertebrates, gnathostomes, amniotes, eutherian mammals, catarrhine primates, hominoids, and hominids etc. (see Figures *12.22–12.24* and see *Box 12.5* for a glossary).

The universality of the genetic code, the huge evolutionary conservation of key biochemical reactions and processes essential for cell function, and the high degree of conservation of some key developmental processes in metazoan cells – these are features which emphasize the close relationship of humans to species that are morphologically quite distinct and evolutionarily distantly related. Of course there are differences too. The more complex metazoans tend to have more complex genomes with more genes and protein domains, and more repetitive DNA. There may also significant differences between different species in various characteristics associated with gene expression. Examples include:

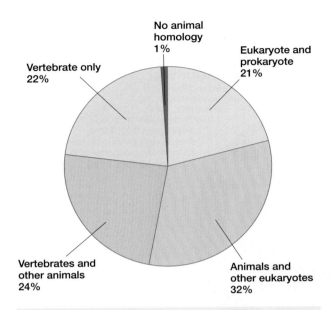

Figure 12.21: Taxonomic distribution of homologs of predicted human and mouse proteins.

Many of our proteins are of ancient evolutionary origin and only less than a quarter originated since vertebrates appeared. Note that sequence databases are still filling up with new sequences and there is expected to be a very small number of truly human-specific genes. Adapted from the International Human Genome Sequencing Consortium (2001) with permission from the Nature Publishing Group.

▶ *operons* – transcription of eukaryotic genes tend to be independently regulated unlike the co-ordinate control in bacterial operons, but a large fraction of *C. elegans* genes are organized in operons which resemble bacterial operons (but are operationally different);

▶ *DNA methylation* (which is far from universal in metazoans – see *Box 9.3*);

▶ *genomic imprinting* (an important feature of mammalian gene expression but does not appear to occur in organisms such as *Drosophila, C. elegans* etc;)

▶ *vertebrate adaptations* e.g. associated with immune systems, hormones etc.

▶ *mammalian adaptations*, e.g. genes associated with the XY sex determination system, X-chromosome inactivation, placentation etc.

Of course, we are also different from other mammals and from our evolutionarily cousins, the great apes. The recent sequencing of the mouse and rat genomes, and the ongoing sequencing of the genomes of chimpanzee (Olson and Varki, 2003) and other primates, will deliver important information on their make up and how they differ from us.

12.4.1 What makes us different from mice?

A recent commentary on the sequencing of the mouse genome suggested that Man's best friend was no longer the dog, but the mouse. Without question, the mouse is the most important mammalian model (for its advantages, see *Box 8.8*), and in some areas we have been heavily reliant on extrapolation from mouse studies (e.g. studying gene expression during development). But of course the mouse is still a *model* and we are increasingly aware of differences between humans and mice. Most of the following observations have been derived from the **M**ouse **G**enome **S**equencing **C**onsortium (2002), which is abbreviated to MGSC (2002) in cross-references below.

General aspects of genome organization

The mouse euchromatic genome size is 2500 Mb, somewhat smaller than the 2900 Mb size of the human euchromatic genome. The difference is mostly due to the greater amount of repetitive DNA (see below). As a result, intron sizes are about 16% shorter in mouse genes on average. Intergenic distances are generally smaller but there is considerable regional variation. Exon sizes and coding DNA sizes (of the order of 500–550 codons on average) are very similar in humans and mice as is the number of exons in orthologous genes. The overall (G + C) content of the mouse genome is 42%, slightly higher than the 41% (G + C) content of the human genome. Nevertheless, the human genome has a significantly higher fraction of very high (G + C) content DNA (MGSC, 2002; Figure 7), and so there are many more CpG islands (~27 000 in repeat-free human DNA) than in mouse (15 000). Regions of conserved human–mouse synteny are on average about 10 Mb long (see *Figure 12.11*).

Interspersed repeats

Much of the human–mouse difference in genome size is due to the higher amount of repetitive DNA in the human genome: transposon-based repeats account for ~45% of the human genome but only ~37% of the mouse genome. The amount of human LINE sequence slightly exceeds that of the mouse LINE sequence. Although LINE sequences have been inherited from the common ancestor of humans and mice, there has been a higher turnover of LINE sequences in the mouse genome. In both genomes only the LINE1 element has been actively transposing but while the clear majority of the current mouse LINE1 sequences underwent transposition following human–mouse divergence the majority of human LINE1 sequences were retained from the common ancestor. LINE-1 repeats are conserved in humans and mouse (and throughout mammals) largely because of conservative selection pressure to maintain the sequence of the large *ORF2* sequence which specifies the reverse transcriptase.

Much of the human–mouse difference in interspersed repeat content is due to the higher amount of SINE DNA in the human genome. None of the SINE elements have been inherited from a common ancestor (unlike LINE elements, LTR elements and DNA transposons – see MGSC, 2002, Tables 5 and 6). Whereas only a single SINE, the Alu repeat, has been active in the human genome, four mouse SINE families have developed, each relying on the LINE1 retrotransposition machinery. Of these, the B2, ID and B4 repeats derived as cDNA copies of tRNA genes but the B1 repeat, like the Alu repeat, derived from 7SL RNA (*Figure 12.25*).

Table 12.5: Comparative occurrence in metazoans of selected protein families, domains and repeats.

Protein family, domain or repeat	H. sapiens (28525 proteins)		M. musculus (20804 proteins)		D. melanogaster (13446 proteins)		C. elegans (20377 proteins)		A. thaliana (25936 proteins)		S. cerevisiae (6202 proteins)	
	Matching proteins	Rank	Matching proteins	Rank	Matching proteins	Rank	Matching proteins	Rank	Matching proteins	Rank	Matching proteins	Rank
Zn-finger, C2H2 type	946	1	482	6	360	1	246	11	191	20	54	10
Immunoglobulin/major histocompatibility complex	883	2	674	3	160	8	120	26	51	110	None	N/A
Rhodopsin-like GPCR superfamily	826	3	1375	1	86	26	403	4	6	836	None	N/A
Immunoglobulin-like	804	4	655	4	144	10	100	35	1	1920	None	N/A
Protein kinase	687	5	486	5	263	2	507	2	1047	1	118	1
Olfactory receptor	477	6	980	2	None	N/A	None	N/A	None	N/A	None	N/A
Serine/Threonine protein kinase	472	7	319	7	205	4	286	7	816	2	113	2
G-protein beta WD-40 repeat	386	8	243	10	188	5	166	17	267	12	101	3
Zn-finger, RING	367	9	222	13	121	15	173	16	466	5	39	16
EGF-like domain	357	10	286	8	101	21	202	13	34	175	None	N/A
RNA-binding region RNP-1 (RNA recognition motif)	342	11	234	12	166	7	160	18	299	10	58	8
Pleckstrin-like	331	12	188	18	79	35	90	46	31	197	29	25
Immunoglobulin subtype	327	13	191	16	79	35	28	161	None	N/A	None	N/A
Proline-rich extensin	324	14	185	19	147	9	150	19	186	21		N/A
Tyrosine protein kinase	314	15	216	15	132	12	192	14	360	7	26	28
KRAB box	314	15	119	32	None	N/A	None	N/A	None	N/A	None	N/A
Calcium-binding EF-hand	298	17	220	14	128	13	134	23	220	16	17	46
Zn-finger, C2H2 subtype	286	18	91	48	2	1116	2	1161	None	N/A	None	N/A
SH3 domain	280	19	189	17	80	33	70	64	4	1075	25	29
Ankyrin	259	20	162	20	94	23	110	27	114	38	19	42
Immunoglobulin C-2 type	259	20	145	26	111	17	67	67	None	N/A	None	N/A
Homeobox	254	22	238	11	107	19	106	30	95	53	8	125
Fibronectin, type III	232	23	157	22	71	42	54	78	4	1075	2	584
Immunoglobulin V-type	223	24	249	9	4	713		N/A	None	N/A	None	N/A
Leucine-rich repeat	212	25	146	25	108	18	69	65	509	4	6	183

The number of different proteins under species names is based on non-redundant proteome sets of SWISS-PROT, TrEMBL and Ensembl entries. The left column lists the top 25 human entries (according to the number of matching proteins) in the InterPro database. Under species names are the number of different matching proteins and the ranking order. Data obtained (March 2003) from the European Bioinformatics Institute via http://www.ebi.ac.uk/proteome.

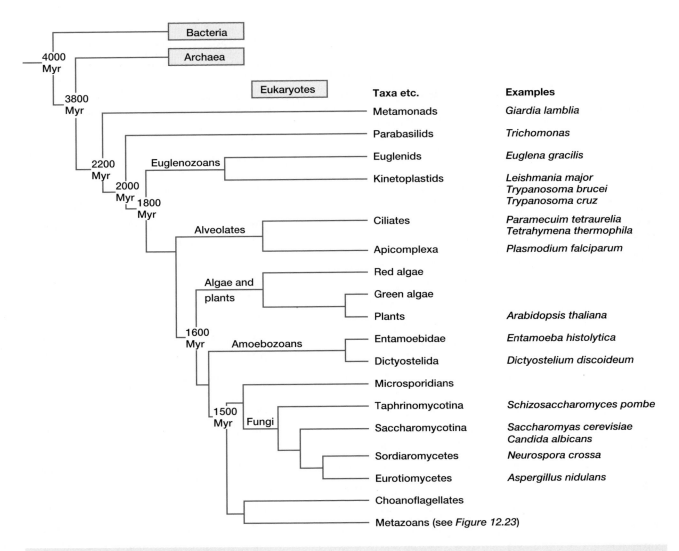

Figure 12.22: A simplified eukaryotic phylogeny.

Note: the choanoflagellates (or collar flagellates) are considered to be the closest living protist relatives of the sponges, the most primitive metazoans (the choanoflagellates are almost identical in shape and function to the choanocytes, or collar cells, of sponges).

The B1 and Alu repeats have, however, significantly diverged in sequence.

Divergence in gene and protein sequence

About 80% of mouse proteins seem to have strict 1:1 orthologs in the human genome, with sequence identities often falling in the 70–100% range. However, perhaps about 10% of orthologous human and mouse proteins show more extreme sequence divergence (*Figure 12.26*). Many of the most rapidly evolving proteins are known to function in host defense/immunity, e.g. MHC genes, or in reproductive processes. Rapidly evolving reproductive proteins include: transition protein 2 (68% amino acid sequence divergence between humans and mice), zona pellucida glycoproteins 2 and 3 (43% and 33% divergence respectively), acrosin (38% divergence), and sperm-specific protamines P15 and P2

(41% and 36% divergence respectively). *Positive (Darwinian)* selection (selection for amino acid replacement to promote diversity) has been identified for several of these kinds of protein (see e.g. Swanson *et al.*, 2001).

Divergence in gene number

The total numbers of human and mouse genes remains to be established but are expected to be broadly similar. Many human and mouse proteins, however, belong to gene families that have undergone differential expansion in at least one of the two genomes, resulting in the lack of a strict 1:1 relationship. Of course, when gene number varies in a gene family, identifying true orthologs may be difficult, especially if there is substantial sequence divergence. A long established example is the major histocompatibility complex (MHC) where orthologs cannot easily be identified among the

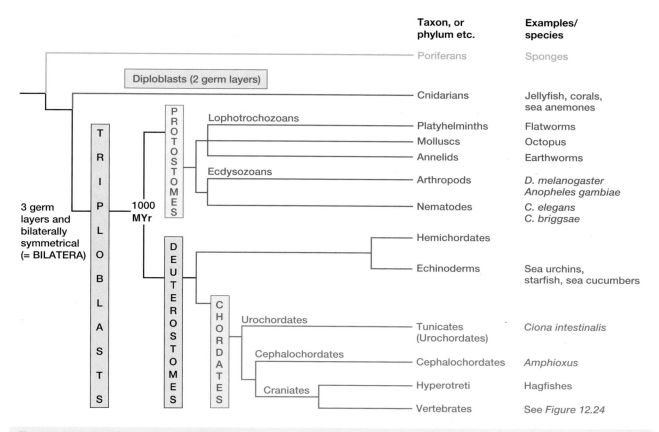

	Taxon, or phylum etc.	Examples/ species
	Poriferans	Sponges
Diploblasts (2 germ layers)	Cnidarians	Jellyfish, corals, sea anemones
Lophotrochozoans	Platyhelminths	Flatworms
	Molluscs	Octopus
	Annelids	Earthworms
Ecdysozoans	Arthropods	*D. melanogaster* *Anopheles gambiae*
	Nematodes	*C. elegans* *C. briggsae*
	Hemichordates	
	Echinoderms	Sea urchins, starfish, sea cucumbers
Urochordates	Tunicates (Urochordates)	*Ciona intestinalis*
Cephalochordates	Cephalochordates	*Amphioxus*
Craniates	Hyperotreti	Hagfishes
	Vertebrates	See *Figure 12.24*

Figure 12.23: A simplified metazoan phylogeny.

Note: the fundamental protostome-deuterostome split which occurred about 1000 MYr ago means that *C. elegans* and *D. melanogaster* are more distantly related from humans than some other invertebrates such as the sea urchin *Strongylocentrotus purpuratus*, the tunicate *Ciona intestinalis*, and the cephalochordate *Amphioxus*.

classical MHC loci (HLA in humans, H-2 in mouse) and where there are considerable differences in the number of nonclassical MHC loci.

An extreme case is the olfactory receptor gene family. The mouse has about 1200 functional genes, more than three times the number of functional human genes (Young *et al.*, 2002), resulting in differing degrees and repertoires of odorant detection between mice and humans. An additional 25 mouse-specific gene clusters are known, 14 containing genes involved in rodent reproduction (MGSC, 2002, Table 16) and five containing genes involved in host defense and immunity. As rodents and primates show some pronounced differences in the physiology of reproduction reproductive traits may have been responsible for powerful evolutionary pressures and the demand for innovation prompted differential gene family expansions.

Gene loss during evolution can also lead to differences in gene number not only in clustered gene families but also in the case of dispersed genes. Some human genes therefore do not appear to have an ortholog in rodent lineages, notably genes in the major pseudoautosomal region and the neighboring sex chromosome-specific region (*Figure 12.15*). Mutation in some of these genes causes disease as

in the pseudoautosomal *SHOX* gene (Leri-Weill syndrome; Langer mesomelic dysplasia) and the Kallman syndrome gene *KAL1* located on Xp (homologs have been identified in *C. elegans* and *D. melanogaster,* but equivalent genes appear to have been deleted from rodent lineages).

Divergence in gene expression

Human–mouse differences in expression of orthologous genes include many examples of differences in RNA processing and the alternative usage of promoters, differences in the pattern of X chromosome inactivation and differences in imprinting. In addition, even when orthologous genes are very highly conserved at the protein level the spatio-temporal patterns of expression not infrequently show significant differences (Fougerousse *et al.*, 2000).

12.4.2 What makes us different from our nearest relatives, the great apes?

Nearly one and a half centuries ago Thomas Huxley correctly identified the chimpanzee and the gorilla as our closest relatives. Ever since, evolutionary geneticists have wrestled with the **trichotomy problem**: which of the two species is our

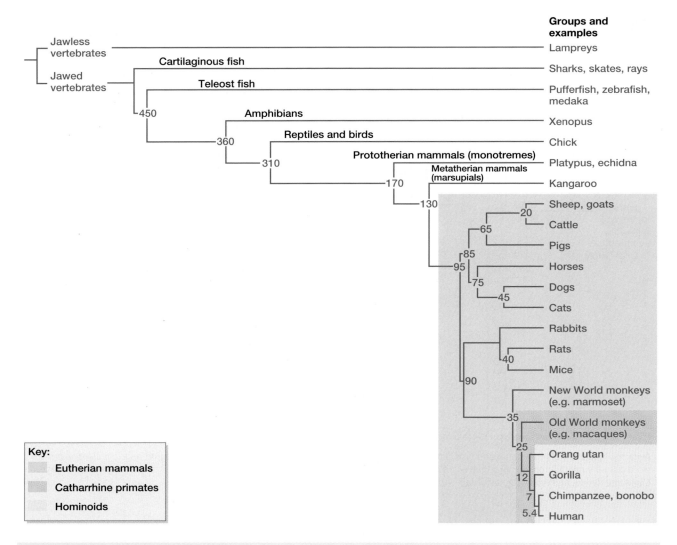

Figure 12.24: A simplified vertebrate phylogeny.

Numbers at nodes show estimated divergence times in millions of years.

closest relative, or has there been simultaneous divergence of the human, chimpanzee and gorilla lineages (*trichotomy*)? Most nucleotide sequence data (e.g. Satta *et al.,* 2000) support a closer relationship between humans and chimpanzees (which includes *Pan troglodytes,* the common chimpanzee, and *Pan paniscus,* the pigmy chimpanzee or bonobo). The *Homo-Pan* clade (= grouping) is also supported by chromosome break-point analyses which show that a 0.1 Mb autosomal fragment must have transposed onto the Y chromosome in the ancestor of humans, common chimpanzees and bonobos.

Divergence of the human–chimpanzee and human–gorilla lineages occurred about 5–6 MYr and ~7 MYr ago respectively leading to a variety of genetic differences (see Gagneux and Varki, 2001; Olson and Varki, 2003). The divergence into separate species (*speciation*) may initially have been driven by small cytogenetic differences, e.g. inversions (which suppress recombination in mammals) and/or mutations in key genes

regulating gamete formation or regulation of early embryonic development.

Genome organization

Classical cytogenetic comparisons emphasize the very strong conservation of *hominoid* (human + great ape) chromosome banding patterns (Yunis and Prakash, 1982). Major structural differences appear limited, including a number of pericentric and paracentric inversions, the recent fusion of two chromosomes to form human chromosome 2, and a reciprocal translocation between the gorilla chromosomes which correspond to human chromosomes 5 and 17 (see *Figure 12.27*). Centromere sequences show very rapid evolution and although an α-satellite sequence is conserved in all human and great ape chromosomes there is very significant sequence divergence most likely as a result of *concerted evolution* of the repeats in individual lineages. More recent multicolor FISH

Box 12.5: A glossary of common metazoan phylogenetic groups and terms.

For more information, see the Tree of Life web page (http://tolweb.org/tree/phylogeny.html) and other phylogenetic sources under Further Reading. See *Figures 12.22–12.24* for simplified eukaryotic, metazoan and vertebrate phylogenies.

Amniotes – vertebrates which develop an amnion. Includes reptiles, birds and mammals but not fishes and amphibians.

Anthropoids – *hominoids* (q.v.) plus monkeys.

Ascidians – sea squirts (e.g. *Ciona intestinalis*), a type of *tunicate* (q.v.).

Bilaterians – bilaterally symmetrical metazoans, including the echinoderms, which have bilaterally symmetrical larvae. They have three germ layers and so are now synonymous with *triploblasts* (q.v.).

Catarrhine primates = *hominoids* (q.v.) plus *old world monkeys*.

Cephalochordates (= *branchiostomes* = *lancets*) – *chordates*, which do not have a skull or vertebral column (e.g. *Amphioxus*).

Chordates – animals which undergo an embryonic stage where they possess a notochord, a dorsal tubular nerve cord and pharyngeal gill pouches; includes *urochordates* (q.v.), *cephalochordates* (q.v.) and *craniates* (q.v.).

Cnidarians (= *coelenterates*) – jellyfish, corals, sea anemones (all radially symmetrical with two germ layers).

Coelomates – organisms with a coelom, or fluid-filled internal body cavity that is completely lined with mesoderm, as opposed to acoelomates (e.g. flatworms) and pseudocoelomates (such as nematodes which have a fluid-filled body cavity but the cavity is not completely lined with mesoderm). Divided into two groups, *protostomes* (q.v.) and *deuterostomes* (q.v.).

Clade – a group of organisms including all the descendants of a last common ancestor (= *monophyletic taxon*).

Craniates – animals with skulls = vertebrates plus hagfish.

Deuterostomes – from the Greek meaning *second mouth* = *chordates* (q.v.) plus *echinoderms* (q.v.). Animals in which the *blastopore* (the opening of the primitive digestive cavity to the exterior of the embryo) is located to the posterior of the embryo and becomes the anus. Later on, the mouth opens opposite the anus. Compare with *protostomes* (q.v.).

Diploblasts – metazoans with two germ layers only: ectoderm and endoderm = *cnidarians* (q.v.) + *ctenophora* (comb jellies). They are radially symmetrical and so this group is often classified as *radiata* (q.v.).

Echinoderms (e.g. sea urchin, starfish etc.) are *radially symmetrical* marine animals but because they have three germ layers they are *triploblasts* and so are classified as *bilaterians* (q.v.).

Eutherian mammals – placental mammals, as opposed to *monotremes* (q.v.) and *metatherian mammals* (q.v.).

Gnathostomes – jawed vertebrates, accounting for the majority of the vertebrates, but lampreys are jawless vertebrates.

Hagfish are closely related to vertebrates but are invertebrate because the notochord does not get converted into a vertebral column.

Hominids – humans and human-like ancestors.

Hominoids = humans plus *great apes* (common chimpanzee, bonobo, gorilla, orang-utan) plus *lesser apes* (gibbons).Compare *anthropoids* (q.v.)

Metatherian mammals (= marsupials).

Monotremes – egg-laying mammals.

New World monkeys (Platyrrhini) – monkeys that are limited to tropical forest environments of southern Mexico, Central, and South America.

Old World monkeys (Cercopithecidae) – monkeys found in a wide variety of environments in South and East Asia, the Middle East, and Africa.

Phylum – a group of species sharing a common body organization.

Platyrrhine primates = *New World monkeys* (q.v.).

Prototherian mammals (= *monotremes* q.v.).

Protostomes – from the Greek meaning *first mouth*. Organisms in which the mouth originates from the *blastopore* (the opening of the primitive digestive cavity to the exterior of the embryo). Later on, the anus will open opposite the mouth. Includes molluscs, annelids and arthropods. Compare with *deuterostomes* (q.v.).

Radiata – radially symmetrical animals with two germ layers only = *diploblasts* (q.v.).

Taxon – a group of organisms recognized at any level of the classification.

Teleost fish – a major grouping of *bony fish* (as opposed to fish with a cartilaginous skeleton, such as sharks, skates and rays). Characterized by a fully moveable upper jaw, rayed fins and a swim bladder.

Triploblasts – metazoans with three germ layers; they are bilaterally symmetrical and hence synonymous with *bilaterians* (q.v).

Tunicates – sea squirts and salps (e.g. *Ciona intestinalis*, the sea squirt).

Urochordates – *chordates* (q.v.), which have a notochord limited to the caudal region (= *tunicates*).

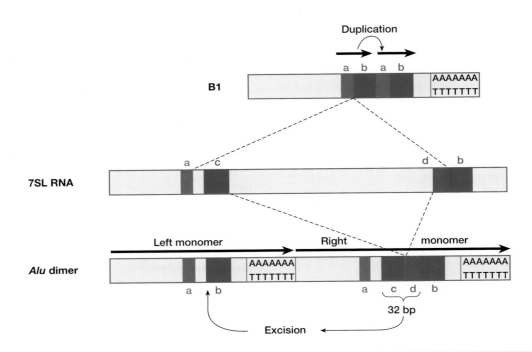

Figure 12.25: The Alu and B1 repeats evolved from processed copies of the 7SL RNA gene.

Extensive homology of the Alu repeat sequences to the ends of the 7SL RNA sequence suggests that a polyadenylated copy of the 7SL RNA gene integrated elsewhere in the genome by a retrotransposition event (see *Figure 9.14* for the general type of mechanism). In some cases, the integrated copies were able to produce RNA transcripts of their own. At a very early stage, an internal segment (between c and d) was lost. Subsequently, a 32-bp central segment containing regions flanking the original deletion (c + d) was deleted to give a related repeat unit. Fusion of the two types of unit resulted in the classical Alu dimeric repeat, with the left (5′) monomer lacking a 32-bp sequence and the right (3′) monomer containing the 32-bp sequence. ***Note*** that multiple copies of Alu monomers are also found in the human genome. In the mouse, a similar process of copying from the 7SL RNA gene appears to have occurred with subsequent deletion of a large internal unit (between a and b), followed by tandem duplication of flanking regions (a + b).

comparative mapping studies (see *Figure 12.28*) have confirmed the strong conservation of synteny, and allowed a proposed ancestral karyotype for hominoids (Muller and Wienberg, 2001).

Sequence samplings suggest that about 95% of the chimpanzee genome can be aligned directly with the corresponding human sequences, and that in these aligned regions sequence divergence averages 1.2%; the nonaligned 5% is due to deletions/insertions (see Olson and Varki, 2003 for references). Despite the very high sequence identity (of noncoding DNA as well as coding DNA), some human-specific Alu subfamilies (Sb1, Sb2), and LTR retrotransposons can be recognized.

Gene differences

When orthologous human and chimpanzee sequences are compared, the coding DNA typically shows 99% or more sequence identity. In some cases, specific alleles of certain human genes are more closely related to orthologs in chimpanzees than they are to other human alleles. For example, at the human HLA-DRβ locus, the alleles *HLA-DRB1*0302* and *HLA-DRB1*0701* are clearly closer in sequence to certain alleles of the orthologous chimpanzee gene *Patr-DRB* than they are to each other (*Figure 12.29*). Such observations are consistent with a comparatively ancient

origin for such divergent alleles, predating man–chimpanzee divergence.

Primate-specific genes are known, including some of recent evolutionary origin and some which appear to have been subject to positive selection for amino acid replacement (Courseaux and Nahon, 2001; Johnson *et al.*, 2001). Human-specific genes are likely to be quite rare, although gene duplication and gene loss – notably in clustered gene families – will result in gene number differences between humans and the great apes. Currently, however, there is little evidence for functional differences between human and chimpanzee genes: the only known biochemical difference is the global depletion in humans of a specific sialic acid, *N*-glycolylneuraminic acid which is expressed (in a developmentally regulated and tissue-specific fashion) in chimpanzees, bonobos and the other great apes. The difference is due to a frameshifting deletion in the CMP *N*-acetylneuraminic acid gene, which occurred in the human lineage about 2 MYr ago, just before the onset of rapid brain expansion (Chou *et al.*, 2002).

The forkhead domain gene *FOXP2* is the first gene to be implicated in the uniquely human characteristics of speech and language. It has been very well conserved during evolution and shows only three amino acid substitutions between humans and mice. However, two of these substitu-

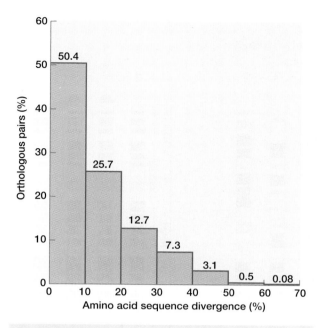

Figure 12.26: Human-rodent protein divergence.

A sample of 1880 human-rodent orthologs (initially reported by Makalowski and Boguski, *Proc. Natl. Acad. Sci. USA* **95**, 9407–9412; 1998) is classified into groups according to the level of amino acid sequence divergence between orthologs. The most highly conserved proteins (e.g. calmodulin, ribosomal proteins, histones etc.) are involved in critical cellular processes. The most rapidly evolving proteins are involved in host defense/immunity or reproduction (see text).

tions are uniquely human (the great apes have the same amino acid as in mouse), and have been suggested to be positively selected during recent human evolution (Enard *et al.*, 2002a). Possibly, they could affect a person's ability to control orofacial movements needed for speaking.

Two *systematic* approaches to identify the molecular basis of our unique characteristics have recently been proposed. The first compares human and primate gene expression on a genomewide scale by microarray-based RNA profiling and two-dimensional electrophoresis–based protein profiling. The preliminary data show that species-specific expression patterns are pronounced in the case of the brain (Enard *et al.*, 2002b). The second approach is to screen the human genome sequence for regions which have experienced strong selection in the human line. Strong selection for a favorable new allele can cause a **selective sweep** (as the new mutant rises in frequency, adjacent chromosomal regions are also swept to fixation) resulting in regions of very low nucleotide diversity. A variety of human genes have been identified in such regions and are targets for identifying genetic differences between modern humans and chimpanzees (Diller *et al.*, 2002).

12.5 Evolution of human populations

The quest to know how human populations evolved goes beyond simple curiosity about our past. It can also unearth evidence for adaptation at the molecular level and can help in interpreting and predicting *linkage disequilibrium*, with consequences for disease-association studies.

Until DNA analyses made their mark, the study of human population evolution was dependent on archeological and paleological studies. The hominid fossil record in Africa begins about 4 MYr ago in the Early Pliocene (the geological epoch from 5.3–1.8 MYr) with representatives of the genus *Australopithecus* from Ethiopia and Tanzania. *Homo erectus* arose more than a million years ago in the Pleistocene (the epoch spanning 1.8 and 0.8 MYr ago), giving rise to our own genus. Some favor the possibility that another species *Homo heidelbergensis* descended from *Homo erectus* and gave rise in turn to *Homo sapiens* and *Homo neanderthalensis* (see Stringer, 2002). Anatomically modern humans (*Homo sapiens sapiens*) began to appear 120 000–100 000 years ago and co-existed with Neanderthals until the latter became extinct about 30 000 years ago.

DNA variation provides an alternative approach to reconstructing the recent history of human populations. Different DNA markers can be used. Mitochondrial DNA and non-recombining Y chromosome sequences have been popular because, unlike other markers, they are not prone to recombination and it is possible to follow lineages directly through the maternal line (mtDNA) or paternal line (Y chromosome). Of course, they are hardly representative of the genome and so autosomal and X-linked sequences have also been studied.

In addition to sampling DNA variation in different populations of modern humans, DNA can be recovered from bones of archeological specimens. Although of short lengths, the recovered DNA sequences can be amplified using PCR and analysed by sequencing or genotyping. The degradation process sets time limits but with great care it is possible to extract **ancient DNA** from specimens up to about 50 000 or more years old. DNA recovered from a 30 000–100 000-year-old Neanderthal specimen had a clearly different mtDNA sequence to that of modern humans: about 3 times the average difference found between different humans, but about half the average difference between humans and chimpanzees (Krings *et al.*, 1997). The data, although necessarily limited, suggest Neanderthals went extinct without contributing any mtDNA to modern humans, and the lineages leading to Neanderthals and modern humans appear to have diverged about 500 000 years or so ago.

12.5.1 Genetic evidence has suggested a recent origin of modern humans from African populations

When applied to modern populations DNA variation studies have clearly suggested a recent origin of modern humans from African populations (see Excoffier, 2002; Templeton, 2002), and fits with the observed greater genetic diversity of African populations (see next Section). An early proposal, the **recent African origin (RAO) model** (also called the uniregional hypothesis), suggested that our species evolved from a small African population that had subsequently colonized the whole world, supplanting former hominids about 100 000–150 000 years or so ago (around the time when anatomically modern humans emerged – see *Figure 12.30*).

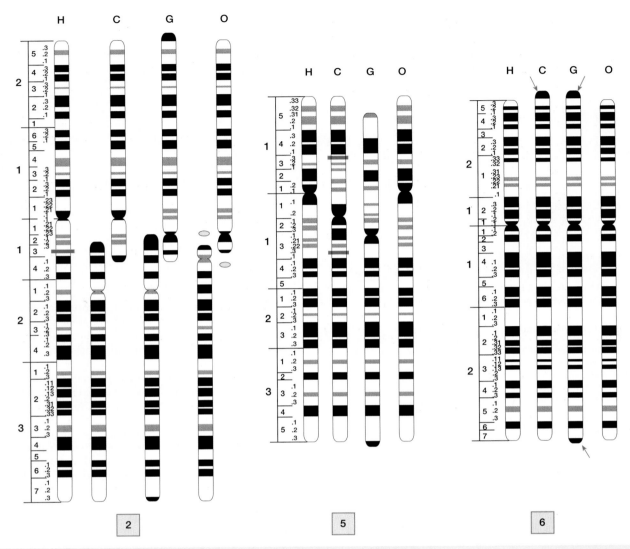

Figure 12.27: Human chromosome banding patterns are very similar to those of the great apes.

The ideograms are from 1000-band late prophase preparations from Human (H), chimpanzee (C), gorilla (G) and orang-utan (O). Human chromosome 2 arose by fusion of two primate chromosomes. Human chromosome 6 is extremely similar to its orthologs – the only readily visible differences are additional telomeric heterochromatin on the short arm of the chimpanzee ortholog, and on both arms of the gorilla ortholog. Orthologs of human chromosome 5 show significant differences. The chimp ortholog has undergone a pericentric inversion (breakpoints corresponding to 5p13 and 5q13), and the gorilla ortholog has undergone a reciprocal translocation with a chromosome corresponding to human chromosome 17). Reprinted with permission from Yunis and Prakash (1982) *Science* **215**, 1525–1530. © 1982, American Association for the Advancement of Science.

The RAO model was based initially on estimates of haplotype variation in human mtDNA. Maternal transmission of mtDNA and absence of recombination means that **coalescence analyses** (*Box 12.6*) can be easily applied to estimate the date of the common ancestor who transmitted the DNA sequence under study to all of the individuals sampled. Using this type of approach, the mtDNA sequences of all modern humans can be traced back to a single individual, the 'mitochondrial Eve'. The analyses show that this individual existed about 100 000–150 000 years ago (5000 to 7000 or so generations) and the data firmly suggest that she lived in East Africa (see below). Of course, the mitochondrial Eve was not the only person living on the planet at that time: there were perhaps about 10 000 individuals living at that time *but unlike Eve, their mtDNA sequences didn't get transmitted to the present human population* (see *Box 12.6*).

The RAO model was controversial because our immediate precursor, *Homo erectus,* is known to have dispersed out of Africa more than a million years ago. The model therefore requires that only an African subset of *Homo erectus* gave rise to modern humans. Although based initially on incorrect estimates, the RAO model has been supported by more reliable

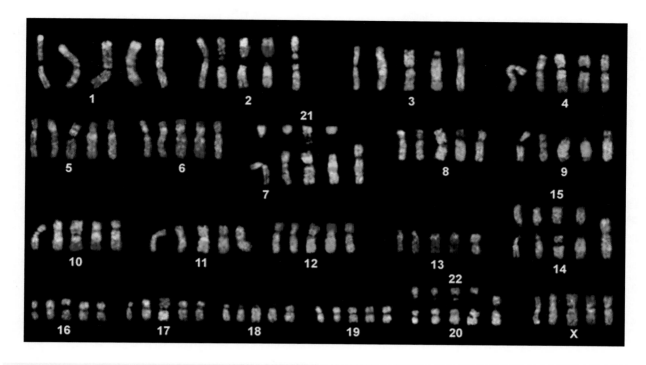

Figure 12.28: Bar-coding of primate chromosomes as a way of revealing structural differences.

Cross-species color banding (*Rx-FISH*) profiles show alignment of orthologous primate chromosomes with human chromosomes 1-22 and X. Chromosome sets (with numbering according to the human homologs) show from left to right: human, chimpanzee, gorilla, orang-utan and macaque. To improve comparisons, chromosomes 2p/2q, 7/21, 14/15 and 20/22, which are single chromosomes in human or in the macaque are shown together with the great ape homologs. This type of analysis has suggested a preliminary ancestral karyotype for human and great apes. Reproduced from Muller and Wienberg (2002) *Hum. Genet.* **109**, 85–94 with permission from Springer-Verlag.

recent data. The mtDNA-based phylogeny shown in *Figure 12.31* provides an example of the general approach. The crucial point is that the *deepest branches* (the ones that diverged earliest from the root, and therefore older in evolutionarily terms) lead exclusively to variants found in the African population strongly suggesting that the mitochondrial Eve lived in Africa. The tree reconstruction methods (e.g the nearest neighbor joining shown in *Figure 12.31*) allow statistical estimates of the time taken for all the currently existing mtDNA sequences to coalesce in the single ancestral mtDNA sequence.

An alternative **multiregional evolution** model has been favored by some paleontologists. Here, modern humans are envisaged to have emerged gradually and simultaneously from dispersed *Homo erectus* populations on different continents, with significant gene flow between the different groups. More recent statistical analyses of haplotype trees have also argued against a simple RAO model. While confirming the dominant role played by Africa in shaping the modern gene pool, Templeton (2002) suggests that humans expanded out of Africa more than once and interbred on a regional basis (see Excoffier, 2002 for various models).

12.5.2 Human genetic diversity is low and is mostly due to variation within populations rather than between them

Studies have consistently shown that genetic diversity in humans is low compared to other species, including mice and apes, suggesting that the human lineage underwent an evolutionarily recent ***population bottleneck***. The pattern of variation is not uniform: non-African populations typically show a subset of the genetic variation found in African populations.

A striking example is the variation seen at the insulin minisatellite, which has been associated with susceptibility to diabetes and other phenotypes. A recent study of three African populations plus three non-African populations identified a total of 22 highly diverged lineages of related alleles. No structural intermediates between these lineages could be found in currently existing populations, suggesting a bottleneck within the ancestry of all humans (Stead and Jeffreys, 2002). The difference between diversity in the African and non-African populations is unusually large: all 22 lineages were identified in Africa but only three lineages were seen in the non-African populations consistent with a common out-of-Africa origin (*Table 12.6*).

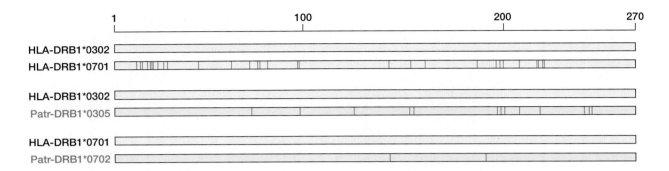

Figure 12.29: Some human alleles show greater sequence divergence than when individually compared with orthologous chimpanzee genes.

From a total of 270 amino acid positions, the *HLA-DRB1*0302* and *HLA-DRB1*0701* alleles show a total of 31 differences (13%). Comparison of either allele with alleles at the orthologous chimpanzee locus (*Patr-DRB1*) identifies more closely related human-chimpanzee pairs, such as *HLA-DRB1*0701* and *Patr-DRB1*0702* (only two amino acid differences out of 270). This suggests that some present-day *HLA* alleles pre-date the human-chimpanzee split. Redrawn from Klein *et al.* (1993), *Scientific American* **269**, 675–680 with permission from Scientific American Inc.

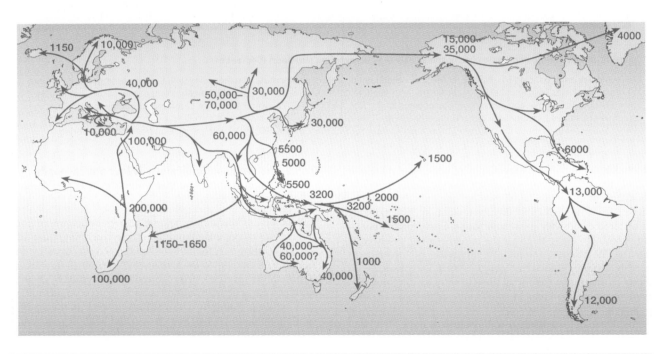

Figure 12.30: The spreading of *Homo sapiens* out of East Africa as postulated by the Recent African Origin model (uniregional hypothesis).

Dates (in years before the present) denote estimated dates of arrival of modern humans at indicated sites as supported by paleontological and archeological records. Dates of migration out of Africa suggested by genetic studies supporting the Recent African Origin model vary in the 50 000–200 000 year range. Figure adapted from Klein and Takahata (2002) *Where do we Come From? The molecular evidence for human descent,* with permission from Springer-Verlag, Berlin.

Box 12.6: Coalescence analyses

In evolutionary genetics **coalescence** is the opposite of divergence. During the normal evolutionary direction, from past to present, genes and DNA sequences diverge from a common ancestral sequence. The point of coalescence analyses is to start from presently existing genetic diversity at some locus and attempt to work backwards though evolution. Of course, it is much easier to work backwards if one follows only the maternal line or the paternal line and so coalescence analyses are comparatively straightforward for mtDNA (maternal line) and nonrecombining Y chromosome sequences (paternal line) because there is no confounding recombination.

After samples have been collected from varied populations coalescence analysis seeks to establish the point when all the different varieties of the modern DNA sequence found in the collected samples *coalesce* into a single ancestral sequence present in the *most recent common ancestor (MRCA)*. Implicit in the concept of coalescence is the inequality of descent: all modern humans and their genes descend from ancestor individuals containing ancestral genes, but not all people who lived in the past had descendants. Different lineages can therefore be connected to different MRCAs (see *Figure*). Ultimately all the genetic variation present at a locus in modern humans can be traced back to a single individual, such as the mitochondrial Eve in the case of mtDNA.

It is important to appreciate that different loci will result in coalescence in different MRCAs. The three billion or so human Y chromosomes on the planet today coalesce into a single Y chromosome of a single individual who lived in the past, the 'Y chromosome Adam'. He may not have lived at the same time as the mitochondrial Eve *because the evolutionary path taken by his Y chromosome was different from that taken by Eve's mitochondrial DNA*. Recombination presents problems for coalescence analysis of autosomal sequences unless specific loci are considered. Recombination does, however, tend to occur in certain areas of the genome rather than others, and our autosomal genome appears to be a mosaic of so-called *haplotype blocks* (segments typically 5 kb–200 kb in length where 3–7 variants account for most of the genetic variation seen in modern humans – see Paabo, 2003). Haplotype blocks tend to be shorter in African populations and so a species-wide mean size for a haplotype block may be of the order of 10 kb or so. Each of the autosomal haplotype blocks will have its own evolutionary history and so thousands of other MRCAs will have contributed to the genetic DNA of modern humans in addition to mitochondrial Eve and the Y chromosome Adam.

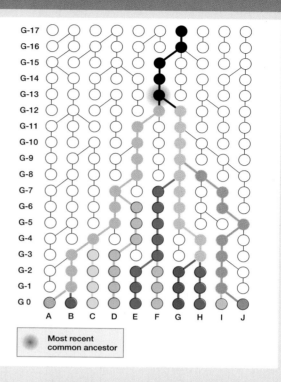

Most recent common ancestor

Coalescence analyses seek to trace back gene or DNA sequence lineagues until they **coalesce** *in a single individual.*

Different lineages of specific genes or DNA sequences in currently existing populations (generation G0, bottom row) can ultimately be traced back to various recent common ancestors in previous generations (G-1 to G-17). The earliest branching lineage is E, which diverged eight generations ago (G-8; from lineages F–H). All lineages coalesce in a most recent common ancestor (MRCA) thirteen generations ago (G-13). All open circles represent cases where the gene/DNA sequence under study has not been transmitted to the present generation.

Despite the differences between populations seen in *Table 12.6*, genetic studies have consistently shown that the great majority of human genetic variation derives from differences *within* populations rather than between them. The most comprehensive such study has found that within-population differences among individuals accounts for 93–95% of human genetic variation while only 3–5% is accounted for by differences between different major population groups (Rosenberg *et al.*, 2002). While every individual person has a unique

biological history, our biological histories are so overlapping it is quite misleading to divide people into biological categories. *The concept of 'race' as a discrete category defined by biology is therefore meaningless because it is not possible to define those categories biologically.*

The study by Rosenberg *et al.* (2002) was a new departure, an effort to define the genetic structure of human populations *without using* a priori *information on the geographic origin of the individuals studied*. Without this information, five major

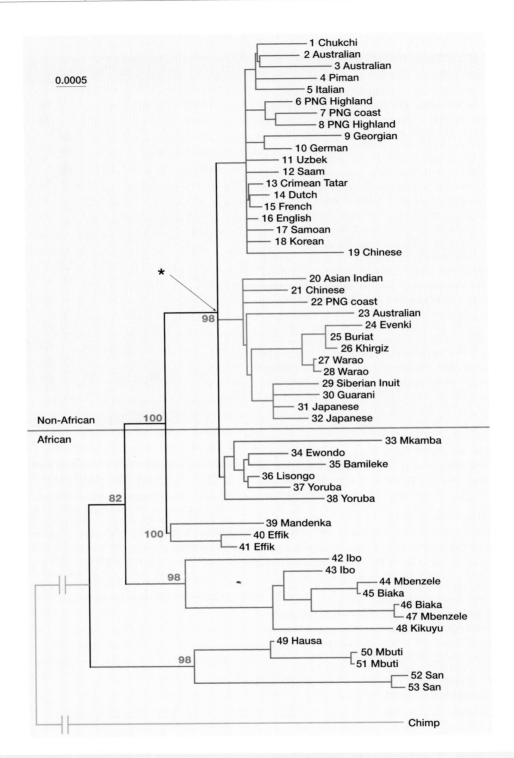

Figure 12.31: mtDNA phylogenies have suggested a recent African origin for modern humans.

A neighbor-joining phylogeny based on essentially complete mtDNA genome sequences (excluding the D-loop) from 53 individuals (population origins given on the right), using the chimpanzee as an outgroup. The data was bootstrapped with 1000 replicates (bootstrap values in red shown on nodes). Individuals of African descent are found below the central horizontal line; non-Africans are above. The node marked with an asterisk refers to the most recent common ancestor of the youngest clade containing both African and non-African individuals. This analysis suggested that modern humans arose from African populations perhaps about 50 000 years or so ago, somewhat more recently than suggested by many other similar analyses. Figure reproduced from Ingman *et al.* (2000) *Nature* **408**, 708–713, with permission from Nature Publishing Group.

Table 12.6: Greater genetic diversity in African populations than in non-African populations at the insulin minisatellite

Lineage	Non-African populations			African populations		
	United Kingdom	Kazakhstan	Japan	Ivory Coast	Zimbabwe	Kenya
I	71.6	85.0	94.9	19.2	18.8	15.5
IIIA	23.0	11.3	4.2	5.8	1.4	7.1
IIIB	5.4	3.8	0.8	3.2	1.4	3.6
F	—	—	—	—	0.7	—
G	—	—	—	1.3	0.7	—
H	—	—	—	—	1.4	3.6
J	—	—	—	9.6	8.0	9.5
K	—	—	—	20.5	17.4	20.2
L	—	—	—	8.3	5.1	3.6
M	—	—	—	3.2	0.7	3.6
N	—	—	—	2.6	5.8	—
O	—	—	—	2.6	—	—
P	—	—	—	—	2.9	2.4
Q	—	—	—	1.9	5.8	9.5
S	—	—	—	4.5	.7	3.6
T	—	—	—	.6	4.3	1.2
U	—	—	—	—	1.4	—
V	—	—	—	1.3	1.4	4.8
W	—	—	—	12.8	13.0	9.5
X	—	—	—	1.9	4.3	2.4
Y	—	—	—	0.6	3.6	—
Z	—	—	—	—	0.7	—

A high resolution system for analyzing the distributions of variant repeats at the *INS VNTR* locus identified 22 highly divergent lineages of related alleles. Lineage frequencies are expressed as a percentage. Only three of the lineages were found in the non-African populations (the symbol — denotes absence of alleles associated with lineage). In marked contrast, all 22 types of lineage were found in Africa where the different populations showed much greater variability. The over-representation of lineage I in non-Africans may be due to positive selection. Reproduced from Stead and Jeffreys (2002) *Am. J. Hum. Genet.* **71**, 1273–1284, with permission from the University of Chicago Press.

genetic clusters were identified which corresponded to five major geographic regions: sub-Saharan Africa; Saharan Africa plus Eurasia (= Europe + Middle East + Central/South Asia); East Asia; the Americas (Amerindians); and Oceania. There is, therefore, a very good agreement between geographic and genetic assignment. Nevertheless, as described before, two genes from two individuals in any one of these geographic regions is *on average* only 4% more similar than two genes selected from individuals belonging to the same region.

Further reading

Carroll SB (2003) Genetics and the making of *Homo sapiens*. *Nature* **422**, 849–857.

Felsenstein J (2003) *Inferring Phylogenies*. Sinauer Associates, Inc., Sunderland, MA.

Holder M, Lewis PO (2003) Phylogeny estimation: traditional and Bayesian approaches. *Nature Rev. Genet.* **4**, 275–284.

Klein J, Takahata N (2002) *Where do we Come From? The molecular evidence for human descent.* Springer-Verlag, Berlin/Heidelberg.

Graur D, Li WH (2000) *Fundamentals of Molecular Evolution*. 2nd Edn. Sinauer, Sunderland, MA.

Meyer A, Van de Peer Y (2003) *Genome Evolution: gene and genome duplications and the origin of novel gene functions*, Kluwer, Dordrecht.

Nei M, Kumar S (2000) *Molecular Evolution and Phylogenetics*. Oxford University Press, Oxford.

Olson S (2002) *Mapping Human History: discovering the past through our genes.* Houghton Mifflin Co., Boston.

Relethford JH (2001) *Genetics and the Search for Modern Human Origins.* Wiley, New York.

Smith JM (1999) *Evolutionary Genetics*. 2nd Edn. Oxford University Press, Oxford.

Electronic reference sources

Clusters of orthologous groups (phylogenetic classification of proteins encoded in complete genomes) – see http://www.ncbi.nlm.nih.gov/COG/

Conservation of human-mouse synteny – see http://www. ensembl.org/Homo_sapiens/syntenyview and http://www. ensembl.org/Mus_musculus/syntenyview.

Proteome analysis server, European Bioinformatic Institute at http://www.ebi.ac.uk/proteome

Tree of Life Web Project at http://tolweb.org/tree/phylogeny.html

References

Abi-Rached L, Gilles A, Shiina T, Pontarotti P, Inoko H (2002) Evidence of *en bloc* duplication in vertebrate genomes. *Nature Genet.* **31**, 100–105.

Bailey JA, Gu Z, Clark RA et al. (2002) Recent segmental duplications in the human genome. *Science* **297**, 1003–1007.

Baxendale S, Abdulla S, Elgar G et al. (1995) Comparative sequence analysis of the human and pufferfish Huntingtons disease genes. *Nature Genet.* **10**, 67–75.

Bonen L, Vogel J (2001) The ins and outs of group II introns. *Trends Genet.* **17**, 322–331.

Brown JR (2003) Ancient horizontal gene transfer. *Nature Rev. Genet.* **4**, 121–132.

Burmester T, Weich B, Reinhardt S, Hankeln T (2000) A vertebrate globin expressed in the brain. *Nature* **407**, 520–523.

Carrel L, Cottle AA, Goglin KC, Willard HF (1999) A first-generation X-inactivation profile of the human X chromosome. *Proc. Natl Acad. Sci. USA* **96**, 14440–14444.

Charchar FJ, Svartman M, El-Mogharbel N (2003) Complex events in the evolution of the human pseudoautosomal region 2 (PAR2). *Genome Res.* **13**, 281–286.

Chou H-H, Hayakawa T, Diaz S et al. (2002) Inactivation of CMP-*N*-acetylneuraminic acid hydroxylase occurred prior to brain expansion during human evolution. *Proc. Natl Acad. Sci. USA* **99**, 11736–11741.

Ciccodicola A, D'Esposito M, Esposito T (2000) Differentially regulated and evolved genes in the fully sequenced Xq/Yq pseudoautosomal region. *Hum. Mol. Genet.* **9**, 395–401.

Courseaux A, Nahon J-L (2001) Birth of two chimeric genes in the *Hominidae* lineage. *Science* **291**, 1293–1297.

Diller KC, Gilbert WA, Kocher TD (2002) Selective sweeps in the human genome: a starting point for identifying genetic differences between modern humans and chimpanzees. *Mol. Biol. Evol.* **19**, 2342–2345.

Doolittle WF (1998) You are what you eat: a gene transfer rachet could account for bacterial genes in eukaryotic nuclear genomes. *Trends Genet.* **14**, 307–311.

Dubchak I, Brudno M, Loots GG et al. (2000) Active conservation of noncoding sequences revealed by 3-way species comparisons. *Genome Res.* **10**, 1304–1306.

Ellis N. (1998) The war of the sex chromosomes. *Nature Genet.* **20**, 9–10.

Enard W, Przeworski M, Fisher SE et al. (2002a) Molecular evolution of FOXP2, a gene involved in speech and language. *Nature* **418**, 869–872.

Enard W, Khaitovich P, Klose J et al. (2002b) Intra- and interspecific variation in primate gene expression patterns. *Science* **296**, 340–343.

Excoffier L (2002) Human demographic history: refining the African model. *Curr. Opin. Genet. Dev.* **12**, 675–682.

Fougerousse F, Bullen P, Herasse M et al. (2000) Human-mouse differences in the embryonic expression patterns of developmental control genes and disease genes. *Hum. Mol. Genet.* **9**, 165–173.

Gagneux P, Varki A (2001) Genetic differences between humans and great apes. *Mol. Phylogenet. Evol.* **18**, 2–13.

Gianfrancesco F, Sanges R, Esposito T (2001) Differential divergence of three human pseudoautosomal genes and their mouse homologs: implications for sex chromosome evolution. *Genome Res.* **11**, 2095–2100.

Graves JA, Wakefield MJ, Toder R (1998) The origin and evolution of the pseudoautosomal regions of human sex chromosomes. *Hum. Mol. Genet.* **7**, 1991–1996.

Ingman M, Kaessmann H, Paabo S, Gyllensten U. (2000) Mitochondrial genome variation and the origin of modern humans. *Nature* **408**, 708–713.

International Human Genome Sequencing Consortium (2001) Initial sequencing and analysis of the human genome. *Nature* **409**, 860–921.

Jeanmougin F, Thompson JD, Gouy M, Higgins DG, Gibson TJ (1998) Multiple sequence alignment with Clustal X. *Trends Biochem Sci.* **23**, 403–405.

Jegalian K, Page DC (1998) A proposed pathway by which genes common to mammalian X and Y chromosomes evolve to become X inactivated. *Nature* **394**, 776–780.

Johnson ME, Viggiano L, Bailey JA et al. (2001) Positive selection of a gene family during the emergence of humans and African apes. *Nature* **413**, 514–519.

Joyce GF (2002) The antiquity of RNA-based evolution. *Nature* **418**, 214–221.

Kaessmann H, Zollner S, Nekrutenko A, Li W-H (2002) Signatures of domain shuffling in the human genome. *Genome Res.* **12**, 1642–1650.

Knight RD, Landweber LF (2000) The early evolution of the genetic code. *Cell* **101**, 569–572.

Koop BF, Hood L (1994) Striking sequence similarity over almost 100 kilobases of human and mouse T cell receptor DNA. *Nature Genet.* **7**, 48–53.

Krings M, Stone A, Schmitz W, Krainitzki H, Stoneking M, Paabo S (1997) Neanderthal DNA sequences and the origins of modern humans. *Cell* **90**, 19–30.

Lahn BT, Page DC (1999) Four evolutionary strata on the human X chromosome. *Science* **286**, 964–967.

Lahn BT, Pearson NM, Jegalian K (2001) The human Y chromosome in the light of evolution. *Nature Rev. Genet.* **2**, 207–216.

Letunic I, Copley RR, Bork P (2002) Common exon duplication in animals and its role in alternative splicing. *Hum. Mol. Genet.* **11**, 1561–1567.

Li W-H, Gu Z, Wang H, Nekrutenko A (2001) Evolutionary analyses of the human genome. *Nature* **409**, 847–849.

Logsdon Jr JM (1998) The recent origins of spliceosomal introns revisited. *Curr. Opin. Genet. Dev.* **8**, 637–648.

Long M (2001) Evolution of novel genes. *Curr. Opin. Genet. Dev.* **11**, 673–680.

Lynch M, Conery JS (2000) The evolutionary fate and consequences of duplicate genes. *Science* **290**, 1151–1155.

Lynch M, Richardson AO (2002) The evolution of spliceosomal introns. *Curr. Opin. Genet. Dev.* **12**, 701–710.

Makalowski W, Boguski MS (1998) Evolutionary parameters of the transcribed mammalian genome: an analysis of 2,820 orthologous rodent and human sequences. *Proc. Natl Acad. Sci. USA* **95**, 9407–9412.

McLysaght A, Hokamp K, Wolfe KH (2002) Extensive genome duplication during early chordate evolution. *Nature Genet.* **31**, 200–204.

Moran JV, DeBerardinis RJ, Kazazian HH Jr (1999) Exon shuffling by L1 retrotransposition. *Science* **283**, 1530–1504.

Mouse Genome Sequencing Consortium (2002) Initial sequencing and comparative analysis of the mouse genome. *Nature* **420**, 520–562.

Muller S, Wienberg J (2001) Bar-coding primate chromosomes: molecular cytogenetic screening for the ancestral hominoid karyotype. *Hum. Genet.* **109**, 85–94.

Needleman SB, Wunsch CD (1970) A general method applicable to the search of similarities in the amino acid sequences of two proteins. *J. Mol. Biol.* **48**, 443–453.

O'Brien SJ, Menotti-Raymond M, Murphy WJ *et al.* (1999) The promise of comparative genomics in mammals. *Science* **286**, 479–481.

Olson MV, Varki A (2003) Sequencing the chimpanzee genome: insights into human evolution and disease. *Nature Rev. Genet.* **4**, 20–28.

Otto SP, Whitton J (2000) Polyploid incidence and evolution. *Annu. Rev. Genet.* **34**, 401–437.

Paabo S (2003) The mosaic that is our genome. *Nature* **421**, 409–412.

Patthy L (1999) Genome evolution and the evolution of exon-shuffling – a review. *Gene* **8**, 103–114.

Perry J, Palmer S, Gabriel A, Ashworth A (2001) A short pseudoautosomal region in laboratory mice. *Genome Res.* **11**, 1826–1832.

Pesce A, Bolognesi M, Bocedi A (2002) Neuroglobin and cytoglobin. *EMBO Reports* **3**, 1146–1151.

Ried K, Rao E, Schiebel K, Rappold GA (1998) Gene duplications as a recurrent theme in the evolution of the human pseudoautosomal region 1: isolation of the gene *ASMTL. Hum. Mol. Genet.* **7**, 1771–1778.

Rosenberg NA, Pritchard JK, Weber JL (2002) Genetic structure of human populations. *Science* **298**, 2381–2385.

Samonte RV, Eichler E (2001) Segmental duplications and the evolution of the primate genome. *Nature Rev. Genet.* **3**, 65–72.

Satta Y, Klein J, Takahata N (2000) DNA archives and our nearest relative: the trichotomy problem revisited. *Mol. Phylogen. Evol.* **14**, 259–275.

Schwartz S, Kent WJ, Smit A (2003) Human-mouse alignments with BLASTZ. *Genome Res.* **13**, 103–107.

Stead JDH, Jeffreys AJ (2002) Structural analysis of insulin minisatellite alleles reveals unusually large differences in diversity between Africans and non-Africans. *Am. J. Hum. Genet.* **71**, 1273–1284.

Stringer C (2002) Modern human origins: progress and prospects. *Proc. Royal Soc. Lond.* B. **357**, 563–579.

Swanson WJ, Yang Z, Wolfner MF, Aquadro CF (2001) Positive Darwinian selection drives the evolution of several female reproductive proteins in mammals. *Proc. Natl Acad. Sci. USA* **98**, 2509–2514.

Tagle DA, Stanhope MJ, Siemieniak DR, Benson P, Goodman M, Slightom JL (1992) The β globin gene cluster of the prosimian primate *Galago crassicaudatus*: nucleotide sequence determination of the 41 kb cluster and comparative sequence analyses. *Genomics* **13**, 741–760.

Templeton AR (2002) Out of Africa again and again. *Nature* **416**, 45–51.

Thomas JW, Schueler MG, Summers TJ *et al.* (2003) Pericentromeric duplications in the laboratory mouse. *Genome Res.* **13**, 55–63.

Ureta-Vidal A, Ettwiller L, Birney E. (2003) Comparative genomics: genome-wide analysis in metazoan eukaryotes. *Nature Rev. Genet.* **4**, 251–262.

Venter JC, Adams MD, Myers EW *et al.* (2001) The sequence of the human genome. *Science* **291**, 1304–1351.

Wang PJ, McCarrey JR, Yang F, Page DC (2001) An abundance of X-linked genes expressed in spermatogonia. *Nature Genet.* **27**, 422–426.

Waterman MS, Smith TF, Beyer WA (1976) Some biological sequence metrics. *Adv. Math.* **20**, 367–387.

Wolfe KH (2001) Yesterday's polyploids and the mystery of diploidization. *Nature Rev. Genet.* **2**, 333–341.

Young JM, Friedman C, Williams EM, Ross JA, Tonnes-Priddy L, Trask BJ (2002) Different evolutionary processes shaped the mouse and human olfactory receptor gene families. *Hum. Mol. Genet.* **11**, 535–546.

Yunis JJ, Prakash O (1982) The origin of man: a chromosomal pictorial legacy. *Science* **215**, 1525–1530.

PART THREE

Mapping and identifying disease genes and mutations

CHAPTER THIRTEEN

Genetic mapping of Mendelian characters

Chapter contents

13.1 Recombinants and nonrecombinants

In principle, genetic mapping in humans is exactly the same as genetic mapping in any other sexually reproducing diploid organism. The aim is to discover how often two loci are separated by meiotic recombination. Consider a person who is heterozygous at two loci (genotype $A_1A_2 B_1B_2$). Suppose the alleles A_1 and B_1 in this person came from one parent, and A_2 and B_2 from the other. Any of that person's gametes that carries one of these parental combinations (A_1B_1 or A_2B_2) is nonrecombinant for those two loci, whereas gametes that carry A_1B_2 or A_2B_1 are recombinant (*Figure 13.1*). The proportion of gametes that are recombinant is the recombination fraction between the two loci A and B.

13.1.1 The recombination fraction is a measure of genetic distance

If two loci are on different chromosomes, they will segregate independently. Considering spermatogenesis in individual II_1 in *Figure 13.1*, at the end of meiosis I, whichever sperm receives allele A_1, there is a 50% chance that it will receive allele B_1 and a 50% chance it will receive B_2. Thus, on average, 50% of the gametes will be recombinant and 50% nonrecombinant. The recombination fraction is 0.5. If the loci are **syntenic**, that is if they lie on the same chromosome, then they might be expected always to segregate together, with no recombinants. However, this simple expectation ignores meiotic crossing-over. During prophase of meiosis I, pairs of homologous chromosomes synapse and exchange segments (*Figure 2.11*). Only two of the four chromatids are involved in any particular crossover. A crossover that occurs between the positions of the two loci will create two recombinant chromatids carrying A_1B_2 and A_2B_1, and leave the two noninvolved chromatids nonrecombinant. Thus one crossover generates 50% recombinants between loci flanking it.

Recombination will rarely separate loci that lie very close together on a chromosome, because only a crossover located precisely in the small space between the two loci will create recombinants. Therefore sets of alleles on the same small chromosomal segment tend to be transmitted as a block through a pedigree. Such a block of alleles is known as a **haplotype**. Haplotypes mark recognizable chromosomal segments that can be tracked through pedigrees and through populations when not broken up by recombination. *Figure 13.9* shows examples.

The further apart two loci are on a chromosome, the more likely it is that a crossover will separate them. Thus the recombination fraction is a measure of the distance between two loci. Recombination fractions define **genetic distance**, which is not the same as **physical distance**. Two loci that show 1% recombination are defined as being 1 **centimorgan (cM)** apart on a genetic map.

13.1.2 Recombination fractions do not exceed 0.5 however great the physical distance

A single recombination event produces two recombinant and two nonrecombinant chromatids. When loci are well separated there may be more than one crossover between them. Double crossovers can involve two, three or four chromatids, but *Figure 13.2* shows that the overall effect, averaged over all double crossovers, is to give 50% recombinants. Loci very far apart on the same chromosome might be separated by three or more crossovers. Again, the overall effect is to give 50% recombinants. Recombination fractions never exceed 0.5, however far apart the loci are.

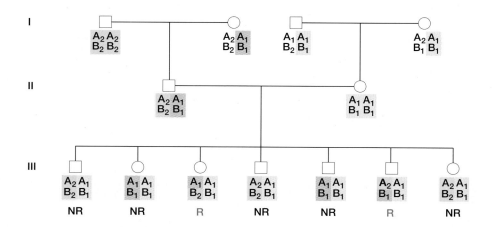

Figure 13.1: Recombinants and nonrecombinants.

Alleles at two loci (locus A, alleles A_1 and A_2; locus B, alleles B_1 and B_2) are segregating in this family. Colored boxes mark combinations of alleles that can be traced through the pedigree. In generation III we can distinguish people who received nonrecombinant (A_1B_1 or A_2B_2) or recombinant (A_1B_2 or A_2B_1) sperm from their father. Because individual II_2 is homozygous at these two loci we cannot identify which individuals in generation III developed from nonrecombinant or recombinant oocytes.

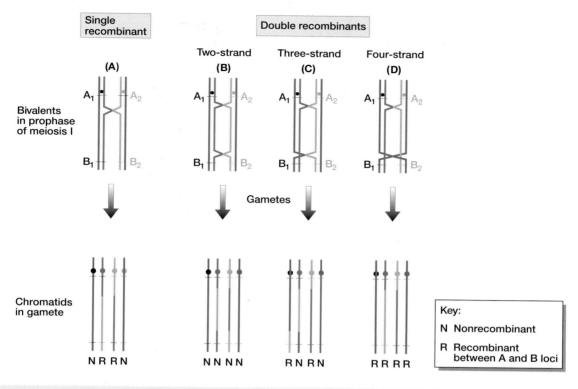

Figure 13.2: Single and double recombinants.

Each crossover involves two of the four chromatids of the two synapsed homologous chromosomes. One chromosome carries alleles A_1 and B_1 at two loci, the other carries alleles A_2 and B_2. **(A)** A single crossover generates two recombinant and two nonrecombinant chromatids (50% recombinants). **(B)** The three types of double crossover occur in random proportions, so the average effect of a double crossover is to give 50% recombinants.

13.1.3 Mapping functions define the relationship between recombination fraction and genetic distance

Because recombination fractions never exceed 0.5, they are not simply additive across a genetic map. If a series of loci, A, B, C, ... are located at 5-cM intervals on a map, locus M may be 60 cM from locus A, but the recombination fraction between A and M will not be 60%. The mathematical relationship between recombination fraction and genetic map distance is described by the **mapping function**. If crossovers occurred at random along a bivalent and had no influence on one another, the appropriate mapping function would be *Haldane's function*:

$$w = -\tfrac{1}{2} \ln (1 - 2\theta)$$
or
$$\theta = \tfrac{1}{2} [1 - \exp (-2w)]$$

where w is the map distance and θ the recombination fraction; as usual ln means logarithm to the base e, and exp means 'e to the power of'. However, we know that the presence of one chiasma inhibits formation of a second chiasma nearby. This phenomenon is called **interference**. A variety of mapping functions exist that allow for varying degrees of interference. A widely used function for human mapping is *Kosambi's function*:

$$w = \tfrac{1}{4} \ln [(1+2\theta) / (1-2\theta)]$$
or
$$\theta = \tfrac{1}{2} [\exp (4w) -1] / [\exp (4w) + 1]$$

A mapping function is needed in multipoint mapping (*Section 13.4*) to convert the raw data on the recombination fraction into a genetic map. Broman and Weber (2000) used a very large set of linkage data to estimate the best mapping function for humans. They estimate that the probability of a true double crossover between markers d cM apart is roughly $(0.0114 d - 0.0154)^4$. For d = 10 cM this works out at only 0.01%. The interested reader should consult Ott's book (see Further Reading) for a fuller discussion of mapping functions.

13.1.4 Chiasma counts and total map length

Each crossover during meiosis produces two recombinant and two nonrecombinant chromatids, giving 50% recombination between flanking markers. One crossover therefore contributes 50 cM to the overall genetic map length. By counting **chiasmata** under the microscope and multiplying the number per cell by 50, we can estimate the total map length in cM. Male meiosis can be studied in testicular biopsies; the average chiasma count is 50.6 per cell (Hultén and Lindsten, 1973; *Figure 13.3A*), giving a map length of 2530 cM. Female meiosis I in humans takes place at 16–24 weeks of fetal life and is very difficult to observe, but a new technique has recently made chiasma counts possible (Tease *et al.*, 2002). Fluorescent antibodies are used to mark the locations of a protein, MLH1, that is part of the recombination machinery (*Figure 13.3B*). Chiasmata are more frequent in female meiosis (exemplifying Haldane's rule that the heterogametic sex has the lower chiasma count). The average autosomal count in ovaries from one aborted female fetus was 70.3, giving a map length of 3515 cM. These cytologically derived map lengths can be compared with the best estimates from genetic mapping of 2590 (male) and 4281 cM (female) (Kong *et al.*, 2002). Uniquely, the Y chromosome, outside the pseudoautosomal region (Section 2.3.3), has no genetic map because it is not subject to synapsis and crossing over in normal meiosis.

13.1.5 Physical vs. genetic maps: the distribution of recombinants

Physical maps show the order of features along the chromosome and their distance in kilobases or megabases; genetic maps show their order and the probability that they will be separated by recombination. While the order of features should be the same on both maps, the distances will only correspond if the probability of recombination per megabase of DNA is constant for all chromosomal locations. In fact, recombination probabilities vary considerably according to sex and chromosomal location. Individuals may vary in the average number of crossovers per meiosis, and if they do, then they will also have different genetic map lengths.

Now that we have the human genome sequence, microsatellite or SNP markers can be physically located in the sequence database by searching for matches to the PCR primers used. Thus the distribution of recombination along a chromosome can be estimated either directly under the microscope (*Figure 13.3*) or by relating genetic map distances to physical separations of markers. Both methods show that *recombination is not random*. There is more recombination towards the telomeres of chromosomes in males, while centromeric regions have recombinants in females but not in males (see *Figure 13.4* and Kong *et al.*, 2002). As mentioned above, interference affects the spacing of double recombinants. As a rough rule of thumb, 1 cM = 1 Mb, but there are recombination 'deserts' up to 5 Mb in length with sex-averaged recombination less than 0.3 cM per Mb, and recombination 'jungles' with > 3 cM per Mb. The most extreme deviation is shown by the pseudoautosomal region at the tip of the short arms of the X and Y chromosomes (see Section 12.2.7). Males have an *obligatory* crossover within this 2.6 Mb region, so that it is 50 cM long. Thus, for this region in males 1 Mb = 19 cM, whereas in females 1 Mb = 2.7 cM.

Recent work also suggests that *recombination is nonrandom at the DNA sequence level*. As described in Section 15.4.3, our chromosomes seem to consist of conserved blocks, typically 20–50 kb long, separated by recombination hotspots. About 95% of all recombination occurs in these 1–2-kb hotspots (*Figure 13.5*; Jeffreys *et al.*, 2001). Moreover, detailed examination of a few such hotspots by A. J. Jeffreys has suggested that

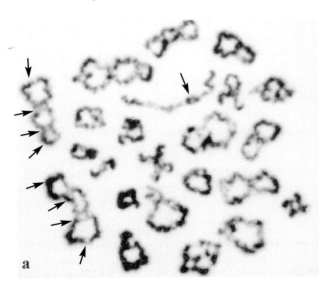

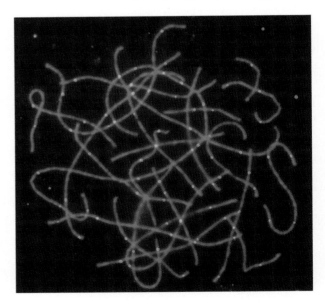

Figure 13.3: Crossovers in male and female meiosis.

(A) Male meiosis: a spermatocyte at metaphase I. *Note* the end-to-end pairing of the X and Y chromosomes. Chiasmata mark the positions of crossovers within each bivalent (arrows). **(B)** Female meiosis at pachytene. The bright dots of fluorescent MLH1 antibody mark positions of crossovers. Photographs courtesy of Professor Maj Hultén, Birmingham. From Hultén and Tease (2003) with permission of Nature Publishing Group.

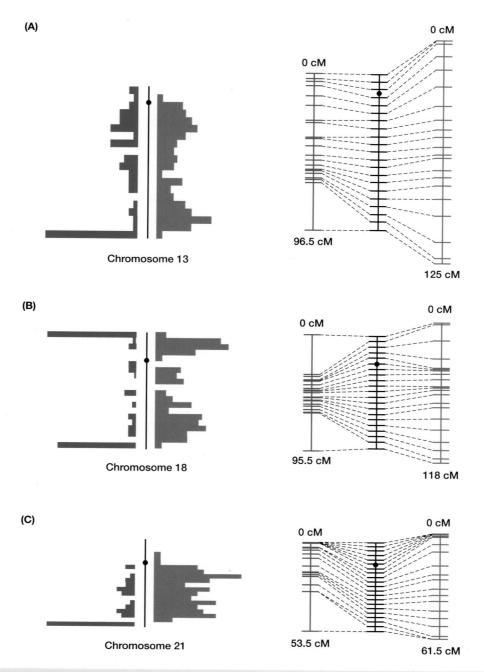

Figure 13.4: The distribution of recombinants is nonrandom and sex specific.

Histograms illustrating the distributions of chiasmata in spermatocytes (blue) and MLH1 foci in oocytes (red) on chromosomes 13, 18 and 21. Each chromosome pair is divided into 5% length intervals. These crossover distribution patterns are also displayed for each chromosome as recombination maps. These maps highlight the different patterns of crossover numbers and distributions in male and female germ cells and the consequent effect on the recombination map. From Hultén and Tease (2003) with permission from Nature Publishing Group.

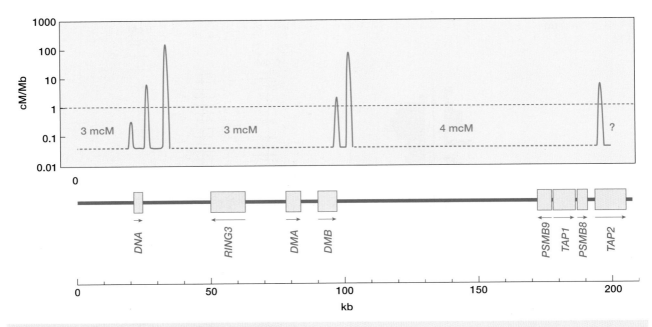

Figure 13.5: Recombination is concentrated in small hotspots.

DNA-level distribution of recombination in males in a region of the major histocompatibility complex. Ninety-five percent of all recombination occurs at highly localized hotspots. Redrawn from Jeffreys *et al.* (2001) *Nat. Genet.* **29**, 217–222, with permission from Nature Publishing Group.

recombination always initiates within 10 bp of the same point. Tantalizingly, the DNA sequence at these points does not have any obvious special feature to explain this behavior.

13.2 Genetic markers

13.2.1 Mapping human disease genes requires genetic markers

Since most human geneticists are interested in diseases, we would like a map to show the order and distance apart of all disease genes. Scoring the recombination fraction between pairs of diseases would be the obvious way to construct such a map – but *disease–disease mapping* is not possible in humans. Genetic mapping, as we have seen (*Figure 13.1*) requires double heterozygotes. People heterozygous for two different diseases are extremely rare. Even if they can be found, they will probably have no children, or be unsuitable for genetic analysis in some other way. For this reason human genetic mapping depends on **markers**. *A marker is any polymorphic Mendelian character that can be used to follow a chromosomal segment through a pedigree.* It helps if the marker can be scored easily and cheaply using readily available material (blood cells rather than a brain biopsy), but the crucial thing is that it should be sufficiently polymorphic that a randomly selected person has a good chance of being heterozygous. *Box 13.1* summarizes the development of human genetic markers, from blood groups through to the present generation of DNA microsatellites and single nucleotide polymorphisms.

Disease-marker mapping, if it is not to be a purely blind exercise, requires *framework maps* of markers. These are generated by *marker–marker mapping*. Although in theory linkage can be detected between loci 40 cM apart, the amount of data required to do this is prohibitive. Ten meioses are sufficient to give evidence of linkage if there are no recombinants, but 85 meioses would be needed to give equally strong evidence of linkage if the recombination fraction was 0.3 (see *Box 13.3* for a guide to these calculations). Obtaining enough family material to test more than 20–30 meioses can be seriously difficult for a rare disease. Thus mapping requires markers spaced at intervals no greater than 10–20 cM across the genome. Given the genome lengths calculated above, and allowing for imperfect informativeness (see below), we need a minimum of several hundred markers. A major achievement in the early stages of the Human Genome Project was to generate upwards of 10 000 highly polymorphic microsatellite markers and place them on framework maps (Broman *et al.*, 1998). These maps made possible the spectacular progress of the 1990s in mapping Mendelian diseases. The newer *disease association studies* (Section 15.4) require very much denser marker maps, and this need has been answered by the development of several million SNP markers.

13.2.2 The heterozygosity or polymorphism information content measure how informative a marker is

For linkage analysis we need **informative meioses** (see *Box 13.2*). The examples in the box show that a meiosis is not informative with a given marker if the parent is homozygous for the marker, and also in half of the cases where both parents have the same heterozygous genotype. For most purposes the mean **heterozygosity** of a marker (the chance that a

Box 13.1: The development of human genetic markers.

Type of marker	No. of loci	Features
Blood groups 1910–1960	~ 20	May need fresh blood, rare antisera Genotype cannot always be inferred from phenotype because of dominance No easy physical localization
Electrophoretic mobility variants of serum proteins 1960–1975	~ 30	May need fresh serum, specialized assays No easy physical localization Often limited polymorphism
HLA tissue types 1970–	1 (haplotype)	One linked set Highly informative Can only test for linkage to 6p21.3
DNA RFLPs (*Figures 7.5A, 7.6*) 1975–	> 10^5 (potentially)	Two allele markers, maximum heterozygosity 0.5 Initially required Southern blotting, now PCR Easy physical localization
DNA VNTRs (minisatellites) (*Figure 7.5B*) 1985–	> 10^4 (potentially)	Many alleles, highly informative Type by Southern blotting Easy physical localization Tend to cluster near ends of chromosomes
DNA VNTRs (microsatellites) (*Figure 7.7*) (di-, tri- and tetra-nucleotide repeats) 1989–	> 10^5 (potentially)	Many alleles, highly informative Can type by automated multiplex PCR Easy physical localization Distributed throughout genome
DNA SNPs (single nucleotide polymorphisms)	> 4×10^6	Less informative than microsatellites Can be typed on a very large scale by automated equipment without gel electrophoresis

VNTR, variable number of tandem repeats.

randomly selected person will be heterozygous) is used as the measure of informativeness. If there are marker alleles $A_1, A_2, A_3...$ with gene frequencies $p_1, p_2, p_3...$, then the proportion of people who are heterozygous is $1 - (p_1^2 + p_2^2 + p_3^2+...)$ (Section 4.5.1). A more sophisticated but seldom used measure, the **polymorphism information content (PIC)**, allows for cases in which someone is heterozygous but uninformative, like pedigree B in *Box 13.2*.

13.2.3 DNA polymorphisms are the basis of all current genetic markers

In the early 1980s **DNA polymorphisms** provided, for the first time, a set of markers that was sufficiently numerous and adequately spaced across the entire genome to allow a whole-genome search for linkage. DNA markers have the additional advantage that they can all be typed by the same technique. Moreover their chromosomal location can be determined using radiation hybrid mapping (*Box 8.4*) or by searching the human genome sequence for a match to the PCR primer used to score the marker. This allows DNA-based genetic maps to be cross-referenced to physical maps, and avoids the

frustrating situation that arose when the long-sought cystic fibrosis gene (*CFTR*) was first mapped. Linkage was established to a protein polymorphism of the enzyme paraoxonase, but the chromosomal location of the paraoxonase gene was not known. The development of DNA markers allowed human gene mapping to start in earnest.

Restriction fragment length polymorphisms (RFLPs)

The first generation of DNA markers were **restriction fragment length polymorphisms (RFLPs)**. RFLPs were initially typed by preparing Southern blots from restriction digests of the test DNA, and hybridizing with radiolabeled probes (see *Figure 7.5*). This technology required plenty of time, money and DNA, and made a whole genome search a heroic undertaking. Nowadays this is less of a problem because RFLPs can usually be typed by PCR. A sequence including the variable restriction site is amplified, the product is incubated with the appropriate restriction enzyme and then run out on a gel to see if it has been cut (*Figure 7.6*). A more fundamental disadvantage is their *low informativeness*. RFLPs have only two alleles: the site is present or it is absent. The maximum heterozygosity is 0.5. Disease mapping using

Box 13.2: Informative and uninformative meioses.

A meiosis is informative for linkage when we can identify whether or not the gamete is recombinant. Consider the male meiosis which produced the paternal contribution to the child in the four pedigrees below. The father has a dominant condition that he inherited along with marker allele A_1. He passes this condition to his daughter – but can we tell whether or not the sperm was recombinant between the gene for the condition and the marker locus?

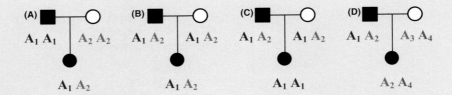

(A) This meiosis is uninformative: the marker alleles in the homozygous father cannot be distinguished. **(B)** This meiosis is uninformative: the child could have inherited A_1 from father and A_2 from mother, or vice versa. **(C)** This meiosis is informative and nonrecombinant: the child inherited A_1 from the father. **(D)** This meiosis is informative and recombinant: the child inherited A_2 from the father.

RFLPs is frustrating because all too often a key meiosis in a family turns out to be uninformative.

Minisatellites

Minisatellite VNTR (variable number tandem repeat) markers were a great improvement. The VNTRs (Section 7.1.3) have many alleles and high heterozygosity. Most meioses are informative. However, the technical problems of Southern blotting and radioactive probes were still an obstacle to easy mapping, and VNTRs are not evenly spread across the genome.

Microsatellites

The advent of PCR finally made mapping relatively quick and easy. Minisatellites are too long to amplify well, and so the standard tools for PCR linkage analysis are **microsatellites**. These are mostly (CA)n repeats. Tri- and tetranucleotide repeats are gradually replacing dinucleotide repeats as the markers of choice because they give cleaner results. Dinucleotide repeat sequences are peculiarly prone to replication slippage during PCR amplification (Section 11.3.1) so that each allele gives a little ladder of 'stutter bands' on a gel, making it hard to read (see *Figure 7.8*). Much effort has been devoted to producing compatible sets of microsatellite markers that can be amplified together in a multiplex PCR reaction to give nonoverlapping allele sizes, so that they can be run in the same gel lane. With fluorescent labeling in several colors, it is possible to score perhaps 10 markers on a sample in a single lane of an automated gel.

Single nucleotide polymorphisms (SNPs)

After 10 years of developing more and more polymorphic markers, it may seem perverse that the newest generation of markers are two-allele **single nucleotide polymorphisms**.

They include the classic RFLPs, but also polymorphisms that do not happen to create or abolish a restriction site. The advantage of SNPs is that they allow ultra-high throughput genotyping and very high marker densities (Wang *et al.*, 1998). Searches for disease susceptibility genes (Chapter 15) require vast numbers of genotypes to be scored with very closely spaced markers. There are not enough microsatellites in the genome to give the required density of one marker per 10 kb or less. Moreover, microsatellites are scored by gel electrophoresis, and this limits throughput. An international consortium of academic and industrial partners has generated a public database of more than 4 million SNPs (dbSNP, accessible through the NCBI website), and methods have been developed for scoring SNPs without gel electrophoresis on an extremely large scale (Section 18.4.2).

13.3 Two-point mapping

13.3.1 Scoring recombinants in human pedigrees is not always simple

Having collected families where a Mendelian disease is segregating, and typed them with an informative marker, how do we know when we have found linkage? There are two aspects to this question:

1. How can we work out the recombination fraction?

2. What statistical test should we use to see if the recombination fraction is significantly different from 0.5, the value expected on the null hypothesis of no linkage?

In some families the first question can be answered very simply by counting recombinants and nonrecombinants. The family shown in *Figure 13.1* is one example. There are two recombinants in seven meioses and the recombination

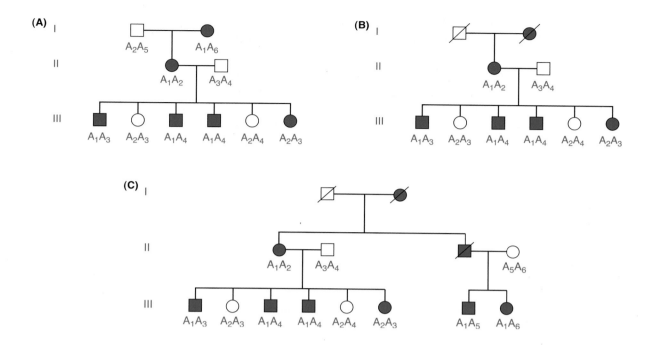

Figure 13.6: Recognizing recombinants.

Three versions of a family with an autosomal dominant disease, typed for a marker A. **(A)** All meioses are phase-known. We can identify III_1–III_5 unambiguously as nonrecombinant and III_6 as recombinant. **(B)** The same family, but phase-unknown. The mother, II_1, could have inherited either marker allele A_1 or A_2 with the disease; thus her phase is unknown. *Either* III_1–III_5 are nonrecombinant and III_6 is recombinant; *or* III_1–III_5 are recombinant and III_6 is nonrecombinant. **(C)** The same family after further tracing of relatives. III_7 and III_8 have also inherited the disease from their father – but we cannot be sure whether their father's allele A_1 is identical by descent to the allele A_1 in his sister II_1. Maybe there are two copies of allele A_1 among the four grandparental marker alleles. The likelihood of this depends on the gene frequency of allele A_1. Thus although this pedigree contains linkage information, extracting it is problematic.

fraction is 0.28. *Figure 13.6A* shows another example. The double heterozygote who is informative for linkage (individual II_1 in both *Figure 13.1* and *Figure 13.6A*), is **phase-known**: we know which alleles were inherited from which parent, and so we can unambiguously score each meiosis as recombinant or nonrecombinant. In *Figure 13.6B*, individual II_1 is again doubly heterozygous, but this time **phase-unknown**. Among her children, either there are five non-recombinants and one recombinant, or else there are five recombinants and one nonrecombinant. We can no longer identify recombinants unambiguously, even if the first alternative seems much more likely than the second. *Figure 13.6C* adds still more complications – yet if this is a family with a rare disease, no researcher would be willing to discard it. Some method is needed to extract the linkage information from a collection of such imperfect families.

13.3.2 Computerized lod score analysis is the best way to analyze complex pedigrees for linkage between Mendelian characters

In the pedigree shown in *Figure 13.6B* it is not possible to identify recombinants unambiguously and count them. It is possible, however, to calculate the overall likelihood of the

pedigree, on the alternative assumptions that the loci are linked (recombination fraction = θ) or not linked (recombination fraction = 0.5). The ratio of these two likelihoods gives the odds of linkage, and the logarithm of the odds is the **lod score**. Morton (1955) demonstrated that lod scores represent the most efficient statistic for evaluating pedigrees for linkage, and derived formulae to give the lod score (as a function of θ) for various standard pedigree structures. *Box 13.3* shows how this is done for simple structures. Being a function of the recombination fraction, lod scores are calculated for a range of θ values. The most likely recombination fraction is the value at which the lod score is maximum. In a set of families, the overall probability of linkage is the product of the probabilities in each individual family, therefore lod scores (being logarithms) can be added up across families.

Calculating the full lod score for the family in *Figure 13.6C* is difficult. To calculate the likelihood that III_7 and III_8 are recombinant or nonrecombinant, we must take likelihoods calculated for each possible genotype of I_1, I_2 and II_3, weighted by the probability of that genotype. For I_1 and I_2, the genotype probabilities depend on both the gene frequencies and the observed genotypes of II_1, III_7 and III_8. Genotype probabilities for II_3 are then calculated by simple Mendelian rules. Human linkage analysis, except in the very simplest

> **Box 13.3: Calculation of lod scores for the families in Figure 13.6.**

▶ Given that the loci are truly linked, with recombination fraction θ, the likelihood of a meiosis being recombinant is θ and the likelihood of it being nonrecombinant is $1-\theta$.

▶ If the loci are in fact unlinked, the likelihood of a meiosis being either recombinant or nonrecombinant is ½.

Family A:

There are five nonrecombinants and one recombinant.

The overall likelihood, given linkage, is $(1-\theta)^5\cdot\theta$.

The likelihood, given no linkage, is $(½)^6$.

The likelihood ratio is $(1-\theta)^5\cdot\theta/(½)^6$.

The lod score, Z, is the logarithm of the likelihood ratio.

θ	0	0.1	0.2	0.3	0.4	0.5
Z	$-$ infinity	0.577	0.623	0.509	0.299	0

Family B:

II$_1$ is phase-unknown.

If she inherited A$_1$ with the disease, there are five nonrecombinants and one recombinant.

If she inherited A$_2$ with the disease, there are five recombinants and one nonrecombinant.

The overall likelihood is $½ [(1-\theta)^5\cdot\theta/(½)^6] + ½[(1-\theta)\cdot\theta^5/(½)^6]$. This allows for either possible phase, with equal prior probability.

The lod score, Z, is the logarithm of the likelihood ratio.

θ	0	0.1	0.2	0.3	0.4	0.5
Z	$-$ infinity	0.276	0.323	0.222	0.076	0

Family C:

At this point nonmasochists turn to the computer.

---

cases, is entirely dependent on computer programs that implement algorithms for handling these branching trees of genotype probabilities, given the pedigree data and a table of gene frequencies.

13.3.3 Lod scores of +3 and −2 are the criteria for linkage and exclusion (for a single test)

The result of linkage analysis is a table of lod scores at various recombination fractions, like the two tables in *Box 13.3*. Positive lods give evidence in favor of linkage and negative lods give evidence against linkage. Note that only recombination fractions between 0 and 0.5 are meaningful, and that all lod scores are zero at $\theta = 0.5$ (because they are then measuring the ratio of two identical probabilities, and $\log_{10} (1) = 0$). The results can be plotted to give curves like those in *Figure 13.7*.

Returning to the two questions posed at the start of this section, we now see that *the most likely recombination fraction is the one at which the lod score is highest*. If there are no recombinants, the lod score will be maximum at $\theta = 0$. If there are recombinants, Z will peak at the most likely recombination fraction ($0.167 = 1/6$ for the family in *Figure 13.6A*, but harder to predict for *Figure 13.6B*).

The second question concerned the threshold of significance. Here the answer is at first sight surprising. *Z = 3.0 is the threshold for accepting linkage*, with a 5% chance of error. Linkage can be rejected if $Z < -2.0$. Values of Z between −2 and +3 are inconclusive. For most statistics $p < 0.05$ is used as the threshold of significance, but Z = 3.0 corresponds to 1000 : 1 odds [$\log_{10}(1000) = 3.0$]. The reason why such a stringent threshold is chosen lies in the inherent improbability that two loci, chosen at random, should be linked. With 22 pairs of autosomes to choose from, it is not likely they would be located on the same chromosome (syntenic) and, even if they were, loci well separated on a chromosome are unlinked. Common sense tells us that if something is inherently improbable, we require strong evidence to convince us that it is true. This common sense can be quantified in a **Bayesian**

calculation (see *Box 13.4*), which shows that 1000 : 1 odds in fact corresponds precisely to the conventional $p = 0.05$ threshold of significance. The same logic suggests a threshold lod of 2.3 for establishing linkage between an X-linked character and an X-chromosome marker (prior probability of linkage $\cong 1/10$).

Confidence intervals are hard to deduce analytically, but a widely accepted support interval extends to recombination fractions at which the lod score is 1 unit below the peak value (the '*lod-1 rule*'). Thus, curve 2 in *Figure 13.7* gives acceptable evidence of linkage ($Z > 3$) with the most likely recombination fraction 0.23 and support interval 0.17–0.32. The curve will be more sharply peaked the greater the amount of data, but in general peaks are quite broad. It is important to remember that distances on human genetic maps are often very imprecise estimates.

Negative lod scores exclude linkage for the region where $Z < -2$. Curve 3 on *Figure 13.7* excludes the disease from 12 cM either side of the marker. While gene mappers hope for a positive lod score, exclusions are not without value. They tell us where the disease is not (**exclusion mapping**). This can exclude a possible candidate gene, and if enough of the genome is excluded, only a few possible locations may remain.

13.3.4 For whole genome searches a genome-wide threshold of significance must be used

In disease studies, families are typed for marker after marker until positive lods are obtained. The appropriate threshold for significance is a lod score such that there is only a 0.05 chance of a false-positive result occurring *anywhere* during a search of the whole genome. As shown in *Box 13.4*, a lod score of 3.0 corresponds to a significance of 0.05 at a single point. But if 50 markers have been used, the chance of a spurious positive result is greater than if only one marker is used. A stringent procedure (Bonferroni correction) would multiply the p value by 50 before testing its significance. The threshold lod score for a study using n markers would be $3 + \log(n)$, that is a lod

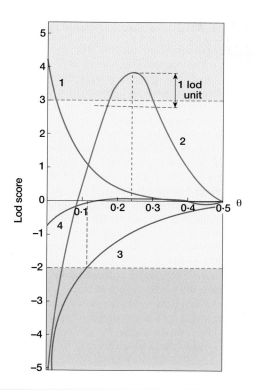

Figure 13.7: Lod score curves.

Graphs of lod score against recombination fraction from a hypothetical set of linkage experiments. Curve 1, evidence of linkage ($Z > 3$) with no recombinants. Curve 2, evidence of linkage ($Z > 3$) with the most likely recombination fraction being 0.23. Curve 3, linkage excluded ($Z < -2$) for recombination fractions below 0.12; inconclusive for larger recombination fractions. Curve 4, inconclusive at all recombination fractions.

score of 4 for 10 markers, 5 for 100, and so on. However, this is over stringent. Linkage data are not independent: if one location is excluded, then the prior probability that the character maps to another location is raised. The threshold for a

genome-wide significance level of 0.05 has been much argued over, but a widely accepted answer for Mendelian characters is 3.3 (Lander and Schork, 1994). For non-Mendelian characters see Section 15.3.4. In practice, lod scores below 5, whether with one marker or many, should be regarded as provisional.

13.4 Multipoint mapping is more efficient than two-point mapping

13.4.1 Multipoint linkage can locate a disease locus on a framework of markers

Linkage analysis can be more efficient if data for more than two loci are analyzed simultaneously. *Multilocus analysis* is particularly useful for establishing the chromosomal order of a set of linked loci. Experimental geneticists have long used *three-point crosses* for this purpose. The rarest recombinant class is that which requires a double recombination. In *Table 13.1*, the gene order A–C–B is immediately apparent. This procedure is more efficient than estimating the recombination fractions for intervals A–B, A–C and B–C separately in a series of two-point crosses. Ideally, in any linkage analysis the whole genome would be screened for linkage, and the full dataset would be used to calculate the likelihood at each location across the genome.

A second advantage of multilocus mapping in humans is that it helps overcome problems caused by the limited informativeness of markers. Some meioses in a family might be informative with marker A, and others uninformative for A but informative with the nearby marker B. Only simultaneous linkage analysis of the disease with markers A and B extracts the full information. This is less important for mapping using highly informative microsatellite markers rather than two-allele RFLPs, but it resurfaces when SNPs are used.

13.4.2 Marker framework maps: the CEPH families

The power of multipoint mapping to order loci is particularly useful for constructing *marker framework maps*. However,

Box 13.4: Bayesian calculation of linkage threshold.

The likelihood that two loci should be linked (the prior probability of linkage) has been argued over, but estimates of about 1 in 50 are widely accepted.

Hypothesis	Loci are linked (recombination fraction = θ)	Loci are not linked (recombination fraction = 0.5)
Prior probability	1/50	49/50
Conditional probability: 1000 : 1 odds of linkage (lod score $Z(\theta) = 3.0$)	1000	1
Joint probability (prior × conditional)	20	~ 1

Because of the low *prior probability* that two randomly chosen loci should be linked, evidence giving 1000 : 1 odds in favor of linkage is required in order to give overall 20 : 1 odds in favor of linkage. This corresponds to the conventional $p = 0.05$ threshold of statistical significance. The calculation is an example of the use of Bayes' formula to combine probabilities (see *Box 18.4* and *Figure 18.15*). See text for description of the lod score.

Table 13.1: Gene ordering by three-point crosses

Class of offspring	Position of recombination (x)	Number
ABC/abc abc/abc	Nonrecombinant	853
ABc/abc abC/abc	(A, B)–x–C	5
Abc/abc aBC/abc	A–x–(B, c)	47
AbC/abc aBc/abc	B–x–(A, C)	95

A cross has been set up between mice heterozygous at three linked loci (*ABC/abc*) and triple homozygotes (*abc/abc*). The rarest class of offspring will be those whose production requires two crossovers. Of the 1000 animals, 142 (95 + 47) are recombinant between A and B, 52 (47 + 5) between A and C, and 100 (95 + 5) between B and C. Only five animals are recombinant between A and C but not between A and B, so these must have double crossovers, A—x—C—x—B. Therefore the map order is A–C–B and the genetic distances are approximately A–(5 cM)–C–(10 cM)–B.

ordering the loci in such maps is not a trivial problem. There are *n*!/2 possible orders for *n* markers, and current maps have hundreds of markers per chromosome. Before the human genome sequence was completed, this was a serious difficulty for map makers. Something more intelligent than brute force computing had to be used to work out the correct order. Even without large-scale sequence data, physical mapping information was immensely helpful. Markers that can be typed by PCR can be used as sequence-tagged sites (STS; *Box 15.4*) and physically localized, either by database searching or experimentally using radiation hybrids (*Box 8.4*). The result is a *physically anchored marker framework*.

Disease–marker mapping suffers from the necessity of using whatever families can be found where the disease of interest is segregating. Such families will rarely have ideal structures. All too often the number of meioses is undesirably small, and some are phase-unknown. Marker–marker mapping can avoid these problems. Markers can be studied in any family, so families can be chosen that have plenty of children and ideal structures for linkage, like the family in *Figure 13.1*. Construction of marker framework maps has benefited greatly from a collection of families (the **CEPH families**) assembled specifically for the purpose by the Centre d'Étude du Polymorphisme Humain (now the Fondation Jean Dausset) in Paris. Immortalized cell lines from every individual ensure a permanent supply of DNA, and sample mix-ups and nonpaternity have long since been ruled out by typing with many markers. As an example, the 1998 CHLC (Cooperative Human Linkage Center) map is based on the results of scoring eight CEPH families with 8325 microsatellites, resulting in over 1 million genotypes (Broman *et al.*, 1998).

13.4.3 Multipoint disease-marker mapping

For disease–marker mapping the starting point is the framework map of markers. This is taken as given, and the aim is to locate the disease gene in one of the intervals of the framework. Programs such as *Linkmap* (part of the Linkage package) or *Genehunter* (Section 13.6.2) can notch the disease locus across the marker framework, calculating the overall likelihood of the pedigree data at each position. The result (*Figure 13.8*) is a curve of lod score against map location. This method is also useful for exclusion mapping: if the curve stays below a lod score of –2 across the region, then the disease locus is excluded from that region.

The apparently quantitative nature of *Figure 13.8* is largely spurious. Peak heights depend crucially on the precise genetic distances between markers, which are usually only known very roughly. Moreover none of the mapping functions (Section 13.1.3) in linkage programs even approximates to the real complexities of chiasma distribution (*Figure 13.4*). However, unless the marker map is radically wrong, it remains true that the highest peak marks the most likely location. If the marker framework is physically anchored, as described above, the stage is then set to search the DNA of the candidate interval and identify the disease gene.

13.5 Fine-mapping using extended pedigrees and ancestral haplotypes

The *resolution* of mapping depends on the number of meioses – the more meioses that are analyzed, the greater is the chance of a recombination event narrowing down the linked region. The small size of most human families severely limits the resolution attainable in family studies. However, sometimes extended family structures can be used for high-resolution mapping. In some societies people are very aware of their clan membership and may see themselves as part of highly extended families. Even in societies where people limit their family feeling to close relatives, ultimately everybody is related, and sometimes one can identify *shared ancestral chromosome segments* among 'unrelated' people. Autosomal recessive diseases lend themselves to such analyses, because a mutated allele can be transmitted for many generations; for most dominant or X-linked diseases the turnover of mutant alleles is too fast to allow sharing over extended families (Section 4.5.2). Assuming one cannot identify carriers of a recessive disease being mapped, the limit of mapping is set by the number of distantly related affected people who have inherited the disease from a common ancestor.

13.5.1 Autozygosity mapping can map recessive conditions efficiently in extended inbred families

Autozygosity is a term used to mean homozygosity for markers *identical by descent*, inherited from a recent common ancestor. People with rare recessive diseases in consanguineous families are likely to be autozygous for markers linked to the disease locus. Suppose the parents are second cousins: they would be expected to share 1/32 of all their genes because of

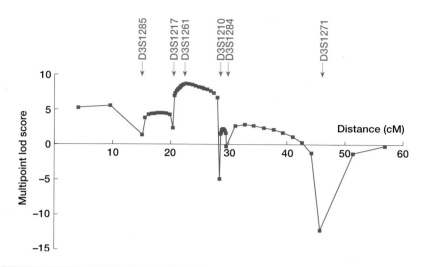

Figure 13.8: Multipoint mapping in man.

The horizontal axis is a framework map of markers and the vertical axis is the lod score from analysis of a family with Waardenburg syndrome. The lod score is calculated for each possible location of the disease locus. Lod scores dip to strongly negative values near the position of loci that show recombinants with the disease. The highest peak marks the most likely location; odds in favor of this location are measured by the degree to which the highest peak overtops its rivals. Redrawn from Hughes *et al.* (1994) *Nat. Genet.* **7**, 509–512 with permission from Nature Publishing Group.

their common ancestry, and a child would be autozygous at only 1/64 of all loci. If that child is homozygous for a particular marker allele, this could be because of autozygosity, or it could be because a second copy of the same allele has entered the family independently. The rarer the allele is in the population, the greater the likelihood that homozygosity represents autozygosity. For an infinitely rare allele, a single homozygous affected child born to second cousins generates a lod score of $\log_{10}(64) = 1.8$. If there are two other affected sibs who are both also homozygous for the same rare allele, the lod score is 3.0 ($\log_{10}(64 \times 4 \times 4)$); the chance that a sib would have inherited the same pair of parental haplotypes even if they are unrelated to the disease is 1 in 4).

Thus quite small inbred families can generate significant lod scores. Autozygosity mapping is an especially powerful tool if families can be found with multiple people affected by the same recessive condition in two or more sibships linked by inbreeding. Suitable families may be found in Middle Eastern countries where inbreeding is common. The method has been applied with great success to locating genes for autosomal recessive hearing loss (Guilford *et al.*, 1994; *Figure 13.9*). Extensive locus heterogeneity (*Figure 4.3*) makes recessive hearing loss impossible to analyze in collections of small nuclear families.

A bold application of autozygosity in a Northern European population enabled Houwen *et al.* (1994) to map the rare recessive condition, benign recurrent intra-hepatic cholestasis, using only four affected individuals (two sibs and two supposedly unrelated people) from an isolated Dutch village. Similar virtuoso applications of autozygosity have been reported from Finland. The more remote the shared ancestor, the smaller is the proportion of the genome that is shared by virtue of that common ancestry, and therefore the greater the significance of demonstrating that the patients share a segment

identical by descent. But at the same time, the remoter the common ancestor, the more chances there are for a second independent allele to enter the family from outside, and so the less likely is it that homozygosity represents autozygosity, either for the disease or for the markers. With remote common ancestry, as in the study of Houwen *et al.*, everything depends on finding people with a very rare recessive condition who are homozygous for a rare marker allele or (more likely) haplotype. The power of Houwen's study seems almost miraculous, but it is important to remember that this methodology applies only to diseases and populations where most affected people are descended from a common ancestor who was a carrier.

13.5.2 Identifying shared ancestral segments allowed high-resolution mapping of the loci for cystic fibrosis and Nijmegen breakage syndrome

Cystic fibrosis (CF) is very rare in the non-European countries where family structures are more likely to allow autozygosity mapping, and so mapping CF depended on rare unfortunate nuclear families with more than one affected child. Using these, CF was mapped to 7q31.2, but after all available recombinants had been used, the candidate region was still dauntingly large (this was in the late 1980s when positional cloning was work for heroes). Arguing that CF mutations might be mostly very old (not only is there no selection against heterozygotes, there is probably positive selection in their favor, see Section 4.5.2), the researchers set out to identify shared ancestral chromosomal segments on CF chromosomes from 'unrelated' patients. Sharing would be indicated by repeatedly finding the same haplotype of marker alleles. The phenomenon is called **linkage disequilibrium** (LD); see Section 15.4 for a fuller description. *Table 13.2*

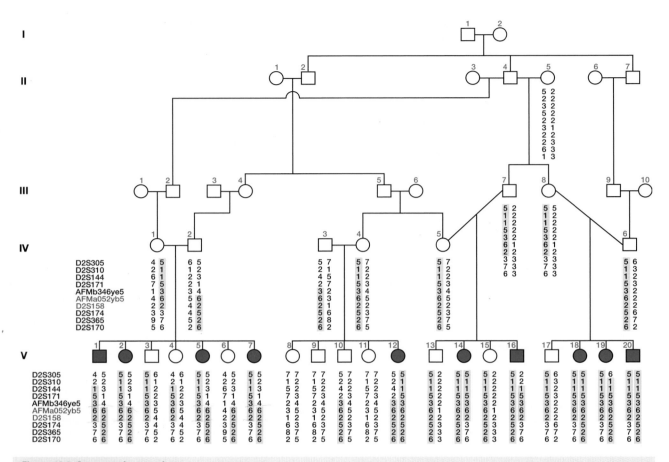

Figure 13.9: Autozygosity mapping.

A large multiply inbred family in which several members suffer from autosomal recessive profound congenital deafness (filled symbols). Color marks a haplotype of markers from chromosome 2 that segregates with the deafness. Markers AFMa052yb5 and *D2S158* are homozygous in all affected people and no unaffected people. The deafness gene should lie somewhere between the two markers that flank these (AFMb346ye5 and *D2S174*). Redrawn from Chaib *et al.* (1996) *Hum. Molec. Genet.* **5**, 155–158 with permission from Oxford University Press.

shows typical data for two markers from within the CF candidate region. Non-CF chromosomes show a random selection of haplotypes, but CF chromosomes tend to carry X_1, K_2. The significance of this is that LD is a very short-range phenomenon (shared ancestral segments are short because of

recurrent recombination), and so it pointed researchers to the exact location of the elusive CF gene.

A more recently cloned gene, governing Nijmegen breakage syndrome (NBS; MIM 251260), shows a more detailed application of the same principle. NBS is a very rare autosomal recessive disease characterized by chromosome breakage, growth retardation, microcephaly, immunodeficiency and a predisposition to cancer. The suspected cause is a defect in DNA repair. Conventional linkage analysis in small nuclear families located the *NBS* locus to chromosome 8p21, but after all recombinants were used the target region still spanned 8 Mb between markers *D8S271* and *D8S270*. Fifty-one apparently unrelated patients and their parents were then typed for a series of microsatellite markers spaced across the candidate region. This generated 102 NBS haplotypes. Of these, 74 looked like derivatives of a common ancestral haplotype, probably of Slav origin (*Figure 13.10*). The most highly conserved region encompassed markers 11 and 12 in *Figure 13.10*, which therefore marked the likely location of the *NBS* gene. Subsequently a gene encoding a novel protein was cloned from this location and shown to carry mutations in

Table 13.2: Allelic association in cystic fibrosis

Marker alleles	CF chromosomes	Normal chromosomes
X_1, K_1	3	49
X_1, K_2	147	19
X_2, K_1	8	70
X_2, K_2	8	25

Data from typing for the RFLP markers XV2.c (alleles X_1 and X_2) and KM19 (alleles K_1 and K_2) in 114 British families with a cystic fibrosis (CF) child. Chromosomes carrying the CF disease mutation tend to carry allele X_1 of XV2.c and allele K_2 of KM19. Data from Ivinson *et al.* (1989).

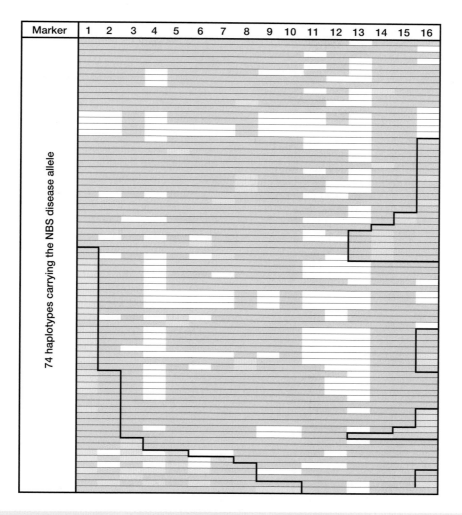

Figure 13.10: An ancestral haplotype in European patients with Nijmegen breakage syndrome.

Apparently unrelated NBS patients often seem to have inherited a chromosome segment at 8p21 from a common ancestor. Haplotypes were defined using 16 markers, shown in chromosomal order across the top of the table. The pink color marks locations with alleles identical to those of the inferred ancestral haplotype. Nonancestral alleles are coded yellow when they differ from the ancestral allele by only 1 or 2 base pairs, and might have been derived from the ancestral allele by mutation; alleles coded gray differ substantially from the ancestral allele, and are likely the result of recombination. Blanks mark loci where there are no data. Only at loci 11 and 12 are there no recombinant (gray) alleles, suggesting that the *NBS* gene maps to this position. Data of Varon *et al.* 1998.

NBS patients. As predicted, patients with the common haplotype all have the same mutation, while those with independent haplotypes had independent mutations (Varon *et al.*, 1998).

LD is a central tool in efforts to identify susceptibility genes for complex diseases, and is discussed in detail in Section 15.4.

13.6 Standard lod score analysis is not without problems

Standard lod score analysis is a tremendously powerful method for scanning the genome in 20-Mb segments to locate a disease gene, but it can run into difficulties. These include:

▸ vulnerability to errors;

▸ computational limits on what pedigrees can be analyzed;

▸ problems with locus heterogeneity;

▸ limits on the ultimate resolution achievable;

▸ the need to specify a precise genetic model, detailing the mode of inheritance, gene frequencies and penetrance of each genotype.

13.6.1 Errors in genotyping and misdiagnoses can generate spurious recombinants

With highly polymorphic markers, common errors such as misread gels, switched samples or nonpaternity will usually result in a child being given a genotype incompatible with the parents. The linkage analysis program will stall until such errors have been corrected. Errors that introduce possible but wrong genotypes are more of a problem, especially misdiagnosis of somebody's disease status. Such errors inflate the length of genetic maps by introducing spurious recombinants: if a child has been assigned the wrong parental allele, it will appear to be a recombinant. Multilocus analysis can help, because *spurious recombinants appear as close double recombinants (Figure 13.11)*. As we saw in Section 13.1.3, interference makes close double recombinants very unlikely. When marker framework maps are made, error-checking routines test the extent to which the map can be shortened by omitting any single test result (see Broman *et al.*, 1998). Results that significantly lengthen the map (i.e. add recombinants) are suspect.

Errors in the order of markers on marker framework maps used to cause headaches (single recombinants could appear to be double), but this problem has receded as genetic maps are cross-checked against physical sequence.

13.6.2 Computational difficulties limit the pedigrees that can be analyzed

As we saw in Section 13.3.2, human linkage analysis depends on computer programs that implement algorithms for handling branching trees of genotype probabilities, given the pedigree data and gene frequencies. *Liped* was the first generally useful program, and *Mlink* (part of a package called Linkage) used the same basic algorithm, the *Elston–Stewart algorithm*, but extended it to multipoint data. The Elston–Stewart algorithm can handle arbitrarily large pedigrees, but the computing time increases exponentially with increasing numbers of possible haplotypes (more alleles and/or more loci). This limits the ability of Mlink to analyze multipoint data. An alternative algorithm, the *Lander–Green algorithm*, is able to handle any number of loci (the computing time increases linearly with the number of loci), but has memory problems with large pedigrees. This algorithm is implemented in the *Genehunter* (Kruglyak *et al.*, 1996) and *Merlin* programs (Abecasis *et al.*, 2002). These programs are particularly good for analyzing whole-genome searches of modest sized pedigrees.

The general theory of linkage analysis is excellently covered in the book by Ott (see Further reading), while the book by

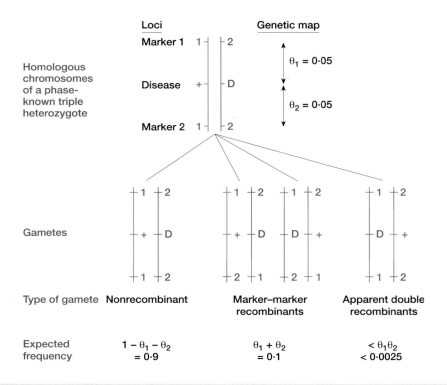

Figure 13.11: Apparent double recombinants suggest errors in the data.

Because of interference (Section 13.1.3), the probability of a true double recombinant with markers 5 cM apart is small, well below $0.05 \times 0.05 = 0.0025$. Apparent double recombinants usually signal an error in typing the markers, a clinical misdiagnosis, or locus heterogeneity such that the disease in this case does not map to locus D but elsewhere in the genome. Mutation in one of the genes or germinal mosaicism are rarer causes.

Terwilliger and Ott (see Further reading) is full of practical advice indispensable to anybody undertaking human linkage analysis.

13.6.3 Locus heterogeneity is always a pitfall in human gene mapping

As we saw in Section 4.1.4, it is common for mutations in several unlinked genes to produce the same clinical phenotype. Even a dominant condition with large families can be hard to map if there is **locus heterogeneity** within the collection of families studied. It took years of collaborative work to show that tuberous sclerosis was caused by mutations at either of two loci, *TSC1* (MIM 191100) at 9q34 and *TSC2* (MIM 191092) at 16p13. With recessive conditions, the difficulty is multiplied by the need to combine many small families. *Autozygosity mapping* (see Section 13.5.1) is the main solution in such cases.

Genehunter or *Homog* and related programs (see Terwilliger and Ott, Further reading) can compare the likelihood of the data on the alternative assumptions of locus homogeneity (all families map to the location under test) and heterogeneity (a proportion α of unlinked families), and give a maximum likelihood estimate of α.

13.6.4 Meiotic mapping has limited resolution

The *resolution of mapping* depends on the number of meioses analyzed. Human families are quite limiting for this purpose – for example, the CEPH family collection (Section 13.4.2) can provide an average resolution of only about 3 Mb. One solution is to use *sperm*. Men may have too few children for high-resolution mapping, but they produce effectively unlimited numbers of sperm. Individual sperm can be scored as recombinant or nonrecombinant for pairs of PCR-amplifiable markers. This method cannot be used to map diseases, but it allows a researcher with the necessary technical skill to carry marker–marker mapping to any resolution desired. Lien *et al.* (2000) describe a typical application. The existence of very localized **recombination hotspots** (*Figure 13.5*) was proved by measuring recombination fractions as low as 0.00001 (Jeffreys *et al.*, 2001). Of course, this only gives information on male recombination.

13.6.5 Characters whose inheritance is not Mendelian are not amenable to mapping by the methods described in this chapter

The methods of lod score analysis described in this chapter require a precise genetic model. The mode of inheritance, gene frequencies and penetrance of each genotype must all be specified. For Mendelian characters, it is not usually a problem to provide plausible figures. Penetrance may require a little thought. If no allowance is made for unaffected people possibly being nonpenetrant gene carriers, or affected people possibly being phenocopies, then such people would be scored as recombinant. On the other hand, if the penetrance is set too low there is a reduction in the power to detect linkage, because a less precise hypothesis is being tested. However, for common complex diseases like diabetes or schizophrenia, the problems are far more intractable. Any genetic model is no more than a hypothesis – we have no real idea of the gene frequencies or penetrance of any susceptibility alleles, or even the mode of inheritance. This makes it very unwise to apply the methods we have described in this chapter to such diseases. Nevertheless, *identifying the genetic components of susceptibility to complex diseases is now a major part of human genetics research.* The ways one can attempt to do this are the subject of Chapter 15.

Further reading

Ott J (1999) *Analysis of Human Genetic Linkage,* 3rd Edn. Johns Hopkins University Press, Baltimore, MD.

Terwilliger J, Ott J (1994) *Handbook for Human Genetic Linkage.* Johns Hopkins University Press, Baltimore, MD.

References

Abecasis GR, Cherny SS, Cookson WO, Cardon LR (2002) Merlin – rapid analysis of dense genetic maps using sparse gene flow trees. *Nature Genet.* **30**, 97–101.

Broman KW, Murray JC, Sheffield VC, White RL, Weber JL (1998) Comprehensive human genetic maps: individual and sex-specific variation in recombination. *Am. J. Hum. Genet.* **63**, 861–869.

Broman KW, Weber JL (2000) Characterization of human crossover interference. *Am. J. Hum. Genet.* **66**, 1911–1926.

Chaib H, Place C, Salem N et al. (1996) A gene responsible for a sensorineural nonsyndromic recessive deafness maps to chromosome 2p22-23. *Hum. Mol. Genet.* **5**, 155–158.

Guilford P, Ben Arab S, Blanchard S, Levilliers J, Weissenbach J, Belkahia A, Petit C (1994) A non-syndrome form of neurosensory, recessive deafness maps to the pericentromeric region of chromosome 13q. *Nature Genet.* **6**, 24–28.

Houwen RHJ, Baharloo S, Blankenship K, Raeymaekers P, Juyn J, Sandkuijl LA, Freimer NB (1994). Genome screening by searching for shared segments: mapping a gene for benign recurrent intrahepatic cholestasis. *Nature Genet.* **8**, 380–386.

Hughes A, Newton VE, Liu XZ, Read AP (1994) A gene for Waardenburg syndrome Type 2 maps close to the human homologue of the microphthalmia gene at chromosome 3p12-p14.1. *Nature Genet.* **7**, 509–512.

Hultén MA, Lindsten J (1973) Cytogenetic aspects of human male meiosis. *Adv. Hum. Genet.* **4**, 327–387.

Hultén MA, Tease C (2003) Genetic maps: direct meiotic analysis. In: Cooper DN (ed) *Encyclopedia of the Human Genome.* Nature Publishing Group, London.

Ivinson AJ, Read AP, Harris R, Super M, Schwarz M, Clayton Smith J, Elles R (1989) Testing for cystic fibrosis using allelic association. *J. Med. Genet.* **26**, 426–430.

Jeffreys AJ, Kauppi L, Neumann R (2001) Intensely punctate meiotic recombination in the class II region of the major histocompatibility complex. *Nature Genet.* **29**, 217–222.

Kong A, Gudbjartsson DF, Sainz J *et al.* (2002) A high-resolution recombination map of the human genome. *Nature Genet.* **31**, 241–247.

Kruglyak L, Daly MJ, Reeve-Daly MP, Lander ES (1996). Parametric and non-parametric linkage analysis: a unified multipoint approach. *Am. J. Hum. Genet.* **58**, 1347–1363.

Lander ES, Schork NJ (1994). Genetic dissection of complex traits. *Science* **265**, 2037–2048.

Lien S, Szyda J, Schechinger B, Rappold G, Arnheim N (2000) Evidence for heterogeneity in recombination in the human pseudoautosomal region: high resolution analysis by sperm typing and radiation-hybrid mapping. *Am. J. Hum. Genet.* **66**, 557–566.

Morton NE (1955) Sequential tests for the detection of linkage. *Am. J. Hum. Genet.* **7**, 277–318.

Tease C, Hartshorne GM, Hultén, MA (2002) Patterns of meiotic recombination in human fetal oocytes. *Am. J. Hum. Genet.* **70**, 1469–1479.

Varon R, Vissinga C, Platzer M *et al.* (1998) Nibrin, a novel DNA double-stranded break repair protein, is mutated in Nijmegen Breakage Syndrome. *Cell* **93**, 467–476.

Wang DG, Fan JB, Siao CJ *et al.* (1998) Large-scale identification, mapping and genotyping of single nucleotide polymorphisms in the human genome. *Science* **280**, 1077–1082.

CHAPTER FOURTEEN

Identifying human disease genes

Chapter contents

A more accurate though less snappy title for this chapter would be 'Identifying genetic determinants of human phenotypes'. The approaches described are equally applicable to identifying determinants of diseases or of normal variations such as red hair or red–green color blindness. Nor would all the determinants that might be identified necessarily be genes, in the sense of protein-coding sequences. By definition they must have an effect on the phenotype, but it might be through some indirect effect on the level of expression of a protein-coding gene or the processing or stability of its mRNA. Understanding why a given DNA sequence variant causes a particular phenotype is the role of molecular pathology (Chapter 16); here we will discuss how to identify the right variant.

It is important not to be misled by phrases like 'the gene for cystic fibrosis', 'the gene for diabetes', and so on. Many human genes were first discovered through research on the diseases caused by mutations in them, and it may take years before their normal function is understood, hence the appeal of this way of naming them. However, you wouldn't describe your domestic freezer as 'a machine for ruining frozen food'. Genes do a job in cells; if the job is not done, or done wrong, the result may be a disease..

Few subjects have moved as fast as human disease gene identification. Before 1980 very few human genes had been identified as disease loci. The few early successes involved a handful of diseases with a known biochemical basis where it was possible to purify the gene product. In the 1980s, advances in recombinant DNA technology allowed a new approach, sometimes given the rather meaningless label 'reverse genetics'. The number of disease genes identified started to increase, but these early successes were hard won, heroic efforts. With the advent of PCR for linkage studies and mutation screening, it all became much easier. Now that the human and other genome projects have made available a vast range of resources, ability to identify a Mendelian disease gene depends almost entirely on having suitable families. Identifying the factors conferring susceptibility to common complex diseases remains, however, exceedingly difficult.

14.1 Principles and strategies in identifying disease genes

There are many different ways of arriving at the final identification (*Figure 14.1*), but all paths converge on a *candidate gene*. One way or another, a candidate gene is identified; the researcher then tests the hypothesis that this is the disease gene by screening it for mutations in patients with the disease.

Candidate genes may be identified without reference to their chromosomal location (Section 14.2) but more commonly, first a candidate chromosomal region is pinpointed, and then candidate genes are identified from within that region (Section 14.3). Now that we have a good (though incomplete) catalog of all the human genes, the task of identifying candidates has become immeasurably easier, though working through a long list of candidates seeking mutations can still be very laborious.

Positional information reduces the list of possible candidates from all 30 000 or so human genes to maybe 10–30 genes in a *candidate region*. This is important because however hard one tries to guess likely candidates, our ability to do so is currently very limited. Over and over again, when a disease gene is finally identified, it remains a complete mystery why mutations should cause that particular disease. Why should loss of function of the FMR1 protein, involved in transporting RNA from nucleus to cytoplasm, cause mental retardation and macro-orchidism (Fragile-X syndrome, MIM 309550), while certain mutations in the TATA binding protein (see Section 1.3.4) cause SCA17 spinocerebellar ataxia (*Table 16.6*)?

14.2 Position-independent strategies for identifying disease genes

Historically, the first disease genes were identified by position-independent methods, simply because no relevant mapping information existed and no techniques were available to generate it. Under those circumstances the candidate must be suggested by knowledge of the gene product: β-globin for sickle cell disease, phenylalanine hydroxylase for phenylketonuria, and so on. Still today, studies coming from a biochemical or cell biology direction may identify protein products of unknown genes. Some method is needed for moving from protein to DNA.

14.2.1 Identifying a disease gene through knowing the protein product

Modern proteomic techniques allow even very tiny quantities of a protein to be identified or partially sequenced by mass spectrometry (Mann *et al.*, 2001 and Section 19.4.2) and chemical microsequencing (Bartlett, 2001). If the sequence of the cDNA encoding those amino acids could be worked out, an oligonucleotide probe could be synthesized and used to screen libraries to recover the cDNA. The problem is the degeneracy of the genetic code – most amino acids can be encoded by any one of several codons. The probe has to be a **degenerate oligonucleotide**, a cocktail of all the possible sequences, chosen to match a part of the amino acid sequence where the number of possible permutations is not hopelessly large. As only one of the oligonucleotides in the mix will correspond to the authentic sequence, it is important to keep the number of different oligonucleotides low so as to increase the chance of identifying the correct target. Tryptophan and methionine are very helpful here, as each has only a single codon. Arginine, leucine and serine, with six codons each, are avoided as far as possible.

Library screening can be tedious when a degenerate probe is used, because the results are greatly influenced by the hybridization conditions. A more rapid alternative is to use partially degenerate oligonucleotides as PCR primers. The number of possible permutations can be reduced by ligating the target cDNA to a vector and using one vector-specific primer and one degenerate protein-specific primer.

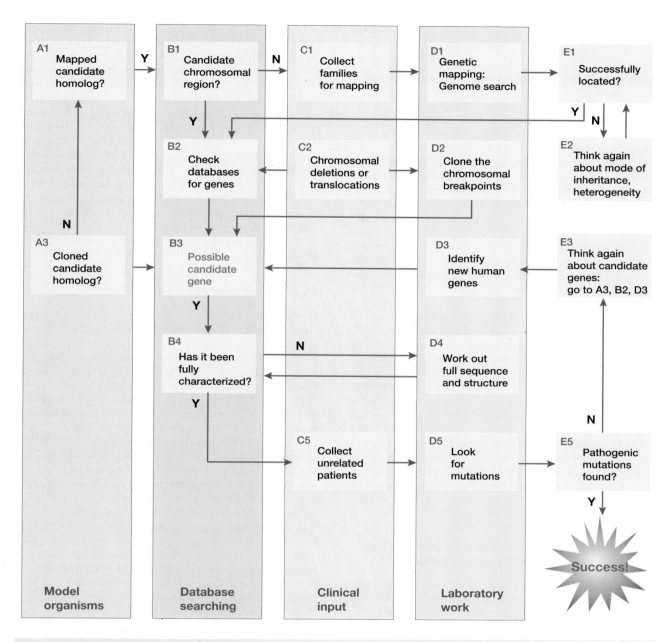

Figure 14.1: How to identify a human disease gene.

There is no single pathway to success, but the key step is to arrive at a plausible candidate gene, which can then be tested for mutations in affected people. Note the interplay between clinical work, laboratory benchwork and computer analysis. Database searching is becoming more and more crucial as information from genome projects accumulates.

However, good luck is required to obtain the desired PCR product, rather than no product or a mess of irrelevant products.

An alternative route, if the protein is available even in minute quantities, is to raise an antibody to the protein and use this to find the gene. Back in 1982 the mRNA encoding phenylalanine hydroxylase was recovered by immunoprecipitation of polysomes that were synthesizing the protein in a cell free system (Robson *et al.*, 1982). Nowadays a **cDNA expression library** would be made by cloning pooled cDNA into an expression vector (Section 5.6.1). Host cells containing clones with the desired gene should produce the protein, or at least parts of the protein, and could be identified by screening colony filters from the library with an appropriate antibody. Everything here depends on the specificity of the antibody, and the hope that the protein is not toxic to the host cell. **Phage display** (Section 5.6.2) would give an alternative approach.

14.2.2 Identifying the disease gene through an animal model

Many human disease genes have been identified with the help of animal models – but nearly always this has been after checking positional information. Maybe a mouse mutant and a phenotypically similar human disease are mapped to chromosomal locations that correspond (using the **Oxford Grid**, see *Figure 14.7*). Then if the mouse gene is cloned, its human homolog becomes a natural candidate. Alternatively a disease gene may be identified in the mouse, and then the human homolog isolated; this can be mapped by fluorescence *in situ* hybridization (Section 2.4.2), and becomes a candidate gene for any relevant disease mapping to that location. This is how the *MITF* gene was identified as a cause of Type 2 Waardenburg syndrome (MIM 193510; Hughes *et al.*, 1994).

It is unusual for a gene identified in an animal model to be tested directly in human patients without any positional confirmation that these are appropriate patients to test, but one such case is *SOX10*. This gene was identified by laborious positional cloning of the mouse *Dominant megacolon* (*Dom*) mutation. *Dom* mice are a long-studied model of human Hirschsprung disease (Section 15.6.2). Patients with a combination of Hirschsprung disease, pigmentary abnormalities and hearing loss (Waardenburg syndrome type IV or WS4, MIM 277580) especially resembled the mice. WS4 is very rare and normally occurs in families too small for mapping, so a panel of WS4 patients was tested for *SOX10* mutations without any prior knowledge of where their disease might map. The gamble paid off when *SOX10* mutations were found, though not in all of the patients (Pingault *et al.*, 1998).

14.2.3 Identification of a disease gene using position-independent DNA sequence knowledge

This most usually arises when the researcher is considering what diseases might be caused by mutations in a particular known gene. Position-independent candidates are also generated by expression array experiments, in which mRNA samples from patients and controls are compared to produce a list of genes whose expression is altered in the disease.

An interesting application of position-independent DNA sequence knowledge is the attempt to clone genes containing novel expanded trinucleotide repeats. As shown in Section 16.6.4, expanded trinucleotide repeats cause several inherited neurological disorders. Often these disorders show **anticipation** – that is, the disease presents at an earlier age and with increased severity in successive generations. If a disease under investigation shows any of these features, it may be worth screening DNA from affected patients for triplet repeat expansions. The *Repeat Expansion Detection* method of Schalling *et al.* (1993) permits detection of expanded repeats in unfractionated genomic DNA of affected patients, and methods have been developed for cloning any expanded repeats detected (Koob *et al.*, 1998). This approach was used in a completely position-independent way to identify a novel repeat expansion involved in a form of spinocerebellar ataxia (SCA8) (Koob *et al.*, 1999).

14.3 Positional cloning

In positional cloning, a disease gene is identified knowing nothing except its approximate chromosomal location. The first successful application was identification of the gene for X-linked chronic granulomatous disease (Royer-Pokora *et al.*, 1985). A major test-bed for positional cloning methods was Duchenne muscular dystrophy (DMD, MIM 310200). Years of careful investigation of the pathological changes in affected muscle had failed to reveal the biochemical basis of DMD. In the early 1980s, several groups competed to clone the DMD gene, using different approaches. The pioneering work of these groups, overcoming formidable technical difficulties to clone an unprecedented gene, was probably the major inspiration for most subsequent positional cloning efforts. This work has been well reviewed by Worton and Thompson (1988).

The successful conclusion of this work in 1986 marked the start of a triumphant new era for human molecular genetics. One after another, the genes underlying important disorders such as cystic fibrosis, Huntington disease, adult polycystic kidney disease, and familial colorectal cancer were isolated. The logic of positional cloning follows the scheme of *Figure 14.2*. However, before the current marker maps, clones and sequence were available, positional cloning could be desperately hard work. By 1995 only about 50 inherited disease genes had been identified by this approach. The frustrating nature of positional cloning was summed up by one researcher in *Figure 14.3*.

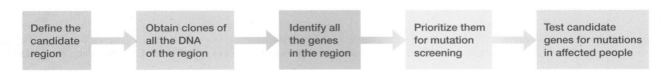

Figure 14.2: The logic of positional cloning.

The figure illustrates the logical progression of positional cloning; however, until the current sequence data and high-resolution marker maps were available, researchers tried all sorts of short-cuts to reduce the labor of pure positional cloning.

Candidate region

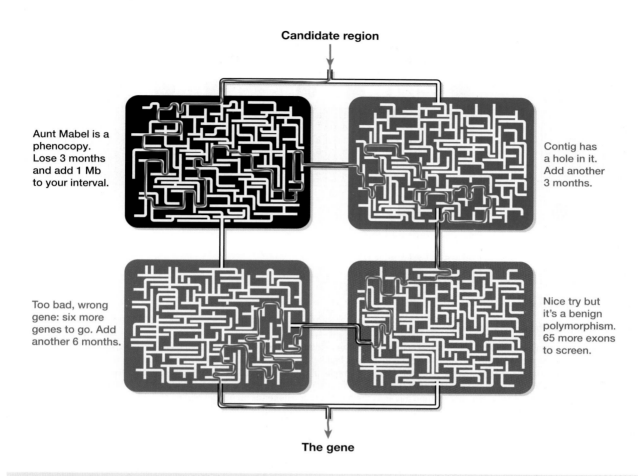

Aunt Mabel is a phenocopy. Lose 3 months and add 1 Mb to your interval.

Contig has a hole in it. Add another 3 months.

Too bad, wrong gene: six more genes to go. Add another 6 months.

Nice try but it's a benign polymorphism. 65 more exons to screen.

The gene

Figure 14.3: The difficult path from candidate region to gene.

One researcher's view of the frustrations of positional cloning. Courtesy of Dr Richard Smith, University of Iowa.

14.3.1 The first step is to define the candidate region as tightly as possible

The difficulty of positional cloning depends very largely on the size of the candidate region, so the first priority is to narrow this down as far as possible. For Mendelian diseases, this is mainly a function of the number of meioses available for study. The **limit of resolution** is reached when the last recombinant has been mapped between closely spaced markers. This is decided by inspecting haplotypes rather than by computer analysis (*Figure 14.4*). Using the rule of thumb that 1 cM = 1 Mb (Section 13.1.5), a family collection with 100 informative meioses might localize a Mendelian disease to a candidate region of around 1 Mb.

When single recombinants define the boundaries of the candidate region, it is important to consider possible sources of error (Section 13.6.1). Meticulous clinical diagnoses are imperative. Key recombinants are more reliable if they occur in unambiguously affected people – an unaffected individual might be a nonpenetrant gene carrier. **Apparent double recombinants** are highly suspect. Sometimes, despite good positive lod scores, there seem to be recombinants with every marker tried. This is usually an indication that one of the families in the study does not map to this region. Alternatively, perhaps the markers are ordered incorrectly on the genetic map.

For non-Mendelian phenotypes, linkage analysis is far less precise and candidate regions are typically 20 cM or more (Chapter 15). That is much too big to search without further clues, hence the emphasis on using **linkage disequilibrium** to narrow the search (Section 15.4). Even for Mendelian diseases, linkage disequilibrium can be a very valuable tool for fine mapping, as we saw for cystic fibrosis and Nijmegen breakage syndrome (Section 13.5.2).

14.3.2 A contig of clones must be established across the candidate region

The techniques of contig construction were described in *Box 8.5*. In the early days, building the contig was a major effort. The paper describing identification of the cystic fibrosis gene (Rommens *et al.*, 1989) summarizes perhaps the most impressive example from this early phase. Using hybridization

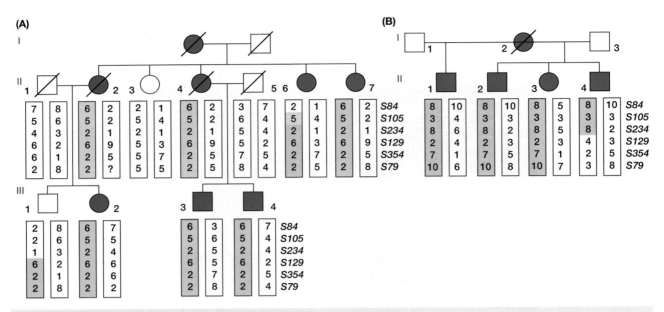

Figure 14.4: Defining the minimal candidate region by inspection of haplotypes.

The two pedigrees show a dominantly inherited skin disorder, Darier–White disease (MIM 124200), which had previously been mapped to 12q. The 12q marker haplotype that segregates with the disease is highlighted. Gray boxes mark inferred haplotypes in dead people. In Pedigree A individual II-6 the recombination maps the disease gene distal to D12584; *D12S105* is uninformative because I-1 was evidently homozygous for allele 5 – compare the genotypes of II-3 and II-7. The recombination shown in III-1 suggests the disease gene maps proximal to *D12S129*, but this requires confirmation because the interpretation depends on the genotypes of II-1 and II-2 being inferred correctly, and on III-1 not being a nonpenetrant gene carrier. The recombination in pedigree B individual II-4 provides the confirmation. The combined data locate the Darier gene to the interval between *D12S84* and *D12S129*. Redrawn from Carter *et al.* (1994) *Genomics* **24**, 378–382. © 1994 with permission from Elsevier.

screening of libraries to identify successive overlapping clones ("**chromosome walking**") was painfully slow with the small-insert phage and cosmid libraries then available, and was supplemented by the now obsolete technique of chromosome jumping (Poustka *et al.*, 1987). Nowadays ready-made contigs can be downloaded from the human genome sequence database, although it is always necessary to check these assemblies before relying on them.

14.3.3 A transcript map defines all genes within the candidate region

Having established a contig, the next step is to catalog all the genes within it. As our knowledge of the human genome improves, it becomes increasingly likely that the cause of any disease under investigation will be mutation in a known gene. A **genome browser** such as Ensembl (http://www.ensembl.org) or the Santa Cruz browser (http://genome.cse.ucsc.edu) is used to display and analyze all the definite and possible genes in the candidate region (*Figure 14.5*; see also Section 8.3.6 and Wolfsberg *et al.*, Further reading). Deeply impressive though these displays are, it is important not to rely totally on them. These extremely sophisticated tools need to be used as adjuncts to thought, and not as replacements for thought. They must be supplemented by first-hand in-depth personal study of the region using a mixture of computer and experimental work. The paper by Reymond *et al.* (2002) illustrates the sorts of extra analysis that can be done.

Computer searches try to wring more data from the databases than the genome browser programs can extract. Gene-finding programs are bad at finding small exons, exons with unusual splice sites or codon biases, or genes with unusually long 5′ or 3′ untranslated regions. Expressed sequence tags (ESTs) that are made of spliced small exons may not get matched to their genomic sequence. Further computer analysis can focus on detecting homologies. Comparison of human and mouse sequences may reveal additional conserved sequences, which would suggest some function. Weak homologies to possible orthologs or paralogs may be missed by the searches used for automatic genome annotation. A more directed, hypothesis-driven search may come up with significant pointers to gene function.

Experimental work focuses on double-checking for mistakes in sequence assembly, such as wrongly ordered subclones, or mistakes in gene assembly, such as missing or spurious exons, split or concatenated genes, and so on. Primers matching different parts of the genomic sequence are used to check that products of the predicted size are produced. Failure to amplify suggests mis-assembly of the genomic sequence. RT–PCR with primers from different putative exons can test whether the predicted product can be amplified at all, and if so, whether it contains the expected intermediate exons. Unsuspected alternative splicing or additional exons may be revealed. 5′-RACE (Section 7.2.3) can be used to try to extend gene sequences, especially if there is no good start

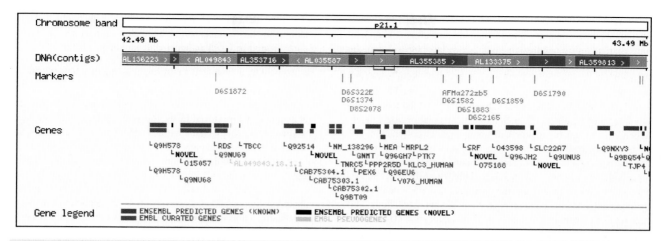

Figure 14.5: Using a genome browser to list the genes in a candidate region.

This partial screendump from the Ensembl genome browser (www.ensembl.org) shows confirmed and predicted genes in a 1-Mb region of chromosome 6p21.1. Clicking on a gene displays information about the gene sequence, structure and homologies. See Section 8.3.6 for more details of this system.

codon (an ATG in a Kozak consensus sequence – see Section 1.5.1) in the furthest upstream exon. Of course, if nothing is known about the expression pattern of the predicted gene, failure to amplify may simply mean that the wrong tissue or cDNA library has been tested. However, if any of the putative gene sequence is present in ESTs, there will be information about which cDNA libraries the EST was isolated from. Where adjacent genes are transcribed in the same direction, an RT–PCR test can be done to see whether they might actually both be part of the same gene.

In addition to these experimental checks on database annotations, **direct searches for transcripts** can be made. The general methods for identifying unknown transcribed sequences from within a contig of genomic clones have been discussed in detail in *Section 7.2*, and are summarized briefly in *Box 14.1*. Until very recently, there were no databases to check, so these methods formed the first line of transcript mapping.

Whenever cDNA libraries are to be screened, the question arises which libraries should be used. Often the pathology of the disease under study suggests a particular investigation. Thus when studying a neuromuscular disease it makes sense to start by screening muscle cDNA libraries. However, the tissue showing the pathology is not necessarily the one with the strongest expression, so if one library fails it is always worth screening others. Fetal brain is a popular choice because it has a particularly high number of expressed sequences.

14.3.4 Genes from the candidate region must be prioritized for mutation testing

From the list of genes that map to the candidate region, one would look for a gene that shows **appropriate expression** and/or **appropriate function**. Alternatively or additionally, as discussed below, one would look for **homology** to some other human or nonhuman gene that is known to have appropriate expression or function, or to have mutants with related phenotypes.

Appropriate expression pattern

A good candidate gene should have an expression pattern consistent with the disease phenotype. Expression need not be restricted to the affected tissue, because there are many examples of widely expressed genes causing a tissue-specific disease (Section 16.7.1), but the candidate should at least be expressed at the time and in the place where the pathology is seen. For example, neural tube defects are likely to involve

---

Box 14.1: Transcript mapping: laboratory methods that supplement database analysis for identifying expressed sequences within genomic clones.

Methods for transcript mapping are described in *Section 7.2*. In summary, these include:

▶ cDNA library screening, using as probes genomic clones from the candidate region;

▶ cDNA selection, for ultra-sensitive detection of cDNAs derived from the candidate region (*Figure 7.11*);

▶ exon trapping, to find genomic sequences flanked by functional splice signals (*Figure 7.10*);

▶ zoo blotting, to seek evolutionarily conserved sequences (*Figure 7.9*)

▶ CpG island identification, to seek the regions of under-methylated DNA that often lie close to genes (*Box 9.3*).

genes that are expressed during the third and fourth weeks of human embryonic development, shortly before or during neurulation. The expression of candidate genes can be tested by RT–PCR, Northern blotting or Serial Analysis of Gene Expression (SAGE, Section 19.3.2). Much of the preliminary work can be done in databases (dbEST or the SAGE database, both accessible through the NCBI homepage www.ncbi.nlm.nih.gov/) rather than in the laboratory. *In situ* hybridization against mRNA in tissue sections (Section 6.3.4) or immunohistochemistry using labeled antibodies provide the most detailed picture of expression patterns. Studies are usually done on mouse tissues, especially for embryonic stages. The common assumption that humans and mice will show similar expression patterns is not always justified, and centralized resources of staged human embryo sections have been established to allow the equivalent analyses to be performed where necessary on human embryos.

Appropriate function

When the function of a gene in the candidate region is known, it may be obvious whether or not it is a good candidate for the disease – rhodopsin and fibrillin (Section 14.6.4) provide examples. For novel genes, sequence analysis will often provide clues to the function: transmembrane domains, tyrosine kinase motifs, and so on can be identified. These may be sufficient to prioritize a gene as a candidate, given the pathology of the disease. For example, ion transport is known to be critical for functioning of the inner ear, so an ion channel gene would be a natural candidate in positional cloning of a deafness gene.

Candidate genes may also be suggested on the basis of a close functional relationship to a gene known to be involved in a similar disease. The genes could be related by encoding a receptor and its ligand, or other interacting components in the same metabolic or developmental pathway. For example, some of the genes implicated in Hirschsprung disease were identified using this logic, as described in Section 15.6.2.

Homology to a relevant paralogous (human) gene

Sometimes a gene in the candidate region turns out to be a close homolog of a known gene (a paralog in humans, or an ortholog in other species). If mutations in the homologous gene cause a related phenotype, the new gene becomes a compelling candidate. For example, after fibrillin was identified as the gene mutated in Marfan syndrome (Section 14.6.4), a paralogous gene, *FBN2*, was mapped to 5q. A related condition,

congenital contractural arachnodactyly (CCA, MIM 121050) mapped to the same region of 5q. *FBN2* mutations were soon demonstrated in CCA patients (Putnam *et al.*, 1995).

Homology to a relevant orthologous (model organism) gene

Over the past decade it has become increasingly clear how far structural and functional homologies extend across even very distantly related species. Virtually every mouse gene has an exact human counterpart, and the same is probably true of other less well explored mammalian species. More surprisingly, extensive homologies can be detected between human genes and genes in zebrafish, *Drosophila*, the nematode worm *Caenorhabditis elegans* and even yeast. A very powerful means of prioritizing candidates from among a set of human genes is therefore to see what is known about homologous genes in these well-studied model organisms, as described in Section 19.2. Such data might include the pattern of expression and the phenotype of mutants. Steinmetz *et al.* (2002) illustrate a systematic screen of yeast mutants for potential human disease genes. Mice are especially useful for such investigations, and their use is considered in more detail below.

Even more than gene sequences, pathways are often highly conserved, so that knowledge of a developmental or control pathway in *Drosophila* or yeast can be used to predict the likely working of human pathways – although mammals often have several parallel paths corresponding to a single path in lower organisms. By contrast, mutant phenotypes are less likely to correspond closely. A striking example is the wingless *apterous* mutant of *Drosophila*. A human gene, *Lhx2*, is able to complement the deficient function of the mutant, so that the flies grow normal wings (*Figure 14.6*). We must have a virtually identical developmental pathway to *Drosophila*, but clearly we use it for a different purpose. Branchio-oto-renal syndrome (Section 14.6.3) provides another example.

14.3.5 The special relevance of mouse mutants

Human–mouse phenotypic homologies provide particularly valuable clues towards identifying human disease genes for several reasons:

▶ programs of systematic mutagenesis are generating very large numbers of mouse mutants (Justice, 2000; Brown and Balling, 2001);

▶ orthologous gene mutations are more likely to produce similar phenotypes in humans and mice than in humans

Figure 14.6: Humans have a gene for making flies grow wings.

The defect in *apterous* mutant flies **(A)** can be corrected either by the wild-type fly gene **(B)** or by the human *LHX2* gene **(C)**. From Rincón-Limas *et al.* (1999) *Proc. Natl Acad. Sci. USA* **96**, 2165–2170 with permission. © 1999 National Academy of Sciences, USA.

and flies or worms. Nevertheless, the similarities may not be as close as one might wish (Section 20.4.6);

▶ mouse phenotypic information often translates readily into positional candidate information. Backcross mapping (see *Box 14.2*) allows quick and accurate mapping in the mouse. Thus most mouse mutants have been mapped, or can easily be mapped. Once a chromosomal location for a gene of interest is known in mouse or humans, it is usually (though not always) possible to predict the location of that gene in the other species. *Figure 14.7* illustrates the general correspondence between mouse and human chromosomal locations, based on orthologous genes that have been mapped in both species. Cross-matching of human and mouse genome sequences provides a very detailed picture of the relationship between human and mouse chromosomes (Gregory *et al.*, 2002; *Figure 14.8*);

▶ exon sequences and exon–intron structures are usually well conserved between orthologous human and mouse genes. This means that once a human or mouse gene is isolated, probes or primers can be designed to screen DNA libraries from the other species in order to identify the orthologous gene;

▶ once a candidate gene has been identified in humans, mouse mutants can be constructed to allow functional analysis. Our ability to make total or conditional knockouts

and to engineer specific mutations in an organism fairly closely related to ourselves makes the mouse a very powerful tool for exploring human gene function.

14.4 Use of chromosomal abnormalities

Chromosomal abnormalities can sometimes provide an alternative method of localizing a disease gene, in place of linkage analysis. For conditions that are normally sporadic, like many severe dominants, chromosome aberrations may provide the only method of arriving at a candidate gene (Section 14.6.1). With luck, they may even point directly to the precise location, rather than defining a candidate region, as with linkage. Balanced abnormalities (translocations or inversions) are particularly useful. Alert clinicians play a crucial role in identifying such patients (*Box 14.3*). Submicroscopic deletions and cryptic translocations are at least as valuable as visible chromosome abnormalities.

14.4.1 Patients with a balanced chromosomal abnormality and an unexplained phenotype are interesting

A balanced translocation or inversion, with nothing extra or missing, would not be expected to have any phenotypic effect on the carrier. If a person with an apparently balanced

---

Box 14.2: Mapping mouse genes.

Several methods are available for easy and rapid mapping of phenotypes or DNA clones in mice. Together with the ability to construct transgenic mice (*Chapter 20*), they make the mouse especially useful for comparisons with humans. Methods include the following.

Interspecific crosses (*Mus musculus / Mus spretus* or *Mus castaneus*)

The species have different alleles at many polymorphic loci, making it easy to recognize the origin of a marker allele. This is exploited in two ways:

▶ constructing marker framework maps. Several laboratories have generated large sets of F$_2$ backcrossed mice. Any marker or cloned gene can be assigned rapidly to a small chromosomal segment defined by two recombination breakpoints in the collection of backcrossed mice. For example, the collaborative European backcross was produced from a *M. spretus / musculus* (C57BL) cross. Five-hundred F$_2$ mice were produced by backcrossing with *spretus*, and 500 by backcrossing with C57BL. All microsatellites in the framework map are scored in every mouse;

▶ mapping a new phenotype. A cross must be set up specifically to do this but, unlike with humans, any number of F$_2$ mice can be bred to map to the desired resolution. *Musculus × castaneus* crosses are easier to breed than *musculus × spretus*.

Recombinant inbred strains

These are obtained by systematic inbreeding of the progeny of a cross, for example the widely used BXD strains are a set of 26 lines derived by over 60 generations of inbreeding from the progeny of a C57BL/6J × DBA/2J cross. They provide unlimited supplies of a panel of chromosomes with fixed recombination points. DNA is available as a public resource, and the strains function rather like the *CEPH families* do for humans (Section 13.4.2). Recombinant inbred strains are particularly suited to mapping quantitative traits (see Section 15.6.8), which can be defined in each parent strain and averaged over a number of animals of each recombinant type. Compared with mice from interspecific crosses, it may be harder to find a marker in a given region that distinguishes the two original strains, and the resolution is lower because of the smaller numbers.

Congenic strains

These are identical except at a specific locus. They are produced by repeated backcrossing, and can be used to explore the effect of changing just one genetic factor on a constant background.

Silver (1995) gives an overview of mouse genetics (see Further reading), and Copeland and Jenkins (1991) describe the uses of interspecific crosses.

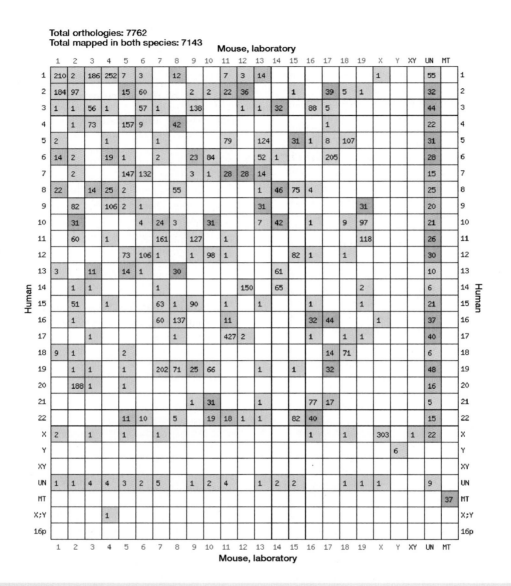

Figure 14.7: Conservation of synteny between human and mouse genetic maps.

The Oxford Grid summarizes the relationship between human and mouse chromosomes. Cells are color coded according to the number of orthologs mapped. The non-random distribution is obvious. The location of a human gene can often be predicted if the mouse location is known, or vice versa. This figure gives an overview; detailed information is contained in a database accessible at http://www.ncbi.nlm.nih.gov/Omim/Homology (DeBry and Seldin, 1996). Figure reproduced with permission from the Mouse Genome Database, Mouse Genome Informatics, The Jackson Laboratory, Bar Harbor, Maine (http://www.informatics.jax.org).

chromosomal abnormality is phenotypically abnormal, there are three possible explanations:

▶ the finding is coincidental;

▶ the rearrangement is not in fact balanced – there is an unnoticed loss or gain of material;

▶ one of the chromosome breakpoints causes the disease.

A chromosomal break can cause a loss-of-function phenotype if it disrupts the coding sequence of a gene, or separates it from a nearby regulatory region. Alternatively, it could cause a gain

of function, for example by splicing exons of two genes together to create a novel chimeric gene (this is rare in inherited disease but common in tumorigenesis, see Chapter 17). In either case, the breakpoint provides a valuable clue to the exact physical location of the disease gene. The precise position of the breakpoint is most easily defined by using FISH (*Figure 14.9*). An example of the power of this approach is the identification of the Sotos syndrome gene (Section 14.6.1). However, the positional clue is not infallible: sometimes breakpoints can alter expression of a gene located on the X chromosome, the woman is left with no active functional copy of the gene. There are about two dozen women worldwide who suffer from DMD because of X-autosome translocations. Each translocation involves a different autosomal breakpoint, but the X-breakpoint is always in Xp21. Study of these women supplemented early linkage work in mapping the DMD gene to Xp21, and one of them, with an X;21 translocation, provided one of the means of cloning the DMD gene (see below).

A rearrangement that places the sought after gene next to a known sequence provides an immediate route to the unknown gene. Many genes in model organisms have been cloned via mutations caused by insertion of a transgene or mobile element. In humans, one attack on the DMD gene used this method. One of the rare women with DMD (see above) had an Xp;21p translocation. Knowing that 21p is occupied by arrays of repeated rRNA genes (Section 9.2.1), Worton's group prepared a genomic library and set out to find clones containing both rDNA and X chromosome sequences. This led to isolation of XJ (X-junction) clones, which turned out to be located within the dystrophin gene, in intron 7 (Worton and Thompson, 1988).

The second possible explanation, above, should not be neglected. Studies of patients who have a *de novo* apparently balanced abnormality and a phenotype have shown that the majority actually have a more complex chromosomal rearrangement, often including a submicroscopic deletion. Loss of even a megabase of DNA would not be visible on standard cytogenetic preparations. Deletions larger than a few kilobases may be detected by FISH (see *Figure 14.12*); smaller ones in heterozygous translocation carriers are best studied by PCR after segregating the derivative chromosome in a somatic cell hybrid (See *Box 8.4*).

14.4.2 Patients with two Mendelian conditions, or a Mendelian condition plus mental retardation, may have a chromosomal deletion

Chromosomal deletions are less valuable for gene identification than balanced abnormalities because the whole deleted region, rather than a specific breakpoint, becomes the focus of search. Nevertheless, deletions have been instrumental in the identification of several major disease genes, including the early landmark triumph, identification of the dystrophin gene.

The starting point here was a boy, 'BB', who had DMD and a cytogenetically visible Xp21 deletion. A technically very difficult **subtraction cloning** procedure was used to isolate clones from normal DNA that corresponded to sequences deleted in BB (Kunkel *et al.*, 1985). Individual DNA clones in the subtraction library were then used as probes in Southern blot hybridization against DNA samples from normal people and DMD patients. One clone, pERT87-8, detected deletions in DNA from about 7% of cytogenetically normal DMD patients. It also detected polymorphisms that were shown by family studies to be tightly linked to DMD. These results showed that pERT87-8 was located much closer to the DMD gene than any previously isolated clones (in fact it was within the gene, in intron 13). Other nearby genomic probes were

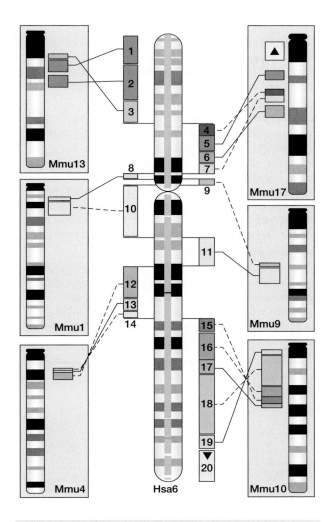

Figure 14.8: Conserved segments between human chromosome 6 (Hsa6) and mouse chromosomes (Mmu).

Twenty blocks of conserved synteny were identified by comparison of the human and mouse genome sequences. Dotted lines indicate blocks inverted relative to the mouse chromosome diagrams. Blue bars in the center of the chromosome ideogram represent the contigs included in the analysis. Reproduced from Gregory *et al.* (2002) *Nature* **418**, 743–750 with permission from Nature Publishing Group.

hundreds of kilobases away by affecting the structure of large-scale chromatin domains (*Box 14.4*).

Even if a translocation breakpoint disrupts a gene, we have lost function of only one of the two copies of the gene. There will be no phenotypic effect unless a 50% reduction in the level of the product causes problems (*haploinsufficiency*, Section 16.4.2). *X-autosome translocations* in females are a special case because of X-inactivation. Inactivation is random, but cells that inactivate the translocated X often suffer lethal genetic imbalances (*Figure 14.10*), so that a female carrier of such a translocation will consist entirely of cells that have inactivated the normal X. If the translocation breakpoint disrupts a gene

Box 14.3: Pointers to the presence of chromosome abnormalities.

Clinicians can make a major contribution to identifying disease genes by finding patients who have causative chromosome abnormalities.

A cytogenetic abnormality in a patient with the standard clinical presentation

If a disease gene has already been mapped to a certain location and then a patient with that disease is found who has a chromosome abnormality affecting that same location, the chromosome abnormality most probably caused the disease.

▶ Patients with balanced translocations or inversions often have breakpoints located within the disease gene, or very close to it. Cloning their breakpoints can provide the quickest route to identifying the disease gene.

▶ With interstitial deletions the breakpoints may be located some distance from the disease gene, but if the deleted segment is smaller than the current candidate region, defining the breakpoints helps localize the gene.

Most such patients will have *de novo* mutations. Some researchers feel that performing chromosome analysis on all patients with *de novo* mutations is a worthwhile expenditure of research effort.

Additional mental retardation

A patient may have a typical Mendelian disease, but in addition be severely mentally retarded. This may be coincidental, but such cases can be caused by deletions that eliminate the disease gene plus additional neighboring genes. Large chromosomal deletions almost always cause severe mental retardation, reflecting the involvement of a high proportion of our genes in fetal brain development. When the patient has a *de novo* mutation, cytogenetic and molecular analysis is warranted.

Contiguous gene syndromes

Very rarely a patient appears to suffer from several different genetic disorders simultaneously. This may be just very bad luck, but sometimes the cause is simultaneous deletion of a contiguous set of genes. Contiguous gene syndromes are described in Section 16.8.1; they are particularly well defined for X-linked diseases.

(A)

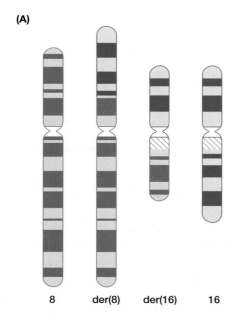

8   der(8)   der(16)   16

(B)

8cen   8pter

A   B   C   D   E   F   G

(C)

Probe	+ve chromosomes
A	8, der(8)
G	8, der(16)
B	8, der(8)
F	8, der(16)
E	8, der(16)
D	8, der(8), der(16)

Figure 14.9: Using fluorescence *in situ* hybridization to define a translocation breakpoint.

(A) Cytogenetically defined translocation t(8;16)(p22;q21). **(B)** Physical map of part of the breakpoint region in a normal chromosome 8, showing approximate locations of seven clones. **(C)** Results of successive FISH experiments. The breakpoint is within the sequence represented in clone D. This result would normally be confirmed using clones from chromosome 16.

Box 14.4: Position effects – a pitfall in disease gene identification.

In general genes appear to be arranged more or less at random on chromosomes, and the exact arrangement or order does not matter. In *Drosophila* however, it is well known that the local megabase-scale chromatin organization can affect gene expression – in particular, genes are silenced if placed within or close to heterochromatin. The same appears to be true in mice and men.

Studies of transgene expression (Section 20.2.3) show that correct tissue-specific gene expression can depend on sequences located hundreds of kilobases away from the coding sequence of a gene. Several human examples are known of translocation breakpoints affecting expression of a gene up to a megabase away. The examples of aniridia (MIM 106210) and the *PAX6* gene, and campomelic dysplasia (MIM 211970) and the *SOX9* gene were mentioned in Section 10.5.1

Thus balanced translocation breakpoints are not necessarily located within, or even very close to, the gene they inactivate, and this reduces their value as tools for cloning disease genes.

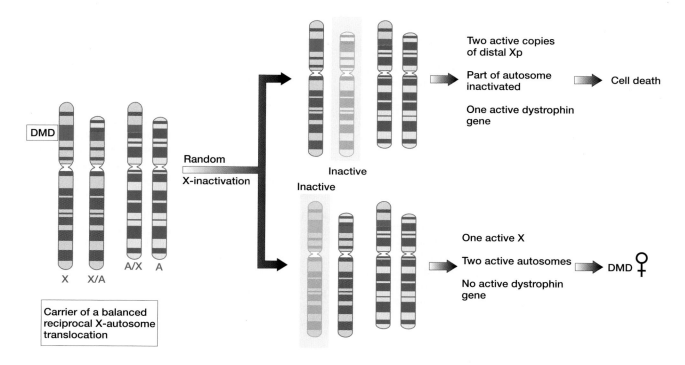

Figure 14.10: Nonrandom X inactivation occurs in female DMD patients with Xp21-autosome translocations.

The translocation is balanced, but the X chromosome breakpoint disrupts the dystrophin gene (red box). X inactivation is random, but cells which inactivate the translocated X die because of lethal genetic imbalance. The embryo develops entirely from cells where the normal X is inactivated, leading to a woman with no functional dystrophin gene. The resulting failure to produce any dystrophin causes DMD.

isolated by chromosome walking, then conserved sequences were sought by zoo blotting and used to screen muscle cDNA libraries. Given the low abundance of dystrophin mRNA and, as we now know, the small size and widely scattered location of the exons, finding cDNA clones was far from easy, but eventually clones were identified, and subsequently the whole remarkable dystrophin gene (see *Figure 10.14*) was characterized.

A more recent example of using deletions concerns the *PHEX* gene that is mutated in the X-linked dominant vitamin D-resistant rickets (MIM 307800). The gene had been mapped to a small interval on Xp, but the demonstration of submicroscopic deletions in four out of 150 affected males allowed attention to be focused on a small part of the candidate region (HYP Consortium, 1995). Deletions are particularly helpful in X-linked conditions because in affected males there is no interference from the normal chromosome.

Microdeletions are generally believed to be the cause of a large number of unexplained genetic syndromes. Because the deleted region is small, they would be especially valuable for identifying the genes involved in the pathology. However, until recently there has been no way to search systematically for them. Hopefully the development of sensitive high-resolution Comparative Genomic Hybridization techniques

(*Box 14.5*) will remedy this. It is important to confirm suspected deletions by FISH (see *Figure 14.12*) or pulsed field gel electrophoresis and Southern blotting – failure of PCR to produce a product may be due to a sequence change in the primer binding site rather than a deletion.

14.5 Confirming a candidate gene

Candidate genes must be tested individually to see if there is good evidence that mutations in them do cause the disease in question. Demonstrating that a candidate gene is likely to be the disease locus can be done by various means.

▶ **Mutation screening.** Screening for patient-specific mutations in the candidate gene is by far the most popular method, because it is generally applicable and comparatively rapid. The reasons why particular mutations may be found in certain diseases are discussed in Chapter 16, and the methods of testing for mutations are described in Chapter 18. Identifying mutations in several unrelated affected individuals strongly suggests that the correct candidate gene has been chosen, but formal proof requires additional evidence.

▶ **Restoration of normal phenotype** *in vitro*. If a mutant phenotype is demonstrable in cells from patients, we can check whether transfection of a normal allele of the candidate gene, cloned into an expression vector (Section 5.6.1), is able to 'rescue' the mutant and restore the normal phenotype. Not all mutant phenotypes are reversible, so a negative result does not necessarily exclude that candidate.

▶ **Production of a mouse model of the disease** (Section 20.4.3). Once a putative disease gene is identified, a transgenic mouse model can be constructed, if no relevant mutant already exists. Loss of function phenotypes can be modeled by knockouts made by gene targeting the mouse germline (Section 20.4.4). For gain of function phenotypes, the disease allele must be introduced into the mouse germline. The mutant mice are expected to show some resemblance to humans with the disease, although this expectation may not always be met even when the correct gene has been identified.

14.5.1 Mutation screening to confirm a candidate gene

Mutation screening is often straightforward for diseases where a good proportion of patients carry independent mutations.

Typically these are severe early onset autosomal dominant or X-linked disorders where the disease phenotype results from loss of function of the gene. As explained in Section 16.3.2, if the correct gene is tested using one or more of the mutation screening procedures described in *Section 18.3*, a panel of samples from unrelated patients will usually show a variety of different mutations. Hopefully these will include some with an obviously deleterious effect on gene expression such as nonsense mutations, frameshifts, and so on. *Figure 16.1* shows an example. Normal controls need to be checked to prove that any changes are not common population variants.

In other circumstances the identification of mutations and the interpretation of mutation screening may be more difficult.

▶ **Unsuspected locus heterogeneity.** Often mutations in several different genes can give almost identical phenotypes, so that a panel of unselected patient samples may have pathogenic mutations in different genes. If the candidate gene being tested is responsible for only a small proportion of cases, most samples will show no mutation in that gene. Ideally, one would use only samples from families with demonstrated linkage to the candidate region, but this may be impracticable – family sizes for recessive and some dominant disorders are often too small for independent linkage analyses, and in some severe dominant disorders most patients present as sporadic cases without a family history.

▶ **Mutational homogeneity.** If most apparently unrelated patients show the same sequence change, this could be the pathogenic mutation or it could be a rare variant in strong linkage disequilibrium with the true mutation. Functional evidence is needed that the change is pathogenic.

▶ **Mutations are not unambiguously pathogenic.** It may be difficult to identify missense mutations as being pathogenic as opposed to being rare neutral variants with no major effect on gene expression. Some guidelines to help decide whether a sequence change is pathogenic are given in *Box 16.4*.

▶ *Mutations may be hard to find.* Apart from the practical problem of screening a large gene with many exons, some mutations are not identifiable by PCR testing of genomic DNA. For example detecting the large inversions that disrupt the Factor VIII gene (Section 11.5.5 and *Figure 11.20*), or the *CFTR* 3849 + 10 kb C > T mutation that

---

Box 14.5: CGH for detecting submicroscopic chromosomal imbalances.

Comparative genomic hybridization (CGH) is used to detect partial monosomies or trisomies, or chromosomal deletions or amplifications. As described in Section 17.3.3, CGH is based on making the test DNA and a control DNA compete to hybridize to a target. The target can be either a spread of normal chromosomes on a microscope slide, using standard methods for fluorescence *in situ* hybridization (Section 6.3.4, or a set of defined BAC clones arrayed on a slide (array-CGH). Chromosomal regions or BAC clones where the copy number is different in the test and control samples show up as areas or spots of different colored fluorescence signal (*Figure 17.3*). Array-CGH has the potential to allow genome-wide screens for microdeletions and microduplications in patients with congenital abnormalities.

activates a cryptic splice site deep within an intron (Section 16.4.1) would require Southern blotting or RT–PCR, respectively.

14.5.2 Once a candidate gene is confirmed, the next step is to understand its function

Identifying the gene involved in a genetic disease opens the way to several lines of investigation. The ability to identify mutations should immediately lead to improved diagnosis and counseling, as described in Chapter 18. Understanding the molecular pathology (why the mutated gene causes the disease; see Chapter 16) may also lead to insight into related diseases, and hopefully eventually to more effective treatment.

A second line of enquiry concerns the **normal function of the gene product**. Until the DMD dystrophy gene was identified we knew nothing about the way the contractile machinery of muscle cells is anchored to the sarcolemma. Analysis of functional domains and motifs and the search for experimentally manipulable homologs in the mouse, fruit fly, nematode and yeast are powerful tools for this work. These large topics are covered in Chapters 19 and 20, respectively; a foretaste of the sort of information that can be generated by database searching can be seen in the following information, taken from the Hereditary Hearing Loss Homepage (http://www.uia.ac.be/dnalab/hhh/), that describes the gene that was identified by positional cloning of an autosomal dominant hearing loss locus (*DFNA1*) in one large Costa Rican family (see OMIM entry 124900):

> The human DFNA1 protein product DIAPH1, mouse *p140mDia*, and *Drosophila diaphanous* are homologs of *Saccharomyces cervisiae* protein *Bni1p*. The proteins are highly conserved overall. The genes encoding these proteins are members of the formin gene family, which also includes the

mouse *limb deformity* gene, *Drosophila cappuccino*, *Aspergillus nidulans* gene *sepA*, and *Schizosaccharomyces pombe* genes *fus1* and *cdc12*. These genes are involved in cytokinesis and establishment of cell polarity. All formins share Rho-binding domains in their N-terminal regions, polyproline stretches in the central region of each sequence, and formin-homology domains in the C-terminal regions.

14.6 Eight examples illustrate various ways disease genes have been identified

14.6.1 Direct identification of a gene through a chromosome abnormality: Sotos syndrome

Sotos syndrome (MIM 117550) is characterized by overgrowth, dysmorphic features and mental retardation. Most cases are sporadic; affected individuals seldom reproduce, and there are no good pedigrees for linkage mapping. A Sotos patient was found with a balanced translocation 46,XX,t(5;8)(q35;q24.1). PAC/BAC and cosmid clones spanning the breakpoint were identified by FISH (as in *Figure 14.9*). Sequencing around the breakpoint revealed a partial genomic sequence homologous to the mouse *Nsd1* gene. The human *NSD1* gene was cloned and characterized, and shown to be disrupted by the translocation (*Figure 14.11*). That could be just coincidental. Proof that *NSD1* was the gene mutated in Sotos syndrome came from demonstrating point mutations in four of 38 independent Sotos patients and microdeletions (*Figure 14.12*) in 20 out of 30 (Kurotaki *et al.*, 2002). It should be noted, however, that genes may be affected by a chromosomal rearrangement even when they are not physically disrupted (*Box 14.4*), so it is not always so straightforward to discover why a translocation causes a disease.

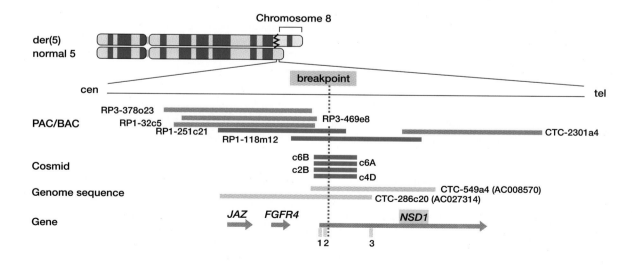

Figure 14.11: A balanced 5;8 translocation disrupts the *NSD1* gene in a patient with Sotos syndrome.

Reproduced from Kurotaki *et al.* (2002) *Nat. Genet.* **30**, 365–366 with permission from Nature Publishing Group.

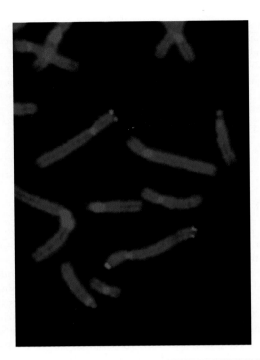

Figure 14.12: *NSD1* **microdeletion demonstrated by fluorescence** *in situ* **hybridization in a patient with Sotos syndrome.**

The two homologs of chromosome 5 are identified by the red FISH probe that recognizes a sequence on 5pter. The green FISH probe is a BAC from 5qter containing the *NSD1* gene; this sequence is lacking on one copy of chromosome 5. Reproduced from Kurotaki *et al.* (2002) *Nat. Genet.* **30**, 365–366 with permission from Nature Publishing Group.

14.6.2 Pure transcript mapping: Treacher Collins syndrome

Treacher Collins syndrome (TCS; MIM 154500) is an autosomal dominant disorder of craniofacial development with a variable phenotype including abnormalities of the external and middle ears, hypoplasia of the mandible and zygomatic complex and cleft palate. Linkage was initially established in 1991 to markers at 5q31-q34. Because the markers in that region known at the time were not very informative, new microsatellites were isolated and used to refine the candidate region to 5q32-33.1. A combined genetic and radiation hybrid map was constructed across this interval, and by 1994 the team had assembled a YAC contig. This was converted to a cosmid contig, and cDNA library screening and exon trapping were used to generate a transcript map. At least seven genes were identified in the critical region. Further rounds of marker isolation and crossover analysis produced a confusing picture of overlapping recombinations, but led eventually to isolation of a candidate incomplete cDNA from a placental library. Northern blotting and zoo blotting showed that the gene was widely expressed and conserved across species, but database searches revealed no significant

homologies. The exon–intron structure was determined, and mutation analysis demonstrated different mutations in five unrelated patients.

Isolation of the *TCOF1* gene (Treacher Collins Syndrome Collaborative Group, 1996) illustrates positional cloning in its purest form. No relevant chromosomal abnormalities were found (there were four patients with TCS who had chromosomal translocations or deletions, but markers from each of the breakpoints showed no linkage to TCS in family studies, so presumably these cases were all coincidental). There is no linkage disequilibrium – this is not surprising since 60% of cases are new mutations. The candidate region is gene-rich, so there were many possible candidates, and the gene eventually identified had no features that made it a particularly promising candidate. The gene product is a nucleolar phosphoprotein, but exactly why mutations should cause TCS is not yet known.

14.6.3 Large-scale sequencing and search for homologs: branchio-oto-renal syndrome

The autosomal dominant branchio-oto-renal syndrome (BOR; MIM 113650: branchial fistulas, malformation of the external and inner ear with hearing loss; hypoplasia or absence of kidneys) was mapped to 8q13 following a clue from an affected patient who had a rearrangement of chromosome 8. The initial interval of 7 cM was refined to 470–650 kb by further mapping and discovery of a submicroscopic chromosomal deletion in the patient mentioned. P1 and PAC clones were isolated by screening genomic libraries with markers within or close to the candidate region, and gaps in the contig were filled by chromosome walking. The minimum tiling path across the candidate region involved three P1 and three PAC clones.

It was decided to identify genes in the contig by large-scale sequencing. Checking the sequence against the EMBL and GenBank protein and nucleic acid databases revealed homology between part of the sequence obtained and the *Drosophila* developmental gene *eyes absent* (*eya*). The genomic sequence was searched for open reading frames, which were translated and compared to the *eya* amino acid sequence. This allowed identification of seven putative exons showing 69% identity and 88% similarity at the amino acid level to the putative *eya* protein. The human cDNA was then isolated from a 9-week total fetal mRNA library, and seven mutations in the gene, named *EYA1* were demonstrated in 42 unrelated BOR patients (Abdelhak *et al.*, 1997).

As genome sequence databases become more complete, the sort of analysis done here is becoming more and more the standard approach. As here, homologies with genes in distantly related organisms may be more apparent in the amino acid sequence than in the DNA sequence or phenotype. At this stage the function of the gene product was not identified, nor was it clear why the *Drosophila* phenotype consists of reduced or absent compound eyes, while humans have no eye problems. As so often with positional cloning, identifying the gene was just the start of understanding the syndrome.

14.6.4 Positional candidates defined by function: rhodopsin and fibrillin

The gene for the human photopigment rhodopsin (*RHO*) was cloned in 1984, and it was mapped to 3q21-qter in 1986. Among disorders involving hereditary retinal degeneration are the various forms of retinitis pigmentosa (RP), which are marked by progressive visual loss resulting from clumping of the retinal pigment. Although *RHO* was a possible candidate gene for some forms of RP, it was only one of many genes encoding proteins known to be involved in phototransduction. However, in 1989, linkage analysis in a large Irish RP family mapped their disease gene to 3q in the neighborhood of *RHO*. This was now a serious candidate gene, and patient-specific mutations were identified within a year (see OMIM entry 180380).

The phenotype of Marfan syndrome (MFS, MIM 154700: excessive growth of long bones; lax joints; dislocation of lenses; liability to aortic aneurysms) suggested some abnormality in connective tissue. Linkage analysis mapped the MFS gene to 15q. When the gene for the connective tissue protein fibrillin was localized to 15q21.1 by *in situ* hybridization, it became an obvious positional candidate. Mutations were soon demonstrated in MFS patients (see McKusick 1991 for discussion of the background).

14.6.5 A positional candidate identified through comparison of the human and mouse maps: *PAX3* and Waardenburg syndrome

Waardenburg syndrome type 1 (WS1, MIM 193500) illustrates the value of human–mouse comparisons. A pedigree of this autosomal dominant but variable condition was shown in *Figure 4.5C*. The characteristic pigmentary abnormalities and hearing loss of WS1 are caused by absence of melanocytes from the affected parts, including the inner ear, where melanocytes are required in the stria vascularis of the cochlea in order for normal hearing to develop. Linkage analysis, aided by the description of a chromosomal abnormality in an affected patient, localized the gene for WS1 to the distal part of 2q. This chromosomal region has strongly conserved synteny to part of mouse chromosome 1. At this point, a likely mouse homolog emerged. The *Splotch* (*Sp*) mouse mutant has pigmentary abnormalities caused by patchy absence of melanocytes. This is probably the result of a defect in the embryonic neural crest. Despite various differences between the two phenotypes, it seemed quite likely that *Sp* and WS1 were caused by mutations in orthologous genes.

Neither gene had been identified, but a positional candidate emerged when the murine *Pax-3* gene was mapped to the vicinity of the *Sp* locus. *Pax-3* encodes a transcription factor that is expressed in mouse embryos in the developing nervous system, including the neural crest. The sequence of murine *Pax-3* was almost identical to limited sequence previously published for an unmapped human genomic clone, HuP2. Such observations prompted mutation screening of *Pax-3* and HuP2 and led to identification of mutations in *Splotch* mice and in humans with WS1 (reviewed by Strachan and Read,

1994). As the underlying genes were clearly orthologs, the *HuP2* gene was re-named *PAX3*.

14.6.6 Inference from function *in vitro*: Fanconi anemia

Fanconi anemia is a recessive disorder with very diverse congenital abnormalities, especially radial aplasia, and a predisposition to bone marrow failure and malignancy, particularly acute myelogenous leukemia. The underlying problem is a defective ability to repair DNA damage. This defect can be observed in cell cultures as hypersensitivity to DNA crosslinking agents such as diepoxybutane. Cell fusion experiments allow Fanconi patients to be divided into at least eight complementation groups (A–H): when fused, cells from patients in different groups complement each other, while those from patients in the same group remain defective. By testing clones from a cDNA library for their ability to cure the diepoxybutane sensitivity of cells from a Group C Fanconi patient, Strathdee *et al.* (1992) were able to isolate the Fanconi anemia group C (*FANCC*; MIM 227645) gene. A similar approach later identified the Group A (*FANCA*) and Group G (*FANCG*) genes. In other applications of functional cloning, the ability of transferred chromosomes or clones to correct the uncontrolled growth of tumor cell lines has been used to help locate and then identify tumor suppressor genes (Section 17.4).

14.6.7 Inference from function *in vivo*: myosin 15 and *DFNB3* deafness

Occasionally a mouse disease gene has been identified by rescuing the mutant phenotype with a transgene of wild-type DNA from a candidate region. This strategy was first used to identify a clock gene (Antoch *et al.*, 1997) and more recently was a crucial step in identifying the human *DFNB3* deafness gene (Probst *et al.* 1998; *Figure 14.13*). Comparative mapping showed that *DFNB3* mapped to a position in humans corresponding to the location in the mouse of the *shaker-2* deafness gene. Transgenic *shaker-2* mice were constructed using wild-type BACs from the *shaker-2* candidate region. A BAC that corrected the phenotype was identified and turned out to contain an unconventional myosin gene, *myo15*. The human *MYO15* gene was then isolated based on its close homology to the mouse gene, its position within the *DFNB3* candidate region was confirmed, and then mutations were demonstrated in *DFNB3* affected people.

14.6.8 Inference from the expression pattern: otoferlin

cDNA libraries of genes specifically expressed in the mouse inner ear have been useful tools for identifying candidate deafness genes. The libraries are produced by *subtractive hybridization* of an inner ear cDNA library against one or more nonspecific libraries, to try to remove genes expressed in a variety of tissues – a tricky technique (Swaroop *et al.*, 1991). We have already seen the linkage analysis that mapped the *DFNB9* nonsyndromic deafness gene (*Figure 13.8*). During subsequent positional cloning (Yasunaga *et al.*, 1999) one partial gene sequence from the critical region showed 90% amino acid identity and 97% similarity to a clone from the mouse inner ear library. The gene, named otoferlin, was fully characterized and mutation screening revealed a nonsense mutation in the family.

An alternative strategy would have been to identify human homologs of the mouse clones and map them by FISH (Section 2.4.2). They would then be positional candidates for any deafness gene mapped to the corresponding position.

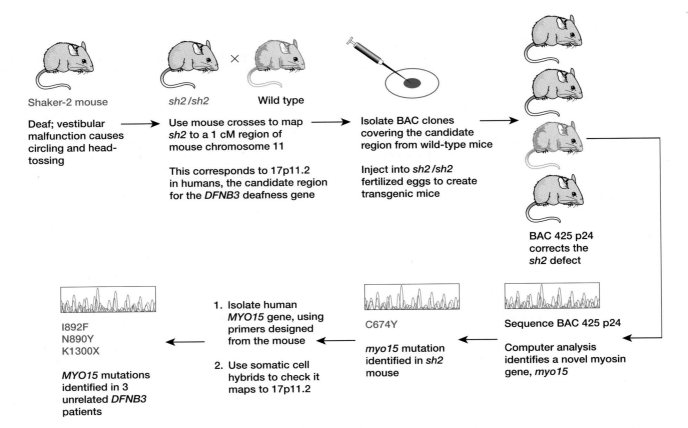

Shaker-2 mouse

Deaf; vestibular malfunction causes circling and head-tossing

sh2/sh2 × Wild type

Use mouse crosses to map *sh2* to a 1 cM region of mouse chromosome 11

This corresponds to 17p11.2 in humans, the candidate region for the *DFNB3* deafness gene

Isolate BAC clones covering the candidate region from wild-type mice

Inject into *sh2/sh2* fertilized eggs to create transgenic mice

BAC 425 p24 corrects the *sh2* defect

I892F
N890Y
K1300X

MYO15 mutations identified in 3 unrelated *DFNB3* patients

1. Isolate human *MYO15* gene, using primers designed from the mouse

2. Use somatic cell hybrids to check it maps to 17p11.2

C674Y

myo15 mutation identified in *sh2* mouse

Sequence BAC 425 p24

Computer analysis identifies a novel myosin gene, *myo15*

Figure 14.13: Functional complementation in transgenic mice as a tool for identifying a human disease gene.

The *shaker-2* mouse mutation was identified by finding a wild-type clone that corrected the defect. Human families with a similar phenotype that mapped to the corresponding chromosomal location proved to have mutations in the orthologous gene.

Further reading

Silver LM (1995) *Mouse Genetics: Concepts and Applications.* Oxford University Press, Oxford.

Wolfsberg T, Wetterstrand K, Guyer M, Collins F, Baxevanis A (2002) A user's guide to the human genome. *Nature Genet.* **32(suppl)**.

References

Abdelhak S, Kalatzis V, Heilig R *et al.* (1997). A human homologue of the *Drosophila eyes absent* gene underlies Branchio-oto-renal (BOR) syndrome and identifies a novel gene family. *Nature Genet.* **15**, 157–164.

Antoch MP, Song E-J, Chang A-M *et al.* (1997) Functional identification of the mouse circadian Clock gene by transgenic BAC rescue. *Cell* **89**, 655–667.

Bartlett SE (2001) Identifying novel proteins in nervous tissue using microsequencing techniques. *Methods Mol. Biol.* **169**, 43–50.

Brown SD, Balling R (2001) Systematic approaches to mouse mutagenesis. *Curr. Opin. Genet. Dev.* **11**, 268–273.

Carter SA, Bryce SD, Munro CS *et al.* (1994) Linkage analyses in British pedigrees suggest a single locus for Darier disease and narrow the location to the interval between *D12S105* and *D12S129 Genomics* **24**, 378–382.

Copeland N, Jenkins NA (1991) Development and applications of a molecular genetic map of the mouse genome. *Trends Genet.* **7**, 113–118.

DeBry RW, Seldin MF (1996) Human/mouse homology relationships. *Genomics*, **33**, 337–351.

Gregory SG, Sekhon M, Schein J *et al.* (2002) A physical map of the mouse genome. *Nature* **418**, 743–750.

Hughes A, Newton VE, Liu XZ, Read AP (1994) A gene for Waardenburg syndrome Type 2 maps close to the human homologue of the *microphthalmia* gene at chromosome 3p12-p14.1. *Nature Genet.* **7**, 509–512.

Hyp Consortium (1995) A gene (PEX) with homologies to endopeptidases is mutated in patients with X-linked hypophosphatemic rickets. *Nature Genet.* **11**, 130–136.

Justice MJ (2000) Capitalizing on large-scale mouse mutagenesis screens. *Nature Rev. Genet.* **1**, 109–115.

Koob MD Benzow KA, Bird TD, Day JW, Moseley ML, Ranum LP (1998) Rapid cloning of expanded trinucleotide repeat sequences from genomic DNA. *Nature Genet.* **18**, 72–75.

Koob MD, Moseley ML, Schut LJ *et al.* (1999) Untranslated CTG expansion causes a novel form of spinocerebellar ataxia (SCA8). *Nature Genet.* **21**, 379–384.

Kunkel LM, Monaco AP, Middlesworth W, Ochs HD, Latt SA (1985) Specific cloning of DNA fragments absent from the DNA of a male patient with an X chromosome deletion. *Proc. Natl Acad. Sci. USA* **82**, 4778–4782.

Kurotaki N, Imaizumi K, Harada N *et al.* (2002) Haploinsufficiency of *NSD1* causes Sotos syndrome. *Nature Genet.* **30**, 365–366.

Mann M, Hendrickson RC, Pandey A (2001) Analysis of proteins and proteomes by mass spectrometry. *Annu. Rev. Biochem.* **70**, 437–473.

McKusick VA (1991) The defect in Marfan syndrome. *Nature* **352**, 279–281.

Pingault V, Bondurand N, Kuhlbrodt K *et al.* (1998) *SOX10* mutations in patients with Waardenburg-Hirschsprung disease. *Nature Genet.* **18**, 171–173.

Poustka A, Pohl TM, Barlow DP, Frischauf AM, Lehrach H (1987) Construction and use of human chromosome jumping libraries from *Not*I-digested DNA. *Nature* **325**, 353–355.

Probst FJ, Fridell RA, Raphael Y *et al.* (1998) Correction of deafness in *shaker-2* mice by an unconventional myosin in a BAC transgene. *Science* **280**, 1444–1447. See also the accompanying paper, **Wang A, Liang Y, Fridell RA** *et al.* (1998) Association of unconventional myosin *MYO15* mutations with human nonsyndromic deafness *DFNB3*. *Science* **280**, 1447–1451.

Putnam EA, Zhang H, Ramirez F, Milewicz DM (1995) Fibrillin-2 (*FBN2*) mutations result in the Marfan-like disorder, congenital contractural arachnodactyly. *Nature Genet.* **11**, 456–458.

Reymond A, Camargo AA, Deutsch S *et al.* (2002) Nineteen additional unpredicted transcripts from human chromosome 21. *Genomics* **79**, 824–832.

Rincón-Limas DE, Lu C-H, Canal I *et al.* (1999) Conservation of the expression and function of *apterous* orthologs in *Drosophila* and mammals. *Proc. Natl Acad. Sci. USA* **96**, 2165–2170.

Robson KJH, Chandra T, MacGillivray RTA, Woo SLC (1982) Polysome immunoprecipitation of phenylalanine hydroxylase mRNA from rat liver and cloning of its cDNA. *Proc. Natl Acad. Sci. USA* **79**, 4701–4705.

Rommens JM, Januzzi MC, Kerem B-S *et al.* (1989) Identification of the cystic fibrosis gene: chromosome walking and jumping. *Science* **245**, 1059–1065.

Royer-Pokora B, Kunkel LM, Monaco AP *et al.* (1985) Cloning the gene for an inherited human disorder – chronic granulomatous disease – on the basis of its chromosomal location. *Nature* **322**, 32–38.

Schalling M, Hudson TJ, Buetow KH, Housman DE (1993) Direct detection of novel expanded trinucleotide repeats in the human genome. *Nature Genet.* **4**, 135–139.

Steinmetz LM, Scharfe C, Deutschbauer AM *et al.* (2002) Systematic screen for human disease genes in yeast. *Nature Genet.* **31**, 400–404.

Strachan T, Read AP (1994) PAX genes. *Curr. Opin. Genet. Dev.* **4**, 427–438.

Strathdee CA, Gavish H, Shannan WR, Buchwald M (1992) Cloning of cDNAs for Fanconi's anemia by functional complementation. *Nature* **356**, 763–767.

Swaroop A, Xu J, Agarwal N, Weissman SM (1991). A simple and efficient cDNA library subtraction procedure: isolation of human retina-specific cDNA clones. *Nucl. Acids Res.* **19**, 1954.

Treacher Collins Syndrome Collaborative Group (1996) Positional cloning of a gene involved in the pathogenesis of Treacher Collins syndrome. *Nature Genet.* **12**, 130–136.

Worton RG, Thompson MW (1988) Genetics of Duchenne muscular dystrophy. *Annu. Rev. Genet.* **22**, 601–629.

Yasunaga S, Grati M, Cohen-Salmon M *et al.* (1999). A mutation in *OTOF*, encoding otoferlin, a FER-1-like protein, causes *DFNB9*, a nonsyndromic form of deafness. *Nature Genet.* **21**, 363–369.

CHAPTER FIFTEEN

Mapping and identifying genes conferring susceptibility to complex diseases

Chapter contents

Throughout the world the main genetic contribution to morbidity and mortality is through the genetic component of common diseases. Thus identifying the genes concerned is a central task for medical research. A logical sequence for research on any complex disease would be:

▶ perform **family, twin or adoption studies** to check that susceptibility is at least partly genetic;

▶ use **segregation analysis** to estimate the type and frequency of susceptibility alleles;

▶ map susceptibility loci by **linkage analysis**, usually of affected sib pairs;

▶ narrow down the candidate region by studying **population associations**;

▶ identify the DNA sequence variants conferring susceptibility, and define their biochemical action.

Below, we go through these steps, and follow this by discussing eight specific diseases to illustrate what can happen in practice. Finally, we discuss the still open question of whether this whole paradigm for identifying susceptibility factors is the key to the future of medical genetics or whether, as some maintain, it is unlikely to be generally successful.

15.1 Deciding whether a non-Mendelian character is genetic: the role of family, twin and adoption studies

15.1.1 The λ value is a measure of familial clustering

Nobody would dispute the involvement of genes in a character that consistently gives Mendelian pedigree patterns or that is associated with a chromosomal abnormality. However, with non-Mendelian characters, whether **continuous (quantitative)** or **discontinuous (dichotomous)**, it is

necessary to prove claims of genetic determination. The obvious way to approach this is to show that the character runs in families. The degree of family clustering of a disease can be expressed by the quantity λ_R, *the risk to relative R of an affected proband compared with the population risk.* Separate values can be calculated for each type of relative, for example λ_s for sibs. The mathematical properties of λ_R are derived by Risch (1990a). *Table 15.1* shows pooled data from a number of studies of schizophrenia. Family clustering is evident from the raised λ values and, as expected, λ values drop back towards 1 for more distant relationships.

15.1.2 The importance of shared family environment

Geneticists must never forget that parents give their children their environment as well as their genes. Many characters run in families because of the *shared family environment* – whether one's native language is English or Chinese, for example. One has therefore always to ask whether shared environment might be the explanation for a familial character. This is especially important for behavioral attributes like IQ or schizophrenia, which depend at least partly on upbringing. Even for physical characters or birth defects it cannot be ignored: a family might share an unusual diet or some traditional medicine that could cause developmental defects. Something more than a familial tendency is necessary to prove that a non-Mendelian character is under genetic control. These reservations are not always as clearly stated in the medical literature as perhaps they should be. *Table 15.5* shows what can happen if shared family environment is ignored.

15.1.3 Twin studies suffer from many limitations

Francis Galton, who laid so much of the foundation of quantitative genetics, pointed out the value of twins for human

Table 15.1: Risk of schizophrenia among relatives of schizophrenics: pooled results of several studies

Relative	No. at risk[a]	Risk, %	λ[b]
Parents	8020	5.6	7
Sibs	9920.7	10.1	12.6
Sibs, one parent affected	623.5	16.7	20.8
Offspring	1577.3	12.8	16
Offspring, both parents affected	134	46.3	58
Half-sibs	499.5	4.2	5.2
Uncles, aunts, nephews, nieces	6386.5	2.8	3.5
Grandchildren	739.5	3.7	4.6
Cousins	1600.5	2.4	3

[a]Numbers at risk are corrected to allow for the fact that some at-risk relatives were below or only just within the age of risk for schizophrenia (say, 15–35 years).
[b]λ Values are calculated assuming a population incidence of 0.8%.
Data assembled by McGuffin (1984).

genetics. *Monozygotic (MZ) twins are genetically identical clones* and will necessarily be **concordant** (both the same) for any genetically inherited character. This is true regardless of the mode of inheritance or number of genes involved; the only exceptions are for characters dependent on postzygotic somatic genetic changes (the pattern of X-inactivation in females, the repertoire of functional immunoglobulin and T-cell receptor genes etc.). *Dizygotic (DZ) twins share half their genes on average,* the same as any pair of sibs. Genetic characters should therefore show a higher concordance in MZ than DZ twins, and many characters do (*Table 15.2*).

A higher concordance in MZ compared to DZ twins does not, however, prove a genetic effect. For a start, half of DZ twins are of unlike sex, whereas all MZ twins are the same sex. Even if the comparison is restricted to same-sex DZ twins (as it is in the studies shown in *Table 15.2*), at least for behavioral traits the argument can be made that MZ twins are more likely to be very similar, to be dressed and treated the same, and thus to share more of their environment than DZ twins.

MZ twins separated at birth and brought up in entirely separate environments would provide an ideal experiment (Francis Crick once made the tongue-in-cheek suggestion that one of each pair of twins born should be donated to science for this purpose). Such separations happened in the past more often than one might expect because the birth of twins was sometimes the last straw for an overburdened mother. Fascinating television programs can be made about twins reunited after 40 years of separation, who discover they have similar jobs, wear similar clothes and like the same music. As research material, however, separated twins have many drawbacks:

▶ any research is necessarily based on small numbers of arguably exceptional people;

▶ the separation was often not total – often they were separated some time after birth, and brought up by relatives;

▶ there is a **bias of ascertainment** – everybody wants to know about strikingly similar separated twins, but separated twins who are very different are not newsworthy;

▶ even in principle, research on separated twins cannot distinguish *intrauterine environmental causes* from genetic causes. This may be important, for example in studies of sexual orientation ('the gay gene'), where some people have suggested that maternal hormones may affect the fetus *in utero* so as to influence its future sexual orientation.

Thus, for all their anecdotal fascination, separated twins have contributed relatively little to human genetic research.

15.1.4 Adoption studies: the gold standard for disentangling genetic and environmental factors

If separating twins is an impractical way of disentangling heredity from family environment, *adoption* is much more promising. Two study designs are possible:

▶ find adopted people who suffer from a particular disease known to run in families, and ask whether it runs in their biological family or their adoptive family;

▶ find affected parents whose children have been adopted away from the family, and ask whether being adopted away saved the children from the family disease.

A celebrated (and controversial) study by Rosenthal and Kety (see Further reading) used the first of these designs to test for genetic factors in schizophrenia. The diagnostic criteria used in this study have been criticized; there have also been claims (disputed) that not all diagnoses were made truly blind. However, an independent re-analysis using DSM-III diagnostic criteria (Kendler *et al.*, 1994) reached substantially the same conclusions. *Table 15.3* shows the results of a later extension of this study (Kety *et al.*, 1994).

The main obstacle in adoption studies is *lack of information about the biological family*, frequently made worse by the undesirability of approaching them with questions. Efficient adoption registers exist in only a few countries. A secondary problem is *selective placement*, where the adoption agency, in the interests of the child, chooses a family likely to resemble the biological family. Adoption studies are unquestionably the gold standard for checking how far a character is genetically determined, but because they are so difficult, they have in the main been performed only for psychiatric conditions, where the nature–nurture arguments are particularly contentious.

15.2 Segregation analysis allows analysis of characters that are anywhere on the spectrum between purely Mendelian and purely polygenic

As we saw in *Figure 4.6*, pure Mendelian and pure polygenic characters represent the opposite ends of a continuum. In between are **oligogenic traits** governed by a few major susceptibility loci, maybe operating against a polygenic background, and maybe subject to major environmental

Table 15.2: Twin studies in schizophrenia

Study	Concordant MZ pairs	Concordant DZ pairs
Kringlen, 1968	14/55 (21/55)	4–10%
Fischer, 1969	5/21 (10/21)	10–19%
Tienari, 1975	3/20 (5/16)	3/42
Farmer, 1987	6/16 (10/20)	1/21 (4/31)
Onstad, 1991	8/24	1/28

The numbers show pairwise concordances, i.e. counts of the number of concordant (+/+) and discordant (+/–) pairs ascertained through an affected proband. Figures in brackets are obtained using a wider definition of affected, including borderline, phenotypes. Concordances can also be calculated probandwise, counting a pair twice if both were probands. This gives higher values for the MZ concordance. Probandwise concordances are thought to be more comparable with other measures of family clustering. Only the studies of Onstad and Farmer use the current standard diagnostic criteria, DSM-III. For references, see Onstad *et al.* (1991) and Fischer *et al.* (1969).

Table 15.3: An adoption study in schizophrenia

	Schizophrenia cases among biological relatives	Schizophrenia cases among adoptive relatives
Index cases (47 chronic schizophrenic adoptees)	44/279 (15.8%)	2/111 (1.8%)
Control adoptees (matched for age, sex, social status of adoptive family and number of years institutionalized)	5/234 (2.1%)	2/117 (1.7%)

The study involved 14 427 adopted persons aged 20–40 years in Denmark, 47 of whom were diagnosed as chronic schizophrenic. The 47 were matched with 47 nonschizophrenic control subjects from the same set of adoptees. Data of Kety *et al.* (1994).

influences. **Segregation analysis** is the main statistical tool for analyzing the inheritance of any character. It can provide evidence for or against a major susceptibility locus and at least partly define its properties. The results can help guide future linkage or association studies.

15.2.1 Bias of ascertainment is often a problem with family data: the example of autosomal recessive conditions

Disease studies rely on collections of cases and families, so a first step is to consider what biases the method of ascertainment may impose on the raw data. Segregation analysis requires large datasets and is very sensitive to subtle biases in the way the data are collected. This can be illustrated by a Mendelian example. Suppose we wish to show that a condition is autosomal recessive. We could collect a set of families and check that the **segregation ratio** (the proportion of affected children) is 1 in 4. At first sight this would seem a trivial task, provided the condition is not too rare. But in fact the expected proportion of affected children in our sample is not 1 in 4. The problem is **bias of ascertainment**.

Assuming there is no independent way of recognizing carriers, the families will be identified through an affected child. Thus the families shown unshaded in *Figure 15.1* will not be ascertained, and the observed segregation ratio in the two-child families collected is not 1/4 but 8/14. Families with three children, ascertained in the same way, would give a different segregation ratio, 48/111. The ratio for any given family size can be estimated from the **truncated binomial distribution**, a binomial expansion of $(\frac{1}{4} + \frac{3}{4})^n$ in which the last term (no affected children) is omitted. Experimental data can be corrected for this bias, most simply by the method of Li and Mantel shown in *Box 15.1*.

The example above presupposes **complete truncate ascertainment**: we collect all families from some defined population who have at least one affected child. But this is not the only possible way of collecting families. We might have ascertained affected children by taking the first 100 to be seen in a busy clinic (so that many more could have been ascertained from the same population by carrying on for longer). Under these conditions, a family with two affected children is twice as likely to be picked up as one with only a single affected child, and one with four affected is four times as likely.

Single selection, where the probability of being ascertained is proportional to the number of affected children in the family, introduces a different bias of ascertainment, and requires a different statistical correction (see *Box 15.1*). We see that *working out a segregation ratio requires data that have been collected in accordance with an explicit scheme of ascertainment*, so that appropriate corrections can be applied.

15.2.2 Complex segregation analysis is a general method for estimating the most likely mix of genetic factors in pooled family data

Analyzing data on the relatives of a large collection of people affected by a familial but non-Mendelian disease is not a simple task. There could be both genetic and environmental factors at work; the genetic factors could be polygenic, oligogenic or Mendelian with any mode of inheritance, or any mixture of these, while the environmental factors may include both familial and nonfamilial variables. In complex segregation analysis a whole range of possible mechanisms, gene frequencies, penetrances, etc., are allowed, and the computer performs a maximum likelihood analysis to find the mix of parameter values that gives the greatest overall likelihood for the observed data. *Table 15.4* shows an example. As with lod score analysis (Chapter 13), the question asked is how much more likely are the observations on one hypothesis compared with another.

In the example of *Table 15.4*, the ability of specific models (sporadic, polygenic, dominant, recessive) to explain the data was compared with the likelihood calculated by a general model ('mixed model'), in which the computer could freely optimize the mixture of single-gene, polygenic and random environmental causes. All models were constrained by overall incidences, sex ratios and probabilities of ascertainment estimated from the collected data. A single-locus dominant model is not significantly worse than the mixed model at explaining the data ($\chi^2 = 2.8$; $p = 0.42$), while models assuming no genetic factors, pure polygenic inheritance or pure recessive inheritance perform very badly. On the argument that simple explanations are preferable to complicated explanations, the analysis suggests the existence of a **major dominant susceptibility** to Hirschsprung disease. Several such factors have now been identified (Section 15.6.2).

However clever the segregation analysis program, it can only maximize the likelihood across the parameters it was

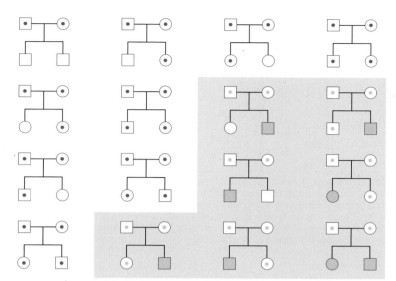

Total children : AA $\frac{8}{32}$ Aa $\frac{16}{32}$ aa $\frac{8}{32}$

Ascertained :   AA $\frac{2}{14}$ Aa $\frac{4}{14}$ aa $\frac{8}{14}$

Correction (Li–Mantel): $p = (R - S)/(T - S) = (8 - 6)/(14 - 6) = \frac{2}{8}$
$= 0.25$

Figure 15.1: Biased ascertainment of families with an autosomal recessive condition (complete truncate ascertainment).

Both parents are carriers of an autosomal recessive condition. Overall, one child in four is affected – but if the families are ascertained through affected children, only the families shown in the shaded area will be picked up, and the proportion of affected children is 8 out of 14. The true ratio is recovered by using the Li-Mantel correction (*Box 15.1*).

given. If a major factor is omitted, the result can be misleading. This was well illustrated by the data of McGuffin and Huckle (*Table 15.5*). They asked their classes of medical students which of their relatives had attended medical school. When they fed the results through a segregation analysis program, it came up with results apparently favoring the existence of a recessive gene for attending medical school. Though amusing, this was not done as a joke, nor to discredit segregation analysis. The authors did not allow the computer to consider the likely true mechanism, shared family environment. The computer's next best alternative was mathematically valid but biologically unrealistic. The serious point McGuffin and Huckle were making was that there are many pitfalls in segregation analysis of human behavioral traits, and incautious analyses can generate spurious genetic effects.

15.3 Linkage analysis of complex characters

15.3.1 Standard lod score analysis is usually inappropriate for non-Mendelian characters

Standard lod score analysis is called **parametric** because it requires a precise genetic model, detailing the mode of inheritance, gene frequencies and penetrance of each genotype. As long as a valid model is available, parametric linkage provides a wonderfully powerful method for scanning the genome in 20-Mb segments to locate a disease gene. For Mendelian characters, specifying an adequate model should be no great problem. Non-Mendelian conditions, however, are much less tractable.

Box 15.1: Correcting the segregation ratio.

Complete truncate ascertainment: $p = (R–S)/(T–S)$

Single selection: $p = (R–N)/(T–N)$

p = true (unbiased) segregation ratio

R = number of affected children

S = number of affected singletons (children who are the only affected child in the family)

T = total number of children

N = number of sibships

Table 15.4: Complex segregation analysis

Model	d	t	q	H	z	x	χ^2	p
Mixed	1.00	7.51	9.6×10^{-6}		0.01	0.15		
Sporadic							334	$<1 \times 10^{-5}$
Polygenic				1.00	1.00		78	$<1 \times 10^{-5}$
Major recessive locus	0.00	8.22	3.8×10^{-3}				35	$<1 \times 10^{-5}$
Major dominant locus	1.00	7.56	1.2×10^{-5}			0.19	2.8	0.42

Data are for families ascertained through a proband with long-segment Hirschsprung disease. Parameters that can be varied are t (the difference in liability between people homozygous for the low-susceptibility and the high-susceptibility alleles of a major susceptibility gene, measured in units of standard deviation of liability), d (the degree of dominance of any major disease allele), q (the gene frequency of any major disease allele), H (the proportion of total variance in liability which is due to polygenic inheritance, in adults), z (the ratio of heritability in children to heritability in adults) and x (the proportion of cases due to new mutation). A single major locus encoding dominant susceptibility explains the data as well as a general model in which a mix of all mechanisms is allowed. Data of Badner *et al.*, 1990.

Table 15.5: A recessive gene for attending medical school?

Model	d	t	q	H	χ^2	p
Mixed	0.087	4.04	0.089	0.008		
Sporadic					163	$<1 \times 10^{-5}$
Polygenic				0.845	14.4	<0.005
Major recessive locus	0.00	7.62	0.88		0.11	N.S.

Data of McGuffin and Huckle (1990) from a survey of medical students and their families. Meaning of symbols as in *Table 15.4*. 'Affected' is defined as attending medical school. The analysis appears to support recessive inheritance, since this accounts for the data equally well as the unrestricted model. The point of this work is to illustrate how analysis of family data can produce spurious results if shared family environment is ignored (see text).

The crucial problem of diagnostic criteria

A major problem is establishing diagnostic criteria that are relevant for genetic analysis. With Mendelian syndromes it is usually fairly obvious which features of a patient form part of the syndrome and which are coincidental. Different features may have different penetrances, but basically the components of the syndrome are those that co-segregate in a Mendelian pattern. No such reality check exists for non-Mendelian conditions. Great efforts are made, especially with psychiatric diseases, to establish diagnostic categories that are *valid*, in the sense that two independent investigators will agree whether or not a certain label applies to a given patient. But a diagnostic label can be valid without being biologically meaningful. Especially for psychiatric and behavioral phenotypes, the diagnostic criteria are often biologically arbitrary. Adhering to them helps make different studies comparable, but does not guarantee that the right genetic question is being asked.

Linkage analysis on 'near-Mendelian' families

Once diagnostic criteria are agreed, one approach to genetic analysis is to look for a subset of families in which the condition segregates in a near-Mendelian manner. Segregation analysis is used to define the parameters of a genetic model in those families, which are then used in a standard (parametric) linkage analysis. Such families may arise in three ways:

▶ any complex disease is likely to be heterogeneous, so the family collection may well include some with Mendelian

conditions phenotypically indistinguishable from the non-Mendelian majority;

▶ the near-Mendelian families may represent cases where, by chance, many determinants of the disease are already present in most people, so that the balance is tipped by the Mendelian segregation of just one of the many susceptibility factors;

▶ the 'near-Mendelian' pattern may be spurious – just chance aggregations of affected people within one family.

In the first case, identifying the Mendelian subset is intrinsically valuable, but does not necessarily cast any light on the causes of the non-Mendelian disease. That was the case with breast cancer (Section 15.6.1) and Alzheimer disease (Section 15.6.3). In the second case, the loci mapped are also susceptibility factors for the common non-Mendelian disease – Hirschsprung disease (Section 15.6.2) provides examples. Finally, early work on schizophrenia exemplified the third case, producing a lod score of 6 that is now generally agreed to have been spurious (see Byerley, 1989). This debâcle was enough to persuade most investigators to switch to nonparametric analysis.

15.3.2 Non-parametric linkage analysis does not require a genetic model

Model-free or **nonparametric** methods of linkage analysis look for alleles or chromosomal segments that are shared by affected individuals. Some of the basic ideas underlying these approaches were set out in three papers by Neil Risch in 1990 (Risch 1990a, b, c). **Shared segment methods** can be used within families (Section 15.3.3), or in whole populations (Section 15.4).

It is important to distinguish segments **identical by descent (IBD)** from those **identical by state (IBS)**. Alleles IBD are *demonstrably* copies of the same ancestral (usually parental) allele. IBS alleles look identical, and may indeed be so, but their common ancestry is not demonstrable, therefore they must be treated mathematically in terms of population frequency rather than Mendelian probability of inheritance from the defined common ancestor. *Figure 15.2* illustrates the difference. For very rare alleles, two independent origins are unlikely, so IBS generally implies IBD, but this is not true for common alleles. Multiallele microsatellites are more efficient than two–allele markers for defining IBD, and multilocus multiallele haplotypes are better still, because any one haplotype is likely to be rare. Shared segment analysis can be conducted using either IBS or IBD data, provided the appropriate analysis is used. IBD is the more powerful, but requires samples from more relatives. In a complex pedigree with several affected people, it is possible to use marker information to calculate the probability that a pair of affected relatives share haplotypes IBD (Arnos *et al.*, 1990).

15.3.3 Shared segment analysis in families: affected sib pair and affected pedigree member analysis

Picking a chromosomal segment at random, pairs of sibs are expected to share 0, 1 or 2 parental haplotypes with frequency ¼, ½ and ¼, respectively. However, if both sibs are affected by a genetic disease, then they are likely to share whichever segment of chromosome carries the disease locus. If everybody with the disease carried a mutant allele at this locus, then they would share at least one parental haplotype if the disease is dominant, and two if the disease is recessive (*Figure 15.3*). This allows a simple form of linkage analysis. **Affected sib pairs (ASP)** are typed for markers, and chromosomal regions sought where the sharing is above the random 1 : 2 : 1 ratios of sharing 2, 1 or 0 haplotypes identical by descent. If the sib pairs are tested only for identity by state, the expected sharing on the null hypothesis must be calculated as a function of the gene frequencies. ASP analysis can be performed without making any assumptions about the genetics of the disease, and it is usually much easier to collect affected sib pairs than extended families. **Multipoint analysis** is preferable to single-point analysis because it more efficiently extracts the information about IBD sharing across the chromosomal region. The MAPMAKER/SIBS program of Kruglyak and Lander (1995) is widely used to analyze multipoint ASP data and produce **nonparametric lod (NPL) scores**.

One drawback of ASP analysis is that candidate regions it identifies are usually impossibly large for positional cloning. Few recombinants separate sibs, so they share large parental chromosome segments, whether by chance or because of a shared susceptibility. Crucially, complex disease analysis has no process analogous to the end-game of Mendelian mapping, where closer and closer markers are tested until there are no more recombinants. If a susceptibility factor is neither necessary nor sufficient for disease, then not all affected sib pairs will share the relevant chromosomal segment. Moreover, sib pairs share many segments by chance. Nevertheless, because of its simplicity and robustness, ASP mapping has been one of the main tools for seeking genes conferring susceptibility to common non-Mendelian diseases (see Section 15.6). The mathematics of ASP analysis are detailed by Sham and Zhao (see Further reading).

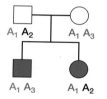
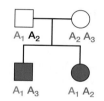

Figure 15.2: Identity by state (IBS) and identity by descent (IBD).

Both sib pairs share allele A₁. The first sib pair have two independent copies of A₁ (IBS but not IBD); the second sib pair share copies of the same paternal A₁ allele (IBD). The difference is only apparent if the parental genotypes are known.

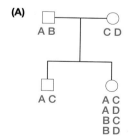
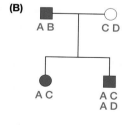
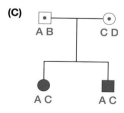

Figure 15.3: Affected sib pair analysis.

(A) By random segregation sib pairs share 0, 1 or 2 parental haplotypes ¼, ½ and ¼ of the time, respectively. **(B)** Pairs of sibs who are both affected by a dominant condition share either one or two copies of the relevant parental chromosomal segment. **(C)** Pairs of sibs who are both affected by a recessive condition necessarily share both parental haplotypes for the relevant chromosomal segment. Above-random haplotype sharing by affected sib pairs identifies chromosomal segments containing susceptibility genes.

Programs such as **GENEHUNTER** (Section 13.6.2) extend shared segment analysis to other relationships. The programs calculate the extent to which affected relatives share alleles identical by descent, and compare the result across all affected pedigree members with the null hypothesis of simple Mendelian segregation (markers should segregate according to Mendelian ratios unless the segregation is distorted by linkage or association). The comparison can be used to compute a nonparametric lod score.

15.3.4 Thresholds of significance are an important consideration in analysis of complex diseases

Whereas most Mendelian genes that have been localized by significant lod scores have subsequently been successfully cloned, the history of complex disease analysis has been marked by a succession of irreproducible results. Candidate regions defined in independent studies of the same disease have all too seldom coincided. Risch and Botstein (1996) outline a typical history, that of manic-depressive psychosis, and Altmüller *et al.* (2001) hammer the message home in a meta-analysis of 101 linkage studies in 31 complex diseases. Whatever the exact cause of these problems in individual cases, a clear common thread is the difficulty of deciding when to call the results significant.

The problems of deciding *appropriate thresholds of significance* are partly technical and partly philosophical. We have already noted the distinction between **pointwise** (or **nominal**) and **genome-wide significance** (Section 13.3.4).

▶ the **pointwise *p* value** of a linkage statistic is the probability of exceeding the observed value at a specified position in the genome, assuming the null hypothesis of no linkage;

▶ the **genome-wide *p* value** is the probability that the observed value will be exceeded anywhere in the genome, assuming the null hypothesis of no linkage.

For a whole-genome study, the appropriate significance threshold is a value where the probability of finding a false positive *anywhere in the genome* is 0.05. Theoretical arguments (Lander and Kruglyak, 1995) suggest genome-wide lod score thresholds of 3.6 for IBD testing of affected sib pairs, and 4.0 for IBS testing. Complex disease studies often estimate their significance thresholds by simulation. Typically, 1000 replicates of the family collection are generated by computer with random marker genotypes, but based on correct allele frequencies, recombination fractions etc. A whole-genome search is conducted in each simulated dataset and the maximum lod score noted. The genome-wide threshold of significance is taken as a score that is exceeded in less than 5% of the replicates.

In response to the frequent failure to replicate claimed localizations of disease susceptibility genes, Lander and Kruglyak (1995) proposed the series of thresholds in *Table 15.6*. Note that a pointwise *p* value of 1×10^{-5} is *not* equivalent to a lod score of 5.0 – the two measures are not the same:

▶ a lod score of 5 means that the data are 10^5 times more likely on the given linkage hypothesis than on the null hypothesis;

▶ a *p* value of 10^{-5} means that the stated lod score will be exceeded only once in 10^5 times, given the null hypothesis.

For some discussion of the Lander and Kruglyak criteria, see the correspondence section of the April 1996 number of *Nature Genetics*.

15.4 Association studies and linkage disequilibrium

Association is not a specifically genetic phenomenon; it is simply a statistical statement about the co-occurrence of alleles or phenotypes. Allele A is associated with disease D if people who have D also have A significantly more often (or maybe less often) than would be predicted from the individual frequencies of D and A in the population. For example, HLA-DR4 is found in 36% of the general UK population but in 78% of people with rheumatoid arthritis.

15.4.1 Why associations happen

A population association can have many possible causes, not all genetic:

Table 15.6: Suggested criteria for reporting linkage (Lander and Kruglyak 1995). The figures for *p* values and lod scores are from Altmüller *et al.* (2001).

Category of linkage	Expected number of occurrences by chance in a whole genome scan	Range of approximate *p* values	Range of approximate lod scores
Suggestive	1	7×10^{-4}–3×10^{-5}	2.2–3.5
Significant	0.05	2×10^{-5}–4×10^{-7}	3.6–5.3
Highly significant	0.001	$\leq 3 \times 10^{-7}$	≥ 5.4
Confirmed	0.01 in a search of a candidate region that gave significant linkage in a previous independent study		

▸ **direct causation:** having allele A makes you susceptible to disease D. Possession of A may be neither necessary nor sufficient for somebody to develop D, but it increases the likelihood;

▸ **natural selection:** people who have disease D might be more likely to survive and have children if they also have allele A;

▸ **population stratification:** the population contains several genetically distinct subsets, and both the disease and allele A happen to be particularly frequent in one subset. Lander and Schork (1994) give the example of the association in the San Francisco Bay area between HLA-A1 and ability to eat with chopsticks. HLA-A1 is more frequent among Chinese than among Caucasians;

▸ **type 1 error:** association studies normally test a large number of markers for association with a disease. Even without any true effect, 5% of results will be significant at the $p = 0.05$ level and 1% at the $p = 0.01$ level. The raw p values need correcting for the number of questions asked (Section 15.4.4). In the past, researchers often applied inadequate corrections, and associations were reported that could not be replicated in subsequent studies;

▸ **linkage disequilibrium (LD):** the goal of association studies in complex disease is to discover associations caused by LD between the marker and disease. The phenomenon of LD is discussed below and *Box 15.2* describes how LD is measured.

15.4.2 Association is in principle quite distinct from linkage, but where the family and the population merge, linkage and association merge

In principle linkage and association are totally different phenomena. Linkage is a relation between *loci*, but association is a relation between *alleles* or *phenotypes*. Linkage is a specifically genetic relationship while association, as mentioned above, is simply a statistical observation that might have various causes.

Linkage does not of itself produce any association in the general population. For example, the *STR45* marker locus is *linked* to the dystrophin locus. Nevertheless, the distribution of *STR45* alleles among a set of unrelated Duchenne dystrophy

patients is just the same as in the general population. However, within a family where a dystrophin mutation is segregating, we would expect affected people to share the same allele of *STR45*, because the loci are tightly linked. Thus *linkage creates associations within families, but not among unrelated people*. However, if two supposedly unrelated people with disease D have actually inherited their disease from a distant common ancestor, they may well also tend to share particular ancestral alleles at loci closely linked to D. Section 13.5.2 showed examples of this phenomenon.

Common ancestors are important because we all have them. All humans are related, if we go back far enough. In so far as a population is one extended family, population-level associations due to LD should exist between ancestral disease susceptibility genes and closely linked markers. A rough calculation suggests that in the UK two 'unrelated' people would typically share common ancestors not more than 22 generations ago. If fully outbred, they would have $2^{22} = 4$ million ancestors each at that time. Twenty-two generations is about 500 years, and in 1500 the population of Britain was around 4 million (*Figure 15.4*).

Suppose the two 'unrelated' people each inherit à disease susceptibility allele from their common ancestor. During the many generations and many meioses that separate them from their common ancestor, repeated recombination will have reduced the shared chromosomal segment to a very small region. Only alleles at loci tightly linked to the disease susceptibility locus will still be shared. For a locus showing recombination fraction θ with the susceptibility locus, a proportion θ of ancestral chromosomes will lose the association each generation, and a proportion $(1-\theta)$ will retain it. After n meioses, a fraction $(1-\theta)^n$ of chromosomes will retain the association. The half-life of LD between loci 1 cM and 2 cM apart is 69 and 34 meioses respectively, since $(0.99)^{69} \cong (0.98)^{34} \cong 0.5$. We calculated above that the ancestry of two 'unrelated' British people completely merges 22 generations back. That calculation was grossly simplified because it assumed the entire British population has been one freely interbreeding unit over the past 500 years. However, it provides a first crude estimate that allelic associations reflecting shared ancestral segments might begin to be noticeable for loci within 1 cM of

---

Box 15.2: Measures of linkage disequilibrium.

If two loci have alleles A,a and B,b with frequencies p_A, p_a, p_B and p_b, there are four possible haplotypes AB, Ab, aB and ab. Let the frequencies of the four haplotypes be p_{AB}, p_{Ab}, p_{aB} and p_{ab}. If there is no LD, $p_{AB} = p_A p_B$ and so on. The degree of departure from this random association can be measured by $D = p_{AB}p_{ab} - p_{Ab}p_{aB}$.

As a measure of LD, D suffers from the property that its maximum absolute value depends on the gene frequencies at the two loci, as well as on the extent of disequilibrium. Among preferred measures are:

▸ $D' = (p_{AB} - p_A p_B) / D_{max}$, where D_{max} is the maximum value of $|p_{AB} - p_A p_B|$ possible with the given allele frequencies

▸ $\Delta^2 = (p_{AB} - p_A p_B)^2 / (p_A p_a p_B p_b)$

D' is the most widely used. It varies between 0 (no LD) and ± 1 (complete association) and is less dependent than D on the allele frequencies. As a rule of thumb, $D' > 0.33$ is often taken as the threshold level of LD above which associations will be apparent in the usual size of dataset. The proliferation of alternative measures suggests that none is ideal (Devlin and Risch, 1995). Particularly, these measures are all developed for pairs of loci, while most whole-genome scans use multipoint analyses. Such data should be inspected for conserved haplotypes, and not simply analyzed for pairwise LD.

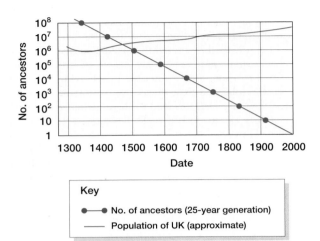

Key

●—● No. of ancestors (25-year generation)

—— Population of UK (approximate)

Figure 15.4: Merging into the gene pool.

A fully-outbred person has 2^n ancestors n generations ago. If the UK population were fully outbred, two 'unrelated' present-day people would share all the same ancestors in 1500. In reality, of course, the population is not fully outbred, and the two people would have strongly overlapping but not identical pools of ancestors in 1500.

each other in the British population. More sophisticated calculations use a Poisson distribution of recombination events and incorporate assumptions about population structure and history (Kruglyak, 1999). A key determinant is the **coalescence time** – the number of generations back to the most recent common ancestor (see *Box 12.6*).

However, the wide stochastic variance and reliance on unknowable details of population history make even the most elaborate calculations unreliable. What is needed is data, and recently increasing amounts of real data have become available.

15.4.3 Many studies show islands of linkage disequilibrium separated by recombination hotspots

The cystic fibrosis gene was identified by climbing up a gradient of LD until a maximum was reached (Section 13.5.2). However, research into other diseases soon showed that smooth gradients of LD were the exception rather than the norm. For Huntington disease (*Figure 15.5*) we can see cases of a strong association with a more distant marker and a weak association with a closer marker. Even more curious, the marker *D4S95*, closely linked to the *HD* locus, detects RFLPs with three enzymes, *Taq*I, *Mbo*I and *Acc*I. Results confirmed in several independent studies showed a strong association with a particular *Acc*I and a particular *Mbo*I allele, but no association with either *Taq*I allele. Such baffling patterns must reflect a complex history, with chance recombination events in small founder populations, and maybe an origin of some marker polymorphisms more recently than some disease mutations.

Recently a number of systematic studies of marker–marker LD across significant chromosomal segments have been reported (Gabriel *et al.*, 2002b and references therein). A common finding is that LD does not decay smoothly with distance. Instead, chromosomes contain a series of islands of relatively long-range LD that are sharply separated from each other (*Figure 15.6*). Within the islands, useful LD may extend for 50 kb (in Europeans; less in Africans), but even very closely

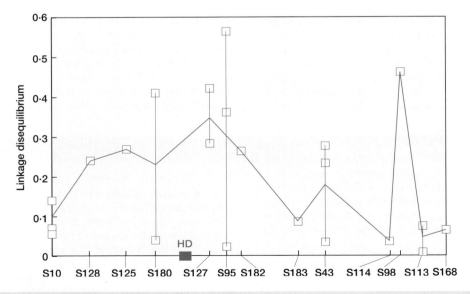

Figure 15.5: Linkage disequilibrium around the Huntington disease locus.

S10, S125 etc. are shorthand for the DNA markers *D4S10, D4S125* etc., shown in their map positions relative to the HD locus. The total distance represented is 2500kb. For some loci, several different RFLPs exist, which sometimes show very different allelic association, for example marker S95 (see text). From Krawczak and Schmidtke (1998) *DNA Fingerprinting,* 2nd edn. BIOS Scientific Publishers, Oxford.

spaced markers in different islands show no LD with each other. Detailed examination of a few regions has confirmed that the island boundaries are indeed recombination hotspots. Presumably this is showing us the mosaic of ancestral chromosomal segments that make up our common heritage. If the LD island structure of a population could be defined across the whole genome, then a set of markers ('hap-SNPs') could be defined to establish haplotypes at each island, that could then be tested for association with any disease. It is suggested that most islands would have only four to six different common haplotypes in any population (Gabriel *et al.*, 2002b). Large-scale efforts to define these structures by extensive marker–marker mapping are now underway (see www.genome.gov/10005336)

15.4.4 Design of association studies

Searching for population associations is an attractive option for identifying disease susceptibility genes. Association studies are easier to conduct than linkage analysis, because no multi-case families or special family structures are needed. Under some circumstances association can also be more powerful than linkage for detecting weak susceptibility alleles (see below). However, it is important to think carefully about the experimental design.

Choice of method to test for association

In any association study the choice of the control group is crucial. However hard one tries to match the controls to the cases, it is impossible to be absolutely certain that nothing has been overlooked. Thus, when an association is found, there is always the worry that it might be caused by inadequately matched controls and not by linkage disequilibrium with a susceptibility locus. The combination of this uncertainty and a

plethora of irreproducible results (especially for HLA–disease associations) caused case–control studies to fall out of favor among human geneticists during the 1980s.

Recently, a clutch of methods has been developed that largely circumvents this problem. Collectively these methods can be called **association studies with internal controls**. The most popular method is the **transmission disequilibrium test** (TDT; Schaid, 1998). The TDT starts with couples who have one or more affected offspring. It is irrelevant whether either parent is affected or not. To test whether marker allele M_1 is associated with the disease, we select those parents who are heterozygous for M_1. The test simply compares the number of cases where such a parent transmits M_1 to the affected offspring with the number when their other allele is transmitted (*Box 15.3*). The result is unaffected by population stratification. An extended TDT (ETDT; Sham and Curtis 1995) has been developed to handle data from multiallelic markers like microsatellites. The TDT can be used when only one parent is available, but this may bias the result (Schaid, 1998). When there are no parents available (a common problem with late-onset diseases) an alternative variant, sib-TDT, looks at differences in marker allele frequencies between affected and unaffected sibs (Spielman and Ewens, 1998).

There has been some argument about whether the TDT is a test of linkage or association. Since it asks questions about alleles and not loci, it is fundamentally a test of association. The associated allele may itself be a susceptibility factor, or it may be in linkage disequilibrium with a susceptibility allele at a nearby locus. The TDT cannot detect linkage if there is no disequilibrium – a point to remember when considering schemes to use the TDT for whole-genome scans.

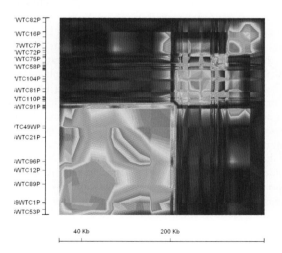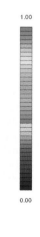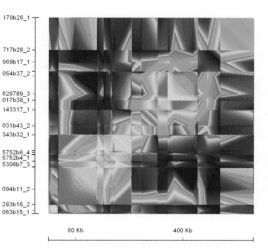

Figure 15.6: Patterns of linkage disequilibrium in two chromosomal regions.

On each square the same set of markers, in chromosomal order, is represented on the X and Y axes. The color at the Cartesian coordinate for each pair of markers shows the strength of pairwise LD according to the scale in the center. The GOLD program fills the space between points by interpolation. If LD were simply a function of distance, each square would show uniform red on the diagonal, shading through to uniform blue for points far away from the diagonal. Note the basic pattern of isolated islands of LD, but with a great deal of complicated detail. Left panel, part of chromosome 2; right panel, part of chromosome 13. Courtesy of Dr. William Cookson, Oxford.

Box 15.3: The transmission disequilibrium test (TDT) to determine whether marker allele M₁ is associated with a disease.

1) Affected probands are ascertained

2) The probands and their parents are typed for the marker

3) Those parents who are heterozygous for marker allele M₁ are selected. They may or may not be affected

Let **a** be the number of times a heterozygous parent transmits M₁ to the affected offspring, and **b** be the number of times the other allele is transmitted.

The TDT test statistic is $(a-b)^2/(a+b)$. This has a χ^2 distribution with 1 degree of freedom, provided the numbers are reasonably large.

As an alternative to TDT, conventional case–control studies are now coming back into favor. Case–control studies need fewer samples than TDT and are easier for late-onset diseases where parents are seldom available. The risk of false associations due to population stratification is felt to have been exaggerated, and data can be checked for possible stratification effects by comparing allele frequencies at a range of unlinked loci in the cases and controls (Pritchard and Rosenberg, 1999). The optimum study design, according to Risch and Teng (1998), is to use affected sib pairs as cases with two unrelated controls.

Whatever association test is used, the problem of multiple testing must be addressed. A full Bonferroni correction (dividing the threshold p value by N, the total number of questions asked) is over-conservative for large values of N. The preferred threshold of significance is $p' = 1-(1-p)^N$ (Emahazion et al., 2001). Some general issues in design of association studies are discussed in nonmathematical detail by Cardon and Bell (2001) – see Further reading.

Selection of markers

Single nucleotide polymorphisms (SNPs, *Box 13.1*) are the markers of choice for association studies, for two reasons:

▶ unlike microsatellites, they are sufficiently numerous (≥ 1 per kb on average) to define LD islands, and can be scored by various ultra-high throughput methods (*Section 18.4.2*).

▶ SNPs are less mutable than microsatellites. If it is true, as often suggested, that susceptibility to common diseases is mainly determined by common ancient DNA variants, one would need to use markers that are stable over a long timescale to identify the ancestral haplotypes.

For chromosomal regions and populations where the pattern of linkage disequilibrium is known (like those illustrated in *Figure 15.6*), one would select markers to define haplotypes in each island of disequilibrium. In the absence of such knowledge, all is guesswork. The older a disease allele is in human history, the higher is the density of SNPs needed to detect it (Kruglyak, 1999; Wright et al., 1999). Some investigators favor SNPs located within genes, and especially those in coding sequences (cSNPs), on the argument that such variants are more likely to be the actual susceptibility determinants. In truth, at present nobody knows the optimum strategy for marker selection. Different diseases in different populations have different histories, and maybe each individual study needs a custom-made strategy.

Choice of population to study

A contentious question is whether association studies will be more fruitful in isolated populations. Populations derived from a small number of founders are expected to show limited haplotype diversity and higher levels of linkage disequilibrium. The belief that disease-causing alleles at susceptibility loci will be more easily identifiable in isolated populations lies behind the DeCode project in Iceland (Gulcher et al., 2001) and similar projects elsewhere. Such populations may well show strong and long-range disequilibrium around loci for the rare Mendelian diseases that characterize the population (e.g. 'Finnish' diseases in Finland), but that disequilibrium exists only on chromosomes carrying the disease allele, which are presumably all derived from a single common ancestor. For more common variants, empirical data do not show strikingly increased disequilibrium (Varilo et al., 2000; Pritchard and Przeworski, 2001). Availability of large numbers of potential subjects with good medical records may be more important than population structure – certainly that is the thinking behind the BioBank project in the UK, which aims to collect medical and lifestyle data and DNA from 500 000 British people aged 45–69 years and follow their health prospectively.

Sub-Saharan African populations show higher genetic diversity than European populations, consistent with the 'out of Africa' hypothesis of human origins, and limited data suggest that linkage disequilibrium is shorter-range in Africans (Reich et al., 2001). On the other hand, populations derived by recent admixture may show very strong linkage disequilibrium (e.g. the Lemba, a Bantu-Semitic hybrid population studied by Wilson and Goldstein, 2001). Theoretically, if the two source populations had widely different incidences of a common disease, the mixed population could be used to map the determinants rather efficiently (like in a mouse cross), but the idea has not been tested in practice. In summary, it is still at present unclear whether any population has special advantages for association studies in complex diseases. For excellent discussions of these issues, see Wright et al. (1999) and Peltonen et al. (2000).

Genotypes or haplotypes?

When individuals rather than families are studied, the raw data consist of genotypes, but association analysis requires haplotypes. Haplotypes are inferred from the genotypes by an expectation-maximization computer analysis (Long et al., 1995), but it is impossible in principle to do this 100% reliably; the only wholly reliable way is to type somatic cell hybrids containing haploid chromosomes (Douglas et al., 2001). Some argue that this is a fundamental flaw in the design of the current large-scale association studies. Optimists believe that the studies will work because most LD islands (Section 15.4.3) have only a small number of common haplotypes, which can be identified adequately reliably from genotype data. It will be interesting to see who is right.

Linkage versus association: playing the numbers game

An important paper by Risch and Merikangas (1996) suggested that association can be more powerful than linkage for detecting weak susceptibility alleles. The authors compared the power of linkage (affected sib pair, ASP) and association (TDT) testing to identify a marker tightly linked to a disease susceptibility locus. They calculated the number of ASPs or TDT trios (affected child and both parents) required to distinguish a genetic effect from the null hypothesis, with a given power and significance level. *Box 15.4* illustrates their method (consult the original paper for more detail), and *Table 15.7* shows typical results of applying their formulae. The conclusion is clear. ASP analysis would require unfeasibly large samples to detect susceptibility loci conferring a relative risk of less than about 3, whereas TDT might detect alleles giving a relative risk below 2 with manageable sample sizes. Susceptibility alleles conferring a relative risk below 1.5 would be very hard to find by either method. Note, however, that their result incorporates various assumptions – in particular it assumes a single ancestral susceptibility allele at the disease locus. Any allelic heterogeneity would rapidly degrade the performance of an association test, while not affecting the power of a linkage test. There are now some data illustrating this (Section 15.6.6).

15.4.5 Linkage and association: complementary techniques

In many ways linkage and association provide complementary data. Linkage operates over a long chromosomal range and can scan the entire genome in a few hundred tests. A typical study of 250 affected sib pairs with 300 markers would require $1.5–3 \times 10^5$ genotypes to be generated (depending whether or not the parents were typed). For a well-automated and well-funded laboratory this is a few weeks' work. On the other hand, we have seen that linkage disequilibrium is a short range phenomenon, with islands of LD typically 20–50 kb in size. A whole-genome TDT scan of 300 trios (affected child and both parents), even testing only one SNP in each island, would require 10^8 genotypes, assuming islands average 25 kb in size. Emerging technologies may soon offer such throughput, but still the cost would be daunting. Thus association studies need to focus on predetermined candidate regions. These may be suggested by reference to animal models or known genes, but alternatively they can be defined by linkage studies. As noted above, candidate regions defined by ASP studies are usually impractically large for positional cloning of susceptibility genes. A natural study design is therefore to start with a genome-wide screen by linkage, probably in affected sib pairs and then, once an initial localization has been achieved, to narrow down the candidate region by linkage disequilibrium mapping.

15.5 Identifying the susceptibility alleles

In Mendelian diseases, finding the right gene may be difficult, but it is usually obvious when one has succeeded. Patients have unambiguous mutations, not present in controls, in a gene whose position has usually been tightly defined by linkage analysis. For complex diseases it is altogether more difficult to distinguish true susceptibility factors from irrelevant DNA polymorphisms. There are three reasons for this:

▶ no single gene mutation is necessary or sufficient to cause the disease, so even a true susceptibility allele will be found in some controls and be absent from some patients.

Box 15.4: Sample sizes needed to find a disease susceptibility locus by a whole genome scan using either affected sib pairs (ASP) or the transmission disequilibrium test (TDT).

Risch and Merikangas (1996) calculated the sample sizes needed to distinguish a genetic effect from the null hypothesis with power (1–β) and significance level α. This Box summarizes their formulae and equations, but the original paper should be consulted for the derivations and for details.

A standard piece of statistics tells us that the sample size M required is given by $(Z_\alpha - \sigma Z_{1-\beta})^2 / \mu^2$, where Z refers to the standard normal deviate. The mean μ and variance σ^2 are calculated as functions of the susceptibility allele frequency (p) and the relative risk γ conferred by one copy of the susceptibility allele. The model assumes that the relative risk for a person carrying two susceptibility alleles is γ^2; that the marker used is always informative; and that there is no recombination with the susceptibility locus.

For ASP, the expected allele sharing at the susceptibility locus is given by $Y = (1+w)/(2+w)$, where $w = [pq(\gamma-1)^2]/(p\gamma + q)$.

$\mu = 2Y-1$ and $\sigma^2 = 4Y(1-Y)$. The genome-wide threshold of significance (probability of a false positive anywhere in the genome = 0.05; testing for sharing IBD) requires a lod score of 3.6, corresponding to $\alpha = 3 \times 10^{-5}$, and $Z_\alpha = 4.014$. For 80% power to detect an effect, $1-\beta = 0.2$ and $Z_{1-\beta} = -0.84$.

For the TDT, the probability that a parent will be heterozygous for the allele in question is $h = pq(\gamma+1)/(p\gamma+q)$. P(trA), the probability that such a heterozygous parent will transmit the high-risk allele to the affected child, is $= \gamma/(1+\gamma)$. $\mu = \sqrt{h(\gamma-1)/(\gamma+1)}$, and $\sigma^2 = 1-[h(\gamma-1)^2/(\gamma+1)^2]$. As discussed above, for an ultimate genome screen involving 1 000 000 tests, $\alpha = 5 \times 10^{-8}$, $Z_\alpha = 5.33$ and, as before, $Z_{1-\beta} = -0.84$.

In Table 15.7 the Z_α, $Z_{1-\beta}$, μ and σ^2 values are used to calculate sample sizes by substituting in the formula $M = (Z_\alpha - \sigma Z_{1-\beta})^2 / \mu^2$. For the TDT, the answer is halved because each parent–child trio allows two tests, one on each parent.

Additionally, the main determinants of susceptibility may be different in different populations;

▸ the patchy nature of LD, with some long-range correlations coexisting with short-range lack of correlation, means that despite the theoretical high resolution of association studies, in reality one can seldom be confident that one is looking in exactly the right place for the susceptibility determinant. Moreover there is no genetic way of identifying the true determinant among a set of alleles that are all in strong disequilibrium with each other;

▸ the genetic variants causing susceptibility to common diseases may not be obvious mutations. Although there are exceptions, Mendelian diseases are usually caused by mutations that completely inactivate a gene, or at least have a major effect on its expression (Chapter 16), and it is usually not too hard to work out whether a candidate DNA sequence variant would do this. Susceptibility to common diseases is more likely to depend on a combination of quite subtle changes in the expression of several genes, no one of which is pathogenic in isolation, and all of which may be quite common in the healthy population. Susceptibility may be better modeled as a quantitative trait locus (QTL, see Section 15.6.8) than a binary (present/absent) factor. Susceptibility factors may be polymorphisms in noncoding DNA that have some small effect on promoter activity, splicing or mRNA stability. For example the UCSNP-43 G allele implicated in susceptibility to Type 2 diabetes (Section 15.6.5) lies deep within an intron of the candidate gene and has a frequency of 0.75 in unaffected controls.

For a thoughtful discussion of the problem (in the context of Type 2 diabetes) see the review by Altshuler, Daly and Kruglyak (2000).

15.6 Eight examples illustrate the varying success of genetic dissection of complex diseases

There is no unified story in genetic analysis of complex diseases. We make no attempt to summarize the current state of play across the whole field – each disease is different. Readers interested in a particular disease should use PubMed to locate a good up-to-date review of their chosen condition. The eight diseases summarized here have been chosen to illustrate some of the recurring themes of complex disease research. We have not concentrated on presenting success stories. In some cases there is little progress to report. Given the large efforts involved in every case, lack of success is almost as interesting as success.

15.6.1 Breast cancer: identifying a Mendelian subset has led to important medical advances, but does not explain the causes of the common sporadic disease

Although the common cancers are usually sporadic, the existence of 'cancer families' has been known for many years. When several relatives suffer the same rare cancer,

Table 15.7: Sample sizes for 80% power to detect significant linkage or association in a genome-wide search

γ	p	ASP analysis		TDT analysis	
		Y	N-ASP	P(trA)	N-TDT
5	0.01	0.534	2530	0.830	747
	0.1	0.634	161	0.830	108
	0.5	0.591	355	0.830	83
3	0.01	0.509	33797	0.750	1960
	0.1	0.556	953	0.750	251
	0.5	0.556	953	0.750	150
2	0.1	0.518	9167	0.667	696
	0.5	0.526	4254	0.667	340
1.5	0.1	0.505	115537	0.600	2219
	0.5	0.510	30660	0.600	950
1.2	0.1	0.501	3951997	0.545	11868
	0.5	0.502	696099	0.545	4606

γ is the relative risk for individuals of genotype Aa compared to aa; p is the frequency of the A susceptibility allele. For affected sib pair (ASP) analysis, Y is the expected allele sharing and N-ASP the number of pairs required for significance, based on IBD testing ($\alpha = 3 \times 10^{-5}$). For transmission disequilibrium testing (TDT), P(trA) is the probability that an Aa parent will transmit A to an affected child, and N-TDT is the number of parent-child trios required for significance. After Risch and Merikangas (1996).

like vestibular schwannomas (see NF2, MIM 101000), a Mendelian syndrome is readily suspected, and investigation of such families has led to the identification of the tumor suppressor genes described in Section 17.4. Breast cancer is common – the lifetime risk for a British woman is 1 in 12 – and so when several relatives have breast cancer, it is less clear whether this is just bad luck or a true cancer family. However, a strong family history of breast cancer is also associated with an unusually early age of onset, with breast plus ovarian cancer, with frequent bilateral tumors and occasionally with affected males. Investigation of such families has led to identification of the *BRCA1* and *BRCA2* genes.

A large-scale segregation analysis of 1500 families (Newman *et al.*, 1988) supported the view that 4–5% of breast cancer, particularly early-onset cases, might be attributable to inherited factors. Families with near-Mendelian pedigree patterns were collected for linkage analysis (*Figure 15.7*; see MIM 113705 for details of the linkage work). In 1990 a susceptibility locus, named *BRCA1*, was mapped to 17q21. The mean age at diagnosis in 17q-linked families was below 45 years. Later-onset families gave negative lod scores. Subsequently in 1994 a linkage search in 15 large families with breast cancer not linked to 17q identified a *BRCA2* locus at 13q12 (see MIM 600185). A hectic race ensued to identify the two mapped genes. *BRCA1* was cloned in 1994 and *BRCA2* in 1995 – see MIM entries 113705 and 600185 for details. It appeared that *BRCA1* might account for 80–90% of families with both breast and ovarian cancer, but a

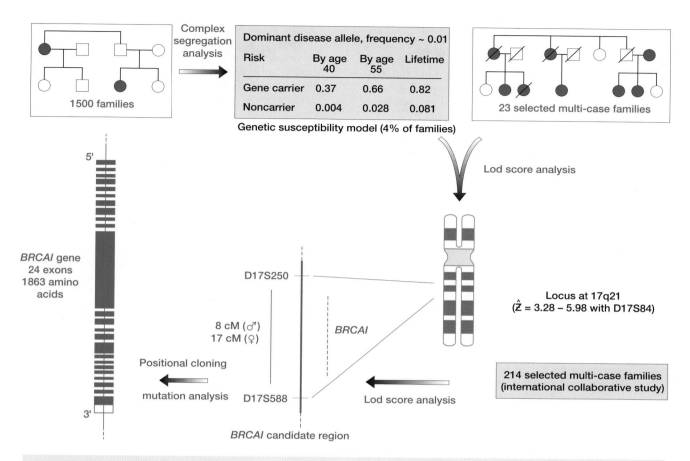

Figure 15.7: How the *BRCA1* gene was found.

Standard positional cloning successfully identified a gene conferring susceptibility to a common disease – but only to a Mendelian subset of the disease. *BRCA1* appears to have little role in the common, sporadic breast cancer.

much smaller proportion of families with breast cancer alone. Male breast cancer was seen mainly in *BRCA2* families.

Both genes encode large novel proteins that eventually turned out to be transcriptional co-activators, with additional roles in DNA repair (see Section 17.5.1). They behave as tumor suppressor genes, in that the inherited mutations cause loss of function (Section 16.3), and tumors from familial cases lose the wild-type allele. However, the story of breast cancer offers a striking contrast to colon cancer. In colon cancer, the *APC* gene, identified through the study of a rare Mendelian form of the disease, is very often mutated in the common sporadic forms as well (Section 17.5.4). *BRCA1* and *BRCA2*, by contrast, are seldom inactivated in sporadic breast cancer. Moreover, mutations at these two loci account for only 20–25% of the familial risk of breast cancer (Pharoah *et al.*, 2002). A survey of 257 families with four or more cases of breast cancer (Ford *et al.*, 1998) suggested that of those with four or five cases of female breast cancer but no ovarian cancer or male breast cancer, 67% probably did not involve *BRCA1/2* mutations. The search for further susceptibility genes is described by Nathanson and Weber (2001). Segregation analysis is consistent with the remaining susceptibility being polygenic, and one common

low-penetrance risk allele has been identified: a mutation in the *CHEK2* cell cycle kinase gene was found in 1.1% of controls but 5% of patients with breast cancer (CHEK2-Breast Cancer Consortium, 2002).

Early data suggested that a woman carrying a *BRCA1* mutation had an 85–90% chance of developing breast cancer, and a 40% risk of ovarian cancer. The risk of breast (but not ovarian) cancer for a woman with a *BRCA2* mutation is similar. However, these analyses were based on women from the large families that had been used for mapping. Later, when mutations were found in women not selected for family history, far more nonpenetrant cases were found among their relatives. In Ashkenazi Jewish women, three particular mutations are rather frequent (185delAG and 5382insC in *BRCA1*, 6174delT in *BRCA2*). A study of relatives of Ashkenazi women with breast cancer, not selected on the basis of family history, suggested a lifetime risk for carriers of these mutations of 36% rather than the 85–90% usually quoted (Fodor *et al.*, 1998). It is a general feature of complex disease research that initial estimates of penetrance, severity and risk to relatives, defined in the family collections used for the original identification of susceptibility loci, exaggerate the risk in the general population (Göring et al., 2001). This has important implications for population screening. Mutations defined in the initial family studies will probably have a much less dramatic effect in cases ascertained by screening, and additionally, low penetrance mutations may well exist that would be missed in the initial studies but revealed by population screening.

15.6.2 Hirschsprung disease: an oligogenic disease

Hirschsprung disease (HSCR) is a congenital absence of ganglia in some or all of the large intestine or colon. The resulting lack of peristaltic action produces a grossly distended megacolon that is lethal neonatally unless the affected segment is removed. Amiel and Lyonnet (2001) review the genetics of HSCR. In 18% of cases HSCR is part of a syndrome, and in a further 12% there are chromosomal anomalies. The remaining 70% have isolated HSCR, a typical 'multifactorial' condition, often familial but non-Mendelian. We have already seen the result of segregation analysis (*Table 15.4*). In HSCR the tools of classical positional cloning have been successful in identifying several loci.

▶ **A chromosomal abnormality:** two HSCR patients were noted to have visible deletions of 10q11q21. The *RET* oncogene lies in the deleted region and encodes a receptor tyrosine kinase that is expressed in the appropriate cells. It turned out that about 50% of familial and 15–35% of sporadic patients with isolated HSCR have a *RET* mutation, but so do some unaffected relatives. The known *RET* variants do not account for the whole of the effect of the *RET* locus on susceptibility; probably noncoding variants also play a role (Gabriel et al., 2002a). See Section 16.6.2 for the interesting molecular pathology of *RET* mutations.

▶ **Linkage and a mouse model:** a second susceptibility locus was mapped to chromosome 13q in a huge multiply inbred Mennonite kindred; the presence of HSCR in several unrelated patients with 13q deletions also pointed to this region. Independently, the mouse endothelin receptor B (*ednrb*) gene had been knocked out as part of an investigation of the role of endothelins in control of vascular tone. Unexpectedly, the knockout turned out to have the phenotype of a well-studied mouse model of HSCR, *piebald-lethal* (s^l). In man, *EDNRB* lay in the HSCR candidate region on 13q, and a mutation was soon demonstrated in the Mennonite kindred. Interestingly, once the family mutation was defined (as W276C) it turned out to be neither necessary nor sufficient. The penetrance was sex-specific and different for each genotype: 0.13 (M), 0.09 (F) for W/W, 0.33 (M), 0.08 (F) for W/C, and 0.85 (M), 0.60 (F) for C/C, illustrating the oligogenic character of Hirschsprung disease. It is well worth reading the reports of this work (Puffenberger et al., 1994a, b) as an illustration of the complexities of gene identification in a relatively common oligogenic disease.

▶ **Direct testing of candidate genes:** since *RET* and *EDNRB* both encode receptors, genes encoding their ligands (*GDNF*, *NTN* and *EDN3*) were natural candidate susceptibility genes. Mutations in each were demonstrated in a small number of families; another family had a mutation in the *ECE1* gene encoding an enzyme necessary for proteolytic maturation of endothelin 3. A further candidate gene, *SOX10*, was identified when the gene underlying another mouse model of HSCR, *dominant megacolon*, was cloned, but it turned out that humans with *SOX10* mutations had complex syndromes rather than typical HSCR. Mutations in *SMAD1P1* on 2q22 are frequent in patients with HSCR and mental retardation.

▶ **Mapping the remaining susceptibility loci:** *RET* is the major gene, but penetrance is incomplete and sex-dependent (65% in males, 45% in females). The other identified genes are important in occasional families, but do not contribute significantly to overall susceptibility. Gabriel et al. (2002a) conducted a systematic linkage analysis, from which they concluded that variations at the *RET* locus, interacting with variation at unknown loci on 3p21 and 19q12, together with a *RET*-dependent modifier at 9q31, can account for all genetic susceptibility to HSCR. However, in the Mennonite kindred mentioned above, susceptibility appears to be determined by an interaction between alleles at the *RET* and *EDNRB* loci, together with an unidentified locus at 16q23 (Carrasquillo et al., 2002). Curiously, there was not even suggestive association with the 3p21 or 19q12 loci, nor with 21q22.3, which had previously been reported as associated within this kindred.

Isolated HSCR emerges as an oligogenic disease (Amiel and Lyonnet, 2001). If positional cloning and functional studies support this analysis, HSCR may well be the first such disease to be fully dissected. An important contributor to this success is the very high value, 187, of λ_s (Gabriel et al., 2002a).

15.6.3 Alzheimer disease: genetic factors are important both in the common late-onset form and in the rare Mendelian early-onset forms, but they are different genes, acting in different ways

Alzheimer disease (AD; MIM 104310) affects about 5% of persons over the age of 65 years, and about 20% over the age of 80 years. There is progressive loss of memory, followed by disturbances of emotional behavior and general cognitive deterioration. Autopsy of the brain reveals loss of neurons with many amyloid-containing plaques. Degenerating neurons contain characteristic neurofibrillary tangles. Rarely, the onset is at a much younger age. The clinical and pathological features are identical in early-onset and late-onset AD, but early-onset disease is sometimes Mendelian and autosomal dominant, whereas late-onset AD is non-Mendelian and shows only modest familial clustering. In dominant early-onset families standard lod score analysis allowed mapping and subsequently cloning of three genes, *APP* at 21q21, presenilin-1 at 14q24 and presenilin-2 at 1q42 (see MIM 104300, 104311 and 600579 respectively). Although mutations at these three loci account for only about 10% of AD cases with onset before age 65 years, they are seen particularly in the rare families afflicted by highly penetrant dominant AD with strikingly early onset. Anybody wanting an insight into what such disease means for a family should read *Hannah's Heirs* by Daniel Pollen (see Further Reading).

Multicase late-onset families showed no evidence of linkage to the loci implicated in early-onset disease, but did show linkage to chromosome 19. Eventually the susceptibility locus was identified as *APOE* (Apolipoprotein E, MIM 107741), located at 19q13.2. There are three alleles with frequencies (in Caucasians) 0.08 (E2), 0.77 (E3) and 0.15 (E4). Both familial and sporadic late-onset AD are strongly associated with the E4 allele, while E2 is associated with resistance to AD. In cross-sectional studies of people over 65, E3/E4 people have about three times the risk, and E4/E4 people about 14 times, compared to E3 homozygotes. ApoE appears to account for about 50% of the susceptibility to late-onset AD. Thus, in contrast to breast cancer, not only the rare Mendelian forms but also the common sporadic form has a substantial degree of genetic determination. A longitudinal study (Meyer *et al.*, 1998) has suggested that E4 may govern the age of onset rather than susceptibility. Once AD has started, its progress is no different in people with or without E4. While the association of E4 with late-onset AD is beyond dispute, the mechanism is unclear. The E2/E3/E4 determinants are amino acid substitutions R112C and R158C; intensive searches have not revealed any 'true' susceptibility determinant in linkage disequilibrium with E4.

Further linkage and association studies have produced a plethora of other candidate susceptibility regions. Emahazion *et al.* (2001) list these and report a large-scale association study using SNPs from 60 candidate regions or genes. The results, after correction of the raw *p* values for multiple testing, barely identified *APOE*, and were negative for all other candidates. The authors use these data to highlight just how difficult it will be to confirm or refute claimed associations in complex diseases. Follow-up studies usually fail to replicate an initial localization – how often is this because the initial report was a Type 1 error, and how often is it because the follow-up study lacked power? Given the statistical tendency of studies to overestimate the effect of any true locus they identify (Göring *et al.*, 2001), it is not easy to decide how powerful a follow-up study would need to be to refute a claimed linkage.

The ApoE story is also very relevant to discussions of the possible social and ethical implications of identifying a susceptibility allele for a common disease (see *Ethics Box 1*).

15.6.4 Type 1 diabetes mellitus: still the geneticist's nightmare?

The first step towards understanding the genetics of diabetes was to distinguish the different types of diabetes (*Table 15.8*). Type 1 and Type 2 are different diseases, with different causes, different natural histories and different genetics. Type 1

diabetes (T1D, insulin-dependent diabetes; see MIM 222100) is caused by autoimmune destruction of pancreatic β-cells, typically affecting young people and requiring lifelong insulin treatment. There is fairly strong familial clustering ($\lambda_s = 15$, MZ twin concordance ca. 30%). An association with certain HLA alleles was already established in the 1970s.

In the UK, about 95% of patients with T1D have the HLA-DR3 and/or DR4 antigens, compared with 45–54% of the general population. HLA accounts for around 40% of the genetic predisposition, but several different haplotypes are associated with susceptibility or resistance. The strong linkage disequilibrium within the major histocompatibility complex makes it difficult to identify the primary determinant of susceptibility. It turns out that haplotypes associated with a low risk of diabetes all carry an allele at the *DQB1* locus in which amino acid 57 is aspartic acid, while high-risk haplotypes have *DQB1* alleles with some other amino acid at this position (Todd *et al.*, 1987). In one study, 96% of diabetics but only 20% of controls were homozygous non-Asp at *DQB1* position 57. Probably a resistance factor is defined by an antigen epitope involving Asp-57 that is shared by several different HLA molecules. The shared epitope idea has been applied also to other HLA-associated autoimmune diseases, with mixed success.

Early work on T1D also identified a second susceptibility locus, *IDDM2*, close to *INS* the structural gene for insulin. Linkage disequilibrium mapping has revealed the actual determinant as a 14-bp minisatellite repeat upstream of the gene. Susceptibility is associated with short repeats (26–63 repeat units). Long repeats (140–210 units) cause the insulin gene to be transcribed at rather higher level in the developing thymus; it is argued that this increases the efficiency of deletion of insulin-reactive T-cell clones during development of the immune system, and so reduces the risk of an autoimmune attack. Both the *HLA-DQB* work and the *INS* findings underscore the likely difference between the type of subtle genetic changes that may underlie susceptibility to a common disease and the gross changes seen in Mendelian diseases.

Several groups have vigorously pursued the strategy of whole genome linkage scans followed by linkage disequilibrium mapping within candidate regions. The European Consortium (2001) paper gives references to the leading linkage studies, and *Table 15.9* lists the main regions implicated to date. In order to prevent the strong HLA effect from masking weaker signals, data are usually stratified for HLA genotype before looking for linkage to other loci. The regions defined by ASP studies are very broad, and moreover the 'true' locus may lie well away from the lod score peak, making it very hard to know how many different susceptibility loci there are on 2q and 6q, for example. In theory the overall contribution of each locus can be measured by expressing its λ_s as a fraction of the overall λ_s for T1D (Risch, 1990a), so that one can know how much there is left to explain. However, the usual practice of using the same dataset to define a candidate region and then estimate λ_s gives very unreliable estimates (Göring *et al.*, 2001). HLA and *INS* probably explain 50% of the susceptibility, but the

Ethics Box 1: Alzheimer disease, ApoE testing and discrimination.

Identifying genotypes susceptible to common diseases raises ethical and social issues which ApoE well illustrates. Will employers and insurers push for widespread use of the tests, and would this lead to unfair discrimination in employment, healthcare and insurance?

Employment. Employers have a legitimate interest in somebody's current ability to do the job, but very little interest in what might happen in 10 years time. Only jobs that require the employer to invest heavily in specialist training are a partial exception to this. In the UK it would be illegal for an employer to discriminate against a healthy person because of their risk of developing a disease at a later date.

Healthcare. In countries with socialized medical systems the issue of discrimination in healthcare does not arise. Where access depends on private insurance, the problem comes back to the question of discrimination in insurance.

Insurance: Commercial insurance rests on the principle of *mutuality*: an underwriter assigns the applicant to a risk pool, and the premium reflects the risk that the applicant brings in to the pool. There will always be a conflict between actuarial fairness and moral fairness in individually underwritten insurance policies. Is it fair that an applicant is penalized because of a genetic constitution that is not under his control? Yet most people have no problems with companies charging different rates for males and females. The obvious solution is to ensure that mutuality-based insurance is not used to provide for the necessities of civilized life, like basic health care, but is limited to products that people can choose whether or not to buy as part of their normal choices in consumer societies.

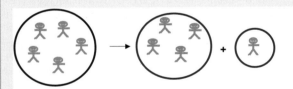

Risk pool A

premium £1.00

Risk pool B

premium £0.50

Risk pool C

premium £3.00

Contrary to popular belief, insurers have very little interest in making people take genetic tests. In the example illustrated, there is no benefit to the insurance company in using genetic testing to split risk pool A into pools B and C. Insurers' interest is in avoiding anti-selection. This arises when proposers use secret knowledge to gain an unfair advantage. Someone who

secretly knows he is at high risk takes out a specially large policy on standard terms, while somebody who knows she is at low risk decides not to bother with insurance. Thus insurers want to know as much as the applicant does about relevant pre-existing tests; the only legitimate reason to make an applicant take extra tests is if they suspect the applicant has in fact already secretly done so.

The criterion of what is 'relevant' is different for clinical testing and testing for insurance. For useful clinical testing, the result must be predictive for each individual. For insurance, in principle it is merely necessary that the test should be able to distinguish groups with significantly different average risks – a much less demanding task. It is the difference between standard deviation and standard error of the mean. *APOE4* is neither necessary nor sufficient for AD. About half of all people with *APOE4* will never develop AD, no matter how long they live, while many AD patients lack *APOE4*. The uncertainty allows neither confirmation of a clinical diagnosis nor useful predictive testing in healthy persons. The American College of Medical Genetics and the American Society of Human Genetics have recommended that ApoE testing should not be used for routine clinical diagnosis or predictive testing in Alzheimer's disease (ACMG/ASHG Working Group, 1995). The only *clinical* indication for ApoE genotyping in Alzheimer disease would be if an expensive drug used to treat Alzheimer patients were effective only for certain ApoE genotypes, or especially if it were harmful to certain genotypes. For insurance, the crucial question is whether ApoE genotypes define pools with risks sufficiently different to make testing important. According to actuaries Macdonald and Pritchard (2001), the basic answer, with a few qualifications, is 'no'.

Although ApoE testing turns out not to raise major public policy issues, other tests may come along that do require social control. The three features of a test that would ring alarm bells are:

▶ it should predict risk that is not already evident in the lifestyle or family history (this excludes, for example, Huntington disease testing);

▶ the condition should be sufficiently common, and the test sufficiently predictive that insurers could not risk ignoring it, but the test should not be available as part of routine clinical service;

▶ confidential private testing should be available and widely used.

overall contribution of the other loci in *Table 15.9* is not known.

Several features of T1D would seem to favor genetic investigation. It is fairly common (three per thousand) so that large patient groups can be assembled for study. Patients are young, and their parents are usually available for TDT analysis. The diagnosis is unambiguous, and there is a good animal model,

the *NOD* (nonobese diabetic) mouse, in which the disease is also polygenic. For all these reasons, and because of its huge financial and personal cost, T1D has been intensively investigated by linkage, candidate gene and model organism studies for decades. Considering the relatively modest progress made to date, perhaps it is time to resurrect the old description of diabetes as the geneticist's nightmare.

Table 15.8: Clinical classification of diabetes

Type 1 diabetes	Type 2 diabetes	MODY
Juvenile onset	Maturity onset (> 40 years)	Juvenile onset
0.4% of UK population	6% of US population	Rare
Requires insulin	Usually controllable by oral hypoglycemics	As type 2 diabetes
No obesity	Strong association with obesity	No obesity

Familial:
  MZ concordance 30%
  sib risk 6–10%
	Familial:
	  MZ twin concordance 40–100%
	  sib risk 30% (maybe subclinical)
		Familial:
		  autosomal
		  dominant?

Associated with HLA-DR3 and DR4	No HLA association	No HLA association

MODY (maturity onset diabetes of the young) is an uncommon Mendelian form for which various genes have been identified (see MIM 600496). Additionally, diabetes can be part of a number of uncommon syndromes.

Table 15.9: Main Type 1 diabetes susceptibility loci suggested by affected sib pair (ASP) or transmission disequilibrium (TDT) analysis

Locus	MIM No.	Location	Status
IDDM1	222100	6p21	λ_s = 3.1; determinant is HLA-DQB
IDDM2	125852	11p15	λ_s = 1.3; determinant is a VNTR upstream of INS gene
IDDM4	600319	11q13	λ_s = 1.6; significant linkage in combined results of three screens (596 families)
IDDM5	600320	6q24–q27	λ_s = 1.2; observed in four studies
IDDM6	601941	18q21	ASP and TDT evidence in one study.
IDDM7	600321	2q31-q33	λ_s = 1.3; seen in three ASP studies. Linkage disequilibrium with candidate gene CTLA4,
IDDM12	600388	2q33	$p = 5 \times 10^{-5}$ but only in some populations.
IDDM8	600883	6q25–q27	λ_s = 1.8; not clearly distinct from IDDM5.
IDDM10	601942	10p11-q11	ASP and TDT data from three studies
IDDM13	601318	2q34	Same as IDDM7 and/or 12?
IDDM15	601666	6q21	Confirmed (though hard to separate from HLA effect)

Data from the OMIM entries and papers cited therein; λ_s values are from Luo *et al.* (1995).

15.6.5 Type 2 diabetes: two susceptibility factors, one so common as to be undetectable by linkage; the other very complex and in certain populations only

Type 2 or noninsulin dependent diabetes mellitus (T2D) is the most common form of this heterogeneous disorder, affecting some 135 million people world-wide. T2D results from a combination of impaired insulin secretion and decreased end-organ responsiveness; known risk factors include age, obesity and lack of exercise. In many developed countries 10–20% of people over 45 years of age are affected, and certain populations have particularly high incidences.

Associations have been reported between T2D and at least 16 different genetic variants (summarized by Altshuler *et al.*, 2000a). A very large study of all these by Altshuler *et al.* replicated only one. A common ($p = 0.15$) allele of the *PPARG* gene was associated with a *decreased* risk of T2D in several cohorts (odds ratio = 0.8). Because the *increased* risk allele is so common ($p = 0.85$) it could never be detected by linkage analysis. *PPARG* is not a surprise candidate gene: it encodes a nuclear hormone receptor that regulates adipogenesis and is a target of the thiazolidinedione drugs used to treat T2D.

Altmüller *et al.* (2001) detail 10 whole genome scans that between them reported significant linkage to 25 loci on 15 different chromosomes. As usual, findings are not well

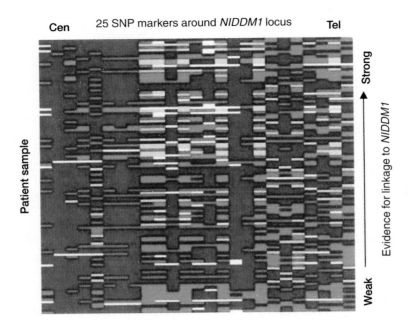

Figure 15.8: Genotypes of Type 2 diabetes patients at the calpain-10 locus.

Each row summarizes results for one patient with 25 SNPs (columns). Blue: homozygous for common allele; red: heterozygous; yellow: homozygous for rare allele; white: no data. Patients are arranged from top to bottom in order of decreasing evidence for linkage to *NIDDM1*. Note how the colors in the central part of the diagram, the location of the *CAPN10* gene, change from top to bottom, but colors in other parts do not. This shows how certain genotypes at the *CAPN10* locus are associated with the evidence for linkage. Adapted from Cox NJ (2001) *Hum. Mol. Genet.* **10**, 2301–2305 with permission from Oxford University Press

replicated between studies. One locus, *NIDDM1* at 2q37, was identified in Mexican Americans and appears to interact with a locus on chromosome 15 to increase susceptibility in this group (see Horikawa *et al.*, 2000 for references). The paper by Horikawa *et al.* describes the identification of a particular combination of SNPs within the calpain 10 (*CAPN10*) gene at 2q37 as the susceptibility determinant in Mexican Americans. To quote an editorial that is well worth reading, 'The study provides an unprecedented look at the cutting edge of positional cloning for complex traits and teaches lessons of lasting importance about just how difficult finding genes for common diseases may turn out to be.' (Altshuler *et al.*, 2000b).

The work started by narrowing the candidate region to a 7-cM region on 2q37. Physically this corresponded to a 1.7-Mb contig (recombination rates tend to be higher near telomeres, Section 13.1.5). A series of polymorphisms from this region were tested, not just for association with T2D, but also for association with the evidence for linkage. The rationale was that only a subset of cases were linked to the 2q37 locus, but these should be the ones carrying the susceptibility determinant. Initial analyses suggested a 66-kb target region, which on examination contained three genes, *CAPN10*, *RNPEPL1* and *GPR35*, and 179 sequence variants (in a panel of 10 Mexican diabetics). This led to identification of a variant, UCSNP-43, where the homozygous G/G genotype showed association with the evidence for linkage and also probably

with diabetes (odds ratio, 1.54; confidence interval, 0.88–2.41). A search was then initiated for haplotypes that were: (a) increased in frequency in the patient groups with greatest evidence of 2q37 linkage; (b) shared by affected sib pairs more often than expected; *and* (c) associated with increased risk of diabetes. A heterozygous combination of two haplotypes (defined by three SNPs within introns of the *CAPN10* gene) fulfilled all the criteria. Neither haplotype in homozygous form was a risk factor. Thus the susceptibility determinant at *NIDDM1* appears to be a particular heterozygous combination of noncoding SNPs interacting with an unidentified factor on chromosome 15. *Figure 15.8* attempts to give a visual impression of what is meant by 'association with the evidence for linkage'.

How confident can we be that this blockbuster project has truly identified the *NIDDM1* susceptibility factor? Follow-up studies have produced conflicting results (summarized by Fullerton *et al.*, 2002); there is certainly no universal association between the proposed susceptibility genotype and T2D worldwide. Calpain-10 was an unexpected candidate; however, an effect of UCSNP-43 on transcription of the gene has been reported, and a possible downstream effect on insulin-stimulated glucose turnover suggested. Overall, this study prompts considerable caution about identifying complex disease susceptibility factors. On the positive side, a novel candidate gene has been identified; on the negative side, it is far from clear that the actual variant causing susceptibility has been identified, nor is it clear how this uncertainty might be resolved. Will other common disease risk factors prove equally obscure and complex? Will they turn out to be specific to particular populations? The heroic labor of Horikawa *et al.* has at the very least provided much food for thought.

15.6.6 Inflammatory bowel disease: a clear-cut susceptibility gene identified

Ulcerative colitis (UC) and Crohn's disease (CD) are forms of inflammatory bowel disease (IBD) which in total affects one to three people per thousand in the Western world. Family and twin studies support a genetic susceptibility for each disease with some shared components. λ_s is 20–30 for CD and 8–15 for UC. Sib pair and family linkage analyses by many groups defined several regions with confirmed or replicated linkage (*IBD1*, 16p12-q13; *IBD2* 12p13; *IBD3* 6p; *IBD4* 14q11-q12; *IBD5* 5q31) and many others of suggestive linkage (for references, see OMIM entry 266600). Linkage to *IBD1* in particular was confirmed with a maximum lod score of 5.79 in CD (but not UC) families in a very large pooled study of 613 families with 1298 CD and/or UC affected sibs (IBD International Genetics Consortium, 2001).

Three groups recently identified the gene involved in *IBD1* (Hugot *et al.*, 2001; Ogura *et al.*, 2001; Hampe *et al.*, 2001). Ogura *et al.* had cloned *CARD15* (*NOD2*) as a regulator of the NF-κB inflammatory response. They tested it in IBD because it mapped to the *IBD1* locus and biologically was a promising IBD candidate. By contrast, Hugot *et al.* pursued a pure linkage-association pathway. The candidate region was narrowed by linkage, and then association studies undertaken. Interestingly, the TDT result which led on to *CARD15* was only borderline significant ($p = 0.05$), and although it was replicated in an independent sample with $p < 0.01$, this time the association was with a different allele of the marker! In fact careful inspection of their data suggest that a random microsatellite-based search for association in a large cohort would not have given positive results. Eventually, stronger though not dramatic associations highlighted a genomic region from which they cloned the then unpublished *CARD15* gene.

The primary susceptibility factor at *IBD1* is a single base insertion, 3020insC, in the C-terminal part of the *CARD15* gene. Hampe *et al.* showed very strong TDT association between CD and this mutation. The insertion has a frequency of .08 in CD patients, .02 in controls and .02 in UC patients, showing that it is a susceptibility factor for CD but not UC. The relative risk of CD is about 3 for heterozygotes and 23 for homozygotes. The insertion results in a protein truncated by 33 amino acids which affects a region that is demonstrably important in activating NF-κB.

This work highlights three points. Primarily, it shows that the enterprise is possible: susceptibility genes can at least sometimes be identified by a genetic approach. Second, it emphasizes the need for very large-scale family studies. And finally, the weak TDT results for linked markers in the study of Hugot *et al.*, compared to the very strong results seen by Hampe *et al.* when they tested the insertion mutation directly,

points an important lesson about association studies. Not surprisingly, other mutations in the *CARD15* gene also contribute to CD susceptibility. Two other cSNPs, G908R and R702W (numbered differently by Hugot *et al.*) are strongly associated with CD, and a whole variety of rare missense changes in the *CARD15* gene are more frequent in CD patients than in controls. Even if linkage disequilibrium is present, each mutation will be associated with its own individual haplotype. Such allelic heterogeneity is entirely to be expected – but it places severe constraints on the chances of localizing a susceptibility gene within a candidate region using an association-based strategy.

15.6.7 Schizophrenia: the special problems of psychiatric or behavioral disorders

Geneticists studying psychiatric illness or antisocial behavior face two additional problems, on top of all the difficulties inherent in studying any complex disease.

(i) *Nature–nurture arguments.* Because shared family environment is such a plausible explanation for the tendency of psychiatric or behavioral problems to run in families, a very high standard of proof is required for claims of genetic factors. For schizophrenia, some of the evidence from family, twin and adoption studies was summarized in *Tables 15.1–15.3*. All too often, nature–nurture arguments are just proxies for political arguments, and are conducted with all the decorum and objectivity of street politics. Both sides base their passion on the false premise that if a condition is genetic, then it cannot be ameliorated by any social intervention. People on the left therefore favor environmental causation, and people on the right favor genetic causes. At present some of the heat has gone out of arguments about psychiatry, but geneticists researching antisocial behavior still attract virulent opposition and unwelcome supporters.

(ii) *Diagnostic criteria.* As we discussed in Section 15.3.1, diagnostic criteria are another very difficult area in psychiatric genetics. 'Schizophrenia should be considered not as a disease, but, like epilepsy, as a syndrome, recognized by a collection of signs and symptoms, which have a diverse pathogenesis' (Trimble, 1996). Agreed criteria, like DSM-IV, are based on the best judgment of experienced psychiatrists about what constitutes the essential core of the condition – but ultimately they are arbitrary. Maybe 'schizoid personality' is one manifestation of the genes that make people susceptible to DSM-IV schizophrenia? Or maybe there is a general genetic susceptibility to psychotic illness, and bipolar and depressive illness should also be included? Genetic analyses often use two or more different sets of criteria, one narrow and one broad, to see which gives the best lod score. Alternatively, researchers may look for linkage to an intermediate phenotype, something that is not schizophrenia but that might form part of the susceptibility and might be more simply genetically determined – some physiological or neuropsychological variant, perhaps.

As a relatively common and intensely distressing familial condition, schizophrenia has long been high on the priority list of genetic researchers. Altmüller *et al.* (2001) list 10 whole genome scans performed since 1994, and there have been many additional studies of individual candidate genes or regions. Typical candidates are genes encoding proteins involved in neurotransmission or affected by drugs used to treat schizophrenia. Large multiply-affected kindreds are not uncommon, and these have often been used for linkage studies. The choice of such families has implications for the analysis: Genehunter, the program most commonly used for complex disease linkage, cannot handle large pedigrees (Section 13.6.2), causing many workers to use parametric lod score analysis (Section 13.3.1), but with a variety of alternative genetic models. Doing this is perfectly valid, but it introduces extra degrees of freedom and therefore reduces the power of the study. Combined with the use of multiple diagnostic criteria (see above) this makes assessing the significance of the results unusually tricky. Nevertheless, multiple models may be biologically realistic – it is entirely plausible that a risk allele at one locus could confer dominant susceptibility to broadly-defined disease, while an allele at another locus could confer recessive susceptibility to just narrowly defined schizophrenia.

Currently one or more studies show evidence for linkage to around a dozen chromosomal regions, but no susceptibility allele has been clearly identified. One promising candidate is the valine allele of a Met158Val polymorphism in the *COMT* (catechol-O-methyl transferase) gene. Biologically this is a plausible candidate, and despite some negative studies, a number of studies have reported positive association and TDT results. Weinberger *et al.* (2001), for example, found a frequency of 52% in controls and 60% in schizophrenics, with an odds ratio of 1.5 for homozygotes. However, even if confirmed, this factor would account for only 4% of the variance in risk. OMIM entry 181500 provides links to discussions of the main candidate regions, while the studies of Gurling *et al.* (2001), Camp *et al.* (2001), Weinberger *et al.* (2001) and Lewis *et al.* (2003) give a good picture of the problems and some of the approaches being used to try and unravel the genetics of this tragic disease. The issues raised here apply equally to a range of other human psychiatric and behavioral phenotypes.

15.6.8 Obesity: genetic analysis of a quantitative trait

As a contentious emotive issue, obesity rivals schizophrenia, with one school of thought, prevalent among thin people, blaming poor self-discipline and moral laxity, and the stout party laying the blame on constitutional predisposition. Obesity is a major public health concern in developed countries, and is of great interest to biotechnology companies, who see the fortune to be made from an effective slimming pill. For our present purposes, the principal interest of obesity is as an example of a quantitative trait (Barsh *et al.*, 2000).

Simply measuring weight does not give a good quantitative variable for analysis because it confuses tall thin people with short fat people. Body mass index (BMI; weight in kg/height in m²), is a much better measure. The mean BMI in the US in 1983 was 22 kg/m². Other investigators have tried to use measures closer to the underlying physiology, for example percent adipose tissue or serum leptin concentration. In each case, some cut-off could be used to define obesity, for example BMI 25–29.9 kg/m² = grade 1, BMI 30–40 kg/m² = grade II and BMI > 40 = grade III, with adjustments for age and sex. However, such an approach is arbitrary and loses data. It is much better to analyze the quantitative variable directly to identify a **quantitative trait locus (QTL)**. Such quantitative measures are the bread and butter of animal and plant breeding. The usual technique for QTL analysis in human pedigrees is variance–component linkage analysis. Almasy and Blangero (1998) describe the mathematical basis of the method. In essence, as with conventional lod score analysis, a lod score is calculated from the ratio of the likelihood of the data on two alternative hypotheses. In this case the data are covariances between relatives for the quantitative phenotype, and the alternative hypotheses are that a QTL does or does not exist at the chromosomal location in question. Covariances can be adjusted for confounding variables such as age and sex. On the hypothesis that a QTL does exist, its effect is predicted from the identity by descent of the chromosomal segment in the pair of relatives, as evidenced by the marker data. The multipoint method described by Almasy and Blangero provides a confidence interval for the location of the QTL and an estimate of its effect. Although we are analyzing BMI as a QTL rather than obesity as a dichotomous trait, it is still preferable to concentrate on people at the extremes of the distribution to maximize the statistical power.

OMIM has 44 entries containing 'obesity' in the title or clinical synopsis, including rare dramatic obesity involving components of the pathways that regulate food intake – the hormone leptin, the leptin receptor, pro-opiomelanocortin and the melanocortin-4 receptor. However most obesity is non-Mendelian. In fact, syndromes with inevitable obesity are of little interest: the aim of obesity research is to discover the mechanisms that make people susceptible or resistant to diet-induced obesity, rather than those that control body weight *per se* – a relevant point when considering animal models. This line of thought suggests that the best research design would consider only people, obese or otherwise, who live on junk food and take no exercise. They could presumably be recruited at a drive-through fast food outlet.

As always, the first task is to check whether there is any genetic component in the causation. Obesity tends to run in families, but this could easily be an effect of shared family environment. However, twin and adoption studies suggest that around 70% of the variance in BMI can be attributed to genetic factors. Segregation analyses have suggested that anything from 20% to 65% of the population variance in BMI might be due to one or two recessive major genes (summarized by Feitosa *et al.*, 2002). Altmüller *et al.* (2001) list five whole genome scans reported before December 2000; two more recent reports are by Feitosa *et al.* (2002) and Deng *et al.* (2002). The latter two both show lod score curves for each chromosome, which illustrate very clearly the general problem of failure to replicate results. As with many other complex diseases, the key question is whether the failure to replicate tells us that the initial report was mistaken, or that the follow-up study lacked power. In the case of obesity, it is also worth reflecting that some of the best data may be hidden in the files of biotechnology companies, not to be revealed until something patentable has been achieved.

15.7 Overview and summary

15.7.1 Why is it so difficult?

Identifying susceptibility factors is proving much more difficult than most people imagined 10 years ago. Weiss and Terwilliger (2000) pertinently ask, 'what percentage of complex disease gene mapping projects, whose grant proposals promised 80% power, have actually been successful in identifying the disease genes assumed to be in the data?' The answer is so far below 80% that something is surely wrong. Their paper is required reading for anybody who feels complex disease research might be easy. Nevertheless, every major venture into the unknown has always been accompanied by compelling scholarly arguments showing why it can never work – which are then proved wrong. Majority opinion is that the failures have been due to lack of power, and can be remedied by larger patient cohorts, more intensive genotyping and a more powerful statistical mill.

The central problem must be heterogeneity

For linkage, affected sib pair analysis is an extremely robust method. The fact that so few significant lod scores are obtained, and that studies so rarely replicate each other (Altmüller et al., 2001) can only be due to locus heterogeneity. Evidently susceptibility to most complex diseases is determined by many minor loci and not a few major loci. Sib pair analysis is limited to detecting fairly strong susceptibility factors. The calculations in *Table 15.7* show that for $\gamma \leq 2$ the excess of allele sharing by affected sib pairs is very small, and detecting it would require huge numbers. Altmüller et al. (2001) compared 101 whole genome scans to see if any lessons could be learned about how to succeed, but were not able to identify any golden rule apart from obvious things like having a large sample to analyze. An additional consideration is that susceptibility in different populations may have different genetic causes. For example, a survey of 483 Japanese patients with Crohn disease (Yamazaki et al., 2002) found little evidence of mutations in *CARD15*, which is a major susceptibility gene among Europeans (see Section 15.6.6).

Association studies, the other main tool for the research, depend on linkage disequilibrium. It is becoming increasingly clear that until we have good disequilibrium maps along the lines of *Figure 15.6*, there is no rational basis for selecting which markers to use. Nevertheless, it is also clear that, contrary to the calculations of Kruglyak (1999), disequilibrium is in many cases extensive enough to give a reasonable chance of success to SNP association studies on current scales. Having chosen appropriate SNPs to test, the major issue is allelic heterogeneity. Inflammatory bowel disease (Section 15.6.6) provides a perfect illustration of the way that the power of association studies dives rapidly as soon as there is more than one common susceptibility allele at a risk locus. The calculations of Risch and Merikangas (1996), so influ-

ential in propelling TDT studies to the forefront of complex disease research, consider a single susceptibility allele at the disease locus. It is important to remember that their calculations apply to that allele, and not to the overall influence of the locus on the disease. The degree of allelic heterogeneity at a susceptibility locus emerges as a key determinant of success or failure. Are most susceptibility alleles ancient common polymorphisms (the 'common disease–common variant' hypothesis), or are they a heterogeneous collection of rare recent mutations, like with most Mendelian diseases? Arguments have been produced for both positions (Reich and Lander, 2001; Pritchard, 2001; Wright et al., 2003). The jury is still out.

15.7.2 If it all works out and we identify susceptibility alleles—then what?

Assuming it can be made to work, identifying susceptibility factors for a disease must advance our understanding of its pathogenesis. That will not automatically lead to improved treatment – a condition may still be incurable even if we understand it well, and purely symptomatic treatment is sometimes very effective – but understanding the pathology should allow more rational and targeted approaches to treatment. Patients with different genotypes may have differences in underlying pathology, and hence respond to different drugs.

Whether identifying risk factors will lead to effective *prevention* is much less clear. Technical developments in high-throughput genotyping will make population screening easier, but one always needs to ask whether such screening would be cost-effective and ethically acceptable. The necessary conditions are discussed in detail in Section 18.3. The most essential single point, more important than any technical issue, is that identifying somebody as at risk should lead to some useful action. In general one is talking about lifestyle changes. Prophylactic drug treatment would be justifiable only if it could be targeted at a small number of high-risk people, and such groups are characteristic of Mendelian rather than complex diseases. Also, Pharoah et al. (2002) make the important point that where high-risk people are identified by multilocus genotyping, interventions that are based on specific mechanisms of predisposition might deal with only a specific small part of their extra risk.

An ability to define genetic susceptibility would greatly assist identification of relevant lifestyle factors. These could be identified far more easily in a cohort who are known to be all genetically susceptible. Opinions differ widely on what impact such knowledge might have. The two extreme positions might be caricatured as the head-in-clouds and head-in-sand positions (*Figure 15.9*). The head-in-clouds position assumes not only that strongly protective lifestyle changes could be identified, but also that people would follow advice to adopt them. Given the current epidemic in most developed countries of entirely preventable disease caused by smoking, obesity and physical inactivity, this looks a little optimistic. Nevertheless one should not collapse into pessimism: if even one common disease can be effectively prevented, the effort will have been worth while.

Figure 15.9: Head in clouds and head in sand – Contrasting views of the impact on 21st century medicine of identifying susceptibility factors for complex disease.

Cartoon by Maya Evans.

Further reading

Cardon LR, Bell JI (2001) Association study designs for complex diseases. *Nature Rev. Genet.* **2**, 91–99.

Falconer DS, Mackay TFC (1996) *Introduction to Quantitative Genetics.* Longman, Harlow.

Kety SS, Rowland LP, Sidman RL, Matthysse SW (eds). (1983) *Genetics of Neurological and Psychiatric Disorders.* Raven Press, New York.

Ott J (1999) *Analysis of Human Genetic Linkage*, 3rd Edn. Johns Hopkins University Press, Baltimore.

Pollen DA (1996) *Hannah's Heir*s (expanded edition). Oxford University Press, Oxford.

Rosenthal D, Kety SS (1968) *The Transmission of Schizophrenia.* Pergamon Press, Oxford.

Sham S, Zhao J (1998) Linkage analysis using affected sib-pairs. In: *Guide to Human Genome Computing*, 2nd Edn (ed. MJ Bishop). Academic Press, San Diego CA.

Terwilliger J, Ott J (1994) *Handbook for Human Genetic Linkage.* Johns Hopkins University Press, Baltimore.

References

ACMG/ASHG Working Group (1995) Use of ApoE testing for Alzheimer disease. *J.A.M.A.* **274**, 1627–1629.

Almasy L, Blangero J (1998) Multipoint quantitative-trait linkage analysis in general pedigrees. *Am. J. Hum. Genet.* **62**, 1198–1211.

Altmüller J, Palmer LJ, Fischer G, Scherb H, Wjst M (2001) Genomewide scans of complex human diseases: true linkage is hard to find. *Am. J. Hum. Genet.* **69**, 936–950.

Altshuler D, Hirschhorn JN, Klannemark M *et al.* (2000a) The common PPARγ Pro12Ala polymorphism is associated with decreased risk of type 2 diabetes. *Nature Genet.* **26**, 76–79.

Altshuler D, Daly M, Kruglyak L (2000b) Guilt by association. *Nature Genet.* **26**, 135–137.

Amiel J, Lyonnet S (2001) Hirschsprung disease, associated syndromes and genetics: a review. *J. Med. Genet.* **38**, 729–739.

Arnos CI, Dawson DV, Elston RC (1990) The probabilistic determination of identity by descent sharing for pairs of relatives from pedigrees. *Am. J. Hum. Genet.* **47**, 842–853.

Badner JA, Sieber WK, Garver KL, Chakravarti A (1990) A genetic study of Hirschsprung disease. *Am. J. Hum. Genet.* **46**, 568–580.

Barsh GS, Farooqi IS, O'Rahilly S (2000) Genetics of body-weight regulation. *Nature* **404**, 644–651.

Byerley WF (1989) Genetic linkage revisited. *Nature* **340**, 340–341.

Camp NJ, Neuhausen SL, Tiobech J *et al.* (2001) Genomewide multipoint linkage analysis of seven extended Palauan pedigrees with schizophrenia, by a Markov-chain Monte Carlo method. *Am. J. Hum. Genet.* **69**, 1278–1289.

Carrasquillo MM, McCalion AS, Puffenberger EG, Kashuk CS, Nouri N, Chakravarti AS (2002) Genome-wide association study and mouse model identify interaction between *RET* and *EDNRB* pathways in Hirschsprung disease. *Nature Genet.* **32**, 237–244.

CHEK2-Breast Cancer Consortium (2002) Low-penetrance susceptibility to breast cancer due to *CHEK2**1100delC in non-carriers of *BRCA1* or *BRCA2* mutations. *Nature Genet.* **31**, 55–59.

Cox NJ (2001) Challenges in identifying genetic variation affecting susceptibility to type 2 diabetes: examples from studies of the calpain-10 gene. *Hum. Mol. Genet.* **10**, 2301–2305.

Deng H-W, Deng H, Liu Y-J *et al.* (2002) A genome-wide linkage scan for quantitative-trait loci for obesity phenotypes. *Am. J. Hum. Genet.* **70**, 1138–1151.

Devlin B, Risch N (1995) A comparison of linkage disequilibrium measures for fine-scale mapping. *Genomics* **29**, 311–322.

Douglas JA, Boehnke M, Gillanders E, Trent JM, Gruber SB (2001) Experimentally-derived haplotypes substantially increase the efficiency of linkage disequilibrium studies. *Nature Genet.* **28**, 361–364.

Emahazion T, Feuk L, Jobs M *et al.* (2001) SNP association studies in Alzheimer's disease highlight problems for complex disease linkage analysis. *Trends Genet.* **17**, 407–413.

European Consortium for IDDM Genome Studies (2001) A genomewide scan for Type 1-diabetes susceptibility in Scandinavian families: identification of new loci with evidence of interactions. *Am. J. Hum. Genet.* **69**, 1301–1313.

Feitosa MF, Borecki IB, Rich SS *et al.* (2002) Quantitative-trait loci influencing body-mass index reside on chromosomes 7 and 13: the National Heart, Lung and Blood Institute family study. *Am. J. Hum. Genet.* **70**, 72–82.

Fischer M, Harvald B, Hauge M (1969) A Danish twin study of schizophrenia. *Br. J. Psychiatr.* **115**, 981–990.

Fisher RA (1918) The correlation between relatives under the supposition of mendelian inheritance. *Trans. R. Soc. Edin.* **52**, 399–433.

Fodor FH, Weston A, Bleiweiss IJ *et al.* (1998) Frequency and carrier risk associated with common *BRCA1* and *BRCA2* mutations in Ashkenazi Jewish breast cancer patients. *Am. J. Hum. Genet.* **63**, 45–51.

Ford D, Easton DF, Stratton M *et al.* (1998). Genetic heterogeneity and penetrance analysis of the *BRCA1* and *BRCA2* genes in breast cancer families. *Am. J. Hum. Genet.* **62**, 676–689.

Fullerton SM, Bartoszewicz A, Ybazeta G *et al.* (2002). Geographic and haplotype structure of candidate Type 2 diabetes-susceptibility variants at the *Calpain-10* locus. *Am. J. Hum. Genet.* **70**, 1096–1106.

Gabriel SB, Salomon R, Pelet A *et al.* (2002a). Segregation at three loci explains familial and population risk in Hirschsprung disease. *Nature Genet.* **31**, 89–93.

Gabriel SB, Schaffner SF, Nguyen H *et al.* (2002b) The structure of haplotype blocks in the human genome. *Science* **296**, 2225–2229.

Göring HHH, Terwilliger JD, Blangero J (2001) Large upward bias in estimation of locus-specific effects from genomewide scans. *Am. J. Hum. Genet.* **69**, 1357–1369.

Gulcher JR, Kong A, Stefansson K (2001) The role of linkage studies for common diseases. *Curr. Opin. Genet. Dev.* **11**, 264–267.

Gurling HMD, Kalsi G, Brynjolfson J *et al.* (2001) Genomewide genetic linkage analysis confirms the presence of susceptibility loci for schizophrenia, on chromosomes 1q32.2, 5q33.2, and 8p21-22 and provides support for linkage to schizophrenia on chromosomes 11q23.3-24 and 20q12.1-11.23. *Am. J. Hum. Genet.* **68**, 661–673.

Hampe J, Cuthbert A, Croucher PJP *et al.* (2001) Association between insertion mutation in *NOD2* gene and Crohn's disease in German and British populations. *Lancet* **357**, 1925–1928.

Horikawa Y, Oda N, Cox NJ *et al.* (2000) Genetic variation in the gene encoding calpain-10 is associated with type 2 diabetes mellitus. *Nature Genet.* **26**, 163–175.

Hugot J-P, Chamaillard M, Zouali H *et al.* (2001) Association of *NOD2* leucine-rich repeat variants with susceptibility to Crohn's disease. *Nature* **411**, 599–603.

IBD International Genetics Consortium (2001) International collaboration provides convincing linkage replication in complex disease through analysis of a large pooled data set: Crohn disease and chromosome 16. *Am. J. Hum. Genet.* **68**, 1165–1171.

Kendler KS, Gruenberg AM, Kinney DK (1994) Independent diagnoses of adoptees and relatives as defined by DSM-III in the provincial and national samples of the Danish Adoption Study of Schizophrenia. *Arch. Gen. Psychiatr.* **51**, 456–468.

Kety SS, Wender PH, Jacobsen B *et al.* (1994) Mental illness in the biological and adoptive relatives of schizophrenic adoptees. Replication of the Copenhagen Study in the rest of Denmark. *Arch. Gen. Psychiatr.* **51**, 442–455.

Krawczak M, Schmidtke J (1994) *DNA Fingerprinting*. BIOS Scientific Publishers, Oxford, p. 650

Kruglyak L, Lander ES (1995) Complete multipoint sib-pair analysis of qualitative and quantitative traits. *Am. J. Hum. Genet.* **57**, 439–454.

Kruglyak L (1999) Prospects for whole-genome linkage disequilibrium mapping of common disease genes. *Nature Genet.* **22**, 139–144.

Lander ES, Kruglyak L (1995) Genetic dissection of complex traits: guidelines for interpreting and reporting linkage results. *Nature Genet.* **11**, 241–247.

Lander ES, Schork N (1994). Genetic dissection of complex traits. *Science* **265**, 2037–2048.

Long JC, Williams RC, Urbanek M (1995) An E-M algorithm and testing strategy for multiple-locus haplotypes. *Am. J. Hum. Genet.* **56**, 799–810.

Lewis CM, Levinson DF, Wise LH *et al.* (2003) Genome scan meta-analysis of schizophrenia and bipolar disorder, Part II: schizophrenia. *Am. J. Hum. Genet.* **73**, 34–48.

Luo D-F, Bui MM, Muir A, Maclaren NK, Thomson G, She J-X. (1995) Affected sib-pair mapping of a novel susceptibility gene to insulin-dependent diabetes mellitus (*IDDM8*) on chromosome 6q25-q27. *Am. J. Hum. Genet.* **57**, 911–919.

Macdonald AS, Pritchard DJ (2001) Genetics, Alzheimer's disease and long-term care insurance. *North American Actuarial J.* **5**, 54–78.

McGuffin P (1984) In: *The Scientific Principles of Psychopathology*, (eds P McGuffin, MF Shanks, RJ Hodgson). Grune and Stratton, London.

McGuffin P, Huckle P (1990) Simulation of Mendelism revisited: the recessive gene for attending medical school. *Am. J. Hum. Genet.* **46**, 994–999.

Meyer MR, Tschanz JT, Norton MC *et al.* (1998) ApoE genotype predicts when – not whether – one is predisposed to develop Alzheimer disease. *Nature Genet.* **19**, 331–332.

Nathanson KL, Weber BL (2001) 'Other' breast cancer susceptibility genes: searching for more holy grail. *Hum. Mol. Genet.* **10**, 715–720.

Newman B, Austin MA, Lee M, King MC (1988) Inheritance of human breast cancer: evidence for autosomal dominant transmission in high-risk families. *Proc. Natl Acad. Sci. USA* **85**, 3044–3048.

Ogura Y, Bonen DK, Inohara N *et al.* (2001) A frameshift mutation in *NOD2* associated with susceptibility to Crohn's disease. *Nature* **411**, 603–606.

Onstad S, Skre I, Torgersen S, Kringlen E (1991) Twin concordance for DSM-III-R schizophrenia. *Acta Psychiatr. Scand.* **83**, 395–401.

Peltonen L, Palotie A, Lange K (2000) Use of population isolates for mapping complex traits. *Nature Rev. Genet.* **1**, 182–190.

Pharoah PD, Antoniou A, Bobrow M, Zimmern RL, Easton DF, Ponder BA (2002) Polygenic susceptibility to breast cancer and implications for prevention. *Nature Genet.* **31**, 33–36.

Pritchard JK (2001) Are rare variants responsible for susceptibility to complex diseases? *Am. J. Hum. Genet.* **69**, 124–137.

Pritchard JK, Przeworski M (2001) Linkage disequilibrium in humans: models and data. *Am. J. Hum. Genet.* **69**, 1–14.

Pritchard JK, Rosenberg NA (1999) Use of unlinked genetic markers to detect population stratification in association studies. *Am. J. Hum. Genet.* **65**, 220–228.

Puffenberger E, Kauffman E, Bolk S *et al.* (1994a) Identity-by-descent and association mapping of a recessive gene for Hirschsprung disease on human chromosome 13q22. *Hum. Mol. Genet.* **3**, 1217–1225.

Puffenberger EG, Hosoda K, Washington SS *et al.* (1994b) A missense mutation of the endothelin-B receptor gene in multigenic Hirschsprung's disease. *Cell* **79**, 1257–1266.

Reich DE, Cargill M, Bolk S *et al.* (2001) Linkage disequilibrium in the human genome. *Nature* **411**, 199–204.

Reich DE, Lander ES (2001) On the allelic spectrum of human disease. *Trends Genet.* **17**, 502–510.

Risch N (1990a) Linkage strategies for genetically complex traits. 1. Multilocus models. *Am. J. Hum. Genet.* **46**, 222–228.

Risch N (1990b) Linkage strategies for genetically complex traits. 2. The power of affected relative pairs. *Am. J. Hum. Genet.* **46**, 229–241.

Risch N (1990c) Linkage strategies for genetically complex traits. 3. The effect of marker polymorphism on analysis of affected relative pairs. *Am. J. Hum. Genet.* **46**, 242–253.

Risch N, Botstein D (1996) A manic depressive history. *Nature Genet.* **12**, 351–353.

Risch N, Merikangas K (1996) The future of genetic studies of complex human diseases. *Science* **273**, 1516–1517 [see also *Science* **275**, 1327–1330 (1997) for discussion].

Risch N, Teng J (1998) The relative power of family-based and case-control designs for linkage disequilibrium studies of complex human diseases. I. DNA pooling. *Genome Res.* **8**, 1273–1288.

Schaid DJ (1998) Transmission disequilibrium, family controls and great expectations. *Am. J. Hum. Genet.* **63**, 935–941.

Sham PC, Curtis D (1995) An extended transmission disequilibrium test (TDT) for multi-allele marker loci. *Ann. Hum. Genet.* **59**, 323–336.

Spielman RS, Ewens WJ (1998) A sibship test for linkage in the presence of association: the sib transmission disequilibrium test. *Am. J. Hum. Genet.* **62**, 450–458.

Todd JA, Bell JI, McDevitt HO (1987) HLA-DQβ gene contributes to susceptibility and resistance to insulin-dependent diabetes mellitus. *Nature* **329**, 599–604.

Trimble MR (1996) *Biological Psychiatry*, 2nd Edn. John Wiley, Chichester, p. 183.

Varilo T, Laan M, Hovatta I, Wiebe V, Terwilliger JD, Peltonen L (2000) Linkage disequilibrium in isolated populations: Finland and a young sub-population of Kuusamo. *Eur. J. Hum. Genet.* **8**, 604–612.

Weinberger DR, Egan MF, Bertolino A *et al.* (2001) Prefrontal neurons and the genetics of schizophrenia. *Biol. Psychiatry* **50**, 825–844.

Weiss KM, Terwilliger JD (2000) How many diseases does it take to map a gene with SNPs? *Nature Genet.* **26**, 151–157.

Wilson JF, Goldstein DB (2001) Consistent long-range linkage disequilibrium generated by admixture in a Bantu-Semitic hybrid population. *Am. J. Hum. Genet.* **67**, 926–935.

Wright AF, Carothers AD, Pirastu M (1999) Population choice in mapping genes for complex diseases. *Nature Genet.* **23**, 397–404.

Wright A, Charlesworth B, Rudan I, Carothers A, Campbell H (2003) A polygenic basis for late-onset disease. *Trends Genet.* **19**, 97–106.

Yamazaki K, Takazoe M, Tanaka T, Kazumori T, Nakamura Y (2002) Absence of mutation in the *NOD2/CARD15* gene among 483 Japanese patients with Crohn's disease. *J. Hum. Genet.* **47**, 469–472.

CHAPTER SIXTEEN

Molecular pathology

Chapter contents

16.1 Introduction

Molecular pathology seeks to explain why a given genetic change should result in a particular clinical phenotype. We have already reviewed the nature and mechanisms of mutations in Chapter 11 (briefly summarized in *Box 16.1*); this chapter is concerned with their effects on the phenotype. Molecular pathology requires us to work out the effect of a mutation on the quantity or function of the gene product, and to explain why the change is or is not pathogenic for any particular cell, tissue or stage of development.

Not surprisingly, given the complexity of genetic interactions, molecular pathology is a very imperfect science. The greatest successes to date have been in understanding cancer, where the phenotype to be explained – uncontrolled cell proliferation – is relatively simple, and in hemoglobinopathies – where the pathology is a very direct result of a globin abnormality. For most genetic diseases the clinical features are the end result of a long chain of causation, and the holy grail of molecular pathology, **genotype–phenotype correlation**, will never be obtained, because in reality even 'simple' Mendelian diseases are not simple at all (Scriver and Waters, 1999; Dipple and McCabe, 2000; Weatherall, 2001). These reviews, especially that by Scriver and Waters, are strongly recommended (see Further reading).

Despite all the difficulties, humans have one great advantage for students of molecular pathology: we know far more about the phenotypes of humans than of any other organism. Not only are we more likely to notice a subtle variant in a human than in a fly or worm, but the healthcare systems world-wide act as a gigantic and continuous mutation screen. Any human phenotype that occurs with a frequency greater than 1 in 10^9 is probably already described somewhere in the literature. Moreover for most identified disease genes, many different mutations are known. We cannot do experiments on humans or breed them to order, but humans provide unique opportunities to observe the phenotypic effects of many different changes in a given gene. As the emphasis of the Human Genome Project moves from cataloging genes to understanding their function, the study of molecular pathology has moved to center stage. Studies of humans generate hypotheses, which must then be tested in animals. Thus investigations of naturally occurring human mutations are complemented by studies of specific natural or engineered mutations in animals (see Chapter 20).

16.2 The convenient nomenclature of *A* and *a* alleles hides a vast diversity of DNA sequences

When we describe the genotype of a cystic fibrosis carrier as *Aa*, what we mean by *a* is any *CFTR* gene sequence mutated so that it does not produce a functioning chloride channel. Over 750 different such alleles have been described. Similarly, *A* means any functioning sequence – the actual DNA sequence of *A* genes in unrelated people will not necessarily be 100% identical. For genetic counseling and pedigree analysis this is the most useful level of description, but for molecular pathology we need to look more closely. *Box 16.1* contains a brief summary of the main types of DNA sequence change seen in pathogenic mutations and *Box 16.2* summarizes the conventions for describing them.

A problem in molecular pathology is that there is currently no comprehensive database listing all known human mutations. The Human Mutation Database (www.hgmd.org) has useful lists for a number of genes but is not comprehensive, while OMIM makes only a rather half-hearted and inconsistent attempt at listing mutant alleles. Such information is not accessible through PubMed because research journals will not normally publish reports of novel mutations in well-studied genes. An attempt is now being made to remedy this through the HUGO Mutation Database Initiative (see www.genomic.unimelb.edu.au/mdi/ for details and many useful links).

16.3 A first classification of mutations is into loss of function vs. gain of function mutations

16.3.1 For molecular pathology, the important thing is not the sequence of a mutant allele but its effect

Knowing the sequence of a mutant allele is important for genetic testing (Chapter 18) but for molecular pathology we need to know what it does. A mutated gene might have all sorts of subtle effects on an organism, but a valuable first question to ask is whether it produces a loss or gain of function.

Box 16.1: The main classes of mutation.

Deletions ranging from 1 bp to megabases

Insertions including duplications

Single base substitutions:

   missense mutations replace one amino acid with another in the gene product;

   nonsense mutations replace an amino acid codon with a stop codon;

   splice site mutations create or destroy signals for exon–intron splicing.

Frameshifts can be produced by deletions, insertions or splicing errors.

Dynamic mutations are tandem repeats that often change size on transmission to children

See *Table 16.1* for some examples, and Chapter 11 for more details and discussion of mechanisms.

Box 16.2: Nomenclature for describing sequence changes.

See www.dmd.nl/mutnomen.html and den Dunnen and Antonarakis (2001) for full details.

Amino acid substitutions

Start with 'p.' to indicate protein if necessary. Use the one-letter codes: A, alanine; C, cysteine; D, aspartic acid; E, glutamic acid; F, phenylalanine; G, glycine; H, histidine; I, isoleucine; K, lysine; M, methionine; N, asparagine; P, proline; Q, glutamine; R, arginine; S, serine; T, threonine; V, valine; W, tryptophan; Y, tyrosine; X means a stop codon. Three-letter codes are also acceptable.

p.R117H or **Arg117His** – replace arginine 117 by histidine (the initiator methionine is codon 1).

p.G542X or **Gly542Stop** – glycine 542 replaced by a stop codon.

Nucleotide substitution

Start with 'g.' (genomic) or 'c.' (cDNA) if necessary. The A of the initiator ATG codon is +1; the immediately preceding base is –1.

There is no zero. Give the nucleotide number followed by the change. For changes within introns, when only the cDNA sequence is known in full, specify the intron number by IVSn or the number of the nearest exon position.

g.1162G→A – replace guanine at position 1162 by adenine.

g.621+1G→T or **IVS4+1G→T** – replace G by T at the first base of intron 4; exon 4 ends at nt 621.

Deletions and insertions

Use del for deletions and ins for insertions. As above, for DNA changes the nucleotide position or interval comes first, for amino acid changes the amino acid symbol comes first.

p.F508del – delete phenylalanine 508.

c.6232_6236del or **c.6232_6236delATAAG** – delete 5 nucleotides (which can be specified) starting with nt 6232 of the cDNA.

g.409_410insC – insert C between nt 409 and 410 of genomic DNA.

▸ In **loss of function mutations** the product has reduced or no function.

▸ In **gain of function mutations** the product does something positively abnormal.

Box 16.3 shows one way of describing these effects.

Loss of function mutations most often produce recessive phenotypes. For most gene products the precise quantity is not crucial, and we can get by on half the normal amount. Thus most inborn errors of metabolism are recessive. For some gene products, however, 50% of the normal level is not sufficient for normal function, and **haploinsufficiency** produces an abnormal phenotype, which is therefore inherited in a dominant manner (see Section 16.4.2). Sometimes also a nonfunctional mutant polypeptide interferes with the function of the normal allele in a heterozygous person, giving a **dominant negative** effect (an antimorph in the terminology of *Box 16.3* – see Section 16.4.3).

Gain of function mutations usually cause dominant phenotypes, because the presence of a normal allele does not prevent the mutant allele from behaving abnormally. Often this involves a control or signaling system behaving inappropriately – signaling when it should not, or failing to switch a process off when it should. Sometimes the gain of function involves the product doing something novel – a protein

containing an expanded polyglutamine repeat forming abnormal aggregates, for example.

Inevitably some mutations cannot easily be classified as either loss or gain of function. Has a permanently open ion channel lost the function of closing or gained the function of inappropriate opening? A dominant negative mutant allele has lost its function but also does something positively abnormal. A mutation may change the balance between several functions of a gene product. Nevertheless, the distinction between loss of function and gain of function is the essential first tool for thinking about molecular pathology.

16.3.2 Loss of function is likely when point mutations in a gene produce the same pathological change as deletions

Purely genetic evidence, without biochemical studies, can often suggest whether a phenotype is caused by loss or gain of function. When a clinical phenotype results from loss of function of a gene, we would expect *any* change that inactivates the gene product to produce the same clinical result. We should be able to find point mutations which have the same effect as mutations that delete or disrupt the gene. Waardenburg syndrome Type 1 (MIM 193500: hearing loss and pigmentary abnormalities) provides an example. As

---

Box 16.3: A nomenclature for describing the effect of an allele.

Null allele or **amorph**: an allele that produces no product

Hypomorph: an allele that produces a reduced amount or activity of product

Hypermorph: an allele that produces increased amount or activity of product

Neomorph: an allele with a novel activity or product

Antimorph: an allele whose activity or product antagonizes the activity of the normal product.

Figure 16.1 shows, causative mutations in the *PAX3* gene include amino acid substitutions, frameshifts, splicing mutations, and in some patients complete deletion of the gene. Since all these events produce the same clinical result, its cause must be loss of function of *PAX3*. Similarly, among diseases caused by unstable trinucleotide repeats (Section 16.6.4), Fragile-X and Friedreich ataxia are occasionally caused by other types of mutation in their respective genes, pointing to loss of function, whereas Huntington disease is never seen with any other type of mutation, suggesting a gain of function.

16.3.3 Gain of function is likely when only a specific mutation in a gene produces a given pathology

Gain of function is likely to require a much more specific change than loss of function. The mutational spectrum in gain-of-function conditions should be correspondingly more restricted, and the same condition should not be produced by deletion or disruption of the gene. Likely examples include Huntington disease (Section 16.6.4), and achondroplasia (MIM 100800: short-limbed dwarfism). Virtually all achondroplastics have the same amino acid change, G380R in the fibroblast growth factor receptor FGFR3. Other substitutions in the same protein produce other syndromes (Section 16.7.3). For unknown reasons, the

mutation rate for G380R is extraordinarily high, so that achondroplasia is one of the commoner genetic abnormalities, despite requiring a very specific DNA sequence change.

Mutational homogeneity is an indicator of a gain of function, but there are other reasons why a single mutation may account for most or all cases of a disease:

▶ diseases where what one observes is very directly related to the gene product itself, rather than a more remote consequence of the genetic change, may be defined in terms of a particular variant product, as in sickle cell disease (see *Box 16.4*);

▶ some specific molecular mechanism may make a certain sequence change in a gene much more likely than any other change – e.g. the CGG expansion in Fragile-X syndrome (Section 16.6.4);

▶ there may be a **founder effect** – for example, certain disease mutations are common among Ashkenazi Jews, presumably reflecting mutations present in a fairly small number of founders of the present Ashkenazi population (Motulsky, 1995);

▶ selection favoring heterozygotes (Section 4.5.3) enhances founder effects and often results in one or a few specific mutations being common in a population.

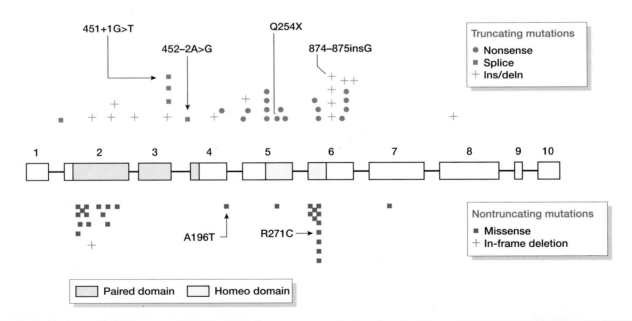

Figure 16.1: Loss of function mutations in the *PAX3* gene.

The 10 exons of the gene are shown as boxes, with the connecting introns not to scale. Shaded areas mark the sequences encoding the two DNA-binding domains of the PAX3 protein. Note that mutations that completely destroy the structure of the PAX3 protein (drawn above the gene diagram) are scattered over at least the first six exons of the gene, but missense mutations (shown below the gene diagram) are concentrated in two regions, the 5′ part of the paired domain and the third helix of the homeodomain. A196T is believed to affect splicing. The other named mutations are mentioned in *Table 16.1*. The 874–875insG mutation introduces a seventh G into a run of six Gs; it has arisen independently several times and illustrates the relatively high frequency of slipped-strand mispairing (Section 11.3.1).

16.3.4 Deciding whether a DNA sequence change is pathogenic can be difficult

Not every sequence variant seen in an affected person is necessarily pathogenic. How can we decide whether a sequence change we have discovered is the sought-for pathogenic mutation or a harmless variant? In descending order of reliability, the criteria are:

1) functional studies showing that the change is pathogenic;

2) precedent: that change has been seen before in patients with this disease (and not in ethnically matched controls);

3) a *de novo* mutation, not present in the parents, in a person with *de novo* disease;

4) a novel sequence change that is absent in a panel of, say, 100 normal controls;

5) the nature of the sequence change (see *Box 16.5*).

Functional studies are the gold standard; none of the other criteria is fully reliable. For gain of function phenotypes, as explained above, the cause is likely to be very specific. Any sequence change different from the standard mutation is probably not pathogenic, at least for the disease in question. Loss of function conditions raise far more questions of interpretation.

16.4 Loss of function mutations

16.4.1 Many different changes to a gene can cause loss of function

Not surprisingly, there are many ways of reducing or abolishing the function of a gene product (*Table 16.1*). Some of these have been discussed in Section 11.4. The hemoglobinopathies (*Box 16.4*) exemplify many of these mechanisms especially well. In fact, the globins could be used to illustrate virtually every process described in this book. Readers are recommended to consult a review such as that by Weatherall *et al.* (2001; see Further reading) to get a parallel narrative on molecular pathology.

Box 16.5 provides some guidelines for considering the likely result of a mutation on the gene product.

Small deletions and insertions have a much more drastic effect on the gene product if they introduce a **frameshift** (that is, if they add or remove a number of nucleotides that is not an exact multiple of three). Deletions in the dystrophin gene provide striking examples (*Figure 16.2*). In general frameshifting deletions produce the severe Duchenne muscular dystrophy, in which no dystrophin is produced, while nonframeshifting mutations cause the milder Becker form, in which dystrophin is present but abnormal.

Nonsense mutations (those that generate **premature termination codons)** normally function as null alleles because they trigger **nonsense-mediated mRNA decay** (Hentze and Kulozik, 1999). It is unusual for the mRNA to be translated to produce a truncated protein. Nonsense-mediated decay is usual whenever a premature termination codon is at least 50 nucleotides upstream of the last splice junction. See Lykke-Andersen *et al.* (2001) for details of the mechanism.

Splicing mutations are often thought of as just mutations that alter the conserved GT…AG sequences at the ends of introns. In fact a much wider variety of sequence changes can affect splicing, but these effects are hard to predict. Thus many mutations affecting splicing go unrecognized unless RT–PCR studies are performed, and this class of mutation is certainly very under-diagnosed. Several examples can be seen among

---

Box 16.4: Hemoglobinopathies.

Hemoglobinopathies occupy a special place in clinical genetics for many reasons. They are by far the most common serious Mendelian diseases on a worldwide scale. Globins illuminate important aspects of evolution of the genome (*Figures 12.4 – 12.6*) and of diseases in populations. Developmental controls are probably better understood for globins than for any other human genes (*Figures 10.22, 10.23*). More mutations and more diseases are described for hemoglobins than for any other gene family. Clinical symptoms follow very directly from malfunction of the protein, which at 15 g per 100 ml of blood is easy to study, so that the relation between molecular and clinical events is clearer for the hemoglobinopathies than for most other diseases.

Hemoglobinopathies are classified into three main groups.

▸ The **thalassemias** are caused by inadequate quantities of the α or β chains. Alleles are classified into those producing no product (α^0, β^0) and those producing reduced amounts of product (α^+, β^+). The underlying defects include examples of all the types detailed here in Section 16.4 (see also *Table 18.5*).

▸ **Abnormal hemoglobins** with amino acid changes cause a variety of problems, of which sickle cell disease is the best known. The E6V mutation replaces a polar by a neutral amino acid on the outer surface of the β-globin molecule. This causes increased intermolecular adhesion, leading to aggregation of deoxyhemoglobin and distortion of the red cell. Sickled red cells have decreased survival time (leading to anemia) and tend to occlude capillaries, leading to ischemia and infarction of organs downstream of the blockage. Other amino acid changes can cause anemia, cyanosis, polycythemia (excessive numbers of red cells), methemoglobinemia (conversion of the iron from the ferrous to the ferric state) etc.

▸ **Hereditary persistence of fetal hemoglobin**, caused by a defect in the normal switch from fetal to adult hemoglobin, can be an important modifier of the clinical effects of the other two classes of mutants.

See Weatherall *et al.* (2001; Further reading) for a comprehensive review.

Table 16.1: Eleven ways to reduce or abolish the production of a functioning gene product

Change	Example
Delete:	
(i) the entire gene	Most α-thalassemia mutations (*Figure 16.3*)
(ii) part of the gene	60% of Duchenne muscular dystrophy (*Figure 16.2*)
Insert a sequence into the gene	Insertion of LINE-1 repetitive sequence (see Section 11.5.6) into *F8* gene in hemophilia A
Disrupt the gene structure:	
(i) by a translocation	X–autosome translocations in women with Duchenne muscular dystrophy (*Figure 14.10*)
(ii) by an inversion	Inversion in *F8* gene (*Figure 11.20*)
Prevent the promoter working:	
(i) by mutation	β-Globin g.-29A→G mutation (*Table 18.5*)
(ii) by methylation	*CDKN2A* gene in many tumors (Section 17.6.1)
Destabilize the mRNA:	
(i) polyadenylation site mutation	α-globin g.AATAAA→AATAGA mutation
(ii) by nonsense-mediated RNA decay	β-Globin p.Q39X
Prevent correct splicing (Section 11.4.3)	
(i) inactivate donor splice site	*PAX3* g.451+1G→T mutation (*Figure 16.1*)
(ii) inactivate acceptor splice site	*PAX3* g.452-2A→G mutation (*Figure 16.1*)
(iii) alter an exonic splicing enhancer	*SMN2* exon 7 g.C6T (Cartegni and Krainer 2002)
(iv) activate a cryptic splice site (maybe deep within an intron)	*LGMD2A* G624G (*Figure 11.12*) β-Globin IVS1–110G→A mutation (*Table 18.5*) *CFTR* 3849+10kb C→T (*Table 18.6*)
Introduce a frameshift in translation	*PAX3* g.874_875insG mutation (*Figure 16.1*)
Convert a codon into a stop codon	*PAX3* p.Q254X mutation (*Figure 16.1*)
Replace an essential amino acid	*PAX3* p.R271C mutation (*Figure 16.1*)
Prevent post-transcriptional processing	Cleavage-resistant collagen N-terminal propeptide in Ehlers Danlos VII syndrome (*Section 16.6.1*).
Prevent correct cellular localization of product	p.F508del mutation in cystic fibrosis

Box 16.5: Guidelines for assessing the significance of a DNA sequence change.

▶ Deletions of the whole gene, nonsense mutations and frameshifts are almost certain to destroy the gene function.

▶ Mutations that change the conserved GT...AG nucleotides flanking most introns affect splicing, and will usually abolish the function of the gene. Many other sequence changes can affect splicing but in ways that are much harder to predict (see below).

▶ A missense mutation is more likely to be pathogenic if it affects a part of the protein known to be functionally important. Computer modeling of the protein structure may help suggest which residues are critical. For example, the missense mutations in *Figure 16.1*, all of which cause loss of function, are concentrated in the key DNA-binding domains of the PAX3 protein.

▶ Changing an amino acid is more likely to affect function if that amino acid is conserved in related genes (orthologs or paralogs).

▶ Amino acid substitutions are more likely to affect function if they are nonconservative (replace a polar by a nonpolar amino acid, or an acidic by a basic one – see *Box 11.3*).

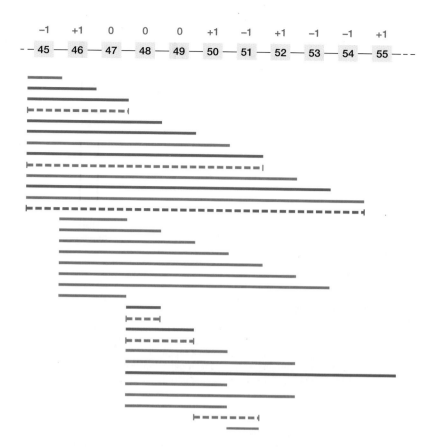

Figure 16.2: Deletions in the central part of the dystrophin gene in patients with Duchenne or Becker muscular dystrophy.

Numbered boxes represent exons 45–55. Numbers above each box show the effect that deleting the exon has on the reading frame (0 no effect, –1 shift 1 nucleotide back, +1 shift 1 nucleotide forward). Red bars show deletions in patients with the lethal DMD, blue bars show deletions in patients with the milder BMD. In general frameshifting deletions cause DMD and frame-neutral ones cause BMD. Exceptions are indicated by dashed lines. Possible reasons for the exceptions include deletions ending within an exon, actions of modifier genes or environmental effects, or maybe clinical or laboratory errors. Data from the Leiden Muscular Dystrophy website www.dmd.nl, which should be consulted for fuller data and background information. *Figure 18.6* shows how these deletions are characterized in the laboratory.

the β-globin mutations listed in *Table 18.5*. Three types of sequence change cause aberrant splicing:

▶ **alterations of normal splice sites:** splicing (Section 11.4.3) requires not just the canonical GT…AG sequence, but also a less rigidly defined sequence context surrounding the site. Changes here may alter the ratio of different splice isoforms rather than abolish splicing altogether – for example the 5T/7T/9T polymorphism in intron 8 of the *CFTR* gene (see MIM 602421) alters the proportion of transcripts that skip exon 9. Such effects are probably frequent contributors to genetic disease, especially perhaps susceptibility to common diseases (Nissim-Rafinia and Kerem, 2002);

▶ **alterations of splicing enhancers** or **silencers:** these important but ill-defined sequences may be located in exons or introns (Nissim-Rafinia and Kerem, 2002). Cartegni and Krainer (2002) provide a particularly clear example of a mutation disrupting an exonic splicing enhancer;

▶ **activation of cryptic splice sites:** apparently innocuous base substitutions may cause a previously inactive sequence to be used as a splice site. The cryptic site might be in an exon or an intron. *Figure 11.12* shows an example. If the site lies deep within an intron, the mutation would be overlooked without RT–PCR studies – for example the *CFTR* mutation 3849 + 10kbC → T activates a cryptic splice site that lies 10 kb inside intron 19.

16.4.2 In haploinsufficiency a 50% reduction in the level of gene function causes an abnormal phenotype

Loss of function mutations tend to be recessive because heterozygotes often function perfectly normally. Sometimes this is because feedback loops at the transcription or protein activity level compensate for the reduced dosage, but in many cases the cell and organism are able to function normally with

molecules into inactive dimers give dominant phenotypes (Hemesath *et al.*, 1994). The ion channels in cell membranes provide another example of multimeric structures that are sensitive to dominant negative effects (Section 16.6.1).

Inherited disease but common in cancer

Making random changes in a gene is quite likely to stop it working, but very unlikely to give it a novel function. The

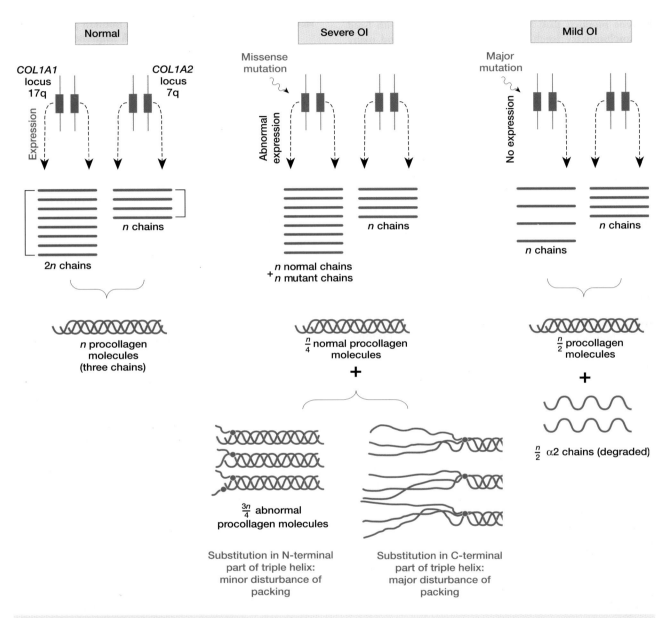

Figure 16.4: Dominant negative effects of collagen gene mutations.

Collagen fibrils are built of arrays of triple-helical procollagen units. The type I procollagen comprises two chains encoded by the *COL1A1* gene and one encoded by *COL1A2*. Null mutations in either gene have a less severe effect than mutations encoding polypeptides that are incorporated into the triple helix and destroy its function.

only mechanism that commonly generates novel functional genes is when a chromosomal rearrangement joins functional exons of two different genes (*Figure 17.4A*). Such exon shuffling was no doubt important in evolution; for molecular pathology, it is most often noticed when it leads to cancer. Many acquired tumor-specific chromosomal rearrangements produce chimeric genes with novel activities that lead to uncontrolled cell proliferation (see *Table 17.3*). A rare case of an inherited point mutation conferring a novel function on a

protein is the Pittsburgh allele at the *PI* locus (MIM 107400; *Figure 16.6*).

16.5.2 Overexpression may be pathogenic

Gross overexpression of certain genes is common in cancer cells. The mechanisms by which somatic genetic changes produce overexpression include massive reduplication of the gene or transposition of a gene normally expressed at low level

▶ gene products that co-operate in interactions with fixed stoichiometry (such as the α and β globins and many structural proteins).

In each case the gene product is titrated against something else in the cell. What matters is not the correct absolute level of product, but the correct relative levels of interacting products. The effects are sensitive to changes in all the interacting partners, thus these dominant conditions often show highly variable expression (see Section 16.6.3). Genes whose products act essentially alone, such as many soluble enzymes of metabolism, seldom show dosage effects.

In addition, there are some cases where a cell with only one working copy of a gene just can't meet the demand for a gene product that is needed in large quantities. An example may be elastin. In people heterozygous for a deletion or loss of function mutation of the elastin gene, tissues that require only modest quantities of elastin (skin, lung) are unaffected, but the aorta, where much more elastin is required, often shows an abnormality (supravalvular aortic stenosis) that can require surgery (see Section 16.8.1).

16.4.3 Mutations in proteins that work as dimers or multimers sometimes produce dominant negative effects

A **dominant negative effect** occurs when a mutant polypeptide not only loses its own function, but also interferes with the product of the normal allele in a heterozygote. Some people would argue that these are gain of function, not loss of function mutations – but that is just an argument about words. Dominant negative mutations cause more severe effects than simple null alleles of the same gene. Structural proteins that build multimeric structures are particularly vulnerable to dominant negative effects. Collagens provide a classic example.

Fibrillar collagens, the major structural proteins of connective tissue, are built of triple helices of polypeptide chains, sometimes homotrimers, sometimes heterotrimers, that are assembled into close-packed crosslinked arrays to form rigid fibrils. In newly synthesized polypeptide chains (preprocollagen), N- and C-terminal propeptides flank a regular repeating sequence (Gly–X–Y)$_n$, where either X or Y is usually proline, and the other is any amino acid. Three preprocollagen chains associate and wind into a triple helix under control of the C-terminal propeptide. After formation of the triple helix, the N- and C-terminal propeptides are cleaved off. A polypeptide that complexes with normal chains, but then wrecks the triple helix can reduce the yield of functional collagen to well below 50% (*Figure 16.4*). The molecular pathology of collagen mutations is very rich, and is discussed below (see Section 16.6.1 and *Table 16.7*).

Nonstructural proteins that dimerize or oligomerize also show dominant negative effects. For example, transcription factors of the b-HLH-Zip family (*Figure 10.9*) bind DNA as dimers. Mutants that cannot dimerize often cause recessive phenotypes, but mutants that are able to sequester functioning molecules into inactive dimers give dominant phenotypes (Hemesath *et al.*, 1994). The ion channels in cell membranes provide another example of multimeric structures that are sensitive to dominant negative effects (Section 16.6.1).

16.4.4 Epigenetic modification can abolish gene function even without a DNA sequence change

Changes that are heritable (from cell to daughter cell, or from parent to child) but that do not depend on changes in DNA sequence are called **epigenetic** (Section 10.1). A set of mini-reviews in the 1 May 1998 issue of *Cell* (Vol 93, pages 301–337) discusses many of the diverse facets of epigenetics. Epigenetic effects are particularly clear in cancer (Jones and Baylin, 2002). For present purposes the main epigenetic mechanism of interest is **DNA methylation**. For fuller detail on this subject, see the review by Bird (2002).

As mentioned earlier (Section 10.4.2), cytosine bases in human DNA that lie next to guanine (a **CpG dinucleotide**) are often methylated at carbon 5, and the pattern of CpG methylation can be stably transmitted when the DNA is replicated (*Figure 16.5*). These methylation patterns are important signals for controlling transcription of genes (Section 10.4.3). X-inactivation depends at least in part on DNA methylation, and epigenetic transmission of the pattern of methylation ensures that the same X is inactivated in mother and daughter cells. Inappropriate methylation can cause a heritable pathogenic loss of function. In many tumors, for example, function of the p16 (*CDKN2A*) tumor suppressor gene is abrogated by methylation of the promoter rather than by mutating its DNA sequence (Section 17.4.3). Conversely, inappropriate demethylation is believed to switch on expression of oncogenes in tumors.

Imprinted genes (Sections 4.3.4, 10.5.4) are a particularly intriguing example of epigenetic modification. Their expression is controlled by patterns of methylation that differ according to the parental origin of the gene. When either the imprinting mechanism malfunctions or the parental origin is not as expected, loss of function or inappropriate expression can occur in intact genes. Imprinted genes occur in clusters that contain both maternally and paternally imprinted genes, some of which are imprinted only in certain tissues. Often both DNA strands can potentially be transcribed, one transcript being a mRNA and the other a large (> 100 kb) untranslatable **antisense RNA**, but transcription of one strand prevents transcription of the other. Thus the molecular pathology of human imprinting diseases is fascinating but exceedingly complicated. *Box 16.6* describes the best known human imprinting diseases, Prader–Willi and Angelman syndromes; for more information on some other diseases see OMIM 130650 (Beckwith–Wiedemann syndrome), OMIM 139320 (*GNAS*) and the reviews by Reik and Walter (2001) and Rougeulle and Heard (2002).

16.5 Gain of function mutations

Table 16.3 lists a number of mechanisms that can produce a gain of function phenotype.

16.5.1 Acquisition of a novel function is rare in inherited disease but common in cancer

Making random changes in a gene is quite likely to stop it working, but very unlikely to give it a novel function. The

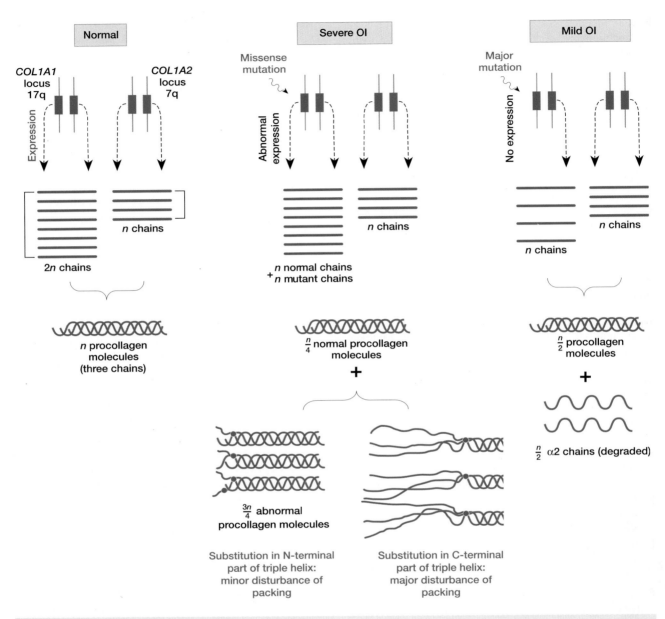

Figure 16.4: Dominant negative effects of collagen gene mutations.

Collagen fibrils are built of arrays of triple-helical procollagen units. The type I procollagen comprises two chains encoded by the *COL1A1* gene and one encoded by *COL1A2*. Null mutations in either gene have a less severe effect than mutations encoding polypeptides that are incorporated into the triple helix and destroy its function.

only mechanism that commonly generates novel functional genes is when a chromosomal rearrangement joins functional exons of two different genes (*Figure 17.4A*). Such exon shuffling was no doubt important in evolution; for molecular pathology, it is most often noticed when it leads to cancer. Many acquired tumor-specific chromosomal rearrangements produce chimeric genes with novel activities that lead to uncontrolled cell proliferation (see *Table 17.3*). A rare case of an inherited point mutation conferring a novel function on a

protein is the Pittsburgh allele at the *PI* locus (MIM 107400; *Figure 16.6*).

16.5.2 Overexpression may be pathogenic

Gross overexpression of certain genes is common in cancer cells. The mechanisms by which somatic genetic changes produce overexpression include massive reduplication of the gene or transposition of a gene normally expressed at low level into a highly active chromatin environment. These are discussed more fully in Section 17.3.3.

Inherited diseases are not often caused by constitutional overexpression of a single gene. Duplication of the *DSS* gene on Xp21.3 causes male to female sex reversal, probably as a direct result of the doubled dosage (Bardoni *et al.,* 1994). For the *PMP22* peripheral myelin protein gene, an increase in gene dosage from two to three copies is enough to produce Charcot–Marie–Tooth disease (see below and *Figure 16.8*). Such modest increases in gene expression are probably seldom pathogenic, although a similar degree of dosage sensitivity of unidentified genes must explain many features of chromosomal trisomies (Section 16.8.2). Over-activity of an abnormal gene product, with normal transcription and translation of the gene, can produce similar effects.

16.5.3 Qualitative changes in a gene product can cause gain of function

Although gains of truly novel functions are very rare in inherited disease, activating mutations that modify cellular signaling responses quite often produce dominant phenotypes. The G-protein coupled hormone receptors provide good examples. Many hormones exert their effects on target cells by binding to the extracellular domains of transmembrane receptors. Binding of ligand causes the cytoplasmic tail of the receptor to catalyze conversion of an inactive (GDP-bound) G-protein into an active (GTP-bound) form, and this relays the signal further by stimulating adenylyl cyclase. Some mutations cause receptors to activate adenylyl cyclase even in the absence of ligand.

▶ Familial male precocious puberty (MIM 176410: onset of puberty by the age of 4 years in affected boys) is found with a constitutively active luteinizing hormone receptor.

▶ Autosomal dominant thyroid hyperplasia can be caused by an activating mutation in the thyroid stimulating hormone receptor (see MIM 275200).

▶ Jansen's metaphyseal chondrodystrophy (MIM 156400: a disorder of bone growth) can be caused by a constitutionally active parathyroid hormone receptor.

▶ A constitutionally active $G_s\alpha$ protein (part of the receptor-coupling G-protein) causes McCune–Albright syndrome or polyostotic fibrous dysplasia (PFD, MIM 174800). PFD is known only as a somatic condition in mosaics – probably constitutional mutations would be lethal. Depending on the tissues carrying the mutant cell line, the result is polyostotic fibrous dysplasia, café au lait spots, sexual precocity and other hyperfunctional endocrinopathies. Loss of function mutations of the same gene often underlie a different disease, Albright's hereditary osteodystrophy (see *Table 16.5*).

16.6 Molecular pathology: from gene to disease

The starting point in thinking about molecular pathology may be either a gene or a disease. These two approaches are considered separately in this section and the next, although of course a full understanding of molecular pathology merges the two.

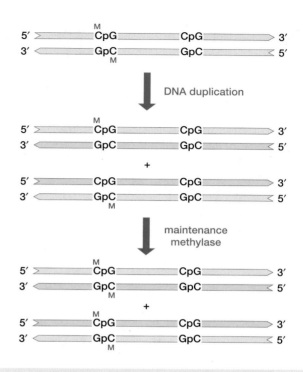

Figure 16.5: Inheritance of the pattern of CpG methylation.

The maintenance methylase recognizes hemi-methylated CpG in newly replicated DNA, allowing the pattern of CpG methylation to be inherited (see Section 10.4.2 and Bird (2002)).

16.6.1 For loss of function mutations the phenotypic effect depends on the residual level of gene function

The DNA sequence changes described in *Table 16.1* can cause varying degrees of loss of function. Many amino acid substitutions have little or no effect, while some mutations will totally abolish the function. A mutation may be present in one or both copies of a gene. People with autosomal recessive conditions are often **compound heterozygotes**, with two different mutations. If both mutations cause loss of function, but to differing degrees, the least severe allele will dictate the level of residual function.

Figure 16.7 represents four possible relations between the level of residual gene function and the clinical phenotype.

A. **A simple recessive condition.** People heterozygous for a mutation that totally abolishes gene function are phenotypically normal, provided their remaining allele is not significantly defective.

B. **A dominant condition caused by haploinsufficiency.** In reality, this simple situation is rare. If a 50% reduction in gene product causes symptoms, a more severe reduction will probably have more severe effects.

C. **A recessive condition with graded severity.** Among many examples are:

Box 16.6: Molecular pathology of Prader–Willi and Angelman syndromes.

Prader–Willi syndrome (PWS) and Angelman syndrome (AS) are both caused by problems with differentially imprinted genes at 15q11-q13. They exemplify the complicated molecular pathology associated with clusters of imprinted genes (Nicholls and Knepper, 2001).

▶ **PWS** (MIM 176260: mental retardation, hypotonia, gross obesity, male hypogenitalism) is caused by lack of function of genes that are expressed only from the paternal chromosome.

▶ **AS** (MIM 105830: mental retardation, lack of speech, growth retardation, hyperactivity, inappropriate laughter) is due to lack of function of a closely linked gene that is expressed only from the maternal chromosome. Some children with AS will have two functional copies of the PWS genes, and vice versa, but this overexpression does not appear to have any phenotypic effect.

As shown in the Table, a variety of events can lead to lack of a paternal (PWS) or a maternal (AS) copy of the relevant chromosome 15 sequences.

▶ *De novo* deletions of 4–4.5 Mb between flanking repeats at 15q11-q13 are the commonest cause. Deletions on the paternal chromosome cause PWS, deletion of the same region on the maternal chromosome causes AS.

▶ **Uniparental disomy** is detected when DNA marker studies show that a person with apparently normal chromosomes has inherited both homologs of a particular pair (No.15 in this case) from one parent. The usual cause is **trisomy rescue**. A trisomy 15 conceptus develops to a multicell stage; normally it would die, but if a chance mitotic nondisjunction (Section 2.5.2) produces a cell with only two copies of chromosome 15 at a sufficiently early stage of development, that cell can go on to make the whole of a surviving baby. Most trisomy 15 conceptuses are $15^M15^M15^P$. One time out of three, random

loss of one chromosome 15 will produce a fetus with maternal uniparental disomy, 15^M15^M. Lacking any paternal 15, the fetus will have Prader–Willi syndrome.

▶ Occasionally something has gone wrong with the mechanism of imprinting. Both chromosome 15 homologs carry the same parent-specific methylation pattern, although marker studies show they originate from different parents. These interesting cases are caused by small deletions that define nonoverlapping PWS and AS **imprinting control elements**.

▶ AS may be caused entirely by lack of expression of the *UBE3A* gene, since some inherited cases have normal chromosome structure and imprinting, but have point mutations in this gene. PWS is more complex. The paternal chromosome encodes a huge transcript (up to 460 kb and 148 exons) with multiple splice forms. The first 10 exons encode two proteins, SNURF and SNRPN, while the downstream exons lack open reading frames but some of the introns encode small nucleolar RNAs (snoRNAs) that form part of the splicing machinery. Absence of the snoRNAs may cause the symptoms of PWS. The downstream part of the transcript overlaps the *UBE3A* gene on the opposite strand, and probably acts as an antisense RNA, preventing transcription of *UBE3A* from the paternal chromosome (Runte *et al.*, 2001).

Origins of Prader–Willi and Angelman syndromes.

Event	Proportion of PWS	Proportion of AS
Deletions	~ 75%	~ 75%
Uniparental disomy	~ 20%	~ 3%
Imprinting errors	~ 2%	~ 5%
Point mutations	Not seen	~ 15% (in *UBE3A* gene)

Table 16.3: Mechanisms of gain of function mutations

Malfunction	Gene	Disease	MIM no.
Overexpression	PMP22	Charcot–Marie–Tooth disease	118200
Receptor permanently 'on'	GNAS	McCune–Albright disease	174800
Acquire new substrate (Pittsburgh allele)	PI	α_1-Antitrypsin deficiency	107400
Ion channel inappropriately open	SCN4A	Paramyotonia congenita	168300
Structurally abnormal multimers	COL2A1	Osteogenesis imperfecta	Various
Protein aggregation	HD	Huntington disease	143100
Chimeric gene	BCR–ABL	Chronic myeloid leukemia	151410

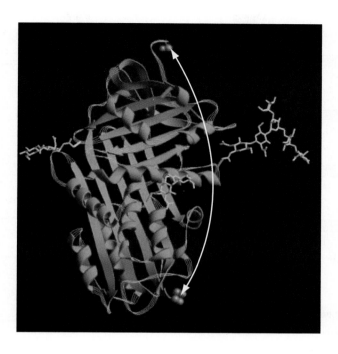

Figure 16.6: An inherited mutation causing a protein to gain a novel function.

Methionine 358 in the reactive center of α_1-antitrypsin acts as a 'bait' for elastase. Elastase cleaves the peptide link between Met358 and Ser359, causing the two residues to spring 65 Å apart, as shown here (green balls). Elastase is trapped and inactivated. The Pittsburgh variant has a missense mutation M358R which replaces the methionine bait with arginine. This destroys affinity for elastase but creates a bait for thrombin. As a novel constitutively active antithrombin, the Pittsburgh variant produces a lethal bleeding disorder. Image from University of Geneva ExPASy molecular biology World Wide Web server.

- mutants in the X-linked hypoxanthine guanine phosphoribosyl transferase (*HPRT*) gene. The extent of residual enzyme activity in mutants correlates well with the clinical phenotype of affected males (*Table 16.4*);

- reduced copy numbers of α-globin genes produce successively more severe effects. As shown in *Figure 16.2*, most people have four copies of the α-globin gene ($\alpha\alpha/\alpha\alpha$). People with three copies ($\alpha\alpha/\alpha-$) are healthy; those with two (whether the phase is $\alpha-/\alpha-$ or $\alpha\alpha/-$) suffer mild α-thalassemia; those with only one gene ($\alpha-/-$) have severe disease, while lack of all α genes ($-/-$) causes lethal hydrops fetalis.

D. **Several phenotypes**. Decreasing residual function of a gene may extend the phenotype, perhaps causing a condition with a different clinical label. Depending on the position of the thresholds, several different situations can arise:

- Several related recessive conditions may be caused by successive reductions in gene function at a single locus. For example, extracellular matrix is rich in sulfated proteoglycans like heparan sulfate and chondroitin sulfate, and defects in sulfate transport interfere with skeletal development. Loss of function mutations in the *DTDST* sulfate transporter cause four related autosomal recessive skeletal dysplasias, diastrophic dysplasia (MIM 226600), multiple epiphyseal dysplasia 4 (MIM 226900),

atelosteogenesis II (MIM 256050) and achrondrogenesis Type 1B (MIM 600972), depending on the extent of loss (Karniski, 2001).

- A 50% loss may have no effect; a somewhat greater loss, caused by dominant negative effects, may produce a dominant condition; while total loss of function produces a more severe recessive condition. Simple loss of function mutations in the *KVLQT1* K^+ channel have no effect in heterozygotes but cause the recessive Jervell and Lange–Nielsen syndrome (MIM 220400: heart problems and hearing loss) in homozygotes. However, a dominant negative mutation in the same gene produces the dominantly inherited Romano–Ward syndrome (MIM 192500: cardiac arrhythmia). In transfected *Xenopus* oocytes Romano–Ward ion channels have about 20% of normal activity, but ion channels from JLN patients totally lack function (Wollnik *et al.*, 1997).

- Mutations in the same gene can produce two or more dominant conditions, the milder one by simple haploinsufficiency, and more severe forms through dominant negative effects. This happens in the *COL1A1* or *COL1A2* genes that encode Type I collagen (*Figure 16.4*). Mutations in these genes usually produce osteogenesis imperfecta (OI; brittle bone disease). Frameshifts and nonsense mutations produce Type 1 OI, the mildest form, while amino acid substitutions in the Gly-X-Y

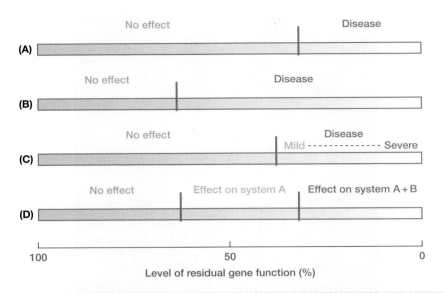

Figure 16.7: Four possible relationships between loss of function and clinical phenotype.

See Section 16.6.1 for discussion.

repeated units are seen in the more severe Types II, III and IV OI. The genotype–phenotype relationship is quite subtle. Substitution of glycine by a bulkier amino acid in the Gly-X-Y unit has a dominant negative effect by disrupting the close packing of the collagen triple helix. The helix is assembled starting at the C-terminal end, and substitution of glycines close to that end has a more severe effect than substitutions nearer the N-terminal end. Skipping of exon 6 (of *COL1A1* or *COL1A2*) has a quite different effect. The site for cleavage of the N-terminal propeptide is lost and abnormal collagen is produced that causes Ehlers Danlos syndrome Type VII (MIM 130060; laxity of skin and joints). A different function has been lost, and a different phenotype results.

16.6.2 Loss of function and gain of function mutations in the same gene will cause different diseases

We have seen that loss of function mutations in the *PAX3* gene cause the developmental abnormality Type 1

Waardenburg syndrome (*Figure 16.1*). A totally different phenotype is seen when an acquired chromosomal translocation creates a novel chimeric gene by fusing *PAX3* to another transcription factor gene, *FKHR* in a somatic cell. The gain of function of this hybrid transcription factor causes the development of the childhood tumor, alveolar rhabdo-myosarcoma (*Table 17.3*).

A striking example concerns the *RET* gene (Manié *et al.*, 2001). *RET* encodes a receptor that straddles the cell membrane. When its ligand (GDNF) binds to the extracellular domain it induces dimerization of the receptors, which then transmit the signal into the cell via tyrosine kinase modules in their cytoplasmic domain. A variety of loss of function mutations – frameshifts, nonsense mutations and amino acid substitutions that interfere with the post-translational maturation of the RET protein – are one cause of Hirschsprung disease (MIM 142623; intractable constipation caused by absence of enteric ganglia in the bowel – see Section 15.6.2). Certain very specific missense mutations in the *RET* gene are seen in a totally different set of diseases, familial medullary thyroid carcinoma and the related but

Table 16.4: Consequences of decreasing function of hypoxanthine guanine phosphoribosyl transferase

HPRT activity (% of normal)	Phenotype
> 60	Normal
8–60	Neurologically normal; hyperuricemia (gout)
1.6–8	Neurological problem (choreoathetosis)
1.4–1.6	Lesch–Nyhan syndrome (choreoathetosis, self-mutilation) but intelligence normal
< 1.4	Classical Lesch–Nyhan syndrome (MIM 308000; choreoathetosis, self-mutilation and mental retardation)

more extensive multiple endocrine neoplasia type 2. These are gain of function mutations, producing receptor that reacts excessively to ligand or is constitutively active and dimerizes even in the absence of ligand. Curiously, some people with missense mutations affecting cysteines 618 or 620 suffer from both thyroid cancer and Hirschsprung disease – simultaneous loss and gain of function. This reminds us that loss of function and gain of function are not always simple scalar quantities; mutations may have different effects in the different cell types in which a gene is expressed.

Table 16.5 lists a number of cases where mutations in a single gene can result in more than one disease. Usually the gain of function mutant produces a qualitatively abnormal protein. Occasionally a simple dosage effect can be pathogenic – the peripheral myelin protein gene *PMP22* is an example. Unequal crossovers between repeat sequences on chromosome 17p11 create duplications or deletions of a 1.5-Mb region that contains the *PMP22* gene (*Figure 16.8*). Heterozygous carriers of the deletion or duplication have one copy or three copies respectively of this gene. People who have only a single copy suffer from hereditary neuropathy with pressure palsies or tomaculous neuropathy (MIM 162500), while as mentioned above, people with three copies have a clinically different neuropathy, Charcot–Marie–Tooth disease 1A (CMT1A; MIM 118220).

16.6.3 Variability within families is evidence of modifier genes or chance effects

Many Mendelian conditions are clinically variable even between affected members of the same family who carry exactly the same mutation. Intrafamilial variability must be caused by some combination of the effects of other unlinked genes (modifier genes) and environmental effects (including chance events). Phenotypes depending on haploinsufficiency are especially sensitive to the effects of modifiers, as discussed above (Section 16.4.2). Waardenburg syndrome is a typical example: *Figure 16.1* shows the evidence that this dominant condition is caused by haploinsufficiency, and *Figure 4.5C* shows typical intrafamilial variation.

Intrafamilial variability is a big problem in genetic counseling because families contemplating childbearing want to know how severely affected a child would be. Thus there is a clinical as well as a scientific motivation to identify modifier genes. Candidate modifier genes may be suggested by knowledge of the biochemical interactions of the primary gene product, or from studies in mice where the necessary genetic analysis is feasible (Nadeau, 2001). The work of Easton *et al.* (1993) on neurofibromatosis type 1 (MIM 162200) shows how statistical analysis of clinical phenotypes within large families can provide evidence for modifier genes. The role of pure chance should also not be ignored, especially in conditions with a patchy phenotype. Examples include patchy depigmentation in Waardenburg syndrome, and the variable numbers of neurofibromata or polyps in neurofibromatosis type 1 and adenomatous polyposis coli (MIM 175100), respectively.

An example of how a modifier might work comes from an interesting family with apparent digenic inheritance of ocular albinism (Morell *et al.*, 1997). Tyrosinase is a key enzyme of melanocytes; deficiency leads to oculocutaneous albinism (MIM 203100). A common variant of the tyrosinase gene,

Table 16.5: Examples of genes responsible for more than one disease

Gene	Location	Diseases	Symbol	MIM no.
PAX3	2q35	Waardenburg syndrome type 1	WS1	193500
		Alveolar rhabdomyosarcoma	RMS2	268220
CFTR	7p31.2	Cystic fibrosis	CF	219700
		Bilateral absence of vas deferens		
RET	10q11.2	Multiple endocrine neoplasia type 2A	MEN2A	171400
		Multiple endocrine neoplasia type 2B	MEN2B	162300
		Medullary thyroid carcinoma	FMTC	155420
		Hirschsprung disease	HSCR	142623
PMP22	17p11.2	Charcot–Marie–Tooth neuropathy type 1A	CMT1A	118220
		Tomaculous neuropathy	HNPP	162500
SCN4A	17q23.1–q25.3	Paramyotonia congenita	PMC	168300
		Hyperkalemic periodic paralysis	HYPP	170500
		Acetazolamide-responsive myotonia congenita		
PRNP	20p12–pter	Creutzfeldt–Jakob disease	CJD	123400
		Familial fatal insomnia	FFI	176640
GNAS	20q13.2	Albright hereditary osteodystrophy	AHO	103580
		McCune–Albright syndrome	PFD	174800
AR	Xcen–q22	Testicular feminization syndrome	TFM	313700
		Spinobulbar muscular atrophy	SBMA	313200

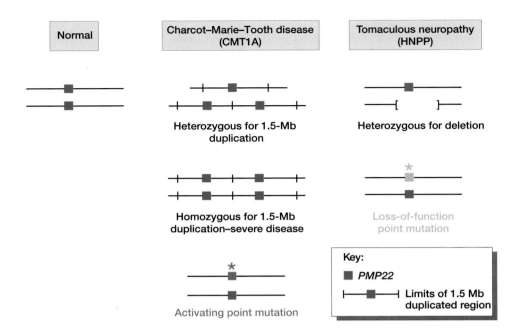

Figure 16.8: Gene dosage effects with the *PMP22* gene.

Most patients with Charcot–Marie–Tooth disease are heterozygous for a 1.5-Mb duplication at 17p11.2, including the gene for peripheral myelin protein, *PMP22* (colored box). A patient homozygous for the duplication had very severe disease. Some patients have only two copies of the *PMP22* gene, but one copy carries an activating mutation. Deletion or loss-of-function mutation of the *PMP22* gene is seen in patients with tomaculous neuropathy (Patel and Lupski, 1994).

R402Q, encodes an enzyme with reduced activity, but the residual activity is sufficiently high for even homozygotes to have normal pigmentation. However, in the family reported by Morell *et al.* people carrying one or two copies of R402Q showed ocular albinism (a mild form of oculocutaneous albinism) when they also carried a mutation in *MITF*, a gene involved in differentiation of melanocytes. Mutations in *MITF* alone do not cause ocular albinism.

16.6.4 Unstable expanding repeats – a novel cause of disease

Unstable expanding nucleotide repeats (**dynamic mutations**) were an entirely novel and unprecedented disease mechanism when first discovered in 1991. The currently known examples are shown in *Table 16.6*. They raise two major questions:

▶ what is the mechanism of the instability and expansion? This was discussed in Section 11.5.2;

▶ why do the expanded repeats make you ill? This is discussed here.

A hallmark of diseases caused by dynamic repeats is **anticipation** – that is, the age of onset is lower and/or the severity worse, in successive generations. See Section 4.3.3 for a cautionary note about anticipation. In some cases intermediate-sized alleles are nonpathogenic but unstable, and readily expand to full mutation alleles (e.g. FRAXA repeats of 50–200 units); in other cases such alleles only very occasionally expand (e.g. HD alleles with 29–35 repeats). Pathogenic expansions fall into two main classes:

▶ highly expanded repeats outside coding sequences;

▶ modest expansions of CAG repeats that encode polyglutamine tracts in the gene product.

Highly expanded repeats outside coding sequences

In Fragile-X syndrome and Friedreich ataxia, an enormously expanded repeat causes a loss of function by abolishing transcription. The same is true for the expanding 12-mer in juvenile myoclonus epilepsy. In each case, the disease is occasionally caused by different, more conventional, loss of function mutations in the gene. Such mutations produce the identical clinical phenotype to expansions, apart (presumably) from not showing anticipation. Other similar highly expanded repeats, such as *FRA16A* (expanded CCG repeat) or *FRA16B* (an expanded 33-bp minisatellite) are nonpathogenic, presumably because no important gene is located nearby. Probably the repeats abolish transcription by altering the local chromatin structure. In the case of fragile-X, this leads to methylation of the promoter.

The large expansion in myotonic dystrophy is different. No other mutation has ever been found in a myotonic dystrophy patient, so there must be something quite specific about the action of the CTG repeat. It has recently become clear that the main pathogenic action (in DM1 and DM2) is through the mRNA binding and sequestering CUG-binding proteins

Table 16.6: Diseases caused by unstable expanding nucleotide repeats

Disease	MIM No.	Mode of inheritance	Location of gene	Location of repeat	Repeat sequence	Stable repeat #	Unstable repeat #
1. Very large expansions of repeats outside coding sequences:							
Fragile-X site A (FRAXA)	309550	X	Xq27.3	5'UTR	(CGG)n	6–54	200–1000+
Fragile-X site E (FRAXE)	309548	X	Xq28	Promoter	(CCG)n	6–25	200+
Friedreich ataxia (FA)	229300	AR	9q13-q21.1	Intron 1	(GAA)n	7–22	200–1700
Myotonic dystrophy (DM1)	160900	AD	19q13	3'UTR	(CTG)n	5–35	50–4000
Myotonic dystrophy 2 (DM2)	602668	AD	3q21	Intron 1	(CCTG)n	12	75–11 000
Spinocerebellar ataxia 8	603680	AD	13q21	Untranslated RNA	(CTG)n	16–37	110–500+
Spinocerebellar ataxia 11	604432	AD	15q14-q21	Intron 9	(ATTCT)n	10–22	Up to 22 kb
Juvenile myoclonus epilepsy (JME)	254800	AR	21q22.3	Promoter	(CCCCGCCCCGCG)n	2–3	40–80
2. Modest expansions of CAG repeats within coding sequences:							
Huntington disease (HD)	143100	AD	4p16.3	Coding	(CAG)n	6–35	36–100+
Kennedy disease	313200	XR	Xq21	Coding	(CAG)n	9–35	38–62
SCA1	164400	AD	6p23	Coding	(CAG)n	6–38	39–83
SCA2	183090	AD	12q24	Coding	(CAG)n	14–31	32–77
Machado-Joseph disease (SCA3)	109150	AD	14q32.1	Coding	(CAG)n	12–39	62–86
SCA6	183086	AD	19p13	Coding	(CAG)n	4–17	21–30
SCA7	164500	AD	3p12-p21.1	Coding	(CAG)n	7–35	37–200
SCA17	607136	AD	6q27	Coding	(CAG)n	25–42	47–63
Dentatorubral-pallidoluysian atrophy (DRPLA)	125370	AD	12p	Coding	(CAG)n	3–35	49–88

SCA, spino-cerebellar ataxia.

that are required for correct splicing of other transcripts. Aberrant splicing leads to loss of the muscle-specific chloride channel, among other transcripts (reviewed by Tapscott and Thornton, 2001). There may be other effects as well, because the site of the expansion forms part of the CpG island of an adjacent gene, *SIX5* (MIM 600963), and the expansion reduces expression of this gene.

The *SCA8* gene (Koob *et al.*, 1999) apparently encodes an untranslated RNA that acts as an antisense regulator of a gene on the other strand of the DNA (similar to the imprinted transcripts mentioned in Section 16.4.4). How the expansion produces the disease is not known – indeed, it is not totally certain that this change is actually pathogenic.

Modest expansions of CAG repeats within coding regions that encode polyglutamine tracts within the gene product

Common features of the eight diseases caused by expansion of an unstable CAG repeat within a gene include:

▸ they are all late-onset neurodegenerative diseases, and except for Kennedy disease, are all dominantly inherited;

▸ no other mutation in the gene has been found that causes the disease;

▸ the expanded allele is transcribed and translated;

▸ the trinucleotide repeat encodes a polyglutamine tract in the protein;

▸ there is a critical threshold repeat size, below which the repeat is nonpathogenic and above which it causes disease;

▸ the larger the repeat, above the threshold, the earlier is the average age of onset (predictions cannot be made for individual patients, but there is a clear statistical correlation).

The androgen receptor mutation in Kennedy disease provides clear evidence that CAG-repeat diseases involve a specific gain of function. Loss of function mutations in this gene are well known and cause androgen insensitivity or testicular feminization syndrome (MIM 300068), a failure of male sexual differentiation. The polyglutamine expansion, by contrast, causes a quite different neurodegenerative disease, although patients often also show minor feminization. The common pathogenic mechanism involves formation of protein aggregates (see below).

In addition to the classic unstable expanded repeats, some diseases can be caused by much shorter and fairly stable trinucleotide repeat expansions. Examples include oropharyngeal muscular dystrophy (*PABP2* gene, MIM 602279), pseudoachondroplasia (*COMP* gene, MIM 600310) and synpolydactyly (*HOXD13* gene, MIM 186000). These are not dynamic mutations, and do not belong with the classic unstable expanded repeats.

Laboratory diagnosis of expanded repeats

A single PCR reaction makes the diagnosis in the polyglutamine repeat diseases. The very large expansions in myotonic dystrophy require Southern blotting. Fragile X is diagnosed by a Southern blotting method that depends on both the size and methylation status of the *FRAXA* gene. For further details and pictures of gels, see Section 18.4.3 and *Figure 18.12*.

16.6.5 Protein aggregation is a common pathogenic mechanism in gain of function diseases

Recently it has become apparent that formation of protein aggregates is a common feature of several adult-onset neurological diseases, including the polyglutamine diseases described above, Alzheimer disease, Parkinson disease, Creutzfeld–Jakob disease and the heterogeneous group known as amyloidoses. The full story is not yet clear, but a common theme is emerging. Globular protein molecules resemble oil drops, with the hydrophobic residues in the interior and polar groups on the outside. Correct folding is a critical and highly specific process, and naturally occurring proteins are selected from among all possible protein sequences partly for their ability to fold correctly. Mutant proteins may be more prone to misfolding. Misfolded molecules with exposed hydrophobic groups can aggregate with each other or with other proteins, and this is somehow toxic to neurons and maybe other cells. Sometimes it appears that a conformational change can propagate through a population of protein molecules, converting them from a stable native conformation into a new form with different properties in a process perhaps analogous to crystallization. The behavior of prion proteins (Prusiner *et al.*, 1998) is the most striking example. Misfolding might start with a chance misfolding of a newly synthesized structurally normal molecule (sporadic cases), a mutant sequence with a greater propensity to misfold (a genetic disease), or a misfolded molecule somehow acquired from the environment (infectious cases). Thus this final common pathway brings together a set of diseases with very disparate origins (Perutz and Windle, 2001; Bucciantini *et al.*, 2002).

16.6.6 For mitochondrial mutations, heteroplasmy and instability complicate the relationship between genotype and phenotype

Point mutations, deletions or duplications that abolish the function of genes in the densely packed mitochondrial genome (*Figure 9.2*) are associated with a broad spectrum of degenerative diseases involving the central nervous system, heart, muscle, endocrine system, kidney and liver. Cells contain many mtDNA molecules. They can be homoplasmic (every mtDNA molecule is the same) or heteroplasmic (a mixed population of normal and mutant mitochondrial DNA). Unlike mosaicism, heteroplasmy can be transmitted from mother to child through a heteroplasmic egg. Both the mutations and the heteroplasmy often seem to evolve with time within an individual. The same individual can carry both deletions and duplications, and the proportion can change with time (Poulton *et al.*, 1993).

The same sequence change is frequently seen in people with different syndromes, and phenotype–genotype correlations are particularly hard to establish. For example, 50% of people with Leber's hereditary optic neuropathy (MIM 535000: sudden irreversible loss of vision) have a G→A substitution at nucleotide 11778 of the mitochondrial genome. Most of these patients are homoplasmic, but about 14%, no less severely affected, are heteroplasmic. Even in homoplasmic families the condition is highly variable; penetrance overall is 33–60%, and 82% of affected individuals are male (Wallace *et al.*, 2001). Possible reasons for the poor correlation include:

▶ heteroplasmy can be tissue-specific, and the tissue that is examined (typically blood or muscle) may not be the critical tissue in the pathogenesis;

▶ mtDNA is much more variable than nuclear DNA, and some syndromes may depend on the combination of the reported mutation with other unidentified variants;

▶ some mitochondrial diseases seem to be of a quantitative nature: small mutational changes accumulate that reduce the energy-generating capacity of the mitochondrion, and at some threshold deficit clinical symptoms appear;

▶ many mitochondrial functions are encoded by nuclear genes (see *Box 9.2*), so that nuclear variation can be an important cause or modifier of mitochondrial phenotype.

The MITOMAP database of mitochondrial mutations (www.mitomap.org) has a good general discussion, plus extensive data tables showing just how great is the challenge of predicting phenotypes.

16.7 Molecular pathology: from disease to gene

Very often the starting point for thinking about molecular pathology is a disease rather than a gene. This approach gives an alternative viewpoint of genotype–phenotype correlations. The overall message is that one must not be naïve when speculating about the gene defect underlying a Mendelian syndrome.

16.7.1 The gene underlying a disease may not be the obvious one

Mutations leading to deficiency of a protein are not necessarily in the structural gene encoding the protein

Agammaglobulinemia (lack of immunoglobulins, leading to clinical immunodeficiency) is often Mendelian. It is natural to assume the cause would be mutations in the immunoglobulin genes. But the immunoglobulin genes are located on chromosomes 2, 14 and 22, and agammaglobulinemias do not map to these locations. Many forms are X-linked. Remembering the many steps needed to turn a newly synthesized polypeptide into a correctly functioning protein (Section 1.5), this lack of one-to-one correspondence between the mutation and the protein structural gene should not come as any great surprise. Failures in immunoglobulin gene processing, in B-cell maturation, or in the overall development of the immune system will all produce immunodeficiency.

One gene defect can sometimes produce multiple enzyme defects

I-cell disease or mucolipidosis II (MIM 252500) is marked by deficiencies of multiple lysosomal enzymes. The primary defect is not in the structural gene for any of these enzymes but in an enzyme, N-acetylglucosamine-1-phosphotransferase, that phosphorylates mannose residues on the glycosylated enzyme molecules. The phosphomannose is a signal that targets the enzymes to lysosomes; in its absence the lysosomes lack a whole series of enzymes.

Mutations often affect only a subset of the tissues in which the gene is expressed

The pattern of tissue-specific expression of a gene is a poor predictor of the clinical effects of mutations. Tissues in which a gene is not expressed are unlikely to suffer primary pathology, but the converse is not true. Usually only a subset of expressing tissues are affected. The *HD* gene is widely expressed, but Huntington disease affects only limited regions of the brain. The retinoblastoma gene (Section 17.6.1) is ubiquitously expressed, but only the retina is commonly affected by inherited mutations. This is also strikingly seen in the lysosomal disorders. Gene expression is required in a single cell type, the macrophage, which is found in many tissues. But not all macrophage-containing tissues are abnormal in affected patients. Explanations are not hard to find:

▶ genes are not necessarily expressed only in the tissues where they are needed. Provided expression does no harm, there may be little selective pressure to switch off expression, even in tissues where expression confers no benefit;

▶ loss of a gene function will affect some tissues much more than others, because of the varying roles and metabolic requirements of different cell types and varying degrees of functional redundancy in the meshwork of interactions within a cell. The 'gatekeeper gene' concept from cancer genetics (Section 17.7.2) is likely to be applicable to many other cell functions and malfunctions, in addition to the cell turnover that goes wrong in cancer;

▶ any gain of function may be pathological for some cell types and harmless for others – see the example of the *RET* gene (Section 16.6.2).

16.7.2 Locus heterogeneity is the rule rather than the exception

Locus heterogeneity describes the situation where the same disease can be caused by mutations in several different genes. It is important to think about the biological role of a gene product, and the molecules with which it interacts, rather than expecting a one-to-one relationship between genes and syndromes. As we saw in Section 4.2.4, clinical syndromes often result from failure or malfunction of a developmental or physiological pathway; equally, many cellular structures and functions depend on multi-component protein aggregates. If the correct functioning of several genes is required, then mutations in any of the genes may cause the same, or a very similar, phenotype.

Once again, the collagens (*Figure 16.4*; Section 16.6.1) provide good examples. We have seen that type I collagen, the major collagen of skin, bone, tendon and ligaments, is built of triple helices comprising two α(1) chains and one α(2). Mutations in either the *COL1A1* or *COL1A2* genes cause the same condition, dominant osteogenesis imperfecta. Type II collagen forms fibrils in cartilage and other tissues including the vitreous of the eye. It is made of homotrimeric helices of COL2A1 chains. Different mutations in the *COL2A1* gene result in an overlapping spectrum of skeletal dysplasias including Stickler syndrome, spondyloepiphyseal dysplasia and Kniest dysplasia. A similar phenotype can result from mutations in the type XI collagen, which is a minor component of the type II fibril. In all these cases, which syndrome is produced depends on the overall effect on the final collagen fibrils, rather than on which gene is mutated.

16.7.3 Mutations in different members of a gene family can produce a series of related or overlapping syndromes

Mutations in members of a gene family with partially overlapping functions can produce a set of partially overlapping phenotypes that are hard to dissect clinically. Mutations affecting fibroblast growth factor receptors illustrate this. The 10 fibroblast growth factors govern important developmental processes through four cell surface receptors, FGFR 1–4. Most tissues express multiple FGFRs, including splice variants of each. The FGFRs are receptor tyrosine kinases that act in a similar manner to the RET protein described above: signal transduction requires receptor dimerization, and this can involve homodimers or heterodimers. FGFR mutants could produce an altered balance of splice forms, change the balance of homo- and heterodimers, reduce signaling by a dominant negative effect or produce constitutively active dimers. Thus there is the potential for complex genetic effects.

Very specific mutations of the receptor genes are responsible for a series of dominant disorders of skeletal growth (*Figure 16.9*). Mutations in FGFR2 on 10q26 are found in

Crouzon, Jackson-Weiss, Pfeiffer and Apert syndromes, while different specific mutations in FGFR3 at 4p16 produce achondroplasia, thanatophoric dysplasia types 1 and 2, hypochondroplasia, Crouzon syndrome with acanthosis nigricans and Muencke's coronal craniosynostosis. Some patients with Pfeiffer syndrome have a mutation in FGFR1. For clinical descriptions, references and an introduction to the molecular pathology of these syndromes, see OMIM and Wilkie (1997). The very specific nature of the mutations suggests a gain of function, and the achondroplasia, thanatophoric dysplasia and Crouzon mutants have been shown to produce receptors with varying degrees of constitutive (ligand independent) activation when transfected into certain types of cells (Naski et al., 1996).

16.7.4 Clinical and molecular classifications are alternative tools for thinking about diseases, and each is valid in its own sphere

The connective tissue disorders caused by collagen gene mutations, which have been a recurring theme in this chapter, illustrate the difference between clinical and molecular classifications of diseases (*Table 16.7*).

▶ All Mendelian diseases can be classified on a molecular basis, first by the locus involved and second by the particular mutant allele at that locus.

▶ Genetic diseases can also be classified clinically according to their symptoms and prognosis. Clinical categories defined in this way may not correspond exactly to a molecular classification, but they may be more useful for counseling and management of patients.

Clinical labels are not simply conventions. They evolve as knowledge of the underlying genetics advances – diseases are grouped together (Duchenne and Becker muscular dystrophy) or split (*BRCA1* breast cancer from sporadic breast cancer). A molecular classification is essential for molecular diagnosis, and it may allow more accurate counseling – for example, only molecular analysis could show that unaffected parents who have more than one child affected with osteogenesis imperfecta are germinal mosaics rather than carriers of a recessive form of OI. However, a full-blown molecular classification is not always clinically useful – for example, although OMIM lists 11 loci causing Usher syndrome (recessive deaf–blindness), clinically it is only useful to distinguish three types, which vary in their severity. Thus a molecular classification illuminates rather than supersedes the clinical classification.

16.8 Molecular pathology of chromosomal disorders

16.8.1 Microdeletion syndromes bridge the gap between single gene and chromosomal syndromes

If our 3000-Mb genome contains 30 000 genes, a deletion of a megabase or so, which is too small to be seen under the

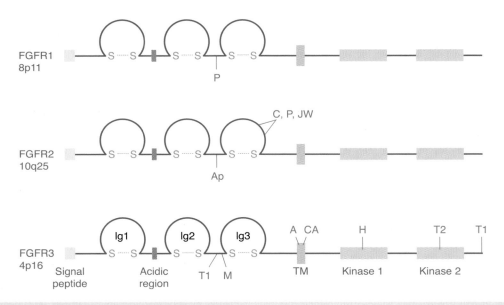

Figure 16.9: Phenotype–genotype correlations in FGFR mutations.

Three of the four highly homologous fibroblast growth factor receptors are shown. Each receptor tyrosine kinase has three immunoglobulin-like extracellular domains (held by S–S bridges), a transmembrane domain (TM), and paired intracellular tyrosine kinase domains. Very specific missense mutations are associated with a series of skeletal dysplasias (achondroplasia A, hypochondroplasia H, thanatophoric dysplasia Types 1 and 2, T1 and T2) and craniosynostosis syndromes (Apert Ap, Crouzon C, Jackson–Weiss JW, Muencke M, Pfeiffer P). Other mutations can cause the Beare–Stevenson cutis gyrata skin disease (CA). Some mutations in the Ig3 domain of FGFR2 are associated in different families with Crouzon, Jackson–Weiss and Pfeiffer syndromes.

microscope, may still involve a dozen or more genes. FISH analysis and array-CGH (Sections 2.4.2, 17.3.3) are revealing increasing numbers of such submicroscopic chromosomal abnormalities (*Table 16.8*). From the point of view of molecular pathology they fall into three classes.

▶ **Single gene syndromes**, where all the phenotypic effects are due to deletion (or sometimes duplication) of a single gene. For example Alagille syndrome (MIM 118450) is seen in patients with a microdeletion at 20p11. However, 93% of Alagille patients have no deletion, but instead have point mutations in the *JAG1* gene located at 20p11. The cause of the syndrome in all cases is haploinsufficiency for *JAG1*.

▶ **Contiguous gene syndromes**, seen primarily in males with X-chromosome deletions. The classic case was the boy 'BB' who suffered from Duchenne muscular dystrophy (MIM 310200), chronic granulomatous disease (MIM 306400) and retinitis pigmentosa (MIM 312600), together with mental retardation (Francke *et al.*, 1985). He had a chromosomal deletion in Xp21 that removed a contiguous set of genes and incidentally provided investigators with the means to clone the genes whose absence caused two of his diseases, DMD and chronic granulomatous disease (Section 14.4.2). Deletions of the tip of Xp are seen in another set of contiguous gene syndromes. Successively larger deletions remove more genes and add more diseases

Table 16.7A: Clinical classification of the connective tissue diseases mentioned in the text and Table 16.7B

Disease	MIM No.	Features
OI type I (divided on dental findings into IA, IB, IC)	166200 (166240)	Mild–moderate bone fragility; blue sclerae; normal stature; hearing loss (50%)
OI type II	166210	Very severe bone fragility; perinatal lethal
OI type III	(166230)	Moderate–severe bone fragility; progressive deformity; very short stature; often hearing loss
OI type IV (divided on dental findings into IVA, IVB)	166220	Mild–moderate bone fragility; normal sclerae; variable stature
SED	183900	Short stature, short neck, spondylo-epiphyseal dysplasia
Stickler syndrome	108300	Mild SED, cleft palate, high myopia, hearing loss
Kniest dysplasia	156550	Disproportionate short stature; short neck; SED, etc.
EDS type VII	130060	Lax joints and skin

OI, Osteogenesis imperfecta (brittle bone disease); SED, spondyloepiphyseal dysplasia; EDS, Ehlers–Danlos syndrome.

Table 16.7B: Molecular classification of the connective tissue diseases mentioned in the text and Table 16.7A.

Gene	Location	Mutations	Syndrome
COL1A1	17q22	Null alleles	OI type I
		Partial deletions; C-terminal substitutions	OI type II
		N-terminal substitutions	OI types I, III or IV
		Deletion of exon 6	EDS type VII
COL1A2	7q22.1	Splice mutations; exon deletions	OI type I
		C-terminal mutations	OI type II, IV
		N-terminal substitutions	OI type III
		Deletion of exon 6	EDS type VII
COL2A1	12q13	Point mutations	SED
		Nonsense mutation	Stickler syndrome
		Defect in conversion	Kniest dysplasia
		Missense	Achondrogenesis II, spondylo-meta-epiphyseal dysplasia
COL11A2	6p21.3	Splicing mutation	Stickler syndrome

to the syndrome (Ballabio and Andria, 1992). Microdeletions are relatively frequent in some parts of the X chromosome (e.g. Xp21, proximal Xq) but rare or unknown in others (e.g. Xp22.1-22.2, Xq28). No doubt deletion of certain individual genes, and visible deletions in gene-rich regions, would be lethal.

▶ **Segmental aneuploidy syndromes.** Autosomal microdeletions are rarely true contiguous gene syndromes. Usually the phenotype in heterozygotes depends on only a subset of the deleted genes, those that are dosage-sensitive (*Figure 16.10*). Several well defined syndromes are produced by recurrent *de novo* microdeletions (Budarf and Emanuel, 1997). The deletion-prone regions are flanked by long repeats, which allow misaligned recombination (*Figure 11.7*). Often the repeats contain transcribed sequences that may make for a more open, recombination-prone chromatin structure. Inversions of some of these megabase-sized regions occur as common nonpathogenic polymorphisms which may predispose to deletion by mispairing of the flanking repeats (Giglio *et al.*, 2001).

Identifying which genes in the deleted region are responsible for which parts of the phenotype of these microdeletion syndromes is proving difficult. Three possible strategies are:

▶ find people who have one component of the syndrome inherited as a Mendelian condition caused by mutation of a single gene within the candidate region;

▶ find people who have smaller than normal microdeletions, and who show only partial features of the syndrome;

▶ delete the corresponding region in the mouse; cross the deleted mice with mice transgenic for individual genes from the region, and correlate the phenotype with the extent of remaining haploinsufficiency.

Williams syndrome (WLS; MIM 194050) provides a good example of the problems. People with WLS have a recognizable face, they are growth retarded, as infants they may have life-threatening hypercalcemia, and they often have supravalvular aortic stenosis (SVAS). Additionally they are usually moderately mentally retarded, to about the same extent as people with Down syndrome, but they have a very distinctive cognitive profile and personality. They are highly sociable, often musical, and talk remarkably well, but have a specific inability in manipulating shapes (visuospatial constructive ability). WLS is caused by a 1.6-Mb deletion, the result of recombination between flanking repeats, especially in people heterozygous for a common inversion of the whole region (Osborne *et al.*, 2001). About 20 genes have been identified in the deleted region (Tassabehji *et al.*, 1999). Identifying the relevant genes from among all those deleted in WLS might provide an entry to identifying genetic determinants of normal human cognition and behavior.

As mentioned above (Section 16.4.1), SVAS can result from deletion or disruption of the elastin gene. This gene lies within the Williams critical region, and haploinsufficiency for elastin is undoubtedly the cause of the SVAS seen in Williams syndrome. Potentially the facial features could also be caused by a deficiency of this connective tissue protein, but this is evidently not the case because people with simple elastin mutations often have SVAS but do not have the

Table 16.8: Syndromes associated with chromosomal microdeletions

The syndromes caused by haploinsufficiency of a single gene are often seen in patients who do not have a microdeletion, but a point mutation in the gene.

Syndrome	MIM No.	Location	Type of anomaly
Wolf–Hirschhorn	194190	4p16.3	Segmental aneuploidy
Cri du chat	123450	5p15.2-p15.3	Segmental aneuploidy
Williams	194050	7q11.23	Segmental aneuploidy
Langer–Giedon	150230	8q24	Contiguous gene (*TRPS1, EXT1*)
WAGR	194072	11p13	Contiguous gene (*PAX6, WT1*)
Prader–Willi	176270	15q11-q13	Segmental aneuploidy (lack of paternal copy)
Angelman	105830	15q11-q13	Lack of maternal *UBE3A*
Rubinstein–Taybi	180849	16p13.3	Haploinsufficiency for *CBP*
Miller–Dieker	247200	17p13.3	Contiguous gene (*LIS1, YWHAE* etc.)
Smith–Magenis	182290	17p11.2	Segmental aneuploidy
Alagille	118450	20p12.1	Haploinsufficiency for *JAG1*
Di George / VCFS	192430	22q11.21	Segmental aneuploidy

WAGR, Wilms tumor, aniridia, genital abnormalities, mental retardation; VCFS, velocardiofacial syndrome.

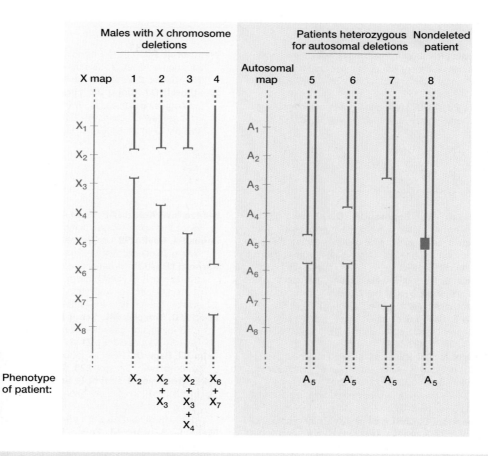

Figure 16.10: X-linked and autosomal microdeletion syndromes.

On the X chromosome, deletion of genes X_1 or X_5 is lethal in males. Patients 1–3 show a nested series of contiguous gene syndromes, while patient 4 has a different contiguous gene syndrome. On the autosome, only gene A_5 is dosage sensitive. Patients 5–7, with different sized deletions, all show the same phenotype as patient 8, who is heterozygous for a loss-of-function point mutation in the A_5 gene.

Williams face. No other features of the syndrome have so far been convincingly attributed to any of the other genes in the region. People with partial deletions of the region seem to have either just SVAS or the full syndrome. Whether the Williams phenotype will be recognizable in mice is an interesting question. Other microdeletion syndromes are also associated with specific behaviors, and so unraveling the constituent genes of these syndromes is an area of great current research interest.

16.8.2 The major effects of chromosomal aneuploidies may be caused by dosage imbalances in a few identifiable genes

Monosomies and trisomies probably owe their characteristic phenotypes to a few major gene effects superimposed on many minor disturbances of development. Genes where a 50% increase of dosage has a major effect must be very uncommon, and so it should be possible to identify the few genes that cause the major features of, for example, Down syndrome

(DS). Studies of patients with translocations show that the DS critical region is in 21q22.2; trisomy of other parts of chromosome 21 does not cause DS. At least two candidate genes for features of DS have been identified from this region: *DYRK*, a gene whose *Drosophila* and mouse homologs (*minibrain*) produces dosage-sensitive learning defects (Altafaj *et al.*, 2001), and *DSCAM*, a cell adhesion molecule expressed in the developing nervous system and heart (Barlow *et al.*, 2001).

X chromosome monosomy and trisomy are particularly interesting because X-inactivation ought to render them asymptomatic in somatic tissues. However, not all X-linked genes are inactivated. The skeletal abnormalities of Turner syndrome are caused by haploinsufficiency for *SHOX*, a homeobox gene in the Xp/Yp pseudoautosomal region (Section 2.3.3, *Figure 12.15*) (Clement-Jones *et al.*, 2000). Other somatic features probably stem from haploinsufficiency of other genes that escape X inactivation and that have a functional Y counterpart (see *Figure 12.14*). Only 18 such genes are known (Lahn and Page, 1997), so the list of potential candidates looks manageable.

Further reading

Dipple KM, McCabe ERB (2000) Phenotypes of patients with 'simple' Mendelian disorders are complex traits: thresholds, modifiers and system dynamics. *Am. J. Hum. Genet.* **66**, 1729–1735.

Scriver CR, Waters PJ (1999) Monogenic traits are not simple: lessons from phenylketonuria. *Trends Genet.* **15**, 267–272.

Weatherall DJ (2001) Phenotype–genotype relationships in monogenic disease: lessons from the thalassaemias. *Nature Rev. Genet.* **2**, 245–255.

Weatherall DJ, Clegg JB, Higgs DR, Wood WG (2001) The hemoglobinopathies. In: *The Metabolic and Molecular Basis of Inherited Disease*, 8th Edn (eds CR Scriver, AL Beaudet, WS Sly, D Valle). McGraw Hill, New York, pp. 4571–4636.

References

Altafaj X, Dierssen M, Baamonde C *et al.* (2001) Neurodevelopmental delay, motor abnormalities and cognitive deficits in transgenic mice overexpressing *Dyrk1A* (*minibrain*), a murine model of Down's syndrome. *Hum. Molec. Genet.* **10**, 1915–1923.

Ballabio A, Andria G (1992) Deletions and translocations involving the distal short arm of the human X chromosome: review and hypotheses. *Hum. Molec. Genet.* **1**, 221–722.

Bardoni B, Zanaria E, Guioli S *et al.* (1994) A dosage sensitive locus at chromosome Xp21 is involved in male to female sex reversal. *Nat. Genet.* **7**, 497–501.

Barlow GM, Chen X-N, Shi ZY *et al.* (2001) Down syndrome congenital heart disease: a narrowed region and a candidate gene. *Genet. Med.* **3**, 91–101.

Bird AC (2002) DNA methylation patterns and epigenetic memory. *Genes Dev.* **16**, 6–21.

Bucciantini M, Giannoni E, Chiti F *et al.* (2002) Inherent cytotoxicity of polypeptide aggregates suggests a common origin of protein misfolding diseases. *Nature* **416**, 507–511.

Budarf ML, Emanuel BS (1997) Progress in the autosomal segmental aneusomy syndromes (SASs): single or multilocus disorders? *Hum. Molec. Genet.* **10**, 1657–1665.

Cartegni L, Krainer AR (2002) Disruption of an SF2/ASF dependent exonic splicing enhancer in *SMN2* causes SMA in the absence of *SMN1*. *Nature Genet.* **30**, 377–384.

Clement-Jones M, Schiller S, Rao E *et al.* (2000) The short stature homeobox gene SHOX is involved in skeletal abnormalities in Turner syndrome. *Hum. Molec. Genet.* **9**, 695–702.

Den Dunnen JT, Antonarakis SE (2001) Nomenclature for the description of human sequence variations. *Hum. Genet.* **109**, 121–124.

Easton DF, Ponder MA, Huson SM, Ponder BAJ (1993) An analysis of variation in expression of neurofibromatosis (NF) type 1 (NF1): evidence for modifying genes. *Am. J. Hum. Genet.* **53**, 305–313.

Francke U, Ochs HD, de Martinville B *et al.* (1985) Minor Xp21 chromosome deletion in a male associated with expression of Duchenne muscular dystrophy, chronic granulomatous disease, retinitis pigmentosa and McLeod syndrome. *Am. J. Hum. Genet.* **37**, 250–267.

Giglio S, Broman KW, Matsumoto N *et al.* (2001) Olfactory receptor-gene clusters, genomic-inversion polymorphisms and common chromosome rearrangements. *Am. J. Hum. Genet.* **68**, 874–883.

Hemesath TJ, Steingrimsson E, McGill G *et al.* (1994) Microphthalmia, a critical factor in melanocyte development, defines a discrete transcription factor family. *Genes Dev.* **8**, 2770–2780.

Hentze MW, Kulozik AE (1999) A perfect message: RNA surveillance and nonsense-mediated decay. *Cell* **96**, 307–310.

Jones PA, Baylin SB (2002) The fundamental role of epigenetic events in cancer. *Nature Rev. Genet.* **3**, 415–428.

Karniski LP (2001) Mutations in the diastrophic dysplasia sulfate transporter (DTDST) gene: correlation between sulfate transport activity and chondrodysplasia phenotype. *Hum. Molec. Genet.* **10**, 1485–1490.

Koob MD, Moseley ML, Schut LJ *et al.* (1999) An untranslated CTG expansion causes a novel form of spinocerebellar ataxia (SCA8). *Nature Genet.* **21**, 379–384.

Lahn BT, Page DC (1997) Functional coherence of the human Y chromosome. *Science* **278**, 675–680.

Lykke-Andersen J, Shu M-D, Steitz JA (2001) Communication of the position of exon–exon junctions to the mRNA surveillance machinery by the protein RNPS1. *Science* **293**, 1836–1839 (see also the preceding paper in that issue).

Mani, S, Santoro M, Fusco A, Billaud M (2001) The RET receptor: function in development and dysfunction in congenital malformation. *Trends Genet.* **17**, 580–589.

Morell R, Spritz RA, Ho L *et al.* (1997) Apparent digenic inheritance of Waardenburg syndrome type 2 (WS2) and autosomal recessive ocular albinism (AROA). *Hum. Mol. Genet.* **6**, 659–664.

Motulsky AG (1995) Jewish diseases and origins. *Nature Genet.* **9**, 99–101.

Nadeau JH (2001) Modifier genes in mice and men. *Nature Rev. Genet.* **2**, 165–174.

Naski MC, Wang Q, Xu J, Ornitz DM (1996) Graded activation of fibroblast growth factor receptor 3 by mutations causing achondroplasia and thanatophoric dysplasia. *Nature Genet.* **13**, 233–237.

Nicholls RD, Knepper JL (2001) Genome organization, function and imprinting in Prader-Willi and Angelman syndromes. *Annu. Rev. Genomics Hum. Genet.* **2**, 153–175.

Nissim-Rafinia M, Kerem B (2002) Splicing regulation as a potential genetic modifier. *Trends Genet.* **18**, 123–127.

Osborne LR, Li M, Pober B *et al.* (2001) A 1.5 million-base pair inversion polymorphism in families with Williams–Beuren syndrome. *Nature Genet.* **29**, 321–325.

Patel PI, Lupski JR (1994) Charcot-Marie-Tooth disease: a new paradigm for the mechanism of inherited disease. *Trends Genet.* **10**, 128–133.

Perutz MF, Windle AH (2001) Cause of neural death in neurodegenerative diseases attributable to expansion of glutamine repeats. *Nature* **412**, 143–144.

Poulton J, Deadman ME, Bindoff L, Morten K, Land J, Brown G (1993) Families of mtDNA re-arrangements can be detected in patients with mtDNA deletions: duplications may be a transient intermediate form. *Hum. Mol. Genet.* **2**, 23–30.

Prusiner SB, Scott MR, DeArmond SJ, Cohen FE (1998) Prion protein biology. *Cell* **93**, 337–348.

Reik W, Walter J (2001) Genomic imprinting: parental influence on the genome. *Nature Rev. Genet.* **2**, 21–32.

Rougeulle C, Heard E (2002) Antisense RNA in imprinting: spreading silence through Air. *Trends Genet.* **18**, 434–437.

Runte M, Huttenhofer A, Gross S, Kiefmann M, Horsthemke B, Buiting K (2001) The IC-SNURF-SNRPN transcript serves as a host for multiple small nucleolar RNA species and as an antisense transcript for *UBE3A*. *Hum. Mol. Genet.* **10**, 2687–2700.

Tapscott SJ, Thornton CA (2001) Reconstructing myotonic dystrophy. *Science* **293**, 816–817.

Tassabehji M, Metcalfe K, Karmiloff-Smith A *et al.* (1999) Williams syndrome: using chromosomal microdeletions as a tool to dissect cognitive and physical phenotypes. *Am. J. Hum. Genet.* **64**, 118–125.

Wallace DC, Lott MT, Brown MD, Kerstann K (2001) Mitochondria and neuro-ophthalmologic diseases. In: *The Metabolic and Molecular Bases of Inherited Disease*, 8th Edn (eds CR Scriver, AL Beaudet, WS Sly, D Valle) McGraw Hill, New York, pp. 2425–2509.

Wilkie AO (1997). Craniosynostosis: genes and mechanisms. *Hum. Mol. Genet.* **6**, 1647–1666.

Wollnik B, Schroeder BC, Kubisch C, Esperer HD, Wieacker H, Jentsch T (1997) Pathophysiological mechanisms of dominant and recessive *KVLQT1* K$^+$ channel mutations found in inherited cardiac arrhythmias. *Hum. Mol. Genet.* **6**, 1943–1949.

CHAPTER SEVENTEEN

Cancer genetics

Chapter contents

17.1 Introduction

Cancer is not so much a disease as the natural end-state of any multicellular organism. We are all familiar with the basic Darwinian idea that a population of organisms which show hereditary variation in reproductive capacity will evolve by natural selection. Genotypes that reproduce faster or more extensively will come to dominate later generations, only to be supplanted in turn by yet more efficient reproducers. The determining factor can be an increased birth rate or a decreased death rate. Exactly the same applies to the population of cells that constitutes a multicellular organism like man. Cellular birth and death are under genetic control, and if *somatic mutation* creates a variant that proliferates faster, the mutant clone will tend to take over the organism. Thus cancer can be seen as a natural evolutionary process.

Cancers are the result of a series of somatic mutations, with in some cases also an inherited predisposition. They can be classified according to the tissue of origin (carcinomas are derived from epithelial cells, leukemias and lymphomas from blood cell precursors, and so on), and by the histology as seen under the microscope (*Figure 17.1*). The recent use of expression arrays to display the overall pattern of gene expression in a tumor (see *Figure 17.18*) promises to refine classification further. These classifications are important for deciding prognosis and management, but they do not explain how the cancer evolved. The aim of cancer genetics is to understand the multi-step mutational and selective pathway that allowed a normal somatic cell to found a population of proliferating and invasive cancer cells. As the key molecular events are revealed, new prognostic indicators become available to the pathologist.

Cancer genetics can be deeply confusing. Every tumor is individual. There are so many different genes that acquire mutations in one or another tumor, and they interact in such complex ways, that it is easy to get lost in a sea of detail. In this chapter we will try to avoid confusion by concentrating on the principles of tumorigenesis, as currently understood, rather than cataloging genes and mutations.

17.2 The evolution of cancer

As described above, cells are under strong selective pressure to evolve into tumor cells. But while tumors are very successful as cells, as organisms they are hopeless failures. They leave no offspring beyond the life of their host. At the level of the whole organism, there is therefore powerful selection for mechanisms that prevent a person dying from a tumor, at least until she has borne and brought up her children. Thus we are ruled by two opposing sets of selective forces. But selection for tumorigenesis is short term, while selection for resistance is long term. The evolution from a normal somatic cell to a malignant tumor takes place within the life of an individual, and has to start afresh with each new individual. But an organism with a good anti-tumor mechanism transmits it to its offspring, where it continues to evolve. A billion years of evolution have endowed us with sophisticated interlocking and overlapping mechanisms to protect us against tumors, at least during our reproductive life. Potential tumor cells are either repaired and brought back into line, or made to kill themselves (apoptosis). No single mutation can circumvent these defenses and convert a normal cell into a malignant one. Long ago, studies of the age dependence of cancer suggested that on average six to seven successive mutations are needed to convert a normal epithelial cell into an invasive carcinoma. In other words, only if half a dozen independent defenses are disabled by mutation can a normal cell convert into a malignant tumor (*Figure 17.2*).

The chance of a single cell undergoing six independent mutations is negligible, suggesting that cancer should be vanishingly rare. However, two general mechanisms exist that can allow the progression to happen (*Box 17.1*). Accumulating all these mutations nevertheless takes time, so that cancer is mainly a disease of post-reproductive life, when there is little selective pressure to improve the defenses still further.

Considering the genes that are the targets of these mutations, two broad categories can be distinguished, although as

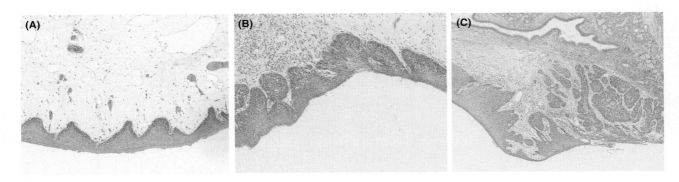

Figure 17.1: Histology of tumors.

Stages in development of a carcinoma. Three histological sections of oral mucosa, stained with hematoxylin-eosin, showing stages in the development of oral cancer. **(A)** Normal epithelium. **(B)** Dysplastic epithelium, which is a potentially premalignant change. The epithelium shows disordered growth and maturation, abnormal cells and increased mitoses. **(C)** Cancer arising from the surface epithelium and invading the underlying connective tissues. The islands of the tumor (carcinoma) show disordered differentiation, abnormal cells and increased and atypical mitoses. Pathologists use such changes in tissue architecture to identify and grade tumors. Courtesy of Dr Nalin Thakker.

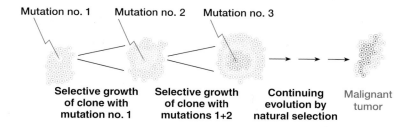

Figure 17.2: Multistage evolution of cancer.

Each successive mutation gives the cell a growth advantage, so that it forms an expanded clone, thus presenting a larger target for the next mutation.

always in biology, they are more tools for thinking about cancer than watertight exclusive classifications.

▸ **Oncogenes** (Section 17.3). These are genes whose normal activity promotes cell proliferation. Gain of function mutations in tumor cells create forms that are excessively or inappropriately active. A single mutant allele may affect the phenotype of the cell. The nonmutant versions are properly called *proto-oncogenes*.

▸ **Tumor suppressor (TS) genes** (Section 17.4). TS gene products inhibit events leading towards cancer. Mutant versions in cancer cells have lost their function. Some TS gene products prevent inappropriate cell cycle progression, some steer deviant cells into apoptosis, while others keep the genome stable and mutation rates low by ensuring accurate replication, repair and segregation of the cell's DNA. Both alleles of a TS gene must be inactivated to change the behavior of the cell.

By analogy with a bus, one can picture the oncogenes as the accelerator and the tumor suppressor genes as the brake. Jamming the accelerator on (a dominant gain of function of an oncogene) or having *all* the brakes fail (a recessive loss of function of a TS gene) will make the bus run out of control.

17.3 Oncogenes

17.3.1 The history of oncogenes

Oncogenes were discovered in the 1960s when it was realized that some animal cancers (especially leukemias and lymphomas) were caused by viruses. Some of the viruses had relatively complicated DNA genomes (SV40 virus, papillomaviruses), but others were *acute transforming retroviruses* that had very simple RNA genomes. The genome of a standard retrovirus has just three transcription units: *gag*, encoding internal proteins; *pol*, encoding the polymerase; and *env*, encoding envelope proteins (each transcript is cleaved to encode several proteins). Great excitement ensued when it was discovered that the transforming properties of acute transforming retroviruses were entirely due to their possession of one extra gene, the *oncogene*. For a short time, some enthusiasts hoped that the whole of cancer might be explained by infection with viruses carrying oncogenes, and that once these were identified their action could be blocked.

Researchers soon discovered that the viral oncogenes were copies of normal cellular genes, the *proto-oncogenes*, that had become accidentally incorporated into the retroviral particles. The viral versions were somehow *activated*, enabling them to *transform* infected cells. Most human cancers do not depend on viruses, but nevertheless their resident proto-oncogenes have become activated. In the late 1970s an assay was developed to detect activated cellular oncogenes. NIH-3T3 mouse fibroblast cells were transfected with random DNA fragments from human tumors. Some of the cells were transformed, and the human DNA responsible could be isolated from the transformants by constructing a phage genomic library and screening for the human-specific *Alu* repeat. The oncogenes identified in the transforming fragments included many which were already known from the viral studies. See Bishop (1983; Further reading) for descriptions of this early work on retroviruses and cellular oncogenes.

Box 17.1: Two ways of making a series of successive mutations more likely.

Turning a normal epithelial cell into a malignant cancer cell requires perhaps six specific mutations in the one cell. If a typical mutation rate is 10^{-7} per gene per cell generation, it is vanishingly unlikely that any one cell should suffer so many mutations (which is why most of us are alive). The probability of this happening to any one of the 10^{13} cells in a person is $10^{13} \times 10^{-42}$, or 1 in 10^{29}. Cancer nevertheless happens because of a combination of two mechanisms:

▸ some mutations enhance cell proliferation, creating an expanded target population of cells for the next mutation (*Figure 17.2*);

▸ some mutations affect the stability of the entire genome, at either the DNA or the chromosomal level, increasing the overall mutation rate.

Because cancers depend on these two mechanisms, they develop in stages, starting with tissue hyperplasia or benign growths, while malignant tumor cells usually advertise their genomic instability by their bizarre karyotypes (*Figure 17.10*).

17.3.2 The functions of oncogenes

Functional understanding of oncogenes began with the discovery in 1983 that the viral oncogene *v-sis* (the v- suffix denotes a viral oncogene) was derived from the normal cellular platelet-derived growth factor B (*PDGFB*) gene. Uncontrolled overexpression of a growth factor would be an obvious cause of cellular hyperproliferation. The roles of many cellular oncogenes (strictly speaking, proto-oncogenes) have now been elucidated (*Table 17.1*). Gratifyingly, they turn out to control exactly the sort of cellular functions that would be predicted to be disturbed in cancer. Five broad classes can be distinguished:

▸ secreted growth factors (e.g. *SIS*);

▸ cell surface receptors (e.g. *ERBB*, *FMS*);

▸ components of intracellular signal transduction systems (e.g. the *RAS* family, *ABL*);

▸ DNA-binding nuclear proteins, including transcription factors (e.g. *MYC*, *JUN*);

▸ components of the network of cyclins, cylin-dependent kinases and kinase inhibitors that govern progress through the cell cycle (e.g. *MDM2*).

If oncogenes are defined as genes that undergo dominant activating mutations in tumors, something over 100 are currently known.

17.3.3 Activation of proto-oncogenes

Some of the best illustrations of molecular pathology in action are furnished by the various ways in which proto-oncogenes can become activated (*Table 17.2*). Activation involves a gain of function. This can be quantitative (an increase in the production of an unaltered product) or qualitative (production of a subtly modified product as a result of a mutation, or a novel product from a chimeric gene created by a chromo-somal rearrangement). These changes are dominant and normally affect only a single allele of the gene. Blume-Jensen and Hunter (2001) give additional examples illustrating the mechanisms described below.

Activation by amplification

Many cancer cells contain multiple copies of structurally normal oncogenes. Breast cancers often amplify *ERBB2* and sometimes *MYC*; a related gene *NMYC* is usually amplified in late-stage neuroblastomas and rhabdomyosarcomas. Hundreds of extra copies may be present. They can exist as small paired chromatin bodies separated from the chromosomes (*double minutes*) or as insertions within the normal chromosomes (*homogeneously staining regions, HSRs*). The genetic events producing these may be quite complex because they usually contain sequences derived from several different chromosomes (reviewed by Pinkel, 1994). Similar gene amplifications are seen in cells exposed to strong artificial selective regimes – for example amplified dihydrofolate reductase genes in cells selected for resistance to methotrexate. In all cases the result is to increase greatly the level of gene expression.

Oncogene amplification in tumors can be studied by *comparative genomic hybridization (CGH)* (Forozan *et al.*, 1997). The analysis also reveals any regions of allele loss or aneuploidy, which may point to tumor suppressor genes (see below). The CGH test uses a mixture of DNA from matched normal and tumor cells in competitive hybridization. The two samples are labeled with red and green fluors, mixed, and used as a hybridization probe. The ratio of red to green hybridization signal is observed. CGH can be done in two ways:

▸ standard CGH (*Figure 17.3A*) hybridizes the mixed samples to spreads of normal chromosomes on a microscope slide (fluorescence *in situ* hybridization, Section 2.4.3). The ratio of red to green FISH signal is plotted along the length of each chromosome, and regions can be picked out where the ratio deviates from the 1 : 1 expec-

---

Table 17.1: Viral and cellular oncogenes

Viral disease	v-onc	c-onc	Location	Function
Simian sarcoma	v-sis	*PDGFB*	22q13.1	Platelet-derived growth factor B subunit
Chicken erythroleukemia	v-erbb	*EGFR*	7p12	Epidermal growth factor receptor
McDonough feline sarcoma	v-fms	*CSF1R*	5q33	Macrophage colony-stimulating factor receptor
Harvey rat sarcoma	v-ras	*HRAS*	11p15	Component of G-protein signal transduction
Abelson mouse leukemia	v-abl	*ABL*	9q34.1	Protein tyrosine kinase
Avian sarcoma 17	v-jun	*JUN*	1p32-p31	AP-1 transcription factor
Avian myelocytomatosis	v-myc	*MYC*	8q24.1	DNA-binding transcription factor
Mouse osteosarcoma	v-fos	*FOS*	14q24.3-q31	DNA-binding transcription factor

The viral genes are sometimes designated v-src, v-myc etc. and their cellular counterparts c-src, c-myc etc. The forms of the c-onc genes in normal cells are properly termed **proto-oncogenes**. Nowadays it is common to ignore these distinctions and simply use the term **oncogenes** for the normal genes. The abnormal versions can be described as activated oncogenes.

Table 17.2: Four ways of activating (proto)-oncogenes

Activation mechanism	Oncogene	Tumor
Amplification	ERBB2	Breast, ovarian, gastric, nonsmall-cell lung, colon cancer
	NMYC	Neuroblastoma
Point mutation	HRAS	Bladder, lung, colon cancer, melanoma.
	KIT	Gastrointestinal stromal tumors, mastocytosis.
Chromosomal rearrangement creating a novel chimeric gene	[Many]	See *Table 17.3*
Translocation to a region of transcriptionally active chromatin	MYC	Translocation to immunoglobulin heavy chain locus by t(8;14) in Burkitt's lymphoma

tation, as measured by a confidence interval. Depending on the direction of deviation, these mark regions of amplification or of allele loss in the tumor. The smallest alteration visible by this technique is around 3 Mb;

▶ array-CGH (*Figure 17.3B*) uses microarrayed DNA instead of chromosomes for the hybridization. The DNA in each spot is made by DOP–PCR (Section 5.2.4) of a BAC clone from a defined chromosomal location. Potentially this offers much higher resolution, limited only by the number of BACs on the array. Current state of the art is 3000 BACs, to give 1 Mb resolution. An added advantage over conventional CGH is that any amplified or deleted sequences are immediately identified at the DNA sequence level, rather than as chromosomal band locations.

Activation by point mutation

The three RAS family genes, *HRAS*, *KRAS* and *NRAS*, that mediate signaling by G-protein coupled receptors (Lowy and Willumsen, 1993), are activated in a great variety of tumors. Binding of ligand to the receptor triggers binding of GTP to the *RAS* protein, and GTP-RAS transmits the signal onwards in the cell (see *Figure 3.5*). RAS proteins have GTPase activity, and GTP-RAS is rapidly converted to the inactive GDP-RAS. Specific activating point mutations in *RAS* genes are frequently found in cells from a variety of tumors including colon, lung, breast and bladder cancers. The mutant RAS protein has reduced GTPase activity, so that the GTP-RAS is inactivated more slowly, leading to excessive cellular response to the signal from the receptor.

Activation by a translocation that creates a novel chimeric gene

This mechanism is rare in carcinomas (epithelial tumors) but common in hematologic tumors and sarcomas. The best known example is the Philadelphia (Ph) chromosome, a small acrocentric chromosome seen in 90% of patients with chronic myeloid leukemia. This chromosome is one product of a balanced reciprocal 9;22 translocation. The breakpoint on chromosome 9 is within an intron of the *ABL* oncogene. The translocation joins the 3′ part of the ABL genomic sequence onto the 5′ part of the *BCR* (*b*reakpoint *c*luster *r*egion) gene

on chromosome 22, creating a novel fusion gene (Chissoe *et al.*, 1995). This chimeric gene is expressed to produce a tyrosine kinase related to the *ABL* product but with abnormal transforming properties (*Figure 17.4A*).

Many tumor-specific breakpoints have now been recognized, and many oncogenes have been identified by cloning them (*Table 17.3*). A database of some 40 000 chromosome aberrations in cancer can be searched at the Cancer Genome Anatomy Project website (Mitelman *et al.*, 2002).

Activation by translocation into a transcriptionally active chromatin region

Burkitt's lymphoma is a childhood tumor common in malarial regions of Central Africa and Papua New Guinea. Mosquitoes and Epstein–Barr Virus are believed to play some part in the etiology, but activation of the *MYC* oncogene is a central event. A characteristic chromosomal translocation, t(8;14) (q24;q32) is seen in 75–85% of patients (*Figure 17.4B*). The remainder have t(2;8)(p12;q24) or t(8;22)(q24;q11). Each of these translocations puts the *MYC* oncogene close to an immunoglobulin locus, *IGH* at 14q32, *IGK* at 2p12 or *IGL* at 22q11. Unlike the tumor-specific translocations shown in *Table 17.3*, the Burkitt's lymphoma translocations do not create novel chimeric genes. Instead, they put the oncogene in an environment of chromatin that is actively transcribed in antibody-producing B-cells. Usually exon 1 (which is non-coding) of the *MYC* gene is not included in the translocated material. Deprived of its normal upstream controls, and placed in an active chromatin domain, *MYC* is expressed at an inappropriately high level. However, that may not be the whole story. These translocations are produced by misdirected action of the special recombinases involved in immunoglobulin V-D-J gene rearrangement (Section 10.6) and the translocated *MYC* gene often contains *de novo* point mutations induced as part of the mechanism for generating antibody diversity.

Many other chromosomal rearrangements produced by the same mechanism put one or another oncogene into the neighborhood of either an immunoglobulin (*IGG*) or a T-cell receptor (*TCR*) gene (Sanchez-García, 1997). Predictably, these rearrangements are characteristic of leukemias and lymphomas, but not solid tumors.

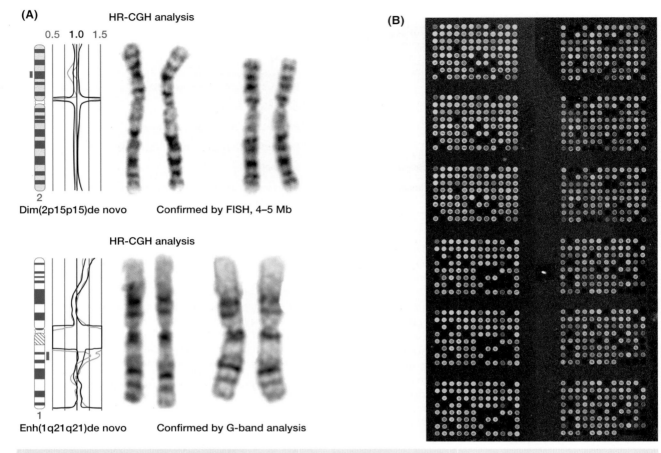

(A) HR-CGH analysis

0.5 1.0 1.5

2

Dim(2p15p15)de novo   Confirmed by FISH, 4–5 Mb

HR-CGH analysis

1

Enh(1q21q21)de novo   Confirmed by G-band analysis

(B)

Figure 17.3: Comparative genomic hybridization.

(A) *Chromosome CGH*. High-resolution CGH analysis of samples from two dysmorphic and mentally retarded boys whose G-banded chromosomes appeared normal. The yellow line is the CGH trace and the black line the 99% confidence interval for 1 : 1 ratio. The first case has a small *de novo* deletion that was confirmed by FISH; the second case has a *de novo* duplication that could in retrospect be seen on the G-banded chromosomes. Images courtesy of Dr. Claes Lundsteen, Copenhagen. **(B)** *Array-CGH*. DOP–PCR fragments from 321 BACs were spotted onto the slide, which was then hybridized to a mixture of DNA from a breast tumor cell line (labeled green) and normal reference DNA (labeled red). BACs containing DNA that is amplified in the tumor show green, those containing DNA deleted in the tumor show red, and those where the tumor is unchanged show yellow. Each panel is spotted in triplicate to allow for variations in hybridization efficiency. Courtesy of Dr. J. Veltman, Nijmegen.

17.4 Tumor suppressor genes

17.4.1 The retinoblastoma paradigm

The background to our understanding of TS genes is described by Stanbridge (1990; see Further reading). An early landmark was Knudson's work in 1971 on retinoblastoma (reviewed by Knudson, 2001). This is an aggressive childhood cancer that develops from retinoblasts, a transient population of rapidly dividing and poorly differentiated cells that, on the bus analogy we used earlier, are already driving dangerously and hence transform relatively easily. About 40% of cases are familial. These are inherited as an incompletely penetrant dominant character (MIM 180200). Familial cases are often bilateral, whereas the sporadic forms are always unilateral. Knudson noted that the age–of–onset distribution of bilateral cases was consistent with a single mutation, while sporadic

cases followed two–hit kinetics. He reasoned that all retinoblastomas involved two 'hits', but that in the familial cases one hit was inherited (*Figure 17.5*). A seminal study by Cavenee *et al.* (1983) both proved Knudson's hypothesis, and established the paradigm for laboratory investigations of TS genes.

Cavenee and colleagues typed surgically removed tumor material from patients with sporadic retinoblastoma, using a series of markers from chromosome 13 (retinoblastoma was known to be sometimes associated with chromosomal changes at 13q14). When they compared the results on blood and tumor samples from the same patients, they noted several cases where the constitutional (blood) DNA was heterozygous for one or more markers, but the tumor cells were apparently homozygous. They reasoned that what they were seeing was one of Knudson's 'hits': loss of one functional

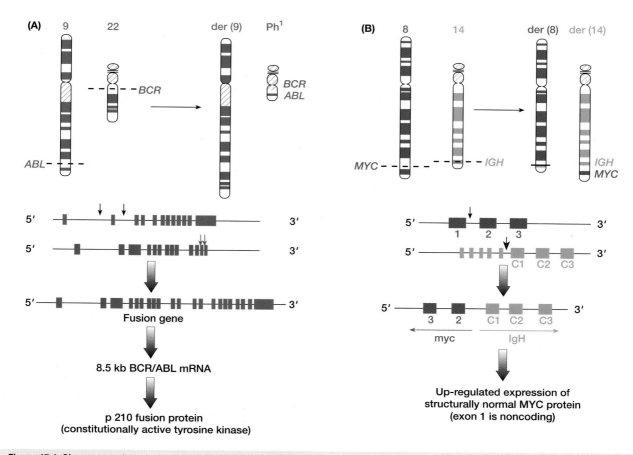

Figure 17.4: Chromosomal rearrangements that activate oncogenes.

(A) Activation by qualitative change in the t(9;22) in chronic myeloid leukemia. The chimeric *BCR-ABL* fusion gene on the Philadelphia chromosome encodes a tyrosine kinase that does not respond to normal controls. See Chissoe *et al.* (1995) for more detail. **(B)** Activation by quantitative change in the t(8;14) in Burkitt lymphoma. The *MYC* gene from chromosome 8 is translocated into the immunoglobulin heavy chain gene. In B-cells this region is actively transcribed, leading to over-expression of *MYC*.

copy of a tumor suppressor gene. Later studies confirmed this interpretation by showing that in inherited cases, it was always the wild-type allele that was lost in this way. Combining cytogenetic analysis with studies of markers from different regions of 13q, Cavenee *et al.* were able to suggest a number of mechanisms for the loss (*Figure 17.6*).

This work had two major implications. First, it suggested that sporadic cancers and familial cancers could share common molecular mechanisms. Second, it suggested two ways of finding TS genes:

▶ mapping and positional cloning of the genes responsible for familial cancers;

▶ scanning tumors for losses of specific chromosomal material.

Table 17.4 lists some of the TS genes that have been identified through studies of rare familial cancers.

Complications to the retinoblastoma paradigm

In retrospect, retinoblastoma is an unusually clear-cut example of a two-hit mechanism, perhaps because, as mentioned above,

retinoblasts are unusual cells. A number of other cancers do closely follow the retinoblastoma example. *APC, NF2,* and *PTC* (*Table 17.4*) are examples where a TS gene identified through investigation of a rare familial cancer turned out to be important in the corresponding sporadic cancer – but this is not always the case.

▶ Some cancers seem to follow different evolutionary paths in familial and sporadic cases. For example, *BRCA1* is inactivated in only 10–15% of sporadic breast cancers, and those tumors form an identifiably distinct molecular subset of breast cancers. Inactivation, when it does happen, is not by the chromosomal mechanisms seen in retinoblastoma (*Figure 17.6*) but by DNA methylation (Section 17.4.3).

▶ Some genes frequently lose function of one allele in tumors, but the retained allele appears fully functional. Sometimes this is because the second allele is inactivated by methylation, which is not detected by some of the experimental methods used, but some cases seem to be genuine one-hit events. It is not unreasonable to

Table 17.3: Chimeric genes produced by cancer-specific chromosomal rearrangements

Tumor	Rearrangement	Chimeric gene	Nature of chimeric product
CML	t(9;22)(q34;q11)	*BCR-ABL*	Tyrosine kinase
Ewing sarcoma	t(11;22)(q24;q12)	*EWS-FLI1*	Transcription factor
Ewing sarcoma (variant)	t(21;22)(q22;q12)	*EWS-ERG*	Transcription factor
Malignant melanoma of soft parts	t(12;22)(q13;q12)	*EWS-ATF1*	Transcription factor
Desmoplastic small round cell tumor	t(11;22)(p13;q12)	*EWS-WT1*	Transcription factor
Liposarcoma	t(12;16)(q13;p11)	*FUS-CHOP*	Transcription factor
AML	t(16;21)(p11;q22)	*FUS-ERG*	Transcription factor
Papillary thyroid carcinoma	inv(1)(q21;q31)	*NTRK1-TPM3* (*TRK* oncogene)	Tyrosine kinase
Pre-B cell ALL	t(1;19)(q23;p13.3)	*E2A-PBX1*	Transcription factor
ALL	t(X;11)(q13;q23)	*MLL-AFX1*	Transcription factor
ALL	T(4;11)(q21;q23)	*MLL-AF4*	Transcription factor
ALL	t(9;11)(q21;q23)	*MLL-AF9*	Transcription factor
ALL	t(11;19)(q23;p13)	*MLL-ENL*	Transcription factor
Acute promyelocytic leukemia	t(15;17)(q22;q12)	*PML-RARA*	Transcription factor + retinoic acid receptor
Alveolar rhabdomyosarcoma	t(2;13)(q35;q14)	*PAX3-FKHR*	Transcription factor

CML, chronic myeloid leukemia; ALL, acute lymphoblastoid leukemia; AML, acute myelogenous leukemia.

Note how the same gene may be involved in several different rearrangements. For further details see Rabbitts (1994).

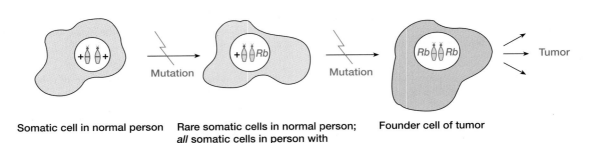

Somatic cell in normal person   Rare somatic cells in normal person; *all* somatic cells in person with familial retinoblastoma   Founder cell of tumor

Figure 17.5: Knudson's two-hit hypothesis.

Suppose there are 1 million target cells and the probability of mutation is 10^{-5} per cell. Sporadic retinoblastoma requires two hits and will affect one person in 10 000 ($10^6 \times 10^{-5} \times 10^{-5} = 10^{-4}$), while the familial form requires only one hit and will be quite highly penetrant, since ($10^6 \times 10^{-5} > 1$). See Knudson (2001) for a more sophisticated treatment.

postulate that there are genes where haploinsufficiency (Section 16.4.2) is enough to provide a growth advantage.

▸ More unusually, there are some cases where three hits are seen. In familial adenomatous polyposis (Section 17.5.3) certain germline mutations are 'weak' and confer an atten-uated phenotype with fewer polyps. Adenomas from such patients often have two somatic mutations in addition to the inherited mutation. One somatic mutation inactivates the wild-type allele. This gives the cells a growth advantage, but they gain a further advantage from a chromosomal deletion that eliminates the partially functional germline *APC* allele.

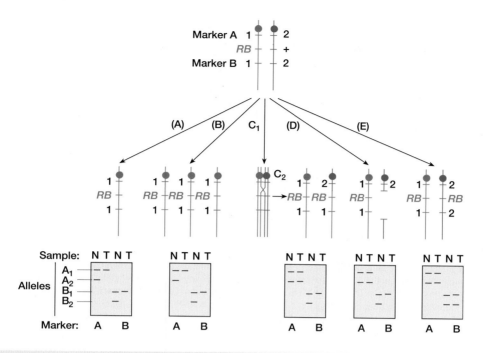

Figure 17.6: Mechanisms of loss of wild-type allele in retinoblastoma.

(A) Loss of a whole chromosome by mitotic nondisjunction. (B) Loss followed by reduplication to give (in this case) three copies of the Rb chromosome. (C) Mitotic recombination proximal to the Rb locus (C₁), followed by segregation of both Rb-bearing chromosomes into one daughter cell (C₂); this was the first demonstration of mitotic recombination in humans, or indeed in mammals. (D) Deletion of the wild-type allele. (E) Pathogenic point mutation of the wild-type allele (adapted from Cavenee *et al.*, 1983). The figures underneath show results of typing normal (N) and tumor (T) DNA for the two markers A and B located as shown. Note the patterns of loss of heterozygosity.

Table 17.4: Rare familial cancers caused by TS gene mutations

Disease	MIM No.	Map location	Gene
Familial adenomatous polyposis coli	175100	5q21	*APC*
Hereditary nonpolyposis colon cancer	120435,12036	2p16, 3p21.3	*MSH2, MLH1*
Breast-ovarian cancer	113705	17q21	*BRCA1*
Breast cancer (early onset)	600185	13q12-q13	*BRCA2*
Li-Fraumeni syndrome	151623	17p13	*TP53*
Gorlin's basal cell nevus syndrome	109400	9q22-q31	*PTC*
Ataxia telangiectasia	208900	11q22-q23	*ATM*
Retinoblastoma	180200	13q14	*RB*
Neurofibromatosis 1 (von Recklinghausen disease)	162200	17q12-q22	*NF1*
Neurofibromatosis 2 (vestibular schwannomas)	101000	22q12.2	*NF2*
Familial melanoma	600160	9p21	*CDKN2A*
von Hippel–Lindau disease	193300	3p25-p26	*VHL*

References to the genes and diseases may be found in OMIM under the number cited. Table 1 of Futreal *et al.* (2001) gives a longer list.

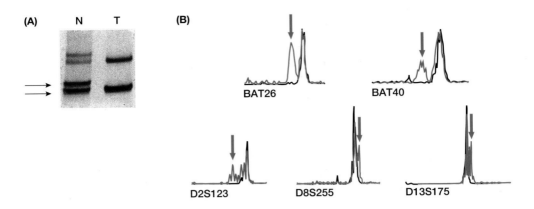

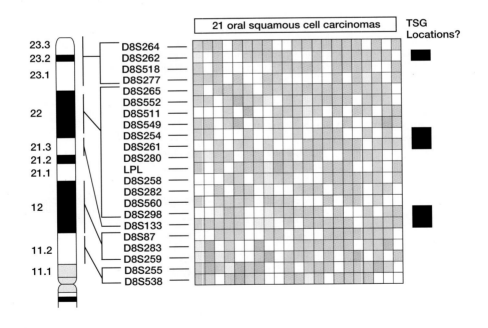

Figure 17.7: Genetic changes in tumors.

(A) *Loss of heterozygosity*. The normal tissue sample (N) is heterozygous for the marker *D8S522* (arrows), while the tumor sample (T) has lost the upper allele. The bands higher up the gel are 'conformation bands', subsidiary bands produced by alternately folded sequences of each allele. Photograph courtesy of Dr. Nalin Thakker, St Mary's Hospital, Manchester, UK. **(B) *Microsatellite instability*** in hereditary non-polyposis colon cancer. Fluorescence sequencer analysis of five microsatellites in HNPCC tumors. Black traces are from constitutional DNA, red from tumor DNA from the same patient. Arrows mark new alleles in the tumors. *BAT26* and *BAT40* are (A)$_n$ runs, *D2S123*, *D8S255* and *D13S175* are (CA)$_n$ dinucleotide repeats. Courtesy of Yvonne Wallis, West Midlands Regional Genetics Laboratory, Birmingham, UK.

Figure 17.8: Possible tumor suppressor genes on chromosome 8p.

The grid shows results of typing constitutional and tumor DNA from a series of patients with oral tumors, using various markers (*D8S264*, etc.) from the locations indicated on chromosome 8p (Wu *et al.*, 1997). Pink marks loss of heterozygosity (LoH), green marks retention, yellow marks tests giving equivocal results, grey marks tests showing microsatellite instability, and clear areas are where markers were uninformative because of constitutional homozygosity. Note the complex pattern of LoH in these tumors, which is typical of such studies. The data suggest the presence of three distinct TS genes on 8p.

17.4.2 Loss of heterozygosity (LoH) screening is widely used for trying to identify TS gene locations

Cavenee *et al.* (1983) saw how somatic genetic changes in retinoblastoma caused LoH at markers close to the *RB* locus. Studies in other cancers confirm the general picture, although the chromosomal mechanisms may be different (Thiagalingam *et al.*, 2001). Thus by screening paired blood and tumor samples (usually sporadic tumors) with markers spaced across the genome, we can hope to discover the locations of TS genes (*Figure 17.7*).

LoH analysis has been a major industry in cancer research over the past decade, but it is fair to ask whether the results have justified the effort. For meaningful results a large panel of tumors must be screened with closely spaced markers. Highly polymorphic microsatellite markers are used in order to minimize the number of uninformative cases where the constitutional DNA is homozygous for the marker, or alternatively array-CGH (*Figure 17.3B*) can be used. Advanced cancer cells may show LoH at as many as one-fourth of all loci, so large samples are needed to tease out the specific changes from the general background. The results suggest the presence of a remarkably large number of TS genes – *Figure 17.8* shows a typical example – yet attempts to follow up such studies by positional cloning have had quite a low success rate.

Figure 17.9 illustrates one pitfall in interpreting LoH data. A more general problem is that tumors typically have grossly abnormal karyotypes with innumerable numerical and structural abnormalities (*Figure 17.10*). There is almost never any information on the karyotypes of the tumors used in LoH studies, but presenting the results as in *Figure 17.8* easily creates an unintended assumption of specific interstitial deletions affecting an otherwise normal chromosome. In fact, LoH is a product of the chromosomal instability and deficient repair of double-strand DNA breaks that are so characteristic of tumor cells. Maybe some of the losses observed are by-products of specific patterns of chromosomal instability, rather than selection for loss of a TS gene.

17.4.3 Tumor suppressor genes are often silenced epigenetically by methylation

TS genes may be silenced by deletion (reflected in loss of heterozygosity) or by point mutations, but a very common third mechanism is methylation of the promoter (see Section 10.4). Overall, tumor cell DNA is hypomethylated compared to the DNA of normal cells, but methylation of specific CpG dinucleotides in the promoters of genes is found in virtually every type of human neoplasm and is associated with inappropriate transcriptional silencing of the gene. Jones and Baylin (2002) list many examples. For some TS genes methylation occurs as a common alternative to point mutation, while in others (for example *RASSF1A* at 3p21, *HIC1* at 17p13.3) methylation is the only known mechanism for tumor-specific loss of function. Standard techniques for mutation screening overlook these changes, so their importance is probably still underestimated.

17.5 Stability of the genome

Genomic instability is an almost universal feature of cancer cells. Instability may be of two types:

▶ **chromosomal instability (CIN)** is the most common form. Tumor cells typically have grossly abnormal karyotypes (*Figure 17.10*), with multiple extra and missing chromosomes, many rearrangements and so on;

▶ **microsatellite instability (MIN)** is a DNA-level instability seen in a few tumors, especially some colon carcinomas (*Figure 17.7B*).

Instability is probably necessary to enable a cell to amass enough mutations (*Box 17.1*). Some have argued that instability is just an incidental by-product of the evolution of a tumor, and that the number of cell divisions in epithelial populations is sufficient for the requisite number of mutations to accumulate, even with standard rates of mutation. However, tumors normally show either CIN or MIN but not both, and this suggests that instability is not a chance feature, but is the result of selection.

17.5.1 Chromosomal instability

The many chromosomal abnormalities seen in tumor cells are mostly random, although they provide material for further selection of faster-growing variants. They probably arise in three ways.

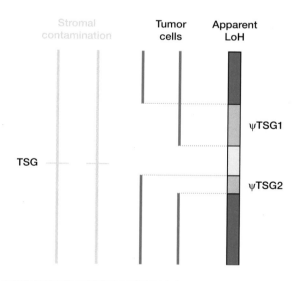

Figure 17.9: A pitfall in interpreting LoH data.

The true event in this tumor is homozygous deletion of TSG. Because of amplification of contaminating stromal material (pale lines), LoH data show retention of heterozygosity at the TSG locus, with LoH at two flanking regions (pink). The pattern suggests the existence of two spurious tumor suppressor loci, ψTSG1 and ψTSG2. The true situation would be revealed by fluorescence *in situ* hybridization, or by immunohistochemical testing for the TSG gene product.

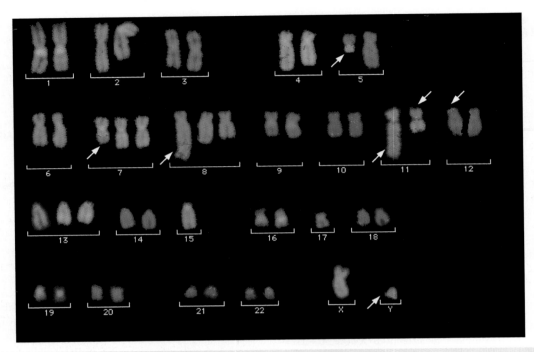

Figure 17.10: Multicolor FISH (M-FISH) karyotype of a human myeloid leukemia-derived cell line.

Note the numerous numerical and structural (arrowheads) abnormalities revealed by 24-color whole chromosome painting. These include t(5;15), der(7)t(7;15), der(8)ins(8;11), der(11)t(8;11), t(11;17) and der(Y)t(Y;12) – the latter shows badly in this cell. Image courtesy of Dr. Lyndal Kearney, Institute of Molecular Medicine, Oxford, UK, from Tosi *et al.* © 1999, reprinted by permission of Wiley-Liss Inc., a subsidiary of John Wiley & Sons, Inc.

▶ Tumor cells lose the *spindle checkpoint* (Jallepalli and Lengauer, 2001; see below). This is probably the main source of the many numerical abnormalities.

▶ Tumor cells are able to proceed through the cell cycle despite having DNA damage. Structural chromosome abnormalities can be a by-product of attempts at DNA replication or mitosis with damaged DNA.

▶ Tumors may also replicate to the point that telomeres become too short to protect chromosome ends, which leads to all sorts of structural abnormalities (see below).

The spindle checkpoint

The spindle checkpoint should prevent chromosome segregation at mitosis until all chromosomes are correctly attached to the spindle fibers. The molecular mechanism is not well understood. A few candidate genes have been identified, but the only one known to be commonly mutated in cancer cells is the *APC* gene that is involved in polyposis coli. *APC* encodes a very large multifunctional protein that is probably involved in a variety of cellular processes. In the colon CIN is observed even in very early adenomas, and $APC^{-/-}$ cells have abnormal mitotic spindles that lead to chromosomal instability.

The DNA damage signaling system

Cells are constantly repairing all sorts of damage to their DNA. The normal response to such damage is to stall the cell cycle until the damage is repaired, and a common feature of many cancers is loss of that control. The bewilderingly complex systems that detect and signal DNA damage are largely conserved from yeast to man (Zhou and Elledge, 2000). A multiprotein machine called the BASC (BRCA1-associated genome surveillance complex), is involved, together with a raft of other proteins. Several well known tumor suppressor genes turn up in this machinery:

▶ **ATM** is a component of the BASC and is an early part of the damage-sensing mechanism. This very large protein relays the signal to a variety of downstream targets. Loss of function of ATM causes *ataxia telangiectasia* (AT, MIM 208900). Affected homozygotes suffer a predisposition to various cancers, immunodeficiency, chromosomal instability and a cerebellar ataxia. Heterozygous carriers of certain specific *ATM* mutations are at increased risk of breast cancer;

▶ **nibrin** complexes with MRE11 and RAD50 proteins. The complex forms part of the BASC and maybe has other functions too. Lack of nibrin causes Nijmegen breakage syndrome (MIM 251260). NBS is clinically rather similar to AT, but includes microcephaly and growth retardation in place of ataxia. Cloning of the *NBS* gene was described in Section 13.5.2;

▶ **BRCA1**, the product of the first gene implicated in familial breast cancer (Section 15.6.1), is another very large protein with multiple functional domains that forms part of the BASC. BRCA1 protein also has functions in recombination, chromatin remodeling, and control of transcription (Scully and Livingston, 2000).

▶ **BRCA2** protein has no structural similarity to BRCA1, but shares many functions. As well as being the cause of some hereditary breast cancer, a particular class of *BRCA2* mutations cause the B/D1 form of Fanconi anemia (see MIM 227650). This is an autosomal recessive syndrome of congenital abnormalities, progressive bone marrow failure, cellular hypersensitivity to DNA damage and predisposition to cancer.

Cells with defects in these proteins seem to be particularly bad at repairing double-strand breaks. It may be that other types of damage do not rely so heavily on detection by this system for their repair.

Telomeres and chromosomal stability

As we saw in Section 2.2.5, the ends of human chromosomes are protected by a repeat sequence $(TTAGGG)_n$, that is maintained by a special RNA-containing enzyme system, *telomerase*. Telomerase is present in the human germline but is absent from most somatic tissues, and telomere length declines by 50–100 bp with each cell generation. In prolonged culture, normal cells reach a point of *senescence*, where they stop dividing. Fibroblasts deficient in p53 and the retinoblastoma protein pRb, or harboring viral oncoproteins, continue beyond senescence and hit *crisis*. Most cells die, but the one in 10^7 or thereabouts that survive have gross chromosomal abnormalities, have acquired telomerase and have become immortal. Probably crisis represents the point where telomeres are so short that they can no longer protect chromosome ends against rearrangement.

One source of the gross chromosomal abnormalities seen in tumor cells may therefore be depletion of telomeres to the point of crisis through excessive cell division (Maser and DePinho, 2002). However, this cannot explain the continuing instability because, one way or another, tumor cells always acquire the ability to maintain their telomeres and replicate indefinitely. 85–90% of full-blown metastatic cancers have telomerase uprated; the remainder use a different mechanism, termed ALT and based on recombination.

17.5.2 DNA repair defects and DNA-level instability

As noted above, cells are constantly repairing damage to their DNA (Hoeijmakers, 2001), and defects in the repair machinery underlie a variety of cancer-prone genetic disorders. The main defects are:

▶ nucleotide excision repair defects – single-strand breaks and crosslinks in DNA that has been damaged by ionizing radiation, UV light or chemical mutagens need to be repaired before the next round of replication (Section 11.6.1). Defects in some of the repair enzymes are seen in several cancer-prone syndromes, particularly the various forms of xeroderma pigmentosum (XP; see MIM 278700). XP patients are homozygous for inherited loss of function mutations, and are unable to repair DNA damage caused

by UV light. They are exceedingly sensitive to sunlight and develop many tumors on exposed skin;

▶ base excision repair defects – defective base excision repair is seldom noted in human cancers, but one form of colon cancer is caused by defects in the MYH repair enzyme (Al-Tassan *et al.*, 2002).

▶ double-strand break repair defects – double-strand breaks are repaired by homologous recombination or nonhomologous end joining (Hoeijmakers, 2001). Both require the ATM-NBS-BRCA1-BRCA2 machine mentioned above;

▶ replication error repair defects – these came to light during studies of colon cancer, as described below.

17.5.3 Hereditary nonpolyposis colon cancer and microsatellite instability

Most colon cancer is sporadic. Familial cases fall into two categories.

▶ **Familial adenomatous polyposis (FAP** or **APC**; MIM 175100) is an autosomal dominant condition in which the colon is carpeted with hundreds or thousands of polyps. The polyps (adenomas) are not malignant, but if left in place, one or more of them is virtually certain to evolve into invasive carcinoma. The cause is an inherited mutation in the *APC* tumor suppressor gene (*Table 17.4*).

▶ **Hereditary nonpolyposis colon cancer (HNPCC**; MIM 120435, 120436) is also autosomal dominant and highly penetrant, but unlike FAP there is no preceding phase of polyposis. HNPCC genes were mapped to two locations, 2p15-p22 and 3p21.3.

Loss of heterozygosity studies on HNPCC tumors produced novel and unexpected results. Rather than lacking alleles present in the constitutional DNA, tumor specimens appeared to contain extra, novel, alleles of the microsatellite markers used. *Figure 17.7B* shows an example. LoH is a property of particular chromosomal regions, but microsatellite instability (MIN) in HNPCC is general. Many tumors show occasional instability of one or a few microsatellites (see for example *Figure 17.8*), but high frequency instability (conveniently defined as > 29% of all markers tested, see Tomlinson *et al.*, 2002) defines a class of MIN$^+$ tumors with distinct clinicopathological features.

In a wonderful example of lateral thinking, Fishel *et al.* (1993) related the MIN$^+$ phenomenon to so-called mutator genes in *E. coli* and yeast. These genes encode an error-correction system that checks newly synthesized DNA for mismatched base pairs or small insertion–deletion loops. Mutations in the *MutHLS* genes that encode the *E. coli* system lead to a 100- to1000-fold general increase in mutation rates. Fishel and colleagues cloned a human homologue of one of these genes, *MutS*, and showed that it mapped to the location on 2p of one of the HNPCC genes, and was constitutionally mutated in some HNPCC families. In all, six homologues of the *E. coli* genes have been implicated in human mismatch repair (Jiricny and Nystrom-Lahti, 2000); see *Table 17.5* and *Figure 17.11*.

Table 17.5: Genes involved in DNA replication error repair

E. coli	Human	Location in man	% of HNPCC
MutS	MSH2	2p16	35%
	MSH3	5q11-q12	0%?
	MSH6	2p16	5%[a]
MutL	MLH1	3p21.3	60%
	MLH3	14q24.3	0%?
	PMS2	7p22	Very low

[a]Atypical late-onset HNPCC and endometrial cancer.

See Jiricny and Nystrom-Lahti (2000) for details.

Patients with HNPCC are constitutionally heterozygous for a loss–of-function mutation, almost always in *MLH1* or *MSH2*. Their normal cells still have a functioning mismatch repair system and do not show the MIN⁺ phenotype. In a tumor, the second copy is lost by one of the mechanisms

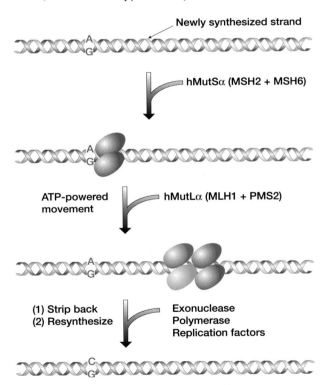

Figure 17.11: Mechanism of mismatch repair.

Replication errors can produce mismatched base pairs or small insertion/deletion loops. These are recognized by hMutSα, a dimer of MSH2/MSH6 proteins, or sometimes by the MSH2/MSH3 dimer hMutSβ. The proteins translocate along the DNA, bind the MLH1/PMS2 dimer hMutLα, then assemble the full 'repairosome' which strips back and re-synthesizes the newly synthesized strand. See Jiricny and Nystrom-Lahti (2000).

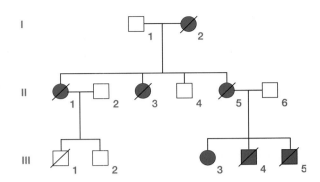

Figure 17.12: A typical pedigree of Li-Fraumeni syndrome.

Malignancies typical of Li-Fraumeni syndrome include bilateral breast cancer diagnosed at age 40 years (I-2); a brain tumor at age 35 years (II-1); soft tissue sarcoma at age 19 years and breast cancer at age 33 years (II-3); breast cancer at age 32 years (II-5); osteosarcoma at age 8 years (III-3); leukemia at age 2 years (III-4); soft tissue sarcoma at age 3 years (III-5). I-1 had cancer of the colon diagnosed at age 59 years of age – this is assumed to be unrelated to the Li-Fraumeni syndrome. Pedigree from Malkin (1994).

shown in *Figure 17.6*. Microsatellite instability is seen in 10–15% of colorectal, endometrial and ovarian carcinomas, but only occasionally in other tumors.

17.5.4 p53 and apoptosis

A major contributor to genomic instability is loss or mutation of *TP53*, the gene encoding the p53 transcription factor. Such loss is probably the commonest single genetic change in cancer. This reflects the central importance of p53, which has been summarized as 'the guardian of the genome' (Vousden, 2000). As mentioned above, cell cycling stalls in cells with damaged DNA. If the damage is not repairable, apoptosis is triggered. p53 has a crucial role in these processes. Normally p53 levels in a cell are low because the protein is rapidly degraded. Signals from a whole range of cellular stress sensors, including the damage sensors, lead to phosphorylation and stabilization of p53. This increases p53-dependent transcription of genes such as p21^WAF1/CIP1 that inhibits cell cycling, and *PUMA* and *PIG3* that control apoptosis. Tumor cells lacking p53 may continue to replicate damaged DNA and do not undergo apoptosis. Note however that loss of p53 is a relatively late event in tumor development (see for example *Figure 17.17*); evidently the earlier stages of tumor evolution are not prevented by p53.

p53 may be knocked out by deletion, by mutation or by the action of an inhibitor such as the *MDM2* gene product (which binds p53 and targets it for degradation; MDM2 also binds pRb, see below) or the E6 protein of papillomavirus. *TP53* maps to 17p12, and this is one of the commonest regions of loss of heterozygosity in a wide range of tumors. Tumors that have not lost *TP53* very often have mutated versions of it. To complete the picture of *TP53* as a TS gene, constitutional mutations in *TP53* are found in families with the dominantly inherited Li-Fraumeni syndrome (MIM 151623). Affected family members suffer multiple primary tumors, typically including soft tissue sarcomas, osteosarcomas, tumors of the breast, brain and adrenal cortex, and leukemia (*Figure 17.12*).

17.6 Control of the cell cycle

Any cell at any time has three choices of behavior: it can remain static, it can divide or it can die (apoptosis). Some cells also have the option of differentiating. Cells select one of these options in response to internal and external signals (*Figure 17.13A*). Oncogenes and tumor suppressor genes play key roles in generating and interpreting these signals.

Life would be very simple if the signal and response were connected by a single linear pathway (*Figure 17.13B*), but this seems never to be the case. Rather, multiple branching, over-lapping and partially redundant pathways control the behavior of the cell (*Figure 17.13C*). Probably such complicated networks are necessary to confer stability and resilience on the extraordinarily complex machinery of a cell. Experimentally, unraveling the precise genetic circuitry of the controls is exceedingly difficult, partly because of their complexity and partly because it is difficult to distinguish direct from indirect effects in transfection or knockout experiments. Hopefully expression microarrays will prove the key to this problem.

For cells such as cancer cells that take the decision to divide, progress through the cycle is subject to several checkpoints (*Figure 17.14*). Three main ones are:

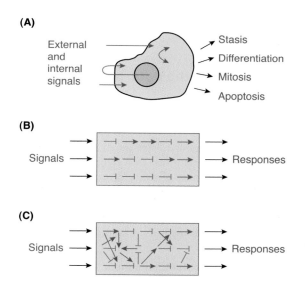

(A)

External and internal signals → Stasis / Differentiation / Mitosis / Apoptosis

(B)

Signals → → Responses

(C)

Signals → → Responses

Figure 17.13: The options open to a cell, and how it chooses.

(A) In response to internal and external signals, a cell chooses between stasis, mitosis, apoptosis and sometimes differentiation. **(B)** An imaginary cell in which signals are linked to responses by linear unbranched pathways of stimulation (→) or inhibition (—|). Human cells do not function like this. **(C)** In real cells signals feed into a complex network of partially redundant interactions, the outcome of which is not easy to predict analytically.

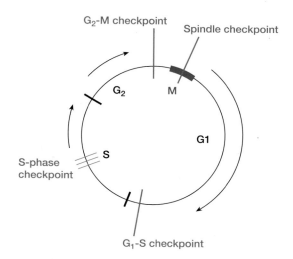

Figure 17.14: Cell cycle checkpoints.

A series of controls prevent the cell from entering S phase or mitosis with unrepaired DNA damage, or starting anaphase of mitosis until all chromosomes are correctly attached to the spindle fibers. There is probably an additional DNA damage checkpoint during S phase.

▸ the G1–S checkpoint – DNA replication is blocked when there is unrepaired DNA damage; irreparable damage leads to apoptosis. There is a probably an independent additional damage checkpoint within S phase;

▸ the G2–M checkpoint – cells are blocked from entering mitosis unless DNA replication and repair of any damage are complete;

▸ the spindle checkpoint – this was described in Section 17.5.1.

17.6.1 The G1–S checkpoint

Figure 17.15 shows part of the circuitry of this checkpoint. It seems to be particularly critical, because it is inactivated in most tumor cells. Three key tumor suppressor genes, *RB*, *TP53* and *CDKN2A*, are central players in carcinogenesis, and are among the most commonly altered genes in tumor cells. Each also has a role in inherited cancers. Probably in fact, all tumor cells need to inactivate both the Rb and p53 arms of the system in order for the cell to by-pass the usual checks on cycling and to avoid triggering apoptosis in response to unrepaired DNA damage or excessive growth signaling.

Function of pRb, the RB gene product

The *RB* gene was identified through its role in retinoblastoma (Section 17.4.1), but it is widely expressed and helps control cycling of all cells. In normal cells the gene product, a 110-kDa nuclear protein, is inactivated by phosphorylation and activated by dephosphorylation. Active (dephosphorylated) pRb binds and inactivates the cellular transcription factor E2F, function of which is required for cell cycle progression (*Figure*

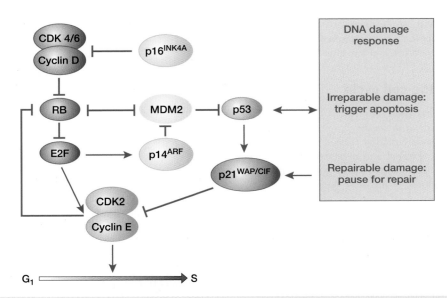

Figure 17.15: Controls on cell cycle progression and genomic integrity mediated by the *RB*, *TP53* and *CDKN2A* (INK4A) gene products.

These controls form at least part of the G1–S cell cycle checkpoint. See Harbour and Dean (2000) for details.

17.15; for fuller detail see Harbour and Dean, 2000). Two to four hours before a cell enters S-phase, pRb is phosphorylated. This releases the inhibition of E2F and allows the cell to proceed to S phase. Phosphorylation is governed by a cascade of cyclins, cyclin-dependent kinases and cyclin kinase inhibitors. Several viral oncoproteins (adenovirus E1A, SV40-T antigen, human papillomavirus E7 protein) bind and sequester or degrade pRb, thus favoring cell cycle progression.

Function of p16^{INK4A} and p14^{ARF}, the CDKN2A gene products

The remarkable *CDKN2A* gene (previously called *MTS1* or *INK4A*) at 9p13 encodes two structurally unrelated proteins (*Figure 17.16*). Exons 1α, 2 and 3 encode the INK4A or p16 protein. A second promoter starts transcription further upstream at exon 1β. Exon 1β is spliced on to exons 2 and 3, but the reading frame is shifted, so that an entirely unrelated protein p14 or ARF (<u>A</u>lternative <u>R</u>eading <u>F</u>rame) is encoded (the mouse homolog is p19).

Both gene products function in cell cycle control (*Figure 17.15*). p16 functions upstream of the RB protein in control of the G1–S cell cycle checkpoint. Cyclin-dependent kinases inactivate pRb by phosphorylation, but p16 inhibits the CDK4/6 kinase. Thus loss of p16 function leads to loss of RB function and inappropriate cell cycling. The other product of the *CDKN2A* gene, p14, mediates G1 arrest by destabilizing MDM2 protein, the product of the *MDM2* oncogene which is amplified in many sarcomas (Pomeranz *et al.*, 1998). MDM2 binds to p53 and induces its degradation. Thus p14 acts to maintain the level of p53. Loss of p14 function leads to excessive levels of MDM2, excessive destruction of p53, and hence loss of cell cycle control.

Inherited *CDKN2A* mutations, usually affecting just p16, are seen in some families with multiple melanoma,

but somatic mutations are very much more frequent. Homozygous deletion of the gene inactivates both the RB and the p53 arms of cell cycle control, and is a very common event in tumorigenesis. Some tumors have mutations that affect p16 but not p14 (e.g. inactivation of the 1α promoter by methylation). Those tumors tend also to have p53 mutations, showing the importance of inactivating both arms of the control system shown in *Figure 17.15*.

17.7 Integrating the data: pathways and capabilities

As we remarked at the start of this chapter, cancer genetics can be deeply confusing. So many genes, so many mutations, such an infinity of combinations…every tumor is unique. But what they all have in common is that each tumor is the product of selection for a specific set of definable capabilities, and that, when we think in terms of pathways rather than individual genes, there is only a limited number of ways of achieving these capabilities. Hahn and Weinberg (2002) attempt to list all these pathways and draw up a 'route map' of cancer.

17.7.1 Pathways in colorectal cancer

Our understanding of tumor evolution is best developed for colorectal cancers, because all the stages of tumor development can be studied in resected colon from patients with familial adenomatous polyposis. *Figure 17.17* shows these steps. Similar but more rudimentary schemes can be made for other tumors where the data are not so rich. The scheme is based on a series of observations:

▶ malignant carcinomas develop from normal colonic epithelium through microscopic *aberrant crypt foci* to benign

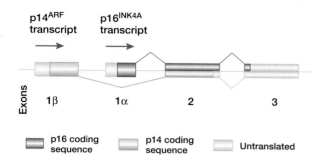

p14^ARF transcript **p16^INK4A transcript**

Exons 1β 1α 2 3

■ p16 coding sequence ■ p14 coding sequence □ Untranslated

Figure 17.16: The two products of the *CDKN2A* gene.

This gene (also known as *MTS* and *INK4A*) encodes two completely unrelated proteins. p16^INK4A is transcribed from exons 1α, 2 and 3, and p14^ARF from exons 1β, 2 and 3 – but with a different reading frame of exons 2 and 3. The two gene products are active in the RB and p53 arms of cell cycle control, respectively, as shown in *Figure 17.15.*

epithelial growths called *adenomas*. Adenomas develop through early (< 1 cm in size), intermediate (> 1 cm but without foci of carcinoma) and late (> 1 cm and with foci of carcinoma) stages, to become *carcinomas*, which eventually *metastasize*;

▶ in FAP, one copy of the *APC* gene on 5q21 is constitutionally mutated. Interestingly, inherited mutations rarely simply inactivate the APC protein; usually they produce a truncated protein that has a dominant negative action (see Section 16.4.3). Evidently simple inactivation of one *APC* allele would not give the necessary growth advantage. The very earliest detectable lesions, aberrant crypt foci, lack all APC expression. About one epithelial cell in 10^6 develops into a polyp, a rate consistent with loss of the second *APC* allele being the determining event;

▶ about 50% of intermediate and late adenomas, but only about 10% of early adenomas, have mutations in the *KRAS* oncogene (a relative of *HRAS*, *Table 17.1*). Thus *KRAS*

mutations may often be required for progression from early to intermediate adenomas;

▶ about 50% of late adenomas and carcinomas show loss of heterozygosity on 18q. This is relatively uncommon in early and intermediate adenomas. It seems likely that the relevant gene is *SMAD4* (also known as *MADH4*) rather than the initial candidate, *DCC* (see White, 1998);

▶ colorectal cancers, but not adenomas, have a very high frequency of *TP53* loss or mutations.

The scheme of *Figure 17.17* is intended to show the most common series of mutations, but not all colon cancers follow this path. Only around 60% of colon cancers have *APC* mutations. However, tumors lacking *APC* mutations often have activating mutations in β-catenin, which is the main downstream target of APC action (Fodde *et al.*, 2001). Further down the progression, FAP tumors tend to lose chromosome 18q, but the tumors of HNPCC are chromosomally stable and do not lose 18q. The selective pressure in both cases is probably loss of growth-inhibitory transforming growth factor (TGF) β signaling. In FAP this happens by loss of the downstream effector *SMAD4* located on 18q. In HNPCC, 90% of MIN^+ tumors have frameshifting mutations in an (A)_8 run in the TGFβ receptor II gene. HNPCC tumors also often have frameshifting mutations in the *AXIN2* gene, that encodes another component of the APC-β-catenin pathway. Thus behind the heterogeneity of molecular events, there is a much more regular picture of pathways that must be inactivated in order for a tumor to develop.

17.7.2 A successful tumor must acquire six specific capabilities

An important and highly recommended review by Hanahan and Weinberg (2000) (see Further reading) suggests another way of looking at the evolution of a tumor. They suggest that the key factors in the transition from a normal cell to a malignant cancer cell are the acquisition of six specific capabilities. The cell must:

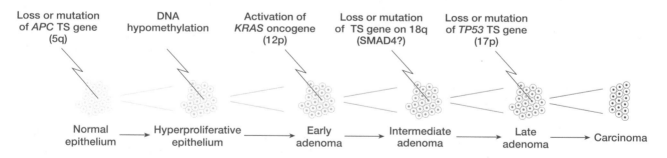

Loss or mutation of *APC* TS gene (5q) → DNA hypomethylation → Activation of *KRAS* oncogene (12p) → Loss or mutation of TS gene on 18q (SMAD4?) → Loss or mutation of *TP53* TS gene (17p)

Normal epithelium → Hyperproliferative epithelium → Early adenoma → Intermediate adenoma → Late adenoma → Carcinoma

Figure 17.17: A model for the multi-step development of colon cancer.

See text for further details. This is primarily a tool for thinking about how tumors develop, rather than a firm description. Every colorectal cancer is likely to have developed through the same histological stages, but the underlying genetic changes are more varied. According to Fearon and Vogelstein (1990) the figure illustrates a particularly common sequence of events. Smith *et al.* (2002) have questioned the validity of this model because in their series of 106 tumors, only seven had mutations in all three of *APC, KRAS* and p53.

▸ become independent of external growth signals;

▸ become insensitive to external anti-growth signals;

▸ become able to avoid apoptosis;

▸ become capable of indefinite replication;

▸ become capable of sustained angiogenesis;

▸ become capable of tissue invasion and metastasis.

Metastasis may not be a specifically selected capability. Ability to metastasize is not clearly advantageous to any cell, and may be an incidental effect of the general genomic disarray of advanced tumor cells. The other five capabilities are the result of specific selection. Relatively little is known about angiogenesis, but we have already seen the roles of activated oncogenes, deranged cell cycle control, p53 mutation and reactivation of telomerase in achieving the first four of these capabilities. Each capability may be acquired by a variety of different genetic changes, but in each case acquiring it requires certain specific pathways to be activated or inactivated. The capabilities are not necessarily acquired in the same order in different tumors, but the requirements for genomic instability and successive clonal expansions at intermediate stages (*Box 17.1*) impose a certain regularity on the process.

The first mutation in the multi-step evolution of a tumor is critical because it should confer some growth advantage on an otherwise normal cell that has all its defenses intact. According to the *gatekeeper hypothesis* (Kinzler and Vogelstein, 1996), in a given renewing cell population one particular gene is responsible for maintaining a constant cell number. Mutation of a gatekeeper leads to a permanent imbalance between cell division and cell death, whereas mutations of other genes have no long-term effect if the gatekeeper is functioning correctly. On this theory, the tumor suppressor genes identified by studies of Mendelian cancers are the gatekeepers for the tissue involved – *NF2* for Schwann cells, *VHL* for kidney cells, and so on. In the case of APC, it is perhaps relevant that APC protein not only regulates the level of β-catenin, a critical controller of cell growth, but may also play an important role in the spindle checkpoint (Section 17.5.1). We have already noted that even very early adenomas show chromosomal instability. It is likely that a single *APC* mutation (of the right type) already confers a growth advantage on colonic epithelium, whereas mutation of a single *BRCA1/2* gene gives no advantage to ductal epithelium. This may explain why *APC* mutation is very common in sporadic colon cancer, but *BRCA1/2* mutations are rare in sporadic breast cancer (Lamlum *et al.*, 1999). How common cancers like lung or prostate cancer that lack Mendelian forms fit the gatekeeper theory is not clear.

17.8 What use is all this knowledge?

In the light of the ideas set out in this chapter, it is easy to see that talk about 'a cure for cancer' is foolish. The new knowledge is however already leading to significant progress in three areas: detection, diagnosis and treatment.

▸ Detection of presymptomatic cancer traditionally relies on physical examination, scanning (mammography etc.), biochemical assays (prostate-specific antigen) or invasive methods (colonoscopy etc.). For reasons that are not entirely clear, tumors shed DNA. Sensitive PCR methods can amplify tumor DNA from urine (for bladder cancer), feces (for colon cancer), saliva or sputum (for oral or lung cancer) or blood (for a variety of cancers) (Sidransky, 2002). Hopefully, testing this DNA for mutations in *TP53*, *APC* and so on, or for specific promoter methylation or microsatellite instability, will allow more effective and noninvasive screening.

▸ As mentioned at the start of this chapter, the traditional histological methods of tumor identification and staging are being increasingly supplemented (but not supplanted) by molecular methods (Lakhani and Ashworth, 2001). This is most advanced for leukemia, where definition of the chromosomal rearrangements provides important pointers to prognosis and management (see, for example, Armstrong *et al.*, 2002). Until recently the complexity of changes in solid tumors defied ready analysis, but **expression profiling** is starting to provide useful classifications of subtypes, again with implications for prognosis and management. *Figure 17.18* shows examples of recent applications of expression arrays in cancer.

▸ The first treatments based on knowledge of specific molecular changes are now established. Herceptin, a monoclonal antibody against the ErbB2 receptor tyrosine kinase, is an effective treatment for breast cancers that have amplification of the *ERBB2* gene, but not for those lacking that feature (de Bono and Rowinsky, 2002). Gleevec (Imatinib, ST-1571) is a specific inhibitor of the BCR–ABL fusion protein (*Figure 17.4A*) and has produced remarkable results in chronic myeloid leukemia (Savage and Antman, 2002). Hopefully these are the first of a new wave of anticancer drugs designed from knowledge of the molecular events in tumors.

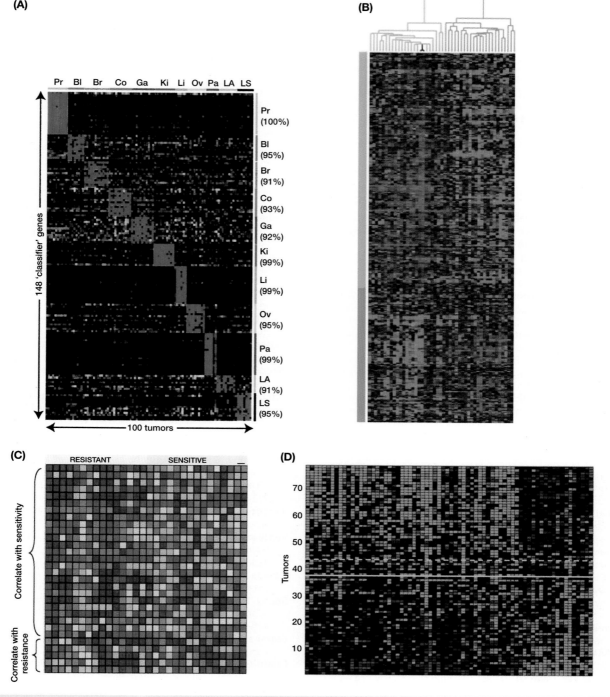

Figure 17.18: Examples of using expression arrays to characterize tumors.

(A) Identifying the tissue of origin. Expression patterns of 148 genes (rows) classify 100 tumors (columns) by origin. Pr, Prostate; Bl, bladder/ureter; Br, breast; Co, colorectal; Ga, gastroesophagus; Ki, kidney; Li, liver; Ov, ovary; Pa, pancreas; LA, lung adenocarcinoma; LS, lung squamous cell carcinoma. Red = increased gene expression, blue = decreased expression. From Su *et al.* (2001). *Cancer Research* **61**, 7388–7393, with permission of The American Association for Cancer Research. **(B) Distinguishing two types of diffuse large B-cell lymphoma.** Clustering of gene expression distinguishes germinal center-like tumors from activated B-like tumors (orange and blue bars). Five-year survival in the two groups is 76% and 16% respectively. Reproduced from Alizadeh *et al.* (2000) with permission from Nature Publishing Group. **(C) Sensitivity to chemotherapy drugs.** Expression profiles of 6817 genes in 60 tumor cell lines were correlated with response to 232 drugs. The figure shows gene expression patterns (rows) in 30 cell lines (columns) that predict sensitivity or resistance to one drug, cytochalasin D. From Staunton *et al.* (2001) with permission from National Academy of Sciences, USA. **(D) Predicting clinical outcome of breast cancer.** Expression patterns of 25 000 genes were scored in 98 primary breast cancers. The figure shows profiles of 70 prognostic marker genes (columns) in 78 cancers (rows). All patients had undergone surgery and radiotherapy; the yellow lines divide those predicted (using two alternative criteria) to require additional chemotherapy (above the line) from those who do not need this unpleasant treatment. Data reproduced from van't Veer *et al.* (2002) with permission from Nature Publishing Group.

Further reading

Bishop JM (1983) Cellular oncogenes and retroviruses. *Annu. Rev. Biochem.* **52**, 301–354.

Hanahan D, Weinberg R (2000) The hallmarks of cancer. *Cell* **100**, 57–70.

Stanbridge EM (1990) Human tumor suppressor genes. *Annu. Rev. Genet.* **24**, 615–657.

References

Alizadeh AA, Eisen MB, Davis RE *et al.* (2000) Distinct types of diffuse large B-cell lymphoma identified by gene expression profiling. *Nature* **403**, 503–511.

Al-Tassan N, Chmiel NH, Maynard J *et al.* (2002) Inherited variants of *MYH* associated with somatic G:C→T:A mutations in colorectal cancer. *Nature Genet.* **30**, 227–232.

Armstrong SA, Staunton JE, Silverman LB *et al.* (2002) MLL translocations specify a distinct gene expression profile that distinguishes a unique leukemia. *Nature Genet.* **30**, 41–47.

Blume-Jensen P, Hunter T (2001) Oncogenic kinase signalling. *Nature* **411**, 355–365.

Cavenee WK, Dryja TP, Phillips RA *et al.* (1983) Expression of recessive alleles by chromosomal mechanisms in retinoblastoma. *Nature* **305**, 779–784.

Chissoe SL, Bodenteich A, Wang Y-F *et al.* (1995) Sequence and analysis of the human *ABL* gene, the *BCR* gene, and regions involved in the Philadelphia chromosomal translocation. *Genomics* **27**, 67–82.

De Bono JS, Rowinsky EK (2002) The ErbB receptor family: a therapeutic target for cancer. *Trends Mol. Med.* **8**(4 Suppl), S18–S26.

Fearon ER, Vogelstein B (1990) A genetic model for colorectal tumorigenesis. *Cell* **61**, 759–767.

Fishel R, Lescoe MK, Rao MRS *et al.* (1993) The human mutator gene homolog *MSH2* and its association with hereditary non-polyposis colon cancer. *Cell* **78**, 539–542.

Fodde R, Smits R, Clevers H (2001) *APC*, signal transduction and genetic instability in colorectal cancer. *Nature Rev. Cancer* **1**, 55–67.

Forozan F, Karhu R, Kononen J *et al.* (1997) Genome screening by comparative genome hybridization. *Trends Genet.* **13**, 405–409.

Futreal PA, Kasprzyk A, Birney E *et al.* (2001) Cancer and genomics. *Nature* **409**, 850–852.

Hahn WC, Weinberg RA (2002) Modelling the molecular circuitry of cancer. *Nature Rev. Cancer* **2**, 331–340.

Harbour JW, Dean DC (2000) The Rb/E2F pathway: expanding roles and emerging paradigms. *Genes Dev.* **14**, 2393–2409.

Hoeijmakers J (2001) Genome maintenance mechanisms for preventing cancer. *Nature* **411**, 366–374.

Jallepalli PV, Lengauer C (2001) Chromosome segregation and cancer: cutting through the mystery. *Nature Rev. Cancer* **1**, 109–117.

Jiricny J, Nystrom-Lahti M (2000) Mismatch repair defects in cancer. *Curr. Opin. Genet. Dev.* **10**, 157–161.

Jones PA, Baylin SB (2002) The fundamental role of epigenetic events in cancer. *Nature Rev. Genet.* **3**, 415–428.

Kinzler KW, Vogelstein B (1996) Lessons from hereditary colorectal cancer. *Cell* **87**, 159–170.

Knudson AG (2001) Two genetic hits (more or less) to cancer. *Nature Rev. Cancer* **1**, 157–162.

Lakhani SR, Ashworth A (2001) Microarray and histopathological analysis of tumours: the future and the past? *Nature Rev. Cancer* **1**, 151–157.

Lamlum H, Ilyas M, Rowan A *et al.* (1999) The type of somatic mutation at APC in familial adenomatous polyposis is determined by the site of the germline mutation: a new facet to Knudson's 'two-hit' hypothesis. *Nature Med.* **5**, 1071–1075.

Lowy DR, Willumsen BM (1993) Function and regulation of RAS. *Annu. Rev. Biochem.* **62**, 851–891.

Malkin D (1994) Germline p53 mutations and heritable cancer. *Annu. Rev. Genet.* **28**, 4443–4465.

Maser RS, DePinho RA (2002) Connecting chromosomes, crisis and cancer. *Science* **297**, 565–569.

Mitelman F, Johansson B, Mertens F (eds) Mitelman Database of Chromosome Aberrations in Cancer (2002). http://cgap.nci.nih.gov/Chromosomes/Mitelman

Pinkel D (1994) Visualizing tumor amplification. *Nature Genet.* **8**, 107–108.

Pomeranz J, Schreiber-Agus N, Liegois NJ *et al.* (1998) The Ink4a tumor suppressor gene product, p19Arf, interacts with MDM2 and neutralizes MDM2's inhibition of p53. *Cell* **92**, 713–723.

Rabbitts TH (1994) Chromosome translocations in human cancer. *Nature* **372**, 143–149.

Sanchez-García I (1997) Consequences of chromosomal abnormalities in tumor development. *Ann. Rev. Genet.* **31**, 429–453.

Savage DG, Antman KH (2002) Imatinib mesylate – a new oral targeted therapy. *New Engl. J. Med.* **346**, 683–693.

Scully R, Livingston DM (2000) In search of the tumour-suppressor functions of *BRCA1* and *BRCA2*. *Nature* **408**, 429–432.

Sidransky D (2002) Emerging molecular markers of cancer. *Nature Rev. Cancer* **2**, 210–219.

Smith G, Carey FA, Beattie J *et al.* (2002) Mutations in APC, Kirsten-ras, and p53-alternative genetic pathways to colorectal cancer. *Proc. Natl Acad. Sci. USA* **99**, 9433–9438.

Staunton JE, Slonim DK, Coller HA *et al.* (2001) Chemosensitivity prediction by transcriptional profiling. *Proc. Natl Acad. Sci. USA* **98**, 10787–10792.

Su AI, Welsh JB, Sapinoso LM *et al.* (2001) Molecular classification of human carcinomas by use of gene expression signatures. *Cancer Res.* **61**, 7388–7393.

Thiagalingam S, Laken S, Willson JKV *et al.* (2001) Mechanisms underlying losses of heterozygosity in human colorectal cancers. *Proc. Natl Acad. Sci. USA* **98**, 2698–2702.

Tomlinson I, Halford S, Aaltonen L *et al.* (2002) Does MSI-low exist? *J. Pathol.* **197**, 6–13.

Tosi S, Giudici G, Rambaldi A *et al.* (1999) Characterization of the human myeloid leukemia-derived cell line GF-D8 by multiplex fluorescence in situ hybridization, subtelomeric probes and comparative genomic hybridization. *Genes Chrom. Cancer* **24**, 231–241.

van't Veer LJ, Dai H, van de Vijver MJ *et al.* (2002) Gene expression profiling predicts clinical outcome of breast cancer. *Nature* **415**, 530–536.

Vousden KH (2000) p53: death star. *Cell* **103**, 691–694.

White RL (1998) Tumor suppressing pathways. *Cell* **92**, 591–592.

Wu CL, Roz L, Sloan P, Read AP *et al.* (1997) Deletion mapping defines three discrete areas of allelic imbalance on chromosome arm 8p in oral and oropharyngeal squamous cell carcinomas. *Genes Chrom. Cancer* **20**, 347–353.

Zhou B-BS, Elledge S (2000) The DNA damage response: putting checkpoints in perspective. *Nature* **408**, 433–439.

CHAPTER EIGHTEEN

Genetic testing in individuals and populations

Chapter contents

18.1 Introduction

Geneticists have no monopoly on DNA-based diagnosis. For microbiologists and virologists, for example, PCR is a central tool for identifying pathogens. Hematologists, oncologists and other pathologists all use DNA testing as a basis for diagnosis. However, for the purposes of this chapter we will define genetic testing as testing for Mendelian factors. The factors may indicate a person's risk of developing or transmitting a disease (which may or may not be Mendelian), or they may be used to identify her or to indicate her relationship to somebody else.

We will use two of the most common Mendelian diseases, cystic fibrosis and Duchenne muscular dystrophy, to illustrate the various testing methods wherever possible. Both involve large genes with extensive allelic heterogeneity, but beyond that, CF and DMD pose rather different sets of problems for DNA diagnosis (*Table 18.1*). Between them, they show many of the issues involved in testing for Mendelian diseases. As always in this book, we concentrate on the principles and not the practical details. The reader interested in specific procedures can find a series of 'best practice' guidelines for laboratory diagnosis of the commoner Mendelian diseases at http//:www.cmgs.org. These have been drawn up at consensus workshops of the UK Clinical Molecular Genetics Society. The book by Elles and Mountford (see Further reading) describes testing methods for a wider range of conditions.

When a clinician brings a sample of a patient's DNA to a laboratory for diagnostic testing, there are three possible questions the laboratory might try to answer.

▸ Does the patient have *any* mutation in *any* gene that would explain his disease? This question is not answerable, now or in the future. Gene testing has to be targeted. Even if it becomes possible to sequence a person's entire genome as a diagnostic procedure, there is no way we could decide which of the several million differences between the DNA sequence of two people might be responsible for a disease, without a specific list of candidates.

▸ Does the patient have *any* mutation in *this particular gene* that might cause his disease? Standard ways of answering this question are considered in Section 18.3, while an alternative approach, gene tracking, is considered in Section 18.5.

▸ Does the patient have a 3-base deletion of the codon for phenylalanine 508 in his *CFTR* gene? The circumstances in which this type of question can be asked, and ways of answering it, are considered in Section 18.4.

18.2 The choice of material to test: DNA, RNA or protein

Genetic testing is almost always done by PCR, applying the methods described in Section 5.2. The few applications of Southern blotting include testing for major gene rearrangements or disruptions and for fragile X and myotonic dystrophy full mutations (Section 16.6.4). The sensitivity of PCR allows us to use a wide range of tissue samples. These can include:

▸ **blood samples** – the most widely used source of DNA from adults;

▸ **mouthwashes or buccal scrapes** – being noninvasive, they are especially favored for population screening programs. Mouthwashes yield sufficient DNA for a few dozen tests, and by using whole genome amplification (Section 5.2.4) more extensive testing of a single sample may be possible;

▸ **chorionic villus biopsy samples** – the best source of fetal DNA (better than amniocentesis specimens);

Table 18.1: The contrasting genetics of cystic fibrosis and Duchenne muscular dystrophy

Cystic fibrosis	Duchenne muscular dystrophy
Autosomal recessive	X-linked recessive
Loss of function mutations	Loss of function mutations

Fairly large gene:
  250-kb genomic DNA
  27 exons
  6.5-kb mRNA
	Giant gene:
	  2400-kb genomic DNA
	  79 exons
	  14-kb mRNA

Almost all mutations are single nucleotide changes	65% of mutations are deletions encompassing one or more complete exons 5% duplications 30% nonsense, splice site etc. mutations Missense mutations are very unusual
New mutations are extremely rare	New mutations are very frequent
Mosaicism is not a problem	Mosaicism is common
Little intragenic recombination	Recombination hotspot (12% between markers at either end of the gene)

Genetic testing in DMD and CF require different sets of approaches.

- **one or two cells removed from eight-cell stage embryos**, for pre-implantation diagnosis after in vitro fertilization;

- **hair, semen**, etc. for criminal investigations;

- **archived pathological specimens**, for typing dead people when no DNA has been stored, or testing tumors for genetic changes. Only short sequences, 250 bp or less, can be reliably amplified from fixed tissue specimens;

- **Guthrie cards** – these are the cards on which a spot of dried blood is sent to a laboratory for neonatal screening for phenylketonuria (PKU) in the UK and elsewhere. Not all of the blood spot is used for the screening test, so if the cards are retained they are a possible source of DNA from a dead child.

RNA has advantages over DNA, but is more difficult to obtain and handle

If a gene has to be scanned for unknown mutations, testing by RT–PCR (see Section 5.2.1) offers several advantages. DNA testing usually involves amplifying and testing each exon separately, and this can be a major chore in a gene with many exons. Most of the mutation-scanning methods (*Table 18.2*) can scan fragments larger than the average sized exon, so that an RT–PCR product can be examined using a smaller number of reactions. Also, only RT–PCR can reliably detect aberrant splicing, which is often hard to predict from a DNA sequence change, or may be caused by activation of a cryptic splice site deep within an intron (see Section 16.4.1). However, RNA is much less convenient to obtain and work with. Samples must be handled with extreme care and processed rapidly to avoid degrading mRNA, and the gene of interest may not be expressed in readily accessible tissues. In addition, many mutations result in unstable mRNA (see sections 11.4.4, 16.4.1), so that the RT–PCR product from a heterozygous person may show only the normal allele.

Functional assays of proteins have a role in genetic testing

A protein-based functional assay might classify the products of a highly heterogeneous allelic series into two simple groups, functional and nonfunctional – which is, after all, the essential question in most diagnosis. The problem with functional assays is that they are specific to a particular protein. DNA technology by contrast is generic. This has obvious advantages for the diagnostic lab, but in addition it encourages technical development, since any new technique can be applied to many problems.

18.3 Scanning a gene for mutations

For the great majority of diseases, as with CF and DMD, there is extensive allelic heterogeneity (see Section 16.3.2). Diagnostic testing therefore usually involves searching for mutations that might be anywhere within or near the relevant gene or genes. *Table 18.2* lists a number of methods that can be used, and these are described briefly below. For laboratory details, see the books by Cotton, Edkins and Forrest or Elles and Mountford (Further reading).

Searching the whole of a candidate gene for sequence variants in a large series of patients will reveal many different variants. Deciding whether or not a variant is pathogenic (and hence represents the sought-after mutation) can be very difficult. *Box 16.5* gives some guidelines for attempting to decide.

18.3.1 Methods based on sequencing

Now that automated fluorescence sequencers are standard laboratory equipment, sequencing becomes more and more attractive as the primary means of mutation scanning (*Figure 18.1*). Other scanning methods serve mainly to reduce the sequencing load by defining which amplicon should be sequenced. As sequencing has become cheaper and easier, the need to reduce the load has diminished, and several previously used alternative methods have fallen out of favor. Alternatives to direct sequencing are now mainly used because they are either quick (denaturing high performance liquid chromatography; dHPLC), cheap (single strand conformation polymorphisms; SSCP) or give some special information (protein truncation test; PTT and quantitative PCR).

Sequencing generates more data than other methods, so the requirement for analysis is greater. Programs are available that automatically report differences between the test sequence and a standard sequence, provided the sequence is of good quality. Quality is critical for avoiding artifacts and reliably detecting base substitutions in heterozygotes.

For sequencing genomic DNA, each exon is normally amplified separately, with maybe 20 bp of flanking intron. This is inefficient because the average exon size in human genes (145 bp) is well below the 500–800 bp that can be read from a good sequencing run. Possible solutions are to use RT–PCR, or to use exon-linking PCR (meta-PCR, Wallace *et al.*, 1999; *Figure 18.2*). Microarray-based resequencing is considered below (Section 18.4.1).

18.3.2 Methods based on detecting mismatches or heteroduplexes

Many tests use the properties of heteroduplexes to detect differences between two sequences. Most mutations occur in heterozygous form (even with autosomal recessive conditions, affected people born to nonconsanguineous parents are likely to be compound heterozygotes, with two different mutations). Heteroduplexes can be formed simply by heating the heterozygous test PCR product to denature it, and then cooling slowly. For homozygous mutations, or X-linked mutations in males, it is necessary to add some reference wild-type DNA. Several properties of heteroduplexes can be exploited:

- heteroduplexes often have abnormal mobility on nondenaturing polyacrylamide gels (*Figure 18.3A*, lower panel). Special gels (Hydrolink™, MDE™) are supposed to improve the resolution. This is a particularly simple method to use. If fragments no more than 200 bp long are tested,

Table 18.2: Methods for scanning a gene for mutations

The table summarizes the advantages and disadvantages of each method for use in a routine diagnostic service. See Section 18.3 for details.

Method	Advantages	Disadvantages
Southern blot, hybridize to cDNA probe	Only way to detect major deletions and rearrangements	Laborious, expensive Needs several μg DNA
Sequencing	Detects all changes Mutations fully characterized	Expensive
Heteroduplex gel mobility	Very simple Cheap	Sequences < 200 bp only Limited sensitivity Does not reveal position of change
dHPLC	Quick, high throughput Quantitative	Expensive equipment Does not reveal position of change
SSCP	Simple, cheap	Sequences < 200 bp only Does not reveal position of change
DGGE	High sensitivity	Choice of primers is critical Expensive primers Does not reveal position of change
Chemical mismatch cleavage	High sensitivity Shows position of change	Toxic chemicals Experimentally difficult
PTT	High sensitivity for chain terminating mutations Shows position of change	Chain terminating mutations only Expensive, difficult technique Usually needs RNA
Quantitative PCR	Detects heterozygous deletions	Expensive
Microarrays	Quick High throughput Might detect and define all changes	Expensive Limited range of genes

dHPLC, Denaturing high performance liquid chromatography; SSCP, single strand conformation polymorphism; DGGE, denaturing gradient gel electrophoresis; PTT, protein truncation test.

insertions, deletions and most but not all single-base substitutions are detectable (Keen *et al.*, 1991);

▶ heteroduplexes have abnormal denaturing profiles. This is exploited in **denaturing high performance liquid chromatography** (dHPLC, *Figure 18.4A*) and **denaturing gradient gel electrophoresis** (DGGE, *Figure 18.3B*). In both cases the mobility of a fragment changes markedly when it denatures. These methods require tailoring to the particular DNA sequence under test, and so are best suited to routine analysis of a given fragment in many samples. dHPLC allows high throughput, which fits well with this use. DGGE requires special primers with a 5′ poly(G;C) extension (a GC clamp) – see Sheffield *et al.* (1992). Once optimized, these methods have high sensitivity;

▶ mismatched bases in heteroduplexes are sensitive to cleavage by chemicals or enzymes. The **chemical cleavage of mismatch** (CCM) method (*Figure 18.4B*) is a sensitive method for mutation detection, with the advantages that quite large fragments (over 1 kb in size) can be analyzed, and the location of the mismatch is pinpointed by

the size of the fragments generated. However, it uses very toxic chemicals, particularly osmium tetroxide (though this can be substituted by potassium permanganate), and is experimentally quite difficult. An alternative is **enzymatic cleavage of mismatches**, which uses enzymes such as T4 phage resolvase or endonuclease VII to achieve the same result without the toxic chemicals. Unfortunately in most people's hands, the quality of the gels produced leaves much to be desired.

Of all these methods, only dHPLC is now widely used in major diagnostic laboratories.

18.3.3 Methods based on single-strand conformation analysis

Single-stranded DNA has a tendency to fold up and form complex structures stabilized by weak intramolecular bonds, notably base-pairing hydrogen bonds. The electrophoretic mobilities of such structures in nondenaturing gels will depend not only on their chain lengths but also on their conformations, which are dictated by the DNA sequence.

(A) Patient

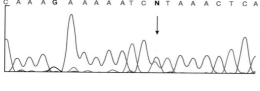

C A A A G A A A A A T C N T A A A C T C A

Control

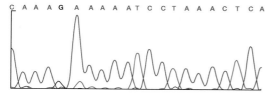

C A A A G A A A A A T C C T A A A C T C A

(B) Control

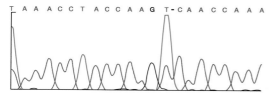

T A A A C C T A C C A A G T - C A A C C A A A

Patient

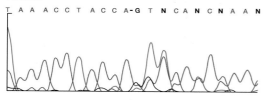

T A A A C C T A C C A - G T N C A N C N A A N

Figure 18.1: Mutation detection by sequencing.

(A) A base substitution in exon 3. The double peak (arrow) shows a heterozygous mutation g.332C→T (p.P67L). **(B)** A single base deletion 3659delC in exon 19. Sequence downstream of the deletion is confused, reflecting overlapping sequence of the two alleles in this heterozygote. The changes would be confirmed by sequencing the reverse strand. Courtesy of Dr Andrew Wallace, St Mary's Hospital, Manchester, UK.

SSCP are detected by amplifying the DNA samples (which may be RT–PCR products), denaturing, snap-cooling and loading on a nondenaturing polyacrylamide gel (*Figure 18.3A*). Primers can be radiolabeled, or unlabeled products can be detected by silver staining. The precise pattern of bands seen is very dependent on details of the conditions. Control samples must be run, so that differences from the wild-type pattern can be noticed. SSCP is very cheap and reasonably sensitive (around 80%) for fragments up to 200 bp long (Sheffield *et al.*, 1993), so it is still widely used. SSCP and heteroduplex analysis can be combined on a single gel, as in *Figure 18.3A*. An elaboration of SSCP, **dideoxy finger-printing**, analyzes each band in a sequencing ladder by SSCP, and is claimed to give 100% sensitivity (Sarkar *et al.*, 1992).

18.3.4 Methods based on translation: the protein truncation test

The PTT (*Figure 18.5*) is a specific test for frameshifts, splice site or nonsense mutations that create a premature termination codon (van der Luijt *et al.*, 1994). The starting material is an RT–PCR product, or occasionally a single large exon in genomic DNA such as the 6.5-kb exon 15 of the *APC* gene or the 3.4-kb exon 10 of the *BRCA1* gene. Nonsense-mediated mRNA decay (Section 16.4.1), that normally prevents production of a truncated protein, does not occur because there is no exon–exon splicing in the assay. Clearly, the strength and weakness of the PTT is that it detects only certain classes of mutation. It would not be useful for cystic fibrosis, where most mutations are nontruncating. But in Duchenne muscular dystrophy, adenomatous polyposis coli or *BRCA1*-related breast cancer, missense mutations are infrequent, and any such change found may well be coincidental and nonpathogenic. For such diseases, the PTT has several advantages. It ignores silent or missense base substitutions, and (like mismatch cleavage methods, but unlike SSCP) it reveals the approximate location of any mutation. Several variants have been developed to give cleaner results, usually by incorporating an immunoprecipitation step – but PTT remains a demanding technique that is not easy to get working well.

18.3.5 Methods for detecting deletions

Homozygous or hemizygous deletions are simple to detect: the deleted sequence will not amplify by PCR. It is important to consider alternative explanations for failure to amplify: maybe there was a technical failure of the PCR or a base substitution in one of the primer binding sites. Deletions can be confirmed by using alternative primers or by Southern blotting. Around 60% of DMD mutations are deletions of one or more exons (see *Figure 16.3*). In affected males, two multiplex PCR reactions (that shown in *Figure 18.6* and one testing exons in the 5′ part of the gene) will reveal 98% of all deletions. Most deletions remove more than one exon. Deletions that appear to affect noncontiguous exons and deletions of just a single exon need confirming.

Testing females to see if they carry a dystrophin deletion is altogether more difficult, and illustrates the special problems posed by heterozygous deletions of one or more whole exons. Such deletions are not detectable when genomic DNA is amplified exon by exon and tested by sequencing, heteroduplex analysis or SSCP, because the mutant allele gives no product. Megabase-sized deletions can be diagnosed by FISH or array-CGH (*Figure 17.3*). For exon-scale deletions alternatives are needed. Sequencing or PTT analysis of RT–PCR products should pick up anything except a deletion of the whole gene, but nonsense-mediated mRNA decay may render the mutant transcript invisible, and RNA-based methods are not always easy to implement in a diagnostic setting.

Several systems for quantitative PCR are available that can detect heterozygous deletions in genomic DNA. The reaction is limited to the early exponential phase (*Figure 5.2*) when the amount of product reflects the amount of template. Product accumulation is measured in real time by fluorescence and

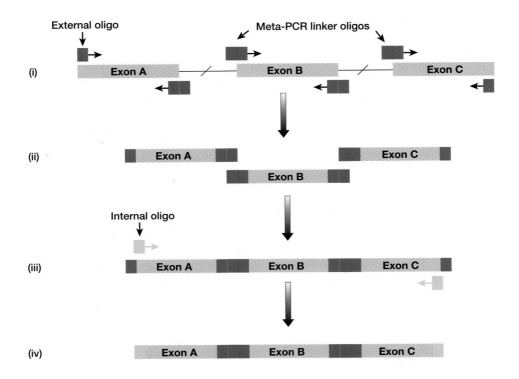

Figure 18.2: Principle of exon-linking PCR (Meta-PCR).

This is a technique for overcoming the disparity between the average size of human exons (145 bp) and the optimum size of product for sequencing (500–800 bp), while retaining the advantages of studying genomic DNA. (i) 2–5 exons (with ca. 20 bp of flanking intron, not shown for clarity) are PCR-amplified using primers that carry specific matching linkers at their 5′ ends. (ii). As primers become exhausted, concatemers form, guided by the linkers (iii). The desired concatemer is amplified in a second round of PCR using specifically designed primers (iv). The product can be designed to have any combination of exons arranged in any order. See Wallace, Wu and Elles (1999).

compared to an internal or parallel standard. The fluorescent signal may be generated by a double-strand DNA-specific binding dye such as SYBR® Green, or by cleavage of a sequence-specific oligonucleotide labeled with a dye and a quencher (TaqMan® assay). An alternative to real-time PCR is the MAPH (Multiplex Amplifiable Probe Hybridization) method of Armour (2000) (*Box 18.1*). MAPH allows a high degree of multiplexing (dosage of at least 50 short sequences can be compared in a single experiment), requires no specially tagged probes or expensive machinery, and is well suited for development in a modest laboratory.

If a deletion is segregating in a family, typing females for microsatellites mapping within the deletion may reveal apparent nonmaternity, where a mother has transmitted no marker allele to her daughter because of the deletion (*Figure 18.7*). In such families 'nonmaternity' proves a woman is a carrier, while heterozygosity (in the daughter or sister of a deletion carrier) proves a woman is not a carrier. Several markers suitable for this purpose have been identified in the introns at deletion hotspots. The method works best in families where there is an affected male in whom the deletion can first be defined.

18.3.6 Methods for detecting DNA methylation patterns

The importance of DNA methylation in the control of gene expression was described in Section 10.4.2. Excessive or deficient methylation of CpG dinucleotides is a common pathogenic mechanism in cancer (Section 17.4.3) and imprinted gene expression (Section 16.4.4). PCR products are always unmethylated, so none of the methods described so far gives information on methylation patterns in a DNA sample. Two main methods are used to check methylation.

▶ Restriction enzyme digestion. *Hpa*II cuts only unmethylated CCGG whereas *Msp*I cuts any CCGG, whether methylated or not. Methylated CpG sites therefore cause the two enzymes to give different restriction fragments. Alternatively, a PCR template containing a CCGG can be digested with *Hpa*II before amplification, if the site is unmethylated the template is cleaved and the PCR will yield no product.

▶ Bisulfite sequencing (Thomassin *et al.*, 1999). When single-stranded DNA is treated with sodium bisulfite, cytosine but not 5-methyl cytosine (5-MeC) is converted to uracil. After bisulfite treatment, the DNA is PCR-amplified (using primers matching the modified sequence) and subjected to

(A)

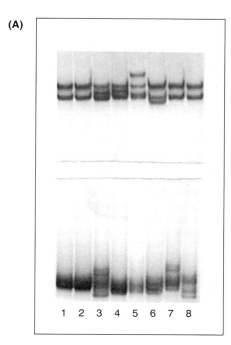

(B)

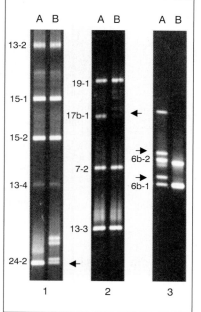

Figure 18.3: Scanning the CFTR gene for mutations.

(A) Heteroduplex and SSCP analysis. Exon 3 was PCR-amplified from genomic DNA. After denaturation, the samples were loaded on to a nondenaturing polyacrylamide gel. Some of each product re-annealed to give double-stranded DNA. This runs faster in the gel and gives the heteroduplex bands seen in the lower panel. The single-stranded DNA runs more slowly in the same gel (upper panel). Lanes 1 and 2 show the pattern typical of the wild-type sequence. Variant patterns can be seen in lanes 3–8. Sequencing revealed mutations G85E, L88S, R75X, P67L, E60X and R75Q, respectively. Courtesy of Dr Andrew Wallace, St Mary's Hospital, Manchester. **(B) Denaturing gradient gel electrophoresis.** Exons of the *CFTR* gene are PCR-amplified in one or more segments and run on 9% polyacrylamide gels containing a gradient of urea–formaldehyde denaturant. The band from any amplicon that contains a heterozygous variant splits into usually four sub-bands (arrows). In the lanes shown, subject A (left lane in each panel) has a variant in amplicon 6 and subject B (right lanes) has variants in amplicons 17 and 24. Characterization of the variants showed that subject B was heterozygous for R1070Q (exon 17). Other variants were nonpathogenic. Courtesy of Dr Hans Scheffer, University of Groningen, The Netherlands.

conventional sequencing. Cytosine and 5-MeC in the original sample show as thymine and cytosine, respectively, in the sequence of the PCR product. This method requires very careful controls to give reliable data.

18.4 Testing for a specified sequence change

Testing for the presence or absence of a known sequence change is a different and much simpler problem than scanning a gene for the presence of *any* mutation. Samples can always be genotyped by sequencing, but conventional sequencing is not an efficient method if only a single nucleotide position is being checked. Some of the main genotyping methods are summarized in *Table 18.3*. Many variants of these and other methods have been developed as kits by biotechnology companies. Typical applications include:

▸ diagnosis of diseases with limited allelic heterogeneity (see *Table 18.4*);

▸ diagnosis within a family. Mutation scanning methods may be needed to define the family mutation, but once it is characterized, other family members normally need be tested only for that particular mutation;

▸ in research, for testing control samples. A common problem in positional cloning is that a patient has a sequence change in a candidate gene. The question then arises, is this change pathogenic (confirming that the candidate gene is the disease gene), or might it be a nonpathogenic polymorphism? One common approach is to screen a panel of normal control samples for the presence of the change. The question of how many controls should be tested is considered by Collins and Schwartz (2002).

▸ SNP genotyping. The aim here is not to find a pathogenic mutation as above, but the problem is identical: to test a DNA sample for a pre-defined sequence variant. SNP typing is the main application of methods for very high throughput genotyping (Section 18.4.2).

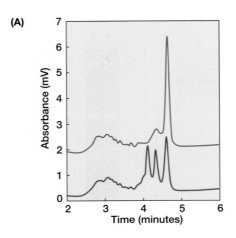

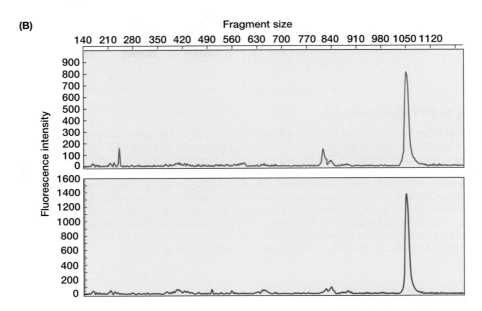

Figure 18.4: Mutation scanning.

(A) DMD mutation scanning by denaturing high-performance liquid chromatography (dHPLC). Exon 6 of the dystrophin gene gives a different pattern in an affected male (blue trace) and a normal control (red trace). Sequencing revealed the splice site mutation 738+1G→T. Because this is an X-linked condition, for males test DNA must be mixed with an equal amount of normal DNA to allow formation of heteroduplexes. Courtesy of Dr Richard Bennett, Children's Hospital, Boston, MA, USA. **(B)** Mutation scanning by chemical cleavage of mismatches. A fluorescently labeled meta-PCR product containing exons 6–10 of the *NF2* gene. Upper track: patient sample; a heterozygous intron 6 splice mutation 600–3c→g is revealed by hydroxylamine cleavage of the 1032-bp meta-PCR product to fragments of 813+239 bp. Lower track: control sample. Courtesy of Dr Andrew Wallace, St Mary's Hospital, Manchester, UK.

18.4.1 Many simple methods are available for genotyping a specified variant

Testing for the presence or absence of a restriction site

When a base substitution mutation creates or abolishes the recognition site of a restriction enzyme, this allows a simple direct PCR test for the mutation (*Figure 7.6*). Although hundreds of restriction enzymes are known, they almost all recognize symmetrical palindromic sites, and many point mutations will not happen to affect such sequences. Also, sites for rare and obscure restriction enzymes are unsuitable for routine diagnostic use because the enzymes are expensive and often of poor quality. Sometimes, however, a diagnostic restriction site can be introduced by a form of PCR mutagenesis (Section 5.5.3) using carefully chosen primers. *Figure 18.8* shows an example.

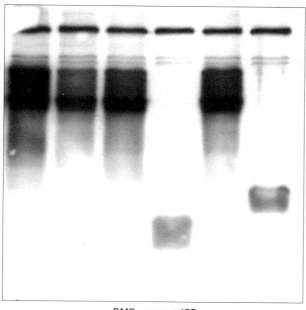

DMD segment 1EF

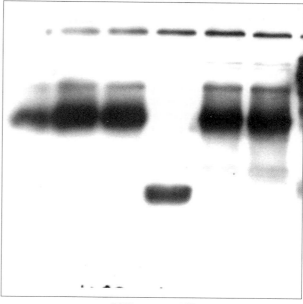

DMD segment 2CD

Figure 18.5: DMD mutation scanning using the protein truncation test (PTT).

A coupled transcription–translation reaction is used to produce labeled polypeptide products encoded by a segment of mRNA. Segments containing premature termination codons produce truncated polypeptides. Here, RT-PCR was used to scan the entire dystrophin gene in ten overlapping segments in a series of DMD patients. The faster running polypeptides identify segments containing termination codons, and the size of the abnormal product indicates the position of the stop codon within the segment. Image courtesy of Dr J.T. den Dunnen and D. Verbove (Leiden, Netherlands); for further details see http://www.dmd.nl

(A)

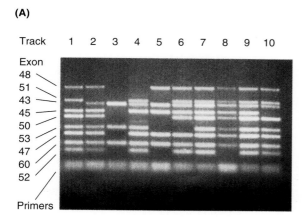

(B)

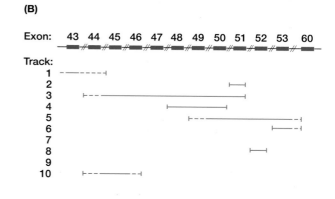

Fig 18.6: Multiplex screen for dystrophin deletions in males.

(A) Products of multiplex PCR amplification of nine exons, using samples from 10 unrelated boys with Duchenne/Becker muscular dystrophy. PCR primers have been designed so that each exon, with some flanking intron sequence, gives a different sized PCR product. Courtesy of Dr R. Mountford, Liverpool Women's Hospital, UK. **(B)** Interpretation: solid lines show exons definitely deleted, dotted lines show possible extent of deletion running into untested exons. No deletion is seen in samples 7 and 9 – these patients may have point mutations, or deletions of exons not examined in this test. Exon sizes and spacing are not to scale. Compare with *Figure 16.3*.

Box 18.1: Multiplex amplifiable probe hybridization (MAPH).

This is a very versatile method of comparing gene dosage for a number of sequences. Genomic DNA from the patient is spotted onto a tiny (2 × 3 mm) nylon filter and hybridized to a mixture of 40 probes, each specific for one exon of the dystrophin gene. After careful washing, the filter is put into a PCR reaction mix that uses a single pair of primers that bind to a sequence present at the end of every probe. One primer can be fluorescently labeled to allow analysis on a gene sequencer.

Probes are designed so that exons can be distinguished by the length of the PCR product, and quantified by the size of the peak on a fluorescence sequencer or by the relative intensities of bands on a manual gel. Because all the probes amplify using the same primer pair, the problems of unequal amplification that plague multiplex PCR are avoided.

For further details of MAPH see Armour *et al.* (2000) and the MAPH website www.nott.ac.uk/~pdzjala/maph/maph.html

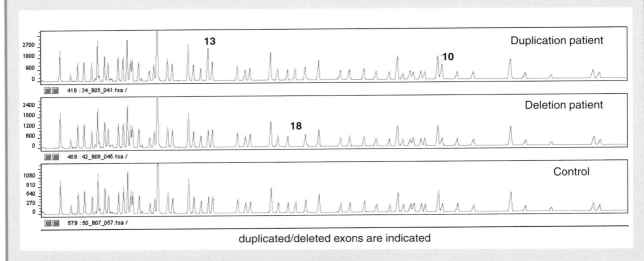

duplicated/deleted exons are indicated

Use of MAPH to detect deletions and duplications of exons of the dystrophin gene.

Comparing peak sizes with the control trace, it is easy to see he duplication of exons 10 and 13 (and presumably 11 and 12, which were not tested in this multiplex) in the top trace, and the deletion of exon 18 in the centre trace. Image courtesy of Dr J.T. den Dunnen and S. White, Leiden, Netherlands; for further details see http://www.dmd.nl

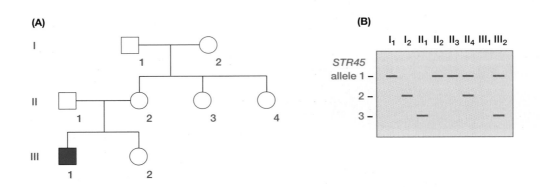

Figure 18.7: Deletion carriers in a DMD family revealed by apparent nonmaternity.

(A) Pedigree. **(B)** Results of typing with the intragenic marker STR45. The affected boy III-1 has a deletion which includes STR45 (lane 7 of the gel is blank). His mother II-2 and his aunt II-3 inherited no allele of STR45 from their mother I-2, showing that the deletion is being transmitted in the family. I-2 is apparently homozygous for this highly polymorphic marker (lane 2), but in fact is hemizygous. The other aunt II-4 and the sister III-2 are heterozygous for the marker, and therefore do not carry the deletion.

Use of allele-specific oligonucleotide (ASO) hybridization

Under suitably stringent hybridization conditions, these short synthetic probes hybridize only to a perfectly matched sequence (Section 6.3.1). *Figure 6.11* demonstrates the use of dot-blot hybridization with ASO probes to detect the single base substitution that causes sickle cell disease. For diagnostic purposes, a reverse dot-blot procedure has often been used. A screen for a series of defined cystic fibrosis mutations, for example, would use a series of ASOs specific for each mutant allele, spotted onto a single membrane which is then hybridized to labeled PCR-amplified test DNA.

The same reverse dot-blot principle is applied on a massively parallel scale in *DNA chips*. Hundred of thousands of 20–25-mer oligonucleotide probes are anchored in defined positions to a solid support (Section 6.4.3). Test DNA is PCR-amplified, fluorescently labeled and hybridized to the array, either alone or (preferably) in competition with a reference wild-type sequence, labeled with a different color. Each individual probe on the chip performs an ASO test for a specific sequence in the test DNA, but overall the chips can scan the entire sequence of a gene and report any base substitution (*Figure 18.9A*). Detection efficiencies are excellent for homozygotes, but less reliable for heterozygotes. Insertions are only detected in so far as oligonucleotides were designed in advance to detect a specific insertion.

Once everything is set up, chip systems are quick and easy to use, though far from cheap. However, it is necessary to define in advance the range of mutations to be detected, and design these into the chip. Thus their main role in mutation detection may be for initial scanning of samples to pick up the common mutations, leaving the difficult cases to be sorted out by other methods.

Allele-specific PCR amplification (ARMS test)

The principle of the ARMS (Amplification Refractory Mutation System) method was shown in *Figure 5.4*. Paired PCR reactions are carried out. One primer (the common primer) is the same in both reactions, the other exists in two slightly different versions, one specific for the normal sequence and the other specific for the mutant sequence. Additional control primers are usually included, to amplify some unrelated sequence from every sample as a check that the PCR reaction has worked. The location of the common primer can be chosen to give different sized products for different mutations, so that the PCR products of multiplexed reactions form a ladder on a gel. With careful primer design the mutation-specific primers can also be made to give distinguishable products. For example, they can be labeled with different fluorescent or other labels, or given 5′ extensions of different sizes. Multiplexed mutation specific PCR is well suited to screening fairly large numbers of samples for a given panel of mutations (*Figure 18.10*).

Oligonucleotide ligation assay (OLA)

In the OLA test for base substitution mutations, two oligonucleotides are constructed that hybridize to adjacent sequences in the target, with the join sited at the position of the mutation. DNA ligase will not covalently join the two oligonucleotides unless they are perfectly hybridized (Nickerson *et al.*, 1990). Various formats for the test are possible, for example as an ELISA or, as in *Figure 18.11*, for analysis on a fluorescence sequencer.

Minisequencing by primer extension

Minisequencing uses the principle of Sanger sequencing (*Figure 7.2*), but adds just a single nucleotide. The 3′ end of the sequencing primer is immediately upstream of the nucleotide that is to be genotyped. The reaction mix contains DNA polymerase plus four differently labeled dideoxy nucleoside triphosphates (ddNTPs). The test DNA acts as template for addition of a single labeled dideoxynucleotide to the primer. One way or another, the nucleotide added to the primer is identified (for example, by running the extended primer on a fluorescence sequencer), allowing the base at the position of interest to be identified.

Minisequencing can readily be adapted to an array-based format (APEX, arrayed primer extension; Tõnisson *et al.*, 2002) to allow the entire sequence of a gene to be checked for base substitutions in a single operation (*Figure 18.9B*).

18.4.2 Methods for high-throughput genotyping

Biotechnology companies are developing numerous systems to meet the challenge of genotyping very large numbers of SNPs for disease association studies (Section 15.4.5). Weaver (2000) has listed a large number of such systems. The underlying genetic method is usually primer extension (minisequencing), allele-specific oligonucleotide hybridization or oligonucleotide ligation, as described above, but formatted to allow automation and very high throughput. Many systems attempt to combine this genetic technology with 'lab on a chip' developments in microfluidics. Examples of other approaches include pyrosequencing and mass spectrometry (*Box 18.2*).

18.4.3 Genetic testing for triplet repeat diseases

The expanded repeats that cause numerous neurological diseases (*Table 16.6*) involve a special set of mutation-specific tests (*Figure 18.12*). For the polyglutamine repeat diseases such as Huntington disease (HD), a single PCR reaction makes the diagnosis. Some of the other expanded repeat diseases are a little more difficult for two reasons. Full mutations may have hundreds or thousands of repeats and do not readily amplify by PCR, especially since most of the repeats have a high GC content. Normal and premutation alleles give clean PCR products, but full mutations may require Southern blotting. Additionally, unlike HD, the mutations mostly cause disease by loss of function, and occasional affected patients have deletions or point mutations that would be missed by testing just the repeat. Myotonic dystrophy is the only one of the 'large expansion' diseases that seems to be completely homogeneous mutationally.

Table 18.3: Methods of testing for a specified mutation

Method	Comments
Restriction digestion of PCR-amplified DNA; check size of products on a gel	Only when the mutation creates or abolishes a natural restriction site, (*Figure 7.6*) or one engineered by use of special PCR primers (*Figure 18.8*)
Hybridize PCR-amplified DNA to allele-specific oligonucleotides (ASO) on a dot-blot or gene chip	General method for specified point mutations; large arrays allow scanning for almost any mutation
PCR using allele-specific primers (ARMS test)	General method for point mutations; primer design critical (*Figure 18.10*) Can be adapted to chip technology Can provide real-time quantitative readout, using TaqMan technology
Oligonucleotide ligation assay (OLA)	General method for specified point mutations (*Figure 18.11*)
PCR with primers located either side of a translocation breakpoint	Successful amplification shows presence of the suspected deletion or specified rearrangement
Check size of expanded repeat	Dynamic repeat diseases (Section 16.6.4) large expansions require Southern blots, smaller ones can be done by PCR only
Pyrosequencing	High-throughput method (*Box 18.2*)
SNAPshot	Mini sequencing by primer extension (Section 18.4.1)
Mass spectrometry	High-throughput method (Box 18.2)

Table 18.4: Examples of diseases that show a limited range of mutations

See Section 16.3 for further discussion of the reasons why some diseases show a limited range of mutations, while others have extensive allelic heterogeneity.

Disease	Cause	Comments
Sickle cell disease	Only this particular mutation produces the sickle-cell phenotype	p.E6V in *HBB* gene See *Figure 6.11*
Achondroplasia	Only G380R produces this particular phenotype; very high mutation rate	Two distinct changes, both causing p.G380R in the *FGFR3* gene (*Figure 16.9*)
Huntington disease, myotonic dystrophy	Gain of function mutations	Unstable expanded repeats See Section 16.6.4
Fragile X	Common molecular mechanism: expansion of an unstable repeat	See Section 16.6.4; other mutations occur, but are rare
Charcot–Marie–Tooth disease (HMSN1)	Common molecular mechanism: recombination between misaligned repeats	Duplication of 1.5 Mb at 17p11.2 (*Figure 16.8*); point mutations also occur
α- and β-Thalassemia	Selection for heterozygotes leads to different ancestral mutations being common in different populations	See *Figure 16.2* (α-thalassemia) and *Table 18.5* (β-thalassemia)
Tay–Sachs disease	Founder effect in Ashkenazi Jews; ancient heterozygote advantage	Two common *HEXA* mutations in Ashkenazim: 4-bp insertion in exon 11 (73%); exon 11 donor splice site G→C (15%)
Cystic fibrosis	Common ancestral mutations in northern European populations, ancient heterozygote advantage	See *Table 18.6* and Section 4.5.3

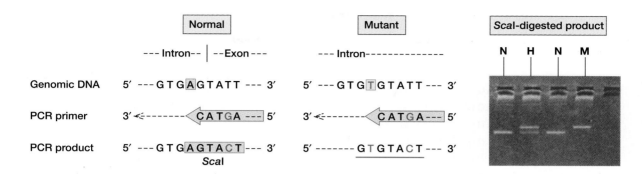

Figure 18.8: Introducing an artificial diagnostic restriction site.

An A→T mutation in the intron 4 splice site of the *FACC* gene does not create or abolish a restriction site. The PCR primer stops short of this altered base, but has a single base mismatch (red G) in a noncritical position which does not prevent it hybridizing to and amplifying both the normal and mutant sequences. The mismatch in the primer introduces an AGTACT restriction site for *Sca*I into the PCR product from the normal sequence. The *Sca*I-digested product from homozygous normal (N), heterozygous (H) and homozygous mutant (M) patients is shown. Courtesy of Dr Rachel Gibson, Guy's Hospital, London, UK.

18.4.4 Geographical origin is an important consideration for some tests

The population genetics of recessive diseases is often dominated by founder effects or the effects of heterozygote advantage. The resulting limited diversity of mutations in a population can make genetic testing much easier. β-thalassemia and cystic fibrosis are good examples. For both these conditions, a very large number of different mutations in the relevant gene have been described, but in each case a handful of mutations account for the majority of cases in any particular population. With β-thalassemia DNA testing is not needed to diagnose carriers or affected people – orthodox hematology does this perfectly well – but it is the method of choice for prenatal diagnosis. Different mutations are predominant in different populations (*Table 18.5*). Provided one has DNA samples from the parents and knows their ethnic origin, the parental mutations can often be found using only a small cocktail of specific tests, after which the fetus can be readily checked.

In cystic fibrosis the F508del mutation is the commonest in all European populations, and is believed to be of ancient origin. However, the proportion of all mutations that are F508del varies, being generally high in the north and west of Europe and lower in the south. Testing for CF mutations divides into two phases. First, a limited number of specified mutations, always including F508del, are sought using the methods described in Section 18.4. As *Table 18.6* shows, there is no obvious natural cut-off in terms of diminishing returns on testing for specific mutations. If this phase fails to reveal the mutations, then if resources allow, a screen for unknown mutations may be instituted, using the methods described in Section 18.3, or alternatively gene tracking (*Figure 18.13*) may be used. The impact of this diversity on proposals for population screening is discussed below (see *Figure 18.18*).

Surprisingly often, when a recessive disease is particularly common in a certain population, it turns out that more than one mutation is responsible. An example is Tay-Sachs disease among Ashkenazi Jews, where there are two common *HEXA* mutations (*Table 18.4*). It is difficult to explain this situation except by assuming there was substantial heterozygote advantage some time when the founding population was small.

18.5 Gene tracking

Gene tracking was historically the first type of DNA diagnostic method to be widely used. It uses knowledge of the map location of the disease locus, but not knowledge about the actual disease gene. Thus it is the only available method for genes that have been mapped but not cloned. Most of the Mendelian diseases that form the bread-and-butter work of diagnostic laboratories went through a phase of gene tracking, then moved on to direct tests once the genes were cloned. Huntington disease, cystic fibrosis and myotonic dystrophy are familiar examples. However, gene tracking may still have a role even when a gene has been cloned. In the setting of a diagnostic laboratory, it is not always cost-effective to search all through a large multi-exon gene to find every mutation. Moreover none of the mutation scanning methods described in Section 18.3 is 100% sensitive, so there are always cases where the mutation cannot be found. In these circumstances, gene tracking using linked markers is the method of choice. The prerequisites for gene tracking are:

1. the disease should be adequately mapped, so that markers can be used that are known to be tightly linked to the disease locus;

2. the pedigree structure and sample availability must allow determination of phase (see below);

3. there must be unequivocally confirmed clinical diagnoses, and no uncertainty about the map location of the disease gene.

18.5.1 Gene tracking involves three logical steps

Box 18.3 illustrates the essential logic of gene tracking. This logic can be applied to diseases with any mode of inheritance. Always there is at least one parent who could have passed on the disease allele to the proband, and who may or may not

(A) Wild-type   A G G T C G T A T C C A T G C C T T A C A G T C C A G G

A>C mutant   A G G T C G T A T C C C T G C C T T A C A G T C C A G G

Cell	Oligo	Wild-type Mismatch	Wild-type Hyb	Mutant Mismatch	Mutant Hyb
10A	A G G T C G T A T a C a T G C C T T A C	1	+	2	−
10G	A G G T C G T A T g C a T G C C T T A C	1	+	2	−
10C	A G G T C G T A T C C a T G C C T T A C	0	++	1	+
10T	A G G T C G T A T t C a T G C C T T A C	1	+	2	−
11A	G G T C G T A T C a a T G C C T T A C A	1	+	2	−
11G	G G T C G T A T C g a T G C C T T A C A	1	+	2	−
11C	G G T C G T A T C C a T G C C T T A C A	0	++	1	+
11T	G G T C G T A T C t a T G C C T T A C A	1	+	2	−
12A	G T C G T A T C C a T G C C T T A C A G	0	++	1	+
12G	G T C G T A T C C g T G C C T T A C A G	1	+	1	+
12C	G T C G T A T C C C T G C C T T A C A G	1	+	0	++
12T	G T C G T A T C C t T G C C T T A C A G	1	+	1	+
13A	T C G T A T C C a a G C C T T A C A G T	1	+	2	−
13G	T C G T A T C C a g G C C T T A C A G T	1	+	2	−
13C	T C G T A T C C a c G C C T T A C A G T	1	+	2	−
13T	T C G T A T C C a T G C C T T A C A G T	0	++	1	+

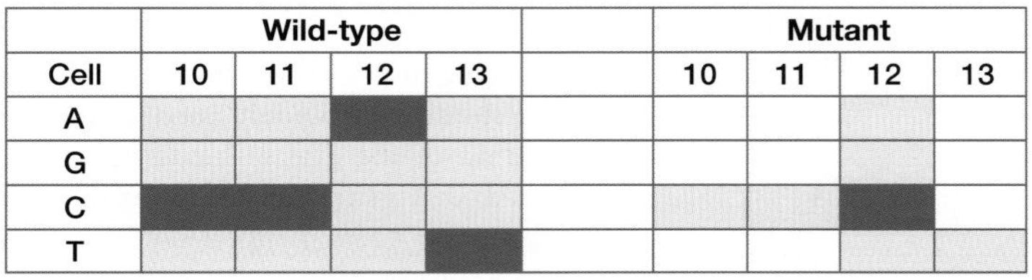

	Wild-type					Mutant			
Cell	10	11	12	13		10	11	12	13
A			■						
G									
C	■	■						■	
T			■						

(B) Wild-type   C T A G T T C G A C G A G G T C G T A T C C A T G C C T T A C A G T C C A G G

Target DNA   C T A G T T C G A C G A G G T C G T A T C C C T G C C T T A C A G T C C A G G

		Nucleotide added
Primer 1	5′ C T A G T T C G A C G A G G T C G T A 3′	ddT-red
Primer 2	5′ T A G T T C G A C G A G G T C G T A T 3′	ddC-blue
Primer 3	5′ A G T T C G A C G A G G T C G T A T C 3′	ddC-blue
Primer 4	5′ G T T C G A C G A G G T C G T A T C C 3′	ddC-blue
Primer 5	5′ T T C G A C G A G G T C G T A T C C A 3′	No label added
Primer 6	5′ T C G A C G A G G T C G T A T C C A T 3′	ddG-yellow (v. weak)
Primer 7	5′ C G A C G A G G T C G T A T C C A T G 3′	ddC-blue (weak)
Primer 8	5′ G A C G A G G T C G T A T C C A T G C 3′	ddC-blue
Primer 9	5′ A C G A G G T C G T A T C C A T G C C 3′	ddT-red
Primer 10	5′ C G A G G T C G T A T C C A T G C C T 3′	ddT-red
Primer 11	5′ G A G G T C G T A T C C A T G C C T T 3′	ddA-green

Figure 18.9: Oligonucleotide arrays for mutation detection.

(A) Principle of mutation detection by hybridization. Oligonucleotides are arrayed in sets of four, with each set corresponding to the four possible bases at a given position (shaded). Mismatches to the wild-type sequence are shown in lower-case red, correct matches in upper case blue. The mutant sequence has an A→C substitution at position 12 (red). When the wild-type or mutant sequences are hybridized to the array the number of mismatches and strength of hybridization are shown on the right. The table illustrates the appearance. *Figure 7.4* shows a real example. **(B)** Principles of a minisequencing array. Each cell of the array contains an oligonucleotide that matches part of the target sequence. Oligos are anchored by their 5′ ends. After hybridization to the target sequence they are used as primers for a single base extension, using color-labeled dideoxyNTPs. A mismatch at the 3′ end prevents extension; mismatches one or two bases in from the end give a weak reaction.

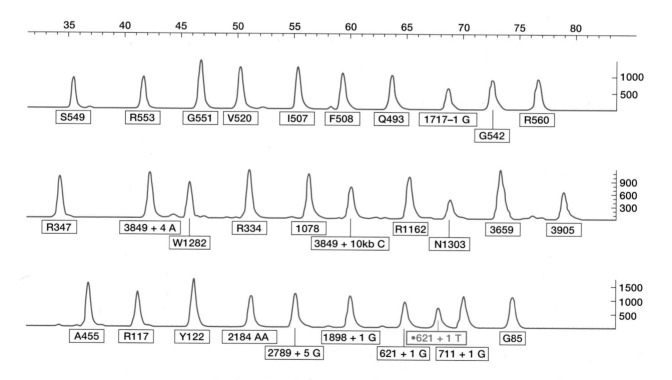

Figure 18.10: Multiplex ARMS test to detect 29 cystic fibrosis mutations.

Each sample is tested with four multiplex mutation-specific PCR reactions (A–D). Within a multiplex each product is a different size. If one of the 29 mutations is present there is an extra band. The identity of the mutation is revealed by the position of the band in the gel, and by which track contains it. Each tube also amplifies two control sequences – these are the top and bottom bands in each gel track; they differ between tracks so that each multiplex carries its own signature pattern. Note that the normal alleles of mutations other than F508del (band in lane B) are not tested for. Samples 1, 2 and 3 show no mutation-specific bands. In sample 4 the extra bands in lanes A and D show the presence of F508del and 1898+1G→A respectively, showing that this DNA comes from a compound heterozygote. Courtesy of Dr Michelle Coleman, St Mary's Hospital, Manchester, UK; data obtained using the Elucigene™ kit from Orchid Biosciences.

have actually done so. The process always follows the same three steps:

1. distinguish the two chromosomes in the relevant parent(s) – i.e. find a closely linked marker for which they are heterozygous;
2. determine phase – i.e. work out which chromosome carries the disease allele;
3. work out which chromosome the consultand received.

Figure 18.13 shows gene tracking for an autosomal recessive disease. The pedigrees emphasize the need for both an appropriate pedigree structure (DNA must be available from the affected child) and informative marker types. Even if the affected child is dead, if the Guthrie card (Section 18.2) can be retrieved, sufficient DNA for PCR typing can usually be extracted from the dried blood spot. Nowadays informativeness of the marker is not a big problem. With over 10 000 highly polymorphic microsatellites mapped across the human genome, it should always be possible to find informative markers that map close to the disease locus.

Figure 18.11: Using the oligonucleotide ligation assay to test for 31 known cystic fibrosis mutations.

After multiplex PCR, a multiple OLA is performed. Ligation oligonucleotides are designed so that products for each mutation and its normal counterpart can be distinguished by size and color of label. A ligation product from the splice site mutation 621+1g→t is seen. The person may be a carrier, or may be a compound heterozygote with a second mutation that is not one of the 31 detected by this kit. Courtesy of Dr Andrew Wallace, St Mary's Hospital, Manchester, UK; data obtained using the kit from ABI Biosystems.

Box 18.2: Two methods for high-throughput genotyping.

Pyrosequencing®. This is a method of examining very short stretches of sequence adjacent to a defined start point. The main use is for SNP typing, where only one or two bases are sequenced. Pyrosequencing uses an ingenious cocktail of enzymes to couple the release of pyrophosphate that occurs when a dNTP is added to a growing DNA chain to light emission by luciferase (Fakhrai-Rad *et al.*, 2002). The method has been developed into a machine that can automatically analyze 10 000 samples a day. Output is quantitative, so that allele frequencies of a SNP can be estimated in a single analysis of a large pooled sample.

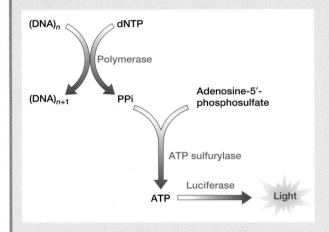

MALDI-TOF MS (matrix-assisted laser desorption/ionization time-of-flight mass spectrometry). Mass spectrometry measures the mass : charge ratio of ions by accelerating them in a vacuum towards a target, and either timing their flight or measuring how far the ions are deflected by a magnetic field (see *Box 19.7* for further details). The problem with using MS for analysis of large molecules like DNA or proteins has been to get free-flying ions. The MALDI technique solves this problem by embedding the macromolecule in a tiny spot of a light-absorbing substance, which is then vaporized by a brief laser pulse (Monforte and Becker, 1997). The time of flight to the target is proportional to the square root of the mass : charge ratio.

Applied to DNA, the technique can measure the mass up to 20 kDa with an accuracy of ± 0.3%. MS can be used as a very much faster alternative to gel electrophoresis for sizing oligonucleotides up to about 100 nucleotides long. Typical applications include analyzing the products of Sanger sequencing reactions or sizing microsatellites. For small oligonucleotides the accuracy is sufficient to deduce the base composition directly from the exact mass. Alternatively, SNPs can be analyzed by primer extension using mass-labeled ddNTPs. The spots of DNA to be analyzed can be arrayed on a plate and the machine will automatically ionize each spot in turn. Current systems can genotype tens of thousands of SNPs per day and, as with Pyrosequencing, samples can be pooled to measure allele frequencies directly.

18.5.2 Recombination sets a fundamental limit on the accuracy of gene tracking

Because the DNA marker used for gene tracking is not the sequence that causes the disease, there is always the possibility of making a wrong prediction if recombination separates the disease and the marker. The recombination fraction, and hence the error rate, can be estimated from family studies by standard linkage analysis (Chapter 13). With almost any disease there should be a good choice of markers showing less than 1% recombination with the disease locus. This follows from the observations that one nucleotide in 300 is polymorphic, and that loci 1 Mb apart show approximately 1% recombination (Section 13.1.5). Ideally one uses an intragenic marker, such as a microsatellite within an intron. The special problem of the Duchenne muscular dystrophy recombination hotspot is mentioned below.

Recombination between marker and disease can never be ruled out, even for very tightly linked markers, but the error rate can be greatly reduced by using two marker loci, situated on opposite sides of the disease locus. With such **flanking** or **bridging markers**, a recombination between either marker and the disease will also produce a marker–marker recombinant, which can be detected (e.g. III-1, *Figure 18.14*). If a marker–marker recombinant is seen in the consultand, then no prediction can be made about inheritance of the disease,

but at least a false prediction has been avoided. Provided no marker–marker recombinant is seen, the only residual risk is that of double recombinants. The true probability of a double recombinant is very low because of interference (Section 13.1.3). Thus the risk of an error due to unnoticed recombination is much smaller than the risk of a wrong prediction due to human error in obtaining and processing the DNA samples. Perhaps a greater risk is unexpected locus heterogeneity, so that the true disease locus in the family is actually different from the locus being tracked.

18.5.3 Calculating risks in gene tracking

Unlike direct mutation testing, gene tracking always involves a calculation. Factors to be taken into account in assessing the final risk include:

▶ the probability of disease–marker and marker–marker recombination;

▶ uncertainty, due to imperfect pedigree structure or limited informativeness of the markers, about who transmitted what marker allele to whom (see *Figure 13.5C* for an example);

▶ uncertainty as to whether somebody in the pedigree carries a newly mutant disease allele (see *Figure 4.8* for an example of this problem in DMD).

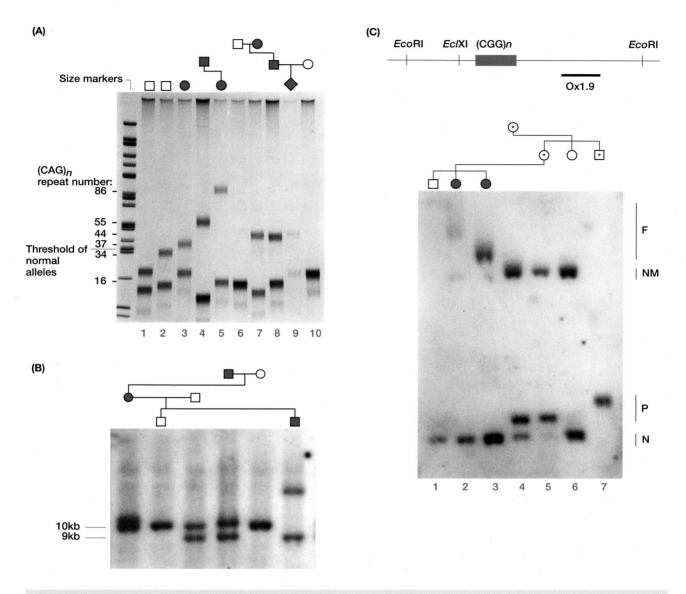

(A) Huntington disease. (B) Myotonic dystrophy. (C) Fragile X.

Figure 18.12: Laboratory diagnosis of trinucleotide repeat diseases.

(A) Huntington disease. A fragment of the gene containing the $(CAG)_n$ repeat has been amplified by PCR and run out on a polyacrylamide gel. Bands are revealed by silver staining. The scale shows numbers of repeats. Lanes 1, 2, 6 and 10 are from unaffected people, lanes 3, 4, 5, 7 and 8 are from affected people. Lane 5 is a juvenile onset case; her father (lane 4) had 45 repeats but she has 86. Lane 9 is an affected fetus, diagnosed prenatally. Courtesy of Dr Alan Dodge, St Mary's Hospital, Manchester, UK. **(B) Myotonic dystrophy.** Southern blot of DNA digested with *Eco*RI. Bands of 9 or 10 kb (arrows) are normal variants. The grandfather has cataracts but no other sign of myotonic dystrophy. His 10-kb band appears to be very slightly expanded, but this is not unambiguous on the evidence of this gel alone. His daughter has one normal and one definitely expanded 10-kb band; she has classical adult onset myotonic dystrophy. Her son has a massive expansion and the severe congenital form of the disease. Courtesy of Dr Simon Ramsden, St Mary's Hospital, Manchester, UK. **(C) Fragile X.** The DNA of the inactivated X in a female, and of any X carrying the full mutation, is methylated. The DNA is digested with a combination of *Eco*RI and the methylation-sensitive enzyme *Ecl*XI, Southern blotted and hybridized to Ox1.9 or a similar probe. The X in a normal male (lane 1) and the active normal X in a female (lanes 2, 3, 4, 6) give a small fragment (labeled N). Unmethylated premutation alleles (P) give a slightly larger band in lanes 4 and 5 (female premutation carriers) and lane 7 (a normal transmitting male). Methylated (inactive) X sequences do not cut with *Ecl*XI and give a much larger band (NM), while the fully expanded and methylated sequence gives a very large smeared band (F) because of somatic mosaicism. Courtesy of Dr Simon Ramsden, St Mary's Hospital, Manchester, UK.

Table 18.5: The main β-thalassemia mutations in different countries

In each country, certain mutations are frequent because of a combination of founder effects and selection favoring heterozygotes.

Population	Mutation	Frequency (%)	Clinical effect
Sardinia	Codon 39 (C→T)	95.7	β^0
	Codon 6 (delA)	2.1	β^0
	Codon 76 (del C)	0.7	β^0
	Intron 1-110 (G→A)	0.5	β^+
	Intron 2-745 (C→G)	0.4	β^+
Greece	Intron 1-110 (G→A)	43.7	β^+
	Codon 39 (C→T)	17.4	β^0
	Intron 1-1 (G→A)	13.6	β^0
	Intron 1-6 (T→C)	7.4	β^+
	Intron 2-745 (C→G)	7.1	β^+
China	Codon 41/42 (delTCTT)	38.6	β^0
	Intron 2-654 (C→T)	15.7	β^0
	Codon 71/72 (insA)	12.4	β^0
	−28 (A→G)	11.6	β^+
	Codon 17 (A→T)	10.5	β^0
Pakistan	Codon 8/9 (insG)	28.9	β^0
	Intron 1-5 (G→C)	26.4	β^+
	619-bp Deletion	23.3	β^+
	Intron 1-1 (G→T)	8.2	β^0
	Codon 41/42 (delTCTT)	7.9	β^0
US black African	−29 (A→G)	60.3	β^+
	−88 (C→T)	21.4	β^+
	Codon 24 (T→A)	7.9	β^+
	Codon 6 (delA)	0.8	β^0

Data courtesy of Dr J. Old, Weatherall Institute of Molecular Medicine, Oxford, UK.

Table 18.6: Distribution of CFTR mutations in 300 CF chromosomes from the North–West of England

Mutation	Exon	Frequency (%)	Cumulative frequency (%)
F508del	10	79.9	79.9
G551D	11	2.6	82.5
G542X	11	1.5	84.0
G85E	3	1.5	85.5
N1303K	21	1.2	86.7
621+1G→T	4	0.9	87.6
1898+1G→A	12	0.9	88.5
W1282X	21	0.9	89.4
Q493X	10	0.6	90.0
1154insTC	7	0.6	90.6
3849+10 kb (C→T)	Intron 19	0.6	91.2
R553X	10	0.3	91.5
V520F	10	0.3	91.8
R117H	4	0.3	92.1
R1283M	20	0.3	92.4
R347P	7	0.3	92.7
E60X	3	0.3	93.0
Unknown/private	–	7.0	100

F508del and a few of the other relatively common mutations are probably ancient and spread through selection favoring heterozygotes; the other mutations are probably recent, rare and highly heterogeneous. CF is more homogeneous in this population than in most others. See *Box 16.2* for nomenclature of mutations. Data courtesy of Dr Andrew Wallace, St Mary's Hospital, Manchester, UK.

Box 18.3: The logic of gene tracking.

Three stages in the investigation of a late-onset autosomal dominant disease where, for one reason or another, direct testing for the mutation is not possible.

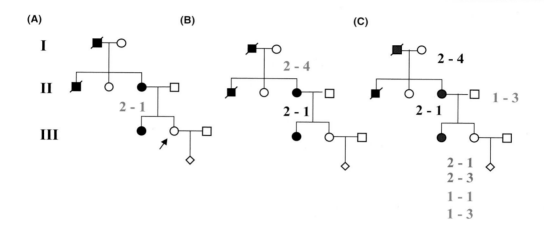

A. III-2 (arrow), who is pregnant, wishes a pre-symptomatic test to show whether she has inherited the disease allele. The first step is to tell her mother's two chromosomes apart. A marker, closely linked to the disease locus, is found for which II-3 is heterozygous.

B. Next we must establish phase – that is, work out which marker allele in II-3 is segregating with the disease allele. I-2 is typed for the marker. II-3 must have inherited marker allele 2 from her mother, which therefore marks her unaffected chromosome. Her affected chromosome, inherited from her dead father, must be the one that carries marker allele 1.

C. By typing III-2 and her father we can work out which marker allele she received from her mother. If she is 2-1 or 2-3, it is good news: she inherited marker allele 2 from her mother, which is the grandmaternal allele. If she types 1-1 or 1-3 it is bad news: she inherited the grandpaternal chromosome, which carries the disease allele.

Note that it is the segregation pattern in the family, and not the actual marker genotype, that is important: if III-2 has the same marker genotype, 2-1, as her affected mother, this is good news, not bad news for her.

Two alternative methods are available for performing the calculation.

Bayesian calculations (see *Box 18.4*)

Bayes' theorem provides a general method for combining probabilities into a final overall probability. The theory and procedure are shown in *Box 18.4*, and a sample calculation is set out in *Figure 18.15*. A very detailed set of calculations covering almost every conceivable situation in DNA diagnostics can be found in the book by Bridge (Further reading), which the interested reader should consult.

For simple pedigrees, Bayesian calculations give a quick answer, but for more complex pedigrees the calculations can get very elaborate. Few people feel fully confident of their ability to work through a complex pedigree correctly, although the attempt is a valuable mental exercise for teasing out the factors contributing to the final risk. An alternative is to use a linkage analysis program.

Using linkage programs for calculating genetic risks

At first sight it may seem surprising that a program designed to calculate lod scores can also calculate genetic risks – but in fact the two are closely related (*Figure 18.16*). Linkage analysis programs are general-purpose engines for calculating the likelihood of a pedigree, given certain data and assumptions. For calculating the likelihood of linkage we calculate the ratio:

$$\frac{\text{likelihood of data}\,|\,\text{linkage, recombination fraction } \theta}{\text{likelihood of data}\,|\,\text{no linkage } (\theta = 0.5)}$$

For estimating the risk that a proband carries a disease gene, we calculate the ratio:

$$\frac{\text{likelihood of data}\,|\,\text{proband is a carrier, recombination fraction } \theta}{\text{likelihood of data}\,|\,\text{proband is not a carrier, recombination fraction } \theta}$$

As in *Box 18.4*, the vertical line | means 'given that'.

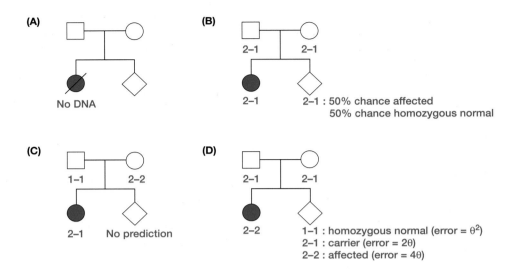

Figure 18.13: Gene tracking for prenatal diagnosis of an autosomal recessive disease.

Four families each have a child affected with a recessive disease. Direct mutation testing is not possible (either because the gene has not been cloned, or because the mutations could not be found). **(A)** No diagnosis is possible if there is no sample from the affected child. **(B)** If everybody has the same heterozygous genotype for the marker, the result is not clinically useful. **(C)** If the parents are homozygous for the marker, no prediction is possible with this marker. **(D)** Successful prediction. The error rates shown are the risk of predicting an unaffected pregnancy when the fetus is affected, or vice versa, if the marker used shows a recombination fraction θ with the disease locus. These examples emphasize the need for both an appropriate pedigree structure (DNA must be available from the affected child) and informative marker types.

18.5.4 The special problems of Duchenne muscular dystrophy

Duchenne dystrophy poses a remarkably wide range of problems for the diagnostic laboratory. Fortunately two-thirds of mutations are deletions, which are easily identified in males (*Figure 18.7*), though very challenging in females. Duplications are hard to spot in either sex and are undoubtedly under-diagnosed. The 30–35% of point mutations pose major problems. Scanning such a large gene (2.4 Mb, 79 exons) for point mutations is a daunting prospect, and therefore gene tracking is often used. However, DMD presents special problems for gene tracking because there is an extremely high recombination frequency across the gene. Even intragenic markers show an average 5% recombination with the disease. Therefore it is prudent to use flanking markers, as in *Figure 18.14*.

The problems do not end here. There is a high frequency of new mutations. The mutation–selection equilibrium calculations in *Box 4.7* show that for any lethal X-linked recessive condition (f = 0), one-third of cases are fresh mutations. Therefore the mother of an isolated DMD boy has only a two in three chance of being a carrier. This has two unfortunate consequences:

▶ it greatly complicates the risk calculations that are necessary for interpreting gene tracking results. The interested reader should consult the book by Bridge (Further reading) for example calculations;

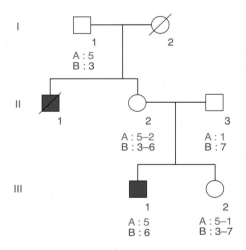

Figure 18.14: Gene tracking in Duchenne muscular dystrophy using flanking markers.

The family has been typed for two polymorphisms A and B that flank the dystrophin locus. III-2 can have inherited DMD only if she has one recombination between marker A and DMD and another between DMD and marker B. If the recombination fractions are θ_A and θ_B respectively, then the probability of a double recombinant is of the order $\theta_A\theta_B$, which typically will be well under 1%. III-1 has a recombination between marker locus A and DMD.

as shown in *Figure 4.8*, the first mutation carrier in a DMD pedigree is very often a mosaic (male or female). This raises yet more problems, both for risk estimation and for interpretation of the results of direct testing.

These factors, together with the particularly distressing clinical course of the disease, the high recurrence risk within families, and the high frequency of DMD in the population mean that DMD remains perhaps the most difficult of all diseases for genetic service providers.

18.6 Population screening

Population screening follows naturally from the ability to test directly for the presence of a mutation. Traditionally a distinction is drawn between screening and diagnosis. A screening test defines a high-risk group, who are then given a definitive diagnostic test. DNA tests are rather different because there are no separate screening and diagnostic tests. However, proposals to introduce any population screening test still need to satisfy the same criteria (*Table 18.7*), regardless of the technology used.

18.6.1 Acceptable screening programs must fit certain criteria

What would screening achieve?

The most important single function of any screening program is to produce some useful outcome. It is quite unacceptable to tell people out of the blue that they are at risk of something unpleasant, unless the knowledge enables them to do something about the risk. Proposals to screen for genes conferring susceptibility to breast cancer or heart attacks must be assessed stringently against this criterion. Predictive testing for Huntington disease might appear to break this rule – but it is offered only to people who know that they are at high risk of HD and who are suffering such agonies of uncertainty that they request a predictive test, and persist despite counseling in which all the disadvantages are pointed out.

Ideally the useful outcome is treatment, as in neonatal screening for phenylketonuria. Increased medical surveillance is a useful outcome only if it greatly improves the prognosis. One of the big risks of over-enthusiastic screening is that it can turn healthy people into ill people. A special case is screening for carrier status, where the outcome is the possibility of avoiding the birth of an affected child. People unwilling to accept prenatal diagnosis and termination of affected pregnancies would not see this as a useful outcome, and in general should not be screened, though there will be some couples who value simply knowing.

An ethical framework for screening

Ethical issues in genetic population screening have been discussed by a committee of distinguished American geneticists, clinicians, lawyers and theologians, and the reader is referred to their report for a very detailed survey (Andrews *et al.*, 1994). It is in the nature of ethical problems that they have no solutions, but certain principles emerge.

▶ Any program must be voluntary, with subjects taking the positive decision to opt in.

Box 18.4: Use of Bayes' theorem for combining probabilities.

A formal statement of Bayes' theorem is:

$$P(H_i|E) = P(H_i).P(E|H_i) / \Sigma \, [P(H_i).P(E|H_i)]$$

$P(H_i)$ means the probability of the ith hypothesis, and the vertical line means 'given', so that $P(E|H_i)$ means the probability of the evidence (E), given hypothesis Hi. An example will probably make this clearer. The steps in performing a Bayesian calculation are:

i   set up a table with one column for each of the alternative hypotheses. Cover all the alternatives;

ii assign a **prior probability** to each alternative. The prior probabilities of all the hypotheses must sum to 1. It is not important at this stage to worry about exactly what information you should use to decide the prior probability, as long as it is consistent across the columns. You will not be using all the information (otherwise there would be no point in doing the calculation because you would already have the answer) and any information not used in the prior probability can be used later;

iii using one item of information not included in the prior probabilities, calculate a **conditional probability** for each hypothesis. The conditional probability is the probability of the information, given the hypothesis, i.e. $P(E|H_i)$ [*not* the probability of the hypothesis given the information, $P(H_i|E)$]. The conditional probabilities for the different hypotheses do not necessarily sum to 1;

iv if there are further items of information not yet included, repeat step (iii) as many times as necessary until all information has been used once and once only. The end result is a number of lines of conditional probabilities in each column;

v   within each column, multiply together the prior and all the conditional probabilities. This gives a joint probability, $P(H_i).P(E|H_i)$. The joint probabilities do not necessarily sum to 1 across the columns;

vi if there are just two columns, the joint probabilities can be used directly as odds. Alternatively the joint probabilities can be scaled to give final probabilities which do sum to 1. This is done by dividing each joint probability by the sum of all the joint probabilities, $\Sigma \, [P(H_i).P(E|H_i)]$.

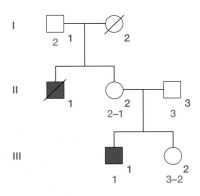

Hypothesis: III$_2$ is	A carrier	Not a carrier
Prior probability	1/2	1/2
Conditional (1): DNA result	0.05	0.95
Conditional (2): CK data	0.7	1
Joint probability	0.0175	0.475
Final probability	0.0175/0.4925	0.475/0.4925
	= 0.036	= 0.964

Figure 18.15: A Bayesian calculation of genetic risk.

III-2 wishes to know her risk of being a carrier of DMD, which affected her brother III-1 and uncle II-1. Serum creatine kinase testing (an indicator of subclinical muscle damage common in DMD carriers) gave carrier : noncarrier odds of 0.7 : 1. A DNA marker that shows on average 5% recombination with DMD gave the types shown. The risk calculation, following the guidelines in *Box 18.4*, gives her overall carrier risk as 3.6%.

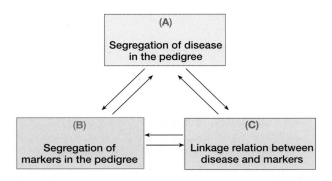

Figure 18.16: Use of linkage analysis programs for calculating genetic risks.

Given information on any two of these subjects, the program can calculate the third. For linkage analysis, the program is given **(A)** and **(B)**, and calculates **(C)**. For calculating genetic risks, the program is given **(B)** and **(C)**, and calculates **(A)**.

▶ Programs must respect the autonomy and privacy of the subject.

▶ People who score positive on the test must not be pressured into any particular course of action. For example, in countries with insurance-based health care systems, it would be unacceptable for insurance companies to put pressure on carrier couples to accept prenatal diagnosis and terminate affected pregnancies.

▶ Information should be confidential. This may seem obvious, but it can be a difficult issue – we like to think that drivers of heavy trucks or jumbo jets have been tested for all possible risks. Societies with insurance-based health care systems have particular problems about the confidentiality of genetic data, since insurance companies will argue that they are penalizing low-risk people by not loading the premiums of high-risk people.

18.6.2 Specificity and sensitivity measure the technical performance of a screening test

Compared with the ethical problems, the technical questions in population screening are fairly simple. The performance of a test can be measured by its sensitivity, specificity and positive predictive value (*Figure 18.17*).

Predictive value of a test

Unexpectedly perhaps, false positive test results can pose a more serious problem than false negatives. Even if they can be filtered out subsequently by a diagnostic test, many people will have been worried unnecessarily by false positives. *Table 18.8* shows that if a test does have a significant false positive rate, then the predictive value is hopelessly low except when testing for very common conditions. In this respect DNA tests are potentially well suited for population screening because, compared with biochemical tests that use arbitrary thresholds, they should generate very few false positives.

Sensitivity of a test

A test must pick up a reasonable proportion of its intended target (i.e. the sensitivity must be high). While the predictive value of DNA tests looks encouraging, the sensitivity usually depends on the degree of allelic heterogeneity. Unless a disease is unusually homogeneous (*Table 18.4*), it is not practicable to test for every conceivable mutation, especially in a large-throughput population screening program. Normally only a subset of mutations will be tested for. *Figure 18.18* shows how the choice of mutations can affect the outcome of a CF carrier screening program.

It is clear that simply testing for the commonest mutation, F508del, would not produce an acceptable program. More affected children would be born to couples negative on the screening program than would be detected by the screening. Whether or not such a program were financially cost-effective, it would surely be socially unacceptable. What constitutes an acceptable program is harder to define. One suggestion focuses on '+/−' couples (i.e. couples with one known carrier and the partner negative on all the tests). The

Table 18.7: Requirements for a population screening program

Requirement	Examples and comments
A positive result must lead to some useful action	▶ Preventive treatment, e.g. special diet for PKU
	▶ Review and choice of reproductive options in CF carrier screening
The whole program must be socially and ethically acceptable	▶ Subjects must opt in with informed consent
	▶ Screening without counseling is unacceptable
	▶ There must be no pressure to terminate affected pregnancies
	▶ Screening must not be seen as discriminatory
The test must have high sensitivity and specificity	▶ Tests with many false negatives undermine confidence in the program
	▶ Tests with many false positives, even if these are subsequently filtered out by a definitive diagnostic test, can create unacceptably high levels of anxiety among normal people
The benefits of the program must outweigh its costs	▶ It is unethical to use limited health care budgets in an inefficient way

partner still might be a carrier of a rare mutation. Ten Kate suggested that an acceptable screening program is one in which the risk for such +/− couples is no higher than the general population risk before screening. For CF screening of Northern Europeans, that would require a sensitivity of about 95%.

18.6.3 Organization of a genetic screening program

Assuming the proposed program looks ethically acceptable and cost-effective, who should be screened? Three examples highlight some of the options.

Neonatal screening: screening for phenylketonuria

All babies in the UK are tested a few days after birth for PKU. A blood spot from a heel-prick is collected on a card (the Guthrie card) during a home visit and sent to a central laboratory. The phenylalanine level in the blood is measured by chromatography or a bacterial growth test. This is the screening test. Babies whose level is above a threshold are called in for a definitive diagnostic test. Only a small proportion eventually turn out to have PKU. The lack of informed consent is justified by the benefit the baby receives from dietary treatment (Smith, 1993).

Prenatal screening: screening for β-thalassemia

Carriers of β-thalassemia can be detected by conventional hematological testing, either before marriage or in the antenatal clinic. Carrier–carrier couples can be offered prenatal diagnosis by DNA analysis. Two ethnic groups in the UK have a high incidence of β-thalassemia: Cypriots and Pakistanis. Screening was quickly accepted by Cypriots in the UK but uptake has been slower among Pakistanis. The comparison illustrates the complex social questions surrounding genetic screening and the relevance of cultural background (Gill and Modell, 1998). Importantly, long-term studies of the Cypriot community show how the success of screening can be measured, not by counts of affected fetuses aborted, but by counts of couples having normal families. Before screening was available, many Cypriot carrier couples opted to have no children; now they are using screening and having normal families (Modell *et al.*, 1984). Pre-implantation diagnosis or fetal stem-cell transplants may one day become alternatives to termination of affected pregnancies, at least for rich people in rich countries.

Population screening for carriers: proposals for cystic fibrosis

It is now technically feasible and financially worthwhile to screen northern European populations to detect CF carriers. Surveys in the UK suggest that most carrier–carrier couples would opt for prenatal diagnosis and would value the opportunity to ensure that they did not have affected children. This view might change if treatment becomes more effective, for example using gene therapy.

If a screening program is to be introduced, two sets of questions must be considered. How many mutations should the laboratory test for, and who should be offered the test? The

	Affected	Not affected
+ve on test	a	b
−ve on test	c	d

Sensitivity of test = a/(a + c)
Specificity of test = d/(b + d)
Positive predictive value = a/(a + b)

Figure 18.17: Sensitivity and specificity of a screening test.

Table 18.8: A test that performs well in the laboratory may be useless for population screening

Prevalence of condition	True positives in population screened	True positives detected by screening	True negatives in population screened	False positives detected by screening	Predictive value of test
1/1000	1000	990	999 000	9990	0.09
1/10000	100	99	999 900	9999	0.0098
1/100000	10	10	999 990	10 000	0.001

In a laboratory trial on a panel of 100 affected and 100 control people, this hypothetical test was 99% accurate: it gave a positive result for 99% of true positives, and a negative result for 99% of true negatives. The table shows results of screening 1 million people. The great majority of all people positive on the test are false positives. Such a test is unlikely to be socially acceptable or financially viable for any Mendelian disease (these typically affect less than 1 person in 1000).

problems raised by allelic heterogeneity have been discussed above (see *Figure 18.18*). On the question of who to screen, *Table 18.9* shows some possibilities considered in the UK. Naturally the way health care delivery is organized in each country will determine the range of possibilities. Preliminary results from controlled pilot studies suggest that none of the methods has had the negative effects (increased anxiety) sometimes predicted.

18.7 DNA profiling can be used for identifying individuals and determining relationships

We use the term *DNA profiling* to refer to the general use of DNA tests to establish identity or relationships. *DNA fingerprinting* is reserved for the technique invented by Jeffreys *et al.* (1985) using multilocus probes. For more detail on this whole area, the reader should consult the book by Evett and Weir (1998; see Further reading).

18.7.1 A variety of different DNA polymorphisms have been used for profiling

DNA fingerprinting using minisatellite probes

These probes contain the common core sequence of a hyper-variable dispersed repetitive sequence GGGCAGGAXG discovered by Jeffreys *et al.* (1985) in the myoglobin gene. The sequence is present in many minisatellites spread around the genome, at each of which the number of tandem repeats varies between individuals. When hybridized to Southern blots the probes give an individual-specific fingerprint of bands (*Figure 18.19*). DNA fingerprinting revolutionized forensic practice, but is now obsolete because of two problems:

▶ the Southern blot procedure requires several micrograms of DNA, corresponding to the content of perhaps a million cells;

▶ it is not possible to tell which pairs of bands in a fingerprint represent alleles. Thus, when comparing two DNA fingerprints, the investigator matches each band

individually by position and intensity. The continuously variable distance along the gel has to be divided into a number of 'bins'. Bands falling within the same bin are deemed to match. Then if, say, 10/10 bands match, the odds that the suspect, rather than a random person from the population, is the source of the sample, are $1 : p^{10}$, where p is the chance that a band in a random person would match a given band (we have simplified by assuming p is the same for every bin and ignoring the need to match on intensity as well as position). Even for $p = 0.2$, p^{10} is only 10^{-7}. It is imperative that the same binning criteria are used for judging matches between two profiles and for calculating p. The criteria can be arbitrary within certain limits, but they must be consistent.

DNA profiling using microsatellite markers

Single-locus profiling uses microsatellite polymorphisms (Section 9.4.3), usually tri- or tetranucleotide repeats, that can be typed by PCR. In theory, even a single cell left at a scene of crime can be typed. Alleles can be defined unambiguously by the precise repeat number, thus avoiding the binning problem. If the gene frequency of each allele in the population is known, an exact calculation can be made of the probability of paternity, of the suspect not being the rapist, etc. The combined genotypes at 10–15 unlinked highly polymorphic loci are usually sufficiently unique to be definitive, one way or the other. Minor variations within repeated units of some microsatellites potentially allow an almost infinite variety of alleles to be discriminated, so that the genotype at a single locus might suffice to identify an individual (Jeffreys *et al.*, 1991), but this 'MVA typing' method remains a research tool.

The use of Y-chromosome and mitochondrial polymorphisms

For tracing relationships to dead persons, Y-chromosome and mitochondrial DNA polymorphisms are especially useful because in each case an individual inherits the complete genotype from a single definable ancestor. An interesting example was the identification of the remains of the Russian

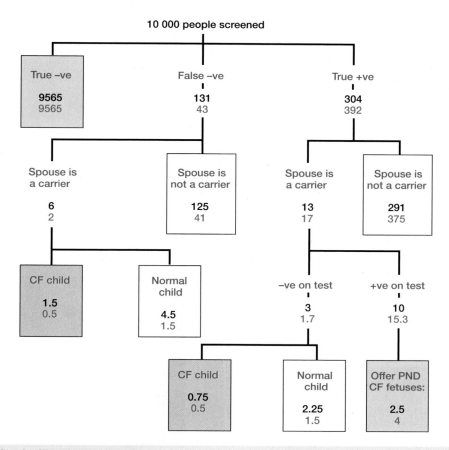

Figure 18.18: Flowchart for CF population screening.

Results of screening 10 000 people, 1 : 23 of whom is a carrier. If a person tests positive, his/her spouse is then tested. Black figures show results using a test which detects 70% of CF mutations (i.e. testing for F508del only); red figures show results for a test with 90% sensitivity. Pink boxes represent cases which would be seen as successes for the screening program (regardless of what action they then take), gray boxes represent failures. PND, Prenatal diagnosis.

Table 18.9: **Possible ways of organizing population screening for carriers of CF**

Group tested	Advantages	Disadvantages
Neonates	▸ Easily organized	▸ No consequences for 20 years ▸ Many families would forget the result ▸ Unethical to test children
School leavers	▸ Easily organized ▸ Inform people before they start relationships	▸ Difficult to conduct ethically ▸ Risk of stigmatization of carriers
Couples from physician's lists	▸ Couple is unit of risk ▸ Stresses physician's role in preventative medicine. ▸ Allows time for decisions	▸ Difficult to control quality of counseling
Women in antenatal clinic	▸ Easily organized ▸ Rapid results	▸ Bombshell effect for carriers ▸ Partner may be unavailable ▸ Time pressure on laboratory
Adult volunteers ('drop-in CF center')	▸ Few ethical problems	▸ Bad framework for counseling ▸ No targeting to suitable users ▸ Inefficient use of resources?

tsar and his family, killed by the Bolsheviks in 1917, by comparing DNA profiles of excavated remains with living distant relatives (Gill *et al.*, 1994).

18.7.2 DNA profiling can be used to determine the zygosity of twins

In studying non–Mendelian characters (Chapter 15), and sometimes in genetic counseling, it is important to know whether a pair of twins are monozygotic (MZ, identical) or dizygotic (DZ, fraternal). Traditional methods depended on an assessment of phenotypic resemblance or on the condition of the membranes at birth (twins contained within a single chorion are always MZ, though the converse is not true). Errors in zygosity determination systematically inflate heritability estimates for non-Mendelian characters, because very similar DZ twins are wrongly counted as MZ, while very different MZ twins are wrongly scored as DZ.

Genetic markers provide a much more reliable test of zygosity. The extensive but now obsolete literature on using blood groups for this purpose is summarized by Race and Sanger (see Further reading). DNA profiling is nowadays the method of choice. The Jeffreys fingerprinting probe provides an immediate impression: samples from MZ twins look like the same sample loaded twice, and samples from DZ twins show differences. When single-locus markers are used, if twins give the same types, then for each locus the probability that DZ twins would type alike is calculated. If the parents have been typed, this follows from Mendelian principles; otherwise the probability of DZ twins typing the same must be calculated for each possible parental mating and weighted by the probability of that mating calculated from population gene frequencies. The resultant probabilities for each (unlinked) locus are multiplied, to give an overall likelihood P_I that DZ twins would give the same results with all the markers used. The probability that the twins are MZ is then:

$$P_m = m \, / \, [m + (1-m)P_I]$$

where m is the proportion of twins in the population who are MZ (about 0.4 for like-sex pairs). Sample calculations are given in Appendix 4 of Vogel and Motulsky (Further reading).

18.7.3 DNA profiling can be used to disprove or establish paternity

Excluding paternity is fairly simple – if the child has a marker allele not present in either the mother or alleged father then, barring new mutations, the alleged father is not the biological father. Proving paternity is in principle impossible – one can never prove that there is not another man in the world who could have given the child that particular set of marker alleles. All one can do is establish a probability of nonpaternity that is low enough to satisfy the courts and, if possible, the putative father.

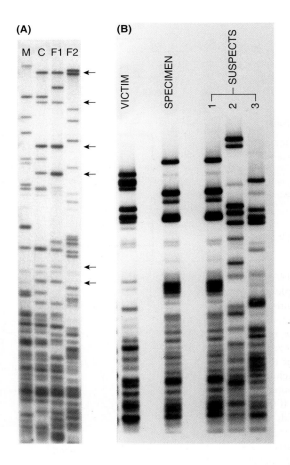

Figure 18.19: Legal and forensic use of DNA fingerprinting.

DNA fingerprinting revolutionized forensic practice, although it has now been superseded by profiling using multiple single-locus polymorphisms. **(A)** A paternity test. Fingerprints are shown from the mother (M), child (C) and two possible fathers (F1, F2). The DNA fingerprint of F1 contains all the paternal bands found in the child, whereas that of F2 contains only one of the paternal bands. **(B)** A rape case. The fingerprint of suspect 1 exactly matches that from the semen sample S on a vaginal swab from the victim. As a result of this evidence, Suspect 1 was charged with rape and found guilty. Photograph courtesy of Cellmark Diagnostics, Abingdon, Oxfordshire, UK.

DNA fingerprinting probes have been widely used for this purpose (*Figure 18.19*). Bands must be binned according to an arbitrary but consistent scheme, as explained above, to decide whether or not each nonmaternal band in the child fits a band in the alleged father. Single-locus microsatellites allow a more explicit calculation of the odds (*Figure 18.20*). A series of 10 unlinked highly polymorphic single-locus markers gives overwhelming odds favoring paternity if all the bands fit.

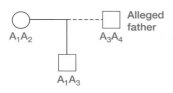

Figure 18.20: Using single-locus markers for a paternity test.

The odds that the alleged father, rather than a random member of the population, is the true father are $1/2 : q_3$, where q_3 is the gene frequency of A3. A series of n unlinked markers would be used and, if paternity were not excluded, the odds would be $(1/2)^n : q_A.q_B.q_C...q_N$.

18.7.4 DNA profiling is a powerful tool for forensic investigations

DNA profiling for forensic purposes follows the same principles as paternity testing. Scene-of-crime material (bloodstains, hairs or a vaginal swab from a rape victim) are typed and matched to a DNA sample from the suspect. One of the most powerful applications of DNA profiling is for preventing miscarriages of justice by proving that a suspect is not the criminal. If the samples don't match, the suspect is excluded, regardless of any circumstantial evidence to the contrary.

If the genotypes do all match, the court needs to know the odds that the criminal is the suspect rather than a random member of the population. Of course, if the alternative were the suspect's brother or his identical twin, the odds would look very different. The fate of DNA evidence in courts provides a fascinating insight into the difference between scientific and legal cultures. Suppose a suspect's DNA profile matches the scene-of-crime sample. There are still at least three obstacles to a rational use of the DNA data.

▶ The jury may simply not believe, or perhaps choose to ignore, the DNA data, as evidently happened in the O.J. Simpson trial [see Weir (1995) for a fascinating account of that case]. Maybe they will decide the incriminating DNA was planted.

▶ An unscrupulous lawyer may try to lead the jury into a false probability argument, the so-called Prosecutor's Fallacy. This consists of confusing the probability the suspect is innocent, given the match, with the probability of a match, given that the suspect is innocent. The jury should consider the first probability, not the second. As *Box 18.5* shows, the two are very different.

▶ Objections may be raised to some of the principles by which DNA-based probabilities are calculated:

(i) The *multiplicative principle*, that the overall probabilities can be obtained by multiplying the individual probabilities for each allele or locus, depends on the assumption that genotypes are independent. If the population actually consisted of reproductively isolated groups, each of whom had genotypes that were fairly constant within a group but very different between groups, the calculation would be misleading. This is serious because it is the multiplicative principle that allows such exceedingly definite likelihoods to be given;

(ii) for single-locus markers, the probability depends on the gene frequencies. DNA profiling laboratories maintain databases of gene frequencies – but were these determined in an appropriate ethnic group for the case being considered?

Box 18.5: The Prosecutor's Fallacy.

The suspect's DNA profile matches a sample from the scene of crime. Does that make him guilty? Consider two different probabilities:

▶ the probability the suspect is innocent, given the match;

▶ the probability of a match, given that the suspect is innocent.

Using Bayesian notation (*Box 18.4*) with M = match, G = suspect is guilty, I = suspect is innocent, the first probability is $P_I|M$, and the second is $P_M|I$. The Prosecutor' Fallacy consists of arguing that the relevant probability is $P_M|I$ when in fact it is $P_I|M$. The calculation below shows how different these two probabilities are.

If the suspect were guilty, the samples would necessarily match: $P_M|G = 1$. Let us suppose that population genetic arguments say there is a 1 in 10^6 chance that a randomly selected person would have the same profile as the crime sample: $P_M|I = 10^{-6}$. Suppose the guilty person could have been any one of 10^7 men in the population. If there is no other evidence to implicate him, he is simply a random member of the population and the prior probability he is guilty (before considering the DNA evidence) is $P_G = 10^{-7}$. The prior probability he is innocent is $P_I = 1 - 10^{-7}, \approx 1$.

Bayes theorem tells us that

$$P_I|M = (P_I . P_M|I) / [(P_I.P_M|I) + (P_G.P_M|G)]$$
$$= 10^{-6} / (10^{-6} + 10^{-7})$$
$$= 1.0 / 1.1$$
$$= 0.9$$

We already saw that

$$P_M|I = 10^{-6} \text{ — quite a difference!}$$

Courts make fools of themselves by ignoring compelling DNA evidence – but this calculation also shows that DNA evidence on its own could not safely convict somebody when there is no other evidence against them, unless $P_M|I$ were well below 10^{-6}. Plans to screen all men in a large town to find a rapist need to take account of this.

Taken to extremes, the argument about the independence of genotypes implies that the DNA evidence might identify the criminal as belonging to a particular ethnic group, but would not show which member of the group it was who committed the crime. These issues have been debated at great length, especially in the American courts. The argument is valid in principle, but the question is whether it makes enough difference in practice to matter. It has become clear that in general it does not. It would be ironic if courts, seeing opposing expert witnesses giving odds of correct identification differing a million-fold ($10^5 : 1$ versus $10^{11} : 1$), were to decide that DNA evidence is hopelessly unreliable, and rely instead on eye-witness identification (odds of correct identification $50 : 50$).

Further reading

Bridge PJ (1997) *The Calculation of Genetic Risks – Worked Examples in DNA Diagnostics*, 2nd Edn. Johns Hopkins University Press, Baltimore, MD.

Clinical Molecular Genetics Society Best Practice Guidelines for Molecular Genetics Services. http://www.cmgs.org

Cotton RGH, Edkins E, Forrest S (eds) (1998) *Mutation Detection: a Practical Approach*. Oxford: IRL Press.

Elles RG, Mountford R (eds) (2003) *Molecular Diagnosis of Genetic Diseases*, 2nd Edn. Humana Press, Totowa, NJ.

Evett IW, Weir BS (1998) *Interpreting DNA Evidence: Statistical Genetics for Forensic Scientists*. Sinauer.

Race RR, Sanger R (1975) *Blood Groups in Man*, 6th Edn. Blackwell, Oxford.

Vogel F, Motulsky AG (1996) *Human Genetics*, 3rd Edn. Springer, Berlin.

References

Andrews LB, Fullarton JE, Holtzman NA, Motulsky AG (1994) *Assessing Genetic Risks – Implications for Health and Social Policy*. National Academy Press, Washington, DC.

Armour JAL, Sismani C, Patsalis PC, Cross G (2000) Measurement of locus copy number by hybridisation with amplifiable probes. *Nucl. Acids Res.* **28**, 605–609.

Collins JS, Schwartz CE (2002) Detecting polymorphisms and mutations in candidate genes. *Am. J. Hum. Genet.* **71**, 1251–1252.

Fakhrai-Rad H, Pourmand N, Ronaghi M (2002) Pyrosequencing: an accurate detection platform for single nucleotide polymorphisms. *Hum. Mutat.* **19**, 479–485.

Gill P, Ivanov PL, Kimpton C et al. (1994) Identification of the remains of the Romanov family by DNA analysis. *Nature Genet.* **6**, 130–135.

Gill PS, Modell B (1998) Thalassaemia in Britain: a tale of two communities. Births are rising among British Asians but falling in Cypriots. *Br. Med. J.* **317**, 761–762.

Jeffreys AJ, Wilson V, Thein LS (1985) Individual-specific fingerprints of human DNA. *Nature* **314**, 67–73.

Jeffreys AJ, MacLeod A, Tamaki K, Neil DL, Monckton DG (1991) Minisatellite repeat coding as a digital approach to DNA typing. *Nature* **354**, 204–209.

Keen J, Lester D, Inglehearn C, Curtis A, Bhattacharya S (1991) Rapid detection of single base mismatches as heteroduplexes on Hydrolink gels. *Trends Genet.* **7**, 5.

Modell B, Petrou M, Ward RH et al. (1984) Effect of fetal diagnostic testing on birth-rate of thalassaemia major in Britain. *Lancet* **ii**, 1383–1386.

Monforte JA, Becker CH (1997) High-throughput DNA analysis by time-of-flight mass spectrometry. *Nature Med.* **3**, 360–362.

Nickerson DA, Kaiser R, Lappin S, Stewart J, Hood L (1990) Automated DNA diagnostics using an ELISA-based oligonucleotide ligation assay. *Proc. Natl Acad. Sci. USA* **87**, 8923–8927.

Sarkar G, Yoon HS, Sommer SS (1992) Dideoxy fingerprinting (ddF): a rapid and efficient screen for the presence of mutations. *Genomics* **13**, 441–443.

Sheffield VC, Beck JS, Nichols B, Cousineau A, Lidral A, Stone EM (1992) Detection of multiallele polymorphisms within gene sequences by GC-clamped denaturing gradient gel electrophoresis. *Am. J. Hum. Genet.* **50**, 567–575.

Sheffield VC, Beck JS, Kwitek AE, Sandstrom DW, Stone EM (1993) The sensitivity of single strand conformational polymorphism analysis for the detection of single base substitutions. *Genomics* **16**, 325–332.

Smith I and MRC Working Party on Phenylketonuria (1993) Phenylketonuria due to phenylalanine hydroxylase deficiency: an unfolding story. *Br. Med. J.* **306**, 115–119.

Thomassin H, Oakeley EJ, Grange T (1999) Identification of 5-methylcytosine in complex genomes. *Methods* **19**, 465–475.

Tõnisson N, Zernant J, Kurg A et al. (2002) Evaluating the arrayed primer extension resequencing assay of TP53 tumor suppressor gene. *Proc. Natl Acad. Sci. USA* **99**, 5503–5508.

van der Luijt R, Khan PM, Vasen H et al. (1994) Rapid detection of translation-terminating mutations at the adenomatous polyposis coli (*APC*) gene by direct protein truncation test. *Genomics* **20**, 1–4.

Wallace AJ, Wu CL, Elles RG (1999) Meta-PCR: a novel method for creating chimeric DNA molecules and increasing the productivity of mutation scanning techniques. *Genet. Test.* **3**, 173–183.

Weaver TA (2000) High-throughput SNP discovery and typing for genome-wide genetic analysis. *New Technologies for the Life Sciences: a Trends Guide* (Supplement issued for *Trends* journals) 36–42.

Weir BS (1995) DNA statistics in the Simpson matter. *Nature Genet.* **11**, 365–368.

White S, Kalf M, Liu Q et al. (2002) Comprehensive detection of genomic duplications and deletions in the DMD gene, by use of multiplex amplifiable probe hybridization. *Am. J. Hum. Genet.* **71**, 365–374.

PART FOUR

New horizons: into the 21st century

CHAPTER NINETEEN

Beyond the genome project: functional genomics, proteomics and bioinformatics

Chapter contents

19.1 An overview of functional genomics

19.1.1 The information obtained from the structural phase of the Human Genome Project is of limited use without functional annotation

The ultimate goal of the structural phase of the Human Genome Project was to provide the complete genome sequence, all three billion base pairs of it! As discussed in Chapter 8, two draft sequences were published in February 2001 (International Human Genome Sequencing Consortium, 2001; Venter *et al.*, 2001) and completed sequences were available by mid 2003. While this was undoubtedly an unparalleled achievement in biology, the sequences by themselves are not particularly informative. Essentially, they are just long strings of the four letters A, C, G and T. In order to find out how this genome sequence helps to construct a functioning human being, the sequence data must be *mined* to extract useful information.

One of the first post-sequencing tasks in any genome project is the identification of all the genes since these represent the major functional units of the genome. In the case of the Human Genome Project, many genes had already been identified in previous studies. Others were predicted on the basis of their structure, their conservation in other genomes, or the fact that they matched expressed sequence tags (ESTs) (Section 8.3.5). The exact number of human genes is still not known, but most estimates now agree on a figure of approximately 30 000. We are therefore a long way from assembling a complete and accurate **gene catalog** of the human genome. However, even if such a resource were available, it would represent nothing more than a list of components. Before we can begin to understand how those components build a human being, we must understand what they do. The next task is therefore to determine the precise functions of each of the genes in the human genome, a process known as **functional annotation**.

19.1.2 The functions of individual genes can be described at the biochemical, cellular and whole-organism levels

The function of a gene is actually the function of its product or products. Most genes encode proteins but a significant majority (see *Figure 9.4*) produce noncoding RNA molecules. The functions of human gene products can be classified at three levels.

▶ The biochemical level. For example, a protein may be described as a kinase or a calcium-binding protein. This reveals little about its wider role in the organism.

▶ The cellular level. This builds in information about intracellular localization and biological pathways. For example, it may be possible to establish that a protein is located in the nucleus and is required for DNA repair even if its precise biochemical function is unknown.

▶ The organism level. This may include information about where and when a gene is expressed and its role in disease. For example, the *HD* gene produces two major transcripts, one of which is enriched in the brain and encodes a protein called huntingtin. Mutations that increase the number of glutamine residues in this protein beyond a certain limit cause Huntington's disease (Section 16.6.4). The precise biochemical and cellular functions of the gene product in healthy people remain to be established.

In order to obtain a complete picture of human gene function, information is required at all three levels. The functional classification of the human gene for glucokinase is shown as an example in *Box 19.1*. A major recent development in functional genomics is the establishment of defined vocabularies to describe gene function across all genomes. An example is the Gene Ontology system (Gene Ontology Consortium, 2000, 2001; see http://www.geneontology.org/). Other useful systems include those employed by the Kyoto Encyclopedia of Genes and Genomes (KEGG; http://www.genome.ad.jp/kegg/) and the oldest and most widely-used system of all, the Enzyme Commission system for enzyme classification.

19.1.3 Functional relationships among genes must be studied at the levels of the transcriptome and proteome

Even if every gene in the genome could be identified and assigned a function, this still does not show how the gene products coordinate the biological activities required to make a living human being. An analogous situation might be the detailed functional description of all the components of a car. This bolt holds the engine to the chassis, this is part of the steering column, this is an electrical switch that operates the indicator lights, but how does the car actually work? In order to appreciate how the functions of thousands of genes combine to generate a human being, it is necessary to study the gene products directly. Arguably, the grand purpose of molecular biology over the last 30 years has been to determine gene functions and link genes into pathways and networks in order to explain how living things work. However, there has been a recent shift in focus from the **reductionist approach** of studying genes and their products one at a time to a **holistic approach** in which many or indeed all gene products are studied simultaneously. This *global* analysis of gene function is the basis of **functional genomics**.

A central concept in functional genomics is the expression of the genome to produce the transcriptome and the proteome. The **transcriptome** is the complete collection of mRNAs in a particular cell, and represents the combined output of transcription, RNA processing and RNA turnover (Velculescu *et al.*, 1997). This is essential in defining the **proteome**, the complete collection of proteins in a particular cell (Wasinger *et al.*, 1995). It is important to realize that the transcriptome and proteome are much more complex than the genome. A single gene can produce many different mRNAs by alternative splicing, alternative promoter or polyadenylation site usage, and special processing strategies like RNA editing (Section 10.3.3). The proteins synthesized from these mRNAs can be modified in various different ways, e.g. by proteolytic cleavage, phosphorylation or glycosylation. Unlike the genome, which is identical in most cells, the transcriptome and

---

Box 19.1: The function of glucokinase.

Gene name: *GCK*

Position in the genome: 7p15

Protein name: glucokinase

Biochemical function: kinase, substrate glucose

Cellular function: glucose metabolism, glycolysis pathway

Organism function: expressed specifically in pancreatic β-cells and hepatocytes, primary regulator of glucose-controlled insulin secretion, loss of function mutations cause diabetes, gain of function mutations can cause hyperinsulinism.

As is the case for many proteins, the biochemical and cellular functions of glucokinase do not reveal its important regulatory role at the whole organism level. Indeed several other enzymes with identical biochemical and cellular activities are encoded in the human genome, but each has a distinct expression pattern and a different higher-level function. Conversely, the expression pattern and disease data indicate the importance of glucokinase in the regulation of insulin production and blood sugar levels, but do not identify its precise biochemical activity.

---

proteome are highly variable. Transcription, RNA processing, protein synthesis and protein modification may all be regulated so that the transcriptome and proteome differ significantly among different cell types and in response to changes in the cell's environment. Analysis at the level of the transcriptome or proteome provides a snapshot of the cell in action, showing the abundance of all the RNAs and proteins under a particular set of circumstances. By understanding the most important features of the transcriptome and proteome in different cell types, and by studying how they change in health and in disease, it becomes possible to build the individual functions of genes into a bigger picture.

19.1.4 High-throughput analysis techniques and bioinformatics are the enabling technologies of functional genomics

Functional genomics has benefited from the development of a whole range of novel experimental strategies to investigate gene function on a global scale. In some cases, existing technology has been adapted for high-throughput analysis. For example, simple gene-by-gene hybridization techniques (Section 7.3.2) have been replaced by DNA arrays and sequence-sampling methods that can be used for global expression profiling. In other cases, entirely new technologies have been invented, such as interaction screening using the yeast two-hybrid system (see below). Because these experiments generate large datasets, bioinformatic support is essential and the functional genomics revolution has been driven as much by the development of new algorithms and the establishment of new databases as it has by experimental advances. Bioinformatics is required for the comparison of sequences and structures, the modeling of structures and interactions, the analysis of global expression data and the sharing of information by means of accessible, user-friendly databases. Existing bioinformatic techniques have been modified for new uses (e.g. clustering algorithms used in phylogenetic analysis have been adapted to mine gene expression data) and new techniques have been developed for specific applications (e.g. algorithms that search protein databases using mass spectrometry data). We discuss these novel technologies and their applications in the remainder of this chapter.

19.2 Functional annotation by sequence comparison

19.2.1 Tentative gene functions can be assigned by sequence comparison

Functional annotation by homology searching is an extension of gene finding

A variety of experimental methods can be used to detect genes in genomic DNA (see Section 7.2). However, because of the large amount of sequence data produced in the genome projects, initial annotation is carried out using computer algorithms that can process the sequence very rapidly. As discussed in Section 8.3.5, such algorithms either predict the existence of genes from first principles or identify sequences with homology to known genes by searching through databases. The methods for doing this are described in more detail in *Box 8.7*.

Homologous genes have similar sequences because they are derived from a common evolutionary ancestor. An evolutionary relationship usually indicates that the two sequences are related not only in structure, but also in function. The simplest way to assign a function to a new gene is therefore to look for *related sequences that have already been annotated*. Generally, comparisons are carried out at the protein level because amino acid sequences are more constrained in evolutionary terms than nucleotide sequences, so protein sequence comparisons are more sensitive.

The robustness of functional annotation based on homology searching depends on many factors, including the degree of similarity between the query sequence and any database hits, the reliability of functional information already in the databases, and the degree to which conserved sequence and structure corresponds to conserved function. A particular query sequence may return a number of matches with different degrees of similarity. In some cases, it may be possible to align matching sequences over their entire lengths, which indicates the sequences have diverged by the accumulation of point mutations alone (*Figure 19.1*). If the sequences are very similar, they might represent homologous genes that carry out identical functions in different species, and which have accumulated mutations due to speciation. As discussed in *Box 12.3*, such genes are known as **orthologs** and an example might be the human and sheep β-globin genes. Functional annotations based on orthologs can be very accurate. A lower degree of similarity might indicate that the genes are homologous but have diverged in functional terms. Such genes are known as **paralogs**, and arise by gene duplication and divergence within a genome (see *Box 12.3*). The human genes for myoglobin and β-globin are examples of paralogs. In such cases, functional predictions may be reliable at the biochemical level (both proteins are oxygen carriers) but their specific cellular and organism-level functions might be very different. It is generally the case that greater structural similarity implies greater functional similarity and that biochemical function is more highly conserved than cellular and organism-level function.

In many cases, database searches do not return hits that match the query sequence over its entire length. Instead, partial alignments are identified in a number of proteins that appear otherwise unrelated (*Figure 19.2*). This reflects the modular nature of proteins and the fact that distinct functions can be carried out by different protein domains (Section 12.1.3). The matching genes have not diverged simply by the accumulation of point mutations, but also by more complex events such as recombination between genes and gene segments leading to exon shuffling (Section 12.1.3). Human proteins involved in blood clotting provide a useful example of this process (Kolkman and Stemmer, 2001; see *Figure 12.3*).

There are several pitfalls to homology-based functional annotation

The principle of functional annotation based on homology searching is that conserved structure always reflects conserved function. However, there are several instances when this assumption is not safe (Orengo *et al.*, 1999):

▶ the presence of **low complexity sequences**, i.e. sequences that are present in many proteins with extremely diverse functions. For example, transmembrane domains, dimerization domains;

▶ **multifunctional sequences**, i.e. sequences that carry out different functions in different proteins. For example, sequences that form an α/β hydrolase fold are found at the catalytic site of six different classes of enzyme as well as in the cell adhesion molecule neurotactin;

▶ **gene recruitment**, i.e. the acquisition of a new function by an existing gene product. For example, many enzymes involved in mundane metabolic processes have been recruited as **crystallins**, the refractive proteins in the lens of the eye. This has been achieved simply by modifying their expression levels but other mechanisms of recruitment include changing the way proteins form complexes and changing their intracellular localization.

Figure 19.2: Similarity searching with a given protein sequence may also identify homologous proteins showing partial alignments with the query sequence.

In this example, the query protein (Q) comprises three distinct domains shown in different colors. The responses may share one, two or all three of these domains, but may also possess additional domains that are not present in the query (red boxes). There may be domain duplications as seen in the first and third responses. The final response shows the special case of permutation, where the order of domains is changed. This generally reflects gene duplication followed by gene fusion and end erosion. Accumulating point mutations relative to the query sequence are shown as white lines.

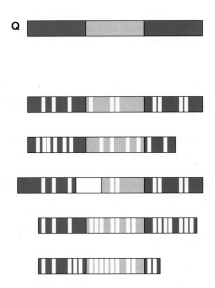

Figure 19.1: Similarity searching with a given protein sequence may identify homologous proteins whose sequences align with the query sequence (Q) over its entire length.

These sequences have diverged by the accumulation of point mutations alone, which may be amino acid substitutions (represented by white lines) or small insertions and deletions. Generally, the more divergent the query and response, the less conserved the functions are likely to be.

Another pitfall of homology searching is that the annotations are necessarily based on other people's experiments and interpretations. All databases, however carefully curated, contain a significant proportion of errors. Basing the function of a new gene on such data may not only be incorrect, but also propagates that error (Brenner, 1999).

19.2.2 Consensus search methods can extend the number of homologous relationships identified

Standard homology search methods employing algorithms of the BLAST family (see *Table 8.2* and *Box 8.7*) are suitable to detect protein sequences that are closely related either over their entire lengths or in one or more domains. However, when the level of sequence similarity falls below 30–40%, these algorithms become less robust and many evolutionary relationships can be missed.

One way in which the performance of homology searching can be improved is to use a consensus search method such as **PSI-BLAST (position–specific iterated BLAST)**, which employs a reiterative search based on *sequence profiles* (Altschul *et al.*, 1997). The principle is shown in *Figure 19.3*. The process begins in the same way as a normal BLAST search but the initial hits are combined into a representative profile which is then used in a second round of searching. Any additional hits are combined with the profile and the process is repeated either for a predetermined number of iterations or until no new hits are identified. This method has been shown to identify three times as many evolutionary relationships as standard homology searching. Further sensitivity can be gained by matching patterns and profiles generated by the alignment of distantly

related proteins and the derivation of short conserved motifs or longer profiles corresponding to functional domains (Eddy, 1998). A number of secondary sequence databases have been created that contain domain motifs and profiles derived from the primary sequence databases (*Table 19.1*). These databases can be searched with a query sequence in order to identify conserved protein domains in very distantly related sequences.

19.2.3 Similarities and differences between genomes indicate conserved and functionally important sequences

From the discussion above, it is clear that closely related genes from different species (orthologs) can be used to assign quite precise functions to uncharacterized genes. Orthologs are not usually identical because separate mutations accumulate in each evolutionary lineage after the speciation event. Therefore, the degree of similarity between orthologs provides a useful measure of evolutionary time and can be used to build phylogenetic trees (Chapter 12). **Comparative genomics** exploits the similarities and differences between genomes to derive structural, functional and evolutionary information (Section 12.3.2). It is based on the principle that genetic similarities between species extend much further than the gene level. Closely related species would be expected to have similar genes *and similar genomes*, the degree of similarity being dependent on the evolutionary divergence of the species (Nadeau and Sankoff, 1998). Similarities are apparent at the sequence level but also at the gross level of genome organization, such that related species often demonstrate **conserved synteny** (a conserved gene order).

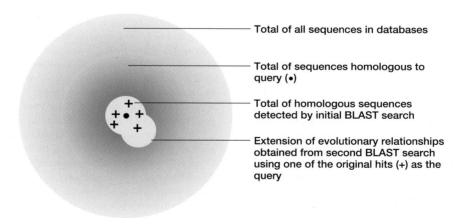

Total of all sequences in databases

Total of sequences homologous to query (•)

Total of homologous sequences detected by initial BLAST search

Extension of evolutionary relationships obtained from second BLAST search using one of the original hits (+) as the query

Figure 19.3: The principle of PSI-BLAST.

The entire contents of the sequence databases are represented by the outer circle. A subset of these sequences will be homologous to the query sequence (dot), although the degree of evolutionary relationship falls the farther away the sequence is from the center. A small inner circle of sequences related to the query will be detected by standard BLAST. In PSI-BLAST, each of the hits in this first search is compiled into a profile which is used in a second search to extend the number of homologous sequences identified. The process can be repeated for a predetermined number of search rounds or until no more hits are found.

Table 19.1: **Secondary databases of protein sequences that can be used to identify conserved elements and protein domains. InterPro is a valuable cross-referencing system that allows each of the databases to be searched with a single query**

Database	Contents	URL
PROSITE	**Sequence patterns** associated with protein families and longer **sequence profiles** representing full protein domains	http://ca.expasy.org/prosite
PRINTS, BLOCKS	Highly conserved regions in multiple alignments of protein families. These are called **motifs** in PRINTS and **blocks** in BLOCKS	http://bioinf.man.ac.uk/dbbrowser/PRINTS http://www.blocks.fhcrc.org
Pfam, SMART, ProDom	Collections of protein domains	http://www.sanger.ac.uk/Software/Pfam http://smart.embl-heidelberg.de/
Interpro	A search facility that integrates the information from other secondary databases	http://www.ebi.ac.uk/interpro/

Synteny is potentially useful for gene mapping and cloning because map information from one species can be used to locate and clone genes from another. In the case of the human genome, the massive effort expended in the construction of high density genetic maps (Chapter 8) is now paying off in more ways than one, since evidence of conserved synteny between humans and other vertebrates is being used to map and clone equivalent genes from other mammals. It is valuable to identify animal orthologs of human disease genes because this may lead to an understanding of why humans are susceptible to certain diseases, and therefore help in the prevention of these diseases and the development of novel drugs. For example, most mammals are not susceptible to HIV. Identifying orthologs of the human genes that confer susceptibility to the disease (e.g. *CCR5*, which encodes the HIV co-receptor displayed on the surface of T cells) might provide insights into novel therapies.

Another application of comparative genomics is the identification of gene regulatory elements. As discussed in Chapter 20, the identification of the promoter and enhancer elements required for normal gene expression is a laborious task, involving artificial expression assays in cultured cells and transgenic animals. A potential shortcut is to compare orthologous genes in related species and look for the most conserved sequence motifs outside the coding region of the gene. Only sequences with a highly conserved function would be represented in both genomes while other sequences would have diverged significantly over evolutionary time (Hardison *et al.*, 2000; Werner, 2003). In this respect, the Japanese puffer fish *Takifugu rubripes* provides an excellent model because it has the smallest known genome of all vertebrates (400 Mb) yet contains approximately the same number of genes as the human genome (Elgar *et al.*, 1996; Aparicio *et al.* 2002).

19.2.4 Comparative genomics can be exploited to identify and characterize human disease genes

The use of homology searching to assign functions to human genes often results in superficial functional assignments such as 'phosphatase' or 'membrane-spanning protein'. Comparative genomics can be exploited to enrich the information provided by homology searching since not only the functions of individual proteins but also of entire pathways, networks and complexes tend to be conserved between related genomes.

Even in the case of distantly-related organisms, such similarities may be expressed at the whole organism level. For example, it is estimated that about 30% of known human disease genes have homologs in yeast. The yeast *SGS1* gene encodes a DNA helicase which is homologous to a human gene, *WRN*, that is defective in **Werner syndrome** (MIM 2777000). This is a disease of premature aging. Individuals with the defective enzyme appear normal in their early years but age dramatically in middle life, developing the wizened appearance of octogenarians by their mid-30s. Common features of the disease include atherosclerosis, osteoporosis, diabetes and cataracts. The symptoms of the disease are thought to be associated in some way with the inability of Werner cells to divide in culture as many times as normal cells, i.e. there is early cell **senescence**. The link between helicase activity and cell senescence is difficult to examine in humans but yeast cells with an inactive *SGS1* gene also show accelerated aging, offering the potential to carry out functional studies of the human disease in yeast cells.

The conservation of gene function between humans and animals is even greater than that between humans and yeast. Over 60% of human disease genes have counterparts in the fly and worm, revealing a core of about 1500 gene families that is conserved in all animals. The insulin signaling pathway is fully conserved between humans and nematodes so mutant worms impaired for insulin signaling are useful models of **type II diabetes**. Due to their microbe-like properties, these nematode mutants can be screened with thousands of potential drugs to identify compounds that return the insulin-insensitive disease physiology to normal. *Caenorhabditis elegans* mutants provide models of many other diseases including neurological disorders, congenital heart disease and kidney disease. We consider the use of animal disease models in more detail in the next chapter.

19.2.5 A stubborn minority of genes resist functional annotation by homology searching

The first genome of a major model organism to be completely sequenced was that of the yeast *Saccharomyces cerevisiae* (Dujon, 1996). Before the sequence became available it was widely believed that most yeast genes had already been identified experimentally and the genome sequence would reveal at most a few hundred additional genes. Scientists therefore got a surprise when they found out that there were over 6000 genes in the yeast genome, only 30% of which were previously known. Homology searching provided functions for another 30% of the genes although the annotations varied considerably in their usefulness. In some cases it was possible to identify both biochemical and cellular functions but for many genes only general biochemical functions could be predicted. This left 40% of the genes with no functional assignments whatsoever. These genes could be divided into two categories (*Figure 19.4*):

▸ **orphan genes**. These are predicted genes that do not match any other sequence in the databases;

▸ **orphan families**. These are predicted genes with homologs in the databases, but the homologs themselves are of unknown function.

For the human genome, the situation is complicated by the uncertainty of gene predictions. In the yeast genome, fewer than 10% of gene predictions were questionable but the figure in the human genome is much higher, perhaps in the order of 25%. Furthermore, while many human genes have been identified, the predicted structure might not be accurate (e.g. there may be missing exons, incorrectly defined exons and phantom exons, or adjacent genes may have been fused). With this taken into account, the outcome of the Human Genome Projects is surprisingly similar to the results obtained from the yeast project. About a third of the listed genes were previously known and were represented in the **RefSeq database**, a highly curated collection of human gene transcripts. Another third was predicted on the basis of homology to ESTs or other sequences and the remaining third were predicted *ab initio*, based on structural criteria alone. Among the latter categories, some of the predictions will be false positives (e.g. matches to pseudogenes and transposable elements) while many genes will have been missed due to the low sensitivity of *ab initio* gene prediction algorithms. Among the previously identified genes and those predicted on the basis of homology, most have been functionally annotated at least at the biochemical level while a minority represent orphan families. Only a small proportion of the genes predicted by *ab initio* methods have been assigned a function. About 60% of human genes are predicted to contain protein domains that are represented in the secondary sequence databases, leaving 40% unassigned. Within this category, the function of perhaps 10% of the genes is known but they do not belong to large protein families, while a third of the genes have no functional assignment at all and are genuine orphans.

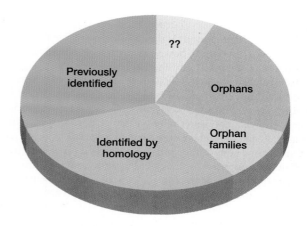

Figure 19.4: Distribution of yeast genes by annotation status in the aftermath of the *Saccharomyces cerevisiae* genome project.

Overall, 30% of the 6000 genes had been identified previously, 30% could be assigned functions on the basis of homology searching, 33% were orphans or members of orphan families and 7%, represented by ??, were questionable open reading frames. The functional annotations based on homology ranged from fully informative (biochemical and physiological function established) or partly informative (likely biochemical function established) to superficial (e.g. probably involved in nitrogen metabolism).

For these orphan genes, homology searching provides no functional information and we must base functional predictions on alternative evidence. The available methods for functional annotation include the following, and are discussed in the rest of this chapter:

▸ analysis of mRNA expression profiles;

▸ analysis of protein expression profiles;

▸ comparison of protein structures;

▸ analysis of protein interactions;

▸ analysis of mutant phenotypes or phenocopies in model organisms.

19.3 Global mRNA profiling (transcriptomics)

19.3.1 Transcriptome analysis reveals how changes in patterns of gene expression coordinate the biochemical activities of the cell in health and disease

Expression patterns provide useful accessory information about the functions of individual genes and can also highlight links between them

Functional annotations based on sequence and structural comparisons can be informative, but in many cases they suggest only a broad biochemical category, such as 'kinase' or 'DNA-binding protein'. Further experiments are required to determine gene functions at the cellular and whole organism levels, and to investigate how the activities of individual gene products are coordinated. The expression profile of a gene can reveal a lot about its role in the body and can also help to identify functional links to other genes. The expression of many genes is restricted to specific cells or developing structures, showing the genes have particular functions in those places. Other genes are expressed in response to external stimuli. For example, they might be switched on or switched off in cells exposed to endogenous signals such as growth factors or environmental molecules such as DNA-damaging chemicals. In such cases, it is not unreasonable to assume that the function of the genes is in some way involved in the perception of such signals or the cell's response to them. Genes with similar expression profiles are likely to be involved in similar processes, and in this way showing that an orphan gene has a similar expression profile to a characterized gene may allow a function to be assigned on the basis of 'guilt by association'. Furthermore, mutating one gene may affect the expression profiles of others, helping to link those genes into functional pathways and networks. This would be the case if the mutated gene encoded a transcription factor, since the abnormal function of that transcription factor would impact on all the genes under its control.

Transcriptome analysis provides more scope for linking genes and their products into functional pathways and networks, and provides new opportunities for drug development

Traditionally, expression analysis has been carried out on a gene-by-gene basis and the methods that are used (e.g. northern blots, RT–PCR, *in situ* hybridization etc.) are optimized for the study of individual genes (see Section 7.3). Some of the techniques are amenable to moderate multiplexing, e.g. it is fairly easy to set up 10 or 20 PCRs, but highly parallel analysis involving hundreds or even thousands of genes is impossible due to the amount of time required to set up individual reactions, and of course the expense of doing so. Even so, the advantages of whole transcriptome analysis are clear. We might wish, for example, to look for genes with altered expression patterns in asthma, or multiple sclerosis, or inflammatory bowel disease. Instead of selecting a few candidate genes for analysis whose expression *might* be affected by the disease, we can look at all genes simultaneously and hence identify every single gene whose expression profile is modified. Some of these genes might be well-characterized already but others could be orphan genes, allowing tentative functions to be assigned. Such analysis also has immediate practical benefits. For example, genes that are upregulated specifically in the disease state might represent useful drug targets, while downregulated genes might encode proteins that could be used therapeutically. The desire for global analysis of gene expression has contributed to the development of novel technologies which allow the simultaneous monitoring of thousands of transcripts and the comparison of

their abundances in different samples. Two major platforms are available. One is sequencing-based and involves the sampling of DNA sequences from representative cDNA populations. This can be thought of as an **open system** because the whole transcriptome is open to analysis. The other is hybridization-based and involves the use of DNA microarrays. This can be thought of as a **closed system** because only those sequences represented on the array can be measured. Before considering these technologies and how they are used it is necessary to dispel a common myth. Although transcriptome analysis methods are generally described as 'transcriptional profiling', it is important to realize that one is not measuring the rate of transcription but the steady state mRNA level, which also takes into account the rate of mRNA decay.

19.3.2 Direct sequence sampling is a statistical method for determining the relative abundances of different transcripts

Global gene expression profiles can be established by sampling cDNA libraries

Probably the most direct way to study the transcriptome is to sequence randomly picked clones from cDNA libraries and count the number of times each sequence comes up. The abundance of each clone represents the abundance of the corresponding transcript in the original sample. If enough clones are sequenced, statistical analysis provides a rough guide to the relative gene expression levels (Audic and Claverie, 1997). This approach has been used to identify differentially-expressed genes but is laborious because large-scale sequencing is required. A potential short cut is to take very short sequence samples, known as **sequence signatures**, and read many of them at the same time. Several techniques have been developed to exploit this (*Box 19.2*) but the one that has had the most impact thus far is serial analysis of gene expression (SAGE), which is discussed in more detail below.

SAGE involves the sampling of short sequence tags that are joined together in a long concatemer

The first high throughput sequence sampling technique to be developed was **serial analysis of gene expression**, or **SAGE** (Velculescu *et al.*, 1995). It is based on two principles:

▸ a short nucleotide sequence tag can uniquely identify the transcript from an individual gene provided it is from a defined position within the transcript. For example, although the total number of human genes is approximately 30 000, a sequence tag of only 9 bp can in principle distinguish 4^9 (262 144) different transcripts. In practice, sequence tags of 12–15 bp are used, which provides even greater discrimination;

▸ concatemerization of the short sequence tags allows the efficient analysis of multiple transcripts in a serial manner. The tags from different transcripts can be covalently linked together within a single clone and the clone can then be sequenced to identify the different tags in that clone. Up to

---

Box 19.2: Sequence sampling techniques for the global analysis of gene expression.

Random sampling of cDNA libraries

Randomly-picked clones are sequenced and searched against databases to identify the corresponding genes. The frequency with which each sequence is represented provides a rough guide to the relative abundances of different mRNAs in the original sample (Audic and Claverie, 1997). This is a very labor-intensive approach, particularly where several cDNA libraries need to be compared.

Analysis of EST databases

ESTs are signatures generated by the single-pass sequencing of random cDNA clones. If EST data are available for a given library, the abundance of different transcripts can be estimated by determining the representation of each sequence in the database (e.g. see Vasmatzis *et al.*, 1998). This is a rapid approach, advantageous because it can be carried out entirely *in silico*, but it relies on the availability of EST data for relevant samples.

Differential display PCR

This procedure was devised for the rapid identification of cDNA sequences that are differentially expressed across two or more samples (Liang and Pardee, 1992). The method has insufficient resolution to cope with the entire transcriptome in one reaction, so populations of labeled cDNA fragments are generated by RT–PCR using one oligo-dT primer with a two-base overhang and one arbitrary primer. The use of different primer combinations produces pools of cDNA fragments representing subfractions of the transcriptome. The equivalent amplification products from two samples (i.e. products amplified using the same primer combination) are then run side by side on a sequencing gel, and differentially expressed cDNAs are revealed by quantitative differences in band intensities. This technique homes in on differentially expressed genes but false

positives are common and other methods must be used to confirm the predicted expression profiles.

Serial analysis of gene expression (SAGE)

In this technique, very short sequence signatures (**sequence tags**) are collected from many cDNAs by exploiting the special properties of type IIs restriction enzymes (see main text and *Figure 19.5*). The tags are ligated together to form long concatemers and these concatemers are sequenced. The representation of each transcript is determined by the number of times a particular tag is counted. Although technically demanding, SAGE is much more efficient that standard cDNA sampling because 50–100 tags can be counted for each sequencing reaction (Velculescu *et al.*, 1995).

Massively parallel signature sequencing (MPSS)

Like SAGE, the MPSS technique uses type IIs restriction enzymes to collect short sequence tags from many cDNAs. However, unlike SAGE (where the tags are cloned in series) MPSS relies on the parallel analysis of thousands of cDNAs attached to microbeads in a flow cell (Brenner *et al.*, 2000). The principle of the method is that a type IIs restriction enzyme is used to expose a four-base overhang on each cDNA. There are 16 possible four-base sequences, which are detected by hybridization of a set of 16 different adaptor oligonucleotides. Each adaptor hybridizes to a different decoder oligonucleotide defined by a specific fluorescent tag. The adaptor contains a site for a type IIs restriction enzyme, allowing another four bases to be exposed, and the process is repeated. By imaging the microbeads after each round of cleavage and hybridization, thousands of cDNA sequences can be read in four-nucleotide chunks. As with SAGE, the number of times each sequence is recorded can be used to determine relative gene expression levels.

---

100 different transcripts can be assayed in one sequencing reaction.

To perform a SAGE analysis, poly(A)+ RNA is first extracted from the source to be investigated and converted to cDNA using a biotinylated oligo(dT) primer. The subsequent events are outlined in *Figure 19.5*. The cDNA is first cleaved with a restriction enzyme such as NlaIII that cuts frequently because it has a 4-bp recognition site (this is termed the **anchoring enzyme**). The liberated 3′ end fragments, which are attached to biotin, are bound to streptavidin beads. The streptavidin-bound cDNA is split into two fractions, and the two fractions are separately ligated *en masse* to two double-stranded oligonucleotide adapters, A and B. Each adapter has an overhang that matches that generated by the anchoring enzyme and, immediately adjacent to this, the 5-bp recognition site for a **type IIs restriction enzyme** such as FokI. Such enzymes have the unusual property of recognizing a specific sequence but cleaving the DNA outside this sequence a particular number of base pairs downstream. Cleavage with

this so-called **tagging enzyme** therefore generates a short signature sequence, defined as a **SAGE tag**, attached to the adapter. Each of the two adapters also contains the annealing site for a different PCR primer. In the next stage of the process, the two pools of adapter-tags are mixed and ligated to produce **ditags** (two SAGE tags joined together) flanked on each side by one of the adapters. The PCR is then used to amplify these molecules. The ditags are released from the amplification products by cleavage with the original anchoring enzyme and the purified ditags are ligated together and cloned. The inserts of the cloning vectors are then sequenced and this allows the tags to be read off in a serial manner. Relative transcript levels are deduced from the frequency with which tag occurs.

In the original experiment, Velculescu *et al.* (1995) reported the recovery of 840 sequence tags from pancreatic cDNA. Of these, 498 represented 77 different transcripts and the most abundant transcripts were all produced from genes known to have pancreatic function (e.g. procarboxypeptidase A1 was represented 64 times and pancreatic trypsinogen 2 was represented 46 times). The particular advantage of SAGE is that the data are digital and therefore it is very easy to compare tag frequencies across different experiments, even if they have been carried out in different laboratories at different times. Several databases have been established for the deposition and comparison of SAGE data. The technique has become even more useful with adaptations that allow very small amounts of starting material to be used (see Velculescu and Volgelstein, 2000).

19.3.3 DNA microarrays use multiplex hybridization assays to measure the abundances of thousands of transcripts simultaneously

The two major types of DNA microarray are made in different ways but the principles of expression analysis are similar for each device

DNA microarrays are miniature devices onto which many different DNA sequences are immobilized in the form of a grid. There are two major types, one made by the mechanical spotting of DNA molecules onto a coated glass slide and one produced by *in situ* oligonucleotide synthesis (Schena *et al.*, 1995; Lockhardt *et al.*, 1996: see Section 6.4.3). Both devices can be described as **microarrays** but a more specific term for the latter is **high density oligonucleotide chip**. There are many commercial sources of spotted arrays and in-house facilities are available to many laboratories. In contrast, high density oligonucleotide chips are produced exclusively by the US biotechnology company Affymetrix Inc. and marketed as **GeneChips**.

Although manufactured in completely different ways, the principles of expression analysis are much the same for the two devices (Harrington *et al.*, 2000). Expression analysis is based on **multiplex hybridization** using a complex population of labeled DNA or RNA molecules (*Figure 19.6*). For both devices, a population of mRNA molecules from a particular source is reverse transcribed *en masse* to form a representative complex cDNA population. In the case of spotted microarrays, a fluorophore-conjugated nucleotide is included in the reaction mix so that the cDNA population is universally labeled. In the case of GeneChips, the unlabeled cDNA is converted into a labeled cRNA (complementary

RNA) population by the incorporation of biotin, which is later detected with fluorophore-conjugated avidin. The complex population of labeled nucleic acids is then applied to the array and allowed to hybridize. Each individual feature or spot on the array contains 10^6–10^9 copies of the same DNA sequence, and is therefore unlikely to be completely saturated in the hybridization reaction. Under these conditions, the intensity of the hybridizing signal at each address on the array is proportional to the relative abundance of that particular cDNA or cRNA in the mixture, which in turn reflects the abundance of the corresponding mRNA in the original source population. Therefore, the relative expression levels of thousands of different transcripts can be monitored in one experiment.

Expression analysis with each type of device is similar in principle, but there are some important practical differences reflecting the nature of the features on the array (cDNAs or oligonucleotides) and the specificity of the hybridization reaction. On spotted microarrays, each feature is represented by a single species of double-stranded cDNA, several hundred base pairs in length, which has to be prepared by PCR from a clone library or similar resource (*Figure 19.6a*). In the case of oligonucleotide chips, each feature is represented by a short single-stranded oligonucleotide, 20–25 nt in length, which is synthesized on the chip during manufacture. There is no need to maintain a clone library because the oligonucleotides can be synthesized based on any desired set of sequences, including those stored in public or proprietary databases (*Figure 19.6b*). One advantage of this is that oligonucleotide sequences can be picked to distinguish between closely related transcripts. With cDNA features, there is a significant risk of cross-hybridization between homologous genes or alternative splice variants. A disadvantage, however, is that the sequences chosen for inclusion on an oligo chip must be known. Conversely, it is possible to generate cDNA microarrays from anonymous clones in cDNA libraries. The specificity of hybridization on cDNA microarrays is high due to the length of the cDNA sequence representing each feature (*Figure 19.6c*). On oligonucleotide chips, however, the specificity of hybridization is lower. Therefore, each gene is represented by 20 different oligos that 'walk' along the sequence (*Figure 19.6d*). There are 20 oligos that match the target sequence perfectly (**perfect match** or **PM oligos**) and 20 that contain a single base mismatch (**mismatch** or

---

Figure 19.5: Multiplex gene expression screening using the SAGE method

The basis of the method is to reduce each cDNA molecule to a representative short sequence tag (about nine nucleotides long). Individual tags are then joined together (*concatemerization*) into a single long DNA clone as shown at the very bottom of the diagram where the numbers above the sequence tags represent a specific cDNA from which the tag was derived. Sequencing of the clone provides information on the different sequence tags which can identify the presence of corresponding mRNA sequences. The mRNA is converted to cDNA using an oligo (dT) primer with an attached biotin group and the biotinylated cDNA is cleaved with a frequently cutting restriction nuclease (the anchoring enzyme, AE; in this example it is *Nla*III which cuts immediately after the G in the 4 bp sequence CATG). The resulting 3′ end fragments which contain a biotin group are then selectively recovered by binding to streptavidin-coated beads, separated into two pools and then individually ligated to one of two double-stranded oligonucleotide linkers, A and B. The two linkers differ in sequence except that they have a 3′ CTAG overhang and immediately adjacent to it, a common recognition site for a type IIs restriction nuclease which will serve as the tagging enzyme (TE). In this example is *Fok*I which recognizes the sequence GGATG (outlined in a box), but cleaves at 9/13 nucleotides downstream. Cleavage with *Fok*I generates a 9 bp sequence tag from each mRNA and fragments from the separate pools can be brought together to form 'ditags' then concatenated as shown.

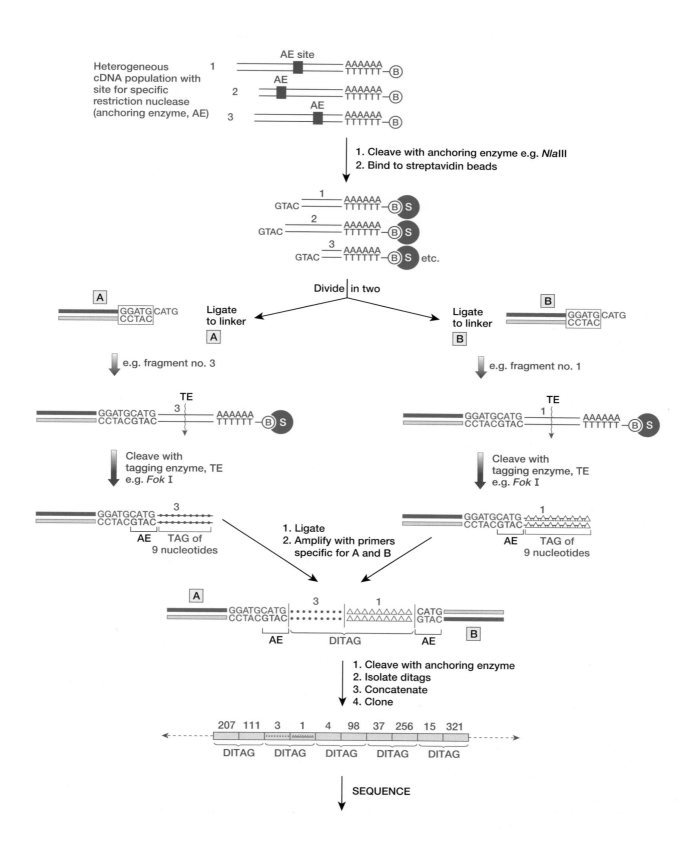

MM oligos) to control for nonspecific hybridization. To determine the signal for a particular gene, the signals of all 20 PM oligos are added together and the signals from all 20 MM oligos are subtracted from the total. Negative control features are also included on cDNA microarrays to normalize for background or nonspecific hybridization. Both types of device also contain positive control features, usually representing constitutively expressed genes such as actin. Alternatively, bacterial genes may be included among the control features and the samples can be 'spiked' with appropriate amounts of bacterial DNA.

Differential gene expression between samples can be demonstrated by comparison across arrays or by dual hybridization to a single array using cDNA populations labeled with different fluorophores

One of the major technical hurdles in any array-based expression analysis experiment is the reproducibility of data between different arrays. This is particularly troublesome if comparative analysis needs to be carried out (e.g. to identify genes that are upregulated in disease) because it becomes unclear whether differences are due to experimental variability or genuine changes in gene expression. This problem is compounded if experiments are carried out in different labs using arrays that have been produced by different facilities. Stringent controls are required to normalize expression data for cross-experimental variation.

One way in which the above problems can be avoided is to hybridize cDNA populations labeled with different fluorophores to the same array simultaneously. As discussed above, under nonsaturating conditions the signal at each feature will represent the relative abundance of each transcript in the sample. If two samples are used, then the ratio of the signals from each fluorophore provides a direct comparison of expression levels between samples fully normalized for variations in signal-to-noise ratio even within the array. The array is scanned at two emission wavelengths and a computer is used to combine the images and render them in false color. Usually, one fluorophore is represented as green and the other as red. Features representing differentially-expressed genes show up as either green or red, while those representing equivalently-expressed genes show up as yellow (*Figure 19.6c*).

19.3.4 The analysis of DNA array data involves the creation of a distance matrix and the clustering of related datapoints using reiterative algorithms

The analysis of gene expression data over many different conditions can show functional links between genes

The raw data from microarray experiments are signal intensities which must be normalized (corrected for background effects and inter-experimental variation, see above) and checked for errors caused by contaminants and extreme outlying values (Yang *et al.*, 2002). The data are summarized as a table of normalized signal intensities where rows on the table represent individual genes and the columns represent different conditions under which gene expression has been measured. In the simplest cases, the table has two columns (e.g. control sample and disease sample) and these may represent the signal intensities from two samples hybridized simultaneously to the array. However, there is no theoretical limit to the number of conditions that can be used. For example, it is possible to test gene expression at a series of developmental time points or at a series of time points after the onset of a viral infection, or when cultured cells have been exposed to a range of drugs and other chemicals.

The next stage of analysis involves the grouping of genes with similar expression profiles (Altman and Raychaudhuri, 2001; Quackenbush, 2001). Generally, the more conditions over which gene expression is tested, the more rigorous the analysis. *Figure 19.7* shows the analysis of three genes over three conditions. Initially, genes A and B appear functionally related because the expression profiles are similar over conditions 1 and 2, while gene C appears different. However, if we consider conditions 2 and 3 in the analysis and omit condition 1, it now appears that genes A and C are functionally related at the expense of gene B. Comparisons over many conditions help to eliminate spurious relationships which may result in false annotations. While the analysis of a simple 3×3 matrix can be carried out by eye, expression data involving thousands of genes and tens of conditions must be mined with the help of computers. Two types of algorithm are used to mine the gene expression data, one in which similar data are clustered in a hierarchy and one in which the clusters are defined in a nonhierarchical manner.

Figure 19.6: Expression analysis with DNA microarrays.

(A) Spotted microarrays are produced by the robotic printing of amplified cDNA molecules onto glass slides. Each spot or feature corresponds to a contiguous gene fragment of several hundred base pairs or more. Pre-synthesized oligonucleotides can also be spotted onto slides (not shown). (B) High-density oligonucleotide chips are manufactured using a process of light-directed combinatorial chemical synthesis to produce thousands of different sequences in a highly ordered array on a small glass chip. Genes are represented by 15–20 different oligonucleotide pairs (PM, perfectly matched and MM, mismatched) on the array. (C) On spotted arrays, comparative expression assays are usually carried out by differentially labeling two mRNA or cDNA samples with different fluorophores. These are hybridized to features on the glass slide and then scanned to detect both fluorophores independently. Colored dots labeled x, y and z at the bottom of the image correspond to hypothetical genes present at increased levels in sample 1 (x), increased levels in sample 2 (y), and similar levels in samples 1 and 2 (z). (D) On Affymetrix GeneChips, biotinylated cRNA is hybridized to the array and stained with a fluorophore conjugated to avidin. The signal is detected by laser scanning. Sets of paired oligonucleotides for hypothetical genes present at increased levels in sample 1 (x), increased levels in sample 2 (y) and similar levels in samples 1 and 2 (z) are shown. Modified from Harrington *et al.* (2000) with permission from Elsevier Science.

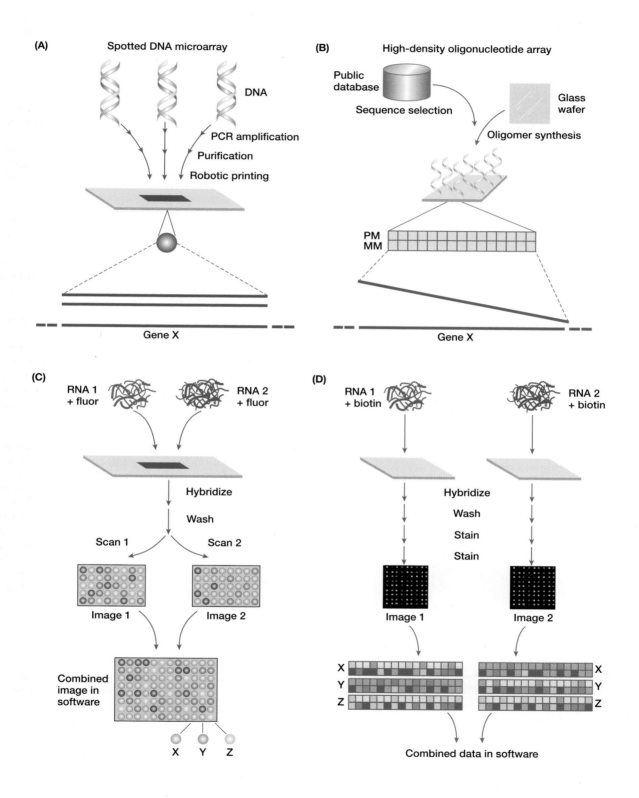

Hierarchical clustering

The general approach in hierarchical clustering is to establish a **distance matrix** which lists the differences in expression levels between each pair of features on the array. Those showing the smallest differences, expressed as the **distance function** d, are then clustered in a progressive manner. **Agglomerative clustering** methods begin with the classification of each gene represented on the array as a *singleton cluster* (i.e. a cluster containing one gene). The distance matrix is searched and the two genes with the most similar expression levels (the smallest distance function) are defined as *neighbors*, and are then merged into a single cluster. The process is repeated until there is only one cluster left. There are variations on how the expression value of the merged cluster is calculated for the purpose of further comparisons.

In the **nearest-neighbor (single linkage) method**, the distance is minimized. That is, where two genes i and j are merged into a single cluster ij, the distance between ij and the next nearest gene k is defined as the lower of the two values $d(i,k)$ and $d(j,k)$. In the **average linkage method**, the average between $d(i,k)$ and $d(j,k)$ is used. In the **farthest-neighbor (complete linkage) method**, the distance is maximized. These methods generate dendograms with different structures (*Figure 19.8*). Less frequently, a *divisive clustering* algorithm may be used in which a single cluster representing all the genes on the array is progressively split into separate clusters.

Nonhierarchical clustering

A disadvantage of hierarchical clustering is that it is time consuming and resource hungry. As an alternative, nonhierarchical methods partition the expression data into a certain predefined number of clusters, therefore speeding up the analysis considerably especially when the dataset is very large. In the **k-means clustering method**, a number of points known

Figure 19.8: Microarray data analysis using the hypothetical expression profiles of four genes (A–D).

Hierarchical clustering methods produce branching diagrams (*dendrograms*) in which genes with the most similar expression profiles are grouped together, but alternative clustering methods produce dendrograms with different topologies. The pattern on the left is typical of the topology produced by nearest neighbor (single linkage) clustering while the pattern on the right is typical of the topology produced by farthest neighbor (complete linkage) clustering.

as **cluster centers** are defined at the beginning of the analysis, and each gene is assigned to the most appropriate cluster center. Based on the membership of each cluster, the means are recalculated, i.e. the cluster centers are repositioned. The analysis is then repeated so that all the genes are assigned to the new cluster centers. This process is reiterated until the membership of the various clusters no longer changes. **Self-organizing maps** are similar in concept but the algorithm is refined through the use of a neural network.

19.3.5 DNA arrays have been used to study global gene expression in human cell lines, tissue biopsies and animal disease models

The comparison of healthy and disease samples can be used to identify disease markers and potential drug targets

DNA arrays have been widely used to characterize the human transcriptome and perhaps the most significant practical application of this technology is the study of disease. The comparison of healthy and disease tissue, represented either by cell lines, tissue biopsies or animal models, allows the identification of genes that are expressed specifically in the disease state or specifically in the nondisease state. Individual genes that are disease-specific can make useful diagnostic markers. Disease-specific genes may encode proteins whose presence or activity contributes to the disease symptoms and these represent potential targets for therapeutic intervention. Similarly, if the products of one or more genes are absent in the disease, the proteins themselves may represent useful therapeutic agents.

The study of McCaffrey *et al.* (2000) is an informative example of the above approach. These investigators used Affymetrix GeneChips to compare the expression profiles of normal arteries and arteries removed from patients with atherosclerotic lesions. They showed that one particular gene, *EGR1*, was upregulated fivefold in the disease state. This gene encodes a transcription factor that is known to control genes encoding growth factors and other signaling molecules, cell adhesion molecules and proteins that regulate blood coagulation. These downstream proteins have roles with obvious

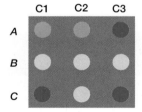

Figure 19.7: This simple example shows the value of monitoring gene expression analysis over several conditions.

The expression profiles of genes A and B appear similar under conditions 1 and 2 (i.e. they both remain static) while gene C appears to be downregulated under condition 2. This suggests genes A and B are functionally linked and C is different. However, if we look at the profiles under conditions 2 and 3, and ignore condition 1, it appears that genes A and C are functionally linked and B is different. Analysis over all three conditions suggests there is no relationship among any of the genes.

implications in the deposition of cholesterol-rich cells on the inner surface of arteries, and EGR1 therefore represents a useful drug target.

Multiplex analysis may provide transcriptional profiles that can be used to distinguish very similar diseases and discover new disease subcategories

Single markers are not useful for the diagnosis of all diseases, particularly those that are closely related (e.g. different types of cancer). In such cases, arrays can be used to derive transcriptional profiles of 50 or 100 genes that offer greater discrimination. Where the disease categories are already known, this procedure is called *class prediction*. The results of the experiment are expected to fit into a certain number of predefined categories and for this reason the data analysis is described as 'supervised'.

An example of this approach is the use of arrays to discriminate between acute myeloid leukemia (AML) and acute lymphoblastoid leukemia (ALL). Although the diseases are similar, they respond to different therapies and a correct diagnosis is therefore essential for successful treatment. Traditionally, a combination of techniques has been used, including the detection of protein markers, differential staining, cytogenetic analysis and the visual inspection of cells in blood smears. None of the tests are 100% accurate and the separate tests may in some cases give conflicting results. Golub *et al.* (1999) used spotted arrays representing about 7000 human genes to examine 38 bone marrow samples. Using the self-organizing maps algorithm discussed above, they correctly identified each sample as representing either AML or ALL based on the profile of about 50 genes.

In some cases, expression profiling of disease samples has identified previously unknown subcategories of the disease (*class discovery*). Because the categories are not defined at the beginning of the experiment, this type of analysis is described as 'unsupervised'. For example, using a cDNA microarray, Alizadeh *et al.* (2000) identified two different subtypes of non-Hodgkin's lymphoma (see *Figure 17.18*). Only 40% of patients with this disease benefit from drug treatment, and until recently there was no way to predict how patients would respond to therapy. However, the two new disease subtypes appear to correspond to the responsive and nonresponsive types of the disease. Similarly, Bittner *et al.* (2000) were able to identify two distinct subtypes of cutaneous melanoma.

Cell lines and animal models can be used to test gene expression profiles under multiple conditions or at multiple time points

While biopsies allow the side-by-side comparison of gene expression profiles in health and disease, cell lines and animal models offer the increased versatility of testing different conditions or different time points. For example, Zhu *et al.* (1998) infected cultured human foreskin fibroblasts with cytomegalovirus and analyzed gene expression profiles at several time points after infection using a cDNA microarray containing 6000 genes. Over 250 genes were shown to be upregulated during the infection, including several that are known to modulate the immune response. Cell lines have also been used to study the effect of growth factors and cytokines (e.g. Der *et al.*, 1998) and transfection with oncogenes (e.g. Khan *et al.*, 1999). An extension of this approach is the use of human cell lines to analyze responses to drugs, chemicals and toxins. Occasionally, such experiments reveal unexpected functional links between genes. For example, Iyer *et al.* (1999) studied the transcriptome of serum-starved cells following the addition of fresh serum and showed that different classes of genes were induced at different time points after refreshment. The first genes to be expressed were proliferation response genes such as *JUN* and *FOS*, but at later time points many of the induced genes were those involved in wound healing, such as *FGF7* and *VEGF*.

19.4 Proteomics

19.4.1 Proteomics encompasses the analysis of protein expression, protein structure and protein interactions

Proteome analysis shows how changes in the abundance of particular proteins coordinate the biochemical activities of the cell

While transcriptome analysis is clearly very useful for the characterization of gene function, it should not be forgotten that the ultimate products of most genes are proteins. Proteomics allows these products to be studied directly, and this is important because it exposes two limitations of transcriptome analysis.

First, partly due to post-transcriptional regulation, not all the mRNAs in the cell are translated, so the transcriptome may include gene products that are not found in the proteome. Similarly, rates of protein synthesis and protein turnover differ among transcripts, therefore the abundance of a transcript does not necessarily correspond to the abundance of the encoded protein (Gygi *et al.*, 1999b). For these reasons, the transcriptome may not accurately represent the proteome either qualitatively or quantitatively.

Second, protein activity often depends on post-translational modifications which are not predictable from the level of the corresponding transcript. Many proteins are present in the cell as inert molecules, which need to be activated by processes such as proteolytic cleavage or phosphorylation. In cases where variations in the abundance of a specific post-translational variant are significant, this means that only proteomics provides the information required to establish a link between gene expression and function. For example, the signaling protein **stathmin** is found at high levels in various cancers, including childhood leukemias, but only the *phosphorylated* form of the protein is a useful marker of the disease. Therefore, it is necessary to study the proteome directly in order to fully understand the functional molecules present in the cell.

The analysis of protein structure and protein interactions can provide important functional information

Like transcriptomics, proteomics can be used to monitor the abundance of different gene products. The expression of all the proteins in the cell can be compared among related samples, allowing proteins with similar expression patterns to be identified, and highlighting important changes in the proteome that occur, for example during disease or the response to particular external stimuli. This is sometimes termed **expression proteomics**. However, another important dimension to proteomics is that functions can often be established by investigating **protein interactions**. Proteins carry out their activities in the cell by interacting with other molecules. Establishing specific interactions between proteins can therefore help to assign individual functions and link proteins into pathways and networks. Further information can be derived from protein interactions with small molecules (which may act as ligands, cofactors, substrates, allosteric modulators etc.) and with nucleic acids. The analysis of protein interactions overlaps with the analysis of protein structure. Knowledge of a protein's three-dimensional structure can help to predict interactions with other proteins and with smaller molecules, which can be very useful in the development of effective drugs. The comparison of protein structures also provides a further way to determine evolutionary links between genes and investigate their functions.

19.4.2 Expression proteomics has flourished through the combination of two major technology platforms: two-dimensional gel electrophoresis (2DGE) and mass spectrometry

The proteome contains tens of thousands of proteins differing in abundance over four or more orders of magnitude. Like nucleic acids, proteins can be detected and identified by specific molecular interactions, in most cases using antibodies or other ligands as probes. However, unlike nucleic acids, there is no procedure for cloning or amplifying rare proteins. Furthermore, the physical and chemical properties of proteins are so diverse that no single, universal methodology analogous to hybridization can be used to study the entire proteome in a single experiment. Therefore, while there has been significant interest in the development of protein chips as direct equivalents of DNA microarrays (*Box 19.3*), alternative technology platforms are currently required for whole proteome analysis.

At the present time, expression proteomics is based on 'separation and display' technology in which complex protein mixtures are separated into their components so that interesting features (e.g. proteins present in a disease sample but absent in a matching healthy sample) can be picked for further characterization. Separation is usually carried out by **two-dimensional gel electrophoresis (2DGE)**, a technique that was developed over 25 years ago and has the power to resolve up to 10 000 proteins on a single gel (Gorg *et al.*, 2000;

Box 19.3: Protein chips.

DNA microarrays can be used for whole genome analysis because all DNA molecules are chemically similar and can be hybridized in one assay. In contrast, proteins are chemically very diverse. Some are acidic while others are basic, some are charged or polar while others are hydrophobic and many undergo post-translational modification that alters their chemical and physical properties still further. For this reason, truly 'universal' protein chips are difficult to produce. However, there have been many technical advances in protein chip technology over the last few years culminating in the creation of a whole proteome chip for the yeast *Saccharomyces cerevisiae* (Zhu *et al.*, 2001). Various different types of protein chip have been described (see Zhu and Snyder, 2003).

▶ **Antibody chips.** These consist of arrayed antibodies and are used to detect and quantify specific proteins in a complex mixture. They can be thought of as miniaturized high-throughput immunoassay devices;

▶ **Antigen chips.** The converse of antibody chips, these devices contain arrayed protein antigens and are used to detect and quantify antibodies in a complex mixture;

▶ **Universal protein chips (functional arrays).** These devices may contain any kind of protein arrayed on the surface and can be used to detect and characterize specific protein–protein and protein–ligand interactions. Various detection methods may be used, including labeling the proteins in solution or detecting changes in the surface properties of the chip, e.g. by surface plasmon resonance. Included within this category are **lectin arrays**, which are used to detect and characterize glycoproteins;

▶ **Protein capture chips.** These devices do not contain arrayed proteins, but other molecules that interact with proteins as broad or specific capture agents. Examples include oligonucleotide aptamers and chips containing molecular imprinted polymers as specific capture agents, or the proprietary protein chips produced by companies such as BIAcore Inc. and Ciphergen Biosystems Inc., which use broad capture agents based on differing surface chemistries to simplify complex protein mixtures;

▶ **Solution arrays.** The latest generation of protein chips are being released from the two-dimensional array format to increase their flexibility and handling capacity. Such devices may for example be based on coded microspheres or barcoded gold nanoparticles.

Herbert *et al.*, 2001). Until a few years ago, the major bottleneck in proteomics was the characterization of the separated proteins, which are represented by anonymous 'spots'. As discussed above, one way in which proteins can be identified and quantified is to use antibodies or other specific probes. However, this approach can be applied only to a small number of proteins in any one experiment and is of course reliant on the availability of such probes. The breakthrough came with the development of annotation techniques based on mass spectrometry, which can be applied to any protein and can be implemented on a large scale (Griffin *et al.*, 2001; Mann *et al.*, 2001). This required innovations in instrument design and new bioinformatic methods for database searching with peptide mass data (Aebersold and Mann, 2003).

2DGE is the major platform for protein separation in proteomics but is not without limitations

There are many different methods for protein separation and all of them exploit particular chemical or physical properties that differ from protein to protein, for example mass, size, charge, solubility and affinity for different ligands. As might be expected, the more properties that are utilized in any separation technique, the greater the resolution. The principle of 2DGE is to separate proteins on the basis of their charge and their mass, with each separation taking place in a different dimension (Gorg *et al.*, 2000; Herbert *et al.*, 2001). A complex protein sample is loaded onto a denaturing polyacrylamide

gel and separated in the first dimension by **isoelectric focusing**. In this technique, the proteins migrate in a pH gradient until they reach their **isoelectric point** (the position at which their charge is neutral with respect to the local pH). The standard procedure is to prepare an **immobilized pH gradient (IPG) gel**, in which the buffering groups are attached to the polyacrylamide matrix, which prevents them drifting and becoming unstable during long gel runs. The gel is then equilibrated in the detergent sodium dodecylsulfate, which binds stoichiometrically to the backbone of denatured proteins and confers a massive negative charge that effectively cancels out any charge differences between individual proteins. Separation in the second dimension is therefore dependent on the mass of the protein, with smaller proteins moving more readily through the pores of the gel. The gel is then stained and the proteins are revealed as a pattern of spots (*Figure 19.9*).

Although 2DGE has a high resolution and is the most widely used technique for protein separation in proteomics, there are several limitations to its usefulness in terms of representation, sensitivity, reproducibility and convenience (particularly in terms of its suitability for automation). Several classes of protein are under-represented on standard gels, including very basic proteins, proteins with poor solubility in aqueous buffers and membrane proteins. It may be necessary to prefractionate the protein sample and use different detergents and buffers to extract different classes of proteins in order to

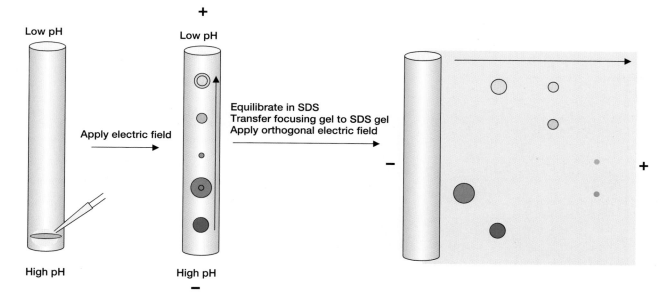

Figure 19.9: Principle of two-dimensional gel electrophoresis.

A complex protein mixture is loaded at the basic end of an isoelectric focusing gel (this is often a tube gel or a pre-cast strip, comprising a polyacrylamide matrix with an immobilized pH gradient). An electric field is applied across the gel and the proteins migrate towards their isoelectric points, where the pI value is equivalent to the surrounding pH and the net charge is zero. This separates the proteins on the basis of their charge (acidic proteins are shown as red and basic proteins as yellow, with shades of orange indicating proteins with intermediate pI values). Although the gel sieves the proteins on the basis of their size, the long gel runs ensure that all proteins reach their isoelectric points and the system achieves equilibrium. The focusing gel is then equilibrated in SDS, which binds with a constant mass ratio to all denatured proteins, and attached to a standard SDS–polyacrylamide gel. Proteins are then separated on the basis of size (indicated by circles of different diameters), with smaller proteins migrating further through the gel than larger ones.

separate them effectively. The sensitivity of 2DGE is dependent on the detection limit for very scarce proteins, and this has been addressed by the development of very sensitive staining reagents known as **SYPRO dyes** which can detect protein spots in the nanogram range. Sensitivity is also influenced by the resolution of the gel, since it is difficult to detect spots representing scarce proteins when they are obscured by those representing abundant proteins. These problems can be addressed by prefractionation, to remove abundant proteins and simplify the initial sample loaded onto the gel, and by increasing the resolution of the separation. In the latter case, the separation distance can be increased through the use of very large format gels, but a more convenient alternative is to use **zoom gels** for isoelectric focusing, that is gels with a very narrow pH range (*Figure 19.10*). For whole proteome analysis, the images obtained from zoom gels can be stitched together using a computer.

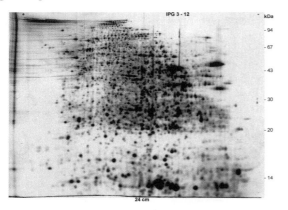

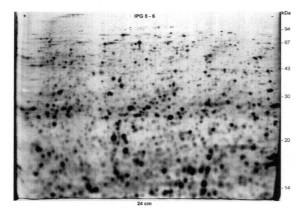

Figure 19.10: The resolving power of narrow pH range gels.

Both images represent mouse liver proteins separated by two-dimensional gel electrophoresis and silver stained to reveal individual protein spots. The top image is a wide pH range gel (pH 3–12) while the bottom image is a narrow pH range gel, which zooms proteins in the pH 5–6 range. Note that in the wider range gel, most proteins are clustered in the middle, reflecting the fact that most proteins have pI values in the 4–7 range. Reproduced from Orengo *et al.* (2002) *Bioinformatics*. Published by BIOS Scientific Publishers.

One of the major limitations of 2DGE is that it is not highly suited for automation, which is required for the high-throughput analysis of many samples. It has been necessary to develop software for spot recognition and quantification, algorithms that can compare protein spots across multiple gels, and robots that can pick interesting spots and process them for MS. Many of the difficulties encountered with 2DGE can be addressed through the development of alternative separation methods based on multi-dimensional high pressure liquid chromatography (HPLC). Such methods are faster, more sensitive, there is no need for protein staining, they allow more accurate quantification, they can resolve proteins that are underrepresented by 2DGE and they are much easier to automate and integrate with downstream analysis. Although lacking the visual aspect of stained gels, it is possible that liquid chromatography methods may eventually displace 2DGE as the major platform for protein separation in proteomics (Lesney, 2001; Wang and Hanash, 2003).

Mass spectrometry is the only universal method for the high throughput annotation of anonymous proteins

Mass spectrometry (MS) is used to determine the accurate masses of molecules in a particular sample or anylate. The principle of protein annotation by MS is the use of accurate molecular masses as query terms to search databases (Mann *et al.*, 2001; Aebersold and Mann, 2003). This can be completed more quickly than the alternative annotation method, the direct sequencing of proteins by Edman degradation, and is easy to automate for high-throughput sample analysis. This is important when one considers it may be necessary to process thousands of spots from a 2D gel or hundreds of HPLC fractions.

Until recently, MS could not be applied to large molecules such as proteins and nucleic acids because these were broken into random fragments during the ionization process. *Soft-ionization* methods such as matrix-assisted laser desorption/ionization (MALDI) and electrospray ionization (ESI) (see *Box 19.4*) now permit the ionization of such molecules without fragmentation (*Figure 19.11*). The masses of proteins, or more usually the peptide fragments derived from them by digestion with proteases, can be used to identify the proteins by correlating the experimentally determined masses with those predicted from database sequences. There are three different ways to annotate a protein by MS (*Figure 19.12*).

▶ **Peptide mass fingerprinting (PMF)**. A simple protein mixture (typically a single spot from a 2D gel) is digested with trypsin to generate a collection of *tryptic peptides*. These are subject to MALDI-MS using a time of flight (TOF) analyzer (*Box 19.4*), which returns a set of mass spectra. These spectra are used as a search query against the SWISS-PROT database. The search algorithm carries out virtual trypsin digests of all the proteins in the database and calculates the masses of the predicted tryptic peptides. It then attempts to match these predicted masses against the experimentally-determined ones.

▶ **Fragment ion searching**. This approach is used if PMF is unsuccessful. The tryptic peptide fragments are analyzed by tandem mass spectroscopy (MS/MS; *Box 19.4*) during which the peptides are broken into random fragments. The mass spectra from these fragments can be used to search against EST databases, which cannot be searched with PMF data because the intact peptides are generally too large. Any EST hits can then be used in a BLAST search to identify putative full length homologs. A dedicated algorithm called MS-BLAST is useful for handling the short sequence signatures obtained from peptide fragments ions.

▶ *De novo* **sequencing of peptide ladders**. In this technique, the peptide fragments generated by MS/MS are arranged into a nested set differing in length by a single amino acid. By comparing the masses of these fragments to standard tables of amino acids, it is possible to deduce the sequence of the peptide fragment *de novo*, even where a precise sequence match is not available in the database. In practice the *de novo* sequencing approach is complicated by the presence of two fragment series, one nested at the N terminus and one nested at the C terminus. The two series can be distinguished by attaching diagnostic *mass tags* to either end of the protein.

In a general approach, PMF is attempted first and, if this is not successful, the other less accurate methods can be tried. PMF is best suited to the analysis of simple proteomes such as the yeast proteome where there are few splice variants and post-translational modifications. Fragment ion analysis is more suited to the analysis of complex proteomes and the algorithm can be modified to take into account the masses of known post-translational modifications. However, it is impossible to account for all variants either at the sequence level (e.g. polymorphisms) or at the protein modification level (e.g. complex glycans). In such cases, *de novo* sequencing may provide sequence signatures that can be used as search queries to identify homologous sequences in the databases.

MS can also be used to analyze differentially expressed proteins

The aim of many proteomics experiments is to identify proteins whose abundance differs significantly across two or more samples. One way in which this can be achieved is to examine 2D gels and identify spots that show quantitative variation, and several software packages are available to aid such comparative investigations. An alternative is to label proteins from two different samples by conjugation with Cy3 and Cy5, and separate them on the same gel (see Patten and Beecham, 2001; Rabilloud, 2002: *Figure 9.13*). This approach, known as **difference gel electrophoresis (DIGE)**, exploits the same principle as differential gene expression with DNA microarrays, as discussed in Section 19.3.3. Yet another way to address the problem is to label proteins from different sources with **isotope coded affinity tags (ICATs**: Gygi *et al.*, 1999a; Sechi and Oda, 2003). A mass spectrometer can easily distinguish and quantify two isotopically labeled forms of the same compound that are chemically identical and can be co-purified. An ICAT method has therefore been developed using a biotinylated iodoacetamide derivative to selectively label protein mixtures at their cysteine residues. The biotin tag allows affinity purification of the tagged cysteine peptides

---

Box 19.4: Mass spectrometry in proteomics.

THE MASS SPECTROMETER

A **mass spectrometer** has three components. The **ionizer** converts the analyte into **gas phase ions** and accelerates them towards the **mass analyzer**, which separates the ions according to their **mass/charge ratio** on their way to the ion detector, which records the impact of individual ions, presenting these as a **mass spectrum** of the analyte.

Soft-ionization methods

The ionization of large molecules without fragmentation and degradation is known as **soft ionization**. Two soft ionization methods are widely used in proteomics. **Matrix-assisted laser desorption/ionization (MALDI)** involves mixing the analyte (the tryptic peptides derived from a particular protein sample) with a light-absorbing matrix compound in an organic solvent. Evaporation of the solvent produces analyte/matrix crystals, which are heated by a short pulse of laser energy. The desorption of laser energy as heat causes expansion of the matrix and analyte into the gas phase. The analyte is then ionized and accelerated towards the detector. In **electrospray ionization (ESI)**, the analyte is dissolved and the solution is pushed through a narrow capillary. A potential difference, applied across the aperture, causes the analyte to emerge as a fine spray of charged particles. The droplets evaporate as the ions enter the mass analyzer.

MASS ANALYZERS

The two simplest types of mass analyzer used in proteomics are the **quadrupole** and **time of flight (TOF)** analyzers. A quadrupole analyzer comprises four metal rods, pairs of which are electrically connected and carry opposing voltages that can be controlled by the operator. Mass spectra are obtained by varying the potential difference applied across the ion stream, allowing ions of different mass/charge ratios to be directed towards the detector. A time of flight analyzer measures the time taken by ions to travel down a flight tube to the detector, a factor that depends on the mass/charge ratio.

TANDEM MASS SPECTROMETRY (MS/MS)

This involves the use of two or more mass analyzers in series. Various MS/MS instruments have been described including triple quadrupole and hybrid quadrupole/time of flight instruments. The mass analyzers are separated by a **collision cell** that contains inert gas and causes ions to dissociate into fragments. The first analyzer selects a particular peptide ion and directs it into the collision cell, where it is fragmented. A mass spectrum for the fragments is then obtained by the second analyzer. These two functions may be combined in the case of more sophisticated instruments, such as the **ion trap** and **Fourier transform ion cyclotron** analyzers.

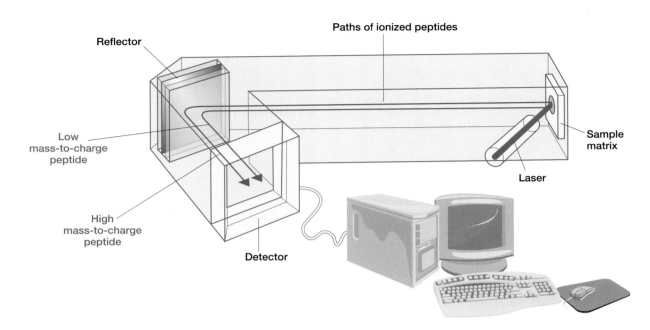

Figure 19.11: Principle of MALDI-TOF MS.

The analyte (usually a collection of tryptic peptide fragments) is mixed with a matrix compound and placed near the source of a laser. The laser heats up the analyte/matrix crystals causing the analyte to expand into the gas phase without significant fragmentation. Ions then travel down a flight tube to a reflector, which focuses the ions onto a detector. The time of flight (the time taken for ions to reach the detector) is dependent on the charge/mass ratio and allows the mass of each molecule in the analyte to be recorded.

after proteolysis with trypsin. The ICAT reagent is available in 'heavy' (d8) and 'light' (d0) isotopically labeled forms, which may be used to differentially label cell pools under different conditions (e.g. health and disease). After labeling, the cells are combined, lysed and the proteins isolated so that purification losses occur equally in both samples. Isotope intensities are compared for peptides as they enter the mass spectrometer. If they are equivalent, then no upregulation or downregulation has occurred, and the protein is of no immediate interest. If the intensities differ, then a change in protein expression has taken place, and the protein is of interest. The amount of the two forms is measured, and the peptide d0-form is fragmented and identified by database searching as discussed in the text.

19.4.3 Expression proteomics has been used to study changes in the proteome associated with disease and toxicity

Proteomics can reveal disease markers and potential drug targets not identified by transcriptome analysis

The varying abundance of specific proteins in healthy and disease samples, or in samples representing the progression of a disease, can help to reveal useful markers and novel drug targets. For example, several proteins have been identified that are expressed at abnormal levels in breast cancer, including PCNA (proliferating cell nuclear antigen, a component of DNA polymerase) and various heat shock proteins (Franzen *et al.*, 1996). Heat shock proteins were also shown to be upregu-

lated in colorectal cancer, while the enzyme cyclooxygenase 2 and a fatty acid binding protein were found to be significantly downregulated (Stulik *et al.*, 1999). Various markers, including different types of keratin, are expressed as bladder cancer progresses from the early transitional epithelium stage to full-blown squamous cell carcinoma. These can be used as markers to assess the degree of differentiation and therefore chart the progression of the disease. Note, however, that care is needed when handling samples because it is easy to contaminate the minute amounts of protein excised from the spots on 2D gels with hair and skin, which are also rich in keratin!

Although the applications of expression proteomics and transcriptomics are similar, proteomics has the advantage that it samples the actual functional molecules of the cell and takes into account post-translational modifications. We have already discussed stathmin as an example – this signaling protein is upregulated in childhood leukemia, but only the phosphorylated form is linked to the disease. Proteomics is also advantageous for the analysis of body fluids, which do not contain mRNA. For example, Celis *et al.* (2000) have shown that a protein called psoriasin is enriched in the urine of bladder cancer patients and can be used as an early diagnostic marker for the disease.

Toxicoproteomics helps to establish the basis of adverse drug responses

As discussed in Chapter 21, individuals have different reactions to drugs due to polymorphic variations in drug receptors and in the enzymes and transport proteins that

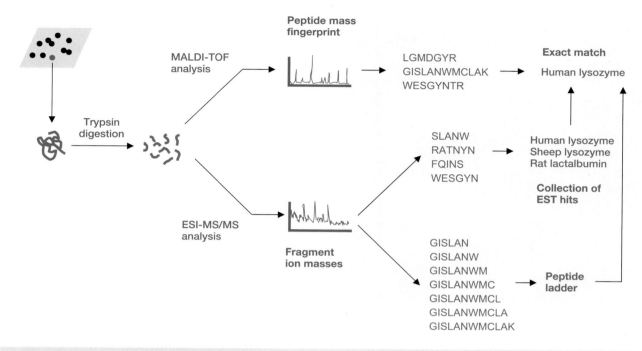

Figure 19.12: Protein annotation by mass spectrometry.

Individual protein samples (e.g. spots from 2D-gels) are digested with trypsin, which cleaves on the C-terminal side of lysine (K) or arginine (R) residues as long as the next residue is not proline. The tryptic peptides can be analyzed as intact molecules by MALDI-TOF, and the masses used as search queries against protein databases. Algorithms are used that take protein sequences, cut them with the same cleavage specificity as trypsin, and compare the theoretical masses of these peptides to the experimental masses obtained by MS. Ideally, the masses of several peptides should identify the same parent protein, in this case human lysozyme. There may be no hits if the protein is not in the database, or more likely, that it has been subject to post-translational modification or artifactual modification during the experiment. Under these circumstances. ESI-tandem mass spectrometry can be used to fragment the ions. The fragment ion masses can be used to search EST databases and obtain partial matches, which may lead eventually to the correct annotation. Alternatively, the masses of peptide ladders can be used to determine protein sequences *de novo*.

determine how drugs are absorbed, metabolized and excreted. Proteomics can play an important role in the prediction and investigation of adverse drug responses by indicating changes in the proteome occurring after drug treatment. Since drugs are molecules that interact directly with proteins and modify their behavior, many drug responses will not affect mRNA abundance and will therefore have no effect on the transcriptome.

As an example, we consider the immunosuppressant drug **cyclosporin A**. This is widely used to prevent rejection following grafts or organ transplants, especially in children. One of the major side effects of cyclosporin A is kidney toxicity, which occurs in nearly 40% of patients. The toxicity is associated with the loss of calcium in the urine and resulting calcification of the kidney tubules. Proteomic analysis of rat, and subsequently human, kidneys from untreated patients and those treated with cyclosporin A showed a striking difference in the level of one particular protein, the calcium-binding protein **calbindin** (Aaicher *et al.*, 1998). This protein was much less abundant in the kidneys of humans and rats treated with cyclosporin A, and immediately suggested the mechanism of cyclosporin A nephrotoxicity. Interestingly, calbindin

is not depleted in monkeys treated with cyclosporin A and consequently they do not suffer the same adverse drug effects as humans. Studying the way in which cyclosporin A is metabolized in monkeys may therefore lead to a mechanism to avoid toxicity in humans.

19.4.4 Protein structures provide important functional information

Protein structure may be conserved even when sequences have diverged to the extent that a homologous relationship can no longer be recognized

At the beginning of this chapter, we showed how similar protein sequences generally have similar structures and therefore similar functions. The function of a protein (or a domain thereof) is dependent on its *tertiary* structure, which is also known as its **fold**. This forms the binding sites, interaction domains and catalytic pockets that actually carry out the biochemical activity of the protein. Within these structures, a small number of amino acid residues may be absolutely critical for specific chemical reactions, e.g. residues within the active

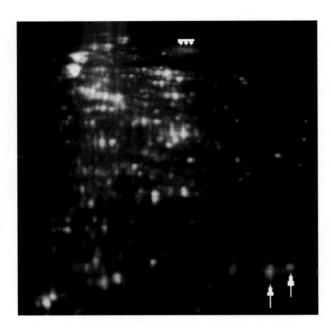

Figure 19.13: Demonstration of the principle of difference gel electrophoresis.

Identical samples of proteins from the bacterium *Erwinia carotovora* were labeled with Cy3 and Cy5, but the Cy5 sample was spiked with eight-fold the normal concentrations of two test proteins, myoglobin and conalbumin. Proteins that are equally abundant in the samples appear yellow, because the Cy3 and Cy5 signals are of equal intensity in the overlay image. The three conalbumin isoforms and two myoglobin isoforms, with higher abundance in the Cy5-labeled population, appear as red spots. Reproduced from Lilley *et al.* (2001) with permission from Elsevier Science.

site of an enzyme that react with the substrate. However, most of the amino acids in a protein have a structural role and help to maintain these critical residues in the correct relative positions. Therefore, evolution acts predominantly on the structure of a protein not its sequence and as long as the fold is conserved, many changes at the sequence level can be tolerated. The overall result is that during evolution protein structure is more highly conserved than sequence.

The practical consequence of structural conservation is that the comparison of protein structures can reveal distant evolutionary relationships and therefore help to assign functions to uncharacterized proteins even when all trace of sequence similarity has vanished. An example is the characterization of AdipoQ, a protein of initially unknown function that is secreted from adipocytes. Structural analysis of this protein shows a clear and unambiguous relationship to the tumor necrosis factor (TNF) family of chemokines and therefore indicates that AdipoQ is also a signaling protein. The structural relationship between AdipoQ and TNFα is shown in *Figure 19.14* along with a multiple sequence alignment of five superfamily members based on the conserved structures (Shapiro and Scherer, 1998).

Algorithms are available for the pairwise comparison of protein structures

The example discussed above suggests that a relatively straightforward way to annotate any orphan gene would be to obtain the structure of the encoded *hypothetical protein* and compare it to known structures in a manner analogous to sequence comparison. In order to achieve this, it would be necessary to obtain the structure of the hypothetical protein (*Box 19.5*) and use bioinformatic resources to compare this structure to protein structures that are already known. The largest repository for protein structures is the **Protein Databank (PDB)** which is maintained by the Research Collaboratory of Structural Biology at Rutgers University (http://www.rcsb.org). The data are stored as flat files with the positional coordinates of each atom in tabulated form. This is known as the **PDB file format**.

Several computer programs are available free over the Internet which convert PDB files into three-dimensional models (e.g. Rasmol, MolScript, Chime). Furthermore, a large number of algorithms has been written to allow protein structures to be compared (Sillitoe and Orengo, 2002). Generally, these work on one of two principles although some more recent programs employ elements of both:

▶ **intermolecular comparison**, the structures of two proteins are superimposed and the algorithm attempts to minimize the distance between superimposed atoms (*Figure 19.15a*). The function used to measure the similarity between structures is generally the **root mean square deviation (RMSD)**, which is the square root of the average squared distance between equivalent atoms (*Figure 19.15a*). The RMSD decreases as protein structures become more similar, and is zero if two identical structures are superimposed. Examples of such algorithms include Comp-3D and ProSup.

▶ **intramolecular comparison**, the structures of two proteins are compared side by side, and the algorithm measures the *internal* distances between equivalent atoms within each structure and identifies alignments in which these internal distances are most closely matched (*Figure 19.15b*). An example of such an algorithm is DALI. Algorithms that employ both methods include COMPARER and VAST.

Structural genomics (structural proteomics) aims to solve the structures of a representative set of proteins covering fold space

There are approximately 30 000 genes in the human genome but many of these can be grouped into families (paralogs) with similar sequences. As discussed above, similar sequence implies similar structure, therefore the different members of each gene family are likely to encode proteins with the same fold. Even gene families that show no significant sequence homology can encode structurally similar proteins, as shown above for AdipoQ and the TNF family. Taking the complication of multiple domain proteins into consideration, it has been estimated that there are fewer that 1000 different protein folds in existence. This means that if one represen-

(A)

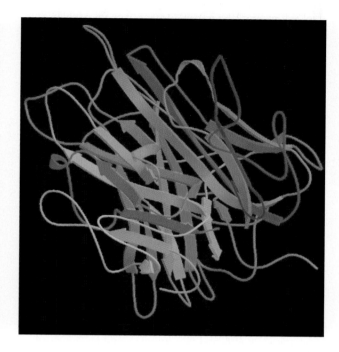

(B)

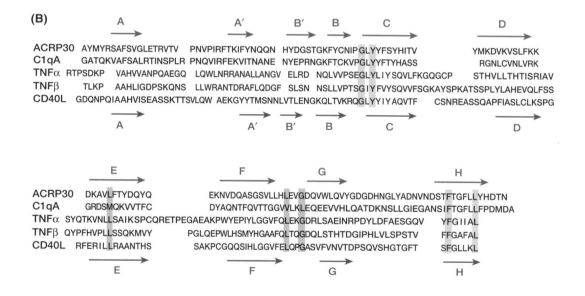

Figure 19.14: Functional annotation on the basis of structural similarity where no sequence homology can be detected.

(A) A ribbon diagram comparison of AdipoQ and TNFα. The structural similarity is equivalent to that within the TNF family. **(B)** Structure-based sequence alignment between several members of the TNF family (CD40L, TNFα and TNFβ) and two members of the C1q family (C1qA and AdipoQ). Highly conserved residues (present in at least four of the proteins) are shaded, and arrows indicate β-strand regions in the proteins. There is little sequence similarity between AdipoQ and the TNF proteins (e.g. 9% identity between AdipoQ and TNFα) so BLAST searches would not identify a relationship. However, the structural alignment shows conserved residue patterns that could be used to recognize novel family members among orphan genes (see Section 19.2.2). Reproduced from Shapiro and Scherer (1998) with permission from Elsevier Science.

tative protein from each fold family could be structurally resolved, it might be possible to annotate all orphan genes using structural data.

Several pilot programs have been initiated around the world to solve protein structures on a global scale in an attempt to cover 'fold space' and obtain the structures of a representative set of proteins (Brenner, 2001; Norin and Sundstrom, 2002; Zhang and Kim, 2003). In most of these projects, high throughput X-ray crystallographic analysis is being carried out using the technique of *multi-wavelength anomalous dispersion (MAD)* (see *Box 19.5* and Heinemann *et al.*, 2001). A common theme in these programs is a funnel effect, where a proportion of proteins is lost at each stage of analysis. For every 25 proteins that are expressed, only five yield crystals and only one generates useful diffraction data. While some of the programs are concentrating on human or pathogen proteins with relevance to disease, most involve proteins from bacteria with small genomes (e.g. *Haemophilus influenzae*) or from thermophiles (e.g. *Methanococcus jannaschii, Methanobacterium thermoautotrophicum*). This is because bacteria with small genomes also have small proteomes but should still contain representatives of all protein families, while proteins from thermophiles should be more stable when expressed in *E. coli*. The output of these pilot programs suggests that about 70% of hypothetical proteins contain known folds and theoretically could be annotated on the basis of structural data, while 30% contain entirely novel folds whose functions are unknown. Complications include the difficulty in providing rigorous definitions of protein structures and the 'Russian doll effect' where there is a continuous range

of intermediate structures between distinct fold types. These problems are discussed in *Box 19.6*.

19.4.5 There are many different ways to study individual protein interactions

The function of a protein may not be obvious even after a comprehensive study of its sequence, structure and expression profile. In such cases, it may be advantageous to identify those proteins with which it interacts specifically, especially if these turn out to be proteins that have been well studied and whose functions are already known. For example, if a new and uncharacterized protein is shown to interact with other proteins that are required for RNA splicing, then the new protein is likely to have a role in the same process. In this way proteins can be linked into functional networks in the cell.

A large number of methods is available for the study of individual protein interactions, including genetic, biochemical and physical techniques (see Phizicky and Fields, 1995). While genetic methods can be applied only to model organisms such as *Drosophila* and yeast, biochemical and physical methods can be applied to human cells directly. A classical biochemical method is **affinity chromatography**, in which a particular bait protein is immobilized on the supporting matrix of a chromatography column, and a cell lysate is added. Proteins that interact with the bait are retained on the column while other proteins are washed through. The interacting proteins can be eluted by increasing the salt concentration of the buffer. Another biochemical approach is **co-immunoprecipitation**,

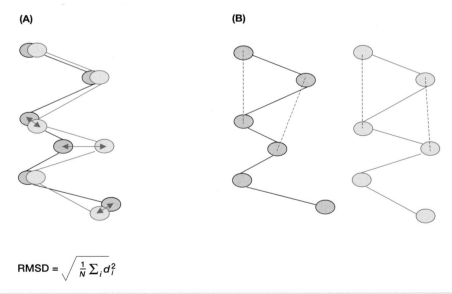

(A) **(B)**

$$\text{RMSD} = \sqrt{\frac{1}{N} \sum_i d_i^2}$$

Figure 19.15: Comparison of protein structures, with circles representing Cα atoms of each amino acid residue and lines representing the path of the polypeptide backbone in space.

(A) Intermolecular comparison involves the superposition of protein structures and the calculation of distances between equivalent atoms in the superimposed structures (shown as bi-directional arrows). These distances are used to calculate the root mean square deviation (RMSD), with the formula shown (R is the RMSD, *d* is the distance between the *i*th pair of superimposed Cα atoms and *N* is the total number of atoms aligned. A small RMSD value computed over many residues is evidence of significantly conserved tertiary structure.
(B) Intermolecular comparison involves side-by-side analysis based on comparative distances between equivalent atoms within each structure (shown as color-coded dotted lines).

Box 19.5: Determination of protein structures.

SOLVING PROTEIN STRUCTURES BY X-RAY CRYSTALLOGRAPHY

X-ray crystallography exploits the fact that X-rays are scattered, or *diffracted*, from protein crystals and the nature of the scattering depends on the organization of the atoms within the crystal. The diffracted X-rays can positively or negatively interfere with each other, generating patterns called **reflections** on a detector.

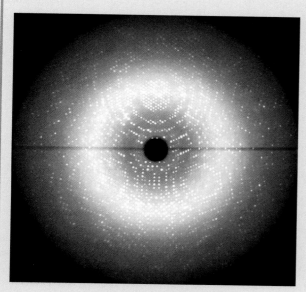

X-ray diffraction image of a protein phosphatase.

Courtesy of Daniela Stock, MRC Laboratory of Molecular Biology, Cambridge, UK.

Electron density maps can be constructed from these data using a mathematical function called the **Fourier transform**. The more data used in the Fourier transform, the more refined and accurate the resulting structural model. Accurate structural determination requires a well-ordered crystal that diffracts X-rays strongly. This can be a significant bottleneck in structural determination because protein crystals can be extremely difficult to grow, even with the availability of automated crystallization stations that allow thousands of reactions to be carried out in parallel under slightly differing conditions. A conventional X-ray source may be used for structural analysis but **synchrotron radiation sources** are favored because they produce X-rays of much greater intensity.

The construction of electron density maps from diffraction data relies on three pieces of information: the *wavelength* of the incident X-rays (which is already known), and the *amplitude* and *phase* of the scattered X-rays. Unfortunately, only the amplitude of scattering can be determined from the pattern of reflections. The phase must be calculated by carrying out further diffraction experiments with *isomorphous crystals*, i.e. crystals of the same structure but which incorporate heavier atoms that produce alternative diffraction patterns. The standard way to achieve this is by soaking the crystals in a heavy metal salt solution so that heavy metal atoms diffuse into the spaces that were originally occupied by solvent. Metal atoms diffract X-rays more strongly than the light atoms normally found in proteins. By comparing the reflections generated by several different isomorphous crystals (a process termed **multiple isomorphous replacement**) the positions of the heavy atoms can be worked out and this allows the phase of diffraction in the unsubstituted crystal to be deduced.

Alternative techniques exploit the phenomenon of *anomalous scattering*, where distinct diffraction patterns are generated when metal atoms in a protein crystal are struck by X-rays whose wavelength is close to their absorption edge. The magnitude of anomalous scattering varies with the wavelength of the incident X-rays, so one type of metal-containing crystal can be bombarded at several different wavelengths and different diffraction patterns obtained from which the phase of scattering can be calculated. This is the basis of techniques such as **SIRAS (single isomorphous replacement with anomalous scattering)** and **MAD (multiple wavelength anomalous scattering)**. In the latter technique, the protein is expressed in bacteria and incorporates the metal-substituted amino acid *selenomethionine*. In each case the differences in the intensities of the reflections caused by anomalous scattering are very small, so synchrotron radiation sources and accurate recording of reflection data are essential. Finally, the electron density map is built into a structural model. This requires one more crucial piece of information – the amino acid sequence – because it is difficult to distinguish amino acid side chains using diffraction data alone.

SOLVING PROTEIN STRUCTURES BY NUCLEAR MAGNETIC RESONANCE SPECTROSCOPY

Some proteins cannot be crystallized, but it may be possible to use **nuclear magnetic resonance (NMR) spectroscopy** to solve their structures if they are relatively small. Until recently, the technique was limited to proteins of 15 kDa or less, but advances in data analysis now allow the proteins up to 100 kDa in size to be studied. The technique exploits the fact that some atomic nuclei have the propensity to switch between magnetic spin states in an applied magnetic field when exposed to radio waves of a certain frequency. When the nuclei flip back to their original orientations, they emit radio waves that can be measured. The frequency of the emitted radio waves depends on the type of atom and its chemical context. For example, the hydrogen nuclei in a methyl group will produce a different signal from those in an aromatic ring. By varying the nature of the applied radio waves, it is also possible to determine how close together two nuclei are in space, even if they are not connected by covalent chemical bonds. This is known as the **nuclear Overhauser effect (NOE)**.

From a two-dimensional NMR spectrum including NOE information, it is possible to work out which nuclei are covalently joined together and which are close together in space. In combination with the amino acid sequence of the protein this allows a list of *distance constraints* to be produced that firstly describes the secondary structural elements of the protein (these have very specific distance relationships, see Section 1.5.5) and then allows models of the tertiary structure to be developed. Generally, there are several variants of the tertiary structure that fit the distance data equally well, so these are deposited as an ensemble of models rather than one precise structure as is the case for X-ray crystallography.

PROTEIN STRUCTURE PREDICTION

Although the technology for solving protein structures has advanced significantly, it remains a labor-intensive and expensive process. An alternative although somewhat less accurate method is to predict protein structures using bioinformatic methods (Jones, 2000). At the present time, it is possible to predict secondary structures in hypothetical proteins quite accurately but tertiary structures require a template structure on which the model can be based.

The secondary structure of a protein can be predicted according to the propensity of certain amino acids to occur in particular secondary structures. Some amino acids, such as glutamate, have a helical propensity (i.e. they are most abundant in α-helices). Others, such as valine, have a strand propensity (i.e. they are most abundant in β-strands and β-sheets). Glycine and proline are unusual in that they are rarely found in secondary structures at all. Indeed they are often found at the ends of helices and strands and appear to act as 'structural terminators'. The appearance of a string of residues with helical or strand propensity strongly indicates the presence of such a structure in the protein. However, secondary structure predictions based on single proteins are unreliable because there are individual examples of all amino acids appearing in all types of secondary

structure. Multiple alignments can remove this uncertainty by identifying conserved blocks of residues that favor the formation of helices or strands. The most sophisticated algorithms, such as PSI-PRED (Jones, 2000), use multiple alignments and also incorporate evolutionary and structural information from the databases to enhance the accuracy of their predictions.

Tertiary structures are much more difficult to predict than secondary structures because there are millions upon millions of different ways in which any given linear chain of amino acids could fold. If enough solvent molecules are incorporated into the model to make it realistic, the system becomes too complex to study without some knowledge of the behavior of known proteins. For this reason, *ab initio* prediction methods are of no practical use. However, if the structure of a closely related protein is available, it can be used as a template to build a structural model of the query sequence. This is known as **comparative modeling** or **homology modeling** and generally works if the two sequences show > 25% identity. More distant relationships can be modeled by *threading*, where the sequence of a hypothetical protein is compared to those in the Protein Data Bank in an attempt to find compatible folds. It is often possible to find folds matching the protein core, which is highly conserved, but the external loops are extremely variable. It may be possible to find similar loops on other proteins using a so-called **spare parts algorithm**. Thus, the structural model is built up as a series of pieces from compatible sequences with known structures. The final step in both threading and comparative modeling is the refinement of the model, which involves moving side chains to avoid clashes and minimizing the overall free energy of the structure.

a technique in which antibodies specific for a particular bait protein are added directly to the cell lysate resulting in the precipitation of the antibody–bait protein complex. Any proteins interacting with the bait are co-immunoprecipitated. Both affinity chromatography and co-immunoprecipitation can be used to isolate entire **protein complexes**, allowing the different components to be identified by MS (Section 19.4.2). This is one of the major technology platforms in interaction proteomics. It has been widely used for the characterization of protein complexes such as ribosomes, the anaphase promoting complex, the nuclear pore complex and signaling complexes, and has recently been applied on a genomic scale (e.g. Gavin *et al.*, 2002; Ho *et al.*, 2002; see *Figure 19.16*). Specific interactions within each complex can be studied by *chemical cross-linking*.

19.4.6 High throughput interaction screening using library-based methods

It would be useful if high-throughput interaction screening methods provided a direct link between the proteins and the

genes encoding them. This has been addressed by the development of library-based methods for interaction screening. In principle, any standard cDNA expression library could be screened with a protein bait instead of a DNA probe. In practice, however, this is laborious because the same library must be screened many times to characterize the interactions of different baits. Two novel technologies have been used to screen for protein interactions and these are discussed below.

Interaction screening by phage display involves panning for interactions between bait proteins immobilized in the wells of microtiter dishes and interacting proteins expressed on the surface of recombinant bacteriophage

Phage display is a form of expression cloning in which foreign DNA fragments are inserted into a bacteriophage coat protein gene (See Section 5.6.2 and Burton, 1995). The recombinant gene can then be expressed as a fusion protein, which is incorporated into the virion and displayed on the surface of the phage (*Figure 19.17*). The fusion phage will bind to any protein that interacts with the foreign component displayed on its surface. Interaction screening is

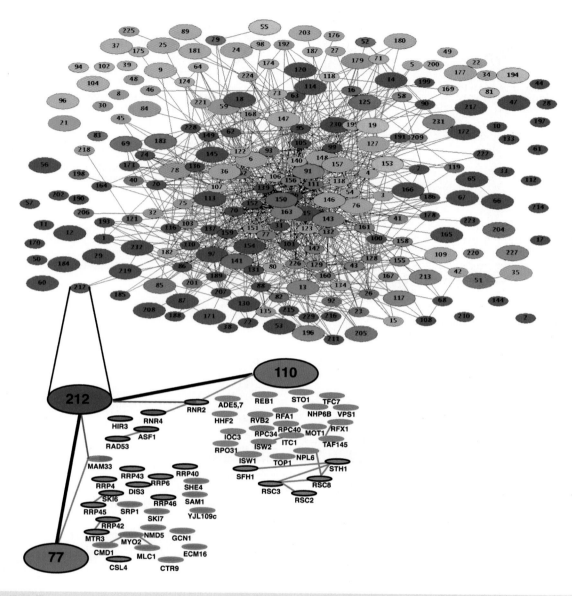

Figure 19.16: The network of protein interactions in yeast discovered by the systematic analysis of protein complexes by mass spectrometry.

A graph representing the yeast protein complex network, established by linking complexes sharing at least one protein (complexes sharing more than nine proteins have been omitted for clarity). In the upper panel, cellular roles of the individual complexes are color coded as follows: red, cell cycle; dark green, signaling; dark blue, transcription, DNA maintenance and chromatin structure; pink, protein and RNA transport; orange, RNA metabolism; light green, protein synthesis and turnover; brown, cell polarity and structure; violet, intermediate and energy metabolism; light blue, membrane biogenesis and traffic. The lower panel is an example of a complex (TAP-C212) linked to two other complexes (TAP-C77 and TAP-C110) by shared components (red lines indicate known physical interactions). After Gavin *et al.* (2002) Functional organization of the yeast proteome by systematic analysis of protein complexes. *Nature* **415**, 141–147 with permission from Nature Publishing Group and Cellzome AG.

carried out by creating a phage display library in which each protein in the proteome is displayed on the phage surface. The wells of microtiter plates are then coated with particular bait proteins of interest, and the phage display library is pipeted into each well. Phage with interacting proteins on their surface will remain bound to the surface of the well while those with noninteracting proteins are washed away. A particular advantage of the technique is that the retained phage, displaying interacting proteins, can be eluted from the wells and used to infect *E. coli*, resulting in massive amplification of the corresponding cDNA sequence, which can then be obtained and used to identify the interacting proteins by

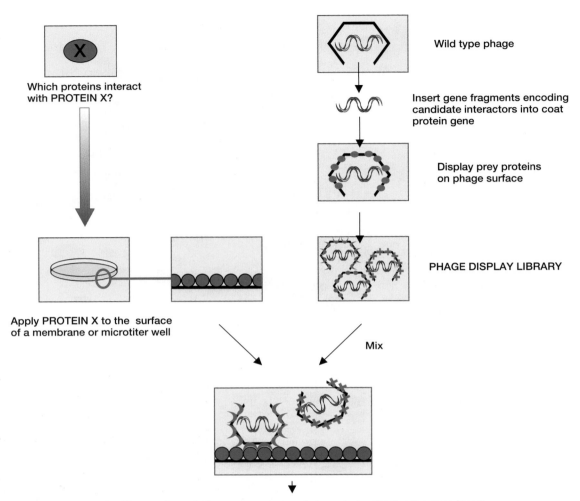

Figure 19.17: The principle of phage display as applied to high-throughout interaction screening (for the general principle, see *Figure 5.22*).

Bait proteins, for which interactors are sought, can be immobilized on the surface of microtiter wells or membranes. All other proteins in the proteome are then expressed on the surface of bacteriophage, by cloning within a phage coat protein gene, to create a phage display library. The wells or membranes are then flooded with the phage library. For any given bait protein (X), phage carrying interacting proteins will be retained while those displaying noninteracting proteins will be washed away. The bound phage can be eluted in a high-salt buffer and used to infect *E. coli*, producing a large amount of phage particles containing the DNA sequence of the interacting protein.

database searching. Disadvantages of phage display include the *in vitro* assay format and the fact that only short peptide sequence can be displayed on the phage surface without disrupting its replication cycle. Both of these features of the technique may prevent specific interactions taking place.

In addition to interaction screening, phage display is also useful for several other applications including:

▶ antibody engineering. Phage display is proving a powerful alternative source of antibodies (including humanized antibodies). It can bypass immunization and even hybridoma technology (Winter *et al.*, 1994; see Section 21.3.4;

▶ general protein engineering. Phage display is a powerful adjunct to random mutagenesis programs as a way of selecting for desired variants from a library of mutants.

The yeast two-hybrid system has the highest throughput of any interaction screening technology and involves the assembly of a functional transcription factor from separate components

The most widely used library method for interaction screening is the yeast two-hybrid system, also called the interaction trap system (Fields and Sternglanz, 1994). As well as identifying proteins that bind to a protein under study, this method can also be used to delineate domains or residues crucial for interaction. Proteins that physically interact are detected by their ability to assemble an active transcription factor and therefore activate a reporter gene and/or selectable marker. The key to the two-hybrid method is the observation that transcription factors comprise two separate domains, a

Box 19.6: Structural classification of proteins.

Functional annotation on the basis of protein structure requires a rigorous and standardized system for the classification of different structures. Several different hierarchical classification schemes are available which divide proteins first into general classes based on the proportion of various secondary structures they contain, then into successively more specialized groups based on how those structures are arranged (Pearl and Orengo, 2002). These schemes are implemented in databases such as SCOP (Structural Classification of Proteins), CATH (Class, Architecture, Topology, Homologous superfamily) and FSSP (Fold classification based on Structure-Structure alignment of Proteins).

These databases differ in the way classifications are achieved. For example, the FSSP system is implemented through fully automated structural comparisons using the DALI program. CATH is semi-automatic, using automatic comparisons that are manually curated. SCOP is a manual classification scheme and is based on evolutionary relationships as well as geometric criteria. Not surprisingly, the same protein may be classified differently when the alternative schemes are used (Hadley and Jones, 1999). There is broad general agreement in the upper levels of the hierarchy, but problems are encountered when more detailed classifications are sought because this depends on the thresholds used to recognize fold groups in the different classification schemes.

Additional problems that lead to confusion in the structural classification of proteins include:

▶ the existence of so-called **superfolds**, which are found in many proteins with diverse tertiary structures. It is necessary to distinguish between *homologous structures* (which are derived from a common evolutionary ancestor) and *analogous structures* (which evolved separately but have converged);

▶ variations in the fold structure between diverse members of the same protein family, which can result in a failure to recognize homologous relationships;

▶ the **Russian doll effect**, which describes a continuum of structures between fold groups. The assignment of a structure to a particular category then becomes very subjective.

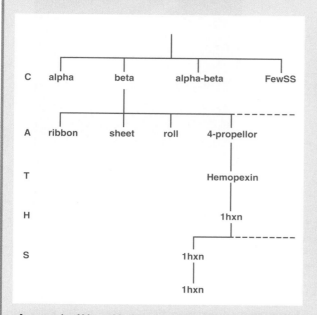

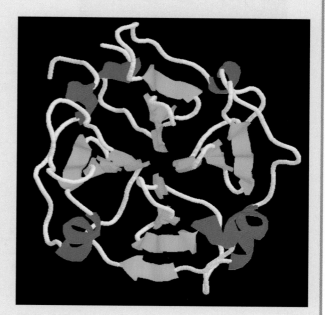

An example of hierarchical structural classification in the CATH database.

continued overleaf

Box 19.6: (continued)

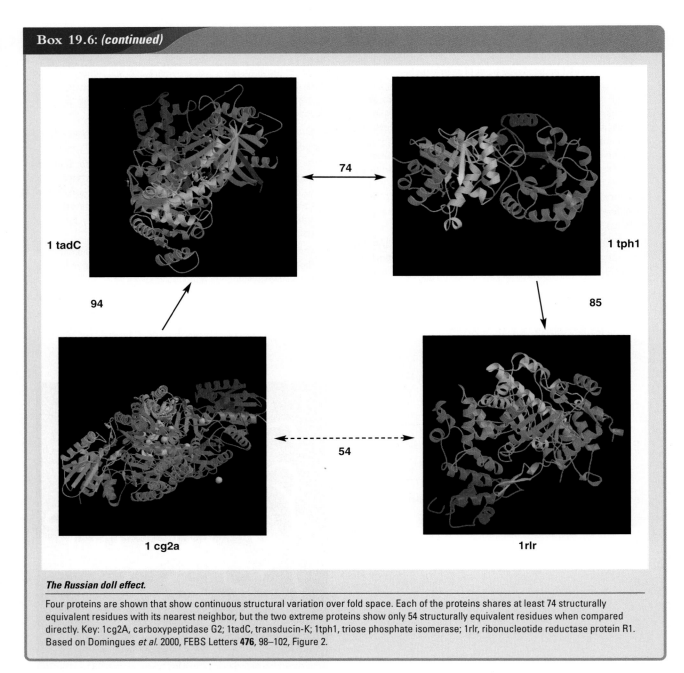

The Russian doll effect.

Four proteins are shown that show continuous structural variation over fold space. Each of the proteins shares at least 74 structurally equivalent residues with its nearest neighbor, but the two extreme proteins show only 54 structurally equivalent residues when compared directly. Key: 1cg2A, carboxypeptidase G2; 1tadC, transducin-K; 1tph1, triose phosphate isomerase; 1rlr, ribonucleotide reductase protein R1. Based on Domingues *et al.* 2000, FEBS Letters **476**, 98–102, Figure 2.

DNA binding domain and a transactivation domain. In most natural transcription factors, the DNA binding and activation domains are part of the same polypeptide (see Section 10.2.4). However, an active transcription factor can also assemble from two interacting proteins carrying separate domains. The object of the two-hybrid system and derivative techniques is to use a target protein as bait for specific recognition by an interacting protein which is fused to a necessary transcription factor component.

To use the two-hybrid system, standard recombinant DNA methods are used to produce a fusion gene which encodes the **bait protein** under study coupled to the DNA-binding

domain of a transcription factor (*Figure 19.18*). Cells transformed with this gene are mated to cells transformed with a library of fusion genes where cDNA sequences are coupled to the coding sequence of the transactivation domain (**prey constructs**). The target cells are also engineered to carry a reporter gene and/or a selectable marker gene which is activated by the assembled transcription factor. Interactions are tested in the resulting diploid yeast cells. If the bait and prey do not interact, the two transcription factor domains remain separate and the marker genes are inactive. However, if the bait and prey do interact, the transcription factor is assembled and the marker genes are then activated facilitating visual identifi-

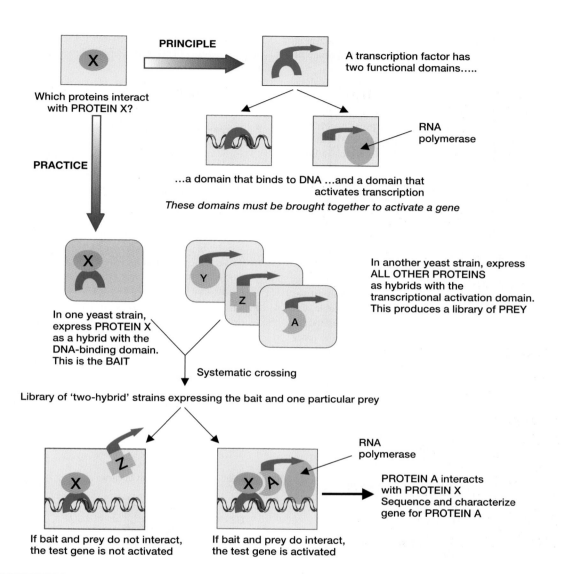

PRINCIPLE

Which proteins interact with PROTEIN X?

A transcription factor has two functional domains.....

RNA polymerase

...a domain that binds to DNA ...and a domain that activates transcription

These domains must be brought together to activate a gene

PRACTICE

In one yeast strain, express PROTEIN X as a hybrid with the DNA-binding domain. This is the BAIT

In another yeast strain, express ALL OTHER PROTEINS as hybrids with the transcriptional activation domain. This produces a library of PREY

Systematic crossing

Library of 'two-hybrid' strains expressing the bait and one particular prey

RNA polymerase

PROTEIN A interacts with PROTEIN X Sequence and characterize gene for PROTEIN A

If bait and prey do not interact, the test gene is not activated

If bait and prey do interact, the test gene is activated

Figure 19.18: Principle and practice of the yeast two-hybrid system.

Transcription factors generally comprise two functionally independent domains, one for DNA binding and one for transcriptional activation. These do not have to be covalently joined together, but can be assembled to form a dimeric protein. This principle is exploited to identify protein interactions. Bait proteins are expressed in one yeast strain as a fusion with a DNA-binding domain and candidate prey are expressed in another strain as fusions with a transactivation domain. When the two strains are mated, functional transcription factors are assembled only if the bait and prey interact. This can be detected by including a reporter gene activated by the hybrid transcription factor. Despite the simplicity of the principle, the technique is prone to reliability and reproducibility problems. False positives arise due to spontaneous autoactivation (the bait or prey can activate the reporter gene on their own), sticky baits and prey (proteins that interact nonspecifically with many others) and irrelevant interactions (chance interactions between proteins that would never encounter each other under normal circumstances, such as those usually found in separate compartments). False negatives may arise due to PCR errors in the construct, and due to nonphysiological conditions (the assay takes place in the nucleus, so proteins from other compartments many not fold or assemble properly).

cation and/or selective propagation of yeast cells containing interacting proteins. From these cells, the cDNA sequence of the prey construct can be identified.

Yeast two-hybrid screening was developed for the analysis of individual proteins (single baits) but in the last few years, the technique has been applied on an increasing scale, culminating with exhaustive studies of all 40 000 000 possible protein interactions in yeast (Uetz *et al.*, 2000; Ito *et al.*, 2000, 2001). Two variants of the technique are used, one in which defined panels of bait and prey are produced and crossed systematically (**matrix method**) and one in which either the prey, or both the bait and prey, are represented by random cDNA fragments (**random library method**) (*Figure 19.19*). The matrix method is exhaustive and systematic, but less reliable than the library method because each protein in the proteome is represented by a single construct that must be produced separately by PCR. The library method involves less work and is advantageous because each prey is represented by multiple, overlapping clones. This reduces the likelihood of false negatives because variations of the same prey construct are available, and also reduces the level of false positives because independent hits with different constructs representing the same prey increase the confidence level attributed to any interactions that are detected. Overall, however, the results of large scale screens must be interpreted with caution since false positive and negative results occur due to a variety of factors (see Legrain *et al.*, 2001 for a review). While most global two-hybrid studies have focused on microbes, a recent pilot program has shown that the technique is equally applicable to mammalian proteomes (Suzuki *et al.*, 2001).

In addition to the basic two-hybrid system, there are derivative methods with more specialized applications. These include the following:

▶ The **one-hybrid system**, which is used to detect interactions between proteins and specific DNA or RNA sequences;

▶ **Three-hybrid systems**, which are used to study more complex protein interactions, including interactions that involve both RNA and proteins (bait and hook);

▶ The **reverse two-hybrid system**, which is used to detect the disruption of protein interactions by transforming yeast with a suicide gene that is activated by the correctly assembled transcription factor.

▶ The **split ubiquitin system**, in which protein interactions bring together the two halves of the protein ubiquitin. The interaction may be monitored by testing for degradation of the bait protein by ubiquitin, or by the release of a functional protein such as a transcription factor.

▶ The **SOS recruitment system**, in which the interacting proteins are recruited to the membrane and complete an essential signaling pathway. The bait protein is tethered to the plasma membrane of a yeast cell that lacks CDC25 activity. Yeast *cdc* 25 mutants are not viable, but they can be rescued by the human ortholog SOS as long as the protein is localized at the membrane. By expressing a library of prey as SOS hybrids, this can be tested. This system is useful for studying the interactions of transcription factors, which would autoactivate the reporter gene in a conventional two-hybrid system.

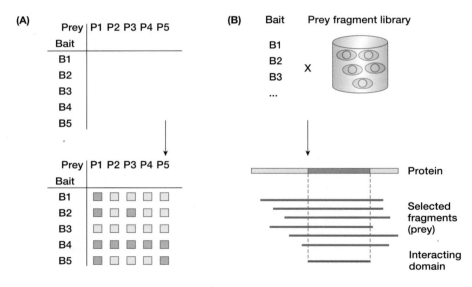

Figure 19.19: The matrix and library screening approaches to build large scale protein interaction maps.

(A) *The matrix approach.* This uses the same collection of proteins (1–5) as bait (B1–B5) and prey (P1–P5). The results can be drawn in a matrix. Autoactivators (for example, B4) and 'sticky' prey proteins (for example, P1 interacts with many baits) are identified and discarded. The final result is summarized as a list of interactions that can be heterodimers (e.g. B2–P3) or homodimers (e.g. B5–P5). **(B)** *The library screening approach.* This identifies the domain of interaction for each prey protein interacting with a given bait. Sticky prey proteins are identified as fragments of proteins that are often selected regardless of the bait protein. Reproduced from Legrain *et al.* (2001) with permission from Elsevier Science.

Table 19.2: **A selection of databases holding information on protein interactions**

Database and comments	URL
Biomolecular Interaction Network Database (BIND) Lists individual protein–protein interactions, pathways and complexes. Also lists protein interactions with other molecules.	http://www.bind.ca
Database of Interacting Proteins (DIP) Lists several thousand protein–protein interactions	http://dip.doe-mbi.ucla.edu
Kyoto Encyclopedia of Genes and Genomes (KEGG). An extensive resource on metabolic and signaling pathways that also has a section devoted to the structure of protein complexes	http://www.genome.ad.jp/kegg/

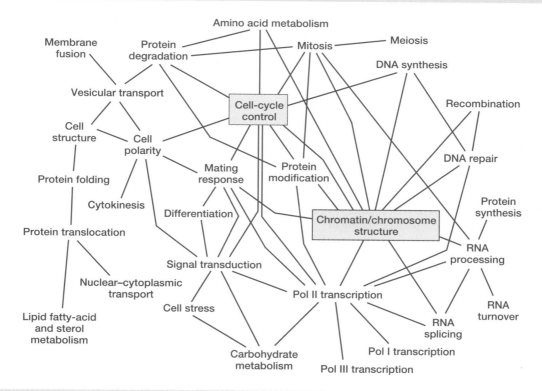

Figure 19.20: **Simplified 'functional group interaction map' of the yeast proteome.**

The data were derived from a more complex map of individual interactions, mostly obtained from two-hybrid screens. Each line indicates that there are 15 or more interactions between proteins of the connected groups. Connections with fewer than 15 interactions are not shown because a small number of interactions occur between almost all groups and often tend to be spurious, i.e. based on false positive results. Only fully annotated yeast proteins are included and many proteins are known to belong to several functional classes. Reproduced from Tucker *et al.* (2001) with permission from Elsevier Science.

▶ Mammalian two-hybrid systems, which can test interactions between proteins with authentic post-translational modifications.

19.4.7 The challenge of interaction proteomics is to assemble a functional interaction map of the cell

Protein interaction data provide useful functional information and, on a proteome-wide scale, potentially allow all the proteins in the cell to be linked into functional pathways and networks. The assimilation and presentation of this data is important and a number of databases have been established with this purpose in mind (*Table 19.2*). These data are mostly derived from large-scale MS and two-hybrid screens, but there is a potentially very large amount of data concerning individual protein interactions 'hidden' in the scientific literature going back many years (see blue lines in *Figure 19.16*). It will be a challenge to extract this information and integrate it with that obtained from recent high-throughput experiments. Interestingly, several bioinformatics tools have been developed to trawl through the literature and identify keywords that indicate protein interactions so that such references can be scrutinized by the human curators of interaction databases (Xenarios and Eisenberg, 2001).

A second problem is that protein interaction networks are not simple to represent, so clear data presentation is a significant challenge (*Figure 19.16*). However, as might be expected, proteins with a similar general function (e.g. membrane transport, DNA repair, amino acid metabolism) tend to interact with each other rather more than with functionally unrelated proteins. Complex maps can therefore be simplified as shown in *Figure 19.20* by grouping those functionally related proteins into foci representing basic cellular processes. Note that these processes themselves are linked in various ways, e.g. proteins that control chromatin structure have more interactions with those involved in DNA repair and recombination than with those involved in amino acid metabolism and protein degradation. This type of presentation allows protein interactions to be presented in a hierarchical manner, and will also provide a benchmark allowing the plausibility of novel interactions to be judged. Three fourths of all protein interactions occur within the same functional protein group, while most others occur with related functional groups. An unexpected interaction between proteins involved in unrelated processes could be regarded with suspicion and tested by rigorous genetic, biochemical and physical assays.

19.4.8 Information about protein interactions with small ligands can improve our understanding of biomolecular processes and provides a rational basis for the design of drugs

As well as protein–protein interactions (see above) and protein–nucleic acid interactions, proteins interact with a large number of small molecules which act either as ligands, substrates, cargoes (in the case of transport proteins), cofactors and allosteric modulators. In all cases, these interactions occur because the surface of the protein and the interacting molecule are complementary in shape and chemical properties, which means that binding leads to a reduction in free energy. The basis of drug design is to identify molecules that interact specifically with target proteins and change their activities in the cell. Most drugs work at the protein level and do so by interacting with the protein and altering its activity in

a manner that is physiologically beneficial. Drug related side effects often reflect nonspecific interactions with other proteins that result in harmful effects. The more we learn about protein structure and protein interactions with small molecules, the more information we can use to design more effective and more specific drugs.

Protein–ligand interactions can be investigated by the high-throughout screening of large numbers of chemical compounds but the amount of work can be reduced by first attempting to model such interactions in order to define suitable **lead compounds**. If the structure of the target protein is available (see above) computer programs known as **docking algorithms** can be used to screen databases of chemical structures and identify potential interacting ligands based on their complementarity. These algorithms attempt to fit small molecules into binding sites using information on steric constraints and bond energies. Examples include AUTODOCK, LIGIN and GRAMM. Chemical databases can be screened not only with a binding site (searching for complementary molecular interactions) but also with another ligand (searching for identical molecular interactions). This process of **rational drug design** has led to the production of several well-known drugs, including Relenza and Captopril.

19.5 Summary

The genome projects produce large amounts of sequence data, which must be mined using computers to identify genes and regulatory elements. Once the genes are available, it is necessary to determine their functions and how the various gene products interact with each other. In this chapter, we have discussed how sequence and structural analysis can be used to provide some information about protein function, and how the development of high-throughput technologies for transcriptome and proteome analysis have helped to link these functions together. One component of functional genomics we have not addressed is the use of mutants to study gene functions. Because this involves the genetic modification of cells and animals, we have deferred the topic until the next chapter, which considers how gene manipulation can be used to study gene function and regulation and to build models of human disease.

Further reading

Eisenberg D, Marcotte EM, Xenarios I, Yeates TO (2000) Protein function in the post-genomic era. *Nature* **405**, 823–827.

Hanash S (2003) Disease proteomics. *Nature* **422**, 226–232.

Lee KH (2001) Proteomics: a technology-driven and technology-limited discovery science. *Trends Biotechnol.* **19**, 217–222.

Lockhart DJ, Winzeler (2000) Genomics, gene expression and DNA arrays. *Nature* **405**, 827–836.

Pandey A, Mann M (2000) Proteomics to study genes and genomes. *Nature* **405**, 837–846.

Primrose SB, Twyman RM (2002) *Principles of Genome Analysis and Genomics*. Blackwell Science Oxford UK.

Various authors (1999) The Chipping Forecast. *Nature Genet.* **21**, (Suppl.) 1–60.

Various authors (2002) The Chipping Forecast II. *Nature Genet.* **32** (Suppl.) 465–551.

Vorm O, King A, Bennett KL, Leber T, Mann M (2000) Protein-interaction mapping for functional proteomics. In *Proteomics: A Trends Guide* 43–47.

References

Aebersold R, Mann M (2003) Mass spectrometry-based proteomics. *Nature* **422**, 198–207.

Aicher L, Wahl D, Arce A, Grenet O, Steiner (1998) New insights into cyclosporine A nephrotoxicity by proteome analysis. *Electrophoresis* **19**, 1998–2003.

Alizadeh AA, Eisen MB, Davis RE, Ma C, Lossos IS, Rosenwald A, Boldrick JC, Sabet H, Tran T, Yu X et al. (2000) Distinct types of diffuse large B-cell lymphoma identified by gene expression profiling. *Nature* **403**, 503–511.

Altman RB, Raychaudhuri S (2001) Whole-genome expression analysis: challenges beyond clustering. *Curr. Opin. Struct. Biol.* **11**, 340–347.

Altschul SF, Madden, TL, Schaeffer AA, Zhang J, Zhang Z, Miller, Lipman DJ (1997) Gapped BLAST and PSI-BLAST: a new generation of protein database search programs. *Nucl. Acids Res.* **25**, 3389–3402.

Aparicio S, Chapman J, Stupka E et al. (2002) Whole-genome shotgun assembly and analysis of the genome of Fugu rubripes. *Science* **297**, 1301–1310.

Audic S, Claverie J (1997) The significance of digital gene expression profiles. *Genome Res.* **7**, 986–995.

Bittner M, Meltzer P, Chen Y, Jiang Y, Seftor E, Hendrix M, Radmacher M, Simon R, Yakhini Z, Ben-Dor A et al. (2000) Molecular classification of cutaneous malignant melanoma by gene expression profiling. *Nature* **406**, 536–540.

Brenner S, Johnson M, Bridgham J, Golda G, Lloyd DH, Johnson D, Luo S, McCurdy S, Foy M, Ewan M et al. (2000) Gene expression analysis by massively parallel signature sequencing (MPSS) on microbead arrays. *Nat. Biotechnol.* **18**, 630–634.

Brenner SE (1999) Errors in genome annotation. *Trends Genet.* **15**, 132–133.

Brenner SE (2001) A tour of structural genomics. *Nature Rev. Genet.* **2**, 801–809.

Burton D (1995) Phage display. *Immunotechnology* **1**, 87–94.

Celis JE, Kruhoffer M, Gromova I, Frederiksenb C et al. (2000) Gene expression profiling: monitoring transcription and translation products using DNA microarrays and proteomics. *FEBS Lett* **480**, 2–16.

Der SD, Zhou A, Williams BR, Silverman RH (1998) Identification of genes differentially regulated by interferon alpha, beta, or gamma using oligonucleotide arrays. *Proc. Natl Acad. Sci.* USA **95**, 15623–15628.

Dujon B (1996) The yeast genome project: what did we learn? *Trends Genet.* **12**, 263–270.

Eddy SR (1998) Multiple alignment and sequence searches. *Bioinformatics, A Trends Guide* **5**, 15–18.

Elgar G, Sandford R, Aparicio S, Macrae A, Vekatesh B, Brenner S (1996) Small is beautiful: comparative genomics with the pufferfish (*Fugu rubripes*). *Trends Genet.* **12**, 145–150.

Fields S, Sternglanz R (1994) The two hybrid system: an assay for protein-protein interactions. *Trends Genet.* **10**, 286–292.

Franzen B, Auer G, Alaiya AA, Eriksson E et al. (1996) Assessment of homogeneity in polypeptide expression in breast carcinomas shows widely variable expression in highly malignant tumors. *Int. J. Cancer* **69**, 408–414.

Gavin AC et al. (2002) Functional organization of the yeast proteome by systematic analysis of protein complexes. *Nature* **415**, 141–147.

Gene Ontology Consortium (2000) Gene ontology: tool for the unification of biology. *Nature Genet.* **25**, 25–29.

Gene Ontology Consortium (2001) Creating the gene ontology resource: design and implementation. *Genome Research* **11**, 1425–1433.

Golub TR, Slonim DK, Tamayo P, Huard C, Gaasenbeek M, Mesirov JP, Coller H, Loh ML, Downing JR, Caligiuri MA et al. (1999) Molecular classification of cancer: class discovery and class prediction by gene expression monitoring. *Science* **286**, 531–537.

Gorg A, Obermaier C, Boguth G, Harder A et al. (2000) The current state of two-dimensional electrophoresis with immobilized pH gradients. *Electrophoresis* **21**, 1037–1053.

Griffen TJ, Goodlet DR, Aebersold R (2001) Advances in proteome analysis by mass spectrometry. *Curr. Opin. Biotechnol.* **12**, 607–612.

Gygi SP, Rist B, Gerber SA, Turecek F, Gelb MH, Aebersold R (1999a) Quantitative analysis of complex protein mixtures using isotope coded affinity tags. *Nature Biotechnol.* **17**, 994–999.

Gygi SP, Rochon Y, Franza BR, Aebersold R (1999b) Correlation between protein and mRNA abundance in yeast. *Mol. Cell Biol.* **19**, 1720–1730.

Hadley C, Jones D (1999) A systematic comparison of protein structure classifications: SCOP, CATH and FSSP *Struct. Fold Des.* **7**, 1099–1112.

Hardison RC (2000) Conserved non-coding sequences are reliable guides to regulatory elements.

Harrington CA, Rosenow C, Retief J (2000) Monitoring gene expression using DNA microarrays. *Curr. Opin. Microbiol.* **3**, 285–291.

Heinemann U, Illing G, Oschkinat H (2001) High-throughput three-dimensional protein structure determination. *Curr. Opin. Biotechnol.* **12**, 348–354.

Herbert BR, Harry JL, Packer NH et al. (2001) What place for polyacrylamide in proteomics? *Trends Biotechnol.* **19**, S3–S9.

Ho Y et al. (2002) Systematic identification of protein complexes in *Saccharomyces cerevisiae* by mass spectrometry. *Nature* **415**, 180–183.

International Human Genome Sequencing Consortium (2001) Initial sequencing and analysis of the human genome. *Nature* **409**, 860–921.

Ito T, Tashiro K, Muta S, Ozawa R, Chiba T, Nishizawa M, Yamamoto K, Kuhara S, Sakaki Y (2000) Toward a protein–protein interaction map of the budding yeast: a comprehensive system to examine two-hybrid interactions in all possible combinations between the yeast proteins. *Proc. Natl Acad. Sci.* USA **97**, 1143–1147.

Ito T, Chiba T, Ozawa R, Yoshida M, Hattori M, Sakaki Y (2001) A comprehensive two-hybrid analysis to explore the yeast protein interactome. *Proc. Natl Acad. Sci.* USA **98**, 4569–4574.

Iyer VR, Eisen MB, Ross DT et al. (1999) The transcriptional program in the response of human fibroblasts to serum. *Science* **283**, 83–87.

Jones DT (2000) Protein structure prediction in the postgenomic era. *Curr. Opin. Struct. Biol.* **10**, 371–379.

Khan J, Bittner ML, Saal LH, Teichmann U, Azorsa DO, Gooden GC, Pavan WJ, Trent JM, Meltzer PS (1999) cDNA microarrays detect activation of a myogenic transcription program by the PAX3-FKHR fusion oncogene. *Proc. Natl Acad. Sci.* USA **96**, 13264–13269.

Kolkman JA, Stemmer WPC (2001) Directed evolution of proteins by exon shuffling. *Nature Biotechnol.* **19**, 423–428.

Legrain P, Wojcik J, Gauthier JM (2001) Protein–protein interaction maps: a lead towards cellular functions. *Trends in Genetics* **17**, 346–352.

Lesney MS (2001) Pathways to the proteome: from 2DE to HPLC. *Modern Drug Discovery*, October issue, pp 33–39.

Liang P, Pardee AB (1992) Differnetial display analysis of eukaryotic messenger RNA by means of the polymerase chain reaction. *Science* **257**, 967–971.

Lilley KS, Razzaq A, Dupree P (2001) Two-dimensional gel electrophoresis: recent advances in sample preparation, detection and quantitation. *Curr. Opin. Chem. Biol.* **6**, 46–50.

Lockhardt DJ, Dong H, Byrne MC et al. (1996) Expression monitoring by hybridization to high-density oligonucleotide arrays. *Nature Biotechnol.* **14**, 1675–1680.

Mann M, Hendrickson RC, Pandey A (2001) Analysis of proteins and proteomes by mass spectrometry. *Annu. Rev. Biochem.* **70**, 437–473.

McCaffrey TA, Fu C, Du B, Eksinar S, Kent KC, Bush H Jr, Kreiger K, Rosengart T, Cybulsky MI, Silverman ES, Collins T (2000) Array-based screening identifies differential expression of heat shock proteins in human atherosclerotic lesions. *J. Clin. Invest.* **105**, 653–662.

Nadeau JH, Sankoff D (1998) Counting on comparative maps. *Trends Genet.* **14**, 495–501.

Norin M, Sundstrom M (2002) Structural proteomics: developments in structure-to-function predictions. *Trends Biotechnol.* **20**, 79–84.

Orengo CA, Jones DT, Thornton JM (2003) *Bioinformatics: Genes, proteins and computers.* BIOS Scientific Publishers, Oxford.

Orengo CA, Todd AE, Thornton JM (1999) From protein structure to function. *Curr. Opin. Struct. Biol.* **9**, 374–382.

Patton WF, Beecham JM (2001) Rainbow's end: the quest for multiplexed fluorescence quantitative analysis in proteomics. *Curr. Opin. Chem. Biol.* **6**, 63–69.

Pearl F, Orengo C (2002) Protein structure classifications. In: Orengo CA, Jones DT, Thornton JM (eds) *Bioinformatics: Genes Proteins and Computers.* BIOS Scientific Publishers, Oxford.

Phizicky EM, Fields S (1995) Protein–protein interactions: methods for detection and analysis. *Microbiol. Rev.* **59**, 94–123.

Quackenbush J (2001) Computational analysis of microarray data. *Nature Rev. Genet.* **2**, 418–427.

Rabilloud T (2002) Two-dimensional gel electrophoresis in proteomics: old, old fashioned, but still it climbs up the mountains. *Proteomics* **2**, 3–10.

Schena M, Shalon D, Davis RW, Brown OP (1995) Quantitative monitoring of gene expression patterns with a complementary DNA microarray. *Science* **270**, 467–470.

Sechi S, Oda Y (2003) Quantitative proteomics using mass spectrometry. *Curr. Opin. Chem. Biol.* **7**, 70–77.

Shapiro L, Scherer PE (1998) The crystal structure of a complement-1q family protein suggests an evolutionary link to tumor necrosis factor. *Curr. Biol.* **8**, 335–338.

Sillitoe I, Orengo C (2002) Protein structure comparison. In: Orengo CA, Jones DT, Thornton JM (eds) *Bioinformatics: Genes Proteins and Computers.* BIOS Scientific Publishers, Oxford UK.

Stulik J, Osterreicher J, Koupilova K, Knizek J et al. (1999) Protein abundance alterations in matched sets of macroscopically normal colon mucosa and colorectal carcinoma *Electrophoresis* **20**, 1047–1054.

Suzuki H, Fukunishi Y, Kagawa I, Saito R et al. (2001) Protein-protein interaction panel using mouse full length cDNAs. *Genome Res.* **11**, 1758–1765.

Tucker CL, Gera JF, Uetz P (2001) Towards an understanding of complex protein networks. *Trends Cell Biol.* **11**, 102–106.

Uetz P, Giot L, Cagney G, Mansfield TA et al. (2000) A comprehensive analysis of protein–protein interactions in Saccharomyces cerevisiae. *Nature* **403**, 623–627.

Vasmatzis G, Essand M, Brinkmann U et al. (1998) Discovery of three genes specifically expressed in human prostate by expressed sequence tag database analysis. *Proc. Natl Acad. Sci. USA* **95**, 300–304.

Velculescu VE, Volgelstein J (2000) Analysing uncharted transcriptomes with SAGE. *Trends Genet.* **16**, 423–425.

Velculescu VE, Zhang L, Vogelstein B, Kinzler KW (1995) Serial analysis of gene expression. *Science* **270**, 484–487.

Velculescu VE, Zhang L, Zhou W, Vogelstein J et al. (1997) Characterization of the yeast transcriptome. *Cell* **88**, 243–251.

Venter JC, Adams MD, Myers EW et al. (2001) The sequence of the human genome. *Science* **291**, 1304–1351.

Wang H, Hanash S (2003) Multi-dimensional liquid phase based separations in proteomics. *J. Chromatography B.* **787**, 11–18.

Wasinger VC, Cordwell SJ, Cerpa-Poljak A et al. (1995) Progress with gene product mapping of the Mollicutes: *Mycoplasma genitalium. Electrophoresis* **16**, 1090–1094.

Werner T (2003) Promoters can contribute to the elucidation of protein function. *Trends Biotechnol.* **21**, 9–13.

Winter G, Griffiths AD, Hawkins RE, Hoogenboom HR (1994) Making antibodies by phage display technology. *Annu. Rev. Immunol.* **12**, 433–455.

Xenarios I, Eisenberg D (2001) Protein interaction databases. *Curr. Opin. Biotechnol.* **12**, 334–339.

Yang YH, Dudoit S, Luu P, Lin DM, Peng V, Ngai J, Speed TP (2002) Normalization for cDNA microarray data: a robust composite method addressing single and multiple slide systematic variation. *Nucleic Acids Res.* **30**, 15.

Zhang C, Kim SH (2003) Overview of structural genomics: from structure to function. *Curr. Opin. Chem. Biol.* **7**, 28–32.

Zhu H, Snyder M (2003) Protein chip technology. *Curr. Opin. Chem. Biol.* **7**, 55–63.

Zhu H, Cong JP, Mamtora G, Gingeras T, Shenk T (1998) Cellular gene expression altered by human cytomegalovirus: global monitoring with oligonucleotide arrays. *Proc. Natl Acad. Sci.* **95**, 14470–14475.

Zhu H, Bilgin M, Bangham R, Hall D et al. (2001) Global analysis of protein activities using proteome chips. *Science* **293**, 2101–2105.

CHAPTER TWENTY

Genetic manipulation of cells and animals

Chapter contents

20.1 An overview of gene transfer technology

Cultured cells and experimental animals have been widely exploited as model systems to study aspects of our biochemistry, physiology and development in health and disease. In recent times, such studies have benefited enormously from **gene transfer technology**, which allows specific DNA sequences to be introduced into the genome of animal cells (**transgenesis**). Before gene transfer techniques were available, the only way in which cells and animals could be genetically altered was by *mutagenesis*, a process that involves the use of radiation or powerful chemicals to generate random modifications in the existing genome.

One of the advantages offered by gene transfer technology is that it allows *new* DNA sequences that have been prepared *in vitro* to be added to the genome. These additional sequences can be genes that provide new functions (*gain of function*) or constructs that inhibit the expression of endogenous genes (*loss of function*). On another level, gene transfer technology allows the existing genome to be modified in a *precise and predetermined way*, so that mutations can be designed and introduced into particular genes *in vivo* (**gene targeting**). Together with sophisticated methods for the regulation of gene expression, gene transfer technology greatly enhances our ability to study and control how genes function, and to construct mouse models of disease.

Gene transfer to cultured mammalian cells was first documented in the early 1960s, when human cells were found to take up and incorporate fragmented genomic DNA from their culture medium (Szybalska and Szybalski, 1962). It took some time for the factors governing DNA transfer to be worked out, but by the early 1970s the introduction of DNA into many cell types was a routine procedure. At about the same time, the first **transgenic animals** were produced. Transgenic animals contain the same additional DNA sequence in every cell and are generated by extending the principles of gene transfer to cells that contribute to the animal germ line. The first such animals were mice containing part of the SV40 (simian virus 40) DNA sequence, created simply by injecting DNA into the blastocele cavities of preimplantation embryos (Jaenisch and Mintz, 1974). Since then, reliable procedures for the genetic modification of mice and many other species have been developed, including invertebrate model organisms (fruit flies and nematodes) as well as fish, frogs, birds, rats and various domestic and livestock mammals. Another experimental approach involving the genetic manipulation of animals has had a major impact more recently. In 1997 a new era in genetics was heralded when the first success in **mammalian cloning** was reported (Wilmut *et al.*, 1997). This involved transfer of the nucleus from an adult cell into an enucleated oocyte (**somatic cell nuclear transfer**) and the technology has subsequently been applied as an alternative route for the production of genetically modified animals.

In the first part of this chapter, we describe the principles and methods of gene transfer, and strategies for the genetic manipulation of animals. In the remainder of this chapter and in Chapter 21, we discuss how these methods are applied, focusing on the following areas of biomedical research.

▶ **The study of gene expression and function** (this chapter). Cultured animal cells have been widely used to investigate gene function and the regulation of gene expression. The reason for this is simple: animal cells provide the correct genetic background and an authentic biochemical context for the study of animal genes. It is not possible to recreate this context accurately with *in vitro* experimental systems. The ability to add genes or to selectively delete or alter specific genes in animals provides a powerful basis for functional analysis in the context of the whole organism.

▶ **Applications in functional genomics** (this chapter). Gene functions can be studied on a global scale through the creation of genome-wide insertional mutant libraries, gene trap libraries and the development of high throughput gene knockdown techniques based on RNA interference (Section 20.2.6). These techniques have been applied to a range of model animals from the nematode *Caenorhabditis elegans* to the mouse, as well as to human cells in culture. They complement the diverse functional genomics strategies discussed in Chapter 19.

▶ **The creation of disease models** (this chapter). Nature has provided some animal models of disease and some have been generated by random mutagenesis, but *not in a predetermined way*. Gene transfer technology allows the creation of specific mutant genotypes in animals thereby increasing their chance of resembling human diseases at the genetic and phenotypic levels. To a more limited extent, pathogenesis can also be modeled in cultured cells.

▶ **The production of recombinant proteins** (Chapter 21). Cultured cells and transgenic animals have been used as 'bioreactors' for the production of recombinant proteins. Mammalian systems are particularly advantageous for the production of therapeutic human proteins because the recombinant proteins undergo authentic post-translational modification.

▶ **Developing the potential for gene therapy** (Chapter 21). Gene transfer to human cells potentially allows the correction or amelioration of genetic defects.

20.2 Principles of gene transfer

20.2.1 Gene transfer can be used to introduce new, functional DNA sequences into cultured animal cells either transiently or stably

Gene transfer is often used to add new and functional genes to animal cells, which generally means that the cells gain the ability to express one or more new proteins. This is described as a **gain of function**. The additional DNA sequence is called a **transgene**, although it may contain one, two or even more actual genes. It is sometimes termed 'foreign DNA' although this is misleading because it is perfectly acceptable to introduce extra copies of an endogenous gene, and indeed this approach can be very useful for studying gene function. The alternative term 'exogenous DNA' can be applied more generally and is preferable. Transgenes can be introduced into cultured cells using a variety of viral and nonviral delivery methods, which are summarized in *Box 20.1*.

The fate of the transgene is important. In some cases it is maintained transiently in the host cells whereas in other cases it becomes a permanent part of the genome. Where virus-mediated gene transfer is used, the fate of the transgene depends on the properties of the virus. Some viruses are only suitable for transient expression while others are capable of latent infections, and long term transgene expression may be possible. Retroviruses are unusual in that a DNA copy of the virus genome integrates into the host genome shortly after infection. Recombinant retroviruses can therefore be used to create stably transformed cell lines.

When DNA is introduced into cultured animal cells by transfection, it is usually in the form of bacterial plasmids that cannot replicate in animals. A large proportion of the cells may initially take up the DNA but it is soon broken down and any transgene expression is transient. The longevity of the plasmid depends on the quality of the DNA and on the host cell line. In most cells, good quality plasmid DNA will persist for 1–2 days but in some cases it may last much longer (in the HEK293 line, for example, plasmid DNA may last for up to 80 h). In a very small proportion of the transfected cells, some of the plasmid DNA integrates into the genome, resulting in **stable transformation**. This is such a rare event that powerful selection strategies must be used to identify the stable transformants. The general strategy is to include a **selectable marker gene** on the plasmid or to cotransfect the cells with two plasmids, one containing the primary transgene and one containing a selectable marker. The selectable marker gene confers a property on the cell that allows it to survive in the presence of a *selective agent*, such as an antibiotic, that kills nontransformed cells (*Box 20.2*). In this way, it is possible to derive stably transformed cell lines.

Alternatively, cells may be transfected with a plasmid vector containing a viral replication origin, in which case the transgene is maintained and replicated episomally. For some vectors, the origin facilitates rapid, runaway plasmid replication, which leads to transient high levels of transgene expression but then kills the cells. This is the case for COS cells transfected with plasmids carrying an SV40 replication origin. Conversely, vectors containing the latent replication origin of Epstein–Barr virus (EBV) are maintained at a moderate copy number and, if appropriate selectable markers are included, can be used to derive stably transformed cell lines.

In summary, both viral and nonviral gene delivery methods can be used to introduce new genes into animal cells either transiently or stably. Stable transformation can be achieved by retroviral integration, the long term episomal maintenance of a replicating vector or the integration of nonreplicating DNA into one of the host cell chromosomes.

20.2.2 The production of transgenic animals requires stable gene transfer to the germ line

The methods discussed above are suitable for the transformation of cells in a culture dish but extending this technology to whole animals is more difficult. It is impossible to introduce DNA uniformly into all the cells of an adult animal. Therefore, in order to produce a fully transgenic animal (an animal containing the same transgene in the same context in every cell) that animal must develop from a *transgenic zygote*. This is achieved by stable gene transfer to the *germ line* and can be carried out in one of two ways (*Figure 20.1*).

▶ Directly, by introducing DNA *selectively* into germ cells, the gametes derived therefrom, or into the egg just after fertilization. If the DNA integrates prior to the first division of the zygote, every cell in the resulting animal will contain the same transgene.

▶ Indirectly, by introducing DNA *nonselectively* into the embryo prior to the formation of the germ line. At this stage, the embryonic cells are totipotent, or at least pluripotent, which means they can form any of the cell types in the developing embryo. Because the DNA integrates after the first division of the zygote, only some of the cells in the embryo will incorporate the transgene. However, if those cells contribute to the germ line of the embryo, transgenic gametes will be produced and the subsequent generation of animals will be transgenic.

The methods for gene transfer to animals are summarized in *Table 20.1* and are discussed in more detail below.

Direct gene transfer to germ line precursor cells is the standard procedure for producing transgenic fruit flies

The introduction of DNA into germ cells is potentially a very useful way to access the germ line because it leads to the production of transgenic gametes. This method is not used in mammals because although mammalian *primordial germ cells* are relatively easy to isolate, culture and transfect, it is difficult to persuade the modified cells to contribute to the germ line when reintroduced into a host animal. However, direct modification of germ cells is the routine method for the production of transgenic fruit flies. In *Drosophila melanogaster*, efficient chromosomal integration is achieved using sequences from a transposable element known as the *P-element*. The transgene is inserted between the two terminal sequences of the P-element and then injected into the *pole plasm* of a young embryo, which is where the nuclei of the *pole cells* (germ line precursor cells) are found. The transposase enzyme required for P-element transposition is provided by a co-injected plasmid, and this facilitates random transgene integration into the genome of one or more of the pole cell nuclei. A single copy of the transgene is usually incorporated.

DNA can be mixed with sperm or sperm nuclei and introduced into unfertilized eggs

The direct modification of gametes is another way to generate transgenic animals. Two very different methods have been devised: sperm-mediated DNA delivery and restriction enzyme-mediated integration.

Sperm-mediated DNA delivery exploits the fact that sperm heads bind spontaneously to DNA *in vitro*. Therefore, sperm are used as delivery vehicles although the *gamete genome itself is not modified*. **ICSI (intracytoplasmic sperm injection)** is an infertility treatment involving the injection

Box 20.1: Methods of gene transfer to animal cells in culture.

There are four general classes of gene transfer methodology that can be applied to cultured animal cells, but transduction and transfection are the most widely used. Variations on these methods can be used to introduce DNA into human cells *in vivo* (Section 21.5).

▶ **TRANSDUCTION**. This is *virus-mediated gene transfer*, i.e. the foreign DNA is packaged inside a virus particle. Animal viruses have evolved various ways to introduce their own DNA or RNA into cells, and many viruses have therefore been exploited as gene transfer vectors. These include:

a) viruses that facilitate *high level transient transgene expression*, e.g. adenovirus, Sindbis virus, Semliki Forest virus, vaccinia virus, baculovirus (in insect cells, which support baculovirus replication);

b) viruses that are maintained in a latent episomal state and facilitate *long term stable transgene expression*, e.g. Epstein–Barr virus, herpes simplex virus, baculovirus (in mammalian cells, which do not support baculovirus replication);

c) viruses that integrate into the genome and can be used for permanent, *stable transformation*, e.g. retroviruses, adeno-associated virus.

▶ **TRANSFECTION**. This involves the use of chemical or physical tricks to persuade cells to take up DNA from the culture medium, the DNA eventually finding its way to the nucleus. Note that, in bacterial systems, transfection refers to the uptake of naked viral (phage) DNA while transformation refers to the uptake of naked plasmid or genomic DNA. In animal cells, transfection refers to the uptake of any naked DNA while transformation, if used in the context of gene transfer, generally refers to a stable and permanent change in genotype. There are numerous different transfection methods including:

a) **chemical transfection**. When DNA is mixed with calcium chloride in the presence of a phosphate buffer, a fine **DNA–calcium phosphate precipitate** is formed. This settles onto the plasma membrane of cultured cells and is taken up by endocytosis. DNA can also form *soluble* complexes with various other chemicals, including DEAE-dextran (diethylaminoethyl-dextran) or the detergent Polybrene. Cells that do not like being coated with calcium phosphate can respond better to these alternatives;

b) **liposome-mediated transfection** (*Figure 21.8*). DNA can be encapsulated within artificial lipid vesicles known as **liposomes** which fuse to the plasma membrane and deposit their cargo into the cytosol. The efficiency of transfection can be increased by deriving the liposome from viral envelopes, which often contain proteins that enhance membrane fusion. Such vehicles are termed *virosomes*. Alternatively, actual cell membranes can be used as the delivery vehicle. This is the basis of *protoplast fusion*, where bacterial protoplasts loaded with DNA are centrifuged onto cultured mammalian cells and induced to fuse with them using polyethylene glycol (PEG). A similar technique employs the hemoglobin-free ghosts of erythrocytes;

c) **lipofection**. Unlike liposome-mediated transfection, where the DNA is encapsulated within a lipid vesicle, *lipofection* involves the formation of a DNA–lipid complex (*lipoplex*) which is taken up efficiently by endocytosis. Lipofection thus has more in common with chemical transfection than liposome-mediated transfection. More recently, cationic polymer-based gene delivery vehicles have become popular (*polyplexes*). These are as efficient as lipoplexes but have the advantage that specific copolymers can be used to modify the physical properties of the delivery vehicle. In some cases it has proven possible to form complexes that have different properties at different temperatures, allowing controlled release of DNA (Yokayama, 2002). This could be particularly useful for *in vivo* gene transfer to particular sites in the body (Chapter 21);

d) **electroporation**. In this method, cells are subjected to a brief electric pulse which causes transient, nanometer-sized pores to appear in the plasma membrane. If DNA is present in the buffer solution at a sufficient concentration, it will be taken up through these pores;

e) **receptor-mediated endocytosis** (*Figure 21.9*). In this method, DNA is attached to the ligand of a cell surface receptor that is recycled by endocytosis. The DNA is released during the processing of the receptor. Under normal circumstances, the endosome containing the receptor– ligand–DNA complex fuses to a lysosome and the DNA is degraded, but this can be prevented by including adenoviral peptides in the transfection complex since these disrupt endosomes. This method can be used to target specific cell types based on their receptor-specificity.

▶ **DIRECT TRANSFER**. This involves the direct physical introduction of DNA into the cell. The obvious example is **microinjection**, which is occasionally used on cultured cells that are recalcitrant to other gene transfer methods but is usually applied to animal eggs, zygotes or early embryos. **Particle bombardment**, which involves the acceleration of high-velocity DNA-coated microprojectiles into cells, is another direct transfer method. This can be used for cultured cells but is the method of choice for the transfection of cells in tissue slices, and can also be used for *in vivo* gene transfer (Section 21.5.4).

▶ **BACTERIAL GENE TRANSFER**. This typically involves the use of live, invasive bacteria that undergo lysis within the animal cell, releasing their cargo of DNA (Higgins and Portnoy, 1998). In the case of *Salmonella* species, lysis occurs in a phagocytic vesicle, while for other species (e.g. *Listeria monocytogenes* and *Shigella flexneri*) lysis occurs after the bacterium has escaped from the vesicle. Bacteria can also attach to the cell surface and transfer DNA through a pilus. *Agrobacterium tumefaciens* uses this method to transfer DNA to plants and can also infect cultured animal cells (Kunik *et al.*, 2001).

Box 20.2: Selectable markers for animal cells.

CATEGORIES OF SELECTABLE MARKER GENE

A **selectable marker gene** confers a property that allows stably transformed cells to survive and grow in the presence of a particular agent that kills or restricts the growth of nontransformed cells. This is known as **positive selection** because cells are being selected on the basis that they carry the marker gene (they are *positive* for the marker gene). Positive selection is essential to isolate the very few cells in a culture dish that are stably transformed following transfection with nonreplicating DNA. There are two major classes of selectable marker gene: endogenous selectable markers and dominant selectable markers.

▶ **Endogenous selectable markers** are genes that are found in the genome of the host cell. Selection can therefore be applied only to cells that *lack a functional copy of that gene*. An example is the *TK* gene, encoding the enzyme **thymidine kinase**. This enzyme converts free thymidine into thymidine monophosphate as part of the **nucleoside salvage pathway**. Under normal circumstances, the salvage pathway is nonessential because thymidine nucleotides can also be synthesized *de novo* from uridine monophosphate. However, if the *de novo pathway* is blocked with the inhibitor **aminopterin**, the cell becomes dependent on TK activity. Therefore, if TK⁻ cells (which lack a functional *TK* gene) are grown on HAT medium (which contains aminopterin and thymidine) only cells stably transformed with a *TK* marker gene are able to survive.

▶ **Dominant selectable markers** are not found in the genome of the host cell and they confer an entirely novel property such as antibiotic resistance. The advantage of such markers is that they can be used in any cell type, i.e. mutants are not required. For example, the *E. coli neo* gene encodes the enzyme neomycin phosphotransferase. This inactivates aminoglycoside antibiotics such as G418, allowing stably transformed cells to be selected simply by adding G418 to the culture medium.

AMPLIFIABLE MARKER GENES

If the chosen selective agent is a *competitive inhibitor* of the marker gene product then it may be possible to use **stepwise selection** to amplify the marker gene and achieve high level transgene expression. This system works because the transgenic locus in stably transformed cells generally contains multiple tandem copies of the selectable marker and primary transgene. Increasing the concentration of the selective agent selects for those cells with the higher copy numbers, since they express the marker at the highest levels. In such cells, recombination events can occur that further increase the copy number and increasing the selection pressure progressively selects for cells with massively amplified transgene arrays. Amplification is a random process and the primary transgene is coamplified along with the marker gene. An example of such a marker is the mouse *dhfr* gene, which encodes the enzyme **dihydrofolate reductase**. This can be amplified by stepwise selection with the competitive inhibitor **methotrexate**.

COUNTERSELECTABLE MARKER GENES

A **counterselectable marker** confers a property that kills stably transformed cells, while allowing nontransformed cells to survive. This is known as **negative selection** because cells are being selected on the basis that they lack the marker gene (they are negative for the marker gene). Such markers are useful for *cell ablation* if they are expressed under the control of restricted promoters. They are also useful for identifying particular types of genetic modification, e.g. they are used to discriminate between gene targeting events and random integration events in ES cells (Section 20.2.4). Some counterselectable markers kill cells directly (e.g. ricin, diphtheria toxin) while others require the presence of a selective agent (e.g. TK can be used for negative selection in the presence of toxic thymidine analogs such as ganciclovir).

---

of sperm heads into the cytoplasm of the egg. ICSI has been used to introduce sperm heads coated with plasmid DNA into mammalian eggs. In the first such experiment, plasmids containing the gene for green fluorescent protein (GFP) were attached to mouse sperm and injected into isolated oocytes. Nearly all of the injected eggs showed GFP activity but only 20% produced transgenic mice, suggesting that in the majority of cases the transgene did not integrate into the cellular genome (Perry *et al.*, 1999). The same technique has been used in rhesus monkeys but although several embryos showed transient GFP activity, none were found to be transgenic.

Restriction enzyme–mediated integration (REMI) is the established method to produce transgenic frogs. Unlike sperm-mediated gene transfer, the gamete genome is modified during the transfer procedure. Sperm nuclei are isolated, decondensed and mixed with plasmid DNA. They are then treated with limiting amounts of a restriction enzyme to introduce nicks. The decondensed nuclei are then transplanted into unfertilized eggs where the nicks are repaired, resulting in the integration of plasmid DNA into the genome. The nuclei are very fragile and the transfer procedure must be completed quickly (Kroll and Amaya, 1996).

Microinjection of DNA into the fertilized egg is an established way to produce transgenic mice, and is used for transient expression assays in fish and amphibians

Transgenic mice have been produced by injecting DNA into the cytoplasm of the fertilized egg but a more efficient technique, and the one that has become established, is to introduce DNA directly into the male pronucleus just after fertilization.

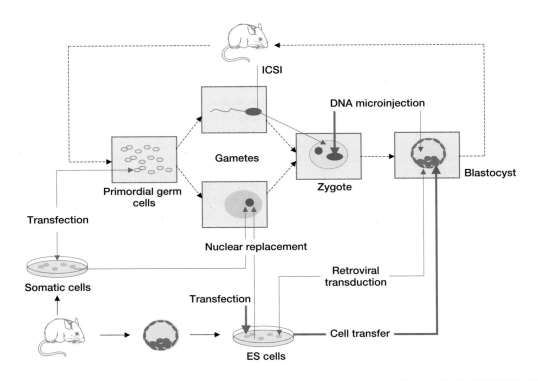

Figure 20.1: The many routes by which genetically modified mice can be produced.

The boxes linked by dotted lines show the mouse life cycle and represent all the stages at which, potentially, genetic modification can be carried out: germ cells, gametes, the zygote, the blastocyst and the adult. The lower part of the figure shows another mouse as a source of donor cells for nuclear transfer and cell transplant procedures. Red arrows show the input of exogenous DNA. Thick arrows represent the most widely used gene transfer methods in mice.

Table 20.1: Gene transfer to animals: Summary of targets and methods for transforming the germ line.

Target cells	Method
Germ cells	Transfection of cultured primordial germ cells (mammals) Injection into embryo at site of germ cell development (*Drosophila*)
Sperm	Attachment of DNA to sperm heads (mammals) Introduction of DNA into decondensed sperm nuclei (*Xenopus*)
Egg/zygote	Microinjection into egg cytoplasm (birds, amphibians, fish, *C. elegans*) Pronuclear microinjection (mammals) Retroviral transfer (mammals, primates) *Nuclear transfer*
Blastocyst	Microinjection into blastocele (mammals) *ES cell transfer* (mice) Retroviral transfer (mammals, birds)
ES cells (*followed by cell transfer*)	Transfection – transgene addition Transfection – gene targeting Retroviral transfer
Somatic cells (*followed by nuclear transfer*)	Transfection – transgene addition Transfection – gene targeting Retroviral transfer

The method is shown in *Figure 20.2*. The microinjected transgene randomly integrates into chromosomal DNA, usually at a single site, and usually as multiple copies (it is not unusual to find 50 or more copies as head-to-tail concatemers). The transgene can integrate immediately, in which case the resulting mouse is transgenic. However, it is more common for the DNA to integrate after one or two cell divisions, in which case the resulting mouse is a **mosaic** containing both transformed and nontransformed cells. Where transformed cells contribute to the germ line, the transgene is passed to the next generation of mice and this can be verified by PCR or Southern blotting or a test for transgene expression. It is possible to achieve germ line transmission in up to 40% of microinjected mouse eggs. Unfortunately, while the technique can be applied to other mammals, the transmission rate is much lower (< 1%). This is partly due to the difficulty in handling eggs, and partly due to the lower survival rates.

Microinjection can also be used to introduce DNA into fish and amphibian eggs but unlike the situation in mammals, this DNA tends to persist in an episomal state and undergoes extensive replication. Injected plasmid DNA may increase 50–100-fold in the early stages of development. This is thought to reflect the fact that there is no transcription in early fish and amphibian development, so there is an extensive stockpile of enzymes and proteins required for DNA replication. The consequence is that fish and frog eggs can be used as transient expression systems. Some of the DNA does integrate, however, and if germ line transmission is established then transgenic lines can be produced. This method is not used in frogs because of the long generation intervals, but is the standard procedure for producing transgenic fish.

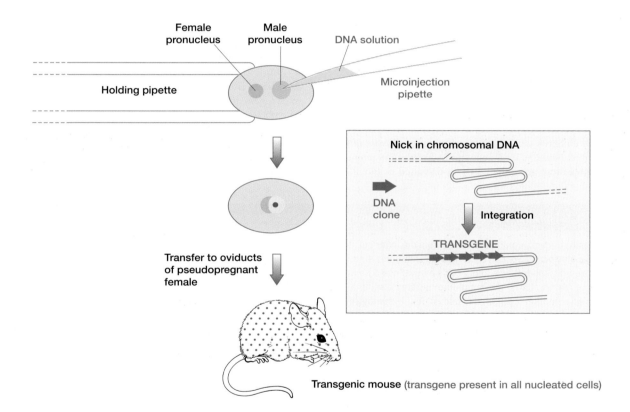

Figure 20.2: Construction of transgenic mice by pronuclear microinjection

Very fine glass pipettes are constructed using specialized equipment: one, a holding pipette, has a bore which can accommodate part of a fertilized oocyte, and thereby hold it in place, while the microinjection pipette has a very fine point which is used to pierce the oocyte and then the *male* pronucleus (because it is bigger). An aqueous solution of the desired DNA is then pipetted directly into the pronucleus. The introduced DNA clones can integrate into chromosomal DNA at nicks, forming transgenes, usually containing multiple head-to-tail copies. Following withdrawal of the micropipette, surviving oocytes are reimplanted into the oviducts of *pseudopregnant* foster females (which have been mated with a vasectomized male; the mating act can initiate physiological changes in the female which stimulate the development of the implanted embryos). Newborn mice resulting from development of the implanted embryos are checked by PCR for the presence of the desired DNA sequence (Gordon, 1992).

Retroviral gene transfer has been used to produce transgenic mammals and birds but the main application of this method is to create developmental mosaics

Genes can be transferred into unselected cells of very early embryos using retroviral vectors because a DNA copy of the vector integrates stably into the host genome following infection. The infection of preimplantation mouse embryos using a recombinant murine retrovirus or injection of the retrovirus into early postimplantation mouse embryos results in the stable transformation of some cells and hence the production of mosaics that may give rise to transgenic offspring. Similar results have been achieved in chickens using avian retroviruses. However, due to several limitations of retroviral gene transfer (the limited amount of foreign DNA that can be incorporated and the tendency for retroviral transgenes to undergo silencing) this technique is not widely used to generate transgenic mice or birds. Instead, it is favored for the production of mosaics that can be used to study gene expression and function in vertebrate development. More recently, the injection of recombinant retroviruses into the perivitelline space of isolated oocytes has allowed the production of transgenic cattle (Chan *et al.*, 1998) and the first ever transgenic primate, a rhesus monkey named ANDi (Chan *et al.*, 2001).

Transgenic mice can be produced by the transfection of embryonic stem cells

Mouse embryonic stem (ES) cells are derived from 3.5–4.5 day postcoitum embryos and arise from the inner cell mass of the blastocyst (see *Box 20.3*). ES cells can be cultured *in vitro* and are easy to transfect using the methods discussed in *Box 20.1*.

However, they remain pluripotent and contribute extensively to all of the tissues of a mouse, including the germ line, when injected back into a host blastocyst and reimplanted in a pseudopregnant foster mother. The developing embryo is a **chimera**, that is it contains two populations of cells derived from different zygotes, those of the blastocyst and the implanted ES cells. This differs from a mosaic in which the cells may be genetically different but are derived from the same zygote (*Figure 4.10*). If the blastocyst and ES cells are derived from mice with different coat colors, chimeric offspring can easily be identified by their patchwork coats. Germ line transmission of the transgene can also be confirmed by screening the offspring of matings between chimeras (usually males) and females with a coat color recessive to that of the strain from which the ES cells were derived (see *Figure 20.3*).

One of the advantages of ES cells is that they can be maintained indefinitely in culture and therefore have all the conveniences associated with cultured cells. Millions of cells can be transformed using simple transfection techniques and the desired genetic modification can be verified at the cell culture stage through the use of selectable markers such as *neo* (*Box 20.2*). In contrast, the other procedures discussed above require the manual processing of individual eggs or embryos, so only a small number of experiments can be carried out. Also, there is no way to select transformed eggs and embryos so transgene integration must be verified in the resulting mice. The biggest advantage of ES cells, however, is their propensity for homologous recombination, which allows genetic modification by gene targeting (Section 20.2.5). Recently, ES cell lines have been derived from chickens and humans (see *Box 20.3* and discussion in Section 21.3.3), but thus far it has been

---

Box 20.3: Isolation and manipulation of mammalian embryonic stem cells.

In mammals, the embryo proper derives from the inner cell mass (ICM) of a blastocyst (Section 3.7.3). Mouse embryonic stem (ES) cells were first isolated from blastocysts by Evans and Kaufman (1981) and Martin (1981). The procedure involves placing a 4.5-day pre-implantation embryo (blastocyst) on a monolayer of feeder cells, which provides a matrix for attachment and also secreted protein factors which inhibit newly formed ES cells from differentiating. Following attachment of the blastocyst, ICM cells proliferate. At an appropriate time, the ICM is physically removed with a micropipette, dispersed into small clumps of cells and seeded onto new feeder cells. Colonies are examined under the microscope for characteristic morphology. They are then picked, dispersed into single cells and reseeded onto feeder layers. Eventually cells with a uniform morphology can be isolated, and cell lines can be established. The term 'ES cell' was introduced to distinguish these embryo-derived stem cells from embryonal carcinoma (EC) cells, which had been derived from teratocarcinomas. In general ES cells have a less restricted developmental potential than EC cells. Both classes of cells are pluripotent and ES cells can give rise to all adult cell types, including germ line cells. They are, however, not totipotent in the same sense as the fertilized oocyte: if implanted in a uterus they

are unable to give rise to an embryo. In some cases EC cells have been reported to contribute to the germ line in chimeras, but when mouse ES cells are injected into an isolated blastocyst from a different strain and implanted in a pseudopregnant foster mother they contribute to chimeras and particularly to the germ line in a more consistent fashion. It is this property which has made mouse ES cells so valuable for research. It should be noted, however, that the ES cells used to form successful germline chimeras are derived from a single mouse strain (129) in combination with a single host embryo strain (C57BL/6). In this combination, the ES cells are vigorous and readily contribute to the germ line.

The search for ES cells in other mammals has occupied researchers for many years, and although ES cells have been isolated from several other species in addition to the mouse, the great success in forming germline chimeras in mouse has not been paralleled. Recently, human embryonic stem cells have been isolated from the blastocyst (Thomson *et al.*, 1998) and have been derived from primordial germ cells (Shamblott *et al.*, 1998). The potential medical applications of human ES cells and the ethical implications of their use are discussed in Chapter 21.

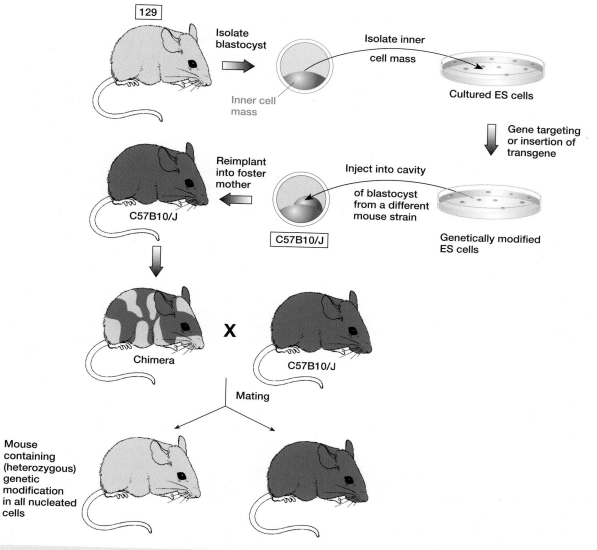

Figure 20.3: Genetically modified ES cells for transferring foreign DNA or specific mutations into the mouse germline

Cells from the *inner cell* mass were cultured following excision of oviducts and isolation of blastocysts from a suitable mouse strain (129). Such embryonic stem (ES) cells retain the capacity to differentiate into, ultimately, the different types of tissue in the adult mouse. ES cells can be genetically modified while in culture by insertion of foreign DNA or by introducing a subtle mutation. The modified ES cells can then be injected into isolated blastocysts of another mouse strain (e.g. C57B10/J which has a black coat color that is recessive to the agouti color of the 129 strain) and then implanted into a pseudopregnant foster mother of the same strain as the blastocyst. Subsequent development of the introduced blastocyst results in a **chimera** containing two populations of cells (including germline cells) which ultimately derive from different zygotes (normally evident by the presence of differently colored coat patches). Backcrossing of chimeras can produce mice that are heterozygous for the genetic modification. Subsequent interbreeding of heterozygous mutants generates homozygotes.

impossible to derive vigorous, reliable ES cell lines from domestic mammals.

Nuclear transfer can be used to produce genetically-modified domestic mammals but it has a very low success rate

The principal techniques used to generate transgenic mice (pronuclear microinjection and ES cell transfection) are either inefficient or impractical in most other mammals. For this reason, alternative methods based on **nuclear transfer technology** have been developed. Nuclear transfer involves the replacement of an oocyte nucleus with the nucleus of a somatic cell, which is then reprogrammed by the oocyte and becomes able to recapitulate the whole of development despite the differentiated state of the donor cell.

The technology itself is not new. It has been used for over 50 years to generate cloned amphibians, and it has been possible to produce cloned mammals from embryonic cells since the late 1980s. In 1995 it was shown for the first time that mammals could be cloned from the nuclei of cultured cells (Campbell *et al.*, 1996) and in 1997 the first mammal cloned from an adult cell was produced (Wilmut *et al.*, 1997). The procedure used by Wilmut and colleagues is summarized in *Figure 20.4*. In the original experiment, only 29 out of a total of 434 oocytes developed to the transferable stage and of these only one developed to term – the famous lamb known as Dolly (see *Figure 20.4B*). The successful cloning of adult animals has forced us to accept that genome modifications once considered irreversible can be reversed and that the genomes of adult cells can be reprogrammed by factors in the oocyte to make them totipotent once again. Research into the control of gene expression during development and basic processes of somatic differentiation, somatic mutation, aging and repair processes will undoubtedly benefit from animal cloning, especially the cloning of mice (Wakayama *et al.*, 1998). We discuss the practical medical benefits and ethical implications of animal and human cloning in Chapter 21, but for now we consider the method solely as another way to generate transgenic animals. The essential point is that if the donor nucleus is taken from a cell that been transfected and contains an additional transgene, then the animal developing from the manipulated oocyte will be transgenic. This was first demonstrated by Schnieke *et al.* (1997) who transfected cultured sheep fibroblasts with the human gene for blood coagulation factor IX and produced a cloned transgenic ewe called Polly producing the recombinant protein in her milk. Similarly, cloned sheep have been produced from cultured somatic cells whose genome has been altered by gene targeting (see below and McCreath *et al.*, 2000).

20.2.3 The control of transgene expression is an important consideration in any gene transfer experiment

The *presence* of a transgene in a cell or transgenic animal is not in itself sufficient to produce a functional product. In order for this to occur, the transgene must also be *expressed*. The control of gene expression is therefore of paramount importance in gene transfer technology. Transgene expression is regulated by sequences present in the expression construct and, when the transgene integrates, also by factors intrinsic to the host genome. Overall, the structure of the promoter is the most important aspect of construct design and we focus on this below. However, robust expression also requires a polyadenylation site, and there may be other considerations, such as optimization of the translational start site and the inclusion of a suitable protein targeting signal.

The transgene promoter defines the basic spatial and temporal pattern of expression

Gene transfer experiments in cell lines often benefit from the use of very active **constitutive promoters**, which cause the transgene to be expressed strongly all the time. Such promoters are generally derived from viruses, which have evolved to express their genes in many different cell types. The regulatory elements most commonly used in mammalian cells include the SV40 early promoter and enhancer, the Rous sarcoma virus long terminal repeat (LTR) promoter and enhancer and the human cytomegalovirus immediate early promoter. These are included in many commercially available expression vectors.

In transgenic animals, it is often desirable to express the transgene in particular tissues or at particular developmental stages. Similarly, in cell lines, it may be necessary to restrict transgene expression to particular stages of the cell cycle or to switch the transgene on only when cells differentiate. By linking the transgene to a suitable **cell- or stage-specific promoter**, the desired expression pattern may be achieved. For example, the neuron-specific enolase gene is expressed only in mature neurons. An upstream regulatory element 1.8 kb in length is sufficient to confer neuron-specific transgene expression in transgenic mice and to upregulate transgene expression specifically in differentiating cultured neuroblasts (Forss-Petter *et al.*, 1990; Sakimura *et al.*, 1995).

Maximum control over transgene expression in both cell lines and animals is provided by **inducible promoters**, which can be switched on and off by controlling the supply of a particular chemical ligand. Typically, the transcription factors that regulate such promoters are structurally modified by this ligand. Several naturally inducible promoters have been utilized including the mouse metallothionein promoter and the mouse mammary tumor virus LTR promoter, both of which are inducible by dexamethasone (a synthetic hormone). In addition, the metallothionein promoter is inducible by heavy metal ions, such as Zn^{2+}. The inducing ligand can be added to the medium of cultured cells and supplied to animals in their drinking water. Generally, the use of such endogenous promoters has been hampered by 'leakiness', i.e. a high background expression level, and by relatively low levels of induction. Also, there may be undesirable effects brought about by the co-activation of endogenous genes that respond to the same ligand, and in the case of transgenic animals there may be differential rates of uptake and elimination of the ligand in different organs. More recently, however, promising systems have been developed that use heterologous components. For example, the *E. coli tet* operon is the basis of tetracycline-regulated inducible expression (*Figure 20.5A*) which allows the production of highly inducible transformed cell lines and transgenic animals (see Saez *et al.*, 1997). Other systems have been devised based on the *E. coli lac* operon and the *Drosophila* hormone ecdysone. Alternatively, inducible expression can be achieved by **chemically-induced dimerization (CID)**, in which a functional transcription factor is assembled from two components in the presence of a divalent ligand (Belshaw *et al.*, 1996). One disadvantage of all these expression systems is that induction occurs at the level of transcription, so there may be a significant delay before a response to induction is seen and a similar delay between the removal of the inductive stimulus and return to the basal state. Where rapid induction and decay is essential, inducible systems that work at the protein level are available. In one such system, the target transgene is expressed

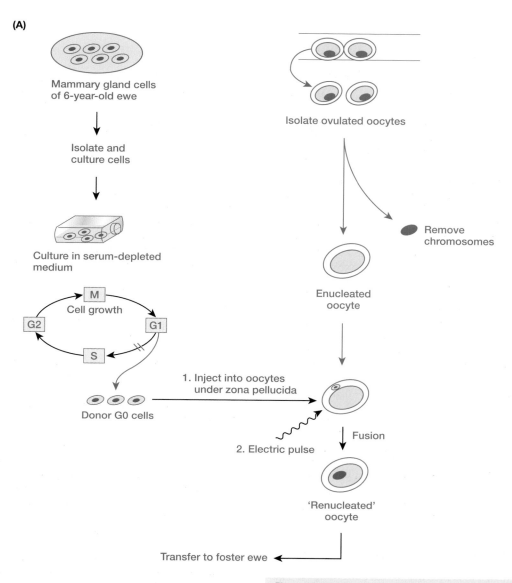

(A)

(B)

Figure 20.4: A sheep called Dolly was the outcome of the first successful attempt at mammalian cloning from adult cells.

(A) The experimental strategy used by Wilmut *et al.* (1997). The donor nuclei were derived from a cell line established from adult mammary gland cells. Nuclear transfer was accomplished by fusing individual somatic cells to enucleated, metaphase II-arrested oocytes. The donor cells were deprived of serum before use, forcing them to exit the cell cycle and enter a quiescent state known as G_0 where only minimal transcription occurs. Since eggs are normally fertilized by transcriptionally inactive sperm whose nuclei are presumably 'programmed' by transcription factors and other chromatin proteins available in the egg, the G_0 nucleus may represent the ideal basal state for reprogramming. Note that in other cloning experiments, different strategies have been used to introduce the donor nucleus into the egg. For example, in the successful cloning of adult mice reported by Wakayama *et al.* (1998), a very fine needle was used to take up the donor cell nucleus with minimal contamination by donor cell cytoplasm. The donor cell was quickly, but very gently, microinjected into the enucleated oocyte. **(B)** Dolly with her first born, Bonnie. Original photo kindly provided by the Roslin Institute.

as a fusion with the estrogen receptor, which is normally sequestered in an inactive complex with Hsp90 (*Figure 20.5B*). In the presence of estrogen, or its analog Tamoxifen, the estrogen receptor is released and the protein to which it is fused can become active (Littlewood *et al.*, 1995).

The expression of integrated transgenes is influenced by position effects and locus structure

A common observation is that independently derived transgenic animals carrying the same transgene construct do not always show the same level or pattern of transgene expression. This is because transgene integration occurs randomly, so both the position and structure of the transgenic locus is variable. Position effects are brought about by the influence of local regulatory elements and chromatin structure. For example, the transgene may integrate next to an enhancer that modifies its expression pattern, or it may integrate within a heterochromatin domain that abolishes expression altogether. The structure of the transgenic locus can also influence expression. For example, if two copies of the transgene happen to be arranged as an inverted repeat, then hairpin RNA can be generated that leads to RNA interference (see Section 20.2.6). Various other aspects of the locus structure may trigger cellular defenses against invasive DNA, leading to transgene silencing by *de novo* DNA methylation. These phenomena are less apparent in transformed cell lines since these are selected on the basis of their ability to express a marker gene strongly.

Position effects can be blocked using dominantly acting regulatory elements and large transgene constructs

Most transgenes are cDNA sequences controlled by short regulatory elements. It is becoming increasingly apparent that while such elements define the minimum requirements for gene expression and function, they do not provide the full complement of sequences required for robust gene expression in a genomic context. There is evidence that certain regulatory elements act as master switches to establish open chromatin domains and protect genes from the influence of regulatory elements and chromatin structure in neighboring domains. These elements are often found within introns and at distant sites. Therefore, position effects can be avoided either by incorporating these elements into transgenes or by using large genomic constructs as transgenes rather than minimal cDNAs.

It has been difficult to define precisely the elements that establish chromatin domains but some, such as boundary elements, matrix attachment regions and locus control regions have been used in transgenes with some success (Section 10.5). In order to study these long range effects more accurately, and to investigate the expression and regulation of human genes in the context of their own *cis* acting regulatory elements, it has been necessary to establish conditions that allow the transfer of large DNA molecules. Major breakthroughs include:

▶ the development of **YAC transgenic mice** (Lamb and Gearhart, 1995). The first report to be published described mice transformed with a 670-kb YAC containing the

human hypoxanthine guanine phosphoribosyltransferase (*HPRT*) gene (Jakobovits *et al.*, 1993). This technique is useful for modeling human diseases caused by large scale dosage imbalance (Section 20.4.5) and has other applications, such as the production of authentic human antibodies in mice (Mendez *et al.*, 1997; for general method, see *Figure 20.6*). YAC transgenics can be produced by cell fusion, microinjection or transfection with liposomes;

▶ the development of **transchromosomic mice** (Tomizuka *et al.*, 1997). These contain human chromosomes or chromosome fragments and are generated by microcell-mediated gene transfer (*Box 8.4*).

20.2.4 Gene transfer can also be used to produce defined mutations and disrupt the expression of endogenous genes

Thus far, we have considered gene transfer technology only from the perspective of *adding* functions to animal cells. Another way in which the same technology can be used, and perhaps its major contribution to biomedical research, is to selectively abolish or alter the functions of endogenous genes. Not only is this a powerful way of investigating gene function (Section 20.3.2) but it also provides a useful route for the creation of disease models that mimic corresponding human diseases precisely (Section 20.4.4). There are three ways in which gene transfer can be used to modify the function of endogenous genes.

▶ **Gene targeting**. In this approach, homologous recombination is used to replace an endogenous gene sequence with a related sequence, therefore introducing a defined mutation at a preselected site in the genome. This is often used simply to disrupt genes with a large insertional cassette, generating null mutations known as **gene knockouts**. However, modified strategies allow more sophisticated forms of genetic manipulation including the introduction of subtle mutations and the replacement of one gene with another.

▶ **Inhibition of gene expression**. In this approach, standard gene transfer methodology is used to add a new DNA sequence to the cell. However, instead of encoding a protein that confers a new function on the cell, the product of this transgene functions solely to inhibit the expression of an endogenous gene. These gene products, which include antisense RNAs, double-stranded RNAs, small interfering RNAs, ribozymes, antibodies and dominantly interfering proteins, can also be introduced directly rather than being expressed from a transgene, although in this case the effects are transient rather than permanent.

▶ **Insertional mutagenesis**. In this approach, the transgene integrates (randomly) into an existing gene and abolishes its function. Like gene targeting this approach alters gene function by introducing a mutation, but in this case the mutation is neither defined nor targeted to a particular site. Mutations in specific genes can be identified by large scale screening (Section 20.3.3).

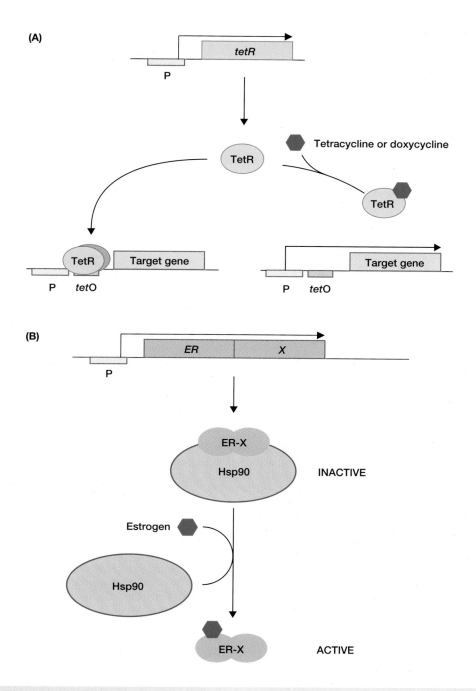

Figure 20.5: Inducible expression systems.

(A) The basic tetracycline-inducible expression system. Constitutive expression of the *E. coli* Tet repressor (TetR) will inactivate any target transgene containing the *tet*O operator sequence, to which the repressor binds. However, in the presence of tetracycline or its analog doxycycline, the structural conformation of the repressor is altered and it can no longer bind the operator, causing the gene to be de-repressed. Note that toxicity effects reflecting high level constitutive expression of the repressor limit the use of this system in some cells. To address this, more sophisticated systems have been developed in which the Tet repressor is converted into an activator (the tTA system) or in which binding is dependent on tetracycline rather than abolished by it (the reverse tTA system). See Saez *et al.* (1997) for a review of these systems. **(B)** The estrogen-inducible expression system. A protein (X) expressed as a fusion with the estrogen receptor (ER) is generally inactive because it is sequestered into a complex with heat shock protein 90 (Hsp90). In the presence of estrogen, however, the fusion protein is released from the complex and the activity of protein X is restored. This is an advantageous system because induction is very rapid, requiring only the dissociation of a protein complex and not transcription followed by protein synthesis.

1. Isolate YACs containing sequences from human IgH (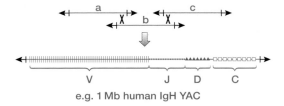) and human IgK chain ( ) loci

2. Allow YACs from each locus to undergo homologous recombination in yeast cells to generate YACs wth large inserts

e.g. 1 Mb human IgH YAC

3. Fuse yeast cell spheroplasts containing large IgYAC to mouse ES cells

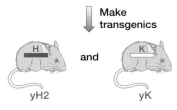

Make transgenics

and

yH2   yK

Mice with single human Ig transloci

4. Cross these mice with the DI strain which has endogenous IgH (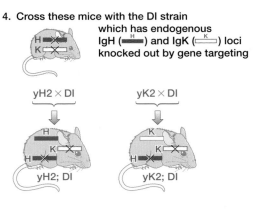) and IgK ( ) loci knocked out by gene targeting

yH2 × DI   yK2 × DI

yH2; DI   yK2; DI

5. Cross yH2; DI to yK2; DI

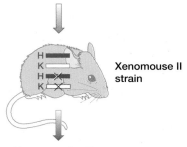

Xenomouse II strain

Human antibodies

Figure 20.6: Use of YAC transgenesis to construct a mouse with a human antibody repertoire

YACs containing Ig sequences are obtained by screening YAC libraries with suitable Ig probes. The recovery of YACs with comparatively small inserts meant that there was a need to artificially construct larger YACs by homologous recombination in yeast cells. Spheroplasts are created by treating the yeast cells so that the outer cell wall is stripped off, making the cells more amenable to cell fusion. See Mendez *et al.* (1997) for further details of the procedures used to construct these mice.

20.2.5 Gene targeting allows the production of animals carrying defined mutations in every cell

Gene targeting involves *in vivo* homologous recombination between an endogenous gene and an exogenous DNA sequence contained in a targeting vector

Interactions between DNA introduced into animal cells and the host genome generally result in random integration of the exogenous DNA at sites of pre-existing chromosome nicks and breaks. However, if the exogenous DNA closely resembles an endogenous gene, a different interaction can occur in which the exogenous and endogenous sequences align and undergo **homologous recombination**. This process, known as **gene targeting**, allows the creation of defined mutations at preselected sites in the host genome. It can therefore be viewed as a form of artificial *in vivo* site-directed mutagenesis (as opposed to the various methods of site-directed *in vitro* mutagenesis described in Sections 5.5.2, 5.5.3.

Homologous recombination is a very rare process in mammalian cells, occurring some 10^4–10^5 times less frequently than random integration. The frequency of homologous recombination depends on both the length of the

homology region (the part of the targeting vector that aligns with the endogenous gene) and the degree of similarity with the target gene. Therefore, the introduced DNA clone, which has been modified to include the desired mutation, is generally a long sequence that is **isogenic** with the genomic DNA of the host cell. Even then, the frequency of genuine homologous recombination events is very low and may be difficult to identify against a sizeable background of random integrations. The strategy used to identify cells in which gene targeting has taken place depends on the construct design (*Figure 20.7*). There are two types of construct.

▸ **Insertion vectors** target the locus of interest by a single reciprocal recombination event, causing insertion of the entire vector (*Figure 20.7A*).

▸ **Replacement vectors** are designed to replace some of the sequence in the chromosomal gene with a homologous sequence from the introduced DNA (*Figure 20.7B*). This can occur as a result of a double reciprocal recombination or by gene conversion.

In both strategies, a large segment of vector DNA containing the *neo* marker gene is introduced into the targeted locus causing disruption and the generation of a null allele (**gene**

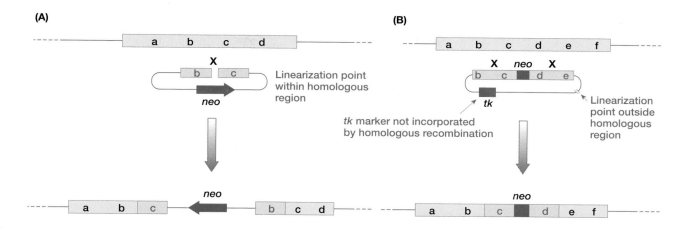

Figure 20.7: Gene targeting by homologous recombination can inactivate genes at a predetermined locus within an intact cell.

(A) *Insertion vector method*. The introduced vector DNA (red) is cut at a unique site within a sequence which is identical or closely related to part of the target endogenous gene (blue). Homologous recombination (X) can occur, inserting the entire vector sequence (including the marker gene *neo*, which confers resistance to the antibiotic G418) into the targeted locus. Genuine targeting events will be rare, however, and in most G418-resistant cells the vector will have integrated randomly. A PCR assay must be used to discriminate between gene targeting and random integration events, using primers designed to anneal strategically in the vector and target gene. **(B) *Replacement vector method*.** In this case, the *neo* gene is contained within the sequence homologous to the endogenous gene, and the vector is cut at a unique location outside the homology region. A double recombination or gene conversion event (X X) can result in replacement of internal sequences within the target gene by homologous sequences from the vector, including *neo*. Again, random integration events are more common but the replacement vector method facilitates a more sophisticated dual selection strategy in which a second marker, this one a counterselectable marker such as *tk* (which confers *sensitivity* to ***ganciclovir***) is placed outside the homology region. The second marker will be incorporated into the genome by random integration but not by homologous recombination. Therefore, cells that are resistant to both G418 and ganciclovir are likely to be correctly targeted. The letters indicate linear order within the gene and do not represent exons.

knockout). However, this may not always be desirable. If a more subtle mutation is required, various two-step recombination techniques can be employed. The *'hit and run' strategy* which is used with insertion vectors and the *'tag and exchange' strategy* which is used with replacement vectors, are shown in *Figure 20.8*.

Genetically modified animals can be produced by gene targeting in ES cells or somatic cells used as donors for nuclear transfer

Gene targeting by homologous recombination was first achieved in cultured human/mouse hybrid somatic cells (Smithies *et al.*, 1985) and has since been demonstrated in somatic cells from a range of mammals. However, the most important application of gene targeting involves mouse ES cells, which are particularly amenable to the technique because homologous recombination occurs at a comparatively high frequency (albeit still much lower than the frequency of random integration). The unusual and highly beneficial combination of three features – ease of culture and transfection, propensity for homologous recombination and pluripotency – means that it is quite straightforward to use ES cells to produce mice containing the same genetic modification in every cell (Capecchi, 1989; Melton, 1994). Once targeted ES cells are available, the general method used to derive the mice is the same as shown in *Figure 20.3*). Such mice are described as 'genetically-modified' or 'targeted' rather than transgenic because they do not necessarily contain any

exogenous DNA. Indeed, it is possible to produce targeted mice containing a single point mutation at a predefined position.

Although the procedure for gene targeting in somatic cells has a generally very low efficiency, fetal fibroblasts with targeted modifications have recently been used as donors for nuclear transfer (Section 20.2.2). This has led to the creation of a range of genetically modified mammals. The initial report described sheep with a new gene introduced at the *COL1A1* locus (McCreath *et al.*, 2000) and more recently two groups have reported the production of pigs with targeted disruptions in the gene for α-1,3-galactosyltransferase (Dai *et al.*, 2002; Lai *et al.*, 2002). This enzyme decorates proteins with carbohydrate groups that are not found in primates, representing one of the major factors responsible for the rejection of organs in pig-to-human organ transplants.

20.2.6 Site-specific recombination allows conditional gene inactivation and chromosome engineering

Site-specific recombination systems are found in several bacteriophages as well as in bacteria and yeast. Each system includes a minimum of two components: a short specific recognition sequence at which recombination occurs and a recombinase enzyme that recognizes this sequence and carries out the recombination reaction when two copies of the sequence are present. The power of site specific recombination is that the recognition sites can be engineered easily into

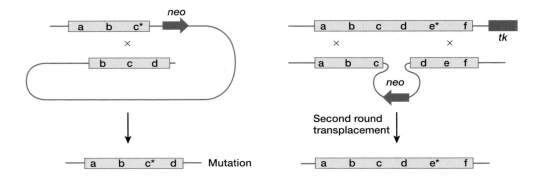

Figure 20.8: Introduction of subtle mutations by gene targeting.

(A) In the *'hit and run' strategy*, used with *insertion vectors*, the subtle mutation is present on the first targeting construct, and intrachromosomal recombination leads to the elimination of the marker gene and vector backbone. **(B)** In the *'tag and exchange' strategy*, used with *replacement vectors*, a second targeting construct is used to replace the mutation introduced by the first. The second construct incorporates a counterselectable marker outside the homology region to avoid random integration.

---

transgenes or targeting vectors and the recombinase enzymes can be supplied conditionally, because the genes encoding these enzymes can be expressed under the control of regulated or inducible promoters. Thus far, the Cre–*lox*P recombination system from bacteriophage P1 has been the most widely used, particularly in genetically-modified mice. The natural function of the **Cre recombinase** (<u>c</u>auses <u>re</u>combination) is to mediate recombination between two *lox*P sequences, which are 34 bp long and comprise two inverted 13-bp repeats separated by a central asymmetric 8-bp spacer (*Figure 20.9*). If the two *lox*P sites are in the same orientation, the intervening sequence between them is excised. If they are in opposite orientations, the intervening sequence is inverted. The Cre-*lox*P system has therefore been applied in a number of different ways, including the site-specific integration of transgenes, the conditional activation and inactivation of transgenes and the deletion of unwanted marker genes. Perhaps the most important applications, however, are conditional gene inactivation and chromosome engineering, which are discussed below (see Lobe and Nagy, 1998).

Conditional gene inactivation

Some genes are critical early in development and simple knock-out experiments are generally not helpful because death ensues at the early embryonic stage. To overcome this problem, methods have been developed to inactivate the target gene in only selected, predetermined cells of the animal, or at a particular stage of development. The animal can therefore survive and the effect of the **conditional knockout mutation** can be studied in a tissue or cell type of interest. An early example of this approach was reported by Gu *et al.* (1994) and involved the conditional knockout of DNA polymerase β, an enzyme that is essential for embryonic development. The gene targeting procedure replaced an essential exon of the endogenous gene with a homologous gene segment flanked by *lox*P sequences. Mice carrying this targeted mutation were then mated with a

strain of mice which carried a *Cre* transgene gene under the control of a T lymphocyte-specific promoter. Offspring of the cross containing both transgenes were identified and survived to adulthood. The Cre product was expressed only in T cells, deleting the essential exon and inactivating the gene (see *Figure 20.10* for the method). A general advantage of this approach is that Cre-transgenic mice can be used over and over again for different experiments. Regardless of which endogenous gene is flanked by *lox*P sites, the Cre transgenic mice described above can be used to disrupt that gene in T cells. Similarly, widely applicable Cre transgenics have been generated in which Cre expression is inducible by tetracycline and in which Cre is constitutively expressed as a fusion to the estrogen receptor and is activated at the protein level by Tamoxifen (Section 20.2.3; Fiel *et al.*, 1996).

Chromosome engineering

Another important recent development is a strategy for chromosome engineering in ES cells which relies on sequential gene targeting and Cre–*lox*P recombination. Gene targeting is used to integrate *lox*P sites at the desired chromosomal locations and, subsequently, transient expression of Cre recombinase is used to mediate a selected chromosomal rearrangement (Ramirez-Solis *et al.*, 1995; Smith *et al.*, 1995; *Figure 20.11*). Chromosome engineering strategies of this type offer the exciting possibility of creating novel mouse

Figure 20.9: Structure of the *lox*P recognition sequence

Note that the central 8-bp sequence which is flanked by the 13-bp inverted repeats is asymmetrical and confers orientation.

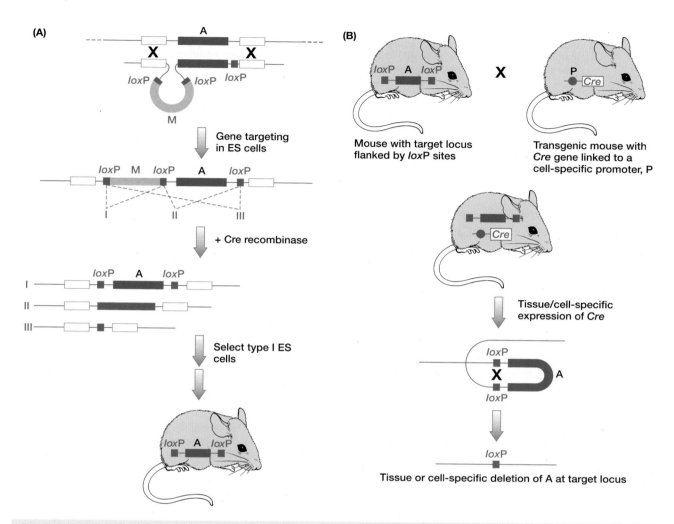

Figure 20.10: Gene targeting with the Cre–*lox*P recombination system can be used to inactivate a gene in a desired cell type

(A) Illustration of a standard homologous recombination method using mouse ES cells, in which three *lox*P sites are introduced along with a marker M at a target locus A (typically a small gene or an internal exon which if deleted would cause a frameshift mutation). Subsequent transfection of a *Cre* recombinase gene and transient expression of this gene results in recombination between the introduced *lox*P sites to give different products. Type 1 recombinants are used to generate mice in which the target locus is flanked by *lox*P sites. Such mice can be mated with previously constructed transgenic mice **(B)** which carry an integrated construct consisting of the *Cre* recombinase gene linked to a tissue-specific promoter. Offspring which contain both the *lox*P-flanked target locus plus the *Cre* gene will express the *Cre* gene in the desired cell type, and the resulting recombination between the *lox*P sites in these cells results in tissue-specific inactivation of the target locus A.

lines with specific chromosomal abnormalities for genetic studies. The multiple targeting and selection steps in ES cells used in the above chromosome engineering methods can be avoided by using the novel approach of Herault *et al.* (1998). This targeted meiotic recombination method takes advantage of the homologous chromosome pairing that occurs naturally during meiosis at the first cell division. A transgene is designed to express Cre recombinase under the control of a *Sycp1* promoter (the *Sycp1* gene encodes part of the synaptonemal complex that facilitates crossing over). As a result Cre recombinase is produced in male spermatocytes during the zygotene to pachytene stages when chromosome pairing occurs.

20.2.7 Transgenic strategies can be used to inhibit endogenous gene function

While gene targeting is undoubtedly the most direct and accurate way to manipulate gene activity in cells and transgenic animals, one major limitation is that the method is only applicable on a routine basis to mice and fruit flies. Therefore, other methods for gene inhibition have been developed that are more generally applicable (*Table 20.2*). These methods are diverse, but they are considered together because they all in some way interfere with gene expression or gene function *without altering the DNA sequence of the target gene*. They are sometimes described as **functional knockout** procedures

(A) 1. Use sequential gene targeting to introduce *lox*P site (▶) plus a marker gene ☐M) into two desired locations on different chromosomes

(B) 1. Use sequential gene targeting to introduce *lox*P site plus a marker gene into two desired locations flanking chromosomal region to be deleted (⌇⌇)

2. Expose *lox*P-containing chromosomes to Cre recombinase

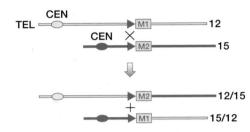

2. Allow to undergo intrachromosomal 'recombination' in presence of Cre recombinase

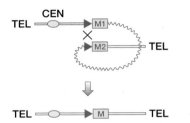

Figure 20.11: Chromosome engineering can be accomplished using Cre–*lox*P systems

(A) Use of targeted insertion of *lox*P sites to facilitate a chromosomal translocation. See Smith *et al.* (1995) for a practical example. **(B)** Use of targeted insertion of *lox*P sites to permit modeling of a microdeletion by intrachromosomal recombination. See Ramirez-Solis *et al.* (1995) for a practical example, and for other examples of intrachromosomal recombinations.

because they generate **phenocopies** of the corresponding mutant phenotype.

Gene expression can be inhibited by targeting specific mRNAs for destruction

RNA is the bridge between genes and proteins, so it should be possible to produce functional knockouts by selectively destroying or inactivating the transcripts of specific genes. There are several ways in which specific mRNAs can be targeted for inactivation, all of which appear to involve the formation of an RNA molecule that is at least partially *double stranded*. Certain genes are regulated by endogenous antisense RNAs known as **small temporal RNAs** (stRNAs, see Section 9.2.3). These appear to work by binding to the tran-

script and inhibiting mRNA processing or protein synthesis. An early gene inhibition strategy was therefore the introduction into cells of **antisense RNA** or **antisense oligonucleotides**. Transient inhibitory effects are seen following the direct introduction of such molecules but permanent inhibition can be achieved by transforming cells, or animals, with an **antisense transgene** that constitutively synthesizes antisense RNA. This was first demonstrated in mice by Katsuki *et al.* (1988) who succeeded in reducing the level of myelin basic protein to 20% of the normal level by expressing an antisense cDNA against the myelin protein gene, thus producing a phenocopy of the *shiverer* mutant. Antisense RNA has also been expressed using an inducible promoter to regulate cell growth in culture (Sklar *et al.*, 1991).

Table 20.2: Summary of methods for interfering with endogenous gene expression without mutating the target gene

Interference at the RNA level	*Interference at the protein level*
Antisense RNA	Dominant negatives
Antisense oligonucleotides	Antibodies, intrabodies
Ribozymes or maxizymes	Aptamers, intramers
Deoxyribozymes	
Sense RNA (cosuppression)	
dsRNA (RNA interference)	
siRNA (RNA interference)	

It was initially assumed that the effect of antisense RNA was stoichiometric (i.e. one antisense molecule is required to block the translation of one transcript and then both are destroyed). More potent inhibition should therefore be possible if the inhibitory molecule is recycled, so that many transcripts are destroyed before the inhibitory molecule is itself degraded. This is the perceived advantage of **ribozymes**, RNA enzymes that cleave RNA molecules catalytically (Section 21.6). Constructs have been designed in which a ribozyme catalytic center is incorporated within an antisense transgene to facilitate the destruction of specific transcripts. Such constructs have been widely used in cell lines, especially to study the inhibition of oncogenes and to combat HIV infection (see Welch *et al.*, 1998) but they have been expressed only infrequently in transgenic mice. One useful example is the targeted inhibition of glucokinase mRNA specifically in pancreatic β-cells by expressing an antisense glucokinase ribozyme transgene under the control of the insulin promoter, creating a model of diabetes (Efrat *et al.*, 1994). Recently, ribozymes whose activity can be controlled by allosteric modulation (so-called **maxizymes**) have been developed and used to inhibit gene expression conditionally (Kuwabara *et al.*, 2000; Famulok and Verma, 2002).

Surprisingly, ribozyme constructs seem to perform no better in most cases than corresponding antisense RNAs lacking the ribozyme catalytic center, suggesting that the antisense RNA may have a more potent effect than would be predicted from stoichiometric binding alone. A clue to the basis of this phenomenon is the ability, in some cases, of *sense RNA* expressed in mammalian cells to inhibit the expression of a corresponding endogenous gene, a phenomenon known as **cosuppression** (Bahramian and Zabl, 1999). Cosuppression has been widely documented in plants and appears to involve the formation of aberrant RNA species that are partially double stranded. Investigations into the ability of antisense RNA and sense RNA to silence genes in the nematode *C. elegans* led to the discovery of a novel phenomenon called **RNA interference** in which the simultaneous introduction of both sense and antisense RNA corresponding to a particular gene led to potent, long lasting and very specific gene silencing (Fire *et al.*, 1998). RNA interference is a highly conserved cellular defense mechanism, which also occurs in mammalian (including human) cells. It is triggered by the presence of **double-stranded RNA (dsRNA)** and causes the degradation of single-stranded mRNAs that have the same sequence as the inducing molecule.

The mechanism of RNA interference is complex, but involves the degradation of the dsRNA molecule into short duplexes, about 21–25 bp in length, by a dsRNA-specific endonuclease called **Dicer** (*Figure 20.12*). The short duplexes are known as **small interfering RNAs (siRNAs)**. These molecules bind to the corresponding mRNA and assemble a sequence-specific RNA endonuclease known as the **RNA induced silencing complex (RISC)**, which is extraordinarily effective and reduces the mRNA of most genes to undetectable levels. Interestingly, the same Dicer enzyme is known to process the small temporal RNAs discussed above that interfere with the expression of endogenous genes.

Similar RNA molecules, which are presumed to have endogenous regulatory functions, have now been identified in many other organisms including humans, and are described as **micro-RNAs (miRNAs**, see Section 9.2.3). It is thought that the different activities of micro-RNAs (blocking translation) and small interfering RNAs (catalytic degradation) may reflect the structure of their precursors and the enzymes that process them prior to Dicer. Micro-RNAs are single stranded and are derived from imperfect duplexes with bulges and loops, while siRNAs are double stranded and are usually derived from perfect duplexes (Pasquinelli, 2002; Voinet, 2002). It is possible there is some cross-talk between these pathways (*Figure 20.12*).

RNA interference can be used in both cells and embryos because it is a systemic phenomenon – the siRNAs appear to be able to move between cells so that dsRNA introduced into one part of the embryo can cause silencing throughout. As well as introducing dsRNA directly into cells (by transfection) or embryos (by injection or other methods), it is possible to express dual transgenes for the sense and antisense RNAs or to express an inverted repeat construct that generates hairpin RNAs that act as substrates for Dicer.

The introduction or expression of long dsRNA molecules is suitable for gene silencing in most animals and in mammalian embryos and embryonic cell lines. However, the results of RNA interference are masked in adult mammalian cells by the **interferon response**, which is a general (not sequence-specific) response to dsRNA molecules over about 30 bp in length. This problem has been circumvented by the direct administration of chemically or enzymatically synthesized siRNAs or the expression of mini-transgenes, generally based on endogenous genes that produce small RNAs (e.g. the U6 snRNA gene). Over the last 3 years, RNA interference has risen from relative obscurity to perhaps the most promising available tool for high throughput functional analysis in cells and animals (see below) and now promises to form the basis of a whole new class of therapeutic agents (Section 21.6). For a useful overview of RNA interference and its applications, see Tuschl and Borkhardt (2002).

Gene function can also be blocked at the protein level

Even if a transcript survives and is translated, gene function can be blocked at the protein level resulting in a loss-of-function phenocopy. If the product of the target endogenous gene functions as a multimer, it may be possible to produce a **dominant negative mutant** version of the gene that can be expressed at high levels in the cell or transgenic animal. In this case, the functional copies of the protein will be sequestered into inactive complexes. This approach has been widely used to block receptor functions since many receptors function as dimers (e.g. Amaya *et al.*, 1991).

Alternatively, it may be possible to express an antibody that recognizes the target protein and neutralizes it. Functional inactivation can be achieved by introducing the antibodies directly into the cell or animal, or by expressing the antibody from a transgene, in which case it is often termed an **intrabody** (Richardson and Marasco, 1995). DNA or RNA oligonucleotides can also bind to and inhibit the activity of

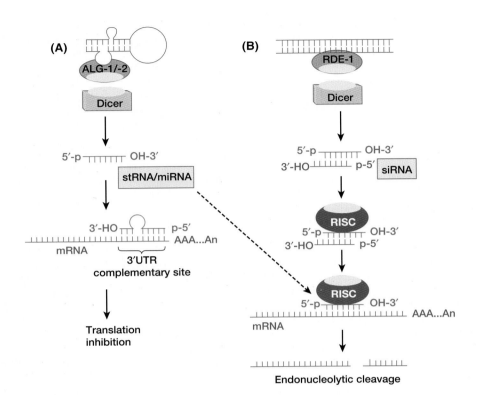

Figure 20.12: Comparison of the stRNA (miRNA) and siRNA pathways in *C. elegans*.

By recognizing specific types of precursor RNAs, related proteins such as ALG-1/ALG-2 and RDE-1 may partly influence processing by DICER. **(A)** The stRNA (miRNA), which is single stranded, is thought to base-pair imperfectly with the 3′ UTR of the target mRNA, repressing the initiation of translation. **(B)** The siRNA, which is double stranded and has 3′ overhangs of two nucleotides, is incorporated into the RNA induced silencing complex (RISC). The siRNA serves as guide for RISC and, upon perfect base pairing, the target mRNA is cleaved in the middle of the duplex formed with the siRNA. It is possible that miRNAs that happen to pair with their target perfectly could be recruited into the RISC. Modified from Voinet (2002) with permission from Elsevier Science.

proteins in a specific manner. These are known as **aptamers**, or if expressed inside the cell, as **intramers** (Famulok *et al.*, 2001).

20.3 Using gene transfer to study gene expression and function

20.3.1 Gene expression and regulation can be investigated using reporter genes

One of the earliest uses for transformed cell lines and trans-genic organisms was the analysis of **gene regulation**, that is the identification of sequences required for specific attributes of gene expression. Gene regulation can be investigated by studying how the deletion of different segments of DNA upstream of the gene, or occasionally in the first intron, affects its expression. Clearly, human cells are the most appropriate system for studying the expression of human genes, and the introduction of constructs containing different amounts of flanking DNA sequence followed by the analysis of gene expression is a logical way to proceed. However, this leaves the issue of how to follow expression of the introduced human

gene in the presence of an endogenous homolog likely to be expressed in the same cells. To address this problem, the presumptive regulatory sequences are cloned in a vector upstream of a **reporter gene**, that is a gene producing a protein that can be detected and quantified using a simple assay (*Box 20.4*).

The general method for mapping regulatory elements in cell lines is shown in *Figure 20.13*. The use of different cell lines, some of which express the corresponding endogenous gene and some of which do not, may provide evidence for promoter elements that permit gene expression in permissive cell lines and prevent gene expression in nonpermissive cell lines. Similarly, elements for inducible gene expression can be located by assaying different promoter constructs in the same cell line either in the presence or absence of induction. A more refined analysis of cell-specific gene expression is possible by studying the reporter expression patterns in transgenic animals, as this can identify elements that control gene expression in particular cell types or developmental stages. However, similar results must be obtained from several, independently-derived transgenic lines to avoid the misinterpretation of expression patterns caused by position effects (see Section 20.2.3).

Box 20.4: Reporter genes for animal cells.

Reporter genes encode proteins that can be detected and quantified using simple and inexpensive assays. They can be used to visualize and measure the efficiency of gene transfer and gene expression, and to investigate intracellular protein localization. Several reporter genes are commonly used in animals.

The *cat* gene from *E. coli* transposon *Tn9* encodes the enzyme **chloramphenicol acetyltransferase**, which transfers acetyl groups from acetyl CoA onto the antibiotic chloramphenicol. A standard *in vitro* assay format is used (Gorman *et al.*, 1982). Cell lysates are mixed with ^{14}C-labeled chloramphenicol, incubated and then separated by thin layer chromatography. CAT activity is determined by measuring the relative amounts of acetylated and nonacetylated chloramphenicol using a phosphorimager or scintillation counter.

The *lacZ* gene from *E. coli* encodes the enzyme **β-galactosidase**, which breaks down lactose and related compounds. A number of specialized derivative substrates are available including **ONPG**, which gives rise to a soluble yellow product and is used for colorimetric *in vitro* assays, and **X-gal**, which gives rise to a blue precipitate and is used for *in situ* assays (Hall *et al.*, 1983). This gene is most widely used with X-gal as a histological marker to show gene expression patterns in transgenic animals. The *E. coli gusA* gene, which encodes the enzyme **β-glucuronidase**, can be used in a similar way.

The *luc* gene from the American firefly (*Photinus pyralis*) encodes the enzyme **luciferase**, which catalyzes the oxidation of luciferin in a reaction requiring oxygen, ATP and magnesium ions (de Wet *et al.*, 1987). The reaction releases a flash of light, the intensity of which is proportional to the level of enzyme activity. This can be detected using a luminometer or a scintillation counter, and the assay is over 100 times more sensitive than conventional assays using CAT, β-galactosidase or β-glucuronidase. The emitted light signal also decays rapidly, so luciferase can be used to monitor rapid changes in gene expression levels. Conversely, CAT, β-galactosidase and β-glucuronidase are very stable proteins, so they can efficiently determine when a gene is switched on, but persist after it is switched off. Luciferase genes from other organisms have slightly different activities and the reactions release light at different wavelengths, allowing the simultaneous monitoring of several genes.

The *gfp* gene from the jellyfish *Aequoria victoria* encodes **green fluorescent protein (GFP)**, a bioluminescent marker that emits bright green fluorescence when exposed to blue or ultraviolet light (Ikawa *et al.*, 1999). GFP activity is measured by quantifying light emission, but unlike luciferase the protein has no substrate requirements and can readily be used to assay cellular processes in real time. GFP fusion proteins are widely used to study protein localization in the cell and protein trafficking within and between cells. A range of variant GFPs can be used for dual labeling and fluorescent proteins of other colors are also available. A mutant fluorescent protein that changes from green to red fluorescence over time can be used to characterize temporal gene expression (Terskikh *et al.*, 2000).

20.3.2 Gene function can be investigated by generating loss-of-function and gain-of-function mutations and phenocopies

Gene functions cannot always be established by disruption or inhibition due to genetic redundancy

Where cellular level studies are possible, RNA interference is becoming the method of choice for generating loss-of-function effects (Section 20.2.6). The functions of a number of genes have already been established by RNA interference in mammalian cell cultures including the gene encoding the vacuolar sorting protein Tsg101, which has been shown to be essential for HIV budding (Garrus *et al.*, 2001). Many other examples are cited in Tuschl and Borkhardt (2002), including the analysis of genes involved in cell division, DNA methylation, cell signaling and membrane trafficking. The most direct way to address the function of a gene in the context of a whole organism is gene targeting (Section 20.2.4). In many cases, such mutations are extremely informative and can reveal a great deal about the function of the gene. So many gene knockout experiments have now been carried out in mice that Internet databases have been established to catalog all the results, categorizing them by phenotype (*Table 20.3*). In a surprisingly large number of cases, however, knockout mutations appear to have very little phenotypic effect. This

may be due to **genetic redundancy**, that is the presence of another gene that is able to carry out the function of the inactivated gene. As a result, double or even triple gene knockouts have been necessary to determine precise functions for some genes.

An informative example concerns the *MyoD* and *Myf*-5 genes, which are expressed early in mouse development. Cloned cDNAs corresponding to both these genes have the profound ability to induce the expression of muscle-specific proteins in many different cell lines. Since both genes encode transcription factors, they appear to be excellent candidates for myogenic regulators. Surprisingly, however, when knockout mutations were produced, in neither case was there a striking phenotype. Muscle development was apparently normal in *MyoD* knockout mice and was slightly delayed in *Myf*-5 knockout mice (Rudnicki *et al.*, 1992). Later, it was shown that the two genes have equivalent functions and each can compensate fully for the absence of the other. Indeed, in normal mice the two transcription factors *repress each other's genes*, so that the knockout of one gene results in a compensatory increase in the expression of the other. The minor phenotype of *myf*-5 knockout mice reflects the slightly earlier onset of expression. In double knockouts, there is a catastrophic absence of muscle development and the mutant mice die just after birth due to asphyxiation.

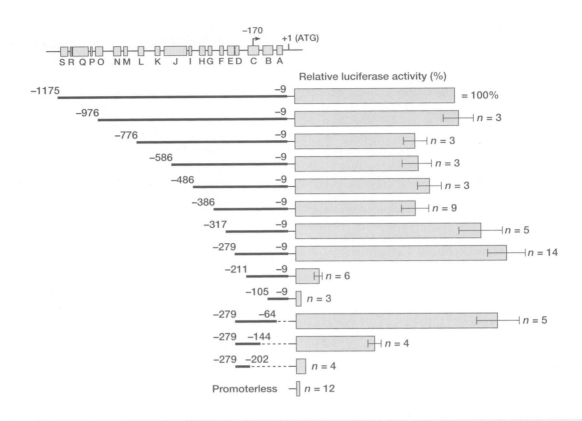

Figure 20.13: Deletion analysis of the human factor VIII gene promoter region.

If all the necessary elements are present for expression of the human gene, high level expression of the reporter gene will result when the construct is transfected into an appropriate type of cultured human cell. A series of progressive deletion constructs can then be made using restriction enzymes or the enzyme exonuclease III from *E. coli* which digests from the 3′ end of double-stranded DNA. Alternatively, a series of PCR amplification products can be designed covering the regions of interest and then cloned into a suitable expression vector. Solid bars to the left indicate variously sized sequences upstream of the human factor VIII gene (*F8*). Boxes at the top indicate upstream sequences which are thought to be protein binding sites. Boxes on the right indicate the level of luciferase activity relative to the intact sequence, based on *n* replicate experiments. On the basis of observed luciferase activity for the different expression clones, the deletion mapping shows that all the necessary elements for maximal promoter activity are located in the region from −279 to −64, including protein binding sites B, C and D. Reproduced from Figueiredo and Brownlee (1995) with permission from The American Society for Biochemistry and Molecular Biology.

Gene functions may be established by gain-of-function experiments if the dosage, activity or distribution of the gene product is critical

If loss-of-function mutations or phenocopies are uninformative, transformed cells and transgenic animals showing gains of function affecting the same gene may be useful. The *MyoD* and *Myf*-5 genes represent a case in point. These genes were known to encode important regulators of muscle development because of their effects on cultured cells. Similarly, many oncogenes have been identified and assigned functions on the basis that they cause cultured cells to proliferate in an uncontrolled manner. The gain of function can be generated either by overexpressing a normal gene product or expressing an overactive mutant version of the gene. In transgenic animals, gain-of-function experiments can be equally useful. A classic example concerns the *Sry* gene, which was shown conclusively to be a major determinant of male development by expression in transgenic mice with two X-chromosomes. These mice, which nature intended to be female, turned out to be male (Koopman *et al.*, 1991). Another type of gain-of-function is **ectopic expression**, where a transgene is expressed outside the normal spatial or temporal domains of the corresponding endogenous gene. There are many useful examples of this approach involving developmental genes. For example, *Hox* genes expressed outside their normal domains disrupt the normal patterning of the embryo and generate limb and skeletal defects (Burke, 2000).

Ectopic expression experiments may also be used to demonstrate genetic redundancy. For example, the mouse *engrailed* genes *En-1* and *En-2* are homeobox-containing genes that are thought to play crucial roles in brain formation. The *En-1* knockout mutant showed serious brain abnormalities as expected but surprisingly *En-2* knockouts have only minor defects. *En-1* expression is switched on 8–10 h before *En-2*, suggesting that perhaps the En-1 protein compensates for the lack of En-2 in *En-2* knockouts. To test for the possibility of functional redundancy, Hanks *et al.* (1995) used

Table 20.3: Internet resources for mammalian transgenesis and mutagenesis, and for large scale animal gene trap and RNAi screens

Resource	URL
Transgenic mouse and targeted mutagenesis resources	
Jackson Laboratory, database of targeted mutant mice	http://jaxmice.jax.org/index.shtml
TBASE, database of transgenic and targeted mutant mice, maintained by the Jackson Laboratory	http://tbase.jax.org
Frontiers in Science gene knockout database	http://www.bioscience.org/knockout/knochome.htm
BioMedNet mouse knockout database	http://biomednet.com/db/mkmd
Mouse ENU mutagenesis resources	
German Human Genome Project ENU mutagenesis	http://www.gsf.de/ieg/groups/enu-mouse.html
MRC mutabase ENU mutagenesis database	http://www.mgu.har.mrc.ac.uk/mutabase
Gene trap resources (for mouse and **Drosophila***)*	
German Gene Trap Consortium	http://tikus.gsf.de
Lexicon Genetics	http://www.lexgen.com/omnibank/omnibank.htm
Berkeley *Drosophila* Genome Project	http://www.fruitfly.org/p_disrupt/index.html
C. elegans *RNA interference resources*	
General information	http://www.wormbase.org
RNAi	http://www.rnai.org

a variant of the knock-out procedure known as **gene knock-in**, in which the targeting construct contained the *En-2* gene within an *En-1* homology region (including the normal *En-1* regulatory elements). The aim was to replace the endogenous *En-1* coding sequence with that of *En-2*, resulting in mice with two *En-2* genes, one expressed in the same manner as *En-1* in normal embryos (see *Figure 20.14*). The resulting *En-1* knock-out mouse had a normal phenotype, demonstrating that the knocked-in *En-2* gene was functionally equivalent to *En-1* (Hanks *et al.*, 1995).

20.3.3 The large scale analysis of gene function by insertional mutagenesis and systematic RNA interference are cornerstones of functional genomics

The experiments described above were designed to investigate gene expression and gene function on an individual basis. As discussed in Chapter 19, however, the vast amount of sequence information and the seemingly endless list of uncharacterized genes resulting from the genome projects now makes it necessary to study gene functions on a *genome wide scale*. Saturation mutagenesis has been used for functional analysis for many years, such studies relying on the use of radiation or potent chemical mutagens such as ethylnitrosourea (ENU) and ethylmethanesulfonate (EMS) to generate populations of mutants representing every gene in the genome. Until recently it was common practice to focus on specific biochemical, physiological or developmental aspects of the species under investigation, and most mutants were discarded.

In the least few years, however, the shift in focus from individual gene studies to functional genomics has turned this idea on its head and genome wide mutagenesis programs are now carried out with the objective of collecting as many mutants affecting as many biological processes as possible. The ultimate aim is to assemble a comprehensive mutant library as a central resource for all investigators. For example, the first genome wide ENU mutagenesis programs in mice involved the screening of a breathtaking 40 000 mouse lines for medically relevant phenotypes, including clinical biochemistry, allergy, immunology, response to physiological stimuli, developmental defects and behavior (Hrabe de Angelis and Balling, 1998, Hrabe de Angelis *et al.*, 2000, Nolan *et al.*, 2000). These programs are ongoing and the results are regularly updated on the websites listed in *Table 20.3*.

Chemical mutagens and radiation tend to induce point mutations. While these may be 'realistic' in that they represent the types of mutations that cause many human diseases, a major problem is that identification of the precise structural changes in the DNA of a single mutant animal often requires a laborious positional cloning approach. The alternative is to use gene transfer as a mutagenesis method. In this case, the mutagen is an insertional DNA sequence, a transgene by any other name, which occasionally integrates into a pre-existing gene and disrupts it. This strategy has one major advantage over irradiation and chemical mutagenesis: it leaves a sequence tag at the locus which is mutated. As a result, rapid molecular characterization of the mutated locus is possible (*Box 20.5*). In *Drosophila*, P-elements are used as insertional mutagens and these are introduced directly into the germ line (Section

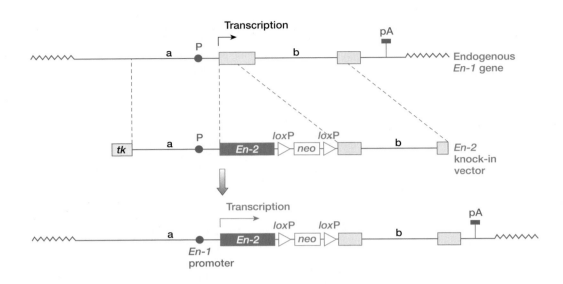

Figure 20.14: The gene knock-in method replaces the activity of one chromosomal gene by that of an introduced gene

The *En-1* gene shown at top has two exons and coding sequences are shown by filled boxes. Its promoter (P) and polyadenylation site (pA) are also shown. The gene targeting vector ('knock-in vector') contains cloned *En-1* gene sequences comprising an upstream sequence *a* which contains the *En-1* promoter, and an internal segment *b* which spans the 3' end of exon 1, the single intron and the 5' coding sequence of exon 2. Separating these two sequences is the coding sequence of the *En-2* gene. Two marker cassettes include a thymidine kinase (*tk*) gene and a neomycin-resistance gene (*neo*) both driven by a phosphoglycerate kinase promoter (chosen because phosphoglycerate kinase is expressed in ES cells). The *neo* gene is flanked by *lox*P sequences. The targeting procedure results in replacement of endogenous sequences *a* and *b* but the 5' coding sequence of the *En-1* gene is deleted. The knocked-in *En-2* gene comes under the conrol of the *En-1* promoter (Hanks *et al.*, 1995). *Note* that the term 'knock-in' has also been applied to any procedure where an endogenous gene is inactivated by insertion of a new gene which is then expressed, even if the latter is only meant to serve as a reporter gene such as *lacZ*.

20.2.2). ES cells provide a way of introducing such mutations into the mouse, since the random integration of DNA sequences can be used to recover insertions in any number of genes. However, because only approximately 3% of the mouse genome is represented by genes, most random insertion events do not result in gene disruption.

In order to increase the likelihood of recovering insertions in genes, the **gene trap** approach was devised (Evans *et al.*, 1997). The underlying principle is that the transgene contains a defective reporter gene or selectable marker that is expressed only if the construct inserts within a gene. For example, the transgene may comprise a reporter gene with an upstream splice acceptor site. If this integrates within a gene, it behaves as an additional exon. When the interrupted gene is transcribed, the reporter gene transcript becomes spliced to the upstream exons and should, when translated, retain its reporter activity. The advantage of this method is that the reporter gene is expressed in the same pattern as the inter-rupted gene, even in the heterozygous state, so information about the gene (its expression profile) can be gained even if the interruption itself is lethal in the homozygous state. A disadvantage is that expression of the reporter relies on the expression of the interrupted gene, so if the gene is not expressed the reporter is not expressed either. This problem has been addressed by expressing the reporter gene from its own (constitutive) promoter, but making polyadenylation

dependent on the surrounding gene (see *Figure 20.15*). This type of approach has been applied on a large scale, with a recent report describing disruption and sequence identifi-cation of 2000 genes in mouse embryonic stem cells (Zambrowicz *et al.*, 1998). Two major initiatives are underway in mice, one organized by the German Gene Trap Consortium, which aims to produce and characterize 20 000 gene trap lines (Wiles *et al.*, 2000) and one organized by the US company Lexicon Genetics. In both cases, databases of insert flanking sequences are maintained. These can be searched by investigators who might identify an interesting gene in their experiments and wish to obtain mouse lines in which that gene is inactivated. A gene trap program is also ongoing in *Drosophila* (Berkeley *Drosophila* Genome Project, see Spradling *et al.*, 1995, 1999), which includes not only P-element mediated gene disruption but also P-element mediated gene *activation*. The latter approach (activation tagging, *Box 20.5*) involves the use of P-elements incorpo-rating a strong, outward-facing promoter. Once integrated, such constructs will activate any adjacent, endogenous gene. This allows insertional mutagenesis to be used to generate gain-of-function mutations as well as loss-of-function ones. Internet resources for all the above programs are listed in *Table 20.3*.

RNA interference (Section 20.2.6) is another method that can be used for high throughput functional annotation.

Box 20.5: Sophisticated vectors used for insertional mutagenesis.

Any DNA sequence can be used as an insertion vector, and as long as the sequence is not already found in the host genome, it can be used as a tag to identify the interrupted gene in simple hybridization or PCR-based assays. However, the inclusion of certain extra features can add functionality to insertion vectors and reveal extra information about the interrupted gene and its product, or even help to clone flanking sequences surrounding the insert.

Gene traps. The insertion vector includes a reporter gene such as *lacZ* downstream of a splice acceptor site, so that the reporter gene is only expressed if the vector integrates within a gene. A variant gene trap vector, sometimes called a promoter trap, comprises a naive reporter gene without a promoter. It is only activated if it inserts downstream of an active promoter (Evans *et al.*, 1997). In both cases, reporter gene expression is dependent on the interrupted gene. The advantage is that the expression pattern of the interrupted gene is mirrored by the reporter gene (which can provide functional information) but the disadvantage is that nonexpressed genes have no reporter expression. This restriction can be removed by expressing the reporter gene from a constitutive promoter but making polyadenylation dependent on the interrupted gene (Zambrowicz *et al.*, 1998).

Enhancer traps. The insertion vector includes a reporter gene such as *lacZ* downstream of a minimal promoter (usually a TATA box) which supports only background transcription on its own. When the construct integrates within the influence of an endogenous enhancer, the reporter gene mimics the expression of the gene controlled by that enhancer (O'Kane and Gehring, 1987). Due to the long distance over which enhancers can act, this strategy is rarely useful for gene identification, but cell specific enhancers can be exploited for other purposes, such as the control of Cre expression (see Section 20.2.5) or the activation of a negative selectable marker for cell ablation (*Box 20.2*).

Activation tags. The insertion vector contains a strong, outward-facing promoter. If the element integrates adjacent to a gene, that gene will be activated by the promoter, possibly generating a gain-of-function phenotype through ectopic expression or overexpression (Rorth *et al.*, 1998).

Plasmid rescue. The insertion vector contains the origin of replication and antibiotic resistance marker from a bacterial plasmid. This means that if the genomic DNA from the transgenic insertion line is digested with a restriction enzyme, diluted and circularized using DNA ligase, the insert and its immediate flanking sequences will form a functional plasmid. If the whole collection of genomic DNA circles is used *en masse* to transform bacteria, only cells containing the plasmid will survive. This is a rapid way to clone and identify the gene sequences surrounding the insertion site (Perucho *et al.*, 1980).

Thus far, this **gene knock-down** method has been applied on a global scale only in the nematode *C. elegans*, but it has shown great potential in mammalian cells and is perhaps the only method suitable for the rapid and direct annotation of human genes, at least those with functions that can be determined at the cellular level. Several large-scale experiments have been carried out in *C. elegans*, involving the synthesis of thousands of dsRNA molecules and their systematic administration to worms either by microinjection, soaking or feeding (Gonczy *et al.*, 2000, Maeda *et al.*, 2001, Fraser *et al.*, 2000). Most recently, a heroic screen was carried out in which nearly 17 000 bacterial strains were generated, each expressing a different dsRNA, representing 86% of the genes in the *C. elegans* genome (Kamath *et al.*, 2003). Over 10% of these worms showed reproducible lethal, sterility, growth or developmental phenocopies.

20.4 Creating disease models using gene transfer and gene targeting technology

Models of human disease are crucially important to medical research. Animal models in particular allow detailed examination of the physiological basis of disease, and they also offer a frontline testing system for studying the efficacy of novel treatments before conducting clinical trials on human subjects. Cell cultures provide an alternative although more limited resource for disease modeling, particularly in terms of investigating the biochemical effects of disease and the response to drugs. Although many individual human disorders do not have a good animal model, animal models exist for some representatives of all the major human disease classes: genetically determined diseases, disease due to infectious agents, sporadic cancers and autoimmune disorders (see Leiter *et al.*, 1987; Darling and Abbott, 1992; Clarke, 1994; Bedell *et al.*, 1997). Some animal models of human disease originated spontaneously; others have been generated artificially by a variety of different routes (*Table 20.4*). Until recently, the great majority of available animal disease models were ones that arose spontaneously or had been artificially induced by random mutagenesis. More recently, gene targeting and transgenic technologies have provided direct ways of obtaining animal and cellular disease models, and targeted mutations in the mouse have been particularly valuable. Interestingly, it has become increasingly clear that disease phenotypes due to comparable mutations in human and mouse orthologs often show considerable differences (Section 20.4.6).

20.4.1 Modeling disease pathogenesis and drug treatment in cell culture

Many aspects of disease pathogenesis have a cellular component which can be effectively studied *ex vivo*. Examples

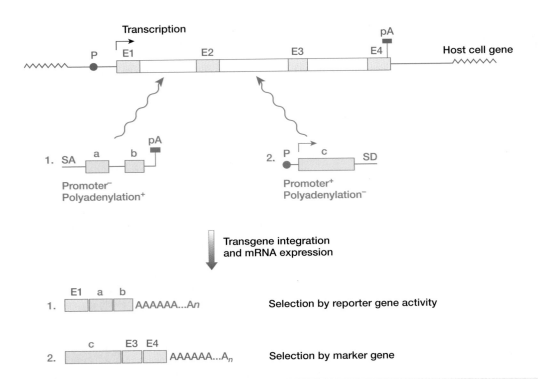

Figure 20.15: Gene trapping uses an expression-defective transgene to select for chromosomal integration events that occur within or close to a gene

A specimen host cell gene with the four exons (E1–E4) is shown at the top together with its promoter (P) and polyadenylation signal (pA). Two possibilities for a gene trap vector are shown. Transgene 1 includes a reporter gene which lacks a promoter. It has two exonic sequences *a* and *b*, a polyadenylation signal (pA) and an upstream sequence which contains a ***splice acceptor sequence (SA)***. In this example, transgene 1 integrates into intron 1 of the host cell gene. During transcription from the endogenous promoter the splice acceptor sequence will help the transgene exons to be spliced to the sequence from the first exon of the endogenous gene, producing a fusion transcript. Selection is based on this transcript having a functional reporter activity (see Evans *et al.*, 1997). Transgene 2 has a marker gene (drug-resistance, e.g. puromycin *N*-acetyltransferase) coupled to a promoter which works in ES cells (usually a phosphoglycerate kinase promoter) and a downstream ***splice donor sequence***, but lacks a polyadenylation signal. Again, integration is intended to permit expression but this time from the transgene's promoter and the splice donor sequence helps the RNA transcript to be spliced to downstream host exons.

include the DNA repair defects of Xeroderma pigmentosum and Fanconi anemia, the premature cell senescence characteristics of Werner syndrome, the cytoskeletal disorganization seen in neurofibromatosis and of course the uncontrolled proliferation of cancer. Where such phenotypes exist, cell models are easier to develop than animal models because gain-of-function defects can be modeled simply by transferring a disease-causing dominant transgene into the appropriate cell type while loss-of-function models can now be efficiently produced by RNA interference. Cell models can be based on human cell lines, and in many cases an artificial model is not required because cells can be isolated from individuals with the corresponding disease. Of course animals are more appropriate because they provide a whole organism context, which is required for the study of diseases with no appreciable cellular phenotype (e.g. essential tremor). For many diseases, however, cell models can be used as a first line of investigation, allowing the screening of many lead compounds and the identification of promising candidate drugs that can be tested on animal models at a later stage.

20.4.2 It may be difficult to identify animal disease models generated spontaneously or induced by random mutagenesis

Spontaneous animal disease models

Mutant human phenotypes, especially those associated with obvious disease symptoms, are subject to intense scrutiny: most individuals who suffer from a disorder seek medical advice. If they present with a previously undescribed phenotype their case may well be referred to experts who often will document the phenotype in the medical literature. Given the motivation of affected individuals and their families, physicians and interested medical researchers, and the large population size for screening (current total global population is over 6 billion individuals), there is a remarkably effective screening process for mutant human phenotypes. In contrast, many animal disease phenotypes will go unrecorded. Only a small percentage of the animal population is in captivity, and recording of spontaneous mutant phenotypes is largely dependent on examination of animal colonies bred for

Table 20.4: Classes of animal models of disease

Process	Examples	Comments
(**A**) Spontaneous	Germline mutation → inherited disorder Somatic mutation → cancer	
(**B**) Artificial intervention or artificially generated	Selective breeding to obtain strains that are genetically susceptible to disease	
	Infect animal strain with relevant microbial pathogen	
	Manipulate environment to induce disease without causing mutation	e.g. injection of pristane, a synthetic adjuvant oil, into the dermis of rats has produced a model of arthritis
	In vivo mutagenesis using a strong mutagen such as X-rays or powerful chemical mutagens such as ethylnitrosourea (ENU)	Large chemical mutagenesis programs have been established to produce mutant mice and zebrafish (Section 20.4.2)
	Genetic modification of fertilized egg cells or cells from the early embryo and subsequent animal breeding (*transgenic* and *gene targeting* technologies)	Sections 20.1–20.3

research purposes and, to a lesser extent, livestock and pet populations. Only mutants with obvious external anomalies are likely to be noticed. Despite the difficulty in identifying spontaneous animal mutants, a number of animal phenotypes have been described as likely models of human diseases (see *Table 20.5* for some examples). In some cases, the animal mutant phenotype closely parallels the corresponding clinical phenotype, but in others there is considerable divergence because of species differences in biochemical and developmental pathways (see Erickson, 1996; Wynshaw-Boris, 1996). Additionally, phenotypic differences may result because of different classes of mutation at orthologous loci.

Random mutagenesis using chemicals and irradiation

Classical methods of producing animal mutants involved controlled exposure to mutagenic chemicals, or to high doses of X-rays. Large numbers of *Drosophila* and mouse mutants have been obtained by this method and, as discussed above, efficient large-scale mutagenesis screens have been conducted on both mice and also zebrafish (which offers some advantages as an animal model; see *Box 20.6*). A major problem with chemically-induced and irradiation-induced mutations, however, is that they are generated essentially at random. In order to identify a mutant phenotype of interest, a laborious screen for mutants needs to be conducted by close examination of the phenotypes following mutagenesis. The mutant phenotypes which have been described in these studies, as well as for spontaneous mutants, show a clear bias towards phenotypes with obvious external abnormalities, simply because of the ease of identifying them. Nevertheless, several important models of human disease have been created using such methods.

Table 20.5: Examples of spontaneous animal mutants

Animal mutant	Phenotypic features and molecular pathogenesis
NOD mouse	Diabetic, but without being obese. Mimics human insulin-dependent diabetes mellitus
mdx mouse	X-linked muscular dystrophy due to mutations in mouse dystrophin gene. The original *mdx* mutant has a nonsense mutation but phenotype is much milder than Duchenne muscular dystrophy (DMD)
Hemophiliac dog	Missense mutation in canine factor IX gene causes complete loss of function. Human homolog is hemophilia B
Watanabe heritable hyperlipidemic (WHHL) rabbit	Hyperlipidemic as a result of a deletion of four codons of the low density lipoprotein receptor gene (*LDLR*); human homolog is familial hypercholesterolemia
Atherosclerotic pigs	Marked hypercholesterolemia. Normal LDL receptor activity, but variant apolipoproteins, including apolipoprotein B
Splotch mouse	Abnormal pigmentation; phenotypic overlap with Waardenburg syndrome suggested that it was an animal model for this disease, and confirmed by identification of mutations in homologous human and murine *PAX* genes
NF damselfish	Extensive neurofibromas suggest that it could be a homolog of human neurofibromatosis type 1

20.4.3 Mice have been widely used as animal models of human disease largely because specific mutations can be created at a predetermined locus

Spontaneous and artificially produced disease phenotypes have been described in a wide range of animal species with differing potentials for modeling human disease. Invertebrates such as *Drosophila* and *C. elegans*, and even yeast cells, can provide some useful disease models, reflecting the existence of certain proteins and pathways that are very strongly conserved throughout evolution (see Section 3.9.2). Evolutionarily distant vertebrates, such as the zebrafish, also offer some advantages as model organisms and have been used to model certain human diseases (*Box 20.6*). Mammals would be expected to provide better disease models but, for a variety of reasons, our closest relatives, the great apes, have not been very useful in providing disease models. Instead, other mammals, notably mice, have been used widely to model human disease (*Box 20.6*). In the case of animal disease models which are artificially induced by exposure to mutagenic chemicals or radiation, or which originate spontaneously, there is little or no artificial control over the resulting phenotype and frequently, the identification of an animal disease model is serendipitous. The great advantage of transgenic/gene targeted mouse models of disease is that *specific disease models can be constructed to order*. Provided that the relevant gene clones are available, including mutant genes in some cases, mice can be generated with a desired alteration in a chosen target gene. All the major classes of disease, inherited disorders, cancers, infectious diseases and autoimmune disorders can be modeled in this way (*Table 20.6*; Smithies, 1993; Clarke, 1994; Bedell *et al.*, 1997). In most cases, the transgenic/gene targeting approaches have been used to model single gene disorders but, increasingly, attempts are being made to produce mouse models of complex genetic diseases, such as Alzheimer's disease, atherosclerosis and essential hypertension effects (Petters and Sommer, 2000).

20.4.4 Loss-of-function mutations can be modeled by gene targeting, and gain-of-function mutations by the expression of dominant mutant genes

Modeling loss-of-function mutations by gene targeting in mice

Many disease phenotypes, including those of essentially all recessively inherited disorders and many dominantly inherited disorders, are thought to result from loss of gene function. The simplest way of modeling the disease for single gene disorders of this type is to make a knockout mouse. The first step is to isolate the orthologous mouse gene and to use a segment of it to knock out the endogenous gene in mouse ES cells using gene targeting (Section 20.2.4). Following injection of the genetically modified ES cells into the blastocyst of a foster mother, and continued development, founder mice are obtained with the targeted mutation in a sizeable proportion of their germ cells. These mice can be interbred and the offspring can be screened for the presence of the desired mutation, and for the presence of the wild-type allele using PCR assays of cells collected from tail bleeds. The gene targeting event is intended to create a null allele (where there is complete absence of gene expression), but sometimes the result may be a **leaky mutation** and the mutant allele retains some gene expression. For example, some of the normal gene product may be obtained by aberrant splicing which skips over the inserted DNA segment. This may explain differences in the severity of the mouse models of cystic fibrosis described by Snouwaert *et al.* (1992) and by Dorin *et al.* (1992). Differences in phenotype may also occur because of *modifier genes* using different mouse strains (see Section 20.4.6).

Table 20.6: **Examples of transgenic or gene-targeted mouse models of human disease**

Human disease or abnormal phenotype	Gene	Method of constructing model
Cystic fibrosis	*CFTR*	Insertional inactivation by gene targeting
β-Thalassemia	*HBB* (β-globin)	Insertional inactivation by gene targeting
Hypercholesterolemia and atherosclerosis	Apolipoprotein genes, e.g. *APOE*	Insertional inactivation by gene targeting
Gaucher's disease	*GBA*	Insertional inactivation by gene targeting
Fragile-X syndrome	*FMR1*	Insertional inactivation by gene targeting
Gerstmann–Sträussler–Scheinker (GSS) syndrome	Prion protein gene (*PRNP*)	Integration of mutant mouse prion gene
Spinocerebellar ataxia type 1 (SCA1)	*SCA1* (ataxin)	Integration of mutant human ataxin gene with expanded triplet repeat
Alzheimer's disease	*APP* (β-amyloid precursor protein)	Integration of mutant full-length *APP* cDNA under control of a platelet-derived growth factor promoter

Box 20.6: The potential of animals for modeling human disease.

Primates *should* provide the best animal models of human disease because they are so closely related to us (see Section 12.4.2). Humans and great apes show extensive developmental, anatomical, biochemical and physiological similarities. Primates are expensive to breed, however, and the population sizes in captivity are very small. Public sensitivity also plays a part. While some are opposed, in principle, to all animal experimentation, those that feel it is justified in the interests of medical research are generally more comfortable with experimentation on small laboratory animals such as mice and rats. Most importantly, primates are not well suited to experimentation: they are comparatively long lived and less fecund than rodents, and so breeding experiments are more difficult to organize. Given that novel therapeutic approaches are rapidly being developed for a range of human disorders (see Chapter 21), the long delay required to perform test experiments on primates has prompted the study of alternative models.

Mice are the most widely used animal models of human disease. They are small and can be maintained in breeding colonies comparatively cheaply. They have a short lifespan (~2–3 years), a short generation time (~3 months) and are prolific (an average female will produce four to eight litters of six to eight pups). Because they can be bred easily, complex breeding programs can be arranged to produce **recombinant inbred strains** and **congenic strains** (Taylor, 1989; see *Box 14.2*) and their short generation time and lifespan means that the effects of transmitting a pathogenic mutation through several generations can be monitored relatively easily. As a result, the genetics of the laboratory mouse have been studied extensively for decades, and the phenotypes of many mutants have been recorded (Lyon and Searle, 1989). Most such mutants have originated spontaneously within breeding colonies. A few have also been produced artificially, initially by X-ray or chemical mutagenesis, but increasingly by gene targeting and insertional mutagenesis. Mapping of the mouse mutants is facilitated by **interspecific back-cross mapping** (Avner *et al.,* 1988; see *Box 14.2*) and by the availability of numerous polymorphic markers (~9000 dinucleotide repeat markers have been mapped). Because syntenic regions of the mouse and humans genomes have been well documented (see *Figure 12.11*), this information is useful in identifying genuinely homologous single gene disorders in mouse and man.

Rats are comparatively large and have been more amenable to physiological, pharmacological and behavioral experiments, especially in cardiovascular and neuropsychiatric studies. They have a longer generation time (11 weeks), and breeding colonies are more expensive. Some classes of human disorders (e.g. hypertension and behavioral disorders) have no good mouse models and instead have relied on rat models. Dense genetic and physical maps of the rat genome have been constructed, and the genome sequence has very recently been obtained (see Section 8.4.4).

The **zebrafish** represents the ideal model of vertebrate development since the fish develop externally and have robust embryos like frogs, have a generation interval similar to mice but produce 40 times more offspring in the same period, and the embryos are transparent like those of *Drosophila* and *C. elegans*. Zebrafish genetics are advanced: large scale mutant screens have been carried out since the mid 1990s, gene transfer technology is routine and dense genetic and physical maps are now available along with EST resources. The zebrafish and human genomes show a moderate amount of synteny. Many of the mutant phenotypes identified in zebrafish genetic screens resemble human disease states, providing useful disease models particularly in the areas of hematopoietic disorders, cardiovascular disease and kidney disease. For example, the *sauternes* gene is the zebrafish ortholog of human *ALAS2*, encoding δ-aminolevulinate synthase, the first enzyme in heme biosynthesis. The zebrafish mutant provided the first animal model of congenital sideroblastic anemia, which is caused by loss of ALAS2 function (Brownlie *et al.,* 1998). Similarly, the *dracula* gene is the zebrafish ortholog of human *FECH*, which encodes ferrochelatase, the last enzyme in heme biosynthesis. The mutation provides a model for erythropoietic protoporphyria (Childs *et al.,* 2000). Other models have been recently reviewed (Dooley and Zon, 2000). The zebrafish and human genomes show certain similarities although, because of the considerable evolutionary divergence between humans and zebrafish, the relevance of zebrafish mutants to human disease will be expected to be confined to disorders affecting pathways that are highly conserved. A number of zebrafish mutants have been produced that are good models of human diseases, and can therefore be used to test candidate drugs. These include models of Alzheimer's disease, disorders of heme biosynthesis, congenital heart disease, polycystic kidney disease and cancer.

Invertebrates and yeast are separated from humans by millions of years of evolution, yet still a core of genes and essential pathways remains conserved. Therefore, simple organisms such as the yeast *Saccharomyces cerevisiae* and the nematode *Caenorhabditis elegans* provide useful models of some diseases, at least as far as providing a test system for drugs, even if the overall phenotype is not particularly relevant. For example, the insulin signaling pathway is entirely conserved between humans and nematodes, so nematodes can be used as models for diabetes. Similarly, diseases affecting very elementary cellular functions (such as the DNA helicase defects in Bloom syndrome) can be modeled in yeast. Importantly, both yeast and *C. elegans* can be handled as microorganisms and can therefore be screened in batches with panels of lead compounds and candidate drugs.

Modeling gain-of-function mutations by expressing a dominant mutant gene

This general experimental design has been used frequently in conjunction with the pronuclear microinjection technique of gene transfer. The disease to be modeled must be one where the presence of an introduced DNA is itself sufficient to induce pathogenesis, and can include inherited gain-of-function mutations and sporadic cancers caused by oncogenes. To model such disorders, it is necessary to clone a mutant gene or, if necessary, design one by *in vitro* mutagenesis. The mutant gene is then simply inserted as a transgene, for example by microinjection into fertilized oocytes. Because there is no requirement for the introduced mutant gene to integrate at a specific location, human mutant genes will suffice although, in some cases, mouse mutant genes have been used. The two examples below illustrate this approach.

▶ An early example was intended to assess whether a leucine substitution found at codon 102 in a prion protein gene in a patient with Gerstmann–Sträussler–Scheinker (GSS) syndrome was pathogenic. An equivalent mutation was artificially designed in a cloned mouse prion protein gene and the mutant gene was then injected into fertilized oocytes to produce transgenic mice. The mice displayed spontaneous neurodegeneration, reminiscent of that found in the human syndrome (Hsiao *et al.*, 1990). A variety of other experiments using prion protein transgenes have been very helpful in understanding prions (Gabizon and Taraboulos, 1997).

▶ Expanded triplet repeats causing neurodegenerative disorders (see Section 16.6.4) constitute another type of gain-of-function mutation. Spinocerebellar ataxia type 1 (SCA1) is a dominantly inherited disorder which results from unstable expansion of a CAG triplet repeat in the ataxin gene. It is characterized by degeneration of cerebellar Purkinje cells, spinocerebellar tracts and some brainstem neurons. Transgenic mice were produced by the introduction of one of two transgenes driven by a Purkinje cell-specific promoter: the normal human ataxin gene (*SCA1*), and a mutant ataxin gene containing an expanded CAG repeat. Both types of transgene were expressed, but only mice with the expanded allele developed ataxia and Purkinje cell degeneration, confirming the gain-of-function hypothesis (Burright *et al.*, 1995).

Modeling human cancers in mice

Considerable effort is also being devoted to constructing mouse models of cancers (see Ghebranious and Donehower, 1998; Macleod and Jacks, 1999).

▶ Gain of function. In this case the disease is due to inappropriate activation of a proto-oncogene and can be modeled by constructing a transgenic mouse. The appropriate oncogene is introduced into the mouse genome by simple transgene integration.

▶ Loss of function. In this case, the disease is due to inactivation of a tumor suppressor gene and can be modeled by constructing knock-out mice through gene targeting. For example, several models have been generated by inactivating the mouse homologs of the *TP53* and *RB1* genes but the phenotypes show only broad similarity to the corresponding human phenotypes, Li–Fraumeni syndrome and retinoblastoma, respectively (see also Section 20.4.6).

20.4.5 Increasing attention is being focused on the use of transgenic animals to model complex disorders

Modeling chromosomal disorders

Existing mouse models for human chromosomal disorders are sparse. In some cases this is due to insufficient conservation of synteny between the two species. Taking the example of Down syndrome (trisomy 21), human chromosome 21 shares a large syntenic region with mouse chromosome 16 (see *Figure 12.11*), but trisomy 16 mice are not good models for Down syndrome because they die *in utero*. Indeed, the trisomy 16 mouse could never be expected to be a good model of human trisomy 21 because the two chromosomes are not absolutely equivalent. Genes in the distal 2–3 Mb of human chromosome 21 have orthologs on mouse chromosomes 17 and 10, and some genes on mouse chromosome 16 have orthologs on chromosomes other than human chromosome 21. In order to produce a better Down syndrome model in the mouse, attention has focused on the Down syndrome critical region at 21q21.3–q22.2 (deduced from observing the phenotypes of rare Down syndrome patients with partial trisomy 21). Within this region, the human *minibrain* gene at 21q22.2 may be an important contributory locus to the associated learning defects. Transgenic mice in which a 180-kb YAC containing the 100-kb human *minibrain* gene but apparently no other gene, develop learning defects (Smith *et al.*, 1997). A segmental trisomy 16 mouse, Ts65Dn, obtained by standard methods of irradiating mice was shown by Reeves *et al.* (1995) to have learning and behavior deficits. Other more recently produced models are discussed by Kola and Hertzog (1998). Future efforts in modeling chromosomal disorders can be expected to take advantage of gene targeting using Cre–*lox*P. As described in Section 20.2.5 this system offers tremendous potential for genome engineering and can be used to engineer chromosome translocations at defined positions on preselected chromosomes. YAC transgenics can also be expected to be important in investigating overexpression of genes in other chromosomal disorders and in disorders resulting from aberrant gene dosage for large regions, such as Charcot–Marie–Tooth disease type 1A, which is due to overexpression as a result of a 1.5-Mb duplication in the *PMP22* gene region (see *Figure 16.8*).

Modeling complex diseases

Increasingly, the focus in human genetics is moving towards understanding the pathogenesis of complex genetic diseases such as atherosclerosis, essential hypertension and diabetes. Such disorders have a complex etiology with multiple genetic and environmental components. Some valuable animal models have been produced for some of these disorders but gene targeting approaches are expected in the future to provide additional badly needed models (Smithies and Maeda,

1995). As long as suitably promising genes can be identified as being involved in the pathogenesis, then breeding experiments can be used to bring different combinations of disease genes together, and the effect of different genetic backgrounds in different strains of mice and of different environmental factors can be assessed. This approach may not be so daunting as it sounds because, increasingly, many complex disease phenotypes are considered to be due mostly to the combination of only a very few major susceptibility genes. For example, a digenic model of spina bifida occulta was generated serendipitously in offspring obtained by crossing a mouse heterozygous for the *Patch* mutation (*Pdgfrb*; platelet-derived growth factor receptor) with a mouse homozygous for the *undulated* mutation (*Pax-1*) (see Helwig *et al.*, 1995). In other cases, such digenic models may suggest possible therapies. For example, crossing *rds* mutant mice (showing retinal degeneration) with transgenic mice expressing *Bcl-2* (which protects cell from apoptosis) led to a slow down in retinal degeneration (Nir *et al.*, 2000).

20.4.6 Mouse models of human disease may be difficult to construct because of a variety of human/mouse differences

It is not uncommon for spontaneous or artificially generated mouse models of disease to show phenotypes that are considerably different from the corresponding human disorders. For example, gene targeting to inactivate several mouse tumor suppressor genes has often produced disappointing mouse models, as in the case of *TP53* and *RB1* (retinoblastoma) knock-outs. There may be problems in achieving the desired type of mutation, for example because of the possibility of *leaky* expression in knockouts discussed above, or the expression of a transgene may be affected by various factors such as the position effects discussed in Section 20.2.3 causing an unexpected phenotype. Setting aside these possibilities, there are several areas where differences between mice and humans could be expected to result in divergent disease phenotypes for mutations in orthologous genes (Erickson, 1989, 1996; Wynshaw-Boris, 1996).

▶ *Differences in biochemical pathways.* Although biochemical pathways in mammals are generally well conserved, some differences are known between the pathways of humans and mice. The human retina appears to depend heavily on the accurate function of the *RB1* gene product, but other vertebrate retinas do not. As a result, spontaneous retinoblastoma mouse mutants have not been described, and retinoblastoma is not a feature of *Rb1* knockout mice. However, it has been shown recently that *Rb1* is required for normal placental development in the mouse. Knockout mice show excessive proliferation of trophoblast cells and a severe disruption of the normal placental labyrinth architecture causing a decrease in vascularization and reduced placental transport (Wu *et al.*, 2003). Another example may be provided by ganglioside degradation pathways. While mutations in the human *HEXA* gene, encoding hexosaminidase, result in the severe lysosomal storage disorder Tay–Sachs disease, inactivation of the mouse ortholog *Hexa* causes the abnormal accumulation of ganglioside in neurons without the motor neuron or learning deficits seen in humans (Wynshaw-Boris, 1996).

▶ *Differences in developmental pathways.* The differences in human and mouse developmental pathways are not well understood but are expected to be significant for some organ systems, such as the brain.

▶ *Absolute time.* Because of the huge difference in the average lifespans of mice and humans, certain human disorders in which the disease is of late onset may possibly be difficult to model in mice.

▶ *Differences in genetic background.* This reflects the importance of *modifier genes*. Most human populations are outbred. Laboratory strains of mice, however, are very inbred. Often a particular phenotype can vary considerably in different strains of mice because of differences in their alleles at other loci (**modifier genes**), which can interact with the locus of interest. A useful example of the importance of genetic background is the ***Min* (multiple intestinal neoplasia) mouse** which was generated by ENU mutagenesis and results from mutations in the mouse *Apc* gene. Mutations in the orthologous human gene, *APC*, cause adenomatous polyposis coli and related colon cancers and the *Min* mouse has been regarded as a good model for such disorders. The phenotype of the *Min* mouse is, however, dramatically modified by the genetic background. For example the number of colonic polyps in mice carrying APC_{Min} is strikingly dependent on the strain of mouse. Similar phenotypic variability is found in human families where different members of the same family may have strikingly different tumor phenotypes although they possess identical mutations in the *APC* gene. Some of the variability could be due to environmental factors, but the involvement of modifier genes had been strongly suspected. The *Min* mouse provides a well-defined genetic system for mapping and identifying modifier genes (Dietrich *et al.*, 1993; MacPhee *et al.*, 1995).

The recent completion of the mouse and rat genome sequences has identified a number of human genes that do not appear to have counterparts in rodents. Examples include the pseudoautosomal *SHOX* gene (deficiencies in which underlie Leri-Weill syndrome and may significantly contribute to Turner syndrome – see Clement-Jones *et al.*, 2000) and the Kalman syndrome gene, *KAL1*.

Further reading

Kuhn R, Schwenk F (1997) Advances in gene targeting methods. *Curr. Opin. Immunol.* **9**, 183–188.

Popko B (ed.) (1998) *Mouse Models of Human Genetic Neurological Disease*. Plenum Press, New York.

Shastry BS (1998) Gene disruption in mice: models of development and disease. *Mol. Cell. Biochem.* **181**, 163–179.

Sikorski R, Peters R (1997) Transgenics on the internet. *Nature Biotechnol.* **15**, 289.

TBASE: a transgenic/targeted mutation database at http://www.gdb.org/Dan/tbase.Html.

Muller U (1999) Ten years of gene targeting: targeted mouse mutants, from vector design to phenotype analysis. *Mechanisms of Development* **82**, 3–21.

References

Amaya E, Musci TJ, Kirschner MW (1991) Expression of a dominant negative mutant of the FGF receptor disrupts mesoderm formation in *Xenopus* embryos. *Cell* **66**, 257–270.

Avner P, Amar L, Dandolo L, Guenet JL (1988) Genetic-analysis of the mouse using interspecific crosses. *Trends Genet.* **4**, 18–23.

Bahramian MB, Zabl H (1999) Transcriptional and posttranscriptional silencing of rodent alpha1 (I) collagen by a homologous transcriptionally self-silenced transgene. *Mol. Cell. Biol.* **19**, 274–283.

Bedell MA, Jenkins NA, Copeland NG (1997) Mouse models of human disease. Part II. Recent progress and future directions. *Genes Dev.* **11**, 11–43.

Belshaw PJ, Ho SN, Crabtree GR, Schreiber SL (1996) Controlling protein association and subcellular localization with a synthetic ligand that induces heterodimerization of proteins. *Proc. Natl Acad. Sci. USA* **93**, 4604–4607.

Brownlie A, Donovan A, Pratt SJ et al. (1998) Positional cloning of the zebrafish sauternes gene: a model for congenital sideroblastic anaemia. *Nature Genet.* **20**, 244–250.

Burke AC (2000) *Hox* genes and the global patterning of the somite mesoderm. *Curr. Top. Dev. Biol.* **47**, 155–181.

Burright EN, Clark HB, Servadio A et al. (1995) SCA1 transgenic mice – a model for neurodegeneration caused by CAG trinucleotide expansion. *Cell* **82**, 937–948.

Campbell KHS, McWhir J, Richie WA, Wilmut I (1996) Sheep cloned by nuclear transfer from a cultured cell line. *Nature* **380**, 64–67.

Capecchi M (1989) The new mouse genetics: altering the genome by gene targeting. *Trends Genet.* **5**, 70–76.

Chan AWS, Homan EJ, Ballou LU, Burns JC, Brennel RD (1998) Transgenic cattle produced by reverse transcribed gene transfer in oocytes. *Proc. Natl Acad. Sci. USA* **95**, 14028–14033.

Chan AWS, Chong KY, Martinovich C, Simerly C, Shatten G (2001) Transgenic monkeys produced by retroviral gene transfer into mature oocytes. *Science* **291**, 309–312.

Childs S, Weinstein BM. Mohideen M-A PK et al. (2000) Zebrafish *dracula* encodes ferrochelatase and its mutation provides a model for erythropoietic protoporphyria. *Current Biol.* **10**, 1001–1004.

Clarke AR (1994) Murine genetic models of human disease. *Curr. Opin. Genet. Dev.* **4**, 453–460.

Clement-Jones M, Schiller S, Rao E, Blaschke RJ, Zuniga A, Zeller R et al. (2000) The short stature homeobox gene *SHOX* is involved in skeletal abnormalities in Turner syndrome. *Human Molecular Genetics* **9**, 695–702.

Dai Y, Vaught TD, Boone J, Chen S-H et al. (2002) Targeted disruption of the α-1,3-galactosyltransferase gene in cloned pigs. *Nature Biotechnol.* **20**, 251–255.

Darling SM, Abbott CM (1992) Mouse models of human single gene disorders. 1. Nontransgenic mice. *BioEssays*, **14**, 359–366.

De Wet JR, Wood KV, De Luca M, Helsinki DR, Subramani S (1987) Firefly luciferase gene: structure and expression in mammalian cells. *Mol. Cell. Biol.* **7**, 725–737.

Dietrich WF, Lander ES, Smith JS et al. (1993) Genetic identification of mom-1, a major modifier locus affecting min-induced intestinal neoplasia in the mouse. *Cell* **75**, 631–639.

Dooley K, Zon LI (2000) Zebrafish: a model system for the study of human disease. *Current Opin. Genet. Dev.* **10**, 252–256.

Dorin JR, Dickinson P, Alton EW et al. (1992) Cystic-fibrosis in the mouse by targeted insertional mutagenesis. *Nature* **359**, 211–215.

Efrat S, Lieser M, Wu Y et al. (1994) Ribozyme-mediated attenuation of pancreatic β-cell glucokinase expression in transgenic mice results in impaired glucose-induced insulin secretion. *Proc. Natl Acad. Sci. USA* **91**, 2051–2055.

Erickson RP (1989) Why isn't a mouse more like a man? *Trends Genet.* **5**, 1–3.

Erickson RP (1996) Mouse models of human genetic disease: which mouse is more like a man? *Bioessays* **18**, 993–998.

Evans MJ, Kaufman MH (1981) Establishment in culture of pluripotential cells from mouse embryos. *Nature* **292**, 154–156.

Evans MJ, Carlton MBL, Russ AP (1997) Gene trapping and functional genomics. *Trends Genet.* **13**, 370–374.

Famulok M, Verma S (2002) In vivo-applied functional RNAs as tools in proteomics and genomics research. *Trends Biotechnol.* **20**, 462–466.

Famulok M et al. (2001) Intramers as promising new tools in functional proteomics. *Chem. Biol.* **8**, 931–939.

Fiel R, Brocard J, Mascrez B, LeMur M, Metzger D, Chambon P (1996) Ligand-activated site-specific recombination in mice. *Proc. Natl Acad. Sci. USA* **93**, 10887–10890.

Figueiredo MS, Brownlee GG (1995) Cis-acting elements and transcription factors involved in the promoter activity of the human factor VIII gene. *J. Biol. Chem.* **270**, 11828–11838.

Fire A, Xu S, Montgomery MK, Kostas SA, Driver SE, Mello CC (1998) Potent and specific genetic interference by double stranded RNA in *Caenorhabditis elegans*. *Nature* **391**, 806–811.

Forss-Petter S, Danielsen PE, Catsicas S, Battenberg E, Price J, Nerenberg M, Sutcliffe JG (1990) Transgenic mice expressing β-galactoisidase in mature neurons under neuron-specific enolase promoter control. *Neuron* **5** (2), 187–197.

Fraser AG, Kamath RS, Zipperlen P, Martinez-Campos M, Sohrmann M, Ahringer J (2000) Functional genomic analysis of *C. elegans* chromosome I by systematic RNA interference. *Nature* **408**, 325–330.

Gabizon R, Taraboulos A (1997) Of mice and (mad) cows – transgenic mice help to understand prions. *Trends Genet.* **13**, 264–269.

Garrus JE, von Schwedler UK, Pornillos OW *et al.* (2001) Tsg101 and the vacuolar protein sorting pathway are essential for HIV-1 budding. *Cell* **107**, 55–65.

Ghebranious N, Donehower LA (1998) Mouse models in tumor suppression. *Oncogene* **17**, 3385–3400.

Gonczy P, Echerverri G, Oegema K, Coulson A *et al.* (2000) Functional genomic analysis of cell division in *C. elegans* using RNAi of genes on chromosome III. *Nature* **408**, 331–336.

Gordon JW (1992) Production of transgenic mice. *Methods. Enzymol.* **225**, 747–771.

Gorman CM, Moffat LF, Howard BH (1982) Recombinant genome which expresses chloramphenicol acetyltransferase in mammalian cells. *Mol. Cell. Biol.* **2**, 1044–1051.

Gu H, Marth JD, Orban PC, Mossmann H, Rajewsky K (1994) Deletion of a DNA-polymerase-beta gene segment in T-cells using cell-type-specific gene targeting. *Science* **265**, 103–107.

Hall CV, Jacob PE, Ringold GM, Lee F (1983) Expression and regulation of *Escherichia coli lacZ* gene fusions in mammalian cells. *J. Mol. Appl. Genet.* **2**, 101–109.

Hanks M, Wurst W, Anson-Cartwright L, Auerbach AB, Joyner AL (1995) Rescue of the *En-1* mutant phenotype by replacement of *En-1* with *En-2*. *Science* **269**, 679–682.

Helwig U, Imai K, Schmahl W, Thomas BE, Varnum DS, Nadeau JH, Balling R (1995) Interaction between *undulated* and *patch* leads to an extreme form of spina-bifida in double-mutant mice. *Nature Genet.* **11**, 60–63.

Herault Y, Rassoulzadegan M, Cuzin F, Duboule D (1998) Engineering chromosomes in mice through targeted meiotic recombination (TAMERE). *Nature Genet.* **20**, 381–384.

Higgins DE, Portnoy DA (1998) Bacterial delivery of DNA evolves. *Nature Biotechnol.* **16**, 138–139.

Hrabe de Angelis M, Balling R (1998) Large scale ENU screens in the mouse. Genetics meets genomics. *Mutations Res.* **400**, 25–32.

Hrabe de Angelis M, Flaswinkel H, Fuchs H, Rathkolb B *et al.* (2000) Genome-wide, large-scale production of mutant mice by ENU mutagenesis. *Nature Genet.* **25**, 444–447.

Hsiao KK, Scott M, Foster D, Groth DF, Dearmond SJ, Prusiner SB (1990) Spontaneous neurodegeneration in transgenic mice with mutant prion protein. *Science* **250**, 1587–1590.

Ikawa M, Yamada S, Nakanishi T, Okabe M (1999) Green fluorescent protein as a vital marker in mammals. *Curr. Top. Dev. Biol.* **44**, 1–20.

Jaenisch R, Mintz B (1974) Simian virus 40 DNA in DNA of healthy adult mice derived from preimplantation blastocysts injected with viral DNA. *Proc. Natl Acad. Sci. USA* **71**, 1250–1254.

Jakobovits A, Moore AL, Green LL *et al.* (1993) Germ-line transmission and expression of a human-derived yeast artificial chromosome. *Nature* **362**, 255–258.

Kamath RS, Fraser AG, Dong Y *et al.* (2003) Systematic functional analysis of the *Caenorhabditis elegans* genome using RNAi. *Nature* **421**, 231–237.

Katsuki M, Sato M, Kimura M, Yokoyama M, Kobayashi K, Nomura T (1988) Conversion of normal behaviour to *shiverer* by myelin basic protein antisense cDNA in transgenic mice. *Science* **241**, 593–595.

Kola I, Hertzog PJ (1998) Down syndrome and mouse models. *Curr. Opin. Genet. Dev.* **8**, 316–321.

Koopman P, Gubbay J, Vivian N, Goodfellow P, Lovell-Badge R (1991) Male development of chromosomally female mice transgenic for Sry. *Nature* **351**, 117–121.

Kroll KL, Amaya E (1996) Transgenic *Xenopus* embryos from sperm nuclear transplantations reveal FGF signaling requirements during gastrulation. *Development* **122**, 3173–3183.

Kunik T, Tzfira T, Kapulnik Y, Gafni Y *et al.* (2001) Genetic transformation of HeLa cells by *Agrobacterium*. *Proc. Natl Acad. Sci. USA* **98**, 1871–1876.

Kuwabara T, Warashina M, Taira K (2000) Allosterically controllable maxizymes cleave mRNA with high efficiency and specificity. *Trends Biotechnol.* **18**, 462–468.

Lai L, Kolber-Simonds D, Park K-W, Cheong H-T *et al.* (2002) Production of α-1,3-galactosyltransferase knockout pigs by nuclear transfer cloning. *Science* **295**, 1089–1092.

Lamb BT, Gearhart JD (1995) YAC transgenics and the study of genetics and human disease. *Curr. Opin. Genet. Dev.* **5**, 342–348.

Leiter EH, Beamer WG, Shultz LD, Barker JE, Lane PW (1987) Mouse models of genetic diseases. *Birth Defects* **23**, 221–257.

Littlewood TD, Hancock DC, Danielian PS, Parker MG, Evan GI (1995) A modified oestrogen receptor ligand-binding domain as an improved switch for the regulation of heterologous proteins. *Nucleic Acids Res.* **23**, 1686–1690.

Lobe CG, Nagy A (1998) Conditional genome alteration in mice. *Bioessays* **20**, 200–208.

Lyon MF, Searle AG (1989) Genetic Variants and Strains of the Laboratory Mouse, 2nd edn. Oxford University Press, Oxford.

Macleod KF, Jacks T (1999) Insights into cancer from transgenic mouse models. *J. Pathol.* **187**, 43–60.

MacPhee M, Chepenik KP, Liddell RA, Nelson KK, Siracusa LD, Buchberg AM (1995) The secretory phospholipase-a2 gene is a candidate for the mom1 locus, a major modifier of apc(min)-induced intestinal neoplasia. *Cell* **81**, 957–966.

Maeda I, Kohara Y, Yamamoto M, Sugimoto A (2001) Large-scale analysis of gene function in *Caenorhabditis elegans* by high-throughput RNAi. *Curr. Biol.* **11**, 171–176.

Martin GR (1981) Isolation of a pluripotent cell line from early mouse embryos cultured in medium conditioned by teratocarcinoma stem cells. *Proc. Natl Acad. Sci. USA* **78**, 7634–7638.

McCreath KJ, Howcroft J, Campbell KHS, Colman A, Schnieke AE, Kind AJ (2000) Production of gene targeted sheep by nuclear transfer from cultured somatic cells. *Nature* **405**, 1066–1069.

Melton DW (1994) Gene targeting in the mouse. *BioEssays* **16**, 633–638.

Mendez MJ, Green LL, Corvalan JRF *et al.* (1997) Functional transplant of megabase human immunoglobulin loci recapitulates human antibody response in mice. *Nature Genet.* **15**, 146–156.

Nir I, Kedzierski W, Chen J, Travis GH (2000) Expression of Bcl protects against photoreceptor degeneration in retinal degeneration slow (rds) mice. *J Neurosci.* **20**, 2150–2154.

Nolan PM, Peters J, Strivens M, Rogers D *et al.* (2000) A systematic, genome-wide, phenotype-driven mutagenesis programme for gene function studies in the mouse. *Nature Genet.* **25**, 440–443.

O'Kane CJ, Gehring WJ (1987) Detection in situ of genetic regulatory elements in *Drosophila*. *Proc. Natl Acad. Sci. USA* **84**, 9123–9127.

Pasquinelli AE (2002) MicroRNAs: deviants no longer. *Trends Genet.* **18**, 171–173.

Perry ACF, Wakayama T, Kishikawa H, Kasai T, Okabe M, Toyoda Y, Yanagimachi R (1999) Mammalian transgenesis by intracytoplasmic sperm injection. *Science* **284**, 1180–1183.

Perucho M, Hanahan D, Lipsich L, Wigler M (1980) Isolation of the chicken thymidine kinase gene by plasmid rescue. *Nature* **285**, 201–210.

Petters RM, Sommer JR (2000) Transgenic animals as models for human disease. *Transgenic Res.* **9**, 347–351.

Ramirez-Solis R, Liu P, Bradley A (1995) Chromosome engineering in mice. *Nature* **378**, 720–724.

Reeves RH, Irving NG, Moran TH *et al.* (1995) A mouse model for Down syndrome exhibits learning and behaviour deficits. *Nature Genet.* **11**, 177–183.

Richardson JH, Marasco WA (1995) Intracellular antibodies: development and therapeutic potential. *Trends Biotechnol.* **13**, 306–310.

Rorth P, Szabo K, Bailey A *et al.* (1998) Systematic gain-of-function genetics in *Drosophila. Development* **125**, 1049–1057.

Rudnicki MA, Braun B, Hinuma S, Jaenisch R (1992) Inactivation of *myoD* in mice leads to upregulation of the myogenic HLH gene *myf-5* and results in apparently normal muscle development. *Nucleic Acids Res.* **18**, 4833–4842.

Saez E, No D, West A, Evans RM (1997) Inducible gene expression in mammalian cells and transgenic mice. *Curr. Opin. Biotechnol.* **8**, 608–616.

Sakimura K, Kushiya E, Ogura A, Kudo Y, Katagiri T, Takahashi Y (1995) Upstream and intron regulatory regions for expression of the rat neuron-specific enolase gene. *Mol. Brain Res.* **28**, 19–28.

Schnieke AE, Kind AJ, Ritchie WA, Mycock K, Scott AR, Ritchie M, Wilmut I, Colman A, Campbell KH (1997) Human factor IX transgenic sheep produced by transfer of nuclei from transfected fetal fibroblasts. *Science* **278**, 2130–2133.

Shamblott MJ, Axelman J, Wang S *et al.* (1998) Derivation of pluripotent stem cells from cultured human primordial germ cells. *Proc. Natl Acad. Sci. USA,* **95**, 13726–13731.

Sklar MD, Thompson E, Welsh MJ *et al.* (1991) Depletion of c-*myc* with specific antisense sequences reverses the transformed phenotype in *ras* oncogene-transformed NIH 3T3 cells. *Mol. Cell. Biol.* **11**, 3699–3710.

Smith AJH, De Sousa MA, Kwabi-Addo B, Heppell-Parton A, Impey H, Rabbitts P (1995) A site-directed chromosomal translocation induced in embryonic stem-cells by Cre-*lox*P recombination. *Nature Genet.* **9**, 376–385.

Smith DJ, Stevens ME, Suclanagunta SP *et al.* (1997) Functional screening of 2 Mb of human chromosome 21q22.2 in transgenic mice implicates *minibrain* in learning defects associated with Down syndrome. *Nature Genet.* **16**, 28–36.

Smithies O, Gregg RG, Boggs SS *et al.* (1985) Insertion of DNA sequences into the human β-globin locus by homologous recombination. *Nature* **317**, 230–234.

Smithies O (1993) Animal models of human genetic diseases. *Trends Genet.* **9**, 112–116.

Smithies O, Maeda N (1995) Gene targeting approaches to complex genetic diseases: Atherosclerosis and essential hypertension. *Proc. Natl Acad. Sci. USA,* **92**, 5266–5272.

Snouwaert JN, Brigman KK, Latour AM *et al.* (1992) An animal-model for cystic-fibrosis made by gene targeting. *Science* **257**, 1083–1088.

Solter D, Gearhart J (1999) Putting stem cells to work. *Science* **283**, 1468–1470.

Spradling AC, Stern DM, Kiss I, Roote J, Laverty T, Rubin GM (1995) Gene disruptions using P transposable elements: an integral component of the *Drosophila* genome project. *Proc. Natl Acad. Sci. USA* **92**, 10824–10830.

Spradling AC, Stern D, Beaton A, Rhem EJ, Laverty T, Mozden N, Misra S, Rubin GM (1999) The BDGP Gene Disruption Project: single P element insertions mutating 25% of vital *Drosophila* genes. *Genetics* **153**, 135–177.

Szybalska EH, Szybalski E (1962) Genetics of human cell lines IV. DNA-mediated heritable transformation of a biochemical trait. *Proc. Natl Acad. Sci. USA* **48**, 2026–2031.

Taylor BA (1989) In: *Genetic Variants and Strains of the Laboratory Mouse,* 2nd edn (eds MF Lyon, AG Searle). Oxford University Press, Oxford, pp. 773–796.

Terskikh A, Fradkov A, Ermakova G *et al.* (2000) 'Fluorescent timer': protein that changes colour with time. *Science* **290**, 1585–1588.

Thomson JA, Itskovitz-Elder J, Shapiro SS *et al.* (1998) Embryonic stem cell lines derived from human blastocysts. *Science* **282**, 1145–1147.

Tomizuka K, Yoshida H, Uejima H *et al.* (1997) Functional expression and germline transmission of a human chromosome fragment in chimaeric mice. *Nature Genet.* **16**, 133–143.

Tuschl T, Borkhardt A (2002) Small interfering RNAs: A revolutionary tool for the analysis of gene function and gene therapy. *Molecular Interventions* **2**, 158–167.

Voinnet O (2002) RNA silencing: small RNAs as ubiquitous regulators of gene expression. *Current Opin. Plant Biol.* **5**, 444–451.

Wakayama T, Perry AC, Zuccotti M, Johnson KR, Yanagimachi R (1998) Full-term development of mice from enucleated oocytes injected with cumulus cell nuclei. *Nature* **394**, 369–374.

Welch PJ, Barber JR, Wong-Staal F (1998) Expression of ribozymes in gene transfer systems to modulate target RNA levels. *Curr. Opin. Biotechnol.* **9**, 486–496.

Wiles MV, Vauti F, Otte J *et al.* (2000) Establishment of a gene-trap sequence tag library to generate mutant mice from embryonic stem cells. *Nature Genet.* **24**, 13–14.

Wilmut I, Schnieke AE, McWhir J, Kind AJ, Campbell KHS (1997) Viable offspring derived from fetal and adult mammalian cells. *Nature* **385**, 810–813.

Wu LZ, de Bruin A, Saavedra HI, Starovic M, Trimboli A, Yang Y *et al.* (2003) Extra-embryonic function of Rb is essential for embryonic development and viability. *Nature* **421**, 942–947.

Wynshaw-Boris A (1996) Model mice and human disease. *Nature Genet.* **13**, 259–260.

Yokoyama M (2002) Gene delivery using temperature-responsive polymeric carriers. *Drug Discovery Today* **7**, 426–432.

Zambrowicz BP, Friedrich GA, Buxton EC, Lilleberg SL, Person C, Sands AT (1998) Disruption and sequence identification of 2,000 genes in mouse embryonic stem cells. *Nature* **392**, 608–611.

CHAPTER TWENTY ONE

New approaches to treating disease

Chapter contents

21.1 Treatment of genetic disease is not the same as genetic treatment of disease

Coming at the end of a book like this, a chapter on treatment might reasonably consider two quite separate matters:

▶ treatment of genetic disease;

▶ genetic treatment of disease.

The latter topic in turn has several major facets:

▶ genotyping individuals to predict their pattern of favorable and adverse responses to drug treatments;

▶ using knowledge of genetics and cell biology to identify new targets for drug development;

▶ using knowledge of genetics and cell biology to develop cell-based therapies;

▶ using genetic techniques to produce drugs, vaccines etc. for treatment of disease;

▶ using genetic techniques directly to treat disease.

Between them, these topics cover most of twenty-first century medical research. It would require several large books, written by teams of experts, to do justice to so vast an area, therefore we will give fairly cursory looks at the earlier topics in the lists, before concentrating on the last topic, using genetic techniques to treat disease. Some important ethical issues raised by these developments are briefly addressed in three Ethics boxes.

21.2 Treatment of genetic disease

It is a popular misconception that if a disease is genetic it must be untreatable. In reality there is no connection at all between cause and treatability of a disease. A profoundly deaf child should be offered hearing aids or a cochlear implant based solely on the child's symptoms and family situation, quite regardless of whether the hearing loss is genetic or not. Orthodox medical treatment aimed at alleviating symptoms of disease is just as applicable to genetic diseases as to any other.

It is, however, true that for many genetic conditions existing treatments are unsatisfactory. A survey 20 years ago (Costa *et al.*, 1983) estimated that treatment improved reproductive capacity in only 11% of Mendelian diseases, improved social adaptation in only 6%, and extended life span to normal in only 15% (of those that reduced longevity). No doubt the figures would be better now, but not dramatically so.

Inborn errors of metabolism are the best candidate subset of Mendelian diseases for conventional treatment. Our detailed biochemical knowledge provides many potential entry points for intervention, including:

▶ **substrate limitation**. Everybody knows the success of dietary treatment of phenylketonuria (although even in successfully treated patients, cognitive development averages half a standard deviation below normal). Several other inborn errors respond equally well to dietary treatment;

▶ **replacement of a deficient product**, such as thyroid hormone for infants with congenital hypothyroidism;

▶ **using alternative pathways to remove toxic metabolites**. Treatments range from simple bleeding as a very effective treatment for hemachromatosis to using benzoate to increase nitrogen excretion in patients with urea-cycle disorders;

▶ **using metabolic inhibitors** such as the drug NTBC that blocks the tyrosine pathway and dramatically improves the prognosis of Type 1 tyrosinemia (Lindstedt *et al.*, 1992).

These and many other examples are discussed in detail by Treacy, Valle and Scriver, whose chapter is recommended to readers interested in this area (see Further reading). It remains true that despite much work, relatively few genetic diseases have wholly satisfactory treatments, hence the emphasis on the novel approaches described below, which apply equally to genetic and nongenetic disease.

21.3 Using genetic knowledge to improve existing treatments and develop new versions of conventional treatments

21.3.1 Pharmacogenetics promises to increase the effectiveness of drugs and reduce dangerous side effects

Drugs are rarely effective in 100% of the patients to whom they are prescribed. For example, among the most commonly prescribed classes of drugs:

▶ 15–35% of patients have an inadequate or no response to beta-blockers;

▶ 7–28% of patients have an inadequate or no response to angiotensin converting enzyme inhibitors;

▶ 9–23% of patients respond inadequately to selective serotonin re-uptake inhibitors;

▶ 20–50% of patients respond inadequately to tricyclic antidepressants.

In some cases (especially psychiatric illness) patients given the same clinical label may actually have different diseases, only one of which responds to a given drug. But for the most part these individual variations in response depend on variations in the absorption, distribution, metabolism and elimination of drugs, and variability of the target receptors.

These individual variations must largely depend on combinations of common polymorphisms in a limited number of genes. Compared to the problems of finding genetic susceptibility factors in common disease (Chapter 15), identifying the variants underlying individual reactions to drugs should be a much more tractable problem. The search can be focused on a definable area of biochemistry, and many hypotheses are amenable to checking *in vitro*. Thus the dream of *individual-specific prescribing* seems attainable. Some sort of chip-based kit would be used in the clinic office to genotype the patient for a few hundred common polymorphisms. The results would determine which drugs would be safe and effective in his case. Cynics might argue that drug companies have no interest in making sure their drug is prescribed only in those cases where it will work – but they have a very strong interest in avoiding prescribing it when it might cause a dangerous side effect. It costs an average of US $800 million, and takes 10–15 years, to develop a compound from initial lead to marketed drug, and many promising compounds fail late in the process because of adverse reactions in a small minority of patients.

Several examples are already known of genetic variants that affect drug responses (*Table 21.1*). These effects can have real clinical importance. Doses of isoniazid that are appropriate for fast acetylators risk causing a peripheral neuropathy in slow acetylators. Patients with the R144C or I359L variants in the *CYP2C9* gene may suffer from excessive anticoagulation and bleeding if they are given the standard maintenance dose of warfarin. A CYP2D6 'poor metabolizer' might get no benefit from codeine-containing painkillers, but if prescribed nortryptilene, might require a daily dose of only 10–20 mg, where an ultra-rapid metabolizer might require 500 mg per day. Wolf *et al.* (2000) quote a report that in patients prescribed psychiatric drugs that are CYP2D6 substrates, adverse drug reactions are observed in every patient with mutations that inactivate the *CYP2D6* gene.

21.3.2 Drug companies have invested heavily in genomics to try to identify new drug targets

It has been claimed that the entire diversity of drugs on the market today act through only about 400 targets. Genomic research has vastly expanded the number of potential targets available for research by drug companies – from a few hundred 10 years ago to maybe 50 000 now. In fact the super-abundance of potential targets is almost an embarrassment to biotech companies. RNA interference (Section 20.2.6) can be used to identify genes where an inhibitory drug might produce a useful effect, before investing in the large scale screening that would be needed to identify possible inhibitors. In the long term completely new classes of drugs will hopefully emerge from this effort.

Genomic or proteomic studies of pathogenic micro-organisms are also important, as guides to the development of new vaccines or treatments. Microbial pathogens have featured strongly on priority lists for genome sequencing (see *Box 8.8*), a notable recent success being the malaria parasite *Plasmodium falciparum* (Gardner *et al.*, 2002). Sequences are analyzed to identify enzymes specific to the pathogen and not the host, which could be targets for inhibitors, or missing enzymes that make the parasite vulnerable to interference with the supply of an essential nutrient. Virulence proteins or gene products expressed early in infection are promising drug targets. They can be identified by expression array studies, and by comparisons of gene content or gene expression in virulent and nonvirulent strains. Proteomics strategies (such as MALDI-TOF MS; see *Box 19.4*) are used to identify proteins on the pathogen cell surface that might be targets for vaccines.

21.3.3 Cell-based treatments promise to transform the potential of transplantation

Cell-based treatments can be seen as natural extensions of present-day transplant methods, but advances in our understanding of stem cells offer the hope of radically extending the present range of options. In theory the possibilities are endless

---

Table 21.1: Examples of genetic variations affecting response to drugs

See Wolf *et al.* (2000) for details.

Enzyme / protein	Variant	Population frequency	Examples of drugs affected
N-acetyl transferase	Slow acetylators	60% (white European) 20% (Oriental)	Isoniazid, procainamide, sulfonamides
Thiopurine methyltransferase	Low activity	(low)	6-mercaptopurine, azathioprine
CYP2D6 P450 cytochrome	Low activity (inactivating mutations)	6% (white European) 1% (Oriental)	Failure to activate: codeine
	Ultra-high activity (tandem gene amplification)	2–7% (white Europeans)	Slow inactivation: psychiatric drugs e.g. nortryptilene, clozapine, haloperidol, imipramine, mianserine; cardiovascular drugs e.g. propanolol
CYP2C9 P450 cytochrome	Low activity	0.2% ?	Ibuprofen, warfarin, tolbutamide
CYP2C19 P450 cytochrome	Low activity	4% (white European) 23% (Oriental)	Mephenytoin, proguanil
β-adrenergic receptor ADRB2	Several SNPs – not yet clear which are important	NA	Variable responses to Albuterol in asthma
ERBB2 receptor	Amplified in some breast cancers	—	Herceptin (effective only when ERBB2 amplified)

NA, Data not available.

– any damaged or worn out tissue or organ might be reno-vated or recreated using appropriate stem cells, and these might be subjected to any sort of genetic manipulation in advance. Daley (2002) gives a judicious assessment of how far these hopes may be realized in the medium term.

Stem cells can be defined as cells that can both self-renew and give rise to differentiated progeny (see Section 3.3). They are probably present at all stages of development and in all tissues. There is a gradient from embryonic stem cells (ES cells), that can potentially produce all germ line and somatic cells of an organism, through to multipotent but tissue-restricted stem cells (*Figure 21.1*). Mouse ES cells have been studied for many years, but the isolation of human ES cells by Thomson *et al.* in1998 propelled stem cell therapy to the fore-front of both medical research and ethical debate. The ethical debate derives from the method of producing ES cells, which inevitably involves destruction of an early human embryo. For a flavor of the arguments, see a series of letters in the *New England Journal of Medicine* **346**, 1619–1622 (2002), which were stimulated by a review by Weissman (2002). Opponents of the use of ES cells often suggest that it is not necessary to use them, because tissue-restricted adult stem cells can do the job equally well. Others contend that few of the available adult stem cell lines are well characterized, and that their growth and differentiation potential is too limited to make them an adequate substitute for ES cells.

Current lines vary widely in how well they grow in culture and in the range of differentiated cells they will produce. An open question is to what extent tissue-restricted stem cells can *transdifferentiate* by migrating into a different tissue and taking on the properties of stem cells of that tissue. Many claims have been made, but few have been rigorously established. As research tools, stem cells of all sorts are immensely important. They may allow identification of the factors that control whether and how cells differentiate. If this led to an ability to control and channel differentiation, stem cells might be used for an almost unlimited range of tissue engineering and tissue repair.

Using stem cells taken from a donor still leaves the problem of transplant rejection. That problem could be solved by using the nuclear transplantation technology that produced Dolly the sheep. The nucleus of an oocyte would be replaced by one from the eventual transplant recipient, and ES cells isolated from the resulting blastocyst (*Figure 21.2*). These cells could in theory be used to produce stem cells, tissues or even whole organs for transplantation with no risk of rejection. This procedure is called *therapeutic cloning*, and is the subject of intense ethical and political controversy (see *Ethics Box 1*).

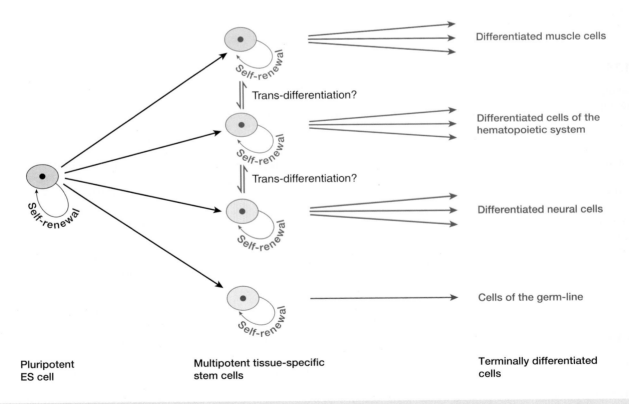

Differentiated muscle cells

Trans-differentiation?

Differentiated cells of the hematopoietic system

Trans-differentiation?

Differentiated neural cells

Cells of the germ-line

Pluripotent
ES cell

Multipotent tissue-specific
stem cells

Terminally differentiated
cells

Figure 21.1: Stem cells.

All stem cells can both self-renew and give rise to more differentiated progeny. Embryonic stem (ES) cells can produce all somatic and germ line cell types of the organism; tissue-restricted stem cells have more limited potential. Whether and if so to what extent tissue-restricted stem cells can trans-differentiate into a cell characteristic of a different tissue is currently controversial.

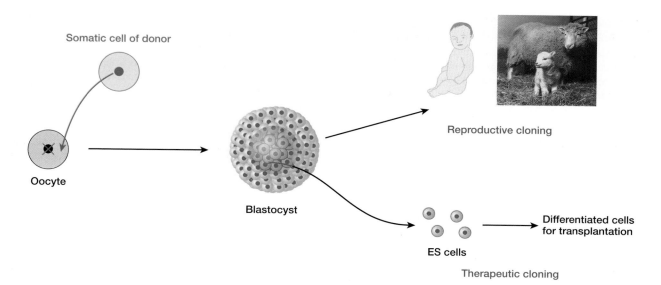

Somatic cell of donor

Oocyte

Blastocyst

Reproductive cloning

ES cells

Differentiated cells for transplantation

Therapeutic cloning

Figure 21.2: Reproductive and therapeutic cloning.

In both procedures a somatic cell nucleus from the donor is transplanted into an enucleated oocyte, which is then stimulated to develop to the blastocyst stage. For reproductive cloning, the blastocyst is implanted in a uterus with the hope that it will develop into a newborn. Human reproductive cloning is banned in most countries. For therapeutic cloning, ES cells are extracted from the inner cell mass of the blastocyst, which is then destroyed. The ES cells are used as a source of cloned donor cells or, potentially, tissues or organs. Human therapeutic cloning is controversial because it involves destruction of a human embryo. The embryo might be a surplus embryo from an *in vitro* fertilization (IVF) clinic, or might be specially created for the purpose.

21.3.4 Recombinant proteins and vaccines

Recombinant proteins can be produced by expression cloning in microorganisms or transgenic livestock

Many therapeutic proteins that were formerly extracted from animal or human sources can be made as recombinant proteins by expression cloning (Russell and Clarke, 1999). The techniques and problems of expression cloning are described in Section 5.6. Recombinant human insulin was first marketed in 1982, and *Table 21.2* lists a number of subsequent examples. Using recombinant proteins avoids many of the safety problems of naturally extracted products. Many hemophiliacs contracted AIDS from HIV-contaminated human Factor VIII, and some children succumbed to Creutzfeld–Jakob disease after injections of growth hormone extracted from human cadaver pituitary glands.

Bacteria are the simplest hosts for expression cloning. Large volumes of culture can be grown, and the media can be defined and controlled to avoid contamination. However, polypeptides produced in bacteria are not usually identical to natural human proteins. Most therapeutic human proteins are glycosylated, and bacteria do not reproduce the human patterns of glycosylation. For this reason there is an increasing interest in mammalian expression systems. These may be cell cultures, but transgenic animals are an attractive alternative. For example a cloned human gene might be fused to a sheep, goat or pig gene expressing a milk protein and inserted into the germ line of the animal. This leads to the idea of 'pharming'. Flocks of transgenic animals would be maintained

that produce a desired human protein at high level in their milk (Velander *et al.*, 1997). Alternatively, transgenic chickens might be made that lay eggs rich in a wanted product.

Transgenic plants are also becoming popular. Plants do not replicate human-specific glycosylation patterns, but they have many advantages for expression cloning in terms of cost and safety. For example, there is considerable interest in producing rice strains engineered to counter the vitamin A deficiency that is a serious public health problem in at least 26 countries in Africa, Asia and Latin America (Ye *et al.*, 2000).

Genetically engineered antibodies

The complicated gene rearrangements in B lymphocytes (Section 10.6) endow each one of us with a huge repertoire of different antibodies as a defense system against innumerable foreign antigens. Antibody molecules function as adapters: they have binding sites for foreign antigen at the variable (V) end, and binding sites for effector molecules at the constant (C) end. Binding of an antibody may by itself be sufficient to neutralize some toxins and viruses, but more usually the bound antibody triggers the complement system and cell-mediated killing.

Artificially produced therapeutic antibodies are designed to be monospecific. Traditionally these were *monoclonal antibodies* (mAbs) secreted by *hybridomas*, immortalized cells produced by fusion of antibody-producing B lymphocytes from an immunized mouse or rat with cells from an immortal mouse B-lymphocyte tumor. Hybridomas are propagated as individual clones, each of which can provide a

---

Ethics Box 1: The ethics of human cloning.

Arguments about the ethics of human cloning need to distinguish *reproductive* from *therapeutic* cloning – even if the final decision is that both types are unethical. Legislatures in many countries have considered bills that either seek to ban every type of cloning, or seek to make a distinction between reproductive (banned) and therapeutic (permitted) cloning.

Reproductive cloning aims to produce a cloned baby. Objections to reproductive cloning are on three grounds: one practical, one principled and one misconceived.

▶ **The practical argument** points to the low success rate of all mammalian cloning experiments, and to the fact that many cloned animals that are born have serious abnormalities, probably because current procedures do not reliably reprogram the epigenetic modifications of the donor cell (Dean *et al.*, 2001). This is simple fact, and clearly any attempt at human reproductive cloning in the present state of knowledge would be grossly unethical. Conceivably advances in knowledge might remove this objection, though it is not clear how we could know without conducting unethical experiments.

▶ **The argument in principle** says that humans should be valued for themselves and not treated as instruments to achieve a purpose. We fully accept this argument, but wonder why it is not applied with equal force to ambitious parents who decide that their 3-year-old will become a tennis champion or a virtuoso violinist.

▶ **The misconceived argument** views clones as not fully individual or maybe not fully human. Much of the anxiety – and also much of the humor – about reproductive cloning is based on a picture of clones as some sort of

programmable zombies. This is very strange, since we all know clones, and are perfectly aware that they are fully human and individual. Engineered clones would be even less identical than MZ twins, since the donor would most likely be an elderly self-obsessed billionaire seeking to achieve a spurious immortality, while the clone would be 60 years younger and born into a totally different environment. To take a hackneyed example, even if somebody were to succeed in cloning Hitler, there is not the slightest reason to imagine the clone would set about exterminating Jews.

Arguments in favor of reproductive cloning are based on scenarios where this is the only option for a couple to have children – for example, if a woman had a serious mitochondrial disorder, she could become pregnant with a donated oocyte (which would supply the mitochondria) into which a nucleus of one of her or her partner's somatic cells had been transplanted.

Therapeutic cloning (*Figure 21.2*) involves the production of embryonic stem cells in which the nucleus derives from a somatic cell of the donor. This is done in order to provide the donor with a source of perfectly matched transplantable cells, tissues or organs. The ethical debate is more complicated here. Many people object to the idea of creating an embryo specifically for this purpose, or even using surplus embryos from *in vitro* fertilization clinics. Other opponents cite a 'slippery slope' argument: if therapeutic cloning is allowed, it might be impossible to prevent unscrupulous operators from diverting cloned embryos into reproductive cloning programs. Against this, proponents argue that if therapeutic cloning is the only way to cure devastating diseases, it would be unethical not to proceed.

---

permanent and stable source of a single mAb. Unfortunately the therapeutic potential of mAbs produced in this way is limited. Although rodent mAbs can be raised against human pathogens and cells, they have a short half-life in human serum and they can elicit production of human anti-rodent antibodies. In addition, only some of the different classes can trigger human effector functions.

These problems can be addressed by **antibody engineering**. Different exons of the immunoglobulin genes encode different domains of the antibody molecule. Therefore exon shuffling at the DNA level allows construction of antibodies with novel combinations of protein domains. Early applications of antibody engineering produced **humanized antibodies** (*Figure 21.3*) in which a lesser or greater part of the rodent mAb is replaced by the human equivalent. Ultimately antibodies were constructed that carried only the *complementarity-determining regions* (*CDRs*) from the rodent mAb – these are the hypervariable sequences of the antigen binding site (*Figure 21.3B*). More recent developments have by-passed hybridoma construction altogether by using phage display technology (Hoogenboom *et al.*, 1998; Section 5.6.2). These systems allow innovative combinations of antibody

domains to be constructed. A very promising class of antibody derivatives are *single chain variable fragments* (scFv; *Figure 21.3C*). These have almost all the binding specificity of a mAb, but in a single nonglycosylated polypeptide which can be made on a large scale in bacterial, yeast or even plant cells (Stöger *et al.*, 2000). Hudson (1999) reviews many of the promising antibody–related engineered molecules that are now starting to enter clinical trials.

Related developments aim to mimic the binding specificity of antibodies in different contexts. Engineered antibody genes can be used to produce nonsecreted intracellular antibodies (**intrabodies**) to bind specific molecules within a cell (Marasco, 1997). Intrabodies can be useful if the simple act of binding inhibits the target. This may be an extremely useful characteristic since, unlike small molecules, intrabodies are not limited to targeting specific classes of protein (kinases, ion channels and so on). Furthermore, intrabodies can be used to carry effector molecules which can activate specific functions when antigen binding occurs. Probably the best example of this is the fusion of caspase 3 to intrabodies (Tse and Rabbitts, 2000). Caspase 3-conjugated intrabodies were prepared against each separate protein moiety of a disease associated

Table 21.2: Examples of pharmaceutical products obtained by expression cloning

Product	For treatment of
Insulin	Diabetes
Growth hormone	Growth hormone deficiency
Blood clotting factor VIII	Hemophilia A
Blood clotting factor IX	Hemophilia B
α-interferon	Hairy-cell leukemia; chronic hepatitis
β-interferon	Multiple sclerosis
γ-interferon	Infections in patients with chronic granulomatous disease
Glucocerebrosidase	Gaucher disease
Tissue plasminogen activator	Thrombotic disorders
Granulocyte-macrophage stimulating hormone	Neutropenia following chemotherapy
Leptin	Obesity
Erythropoietin	Anemia

fusion protein such as bcr–abl. In cancer cells both intrabodies bind to the fusion protein; this allows the caspase 3 to dimerize and thus trigger selective apoptosis of cells containing the fusion protein.

An alternative to recombinant antibodies is the use of short peptides as binding partners for specific target proteins (Hoppe–Seyler and Butz, 2000). Indeed, protein–protein interactions can be abandoned altogether in favor of DNA or RNA molecules (**'aptamers'**). The ability to amplify DNA, and to produce the desired molecule on a large scale, are the key advantages of using DNA or RNA. Starting with a large pool of random oligonucleotides, repeated rounds of selection and amplification are used to isolate molecules that bind and inhibit a target protein with affinity similar to conventional antibodies (White *et al.*, 2000).

Genetically engineered vaccines

Recombinant DNA technology can be applied to antigens as well as antibodies. Pathogenic microorganisms may be disabled by genetic modification so that an attenuated live vaccine can be used safely. Genetically modified plants might be used to produce edible vaccines. Changes can be made to an antigen to improve its visibility to the immune system, so as to produce an enhanced response. Many DNA or RNA

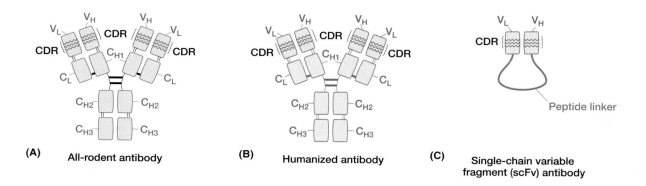

(A) All-rodent antibody **(B)** Humanized antibody **(C)** Single-chain variable fragment (scFv) antibody

Figure 21.3: Antibody engineering.

(A) Classic monoclonal antibodies (mAbs) are monospecific rodent antibodies synthesized by hybridomas made by fusing B-cells from immunized mice or rats with cells from an immortal mouse B-lymphocyte tumor. Humans may produce anti-rodent antibodies that neutralize mAbs. **(B)** This problem can be minimized by engineering mAbs so that all the molecule is of human origin except for the hypervariable complementarity-determining regions (CDRs) that determine the specificity. **(C)** scFvs are engineered single polypeptide chains. Depending on the length of the linker, they bind their target as monomers, dimers or trimers. Multimers bind their target more strongly than monomers.

vaccines are now in clinical trials. These are typically plasmids designed to express an antigen of a pathogen or tumor at high level after delivery by intramuscular injection (Reyes-Sandoval and Ertl, 2001).

This section has provided a necessarily brief overview of the very wide range of applications of genetic knowledge and techniques in medical and pharmaceutical research; we will now look in greater detail at the prospects for direct genetic intervention in disease processes by gene therapy.

21.4 Principles of gene therapy

Gene therapy involves the direct genetic modification of cells of the patient in order to achieve a therapeutic goal. There are basic distinctions in the types of cells modified, and the type of modification effected.

▶ **Germ-line gene therapy** produces a permanent transmissible modification. This might be achieved by modification of a gamete, a zygote or an early embryo. Germ-line therapy is banned in many countries for ethical reasons (see *Ethics Box 2*).

▶ **Somatic cell gene therapy** aims to modify specific cells or tissues of the patient in a way that is confined to that patient. All current gene therapy trials and protocols are for somatic cell therapy.

Somatic cells might be modified in a number of different ways (*Figure 21.4*).

▶ **Gene supplementation** (also called gene augmentation) aims to supply a functioning copy of a defective gene. This would be used to treat loss-of-function conditions (Section 16.4) where the disease process is the result of a gene not functioning here and now. Cystic fibrosis would be a typical candidate. It would not be suitable for loss-of-function conditions where irreversible damage has already been done, for example through some failure in embryonic development. Cancer therapy could involve gene supplementation to increase the immune response against a tumor or to replace a defective tumor suppressor gene.

▶ **Gene replacement** is more ambitious: the aim is to replace a mutant gene by a correctly functioning copy, or to correct a mutation *in situ*. Gene replacement would be required for gain-of-function diseases where the resident mutant gene is doing something positively bad.

▶ **Targeted inhibition of gene expression** is especially relevant in infectious disease, where essential functions of the pathogen are targeted. It could also be used to silence activated oncogenes in cancer, to damp down unwanted responses in autoimmune disease and maybe to silence a gain-of-function mutant allele in inherited disease.

▶ **Targeted killing of specific cells** is particularly applicable to cancer treatment.

Expectations about gene therapy have followed a manic-depressive course over the past 15 years as cycles of over-optimism are followed by bouts of excessive pessimism. An important report for the US National Institutes of Health in 1995 attempted to inject some reality (*Box 21.1*). Since then we have seen optimism grow, only to be derailed by the death of Jesse Gelsinger, raised again by the successful treatment of children with severe combined immunodeficiency, then once again punctured by the news that first one and then a second of the children has developed leukemia, almost certainly as a result of the treatment (see below for details of all these events). Perhaps one cause of these exaggerated reactions is confusion over the natural timescales of such work. Because diagnostic testing can often start within weeks of a gene being cloned, people perhaps expect therapy to be not far behind, whereas really this is drug development, which runs on a timescale of decades.

Developing practical gene therapy is a long-haul process; additionally, reports of successful progress trigger ethical concerns centered around the idea of 'designer babies' (see *Ethics Box 3*). Nevertheless, many academic and commercial laboratories are working hard in this area, and over 600 trial protocols related to gene therapy have been approved. *Figure 21.5* shows statistics from the database of trials maintained at www.wiley.co.uk/genetherapy/clinical. The book by Templeton and Lasic (Further reading) provides more depth on many of the topics covered in the rest of this chapter.

21.5 Methods for inserting and expressing a gene in a target cell or tissue

21.5.1 Genes can be transferred to the recipient cells in the laboratory (*ex vivo*) or within the patient's body (*in vivo*)

Ex vivo gene transfer involves transfer of cloned genes into cells grown in culture. Those cells that have been transformed successfully are selected, expanded by cell culture *in vitro*, then replaced in the patient. To avoid rejection by the immune system, the patient's own cells (*autologous cells*) are used whenever possible (*Figure 21.6*). This approach is used for cells that are accessible for initial removal and that can be induced to engraft and survive for a long time after replacement. Examples include cells of the hematopoietic system, skin cells, etc.

In vivo gene transfer is the only option in tissues where the recipient cells cannot be cultured *in vitro* in sufficient numbers (e.g. brain cells) or where cultured cells cannot be re-implanted efficiently in patients. Tissue targeting is an important consideration. The gene transfer construct may be emplaced directly into the target tissue, or it may be injected into the general circulation but designed in some way so as to be taken up only by the desired cell type. As there is no way of selecting and amplifying cells that have taken up and expressed the foreign gene, the success of this approach is crucially dependent on the general efficiency of gene transfer and expression.

21.5.2 Constructs may be designed to integrate into the host cell chromosomes or to remain as episomes

For achieving long term gene expression it would seem desirable to integrate the foreign gene into a chromosome of the host cell – preferably a stem cell. Then the construct is replicated whenever the host cell or its daughters divide.

Ethics Box 2: Germ line versus somatic gene therapy.

Germ line gene therapy involves making a genetic change that can be transmitted down the generations. This would most likely be done by genetic manipulation of a pre-implantation embryo, but it might occur as a by-product of a treatment aimed at somatic cells that incidentally affected the patient's germ cells. Somatic cell therapy treats just certain body cells of the patient without having any effect on the germ line. For purely technical reasons, germ line therapy is not currently a realistic option, but there are ethical issues that will still be there when the technical problems are solved, and genetic manipulation of the germ line is prohibited by law in many countries.

▶ The argument in favor of germ line therapy is that it solves the problem once and for all. Why leave the patient's descendants at risk of a disease if you could equally well eliminate the risk?

▶ The main argument against germ line therapy is that these treatments are necessarily experimental. We cannot foresee every consequence, and the risk is minimized by ensuring that its effects are confined to the patient we are treating. This would imply that once we have enough experience of somatic therapy, it will be ethical to proceed to germ line therapy. However, hopefully the initial treatment was done with informed consent, but later generations are given no choice. This leads to the view that we have a responsibility not to inflict our ideas or products on future generations so that, on this argument, germ line therapy will always be unethical.

The argument in favor needs to be set against the population genetic background. The equations set out in Section 4.5 are highly relevant here; in addition there is a strong practical argument that germ line therapy is unnecessary.

▶ For recessive conditions only a very small proportion of the disease genes are carried by affected people; the great majority are in healthy heterozygotes. The Hardy–Weinberg equation gives the ratio of carriers to affected in a population as $2pq : q^2$ where q is the frequency of the disease allele and $p = 1 - q$. Since each carrier has one copy of the disease allele while each affected person has two, the proportion of disease alleles present in affected people is $2q^2/(2pq + 2q^2)$ which simplifies to just q. So for a recessive disease affecting one person in 10 000 ($q^2 = 1/10\,000$, $q = 0.01$) only 1% of disease alleles are in affected people. Whether or not we stop affected people transmitting their disease genes (either by germ line therapy or by the cruder option of sterilizing them as the price of treatment) has very little effect on the frequency of the disease in future generations.

▶ For fully penetrant dominant conditions all the disease alleles are carried by affected people, and for X-linked recessives the proportion is 1/3. But the dream of eliminating such diseases once and for all from the population falls down because the equations in Box 4.7 show that most serious dominant or X-linked diseases are maintained in the population largely by recurrent mutation.

▶ A third, and cogent, objection is that germ line therapy is not necessary. Candidate couples would most likely have dominant or recessive Mendelian disorders (recurrence risk 50% and 25% respectively). Given a dish containing half a dozen IVF embryos from the couple, it would seem crazy to select the affected ones and subject them to an uncertain procedure, rather than simply to select the 50% or 75% of unaffected ones for re-implantation.

The argument that somatic therapy is less risky than germ line therapy seems incontrovertible. In particular, the safer nonintegrating vectors could not be used for germ line therapy. We are less convinced by the general argument that it is unethical to impose our choices on future generations. Maybe it is, but in fact we do so all the time. It would be easier to take this argument seriously if governments showed an equal concern not to inflict climate change or massive over-population on future generations.

However, integration carries certain problems and risks. Integration of most constructs occurs at random sites, and will be different in different cells of the patient. The local chromosomal environment can have unpredictable effects on expression of the construct – it may never be expressed, be expressed at an undesirably low level, or may be expressed for a short time and then irreversibly silenced. Worse, the integration may alter expression of endogenous genes. The insertion point might be within the sequence of an endogenous gene, leading to insertional inactivation of that gene. The greatest worry is that insertion of a highly expressed construct may activate an adjacent oncogene, similar to the activation of *MYC* in Burkitt's lymphoma (*Figure 17.4B*) – and indeed, this is precisely what seems to have happened in two of the children successfully treated for severe combined immunodeficiency (Check, 2002, 2003). Apparently in at least one of the 10^6 modified T cells in each of the two children, a random retroviral insertion had activated the *LMO2*

oncogene. Early on after treatment there were 50 different insertion sites among the patient's T cells, but eventually this one clone outgrew all others, leading to a novel form of T-cell leukemia. Unless this turns out to be the result of some avoidable aspect of the vector or protocol used in these particular cases, it is likely that this experience will lead to a general rejection of randomly integrating vectors as tools for gene therapy. As *Figure 21.5B* shows, this would be a major setback for the whole field.

For all these reasons, vectors that remain as extrachromosomal episomes seem likely to become the mainstream gene therapy tools. Their disadvantage is the limited duration of gene expression. If the target cells are actively dividing, the episomes will tend to be diluted out as the cell population grows. Thus there is no possibility of achieving a permanent cure, and repeated treatments may be necessary. For some purposes, for example killing cancer cells or combating an acute infection, this is not a problem because there is no

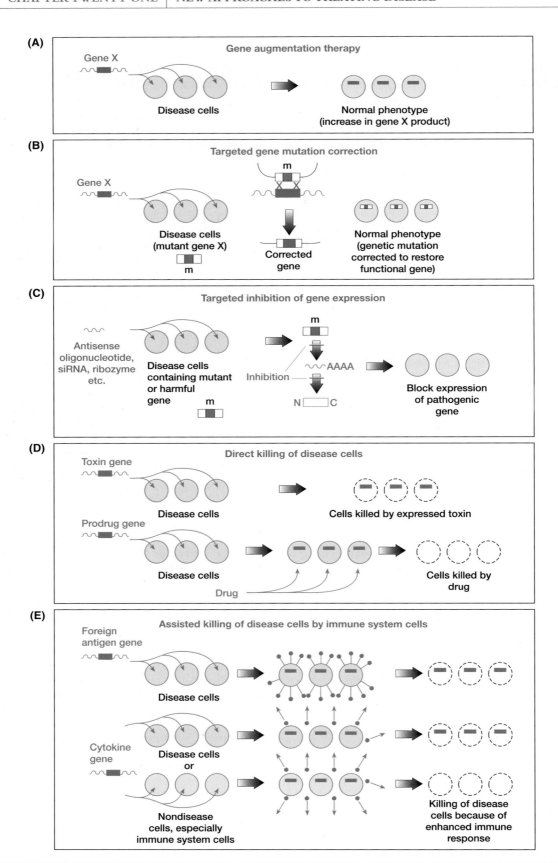

Figure 21.4: Strategies for gene therapy.

See text for details. SiRNA, small interfering RNA, see Section 21.6.

Box 21.1: 1995 NIH Panel report on Gene Therapy (Orkin-Motulsky report).

A panel was convened by the NIH to assess the current status and promise of gene therapy, and provide recommendations regarding future NIH-sponsored research in this area. It reported in December 1995 (www4.od.nih.gov/oba/rac/panelrep.htm). Among the findings were:

▶ somatic gene therapy is a logical and natural progression in the application of fundamental biomedical science to medicine and offers extraordinary potential, in the long term, for the management and correction of human disease, including inherited and acquired disorders, cancer and AIDS…;

▶ while the expectations and the promise of gene therapy are great, clinical efficacy has not been definitively demonstrated at this time in any gene therapy protocol, despite anecdotal claims of successful therapy and the initiation of more than 100 approved protocols;

▶ significant problems remain in all basic aspects of gene therapy. Major difficulties at the basic level include shortcomings in all current gene transfer vectors and an inadequate understanding of the biological interaction of these vectors with the host;

▶ overselling of the results of laboratory and clinical studies by investigators and their sponsors – be they academic, federal or industrial – has led to the mistaken and widespread perception that gene therapy is further developed and more successful than it actually is. Such inaccurate portrayals threaten confidence in the integrity of the field and may ultimately hinder progress toward successful application of gene therapy to human disease.

Most of the observations remain pertinent today, although there has been one definite cure (see Section 21.7.1)

Ethics Box 3: Designer babies.

The 'designer baby' catchphrase encapsulates two sets of worries:

▶ People will use *in vitro* fertilization and pre-implantation diagnosis to select embryos with certain desired qualities and reject the rest even though they are normal. This contrasts with the use of the same procedure to avoid the birth of a baby with a serious disease.

▶ People will use the therapeutic technologies described in this chapter, not to treat disease but for *genetic enhancement*, i.e. endowing genetically normal people with superior qualities.

The first scenario is already with us in the form of pre-implantation sex selection and a few highly-publicized cases where a couple have sought to ensure that their next child can provide a perfectly matched transplant to save the life of a sick child. Current cases involve a transplant of stem cells from cord blood. That would do no harm to the baby – it would be different if it was proposed to take a kidney. It is often suggested that these cases are the start of a slippery slope that leads inevitably to demands for very extensive specification of the genotype of the baby – as envisaged in the phrase 'designer baby'. This is wrong. Simply selecting for HLA-compatibility means only one in four embryos is selected. Most IVF procedures produce only a handful of embryos, and usually two to three are implanted to maximize the chance of success. Selection on multiple criteria is simply not compatible with having enough embryos to implant. More generally, nature has endowed us with a simple and highly agreeable method of making babies that is very effective for the large majority of couples, and it is hard to imagine most people abandoning this in favor of a long drawn out, unpleasant and highly invasive procedure that costs a fortune and has a low success rate.

Genetic enhancement is a difficult question – or it will become one, once we have any idea which genes to enhance. On the one hand, parents are supposed to do what they can to give their children a good start in life; on the other hand, options available only to the rich are widely seen as buying an unfair advantage, at least in Britain. Perhaps fortunately, we are a long way from identifying suitable genes, even if the techniques for using them were available. Attempts to produce genetically enhanced animals have not been a success and in some cases have been spectacular failures (see Gordon, 1999). Long term, the possibilities must be immense, and there will surely be very difficult ethical issues to confront. A sign that people are not yet thinking realistically about it is that they always put intelligence at the top of their list of desirable attributes – have these people never compared the lifestyles of professors and footballers?

---

requirement for long term expression. Moreover, if something does go wrong, a noninserted gene is self-limiting in a way that a gene inserted into a chromosome is not.

21.5.3 Viruses are the most commonly used vectors for gene therapy

No one gene transfer system is ideal; each has its limitations and advantages. However, mammalian viruses have been the most commonly used vectors for gene transfer because of their high efficiency of transduction into human cells. *Figure 21.5B* shows that about 70% of approved protocols use viral vectors. A number of different viral systems have been developed (reviewed by Kay *et al.*, 2001).

Oncoretroviral vectors

Retroviruses are RNA viruses that possess a reverse transcriptase, enabling them to synthesize a cDNA copy of their

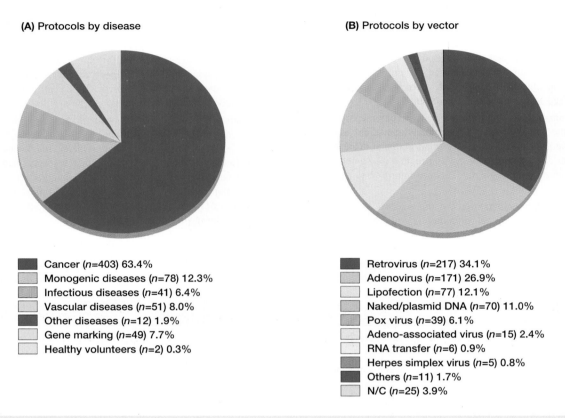

(A) Protocols by disease

- Cancer (*n*=403) 63.4%
- Monogenic diseases (*n*=78) 12.3%
- Infectious diseases (*n*=41) 6.4%
- Vascular diseases (*n*=51) 8.0%
- Other diseases (*n*=12) 1.9%
- Gene marking (*n*=49) 7.7%
- Healthy volunteers (*n*=2) 0.3%

(B) Protocols by vector

- Retrovirus (*n*=217) 34.1%
- Adenovirus (*n*=171) 26.9%
- Lipofection (*n*=77) 12.1%
- Naked/plasmid DNA (*n*=70) 11.0%
- Pox virus (*n*=39) 6.1%
- Adeno-associated virus (*n*=15) 2.4%
- RNA transfer (*n*=6) 0.9%
- Herpes simplex virus (*n*=5) 0.8%
- Others (*n*=11) 1.7%
- N/C (*n*=25) 3.9%

Figure 21.5: Gene therapy trial protocols.

(A) Distribution by disease. **(B)** Distribution by vector. The figures include all approved protocols for completed, ongoing or pending trials listed in December 2002. Reproduced from www.wiley.co.uk/genetherapy/clinical with permission.

genome. Retroviruses deliver a nucleoprotein complex (preintegration complex) into the cytoplasm of infected cells. This complex reverse transcribes the viral RNA genome and then integrates the resulting cDNA into a single random site in a host cell chromosome. Integration requires the retroviral cDNA to gain access to the host chromosomes, and it is only able to do this is when the nuclear membrane dissolves during cell division. Because of this, these retroviruses can only infect dividing cells. This limits potential target cells – some important cell targets, such as mature neurons, never divide. The property of transducing only dividing cells can, however, be turned to advantage in cancer treatment. Actively dividing cancer cells in a normally nondividing tissue like brain can be selectively infected and killed without major risk to the normal cells.

Given the ability of native retroviruses to transform cells, it is clearly crucial to engineer gene therapy vectors so as to eliminate this possibility. Much ingenuity has been expended in designing systems that can produce only permanently disabled viruses. Retroviruses normally have three transcription units, *gag*, *pol* and *env*, and a *cis*-acting RNA element ψ recognized by viral proteins that package the RNA into infectious particles. In the vector, *gag*, *pol* and *env* are replaced by the therapeutic gene, with a maximum cloning capacity of 8 kb. This construct is packaged in a special cell that can contribute the necessary *gag*, *pol* and *env* functions but does not contain an intact retroviral genome (*Figure 21.7*).

Retroviruses are very efficient at transferring DNA into cells, and the majority of early trials of gene therapy have used retroviral vectors. However, increasing concern about the risk of insertional mutagenesis is moving the main emphasis towards nonintegrating vectors.

Adenoviral vectors

Adenoviruses are DNA viruses that cause benign infections of the upper respiratory tract in humans. They can be produced in high titers (much higher than retroviruses) and they efficiently transduce both dividing and nondividing cells. The linear double-stranded DNA genome remains nonintegrated as an episome within the cell nucleus. As explained above, this has advantages in safety and disadvantages in the short term expression obtained. As with retroviral vectors, adenoviral vectors are disabled and rely on a packaging cell to provide vital functions. The adenovirus genome is relatively large and different constructs have

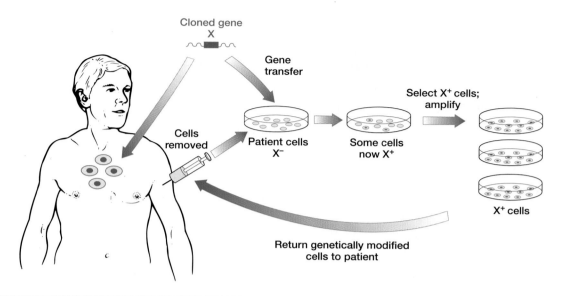

Figure 21.6: *In vivo* and *ex vivo* gene therapy.

Where possible, cells are removed from the patient, modified in the laboratory and returned to the patient (*ex vivo* gene therapy; green arrows). This allows just the appropriate cells to be treated, and the cells can be checked before they are replaced to make sure that the desired change has been achieved. For many tissues this is not possible and the cells must be modified within the patient's body (*in vivo* gene therapy; blue arrow).

varying sections of the genome deleted (Yeh and Perricaudet, 1997). 'Gutless' vectors have all viral genes deleted, and can accommodate up to 35 kb of therapeutic DNA.

The big problem with adenoviral vectors is their immunogenicity (Kafri *et al.*, 1998). Even though a live replication competent adenovirus vaccine has been safely administered to several million US army recruits over several decades (as protection against natural adenoviral infections), unwanted immune reactions have been a problem in several gene therapy trials. Jesse Gelsinger died in September 1999 2 days after receiving 6×10^{13} recombinant adenoviral particles by intra-hepatic injection in a Phase I trial of gene therapy for ornithine transcarbamylase deficiency. Patients in other trials have suffered lesser but significant inflammatory responses. Moreover, because these vectors are noninte-grating, gene expression is short term. The first generation recombinant adenoviruses used in cystic fibrosis gene therapy trials showed that transgene expression declined after about 2 weeks and was negligible by 4 weeks. Repeated administration would be necessary for sustained expression, which could only exacerbate the immune response. Maybe adenoviral vectors will find their main application where high level but transient expression is needed, for example for killing cancer cells.

Adeno-associated virus vectors

Adeno-associated viruses (AAVs) are nonpathogenic single-stranded DNA viruses that rely on co-infection by an adeno or herpes helper virus to replicate. Unmodified human AAV integrates into chromosomal DNA at a specific site on 19q13.3–qter. This is a highly desirable property, providing the advantage of long term expression without the risk of inser-tional mutagenesis that is such a problem with retroviruses. Unfortunately the specificity of integration is provided by the viral *rep* protein, and the *rep* gene is deleted in the constructs used for gene transfer. As with other systems, the functions necessary for production of virus particles (including *rep*) are provided by a packaging cell. Ninety-six percent of the AAV genome has been deleted in AAV-based vectors. This provides a high degree of safety because the recombinant vectors contain no viral genes. However, the AAV genome is very small, and even these highly deleted vectors can only accom-modate inserts up to 4.5 kb.

Lentiviruses

These are specialized retroviruses that have the useful attribute of infecting nondividing cells (Vigna and Naldini, 2000). Like other retroviruses, they integrate into the host chromosomes at random, giving the possibility of long term gene expression but with all the safety implications that integration carries. Human HIV is the basis of most lentiviral vectors, which understandably provokes nervousness about the risk of inad-vertently generating replication competent virus. The HIV genome is more complex than the *gag*, *pol* and *env* of standard retroviruses (see *Figure 21.13*), and much work has been devoted to eliminating unnecessary genes and generating safe packaging lines while retaining the ability to infect non-dividing cells. Self-inactivating vectors provide an additional layer of safety (Miyoshi *et al.*, 1998).

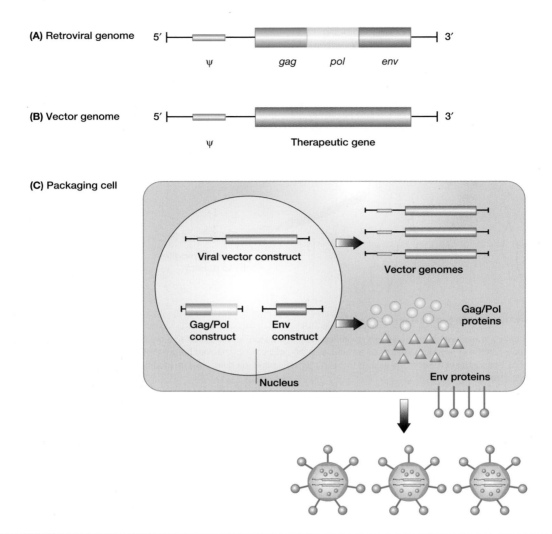

Figure 21.7: Construction and packaging of a retroviral gene therapy vector.

The *gag*, *pol* and *env* genes are deleted from the retroviral genome and replaced by the therapeutic gene. The ψ sequence is retained; this is recognized by viral proteins for assembly of the RNA into a virus particle. *Gag, pol* and *env* functions are supplied by the packaging cell, but the genes are on physically separate molecules and there is no ψ sequence. This is done in order to minimize the risk of producing replication competent viruses. Recombinant viral genomes are packaged into infective but replication deficient virus particles which bud off from the cell and are recovered from the supernatant. Reprinted by permission from Somia and Verma (2000) *Nature Review Genetics*, copyright 2000 Macmillan Magazines Ltd.

Herpes simplex virus vectors

HSV vectors are tropic for the central nervous system (CNS). These are complex viruses with a double-stranded DNA genome of 152 kb containing at least 80 genes. They can establish lifelong latent infections in sensory ganglia in which they exist as nonintegrated extrachromosomal elements. The latency mechanism might be exploitable to allow long term expression of transferred genes, hopefully spreading through a synaptic network. Their major applications would be in delivering genes into neurons for the treatment of diseases such as Parkinson's disease and CNS tumors. Insert capacity is at least 30 kb. Practical vectors are still at an early stage of development (Fink and Glorioso, 1997).

21.5.4 Nonviral vector systems avoid many of the safety problems of recombinant viruses, but gene transfer rates are generally low

In the laboratory it is relatively easy to get foreign DNA into cells, and some of the methods have potential in gene therapy if the safety problems of viral systems prove intractable. However, at present all suffer from a low rate of gene transfer, and a short duration of expression.

Liposomes

Liposomes are synthetic vesicles that form spontaneously when certain lipids are mixed in aqueous solution. Phospholipids for example can form bilayered vesicles that mimic the structure of biological membranes, with the hydrophilic phosphate groups on the outside and the hydrophobic lipid tails in the inside. The DNA to be transferred is packaged *in vitro* with the liposomes and used directly for gene transfer to a target tissue *in vivo* (*Figure 21.8*). Cationic liposomes have a positive surface charge and bind DNA on the outside; anionic liposomes have a negative surface charge, and form with the DNA on the inside. The lipid coating allows the DNA to survive *in vivo*, bind to cells and be endocytosed into the cells. Cationic liposomes have been the most commonly used in gene transfer experiments (see Huang and Li, 1997 for references). Unlike viral vectors, the DNA–lipid complexes are easy to prepare and there is no limit to the size of DNA that is transferred. However, the efficiency of gene transfer is low, and the introduced DNA is not designed to integrate into chromosomal DNA. As a result, any expression of the inserted genes is transient.

Direct injection or particle bombardment

In some cases, DNA can be injected directly with a syringe and needle into a target tissue such as muscle. This approach has been considered, for example, for Duchenne muscular dystrophy. Early studies investigated intramuscular injection of a dystrophin gene into a mouse model, *mdx* (Acsadi *et al.*, 1991). Because of the huge size of the native dystrophin gene, a minigene was used, comprising the dystrophin cDNA fitted with regulatory sequences for ensuring high level expression, such as a powerful viral promoter. An alternative direct injection approach uses particle bombardment (biolistic or 'gene gun') techniques: DNA is coated on to metal pellets and fired from a special gun into cells. Successful gene transfer into a number of different tissues has been obtained using this simple and comparatively safe method. However, with any of these direct injection methods the efficiency of gene transfer is very poor, and the injected DNA is not stably integrated. This may be less of a problem in tissues such as muscle which do not regularly proliferate, and in which the injected DNA may continue to be expressed for several months.

Receptor-mediated endocytosis

In this method the DNA to be transferred is coupled to a targeting molecule that can bind to a specific cell surface receptor, inducing endocytosis and transfer of the DNA into the cell. For example, hepatocytes clear asialoglycoproteins from the serum via receptors on their cell surface. An asialoglycoprotein that is covalently linked to polylysine will reversibly bind DNA by an electrostatic interaction between the positively charged polylysine and the negatively charged DNA. If the complex is infused into the liver via the biliary tract or vascular bed, it is selectively taken up by hepatocytes. A more general approach uses the transferrin receptor which is expressed in many cell types, but is relatively enriched in proliferating cells and hematopoietic cells (*Figure 21.9*).

Normally substances internalized in this way are delivered to lysosomes, and it is necessary to provide some escape mechanism if the DNA is to reach the cell nucleus. One possibility is to co-transfer adenovirus or adenovirus proteins. These specifically disrupt the early endosome, allowing the contents to escape. Gene transfer efficiency may be high, but the method is not designed to allow integration of the transferred genes.

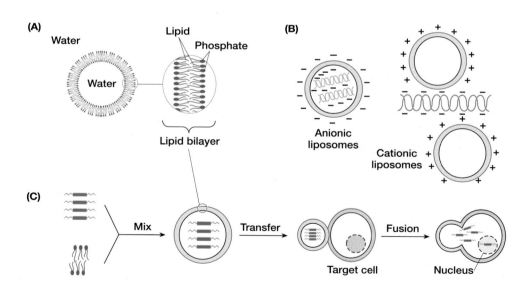

Figure 21.8: Use of liposomes for gene delivery *in vivo*.

(A) Liposomes are synthetic vesicles that form spontaneously in aqueous solution when certain lipids are mixed. They can carry a positive (cationic) or negative (anionic) surface charge, depending on the chemistry of the lipids used. **(B)** The DNA cargo is transported inside anionic liposomes or bound to the surface of cationic liposomes. **(C)** DNA delivery occurs when a liposome fuses with the plasma membrane of a cell.

21.6 Methods for repairing or inactivating a pathogenic gene in a cell or tissue

The methods described in the last section all aim, one way or another, to insert a desired gene into a target cell or tissue and have it expressed at a level and for a time that are appropriate for the particular clinical problem. However, some problems require a different approach. Conditions caused by a *gain of function* of a gene (Section 16.5) or a *dominant negative effect* (Section 16.4.3) will not benefit from this form of treatment. If the problem is a resident gene that is doing something positively harmful, the only solution is to remove or inactivate the offending gene. Typical examples are dominant Mendelian conditions (other than those caused by haploinsufficiency, Section 16.4.2), activated oncogenes in cancer cells, and inappropriate immune reactions in autoimmune diseases. Infectious diseases might also be treated by inhibiting a pathogen-specific gene or gene product.

Various strategies can be tried to achieve this. The offending gene might be physically destroyed, or expression could be prevented by downregulating transcription, destroying the transcript or inhibiting the protein product. Whatever method is used, some sort of agent or construct must be got into a target cell and made to function there. In that respect, the problems of efficient delivery and expression are similar to those described in the previous section. In the case of dominant diseases or activated oncogenes, there is the additional problem of designing an agent that will selectively attack the mutant allele without affecting the normal allele. That area is usually seen as the second wave of gene therapy. Current studies here aim at proof of principle: practical treat-

ments will probably only be developed once gene augmentation therapies are seen to work successfully.

21.6.1 Repairing a mutant allele by homologous recombination

Many standard genetic engineering techniques in lower organisms use homologous recombination to replace one sequence with another, so it is natural to consider using homologous recombination to repair a pathogenic mutant gene. In Small Fragment Homologous Recombination, one of the standard methods is used to deliver a 400–800-bp DNA duplex containing the wild type sequence into the target cell (Goncz *et al.*, 2001). Proof of principle has been reported, but it has proved difficult to achieve efficiencies high enough to be therapeutically useful.

21.6.2 Inhibition of translation by antisense oligonucleotides

Antisense oligonucleotides, typically 12–30 nucleotides long, can inhibit translation of a complementary mRNA by forming a dsRNA or DNA–RNA double helix (Galderisi *et al.*, 1999). In some organisms, such as the nematode *C. elegans*, dsRNA is efficiently converted into small interfering RNA molecules (siRNA), which inactivate homologous transcripts by the poorly understood but highly efficient RNAi mechanism (see below). In humans this apparently does not happen. Human cells have little or none of the Dicer enzyme that is needed to make siRNAs. Instead, dsRNA or DNA–RNA is broken down by RNAse H. This

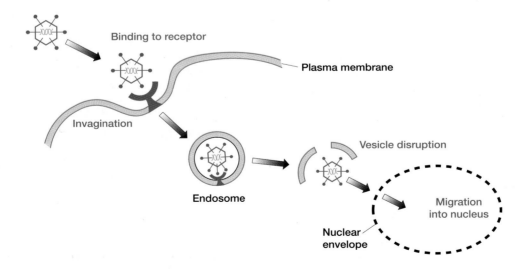

Figure 21.9: Gene transfer via receptor-mediated endocytosis.

After binding the ligand, the plasma membrane invaginates then pinches off, leaving the receptor–ligand complex in an intracellular vesicle, the endosome. The ligand might be an adenovirus carrying a therapeutic gene, as shown here, or it could be the therapeutic DNA bound to some other molecule for which the target cell has a specific receptor. Endosomes are normally targeted to lysosomes for degradation. In order to be expressed, the foreign DNA must somehow escape from the endosome before this happens, and reach the nucleus. Adenoviruses specifically disrupt endosomes, allowing efficient escape.

enzyme degrades the RNA but not the DNA strand of an RNA–DNA duplex, thus allowing DNA antisense molecules to behave semi-catalytically. The antisense molecule is usually chemically modified to make it more resistant to cellular nucleases. Phosphorothioate bonds replace phosphodiester links, or peptide nucleic acids are used where the entire sugar phosphate backbone is replaced with a nuclease resistant synthetic peptide framework. This carries the disadvantage of nonspecific toxicity, also the antisense construct must be synthesized chemically and exogenously, whereas unmodified antisense oligos can be produced inside the cell from a suitable plasmid. Antisense oligos have rather unpredictable effects. In some cases the target gene is silenced efficiently and specifically, but often there is little effect or nonspecific effects. DsRNAs more than about 30 bp long often induce production of interferon, leading to a general shutdown of translation. In 1998 Vivatrene, a phosphorothioate modified DNA for treatment of cytomegalovirus infections in AIDS patients, became the first antisense drug to reach the market.

21.6.3 Selective destruction or repair of mRNA by a ribozyme

Over the years a growing number of instances have been discovered where RNA molecules function as enzymes (*ribozymes*, see Doudna and Cech, 2002). For example, catalytic RNAs are involved in telomerase (acting on DNA), in intron–exon splicing (acting on RNA), and as the peptidyl transferase of ribosomes (acting on polypeptides). RNA molecules have many properties that fit them to act as enzymes. They form a variety of three-dimensional structures by virtue of which they can bind specifically to DNA, RNA or protein molecules, and they have many potentially reactive hydroxyl and amino groups.

Gene therapists have been particularly interested in ribozymes that cleave target RNAs in a sequence-specific manner. In nature self-cleaving RNAs generate genome-length strands by site-specific cutting of concatemers in RNA viruses that replicate by a rolling circle mechanism. Natural examples include the hammerhead (40 nt), hairpin (70 nt), HDV (human delta virus; 90 nt) and VS (Varkud satellite; 160 nt) ribozymes (Doudna and Cech, 2002). Hammerhead ribozymes can be converted by genetic engineering into very versatile trans-cleaving enzymes, whose target RNA sequence can be specified simply by fitting the ribozyme with complementary sequences (*Figure 21.10*). As with antisense oligonucleotides (see above), the ribozyme can be made resistant to nucleases in the cell by chemical modification, but it must then be delivered exogenously rather than being produced by natural transcription within a transfected target cell. The first phase I clinical trials of ribozymes began in 1998, and phase II trials are under way using ribozymes that target the mRNA of the VEGF receptor to inhibit angiogenesis in breast and colorectal tumors (Sullenger and Gilboa, 2002).

Because the sequence specificity of ribozymes is controlled by well understood base pairing, they can readily be designed to specifically degrade a mutant mRNA in a dominant disease, while leaving the wild type transcript intact. More ambitiously, ribozymes have been designed to *repair* mutant

transcripts by trans-splicing reactions in which a predetermined mutant sequence is cut out of the mRNA and replaced by the wild type sequence (Sullenger and Gilboa, 2002). All these ideas have been shown to work in experimental systems, but whether they can be developed into practical therapies remains to be seen.

21.6.4 Selective inhibition of the mutant allele by RNA Interference (RNAi)

This is the technique to which the most excitement currently attaches. As described in Section 20.2.6 specific and very efficient inhibition of gene expression can be obtained in a wide range of organisms by the use of small interfering RNAs (siRNAs). These are 21–23-nt double-stranded RNAs with 5′ phosphorylated termini and 2-nt 3′ overhangs. In many organisms longer dsRNAs are efficiently cleaved to siRNAs by the Dicer enzyme. This does not work in human cells, but siRNAs can be generated within human cells by transcription from suitable DNA vectors. The transcripts are designed so that they can snap back to form stem–loop hairpins. The stem is 19–29 bp, designed to match the target, and the loop is 6–9 nt. At present RNAi is a highly efficient laboratory tool for creating loss of function phenotypes. Whether it can also be developed into a practical means of therapy is less clear, though the potential is exciting.

21.7 Some examples of attempts at human gene therapy

Figure 21.5A shows that of the 600 or so gene therapy protocols that have been approved, 63% have been for cancer and only 12% for monogenic disorders. Another 6% were for infectious diseases and 8% for vascular diseases. Here we give brief overviews, and some references, of the progress of gene therapy in a number of important areas. Despite the limited number of trials, monogenic diseases have always been high on the gene therapy agenda, and the first definitive success has been in that area.

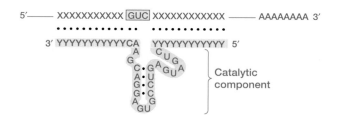

Figure 21.10: Using a trans-cleaving hammerhead ribozyme to destroy a mutant mRNA.

Natural hammerhead ribozymes are self-cleaving. For gene therapy purposes the strand containing the cleavage site (target) is separate from the catalytic strand (the engineered ribozyme). By designing the ribozyme with sequence (YYYY….) complementary to the target (XXXX….), almost any desired RNA can be specifically targeted.

21.7.1 The first definite success: a cure for X-linked severe combined immunodeficiency

Back in 1990 an attempt to treat autosomal recessive severe combined immunodeficiency (SCID) was hailed as a success in a blaze of publicity. T lymphocytes from an affected child with adenosine deaminase deficiency were transfected *ex vivo* with a retroviral vector containing a functioning ADA gene, expanded in culture and replaced in the patient. Later other patients received similar treatment, with up to 10–12 treatments per patient. Several patients reported dramatic clinical improvements. However, all the patients continued also to receive treatment with an enzyme preparation, making it uncertain how much of their improvement was due to the gene therapy.

The first unambiguous success came for a related condition, X-linked SCID (Cavazzana-Calvo *et al.*, 2000; see *Figure 21.11*). Again, the treatment was *ex vivo*, using a retroviral vector encoding the γc chain of the cytokine receptor gene *IL2R*. Bone marrow cells expressing CD34, a marker of hematopoietic stem cells, were incubated for 3 days with the retroviral vector, during which time they increased in number five- to eightfold, and then returned to the patient (*Figure 21.11*). Nine of 11 treated patients were cured and enabled to lead a normal life (Hacein-Bey-Abini *et al.*, 2002). However, as mentioned above, two of them have subsequently developed a leukemia almost certainly as a result of insertional activation of

the *LMO2* oncogene (Check, 2002, 2003). All trials involving retroviral transduction of large pools of lymphocytes were quickly suspended worldwide, pending full understanding of these tragic events, and at the time of writing it is not clear how many will ever be reinstated.

21.7.2 Attempts at gene therapy for cystic fibrosis

Among Mendelian diseases, cystic fibrosis should be one of the more amenable to gene therapy. The problem is caused by lack of the *CFTR*-encoded chloride channel, mainly in airway epithelium. Studies of patients with partially active *CFTR* alleles suggest that 5–10% of the normal level would be sufficient to produce a good clinical response. This would at least prevent progression of the disease, and reverse some secondary effects. Tissue-specific gene delivery might be accomplished by using an aerosol inhaler. At least 18 clinical trials of gene therapy for CF have been reported (Davies *et al.*, 2001). Initial trials used adenoviruses, which naturally infect nondividing lung cells. At high doses these sometimes provoked inflammatory reactions, and especially after the adenovirus-triggered death of Jesse Gelsinger (see above), are now viewed with some suspicion. Four of the five most recent trials used liposomes or AAV. Several trials have demonstrated gene transfer and short lived expression, but it is clear that all these trials fall well short of clinical applicability.

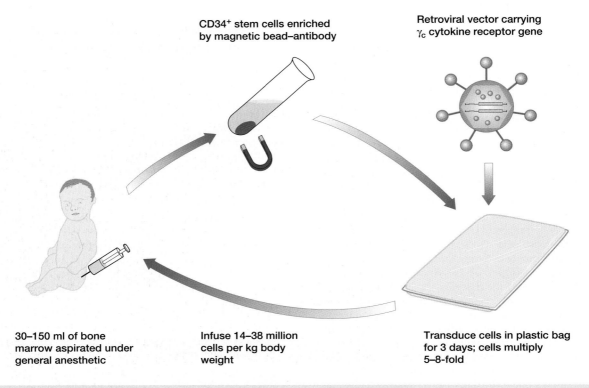

CD34⁺ stem cells enriched by magnetic bead–antibody

Retroviral vector carrying γc cytokine receptor gene

30–150 ml of bone marrow aspirated under general anesthetic

Infuse 14–38 million cells per kg body weight

Transduce cells in plastic bag for 3 days; cells multiply 5–8-fold

Figure 21.11: Gene therapy of X-linked severe combined immunodeficiency disease (X-SCID).

This is the first clear success of gene therapy. Of 11 boys aged 1–11 months treated at the Necker-Enfants Malades Hospital, Paris, nine were cured. See Hacein-Bey-Abini *et al.* (2002). Two of the nine unfortunately later developed a form of leukemia, almost certainly as a result of activation of the *LMO2* oncogene by nearby insertion of the retroviral vector.

A significant problem is the physical barrier of mucus and glycocalyx that covers lung airway epithelial cells, especially in the infected lungs of CF patients. Gene therapy agents can be delivered into the airways, but more sophisticated vehicles will be necessary to allow efficient transduction of epithelial cells. Transfecting the right cells is also a problem. Adenoviruses tend to infect basal cells of the epithelium, whereas the highest levels of natural CFTR expression are seen in submucosal glands. Ideally stem cells should be targeted, because surface epithelial cells have a lifespan of only about 120 days, so that repeated administration will be necessary, with all the attendant problems of the immune response.

21.7.3 Attempts at gene therapy for Duchenne muscular dystrophy

Studies of female carriers of DMD, and of patients with the milder Becker muscular dystrophy, show that restoring about 20% of normal dystrophin gene expression in muscle would benefit DMD patients. However, gene therapy for DMD faces the twin problems of the huge size of the dystrophin gene (2.4 Mb), and of delivering it into both skeletal and cardiac muscle cells. See Chamberlain (2002) for a review of progress. Even the cDNA is too large, at 14 kb, for many vectors. Based on the observation of a patient who had a deletion of exons 17–48 of the dystrophin gene (46% of the coding sequence) but suffered only the mild Becker dystrophy, a 6.3-kb minigene has been constructed that can be accommodated in adenoviral and retroviral vectors. It is important to avoid triggering cell-mediated immune responses; using a muscle-specific promoter to drive the transgene helps minimize these. Adeno-associated viruses have shown good persistence in muscle, at least in healthy mice, and plasmids are also promising vehicles. Delivery of the constructs, not just into skeletal muscle but also into heart and diaphragm is necessary for successful treatment, and so far this problem has not been solved. An elegant alternative approach is to try to up-regulate expression of utrophin (MIM 128240), a dystrophin-like molecule that normally binds the actomyosin complex to the muscle cell membrane just at the neuromuscular junction. Utrophin is encoded by a separate autosomal locus that remains intact in DMD boys.

Cell therapy has also been attempted in DMD by transplanting myoblasts into muscles of affected boys. The occasional patient in some studies showed some dystrophin positive muscle fibers, but no clinical improvement was seen in any patient. It has been claimed that stem cells delivered by bone marrow transplantation may be able to migrate to muscle and act as muscle stem cells (Ferrari et al., 1998). If confirmed, this offers real hope of treating DMD by the methods that cured SCID (see above).

Table 21.3: Examples of cancer gene therapy trials

These are mostly Phase I (basic safety) trials; at this stage of development, most trials do not aim to benefit patients clinically. See the NIH clinical trials database (www4.od.nih.gov/oba/rac/clinicaltrial.htm) for a comprehensive survey.

Disorder	Cells altered	Gene therapy strategy
Ovarian cancer	Tumor cells	Intraperitoneal injection of retrovirus or adenovirus encoding full-length p53 or BRCA1 cDNA, with the hope of restoring cell cycle control
Ovarian cancer	Tumor cells	Inject adenovirus encoding a scFv antibody to ErbB2. Hope to inactivate a growth signal
Malignant melanoma	Tumor-infiltrating lymphocytes	Extract TILs from surgically removed tumor and expand in culture. Infect TILs *ex vivo* with a retroviral vector expressing tumor necrosis factor-α, infuse into patient. Hope that TILs will target remaining tumor cells, and the TNFα will kill them. See *Figure 21.4E* for the principle
Various tumors	Tumor cells	Transfect tumor cells with a retrovirus expressing a cell surface antigen e.g. HLA-B7, or a cytokine, e.g. IL-12, IL-4, GM-CSF or IFN-γ. Hope this enhances the immunogenicity of the tumor, so that the host immune system destroys it. Often done *ex vivo* using lethally irradiated tumor cells. See *Figure 21.4E* for the principle
Prostate cancer	Dendritic cells	Treat autologous dendritic cells with a tumor antigen or cDNA expressing the antigen, to prime them to mount an enhanced immune response to the tumor cells. See *Figure 21.4E* for the principle
Malignant glioma (brain tumor)	Tumor cells	Inject a retrovirus expressing thymidine kinase (TK) or cytosine deaminase (CDA) into the tumor. Only the dividing tumor cells, not the surrounding nondividing brain cells, are infected. Then treat with gancyclovir (TK-positive cells convert this to the toxic gcv phosphate) or 5-fluorocytosine (CDA-positive cells convert it to the toxic 5-fluorouracil). Virally-infected (dividing) cells are selectively killed. See *Figure 21.12*
Head and neck tumors	Tumor cells	Inject ONYX-015 engineered adenovirus into tumor. The virus can only replicate in p53-deficient cells, so selectively lyses tumor cells. This treatment was effective when combined with systemic chemotherapy

TIL, Tumor-infiltrating lymphocytes; TNF, tumor necrosis factor; IL, interleukin; GM-CSF, granulocyte-macrophage stimulating factor; IFN, interferon.

21.7.4 Gene therapy for cancer

Over 60% of all approved gene therapy trial protocols have been for cancer (*Figure 21.5*). *Table 21.3* lists a number of examples, chosen to illustrate the range of approaches. These include:

▶ gene supplementation to restore tumor suppressor gene function;

▶ gene inactivation to prevent expression of an activated oncogene;

▶ genetic manipulation of tumor cells to trigger apoptosis;

▶ modification of tumor cells to make them more antigenic, so that the immune system destroys the tumor;

▶ modification of dendritic cells to increase a tumor-specific immune reaction;

▶ use of oncolytic viruses that are engineered to selectively kill tumor cells;

▶ genetic modification of tumor cells so that they, but not surrounding nontumor cells, convert a nontoxic prodrug into a toxic compound that kills them (*Figure 21.12*).

21.7.5 Gene therapy for infectious disease: HIV

As the most important viral pathogen of humans, HIV-1 is the target of huge research efforts in every relevant branch of medical science. Genetic manipulation has been important in two areas. Much work has gone into attempts to develop genetically engineered vaccines; in addition many researchers have considered genetic manipulation of host cells to make them resistant to HIV. The NIH database of clinical trials (www4.od.nih.gov/oba/rac/clinicaltrial.htm) lists almost 40 trial protocols involving gene transfer submitted in 1992–2001. Almost all of these are similar in principle to the method that cured X-SCID (see above). Hematopoietic stem cells are transfected with a gene that will hopefully inhibit HIV replication, usually in a retroviral vector, and the treated cells are returned to the patient. Since the major pathology of AIDS is the infection and destruction of lymphocytes, this is a natural way to try to prevent HIV infection from developing into AIDS.

HIV is a retrovirus. The mature viral particle consists of two identical copies of the single-stranded RNA genome, plus some core proteins, contained in an envelope made of viral glycoproteins and lipids picked up from the host cell membrane when the virus buds off (*Figure 21.13*). As well as the *gag*, *pol* and *env* genes shared by all retroviruses, HIV replication requires the Tat and Rev regulatory proteins. These, and the viral RNA sequences to which they bind (TAR and RRE) are the targets of choice for many of the attempts at making lymphocytes resistant to HIV. Strategies include:

▶ **use of antisense RNAs**. Retroviruses have been constructed that encode antisense RNAs to TAR, to the overlapping *tat* and *rev* mRNAs, and to the *pol* and *env* mRNAs;

▶ **use of decoy RNAs**. A retrovirus directing high-level expression of a transcript containing the RRE sequence might be able to sequester all the Rev protein and prevent HIV replication;

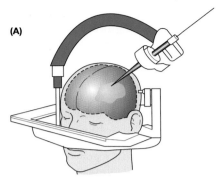

(A)

MRI-guided stereotactic implantation of vector producer cells (VPC) into CNS tumors *in situ*

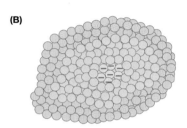

(B)

Vector producing cells inside the tumor

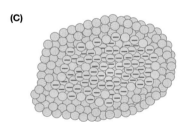

(C)

Retroviruses infect tumor cells but not normal cells

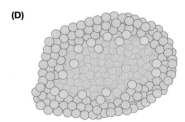

(D)

Gancyclovir kills the infected cells

Figure 21.12: *In vivo* gene therapy for brain tumors.

A retrovirus is engineered to produce the herpes simplex virus thymidine kinase (HSV-TK). Vector-producing cells (VPC; blue) are injected into the brain tumor. Because retroviruses infect only dividing cells, they infect the tumor cells (pink) but not the surrounding normal brain tissue (green). The nontoxic prodrug gancyclovir (gcv) is given intravenously. In TK+ cells gcv is converted to the highly toxic gcv-triphosphate and the cell is killed.

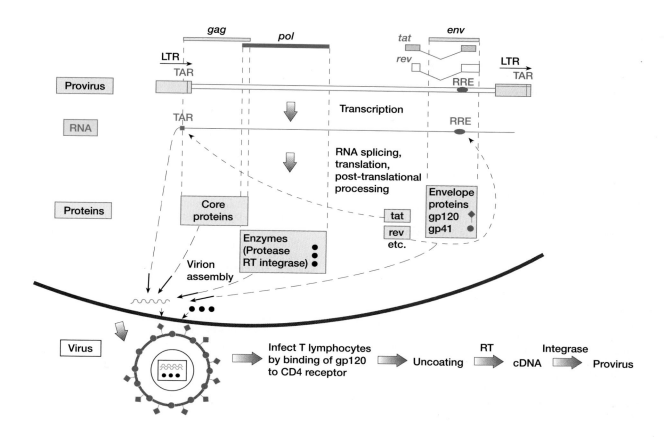

Figure 21.13: The HIV-1 virus life-cycle.

HIV-1 is a retrovirus with an RNA genome. As with other retroviruses, there are *gag, pol* and *env* genes. A reverse transcriptase makes a DNA copy that integrates as a provirus into the host chromosomes. Viral mRNA and proteins are packaged into virus particles that bud off from the host cell membrane. Unlike the simple retroviruses, HIV-1 encodes two regulatory proteins, Tat and Rev, that bind TAR and RRE sequences, respectively, in the viral genome and are essential for viral replication. These are the main targets for gene therapy of HIV.

▶ **use of dominant negative mutants**. Some retroviral constructs encode a mutant Rev protein, RevM10. This binds the RRE but then will not assemble the multi-protein complex required to export the RNA from the nucleus;

▶ **use of ribozymes**. Several groups have made retroviruses that encode ribozymes. RRz2 is a hammerhead ribozyme directed against the *tat* regulatory region; another type, the hairpin ribozyme, has also been used to cleave the HIV genome;

▶ **use of intrabodies**. Retroviruses encoding scFv intracellular antibodies (see Section 21.3.4) have been used to try to inactivate the Tat or Rev regulatory proteins, or the gp160 coat glycoprotein.

In the early stages of HIV infection there is a massive turnover of lymphocytes, as the immune system struggles to destroy infected cells. If the immune response were rather more

effective, maybe the virus could be contained at this stage. Apart from efforts at developing vaccines, work has also been devoted to modifying T cells so that they kill infected cells more efficiently. Retroviruses have been designed to make CD8[+] T cells express a chimeric T cell receptor that targets their cytotoxic response to HIV infected cells.

Details of all these approaches can be found in the NIH database cited under 'Infectious Diseases' (click on the Scientific Abstract for any protocol of interest). *In vitro* the manipulated cells often show high resistance to HIV infection, and *in vivo* long term (several months up to 1 or 2 years) bone marrow engraftment has been demonstrated in a few trials. The open question is whether engraftment can be made to occur at a sufficiently high level to provide a clinically useful pool of HIV resistant lymphocytes. High level engraftment might be achieved by first destroying the patient's existing marrow with cytotoxic chemicals and radiation – but doing this to a patient with AIDS would be a desperate measure.

Further reading

NIH database of gene therapy trials: www4.od.nih.gov/oba/rac/clinicaltrial.htm

Wiley database of approved gene therapy protocols: www.wiley.co.uk/genetherapy/clinical

Templeton NS, Lasic DD (eds) (2000) *Gene Therapy: Therapeutic Mechanisms and Strategies.* Marcel Dekker, New York.

Treacy EP, Valle D, Scriver CR (2001) Treatment of genetic Disease. In: *Metabolic and Molecular Basis of Inherited Disease*, 8th Edn (eds CR Scriver, AL Beaudet, WS Sly, MD Valle). McGraw-Hill, New York

References

Acsadi G, Dickson G, Love DR et al. (1991) Human dystrophin expression in mdx mice after intramuscular injection of DNA constructs. *Nature* **352**, 815–818.

Cavazzana-Calvo M, Hacein-Bey S, de Saint Basile G et al. (2000) Gene therapy of human severe combined immunodeficiency (SCID)-X1 disease. *Science* **288**, 669–672.

Chamberlain JS (2002) Gene therapy of muscular dystrophy. *Hum. Mol. Genet.* **11**, 2355–2362.

Check E (2002) A tragic setback (News Feature). *Nature* **420**, 116–118.

Check E (2003) Second cancer case halts gene-therapy trials (News Feature). *Nature* **421**, 305.

Costa T, Scriver CR, Childs B (1983) The effect of mendelian disease on human health: a measurement. *Am. J. Med. Genet.* **21**, 231–242.

Daley GQ (2002) Prospects for stem cell therapeutics: myths and medicines. *Curr. Opin. Genet. Dev.* **12**, 607–613.

Davies JC, Geddes DM, Alton EWFW (2001) Gene therapy for cystic fibrosis. *J. Gene Med.* **3**, 409–417.

Dean W, Santos F, Stojkovic M et al. (2001) Conservation of methylation reprogramming in mammalian development: Aberrant reprogramming in cloned embryos. *Proc. Natl Acad. Sci. USA* **98**, 13734–13738.

Doudna JA, Cech TR (2002) The chemical repertoire of natural ribozymes. *Nature* **418**, 222–228.

Ferrari G, Gusella-De Angelis G, Coletta M et al. (1998) Muscle regeneration by bone marrow-derived myogenic progenitors. *Science* **279**, 1528–1530.

Fink DJ, Glorioso JC (1997) Engineering herpes simplex virus vectors for gene transfer to neurons. *Nature Med.* **3**, 357–359.

Galderisi U, Cascino A, Giordano A (1999) Antisense oligonucleotides as therapeutic agents. *J. Cell. Physiol.* **181**, 251–257.

Gardner MJ, Shallom SJ, Carlton JM et al. (2002) Genome sequence of the human malaria parasite *Plasmodium falciparum. Nature* **419**, 498–511.

Goncz KK, Colosimo A, Dallapiccola B et al. (2001) Expression of F508 CFTR in normal mouse lung after site-specific modification of CFTR sequences by SFH. *Gene Ther.* **8**, 961–965.

Gordon JW (1999) Genetic enhancement in humans. *Science* **283**, 2023–2024.

Hacein-Bey-Abina S, Le Deist F, Carlier F et al. (2002) Sustained correction of X-linked severe combined immunodeficiency by *ex vivo* gene therapy. *New Engl. J. Med.* **346**, 1185–1193.

Hoogenboom HR, De Bruine AP, Hufton SE, Hoet RM, Arends J-W, Roovers RC (1998) Antibody phage display technology and its applications. *Immunotechnology* **4**, 1–20.

Hoppe-Seyler F, Butz K (2000) Peptide aptamers: powerful new tools for molecular medicine. *J. Mol. Med.* **78**, 426–430.

Huang L, Li S (1997) Liposomal gene delivery: a complex package. *Nature Biotechnol.* **15**, 620–621.

Hudson PJ (1999) Recombinant antibody constructs in cancer therapy. *Curr. Opin. Immunol.* **11**, 548–557.

Kafri T, Morgan D, Krahl T et al. (1998) Cellular immune response to adenoviral vector infected cells does not require *de novo* viral gene expression: implications for gene therapy. *Proc. Natl Acad. Sci. USA* **95**, 11377–11382.

Kay MA, Glorioso JC, Naldini L (2001) Viral vectors for gene therapy: the art of turning infectious agents into vehicles of therapeutics. *Nature Med.* **7**, 33–40.

Lindstedt S, Holme E, Locke EA et al. (1992) Treatment of hereditary tyrosinaemia type 1 by inhibition of 4-hydroxyphenylpyruvate dioxygenase. *Lancet* **340**, 813–817.

Marasco WA (1997) Intrabodies: turning the humoral immune system inside out for intracellular immunization. *Gene Ther.* **4**, 11–15.

Miyoshi H, Blomer U, Takahashi M, Gage FH, Verma IM (1998) Development of a self-inactivating lentivirus vector. *J. Virol.* **72**, 8150–8157.

Reyes-Sandoval A, Ertl HC (2001) DNA vaccines. *Curr. Mol. Med.* **1**, 217–243.

Russell CS, Clarke LA (1999) Recombinant proteins for genetic disease. *Clin. Genet.* **55**, 389–394.

Somia N, Verma IM (2000) Gene therapy: trials and tribulations. *Nat. Rev. Genet.* **1**, 91–99.

Stöger E, Vaquero C, Torres E et al. (2000) Cereal crops as viable production and storage systems for pharmaceutical scFv antibodies. *Plant Mol. Biol.* **42**, 583–590.

Sullenger BA, Gilboa E (2002) Emerging clinical applications of RNA. *Nature* **418**, 252–258.

Thomson JA, Itskovitz-Eldor J, Shapiro SS et al. (1998) Embryonic stem cell lines derived from human blastocysts. *Science* **282**, 1145–1147.

Tse E, Rabbitts TH (2000) Intracellular antibody-caspase-mediated cell killing: an approach for application in cancer therapy. *Proc. Natl Acad. Sci. USA* **97**, 12266–12271.

Velander WH, Lubon H, Drohan WN (1997) Transgenic livestock as drug factories. *Sci. American* **276**, 70–74.

Vigna E, Naldini L (2000) Lentiviral vectors: excellent tools for experimental gene transfer and promising candidates for gene therapy. *J. Gene Med.* **5**, 308–316.

Weissman IL (2002) Stem cells – scientific, medical and political issues. *New Engl. J. Med.* **346**, 1576–1579.

White RR, Sullenger BA, Rusconi CP (2000) Developing aptamers into therapeutics. *J. Clin. Invest.* **106**, 929–934.

Wolf CR, Smith G, Smith RL (2000) Pharmacogenomics. *Br. Med. J.* **320**, 987–990.

Ye X, Al-Babili S, Kloti A et al. (2000) Engineering the provitamin A (beta-carotene) biosynthetic pathway into (carotenoid-free) rice endosperm. *Science* **287**, 303–305.

Yeh P, Perricaudet M (1997) Advances in adenoviral vectors: from genetic engineering to their biology. *FASEB J.* **11**, 615–623.

Glossary

Note. In addition to this general glossary, there are also four specialized glossaries:

▶ Glossary of PCR methods, *Box 5.1*, p. 124

▶ Glossary of nucleic acid hybridisation, *Box 6.3*, p. 166

▶ Glossary of genomics, *Box 8.1*, p. 209

▶ Glossary of metazoan phylogenetic groups, *Box 12.5*, p. 383

Where possible, reference is made to a figure, box or section of text that illustrates or expands the topic. Words in *italics* refer to further glossary entries or to figure and box references.

Affected sib pair (ASP) analysis: a form of *nonparametric linkage analysis* based on measuring *haplotype* sharing by *sibs* who both have the same disease. See *Figure 15.3*.

Alleles: alternative forms of the same gene.

Allele-specific oligonucleotide (ASO): a synthetic oligonucleotide, typically ca. 20 nt long, whose hybridization to its target sequence can be disrupted by a single base-pair mismatch under suitable conditions. ASOs are used as allele-specific hybridization probes (see *Figure 6.11*), or as allele-specific primers in PCR. See *ARMS*.

Allelic association: see *Linkage disequilibrium*.

Allelic exclusion: the mechanism whereby only one of the two immunoglobulin alleles in B lymphocytes, or T-cell receptor alleles in T lymphocytes, is expressed. More generally, any naturally occurring mechanism that causes only one allele to be expressed. See *Box 10.4*.

Allelic heterogeneity: the existence of many different disease-causing alleles at a locus. The normal situation for diseases caused by loss of function of a gene – see, for example, *Figure 16.1*.

Alternative splicing: the natural use of different sets of splice junction sequences, to produce more than one product from a single gene. See Section 10.3.2, *Box 10.4*.

Alu repeat: a highly repetitive non-coding DNA sequence found in primate genomes.

Alu-PCR: see *Box 5.1*.

Amnion: one of the four extra-embryonic membranes of mammals. See *Box 3.8*.

Aneuploidy: a chromosome constitution with one or more chromosomes extra or missing from a full euploid set.

Anneal: see *Box 6.3*.

Anticipation: the tendency for the severity of a condition to increase in successive generations. Commonly due to *bias of ascertainment* (see Section 4.3.3), but seen for real with *dynamic mutations* (Section 16.6.4).

Anticodon: the 3-base sequence in a tRNA molecule that base-pairs with the codon in mRNA. See *Figures 1.7B* and *1.20*.

Antimorph: an allele with a *dominant negative* effect.

Antisense RNA: a transcript complementary to a normal mRNA, made using the non-template strand of a gene. Naturally occurring antisense RNAs are important regulators of gene expression.

Antisense strand (template strand): the DNA strand of a gene, which, during transcription, is used as a template by RNA polymerase for synthesis of mRNA. See *Figure 1.12*.

Apopotosis: programmed cell death.

Archaea: single-celled prokaryotes superficially resembling bacteria, but with molecular features indicative of a third kingdom of life. See *Box 12.4*.

ARMS (Amplification refractory mutation system): allele-specific PCR. See *Figures 5.4* and *18.10*.

Association: a tendency of two characters (diseases, marker alleles etc.) to occur together at non-random frequencies. Association is a simple statistical observation, not a genetic phenomenon, but can sometimes be caused by *linkage disequilibrium*. See Section 15.4.1.

Assortative mating: marriage between people of similar phenotype or genotype (e.g. tall people tend to marry tall people, deaf people tend to marry deaf people; some people prefer to marry relatives). Assortative mating can produce a non-Hardy-Weinberg distribution of genotypes in a population.

Autosome: any chromosome other than the sex chromosomes, X and Y.

Autozygosity: in an inbred person, *homozygosity* for alleles identical by descent.

Autozygosity mapping: for autosomal recessive disorders involves searching large inbred kindreds for loci where all affected individuals are autozygous for the same allele. See *Figure 13.8*.

BAC (bacterial artificial chromosome): a recombinant plasmid in which inserts up to 300 kb long can be propagated in bacterial cells. See Section 5.4.3.

Base complementarity: see *Box 6.3.*

Bayesian statistics: a branch of statistics that forms the basis of much genetic risk estimation. See *Box 18.4, Figure 18.15.*

Bias of ascertainment: distorted proportions of phenotypes in a dataset caused by the way cases are collected. See Section 15.2.1.

Biometrics: the statistical study of quantitative characters.

Bisulfite sequencing: a method for detecting patterns of DNA methylation. See Section 18.3.6.

Bivalent: the four-stranded structure seen in prophase I of meiosis, comprising two synapsed homologous chromosomes. See *Figure 2.11.*

BLAST: a family of programs that search sequence databases for matches to a query sequence. See *Box 7.3.*

Blastocyst: a very early stage in embryonic development when the embryo consists of a hollow ball of cells with a fluid-filled internal compartment, the blastocele.

Blunt-ended: of a DNA fragment, having no single-stranded extensions.

Bootstrapping: a statistical method designed to check the accuracy of an evolutionary tree constructed from comparative sequence analysis. The method involves numerous cycles of replacing some of the original sequence data by random subsamples taken from the original data and re-calculating at each cycle the likelihood of the original tree being correct. See *Figure 12.18.*

Boundary elements: sequences that define the boundaries of co-ordinately regulated chromatin domains in chromosomes. See *Box 8.2.*

Branch site: in mRNA processing, a rather poorly defined sequence (consensus CTRAY; R = purine, Y = pyrimidine) located 10–50 bases upstream of the splice acceptor, containing the adenosine at which the lariat splicing intermediate is formed. See *Figure 1.15.*

C value paradox: the lack of a direct relationship between the content of DNA in the cells of an organism (the C value) and the complexity of the organism.

Candidate gene: in positional cloning, a gene from the appropriate chromosomal location that is suspected of being the disease gene. The suspicion would be tested by seeking mutations in patients.

Cap: a specialized chemical group that cells add to block the 5' end of mRNA. See *Figure 1.17.*

cDNA (complementary DNA): DNA synthesized by the enzyme reverse transcriptase using mRNA as a template, either experimentally (see *Figures 4.8* and *Figure 6.5*) or *in vivo* (see *Figures 7.13* and *14.18*).

cDNA selection: a hybridization-based method for retrieving genomic clones that have counterparts in a cDNA library. See *Figure 7.11.*

CentiMorgan (cM): the unit of genetic distance. Loci 1 cM apart have a 1% probability of recombination during meiosis. See *Figure 13.4* for the relation between genetic and physical distances.

centiRay (cR): see *Box 8.1.*

Centric fusion: see *Robertsonian fusion.*

Centromere: the primary constriction of a chromosome, separating the short arm from the long arm, and the point at which spindle fibers attach to pull chromatids apart during cell division. See Section 2.3.2.

CEPH families: a set of families assembled by the Centre d'Etude du Polymorphisme Humain in Paris to assist the production of marker-marker framework maps.

Chemical cleavage of mismatches (CCM): a method of scanning kilobase-sized DNA fragments for mutations. See *Table 18.2, Figure 18.4B.*

Chiasma (plural: chiasmata): the physical manifestation of meiotic recombination, as seen under the microscope. See *Figure 13.3.*

Chimera: an organism derived from more than one zygote. See *Figure 4.10.*

Chordate: see *Box 12.5.*

Chorion: one of the four extra-embryonic membranes of mammals. See *Box 3.8.*

Chromatid: from the end of S phase of the cell cycle (*Figure 2.1*) until anaphase of cell division, chromosomes consist of two sister chromatids. Each contains a complete double helix and the two are exact copies of each other.

Chromatin fiber: the 30 nm coiled coil of DNA and histones that is believed to be the basic conformation of chromatin. See *Figure 2.3.*

Chromosome painting: fluorescence labeling of a whole chromosome by a *FISH* procedure in which the probe is a cocktail of many different DNA sequences from a single chromosome.

Chromosome walking: isolating sequences adjacent on the chromosome to a characterized clone by screening genomic libraries for clones that partially overlap it.

Cis-acting (of a regulatory factor): controlling the activity of a gene only when it is part of the same DNA molecule or chromosome as the regulatory factor. Compare *trans-acting* regulatory factors which can control their target sequences irrespective of their chromosome location.

Clone: see *Box 8.1.*

Coalescence time: in population genetics, the number of generations back to the most recent common ancestor. See *Box 12.6.*

Coding DNA: DNA that encodes the amino acid sequence of a polypeptide (or occasionally a functional mature RNA that does not specify a polypeptide).

Codon: a nucleotide triplet (strictly in mRNA, but by extension, in genomic coding DNA) that specifies an amino acid or a translation stop signal.

Coefficient of selection: the chance of reproductive failure for a certain genotype, relative to the most successful genotype. See Section 4.5.2.

Comparative genomic hybridization (CGH): use of competitive fluorescence *in situ* hybridization to detect chromosomal regions that are amplified or deleted, especially in tumors. See *Figure 17.3.*

Complementary strands: two nucleic acid strands are said to be complementary in sequence if they can form sufficient base pairs so as to generate a stable double-stranded structure.

Complementation: two alleles complement if in combination they restore the wild-type phenotype (see *Box 4.2*). Normally alleles complement only if they are at different loci, although some cases of *interallelic complementation* occur.

Complexity: the complexity of a genome is the total length or proportion of unique sequences. Low-complexity sequences are present many times in a genome or sample.

Compound heterozygote: a person with two different mutant alleles at a locus.

Conditional knock-out: an engineered mutant that causes loss of gene function under some circumstances (for example, raised temperature), or in some cells, but not others. See Section 20.2.6.

Conservation of synteny: when genetic maps of two organisms are compared, synteny is conserved if two or more loci that are located on the same chromosome in one organism are also located together on a single chromosome in the other.

Conservative substitution: a mutation causing a codon to be replaced by another codon that specifies a different amino acid, but one which is related in chemical properties to the original amino acid.

Constitutional: a genotype, abnormality or mutation that was present in the fertilized egg, and is therefore present in all cells of a person, as distinct from a *somatic* change.

Constitutive expression: a state where a gene is permanently active. Mutations that result in inappropriate constitutive expression are often pathogenic.

Contig: a list or diagram showing an ordered arrangement of cloned overlapping fragments that collectively contain the sequence of an originally continuous DNA strand.

Contiguous gene (segmental aneuploidy) syndrome: a syndrome caused by deletion of a contiguous set of genes, several or all of which contribute to the phenotype. See Section 16.8.1.

Continuous character: a character like height, which everybody has, but to differing degree – as compared with a *dichotomous character* like polydactyly, which some people have and others do not.

CpG dinucleotide: the sequence 5' CG 3' within a longer DNA molecule. CpG dinucleotides are targets of a specific *DNA methylation* system in mammals that is important in control of gene expression.

CpG island: short stretch of DNA, often < 1 kb, containing frequent unmethylated *CpG dinucleotides*. CpG islands tend to mark the 5' ends of genes. See *Box 9.3*.

Cre-lox system: a technique for generating predefined chromosomal deletions. *Cre* is a bacteriophage P1 gene whose product facilitates recombination between *lox*P sequences. So called because it *cre*ates *re*combination. See *Figures 20.10, 20.11*.

Cryptic splice site: a sequence in pre-mRNA with some homology to a splice site. Cryptic splice sites may be used as splice sites when splicing is disturbed or after a base substitution mutation that increases the resemblance to a normal splice site. See *Figures 11.12, 11.13*.

Cytoskeleton: an internal scaffold of protein filaments which occurs within cells and has a crucially important functions in regulating cell shape, cell movement, and intracellular transport. See *Box 3.2*.

Degenerate oligonucleotides: a series of oligonucleotides synthesized in parallel by allowing nucleotide flexibility at certain nucleotide positions. When used as hybridization probes or as PCR primers they can hybridize to, or amplify a family of sequences.

Denaturation: dissociation of complementary strands to give single-stranded DNA and/or RNA.

Deuterostomes: see *Box 12.5*.

DGGE (denaturing gradient gel electrophoresis): a method of mutation detection by electrophoresis of PCR products through gels containing a gradient of a denaturant. See *Figure 18.3B*.

dHPLC (denaturing high-performance liquid chromatography): a method for mutation detection based on the properties of *heteroduplexes*, capable of high sensitivity and throughput. See *Figure 18.4*.

Dichotomous character: a character like polydactyly, which some people have and others do not have – as compared to a *continuous character* like height, which everybody has, but to differing degree.

Differential display: see *Box 5.1*.

Differentiation: the process whereby cells become specialized and committed to form ultimately mature specific cell types.

Diploid: having two copies of each type of chromosome; the normal constitution of most human somatic cells.

Distal (of chromosomes): positioned comparatively distant from the centromere.

DNA chip: see *Microarray*.

DNA fingerprinting: a now obsolete method of identifying a person for legal or forensic purposes based on probing Southern blots with a hypervariable minisatellite probe. See *Figure 18.19*.

DNA library: see *Box 8.1*.

DNA marker: see *Marker (genetic)*.

DNA methylation: in the context of human DNA, this almost always means conversion of cytosine (usually in a *CpG dinucleotide*) into 5-methyl cytosine.

DNA profiling: using genotypes at a series of polymorphic loci to recognize a person, usually for legal or forensic purposes. See Section 18.7.

DNAse I-hypersensitive sites: regions of chromatin that are rapidly digested by DNAse I. They are believed to be important long-range control sequences. See Section 10.5.2.

Dominant negative (antimorph): a mutant gene whose product can inhibit the function of the wild-type gene product in heterozygotes. See, for example, *Figure 16.4*.

Dominant: in human genetics, any trait that is expressed in a heterozygote. See also *semi-dominant*.

Dosage compensation: any system that equalizes the amount of product produced by genes present in different numbers. In mammals, it describes the X-inactivation mechanism that ensures equal amounts of X-encoded gene products in XX and XY cells. See *Figure 10.26*.

Dosage sensitivity: property of a gene where a change in the copy number produces an abnormal phenotype.

Duplicative (or **copy** or **replicative**) **transposition:** transposition of a copy of a DNA sequence, while the original remains in place.

Dynamic mutation: an unstable expanded repeat that changes size between parent and child. See Section 16.6.4.

Ectoderm: one of the three germ layers of the embryo. It is formed during *gastrulation* from cells of the *epiblast* (see *Figure 3.15*) and gives rise to the nervous system and outer epithelia (see *Box 3.5*).

Electroporation: a method of transferring DNA into cells *in vitro* by use of a brief high-voltage pulse.

Embryonic stem cells: see *ES cells*.

Empiric risks: risks calculated from survey data rather than from genetic theory. Genetic counseling in most non-mendelian conditions is based on empiric risks. See Section 4.4.4.

Endoderm: one of the three germ layers of the embryo. It is formed during *gastrulation* from cells migrating out of the *epiblast* layer (*Figure 3.15*). See *Box 3.5* for derivatives of embryonic endoderm.

Enhancer trap: a technique for identifying strongly expressed genes in an organism.

Enhancer: a set of short sequence elements which stimulate transcription of a gene and whose function is not critically dependent on their precise position or orientation. See *Box 10.2*.

Epiblast: the layer of cells in the pregastrulation embryo which will give rise to all three germ layers of the embryo proper, plus the extraembryonic ectoderm and mesoderm. Compare *hypoblast*

Epigenetic: heritable (from mother cell to daughter cell, or sometimes from parent to child), but not produced by a change in DNA sequence. *DNA methylation* is the best understood epigenetic mechanism.

Episome: any DNA sequence that can exist in an autonomous extra-chromosomal form in the cell. Often used to describe self-replicating and extra-chromosomal forms of DNA.

Epitope: a part of an antigen with which a particular antibody reacts.

ES (embryonic stem) cells: undifferentiated, *pluripotent* cells derived from an embryo. A key tool for genetic manipulation. See Section 3.4.5, *Figures 21.1, 20.3*.

EST (expressed sequence tag): see *Box 8.1*.

Euchromatin: the fraction of the nuclear genome which contains transcriptionally active DNA and which, unlike *heterochromatin*, adopts a relatively extended conformation.

Eugenics: "improving" a population by selective breeding from the "best" types (positive eugenics) or preventing "undesirable" types from breeding (negative eugenics).

Euploidy: the state of having one or more complete sets of chromosomes with none extra or missing; the opposite of aneuploidy.

Exclusion mapping: genetic mapping with negative results, showing that the locus in question does not map to a particular location. Particularly useful for excluding a possible *candidate gene* without the labor of mutation screening.

Exon trapping: a technique for detecting sequences within a cloned genomic DNA that are capable of splicing to exons within a specialized vector. See *Figure 7.10*.

Exon: a segment of a gene that is represented in the mature RNA product. Individual exons may contain coding DNA and/or noncoding DNA (untranslated sequences). See *Figures 1.14, 1.19*.

Expression library: a library of cDNAs cloned in a vector that allows them to be expressed. See Section 5.6.

Expression profiling: obtaining a genome-wide picture of mRNA levels, normally by microarray analysis of total cellular cDNA (*Figures 17.18, 19.8*). See also *SAGE*.

Extracellular matrix: a meshwork-like substance found within the extracellular space and in association with the basement membrane of the cell surface. It provides a scaffold to which cells adhere and serves to promote cellular proliferation.

Fitness (f): in population genetics, a measure of the success in transmitting genotypes to the next generation. Also called biological fitness or reproductive fitness. f always lies between 0 and 1.

Fluorescence *in situ* hybridization (FISH): *in situ hybridization* using a fluorescently labeled DNA or RNA probe. A key technique in modern molecular genetics – see *Figures 2.17, 2.18*.

Folds: modules of three-dimensional protein structures. Most protein structures are built from a limited repertoire of folds that are shared among many proteins. See Section 19.4.4.

Founder effect: high frequency of a particular allele in a population because the population is derived from a small number of founders, one or more of whom carried that allele.

Frameshift mutation: a mutation that alters the normal translational *reading frame* of a mRNA by adding or deleting a number of bases that is not a multiple of three.

Functional genomics: analysis of gene function on a large scale, by conducting parallel analyses of gene expression/function for large numbers of genes, even all genes in a genome.

Fusion (hybrid) gene: a gene containing coding sequence from two different genes, usually created by unequal crossover (*Figure 9.17*) or chromosomal translocations (*Figure 18.7A*).

Fusion protein: the product of a natural or engineered *fusion gene*: a single polypeptide chain containing amino acid sequences that are normally part of two or more separate polypeptides. See *Box 20.2*.

Gastrulation: a highly dynamic process involving conversion of the two-layer pre-gastrulation embryo (consisting of *epiblast* plus *hypoblast*) to one that contains the three germ layers: *ectoderm, mesoderm* and *endoderm*.

Gene conversion: a naturally occurring nonreciprocal genetic exchange in which a sequence of one DNA strand is altered so as to become identical to the sequence of another DNA strand. See *Figure 11.10*.

Gene frequency: the proportion of all alleles at a locus that are the allele in question. See *Section 4.5*. Really we mean allele frequency, but the use of gene frequency is too well established now to change.

Gene replacement: gene therapy by replacing an endogenous faulty gene with a correctly functioning version. See *Figure 21.4*.

Gene supplementation (or **augmentation**): gene therapy by introducing a functional gene into the patient's cells without manipulating any of the endogenous genes. Suitable for correcting loss of function phenotypes. See *Figure 21.4*.

Gene targeting: targeted modification of a gene in a cell or organism. See *Figures 20.7, 20.8, 20.10*.

Gene tracking: predicting genotypes within pedigrees by using linked markers to follow a chromosomal segment. Sometimes called Indirect Testing. See *Box 18.3*.

Gene trap: a method of selecting *transgene* insertions that have occurred into a gene.

Genetic distance: distance on a genetic map, defined by recombination fractions and the *mapping function*, and measured in centiMorgans. See *Section 13.1*.

Genetic enhancement: the possible application of molecular genetic technologies to alter a normal human phenotype in some way that is considered to be beneficial. See *Ethics Box 3* in Chapter 21.

Genetic map: see *Box 8.1*.

Genetic redundancy: performance of the same function in parallel by genes at more than one locus, so that loss of function mutations at one locus do not cause overall loss of function

Genome browser: a program that provides a graphical interface for interrogating genome databases. See *Section 8.3.6*.

Genome: the total set of *different* DNA molecules of an organelle, cell or organism. The human genome consists of 25 different DNA molecules, the mitochondrial DNA molecule plus the 24 different chromosomal DNA molecules. Cf. *transcriptome, proteome*.

Genome-wide *p* value: in testing for linkage or association, the probability on the null hypothesis of observing the statistic in question anywhere in a screen of the whole genome (cf *pointwise p value*). See *Section 15.3.4*.

Genotype: the genetic constitution of an individual, either overall or at a specific locus.

Germ cells (or **gametes**): sperm cells and egg cells.

Germinal (gonadal, or gonosomal) mosaic: an individual who has a subset of germline cells carrying a mutation that is not found in other germline cells. See *Figure 4.9*.

Germ-line: the germ cells and those cells which give rise to them; other cells of the body constitute the *soma*.

Haploid: describing a cell (typically a gamete) which has only a single copy of each chromosome (e.g. the 23 chromosomes in human sperm and eggs).

Haploinsufficiency: a locus shows haploinsufficiency if producing a normal phenotype requires more gene product than the amount produced by a single copy. See *Section 16.4.2, Table 16.2*.

Haplotype: a series of alleles found at linked loci on a single chromosome.

Haplotype block: a particular variant of a polymorphic extended segment of chromosomal DNA (typically 10–100 kb) that is present in many members of a population and is assumed to represent an ancestral variant. See *Box 12.6* and Section 15.4.3.

Hardy–Weinberg distribution: the simple relationship between gene frequencies and genotype frequencies that is found in a population under certain conditions. See Section 4.5.

Hemizygous: having only one copy of a gene or DNA sequence in diploid cells. Males are hemizygous for most genes on the sex chromosomes. Deletions occurring on one autosome produce hemizygosity in males and in females.

Heritability: the proportion of the causation of a character that is due to genetic causes. See *Box 4.4*.

Heterochromatin: a chromosomal region that remains highly condensed throughout the cell cycle and shows little or no evidence of active gene expression. Constitutive heterochromatin is found at centromeres plus some other regions (for which, see *Figure 2.15*).

Heteroduplex: double-stranded DNA in which there is some mismatch between the two strands. Important in mutation detection – see Section 18.3.2.

Heteroplasmy: mosaicism, usually within a single cell, for mitochondrial DNA variants. See Sections 4.2.5, 11.4.2.

Heterozygote advantage: the situation when somebody heterozygous for a mutation has a reproductive advantage over both homozygotes. Sometimes called overdominance. Heterozygote advantage is the reason why several severe recessive diseases remain common. See *Box 4.8*.

Heterozygote: an individual having two different alleles at a particular locus.

Homologs (chromosomes): the two copies of a chromosome in a diploid cell. Unlike sister chromatids, homologous chromosomes are not copies of each other; one was inherited from the father and the other from the mother.

Homologs (genes): two or more genes whose sequences are significantly related because of a close evolutionary relationship, either between species (*orthologs*) or within a species (*paralogs*).

Homoplasmy: of a cell or organism, having all copies of the mitochondrial DNA identical. Cf. *heteroplasmy*.

Homozygote: an individual having two identical alleles at a particular locus. For clinical purposes a person is often described as homozygous AA if they have two normally-functioning alleles, or homozygous aa if they have two pathogenic alleles at a locus, regardless whether the alleles are in fact completely identical at the DNA sequence level. Homozygosity for alleles *identical by descent* is called *autozygosity*.

Hotspot: a sequence associated with an abnormally high frequency of recombination or mutation.

Hox genes: a subset of homeobox genes which are organized in clusters and have important roles in anterior-posterior patterning. See *Figures 3.10* and *3.12*.

Hybrid cell panel: a collection of *somatic cell hybrids* or *radiation hybrids* used for physical mapping.

Hypermorph: an allele that produces an increased amount or activity of product.

Hypoblast: the layer of cells in the pre-gastrulation embryo which gives rise to extraembryonic endoderm

Hypomorph: an allele that produces a reduced amount or activity of product.

Identity by descent (IBD): alleles in an individual or in two people that are known to be identical because they have both been inherited from a demonstrable common ancestor.

Identity by state (IBS): alleles that appear identical, but may or may not be *identical by descent* because there is no demonstrable common source. See *Figure 15.2*.

***In situ* hybridization:** a form of molecular hybridization in which the target nucleic acid is denatured DNA within chromosome preparations (**chromosome *in situ* hybridization**) or RNA within cells of tissue sections immobilized on a microscope slide (**tissue *in situ* hybridization** – see *Figure 6.15*) or within whole embryos (**whole mount *in situ* hybridization** – see *Figure 7.15*) .

Imprinting: determination of the expression of a gene by its parental origin. See Section 10.5.4 and *Box 16.6*.

Inbreeding: marrying a blood relative. The term is comparative, since ultimately everybody is related. The *coefficient of inbreeding* is the proportion of a person's genes that are *identical by descent*.

Inducible promoter: a *promoter* whose activity can be switched on and off by an external agent. See *Figure 20.5A*.

Informative meiosis: in linkage analysis, a meiosis is informative if the genotypes in the pedigree allow us to decide whether it is recombinant or not (for a given pair of loci). See *Box 13.2*.

Inner cell mass: a group of cells located internally within the *blastocyst* which will give rise to the embryo proper. See *Figure 3.13*.

Insertional mutagenesis: mutation (usually abolition of function) of a gene by insertion of an unrelated DNA sequence within the gene.

Interference: in meiosis, the tendency of one crossover to inhibit further crossing over within the same region of the chromosomes. See Section 13.1.3.

Interphase: all the time in the cell cycle when a cell is not dividing.

Intron: noncoding DNA which separates neighboring exons in a gene. During gene expression introns are transcribed into RNA but then the intron sequences are removed from the pre-mRNA by splicing. See *Figure 1.14*. Can be classified according to the mechanism of splicing (See *Box 12.1*) or, if separating coding DNA sequences, by their precise location within codons (See *Box 12.2*).

Isochromosome: an abnormal symmetrical chromosome, consisting of two identical arms, which are normally either the short arm or the long arm of a normal chromosome.

Isoforms/isozymes: alternative forms of a protein/enzyme.

Isogenic: two or more organisms or cells having identical genotypes. For example, different mice belonging to a particular inbred strain, e.g. C57B10/J, are isogenic.

Karyotype: strictly, this means a summary of the chromosome constitution of a cell or person, such as 46,XY. However, the term is often loosely used to mean an image showing the chromosomes of a cell sorted in order and arranged in pairs, such as *Figure 2.14*.

Knock-down: targeted inhibition of expression of a gene by, for example, using specific *antisense RNA* or *siRNA* to bind to RNA transcripts.

Knock-in mutation: a targeted mutation that replaces activity of one gene by that of an introduced gene (usually an allele). See *Figure 20.14*.

Knock-out mutation: the targeted inactivation of a gene within an intact cell.

Lagging strand: in DNA replication, the strand that is synthesized as Okazaki fragments (see *Figure 1.9*).

Leading strand: in DNA replication, the strand that is synthesized continuously (see *Figure 1.9*).

Ligation: formation of a 3'–5' phosphodiester bond between nucleotides at the ends of two molecules (*intermolecular ligation*) or the two ends of the same molecule (*intramolecular ligation, cyclization*).

LINE (*l*ong *i*nterspersed *n*uclear *e*lement): a class of repetitive DNA sequences that make up about 20% of the human genome (*Table 9.15*). Some are active transposable elements. See *Figures 9.17, 9.18*.

Lineage: a series of cells originating from a progenitor cell.

Linkage disequilibrium: a statistical association between particular alleles at separate but linked loci, normally the result of a particular ancestral *haplotype* being common in the population studied. An important tool for high resolution mapping. See *Figures 13.10, 15.6* and *Box 15.2*.

Linkage: the tendency of characters (phenotypes, marker alleles etc) to co-segregate in a pedigree because their determinants lie close together on a particular chromosome.

Linker (adapter) oligonucleotide: a double-stranded oligonucleotide which can be ligated to a DNA molecule of interest and which has been designed to contain some desirable characteristic, e.g. a favorable restriction site.

Liposome: a synthetic lipid vesicle used to transport a molecule of interest into a cell. See *Figure 21.8*.

Locus control region (LCR): a stretch of DNA containing regulatory elements which control the expression of genes in a gene cluster that may be located tens of kilobases away. See *Figure 10.23*.

Locus heterogeneity: determination of the same disease or phenotype by mutations at different loci. A common problem in mapping genetic diseases. See Sections 4.2.4, 16.7.2.

Locus: a unique chromosomal location defining the position of an individual gene or DNA sequence.

Lod score (z): a measure of the likelihood of genetic linkage between loci. The log (base 10) of the odds that the loci are linked (with recombination fraction θ) rather than unlinked. For mendelian characters a lod score greater than +3 is evidence of linkage; one that is less than –2 is evidence against linkage. See *Box 13.3, Figure 13.7*.

Loss of heterozygosity (LoH): homozygosity or hemizygosity in a tumor or other somatic cell when the *constitutional* genotype is heterozygous. Evidence of a somatic genetic change. See *Figure 17.6, 17.7*.

MALDI (matrix-assisted laser desorption/ionization): a mass spectrometric method commonly used for identifying DNA or protein molecules. See *Box 19.4*.

Manifesting heterozygote: a female carrier of an X-linked recessive condition who shows some clinical symptoms, presumably because of skewed *X-inactivation*. See Section 4.2.2.

MAPH (multiplex amplifiable probe hybridization): a method of detecting deletions or duplications of exons. See *Box 18.1*.

Mapping function: a mathematical equation describing the relation between recombination fraction and *genetic distance*. The mapping function depends on the extent to which *interference* prevents close double recombinants. See Section 13.1.3.

Marker chromosome: an extra abnormal chromosome of unidentified origin.

Marker (genetic): any polymorphic *mendelian* character that can be used to follow a chromosomal segment through a pedigree. Genetic markers are usually DNA polymorphisms. See *Box 13.1*.

Marker (chromosome): an extra chromosome of unidentified origin.

Matrilineal inheritance: transmission from just the mother, but to children of either sex; the pattern of mitochondrial inheritance. See *Figure 4.4*.

Mean heterozygosity: of a marker, the likelihood that a randomly selected person will be heterozygous. A measure of the usefulness of the marker for linkage analysis (see Section 11.2.2).

Melting temperature (T_m): see *Box 6.3*.

Mendelian: of a pedigree pattern, conforming to one of the archetypal patterns shown in *Figure 4.2*. A character will give a mendelian pedigree pattern if it is determined at a single chromosomal location, regardless whether or not the determinant is a gene in the molecular geneticist's sense.

Mesoderm: one of the three germ layers of the embryo. It is formed during *gastrulation* by cells migrating out of the *epiblast*. See *Figure 3.15* and see *Box 3.5 for derivatives of embryonic mesoderm*.

Metaphase: the stage of cell division (mitosis or meiosis) when chromosomes are maximally contracted and lined up on the equatorial plane (metaphase plate) of a cell. See *Figures 2.10, 2.11, 2.15*.

Metazoans: multicellular animals as opposed to unicellular *protozoa*.

Microarray: a miniature array of different DNA or oligonucleotide sequences on a glass surface that is intended to be used in a hybridization assay. The sequences may be pre-formed DNA molecules that have been deposited using automation or oligonucleotide sequences which are synthesized *in situ* to make a **DNA chip**.

Microdeletion: a chromosomal deletion that is too small to be seen under the microscope (typically <3 Mb).

Micro-RNAs (miRNA): short (22 nt) RNA molecules encoded within normal genomes that have a role in regulation of gene expression and maybe also of chromatin structure. Sometimes called small temporal RNA (stRNA). See *Figures 9.6, 20.12*.

Microsatellite: small run (usually less than 0.1kb) of tandem repeats of a very simple DNA sequence, usually 1–4 bp, for example $(CA)_n$. Often polymorphic, providing the primary tool for genetic mapping during the 1990s. Sometimes also described as STR (simple tandem repeat) or SSR (simple sequence repeat) polymorphism. See *Figures 7.7, 7.8*.

Microsatellite instability: a phenomenon characteristic of certain tumor cells, where during DNA replication the repeat copy number of microsatellites is subject to random changes. Abbreviated to **MIN, MSI** or **RER** (replication error). See *Figure 17.7*.

Microtubules: long hollow cylinders constructed from tubulin polymers which form part of the *cytoskeleton* (see *Box 3.2*).

MIM number: the catalog number for a gene or mendelian character, as listed in the *OMIM* database.

Minisatellite DNA: an intermediate size array (typically 0.1–20 kb long) of short tandemly repeated DNA sequences. See *Box 7.2*. Hypervariable minisatellite DNA is the basis of DNA *fingerprinting* and many *VNTR* markers.

Minisequencing: a method of detecting sequence variants at a predefined position by sequencing just one or two nucleotides downstream of a primer. Often used in a microarray format. See *Figure 18.9B*.

Mismatch repair: a natural enzymic process that replaces a mispaired nucleotide in a DNA duplex (most likely present because of an error in DNA replication) to obtain perfect Watson–Crick base pairing.

Missense mutation: a nucleotide substitution that results in an amino acid change. See *Box 11.3*.

Modifier gene: a gene whose expression can influence a phenotype resulting from mutation at another locus. See Section 16.6.3.

Monoallelic expression: expression of only one of the two copies of a gene in a cell, because of X-inactivation, imprinting or other epigenetic change, or because of the gene rearrangements that take place with immunoglobulin and T-cell receptor genes. See *Box 10.4* for examples.

Monoclonal antibody (mAb): pure antibodies with a single specificity, produced by hybridoma technology, as distinct from *polyclonal antibodies* that are raised by immunization. See *Box 7.4*.

Mosaic: an individual who has two or more genetically different cell lines derived from a single zygote. The differences may be point mutations, chromosomal changes, etc. See *Figure 4.10*.

Multifactorial: a character that is determined by some unspecified combination of genetic and environmental factors. Cf. *polygenic*.

Multigene family: a set of evolutionarily related loci within a genome, at least one of which can encode a functional product. See Section 9.3.

Neomorph: an allele with a novel activity or product.

Neural crest: a highly versatile group of cells which give rise to part of the peripheral nervous system, melanocytes, some bone and muscle, the retina and other structures. Neural crest cells form during neurulation as a specific population of cells, which arise along the lateral margins of the neural folds then detach from the neural plate and migrate to many specific locations within the body. See *Figure 3.16A.*

Nondisjunction: failure of chromosomes (sister chromatids in mitosis or meiosis II; paired homologs in meiosis I) to separate (disjoin) at anaphase. The major cause of numerical chromosome abnormalities. See Section 2.5.2.

Nonhomologous recombination: recombination between sequences that either have no homology or have limited local homology. A major cause of insertions and deletions, at the genetic or chromosomal level. See for example *Figures 11.7, 16.2.*

Nonparametric: in linkage analysis, a method that does not depend on a specific genetic model, such as *affected sib pair analysis.*

Nonpenetrance: the situation when somebody carrying an allele that normally causes a dominant phenotype does not show that phenotype. An effect of other genetic loci or of the environment. A pitfall in genetic counseling. *Figure 4.5B* shows an example.

Nonsense mutation: a mutation that occurs within a codon and changes it to a stop codon. See *Box 11.3.*

Nonsense-mediated mRNA decay: a cellular mechanism that degrades mRNA molecules that contain a premature termination codon (>50 nt upstream of the last splice junction). See Section 11.4.4.

Northern blot: a membrane bearing RNA molecules that have been size-fractionated by gel electrophoresis, used as a target for a hybridization assay. Used to detect the sizes of transcripts of a gene of interest in a collection of adult or fetal tissues. See *Figure 5.13.*

Notochord: a flexible rod-like structure that forms the supporting axis of the body in simple chordates and simple vertebrates, and in embryos of the more complex vertebrates.

Nucleolar organizer region (NOR): the satellite stalks of human chromosomes 13, 14, 15, 21 and 22. NORs contain arrays of ribosomal DNA genes and can be selectively stained with silver. Each NOR forms a nucleolus in telophase of cell division; the nucleoli fuse in interphase.

Nucleosome: a structural unit of chromatin. See *Figure 2.3.*

Null allele: a mutant allele that produces no product.

Oligogenic: a character that is determined by a small number of genes acting together.

Oligonucleotide ligation assay (OLA): a method for detecting a predefined sequence change. See *Figure 18.11.*

OMIM: On-line *M*endelian *I*nheritance in *M*an, the central database of human genes and mendelian characters (http:

//www3.ncbi.nlm.nih.gov/omim/. *MIM numbers* are the index numbers for entries in OMIM.

Oncogene: a gene involved in control of cell proliferation which, when overactive can help to transform a normal cell into a tumor cell. See *Table 17.1.* Originally the word was used only for the activated forms of the gene, and the normal cellular gene was called a *proto-oncogene*, but this distinction is now widely ignored.

One gene-one enzyme hypothesis: the hypothesis advanced by Beadle and Tatum in 1941 that the primary action of each gene was to specify the structure of an enzyme. Historically very important, but now seen to be only part of the range of gene functions.

Open reading frame (ORF): a significantly long sequence of DNA in which there are no termination codons in at least one of the possible reading frames. Six reading frames are possible for a DNA duplex because each strand can have three reading frames.

Ortholog: one of a set of homologous genes in different species (e.g. *PAX3* in humans and *Pax3* in mice). See *Box 12.3.*

Overdominant: phenotypes showing *heterozygote advantage.* A term used in population genetics.

Palindrome: a DNA sequence such as ATCGAT that reads the same when read in the 5'→3' direction on each strand. DNA-protein recognition, for example by restriction enzymes, often relies on palindromic sequences.

Paracentric inversion: inversion of a chromosomal segment that does not include the centromere. See *Figure 2.20.*

Paralog: one of a set of homologous genes within a single species. See *Box 12.3.*

Parametric: in linkage analysis, a method such as standard *lod score* analysis, that requires a tightly specified genetic model.

Partial digestion: digestion, usually of DNA by a restriction enzyme, that is stopped before all target sequences have been cut. The object is to produce overlapping fragments. See *Figure 5.9.*

Penetrance: the frequency with which a genotype manifests itself in a given phenotype.

Pericentric inversion: inversion of a chromosomal segment that includes the centromere. See *Figure 2.20.*

Phage display: an expression cloning method in which foreign genes are inserted into a phage vector and are expressed to give polypeptides that are displayed on the surface (protein coat) of the phage.

Pharmacogenetics: the study of the influence of individual genes or alleles on the metabolism or function of drugs.

Pharmacogenomics: the use of genome resources (genome sequences, expression profiles etc.) to identify new drug targets.

Phase (of linked markers): the relation (coupling or repulsion) between alleles at two linked loci. If allele A1 is on the same physical chromosome as allele B1, they are in coupling; if they are on different parental homologs they are in repulsion. See *Figure 13.6.*

Phase (of the cell cycle): G1, S, G2, M and Go phases (see *Figure 2.1*).

Phase (of an intron): a term used to classify introns in coding sequences according to the position at which they interrupt the message (see *Box 12.2*).

Phenotype: the observable characteristics of a cell or organism, including the result of any test that is not a direct test of the *genotype*.

Phylogeny: classification of organisms according to perceived evolutionary relatedness. See *Figures 12.22–12.24*.

Plasticity: see *Transdifferentiation*.

Pluripotent: strictly, this means the capacity to give rise to very many, but not all of the different cell types to which the zygote can give rise. The zygote and the immediate cells that it gives rise to are said to be totipotent, but differentiation by the blastocyst stage means that the cells of the inner cell mass are pluripotent rather than totipotent.

Point mutation: a mutation causing a small alteration in the DNA sequence at a locus. The meaning is a little imprecise: when being compared to chromosomal mutations, the term point mutation might be used to cover quite large (but submicroscopic) changes within a single gene, whereas when mutations at a single locus are being discussed, point mutations would normally mean the substitution, insertion or deletion of just a single nucleotide.

Pointwise *p* value: in linkage analysis, the probability on the null hypothesis of exceeding the observed value of the statistic at one given position in the genome. Cf. *genome-wide p value*. See Section 15.3.4.

Polyadenylation: addition of typically 200 A residues to the 3' end of a mRNA. The poly(A) tail is important for stabilizing mRNA. See *Figure 1.18*.

Polygenic: a character determined by the combined action of a number of genetic loci. Mathematical polygenic theory (see Section 4.4) assumes there are very many loci, each with a small effect.

Polymorphic markers: see *Box 8.1*.

Polymorphism: strictly, the existence of two or more variants (alleles, phenotypes, sequence variants, chromosomal structure variants) at significant frequencies in the population. Looser usages among molecular geneticists include (1) any sequence variant present at a frequency >1% in a population (2) any non-pathogenic sequence variant, regardless of frequency.

Polyploid: having multiple chromosome sets as a result of a genetic event that is abnormal (e.g. constitutional or mosaic triploidy, tetraploidy, etc.), or programmed (e.g. some plants and certain human body cells are naturally polyploid).

Positional cloning: cloning a gene knowing only its chromosomal location.

Primer: a short oligonucleotide, often 15–25 bases long, which base-pairs specifically to a target sequence to allow a polymerase to initiate synthesis of a complementary strand.

Primordial germ cells: cells in the embryo and fetus which will ultimately give rise to germ-line cells

Probe: a known DNA or RNA fragment (or a collection of different known fragments) which is used in a hybridization assay to identify closely related DNA or RNA sequences within a complex, poorly understood mixture of nucleic acids. In standard hybridization assays, the probe is labeled but in reverse hybridization assays the target is labeled (see *Box 6.4, p. 169*).

Promoter: a combination of short sequence elements, normally just upstream of a gene, to which RNA polymerase binds in order to initiate transcription of the gene. See *Figure 1.13*.

Proofreading: an enzymic mechanism by which DNA replication errors are identified and corrected.

Protein truncation test (PTT): a method of screening for chain-terminating mutations by artificially expressing a mutant allele in a coupled transcription–translation system. See *Figure 18.5*.

Protein truncation test: a method of screening for chain-terminating mutations by artificially expressing a mutant allele in a coupled transcription–translation system. See *Figure 17.9*.

Proteome: the total set of different proteins in a cell, tissue or organism. Cf. *genome, transcriptome*

Proto-oncogene: see *Oncogene*.

Protostomes: see *Box 12.5*.

Proximal (of chromosomes): positioned comparatively close to the centromere.

Pseudoautosomal regions: regions at each tip of the X and Y chromosomes containing X-Y homologous genes. Because of X-Y recombination, alleles in these regions show an apparently autosomal mode of inheritance. See *Figure 12.15*.

Pseudogene: a DNA sequence which shows a high degree of sequence homology to a nonallelic functional gene but which is itself nonfunctional.

Pyrosequencing: a proprietary method for checking sequence very close to a predefined start point. See *Box 18.2*.

Quantitative trait locus (QTL): a locus important in determining the phenotype of a continuous character. Section 15.6.8 discusses the search for QTL's underlying human obesity.

RACE-PCR: *see Box 5.1, p. 124.*

Radiation hybrid: in human physical mapping, a rodent cell that contains numerous small fragments of human chromosomes. Produced by fusion with a lethally irradiated human cell. Radiation hybrid panels allow very rapid mapping of *STSs*. See *Figure 10.4*.

Rare cutter: a restriction nuclease which cuts DNA infrequently because the sequence it recognizes is large and/or contains one or more CpGs. Examples are *Not*I, *Sac*II and *Bss*HII. See *Table 4.1*.

Reading frame: during translation, the way the continuous sequence of the mRNA is read as a series of triplet codons. There are three possible reading frames for any mRNA, and the correct reading frame is set by correct recognition of the AUG initiation codon.

Real-time PCR: *see Box 5.1, p. 124.*

Recessive: a character is recessive if it is manifest only in the homozygote.

Recombinant (linkage analysis): a person who inherits from a parent a combination of alleles that is the result of a crossover during meiosis. See *Figure 13.1*.

Recombinant DNA: an artificially constructed hybrid DNA containing covalently linked sequences with different origins, for example a vector with an insert.

Recombination fraction: for a given pair of loci, the proportion of meioses in which they are separated by recombination. Usually signified as θ. θ, values vary between 0 and 0.5. See *Section 13.1*.

Repetitive DNA: a DNA sequence that is present in many identical or similar copies in the genome. The copies can be tandemly repeated or dispersed.

Replication slippage: a mistake in replication of a tandemly repeated DNA sequence, that results in the newly synthesized strand having extra or missing repeat units compared to the template. See *Figure 11.5*.

Replicon: any nucleic acid that is capable of self-replication. Many cloning vectors use extrachromosomal replicons (as in the case of plasmids). Others use chromosomal replicons, either directly (as in the case of yeast artificial chromosome vectors), or indirectly, by allowing integration into chromosomal DNA.

Reporter gene: a gene used to test the ability of an upstream sequence joined on to it to cause its expression. Putative cis-acting regulatory sequences can be coupled to a reporter gene and transfected into suitable cells to study their function. Alternatively, transgenic animals (and other organisms) are often made with a promotorless reporter gene integrated at random into the chromosomes, so that expression of the reporter marks the presence of an efficient promoter. See *Box 20.4*.

Reporter molecule: a molecule whose presence is readily detected (for example, a fluorescent molecule) that is attached to a DNA sequence we wish to monitor. See, for example *Figure 6.7*.

Response elements: sequence usually located a short distance upstream of promoters that makes gene expression responsive to some chemical in the cellular environment. *Table 10.4* lists some examples.

Restriction fragment length polymorphism (RFLP): a genetic marker consisting of variable sizes of allelic restriction fragments resulting from a DNA sequence polymorphism. See *Box 7.2*. RFLPs were originally assayed by *Southern blotting* (*Figure 7.5A*) but now usually by PCR (*Figure 7.6*).

Retrotransposon (retroposon): a transposable DNA element that transposes by means of an RNA intermediate. Retroposons encode a *reverse transcriptase* that acts on the RNA transcript to make a cDNA copy, which then integrates into chromosomal DNA at a different location. See *Box 7.4*, *Figures 7.17, 7.18*.

Retrovirus: an RNA virus with a reverse transcriptase function, enabling the RNA genome to be copied into cDNA prior to integration into the chromosomes of a host cell.

Reverse transcriptase: an enzyme that can make a DNA strand using an RNA template. Used to make cDNA libraries (*Figure 4.8*) and for RT-PCR (*Section 20.2.4*). Reverse transcription is an essential part of the retroviral life cycle (*Figure 18.2*) but not, as far as is known, of normal cell metabolism.

RFLP: *see Restriction Fragment Length Polymorphism*

Ribozyme: a natural or synthetic catalytic RNA molecule. See *Figure 21.10*.

RNA editing: a natural process in which specific changes occur post-transcriptionally in the base sequence of an RNA molecule. Occurs rarely in human genes. See *Figure 8.16*.

RNA splicing: see *Splicing*.

RNAi (RNA interference): the use of *siRNAs* to *knock down* (but rarely completely abolish) expression of specified genes. A powerful tool for studying gene function.

Robertsonian fusion: a chromosomal rearrangement that converts two acrocentric chromosomes into one metacentric or submetacentric. See *Figure 2.21*. Sometimes called *centric fusion*, although the point of exchange is actually in the proximal short arm, and not at the centromere.

RT-PCR (reverse transcriptase PCR): see *Box 5.1*, p. 124.

SAGE (serial analysis of gene expression): a method of *expression profiling* based on sequencing. See *Figure 19.5*.

Satellite (on chromosome): stalked projection variably present on the short arms of human acrocentric chromosomes (13, 14, 15, 21, 22).

Satellite DNA: originally described a DNA fraction that forms separate minor bands on density gradient centrifugation because of its unusual base composition. The DNA is composed of very long arrays of tandemly repeated DNA sequences. See *Section 9.4.1*.

Secondary structure: regions of a single stranded nucleic acid or protein/polypeptide molecule where chemical bonding occurs between distantly spaced nucleotides or amino acids resulting in complex structures. Secondary structure is often due to intra-strand hydrogen bonding (*Figures 1.7, 1.24*).

Segmental aneuploidy (or aneusomy) syndrome: see *Contiguous gene syndrome*.

Segmental duplication: the existence of very highly related DNA sequence blocks on different chromosomes or at more than one location within a chromosome. See *Figures 12.12, 12.13*.

Segregation analysis: the statistical methodology for inferring modes of inheritance.

Segregation ratio: the proportion of offspring who inherit a given gene or character from a parent.

Semi-dominant: an allele that, in the heterozygote, produces a phenotype intermediate (but not necessarily halfway) between the wild-type and the homozygote. A term widely used in mouse genetics, but better avoided, at least in human genetics, since dominance is a property of a character and not of an allele.

Sense strand: the DNA strand of a gene that is complementary in sequence to the template (antisense) strand, and identical to the transcribed RNA sequence (except that DNA contains T where RNA has U). Quoted gene sequences are always of the sense strand, in the 5'→3' direction. See *Figure 1.12*.

Sequence homology: a measure of the similarity in the sequences of two nucleic acids or two polypeptides.

Sequence *tagged site* (STS): any unique piece of DNA for which a specific PCR assay has been designed, so that any DNA sample can be easily tested for its presence or absence. See *Box 10.3*.

Shotgun sequencing: sequencing of DNA fragments that have been randomly generated from a large clone or a whole genome. See *Figure 8.3*.

Sib-pair analysis: see *Affected sib-pair analysis*.

Sibs: brothers or sisters.

Signal sequence (leader sequence): a sequence of about 20 amino acids at the N-terminus of a polypeptide that controls its destination within or outside the cell. See Section 1.5.4.

Silencer: combination of short DNA sequence elements which suppress transcription of a gene. See *Box 10.2*.

Silent (synonymous) mutation: a mutation that changes a codon but does not alter the aminoacid encoded. See *Box 11.3*. Such mutations may still have effects on mRNA splicing or stability.

SINE (short *interspersed nuclear element*): a class of moderate to highly repetitive DNA sequence families, of which the best known in humans is the *Alu repeat* family. See *Table 9.15* and *Figures 9.17, 9.18*.

siRNA (small interfering RNA): A 21–23 nt double stranded RNA that specifically abolishes the function of a mRNA to which is it homologous. See Section 20.2.7, *Figure 20.12*.

Simple sequence repeat polymorphism: see *Microsatellite*

Sister chromatid exchange (SCE): a recombination event involving sister chromatids. Since sister chromatids are duplicates of each other, such exchanges should have no effect unless they are *unequal*. However, an increased frequency of SCEs is evidence of DNA damage.

Sister chromatid: two chromatids present within a single chromosome and joined by a centromere. Nonsister chromatids are present on different but homologous chromosomes.

Site-directed mutagenesis: production of a specific predetermined change in a DNA sequence. Can be done *in vitro* on cloned DNA (Sections 5.5.2 and 5.5.3) or *in vivo* by homologous recombination (Section 20.2.6).

Slipped strand mispairing: see *replication slippage*

SNP (single nucleotide polymorphism): any polymorphic variation at a single nucleotide. SNPs include *RFLPs*, but also other polymorphisms that do not alter any restriction site. Although less informative than *microsatellites*, SNPs are more amenable to large-scale automated scoring.

Somatic cell hybrid: an artificially constructed cell formed by fusing two different types of somatic cell, notably cells from different species. Human-rodent hybrid cells have been valuable mapping tools. See *Box 8.4*.

Somatic cell: any cell in the body except the gametes.

Somites: discrete blocks of segmental mesoderm that will establish the segmental organization of the body by giving rise to most of the axial skeleton (including the vertebral column), the voluntary muscles and part of the dermis of the skin.

Southern blot: transfer of DNA fragments from an electrophoretic gel to a nylon or nitrocellulose membrane (filter), in preparation for a hybridization assay. See *Figure 5.12*.

Specificity: in testing, a measure of the performance of a test. Specificity = (1 − false positive rate), see *Figure 18.17*.

Splice acceptor site: the junction between the 3' end of an intron and the start of the next exon. Consensus sequence $y_{11}nyagR$ (y = pyrimidine, R = purine; upper case = exon). See *Figure 1.15*.

Splice donor site: the junction between the end of an exon and the start (5' end) of the downstream intron. Consensus sequence (C/A)AG*gt*ragt (r = purine; upper case = exon). See *Figure 1.15*.

Spliceosome: a ribonucleoprotein complex used in RNA *splicing*.

Splicing: normally **RNA splicing**, in which RNA sequences transcribed from introns are excised from a primary transcript and those transcribed from exons are spliced together in the same linear order as the exons (see *Figures 1.14, 1.16*). A form of **DNA splicing** is important for assembling productive immunoglobulin and T cell receptor genes in B and T lymphocytes respectively (see *Figures 10.28, 10.29*).

Splicing enhancer: a sequence (which can be exonic or intronic) that increases the probability that a nearby potential splice site will actually be used. See Sections 11.4.3 and 16.4.1.

SSCP or SSCA: *single* *stranded* *conformation* *polymorphism or analysis*, a commonly used method for point mutation screening. See *Figure 18.3A*.

Stem cell: a cell which can act as a precursor to differentiated cells but which retains the capacity for self-renewal. See Section 3.4.3.

Sticky ends (cohesive termini): short single-stranded projections from a double-stranded DNA molecule, typically formed by digestion with certain restriction enzymes. Can Molecules with complementary sticky ends can associate, and can then be covalently joined using *DNA ligase* to form recombinant DNA molecules. See *Figures 4.3, 4.4*.

Stratification: the existence of genetically different groups within a population supposed to be homogeneous.

STR polymorphism (short tandem repeat polymorphism): see *Microsatellite*.

STS: see *Box 8.1*.

Subtraction cloning: a method of cloning DNA sequences that are present in one DNA sample and absent in a second, generally similar, sample. Used to select tissue-specific cDNAs by subtraction against a library of ubiquitously expressed cDNAs, or to clone a gene that is deleted in a patient with a disease by subtraction against normal DNA.

Suppressor tRNA: a mutant transfer RNA molecule with a nucleotide substitution in the anticodon. Shows an altered coding specificity and is able to translate a nonsense (or missense) codon. See *Box 5.3*.

Syncytium: a cell that contains multiple nuclei as a result of cycles of DNA replication without cell division, or as a result of fusion of multiple cells (as in the case of muscle fiber cells).

Synonymous (silent) substitution: a substitution that replaces one codon by another that encodes the same amino acid. See *Box 9.2*.

Synteny: loci are syntenic if on the same chromosome. Syntenic loci are not necessarily linked: loci sufficiently far apart on the chromosome assort at random, with 50% recombinants.

Telomere: a specialized structure at the tips of chromosomes. It consists of an array of short tandem repeats, (TTAGGG)$_n$ in humans, which form a closed loop and protect the chromosome end.

Template strand: in transcription, the DNA strand that base-pairs with the nascent RNA transcript. See *Figure 1.12*.

Termination codon: a UAG (*amber*), UAA (*ochre*), or UGA (*opal*) codon in a mRNA (and by extension, in a gene) that signals the end of a polypeptide.

Therapeutic cloning: a proposed method of treating disease using cells made by transplanting nuclei from cells of the patient into human *ES cells*. See *Figure 21.2*.

Trans-acting: (of a regulatory factor) affecting expression of all copies of the target gene, irrespective of chromosomal location. Trans-acting regulatory factors are usually proteins which can diffuse to their target sites.

Transcription unit: a stretch of DNA that is naturally transcribed in a single operation to produce a single primary transcript. In straightforward cases, a transcription unit is the same thing as a gene.

Transcriptome: the total set of different RNA transcripts in a cell or tissue. Cf. *genome*, *proteome*.

Transdifferentiation (or **plasticity**): the possible ability for cells that appear to be committed to a particular type of differentiation to be diverted into another differentiation pathway.

Transduction: virus-mediated gene transfer. See *Box 19.1* and *Ethics Box 1* in Chapter 21.

Transfection: uptake of DNA by eukaryotic cells (the equivalent of *transformation* in bacteria, but that word has a different meaning for eukaryotic cells). Also uptake of plasmid DNA by bacterial cells. See *Box 19.1*.

Transformation (of a cell): 1. uptake by a *competent* bacterial cell of naked high molecular weight DNA from the environment; 2. alteration of the growth properties of a normal eukaryotic cell as a step towards evolving into a tumor cell.

Transgene: an exogenous gene that has been transfected into cells of an animal or plant. It may be present in some tissues (as in human gene therapies) or in all tissues (as in germ-line engineering, e.g. in mouse). Introduced transgenes may be episomal and be transiently expressed, or can be integrated into host cell chromosomes.

Transgenic animal: an animal in which artificially introduced foreign DNA (a **transgene**) becomes stably incorporated into the germline. See *Figures 20.2, 20.3*.

Transition: G⇔A (purine for purine) or C⇔T (pyrimidine for pyrimidine) nucleotide substitution.

Translocation: transfer of chromosomal regions between non-homologous chromosomes. See *Figure 2.21*.

Transmission disequilibrium test (TDT): a statistical test of allelic association. See *Box 15.3*.

Transposon: a mobile genetic element – see *Figure 9.17*.

Transversion: a nucleotide substitution of purine for pyrimidine or vice versa.

Triploid: of a cell, having three copies of the genome; of an organism, being made of triploid cells.

Trisomy rescue: survival of an initally trisomic embryo because a chance mitotic non-disjunction produces a disomic cell that becomes the progenitor of the whole fetus. Can result in *uniparental disomy*.

Trisomy: having three copies of a particular chromosome, e.g. trisomy 21.

Trophoblast (=trophectoderm): outer layer of polarized cells in the *blastocyst* (see *Figure 3.13*) which will go on to form the *chorion*, the embryonic component of the placenta.

Tumor-suppressor gene (TSG): a gene whose normal function is to inhibit or control cell division. TSG are typically inactivated in tumors. See Section 17.4.

Two-hit hypothesis: Knudson's theory that hereditary cancers require two successive mutations to affect a single cell. See *Figure 17.5*.

Two-hybrid system: see *Yeast two-hybrid system*.

Unequal crossover (UEC): recombination between nonallelic sequences on nonsister chromatids of homologous chromosomes. See *Figure 11.7*.

Unequal sister chromatid exchange (UESCE): recombination between nonallelic sequences on sister chromatids of a single chromosome. See *Figure 11.7*.

Uniparental disomy: a cell or organism in which both copies of one particular chromosome pair are derived from one parent. Depending on the chromosome involved, this may or may not be pathogenic. See Section 2.5.4, *Box 16.6*.

Untranslated regions (5' UTR, 3'UTR): regions at the 5' end of mRNA before the AUG translation start codon, or at the 3' end after the UAG, UAA or UGA stop codon. See *Figure 1.19*.

Variable expression: Variable extent or intensity of phenotypic signs among people with a given genotype. See for example, *Figure 4.5C*.

Variable number tandem repeat (VNTR) polymorphism: *microsatellites*, *minisatellites* and *satellite DNAs* are arrays of tandemly repeated sequences that often vary between people in the number of repeat units. See *Box 7.2*. The term VNTR is often used to mean specifically minisatellites.

Vector: a nucleic acid that is able to replicate and maintain itself within a host cell, and that can be used to confer similar properties on any sequence covalently linked to it.

Western blotting: a process in which proteins are size-fractionated in a polyacrylamide gel, then transferred to a nitrocellulose membrane for probing with an antibody. See *Figure 20.9*.

Whole genome amplification: a PCR method using highly degenerate primers that can amplify a very large number of random sequences spread across the genome. Can be used to allow repeated testing of DNA from a single cell, for example in typing single sperm (Section 11.5.4).

Whole mount *in situ* hybridization: see *In situ hybridization.*

X inactivation (lyonization): the inactivation of one of the two X chromosomes in the cells of female mammals by a specialized form of genetic *imprinting*. See Section 10.5.6.

YAC (yeast artificial chromosome): an vector able to propagate inserts of a megabase or more in yeast cells. See *Figure 5.17.*

YAC transgenic: a *transgenic* mouse in which the transgene is a complete *YAC*, allowing studies of the regulatory effects of sequences surrounding the gene involved. See Section 20.2.3.

Yeast two-hybrid system: an important system for identifying and purifying proteins that bind to a protein of interest. See *Figure 19.18.*

Zinc finger: a polypeptide motif which is stabilized by binding a zinc atom and confers on proteins an ability to bind specifically to DNA sequences. Commonly found in transcription factors. See *Figure 10.8.*

Zoo-blot: a Southern blot containing DNA samples from a range of different species. See *Figure 7.9.*

Zygote: the fertilized egg cell.